OXYGEN TRANSPORT
TO TISSUE XII

ADVANCES IN EXPERIMENTAL MEDICINE AND BIOLOGY

Editorial Board:

NATHAN BACK, *State University of New York at Buffalo*

IRUN R. COHEN, *The Weizmann Institute of Science*

DAVID KRITCHEVSKY, *Wistar Institute*

ABEL LAJTHA, *N. S. Kline Institute for Psychiatric Research*

RODOLFO PAOLETTI, *University of Milan*

Recent Volumes in this Series

A Continuation Order Plan is available for this series. A continuation order will bring delivery of each new volume immediately upon publication. Volumes are billed only upon actual shipment. For further information please contact the publisher.

OXYGEN TRANSPORT TO TISSUE XII

Edited by

Johannes Piiper

Max-Planck-Institute for Experimental Medicine
Goettingen, Federal Republic of Germany

Thomas K. Goldstick

Northwestern University
Evanston, Illinois

and

Michael Meyer

Max-Planck-Institute for Experimental Medicine
Goettingen, Federal Republic of Germany

PLENUM PRESS • NEW YORK AND LONDON

Library of Congress Cataloging-in-Publication Data

International Society on Oxygen Transport to Tissue. Meeting (17th :
 1989 : Göttingen, Germany)
   Oxygen transport to tissue XII / edited by Johannes Piiper, Thomas
 K. Goldstick, and Michael Meyer.
       p.   cm. -- (Advances in experimental medicine and biology ; v.
 277)
   "Proceedings of the Seventeenth Annual Meeting of the
 International Society on Oxygen Transport to Tissue, held July
 21-24, 1989, in Goettingen, Federal Republic of Germany"--T.p.
 versos.
   Includes bibliographical references.
   Includes indexes.

   1. Tissue respiration--Congresses.   I. Piiper, Johannes.
 II. Goldstick, Thomas K.   III. Meyer, M. (Michael) IV. Title.
 V. Title: Oxygen transport to tissue 12. VI. Title: Oxygen
 transport to tissue twelve. VII. Series.
   [DNLM: 1. Biological Transport--congresses. 2. Oxygen--blood-
 -congresses. 3. Oxygen Consumption--congresses.
 4. Spectrophotometry, Infrared--congresses.   W1 AD559 v. 277 / WF
 110 I61o 1989]
 QP177.I56 1989
 599'.012--dc20
 DNLM/DLC
 for Library of Congress   90-7835
                                                         CIP

Proceedings of the Seventeenth Annual Meeting of the
International Society on Oxygen Transport to Tissue,
held July 21-24, 1989, in Goettingen, Federal Republic of Germany

ISBN 978-1-4684-8183-9   ISBN 978-1-4684-8181-5 (eBook)
DOI 10.1007/978-1-4684-8181-5

INTERNATIONAL SOCIETY ON OXYGEN TRANSPORT

TO TISSUE 1988-89

Officers:

President: J. Piiper, West Germany
President-Elect: M. McCabe, Australia
Past President: K. Rakusan, Canada
Secretary: N.S. Faithful, England
Treasurer: J. Grote, West Germany
Treasurer-Elect: S.M. Cain, USA

Executive Committee:

D.T. Delpy, England   A. Mayevsky, Israel
W. Erdmann, The Netherlands E.M. Nemoto, USA
T.K. Goldstick, USA   Z. Turek, The Netherlands
T. Koyama, Japan   P. Vaupel, West Germany
I.S. Longmuir, USA

GOETTINGEN MEETING, July 21-24, 1989

Local Organizing Committee:

J. Piiper (chairman)   M. Meyer

Sponsors:

We are grateful for the generous financial support
for the 1989 ISOTT meeting received from the following:

GHG Medizin-Elektronik AG, Zurich
Keithley Instruments GmbH, Munich
Gould Electronics GmbH, Seligenstadt
MKS-Instruments GmbH, Munich
Stemmer Software GmbH, Puchheim
MH-Ges. f. Hardware/Software mbH, Erftsadt
Siemens AG, Erlangen
National Science Foundation, Washington, D.C.

PREFACE

The International Society on Oxygen Transport to Tissue (ISOTT) was founded in 1973 "to facilitate the exchange of scientific information among those interested in any aspect of the transport and/or utilization of oxygen in tissues". Its members span virtually all disciplines, extending from various branches of clinical medicine such as anesthesiology, ophthalmology and surgery through the basic medical sciences of physiology and biochemistry to most branches of the physical sciences and engineering.

The seventeenth annual meeting of ISOTT was held in 1989 for four days, from July 21 to 24, at the Max Planck Institute for Experimental Medicine and the adjoining University Hospital (Klinikum), in Goettingen, Federal Republic of Germany. It attracted 147 active registrants and approximately 40 accompanying persons. The very successful format originated by Dr. Ian Longmuir in 1985, consisting of posters accompanied by an abbreviated oral summary, was continued with slight modification. Virtually all of the presentations utilized this format, with each poster session preceded by a formal discussion during which the presenter briefly reviewed the poster aided by a few slides. All posters remained in place for the entire four days of the meeting. Simultaneous sessions were not utilized. The theme of this meeting emphasized respiration but essentially all aspects of physiological oxygen transport were covered, as the 105 manuscripts comprising this volume demonstrate. The organizing committee is most grateful to Frau Irmgard Barteczko and Frau Helgard Rinnert for their assistance in arranging the many details of the meeting and to Reiner Schubert for his photographic and audiovisual assistance, particularly in taking the group photograph. The committee also acknowledges the assistance of the many other personnel from the Physiology Department at M. P. I. Goettingen including scientists, technicians, and housekeeping staff who assisted with many facets of the meeting. All contributed to the success of the meeting.

The editors reviewed all manuscripts. Extensive revisions were made in about 15% and modest revision in about another 30%. Minor errors in format and some typos were not corrected. Except for some revisions and retyping of a few manuscripts by the editorial staff, all of the camera-ready manuscripts in this volume were prepared by the authors themselves and we greatly appreciate their cooperation. We also wish to acknowledge the skillful, patient and careful work involved in the preparation of this volume by Frau Renate Hahn of Goettingen and Rod D. Braun of Evanston.

For the editors

Thomas K. Goldstick

February, 1990

CONTENTS

MATHEMATICAL MODELS

METHODS AND INSTRUMENTATION

RESPIRATORY SYSTEM

WHOLE BODY

TUMORS

MATHEMATICAL MODELS

COUNTER-CURRENT BLOOD FLOW IN TISSUES:

PROTECTION AGAINST ADVERSE EFFECTS

Hirosuke Kobayashi[1], Bernd Pelster[2], Johannes Piiper[3] and Peter Scheid[2]

[1] Department of Medicine, School of Medicine, Keio University, Tokyo 160, Japan
[2] Institut für Physiologie, Ruhr-Universität Bochum 4630 Bochum, F.R.G.
[3] Abteilung Physiologie, Max-Planck-Institut für experimentelle Medizin, 3400 Göttingen, F.R.G.

INTRODUCTION

Histological studies indicate parallel arrangements of small arteries and veins as well as arterioles and venules in a number of tissues, such as kidney (cf. Solez and Heptinstall, 1980), skin (Sparks, 1978), placenta (Barcroft and Barron, 1946), heart (Hutchins et al., 1986), and intestine (cf. Jodal and Lundgren, 1986). There exists also functional evidence for inert gas back-diffusion in counter-current blood flow in several tissues, such as muscle (Piiper et al., 1984; cf. Wagner, 1987), brain (Stosseck, 1970), and heart (Roth and Feigl, 1981). Effros et al. (1984) also observed counter-current diffusion of labeled n-butanol in the rabbit renal cortex.

In spite of this evidence for counter-current blood flow in tissue, little is known on the role of counter-current O_2 and CO_2 gas exchange in tissue oxygenation. Counter-current gas exchange has been considered to be detrimental for oxygen supply to the tissue, since part of the oxygen in the arterial blood would diffuse into the venous blood, thus forming an oxygen diffusion shunt (Lübbers, 1968; Piiper et al., 1984).

Even less is known about the effect of CO_2 on O_2 in tissues with counter-current blood flow. Grossmann (1982) incorporated the O_2/CO_2 interaction (Bohr/Haldane effect) in the counter-current model of skin circulation, but his interest was focused on the function of surface P_{O_2} and P_{CO_2} electrodes. Roth and Wade (1986) also accounted for the O_2/CO_2 interaction in their counter-current model in skeletal muscle, but they did not estimate the influence of the Bohr/Haldane effect.

Oxygen Transport to Tissue XII, Edited by J. Piiper et al.
Plenum Press, New York, 1990

In the present study, a simple counter-current gas exchange model including O_2/CO_2 interaction was evaluated in an attempt to investigate the physiological significance of the counter-current system as well as the Bohr effect for tissue oxygenation. A model excluding the Bohr effect was used as a control. Situations investigated with this model are (1) hypoxia with and without acid production in the tissue, (2) normoxia, and (3) hyperoxia, at resting condition.

MODEL

The model consists of a counter-current gas exchange unit and a blood/tissue gas exchange unit (Figure 1). Assumptions and basic parameters made in this model are given elsewhere (Kobayashi et al., 1989). Briefly, O_2 and CO_2 reaction in blood is considered to take place instantaneously, the counter-current system being composed of a distributed diffusion barrier rather than a lumped one, as in the model of Roth and Wade (1986). This barrier is assumed to be permeable to O_2 and CO_2, but impermeable to H^+ or HCO_3^-.

The symbols employed are as follows:

D is the diffusing capacity of the exchange barrier, mmol gas$\cdot$min$^{-1}\cdot$torr^{-1};

D_{O_2}, D_{CO_2} are the diffusing capacity to O_2 and CO_2 of the exchange barrier;

L is the length of the transfer region, cm;

$\dot{Q}$ is the blood flow rate, l$\cdot$min^{-1};

$\dot{M}$ is the (diffusive) transfer rate across the barrier, mmol gas$\cdot$min^{-1};

$\dot{M}_{O_2}$, $\dot{M}_{CO_2}$ are the O_2 uptake and CO_2 production rate of tissue.

D is assumed to be independent of x, so that the diffusing capacity of any length element, dx, of the exchange barrier is $DdxL^{-1}$. The value of $\dot{M}$ of this element is $\dot{M}(x)dxL^{-1}$.

The following basic values were assumed in the calculations: $\dot{M}_{O_2}/\dot{Q} = 2.0$ mmol$\cdot$l^{-1}, $\dot{M}_{CO_2}/\dot{Q} = 1.6$ mmol$\cdot$l^{-1}, $D_{CO_2}/D_{O_2} = 20$, $Pa_{CO_2} = 40$ torr. Pa_{O_2} was assumed to be 40 torr in hypoxia, 90 torr in normoxia, and variable, between 100 and 700 torr, in hyperoxia. For tissue acid production, base excess values in the blood were assumed as -1 or -10 mEq$\cdot$l^{-1}.

Basic equations

The diffusive transfer rate, $\dot{M}(x)\cdot dx\cdot L^{-1}$, of a gas across the element of diffusing capacity, $D\cdot dx\cdot L^{-1}$, in the counter-current compartment at site x (Figure 1) equals:

$$\dot{M}(x) = D \cdot [Pa(x) - Pv(x)] \tag{1}$$

4

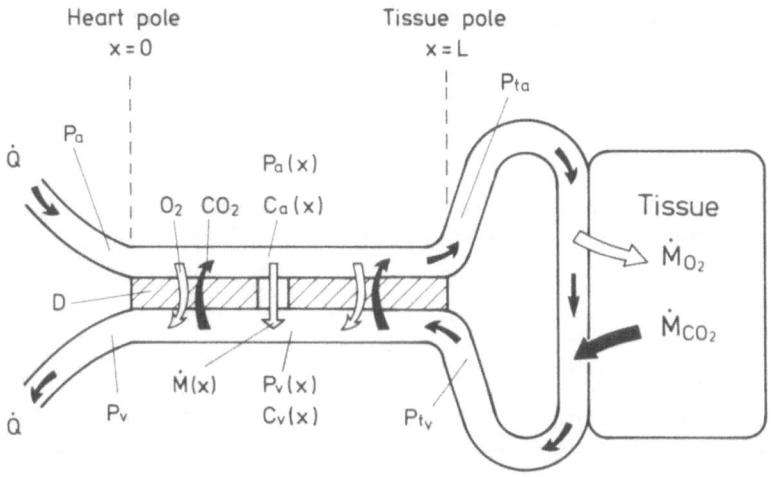

Counter-current gas exchange   Blood/tissue gas exchange

Fig. 1 Counter-current model. The heart pole (x = 0) indicates the inflow of arterial vessels, a, into the counter-current compartment, and also the outflow of venous vessels, v. The tissue pole (x = L) represents the outflow of the arterial vessels, ta, and the inflow of venous vessels, tv. D, diffusing capacity.

in which a and v refer to arterial and venous blood.

This gas transfer leads to a change of gas content, C(x), in both arterial and venous blood:

$$\dot{M}(x) = -\dot{Q}L \frac{dCa}{dx} \tag{2}$$

$$\dot{M}(x) = -\dot{Q}L \frac{dCv}{dx} \tag{3}$$

Combining equations (2) or (3) with (1) yields

$$dC = -\frac{1}{\dot{Q}} \frac{Ddx}{L} [Pa(x) - Pv(x)] \tag{4}$$

which is valid for any gas including O_2 and CO_2 and for arterial and venous blood. This equation was integrated for Ca from x = 0 to L, an exchange was allowed with the tissue effected at L, and then the equation was integrated for Cv from x = L to 0.

With the initial condition at x = 0: Ca(x) = Ca, Pa(x) = Pa, $Cv_{O_2} = Ca_{O_2} - \dot{M}_{O_2}/\dot{Q}$, and $Cv_{CO_2} = Ca_{CO_2} + \dot{M}_{CO_2}/\dot{Q}$ (5)

eq. (4) was solved for O_2 and CO_2, using the Runge-Kutta integration and an iterative method (*cf.* West and Wagner,

5

1977). This yields the corresponding values at x = L, Cta, Ctv, Pta, and Ptv.

The dissociation curve of human blood including the Bohr effect (Kelman, 1966) was used in the calculation. The standard dissociation curve without the Bohr effect, i.e., with Pa_{CO_2} fixed at 40 torr and pH at 7.40, was also used in the calculation as a control in order to investigate the specific role of the Bohr effect in the tissue oxygenation.

RESULTS

Hypoxia, with and without acid production in the tissue

Figure 2 shows that, with increasing effectiveness of the counter-current exchange, i.e. with increasing D/$\dot{Q}$,

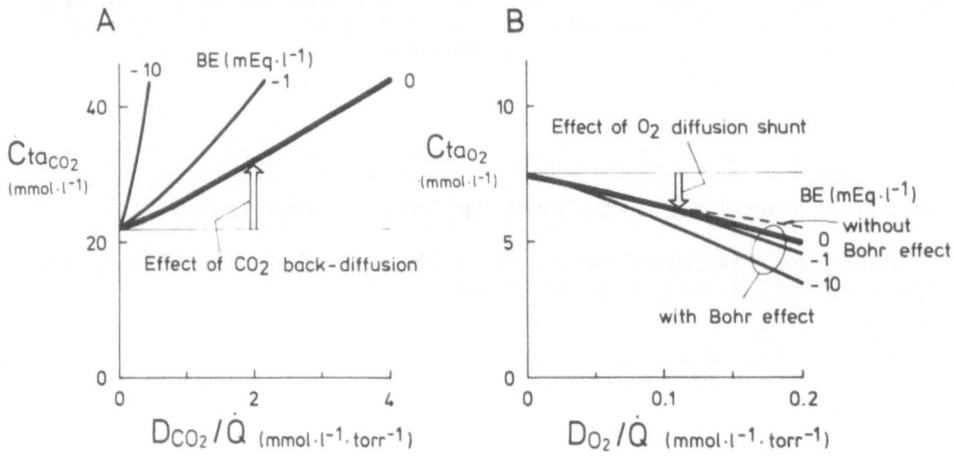

Fig. 2 Hypoxia. A: accumulation of CO_2 in the counter-current system. B: diffusion shunt of O_2. BE: Base excess in the venous blood. (see text)

Cta_{CO_2} is elevated (A) and Cta_{O_2} is decreased (B). These effects originate from trapping CO_2 by veno-arterial back-diffusion and from shunting O_2 by an arterio-venous diffusion shunt of O_2.

These effects are further enhanced when fixed acid is released into the blood from the tissue (negative base excess in venous blood, BE), since H^+ and HCO_3^- are combined to form CO_2 molecules, thus increasing venous P_{CO_2} and enhancing CO_2 back-diffusion. The O_2 diffusion shunt is also enhanced, since Pa_{O_2} is elevated by CO_2 and H^+ *via* the Bohr effect, thus increasing the arterio-venous pressure difference.

6

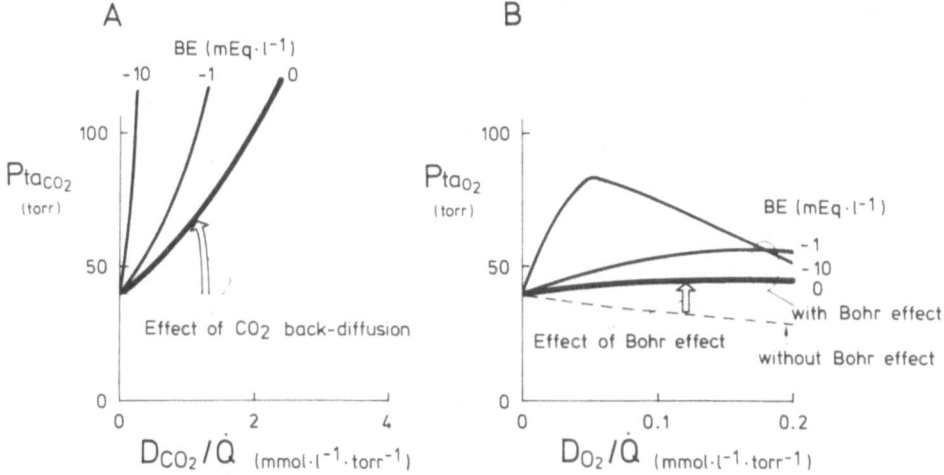

Fig. 3 Hypoxia. A: Pta_{CO_2} enhancement due to CO_2 back-diffusion. B: Pta_{O_2} elevation due to the Bohr effect.(see text)

Figure 3A shows that P_{CO_2} in the blood at the tissue pole (Pta_{CO_2}) is enhanced, corresponding to the CO_2 trapping shown in fig. 2A. This effect is further enhanced when fixed acid is released into the blood. Pta_{O_2} is likewise decreased (fig. 3B) when there is no Bohr effect. With Bohr effect, however, Pta_{O_2} is elevated with increasing $D/\dot{Q}$. Also the increment in Pta_{O_2} in the presence of the Bohr effect is further enhanced when fixed acid is released into the blood from the tissue (fig. 3B).

Normoxia

Without the counter-current flow, Pta_{O_2} is independent of $D_{O_2}/\dot{Q}$, and is kept at the same level as Pa_{O_2} (fig. 4). With the counter-current flow, Pta_{O_2} is reduced with increasing $D_{O_2}/\dot{Q}$, but, with the Bohr effect, Pta_{O_2} is higher than the level without the Bohr effect, and remains almost constant at a higher $D_{O_2}/\dot{Q}$.

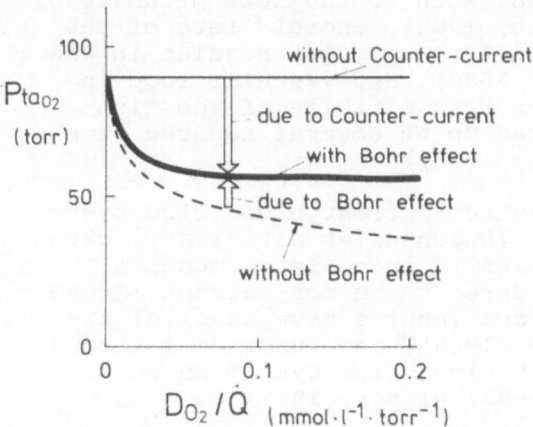

Fig. 4 Normoxia. Pta_{O_2} decreases with increasing $D_{O_2}/\dot{Q}$. This decrement is partially counteracted with the Bohr effect.

7

Hyperoxia

Fig. 5 shows that, with the counter-current flow, Pta_{O_2} under hyperoxia becomes almost constant at a lower Pta_{O_2} level, irrespective of Pa_{O_2}. Pta_{O_2} with the Bohr effect is higher than the level without the Bohr effect at any Pa_{O_2}.

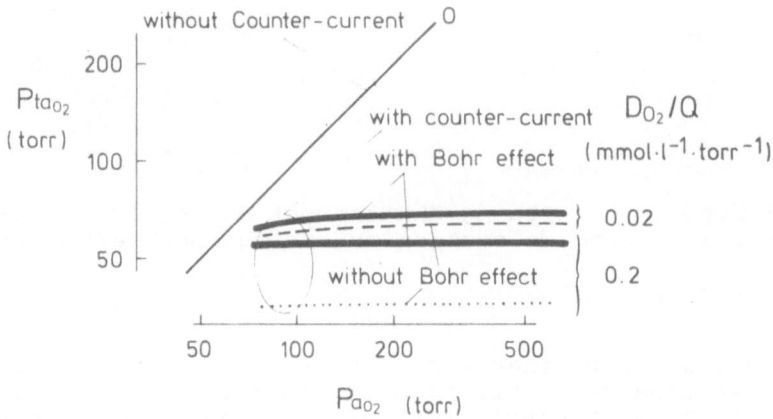

Fig. 5 Hyperoxia. With counter-current blood flow system, Pta_{O_2} remains almost constant. Pta_{O_2} with the Bohr effect is higher than the level without the Bohr effect. The larger the $D_{O_2}/\dot{Q}$ is, the smaller the Pta_{O_2} becomes.

DISCUSSION

Counter-current blood flow systems in the tissue

Both anatomical and functional evidence suggests the existence of counter-current gas exchange between pre-capillary and post-capillary vessels in various tissues. Well-developed counter-current blood flow systems are known to exist in several animal tissues, such as the rete mirabile of the fish swimbladder (*cf.* Fänge, 1983), choroid rete of the fish eyes (Wittenberg *et al.*, 1974), vascular bundles in the kidney (Solez and Heptinstall, 1980), and vascular loop in the skin (Sparks, 1978). In the rete mirabile of the fish swimbladder, P_{O_2} can be elevated up to several hundred atmospheres (Scholander and Van Dam, 1953).

A moderately developed counter-current blood flow system is found to exist in the heart (Hutchins *et al.*, 1986), particularly in the horse (Kiss, 1957). In skeletal muscles, capillary circulation is considered to be con-current (Groom *et al.*, 1976), but arterioles and venules have parallel arrangements (McKay and Lipowsky, 1988), and functional studies also suggests a counter-current blood flow system in the skeletal muscle (*cf.* Wagner, 1987; Piiper, 1987).

The role of the Bohr effect with counter-current system in tissue oxygenation

The counter-current diffusion shunt of O_2 would, in fact, reduce capillary oxygen supply. In the presence of the Bohr effect, however, $Ptao_2$ can be elevated compared with the level without the Bohr effect. The effect is particularly prominent in hypoxia and may be further pronounced when lactic acid production is stimulated with tissue hypoxia. The elevated Pao_2 in the capillary blood enables enhanced O_2 transport to the tissue in spite of the reduced amount of O_2 in the capillary blood due to O_2 diffusion shunt.

During hyperoxia, the Bohr effect exerts no major role in Po_2 enhancement at the entrance to the counter-currrent system since hemoglobin is fully saturated with O_2, whereby a shift in the O_2 dissociation curve is not effective in Po_2 elevation. In this situation, the counter-current acts as an O_2 diffusion shunt, thus reducing Po_2 at the tissue and preventing toxic effects of hyperoxia.

SUMMARY

In hypoxia, the tissue counter-current can thus, by virtue of the Bohr effect, increase tissue Po_2 and thus tissue oxygenation. In hyperoxia, on the other hand, the counter-current system, acting as a diffusion shunt, can protect the tissue against adverse O_2-toxic effects. It thus appears, that the counter-current system is advantageous for O_2 supply to tissues.

REFERENCES

Barcroft, J., and Barron, D. H., 1946, Observations on the form and relations of the maternal and fetal vessels in the placenta of the sheep, *Ant. Rec.*, 94: 569-595.

Effros, R. M., Taki, K. Reid, and E., Silverman, P., 1984, Countercurrent diffusion in the renal cortex of the rabbit, *Circ. Res.*, 55: 463-467.

Fänge, R., 1983, Gas exchange in fish swim bladder, *Rev. Physiol. Biochem. Pharmacol.*, 97: 111-158.

Groom, A. C., Plyley, M. J., and Sutherland, G., 1976, Oxygen transport in skeletal muscle: capillary geometry in longitudinal section, *in*: "Oxygen Transport to Tissue II", J. Grote, D. Reneau and G. Thews, eds., Plenum Press, New York, pp. 685-692.

Grossman, U., 1982, Simulation of combined transfer of oxygen and heat through the skin using a capillary-loop model, *Mathematical Bioscience*, 61: 205-236.

Hutchins, G. M., More, G. W., and Hatton, E. V., 1986, Arterial-venous relationships in the human left ventricular myocardium: anatomic basis for countercurrent regulation of blood flow, *Circulation*, 74: 1195-1202.

Jodal, M., and Lundgren, O., 1986, Countercurrent mechanisms in the mammalian gastrointestinal tract, *Gastroenterology*, 91: 225-241.

Kelman, G. R., 1966, Digital computer subroutine for the conversion of oxygen tension into saturation, *J. Appl. Physiol.*, 21: 1375-1376.

Kiss, F., 1957, Funktionell-Morphologische Angaben zum venösen Kreislauf, *Acta Anat.*, 30: 358-370.

Kobayashi, H., Pelster, B., Piiper, J., and Scheid, P., 1989, Significance of the Bohr effect for tissue oxygenation in a model with counter-current blood flow, *Respir. Physiol.*, 76: 277-288.

Lübbers, D. W., 1968, The oxygen pressure field of the brain and its significance for the normal and critical oxygen supply of the brain, *in*: "Oxygen Transport in Blood and Tissue", D. W. Lübbers, U. C. Luft, G. Thews, and E. Witzleb, eds, Georg Thieme, Stuttgart, pp. 124-139.

McKay, C. B., and Lipowsky, H. H., 1988, Arteriovenous distribution of transit times in cremaster muscle of the rat, *Microvasc. Res.*, 36: 75-91.

Pelster, B., Kobayashi, H., and Scheid, P., 1989, Metabolism of the pefused swimbladder of the European eel: oxygen, carbon dioxide, glucose and lactate balance, *J. Exp. Biol.*, (in press).

Piiper, J., 1987, Role of diffusion shunt in transfer of inert gases and O_2 in muscle, *in*: "Oxygen Transport to Tissue X", M. Mochizuki, C. R. Honig, T. Koyama, T. K. Goldstick, and D. F. Bruley, eds., Plenum Press, New York and London, pp. 55-61.

Piiper, J., Meyer, M., and Scheid, P., 1984, Dual role of diffusion in tissue gas exchange: blood-tissue equilibration and diffusion shunt, *Respir. Physiol.*, 56: 131-144.

Roth, A. C., and Feigl, E. O., 1981, Diffusional shunting in the canine myocardium, *Circ. Res.*, 48: 470-480.

Roth, A. C., and Wade, K., 1986, The effects of transmural transport in the microcirculation: a two gas species model, *Microvasc. Res.*, 32: 64-83.

Scholander, P. F., and Van Dam, L., 1953, Composition of the swimbladder gas in deep sea fishes, *Biol. Bull.*, 104: 75-86.

Solez, K., and Heptinstall, R. H., 1980, The anatomy of the renal circulation, *in*: "Structure and Function of the Circulation", Vol. 1, C. J. Schwartz, N. T. Werthessen, and S. Wolf, eds., Plenum Press, New York and London, pp. 631-660.

Sparks, H. V., 1978, Skin and muscle, *in*: "Peripheral Circulation", D. C. Johnson, ed., John Wiley & Sons, New York, pp. 193-230.

Stosseck, K., 1970, Hydrogen exchange through the pial vessel wall and its meaning for the determination of the local cerebral blood flow, *Pflügers Arch.*, 320: 111-119.

Wagner, P. D., 1987, Peripheral inert-gas exchange, *in*: "Handbook of Physiology", Section 3, The Respiratory System Vol. IV, Gas Exchange, L. E. Farhi and S. M. Tenney, eds., American Physiological Society, Bethesda, pp. 257-281.

West, J. B., and Wagner, P. D., 1977, Pulmonary gas exchange, *in*: "Bioengineering Aspects of the Lung", J. B. West, ed., Marcel Dekker, New York, pp. 361-457.

Wittenberg, J. B., and Haedrich, R. L., 1974, The choroid rete mirabile of the fish eye. II. Distribution and relation to the pseudobranch and to the swimbladder rete mirabile, *Biol. Bull.*, 146: 137-156.

CONCENTRIC OXYGEN DIFFUSION IN TISSUE WITH HETEROGENEOUS PERMEABILITY AND CONSUMPTION

Louis Hoofd, Zdenek Turek, and Stuart Egginton*

Dept. Physiology, University of Nijmegen, Nijmegen, The Netherlands.
*Dept. Physiology, The University of Birmingham, Medical School, Birmingham, U.K.

INTRODUCTION

Most modelling of oxygen diffusion in tissue has considered the properties of the tissue to be homogeneous, not dependening on location. However, in tissue different portions can be discerned, resulting in heterogeneity with respect to oxygen diffusion and consumption. Such tissue heterogeneity was considered in a concentric diffusion model (Rakusan et al., 1984), the oxygen diffusing inward into a circular region consisting of two zones: a central one and a ring-shaped outer one, each having different oxygen consumptions. Capillaries were located at the outer border of the circular region, regularly spaced and all having the same capillary radius and oxygen pressure, while oxygen permeability had to be the same in both zones.

In the model presented here these latter constraints are removed, and new possibilities added. Capillaries still have to be at the outer border but may vary in location, radius and oxygen pressure, while diffusive properties of the two zones may be different. This permits a more thorough investigation of the effects of tissue diffusion and consumption heterogeneity. For instance, the possiblity of "guiding" the oxygen through a highly permeable outer (lipid-like) zone can be investigated; and also, by assigning to neighbouring capillaries different P_{O_2}'s and different intercapillary distances, the effect of "shunting" the oxygen from one capillary to the other can be analyzed. The consequences of differing capillary-to-fibre ratios can also be modelled by varying the number of capillaries surrounding the circular region ("fibre").

The model was based on combination of two other model descriptions: the concentric double-zone model mentioned above (Rakusan et al., 1984) and a general descriptive model in homogeneous tissue (Hoofd et al., 1989). The latter model also allows for the incorporation of two other phenomena: facilitated diffusion by myoglobin and a difference in O_2 partial pressure between "inside" and "outside" the capillary, referred to here as "Extraction Pressure" (Hoofd et al., this volume). While these features are also incorporated here, facilitation can be accounted for only in special cases.

MATHEMATICAL MODEL

The model presented here describes diffusion into a circular region from capillaries located at the outer border, combining features of two other models, the concentric model (Rakusan et al., 1984) and the generalized multi-capillary flat-plane model (Hoofd et al., 1989), and extended to account for different oxygen permeabilities in an outer and an inner zone. The outline is shown in the right panel of figure 1. Locations in the area are given in terms of the coordinate pair $\underline{r}=(r,\phi)$ of distance from the centre and angle. The inner and outer zones (numbered 0 and 1 here) are concentric and range from r=0 to r=tR and from r=tR to r=R respectively and can have different oxygen consumption M_o, M_1 and oxygen permeabilities (the product of diffusion coefficient and solubility) P_o, P_1. Capillaries are located anywhere on the outer border, characterized by the angle ϕ_k. Each capillary k can have a different radius r_{ck} and capillary oxygen pressure P_{ck}.

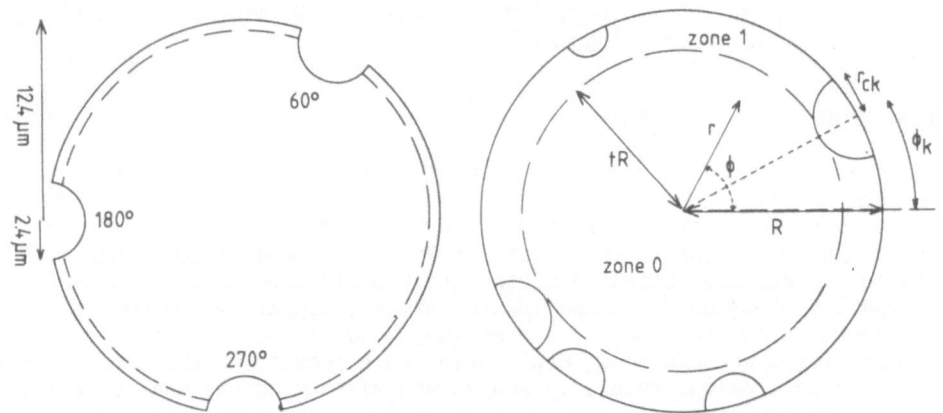

Fig. 1. Outline of the concentric heterogeneous model (right panel) and arrangement used in the first set of calculations (left panel). Capillaries are located on the outer border of a circular field consisting of two concentric zones.

Mathematical description

In principle, the mathematical solution of this system is derived from the general solution in a flat plane (Hoofd et al., 1989) as a combination of a field term and source terms, where the source terms coincided with the capillary locations. However, this cannot be maintained here due to the different oxygen permeabilities that the two zones may have. Mathematically, the boundary between these two zones acts as a "mirror", creating virtual sources at their "mirrored" locations also. For example, given a capillary located at $\underline{R}_k=(R,\phi_k)$, the zone boundary at r=tR will create a virtual image at (t^2R,ϕ_k). This process will be repeated at the outer border r=R, creating a second image at $(R/t^2,\phi_k)$, which in turn will be reflected at the inner border in an image at (t^4R,ϕ_k) and so on. An infinite series of images results in order to representat this one capillary and the same holds for the other ones.

The field term was the basic solution of the governing differential equation $\nabla^2 P=M$, where P is oxygen partial pressure, and was equal to $Mr^2/4P$ for a circular area. Here it still can be solved as radially symmetric, but

14

due to the different properties for both zones the full radial solution of the differential equation has to include a logarithmic term:

$$P = C_{Pz} + \frac{M_z}{4P_z} \left[r^2 + B_z \ln(r^2) - \sum_{i=1}^{N} \frac{A_{zi}}{\pi} F_{zi}(\underline{r}) \right] \tag{1}$$

where the index z denotes zone z, N is the number of capillaries and C_{Pz}, B_z and A_{zi} are constants to be solved from the boundary conditions. The functions F_{zi} represent the sum of mirrored source terms for the ith capillary in zone z:

$$F_{zi}(\underline{r}) = \sum_{j=n_z}^{\infty} \left[\frac{P_1 - P_0}{P_1 + P_0} \right]^{|j|} \ln(|\underline{r} - t^{2j}\underline{R}_i|^2) \tag{2}$$

where $n_z=0$ for z=0, the inner zone, and $-\infty$ for z=1, the outer ring. Note, that the infinite series is due to the difference in permeabilities P_z and that only the term j=0 remains if these are equal. The terms in eq.(2) arise from the constraints that both oxygen pressure P and oxygen flux $\underline{J}=-P_z\nabla P$ have to be continuous and finite (except at $\underline{R}_i$) in both zones and also across their boundary, at r=tR, and that there is no flux across the outer border, at r=R. Also, the following solutions and relationships are found:

$$B_0 = 0; \quad B_1 = R^2 \left[\frac{P_1}{P_0} \left\{ 1 - t^2(1 - \frac{M_0}{M_1}) \right\} - 1 \right] \tag{3}$$

$$\frac{M_0}{P_0} A_{0i} = \frac{2M_1}{P_0 + P_1} A_{1i} \tag{4}$$

$$\sum_{i=1}^{N} A_{1i} = 2\pi R^2 \left\{ 1 - t^2(1 - \frac{M_0}{M_1}) \right\} \tag{5}$$

and there is an equation relating C_{P0} to C_{P1}. The usual interpretation of the terms A_{zi} is that these are the oxygen supply areas of the respective capillaries, but here the interpretation is somewhat more difficult because of the infinite series of logarithmic terms. However, eq.(5) is independent of P_z and thus also holds for equal permeabilities where there is only one term instead of an infinite series. The right-hand term of eq.(5) is seen to be equal to twice the total consumption area if the inner zone is weighted accordingly; this is consistent with the fact that only half of each capillary lays in the defined circular region. Obviously, the virtual capillaries also have virtual supply areas, not contributing to the supply areas of the real capillaries.

The final solution of the terms C_{Pz} and A_{zi} comes from the boundary conditions that P must equal the capillary oxygen pressure P_{ck} for each capillary k, with radius r_{ck}, at the capillary border (so, strictly speaking P_{ck} is the oxygen partial pressure just outside the capillary). However, eq.(1) will show some variation in P along the outline of the capillary. There are several ways to deal with this problem in taking an "average" value; here, we adopt the same method as in the general flat-plane model by calculating the mean value of P over the full capillary outline, leading to:

$$P_{ck} = C_{P1} + \frac{M_1}{4P_1} \left[R^2 + r_{ck}^2 + B_1 \ln(R^2) - \sum_{i=1}^{N} \frac{A_{1i}}{\pi} F_{1i}(\underline{R}_k') \right] \tag{6}$$

15

where $\underline{R}_k'=\underline{R}_k$ for $k\neq i$ or $j\neq 0$ and $\underline{R}_k'=\underline{R}_k(1-r_{ck}/R)$ for $k=i,j=0$ (resulting in a value $\ln(r_{ck}^2)$ there instead of the impossible value $\ln(0)$).

With the above set of equations, any situation, for any combination of capillary pressures and capillary radii, can be solved.

Implementation

A computer program was written that calculates P for any rectangularly spaced grid within the circular region (TWOCON). The features of this program are:
- The input data of consumption and permeability are entered as a mean value of M/P for the whole region, and as ratios M_1/M_0 and P_1/P_0. This facilitates a correct comparison between different situations, since the mean value should reflect the overall measured situation and be equal for any set of different ratios of M and P. The program calculates the individual values M_0, M_1, P_0 and P_1, taking into account that the capillaries do not consume oxygen;
- Capillary pressures can be entered as "inside" values and the individual "outside" values P_{ck} are calculated from the mean difference for all capillaries. This difference was defined earlier as "Capillary Barrier" (Hoofd et al., 1989) or as "Equivalent Pressure Difference" (Turek et al., 1989) but a better name may be "Extraction Pressure" (Hoofd et al., this volume) since it is the driving force needed to get the oxygen from inside to just outside of the capillary. A mean value of this must be given;
- A facilitation pressure can be given, P_F (de Koning et al., 1981), to account for myoglobin-facilitated oxygen diffusion. However, this is only valid if either P_F is the same for both zones (which is unlikely) or myoglobin saturation at the interface between the two zones is close to 1. Otherwise, the boundary conditions for P at r=tR can no longer be satisfied.

The program is written in C and runs on an IBM compatible or Atari ST computer.

RESULTS

The most important feature of the present method is that it allows an investigation of the effects of tissue heterogeneity on oxygen transport in tissue in a simplified model. Contrary to former concentric diffusion models, where capillaries had to be equal and evenly spaced, differing capillary pressures and distances are possible here. The simplest situation is with 3 capillaries, here located at 60°, 180° and 270° in order to have different intercapillary distances; see figure 1, left panel. Different capillary oxygen pressures were chosen to investigate the possible effects of oxygen diffusion shunting: the capillary at 270° was 30 mmHg lower. Region radius and capillary radius were 12.4 μm (10.1x√1.5, for 3x½ capillary) and 2.4 μm respectively (Turek et al., 1986), and M/P was 0.19 mmHg/μm² (M of Bourdeau-Martini et al., 1974; P of Krogh, 1919). An outer zone comprising 8% of the total area could have a permeability 4.4 times that of the inner zone (Egginton and Sidell, 1988).

The situations with equal and different permeabilities ratios are compared in figure 2, for equal consumption in both layers and for capillary P_{O_2}'s of 50 and 20 mmHg, respectively, in two different representations; in the top panels, P_{O_2} is shown vertically for a grid of equidistant points in the circular area (axonometric plot) and in the bottom panels lines are drawn connecting points of equal P_{O_2} (isopleths). In the axonometric plots, capillaries are shown as the shaded (semi-)circular areas, for the respective capillary P_{O_2} values; in the isopleths these capillaries are at the

16

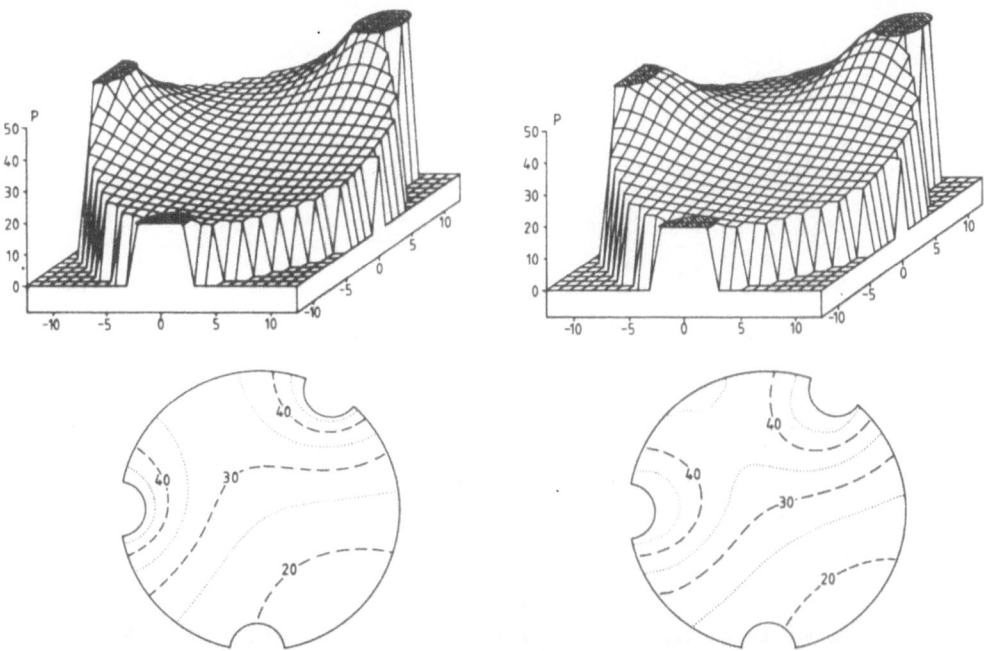

Fig. 2. Oxygen partial pressure for the 3-capillary situation (see text) with $P_1/P_o = 1$ (left panels) and 4.4 (right panels), without changing overall permeability. Top panels: axonometric plot of P_{O_2} against location; bottom panels: P_{O_2} isopleths.

isopleth values of 50 and 20 mmHg respectively. The left panels show the situation for a homogeneous diffusivity over the whole area, $P_1/P_o = 1$; the right panels for $P_1/P_o = 4.4$. In both situations consumption was taken equal in both zones, $M_1/M_o = 1$. The differences are better depicted in the isopleth plots but are marked especially when one realizes that the **overall** permeability was the same for both situations. The thin (8% area; compare left panel of figure 1) outer high-permeable zone has the effect of guiding O_2, into the large inner zone, thereby raising the P_{O_2} at almost all locations. This rise is best seen in the axonometric plot close to the capillary, where the slope of P_{O_2} is less steep and thus the O_2 flux is reduced towards the inner zone, which is compensated by a higher flux (less steep slope but much higher permeability) into the outer zone, and later redistributed to the inner zone. Although the inner zone permeability is only 79% of the overall value and thus 21% lower than in the homogeneous case, this is more than compensated by the guiding effect, so that P_{O_2} in the centre is about 5 mmHg higher than in the homogeneous case.

Compared to this, the influence of heterogeneity in oxygen consumption was almost absent. Calculation of situations where M_1/M_o was 0.05 instead of 1 showed almost no difference compared to the above cases. The oxygen is hardly consumed in the outer zone, but this only slightly raises P_{O_2} there since most of the gradient is needed for the transport to the large inner zone; there it is consumed at a somewhat higher rate (ca. 8%), decreasing P_{O_2} and resulting in almost the same values in the inner zone.

It is often suggested that concentric diffusion, from sources outside inward into a muscle fibre, might be beneficial as compared to eccentric diffusion, from a central capillary into the surrounding area like in the Krogh model (Groom et al., 1984; Piiper and Scheid, 1986), because the drop in oxygen partial pressure from the capillary to the most distant locations

17

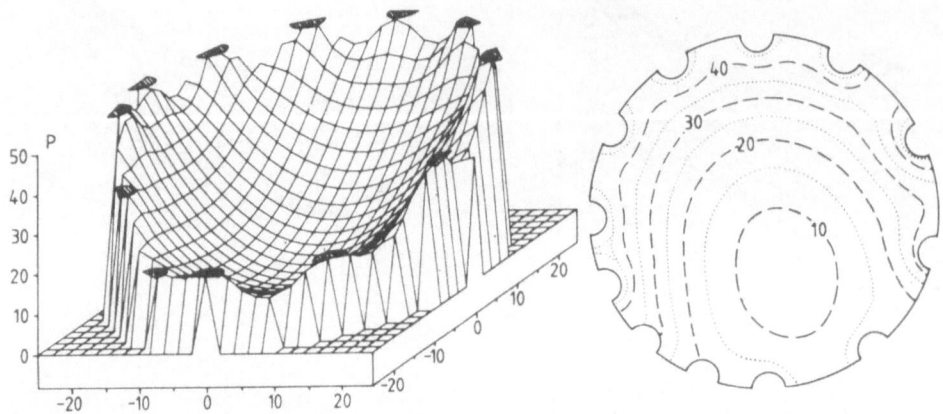

Fig. 3. Po_2 calculated for an area of 24.8 µm radius surrounded by 12 capillaries. Left panel: axonometric plot of Po_2 against location in the plane; right panel: Po_2 isopleths.

would be less. However, one must realize that the tissue properties are a given and that correct comparison between different models should be performed for the same overall data. So, in the above examples the overall values of M and P were taken identical; here, we may compare different numbers of capillaries on the surrounding outer border, but with the same number of capillaries per tissue region, i.e., the same capillary density.

In figure 3, a situation is shown with a fourfold number of capillaries compared to the above case, 12 instead of 3, regularly spaced in between these 3, i.e., at $37\frac{1}{2}°$, 60°, 90°, 120°, 150°, 180°, $202\frac{1}{2}°$, 225°, $247\frac{1}{2}°$, 270°, $307\frac{1}{2}°$ and 345°. Capillary P_{O_2}'s were chosen equal to the neighbouring capillary or half-way for the capillaries added in between; so, the first seven capillary P_{O_2}'s are 50 mmHg, those at 225° and 345° are 35 mmHg and the remaining three are 20 mmHg. In order to have the same capillary density the area also must be four times larger resulting in a doubled radius of 24.8 µm. Again, P_{O_2} is shown in an axonometric and an isopleth plot, with the capillaries represented as shaded semicircles in the axonometric plot and delineated by their respective P_{O_2} lines in the isopleth plot. The most striking difference against the 3-capillary case is, that P_{O_2} falls to lower values and that these values are reached inside the area instead of at the border. So, increasing muscle fibre radius while keeping capillary density constant (as it might occur in contracting muscle) is not advantageous for the tissue.

DISCUSSION

The most important feature of model presented here is, that it offers the opportunity to study heterogeneity of permeability in tissue. Although a Krogh cylinder could also be modelled with different layers, this would be concentric around the capillary. The situation here, where, e.g., the inner zone could resemble a muscle fibre and the outer zone connective tissue and intercellular space, seems more adequate. Heterogeneity of oxygen consumption in a similar model was presented earlier (Rakusan et al., 1984) and found to have little influence on tissue P_{O_2}, leading to almost the same P_{O_2} distribution as with the Krogh model.

Another interesting feature is the possibility to have different cap-

18

illary P_{O_2}'s and irregular capillary spacing. In the 3-capillary model, the capillary at 270° had a lower P_{O_2} and was closer to the 180°-capillary than to the 60°-capillary (distances of 17.54 and 23.95 µm respectively). This results in an inward flux from the 180°-capillary, as seen clearly from the slope at the left in the axonometric plots of figure 2; this might be compared with oxygen shunting from one capillary to another. More oxygen flows out at the other sides, so that the capillary still has a net oxygen supply area of 2.4% of the total area. It is remarkable that the shunt is almost independent of the permeability of the connecting region; it makes hardly any difference if the zone 1, in between, has a 4.4 times higher permeability than the inner zone or not. Obviously, most of the oxygen goes to the inner zone and the shunt is only a side-effect.

The model also allows for incorporation of an extraction pressure (Hoofd et al., this volume). This is defined as the pressure difference between the O_2 supplying item (e.g., the RBC in the capillary) and the rim of the capillary, where diffsuion into the tissue starts (P_{ck}). For a mean extraction pressure of 10 mmHg and capillary P_{O_2}'s of 65 mmHg (60°, 180°) and 20 mmHg (270°) we get approximately the same P_{ck}'s and also similar isopleths as in figure 2.

In our opinion, the way the 12-capillary case was constructed offers the correct comparison for different capillary-to-fibre ratios. Although inward diffusion is easier, with less drop in P_{O_2} (Piiper and Scheid, 1986), the increasing distance will lead to an ever larger drop in P_{O_2} so that the net effect is disadvantageous. A regular 3-capillary case is consistent with a hexagonal arrangement and thus comparable with a Krogh model (Rakusan et al., 1984).

The model presented here has to much features to be covered here, so that only some representative situations were explicitly handled. These were chosen to elucidate the findings as mentioned above. In the computer program, besides the capillary data only values of M/P, R and ratios of M, P and R are requested so that it is easy to use.

The computer program runs on IBM PC and on Atari ST and is available on request.

REFERENCES

Bourdeau-Martini, J., Odoroff, C.L., and Honig, C.R., 1974, Dual effect of oxygen on magnitude and uniformity of coronary intercapillary distance, Am. J. Physiol., 226: 800-810.

Egginton, S., Sidell, B.D., 1989, Thermal acclimation induces adaptive changes in subcellular structure of fish skeletal muscle, Am. J. Physiol., 256: R1-R9.

Groom, A.C., Ellis, C.G., and Potter, R.F., 1984, Microvascular geometry in relation to modeling oxygen transport in contracted muscle, Am. Rev. Respir. Dis., 129: S6-S9.

Hoofd, L., Turek, Z., and Olders, J., 1989, Calculation of oxygen pressures and fluxes in a flat plane perpendicular to any capillary distribution, in: "Oxygen Transport to Tissue XI," K.Rakusan, G.Biro, T.K.Goldstick, and Z.Turek, eds., Plenum Press, New York and London, pp. 187-196.

Hoofd, L., Olders, J., and Turek, Z., this volume, Oxygen pressures calculated in a tissue volume with parallel capillaries.

de Koning, J., Hoofd, L.J.C., and Kreuzer, F., 1981, Oxygen transport and the function of myoglobin: theoretical model and experiments in chicken gizzard smooth muscle, Pflügers Arch., 389: 211-217.

Krogh, A., 1919, The number and distribution of capillaries in muscles with calculations of the oxygen pressure head necessary for supplying the tissue, J. Physiol., 52: 409-415.

Piiper, J., and Scheid, P., 1986, Cross-sectional P_{O_2} distributions in Krogh cylinder and solid cylinder models. Respir. Physiol., 64: 241-251.

Rakusan, K., Hoofd, L., and Turek, Z., 1984, The effect of cell size and capillary spacing on myocardial oxygen supply, in: "Oxygen Transport to Tissue VI," D.F.Bruley, H.I.Bicher, and D.Reneau, eds., Plenum Press, New York and London, pp. 463-477.

Turek, Z., Olders, J., Hoofd, L., Egginton, S., Kreuzer, F., and Rakusan, K., 1989, P_{O_2} histograms in various models of tissue oxygenation in skeletal muscle, in: "Oxygen Transport to Tissue XI," K.Rakusan, G.Biro, T.K.Goldstick, and Z.Turek, eds., Plenum Press, New York and London, pp. 221-237.

Turek, Z., Hoofd, L., and Rakusan, K., 1986, Myocardial capillaries and tissue oxygenation, Can. J. Cardiol., 2: 98-103.

OXYGEN PRESSURES CALCULATED IN A TISSUE VOLUME WITH PARALLEL CAPILLARIES

Louis Hoofd, Jos Olders, and Zdenek Turek

Dept. Physiology, University of Nijmegen, Nijmegen, The Netherlands

INTRODUCTION

Recently, we presented a model of oxygen diffusion in a flat plane with arbitrary capillary distribution (Hoofd et al., 1989). Capillary locations were read in from a tissue photomicrograph, and P_{O_2} was calculated in the flat plane of the cross section, for a given set of input parameters, assuming only two-dimensional diffusion.

Muscle tissue with its highly parallel capillarization can be sliced perpendicular to the capillaries, where the prominent P_{O_2} gradients are supposed to be in these slices and not along the capillaries. This allows for calculation of P_{O_2} in such a muscle tissue by consecutive planes, where the capillary P_{O_2} in the next plane follows from the amount extracted in each plane. This model is worked out here.

The model assumes homogeneous oxygen consumption, following zero-order kinetics, uniform permeability of the tissue and parallel capillaries. All other parameters, including entrance and end of the capillary, can be chosen. Allowance can be made in a formal way for a difference in P_{O_2} between inside the capillary and just outside, where the diffusion into the tissue commences; this was formerly called "Capillary Barrier" (Hoofd et al., 1989) or "Equivalent Pressure Difference" (Turek et al., 1989); here we propose "Extraction Pressure" as a better term. Facilitated diffusion by myoglobin in the tissue is also incorporated.

MODELLING

The model presented here is for a piece of tissue with parallel capillaries, a cylinder or a rectangular block, with the axis parallel to the capillaries; so, cross sections perpendicular to the axis have an identical capillary arrangement, which may be chosen arbitrarily. Oxygen is transported in the tissue by diffusion, with oxygen permeability P (the product of diffusion coefficient D_{O_2} and solubility α), and consumed by zero-order reaction, with reaction rate M; both P and M do not depend on location (homogeneous tissue). Oxygen comes from the (blood in the) capillaries which may have their begin and end anywhere and have different O_2 properties (initial oxygen pressure, hematocrit, etc.). A similar rectangular arrangement was calculated by Ellsworth et al. (1988) by numerical tech-

Oxygen Transport to Tissue XII, Edited by J. Piiper *et al.*
Plenum Press, New York, 1990

niques, for a slightly different set of describing equations. Here, an analytical approximate solution is presented under the assumptions that: 1) axial diffusion is of negligible influence and 2) capillaries in each cross sectional plane can be represented ("seen" from outside the capillary) as point sources of oxygen. Facilitated diffusion by myoglobin can be incorporated under the additional assumption that myoglobin saturation is close to equilibrium with local oxygen partial pressure P_{O_2} and that myoglobin distribution is uniform.

Two-dimensional solution

The basis for the mathematical modelling is the generalized approximate solution for a homogeneous flat plane with arbitrary capillary distribution derived earlier (Hoofd et al., 1989). At any location $\underline{r}$ in the plane, the equation from which P_{O_2} is obtained can be written as:

$$P_{O_2} + P_F S_{MbO2} = C_P + \frac{M}{4P} \left[Fld(\underline{r}) - \sum_{i=1}^{N} \frac{A_i}{\pi} \ln(|\underline{r}-\underline{R}_i|^2) \right] \tag{1}$$

where $P_F S_{MbO2}$ is the myoglobin contribution to oxygen diffusion (facilitated diffusion), P_F is facilitation pressure (de Koning et al., 1981), S_{MbO2} is myoglobin oxygen saturation, N is the number of capillaries and $\underline{R}_i$ is the ith capillary location. C_P and A_i are constants where A_i is the oxygen supply area of the ith capillary; these constants have to be solved from the boundary conditions being: 1) the total area A of the plane is equal to the sum of all supply areas A_i and 2) each capillary oxygen pressure P_{ci} is equal to the average value of eq.(1) over the capillary outline. The equation was derived for a circular region, in where the "background term" $Fld(\underline{r})$ was solved as:

$$Fld(\underline{r}) = |\underline{r}|^2 \tag{2}$$

In addition to the original solution, this term also can be derived for other geometries; for a rectangular plane with coordinates $\underline{r}=(x,y)$, the origin in the centre and width, length (w,l):

$$Fld(\underline{r}) = x^2+y^2 + \frac{1}{\pi} \sum_u \sum_v \left[uv \cdot \ln(u^2+v^2) - \frac{u^2-v^2}{2} \, arctg\left[\frac{u^2-v^2}{2uv} \right] \right] \tag{3}$$

where the summations go over $u=\frac{1}{2}w-x$, $\frac{1}{2}w+x$ and $v=\frac{1}{2}l-y$, $\frac{1}{2}l+y$.

Three-dimensional extension

The two-dimensional model can be extended into the third dimension, coordinate z, if diffusion perpendicular to the plane is unimportant, in the same way as derived by Kety (1957) for the classical Krogh cylinder. A necessary condition for this is, that the capillaries run in parallel perpendicular to the plane. Then, the planes can be piled to form a cylinder (for the circular region) or a block (for the rectangular one). Blood flows in from one side, at z=0, and in each consecutive slab of thickness δz the amount of oxygen fetched from the capillary is subtracted from its oxygen content so that the capillary P_{O_2}'s of the following slab can be calculated. With a capillary radius r_{ci}, the volume of blood in this part of the capillary is $V_{ci}=\pi r_{ci}^2 \delta z$ and the amount of oxygen available is $C_{O_2}V_{ci}$ where C_{O_2} is the total oxygen concentration (bound+free). This amount of oxygen is available during the time of passage δt through the slab, which is equal to V_{ci}/F_i, where F_i is the capillary blood flow, and in this time must supply a tissue volume $(A_i-\pi r_{ci}^2)\delta z$ (total minus capillary volume). So, the change in oxygen concentration is found by subtracting this amount:

22

$$\delta C_{O_2} \cdot V_{ci} = -M \cdot (A_i - \pi r_{ci}^2) \cdot \delta z \cdot \delta t \tag{4}$$

which, with the above substitutions, can be worked out as:

$$\frac{\delta C_{O_2}}{\delta z} = -\frac{M}{F_i}(A_i - \pi r_{ci}^2) \tag{5}$$

Total oxygen concentration is the sum of free dissolved gas and oxygen bound to the hemoglobin in the blood:

$$C_{O_2} = C_{Hm}S_{HbO2} + \propto P_{O_2} \tag{6}$$

where C_{Hm} is the heme concentration (four times the hemoglobin concentration) and S_{HbO2} is the hemoglobin oxygen saturation. Note that $\propto$ and P_{O_2} here are of the capillary blood.

Additional considerations

To obtain P_{O_2} from eq.(6), we need an expression for the blood oxygen saturation curve (S_{HbO2} against P_{O_2}). The oldest and simplest relation is that of the Hill equation:

$$S_{HbO2} = \frac{(P_{O_2}/P_{50})^n}{1 + (P_{O_2}/P_{50})^n} \tag{7a}$$

where P_{50} is the half-saturation pressure and n is the Hill coefficient. This description is adequate down to ca. 10% saturation but below this the saturation is underestimated. Therefore, for the lower part we use a modified Pauling model (Bouwer, 1987):

$$S_{HbO2} = \frac{t^3(P_{O_2}/P_{50})+3t^4(P_{O_2}/P_{50})^2+3t^3(P_{O_2}/P_{50})^3+(P_{O_2}/P_{50})^4}{1+4t^3(P_{O_2}/P_{50})+6t^4(P_{O_2}/P_{50})^2+4t^3(P_{O_2}/P_{50})^3+(P_{O_2}/P_{50})^4} \tag{7b}$$

with the parameter t so that the slopes are equal for $P_{O_2}=P_{50}$, resulting in $n=4(1+t^3)/(1+4t^3+3t^4)$. By applying eq.(7a) for $P_{O_2}>P_{50}$ and eq.(7b) for $P_{O_2}<P_{50}$ we have, with only two parameters, a good description of the whole saturation curve.

Another concern is the oxygen consumption M. This value is determined experimentally for a whole tissue including the capillaries (M_{av}), but in the description here it is the consumption outside of the capillaries; as is most clearly seen from eq.(5). So, the consumption term M in eqs.(1), (4) and (5) is taken higher such that the overall value equals the experimental given one:

$$M = \frac{M_{av}}{1 - \frac{1}{A}\sum_{i=1}^{N}\pi r_{ci}^2} \tag{8}$$

RESULTS

The above mathematical description can be applied to a great number of different situations, since any capillary parameter can be specified separately. In this paper, we were mainly interested in capillary heterogeneities, so two exemplary situations were calculated as shown in figure 1.

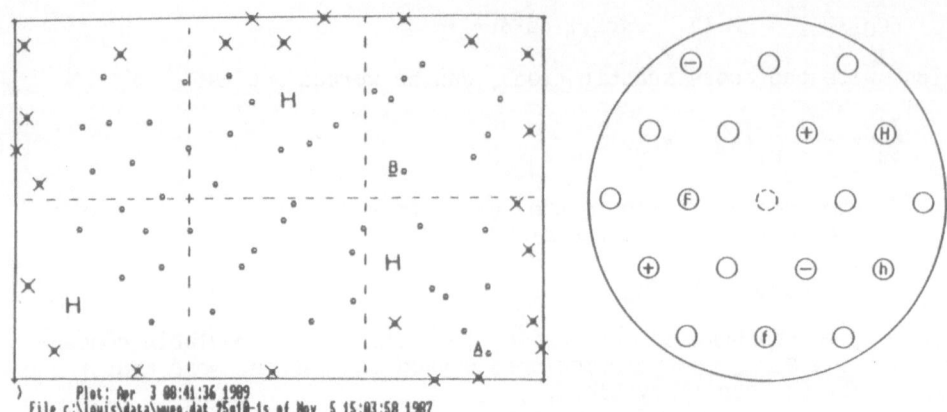

Plot: Apr 3 08:41:36 1989
File c:\louis\data\wupo.dat 25g18-1s of Nov 5 15:03:58 1987

Fig. 1. Cross sections of the two situations calculated: rat heart tissue (computer replot; left panel), cross section of a tissue block with heterogeneous capillarity, and regular hexagonal arrangement (right panel), cross section of a tissue cylinder. Capillaries are indicated as circles. About the other indications in the panels see text.

The left panel shows a computer representation of capillary locations in a cross section of rat heart tissue; the right one shows 19 capillaries filling a circular field in a regular hexagonal arrangement. In the first case, heterogeneity in capillary distribution was studied, with all the other parameters equal; in the second case with equal intercapillary distances, the heterogeneity in capillary length, flux, hematocrit and (initial) P_{O_2} were studied. The calculation step length δz was 5 µm in all cases; checks against other step lengths (10 µm and 20 µm) indicated that this step was short enough.

The rat heart tissue was 177.5 µm (width) x 140.4 µm (length) x 500 µm (depth), covering 71 capillaries and selected to adequately represent mean and distribution of capillary distances of the whole tissue (Turek et al., 1986). M/P was taken 0.19 mmHg/µm² (M of Bourdeau-Martini et al., 1974; P of Krogh, 1919). Myoglobin data (all data for 37°C) were P_m= 14 mmHg (Mb concentration of Turek et al., 1973; D_{O_2}/D_{Mb} of Federspiel, 1986) and P_{50}= 5.3 mmHg (Gayeski, 1981). $\propto F/P$ was 10.56 µm (oxygen solubility $\propto$ of Handbook of Chemistry and Physics; F of Rakusan and Blahitka, 1974). Blood data were $C_{Hm}/\propto$= 6333 mmHg (O_2 capacity of 20ml/100ml) and P_{50}= 37 mmHg, n= 2.7 (Turek et al., 1973). Capillary radius was 2.4 µm (Turek et al., 1986); all capillaries had the same flow and hematocrit, started at the same capillary P_{O_2} of 100 mmHg and followed the full axis length of 500 µm. Results are shown in figure 2. Mean values and variance were calculated excluding the 24 capillaries close to the border of the picture (determined by their domains (Hoofd et al., 1985) touching the border), crossed out in figure 1. Due to the different intercapillary spacing, supply areas are different; this causes capillaries with larger areas to loose more oxygen, resulting in a lower capillary P_{O_2} so in a smaller supply area in the next plane, neighbouring capillaries taking over some of the supply. This effect, however, is limited; after the first 30 µm, supply areas hardly change. Two extreme capillaries were also followed, indicated A (small supply area, so high P_{O_2}) and B (large area, low P_{O_2}) in figure 1, left panel. Towards the capillary ends, supply areas become more homogeneous and capillary P_{O_2}'s more heterogeneous but only slightly; e.g., variance in P_{O_2} is 4.0 mmHg at 250 µm and 4.8 mmHg at 500 µm.

24

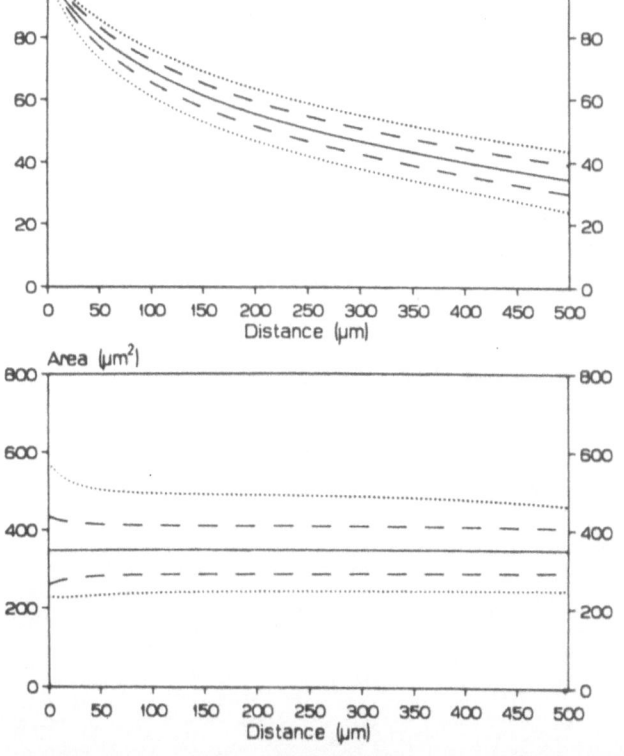

Fig. 2. Capillary P_{O_2} (top) and capillary O_2 supply areas (bottom) followed along the capillary length for the rat heart example where all capillaries start at a P_{O_2} of 100 mmHg. Solid lines: mean value; dashed: ±variance; dotted: extremes.

The choice of all capillaries starting at a P_{O_2} of 100 mmHg seems not to be realistic and therefore also a situation was calculated dividing the tissue area in six parts of equal size, as shown by the dashed lines in the left panel of figure 1; the capillaries in the parts indicated with an H now started at a P_{O_2} of 100 mmHg whereas the other ones were given P_{O_2}'s according to the distribution that the other capillaries reached at 250 μm ("half-way"). This simulates staggering, where groups of capillaries have their arterial supply at different levels. The results, again mean and variance for the non-border capillaries, are shown in figure 3. The two groups clearly differ also in supply areas; these are quite separate in the beginning but tend to become more homogeneous towards the capillary ends. Most remarkable, however, is how the course of capillary P_{O_2}'s resembles that of the former case, figure 2; in spite of the different supply areas mean P_{O_2}'s at 250 μm were quite close, 46.4±5.0 mmHg here and 50.6±4.0 mmHg in the former case.

In the hexagonal arrangement, right panel of figure 1, capillary separation was identical, 20 μm, but the other factors varied. Cylinder radius was 45.77 μm (19 capillaries); the other data were the same as above. The capillaries indicated with a + started at 100 mmHg, those with a - at 50 mmHg and the rest at 75 mmHg; F means a 40% higher flow, f a 30% lower

25

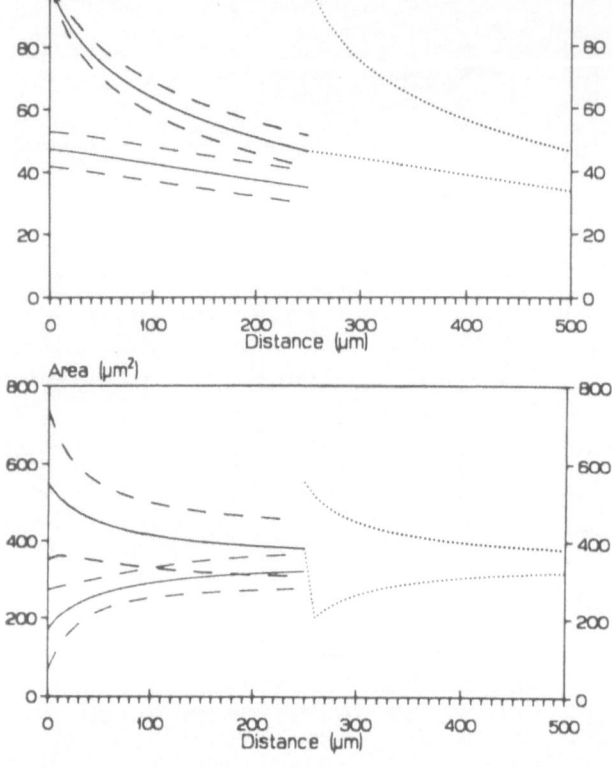

Fig. 3. Effect of staggering on capillary P_{O_2} (top) and capillary O_2 supply areas (bottom) for the rat heart example where the capillaries in three of the six areas of figure 1 start at a P_{O_2} of 100 mmHg and the rest is already "half-way". Solid lines: mean values; dashed: ±variance; dotted: extended mean values after 250 μm, where the low-P_{O_2} group ends and is replaced by new capillaries, with P_{O_2}= 100 mmHg.

flow; H means a 40% higher hematocrit, h a 30% lower hematocrit; the central capillary not runs over the full axis length of 500 μm but only between 100 μm and 350 μm. The results are shown in figure 4. Capillaries with equal flow and hematocrit tend towards the same P_{O_2} and supply area, especially for the high-P_{O_2} capillaries, but divergence results if flow or hematocrit are different (the H and h cases are not shown separately because their P_{O_2} is almost equal to the F and f capillaries respectively). Note that at 100 μm, where the central capillary comes in, supply areas are decreased but almost not for the capillaries in the outer ring (e.g., capillary f, lower dotted line); obviously, these are "shielded off" from the centre quite effectively by the inner ring.

DISCUSSION

From model calculations considered so far, the following conclusions can be drawn:

Capillaries with large supply areas or low flow (or hematocrit) loose much oxygen, resulting in a low P_{O_2} as compared to the other capillaries. This decreases their O_2 supply areas, neighbouring capillaries taking over

26

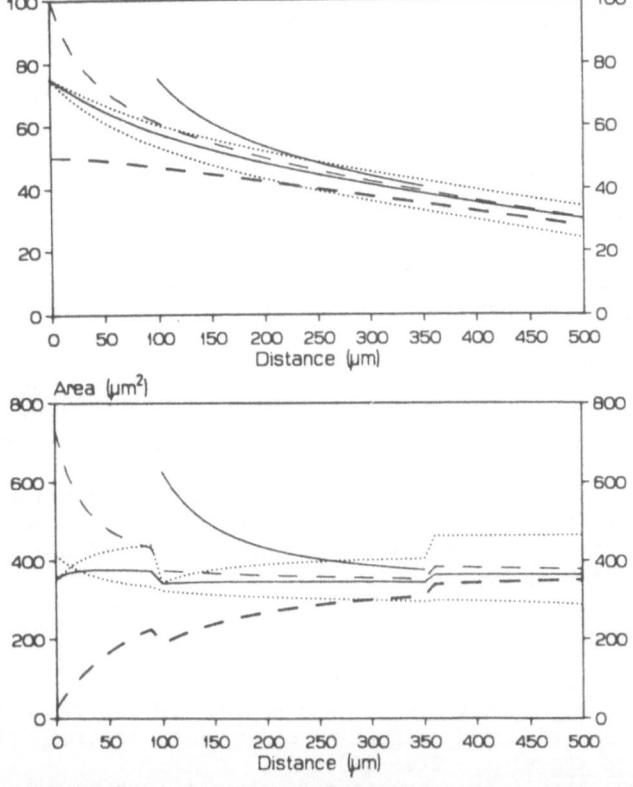

Fig. 4. Capillary P_{O_2} (top) and capillary O_2 supply areas (bottom) followed along the capillary length for the hexagonal arrangement. Solid lines: central capillary (from 100 - 350 μm) and mean value for the "normal" capillaries; dashed: different initial P_{O_2}; dotted: different flow.

part of the supply. Consequently, they loose less O_2 further on and capillary P_{O_2} decrease becomes some more homogeneous.

This phenomenon only works with nearby capillaries. If something happens a bit farther away, this is hardly "seen"; in the hexagonal arrangement, adding a new capillary at high P_{O_2} only 1.7 times the mean intercapillary distance away had little effect on O_2 supply area and thus on O_2 loss. Consequently also, on a larger scale, regions of several capillaries that are poorly supplied will go down in oxygen more rapidly, not or hardly compensated by other well-supplied regions.

For what happens in the tissue a bit more consideration is needed. In each cross-sectional plane, tissue P_{O_2} can be calculated also but we will have to account for a difference in P_{O_2} inside and just outside the capillary, where diffusion into the tissue starts. In this small region, and especially near the red blood cells, O_2 delivery to the tissue commences and there is a quite large pressure gradient needed to drive this initial transport. The resulting pressure difference was accounted for in the flat-plane model (Hoofd et al., 1989) by defining a mean value for all capillaries and calculating the individual drop from the supply area. The P_{O_2} drop was called "Capillary Barrier" but here we propose "Extraction Pressure" as a better term, since it is the pressure needed already to get the oxygen out of the capillary anyway; after that, P_{O_2} will decrease further into the supply area. The formal representation by an "Extraction Pressure" leaves its value to be determined by separate other methods (Popel, 1989). The tissue P_{O_2} of course will depend critically on the value of the extraction pressure, as shown earlier (Hoofd et al., 1989), but has only little effect on the course of the capillary P_{O_2}'s; recalculating the above cases with a mean extraction pressure of 10 mmHg only gave slightly different results, although the supply areas were somewhat more different, consistent with earlier findings (Hoofd et al., 1989).

The relative importance of neglecting axial diffusion can also be estimated. The describing differential equation, $P\nabla^2 P_{O_2}=M$, is solved here for a two-dimensional ∇^2, neglecting the term $\partial^2/\partial z^2$ of the three-dimensional ∇^2. So, the contribution of axial diffusion can be estimated by comparing $\partial^2 P_{O_2}/\partial z^2$, or here $\delta P_{O_2}/\delta z^2$, with M/P. This value is largest at high P_{O_2} where hemoglobin is only little and alinearly desaturated; for the rat-heart case with uniform $P_{O_2}=100$ mmHg we calculate 0.0084 mmHg/μm^2 in the first slab, 4.4% of M/P. This value decreases rapidly; at z=250 μm a value of 0.0003 remains, only 0.15% of M/P. So, axial diffusion indeed seems unimportant, except maybe in special situations, e.g., when a new capillary enters the system.

We believe, that the above model can give a good description of striated muscle tissue P_{O_2}, provided a correct description of capillary entrance points, hematocrits and flow distribution is available as well as a good estimate of the extraction pressure.

REFERENCES

Bourdeau-Martini, J., Odoroff, C.L., and Honig, C.R., 1974, Dual effect of oxygen on magnitude and uniformity of coronary intercapillary distance, Am. J. Physiol., 226: 800-810.

Bouwer, S., 1987, Facilitated oxygen diffusion through hemoglobin solutions. Dissertation Thesis, Univ. Nijmegen, The Netherlands.

Ellsworth, M.L., Popel, A.S., and Pittman, R.N., 1988, Assessment and impact of heterogeneities of convective oxygen transport parameters in capillaries of striated muscle: experimental and theoretical, Microvasc. Res., 35: 341-362.

Federspiel, W.J., 1986, A model study of intracellular oxygen gradients in a myoglobin-containing skeletal muscle fiber, Biophys. J., 49: 857-868.

Gayeski, T.E.J., 1981, A cryogenic microspectrophotometric method for measuring myoglobin saturation in subcellular volumes; application to resting dog gracilis muscle. Thesis, Univ. Rochester, NY, USA.

Handbook of Chemistry and Physics, ed. 44, 1963, The Chemical Rubber Co., Cleveland, Ohio, USA.

Hoofd, L., Turek Z., Kubat K., Ringnalda B.E.M., Kazda S., 1985, Variability of intercapillary distance estimated on histological sections of rat heart. in: "Oxygen Transport to Tissue VII," F.Kreuzer, S.M.Cain, Z.Turek, and T.K.Goldstick, eds., Plenum Press, New York and London, pp. 239-247.

Hoofd, L., Turek, Z., and Olders, J., 1989, Calculation of oxygen pressures and fluxes in a flat plane perpendicular to any capillary distribution, in: "Oxygen Transport to Tissue XI," K.Rakusan, G.Biro, T.K.Goldstick, and Z.Turek, eds., Plenum Press, New York and London, pp. 187-196.

Kety, S.S., 1957, Determinants of tissue oxygen tension, Fed. Proc., 16: 666-670.

de Koning, J., Hoofd, L.J.C., and Kreuzer, F., 1981, Oxygen transport and the function of myoglobin: theoretical model and experiments in chicken gizzard smooth muscle, Pflügers Arch., 389: 211-217.

Krogh, A., 1919, The number and distribution of capillaries in muscles with calculations of the oxygen pressure head necessary for supplying the tissue, J. Physiol., 52: 409-415.

Popel, A.S., 1989, Theory of oxygen transport to tissue, Crit. Revs. Biomed. Engng., 17: 257-321.

Rakusan, K., Blahitka, J., 1974, Cardiac output distribution in rats measured by injection of radioactive microspheres via cardiac puncture. Can. J. Physiol. Pharmacol., 52: 230-235.

Turek, Z., Ringnalda, B.E.M., Grandtner, M., and Kreuzer, F., 1973, Myoglobin distribution in the heart of growing rats exposed to a simulated altitude of 3500 m in their youth or born in the low pressure chamber. Pflügers Arch., 340: 1-10.

Turek Z., Hoofd L., Rakusan K., 1986, Myocardial capillaries and tissue oxygenation. Can. J. Cardiol., 2: 98-103.

Turek Z., Olders J., Hoofd L., Egginton S., Kreuzer F., Rakusan K., 1989, PO_2 histograms in various models of tissue oxygenation in skeletal muscle. in: "Oxygen Transport to Tissue XI," K.Rakusan, G.Biro, T.K.Goldstick, and Z.Turek, eds., Plenum Press, New York and London, pp. 227-237.

DEPENDENCE OF CEREBRAL CAPILLARY HEMATOCRIT ON RED CELL FLOW

SEPARATION AT BIFURCATIONS: A COMPUTER SIMULATION STUDY

Antal G. Hudetz

Department of Physiology
Medical College of Wisconsin
Milwaukee, Wisconsin, USA

INTRODUCTION

Previously, we have attempted to estimate hemodynamic parameters of the cerebrocortical microcirculation based on complete geometrical and topological information of a cortical microvascular network (Hudetz et al, 1989). An important element of the calculations is the model of partitioning of red blood cells at vascular bifurcations which primarily determines local vessel hematocrit. The preference of erythrocytes to enter branches with higher flow has been reported and mathematically modeled (Schmid-Schoenbein et al, 1980; Papenfuss and Gross, 1981; Klitzman and Johnson, 1982; Fenton et al, 1985; Levin et al, 1986; Secomb et al, 1989; etc.). The influence of this effect on the calculated hematocrit distribution has not been evaluated systematically, except the recent abstract by Hsu and Cokelet (1989).

The objective of the present work was to evaluate the significance of preferential red cell flow parameter, B, as defined by Klitzman and Johnson (1982) on the hematocrit distribution in cerebral microvascular networks. The nature of capillary hematocrit distribution and its dependence on vascular architecture and hemodynamic conditions is especially important in the brain, in light of the precisely controlled but often vulnerable cerebral oxygen supply-demand relationship. A new, larger cerebrocortical network was used in this study.

EXPERIMENTAL METHODS

The cerebrocortical microvascular network was mapped as described previously (Hudetz et al, 1989). Briefly, a vascular corrosion cast of the rat brain was prepared using Batson's #17 compound following perfusion fixation. 0.5 to 1 mm thick coronal sections of the tissue were cut. Following maceration of the cerebral tissue, the cast was dehydrated and was embedded in epoxy. The topology and geometry of the cerebrocortical microvascular network was reconstructed using a computer-aided optical sectioning video-microscope system. In essence, the 3-dimensional course of vessels was traced on the video image of the cast by a superimposed cursor and by adjusting the elevation of the microscope stage to keep the vessel segment at the cursor in focus. Vessel diameters were also measured off the video image at several locations along each vessel.

Oxygen Transport to Tissue XII, Edited by J. Piiper *et al.*
Plenum Press, New York, 1990

MATHEMATICAL METHODS

A first approximation of blood flow distribution in the cerebrocortical microvascular network was calculated assuming constant blood viscosity. The hydraulic resistance of each capillary was estimated from its mean diameter and length based on Poiseuille flow. Equations for flow were written for each capillary segment, and at each branch point the flow balance was written as a second equation. At the arterial and venous ends constant pressure boundary conditions were used. At broken ends of capillaries, estimated mean capillary pressure prescribed as a boundary condition. The resulting system of equations was solved by matrix inversion. The effect of local variations in apparent blood viscosity were accounted for in an iterative manner. A semi-empirical mathematical model of apparent relative viscosity as a function of vessel diameter and vessel discharge hematocrit was used for this purpose (Hudetz et al, 1989). The partitioning of red cells at bifurcations was determined by the following formula of Klitzman and Johnson (1982):

$$[(1 - f)/f] = [(1-q)/q]^B$$

where $q = Q_1/Q$ and $f = F_1/F$ are fractional blood flow and red cell flux in branch 1 with respect to those in the parent vessel. Various values of B between 1 and 2 were used in the simulation. In one case, B was allowed to vary with local discharge hematocrit, as suggested by Dellimore et al. (1983).

Based on the previously calculated flow values and on the bifurcation rule, hematocrit was calculated in all vessels. From vessel diameter and the obtained hematocrit, vascular resistance was updated and the blood flow was recalculated. This iterative procedure was continued until convergence in the flow values was reached. The calculations were performed on the Apollo 10000 supercomputer.

RESULTS

The present calculations were carried out in a cerebrocortical network containing 116 internal vessel segments. Vascular diameters ranged between 3.6 and 20.4 μm (lognormal distribution, mean: 6.8 ± 2.2, SD). Segment lengths varied between 9.8 and 302.4 μm (lognormal distribution, mean: 74.4 ± 62.9, SD).

The results summarizing calculated microvessel hematocrit (Ht_d) at various cell partition parameter B are displayed in <u>Figure 1</u>. At B < 1.3 the Ht_d values were distributed close to the feed hematocrit. At B = 1.3 to 1.5 the Ht_d values became more heterogeneous. For B > 1.7 the Ht_d distribution was obviously skewed to the left. When B was made to be hematocrit-dependent, as suggested by Dellimore et al (1983), bimodal Ht_d distribution was obtained.

DISCUSSION

The results suggest that minor changes in the preferential red cell flow at bifurcations influence strongly the distribution of microvessel discharge hematocrit. This emphasizes the need for precise measurement of the phase separation parameter B, if reliable computer-predictions of microvessel hematocrit are to be obtained. A recent study of Secomb et al (1989) applies a more accurate, multi-parameter model of phase separation at diverging bifurcations. The parameters of their model are functions of the diameters of the parent and daughter vessels as well as of the local

32

discharge hematocrit. The predicted hematocrit values in the rat mesenteric microvascular network show similar distribution to that of the measured ones. In contrast to our results for the cerebral network, their hematocrit distribution is not bimodal, despite the hematocrit-dependence of the model parameter B. Unfortunately, a comparison of predicted and measured hematocrit distributions in the cerebral cortex cannot be made at this time, due to the lack of experimental data.

The present results suggest also that preferential red cell flow may be an important determinant of hematocrit distribution in the cerebral cortex. Should red cell flow separation be enhanced for any reason, local

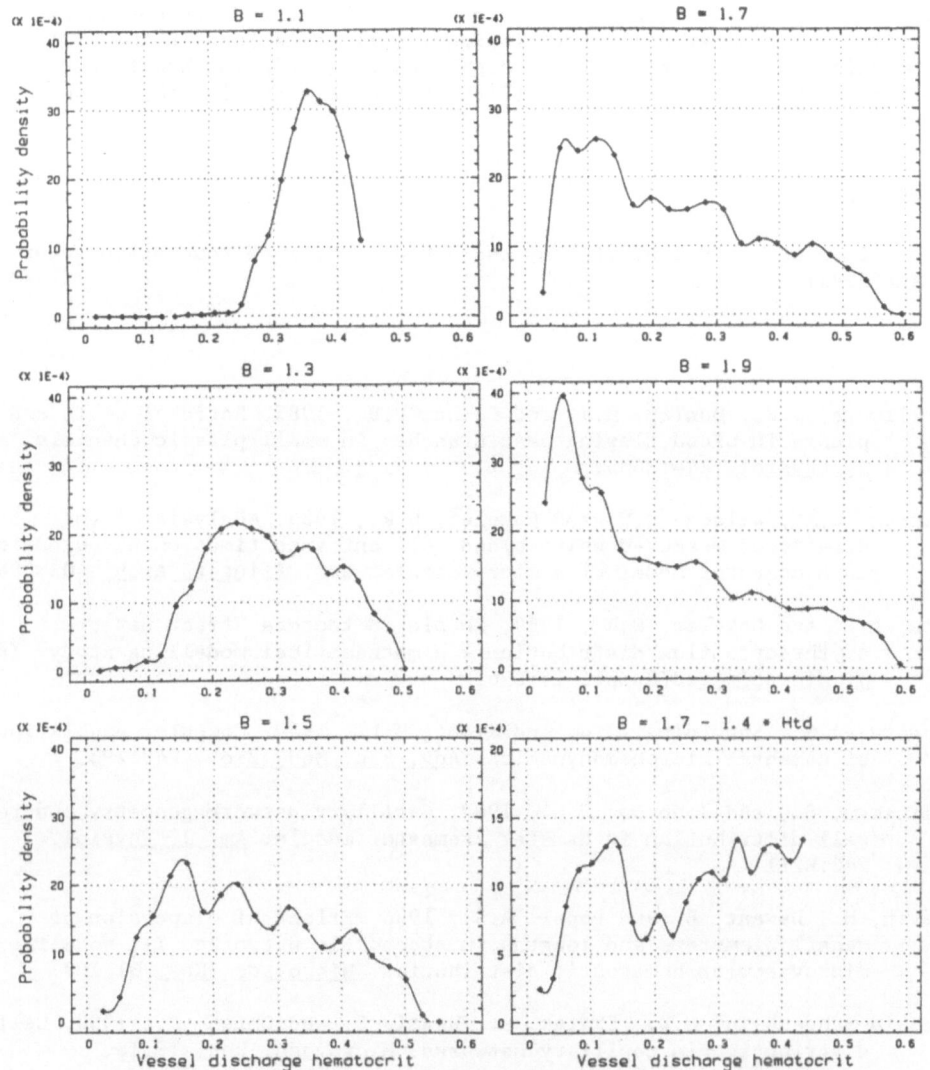

Fig. 1. Probability density of calculated hematocrit distribution in a cerebral microvascular network. With increasing red cell flow separation associated with the increasing value of parameter B, more and more vessels have low hematocrit. In case of hematocrit dependent flow separation (bottom right panel), the hematocrit distribution becomes bimodal.

hematocrit may be severely reduced in a large number of capillaries. Similar conclusions were reached in a recent study by Hsu and Cokelet (1989) by modeling the erythrocyte flux distribution in striated muscle. The strong dependence of erythrocyte flux distribution on the partitioning of cells and plasma at bifurcations implies its importance in physiologic regulation of the microcirculation and in the development of pathological flow situations.

SUMMARY

The influence of preferential red cell entry into microvascular branches with higher flow on microvessel hematocrit distribution was studied by mathematical modeling in a reconstructed cerebrocortical microvascular network. More heterogeneous hematocrit distribution was obtained at stronger cell partitioning. Small variations in the cell separation parameter resulted in significantly different hematocrit distributions. The significance of these findings in vivo should be further evaluated.

ACKNOWLEDGEMENT

This work was supported by the National Science Foundation grant CBT-8822851.

REFERENCES

Dellimore, J.W., Dunlop, M.J. and Canham P.B., 1983, Ratio of cells and plasma in blood flowing past branches in small plastic channels. Am. J. Physiol. 244:H635.

Fenton, B.M., Wilson, D.W. and Cokelet, G.R., 1985, Analysis of the effects of measured white blood cell entrance times on hemodynamics in a computer model of a microvascular bed. Pflugers Arch. 403:396.

Hsu, L.L. and Cokelet, G.R., 1989, Rheologic factors affect network erythrocyte flux distribution - a mathematical modelling study. Int. J. Microcirc. 8(Suppl. 1):S26.

Hudetz, A.G., Spaulding, J.G. and Kiani, M.F., 1989, Computer simulation of cerebral microhemodynamics. Adv. Exp. Med. Biol. 248:293.

Klitzman, B., and Johnson, P.C., 1982, Capillary network geometry and red cell distribution in hamster cremaster muscle. Am. J. Physiol. 242:H211.

Levin, M., Dawant, B. and Popel, A.S., 1986, Effect of dispersion of vessel diameters and lengths in stochastic networks. II. Modeling of microvascular hematocrit distribution. Microvasc. Res. 31:202.

Schmid-Schoenbein, G.W., Skalak, R., Usami, S. and Chien, S., 1980, Cell distribution in capillary networks. Microvasc. Res. 19:18.

Secomb, T.W., Pries, A.R., Gaehtgens, P., Gross, J.F., 1989, In: "Frontiers of Cardiovascular Mechanics," J.S. Lee and T.C. Skalak, eds., Springer, New York, pp. 39-49.

Papenfuss, H.-D. and Gross, J.F., 1981, Microhemodynamics of capillary networks, Biorheology 18:673.

MEMBRANE RESISTANCE TO OXYGEN TRANSPORT

INSIDE HYBRIDOMA CELLS IN SUSPENSION CULTURE

Kyung A. Kang and Dewey D.Y. Ryu

Department of Chemical Engineering
University of California at Davis
California 95616, USA

INTRODUCTION

Monoclonal antibodies (MAb) have been used as valuable analytical, diagnostic, and therapeutic tools for modern bioscience and medicine. Because of their high specificity, monoclonal antibodies have also been used for the purification of biomaterials which need high purity. Due to the high demand for monoclonal antibodies hybridoma cell culture, which enables one to produce monoclonal antibodies in large quantity, has become one of the most interesting topics and has been studied by many biochemical engineers.

We have been investigating murine hybridoma cell culture for the mass production of monoclonal antibody which has been used for the immunoaffinity purification of an enzyme. Until now we have studied basic growth kinetic parameters using a small scale batch reactor. We are in the process of scaling up to large scale reactors for continuous suspension culture.

Oxygen has been always one of the most important operational parameters that has to be controlled for the maximum productivity in cell culture. For understanding cell growth and MAb production it is very important and necessary to study the single cell responses to different environmental conditions. However, oxygen transport in a single hybridoma cell has not been well studied because of the technical difficulties in measuring oxygen concentration in a single cell. One of the primary operational parameters which are frequently measured during hybridoma cell culture is the dissolved oxygen concentration (DOC) in media correlated with cell growth and monoclonal antibody production.

The effect of oxygen on mammalian cell growth has been studied by many researchers (Cooper, et al., 1958; Rueckert and Mueller, 1960; Haugaard, 1968; Mizrahi, et al., 1972; Suleiman and Stevens, 1987; Dong, et al., 1987). Some of them have also studied the cytotoxic effect of hypertensive or hyperoxic oxygen supply (Cooper, et al., 1958; Rueckert and Mueller, 1960; Haugaard, 1968; Mizrahi, et al., 1972; Dong, et al., 1987). However, it is difficult to judge whether these studies represent the actual effect of concentrations on cell growth since the DOC in the media or the oxygen transported to the cell would be significantly different from the measurements of oxygen tension or oxygen content in the supply gas. Recently, some of researchers examined the effect of DOC on hybridoma cell culture (Boraston, et al., 1982; Reuveny, et al., 1986; Miller, et al., 1987). Nevertheless, the optimum DOC's on the cell growth and MAb production depend on the types of reactors or cell lines used.

One of the difficulties in understanding the effect of DOC on the growth and MAb production of hybridoma cell is that it is very difficult to measure, if not impossible, the oxygen concentration inside a living cell. Our research group has been trying to develop a methodology for obtaining indicative values of the intracellular oxidative level by measuring intracellular NADH concentration in suspension culture (Chillakuru, 1989). However, it is not easy to obtain meaningful values from this type of measurements because it requires very high cell concentration and can give only averaged values of the oxidative state of cells instead of actual intracellular oxygen concentration. Therefore, the effect of the DOC in the media may be the only indirect index to correlate the effect of environmental oxygen supply conditions to cell growth and MAb production, at the present time.

In this paper, we have simulated the oxygen transport from the media into a single hybridoma cell by using currently available experimental parameters and have attempted to interpret simulation results.

DESCRIPTION OF SIMULATED SYSTEM AND PARAMETER USED

Physiological and cytological data for a single hybridoma cell were obtained from the paper of Renau-Piqueras et al. (1983). Experimental system and DOC data were from the research of Miller et al (1987). In Miller's experiment, suspension culture in a continuous reactor of working volume of one liter was used.

Both groups of researchers used hybridoma derived from the Sp2 myeloma. The simulated oxygen transport system is a single hybridoma cell growing in a continuous suspension culture system (Figure 1).

(1) Geometry of the system simulated: According to Renau-Piqueras' measurements on volume and surface area of a single hybridoma cell (1983), the cells seemed to be spherical. From their measurement on volume and surface area, and the radius was calculated with the assumption that the cells were spherical. The radius of the cell was approximately 5.2 μm. (Figure 1).

(2) Volume of the hybridoma: $589.0.\mu m^3$ (Renau-Piqueras et al., 1983).

(3) Surface of the hybridoma: $340.0\ \mu m^2$ (Renau-Piqueras et al., 1983).

(4) Oxygen diffusion coefficient, D: $1.75 \times 10^{-5}\ cm^2/s$ (from the data for DS-Carcinosarcoma, Grote et al., 1977).

(5) Oxygen solubility in hybridoma cell, S: 1.79×10^{-2} ml-O_2 /ml-atm (Grote et al., 1977).

(6) Mass transfer coefficient, k_s: 0.065 cm/s (Harriot, 1962).

(7) Dissolved oxygen concentration in the media: In Miller's experiments, the lowest DOC they used was 0.1 % of air saturation. The maximum cell viability occurred at DOC of 0.5 %. Oxygen was the growth limiting factor until DOC became 10 % of air saturation. MAb production was maximum at 50 %. Cell viability at DOC between 30 to 100 % of air saturation was almost constant (see Figure 2).

We have simulated five different intracellular oxygen concentration profiles when DOC's of media were as follows.

36

(a) the lowest DOC of their measurement: 0.1 %,
(b) DOC at maximum cell growth: 0.5 %,
(c) DOC at which oxygen is no longer a growth limiting factor:
    10.0 %,
(d) DOC at maximum MAb production: 50.0 %, and
(e) DOC at 100.0 % of air saturation

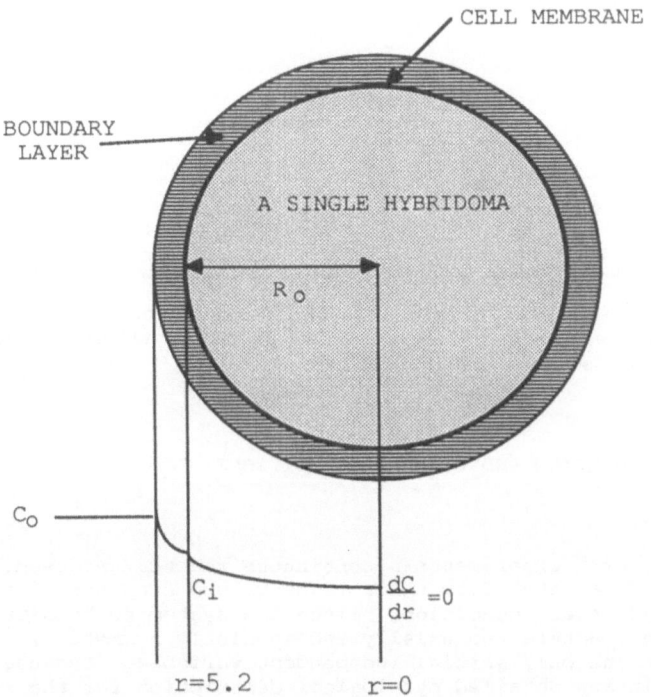

Figure 1. System Simulated. A Single Hybridoma Cell (unit = μm).

(8) Oxygen consumption rate: From Miller's experiments, a single
    hybridoma cell consumes at the rate of 0.19×10^{-9} mmol/cell-hr when
    DOC was above 10 % of air saturation. Oxygen consumption decreased
    to 0.08×10^{-9} mmol/cell-hr with decreasing DOC until 0.1 % of air
    saturation (Figure 2).

    For this simulation, oxygen consumption rate was assumed to be
    zeroth order and the values of consumption rates for different DOC's
    were

    (a) Oxygen consumption rate at DOC of 0.1 % was 0.08×10^{-9}
        mmole-O_2/cell-hr,
    (b) Oxygen consumption rate at DOC of 0.5 % was 0.14×10^{-9}
        mmole-O_2/cell-hr,
    (c), (d), and (e) Oxygen consumption rates at DOC of 10 %, 50 %, and
        100 % were 0.19×10^{-9} mmole-O_2/cell-hr (see Figure 2).

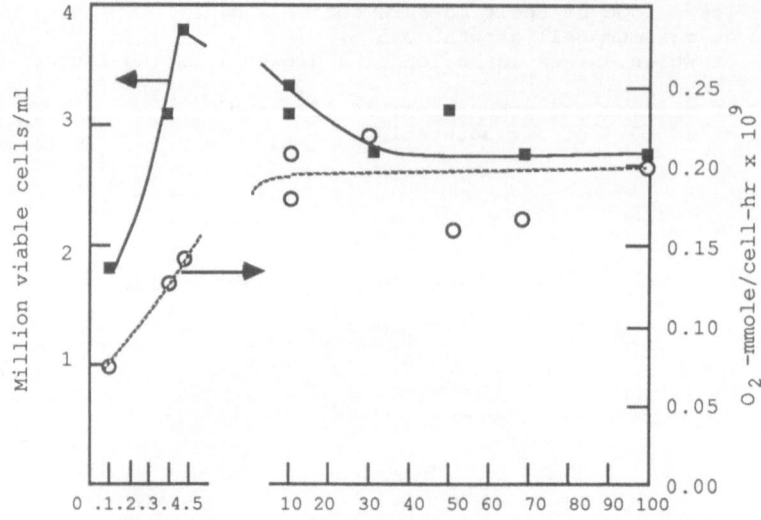

Dissolved oxygen concentraion (% of air saturation)

Figure 2. Cell Viability and Oxygen Consumption Rate with changing Dissolved Oxygen Concentration in Media (from the paper of Miller et al., 1987).

ILLUSTRATIVE EQUATION AND BOUNDARY CONDITIONS

(1) Equations

For Miller's experiments a continuous reactor was used. Therefore, experimental data taken from their reactor were supposed to represent those of steady state condition. Since the system to be simulated was assumed to be a sphere and axially and angularly symmetric, radial position r is the only spatial independent variable. Because we were not able to obtain any detailed cytological description for the intracellular materials and their corresponding parameters of oxygen transport in hybridoma cells, we assumed the cell as a lumped parameter system.

The governing equation of oxygen transport in the cell were by diffusion and zeroth order reaction.

$$D\left[\frac{d^2 C}{dr^2} + \frac{2}{r}\frac{dC}{dr}\right] - \frac{R_x}{S} = 0.0 \tag{1}$$

where D is oxygen diffusion coefficient, C is oxygen concentration in the cell, r is radial position, R_x is zeroth order oxygen consumption rate, and S is oxygen solubility of hybridoma cells.

(2) Boundary Conditions

At the center of the hybridoma cell, the flux of oxygen is zero, because of the axial and angular symmetry assumption.

$$\frac{dC}{dr} = 0.0 \qquad\qquad \text{at } r = 0.0 \tag{2}$$

At the hybridoma cell membrane, mass transfer rate from outside to the membrane is the same as diffusion rate from the membrane to the inside cell.

38

$$k_s (C_o - C_i) = D \frac{dC_i}{dr} \qquad\qquad at \ r = r_o \qquad (3)$$

where k_s is mass transfer coefficient in the boundary layer and C_o, and C_i are dissolved oxygen concentration in the media and at the cell membrane, respectively.

Equation (1) may be solved very easily by replacing $y = C \cdot r$ and using boundary conditions (2) and (3). The analytical solution obtained for the governing equation (1) becomes

$$C = C_o + \frac{R_x}{6D}(r^2 - r_o^2) - \frac{R_x r_o}{3 k_s} \qquad (4)$$

RESULTS AND DISCUSSION

Figure 3 is the simulation result of the intracellular oxygen concentration profiles for five different DOC's

(a) Media DOC of 0.1 % of air saturation (0.16 mmHg) and oxygen consumption rate of 0.08 x 10^{-9} mmole-O_2/cell-hr,

(b) Media DOC of 0.5 % (0.8 mmHg), at which hybridomas had the highest cell viability, and oxygen consumption rate of 0.14 x 10^{-9} mmole-O_2/cell-hr,

(c) Media DOC of 10 % (16 mmHg), at which oxygen was no longer growth limiting, and oxygen consumption rate of 0.19 x 10^{-9} mmole-O_2/cell-hr,

(d) Media DOC of 50 % (80 mmHg), at which MAb production was the maximum, and oxygen consumption rate of 0.19 x 10^{-9} mmole-O_2/cell-hr, and

(e) Media DOC of 100 % of air saturation (160 mm Hg) and oxygen consumption rate of 0.19 x 10^{-9} mmole-O_2/cell-hr.

Figure 3 shows the simulated oxygen profiles of the five different conditions above in terms of absolute values of oxygen partial pressure. From these five profiles, (a)-(b), it can be seen that mass transfer resistance in the boundary layer to the cell membrane is not significant. It appears that oxygen transport is intracellular diffusion or reaction limited.

Figure 4 shows that intracellular concentration profiles normalized by media DOC's (C_o's). From this figure we can see that as DOC in the media increases, the gradient of oxygen concentration decreases. Nevertheless, the gradients for all five cases are low. For all five cases, the oxygen concentration in the cell is always more than 93 % of outside media. Even for the case of 10 % air saturation, at which cells begin to have their full oxygen consumption rate (i.e. oxygen is no longer a growth limiting factor), the oxygen concentration at the center of the cell is almost the same as media DOC (99.8 % of media DOC). In the cases of 50 and 100 % of air saturation there was almost no concentration gradient. This low gradient is mainly due to the low oxygen consumption rate of hybridoma cells and no intracellular diffusion resistance to oxygen transport. According to Miller's experiments, cell viability decreased as the media DOC increased (see figure 2) and they presumed that this was due to the oxidative damage. However, cell viability did not decrease further when DOC increased to higher than 30 % of air saturation. Moreover oxygen consumption rate did not increase beyond a media DOC of 10 % of air saturation. In the case of 100 % air saturation, entire cell is saturated by air, which means the entire cell has as high as 160 mmHg of oxygen concentration. Most of the time, air saturation conditions for hybridoma cell reactor can be obtained by

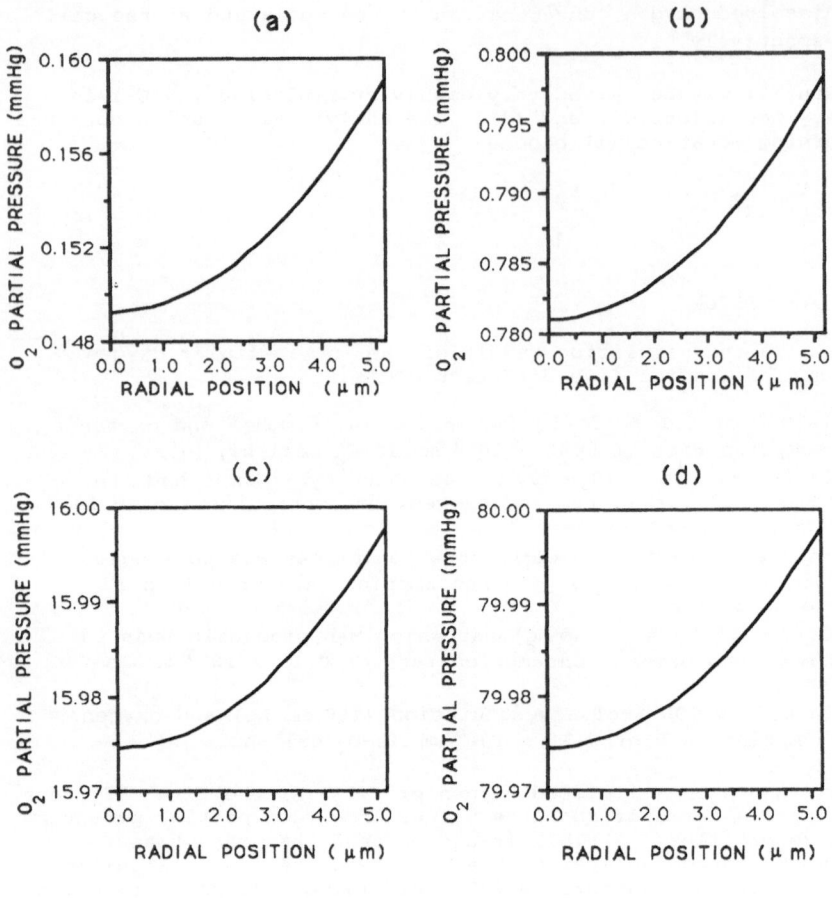

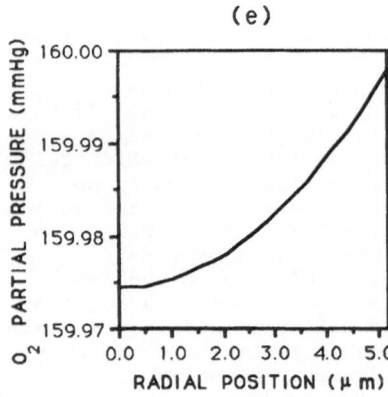

Figure 3. Intracellular Oxygen Concentration Profiles in a Single Hybridoma Cell, (a) When Media DOC is 0.1 % of Air Saturation, (b) When Media DOC is 0.5 % of Air Saturation, (c) When Media DOC is 10 % of Air Saturation, (d) When Media DOC is 50 % of Air Saturation, and (e) When Media DOC is 100 % of Air Saturation.

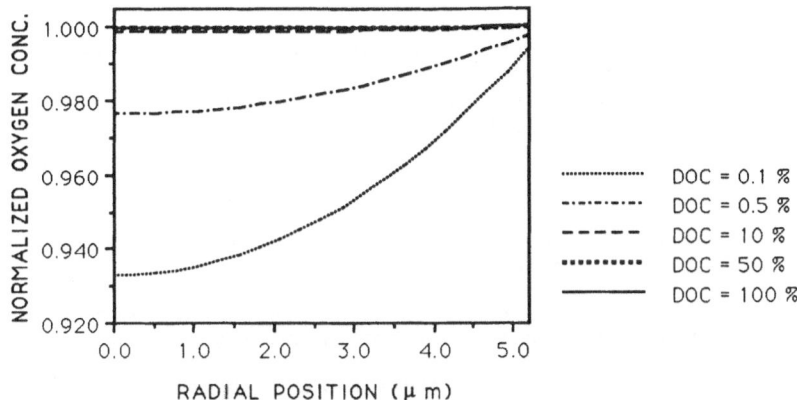

Figure 4. Normalized Intracellular Oxygen Concentration Profiles of Five Different DOC's in the Media (Normalized by DOC's in the Media).

supplying a gas mixture of oxygen and air, which means hyperoxic oxygen supply (in Miller's experiments, also). There has been much research on cytotoxic effect of hypertension or hyperoxia of supply gas on animal cells or animal organs, especially, pulmonary organs (Balentine, 1982). However, these studies were done by changing oxygen content or tension in the supply gas, and not in terms of DOC. We have not been able to obtain any relevant information relating the level of cytotoxicity to intracellular oxygen concentration. Therefore, it is difficult to utilize the data from these papers how the oxygen concentration in the media actually affects cell viability and MAb production during *in vitro* cell cultivation.

There could be two plausible explanations for the relatively high intracellular oxygen concentration obtained from the simulation.

One possible explanation might be that hybridoma cells may be able to survive high intracellular oxygen concentration. Usually, the cytotoxicity of oxygen in living cells is due to oxygen induced radicals such as superoxide radical, hydrogen peroxide, hydroxyl radical, singlet oxygen, or peroxide radical. These species may be scavenged by intracellular enzymes (catalase, superoxide dismutase, or peroxidase) and some other organic compounds (Vitamin E, β-carotene, Ascorbate, Glutathione, etc.; Frank, 1985) It has been known for some time that tumor cells have low level of these enzymes which can eliminate cytotoxic radicals. This could imply that tumor cells are more susceptible to these radicals (Oberley, 1982; Balentine, 1982). Since hybridoma cells are derived from myeloma, there is a strong possibility that they are also low in radical scavenging enzymes. A plausible explanation for the high viability of hybridoma cell with high intracellular oxygen concentration and lower level of radical scavenging enzymes could be that these cells have lowered capacity to produce cytotoxic radicals. This has been suggested by some biochemists (Oberley, 1982), though it has not yet been confirmed with definite evidences, yet.

Another likely explanation could be that there might be a certain cellular function that limits oxygen transport to the inside of the cell, possibly at the cell membrane, which is not included in our mathematical model. With the assumption of low oxygen consumption rate and negligible intracellular mass transport resistance, cells become rapidly saturated

41

with oxygen transported from outside. If we assume mass transfer resistance at the membrane, with a saturation limited uptake mechanism (i.e. a type of Michaelis-Menton kinetics), then the cells would take up only limited amounts of oxygen. Thus, oxygen would be transported to the cells without much resistance when DOC in the media is low. However, at high DOC, cells would be able to limit oxygen transport to avoid cytotoxic effects of high oxygen concentration.

CONCLUSION

From our study on the simulation of intracellular oxygen concentration in hybridoma cells, we conclude that

(1) Mass transfer resistance of the boundary layer outside the hybridoma cell in continuous suspension culture is not significant.

(2) In all simulations for five different DOC's in the media and their corresponding oxygen consumption rates, intracellular oxygen concentrations were not significantly different from those of media. The reasons for this are, mainly, because of the low oxygen consumption rate of hybridoma cells and rapid oxygen transport with no other resistance except intracellular diffusion. From these results we suggest that hybridoma cells seem to have either

    (a) tolerance to the high intracellular oxygen concentration or
    (b) some other mechanism to limit oxygen transport into the cell, possibly, at the cell membrane.

(3) In order to elucidate the actual mechanism of intracellular oxygen transport it is necessary to develop proper techniques for measuring the intracellular oxygen concentration in the cell in terms of, not only oxidative state of cells (such as NADH or ATP level), but also actual intracellular oxygen concentration.

SUMMARY

We have examined oxygen transport from media to a single hybridoma cell for different dissolved oxygen concentrations in media and oxygen consumption rates of cells. Intracellular oxygen concentration profiles inside the cell were simulated using currently available parameters.

From the simulation, it was found that mass transfer resistance at the boundary layer does not seem to be significant. Intracellular oxygen concentrations did not differ very much from outside media concentration and the concentration gradients were relatively low for all five different media DOC's (from 0.1 to 100 % air saturation). From these results, it is suspected that either hybridoma cells have better tolerance to high intracellular oxygen concentrations than normal cells or that the cell membrane has a certain mechanism for limiting oxygen transport into the cell to avoid the oxidative damage.

REFFERENCES

Balentine, J. D., Pathology of Oxygen Toxicity, Academic Press, New York, 1982.

Chillakuru, R., 1989, Personal Communication.

Cooper, P. D., Burt, A. M., and Wilson, J. N., 1958, "Critical Effect of Oxygen Tension on Rate of Growth of Animal Cells in Continuous Suspended Culture", Nature, 182, 1508-1509.

Dong, G., Hu, S., Zhao, G., Gao, S., and Wu, L., 1987, "Experimental Study on Cytotoxic Effects of Hyperbaric Oxygen and Photodynamic Therapy on Mouse Transplanted Tumor", Chinese Medical Journal, 100 (9), 697-702.

Grote, J., Susskind, R., and Vaupel, P., 1977, "Oxygen Diffusivity in Tumor Tissue (DS-Carcinoma) under Temperature Conditions within the Range of 20-40 ^{O}C, Pflugers Arch., 371, 37-42.

Harriot, P., 1962, "Mass Transfer to Particles", 8 (1), AIChE Journal, 93-102.

Haugaard, N., 1968, "Cellular Mechanisms of Oxygen Toxicity", 48(2), Physiological Reviews, 311-373.

Mizrahi, A., Vosseller, G. V., Yagi, Y., and Moore, G. E., 1972, "the Effect of Dissolved Oxygen Partial Pressure on Growth, Metabolism and Immunoglobulin Production in a Permanent Human Lymphocyte Cell Line Culture", 139, P.S.E.B.M., 118-122.

Miller, W. M., Wilke, C. R., and Blanch, H. W., 1987, "Effect of Dissolved Oxygen Concentration on Hybridoma Growth and Metabolism in Continuous Culture", Journal of Cellular Physiology, 132, 524-530.

Oberley, L. W., Superoxide Dismutase and Cancer, 1982, in: "Superoxide Dismutase", Vol II, edited by Oberley, L. W., CRC Press, Boca Raton, Florida.

Frank, L., Oxygen Toxicity in Eukaryotes, 1985, in: "Superoxide Dismutase", Vol III, edited by Oberley, L.W., CRC Press, Boca Raton, Florida.

Renau-Piqueras, J., Perez-Serrano, M. D., and Martinez-Ramon, A., 1983, "Stereological Study of Murine Myeloma and Hybridoma Cells in vitro and in vivo, J. Submicrosc. Cytol., 15 (3), 607-618.

Reuveny, S., Velez, D., Macmillan, J. D., and Miller, L., "Factors affecting Cell Growth and Monoclonal Antibody Production in Stirred Reactors", 1986, Journal of Immunological Method, 86, 53-59.

Satterfield, C. N., 1970, "Mass Transfer in Heterogeneous Catalysis", MIT press, Massachusetts.

Suleiman, A. S. and Stevens, J. B. , 1987, "The Effect of Oxygen Tension on Rat Hepatocytes in Short-Term Culture", 23 (5) in vitro Cellular & Developmental Biology, 332-338.

METHODS AND INSTRUMENTATION

DEVELOPMENT OF A MICRO TRANSMISSION CELL FOR IN VIVO

MEASUREMENT OF SaO$_2$ AND Hb

J.A.H.Bos, W.Schelter*, W.Gumbrecht*, B.Montag*, E.P.Eijking,
S.Armbruster, W.Erdmann and B.Lachmann

Dept. of Anesthesiology, Erasmus University Rotterdam, The
Netherlands and *Siemens AG, Erlangen, West Germany

INTRODUCTION

Oxygen transport in blood depends on oxygen partial pressure (PaO$_2$), oxygen saturation (SaO$_2$), hemoglobin content (Hb) and cardiac output (CO). At least SaO$_2$ and Hb-content as well as CO should be known to achieve reliable insights into tissue oxygen supply of patients in intensive care and during anesthesia. Methods for cardiac output and saturation measurements have been available for some time but there exists so far no method to measure the Hb-content continuously in vivo as well as in vitro.

Since a few years, pulse oximetry has received extensive use in routine medical practice (for review see: Parker, 1987). Recently fiberoptic catheters have been developed for continuous measurement of mixed venous oxygen saturation. The principle of this technique is based on two- and three wavelength spectrophotometric measurements. These systems have a high accuracy and stability. However the major disadvantage of these invasive systems is that they need in vivo calibration and are still subject to drift especially if two-wavelength catheters are used (Gettinger, Detraglia, Glass, 1987). To avoid these shortcomings a Micro Transmission cell has been developed capable of in vivo on line measurements of SaO$_2$ and Hb-content.

METHODS

The development of the Micro Transmission Cell was based on the principle that the uptake of oxygen by desoxyhemoglobin forming oxyhemoglobin causes a change in extinction at wavelengths of 660 nm and 950 nm when analyzed with spectrophotometry. The extinction coefficient decreases at 660 nm and increases at 950 nm wavelength when the ratio of oxyhemoglobin versus desoxyhemoglobin increases (Fig 1). This method, called oximetry, was already developed in the early forties. For SaO$_2$ measurements hemolyzed blood was used and therefore Lambert-Beer's law could be applied.

The main problem to overcome while applying this newly developed technique was the fact that transmission through whole blood as compared to hemolyzed blood, produces scattering and absorption through intact erythrocytes. The hematocrit (Ht), fibrinogen

Oxygen Transport to Tissue XII, Edited by J. Piiper *et al.*
Plenum Press, New York, 1990

content (and other constituents) in blood also influence extinction measurements for which must be corrected in order to obtain an accurate and reliable value of the SaO_2 and Hb-content.

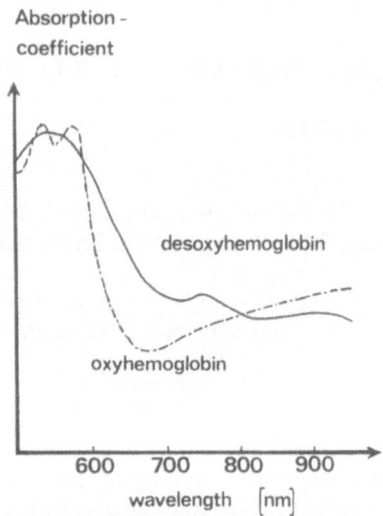

Figure 1. Absorption coefficient of desoxyhemoglobin and oxyhemoglobin versus the measured wavelength.

Assuming blood plasma to be transparant, the attenuation coefficient μ_δ can be written as:

$$\mu_\delta = a_{\delta,ox}\, c_{ox}\, l_{\delta,ox} + a_{\delta,Dox}\, c_{Dox}\, l_{\delta,Dox} + s_{\delta,SaO2,Ht} \tag{1}$$

where δ (subscript) describes wavelength, "a" is the absorption coefficient, "c" is the concentration of oxy- or desoxyhemoglobin, "l" is optical path, subscripts ox and Dox stand for oxyhemoglobin and desoxyhemoglobin respectively. "S" denotes the component of scattering.

Figure 2 shows the attenuation coefficient μ_{660} of 660 nm light plotted against the attenuation coefficient μ_{950} of 950 nm light for pig's blood mixed with Alsever's solution (2,05% Glucose, 0.42% Sodium chloride, 0,80% Tri-sodium citrate (Dihydrate), pH 6,1). The upper and lower solid lines indicate zero and 700 Torr pO_2 respectively. For zero pO_2, μ_{660} is proportional to μ_{950} with increasing c_{Hb} but for higher pO_2 values a non-linear relationship exists between both attenuation coefficients.

For constant Hb the dashed curves show linear, i.e. Lambert-Beer behaviour. All dashed lines have the same slope, as described in the following equation:

$$\frac{\mu_{660}(c_{ox}=0) - \mu_{660}(c_{Dox}=0)}{\mu_{950}(c_{ox}=0) - \mu_{950}(c_{Dox}=0)} = \text{constant} \tag{2}$$

48

$\mu_{660}(c_{ox}=0)$ denotes attenuation at 660 nm with $c_{ox} = 0$ g ml^{-1}, etc.

Inserting equation 1 into eqn. 2 leads to the following equation:

$$\frac{a_{660,ox}\ c_{Hb}\ l_{660,ox} + s_{660,ox,Ht} - a_{660,Dox}\ c_{Hb}\ l_{660,Dox} - s_{660,Dox,Ht}}{a_{950,ox}\ c_{Hb}\ l_{950,ox} + s_{950,ox,Ht} - a_{950,Dox}\ c_{Hb}\ l_{950,Dox} - s_{950,Dox,hct}}$$

$$= \text{constant} \tag{3}$$

Equation 3 is only valid if the folowing conditions are fulfilled:

a. $l_{\delta,ox} = l_{\delta,Dox} = l_{\delta}$
b. l_{δ} is independent of c_{ox} and therefore of SaO_2
c. s_{δ} is independent of SaO_2
d. l_{660}/l_{950} = constant

Thus equation 3 can also be written in a more simple way:

$$\frac{l_{660}}{l_{950}} \cdot \frac{a_{660,ox} - a_{660,Dox}}{a_{950,ox} - a_{950,Dox}} = m = \text{constant} \tag{4}$$

m denotes the slope of the dashed curves in figure 2. According to fig. 1 a pseudoisosbestic attenuation μ_{iso}, i.e. an attenuation which is independent of SaO_2 can be calculated :

$$\mu_{iso} = \mu_{660}\ \frac{m}{1-m} - \mu_{950}\ \frac{1}{m-1} \tag{5}$$

μ_{iso} is then only a function of C_{Hb}
Knowledge of the attenuation coefficients at the two wavelengths and the correction factor for the scattering by erythrocytes enables the development of a photometer and accordingly calculation of SaO_2 and Hb.

CALCULATION OF HEMOGLOBIN CONCENTRATION C_{Hb} AND SaO_2

The relationship of the C_{Hb} to red and infrared attenuation coefficients can be sufficiently approximated by a second order polynomial of μ_{iso}, according to equation 5. SaO_2 is determined by drawing a line with slope m through the crossingpoint of μ_{950} and μ_{660} attenuation for the same Hb-content values. This line crosses with the lower and upper solid lines in Fig. 2. which reflects the 0 saturation respectively the 100% saturation. The distance of the measurement point to the upper point of intersection divided by the

49

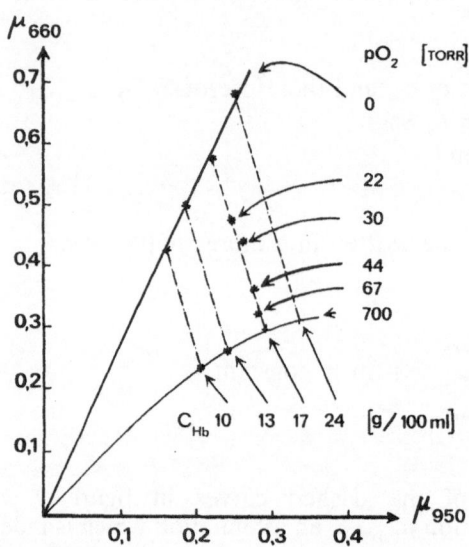

Figure 2. Attenuation coefficient of red light versus
extinction coefficient of infrared light
for different values of Hb-content and SaO_2. Dashed
lines mark values for constant Hb-content. For more
details see text.

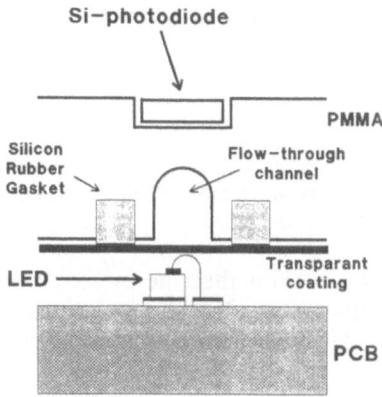

Figure 3. Scheme of the Micro Transmission Cell

distance between 0 and 100% SaO_2 intersection yields SaO_2 according to:

$$SaO_2 = \frac{c_{ox}}{c_{ox} + c_{Dox}}$$

DESCRIPTION OF THE CELL AND ITS PROCEDURES

The Micro Transmission Cell consists of two 660 nm LED's (Light Emitting Diode), two 950 nm LED's in one row and a silicon photodiode altogether mounted in a flow through cell (Fig. 3). The design of such a cell has been described earlier for the chemFET cell (Gumbrecht, Schelter, Montag, 1987).

A two-syringe-pump system is connected to the cell to perform sampling. Samples are withdrawn through a double-lumen catheter inserted into a vessel (arterial or venous) of which the inner lumen is connected to the cell. A syringe-pump (waste pump) draws fluid (calibration fluid or blood) through the cell with a constant flow rate of approximately 30-40 μl min^{-1}. The outer lumen, which exceeds the length of the inner catheter, is connected to a pump which infuses calibration fluid intermittently at a rate of approximately 120 μl min^{-1} thus decisively exceeding the drawing rate of the so-called waste pump. For calibration fluid, heparinized saline is used to minimize the clot formation in the catheter and in the micro transmission cell.

During the calibration phase, the calibration fluid is pumped through the outer lumen of the double-lumen catheter in excess and part of it drawn via the inner lumen into the cell by the waste pump while the excess fluid flushes the catheter entrance. The cell can now recalibrate and clean itself before taking another blood sample.

Due to the intermittent stops of the calibration pump, during the pauses blood is drawn through the inner lumen into the cell and a measurement sample is taken. The signals of the cell are sent to a computer and processed.

The values are available 30 seconds after the sample is withdrawn. The whole system then automatically recalibrates itself and the cycle can start all over again.

The volume of the blood sample is approximately 10 μl per measurement cycle. Thus with a measurement cycle of 2 minutes, not more than 7.2 ml of blood have to be withdrawn during 24 hours.

To achieve correct software adjustment, in vitro calibration of the system was performed with pig blood mixed with a 50 % Alsever's solution to get samples with undamaged erythrocytes and different values of SaO_2 were established by variation of oxygen partial pressure (pO_2).

In vivo studies in pigs have proven that the system works accurately and without failure during long time application. Necessary adjustments of the software for measurements under unphysiological low hemoglobin conditions have still to be completed (Eijking, Bos et al, 1989)

CONCLUSION

Results show that the newly developed Micro Transmission Cell is capable to monitor the SaO_2 and Hb-content in arterial and venous blood in the physiological range with high accuracy. The use of heparinized calibration solution allows long time stability (> 24 hrs) and thereby satisfies intensive care requirements.

In the near future it should be possible to integrate this saturation measurement method in a chemFET cell for on line measurement of electrolytes and thereby combining the advantages of both systems (Gumbrecht, Schelter, Montag, 1987)

REFERENCES

Eijking, E.P., Bos, J.A.H., Armbruster, S., Schelter, W., Gumbrecht, W., Montag, B., Erdmann, W., Lachmann, B. ,1990, Online measurements of SaO_2, Ht and Hb using a micro transmission cell, This volume.

Gettinger, A., DeTraglia, M.C., Glass, D.D ,1987, In vivo comparison of two mixed venous saturation catheters. Anesthesiology. 66, 373-375

Gumbrecht, W, Schelter, W, Montag, B. ,1987, A chemFet microcell system for medical and biotechnological online electrolyte monitoring. Third conference on sensors and their applications, Cavendish laboratory, Cambridge UK (conference report), 155-156

Parker, D., 1987, Sensors for monitoring blood gases in intensive care, J. Phys.E: Sci. Instr. 20; 1103-1112

ONLINE MEASUREMENTS OF SaO₂, Ht AND Hb USING A MICRO TRANSMISSION CELL

E.P.Eijking, J.A.H.Bos, S.Armbruster, W. Schelter*,
W. Gumbrecht*, W.Erdmann and B.Lachmann

Dept. of Anesthesiology, Erasmus University Rotterdam, The
Netherlands and *Siemens AG, Erlangen, West Germany

INTRODUCTION

Oxygen saturation (SaO_2) and hemoglobin concentration (Hb) are important parameters for evaluation of oxygen transport in patient monitoring. Before the advent of noninvasive monitoring of SaO_2 with pulse oximetry (Yoshiya et al, 1980; Severinghaus and Honda, 1987) intermittent sampling of blood gases was the only reliable method of assessing arterial oxygenation and probably still is the most frequently used method. While intermittent sampling is satisfactory under most clinical conditions, this method does not give any indication of the oxygenation status between two samples (Lam, 1987). As shown by Thorson and colleagues (1983), considerable spontaneous variation in PaO_2 may occur in patients in the intensive care unit. Arterial hypoxaemia is often the cause of complications during anesthesia, resulting in for example neurological impairment and cardiac ischemia. The majority of incidents of severe hypoxaemia during anesthesia is the result of failure of oxygen delivery to the arterial blood. The common causes are mishaps reducing the inspired oxygen concentration, upper airway obstruction, misplacement of an endotracheal tube and hypoventilation (Knill, 1985). To prevent complications due to a failure of arterial oxygenation there is a strong need for continuous measurement of the arterial oxygenation status of the patient.

Recently a Micro Transmission Cell based on spectrophotometry has been developed for simultaneous in vivo monitoring of SaO_2 and Hb. Briefly explained, the principle of this cell is based on the fact that uptake of oxygen by hemoglobin results in a decrease in extinction coefficient at a wave length of 660 nm and an increase in extinction coefficient at 950 nm. Knowing the two different extinction coefficients SaO_2, Hb and hematocrit (Ht) can be calculated. Via a double lumen catheter specifically designed to feed the measuring Micro Transmission Cell by means of a computerized pump system, blood samples of 10 μl are automatically taken every 2 minutes and SaO_2, Hb and Ht are measured. Before each sampling the cell is calibrated with a heparinized calibration solution (Bos, Schelter and Gumbrecht, this volume).

The purpose of this study was to compare the values measured with this newly developed Micro Transmission Cell with the values obtained by standard methods for measuring SaO_2, Hb (OSM 2 Hemoximeter, Radiometer, Denmark) and Ht (Ht centrifuge) at different O_2 saturations and Hb concentrations.

Oxygen Transport to Tissue XII, Edited by J. Piiper *et al.*
Plenum Press, New York, 1990

METHODS

Pigs (approximately 12 kg BW) were anesthetized with Dormicum® 0.5 mg/kg i.m., Ketamine® 10 mg/kg i.m. and Pentothal® 10 mg/kg i.v. During the experiments Pavulon® 0.5-1.0 mg/kg/h and Dormicum® 1-3 mg/kg/h were used. The animals were ventilated with a Servo Ventilator 900C (Siemens-Elema, Sweden) at a frequency of 15/min. Tidal volume was set to get a $PaCO_2$ between 35 and 40 mmHg. The catheter leading to the Micro Transmission Cell was placed in the right femoral artery. In the left femoral artery an arterial line was placed from which the blood samples were taken in duplo for comparison.

The following steps were taken to alter the O_2 saturation and the Hb concentration:

1. Normoventilation with 100% oxygen
2. Normoventilation with 21% oxygen
3. Hypoventilation with 21% oxygen (hypoventilation was induced by decreasing the tidal volume, resulting in a $PaCO_2 > 60$ mmHg)
4. After normoventilation with 100% oxygen, hemodilution was produced by withdrawing 10% of the animals' blood volume and replacing it with a plasma expander (Isodex®)

This stepwise hemodilution procedure was repeated about 6 to 7 times, until the animals died of anemia.

RESULTS

The above described measurements were performed in 4 pigs during a 5 hour monitoring period each. Figure 1 shows the O_2 saturation values from the Micro Transmission Cell plotted against the O_2 saturation values obtained with the OSM 2 Hemoximeter. Altogether 227 paired values were acquired. Regression analysis shows a high correlation. Hematocrit values obtained with the Micro Transmission Cell plotted against Ht values obtained with the Ht-centrifuge are shown in figure 2. Altogether 108 paired samples were measured. Regression analysis again shows a high correlation. Figure 3 demonstrates the results of a 5 hour control period to investigate the stability of the Micro Transmission Cell. This is an individual example of measurements in one pig in which the steps to alter the O_2 saturation and the Hb concentration were repeated 7 times. In this figure a solid line is drawn, which clearly reveals that adjustment of the software to lower Hb concentrations (<8 gr/100 ml) is still lacking.

DISCUSSION

The results obtained from the animal studies show that the Micro Transmission Cell is capable of following the trends of SaO_2, Ht and Hb at different Hb concentrations and/or O_2 saturations. To acquire more accurate values for Hb concentrations below 8 gr/100 ml the software still has to be adjusted. Using a measurement cycle of 2 minutes, only 7.2 ml are withdrawn during 24-hours monitoring. The use of a heparinized calibration solution permits long term stability by preventing deposition of fibrine in the catheter and continuous calibration of the cell, which is one of the greatest advantages of this system. Continuous assessment of the oxygenation status and future integration of the cell with an already developed system for online measurement of electrolytes (Gumbrecht and Schelter, 1987), will be an improvement of patient monitoring during anesthesia and intensive care monitoring.

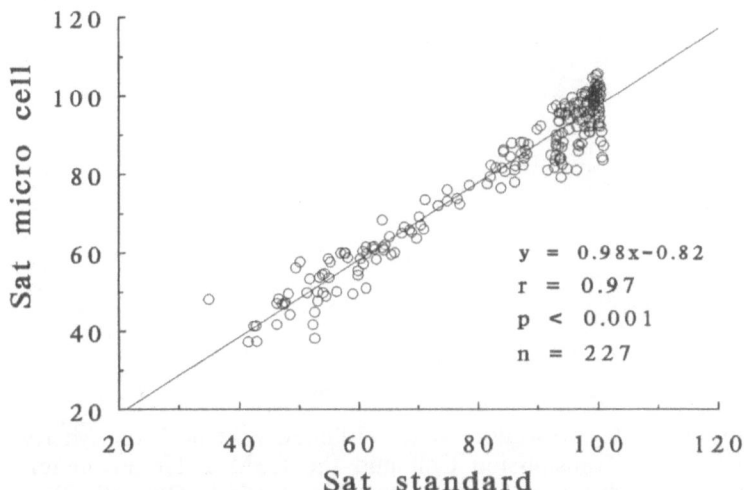

Arterial Saturation %

$y = 0.98x - 0.82$
$r = 0.97$
$p < 0.001$
$n = 227$

Figure 1.   Relationship between O_2 saturation measured with the Micro Transmission Cell and O_2 saturation measured with the OSM 2 Hemoximeter.

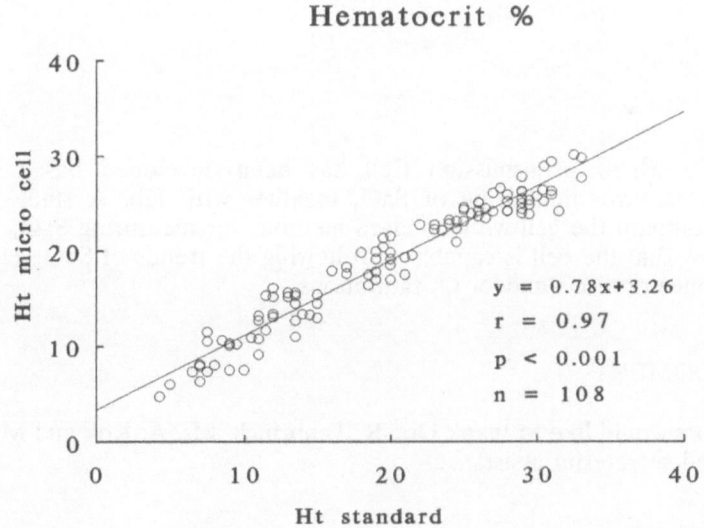

Hematocrit %

$y = 0.78x + 3.26$
$r = 0.97$
$p < 0.001$
$n = 108$

Figure 2.   Relationship between Ht measured with the Micro Transmission Cell and Ht measured with the Ht centrifuge.

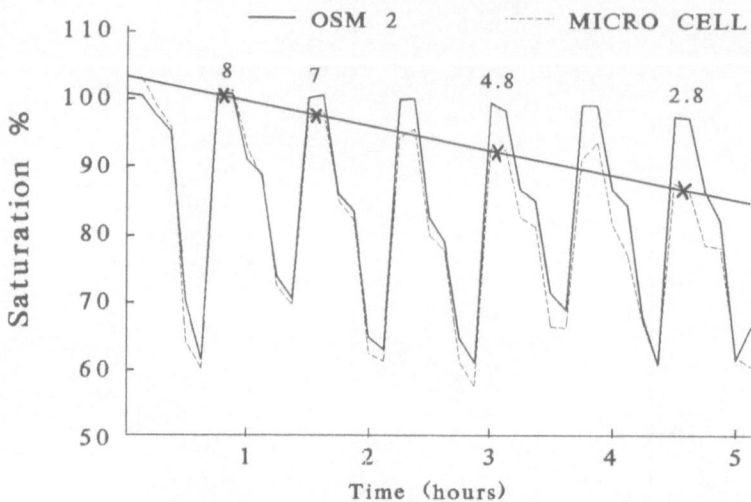

Arterial Saturation %

Figure 3. O_2 saturation values obtained with both the Micro Transmission Cell and the OSM 2 Hemoximeter during a 5 hour monitoring period. The solid line represents a non-adjustment of the software to the lower Hb concentrations (< 8gr/100 ml). The crosses in the figure represent Hb concentrations (in gr/100 ml) after hemodilution.

SUMMARY

Recently a Micro Transmission Cell has been developed based on spectrophotometry for in vivo monitoring of SaO_2 together with Hb. A study in pigs was performed to compare the cell with standard methods for measuring SaO_2, Hb and Ht. The results show that the cell is capable of following the trends of SaO_2, Hb and Ht at different Hb concentrations and/or O_2 saturations.

ACKNOWLEDGMENT

The authors would like to thank Drs. R. Tenbrinck, Mr. A. Kok and Mrs. L. Visser for technical and secreterial assistance.

REFERENCES

Bos, J.A.H., Schelter, W., Gumbrecht, W., Montag, B., Eijking, E.P., Armbruster, S., Erdmann, W. and Lachmann, B., 1990, Development of a Micro Transmission Cell for in vivo measurement of SaO_2 and Hb, This volume.

Gumbrecht, W., Schelter, W. and Montag, B., 1987, A chemFET microcell system for medical and biotechnical online electrolyte monitoring, Third conference on sensors and their applications, Cavendisch Laboratory, Cambridge, UK: 155-156.

Knill, R.L., 1985, Evaluation of arterial oxygenation during anaesthesia, Can. Anaesth. Soc. J., 32:3: S16-S19.

Lam, A.M., 1987, Continuous arterial PO_2 monitoring, Can. J. Anaesth., 34:1: 58-59.

Severinghaus, J.W. and Honda, Y., 1987, History of blood gas analysis. VII. Pulse Oximetry, J. Clin. Monit., 3(2): 135-138.

Thorson, S.H., Marini, J.J., Pierson, D.J. and Hudson, L.D., 1983, Variability of arterial blood gas values in stable patients in the ICU, Chest, 84: 14-18.

Yoshiya, I., Shimada, Y. and Tanaka, K., 1980, Spectrophotometric monitoring of arterial oxygen saturation in the fingertip, Med. Biol. Eng. Comput., 18: 27-32.

KRYPTON FILLED FLASHLAMP: A POSSIBLE NEW LIGHT SOURCE FOR NEAR INFRARED SPECTROSCOPY IN VIVO

M. Essenpreis, J. Spahn, W. Waidelich*, and H.T. Versmold†

* Institut für Medizinische Optik der Universität München
  Barbarastr. 16, D–8000 München 40
† Klinikum Großhadern der Universität München, Neonatologie der
  Frauenklinik, Marchioninistr. 15, D–8000 München 70

INTRODUCTION

Hypoxic brain damage of newborn infants is the prominent problem of perinatal medicine. Research in this field is limited, because new techniques for measuring the oxygen supply to brain cells are still in development. There is a need for further improvement of existing techniques and for a simplification of the measuring systems and their applicability. Near infrared (NIR) spectroscopy in vivo has been shown to be a powerful tool for non–invasive monitoring of several cerebral parameters, e.g. changes in blood oxygenation or blood volume. Since Jöbsis' (1977) first description of this technique marked improvements could be achieved by a number of workers. Recent progress by Cope and Delpy (1988) lead to new clinical applications as the quantitative estimation of cerebral blood flow.

A considerable restriction of the few actually working instruments is the instability of the emitted light energy and the lack of powerful laser diodes for the far red region (750 to 780 nm). Using a krypton filled flashlamp we introduced a cheap and powerful light source which could be shown to have a very constant light output. The purpose of this paper is to describe the realization and the applicability of a highly sensitive spectral photometer system for NIR spectroscopy in vivo, based on a krypton flashlamp. In first in vivo tests on human tissues we could obtain results equivalent to those with an extensive laser diode system.

SYSTEM DESCRIPTION

Our photometer system consists of three main components: a krypton flashlamp as the light source (EG&G FK 272), a highly sensitive GaAs photomultiplier tube for light detection (RCA C31034A), and a microcomputer (IBM AT–02) for system control and data processing.

In contrast to the xenon filled flashlamp used by Ferrari et al. (1985), the krypton type provides several suitable emission peaks in the red and NIR (750 to 900 nm, see Fig. 1). Four of these peaks are selected by interference filters with a half bandwidth of 10 to 16 nm and a transmission of more than 50% (see Fig. 2). Transmission maximum of each filter is centered at an emission peak of the flashlamp: 762 nm, 814 nm, 850 nm and 885 nm. The half bandwidth of the light

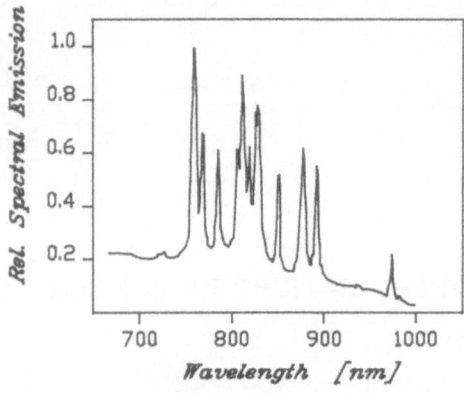

Fig. 1.   Relative spectral emission of the krypton flash lamp in the NIR (maximum output of whole spectrum, visible and NIR set, to 1)

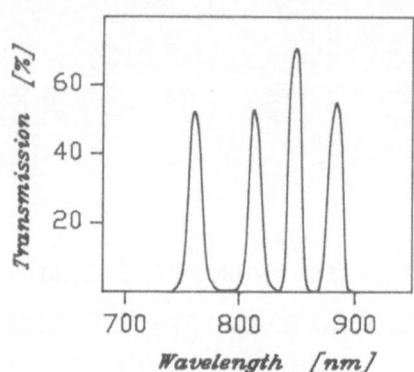

Fig. 2.   Spectral transmission of four interference filters used in the system.

penetrating the tissue is due to the half bandwidth of the emission peaks, which are 5 to 10 nm. Therefore systematic errors due to thermal or angular wavelength drift of the interference filters are minimal. During operation the flash lamp is run with 1 J flash energy (electrical) and 50 Hz repetition rate. Light pulse duration is $10\mu s$ (1/3 of peak). After 100 flashes the wavelength is changed by changing the interference filter in the optical path. The light arc of the flash emission is focused on either the sample directly or a fiber optic bundle, which guides the light to the tissue. Within that optical path the interference filters are positioned close to the entrance of the

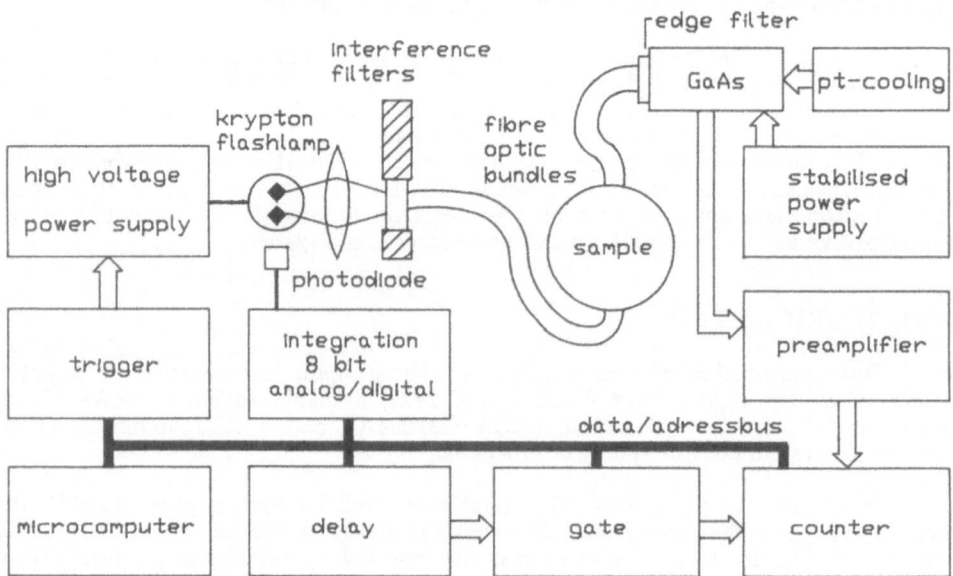

Fig. 3.   Schematic diagram of the spectrophotometric system based on a krypton flashlamp.

fiber bundle and changed by turning a filter wheel. In order to reduce the energy load of tissue, we select the wavelengths before illuminating the sample and not after transmission (Ferrari et al., 1985).

The light emission of the flash lamp is monitored by a photodiode giving a reference to the mircocomputer via an integrating circuit and an analog/digital converter. It can be shown, that the emitted light energy is almost constant and varies only in a range of 0.5% from flash to flash. This is negligible compared to the statistical error in light detection.

Transmitted light is detected by a photomultiplier (PMT), cooled down to −30 ºC by Peltier−cooling (see Fig. 3). The GaAs PMT−tube shows a high absolute responsivity up to 900nm, although the quantum efficiency decreases with increasing wavelength. To achieve highest sensibility, the PMT is run in photon counting mode, the most effective way of light detection. Light is collected by a second fiber bundle from the tissue, filtered by an edge filter (transmission wavelength > 730 nm) to reduce noise caused by scattered room light and focused on the GaAs photocathode of the multiplier. The PMT signal is preamplified and a 20 MHz counter, controlled by the mircocomputer, counts the pulses. Non−linearity effects like dead−time effect and dark pulses, known for photon counting, are taken into account. Absorption measurements on neutral density filters of defined absorption show linearity over a range from 7 to 10 optical densities (OD). At 8.5 OD the resolution is 0.02 OD, based on a time resolution of 10 seconds. Higher resolution in time results in less resolution in optical density, and vice versa. The drift of the whole system is less than 0.009 OD/h after preheating, which we considered acceptable for our purposes.

EXPERIMENTAL RESULTS

First in vivo results are shown in Fig. 4. After 5 minutes of initial observation, arterial occlusion of the forearm of an adult volunteer was induced, using a cuff around the upper arm. After 10 minutes of occlusion the cuff was released, followed by another 15 minutes of observation. Absorption changes of the forearm at three different wavelengths can be seen (Fig. 4). These absorption changes agree with those described by Cope and Delpy (1988) using their laser diode system in similar experiments.

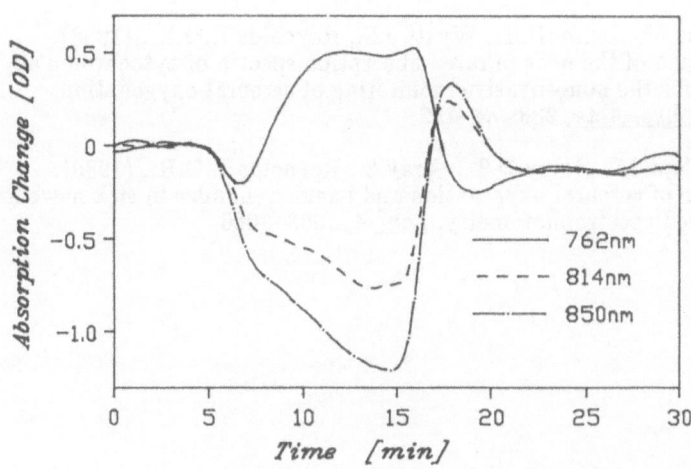

Fig. 4.   Absorption changes of an adult forearm during arterial occlusion.

The rapid decrease of absorption after occlusion for all three displayed wavelengths indicates the decrease in blood volume within the observed tissue volume. It appears that this is caused by continuously pressing the ends of the fiber bundles onto the tissue. Instant effects of hypoxia can be seen and dominate the further development of the absorption changes. The absorption change at 762 nm shows a different behavior compared to the others. This is mainly due to the concentration changes of hemoglobin (Hb) and oxyhemoglobin (HbO_2). Corresponding to the wellknown absorption characteristics of Hb and HbO_2 (Wray et al., 1988), absorption at 762 nm increases when oxygenation falls, while at wavelengths longer than 798 nm, the isosbestic wavelength (Wray et al., 1988), absorption is reduced. After releasing the cuff an overshoot compared to the initial values is recorded. It indicates a higher oxyhemoglobin concentration within the observed tissue volume than in the first five minutes of measurement. Ten minutes after release of the occlusion these absorption changes return to zero.

SUMMARY

We have designed a reliable and flexible low cost instrument for NIR spectroscopy. A krypton filled flashlamp was used as inexpensive light source. Providing a number of suitable emission peaks in the NIR, this flashlamp is ideal for NIR spectroscopy. Application to other NIR spectroscopy systems, e.g. to CCD (Charge Coupled Device) spectrophotometers, should be possible. The introduced system was tested on human arm tissue during arterial occlusion. Results equivalent to those described by other authors could be obtained by this new technical approach.

REFERENCES

Cope M., Delpy D.T., (1988), System for long—term measurement of cerebral blood and tissue oxygenation on newborn infants by near infra—red transillumination, Med. & Biol. Eng. & Comput., 26:289—294

Ferrari M., Giannini I., Sideri G., Zanette E., (1985), Continuous non invasive monitoring of human brain by near infrared spectroscopy, in: "Oxygen transport to tissue VII", Kreuzer et al., eds., Plenum Press, New York

Jöbsis F.F., (1977), Noninvasive, infrared monitoring of cerebral and myocardial oxygen deficiency and circulatory parameters, Science, 198:1264—1267

Wray S., Cope M., Delpy D.T., Wyatt J.S., Reynolds E.O.R., (1988), Characterization of the near infrared absorption spectra of cytochrome aa_3 and haemoglobin for the non—invasive monitoring of cerebral oxygenation, Biochim. Biophys. Acta, 933:184—192

Wyatt J.S., Cope M., Delpy D.T., Wray S., Reynolds E.O.R., (1986), Quantification of cerebral oxygenation and haemodynamics in sick newborn infants by near infrared spectrophotometry, Lancet, 1063—1066

A CONTINUOUS WAVE SPECTROSCOPIC (CWS) STUDY OF HEMOPROTEIN AND OTHER MOLECULES IN MITOCHONDRIAL SUSPENSION, CELL SUSPENSION AND TISSUE

S. Nioka, K.S. Reddy, A. Tanaka, and B. Chance

Department of Biochemistry and Biophysics
University of Pennsylvania
Philadelphia, PA 19104 USA

INTRODUCTION

Optical spectroscopy is the simplest technique for quantitative determination of hemoglobin, myoglobin, and other mitochondrial enzymes. This technique, however, is restricted to transparent and clear solutions in which optical density is a linear function of the absorber concentration. In scattering media, tissues, cells and cell organelles such as mitochondria, the dual wavelength, double beam technique has been extensively and successfully used to observe changes of the redox state in the mitochondrial chain, as well as changes in oxygen saturation of hemoglobin (Chance, 1951). This technique has also been used for quantitative determination of cytochromes in mitochondria, yeast, and E-coli cell suspensions. In most of these experiments, the effect of scattering on enzyme concentration was not quantified (Jöbsis, 1977) and only a few studies have focused in the scattering factor on attenuation spectra (Hoffman and Lübbers, 1985). In order to evaluate the scattering effect on the absorption spectra of tissues, the spectral characters of hemoglobin, cytochrome c, and cytochrome oxidase were studied. To make the comparison valid, transmittance spectra were collected along a straight path at a distance of 1 cm. The extinction coefficient was determined from continuous wave spectroscopy (CWS) in clear media. These studies provide important insights with respect to the effect of scattering on tissue spectra.

METHOD

Preparation of Hemoglobin: Human blood was freshly acquired and washed three times with a saline phosphate buffer mixture at pH of 7.2. A few drops of toluene were used to lyse the red cells and after vigorous shaking, the sample was centrifuged for use in hemoglobin spectral studies. Oxyhemoglobin and deoxyhemoglobin were made by bubbling diluted samples with oxygen and 100% argon respectively.

Preparation of Cytochromes: Cytochrome c was purchased from Sigma Chemical Company and cytochrome aa_3 was extracted and purified by the procedure of Li et al. (1987). The oxidized state of the enzymes was then reduced by sodium dithionate.

Preparation of Scattering Mediums

A) Mitochondrial Suspensions: Rats were anesthetized with 50 mg/kg of sodium pentobarbital and an incision was made on the abdominal wall to expose the liver. A teflon cannula (20 G) was inserted into the portal vein and the liver was perfused with 200 ml Ringer solution for 5 min. The liver was excised and minced in the phosphate buffer at pH of 7.2. The isolation procedure of Chance and Williams (1955) was used to prepare the mitochondria. The mitochondrial solutions were thinned by using 2% cholate in phosphate buffer.

B) <u>Yeast Suspensions</u>: A Baker's yeast solution (6-8% yeast/phosphate buffer), which had 3 to 4 light attenuation characters, was used for all the studies.

C) <u>Brain Homogenates</u>: Dogs and cats were anesthetized with an overdose of pentabarbiturate. The brain from each animal was perfused with Ringer-phosphate buffer at pH of 7.2 through carotid arteries and were frozen in liquid nitrogen. The brain was then minced and homogenized and used for the CWS study.

<u>Continuous Wave Spectroscopy (CWS) and Spectra Analysis</u>

The optical spectra of clear samples and of the samples which are attenuated by an absorber-scatterer mixture were measured using a multi-channel photodiode array system (Otsuka Electronics MCPD 1000), scanning over the wavelength range of 500 - 1100 nm. An incident light source, a 120 watt Tungsten lamp, and the MCPD monochromator having a 0.2 mm slit and providing a 2 nm resolution were coupled by glass fiber optical light guides. Scans, collected for 25 milliseconds, were compiled over 1 second intervals yielding 40 spectra. These spectra were averaged to improve the signal to noise ratio. Standard white paper and water were used for the reference sample. The reference spectra light intensity was controlled so that it was approximately equal to the transmission intensity of the sample. All averaged spectra were subtracted from the reference spectra to give an absolute spectrum of the absorber.

<u>Multicomponent Curve Fitting Analysis</u>

The concentrations of spectral components representing absorbing molecules (hemoproteins) were calculated from optical densities and extinction coefficients of the cytochromes using linear equations which follow Beer-Lambert's law.

$$\text{Optical Density (calculated)} = \sum_{i=1}^{n} \epsilon_i [C]_i$$

where n includes cytochrome aa_3, oxidized and reduced cytochrome c, and the difference between reduced cytochrome b and oxidized cytochrome b (n=5). For the near infrared region, water was added to the calculation as an extra component of the system (n=6).

RESULTS AND DISCUSSION

Figures 1A, 1B, 1C, and 1D show the optical spectra of purified hemoprotein; Hb and HbO_2, cytochrome c oxidized and reduced, cytochrome aa_3 oxidized and reduced, and the difference spectra of oxidized-reduced cytochrome b. Water and fat spectra are shown in Figures 1E and 1F. When the spectra are taken in scattered medium even the weak absorption bands such as the 760 nm deoxyhemoglobin band and the 655 nm cytochrome aa_3 band could easily be seen, indicating the inherent resolution and good signal to noise ratio obtained with the MCPD monochromator. Since the path length is longer than the distance between the input and output lightguides, water peaks may appear in the spectra (1E) and fat can be seen in cells and tissues (1F).

Attenuation spectra of mitochondrial suspensions, yeast suspensions and brain tissue are presented in Figures 2A, 2B and 2C. Each spectrum has identifiable hemoprotein peaks in oxygenated and deoxygenated states. Some similarities are found in the three spectra. All have typical cytochrome peaks (c at 550 nm, b at 560 nm and aa_3 at 605 nm) in the deoxygenated state. Copper absorption of cytochrome aa_3 at 830 nm is obscured in all spectra due to an overlapping water peak (840 nm). Furthermore, the amount of water varied during the experiment due to scattering (Figure 2C). Some differences are observed in these three different scattering materials. No water peaks appeared in the cholic acid-mitochondrial preparation while the other two spectra had distinct water peaks suggesting that light path lengths are more than 1 cm (water light path reference = 1 cm). When comparing brain spectra of the differing treatments the varying amounts of water suggest that during a transition from oxygenated brain tissue to deoxygenated brain tissue, the light path length was changed.

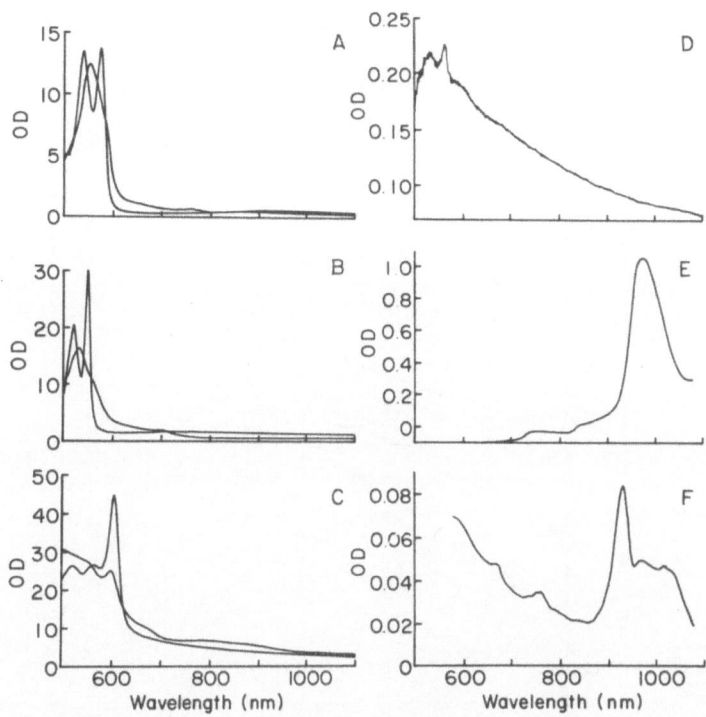

FIGURE 1. Optical spectra of purified hemoproteins Hb, HbO_2 (1A), oxidized and reduced cytochrome c (1B), oxidized and reduced cytochrome aa_3 (1C), difference spectrum of oxidized-reduced cytochrome b from antimycin treated mitochondria (1D), water (1E) and fat [butter] (1F).

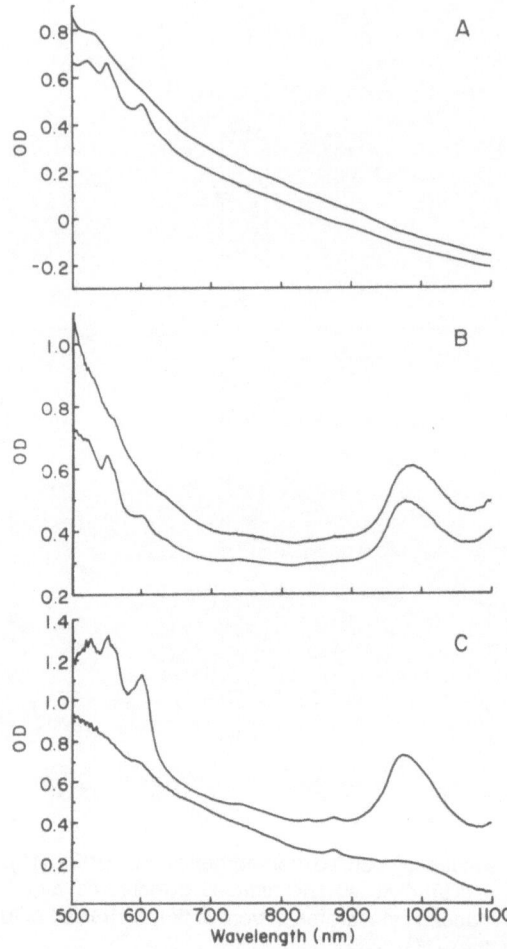

FIGURE 2. Attenuation spectra of mitochondrial suspensions (2A), yeast suspensions (2B), brain homogenates (2C). In each spectra, samples were oxygenated and deoxygenated. Optical density was offset by about 2 O.D. for yeast and brain samples. In 2C the two offset spectra show the oxidized and reduced states.

When cytochromes were reduced, attenuation spectra of these three samples were fitted according to a linear relationship shown in Figures 3A, 3B and 3C. Two fits were calculated using wavelength ranges between 500 nm and 650 nm and between 700 nm and 1100 nm. Sharp cytochrome c peaks at 550 nm in these scattering mediums were attenuated and flattened. The ratios of [cytochrome c] / [cytochrome aa_3] calculated from values obtained in equation 1, for mitochondria, yeast and brain tissue are 3, 6 and 1 respectively (500 to 650 nm wavelengths). The fitting of near infrared wavelengths were not acceptable. Copper bands at 830 nm are too small to detect and water peak disturbances, such as shifting peak position make fitting difficult. In addition, there are some peaks which have not been identified, for example, the 870 nm in brain and yeast spectra.

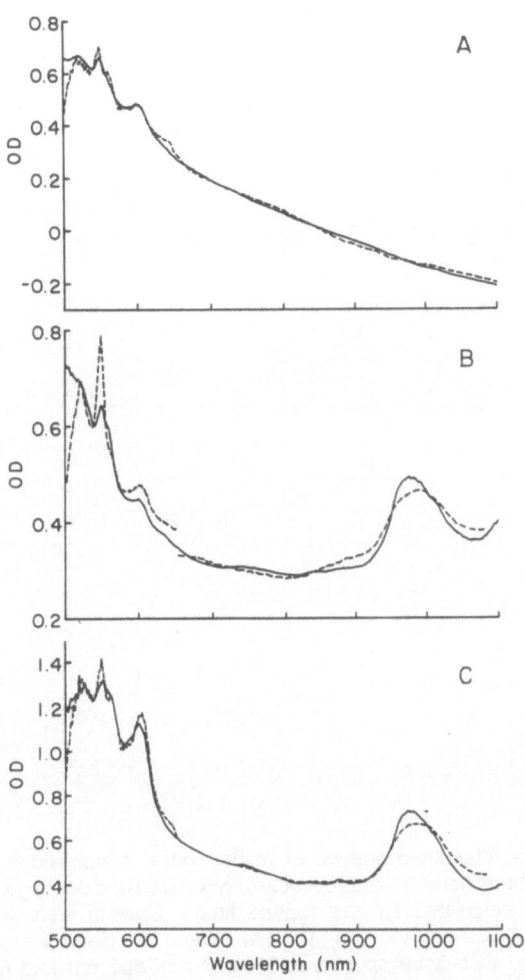

FIGURE 3. Attenuation and calculated (theoretical) spectra of mitochondrial suspensions (3A), yeast suspensions (3B), and brain homogenates (3C) were analyzed between 500 nm to 650 nm and 700 nm to 1100 nm. In the wavelength range 500-650 nm, cytochrome absorption peaks were attenuated by scattering. Unknown factors affected the shape of the spectra in the 700-1100 nm range, and thus the calculated spectra were not a good fit. A peak typically appeared in the homogenated brain at 870 nm. The water peak shifted from 980 nm in phosphate buffer to 970 nm in brain suspensions.

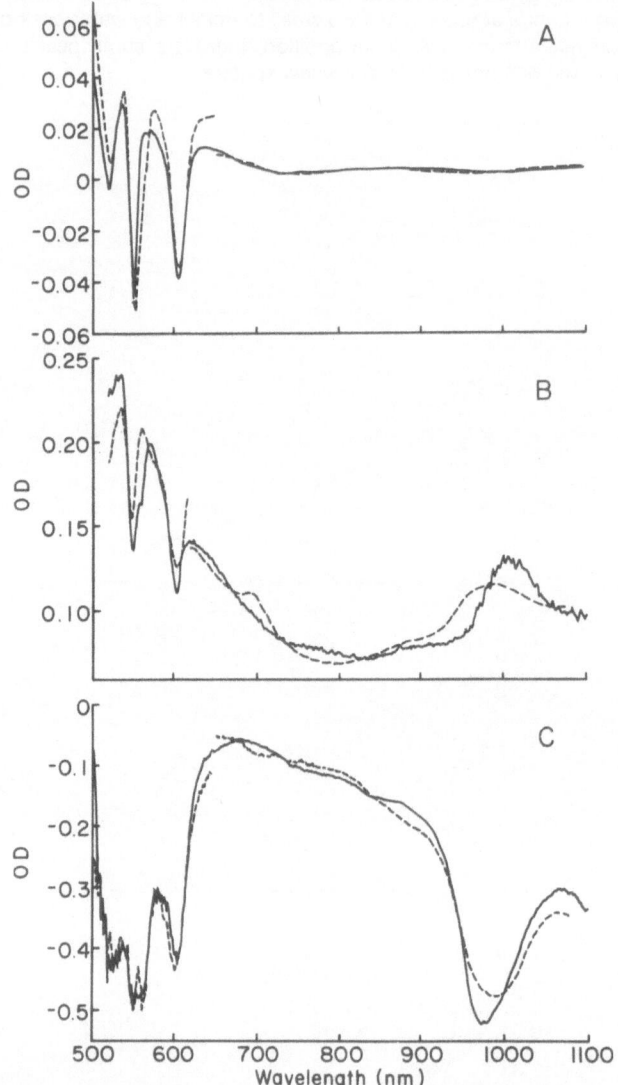

FIGURE 4. Difference attenuated spectra of mitochondrial suspensions (4A), yeast suspensions (4B),and brain homogenates between oxygenated and deoxygenated states (4C) are shown with calculated spectra (dotted line). Spectra were fitted in two wavelength ranges: 500-650 nm and 700-1100 nm. There are discrepancies between attenuated spectra and calculated spectra at 500-520 nm, 560 nm and 650 nm. In infrared (650-1100 nm) spectra, the water peak appears in a range different from the range found in visual (500-650 nm) spectra.

To eliminate the effects of scattering and the presence of unknown absorbers, which produce a non-linear baseline in spectra, it is convenient to use a "difference spectra", where values are obtained by taking the difference between the deoxygenated and oxygenated mitochondrial states. Reduced-oxidized difference spectra are presented in Figures 4A, 4B, and 4C.

Antimycin treated reduced spectra do not include oxidized cytochrome b, therefore the difference spectra between reduced mitochondria and antimycin treated oxidized mitochondria do not have cytochrome b. These spectra can be calculated from the multicomponent factor analysis with cytochromes c and aa_3 as shown in Figures 5A, 5B and 5C.

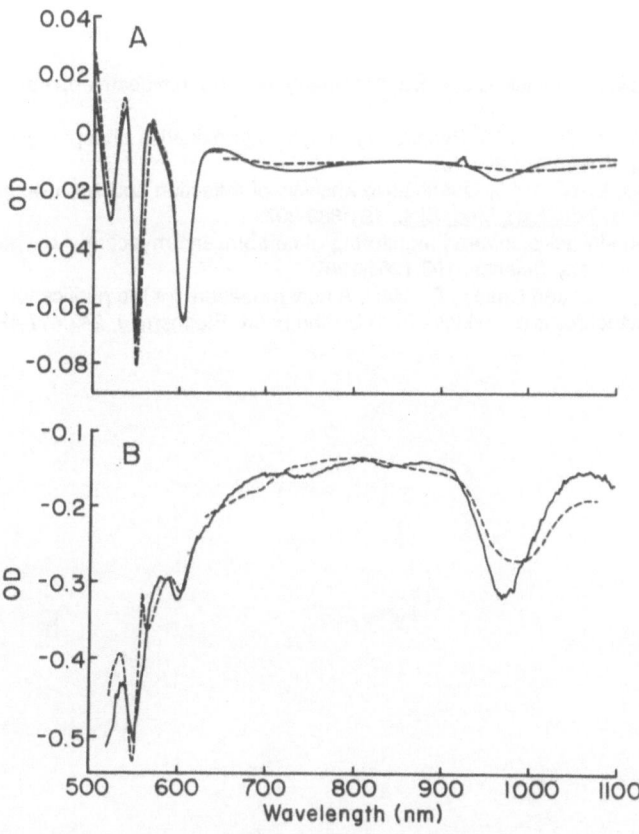

FIGURE 5. Antimycin treated difference spectra of mitochondrial suspensions (5A), yeast suspensions (5B) and brain homogenates between oxygenated and deoxygenated states (5C) are shown with calculated spectra (dotted line). The mitochondrial samples yielded better calculated values than yeast samples in the 500-650 nm range. Satisfactory calculations were not obtained in the infrared region due to poor fitting of the water peaks.

69

CONCLUSIONS

The existence of charge transfer bands of hemoproteins in the visual to infrared regions (500 to 1100 nm) are manifold. Furthermore, they are redox dependent. From this study we can conclude the following:

1) Hemoproteins are in sufficient quantity to be identified in the visual (500-650 nm) wavelength but not in the infrared region (700-1100 nm).

2) The relationship between O.D. (light attenuation) and the theoretical extinction of each hemoprotein component is non-linear. In the wavelength region from 500nm to 650 nm and at wavelengths higher than 930 nm, this non-linearity appears to be much more profound.

3) In the wavelength region from 650 nm to 930 nm, the attenuated spectra of hemoproteins are identifiable. The 1 cm spacing between light guide input and output, however, is not sufficient for an accurate analysis. Furthermore, when water peaks are superimposed on the hemoprotein absorption bands, it is difficult to accurately fit hemoprotein peaks. Additionally, unknown factors make it difficult to calculate absorption of hemoproteins.

ACKNOWLEDGEMENT

This work was supported by National Institutes of Health Grant NS 22881.

REFERENCES

Chance, B., 1951, Rapid and sensitive spectrophotometry. III A double beam apparatus, Rev. Sci. Instru. 22:634-638.

Chance, B., and Williams, E.R., 1955, Respiratory enzymes in oxidative phosphorylation. II Difference spectra, J. Biol. Chem, 217:395-407.

Hoffmann, J., Lübbers, D. W., 1985, Quantitative analysis of reflection spectra: evaluation of simulated reflection spectra, Adv. Exp. Med. Biol., 191:889-897.

Jöbsis, F. F., 1977, Non-invasive, infrared monitoring of cerebral and myocardial oxygen sufficiency and circulatory parameters, Science, 198:1264-1267.

Li, Y., Naqui, A., Frey, T. G., and Chance B., 1987, A new procedure for the purification of monodisperse highly active cytochrome c oxidase from bovine heart, Biochem. J. 242:417-423.

A TIME RESOLVED SPECTROSCOPIC (TRS) STUDY OF MIGRATION OF VISUAL TO INFRARED WAVES IN BRAIN TISSUE IN RELATION TO ABSORPTION OF HEMOPROTEINS

S. Nioka, G. Holtom, H. Miyake, M. Maris, and B. Chance

Department of Biochemistry/Biophysics and Department of Chemistry
University of Pennsylvania
Philadelphia, PA 19104 USA

INTRODUCTION

Previous optical spectroscopic studies of brain tissue revealed that only short photon migration distributions yeilds absorption spectra in the visual wavelength range (Heinrich et al. 1985), while longer wavelengths penetrated over greater distances (Brazy et al., 1985). The alpha and beta bands of hemoproteins in brain tissue have extremely high absorption coefficients which reduce the length of light migration distributions. This phenomena can be explained by photon migration theory. As shown in Figure 1A, in scattering media such as a brain tissue, light path lengths have a distribution function. In this figure, the Y-axis represents the log of photon intensity (I) or number of photons, and the X-axis represents time in nanoseconds. Migrating light path lengths (L), can be represented by use of the conversion factor: 1 ns = 23 cm (light path length in water).

Photon intensity decay is exponential having the following expression:

$$I = I_o \exp^{(-\epsilon [C] + S)L} \tag{1}$$

Photon migration distributions are primarily caused by a scattering factor (S). Photon absorption also affects the distribution pattern. Equation (1) only expresses the effect of absorption of the photon migration pattern in scattered medium. When the number of molecules which absorb photons changes without altering the scattering factor (S), then the difference in the slope is equal to $\epsilon \Delta[C]$.

$$I = \Delta I_o \exp^{-\epsilon [C]L} \tag{2}$$

Equation (2) implies that a change of slope is attributed to a change in absorber concentration [C]. In Figure 1A, two photon migration distributions in the brain which have two different hemoglobin concentrations are superimposed. The brain containing low hemoglobin concentration (Figure 1A) exhibits slower decay of the photon distribution curve as equation (2) suggested.

Changes in photon migration path length is always coupled with a change in slope. If a slope becomes steeper because of high deoxyhemoglobin concentration during hypoxia, then the mean path length must decrease as can be seen in Figure 1B. According to the random walk theory, this phenomena can be explained as follows: Photons which have a longer photon migration pathway have a greater probability of being absorbed. Mean photon path length is shortened by high amounts of an absorber. This principle was tested by using a yeast-hemoglobin solution to model brain tissue and brain homogenates. We utilize time resolved spectroscopy (TRS) to study the properties of photon migration as related to the absorber concentration and wavelength dependent patterns of photon migration at wavelengths from 500 to 900 nm.

Oxygen Transport to Tissue XII, Edited by J. Piiper *et al.*
Plenum Press, New York, 1990

MATERIALS AND METHODS

Time Resolved Spectroscopy

A mode-locked laser emits a train of pulses at 532 nm at a repetition rate of 76 MHz, which activates a cavity-dumped liquid dye laser with output pulses at 3.8 MHz. The liquid dyes of Rhodamine 590 and LDS 751 create tunable wavelengths ranging from 565 to 630 and 730 to 815 nm. Light is emitted in a pulse train of approximately 100 ps using Ar^+ laser for LDS 751, impinging upon a sample. Glass fiber bundles (3 mm) are used to guide the light through the sample chamber. Emergent light is coupled to the photomultiplier tube (PMT). Typical spectra are shown in Figure 1.

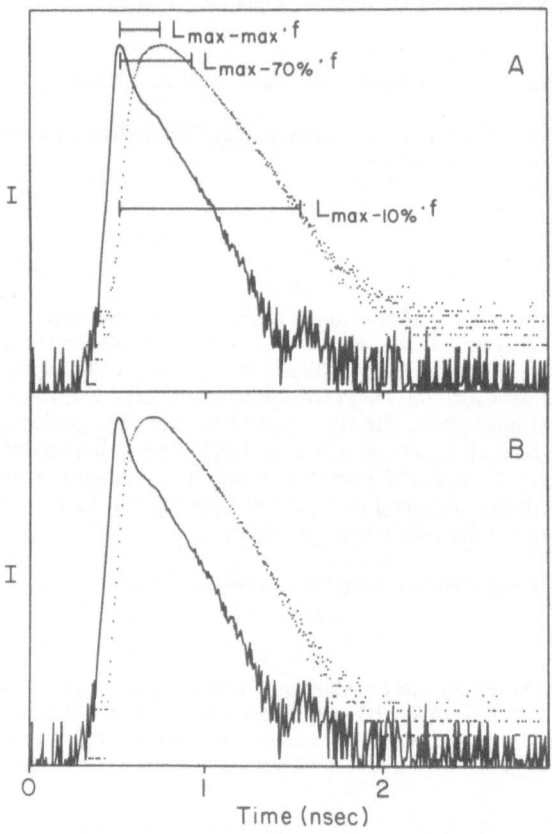

FIGURE 1. Time resolved spectra of cat homogenized perfused brains; the distance between light guides is 1 cm. Light intensity is expressed logarithmically without hemoglobin (1A) and with a residual amount of hemoglobin (1B). The curve to the left indicates the incident light distribution while the curve to the right indicates the emergent light distribution. Note that emergent light decay is higher within the hemoglobin-containing brain. Three time values were selected from the emergent light decay curve: time at which maximum, 70%, and 10% emergent light intensity was reached. These values were subtracted from the time at which the maximum incident light occurred (Lmax-max, Lmax-70%, Lmax-10%). These time differences were then converted into light path length by using the light velocity factor (f) in water.

Subjects and Experimental Protocol

Cat and dog brains, some perfused with saline-phosphate buffer at pH 7.4 and others not perfused, were Funnel frozen and homogenized. A 6% ratio of dry Baker's Yeast to phosphate buffer (pH 7.2) was used to model the migration patterns in brain tissue. Human hemoglobin was added to yeast solutions and yeast concentration was kept constant while hemoglobin concentrations were varied from $1\,\mu$M to 1 mM. These samples were placed in a 30 ml cylindrical container where light guides were placed 1 cm apart at a 180 degree angle. Since the light guides were directly facing the solutions and the container wall was located far from the core, the container had little influence on light migration distribution spectra.

ANALYSIS

Light Path Length Analysis

Light path lengths at wavelengths of 500 to 800 nm were measured in the 6% dry yeast/water solution. The yeast model shows a distinct light migration distribution. Figure 2 illustrates the effect of hemoglobin on the emergent Lmax-10% path length. As the amount of absorber (deoxyhemoglobin) increases, light path length decreases. Three time values were selected from lightpath distributions of various wavelengths which corresponded to maximum, 70% and 10% emergent light intensity. Each of these values was subtracted from the time at which the incident light reached a maximum (Lmax-max, Lmax-70%, Lmax-10%) and converted into path lengths using the velocity of light in water (Figure 1).

Light Intensity Decay

Light intensity decay (μ) was fitted exponentially between 40% and 5% of the maximum intensity decay curve.

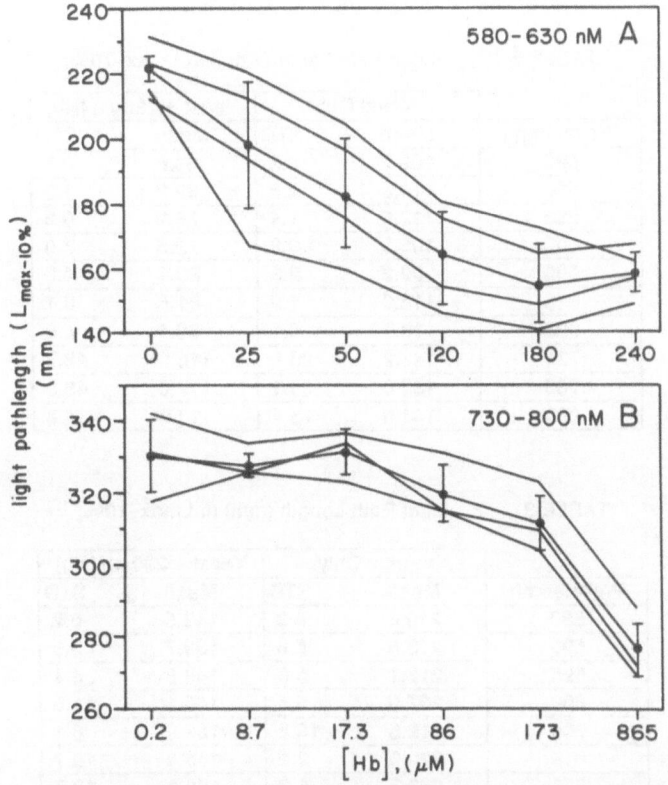

FIGURE 2. Light path lengths (Lmax-10%) as a function of [Hb] in a yeast-hemoglobin model for wavelengthd in the range of 580 nm to 630 nm (2A) and for wavelengths in the range of 730 to 800 nm (2B).

RESULTS AND DISCUSSION

Light Path Length Analysis

As shown in Tables 1-3, the 10% distribution path lengths demonstrated the greatest sensitivity to hemoglobin concentration. Photons of the 10% distribution component migrate over longer distances. These distances are shortened by 5-8 cm with the addition of hemoglobin to the yeast solution.

Lmax-max values at various wavelengths were plotted against [Hb] as shown in Figure 3. Low sensitivity to [Hb] implies that the maximum emergent light intensity is more significantly influenced by scattering than by the amount of absorber (Patterson et al., 1989).

TABLE 1. Light Path Length (mm) at Lmax–max

Wavelength	Yeast Only		Yeast + 230μm [Hb]	
	Mean $\pm$	STD	Mean $\pm$	STD
580	69.8	0.7	36.5	7.3
590	69.6	3.3	31.7	0.1
595	63.3	2.3	45.8	2.0
600	64.7	0.9	39.8	10.6
605	76.1	1.7	53.5	4.9
615	79.9	0.3	49.7	6.0
625	74.9	2.0	48.4	4.7
735	96.6	19.6	67.8	27.8
760	76.5	22.7	71.0	31.4
800	90.7	10.0	67.6	28.7

TABLE 2. Light Path Length (mm) at Lmax–70%

Wavelength	Yeast Only		Yeast + 230μm [Hb]	
	Mean $\pm$	STD	Mean $\pm$	STD
580	109.2	2.3	64.6	3.5
590	114.3	2.5	62.2	1.5
595	112.9	1.4	76.2	0.8
600	105.3	0.9	72.6	7.0
605	109.2	2.3	80.1	4.8
615	113.0	1.2	81.5	6.1
625	116.7	0.1	80.1	4.8
735	147.2	29.5	116.9	42.7
760	132.0	35.1	117.6	43.8
800	144.0	13.4	113.0	38.8

TABLE 3. Light Path Length (mm) at Lmax–10%

Wavelength	Yeast Only		Yeast + 230μm [Hb]	
	Mean $\pm$	STD	Mean $\pm$	STD
580	210.8	4.3	148.5	0.2
590	213.5	6.9	139.7	3.6
595	212.1	5.6	149.9	4.1
600	207.0	5.4	145.1	11.5
605	212.5	13.2	153.9	8.1
615	220.9	2.2	156.5	10.7
625	233.7	5.3	159.1	10.8
735	276.3	56.1	232.5	74.1
760	263.6	62.1	230.2	74.7
800	275.5	36.3	228.4	68.0

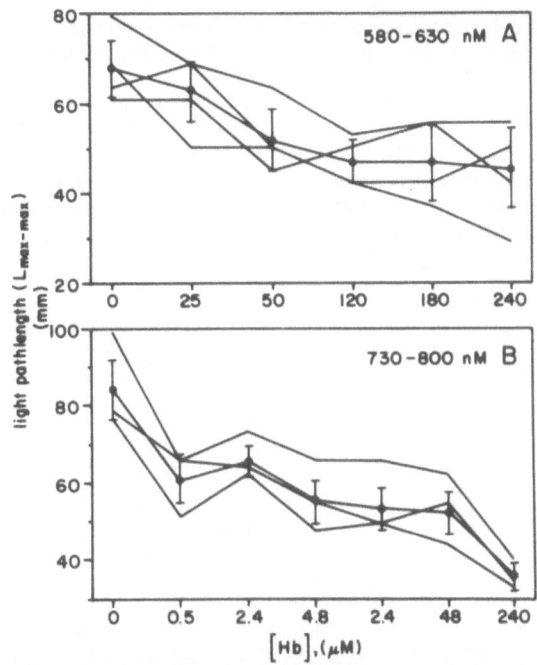

FIGURE 3. Light path lengths (Lmax-max) as a function of [Hb] in yeast hemoglobin model for wavelengths in the range of 580 nm to 630 nm (3A) and for wavelengths in the range of 730 to 800 nm (3B).

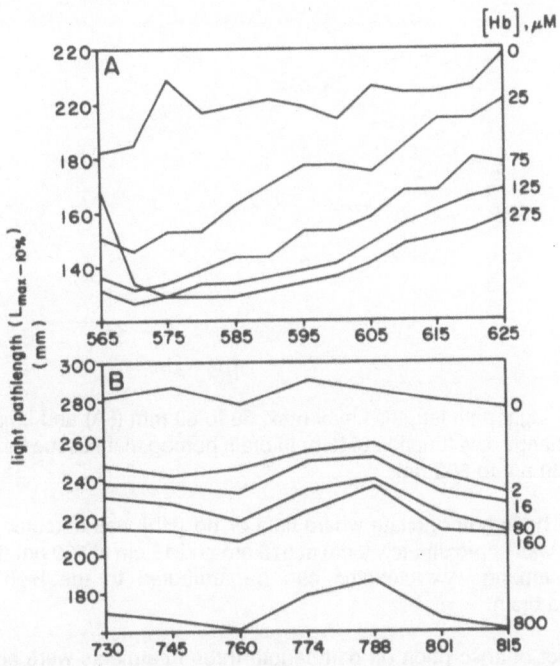

FIGURE 4. Light path lengths (Lmax-10%) as a function of wavelengths and [Hb] in yeast hemoglobin model for wavelengths in the range of 565 nm to 625 nm (4A) and for wavelengths in the range 730 to 815 nm (4B). These spectra show deoxyhemoglobin peaks at 565nm (4A) and at 760 nm (4B).

The relationship of light path length to [Hb] in two wavelength ranges are displayed in Figure 4 (Lmax-10%). Extinction maxima of deoxyhemoglobin can be clearly detected at 565 and 760 nm.

In homogenated brain samples, similar characteristics between path length and [Hb] are observed. In these [Hb] containing adult brains, hemoglobin concentrations were significant and produced a Lmax-max of 3 cm and Lmax-10% of 13 cm in the 565 to 630 nm range. There was no change in light path length as [Hb] was added in the homogenated brain sample. Analysis of the wavelength range from 700 - 800 nm, however, shows that path lengths are shortened by [Hb] (Figure 5). At physiological [Hb], the Lmax-max and Lmax-10% were equivalent to 5 cm and 12 cm respectively.

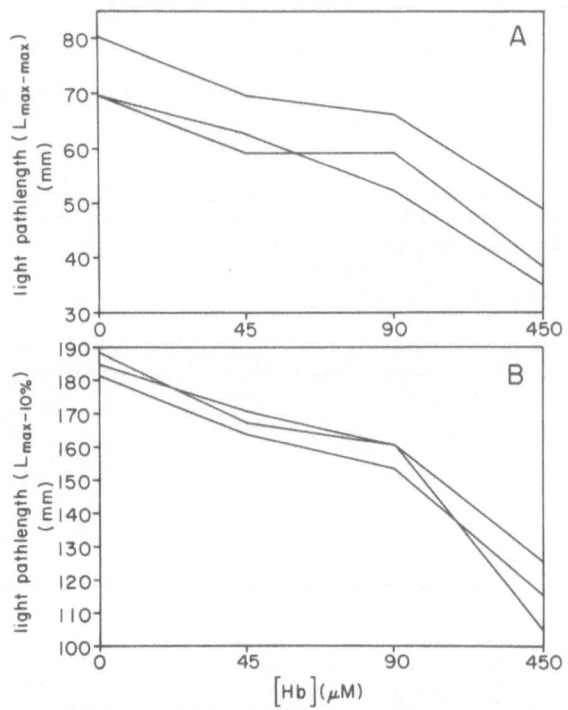

FIGURE 5. Light path lengths Lmax-max, 30 to 80 mm (5A) and Lmax-10%, 110 to 190 mm (5B) change as a function of [Hb] in brain homogenate tissue for wavelengths in the range of 730 nm to 800 nm.

Using perfused brain homogenate where little or no [Hb] was detected in the wavelength domain spectra, Lmax-max was approximately 9 cm at 576 nm and 15 cm at 630 nm (Figure 6). These small path length differences among wavelengths can be attributed to the high extinction coefficients of hemoproteins in the brain.

To test the effect of absorption on path length three treatments were administered to brain tissue. Extinction coefficients were altered by changing molecular states. Oxygen was used to produce oxyhemoglobin and an oxidized redox state. Potassium cyanide was used to produce oxyhemoglobin and a reduced redox state and sodium dithionide was used to produce deoxyhemoglobin and a reduced redox state. The differences in light path lengths, exhibited by the treated samples, were within 2 cm of each other.

76

Exponential Decay (μ) of the Lightpath Distribution in Yeast-Hemoglobin Model

We tested if, in equation (2), $\epsilon \Delta[C]$ adequately represents light intensity decay by plotting $\Delta[C]$ versus light intensity decay (Figure 7). Light intensity decay is proportional to a wide range of [Hb] values at 700-800 nm, while at 550-630 nm, there is a small range of [Hb] values that show this relationship. At higher hemoglobin concentrations, there are no correlations between [Hb] and light intensity decay. The transition values of [Hb], where correlations with light intensity decay cease are 25 μM at 570 nm and 45 μM at 600 nm (Figure 7A) and 450 μM at 630 nm (Figure 7B). If the hemoglobin concentration in the brain exceeds this transition level, it would be impossible to estimate light path length from the concentration. These non-linear relationships are probably due to artifacts caused by stray light. This effect may be due to the high extinction of hemoproteins or due to a consequence of the strong absorption of light in the visible region.

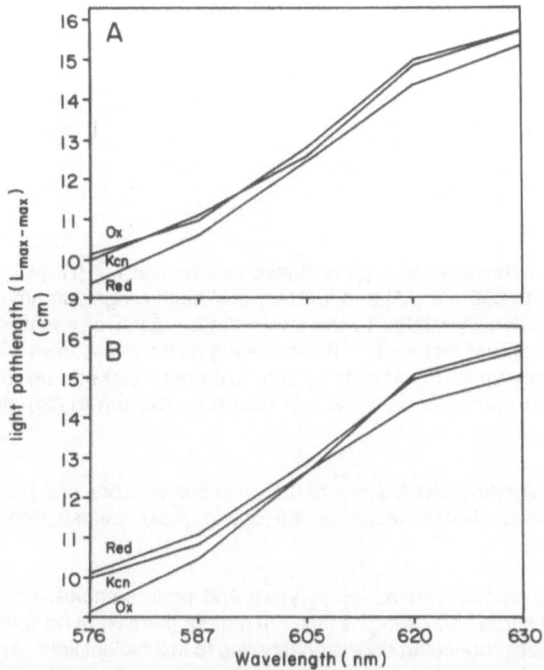

FIGURE 6. Light path lengths as a function of wavelength, 576 nm to 630 nm, in perfused brain without hemoglobin (6A) and with residual hemoglobin (6B). Results from treated samples (oxygen, KCN and dithionite treatments) in both figures show only slight differences in path lengths. Higher extinction coefficients and increased hemoglobin concentration reduce path lengths by values within 1 cm.

CONCLUSION

Time resolved spectroscopy provides photon migration distributions as predicted in the theoretical calulation of equation (2). Light path lengths, defined as Lmax-max, Lmax-70%, and Lmax-10%, were

77

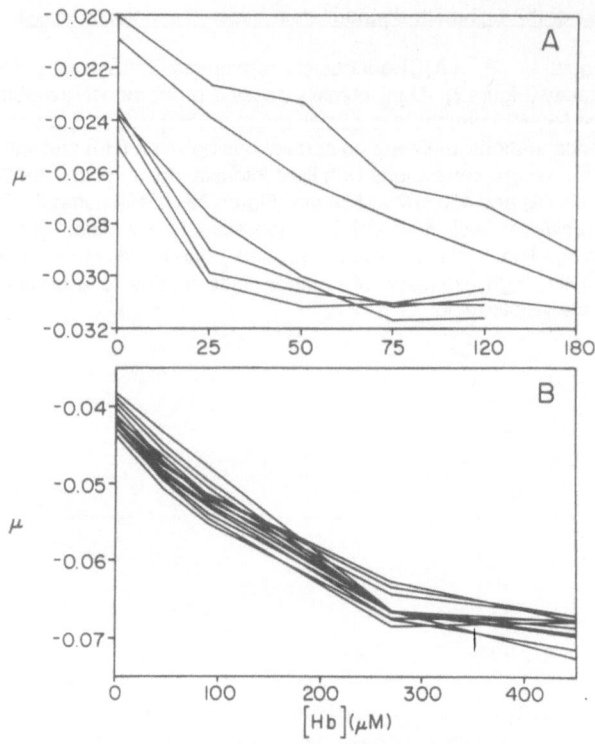

FIGURE 7. Light intensity decay (μ) is plotted as a function of [Hb] for the wavelength range of 570 nm to 630 nm (7A) and for the wavelength range 730 nm to 790 nm (7B). In 7B, μ is proportionally related to the concentration Δ[C] of equation (2). In 7A, μ values at [Hb] of 25 μM increase as the wavelength decreases from 630 to 570. This principle holds only up to 75 μM of [Hb] from 570 nm to 590 nm, up to 75 μM of [Hb] from 600 nm to 610 nm, and up to 180 μM [Hb] from 620 nm to 630 nm.

measured from photon migration distributions of 6% yeast suspensions and brain homogenates. Using samples 1 cm in thickness yielded values of 7-9 cm in yeast suspensions and 7-16 cm in brain homogenate preparations.

When 230 μM of hemoglobin was added to yeast and brain samples, the light path lengths were shortened by 30-50%. At shorter wavelengths (500 nm region) there were no linear relationships between path lengths and hemoglobin concentration, probably due to insufficient light. The effective path lengths decrease as hemoglobin increases due to greater photon absorption. Therefore, the total volume of tissue through which photons migrate is reduced.

ACKNOWLEDGEMENT

This work was supported by National Institutes of Health Grants NS 22881 and RR 01348.

REFERENCES

Brazy, J. E., Lewis, D. V., Mitnick, M. H., and Jöbsis, F. F., 1989, Noninvasive monitoring of cerebral oxygenation in preterm infants: preliminary observations, Pediatrics, 75:217-225.

Heinrich, U., Hoffmann, J., Baumgärtl, H., Yu, B., Lübbers, D. W., 1985. Oxygen supply of the blood-free perfused buinea pig brain at three different temperatures, Adv. Exp. Med. Biol., 191:77-90.

Patterson, M. S., Chance, B., and Wilson, B. C., 1989, Time resolved reflectance and transmittance for the non-invasive measurement of tissue optical properties, Applied Optics, 28:2331-2336.

THE EFFECT OF OPTODE POSITIONING ON OPTICAL PATHLENGTH IN NEAR INFRARED SPECTROSCOPY OF BRAIN

P. van der Zee, S.R. Arridge, M. Cope, D.T. Delpy

Department of Medical Physics & Bioengineering
University College London
1st Floor Shropshire House
11-20 Capper Street
London WC1E 6JA

INTRODUCTION

The use of optical spectroscopy for the non-invasive monitoring of tissue oxygenation and metabolism is well established (Chance et al., 1975). Historically because of the high absorption by tissue of light in the visible range, optical monitoring was often restricted to measurements of reflected light (Jöbsis et al., 1977). Subsequently Jöbsis showed that by using near infrared light (NIR), tissue absorption became sufficiently low to make transillumination of the cat head possible (Jöbsis, 1977). In the near infrared region (700-1300 nm) there is sufficient spectral information available to permit changes in the concentration of haemoglobin and cytochrome aa_3 to be calculated, and hence changes in the oxygenation state of the brain (Brazy et al., 1985, 1986; Ferrari et al., 1986; Fox et al., 1985). This technique is now used routinely to monitor cerebral oxygenation and haemodynamics in the human newborn infant (Wyatt et al., 1986; Edwards et al., 1988), using an instrument designed to transilluminate the heads of most newborn infants (Cope and Delpy, 1988). This instrument allows for measurements through heads up to 8-9 cm in diameter.

In order to quantitate NIRS haemoglobin and cytochrome aa_3 data, it is normally necessary to know the optical pathlength of the light in the tissues. Due to the high amount of light scattering in tissues this is appreciably larger than the interoptode spacing. For the case of transillumination through a slab of tissue we have shown (Delpy et al., 1988) that the effective optical pathlength through the tissue is equal to the interoptode spacing multiplied by a near constant factor (the differential pathlength factor or DPF) which is dependent upon the optical characteristics of the tissue. The DPF, which relates a change in measured light transmission to a change in true absorption, can be derived from a measurement of the time of flight for ultrashort optical pulses through the tissue. For the above slab geometry, using both a Monte Carlo model and measurements on a phantom, we were able to show that the DPF corresponded to the mean transit time of the photons emerging from the tissue. By applying this time of flight technique to the case of the transilluminated heads of pre term infants, the DPF has been found to be 4.4 (Wyatt et al., 1989). In a further study in animals, the change of the DPF with absorption in the brain was investigated (Delpy et al., 1989). The DPF was found to change by 30%/OD/cm, the change being approximately linear over the whole physiological range from $FiO_2 = 100\%$ to death. For the normal physiological range of oxygenation and blood volume, the maximum

Oxygen Transport to Tissue XII, Edited by J. Piiper et al.
Plenum Press, New York, 1990

absorption change observed in the brain, both in animals and in human studies, is typically 0.2 OD/cm, resulting in a DPF change of less than 10%.

Although the DPF, as determined by the time of flight (TOF) measurements, has already been applied to data from rat and human infant brain, the validity of the mean time of flight as a measure of pathlength has only been verified so far for a slab geometry. Also it is not always possible to transilluminate the head of larger infants, and in these cases the optodes must be positioned such that they only partially illuminate the tissues. In the case of large full term infants it is often necessary to position one optode over the fontanelle with the other on the side of the head. In such cases, the angle between the optodes may be 90° or less. The questions that must be asked in these circumstances are firstly whether the DPF as determined from optical time of flight measurements applies to a spherical geometry, which more closely resembles the shape of the head, and secondly whether the calculated DPF is valid for arbitrary optode positioning on the (approximately) spherical head.

MODELLING OF LIGHT TRANSPORT IN TISSUE.

To investigate the effects of optode geometry on the effective pathlength for light going through highly scattering tissues, a Monte Carlo model for light transport in tissue was used (van der Zee and Delpy, 1987). The model gives a full three dimensional simulation of light transport. The scattering phase function of the tissue used in the model is based on an experimentally measured phase function for rat brain at 783 nm (van der Zee and Delpy, 1988). Specular reflection and refraction at the tissue boundary, due to differences in refractive index were taken into account.

In this simulation, a perfectly collimated, infinitesimally narrow and infinitely short pulse of light was incident at (0,0,-r), on a sphere of radius r (Figure 1). The model keeps track of the photons as they pass through the sphere. The angular position on the sphere, Θ, and the transit time t are recorded for each photon that emerges. The data is collected into ten annular rings for values of Θ in 20° intervals centred on: 0°, 20°, 40°, 60°, 80°, 100°, 120°, 140°, 160° and 180°. Within each of these annular rings a distribution of transit times is obtained, and the mean of this time (t_{mean}) is determined.

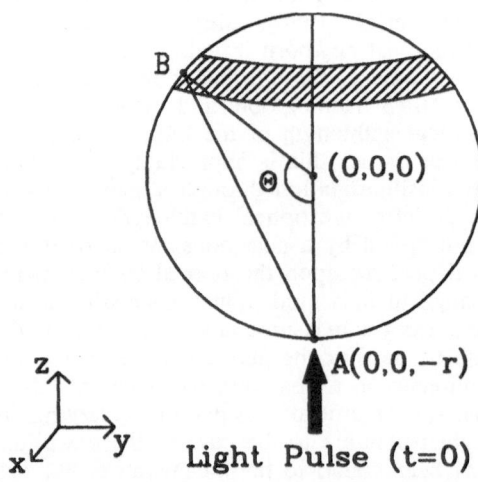

Figure 1. Geometry for the model. Light enters at the bottom of the sphere. The exit position is specified by the angle Θ. The optode spacing equals AB.

80

Also the intensity of the exiting light in each ring, normalised for the surface area of that ring is calculated. For the interoptode spacing, the cord length AB from input to output site is used ($AB = 2r*sin(\Theta/2)$). The DPF for a particular angle is obtained as follows :

$DPF(\Theta) = (t_{mean}(\Theta)*c/n)/AB$.

Here c is the velocity of light in vacuo and n is the tissue refractive index, taken to be 1.4 (Gahm et al., 1986). For each ring the change in attenuation per unit change in absorption, normalised for the chordlength AB, is the incremental Beer Lambert pathlength factor for that angle (ie $(\delta OD(\Theta)/\delta\mu_a)/AB$). The parameters for which the model was run were: sphere radius r=15mm, scattering coefficient μ_s=10mm^{-1} (1/e) and for a range of absorption coefficients : μ_a=0.01, 0.02, 0.04, 0.06, 0.08, 0.10mm^{-1} (1/e). The total number of detected photons for each absorption coefficient was 1.2 million.

RESULTS

Figures 2a, b, c show the calculated differential pathlength factor derived from the mean of the time of flight plotted against the incremental Beer Lambert pathlength factor determined from $(\delta OD/\delta\mu_a)/AB$, for exit angles of respectively 130-180°, 70-130° and 10-70°. The line if identity is also shown.

In Figures 3a and 3b, the DPF is plotted as a function of Θ for absorption coefficients μ_a of 0.01 to 0.10mm^{-1}. The DPF can be seen to decrease with increasing absorption, as reported previously (Delpy et al., 1989), furthermore there is a variation with Θ.

In Figure 4, the DPF, normalised to unity at μ_a = 0.010 is plotted as function of absorption coefficient, μ_a, for different values of Θ.

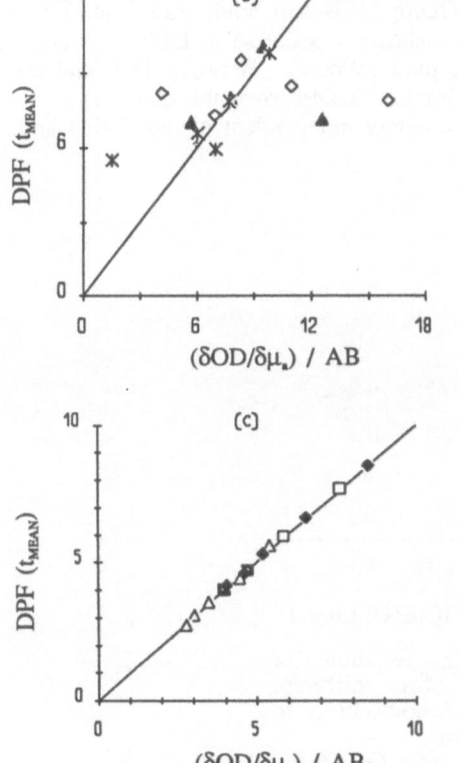

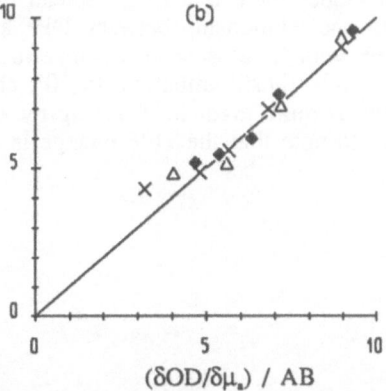

Figure 2. DPF calculated from time of flight measurement versus incremental Beer Lambert pathlength factor, for different exit angles Θ.
(a): (◊)=180°, (△)=160°, (*)=140°.
(b): (♦)=120°, (△)=100°, (×)=80°.
(c): (♦)=60°, (□)=40°, (△)=20°.

81

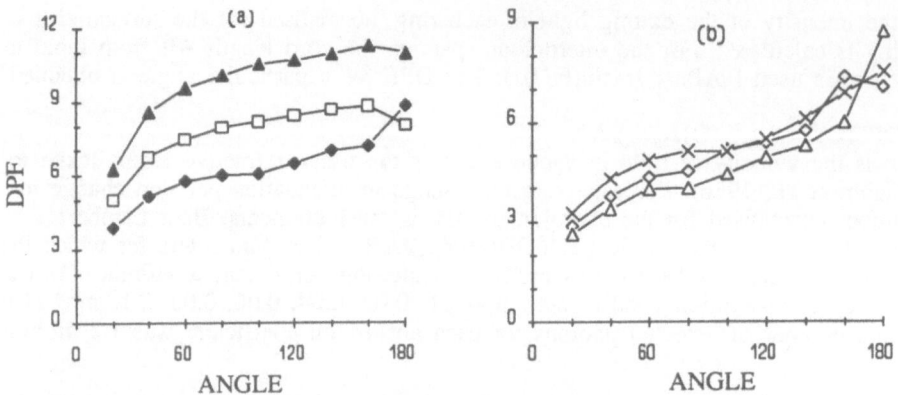

Figure 3. Differential pathlength factor (DPF) as a function of angle for different absorption coefficients, μ_a. (a): $(\triangle)$=0.01, $(\square)$=0.02, $(\blacklozenge)$=0.04. (b): $(\times)$=0.06, $(\lozenge)$=0.08, $(\triangle)$=0.10 mm^{-1}.

DISCUSSION

The first question asked concerned the validity for a spherical geometry of the use of the mean of the optical time of flight to determine the differential pathlength factor (DPF) relating variations in attenuation to true absorption changes. The data in Figures 2a,b & c show a good linear relationship with unity slope and zero intercept between the DPF derived from the mean transit time and the incremental Beer Lambert pathlength factor, $(\delta OD/\delta\mu_a)/AB$. This is seen clearly for the smaller exit angles. For the larger angles there is a larger spread in the data points due to lack in statistics, but the one to one relationship between DPF and $(\delta OD/\delta\mu_a)/AB$ still holds true. The change in the DPF with Θ, as seen in Figures 3a and 3b, shows a decrease in DPF in going from 180°, full transillumination, to 20°, close to pure reflection. Between 180° and 60° this change is quite gradual, changing by approximately 2 DPF over the range. It is interesting to note that the DPF change is approximately independent of absorption and

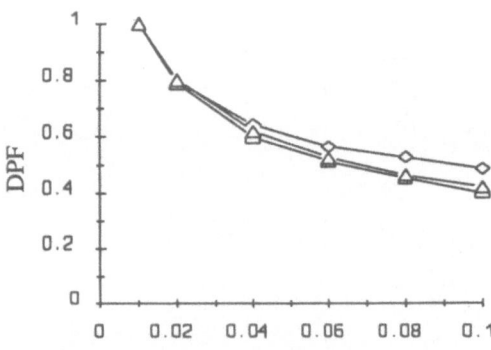

ABSORPTION COEFFICIENT (mm^{-1})

Figure 4. DPF as a function of absorption coefficient for different angles, Θ. The data are normalised to the DPF's at μ_a=0.01 mm^{-1}. $(\lozenge)$: Θ=140°, $(\square)$: Θ=80°, $(\triangle)$: Θ=20°.

82

thus the percentage change will be greater for the higher absorption values. The typical value for blood perfused brain would be 2.7%/10°. At Θ less than 60° there is a more rapid decrease in DPF with angle, particularly for the case of low absorption. This could lead to large errors in the calculation of chromophore concentration from measurements made in reflection mode. In particular, reflection measurements made on tissues of low absorption could be prone to error due to the optical pathlength changing substantially with variations in tissue absorption (eg fluorocarbon perfused rat brain, where the total absorption coefficient is only $0.018mm^{-1}$ at 780nm, for 20μMolar of oxygenated cytochrome aa_3 and 80% water content). A possible explanation for the lower DPF values at decreasing angle is that photons are lost from the sphere. With decreasing angle Θ (Figure 1), photons are unable to take long paths on the outside of the sphere and are therefore lost to free space. This leads to a shortening of the mean time relative to the chord length AB. This argument is consistent with the rapid decrease in DPF at very small angles.

The relative change in the DPF with absorption, as shown in figure 4, is seen to be largest at low absorption, and for smaller angles, with remarkably little difference between the results for $\Theta=80°$ and $\Theta=20°$.

CONCLUSIONS

Using a computer simulation for light transport in tissue it has been shown that the mean of the time of flight for a short pulse of light through a sphere of tissue is an accurate measure of the DPF, the factor used in converting changes in attenuation to true concentration changes. This DPF changes with angular position on the sphere. For the range of absorption coefficients found in normally perfused brain, and for optode geometries varying from 180° to 60°, use of the same DPF as a multiplier of the direct interoptode spacing leads to an error in the calculated concentration changes of up to 35%. Therefore for more precise calculations, the DPF should be varied according to the optode geometry employed in the measurements. For angles of less than 60° there is a more rapid change in DPF with both angle and tissue absorption coefficient precluding the use of a simple constant pathlength factor. This leads to a difficulty in the interpretation of reflection type measurements in the near infrared region. These problems could of course be overcome by incorporating a time of flight measurement facility into the tissue spectrometer.

ACKNOWLEDGEMENTS

This work was carried out with funding provided by the SERC, the Wellcome Trust and Hamamatsu Photonics K.K.

REFERENCES

Brazy, J.E., Lewis, D.V., Mitnick, M.H., Jöbsis, F.F., 1985, Noninvasive monitoring of cerebral oxygenation in preterm infants: preliminary observations. Pediatrics, 75, 217-225

Brazy, J.E., Lewis, D.V., 1986, Changes in cerebral blood volume and cytochrome aa3 during hypertensive peaks in preterm infants. Pediatrics, 108, 983-987

Chance, B., Legallis, V., Sorge, J., Graham, N., 1975, A versatile time sharing multichannel spectrophotometer, Reflectometer and Fluorometer. Anal. Biochem., 66, 498-514.

Cope, M., and Delpy, D.T., 1988, System for long term measurement of cerebral blood and tissue oxygenation on newborn infants by near infrared transillumination, Med. Biol. Eng. Comp. 26, 3, 289-294.

Delpy, D.T., Cope, M., van der Zee, P., Arridge, S.R., Wray, S. and Wyatt, J.S., 1988, Estimation of optical pathlength through tissue from direct time of flight measurement, Phys. Med. Biol. 33, 12, 1433-1442.

Delpy, D.T., Arridge, S.R., Cope, M., Edwards, D., Reynolds, E.O.R., Richardson, C.E., Wray, S., Wyatt, J.S. and van der Zee, P., 1989, Quantitation of pathlength in optical spectroscopy. Adv. Exp. Med. & Biol., 247, 41-46

Edwards, A.D., Wyatt, J.S., Richardson, C., Cope, M., Delpy, D.T., Reynolds, E.O.R., 1988, Cotside measurement of cerebral blood flow in ill newborn infants by near infrared spectroscopy, Lancet, ii, 770-771.

Ferrari, M., De Marchis., Giannini., Nicola, A., Agostino, R., Nodari, S., Bucci, G., 1986, Cerebral blood volume and haemoglobin oxygen saturation monitoring in neonatal brain by near infrared spectroscopy. Adv. Exp. Med. Biol., 200, 203-212

Fox, E., Jöbsis, F.F., Mitnick, M.H., 1985, Monitoring cerebral oxygen sufficiency in anaesthesia and surgery. Adv. Exp. ed. & Biol., 191, 849-854

Gahm, T., and Witte, S., 1986, Measurement of the optical thickness of transparent tissue layers. J. Microscopy. 141, 101-110

Jöbsis, F.F., Keizer, J.H., LaManna, J.C., Rosenthal, M., 1977, Reflectance spectrophotometry of cytochrome aa$_3$ in vivo. J. Appl. Physiol. 43, 858-872.

Jöbsis, F.F., 1977, Non invasive, infrared monitoring of cerebral and myocardial oxygen sufficiency and circulatory parameters, Science, 198, 1264-1267.

van der Zee, P. and Delpy, D.T., 1987, Simulation of the point spread function for light in tissue by a Monte Carlo model, Adv. Exp. Med. & Biol., 215, 179-192.

van der Zee, P. and Delpy, D.T., 1988, Computed point spread functions for light in tissue using a measured volume scattering function, Adv. Exp. Med. & Biol., 222, 191-198.

Wyatt, J.S., Cope, M., Delpy, D.T., Wray, S. and Reynolds, E.O.R., 1986, Quantitation of cerebral oxygenation and haemodynamics in sick newborn infants by near infrared spectroscopy, Lancet, 8515, 1063-1066.

Wyatt, J.S., Cope, M., Delpy, D.T., Richardson, C.E., Edwards, A.D., Wray, S.C., Reynolds, E.O.R., 1989, Quantitation of cerebral blood volume in newborn infants by near infrared spectroscopy. J. Appl. Physiol. (submitted).

CEREBRAL CYTOCHROME-C-OXIDASE COPPER BAND QUANTIFICATION IN PERFLUOROCARBON EXCHANGE TRANSFUSED CATS.

M. Ferrari*, Daniel F. Hanley°, David A. Wilson°, Richard J. Traystman°

*Department of Biomedical Sciences and Technology, and Biometrics University of L'Aquila, 67100 L'Aquila, Italy and Physiopathology Laboratory, Istituto Superiore di Sanità, Viale Regina Elena 299, 00161 Rome, Italy
°Department of Anesthesiology and Critical Care, Johns Hopkins University, 21205 Baltimore, MD USA

INTRODUCTION

Near infrared (IR) spectroscopy (NIRS) offers physicians and associate health care persons a potential means of monitoring changes in brain hemoglobin content and cerebral oxygenation noninvasively (Jobsis, 1977; Ferrari et al., 1989). The principles of NIRS have been explained previously (Jobsis, 1977; 1985). Briefly, IR light in the 700-1000 nm region is transmitted through soft and hard tissues of the head to provide a signal for spectrophotometric purposes. The intensity of the transmitted reflected light depends primarily on the wavelength, scattering effects, and upon NIRS absorption by the most abundant chromophore, i.e. the heme of hemoglobin. It has been suggested that cytochrome-c-oxidase (Cyt a,a3) redox band, previously described "in vitro" (Wharton and Tzagoloff, 1964), contributes to the absorption band providing a means for "in vivo" quantification (Jobsis, 1977; Brazy and Lewis, 1986; Wyatt et al. 1986).

Two copper centers confer on the oxidized Cyt a,a3 the capacity to absorb ligth in the range 800-850 nm, with a maximum at approximately 830 nm. The most notable is the visible copper ($CuA2^+$). Recognizing the interference of the hemoglobin presence and its content changes, the band of the oxidized Cyt a,a3 was evidenced in hemoglobin free rats (Ferrari et al., 1983; Kariman and Burkhart, 1985; Kaizer et al., 1985; Hazeki et al., 1987; Jobsis et al., 1988; Wray et al., 1988). These studies have different pitfalls: a) animal models were not appropiate (Kariman and Burkhart, 1985; Keizer et al., 1985); b) cerebral blood flow (CBF) and the brain electrical activity were not measured or quantified (Ferrari et al., 1983; Kariman and Burkhart, 1985; Kaizer et al., 1985; Hazeki et al., 1987; Jobsis et al., 1988; Wray et al., 1988); c) Cyt a,a3 spectral changes were calculated by 2 and 3 wavelength algorithms that are poorly documented; and d) the spectra had poor signal to noise ratio.

The purpose of this study was to test the hypothesis that cerebral Cyt a,a 3 oxidized copper band can be detected by NIRS on PFC exchange-transfused cats with intact brain hemodynamics as well as electrical activity. The spectra were carried out using a high resolution fast scanning spectrophotometer on hemoglobin free brains of PFC exchange-transfused cats. Cyt a,a3 redox state was evaluated during 5% and 8% CO2 (balance O2) respiration or terminal anoxic anoxia. Brain function was assessed by CBF measurement and somatosensory evoked potential (SEP) measurements (CBF, cerebral blood flow).

METHODS

Seven cats were anesthetized, intubated and paralyzed. Tidal volume and respiratory rate were adjusted to maintain end-tidal carbon dioxide near 4%. The left brachial artery was cannulated to obtain arterial samples and blood pressure recording. The microsphere reference sample was also collected from this catheter. Radiolabelled microsphere injections were made into the left ventricle via a catheter inserted through a femoral artery. The other femoral artery was catheterized for animal bleeding and to obtain blood pressure recordings. Two femoral veins catheters were used for drugs and perfluorochemical emulsions (PFC) infusions. The head was positioned in a stereotaxic apparatus and, after a midline scalp incision and retraction of the temporalis muscles, the skull surface was throughly scraped.

Arterial blood pressure was measured by a pressure transducer. Blood samples were analyzed immediately with an Instrumentation Laboratory Model 282 CO-Oxymeter (Lexington, MA) and a Radiometer ABL 3 gas analyzer (Copenhagen, Denmark).
Regional blood flow was measured with 15 μm diameter, radiolabelled microspheres (^{153}Gd, ^{113}Sn, ^{103}Ru, ^{95}Nb, ^{46}Sc; Du Pont-NEN Products, Boston, MA) using the reference sample method (Marcus 1976). Approximately 7×10^5 microspheres in .3 ml were injected into the left ventricle over 20 sec, followed by a 15 sec flush of 3.5 ml saline. The reference sample was withdrawn from the proximal aorta, using a Harvard dual-syringe pump. The withdrawal rate was 1.94 ml/min starting 30 sec prior to the injection and continuing for 2 min after the last flush of the catheter. At the conclusion of the experiment the brain was removed and placed in 10% buffered formalin, where it was hardened for 3-5 days before dissection. A brain tissue plug, roughly corresponding to the brain tissue volume underneath the two optic fibers (brain optical field) was cut by a well demarcated cylinder and was sliced in 5 discs of about 2 mm thickness The remaining cerebral tissues were pooled for measurement of total CBF. After weighing, the samples were placed in 15-ml vials and counted on a scintillation spectrometer (Minaxi gamma, United Technologies Packard, Santa Ana, CA). The following energy window settings (in KeV) were used: ^{153}Gd, 68-170; ^{113}Sn, 360-440; ^{103}Ru, 450-560; ^{95}Nb, 690-820; ^{46}Sc, 830-1200. The overlap of isotope activity among the five windows was substracted by solving simultaneous equations using overlap coefficients determined from pure isotope spectra (Heymann et al.. 1977). Blood flow was calculated as the product of the overlap-corrected counts in the tissue times the arterial reference withdrawal rate divided by the counts in the arterial reference sample.

Spectral measurements were made using a prototype spectrophotometer built by NIRSystem of Silver Spring, MD previously described (Ferrari et al., 1989). Briefly, the instrument is a microprocessor controlled scanning spectrometer capable of producing continuous spectra in the 730-960 nm region at a rate of 3 Hz. A fiber optic bundle (5 mm active diameter) was used to illuminate the skull overlying the right parietal cortex. The same sized optic fiber bundle, forming an angle of 50-60 degrees with the first, was placed symmetrically on the contralateral hemisphere. Both fibers, roughly at 13 mm from the midline of the animal's head, were held by a stereotaxic frame to avoid movement during the experiment (Fig. 1). Photons passing through the bone (a very low IR absorbing material) are partially absorbed and/or back scattered by gray matter and the circulating hemoglobin present in the illuminated vasculature. Depth of the incident photons into the brain can be estimated by taking into account optical properties of the near IR transilluminated skull, brain tissue and meninges (Wilson and Patterson, 1986; Eggert and Blazek, 1987). The intensity of the collected light, defined as diffuse transmittance (T/d) (Williams and Norris, 1987), was measured by a silicon photodetector. Collected spectra are numerically averaged spectra and sampled over successive sampling periods. The averaging process tends to reduce the random noise that is present in all spectrophotometric measurements. Approximately thirty scans were sufficient to produce acceptable spectra. Optical density measurements, expressed as log (1/Td), were obtained by subtracting a reference spectrum from the brain spectra. The conversion of T/d to log (1/Td) provides a signal which varies linearly with concentration (Williams and Norris, 1987).

Plotting log (1/Td) as a function of wavelength gives a spectrum that depends on the oxy and deoxyhemoglobin contents, Cyt a,a3 redox state and scattering effects of the optical field.

86

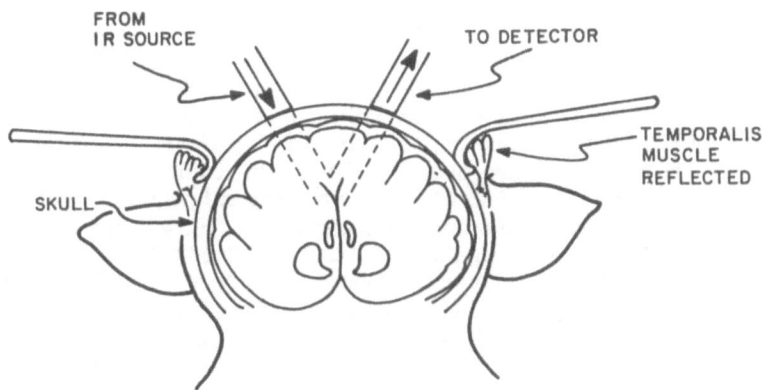

Fig. 1. Experimental lay-out.

The stored spectra can be analyzed and edited with software supplied by the manufacturer. Hemoglobin clearance and Cyt a,a3 redox changes were evaluated by difference in spectra between the different steady state conditions.

Stimulating needle electrodes were placed percutaneosly on palmar surface of the left forelimb in a location to cause a distinct digital twitch. Thereafter muscle paralysis was maintained with pancuronium bromide. Electrode holes were drilled into the skull over the right sensory cortex. The holes were filled with conductive gel, and 2 mm silver ball electrodes were placed and secured. Active electrode locations were referenced to the frontal sinus electrode. SEP were evaluated utilizing a Nicolet Model CA-1000 signal averager. Stimuli were delivered at a rate of 5.9/s and averaged. A stimulus intensity twice that necessary to cause a muscle twitch was utilized throughout the experiment. Waves occuring within 80 ms of stimulus were assessed. A consistent waveform is found in the cat consisting of a positive-negative-positive complex occurring 11-40 ms after the stimulus (Allison and Hume, 1981). The latency of P1 (first positive wave, about 14 ms) was measured at the maximum deflection of the wave. The amplitude of P1N1 wave was measured from the first positive defletion (P1) to the maximum deflection of the initial negative deflection (N1).

All preparations were allowed 60 min to stabilize following surgery. Control measurements of hematocrit were made and heparin administered. The blood-PFC exchange was performed in two steps. Lactate Ringer solution was infused intravenously while withdrawing equal volumes of whole blood from the femoral artery. Infusion was continued until peripheral hematocrit was reduced to approximately 65% of the initial value. At this point, Ringer solution was replaced with FC-43 emulsion. The infusion was continued for 150-180 min, until the desired hemoglobin washout was obtained. Arterial pressure was allowed to vary spontaneously and averaged around 90 mmHg. The residual hemoglobin was converted to carboxyhemoglobin, which does not absorb in the near IR range, by ventilation for 10 min with 4% carbon monoxide (CO) in 96% O_2 gas mixture.

At the end of the exchange-transfusion the animals were challenged for 10 min with 5% and 8% CO_2 in O_2. During the experiment, spectra were obtained for 7 separate conditions: 1) during 100% O_2 respiration; 2) following partial washout of hemoglobin by Lactate; 3) during and following complete blood-PFC exchange; 4) before, during and after CO challenge; 5) before, during and after CO_2 inhalation; 6) before and during ventilation with 100% N_2; and 7) post mortem. Animal death was verified by flattening of the SEP and levelling of the arterial pressure at 0 mmHg. SEP and arterial blood samples were collected for each saved spectrum. Flow measurements were carried out: a) during 100% O_2 respiration before the exchange transfusion began; b) following partial washout of hemoglobin by Lactate, c) and following complete blood-PFC-exchange transfusion; and d) after 5% CO_2 and 8% CO_2 ventilation.

Data in text, table, and figures are represented as means $\pm$ SE. Analysis of variance (ANOVA) for repeated measures was used.

87

RESULTS

Physiological data are shown in Table 1. Mean arterial pressure significantly decreased during the exchange transfusion from a control value of 114 ± 6 mmHg to 97 ± 8, 89 ± 7, 83 ± 5 and 70 ± 6 after Lactate hemodilution, complete blood-PFC exchange, 5% CO_2, and 8% CO_2. Hemoglobin dropped from $10.1 \pm .2$ g/100 ml in control conditions to $6.4 \pm .3$ after Lactate hemodilution, and $.4 \pm .03$ following complete blood-PFC exchange. $PaCO_2$ was stable during the FC-43 transfusion procedure. In contrast, PaO_2 and pH both significantly increased at the end of exchange transfusion. Flow increased significantly at the end of exchange transfusion in the region that corresponds to the brain optical field. CO_2 respiration provoked a severe pH decrease, that dropped from $7.42 \pm .02$ at end of PFC-transfusion to $7.09 \pm .01$ at the end of 8% CO_2 respiration. Blood flow did not significantly increase in comparison to end of PFC transfusion during 5% as well as 8% CO_2. The stability of the cat's cerebral electrical function through the protocol was highlighted by SEP amplitude and latency.

Table 1. Arterial pressure, PaO_2, pH, somatosensory evoked potentials (SEP) amplitude and latency, and blood flow during control, after Lactate Ringer's hemodilution (HD), after the end of blood-FC 43 exchange transfusion (ET), after 10 min 5% CO_2 - 95% O_2, and after 10 min 8% CO_2 - 92% O_2.

	Control	HD	ET	5 % CO_2	8 % CO_2
Arterial Pressure (mmHg)	114 ± 6	97 ± 8	89 ± 7*	83 ± 5	70 ± 6
PaO_2	523 ± 9	539 ± 22	627 ± 11*	602 ± 18*	583 ± 18*
pH	$7.34 \pm .01$	$7.35 \pm .02+$	$7.42 \pm .02$*	$7.18 \pm .01+$	$7.09 \pm 01+$
SEP amplitude (%)	100	98.1 ± 5.3	93.9 ± 6.7	93 ± 7	$83.2 + 3.4$
SEP latency (ms)	$14.1 \pm .9$	$14.1 \pm .9$	$13.2 \pm .6$	$14.4 \pm .6$	$15.3 \pm .7+$
Flow in the optical field (ml/min/100g)	118 ± 13	$161 \pm 15+$	263 ± 22*	280 ± 38*	259 ± 39*

Values are means$\pm$SE. N=7. * p<.01 compared with control; + p<.01 compared with ET.

The spectra obtained during 100% O_2 respiration before the exchange transfusion beginning, showed an absorption band centered around 755 nm referable mostly to the deoxyhemoglobin content of the optical field (Jobsis, 1977). At higher wavelengths, the most relevant feature is the left shoulder of a large peak, centered around 960 nm, attributable to water (Williams and Norris, 1987). Partial washout of hemoglobin by Lactate provoked a parallel decrease of the log (1/Td). A further progressive decrease of log (1/Td) at every wavelength was the time effect of the PFC exchange-transfusion. After complete blood-PFC exchange, the 755 nm peak was flattened and almost disappeared.

Although the extensive PFC exchange-transfusion virtually eliminated circulating hemoglobin, red blood cells from bone marrow stores might be continuosly released. The residual hemoglobin contribution could be responsible of spectra tiltings and shifts. Anoxic anoxia, in fact, could trigger vasodilation and blood volume increase as well as trasform all the residual hemoglobin into deoxyhemoglobin. Therefore a 10 min 4% CO-96 % O_2 challenge was performed just before the end of the PFC exchange-transfusion in order to convert the residual hemoglobin to carboxyhemoglobin. SEP amplitude and latency as well as mean blood pressure were unaffected during the CO respiration. 1-2 minutes CO respiration was sufficient to get a parallel decrease of log (1/Td) in all the wavelength range. No further spectral changes were observed in any animal during the following 8-9 min CO respiration. Figure 2 shows the average of the differences between control and the spectrum following complete blood-PFC exchange. Each spectrum is the sum of the spectral effects due to 1) the partial hemoglobin washout by Lactate, 2) the blood-PFC exchange

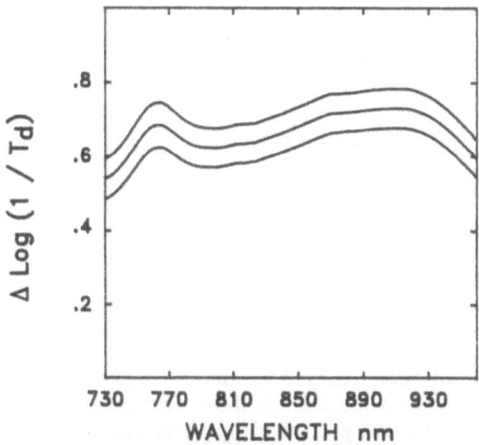

Fig. 2. Cerebral near IR spectral effects (mean ± standard error) of the blood-FC-43 exchange-transfusion obtained from 6 cats. Each spectrum is the difference between an hyperoxic status (100% O_2) before the transfusion beginnning, and an hyperoxic status at the end of the transfusion (arterial hemoglobin lower than 4% of control). The shapes reflect the ratio oxy/deoxyhemoglobin; the absolute values reflect the total hemoglobin content cleared from the brain optical field. Td, diffuse transmittance.

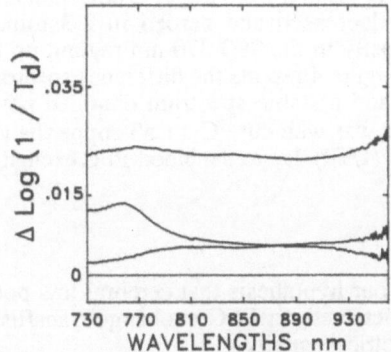

Fig. 3. Cerebral near IR difference spectra from 3 FC-43 exchange-transfused cats. Each spectrum is the difference between an hypercapnic status (5%CO_2-95%O_2) and an hyperoxic status (100% O_2).Delta Td, diffuse transmittance difference.

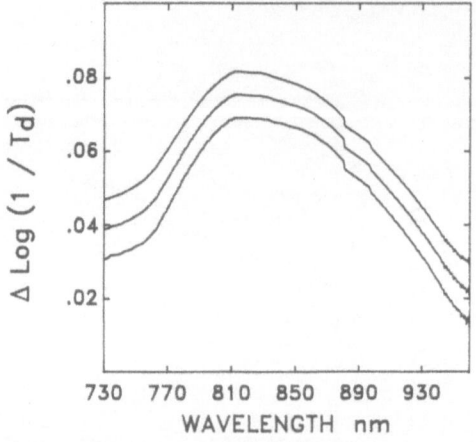

Fig. 4. Cerebral near infrared difference spectra (mean ± standard error) from 6 FC-43 exchange transfused cats. Each spectrum is the difference between a hypercapnic status (8% CO_2-92%O_2) and post mortem produced by ventilation with pure nitrogen. Td, diffuse transmittance.

transfusion, and 3) the CO challenge. The shape of this spectrum reflects mostly the ratio oxy/deoxyhemoglobin in the six animals. On the same animals 10 min 5% CO_2-95% O_2 respiration did not provoke a statistically significant increase of flow in the brain optical field as well as in all tested areas (Table 1). Figure 3 reports the corresponding difference spectra from 3 separate animals. Each spectrum is the difference between the one obtained at the end of the 5% $FiCO_2$ respiration and the one recorded during 100% O_2 FiO_2, just before CO_2 challenge beginning. The spectra had a parallel increase in all the explored range and did not highlight any feature referable to the Cyt a,a3 copper band. A further shift up of all the spectrum was observed increasing the CO_2 in the inspired gas mixture (8% CO_2-92% O_2) without any spectral features attributable to Cyt a,a3.

The Cyt a,a3 copper band could be clearly detected when the animals started to breath nitrogen. Oxygen was quickly cleared from the PFC emulsion and arterial pressure as well as SEP amplitude rapidly decreased and zeroed in 2-3 min. The log (1/Td) spectra progressively decreased, mostly in the 790-870 nm region, up to the animal death when they remained unchanged. Figure 4 reports the difference spectra between the hypercapnic status (8% CO_2-92% O_2) and a stable spectrum obtained post mortem. A broad band centered around 820-845 nm was well cut. Cyt a,a3 copper band amplitude corresponded to about 11% of the delta log (1/Td) due to the blood-PFC exchange transfusion.

DISCUSSION

These results confirm our hypothesis that cerebral low potential copper band of the oxidized Cyt a,a3 can be detected on PFC exchange-transfused cats with intact brain hemodynamics as well as electrical activity.

There is a vast literature on the chemistry and "in vitro" biochemistry of Cyt a,a3, the enzyme that catalyzes the final step of the mitochondrial electron transport chain in which electrons, derived from the oxidation of ferrocytochrome c, are consumed in the reduction of oxygen to water. Each molecule of the enzyme contains two copper ions and two hemes. Three of these four metal centers are known to play a critical role in the catalytic cycle: one of these hemes (referred as Fea) is the primary electron acceptor from cytochrome c, and its

90

location within the enzyme is presumed to be near that of the cytochrome c binding site; the other heme (Fea3) and the copper ion (CuB) together make up the binuclear site that binds O_2 while they undergo a complex sequence of chemical conversions that ultimately result in the synthesis of water molecules. The role played by the fourth metal center, the low potential copper ion site (CuA), is less well understood. The CuA site exhibits unique spectral properties i.e. a broad absorption band in the near IR region centered at 830 nm (Wharton and Tzagoloff, 1964) that could be used to non invasively monitor the redox state of the enzyme "in vivo". The band arises from the CuA^{2+} site and has been assigned to a charge-transfer transition between the cupric ion and a cystein ligand. The extinction coefficient of this absorption is approximately 2000 M^{-1} cm^{-1}, which makes monitoring the CuA site by this method more difficult and less precise than monitoring the heme sites, which have much stronger absorptions in the visible region (Greenwood et al., 1974). The IR band is completely abolished when the enzyme is reduced or when the CuA is substituted with sodium p-hydroxy-mercuri-benzoate (Gelles and Chan, 1985). The fact that the enzyme with this substitution mantains at least 20% of the catalytic activity supports the hypothesis that CuA is not a prerequisite for competent catalysis of electron transfer by the enzyme. In addition available kinetic data on CuA and Fea may be interpreted as CuA can accept electrons directly from cytochrome c. Still unknown is the role of the extra copper which has been recently evidenced in the Cyt a,a3 from different sources (Steffens et al., 1987).

In our "in vivo" experimental animal model the CuA band could be identified on the possibility of carrying out fast near IR spectra of brain tissues which are semi-trasparent to near IR light. Near IR log (1/Td) brain spectra are characterized by broad overlapping absorption bands. The absolute value of the signal is a function of the tissue optical scattering properties, the optic fibers geometry, and the absorption of the brain chromophores. "In vivo", there are a multiplicity of scattering interfaces within the tissue, the optical path length is unknown and could change, and scattering is much greater than absorption. The large decrease in log (1/Td) observed during the Lactate hemodilution as well as during the blood-PFC exchange transfusion confirms that the absorption in this spectral region is strongly dependent on the content of oxy and deoxyhemoglobin in the optical field. The present, as well as previous studies, were performed on animals transfused with FC-43, which has excellent oxygen and carbon dioxide carrying capacity (Geyger, 1975). To fully utilize this capacity, high partial pressures of oxygen were used. Unlike hemoglobin, where oxygen is chelated to the molecule, the solvent action of the PFC for oxygen or any other gases do not involve any kind of chemical or chelating process. The amount of gas carried by the PFC increases linearly with the partial pressure. In addition they possess particular characteristics, such as smaller particle size and lower viscosity as compared to blood, which may contribute to an increase in flow through microvessels. A two fold CBF increase, following complete blood-PFC exchange, has been described (Lee et al., 1988) and evidenced also in this study. Five and 8 % CO_2 (balance oxygen), a well known cerebral vasodilator on intact animals, did not increase cerebral flow probably because the cerebrovascular tree had already reached the complete vasodilation at the end of the PFC exchange transfusion.

In this study, cat Cyt a,a3 IR band was evidenced around 820-845 nm and had a similar shape and wavelength range as rat Cyt a,a3 IR band reported by different authors using different experimental setup. The rat Cyt a,a3 band was found to be approximately 10% of the absorption change which occurred upon replacement of the hemoblogin with PFC; therefore very close to the 11% reported in our study performed on a different animal species. The only discrepancy with the Cyt a,a3 rat spectrum of the most recent study was found in the 730-770 nm region and could be easily attributed to some residual deoxy-hemoglobin and/or deoxyhemoglobin still present in the rat brain optical field (Wray et al., 1988).

Unfortunately, the spectral data of the present study could not address the debated issue of the "in vivo" partial reduction of Cyt a,a3 (La Manna et al., 1987). The IR band amplitute, measured by difference spectra, could not be increased by CO_2 respiration. The high flow values suggest the presence of a maximal oxygen supply to the brain and, therefore, supposedly a complete Cyt a,a3 oxidation.

In conclusion, the fast scanning spectrometer could give an accurate measurement of the Cyt a,a3 IR band amplitude. The band was about 11% of the delta log (1/Td) due to FC-43 exchange-transfusion and about 0.5% of log (1/Td) measured before the exchange transfusion. These data suggest that instruments with higher accuracy than most of the presently available prototypes have to be employed for quantifing the band. The advantage of being able to follow metabolic changes in the intact animal makes Cyt a,a3 near IR copper band a valuable tool which should soon be further explored in different experimental, as well as clinical settings.

SUMMARY

Rapid scanning near infrared spectroscopy (730-960 nm) was utilized to determine cat brain cytochrome-c-oxidase after blood-perfluorocarbon (FC-43) exchange. Spectra were carried out before, during, and after the exchange transfusion on cats with preserved somatosensory evoked potentials and microsphere determined cerebral blood flow (CBF). Carbon dioxide (8%) did not produce an increase in CBF. Difference spectra between the hypercapnic state (8% CO_2-92% O_2) and post mortem, demonstrated the presence of a broad absorption band centered around 820-845 nm which could be attributed to the oxidized low potential copper ion of cytochrome-c-oxidase. We were unable to further oxidize this band by adding carbon dioxide to the inspired gas mixture. This may be due to the near maximal CBF levels present in this preparation. The spectral data support the hypothesis that near infrared spectroscopy can quantify changes in the redox state of cytocrome-c-oxidase.

REFERENCES

Allison, T., and Hume, A.L., 1981, A comparative analysis of short-latency somatosensory evoked potentials in man, monkey, cat, and rat. Exp. Neurol. 72: 592.

Brazy, J.E., and Lewis, D., V., 1986, Changes in cerebral blood volume and cytochrome a,a3 during hypertensive peaks in preterm infants. J. Pediatr. 108:983.

Eggert, H. R., and Blazek, V., 1987, Optical properties of human brain tissue, meninges, and brain tumors in spectral range of 200 to 900 nm. Neurosurgery. 21: 459.

Ferrari, M., Giannini, I., Carpi, A., and Fasella, P., 1983, Non-invasive near infrared spectroscopy of brain in fluorocarbon exchange-transfused rats. Physiol. Chem. Phys. Med. NMR. 5: 107.

Ferrari, M., Wilson, D.A., Hanley, D.F., Hartman, J.F., Traystman, R.J., Rogers, M.C., 1989, Non invasive determination of cerebral venous hemoglobin saturation in the dog by derivative near infrared spectroscopy. Am.J Physiol. 256: H149.

Gelles, J., and Chan, S.I., 1985, Chemical modification of the CuA center in Cytochrome c Oxidase by sodium p-(hydroxymercuri)benzoate. Biochemistry. 24: 3963.

Geyer, R.P., 1975, Bloodless rats through the use of artificial blood substitutes. Fed. Proc. 34:1499.

Greenwood, C., Wilson, M.T., and Brunori, M., 1974, Studies on partially reduced mammalian citochrome oxidase: reactions with carbon monoxide and oxygen, Biochem. J., 137:205.

Hazeki, O., Seiyama, A., and Tamura, M., 1987, Near-infrared spectrophotometric monitoring of haemoglobin and cytochrome a,a3 in situ. Adv. Exp. Med. Biol.215: 283.

Heymann, M.A., Payne, B.D., Hoffman, J.I.E.E.and Rudolph, A.M., 1977,Blood flow measurements with radionuclide-labeled particles. Prog. Cardio-vasc. Dis. 20:55.

Jobsis, F.F., 1977, Noninvasive, infrared monitoring of cerebral and myocardial oxygen sufficiency and circulatory parameters. Science. 198:1264.

Jobsis-Vander Vliet, F.F., 1985,Non-invasive, near infrared monitoring of cellular oxygen sufficiency in vivo. Adv. Exp. Med. Biol. 191: 833.

Jobsis, F.F., Piantadosi, C.A., Sylvia, A.L., Lucas, S.K., and Keizer, H.H., 1988, Near-infrared monitoring of cerebral oxygen sufficiency. 1. Spectra of cytochrome c oxidase. Neurol. Res. 10: 7.

Kariman, K. and Burkhart, D.S., 1985, Heme-copper relationship of cytochrome oxidase in rat brain in situ. Biochem. Biophys. Res. Comm. 126: 1022.

Kaizer, H.H., Jobsis-Vander Vliet, F.F, Lucas, S.S, Piantadosi, C.A., and Sylvia, A.L., 1985, The near infrared band of cytochrome a,a3 in purified enzyme, isolated mitochondria and in the intact brain in situ. Adv. Exp. Med. Biol. 191: 823.

La Manna, J.C.,Sick, T.J., Pikaarsky, S.P, and Rosenthal, M., 1987, Detection of an oxidable fraction of cytochrome oxidase in intact rat brain. Am. J. Physiol. 253:C477.

Lee, P.A., Sylvia, A.L., and Piantadosi, C.A., 1988, Effect of fluorocarbon-for-blood exchange on regional cerebral blood flow in rats. Am. J. Physiol. 254:H719.

Marcus, M.L., Heistad, D.D, Ehrhardt, J.C., and Abbout, F.M., 1976, Total and regional cerebral blood flow measurement with 7-, 10-, 15-, 25-, and 50- um microspheres. J. App. Physiol. 40: 501.

Steffens, G., Biewald, C.R., and Buse, G., 1987, Cytochrome c oxidase is a three-copper, two-heme-A protein. Europ. J. Biochem., 164:295.

Wharton, D.C, and Tzagoloff, A., 1964, Studies on the lectron transfer system. LVII. The near infrared absorption band of cytochrome oxidase. J. Biol. Chem. 239: 2036.

Williams P. and Norris, K, 1987, Near-Infrared technology in the agricultural and food industries. St. Paul, Min: American Association of Cereal Chemists.

Wilson, B.C. and Patterson, M.S., 1986, The physics of photodynamic therapy. Phys. Med. Biol. 31: 327.

Wray, S., Cope, M., Delpy, D.T., Wyatt, J.S., Reynolds, E.O.R., 1988, Characterization of the near infrared absorption spectra of cytochrome a, a3 and hemoglobin for the non-invasive monitoring of cerebral oxygenation. Biochem. Biophis. Acta 933: 184.

Wyatt, J. S., Cope, M., Delpy, D.T., Wray, S., Reynolds, E.O.R., 1986, Quantification of cerebral oxygenation and hemodynamics in sick newborn infants by near infrared spectrophotometry. Lancet II:1063.

INTERACTION OF OXYGEN PARTIAL PRESSURE AND ENERGY METABOLISM WITH THE RELAXATION RATE OF INORGANIC PHOSPHATE: A ^{31}P NMR STUDY[*]

P. Okunieff[1], T. Tokuhiro[2], P. Vaupel[3], and L.J. Neuringer[2]

[1]Dept. Radiation Medicine, Massachusetts General Hospital Cancer Center, Harvard Medical School, Boston MA 02114, USA
[2]Francis Bitter Magnet Laboratory, Massachusetts Institute of Technology, Cambridge, MA 02139, USA
[3]Institute of Physiology & Pathophysiology, Pathophysiology Division, University of Mainz, 6500 Mainz, FRG

INTRODUCTION

It is well known that oxygen molecules present in liquid or solid samples can shorten NMR spin-lattice relaxation times (T_1) of the nucleus under investigation. Several studies have shown large decrements of ^{1}H, ^{13}C, and ^{19}F relaxation times mediated by the O_2 molecule (Lees and Muller, 1961; Ohuchi et al., 1979; Fishman et al., 1989). The positions of the ^{31}P atoms in phosphate compounds are stereometrically similar to some ^{13}C atoms in organic compounds, and thus an effect of oxygen on the ^{31}P T_1 is expected. Recently, we have shown (Okunieff et al., 1988) that oxygen breathing can significantly reduce the T_1 of ^{31}P in the inorganic phosphate molecule (P_i). The degree to which this change was mediated by the O_2 molecule compared to secondary enzymatic processes that are also augmented when 100% oxygen is inspired (Okunieff et al., 1989), however is unknown.

Use of ^{1}H relaxation rates to estimate tissue oxygenation has proven difficult. This is due to an overlapping modification of

---

[*]This work supported in part by NIH grants CA48096, RR00995, and CA13311, and by the American Cancer Society Career Development Award.

Oxygen Transport to Tissue XII, Edited by J. Piiper et al.
Plenum Press, New York, 1990

95

the apparent T, induced by blood flow and associated water motion into and out of the volume of interest (Okunieff, 1987). The obvious utility of developing an endogenous non-invasive marker of tissue hypoxia led us to explore the possibility of using ^{31}P relaxation kinetics as a marker of tissue oxygenation states.

MATERIALS AND METHODS

Solution Studies

The tumor high energy phosphate cytoplasmic milieu was modeled using the following set of reagents[**]: AMP, phosphocreatine (PCr), ATP, glycerophosphocholine and P_i. $MgCl_2$ was added in sufficient concentration to assure near complete PCr and ATP binding, and the pH was adjusted to 7.4 using KOH or HCl as needed. Three samples were prepared. Sample A was equilibrated with oxygen, and was 99.7% D_2O. Sample B was equilibrated with purified N_2 (<5ppm O_2) and also had a 99.7% D_2O content. The final sample (C) was equilibrated with purified N_2 and had a 9.4% D_2O content. The degassing procedure included eight freeze-pump-thaw cycles. Specifically, the sample was frozen in a liquid nitrogen bath during vacuum suction, and then thawed under continued vacuum suction. The sample was then regassed with commercially available pure N_2 gas (<5ppm O_2) which was further purified over two stages of strong alkaline solution and pyrogallol. After repeating this procedure 8 times, the glass NMR tube was sealed by melting its mouth, maintaining the oxygen content in the solution below 5 ppm.

T_1 measurements were performed in an 8.5 T magnet operating at 145.6 MHz for ^{31}P. The inversion recovery pulse sequence was used to determine T_1 relaxation rates. Sample temperature was maintained at 37.0 ± 0.3 °C (mean ± range).

In-Vivo Tumor Studies

C3H mice with 6-7 mm diameter FSaII tumors implanted in the hind foot dorsum were used. T_1 measurements were done using the same 8.5 T spectrometer described above. The progressive saturation pulse sequence (90°-τ-90°-τ) was used (Okunieff et al., 1988). Respiratory gas was modified external to the magnet and probe, and ischemia was similarly induced by a tourniquet applied using a trocar and suture loop mechanism which extended outside the magnet

---

[**] Reagents purchased from Sigma Chemical Co., St. Louis, MO, USA.

Table 1. T_1 Relaxation Times of Various Phosphates in Aqueous Solution at 37°C (sec)[1]

	ATP		PCr	PDE	P_i	PME
	$\alpha-P$	$\beta-P$ $\gamma-P$				
Sample A O_2 filled $\alpha = 0.003$[2]	0.95	0.99 1.50 (±0.10)	3.86 (±0.31)	4.22	8.01	3.16
Sample B N_2 filled $\alpha = 0.003$	0.99	1.08 1.53	5.04	6.15	15.3	3.82
Sample C N_2 filled $\alpha = 0.906$	0.82	0.97 1.09	3.90	5.31	9.65	2.92

[1]Standard deviations, unless specifically reported were less than 8% upon repeated measurements.
[2]α = mole fraction of H_2O in a mixture of H_2O and D_2O.

and NMR probe. Radiation of tumors to a homogeneous dose (range of dose in the irradiated volume ±3%) in a single fraction of 30 Gy (3000 rad) was delivered via parallel opposed ^{137}Cs sources when needed, and water bath hyperthermia (43.5°C x 60 min) of the tumor was performed on a subset of tumors. For each experiment pair-ed control and pO_2 modified tumor ^{31}P relaxation rates were measur-ed in the same animal, except for the hyperthermia treated tumors in which separate control animals were required.

RESULTS AND DISCUSSION

Solution Studies

Table 1 summarizes the ^{31}P T_1 values for the three samples A, B, and C, and Figure 1 shows a typical spectrum with the resonance peaks identified. All peaks refer to those identified in Fig. 1. AMP was used to represent phosphomonoesters (PME), glycerophosphocholine was used to represent the phosphodiesters (PDE), and ATP was used to represent the nucleoside triphosphates (NTP) detected in-vivo in tumors. Three points can be made from the data in Table 1: (1) There is virtually no difference in the T_1 values for the three ATP peaks between samples A and B; (2) the oxygen molecule did moderately affect the AMP and diester ^{31}P relaxation rate as evidenced by the differences seen between samples A and B; and (3) a significantly shorter T_1 for inorganic phosphate ion can be seen in sample A (oxygen filled) than sample B (nitrogen filled).

Using the following five equations and the experimental results in Table 1 it is possible to distinguish the components of ^{31}P relaxation rate ($1/T_1$) due to dipole-dipole interactions with water protons [$\underline{P}$-$\underline{H}_2O$], the oxygen molecule [$\underline{P}$-O_2], deuterium [$\underline{P}$-$\underline{D}_2O$], and lumped other interactions [$\underline{P}$-$\underline{X}$].

$$(1/T_1)_A = (1/T_1)_{[\underline{P}-\underline{O}_2]} + \alpha \cdot (1/T_1)_{[\underline{P}-\underline{H}_2O]} + (1-\alpha) \cdot (1/T_1)_{[\underline{P}-\underline{D}_2O]}$$
$$+ (1/T_1)_{[\underline{P}-\underline{X}]} \hspace{2cm} \text{(Equ. 1)}$$

$$(1/T_1)_B = \alpha \cdot (1/T_1)_{[\underline{P}-\underline{H}_2O]} + (1-\alpha) \cdot (1/T_1)_{[\underline{P}-\underline{D}_2O]} + (1/T_1)_{[\underline{P}-\underline{X}]}$$
$$\text{(Equ. 2)}.$$

$$(1/T_1)_{[\underline{P}-\underline{O}_2]} = (1/T_1)_A - (1/T_1)_B \hspace{2cm} \text{(Equ. 3)}$$

98

Table 2. Various Contributions to ^{31}P $(1/T_1)$ in Aqueous Solutions of Phosphates at 37°C (sec^{-1})

Resonance	$(1/T_1)$ [P-X]	$(1/T_1)$ [P-H$_2$O]	$(1/T_1)$ [P-O$_2$]	$(1/T_1)$Total*
α- ATP	0.994	0.248	0.043	1.285
β- ATP	0.91	0.125	0.084	1.119
γ- ATP	0.633	0.313	0.013	0.959
PCr	0.194	0.0688	0.061	0.324
PDE	0.161	0.0305	0.074	0.265
Pi	0.0624	0.0454	0.060	0.168
PME	0.256	0.0957	0.055	0.407

*Total relaxation rate that would be observed in an aqeous (H$_2$O) solution equililbrated with 100% oxygen at 1 atmosphere pressure.

$$(1/T_1)_{[B \text{ or } C]} = (\alpha + (1-\alpha)F) \cdot (1/T_1)_{[P-H_2O]} + (1/T_1)_{[P-X]} \qquad \text{(Equ. 4)}$$

$$(1/T_1)_{[P-H_2O]} = ((1/T_1)_C - (1/T_1)_B)/(1-F)\beta \qquad \text{(Equ. 5)}$$

Here $(1/T_1)_{[A,B, \text{or } C]}$ represents the experimental measurement made of samples A, B, or C respectively. "α" is the ^1H mole fraction in H_2O or HOD and is 0.003 for sample B and 0.906 for

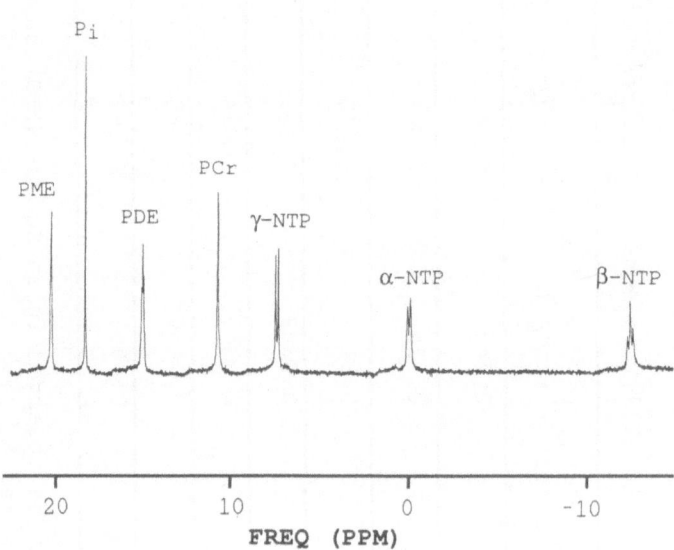

Figure 1. ^{31}P NMR spectrum of solution A.

sample C. The constant F relates the differential effects of the proton and phosphorus nucleus on ^{31}P relaxation using the gyromagnetic constants (γ) for ^{31}P and ^1H and equals $(8/3) \cdot (\gamma_P/\gamma_H)^2 = 0.0623$. And lastly, β is the difference in ^1H mole fractions (α) between samples C and B ($\beta=0.903$). The results of these calculations are given in Table 2. An inspection of Table 2 reveals: (1) in a sample under the 100% oxygen atmosphere, the contributions of oxygen to inorganic phosphate relaxation were large, for other phosphates the contribution was smaller, and for the ^{31}P nuclei of ATP the effect was negligible; and (2) except for inorganic

100

Table 3. $(1/T_1)$ of Various ^{31}P Resonances Measured In-Vivo
Before and After 100% Oxygen Breathing $(sec^{-1})*$

Resonance	Treatment Group	Air Breathing	100% Oxygen Breathing
PME	Control	0.288	0.254
	Post-XRT	0.330	0.231
	Post-XRT	0.169	0.192
	Post-XRT	0.291	0.236
P_i	Control	0.189	0.331
	Post-XRT	0.251	0.301
	Post-XRT	0.203	0.405
	Post-XRT	0.272	0.350
PCr	Control	0.366	0.377
	Post-XRT	0.338	0.429
	Post-XRT	0.254	0.382
	Post-XRT	0.277	0.220
γ-NTP	Control	0.649	0.669
	Post-XRT	0.813	0.775
	Post-XRT	0.690	1.010
	Post-XRT	0.725	0.794
α-NTP	Control	0.775	0.885
	Post-XRT	1.136	0.855
	Post-XRT	0.847	1.266
	Post-XRT	0.870	0.990
β-NTP	Control	0.943	0.800
	Post-XRT	1.053	1.190
	Post-XRT	0.917	0.855
	Post-XRT	0.800	0.980

*The results from 4 tumors in 4 different animals is presented.
One tumor was untreated (Control), and the other 3 tumors were
studied 36-48 hr after a radiation dose of 30 Gy (Post-XRT).

phosphate, the greatest contribution to $1/T_1$ arose from the dipole–dipole interactions between the ^{31}P nucleus under consideration and magnetic nuclei other than water protons.

In-Vivo Tumor Studies

Two sets of measurements were made using a system designed to augment tumor tissue oxygenation (Table 3). The method is simple, and involves replacing air with 100% O_2 as the respiratory gas. Since the P_{50} of hemoglobin in mouse is $\approx 40 - 50$ mmHg, the normal saturation of hemoglobin during air breathing could be under 75% at the capillary level. Nevertheless, the increase in average tissue pO_2 and therefore tissue oxygen content is likely to be modest. The data in Table 3 show that despite the expected small change in tissue oxygen content that occurs after breathing oxygen at one atmosphere, the T_1 of inorganic phosphate was extremely sensitive to the respiration of oxygen while the T_1s of other ^{31}P nuclei were insensitive to oxygen respiration. This pattern of sensitivity is consistent with the solution studies however the magnitude of the O_2 dependence of P_i relaxation was greater then predicted. This experimental observation is likely to have been due to apparent relaxation processes (Okunieff et al., 1989) resulting from enhanced ATPase kinetics. Specifically increased magnetization transferred between the γ–phosphate of ATP and P_i (ATP $\leftrightarrow$ ADP + P_i), due to the large difference in their relaxation rates, could result in an apparent reduction of the P_i T_1.

The 30 Gy given 36 – 48 hours before spectroscopy had no significant effect on T_1 (Table 3) consistent with reoxygenation to the base-line level with no significant change in tumor oxygenation and metabolic rate (Suit et al., 1985).

Table 4. Effect of Ischemia on the In-Vivo ^{31}P Spin-Lattice Relaxation ($1/T_1$) of Inorganic Phosphate (sec^{-1})

Before Ischemia	During Ischemia
0.246	0.224
0.259	0.169
0.256	0.166

A decrease in tissue oxygenation was mediated by tourniquet occlusion of blood flow. Using this technique, T_1 measurement could be made just before, and then after production of ischemia. Tumor temperature was maintained using warmed air throughout the experiment ($T = 32°C$). The spectra for measurement of ^{31}P relaxation were begun after at least 60 minutes of blood flow occlusion to assure a stable metabolic state during the measurement procedure. The data are summarized in Table 4. Induction of hypoxia by vascular occlusion significantly prolonged the T_1 of inorganic phosphate. As with oxygen breathing the degree of change was greater than that expected on the basis of the solution studies and suggests an associated decreased chemical exchange between the γ-ATP phosphate and P_i.

Table 5. Effect of Hyperthermia (43.5°C x 60 min) on the In-Vivo ^{31}P Spin-Lattice Relaxation ($1/T_1$) of Inorganic Phosphate (sec^{-1})

Control (untreated)	Treated (12 to 18 hr after hyperthermia)
0.221	0.595
0.259	0.704
0.246	0.565
-----	0.375

Direct dipole interactions superimposed on indirect chemical reaction kinetics enhances the utility of apparent P_i relaxation rates for detection of changes in tissue oxygen content after oxygen breathing and after induction of ischemia. However, superimposed interactions can also yield unwanted artifacts that make interpretation of T_1 relaxation times difficult. Treatment of tumors with hyperthermia, for example, causes protein coagulation, tissue necrosis and hemorrhage, as well as vascular shut down. Since coagulation is likely to have a large effect on ^{31}P nuclear mobility, and Fe^{++} released into hemorrhagic tissue will have a similar effect on ^{31}P relaxation, changes in T_1 after hyperthermia must be interpreted with caution. The reduction in T_1 therefore seen after 43.5°C x 60 min of heat treatment could (but may not) mean that pO_2 has increased in this nearly devitalized non-metabolizing tissue (Table 5).

CONCLUSIONS

In aqueous solution, at 37°C, the oxygen molecule has an important role in determining the relaxation rate of ^{31}P nuclei of inorganic phosphate. In-vivo, the apparent T_1 of inorganic phosphate is extremely sensitive to tissue oxygen content. Since in-vivo the magnitude of the apparent spin-lattice relaxation interaction with O_2 is greater than would be predicted on the basis of solution studies, and since we have previously shown the high rate of ATPase catalyzed chemical activity in tumors (Okunieff et al., 1989), it is likely that this second process contributes significantly to the direct interactions of the O_2 molecule. In conclusion, the endogenous ^{31}P relaxation is in many cases a good and potentially useful indicator of changes in tissue oxygen content. At present, it cannot however be used to quantify tissue pO_2 or oxygen content, and is not sufficiently specific to allow for estimation of tissue oxygenation status under all circumstances.

REFERENCES

Fishman, J.E., Joseph, P.M., Carvlin, M.J., Saadi-Elmandjra, M., Mukherji, B., and Sloviter, H.A., 1989, In-vivo measurements of vascular oxygen tension in tumors using MRI of a fluorinated blood substitute, Invest. Radiol. 24:65.

Lees, J., and Muller, B.H., 1961, Degassing of liquids for nuclear magnetic spin-lattice relaxation studies, J. Chem. Phys. 34:341.

Ohuchi, M., Fujito, T., and Imanari, M., 1979, Spin-lattice relaxation of C-13 in the rotating frame, J. Mag. Reson. 35:415.

Okunieff, P., Ramsay, J., Tokuhiro, T., Hitzig, B.M., Rummeny, E., Neuringer, L.J., and Suit, H.D., 1988, Estimation of tumor oxygen and metabolic rate using ^{31}P MRS: Correlation of spin-lattice relaxation with tumor growth rate and DNA synthesis, Int. J. Radiat. Oncol. Biol. Phys. 14:1185.

Okunieff, P., Vaupel, P., and Neuringer, L.J., 1989, ^{31}P nuclear magnetic resonance measurements of ATPase kinetics in malignant tumors, Adv. Exptl. Med. Biol. 248:819.

Okunieff, P., 1987, Unpublished data.

Suit, H.D., Sedlacek, R.S., Silver, G., and Dosoretz, D., 1985, Pentobarbital anesthesia and the response of tumor and normal tissue in the C3Hf/Sed mouse to radiation, Radiat. Res. 104:47.

PH CHANGES IN FRONT OF THE HYDROGEN GENERATING ELECTRODE DURING MEASURE-

MENTS WITH AN ELECTROLYTIC HYDROGEN CLEARANCE SENSOR

H. Baumgärtl, W. Zimelka, D.W. Lübbers

Max-Planck-Institut fuer Systemphysiologie

Rheinlanddamm 201, D-4600 Dortmund 1, F.R.G.

INTRODUCTION

The electrolytic hydrogen clearance according to Lübbers and Stosseck (1970) uses a sensor which combines an electrode generating electrochemically molecular hydrogen with another electrode which measures hydrogen polarographically (pH_2/H_2 sensor). Although a separating membrane could be applied, up to now the sensor is used only in direct contact with the tissue so that the measurements may directly influence local tissue. Using a transparent sensor Stosseck et al. (1974) found that generating currents of $0.2 - 2\,\mu A$ of a duration of $1 - 2$ s did not show any change of the vascular diameters (brain cortex), but that larger currents of $3 - 4\,\mu A$ of the same duration caused local vasoconstriction and produced visible gas bubbles. For explanation, a direct effect of the electrical current on the smooth muscle was discussed. However, since during the generation of molecular hydrogen hydroxyl ions are produced, possible pH changes and their effect on microcirculation have to be considered, especially if large generating currents are recommended, as in recent publications (for example, $100\,\mu A$ during $2 - 6$ s (Koshu et al., 1982), to guarantee better results.

To obtain information about the amount and the distribution of hydroxyl ions during the H_2 generation, as a model we investigated the pH distribution in front of the Pt wire of the pH_2/H_2 sensor covered with a $2 - 3$ mm layer of agar.

METHOD

Fig. 1 demonstrates the schematic view of our experimental setup. The pH_2/H_2 sensor forms the bottom of a circular vessel filled with agar (A: 1% agarose in water). The thickness of the agar layer (d) is about $2 - 3$ mm. The H_2 generating Pt wire (G) is surrounded by 4 platinized pH_2-measuring wires (only two wires M_1, M_2 are shown) insulated by glass (I). The distance between G and M (D_{GE}) amounts to about $300 - 350\,\mu m$ and between M_1 and M_2 (D_{EE}) to about $500\,\mu m$. The generating wire has a reference electrode of platinum (R_1), the pH_2-measuring electrodes one of Ag/AgCl (R_2). The reference electrodes are in contact with the agar. The local distribution of OH^- ions, pH_2, and the pO_2 in the agar is measured by microelectrodes:

Oxygen Transport to Tissue XII, Edited by J. Piiper *et al.*
Plenum Press, New York, 1990

107

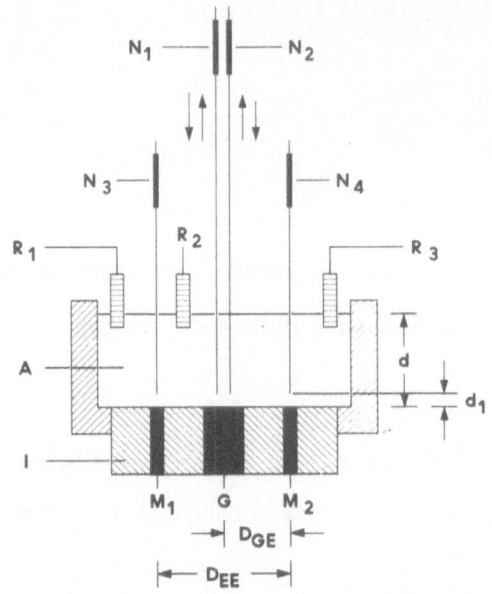

Fig. 1. Schematic view of the experimental setup.

1. pH microelectrodes (N_1, N_3). The pH-sensitive cone of the electrode (Saito et al., 1976) has a length of ca. 50 μm and a tip diameter of about 2 μm. The upper end diameter amounts to ca. 10 μm. The electrode used with an Ag/AgCl reference electrode (R_2) is calibrated at 23°C in 3 standard buffer solutions. The sensitivity was found to be 53 - 55 mV/pH. The drift is less than 0.5 mV/h, the response time less than 3 s.

2. pO_2 microelectrode (N_4). The glass tip of the microelectrode (Baumgärtl and Lübbers, 1983) has a diameter of 0.7 - 1.5 μm, the platinum wire of 0.5 - 1 μm. The electrode has a small recess of a depth of ca. 5 μm which is membranized by collodion and polystyrene. The reference electrode is formed by a thin layer of tantalum, platinum, and silver, directly sputtered on the shaft of the electrode. The reduction current at air amounts to about 0.095 nA/150 Torr and the drift about 1 %/h.

3. pH_2 microelectrode (N_2). The construction of the pH_2 microelectrode equals that of the pO_2 microelectrode, but the Pt surface in the recess is platinized to improve the stability and the sensitivity of the electrode current. In 100 % H_2 the current amounts to about 0.25 nA, the response time is 2 - 4 s, the drift to about 3 %/h.

As standard conditions for H_2 generation we use in all experiments (except Fig. 2) a voltage of 0.9 V, a current of 0.1 μA, and a pH_2/H_2 sensor which has a H_2-generating platinum wire of r = 100 μm and a pH_2-measuring wire of r = 50 μm. The temperature is 22° -23°C.

RESULTS AND DISCUSSION

To find out which voltage is needed to generate hydrogen, steps of negative voltage (0.1 V, 1 min) are applied between the generating electrode (G, diameter 100 μm) and the platinum reference electrode and the current is measured (Fig. 2). To control the chemical reaction pH, pO2, and pH2 microelectrodes are positioned by a micromanipulator (arrow), e.g. in front of

108

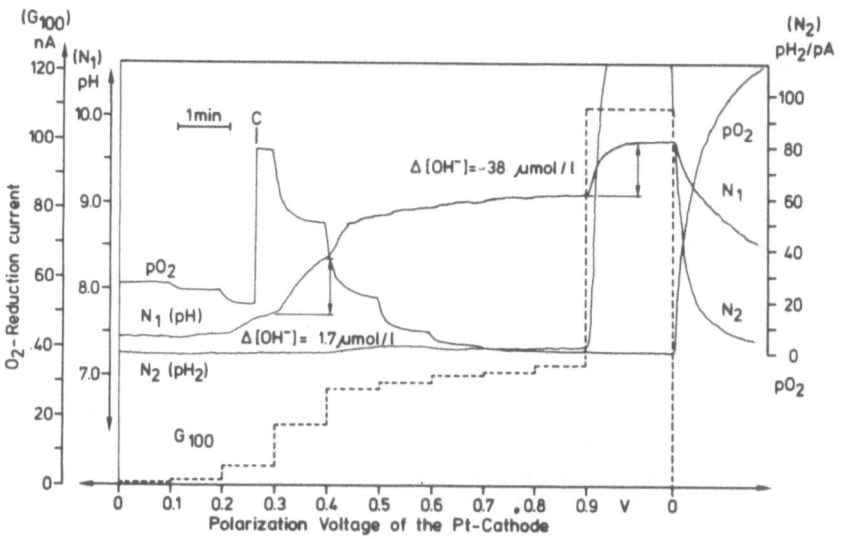

Fig. 2. Electrochemical generation of hydrogen in an air-equilibrated agar
layer and the accompanying pH, pH_2, and pO_2 changes.

the generating wire in a distance of about 150 μm. As expected, up to a
voltage of 0.8 V the current increases with increasing voltage and forms
a polarogram indicating oxygen reduction

$$O_2 + 2H_2O + 4e^- = 4OH^-.$$

The pO_2 microelectrode measures a decrease in local pO_2 because a pO_2
gradient develops in front of the O_2-reducing electrode. Local pO_2 approaches zero Torr (C: change of sensitivity). The pH electrode shows the
formation of hydroxyl ions by an increase in pH: for example, by switching
from 0.3 - 0.4 V pH rises from 7.7 - 8.35, which corresponds to an increase
in hydroxyl ion concentration of 1.7 μmol/l. Changing to 0.9 V, the current
increases three-fold (35 - 110 nA) whilst the local pO_2 remains close to
zero. The pH_2 increases immediately showing that hydrogen is generated at
this voltage. The pH increases further, since also during hydrogen generation, hydroxyl ions are produced:

$$2H_2O + 2e^- = H_2 + 2OH^-.$$

After one minute, the steady state concentration of hydroxyl ions is reached
(pH = 9.7). The increase between 0.8 and 0.9 V corresponds to about 38 μmol
OH^-/l. After 100 s, voltage is switched off and pH, pH_2, and pO_2 reach
their initial values later as a result of diffusion.

Fig. 3 shows development and equalization of the diffusion field of
the hydroxyl ions on the onset and after the end of H_2 generation measured
about 150 μm in front of the H_2-generating wire. At an initial pH of 7.4,
it takes about 5 s before hydroxyl ions reach the electrode. At first,
the pH increases steeply, but with increasing pH, the increase becomes
smaller. This is partly due to the logarithmic pH scale - the same pH change
produces at a pH of 9.4 a 100 times larger change in OH^- ion concentration
than at a pH of 7.4 - but also due to the fact that with increasing time,
hydroxyl ions diffuse into a larger volume. Because of the expansion by
diffusion, the first half of the OH^- concentration change (t_{50}) occurs in
37 s and the second half in 67 s. After the end of H_2 generation, the OH^-
concentration decreases by 50 % in 24 s and reaches the initial value after
about 12 min.

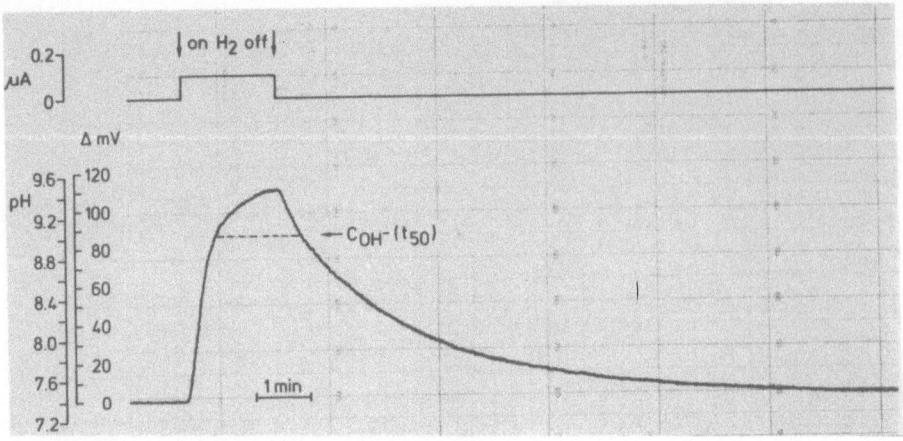

Fig. 3. Temporal changes of pH on the onset and after the end of H_2 generation.

With longer periods of H_2 generation pH changes become smaller and after 8 - 9 min, a quasi steady state field of pH is attained. This is demonstrated in Figs. 4 and 5. In Fig. 4 the development of the quasi steady state in 150 μm and 500 μm distance of the H_2-generating wire is compared. The distant pH electrode B (1) needs a longer time to respond than the nearer one (A) (B: 25 - 26 s, A: 5 - 6 s), (2) reacts more slowly (t_{50} of the OH^- ion concentration change: B: 3 min 15 s, A: 1 min 25 s), and (3) reaches a lower final value (B: pH 8.80, A: pH 9.56). Between the final values there is a concentration gradient of 59 μmol OH^-/l over a distance of 500 μm. Fig. 5 shows the pH distribution over a distance of 2.4 mm under quasi steady state conditions. The distribution is measured by withdrawing the electrode in 50 μm steps and waiting for 20 s. The original record in the upper part demonstrates that local pH is well measurable and sufficiently stable, even at lower pH values with their higher sensitivity against small changes in hydroxyl ion concentration. In Fig. 5 all measurements are collected including the data of the original record (a to b). Between 0.15 and 1.1 mm pH decreases almost linearly from pH 9.82 to pH 8.2 with a slope of ΔpH 0.17/100 μm. At larger distances pH decreases less and reaches its initial value at a distance of 2.3 - 2.4 mm from the surface of the H_2-generating wire. This means that the OH^- concentration gradient, which is responsible for the transport of the hydroxyl ions, changes exponentially; for example, the pH change from 9.82 to 9.65 (between 0.15 and 0.25 mm) corresponds to an OH^- concentration gradient of 21.4 μmol/(0.1 mm$\cdot$l), whereas the OH^- concentration gradient between 1.1 and 1.2 mm (pH 8.20 to 8.02) only amounts to 0.51 μmol/(0.1 mm$\cdot$l). Under quasi steady state conditions the OH^- ion flux is about 8.8 μmol OH^- calculated by the product [hemispherical area (in mm)] $\cdot$ [OH^- ion concentration gradient (in mol/l/min)]. It decreases towards the periphery. All these experiments demonstrate clearly that during H_2 generation the production of hydroxyl ions is, indeed, changing local pH.

The actual observable pH change however depends, apart from the current intensity, on the special measuring conditions: the local buffer capacity and the transport conditions in the medium. To test this influence we investigated the change of the local OH^- concentration during constant H_2 generation (standard conditions) in agar, in a 0.9 % NaCl solution, in a bicarbonate-phosphate buffer (Renooij et al., 1983), and on the surface of the tongue of an anaesthetized frog.

110

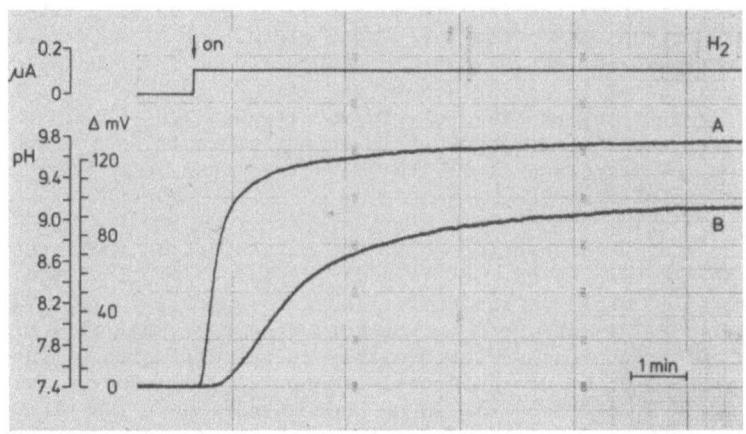

Fig. 4. Spreading of hydroxyl ions in the agar layer during constant electrochemical H_2 generation measured at a distance of 150 μm (A) and 500 μm (B) in front of the H_2-generating wire.

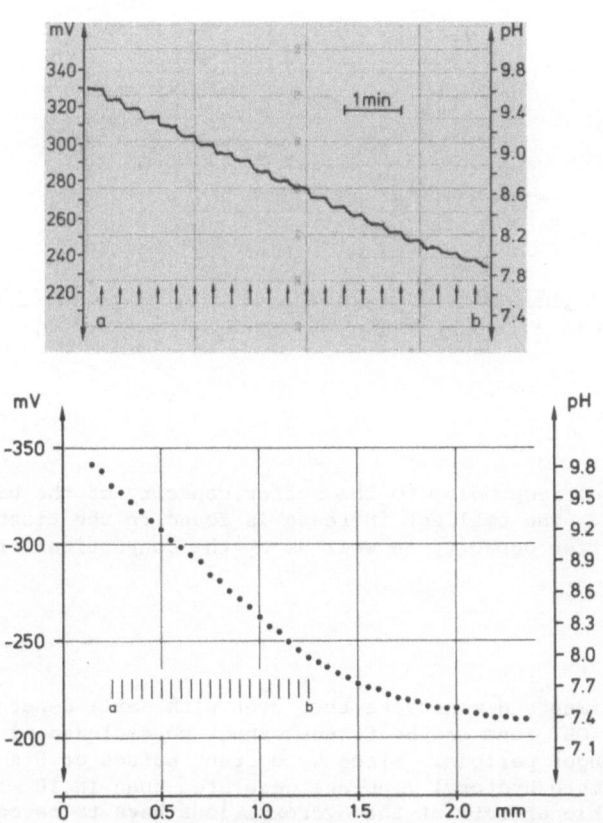

Fig. 5. Local distribution of pH in front of a H_2-generating platinum wire in an air-equilibrated agar layer under quasi steady state conditions (upper part: original record (a - b) steps of 50 μm).

The measurement on the frog tongue (Fig. 6) was carried out in the following way: the pH_2/H_2 sensor is placed close to a blood vessel and measures the tissue pH in a distance of 150 - 250 μm from the H_2-generating wire. The initial pH is rather stable and amounts to 6.8 - 6.9. After the onset of H_2 generation, pH rises slowly and reaches 7.2 - 7.3. This increase is much smaller than in agar, but still well measurable. After the end of H_2 generation, pH decreases slowly to 7.1 - 7.2; then, a movement of the tongue broke the electrode.

With the same H_2 generation local OH^- concentration increases in the following way: In agar by 60 μmol/(1·4 min) (Fig. 4, A), in 0.9 % NaCl by ca. 12.5 μmol/(1·4 min), in the buffer (buffer capacity = 0.31 μVal/1 pH· ml) by 0.24 μmol/(1·4 min), and in the tongue by 0.11 μmol/(1·4 min). In agar the OH^- distribution is caused mainly by the diffusion of the OH^- ions. In a 0.9 % NaCl solution an additional transport by convection is demonstrated by a 4 - 5-fold decrease in OH^- concentrations. The strong reduction in the buffer (250 times smaller than in agar) is caused by the buffering

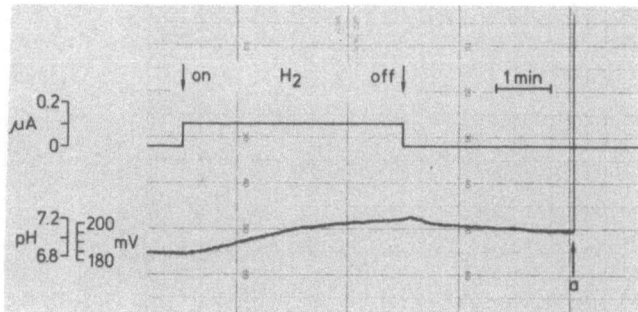

Fig. 6. Local pH changes on the surface of a blood-perfused frog tongue covered by a pH_2/H_2 sensor during H_2 generation.

of the OH^- ions corresponding to the buffer capacity of the bicarbonate-phosphate buffer. The smallest increase is found in the tissue; this is caused by its buffer capacity as well as by the convectional transport (blood flow).

CONCLUSION

These experiments demonstrate that even with small generating currents local effects of OH^- ions on the tissue cannot be excluded if such currents are used over longer periods. Since by current pulses of 6 s and 100 μA about ten times more hydroxyl ions are generated than in 10 min by a current of 0.1 μA, possible effects of the hydroxyl ions have to be considered. Additionally, Fig. 2 demonstrates that during the H_2 generation at 0.9 V oxygen is reduced, whereby local tissue hypoxia may be initiated. Under such conditions the use of a thin membrane would be advisable to protect the tissue.

REFERENCES

Baumgärtl, H., and Lübbers, D. W., 1983, Microcoaxial needle sensor for polarographic measurement of local O_2 pressure in the cellular range of living tissue. Its construction and properties, in: "Polarographic Oxygen Sensors", E. Gnaiger, H. Forstner, eds., Springer-Verlag, Berlin-Heidelberg-New York.

Koshu, K., Kamiyama, K., Oka, N., Endo, S., Takaku, A., and Saito, T., 1982, Measurement of regional blood flow using hydrogen gas generated by electrolysis, Stroke, 13:483.

Lübbers, D. W., and Stosseck, K., 1974, Quantitative Bestimmung der lokalen Durchblutung durch elektrochemisch im Gewebe erzeugten Wasserstoff, Naturwissenschaften, 57:311.

Renooij, W., Janssen, L. W. M., Akkermans, L. M. A., Lagey, C. L. R. S., and Wittebol, P., 1983, Electrode oxygen consumption and its effect on tissue oxygen tension, Clin. Orthop., 173:239.

Saito, J., Baumgärtl, H., and Lübbers, D. W., 1976, The RF sputtering technique as a method for manufacturing needle-shaped pH microelectrodes, in: "Ion-selective Electrodes and Enzyme Electrodes in Biology and Medicine", M. Kessler, L. C. Clark jr., D. W. Lübbers, I. A. Silver, W. Simon, eds. Urban & Schwarzenberg, München-Berlin-Wien.

Stosseck, K., Lübbers, D. W., and Cottin, N., 1974, Determination of local blood flow (microflow) by electrochemically generated hydrogen, Pflügers Arch., 348:225.

DIFFUSION OF OXYGEN AND HYDROGEN GAS IS

FASTER THROUGH A LAYER OF SUSPENDED CULTURED

C_6 CELLS THAN THROUGH THE MEDIUM

Minoru Tomita, Fumio Gotoh, Norio Tanahashi, Masahiro Kobari, Tamotsu Shinohara, Yasuo Terayama, Ban Mihara, Kouichi Ohta

Department of Neurology, School of Medicine, Keio University, Tokyo 160, Japan

INTRODUCTION

The gas diffusion in the living cerebral cortex has been reported by Gotoh et al. to be faster than that in the dead cortex (Gotoh, Tazaki and Meyer, 1961). This was attributed by them to a mixing effect due to the vasomotor action of the microvasculature in the tissue, by which gas molecules could be transported to far distant areas. The present study examined whether or not such a facilitated process persisted even in a cell suspension where the microcirculation was deprived.

METHOD

Glioma C_6 cells (provided via the following route: Nierenberg Inst., NIH - Prof. Haruhiro Higashida, Department of Biophysics, Neuroinformation, Research Institute, Kanazawa University School of Medicine - Prof. Yutaka Nagata, Department of Physiology, Fujita-Gakuen University School of Medicine) were inoculated into flasks containing 10% Dubecco's modified Eagle's medium plus 10% Gibco newborn calf serum (DMEM). The cells were then incubated at 37°C in 5% CO_2 and oxygen. Immediately before the experiment, cells were harvested, centrifuged and then diluted with DMEM so that the cell volume became 50%. This cell suspension was prepared each time for the following experiment. In order to measure the gas diffusion, a modified platinum PO_2 electrode of the recessed type (Davies and Brink, 1942) was used. As shown in Fig. 1, a 250 μm platinum wire was sealed in a glass tube of 3.0 mm in outer diameter, and only the cut surface was exposed. The glass tube was inserted into a glass sheath of 4.5 mm in inner diameter together with an Ag-AgCl reference electrode. In front of the platinum surface, there was a recess of 1.3 mm in depth and 4.5 mm in diameter, which was filled with fluid and covered with a thin Teflon membrane. Two fluids were used in the recess: first, medium alone, i.e. DMEM; and then, a suspension of cultured cells in DMEM. No gas was equilibrated with the cell suspension in the recess in the preparation period lasting approximately 10 min. With one of the respective fluids in the recess, two gases were blown towards the electrode for 20 s at room temperature (approximately 25°C): oxygen gas with a parallel resistance of 68 kΩ and an applied negative potential of -0.5 volts to the platinum surface, and hydrogen gas with a 680 kΩ resistance and an applied potential of +0.5 volts, alternately. The diffusion of the oxygen gas and of the hydrogen gas through the fluid layer in the recess was estimated from response curves to the respective gases. From each response curve (C(t)), the appearance time (AT), mode transit time or peak time (PT), mean transit time (MT) calculated as MT $= \int_0^\infty t \cdot C(t)dt / \int_0^\infty C(t)dt$, and height (H) for O_2 and H_2 were measured.

Oxygen Transport to Tissue XII, Edited by J. Piiper *et al.*
Plenum Press, New York, 1990

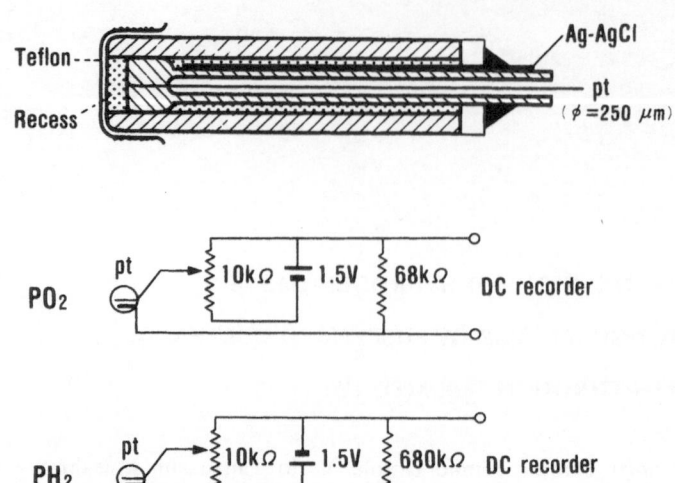

Fig. 1. Platinum electrode and diagram of the electric circuit for determining the PO_2 gas tension. When the electrode was used for hydrogen gas, the parallel resistance of 68 kΩ was changed to 680 kΩ with the polarity of the DC battery reversed.

RESULTS

When gas was introduced towards the electrode, the response curve built up after a short lag time: a negative curve for oxygen gas (Fig. 2A and 2B) and a positive curve for hydrogen gas (Fig. 3A and 3B). The curve in general was smooth, but with a rather rapid rise, single-peaked, and gradual in the descending part. We found that the presence of cells in DMEM shortened both PT and MT, and decreased the height of the curve responding to O_2. The response curve to H_2 revealed similar changes except that the height was somewhat increased with cells. A summary of the experiment is given in Table 1. AT for H_2 was 9.0 ± 0.4 s in DMEM, and 8.3 ± 0.4 s in the cell suspension. On the other hand, AT for O_2 was 14.3 ± 2.0 s in DMEM, and 16.3 ± 3.2 s in the cell suspension. MT for H_2 was 164.2 ± 31.2 s in DMEM, and 126.4 ± 6.2 s in the cell suspension. The difference was statistically significant (p < 0.05). The shortening of MT for O_2 was more marked in the presence of cells, from 125.0 ± 21.2 s to 79.0 ± 9.2 s (p < 0.01).

Table 1.   Summary of Experiment

	DMEM	Cell Suspension	N	p value
Appearance time for H_2	9.0 ± 0.4 s	8.3 ± 0.4 s	5	N.S.
Peak time for H_2	116.0 ± 24.1 s	65.1 ± 5.8 s	6	p<0.05
Mean transit time for H_2	164.2 ± 31.2 s	126.4 ± 6.2 s	6	p<0.05
Peak height for H_2	4.2 ± 0.3	10.6 ± 4.7	6	p<0.05
Appearance time for O_2	14.3 ± 2.2 s	16.3 ± 3.2 s	6	N.S.
Peak time for O_2	229.4 ± 16.9 s	102.2 ± 38.1 s	8	p<0.01
Mean transit time for O_2	125.0 ± 21.2 s	79.0 ± 9.2 s	8	p<0.01
Peak height for O_2	6.4 ± 1.3	2.5 ± 0.8	8	p<0.01

Units for the height (H) are arbitrary units.

116

−0.5 V, 68kΩ, 25°C

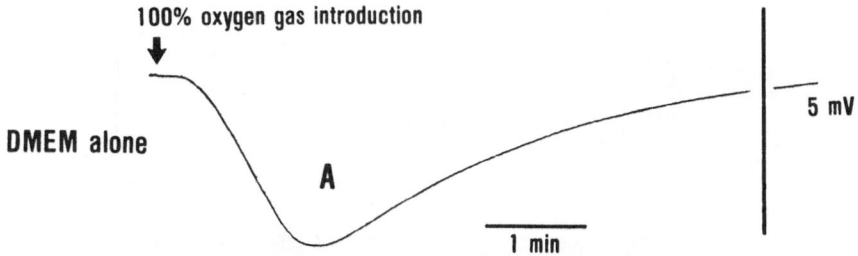

Fig. 2 A. Response curve to oxygen with DMEM in the recess. The time scale and calibration for the sensitivity of the electrode are indicated.

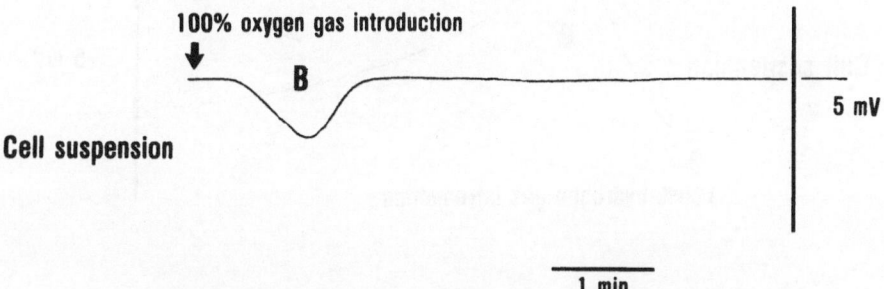

Fig. 2 B. With cell suspension in the recess, the response curve to oxygen was markedly shortened in time and the height of the curve was greatly suppressed.

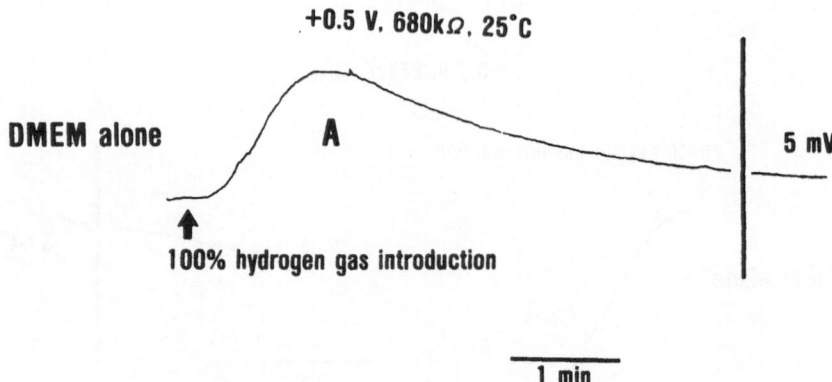

+0.5 V, 680kΩ, 25°C

DMEM alone

A

5 mV

↑
100% hydrogen gas introduction

1 min

Fig. 3 A. Response curve to hydrogen with DMEM in the recess. It should be noted that the polarity of the curve was opposite to that for oxygen gas.

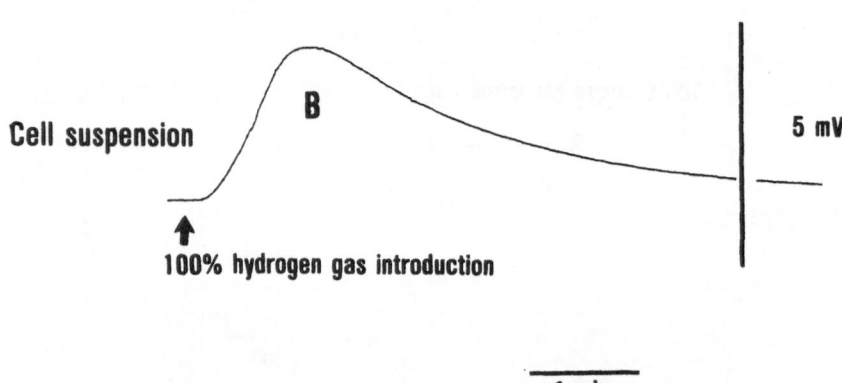

Cell suspension

B

5 mV

↑
100% hydrogen gas introduction

1 min

Fig. 3 B. With the cell suspension, the response curve to hydrogen gas was shortened, but the height of the curve was somewhat increased, giving rise to a contrast with the case of oxygen gas shown in Fig. 2.

DISCUSSION

The above data indicate that the diffusion of gases through a layer of living cells in suspension was faster than that through the medium alone, and the acceleration appeared to be more marked for oxygen than for hydrogen diffusion. This acceleration cannot be explained by oxygen transport mediated by hemoglobin or myoglobin (Kreuzer and Hoofd, 1984) since C_6 cells lack these substances. Hook et al. (1988), based on their model analysis of the O_2 transfer mechanism, concluded that diffusion was the main process limiting O_2 uptake and release by RBC, while the finite reaction kinetics of O_2 with hemoglobin exerted a smaller limiting effect.

The diffusion of oxygen gas estimated from the mean transit time of the response curve could be exaggerated by oxygen consumption by the cells. The diminution of the curve height by half with 50% cells as shown in the present data, or even the disappearance of the response curve with 80% cells which was observed in a separate series of experiments, lends support to this possibility.

Although the actual diffusion of oxygen as estimated from the response curve was apparently modified, in terms of the transit time, by the cell metabolism, the above conclusion concerning facilitated diffusion of gases through a layer of living cells in suspension appeared to be valid based on the data for hydrogen gas. The actual mechanism of the acceleration of gases through the cell suspension remains unknown. One possibility is a mixing effect induced by "water metabolism" of the cells. It is well known that small ions move unremittingly in and out through the cell membrane. Such ionic movement must be accompanied by water movement (Nagasawa et al., 1986), which may be termed the water metabolism. When the influx of ions and efflux of ions are the same, no net volume change occurs, yet a mixing effect in the environmental fluid ensues. In addition, long-span observation of living cells under a fast-speed video display has indicated that they move intermittently pendulating, twisting, swelling and shrinking. Coupled transport of water and ions through membranes with or without cell motion imparted by osmotic work across the cell membranes (Tomita, Gotoh and Kobari, 1988) may contribute to mixing of the extracellular fluid, accelerating gas diffusion through the tissue.

CONCLUSION

Based on the above data, we conclude that the diffusion of gases was faster in the presence of cultured glioma cells than in DMEM alone. One possible mechanism could involve intermittent cell motion yielding a mixing effect in the medium. However, the actual diffusion of oxygen as estimated from the response curve through the living cell suspension appeared to be attenuated in height and modified in its transit time by the cell metabolism.

ACKNOWLEDGEMENT

The authors would like to thank Prof. Akimichi Kaneko, National Institute for Physiological Sciences, Okazaki, for his helpful advice and his courtesy in allowing us to utilize his facilities at the National Institute for Physiological Sciences, Okazaki 444, Japan.

REFERENCES

Davies, P.W., and Brink, F., Jr., 1942, Microelectrodes for measuring local oxygen tension in animal tissues. Rev. Sci. Instr., 13:524.

Gotoh, F., Tazaki, Y., and Meyer, J.S., 1961, Transport of gases through brain and their extravascular vasomotor action. Exp. Neurol., 4:48.

Hook, C., Yamaguchi, K., Scheid, P., and Piiper, J., 1988, Oxygen transfer of red blood cells: experimental data and model analysis. Resp. Physiol., 72:65.

Kreuzer, F., and Hoofd, L., 1984, Facilitated diffusion of oxygen: possible significance in blood and muscle. Adv. Exp. Med. Biol., 169:3.

Nagasawa, M., Tasaka, M., and Tomita, M., 1986, Coupled transport of water and ions through membranes as a possible cause of cytotoxic edema. Neurosci. Lett., 66:19.

Tomita, M., Gotoh, F., and Kobari, M., 1988, Colloid osmotic pressure of cat brain homogenate separated from autogenous CSF by a copper ferrocyanide membrane. Brain Res., 474:165.

A METHOD FOR MEASURING THE RATE OF OXYGEN RELEASE FROM FLOWING ERYTHROCYTES IN MICROVESSELS

N. Tateishi, N. Maeda and T. Shiga[*]

Department Physiology, School of Medicine, Ehime University
Shigenobu, Onsen-gun, Ehime 791-02, Japan
[*]Department Physiology, School of Medicine, Osaka
University, Nakanoshima, Kita-ku, Osaka 530, Japan

INTRODUCTION

Physiologically, dynamics of oxygen release from erythrocytes in capillary to tissues is one of the most important phenomena. The major determinants of oxygen supply to tissues are blood flow to tissues, oxygen content in blood and the density of capillaries. The importance of capillary density has been well examined since the pioneering study of Krogh[1].

For the quantitative study of the phenomenon, the changes of oxygen saturation of flowing erythrocytes in microvessels, the oxygen tension in both tissues and microvessels, and the dimension of microvessels must be measured. In the present study, we combine a scanning spectrophotometer with an inverted microscope through a light guide, in order to record the absorption spectrum of a single capillary in living tissue (such as mesentery). From the absorption spectra, the hemoglobin concentration and the oxygen saturation are calculated. The flow velocity of erythrocytes in microvessels is determined by a method of dual-spots cross-correlation, and the diameter of the vessel is measured on the digitized video-image with an image processor. On the basis of data obtained at two points of a microvessel, the rate of oxygen release from flowing erythrocytes can be estimated.

EXPERIMENTAL PROCEDURE

Whole schematic diagram of apparatus used in the present experiment is shown in Fig.1. Under pentobarbital anesthesia (i.p., 50 µg/g body weight), Wister rats (6 - 9 weeks old, male, 180-250g) were used for the experiments.

Microscopic observation of microvessel in rat mesentery: A portion of mesentery of rat was put between a cylindrical acrylate platform (diameter of 10 mm) and a piece of acrylate cover plate. Isotonic sodium phosphate-buffered saline of 37°C was poured between mesentery and a cover plate, in order to overperfuse the mesentery (for the measurement of the rate of oxygen release from flowing erythrocytes, 5.7mM sodium hydrosulfite was added to the saline). During measurement, the rectal temperature was always monitored.

Oxygen Transport to Tissue XII. Edited by J. Piiper *et al.*
Plenum Press, New York, 1990

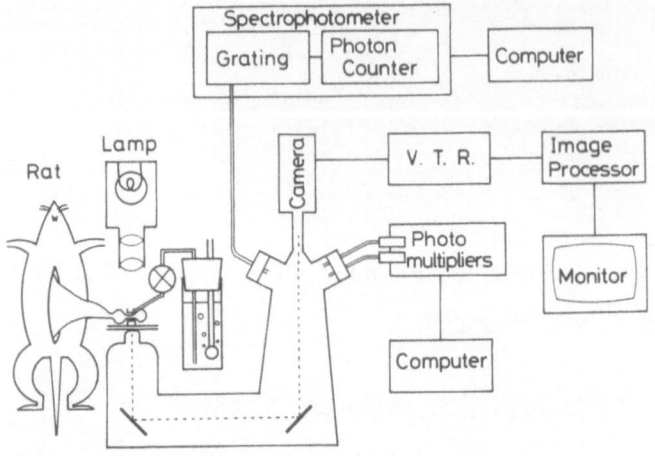

Fig. 1 Schematic diagram of the apparatus

<u>Measurement of oxygen saturation of hemoglobin in flowing erythrocytes</u>: A scanning-grating spectrophotometer (USP-410, Unisoku Co., Osaka) with a sensitive photon counting detector (10^3 - 3×10^6 photons/sec) was used to record a visible absorption spectrum of flowing erythrocytes[2,3]. The spectrophotometer was connected to an eyepiece of the microscope (IMT-2, Olympus Optics Co., Tokyo) by a thin light-guide (ϕ = 0.4mm). By using an objective lens of 40×, visible spectrum in the wavelength range of 500–600 nm was obtained from a spot of 5 µm diameter on a microvessel: the grating (250 steps) and the gate time (100 msec/step) were controlled by a computer(PC-9801, Nippon Electr. Co., Tokyo).

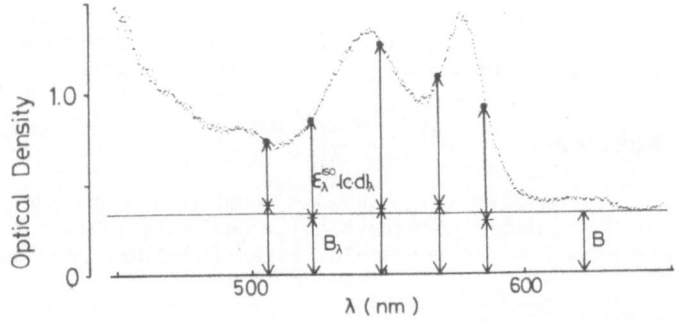

Fig. 2 Absorption spectrum of flowing erythrocytes in a
        microvessel measured by a scanning-grating
        spectrophotometer. Large dots with arrow show
        isosbestic points.

An optical density of flowing erythrocytes in a microvessel at a wavelength (OD) is given by the following equation,

$$OD = (\varepsilon^{oxy} \cdot F + \varepsilon^{deoxy} \cdot (1 - F)) [c \cdot d] + B$$

, where ε^{oxy} and ε^{deoxy} are extinction coefficients of oxy- and deoxy-hemoglobins[4], respectively. F is the fraction of oxyhemoglobin; c, hemoglobin concentration; d, optical path length; B, light scattering.

At isobestic points between oxy- and deoxy-hemoglobin (506.5, 522, 548, 569 and 586 nm),

$$OD = \varepsilon \cdot [c \cdot d] + B$$

, since $\varepsilon^{oxy} = \varepsilon^{deoxy} = \varepsilon$. Plotting OD against ε, the slope gives $[c \cdot d]$ and the intercept gives B.

A typical spectrum obtained for flowing erythrocytes in a single microvessel is shown in Fig.2. Actually, light scattering, B, was independent of wavelength (in 500-600 nm), in the present apparatus. The oxygen saturation was calculated from OD at additional seven wavelengths and was averaged.

Measurement of flow velocity of erythrocytes: Ten fine light guides (ϕ = 0.2 mm) were linearly aligned, and the set was inserted into a block of eyepiece. In order to monitor the changes of light intensity at two separate spots on an erythrocyte flow in a single microvessel (see Fig.3), two appropriate guides were selected and connected to a pair of photomultiplier (of special-made, Unisoku Co., Osaka). The output current was digitized by a high speed multichannel A/D converter (ADM-1198BPC, Micro Science Co., Tokyo). The cross-correlation between the changes of light intensity obtained from two channels (at up- and down-streams) was calculated with a microcomputer (PC-9801, Nippon Electr. Co., Tokyo), varying the delay time of 2nd channel situated at down-stream, as shown in Fig.3.

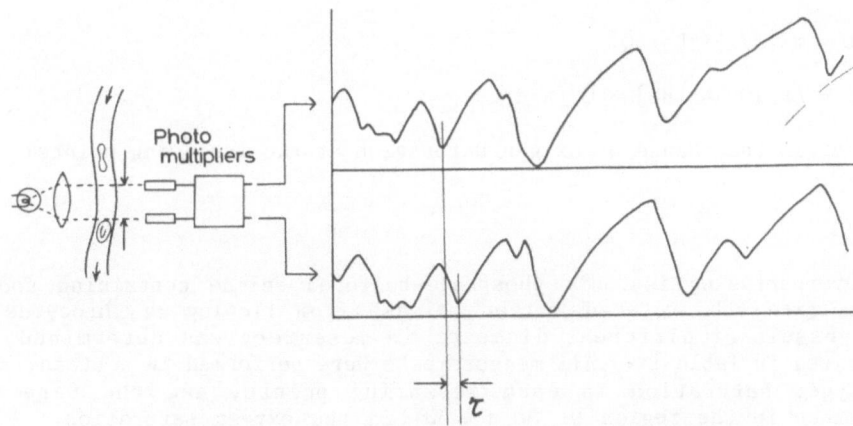

Fig. 3 Measurement of flow velocity of erythrocytes. Light intensity (in ordinate) is monitored at two separate points on a microvessel (with the distance between arrows, 1). The changes in lower figure (observed at down-stream) is delayed by τ in time from those in upper figure (observed at up-stream).

The velocity of erythrocyte flow (V) is estimated by the equation,

$$V = 1/\tau,$$

, where 1 is the distance between two spots and τ is the delay time giving the highest correlation coefficient.

Measurement of diameter of microvessel: The diameter of microvessel was measured on the video-image. A pointer of an image processor (PIP-4000, ADS Co., Nara) was set on one side of vessel wall as an origin, then the pointer was moved successively to two points (A and B) on the other side of wall (to form a triangle). A perpendicular from the origin on the line AB gives an estimate of vessel diameter.

Rate of oxygen release from flowing erythrocytes in microvessels : In order to estimate the rate of oxygen release, the following values at two points with an interval (L, μm) on a microvessel are measured (L > 100×1) : (1) the optical density, thus [c·d] (2) the erythrocyte velocity (V, μm/sec), (3) the oxygen saturation of hemoglobin in erythrocytes (F, %) and (4) the diameter of microvessel (d, μm).

The hemoglobin concentration in the microvessel ([Hb], mM) is calculated by the equation,

$$[Hb] = [c \cdot d] \ / \ d,$$

and the flow volume (Q) is calculated by the equation,

$$Q = \pi \cdot (d/2)^2 \cdot V.$$

Thus, the flow of hemoglobin is [Hb] $\times$ Q. Each averaged values of [Hb] and Q at two measuring points ([$\overline{Hb}$] and $\overline{Q}$, respectively) were used for the calculation of the transit time through the microvessel section of length, L, (t, sec) and the rate of oxygen release (R, mM/μm^2/sec) in the microvessel section, as follows:

$$t = \pi \cdot (d/2)^2 \cdot L \ / \ \overline{Q}$$

$$R = \Delta F/100 \times [\overline{Hb}] \times \overline{Q}/(\pi \cdot d \cdot L)$$

where ΔF is the change of oxygen saturation at two measuring points.

RESULTS

Overperfusing isotonic phosphate-buffered saline containing sodium hydrosulfite, the rate of oxygen release from flowing erythrocytes in microvessels of different diameter in mesentery was determined, as summarized in Table 1. All measurements were performed in a steady sate of oxygen saturation in each measuring points, and the rate was determined in the region of 80 to 30% in the oxygen saturation. The results clearly showed that the rate of oxygen release was the largest in capillary range, and that the rate in arterial side (the diameter, 13 to 20 μm) was significantly larger than that in venous side (the diameter, 15 to 30 μm).

Discussion

The dynamic analyses of the motion, the dimension, and the changes in light intensity within a microscopic field is important for

124

Table 1 Rate of oxygen release measured in microvessels of rat mesentery

Microvessels	Diameter (μm)	Rate ($mM \cdot sec^{-1} \cdot mm^{-2}$)*
Arterial side	13–20	83.0 ± 42.3 (n=3)
Capillary	< 13	132.7 ± 26.1 (n=6)
Venous side	15–30	13.2 ± 4.7 (n=8)

*Values are expressed as mean $\pm$ S.E.

understanding the function of microcirculation, especially the oxygen transport to tissues. Some methods have been developed for the measurements of these parameters with advantages and disadvantages. The apparatus used in the present study has the following characteristics in the measurements of oxygen saturation of flowing erythrocytes and the flow velocity of erythrocytes in microvessels. The usefulness of the apparatus was evaluated by using microvessels in rat mesentery.

1. On the measurement of hemoglobin concentration and oxygen saturation:
     Pittman and Duling[5] and Duling et al[6] have reported a spectrophotometric method for the determination of oxygenation state of blood in a microvessel, by measuring the optical density at three wavelengths. Hester and Duling[7] have measured the oxygen saturation in microvessels up to 15 μm in diameter. In the present study, the use of photon counting system increased the sensitivity of measurement (a numerical filtering was applied, in order to minimize the signal-to-noise ratio). The analysis of spectrum at five wavelengths at isosbestic points (for deoxy- and oxy-hemoglobins) and at additional seven wavelengths in the range of 500 – 600 nm increased the accuracy for the determination of hemoglobin concentration and the oxygen saturation in flowing erythrocytes, respectively. Thus the present method was extended to the measurement in narrower microvessels (up to 8 μm).

2. On the flow velocity of erythrocytes in microvessels:
     For the measurement of the velocity of flowing erythrocytes, the dual-slit photometric method[8] and the dual window television method[9] are used. However, the dual window television method cannot measure the velocity higher than 2.5 mm/sec, because of the limitation of the scanning rate of video image. The use of high speed analog-digital converter, in the present study, made possible to measure the velocity up to 50 mm/sec. Furthermore, since the measurement can be carried out within the short length of microvessel (up to 5 μm), the selection of microvessels, such as straight vessels, is not always needed for the measurement.

     The present apparatus was evaluated for the microvessels of rat mesentery. The oxygen release was compelled to flowing erythrocytes in a microvessel by overperfusing the fully deoxygenated saline. In the present study, the rate of oxygen release is estimated by the deoxygenation rate of hemoglobin. Since the oxygen content released from flowing erythrocytes by the gradient of oxygen tension is greatly affected not only by the nature of vessel wall (especially, the thickness), but also by the oxygen equilibrium characteristics of erythrocytes. Thus, the measurement was performed in the near linear range of the change of oxygen saturation to oxygen tension (30-80 %). The present study showed that the rate of oxygen release in capillaries is the largest, and that the rate in arterial side is larger than that in venous side. The latter result is similar to that observed for microvessels of hamster cheek pouch retractor muscle[10].

ACKNOWLEDGEMENT

The work was supported in part by grants from the Ministry of Education, Science and Culture of Japan and from the Ehime Health Foundation.

REFERENCE

1. A. Krogh, The number and distribution of capillaries in muscles with calculations of the oxygen pressure head necessary for supplying the tissue, J. Physiol. Lond. 52:409-415 (1919).

2. T. Shiga, N. Maeda and N. Tateishi, Kinetics of blood cell aggregation by video-image processing, in:"Microcirculation an Update", Vol.1, M. Tsuchiya, M. Asano, Y. Mishima, and M. Oda, ed., pp.433-436, Elsevier, Amsterdam (1987).

3. N.Tateishi, N. Maeda and T. Shiga, A new method for spectral analysis of flowing blood in microvessel, in:"Microcirculation Annual 1988", M. Tsuchiya, M. Asano and S. Matsuyama, ed., pp.121-122, Nihon-igakukan, Tokyo (1989).

4. O. W. Van Assendelft, "Spectrophotometry of Haemoblobin Derivatives." Royal VanGorcum. Assen, The Netherland (1970).

5. R. N. Pittman, and B. R. Duling, A new method for the measurement of percent oxyhemoglobin, J. Appl. Physiol. 38:315-320 (1975).

6. B. R. Duling, D. N. Damon, S. R. Donaldson, and R. N. Pittman, A computerized system for densitometric analysis of the microcirculation, J.Appl. Physiol., 55:642-651 (1983).

7. R. L. Hester, and B. R. Duling, Red cell velocity during functional hyperemia:implications for rheology and oxygen transport, Am. J. Physiol., 255:H236-H244 (1988).

8. H. Wayland, and P. C. Johnson, Erythrocyte velocity measurement in microvessels by a two slit photometric method, J. Appl. Physiol., 22:333-337 (1967).

9. M. Intaglietta, N. R. Silverman, and W. R. Tompkins, Capillary flow velocity measurements in vivo and in situ by television methods, Microvasc. Res., 10:165-179 (1975)

10. D. P. Swain, and R. N. Pittman, Oxygen exchange in the microcirculation of hamster retractor muscle, Am. J. Physiol., 256:H247-H255 (1989).

SPATIAL VARIATION OF THE LOCAL TISSUE OXYGEN DIFFUSION

COEFFICIENT MEASURED IN SITU IN THE CAT RETINA AND CORNEA

Hang-Duk Roh, Thomas K. Goldstick and Robert A. Linsenmeier

Departments of Chemical and Biomedical Engineering
Northwestern University
Evanston, Illinois 60208 USA

INTRODUCTION

Precise knowledge of the oxygen diffusion coefficient (D) in tissue is essential for the analysis of oxygen transport. The PO_2 profile in a tissue can be measured with a deeply recessed (L/d > 5), polarographic microelectrode (Linsenmeier, 1986; Haugh et al., 1990). By contrast, one with a shallow recess (L/d < 1) is affected by D in the tissue. It is possible to utilize this feature to determine the tissue D from the polarization transient when the electrode is turned on (i.e., the imposed voltage is changed). The faster the approach to steady state, the larger the value of D must be. Previous attempts to utilize a microelectrode turn-on transient to give local values of D in tissue (Erdmann and Krell, 1976; Buerk, 1980; Buerk and Goldstick, 1990) were plagued by the problem that initially the extremely large current saturated the amplifier. In the present study, the amplifier saturation was eliminated by open-circuiting the amplifier for a few ms initially. All of the data recorded after the initial delay, including those at very early times, were used for the calculation of D by a nonlinear regression analysis. Apparently there have been no reliable previous measurements of D in the mammalian retina or cornea in situ. The usual previous approach has been to make the measurement on a slice of excised tissue in vitro. The measurements in the present study are therefore novel because they are local and made in situ in the intact tissue of a living animal.

THEORY

The turn-on transient current depends on both the electrode geometry and the physical properties of the tissue. For both spherical and planar disk electrodes of radius r_o in homogeneous, nonconsuming media, the theoretical transient equation is exact (Laitinen and Kolthoff, 1939 ; Soo and Lingane, 1964) and may be written in the general form:

$$I(t) = I_\infty \left\{ 1 + \sqrt{tc/t} \right\} \tag{1}$$

where tc is the characteristic time; for a spherical electrode $tc = r_o^2/\pi D$ and for a disk $tc = \pi r_o^2/16D$. For these simple shapes, tc is only dependent on the electrode geometry and D. Thus, the ratio of D measured by the same electrode in any two media (e.g., tissue and isotonic saline) with different D can be calculated from the relationship:

$$D_1/D_2 = tc_2/tc_1 \tag{2}$$

Although equation 1 would be adequate for an electrode of simple and known geometry, it

Oxygen Transport to Tissue XII, Edited by J. Piiper et al.
Plenum Press, New York, 1990

127

cannot describe the transients generated by most practical microelectrodes with their complex, usually unknown geometries. In an attempt to compensate for the geometrical factors empirically, Buerk and Goldstick (1990), among others (Kakihana et al., 1981), modified equation 1. Departures from equation 1 were especially observed during the initial part of the transient response. For example, Kakihana et al. (1981), who estimated D for ions in solutions using bare, planar disk electrodes, added an empirical, compensating, exponential term whose parameters were determined by an approximation method. Other investigators have employed equation 1 unmodified. Parker and Winlove (1983) used a 20 µm bare planar electrode in various connective tissues and fitted the measured data using simple linear regression. Winlove and Parker (1984) also tried to measure D and oxygen concentration from the slope and intercept of equation 1 using a planar disk electrode in various aqueous solutions. All previous modifications along with several other similar, semi-empirical equations and equation 1 (a total of seven different equations) were evaluated by Roh (1989). He tested how well they fitted actual experimental data generated by a shallowly recessed oxygen microelectrode. As expected, the deviation from equation 1 occurred in the initial part of the transient which is dependent on D inside the recess. The best fitting equation in terms of consistency and accurate representation of the data for a recessed microelectrode was found to be:

$$I(t) = I_\infty \left\{ 1 + \sqrt{tc/t} + A \exp\left[-\sqrt{t/tc} \right] \right\} \tag{3}$$

Equation 3 was fitted to the transient experimental data by nonlinear regression analysis. For the bare needle microelectrodes, the fitting constant A turned out to be negligible (giving equation 1), but for the recessed microelectrodes A was significant.

MATERIALS AND METHOD

The methods for animal preparation and visual stimulation were essentially the same as those previously described by Yancey and Linsenmeier (1989) and Linsenmeier, Goldstick and Zhang (1989). A conditioned adult cat was initially anesthetized with 20 mg/kg sodium thiamylal injected intravenously. Arterial, venous and tracheal cannulas were inserted. After the cannulation, the animal was moved to a shielded cage where its head was mounted in a head holder. The cat was electrically insulated from the cage so that it was isolated from ground. During surgery, the long term anesthetic urethane was infused at about 200 mg/hr until a loading dose of 200 mg/kg had been given. During the rest of the experiment urethane was given at 20 to 40 mg/kg·hr. During the eye surgery, sodium thiamylal continued to be given as needed. The temporal side of the right eye was surgically exposed to allow enough room for the apparatus used to fix the eye and insert the electrode. After most of the surgery had been finished, the cat was paralyzed with sufficient pancuronium bromide, 0.2 - 0.3 mg/kg·hr, to stop respiration and it was artificially respirated. Body temperature was kept at $37 - 38^\circ C$ with a regulated heating pad. Arterial PO_2, PCO_2, and pH were measured with a blood gas analyzer (model 158, Corning Medical and Scientific, Medfield, MA) and were maintained within the normal range for cats ($P_aO_2 \simeq 100$ mm Hg, $P_aCO_2 \simeq 30$ mm Hg and $pH_a \simeq 7.40$) by respirator adjustment and addition of a small amount of 100 % oxygen to the inspired gas.

Figure 1 shows the circuit used to make the present measurements. Microelectrode turn-on transient currents were measured by closing the first micro reed relay, which imposed a voltage step (0 initially to -0.7 V) between the oxygen microelectrode cathode and the chlorided silver anode reference. An isolated Schmidt trigger closed this first micro relay and also readied the computer for the data acquisition. After a 1 to 25 ms delay, to eliminate initial amplifier saturation, the second relay was switched to include the picoammeter in the circuit and to begin data acquisition. To insure equilibrium prior to each transient measurement, the electrode was not polarized for at least 30 s prior to initiating a measurement. In making measurements, 3 to 6 consecutive transients were measured and analyzed and the parameters (I_∞, tc, and A) were averaged. An extremely high speed picoammeter (Model 417, Keithley Instruments, Cleveland, Ohio, USA) was essential as the amplifier. A microcomputer (Model PC, IBM, Boca-Raton, Florida, USA) was used to collect the data. Depending on the electrode size, the data were recorded for either 1 or 5 s following the voltage step, and the data acquisition rate was set to either 1000 Hz or 200 Hz, respectively. The data were digitized (Model DT2801, Data Translation, Marlboro, MA, USA), stored on a floppy disk, and sent to the university mainframe (CDC Cyber 845) offline for nonlinear regression analysis. The nonlinear regression analysis fitted either the semi-empirical model (equation 3) to

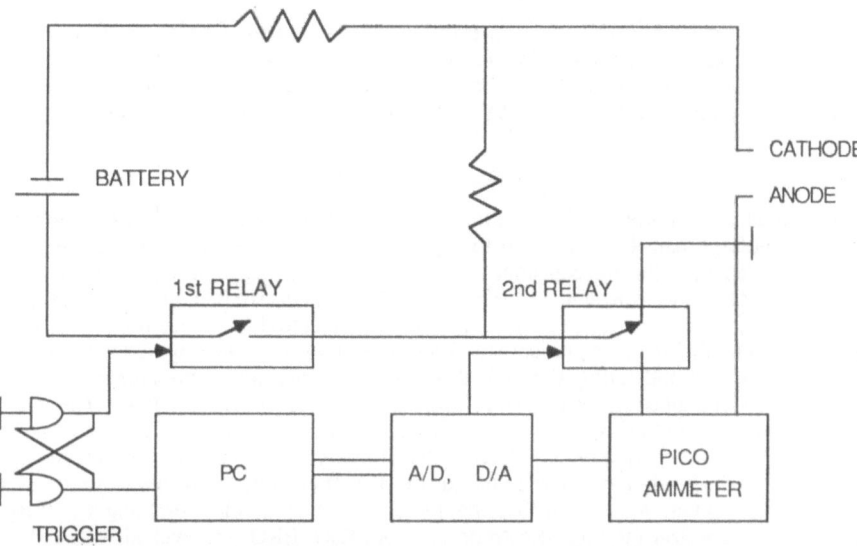

Figure 1. Measuring circuit for determining the oxygen diffusion coefficient from a microelectrode turn-on transient, including amplifier and data acquisition system.

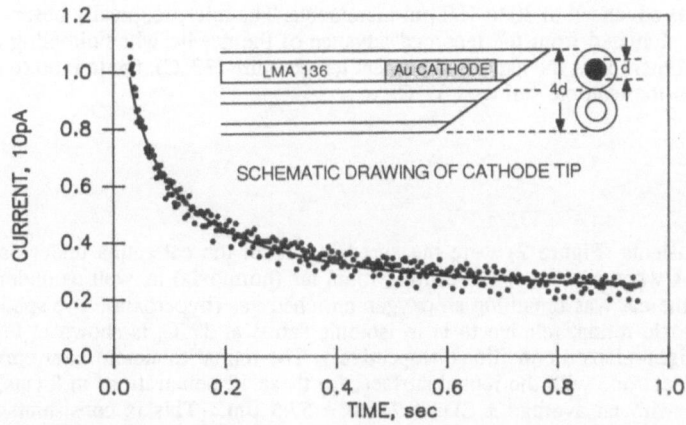

Figure 2. Typical retinal turn-on transient (in this case 233 μm from the vitreous) and schematic of recessed oxygen microelectrode used. Data were taken at 1000 Hz but for clarity only every third point is shown.

the recessed microelectrode data, or the theoretical model (equation 1) to the needle microelectrode data. A typical example is shown in Figure 2.

For calibration, microelectrode turn-on transients were measured before and after the tissue measurements using isotonic saline equilibrated with either 4,8 or 21 percent oxygen (depending on the tissue PO_2 to be measured), at $37^{\circ}C$. To validate the method D was measured in glycerine solutions at various concentrations (Roh, 1989). These agreed with literature values.

Retina

For the retinal measurements, glass microelectrodes with gold plated cathodes < 1 μm diameter and an attached voltage barrel (total size 4 to 5 μm) were constructed following the technique of Linsenmeier and Yancey (1987) except that the recess depth was extremely short (L/d < 1) and a different low melting alloy, Cerrolow-136 (Cerro Metal Products, Bellefonte, Pennsylvania, USA), was used as a filling metal instead of their LMA-117. The electrode was positioned in the vitreous and advanced toward the retina continuously (Linsenmeier et al., 1989) until a negative voltage deflection was recorded indicating the retinal surface (Yancey and Linsenmeier, 1989). The electrode was then advanced in 3 μm steps for measurements every 30 μm. The retina was penetrated near the area centralis far from major blood vessels. All measurements were made during penetration. Each complete penetration took more than 2 hours since 3 to 6 transients were measured at each location and 3 minutes were required for the transition from light- to dark-adapted conditions and vice versa. At each measuring location, the local electroretinogram (ERG) was recorded in order to confirm the microelectrode location. Each ERG was evoked with a short flash of diffuse light of an illumination just sufficient to saturate the rod receptors. In cases where light-adaptation is referred to below, this same illumination was used.

Cornea

For the corneal measurements, commercially available, 0.5 mm shaft diameter, bare, tungsten, needle electrodes with their tips sharpened to a 5° half angle and < 1 μm tip diameter (Model 912, Diamond General, Ann Arbor, Michigan, USA) were utilized. These were lacquer-insulated here, except for the 10 μm exposed tip which was gold-plated under microscopic control. No membrane was applied. The cat cornea could not be penetrated by the recessed glass microelectrodes, because of their fragility and the toughness of the cornea. Even the tungsten microelectrodes occasionally caused dimpling of the cornea. Throughout the measurements the epithelial surface of the cornea was kept moist by a saline drip. Corneal measurements were made under microscopic control in order to ensure that the microelectrode did not bend. The microelectrode penetrated the cornea centrally, at an angle of approximately 60° with its surface. Measurements were made as the needle microelectrode was advanced in 30 to 100 μm increments. The microscopically observed penetration agreed with that calculated from the recorded advance of the needle when dimpling did not occur. To calculate the corneal D at the estimated corneal temperature ($33^{\circ}C$), the tc ratio (equation 2) was multiplied by D in the isotonic saline at $37^{\circ}C$.

RESULTS

Retina

Turn-on transients (Figure 2) were measured in situ in the cat retina under dark- and light-adapted conditions when the cat was breathing room air (normoxia) as well as under dark-adapted conditions when the cat was breathing an oxygen-enriched gas (hyperoxia). The spatial distribution of D in the normoxic retina, relative to D in isotonic saline at $37^{\circ}C$, is shown in Figures 3 and 4 under dark- and light-adapted conditions, respectively. The retinal thickness after correction for the 45° angle of the electrode with the retinal surface, for these 12 penetrations in 8 cats, was between 148 and 329 μm with an average ± SD of 238.8 ± 57.5 μm. This is consistent with previous, similarly corrected measurements made in cat retinas in this laboratory (Haugh et al., 1990).

The average D value in the vitreous humor was found to be 71.8 ± 10.9 (n=8) percent of that in isotonic saline at $37^{\circ}C$.

Ignoring the four very high points in Figure 3 (see Discussion), the local D across the cat retina for 8 dark-adapted penetrations averaged 70.6 ± 3.3 percent of that in isotonic saline at $37^{\circ}C$. Ignoring the two very high points in Figure 4, it averaged 71.7 ± 5.7 percent of that in isotonic

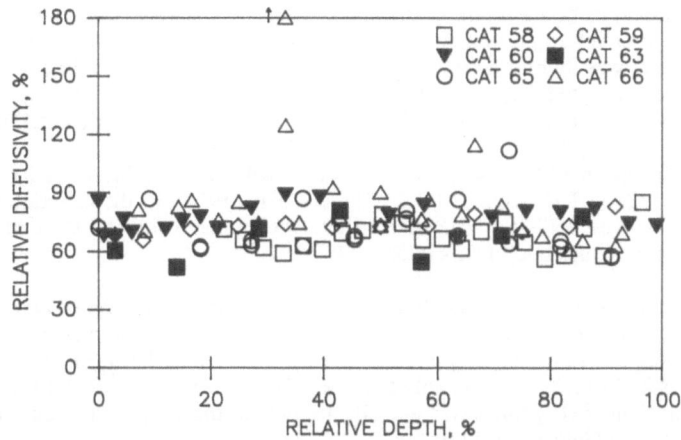

Figure 3. Spatial distribution of D in six cat retinas in situ, under normoxic, dark-adapted conditions, relative to that in isotonic saline at 37°C. The abscissa gives relative distance in percent; 0 % is the vitreous-retinal interface and 100 % is the choroidal retinal interface. Different symbols represent data on different cats, each cat with a different shallowly recessed microelectrode of cathode diameter ca 1 μm. There were a total of eight penetrations in the six cats.

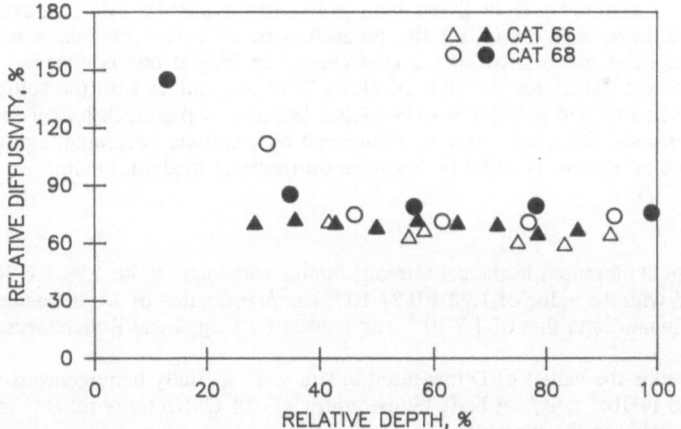

Figure 4. Spatial distribution of D in two cat retinas in situ, under normoxic, light-adapted conditions, relative to that in isotonic saline at 37°C. The abscissa gives relative distance in percent; 0 % is the vitreous-retinal interface and 100 % is the choroidal-retinal interface. In this case, a different symbol represents a different penetration (two penetrations in each of the two cats). A different shallowly recessed microelectrode of cathode diameter ca 1 μm was used in each cat.

saline at $37^{\circ}C$ for the 4 light-adapted penetrations. Systematic light-adapted measurements were not made in the innermost retina near the vitreous (Figure 4) because a few values of D measured there (not shown) were found to be the same as those under the dark-adapted condition. The dark- and light-adapted average D values are not statistically significantly different. The mean retinal D (12 penetrations in 8 cats) was therefore 70.9 ± 4.0 percent of that in isotonic saline at $37^{\circ}C$. Figure 5 shows that, under dark-adapted, hyperoxic conditions (approximately 100 % O_2 inspired), D appears to be approximately the same as that under dark- and light-adapted normoxic conditions.

The above percentage (70.9 ± 4.0) can be used to calculate a value for D. Correcting D in isotonic saline at $25^{\circ}C$, $2.07 \cdot 10^{-5}$ cm^2/s (Goldstick and Fatt, 1970), to the calibration temperature, $37^{\circ}C$, at 2.5 %/$^{\circ}C$, gives $2.78 \cdot 10^{-5}$ cm^2/s. The average retinal value of D is therefore $1.97 \pm 0.11 \cdot 10^{-5}$ cm^2/s at its normal body temperature, $37 - 38^{\circ}C$.

Steady state currents (I_∞) were also obtained under dark- and light-adapted conditions from the nonlinear regression fitting of the turn-on transients. In the outer retina, these agreed in general form with the PO_2 profiles measured by Linsenmeier (1986). There was, however, considerable scatter between I_∞ profiles in different cats probably because of differences in choroidal PO_2. Since neither this nor the simultaneous PO_2 profile was actually measured, the I_∞ profiles could not be used to calculate Dk (Buerk and Goldstick, 1990).

Cornea

Figure 6 shows local D measured in the cat corneal stroma in situ, at approximately $33^{\circ}C$, relative to isotonic saline at $37^{\circ}C$, for those values obtained with intact, unbent tungsten microelectrodes. The stromal D (7 sites in 3 cats) averaged 42.8 ± 7.3 percent of that in saline. Correcting D in isotonic saline at $25^{\circ}C$, $2.07 \cdot 10^{-5}$ cm^2/s (Goldstick and Fatt, 1970), to the calibration temperature, $37^{\circ}C$, gives $2.78 \cdot 10^{-5}$ cm^2/s. The average value of D is therefore $1.19 \pm 0.20 \cdot 10^{-5}$ cm^2/s for the cat corneal stroma at its normal in situ temperature of approximately $33^{\circ}C$.

DISCUSSION

The present study provides a method for the measurement of local D using oxygen microelectrode turn-on transients and also gives the first D profiles across the intact cat retina and cornea in situ. This method has the distinct advantage that it provides more detailed information about the spatial variation of D in tissue than previously available. For measuring D in tissue, previous methods have been based on the measurement of either the flux across, or the PO_2 transient in, an excised tissue slice after a step change in PO_2 at one boundary. These methods give only the average values for the slice of either D or its product with the solubility (k). They also cause tissue trauma and possible anoxia. Also, because of the much longer times involved in both previous methods, the results may be influenced by electrode poisoning, aging and drift. The present study uses an extremely rapid (< 5 s) turn-on transient method, in situ.

Retina

The average D measured in the cat vitreous humor was found to be $1.99 \pm 0.30 \cdot 10^{-5}$ cm^2/s. It can be compared with the value of $1.92 \pm 0.27 \cdot 10^{-5}$ cm^2/s estimated by Linsenmeier et al. (1981) in the cat vitreous humor, and that of $1.7 \cdot 10^{-5}$ cm^2/s used by Briggs and Rodenhaeuser (1973).

In the cat retina the values of D measured in situ were spatially homogeneous with an average value of $1.97 \pm 0.11 \cdot 10^{-5}$ cm^2/s at body temperature, $37 - 38^{\circ}C$. No other reliable measurements of retinal D are available in the literature.

The validity of the present method for the determination of local retinal D depends on the ability of the semi-empirical model to account for the microelectrode recess. A computer simulation of the turn-on transient of a recessed microelectrode, and experimental measurements in glycerine solutions of various concentration, have demonstrated that the computed tc is inversely proportional to the tissue D when L/d < 1 and the difference between the tissue and saline values of D is not large (Roh, 1989). In the retinal measurements, both of these conditions were achieved.

Several very large values of D were observed under normoxic, dark-adapted conditions (Figure 3). In the outer retina (50 to 100 % relative depth) these are attributed to the dependence of the measured D on PO_2, at very low PO_2. In the inner retina (0 to 50 % relative depth) these are

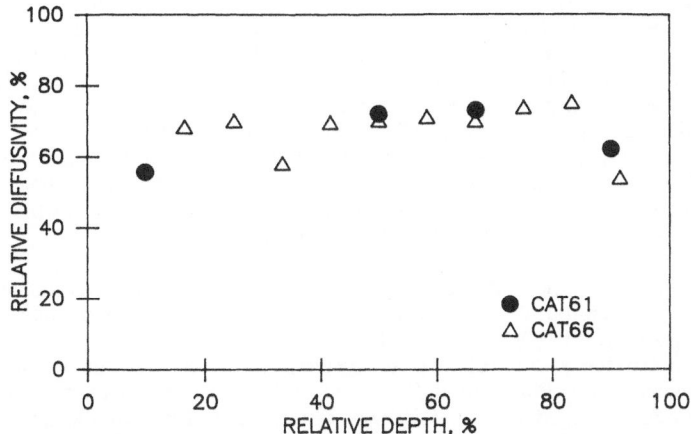

Figure 5. Spatial distribution of D in two cat retinas in situ, under hyperoxic, dark-adapted conditions, relative to that in isotonic saline at 37°C. The abscissa gives relative distance in percent; 0 % is the vitreous-retinal interface and 100 % is the choroidal-retinal interface. Each symbol represents data from a single penetration in a single cat retina, each cat with a different shallowly recessed microelectrode of cathode diameter ca 1 μm.

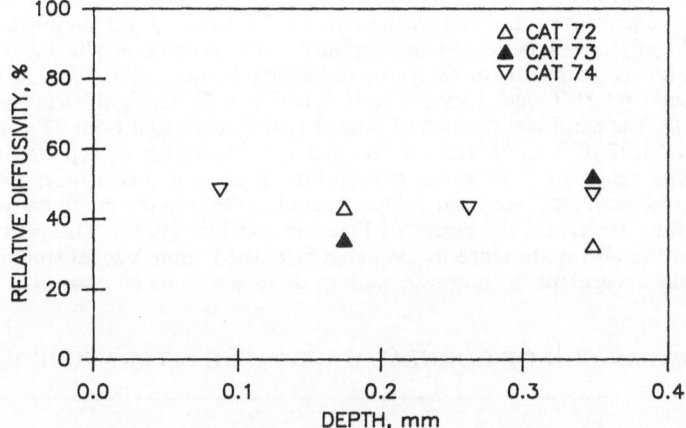

Figure 6. Spatial distribution of D in three cat corneal stromas, in situ, at their normal body temperature of 33°C, relative to that in isotonic saline at 37°C. The abscissa gives the approximate perpendicular distance from the epithelial surface. The endothelial- aqueous interface is at approximately 0.6 mm. Each symbol represents data from a single penetration in a single cat cornea. Each penetration was made with a different tungsten needle microelectrode of exposed tip length ca 10 μm. Reliable measurements could not be made beyond about 0.4 mm because of bending of the tungsten microelectrodes.

attributed to the retinal circulation. The transretinal PO_2 profile measured by Linsenmeier (1986) shows the presence of a minimum PO_2, close to 0 mm Hg, in the middle of the outer retina under normoxic, dark-adapted conditions, but not under light-adapted or hyperoxic, dark-adapted conditions. The dependence of the measured D on PO_2 was tested in isotonic saline equilibrated with 0, 4, 8 and 21 % oxygen and D was fairly constant with PO_2 except at zero PO_2. Much higher measured D values were obtained close to zero, apparently because of other electrode processes not involving oxygen. Because tissue transients in the absence of oxygen were obviously unavailable to correct the experimental, in situ, tissue transients, no account was taken of these. Consequently, values for D were overestimated at very low PO_2 in the outer retina under normoxic, dark-adapted conditions. High values of D in the outer retina were not seen under light-adapted (Figure 4) or hyperoxic (Figure 5) conditions. On the other hand, the high values of D in the inner retina, at about 25% depth, were observed under all conditions in the same cat.

Although with their much shorter recess and absence of any membrane material, the microelectrodes used here might have been expected to be more subject to poisoning effects than the deeply recessed ones used for PO_2 measurement, this did not seem to be the case. Poisoning would have shown up as a difference in tc between successive penetrations, but this was only 3 - 6 %, well within the experimental error. Poisoning did not seem to influence the measurements either. Tissue damage along the penetration tract was minimized (Linsenmeier, 1986) by the tiny microelectrodes employed. An intracellular voltage response to light was never observed.

Cornea

The average D measured in situ in the cat corneal stroma at its body temperature, approximately $33^{\circ}C$, was found to be $1.19 \pm 0.20 \cdot 10^{-5}$ cm^2/s. Although no other values are available for cat corneas, the present in situ value may be compared with D measured directly in the excised rabbit corneal stroma (Table 1), $0.67 \cdot 10^{-5}$ cm^2/s at $33^{\circ}C$ (Takahashi and Fatt, 1965) and $0.61 \cdot 10^{-5}$ cm^2/s (Takahashi et al., 1966), after correcting the original value from $25^{\circ}C$ to $33^{\circ}C$ at 2.5 %/$^{\circ}C$. In addition, Freeman and Fatt (1972) measured the product Dk in the excised rabbit corneal stroma at $23.5^{\circ}C$ to be $3.0 \cdot 10^{-10}$ (ml $O_2 \cdot cm^2$)/(ml·s·torr). Takahashi et al. (1966) estimated the corneal k at $25^{\circ}C$, to be $2.80 \cdot 10^{-5}$ ml O_2/(ml·torr). Correcting it from $25^{\circ}C$ to $23.5^{\circ}C$, at-1.0 %/$^{\circ}C$, gives $2.84 \cdot 10^{-5}$ ml O_2/(ml·torr). The Freeman and Fatt value of D, calculated by dividing by k and then correcting to $33^{\circ}C$, becomes $1.33 \cdot 10^{-5}$ cm^2/s. The rabbit corneal D may be similarly calculated from the data reported by Heald and Langham (1956), after temperature correction from $37.5^{\circ}C$ to $33^{\circ}C$ and again using the temperature corrected k value of Takahashi et al. (1966). It becomes $0.25 \cdot 10^{-5}$ cm^2/s, well below all other values. The present in situ value may also be compared with the bovine corneal D measured by Grote and Zander (1976). It was reported to be $1.70 \pm 0.24 \cdot 10^{-5}$ cm^2/s at $30^{\circ}C$ and $1.95 \pm 0.32 \cdot 10^{-5}$ cm^2/s at $37^{\circ}C$, significantly above all other values for corneal D. The empirical formula of Vaupel (1976), corrected from $37^{\circ}C$ to $33^{\circ}C$, gives, an estimate of D of $1.07 \cdot 10^{-5}$ cm^2/s for a tissue that is 77 % water, as reported for the cornea (Maurice, 1984). The values of D measured here in the cat corneal stroma were twice the values obtained by Takahashi and Fatt (1965) and Takahashi et al. (1966) in the rabbit but relatively close to that calculated from the rabbit Dk results of Freeman and Fatt (1972). The cat corneal stroma value measured here is also quite close to the value calculated from Vaupel's empirical formula. None of the previous measurements, however, were made in situ in intact corneas.

TABLE 1 COMPARISON OF CORNEAL D WITH THE LITERATURE VALUES

Corneal D at $33^{\circ}C$ ($\times 10^{-5}$ cm^2/s)	Study	Measurements
0.25	Heald and Langham (1956)	Rabbit whole cornea
0.61	Takahashi et al. (1966)	Rabbit stroma
0.67	Takahashi and Fatt (1965)	Rabbit stroma
1.07	Vaupel (1976)	Est. for 77 % H_2O
1.19	This study (1990)	Cat stroma
1.33	Freeman and Fatt (1972)	Rabbit stroma
1.80	Grote and Zander (1976)	Bovine cornea

134

SUMMARY

A method for measuring the local oxygen diffusion coefficient (D) in an intact tissue, in situ, in a living cat is described. Values of D were calculated from nonlinear regression analysis of the polarographic (turn-on) transients using a semi-empirical model for the retina and a theoretical one for the cornea. Two types of microelectrodes were employed: in the retina, ones with extremely short recesses; and in the cornea, bare metal needles. The local D in the cat retina was practically homogeneous with a mean of $1.97 \pm 0.11 \cdot 10^{-5}$ cm^2/s, at its body temperature of $37\text{-}38^\circ$C, 70.6 ± 3.3 percent of that in isotonic saline at 37°C. In the cat corneal stroma, at its normal temperature in situ of 33°C, D was also virtually homogeneous with a mean of $1.19 \pm 0.20 \cdot 10^{-5}$ cm^2/s, 42.8 ± 7.3 percent of that in isotonic saline at 37°C.

SYMBOLS

$A =$	dimensionless constant
$d =$	cathode diameter, μm
$D =$	diffusion coefficient of oxygen, cm^2/s
$I(t) =$	transient current, Amps
$I_\infty =$	steady state current, Amps
$k =$	solubility of oxygen, ml O_2(STP)/(ml·torr)
$L =$	recess depth, μm
$P, PO_2 =$	oxygen tension, torr
$Q =$	oxygen consumption rate, ml O_2(STP)/(ml·s)
$r_o =$	cathode radius, μm
$t =$	time, ms
$tc =$	characteristic time, ms

ACKNOWLEDGEMENTS

We gratefully acknowledge the technical skill in animal preparation kindly provided by Rod D. Braun and the assistance in making the needle electrodes provided by Dr. David Ferster. We also acknowledge the support received from the U.S. Public Health Service National Eye Institute grants EY-05034 and EY-07041, and from the Whitaker Foundation.

REFERENCES

Briggs, D. and Rodenhaeuser, J.-H., 1973, Distribution and consumption of oxygen in the vitreous body of cats, *in*: "Oxygen Supply: Theoretical and Practical Aspects of Oxygen Supply and Microcirculation of Tissue", M. Kessler, D.F. Bruley, L.C.Clark, Jr., D.W. Luebbers, I.A. Silver, J. Strauss, eds., Urban and Schwarzenberg, Munich, pp. 265-269.

Buerk, D.G., 1980, "Hypoxia in the Walls of Large Blood Vessels" (PhD thesis), Northwestern University, Evanston, IL, USA.

Buerk, D.G. and Goldstick, T.K., 1990, Oxygen diffusion coefficient determined from polarographic oxygen electrode turn-on transients: Spatial variation of diffusivity in the aortic wall, In preparation.

Erdmann, W. and Krell, W., 1976, Measurements of diffusion with noble metal electrodes, *Adv. Exp. Med. Biol.* 75 : 225-228.

Freeman, R.D. and Fatt, I., 1972, Oxygen permeability of the limiting layers of the cornea, *Biophys. J., 12* : 237-247.

Goldstick, T.K. and Fatt, I., 1970, Diffusion of oxygen in solutions of blood proteins, *Chem. Eng. Prog. Symp. Ser. No. 99, 66* : 101- 113.

Grote, J. and Zander, R., 1976, Corneal oxygen supply conditions, *Adv. Exp. Med. Biol., 75* : 449-455.

Haugh, L.M., Linsenmeier, R.A., and Goldstick, T.K., 1990, Mathematical models of the spatial distribution of retinal oxygen tension and consumption, including changes upon illumination, *Annals Biomed. Eng., 18* :19-36.

Heald, K. and Langham, M.E., 1956, Permeability of the cornea and blood-aqueous barrier to oxygen, *Brit. J. Ophthal., 40* : 705-720.

Kakihana, M., Ikeuchi, H., Sato, G.P., and Tokuda, K., 1981, Diffusion current at microdisk electrodes - Application to accurate measurement of diffusion coefficients, *J. Electroanal. Chem., 117* : 201-211.

Laitinen, H.A. and Kolthoff, I.M., 1939, A study of diffusion processes by electrolysis with microelectrodes, *J. Am. Chem. Soc., 61* : 3344-3349.

Linsenmeier, R.A., 1986, Effects of light and darkness on oxygen distribution and consumption in the cat retina, *J. Gen. Physiol., 88* : 521-542.

Linsenmeier, R.A., Goldstick, T.K.,Blum, R.S., and Enroth-Cugell, C., 1981, Estimation of retinal oxygen transients from measurements made in the vitreous humor, *Exp. Eye Res., 32* : 369-379.

Linsenmeier, R.A., Goldstick, T.K., and Zhang, S.-L., 1989, Chinese herbal medicine increases tissue oxygen tension, *Adv. Exp. Med. Biol., 248* : 795-801.

Linsenmeier, R.A. and Yancey, C.M., 1987, Improved fabrication of double-barreled recessed cathode O_2 microelectrodes, *J. Appl. Physiol., 63* : 2554-2557.

Maurice, D.M., 1984, The cornea and sclera, *in*: "The Eye", H. Davson, ed., 3rd ed., vol 1b, Academic Press, New York, p. 12.

Parker, K.H. and Winlove, C.P., 1983, The measurement of oxygen diffusivity and solubility in tissue using a transient polarographic technique, *J. Physiol. (London), 341* : 45P-46P.

Roh, H.-D., 1989, "Local Oxygen Diffusion Coefficients in the Cat Retina and Cornea" (PhD thesis), Northwestern University, Evanston, IL, USA.

Soo, Z.G. and Lingane, J., 1964, Derivation of the chronoamperometric constant for unshielded, circular, planar electrodes, *J. Phys. Chem., 68* : 3821-3828.

Takahashi, G.H. and Fatt, I., 1965, The diffusion of oxygen in the cornea, *Exp Eye Res., 4* : 4-12.

Takahashi, G.H., Fatt, I., and Goldstick, T.K., 1966, Oxygen consumption rate of tissue measured by a micropolarographic method, *J. Gen. Physiol., 50* : 317-335.

Vaupel, P., 1976, Effect of percentual water content in tissue and liquids on the diffusion coefficient of O_2, CO_2, and H_2, *Pfluegers Arch., 361* :201-204.

Winlove, C.P. and Parker, K.H., 1984, The measurement of oxygen diffusivity and concentration by chronoamperometry using microelectrodes, *J. Electroanal. Chem., 170* : 293-304.

Yancey, C.M. and Linsenmeier, R.A., 1989, Oxygen distribution and consumption in the cat retina at increased intraocular pressure, *Invest. Ophthalmol. Visual. Sci., 48* : 600-611.

A METHOD TO MEASURE THE DIFFUSION COEFFICIENT OF MYOGLOBIN IN INTACT SKELETAL MUSCLE CELLS.

Klaus D. Jürgens, Thomas Peters and Gerolf Gros

Zentrum Physiologie, Medizinische Hochschule
Hannover, FRG

INTRODUCTION

A new method was developed to measure the diffusion coefficient of myoglobin in intact muscle cells, a quantity that allows one to estimate the contribution of myoglobin to facilitated oxygen transport. So far, the diffusion coefficient of myoglobin in mammalian muscle cells has been estimated from results obtained from muscle homogenates (Moll, 1968) or from measurements performed on frog muscle cells, into which myoglobin was injected (Baylor and Pape, 1988). With these data no reliable estimate can be made of the extent to which myoglobin diffusion facilitates oxygen transport within mammalian muscle cells.

METHOD

The basic idea of this method is to instantaneously convert a certain amount of oxymyoglobin into metmyoglobin within geometrically defined areas of red muscle cells and to record the subsequent diffusion process of metmyoglobin out of these areas (and oxymyoglobin from neighboring parts of the muscle fibers into these areas) by measuring the related change of light absorbance in the Soret band. The wavelength in this band is chosen so that there is a maximal difference in the extinction coefficients of oxy- and metmyoglobin. Fig. 1 shows the difference spectrum of human hemoglobin in the oxygenated and the completely oxidized state of the heme groups, which is very similar to the corresponding myoglobin difference spectrum.

Fig.2 shows a schematic diagram of the experimental setup. The oxidation of the heme groups of myoglobin is achieved by pulse irradiation with ultraviolet light (Demma and Salhany, 1977) from a super pressure mercury lamp, which is guided through a quartz optical system onto the sample. After generating the rapid change in metmyoglobin concentration, the recovery of the light absorbance due to diffusion of met- and oxymyoglobin is measured with the microscope photometer, usually at a monochromator wavelength of 420 nm. Duration of the UV light pulse, selection of the wavelength, and averaging of

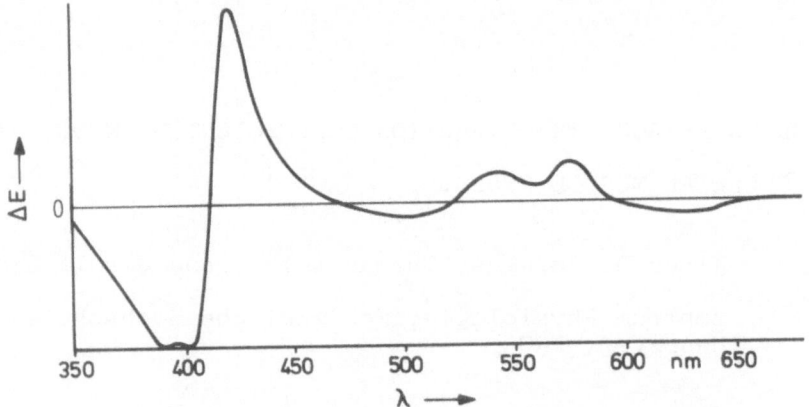

Fig. 1. Difference absorbance spectrum of human
oxyhemoglobin and methemoglobin

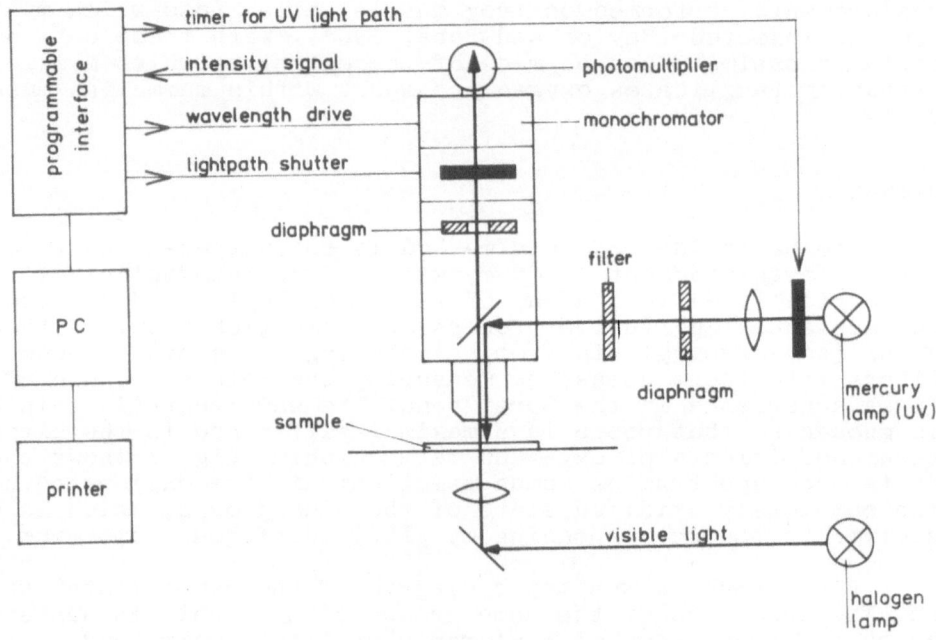

Fig. 2. Diagram of the experimental set-up

the photomultiplier signal are controlled by a personal computer via a programmable interface. It is possible either to record an absorbance spectrum within a given wavelength range or, with a multiplex technique, absorbance kinetics at up to four wavelengths. The time resolution is 1 ms, the wavelength resolution 0.5 nm.

Two different shapes of the UV irradiated field have been used, circular and rectangular. Circular fields with a radius (a) between 16 and 160 µm were chosen for measurements on samples consisting of gelatin slices impregnated with hemoglobin or myoglobin (gelatin was soaked in 1 mM hemoglobin or myoglobin solution for some hours, rinsed in water, and placed between two quartz glass-plates). In this case the absorbance change was measured in a circular area with radius b = 2.5 µm in the center of the UV field. The diffusion process for this geometrical arrangement following an initial absorbance change δE_0 is described by the solution of the differential equation in cylindric coordinates. As long as b<<a, the process can be approximated by the solution for the center of the field (r=0) (Crank, 1956), which is

$$E(t) = E_0 - \delta E_0(1 - e^{-(a^2/4Dt)}), \tag{1}$$

where

$$\delta E_0 = (\epsilon_{met} - \epsilon_{oxy}) \cdot \delta C_{met} \cdot d \tag{2}$$

(D - diffusion coefficient, $\epsilon_{met}, \epsilon_{oxy}$ - extinction coefficients of met- and oxy-state, δC_{met} - initial change in methemoglobin or metmyoglobin concentration, d - sample thickness).

For measuring the myoglobin diffusion coefficient in muscle cells, diaphragms, which constitute a thin layer of red muscle, are dissected out of rats after perfusing the animals with Ringer solution to remove red blood cells from the tissues. The muscle sample is mounted onto a stainless steel holder and placed in a chamber, which is continuously perfused at controlled temperature with Ringer solution equilibrated with a mixture of O_2, N_2, and CO_2. The chamber is placed in the microscope spectrophotometer. For measuring the absorbance of the sample, the chamber is transluminated through two windows by visible light. The window on the top side is made of quartz glass and the one on the bottom made of normal glass (Fig.3).

In the case of measuring diffusion in muscle tissue, the sample is irradiated with UV light in a rectangular field (height 2h), oriented with its longer side perpendicular to the muscle cell axis and extending over 3 muscle cells. The subsequent absorbance change is recorded within a small rectangular field (height 2w) in the middle of the UV field. With this field configuration (Fig.4), generated by rectangular diaphragms variable in width, the myoglobin diffusion along the longitudinal cell axis is observed.

Since w<<h, the solution of the one-dimensional diffusion equation for an initial metmyoglobin concentration in the range -h<x<h can be approximated by the equation for x=0 (Crank,1956)

$$E(t) = E_0 - \delta E_0 \, erf(h/2\sqrt{Dt}), \tag{3}$$

where δE_0 is defined by eq.2.

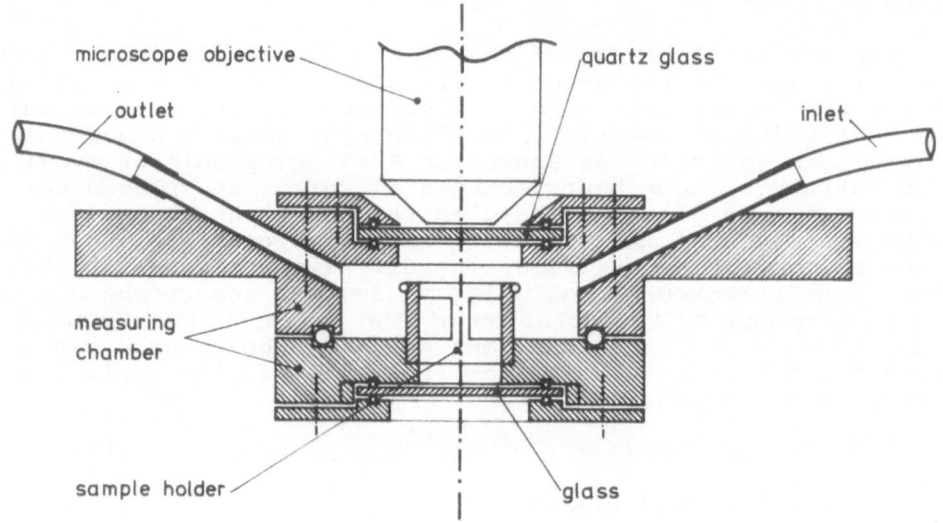

Fig. 3. Scheme of the measuring chamber

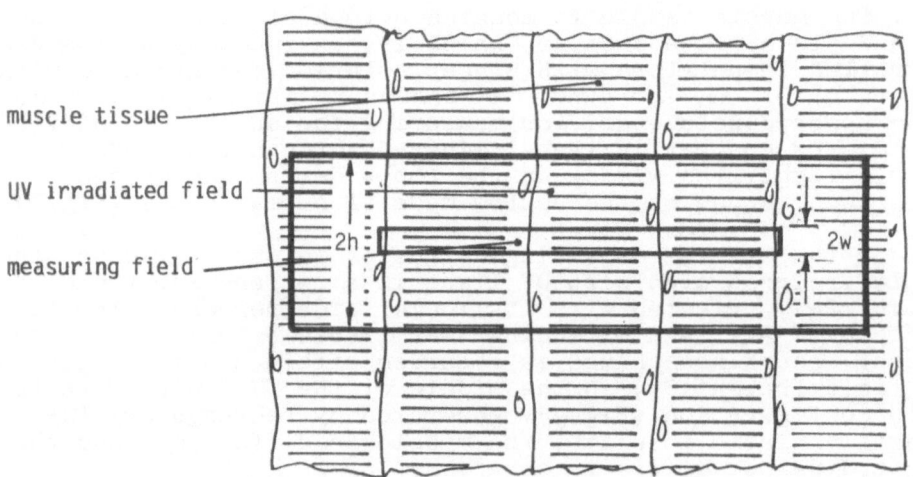

Fig. 4. Microscope diaphragm arrangement for diffusion measurements in muscle tissue

RESULTS AND DISCUSSION

The experimental set-up was tested with a model system, gelatine slices soaked in hemoglobin or myoglobin solution. Fig.5 shows the absorbance spectrum of hemoglobin in gelatin before and immediately after UV irradiation. It can be seen that the largest absorbance changes occur in the Soret band.

The kinetic measurements were made at 420 nm. Fig.6 shows the kinetics of the diffusion process of hemoglobin in gelatin at two different diaphragm diameters, 25 and 40 µm. As expected from eq.1, the recovery is slower when the UV field is larger. Evaluation of 10 recordings of the absorbance kinetics at three different radii of the UV irradiated field yielded the result $D_{hem}=(0.44\pm0.06)10^{-7}$ cm^2/s (T=25°C). Corresponding calculations for myoglobin in gelatin led to the result (T=25°C) $D_{myo}=(0.60\pm0.05)10^{-7}$ cm^2/s. The latter value was confirmed by diffusion measurements using radioactively labelled myoglobin in gelatin, from which a value of $D_{myo}=(0.58\pm0.07)10^{-7}$ cm^2/s was obtained.

Before diffusion kinetics of myoglobin in rat diaphragm muscles was recorded, the absorbance spectrum before and immediately after exposure to the UV light was measured in order to check if the myoglobin was completely oxygenated and to what extent metmyoglobin was formed by a given irradiation time. Fig.7 shows the absorbance spectrum of a diaphragm muscle equilibrated with carbogen (a) and nitrogen (c), respectively. Only in the oxygenated state UV light causes an absorbance change due to metmyoglobin formation (b).

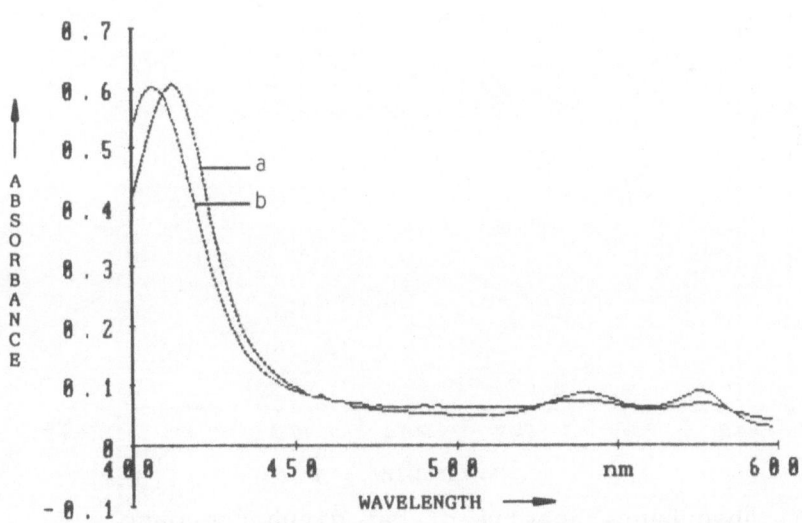

Fig. 5. Absorbance spectrum of human hemoglobin in gelatin before (a) and after (b) UV irradiation

141

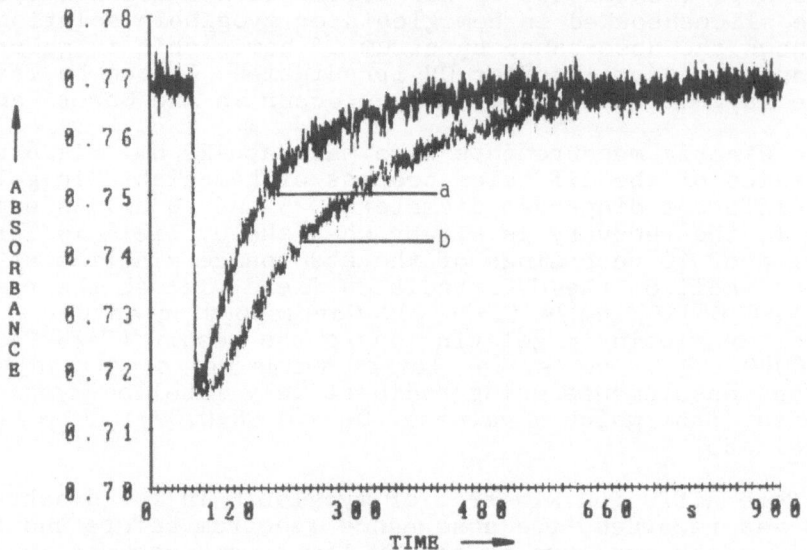

Fig. 6. Absorbance as function of time of UV irradiated hemoglobin in gelatin, a: UV field radius 25 µm, b: UV field radius 40 µm.

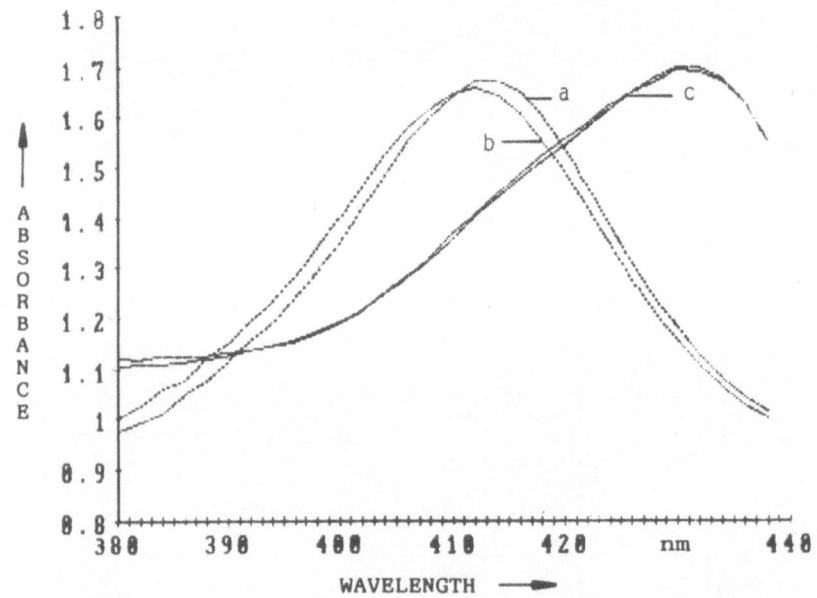

Fig. 7. Absorbance spectra of rat diaphragm muscle, a: oxygenated sample, b: oxygenated sample after UV irradiation, c: deoxygenated sample before and after UV irradiation

142

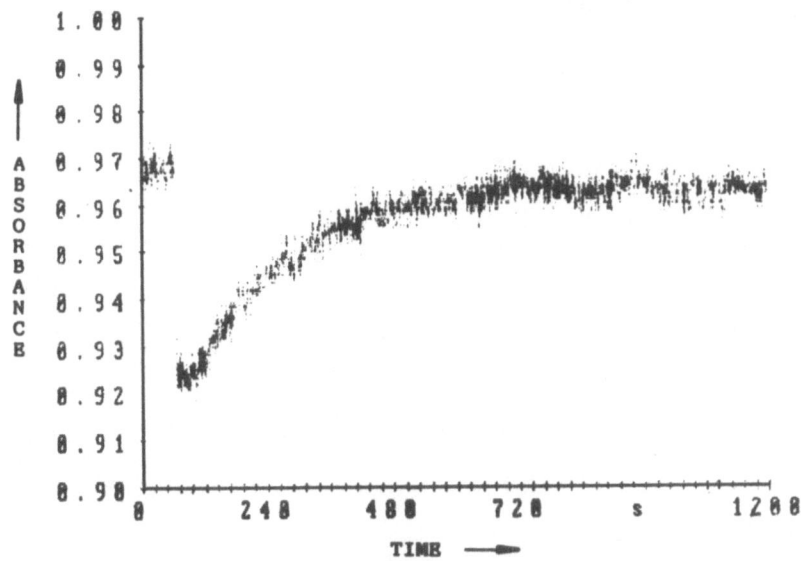

Fig. 8. Absorbance as function of time for UV irradiated
rat diaphragm muscle, UV field height 142 μm.

Fig.8 gives an example for the time course of the absorbance change in the diaphragm muscle after an initial metmyoglobin formation in a field with h = 71 μm. A preliminary result for the diffusion coefficient of myoglobin in rat diaphragm obtained from 8 kinetics (T=25°C) is $D_{myo}=(4.9\pm3.0)10^{-7}$ cm^2/s. This value is in the range assumed for myoglobin from measurements in solutions containing 18% protein (Riveros-Moreno and Wittenberg, 1972). It is about three times larger than the values calculated for muscle homogenate (Moll, 1968) and frog muscle (Baylor and Pape, 1988), but it may become necessary to correct this value to some extent, since in our calculation the spontaneous enzymatic reduction of metmyoglobin and the influence of cytochromes on the measured signal are not taken into account.

REFERENCES

Baylor, S.M. and Pape, P.C., 1988, Measurement of myoglobin
    diffusivity in the myoplasm of frog skeletal muscle
    fibers, J. Physiol., 406:247.
Crank, J., 1956, "The Mathematics of diffusion", Oxford
    University Press, London.
Demma, L.S. ,Salhany, J.M., 1977, Direct generation of
    superoxide anions by flash photolysis of human
    oxyhemoglobin, J. Biol. Chem., 252:1226.
Moll, W., 1968, The diffusion coefficient of myoglobin in
    muscle homogenate, Pflüg. Arch., 299:247.
Riveros-Moreno, V. and Wittenberg, J.B., 1972, The self-
    diffusion coefficients of myoglobin and hemoglobin in
    concentrated solutions, J. Biol. Chem., 247:895.

A NEW MODEL FOR LONG-TERM INVESTIGATIONS OF CEREBRAL OXYGEN SUPPLY IN RATS

A. Hagendorff, K. Zimmer, and J. Grote

Department of Physiology I, University of Bonn
D-5300 Bonn, FRG

INTRODUCTION

Micro-electrode as well as micro-application studies of the brain cortex require in most cases the exposure of the cerebral surface. Since trepanation of the skull and opening of the dura cause a drastic reduction of intracranial pressure, especially in small animals, brain swelling and local ischemia are often to be observed (Maekawa et al., 1979; Morii et al., 1986). In order to prevent brain herniation with subsequent insufficient tissue oxygen supply in the course of the experiments, the conventional open cranial window technique was modified by covering the exposed cortical area with an oil layer, by means of which the level of hydrostatic pressure was maintained close to that of the intracranial pressure (Morii et al., 1986). The efficiency of the modified model was evaluated by long-term investigations of the tissue oxygen supply in the brain cortex measuring the tissue oxygen tension distribution over a period of up to 6 hours.

METHODS

The experiments were performed on 28 male Sprague-Dawley rats (ZFV Hannover, FRG) with an average body weight of 330 g. The spontaneously breathing animals were anesthetized with thiobarbital (Inactin-Byk, 100 mg/kg body weight, i.p.). After tracheotomy, one femoral vein and one femoral artery were cannulated for volume and drug substitution as well as continuous blood pressure recording and intermittent blood sampling. In addition to arterial blood gas tension and pH measurements, hematocrit and hemoglobin concentration were determined. For brain surface preparation the rats were fixed in a stereotactic frame. Following midline incision of the skin and removal of parts of the temporal muscle, a craniotomy was performed over the right parietal cortex.

In a first series of 12 experiments the tissue oxygen tension distribution of the brain cortex was determined 20, 40, 80, 100, 140 and 160 minutes after opening of the dura. At each measuring period about 160 to 200 tissue oxygen tension values were taken. Mean histograms summarize the data of all experiments determined under comparable conditions. In a second series of experiments tissue oxygen tension measurements were performed after covering the exposed brain surface with oil. As shown in Figure 1 a plastic tube was fixed above the brain area under investigation and

Oxygen Transport to Tissue XII, Edited by J. Piiper *et al.*
Plenum Press, New York, 1990

Figure 1. Scheme of the modified open cranial window model

filled with high-viscosity paraffin oil (Merck, Darmstadt; spec. gravity 0.88 kg·l^{-1}) up to a hydrostatic pressure which equaled the intra-cranial pressure of 3 to 5 mmHg (Morii et al., 1986). The oil was heated to 37 °C. When applying the modified open cranial window technique, brain herniation was observed in none of the 16 experiments. Tissue oxygen tension was measured in the superficial cell layers of the exposed cortical region using multiwire surface microelectrodes (Kessler and Grunewald, 1969; Grote et al., 1981). Calibration was performed at brain surface temperature with saline solutions of different oxygen tensionsand a carbon dioxide tension of about 40 mmHg. The electrodes were mounted on a counterbalanced system, which prevented blood flow restriction in the measuring field due to the weight of the electrodes. Calibration was repeated every hour during the investigations. Body temperature and brain surface temperature were continuously monitored and kept in normal range. To exclude the formation of local brain edema in the measuring field, tissue water content was determined at the end of each expermiment.

RESULTS and DISCUSSION

Applying the conventional open cranial window technique, the tissue oxygen tension measurements resulted in histograms summarized in Figure 2. In the first measuring period, 10 to 25 minutes after dura opening, the mean oxygen tension distribution of the investigated brain region (n = 2295) ranged from very few low oxygen tension values between 0 and 5 mmHg to maximal values between 60 and 65 mmHg. The median of the cortical oxygen tensions was found to be 31.3 mmHg. At the same time mean values of 86 and 41 mmHg were determined for arterial oxygen and carbon dioxide tension, respectively. Comparable data were obtained in spontaneously breathing rats by other authors (Schröck and Kuschinsky, 1988). The mean arterial blood pressure was 130 mmHg. The derived tissue oxygen tension histogram of the brain cortex of rats as given in Figure 2 is comparable to those found by several investigators in rats as well as in other mammals under normal conditions (Skolasinska et al., 1984; Feng et al., 1986; Hagendorff et al., 1986). As time progressed, however, the cerebral tissue oxygen tensions decreased, indicating hypoxia with anoxia in single cortical cells, despite respiratory gas tensions of arterial blood and mean arterial blood pressure remaining in normal range. The corresponding histograms (s. Fig. 2.) show a pronounced shift to lower values. The medians of tissue oxygen tension were found to be below 20 mmHg (s. Tab. 1). The

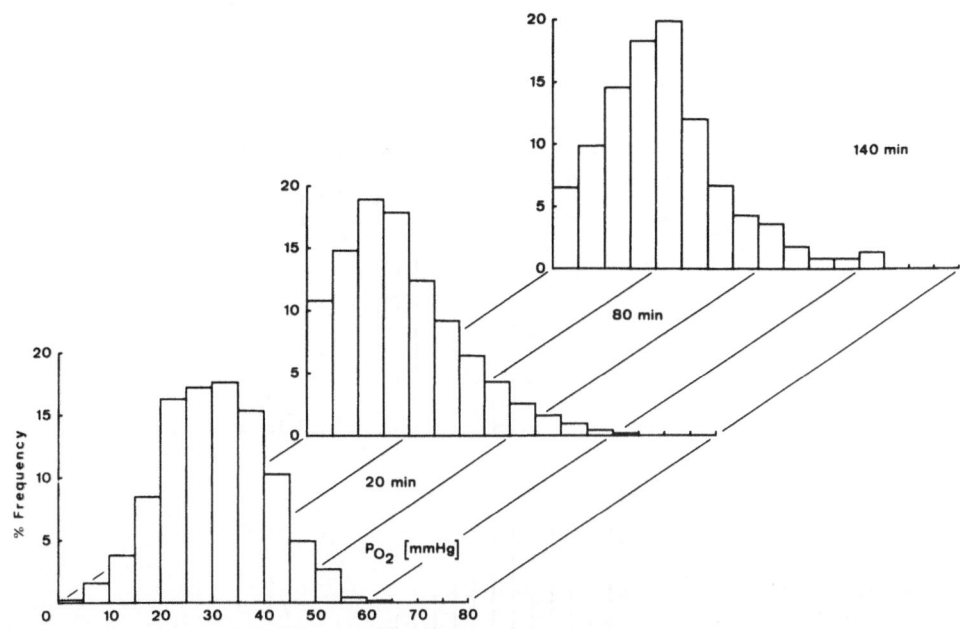

Figure 2: Mean tissue oxygen tension distributions of the brain cortex of rats determined during arterial normoxia and normocapnia as well as normal arterial blood pressure in the course of 12 experiments when applying the conventional open cranial window technique

pronounced decrease in tissue oxygen tension may be caused by a restriction in regional cortical oxygen supply as can be concluded from changes in the diameter of the supply vessels. Preliminary measurements of the outer diameter of pial vessels showed a decrease in the diameter of the arteries and a simultaneous increase in the diameter of the veins. In addition, at the end of the experiments, tissue water content was raised as compared to controls (3.85 vs. 3.76 ml·g dry weight^{-1}.

Table 1. Medians of cortical tissue oxygen tension (means $\pm$ SD) determined in rats during arterial normoxia and normocapnia when applying the conventional open cranial window technique (A) or the modified open cranial window technique (B)

Time after dura opening (min)	Median of tissue oxygen tension (mmHg)	
	A	B
20	31.3 ± 6.5	31.8 ± 5.8
40	18.6 ± 4.9	25.3 ± 5.8
80	16.0 ± 3.8	23.1 ± 5.1
100	15.6 ± 3.3	25.2 ± 4.4
140	18.4 ± 4.9	25.9 ± 3.0
160	17.3 ± 2.3	27.6 ± 5.2
200	–	28.1 ± 3.1
220	–	29.9 ± 4.0
260	–	33.9 ± 6.9
280	–	35.5 ± 6.4

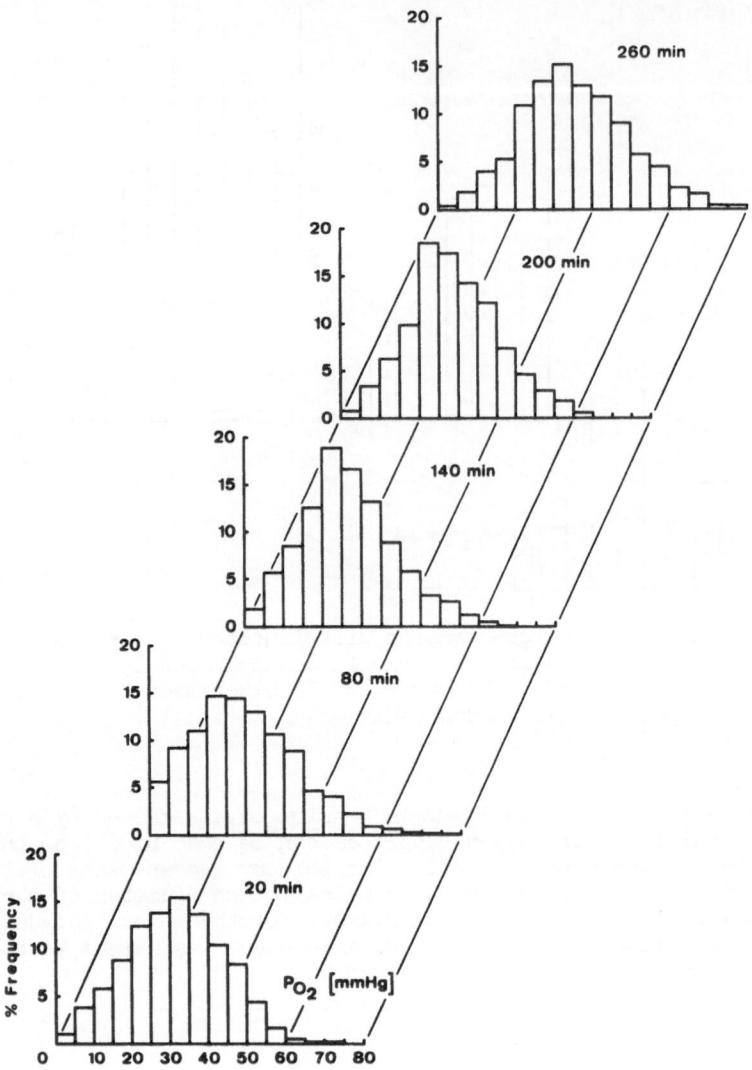

Figure 3: Mean tissue oxygen tension distributions of the brain cortex of rats determined during arterial normoxia and normocapnia as well as normal arterial blood pressure in the course of 16 experiments when applying the modified open cranial window technique

The second group of experiments performed using the modified open window model resulted in mean tissue oxygen tension histograms as shown in Figure 3. The respiratory gas tensions of arterial blood as well as the arterial blood pressure remained constant within normal range during all 16 experiments. The initial series of oxygen tension measurements were completed about 25 minutes after opening of the dura, nine further series followed at intervals of 20 or 40 minutes (s. Tab. 1). The mean tissue oxygen tension distribution observed at the beginning of the investigations was normal with a median of 31.8 mmHg. The derived histogram which summarized 2833 oxygen tension values is comparable to those found in the brain cortex of other species at arterial normoxia and normocapnia (Grote et al., 1981; Hagendorff et al., 1986; Gaab et al., 1987). The histograms determined during the following measuring periods show, in contrast to the first group of experiments, a stable tissue oxygenation without signs of an insufficient cortical oxygen supply. Even among the relatively large number of tissue oxygen tensions found during the third measuring period 80 minutes after opening of the dura in the lowest tissue oxygen tension class, less than 0.5% of the values were found to be below 2 mmHg. The corresponding medians of the tissue oxygen tensions showed a slight transient decrease with mean values of about 25 mmHg. In the second half of the experiments the medians of tissue oxygen tension increased reaching values above the initial level as summarized in Table 1. Since the hematocrit as well as the blood hemoglobin concentration decreased simultaneously from 47.3 % and 14.3 $g \cdot dl^{-1}$ to 38.7 % and 11.7 $g \cdot dl^{-1}$, respectively, the observed improvement of tissue oxygenation may in part be due to hemodilution caused by blood substitution following blood sampling (Hudak et al., 1986; Desjardins and Duling, 1987). The diameters of the pial arteries did not change significantly during the experiments whereas the diameters of the veins slightly increased. The measurements of tissue water content resulted in normal values, indicating that no local edema was present in any of the experiments.

When summarizing the results of tissue oxygen tension measurements in the brain cortex of rats, it can be concluded that the modified open cranial window model is suitable for short as well as long-term investigations of tissue respiration and metabolism or tissue functions in the exposed brain cortex without pathophysiological implications caused by brain swelling and brain ischemia.

REFERENCES

Desjardins, C. and Duling, B.R., 1987, Microvessel hematocrit: measurement and implications for capillary oxygen transport, Am. J. Physiol., 252: H494.

Feng, Z., Roberts, E.L., Thomas, J.S., and Rosenthal, M., 1988, Depth profile of local oxygen tension and blood flow in rat cerebral cortex, white matter and hippocampus, Brain Research, 445: 280.

Gaab, M.R., Poch, B., and Heller, V., 1987, Oxygen metabolism and microcirculation of the brain in various conditions: experimental investigations, in: "Adv. in Neurosurg. 15: 35", R. Wüllenweber, M. Klinger, and M. Brock, eds., Springer, Berlin, Heidelberg, New York.

Grote, J., Zimmer, K., and Schubert,R., 1981, Effects of severe arterial hypocapnia on regional blood flow regulation, tissue PO_2 and metabolism in the brain cortex of cats, Pflügers Arch., 391: 195.

Hagendorff, A., Haller, C., and Grote, J., 1986, Der Einfluß einer postischämischen Hypokapnie auf die cerebrale O2-Versorgung, in: "Physiologie und Pathophysiologie des Gefäßsystems, Funktionsanalyse biologischer Systeme 17: 165", J. Grote, and E. Witzleb, eds., Fischer, Stuttgart, New York.

Hudak, M.L., Koehler, R.C., Rosenberg, A.A., Traystman, R.J., and Jones, M.D., 1986, Effect of hematocrit on cerebral blood flow, Am. J. Physiol., 251: H63.

Kessler, M., and Grunewald, W., 1969, Possibilities of measuring oxygen pressure fields in tissue by multiwire platinum electrodes, Prog. Resp. Res., 3: 147.

Maekawa, T., McDowall, D.G., and Okuda, Y., 1979, Brain-surface oxygen tension and cerebral cortical blood flow during hemorrhagic and drug- induced hypotension in the cat, Anesthesiol., 51: 313.

Morii, S., Ngai, A.C., and Winn, H.R., 1986, Reactivity of rat pial arterioles and venules to adenosine and carbon dioxide: with detailed de scription of the closed cranial window technique in rats, J.Cereb.Blood Flow Metabol., 6:34.

Schröck, H., and Kuschinsky, W., 1988, Cerebral blood flow, glucose use, and CSF ionic regulation in potassium-depleted rats,, Am. J. Physiol., 254: H250.

Skolasinska, K., Günther, H., Höper, J., and Funk, R., 1984, Oxygen supply to the brain cortex in SHR and normotensive rats, in:" Adv. Exp. Med. Biol. 169: 271", Plenum, Ney York.

AN EXPERIMENTAL SET-UP FOR THE BLOOD PERFUSED WORKING ISOLATED RAT HEART

J. Olders, T. Boumans, J. Evers, and Z. Turek

Dept. Physiology, University of Nijmegen, Nijmegen, The Netherlands

INTRODUCTION

The isolated working heart preparation, first described by Neely et al. (1967), has now been used for over two decades to assess myocardial performance and left ventricular function in response to various disease states, drug treatments etc. However, usually a well oxygenated electrolyte solution is used as perfusion medium, which makes the oxygenation of these hearts very unphysiological. The present study reports on an experimental set-up for the isolated working heart model that can be used to study myocardial oxygenation. The perfusion system is a modification of that described by Duvelleroy et al. (1976) for perfusion of the heart with an erythrocyte suspension in buffer. A microcomputer data acquisition and analysis system was developed to avoid time consuming data evaluation by hand. With this set-up, important hemodynamic variables like myocardial oxygen consumption, cardiac work and maximal positive and negative first derivative of the left ventricular pressure can be calculated on line. Our aim is to use this set-up for the evaluation of the mathematical modeling work of our group on tissue oxygenation (e.g. Turek et al., in press; Hoofd et al, this volume) by varying the various input parameters concerning the oxygen supply of the hearts.

PERFUSION SYSTEM

Washed fresh pig erythrocytes suspended in a Krebs-Henseleit bicarbonate buffer are used as perfusion medium ("blood"). The Krebs-Henseleit buffer consists of (mmol/l): NaCl (120), KCl (4.7), $CaCl_2$ (2.5), MgSO4 (1.2), KH_2PO4 (1.2), $NaHCO_3$ (30.0) and glucose (5.0), modified with calcium EDTA (0.5) and pyruvate (2.0). Bovine serum albumine (fraction V) is added to a final concentration of 15 g/l. Fresh pig blood is collected in reservoirs containing heparin (50 mg/l), filtered (pore size 45 µ) and centrifuged 3 min at 3,500 rpm. The plasma is discarded and the erythrocytes are washed three times with equal volumes of saline and finally mixed with the electrolyte solution to the desired hematocrit.

In Fig. 1 a schematic representation of the perfusion system is shown. It is based on that described by Duvelleroy et al. (1976). The perfusion apparatus is installed in in a cage in which the environmental temperature can be kept at 37 °C. Parts of the apparatus are additionally thermostated and variations of the temperature of the perfusate entering the heart are found to be less then 0.2 °C. The left atrium of the heart is perfused via an arterial 25 μ filter (FA) from three identical but separate circuits containing different perfusates (only one circuit is shown, indicated by I). Each circuit consists of a reservoir and an oxygenator (OX) through

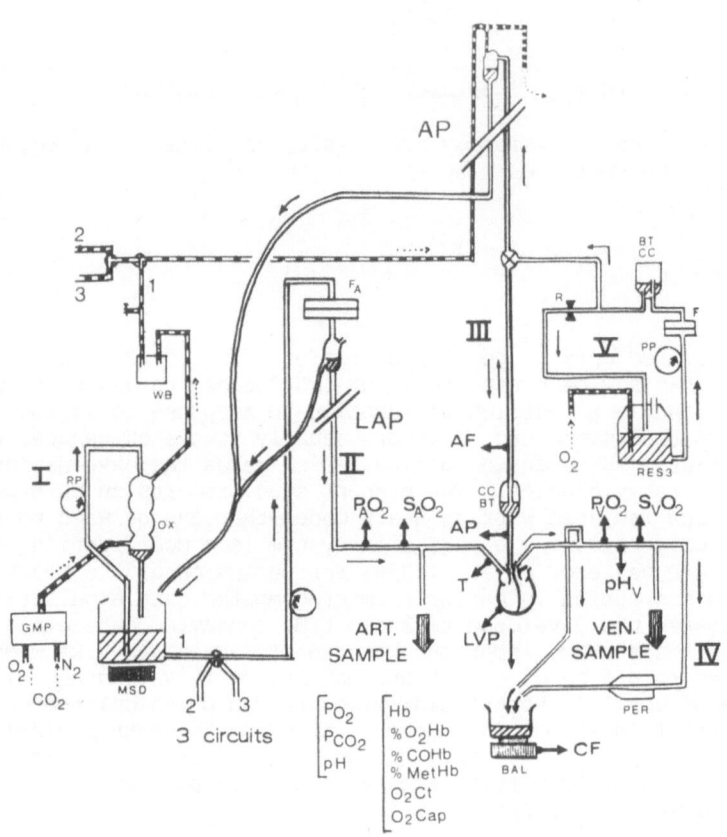

Fig. 1. Schematic representation of the perfusion apparatus. See text for detailed explanation.

which the perfusate is continuously circulated by a roller pump (RP). The oxygenators are exposed to various gas mixtures made by a gas mixing pump (GMP) by mixing 95 % O_2/ 5% CO_2 with 95 % N_2/ 5% CO_2. This in order to control P_{O_2}, P_{CO_2} and S_{O_2} of the perfusates. A magnetic stirring device (MSD) ensures homogeneity of hematocrit. Rapid changes in the perfusate introduced into the left atrium (atrial line indicated by II) can be made by conversion from one circuit to another. The left atrial filling

152

pressure (LAP) or preload (PL) can be changed by varying the height of an overflow before the heart. The perfusate is pumped back to the reservoir by the left ventricle against a hydrostatic pressure column with a controlable height, determining the aortic pressure (AP) or afterload (AL) (the aortic line is indicated by III). A compression chamber (CC) provides adequate compliance for the aortic cannula. Venous coronary blood is collected through a catheter in the pulmonary artery (the coronary line is indicated by IV). A perfusator (PER) leads a constant part of the coronary flow along the measurings probes. The total coronary flow, including a small part dripping off from the heart, is collected and measured by a balance (BAL).

After removal of the heart from the thorax of the anesthetized rat (ether), it is immersed in ice-chilled buffer and remnant lung tissue and fat are removed. The heart is then quickly attached to the aortic cannula of the perfusion apparatus and retrograde perfusion according to Langendorff is started, with a buffer solution without red cells . The circuit for the retrograde perfusion (indicated by V) consist of a reservoir bubbled with 95 % O_2/ 5% CO_2 from which the buffer is pumped through a filter (1.2 μ) and a bottle, inserted for bubble trapping and reduction of peristaltic pressure pulses. The buffer partly enters the aortic line and the perfusion pressure can be regulated by varying the resistance (R). During the retrograde perfusion the left atrium and pulmonary artery are cannulated and electrodes for pacing can be attached to the right atrium. After opening of the atrial line the heart starts to eject and the retrograde line is closed.

DATA ACQUISITION AND ANALYSIS

A microcomputer data-acquisition and analysis system for on line measurements of haemodynamics and oxygen consumption of the heart was developed. A schematic representation of this set-up is shown in Fig. 2. Aortic flow (AF) and mean aortic pressure (AP) are measured by means of an electromagnetic flow probe (Skalar) and a pressure transducer (Gould) in the aortic line. The P_{O_2} and S_{O_2} values of the erythrocyte suspension in the atrial and coronary lines are monitored by means of P_{O_2} electrodes and S_{O_2} probes (home built). These signals are analog sampled at minimal intervals of 5 s. The coronary flow is measured by means of a balance (mettler PS 1200) which is sampled digitally. The left ventricular pressure can be measured by means of a catheter inserted into the left ventricle via the apex of the heart and connected to a pressure transducer (Gould). The catheter is flexible and thin to avoid negative effects on the contraction patterns of the left ventricle. Before the experiment is started the oxygen dissociation curve (ODC) of the blood is established by measuring and calculation of the P50 and Hill's n. During an experiment, several blood samples are taken for measurements of P_{O_2}, P_{CO_2}, pH and Hb-derivatives (IL 1312 and 482), both in the atrial and the coronary line. The P_{O_2} and S_{O_2} values of the samples are used for the calibration of the P_{O2} electrodes and S_{O_2} probes.

The automatization set-up consists of an interface (home built) with 8 signal-conditioning units for amplification and filtering of the signals and a Dash-8 interface chart for communication with the Dash-8 AD chart in the microcomputer. An analog multiplexer allows sampling of more than 8 signals. A digital multiplexer chart and a Pio-12 interface chart were implemented for sampling of the digital signal of the balance and for communication with the Pio-12 chart in the microcomputer. The main sampling and analysing program was written in Turbo Pascal version 3.01 A, using

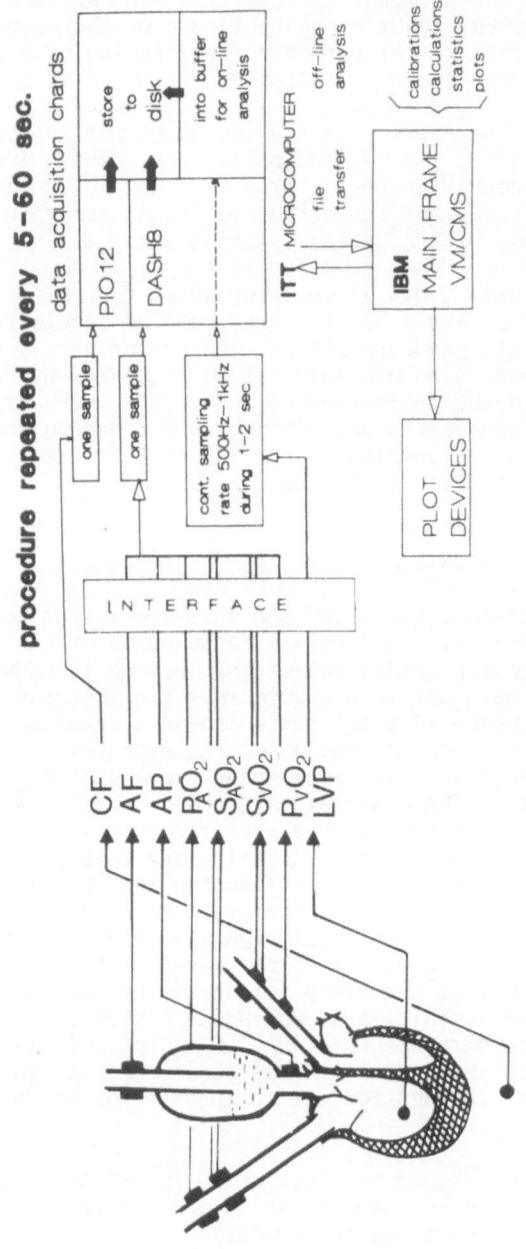

Fig. 2. Schematic representation of the automated set-up for data acquisition and analysis.

tools subroutines (Metrabyte) and was running on an XT-microcomputer (ITT). An extensive description of the hard and soft ware is given in an internal report (Boumans, 1988)

At each sample point the following variables are calculated: cardiac output (CO) is calculated as the sum of aortic and coronary flow. External cardiac work (ECW) is computed as the product of the mean aortic pressure, cardiac output and a coefficient (1.333222 x 10-4) which expresses work in J/min. Myocardial oxygen consumption (MV_{O_2}), ml/min, was calculated as the product of coronary flow times the arteriovenous O_2 content difference. Efficiency (EFF) is computed as ECW/(MV_{O_2} x 20.9) %. Calculations can be done on line if the calibration parameters of the various probes are known. S_{O_2} values can be calculated from P_{O_2} values (or vise versa) by means of the pH, P_{CO_2} and ODC, in case data of a P_{O_2} electrode or S_{O_2} probe are lacking. The interval time for sampling can be choosen between 5 and 60 s. When the left ventrical pressure is also measured, the minimal interval for sampling has to be extended to 20 s to allow on-line analysis of this signal.

ANALYSIS LEFT VENTRICULAR PRESSURE

At each sample point, the left ventricular pressure (LVP) can be sampled by a single key depression, with a sample period of 1 or 2 s. and a sample frequency of 1 kHz or 500 Hz (1000 data points). The hearts are either paced with a frequency of 5 Hz or beat freely, usually with a heart rate of about 240 beats/min. This means that 3-10 complete pressure pulses are collected and stored for analysis. A newly written computer program is used for on-line analysis of the left ventricular pressure; the flow diagram of this program is shown in Fig. 3. First the maximum first derivative (dLVP/dt max) of the signal is determined by calculating areas of maximum change over two points separated by several samples. From this the threshold times of each pressure pulse (at 0.7 x (dLVP/dt max)) are identified and the mean cycle time and heart rate calculated. The number of complete cycles in the signal is determined, with their times of start and end. These cycles are sequentially loaded into a buffer for further analysis. In Fig. 4 this analysis is illustrated. For each cycle the maximum and minimum pulse pressures are found and from these 10 % and 90 % pressure values are calculated. The start and finish times of a pulse are defined as the time at which the pressure was 1 % of the maximum. These times are found by fitting an exponential equation to the beginning and ending of the curve (or, more easily, by determining the times at which the value of 0.2 x (dLVP/dt max) are exceeded). Similarly the time of peak-pressure is located by fitting a cubic equation to the top 10 % of the pressure curve (indicated in the figure by the dashed line). The maximum positive and negative slopes of the pressure curve are calculated (pos. and neg. dLVP/dt). The area under the contraction curve is calculated by use of Simpson's Rule, that is, by fitting quadratic equations to each set of three consecutive data points. This is done as well for the area from baseline (diastolic LVP) to peak LVP as for the area from peak LVP back to the baseline. The above calculations are done on each complete cycle in the signal and the averaged results with their standard deviations are stored to disk.

RESULTS AND DISCUSSION

An example of an experiment is shown in Fig. 5. The horizontal axes represent the time scale. When the pre- and afterload pressures are changed this is indicated in the bottom bar and by the thin vertical dashed lines.

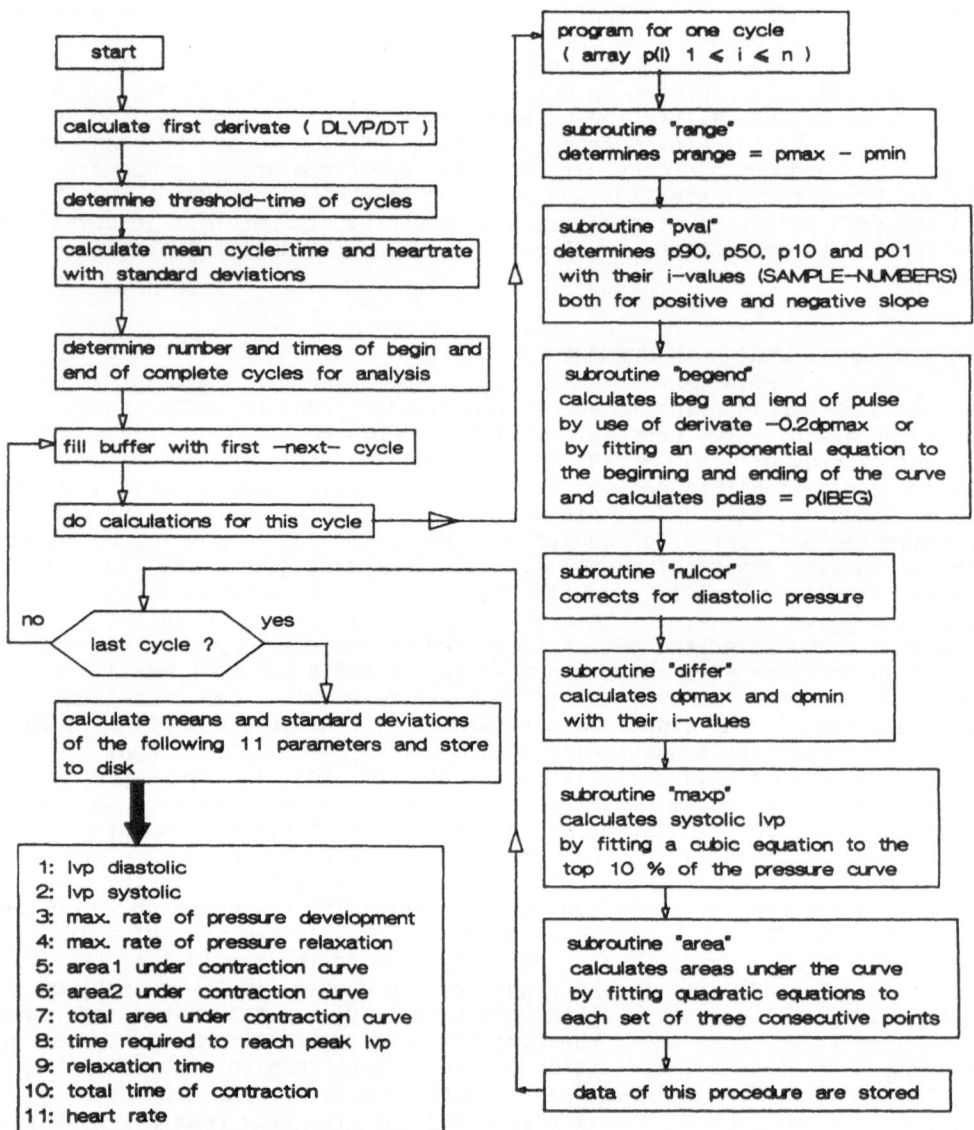

Fig. 3. Flow diagram of the computer program for the analysis of the left ventricular pressure curve.

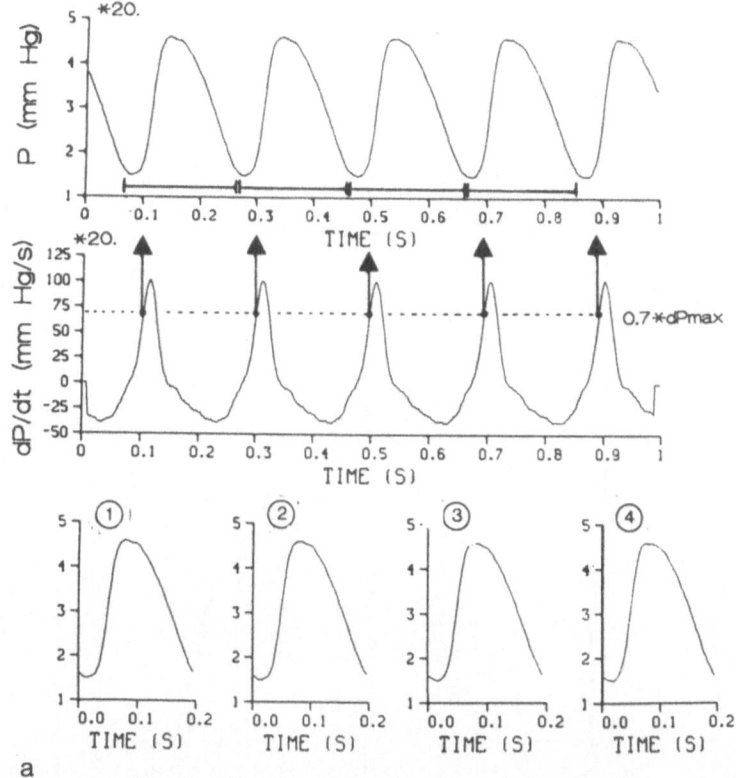

a

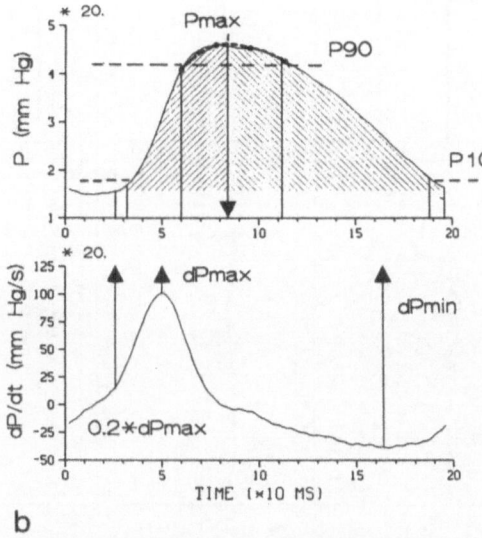

b

Fig. 4. Analysis of the left ventricular pressure curve.
   a. Left ventricular pressure curve (upper), and its first
      derivative (middle), used to select complete pressure pulses
      out of the signal (lower).
   b. Analysis of the first pressure pulse.

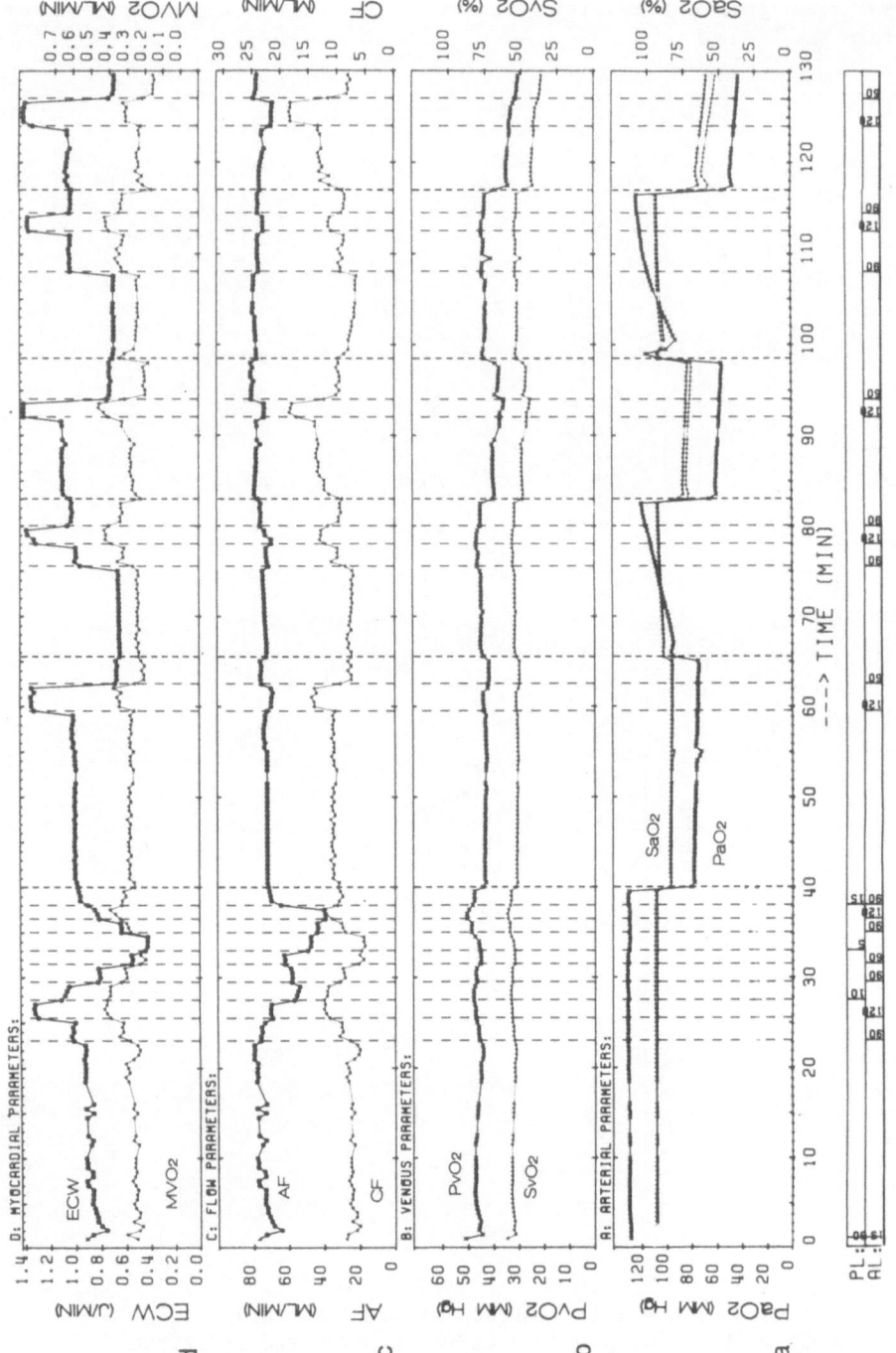

Fig. 5. Example of an experiment with the blood perfused isolated rat heart. See text for explanation.

In panel a the values of the P_{O_2} and S_{O_2} in the arterial line are shown. Both the measured and the calculated S_{O_2} (from P_{O_2}, pH and ODC) are plotted. Panel b shows the PO_2 and measured SO_2 values of the coronary line. In panel c the values of the aortic and coronary flow are represented, whereas the upper panel (d) shows the calculated values of the external heartwork and the myocardial oxygen consumption. From the 20th till the 40th min of the experiment the left ventricular function at a preload of 10, 15 and 20 mm Hg and (at each preload value) an afterload of appr. 65, 90 and 120 mm Hg was established. After this period hypoxia was induced at a preload of 15 mm Hg and an afterload of 90 mm Hg by decreasing the arterial SO_2 in steps by changing from one perfusate to another (indicated by the thick vertical dashed lines). At the end of each saturation step three afterload steps were made. It can be seen that a drop in arterial S_{O_2} is accompanied by a drop in venous P_{O_2} and an increase in coronary flow, so that myocardial consumption remains almost stable. When the arterial S_{O_2} decreases below 60 %, coronary vasodilatation is maximal and the aortic flow starts to decrease slightly.

The preparation described is stable for at least 2 hours and the measured and calculated hemodynamic variables agree well with those reported by Duvelleroy et al. (1976). The results of a first series of experiments are presented elsewhere (Olders et al., this volume). The main differences between our perfusion apparatus and that described by Duvelleroy et al. are the addition of the retrograde circuit, the extension to three supply circuits (needed when more than two saturation steps are made), the way of oxygenation (without the expensive membran oxygenators) and the variable afterload. By varying the oxygen supply parameters of the perfusate (i.e. saturation, hematocrit or oxygen affinity and combinations of those), this preparation offers an opportunity to study changes in myocardial oxygenation per se as well as the effects on cardiac performance.

The data acquisition and analysis system not only saves time, but it offers also the advantage of a more detailed insight into the experimental process during the experiment itself. Snoeckx et al. (1986) described an automated system for hemodynamic measurements in isolated buffer-perfused working rat hearts. They did not measure MV_{O_2} and they used the electrogram (maximal ventricular depolarization) of the heart to detect left ventricular pressure pulses. Our analysing program can be used without measuring the electrogram and avoids difficulties in identifying ventricular depolarization pulses. The program for the analysis of one pressure pulse in the left ventricular signal is based on a description by Harris et al. (1983). A great advantage of our extension of the program is that the analysis of a complete signal sequence is carried out on line without the necessity to select the pressure pulses by hand.

ACKNOWLEDGEMENTS

Thanks are due to the Department of Electronics, Faculty of Medicine, University Nijmegen for support in developing the data acquisition device. This work was in part supported by NWO, the Netherlands Organization for Scientific Research.

REFERENCES

Boumans, T., 1989, Data-acquisitie systeem t.b.v. onderzoek naar de fysiologie van het rattehart, Internal Report, Dept. Electronics, University Nijmegen.

Duvelleroy, M.A., M. Duruble, J.L. Martin, B. Teisseire, J. Droulez, and M.Cain, 1976, Blood-perfused working isolated rat heart, J. Appl. Physiol., 41: 603-607.

Harris, D.P., Marriot, M.L., and McNeill, J.H., Microcomputer acquisition and analysis system for the isolated working heart preparation, 1983, J. Pharm. Meth., 10: 65-73.

Hoofd, L., Olders, J., and Turek, Z., this volume, Oxygen pressures calculated in a tissue volume with parallel capillaries.

Olders, J., Turek, Z., Evers, J., Hoofd, L., Oeseburg, B., and Kreuzer, F., this volume, Comparison of Tyrode and blood perfused working isolated rat hearts,

Snoeckx, L.H., Schrijen, J.J., van Bilsen, M., Lammers, W.J., van der Nagel, T., van der Vusse, G.J., and Reneman, R.S., 1986, A microcomputer system for hemodynamic measurements in isolated, working rat hearts, Comp. Biol. Med. 4: 301-309.

Turek, Z., Rakusan, K., Hoofd, L., Olders, J. and Kreuzer, F., in press, Various models of myocardial oxygenation: the effect on calculated PO_2 histograms.

COMPARATIVE DISTRIBUTIONS OF NUMERICAL AND AREAL INDICES OF TISSUE

CAPILLARITY

Stuart Egginton and Zdenek Turek[*]

Departments of Physiology, The University of Birmingham Medical School, Birmingham B15 2TJ UK and [*] University of Nijmegen, Nijmegen, The Netherlands

INTRODUCTION

Indices of tissue capillarity based on numerical ratios have many limitations when used other than for global descriptions. This is in part a result of the most common approaches to the problem of quantification being implicitly based on the Krogh cylinder model. Not only does this use unrealistic (imposed) boundaries but utility of the derived indices are essentially limited to gross averages, being unable to account for the variation in fibre size and/or spatial heterogeneity of fibre types which is found in most striated muscle.

We have explored the utility of indices derived from a planar analysis of capillarity, based on area rather than number distribution, that uses natural boundaries which accounts for local interactions between individual capillaries and fibres (Egginton, Turek and Hoofd, 1987; Egginton and Ross, 1989a,b). We present a comparison of these different approaches in quantifying capillarity from histological transverse sections of skeletal muscle with varying metabolic capacity, from representative mammalian (rat) and fish (eel) species.

MATERIAL AND METHODS

Sample preparation and analytical approach have been described previously (Egginton, Turek and Hoofd, 1987). Indices of the inter-relationship between capillaries and muscle fibres based on numerical distributions which were considered are 1) the number of capillaries around a fibre [CAF or, in morphometric notation, $N(c,f)$] and 2) the number of fibres surrounding a capillary, the 'sharing factor' of Plyley and Groom (1975) [SF, or $N(f,c)$]. The corresponding newly developed indices based on areal distributions are 1) the number of capillary domains intersecting a muscle fibre profile (or domain:fibre ratio, DFR) and 2) the number of fibres overlapping a domain (or fibre:domain ratio, FDR) (see Fig. 1).

In addition, we also analysed the distribution of domain areas, muscle fibre cross sectional areas, and the local capillary to fibre ratio (LCFR; the cumulative fraction of domains intersecting a muscle fibre profile). The goodness of fit with respect to normal and logarithmic normal distributions were estimated by residual mean squares

Oxygen Transport to Tissue XII, Edited by J. Piiper *et al.*
Plenum Press, New York, 1990

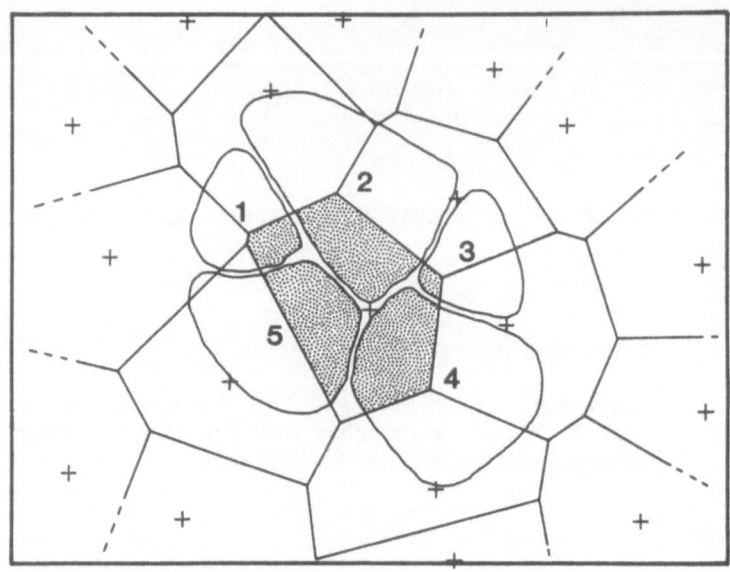

Fig 1   Geometrical relationship between domains and muscle fibre profiles.   One domain (stippled) intersects five different fibres, including one that lacks direct capillary contact (FDR=5).   Conversely, the number of domains intersecting one fibre gives the DFR, while the sum of all fractions of domains that intersect an individual fibre gives the LCFR (from Egginton, 1989).

Table 1   Mean squares of differences with respect to normal or logarithmic normal distribution

Muscle		Domain area	Fibre area	Local capillary to fibre ratio
Mammalian slow	normal lognorm	0.022 0.004	0.131 0.060	0.084 0.080
Mammalian fast	normal lognorm	0.113 0.010	0.332 0.090	0.100 0.082
Fish slow	normal lognorm	0.149 0.035	0.577 0.212	0.457 0.297
Fish fast	normal lognorm	1.792 1.880	1.128 0.093	1.119 0.084

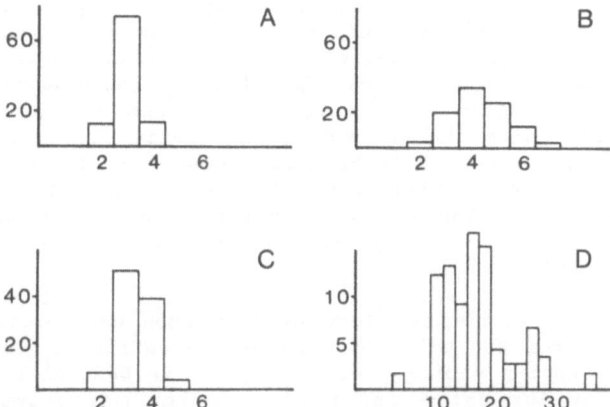

Fig. 2. Comparison of numerical vs. areal indices of capillary supply to individual muscle fibres. Number of capillaries surrounding (contiguous with) a fibre - N(c,f) - and number of capillary domains intersecting (overlapping) individual fibre profiles - DFR - for mammalian slow (A,B) and fast (C,D) muscles, respectively. N(c,f) = A,C; DFR = B,D.

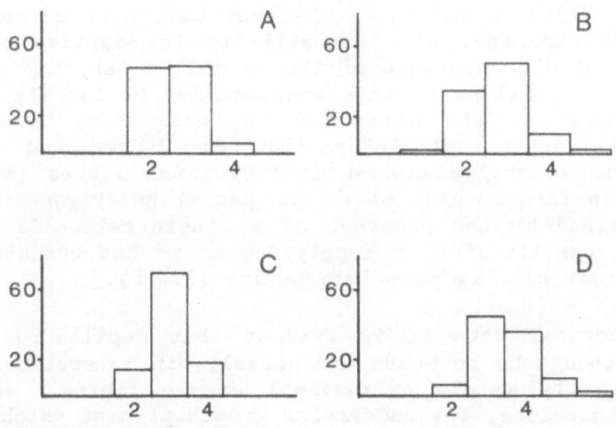

Fig. 3. As Fig. 2, for fish slow (A,B) and fast (C,D,) muscles, respectively.

for the four different types of skeletal muscle (see Table 1).

RESULTS AND DISCUSSION

When the local numerical relationship between capillaries and fibres, N(c,f) and N(f,c), are compared with the corresponding indices based on the area of tissue supplied by an individual capillary, DFR and FDR, one difference is most clear. In all cases the distribution of the area-based index is displaced to the right with a greater spread of values (Figs 2-5), being therefore more informative and potentially more sensitive to adaptive changes in capillary supply. This is particularly evident in fish muscle (Figs 3 and 5), where the sparse capillary network highlights the inadequate descriptive power of the numerical indices, but may also be of significance when quantifying capillarity of mammalian fast muscle.

The discrepancy is perhaps less evident when considering the local capillary supply from the point of view of concentric oxygen delivery to fibres from surrounding capillaries, ie N(c,f) vs DFR, although the area based index provides extra information which may prove helpful in modelling small scale (local) transport of oxygen. For example, while mammalian slow and fast muscle have a minimum of 2 and 1 contiguous capillaries, the DFR shows that such poorly oxygenated fibres are potentially influenced (=supplied?) by 3 and 2 capillaries, respectively (Fig 2). The poor descriptive power of numerical indices is emphasised when fish muscle is considered; even the highly oxidative slow muscle has fibres that lack direct capillary contact, a situation in common with a staggering 60% of fast fibres. Clearly this has profound implications for oxygen transport, which the DFR indicates is likely to involve primarily one, sub-adjacent capillary (Fig 3).

The complimentary anatomical indices, ie N(f,c) vs FDR, may be more appropriate if one considers the capacity for oxygen transport to be limited by eccentric diffusion from capillaries to surrounding fibres. This reveals a significant increase in spread of values for the area based index compared to the rather coarse distribution offered by the numerical index (Figs 4 and 5). The resultant poor correlation with muscle oxidative capacity, and insensitivity to adaptive changes, has led to the gradual disappearance of the so called 'sharing factor' from the literature. Objections to this approach may be largely overcome by the FDR, as this has the potential of being more informative by partitioning the anatomical interaction of fibres and capillaries according to the metabolic demand of individual fibres (Egginton and Ross, 1989b). In fish muscle, where the patial heterogeneity of oxygen demand is minimised by the presence of a single metabolic fibre type, this analytical partitioning of supply/demand is not essential and the area based index is clearly more informative (Fig 5).

From histological data it is evident that capillaries are found predominately, though by no means exclusively, in interstices formed by close-packing of the mainly cylindrical muscle fibres. As a first approximation, therefore, the underlying growth process which results in a given spread of fibre size may determine the subsequent pattern of capillary distribution, since the network of interstices will offer a pathway of least resistance for growth of new vessels during development, or during adaptation to increased oxidative demand such as sustained activity. In most cases the cross sectional area of both domain and fibre profiles showed a closer fit to logarithmic normal,

164

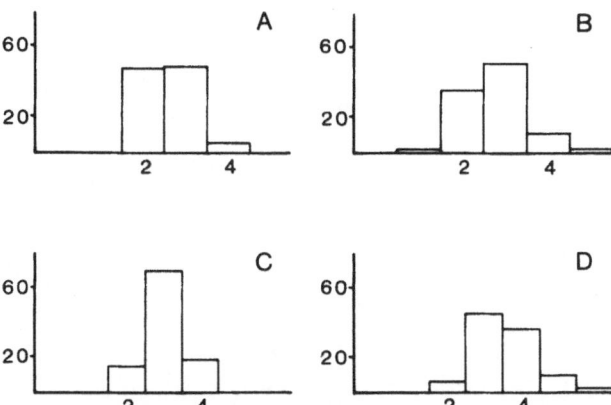

Fig. 4. Comparison of numerical vs. areal indices of fibre packing around capillaries. Number of fibres surrounding (contiguous with) a capillary - N(f,c) - and number of fibre profiles intersecting (overlapping) individual domains - FDR - for for mammalian slow (A,B) and fast (C,D) muscles, respectively. N(c,f) = A,C; FDR = B,D.

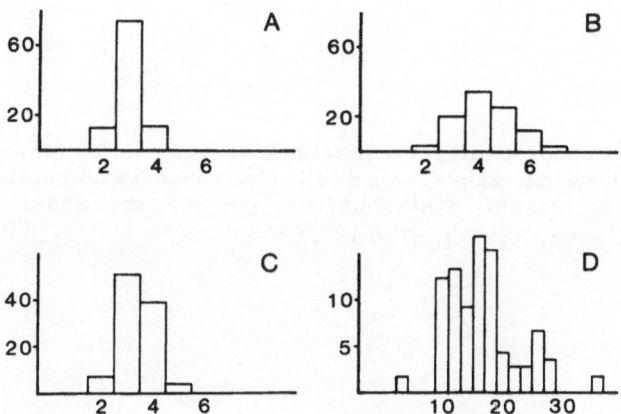

Fig. 5. As Fig. 4, for fish slow (A,B) and fast (C,D) muscles, respectively.

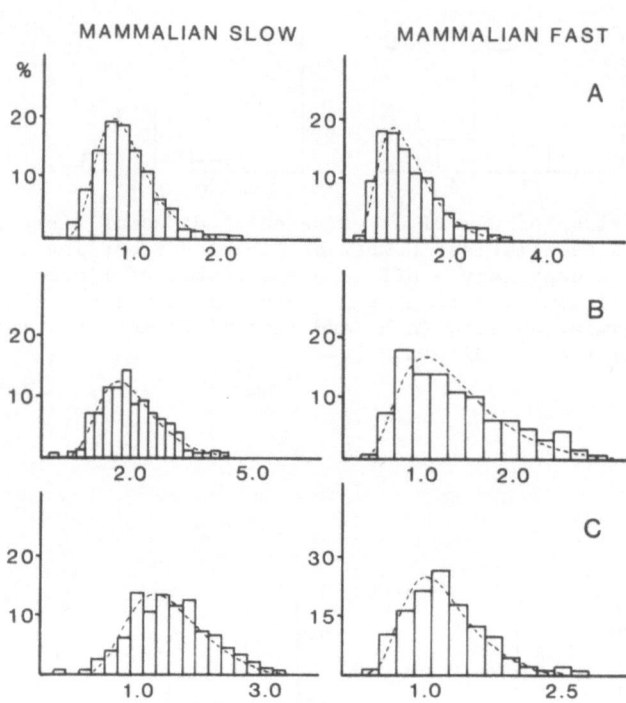

Fig 6 Distributions of capillary domain area (A) muscle fibre area (B), and local capillary to fibre ratio (C) in mammaliam skeletal muscle. Fitted logarithmic normal distributions (-----) are shown. Y axis, frequency (%); X axis, area ($m^2 \times 10^3$).

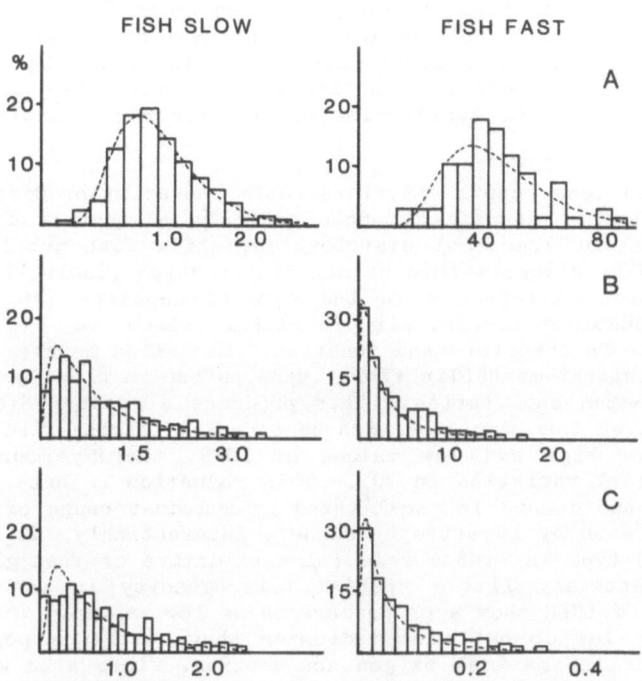

Fig 7 As Fig 6, for fish skeletal muscle.

rather than normal (Gaussian) distributions (Table 1; Fig 6). As this distribution is often found in nature reflecting normal, ontogenetic growth (eg size of vesicles, capillaries, nerves etc) it is possible that the similar form of spatial heterogeneity of capillaries and fibres simply reflects physical constraints on the system. Alternatively, the geometry of the capillary bed may undergo a continual reordering to maintain this distribution, where individual elements are separated by fibre hypertrophy. In order to distinguish between these two possibilities (passive v active co-distributions) we need to explore further the distribution of a complimentary tesselation to that formed by capillaries or fibres, and examine the constraints on the process of growth that would permit parallel development.

The exception to this similarity, Guassian distribution of domain area in fish fast muscle (Fig 7), may be explained by the extremely sparse distribution of capillaries (CD around 20 mm^{-2}) and large spread of fibre area (as this muscle undergoes continual hyperplasia). In this case, control of capillary distribution is unlikely to be affected by local variations in muscle composition, and presumably maintains an evenly spaced network of capillaries to minimise regional variations in oxygen tension.

The derived local capillary:fibre ratio showed minor differences at the species level. Mammalian muscle could be adequately described by either a normal or lognormal distribution, while fish muscle LCFR was best described by a logarithmic normal distribution (Table 1). Although this may reflect differences in the aerobic capacity (VO_2) of these muscles in absolute terms, it is likely also to reflect other heterogeneities in structure and function. Mammalian muscle consists of two or more distinct metabolic fibre types, often in close proximity and supplied by common capillaries. This produces a greater $N(c,f)$ around glycolytic fibres than may otherwise be expected, giving rise to an even distribution of high and low values of LCFR, thereby modulating the impact of spatial variation in VO_2. This reduction in heterogeneity of oxygen supply and demand is facilitated by a modest range of fibre size (reflecting growth by hypertrophy alone). Interestingly, while there is only one fibre type in fish muscle (slow oxidative or fast glycolytic), and hence relatively little spatial heterogeneity in oxygen demand, distributions of LCFR show a preponderance of low values. This reflects the relatively low CD but also indicates that, within a population of fibres with similar maximal oxygen consumption, fibre area may play an important role in determining capillarity (Egginton and Ross, 1989b).

In conclusion, indices of capillarity based on area rather than number distributions are right-shifted with a greater spread of values, being therefore more informative and potentially more sensitive to adaptive changes in capillary supply. In most cases domain area, fibre area and LCFR are best described by logarithmic normal distributions. The exceptions, domains of fish fast muscle and mammalian LCFR, may reflect the sparse capillary network and the close proximity of different fibre types, respectively.

REFERENCES

Egginton, S., 1989, Morphometric analysis of tissue capillary supply, In: 'Vertebrate Gas Exchange from Environment to Cell', Boutilier, R.G., ed., Springer Verlag (in press).

Egginton, S. and Ross, H.F., 1989a, Quantifying capillary distribution in four dimensions. In: 'Oxygen Transport to Tissue XI', Rakusan, K., Biro, G.P., Goldstick, T.K. and Turek, Z., eds., Plenum Press, New York and London, pp. 271-280.

Egginton, S. and Ross, H.F., 1989b, Influence of muscle phenotype on local capillary supply. In: 'Oxygen Transport to Tissue XI', Rakusan, K., Biro, G.P., Goldstick, T.K. and Turek, Z., eds., Plenum Press, New York and London, pp. 281-291.

Egginton, S., Turek, Z. and Hoofd, L., 1987, Morphometric analysis of sparse capillary networks. In: 'Oxygen Transport Tissue IX', Silver, I.A. and Silver, A., eds., Plenum Press, New York and London, pp. 1-12.

Plyley, M.J. and Groom, A.C., 1975, Geometrical distribution of capillaries in mammalian striated muscle, Am. J. Physiol., 228:1376-1383.

BLOOD AND BLOOD SUBSTITUTES

HYPERCHYLOMICRONEMIA, OXYGEN AFFINITY AND PROTON PASSAGE

ACROSS THE RED CELL MEMBRANE

M. J. Poss, I. S. Longmuir and E. T. Moser

Department of Biochemistry
North Carolina State University
Raleigh, NC 27695-7622 U.S.A.

INTRODUCTION

Diminished myocardial oxygen supply is clearly an important factor in the incidence and progress of cardiovascular disease (Selwyn and Ganz, 1988). However the pathophysiology of the various molecular events which cause acute cardiac dysfunction remain unclear.

Our experiments were designed to explore the possibility that one type of hyperlipidemia, hyperchylomicronemia (HC), might affect the affinity of red blood cells for oxygen. If the affinity of red cells for oxygen was increased, then the supply of oxygen to the myocardium would be reduced.

Previously, Neville and Clemmer (1977) showed a positive relationship between increased oxygen affinity and the role of myocardial hypoxia in the incidence of cardiovascular disease.

Our research follows earlier investigations presented by Ditzel and Dyerburg (1977). They showed that normal red cells exposed to high concentrations of chylomicra exhibit an increased affinity for oxygen. Furthermore, they suggested that this might be due to an increase in intracellular pH brought about by an increase in membrane permeability due to a decreased membrane cholesterol content.

Recently (Poss and Longmuir, in press) we presented evidence in support of the claim that normal red cells exposed to 6% chylomicra exhibit an increased affinity for oxygen, a leftward shift in the oxygen dissociation curve. We found that in these samples the intracellular pH rose to approximate the extracellular pH, so that the proton gradient across the red cell membrane was largely abolished.

In this paper we present additional data to support the idea that HC affects the affinity of the red cell for oxygen by abolishing the proton gradient across the red cell membrane. An oxidative phosphorylation uncoupling agent appears to produce a similar increased affinity for oxygen by immediately abolishing the proton gradient across the red cell membrane.

We found no evidence to support the suggestion that HC produces a decrease in the amount of cholesterol contained in the red cell membrane. Therefore the nature of the molecular mechanism which abolishes the proton gradient remains unclear.

MATERIALS AND METHODS

The principal of the method used in determining the shift in the oxygen dissociation curve is the same as explained in Poss and Longmuir (in press). The only exception being that we used the Yellow Springs Instruments Model 5300 Biological Oxygen Monitor. The advantage of this instrument was that it facilitated parallel experiments using normal and treated blood samples.

The apparatus consisted of two water-jacketed glass cells, each fitted with rubber stoppers through which pO_2 and pH electrodes were inserted. Inlet and outlet needles allowed for the addition of solutions and a stirrer bar permitted mixing.

Blood was obtained from an antecubital vein of the same subject (project approved by NCSU's IRB) and coagulation was prevented by adding K-EDTA (1.5 mg/ml). Heavy whipping cream (MacFarlane et al., 1941; MacLagen and Billimoria, 1956) or emulsified coconut oil (Shafiroff and Frank, 1947) was added to approximate an acute clinically recognized level of HC (60 mg/ml). The oxidative phosphorylation uncoupling agent, carbonyl cyanide m-chlorophenyl hydrazone (CCCP), was added (1.0μM) to abolish the proton gradient across the red cell membrane. The effect of these three agents on the pO_2 of blood was compared.

The pO_2 was continuously recorded. The difference in slopes between normal blood and treated blood revealed a decreased pO_2 in treated blood. A corresponding leftward shift in the oxygen dissociation curve was determined by the method of Longmuir and Chow (1970).

Intracellular pH was determined by the freeze/thaw method of Hilpert et al. (1963). Intracellular and extracellular pH were measured at the completion of the experiment for both normal and treated blood.

Total red cell membrane cholesterol was measured by the method of Lubin et al. (1988). Total membrane cholesterol concentrations were determined after 2 and 8 hours of incubation in normal blood + 6% coconut oil emulsion.

RESULTS

The pO_2 of blood slowly decreased when 6% lipid emulsions were added. A similar but immediate decrease was observed when 1.0μM CCCP was added (Fig. 1)

The ΔpO_2 at varying percent saturations reflected a 6.0 mm Hg decrease in P-50 for both normal blood + 6% lipid emulsions and normal blood + 1.0 μM CCCP (Fig. 2). This leftward shift in the oxygen dissociation curves revealed a significant increased in the affinity of hemoglobin for oxygen.

In both cases the pH gradient across the red cell membrance was largely abolished. In parallel samples of normal blood, the pH gradient remained unchanged (Table 1).

No significant change in the total mebrance cholesterol was observed during the length of time that a ΔpO_2 was observed.

174

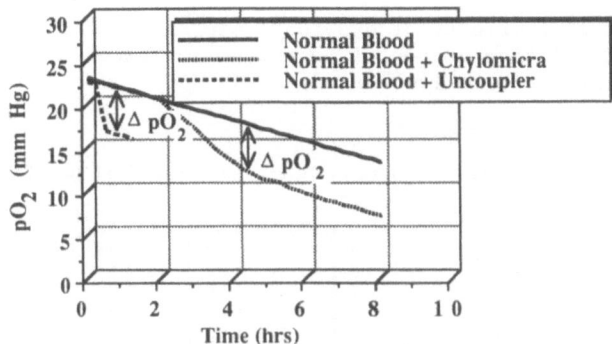

Fig. 1. A model experiment showing the recorder tracing for normal blood, normal blood +6% chylomicra (coconut oil emulsion), and $1\mu\underline{M}$ oxidative phosphorylation uncoupler (CCCP).

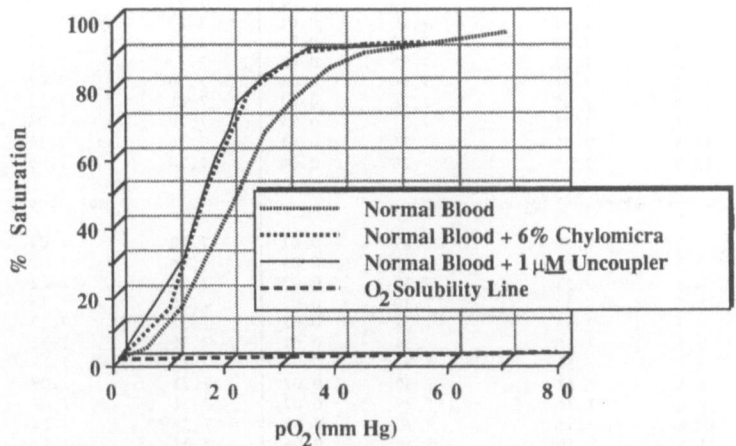

Fig. 2. An oxygen dissociation curve for normal blood is compared to those for normal blood + 6% chylomicra (coconut oil emulsion) and normal blood + $1\,\mu\underline{M}$ oxidative phosphorylation uncoupler.

175

Table 1. The similar effects of hyperchylomicronemia and an oxidative phosphorylation uncoupler on the pO_2 of blood and the pH gradient across the red cell membrane.

A.   Normal Blood + 6% Cream   Normal Blood

Initial pO_2 (mm Hg)	ΔpO_2 (mmHg)	Extracellular pH	Intracellular pH	ΔpH	Extracellular pH	Intracellular pH	ΔpH
10.6	1.3	7.15	7.14	0.01	7.16	7.00	0.16
16.1	3.5	7.14	7.11	0.03	7.15	6.95	0.20
20.4	5.5	7.31	7.25	0.06	7.26	7.03	0.23
22.4	5.9	7.28	7.27	0.01	7.29	7.11	0.18
22.9	6.1	7.35	7.32	0.03	7.36	7.14	0.22
23.5	6.8	7.22	7.19	0.03	7.22	7.00	0.22
27.3	7.3	7.31	7.30	0.01	7.30	7.04	0.26
29.0	8.2	7.18	7.16	0.02	7.20	6.99	0.21
31.8	9.2	7.21	7.16	0.05	7.21	7.02	0.19
33.0	9.6	7.28	7.25	0.03	7.26	7.08	0.18
44.2	11.2	7.29	7.25	0.04	7.29	7.07	0.22
48.1	11.4	7.33	7.30	0.03	7.34	7.15	0.19
53.0	12.8	7.32	7.28	0.04	7.33	7.11	0.22

B.   Normal Blood + 6% Coconut Oil Emulsion   Normal Blood

9.5	1.1	7.18	7.16	0.02	7.18	6.99	0.19
11.7	2.6	7.22	7.21	0.01	7.14	6.96	0.18
20.7	5.6	7.26	7.22	0.04	7.24	7.00	0.24
22.1	6.0	7.29	7.25	0.04	7.31	7.11	0.20
24.1	7.1	7.26	7.25	0.01	7.28	7.07	0.21
27.6	7.7	7.31	7.30	0.01	7.26	7.06	0.20
32.2	9.8	7.32	7.30	0.02	7.22	7.00	0.22
33.3	9.9	7.21	7.20	0.01	7.23	7.04	0.19
36.1	10.1	7.33	7.32	0.01	7.31	7.11	0.20
38.7	10.7	7.31	7.28	0.03	7.33	7.11	0.23
43.4	11.1	7.27	7.26	0.01	7.30	7.07	0.23
45.5	11.8	7.29	7.26	0.03	7.31	7.13	0.18
47.1	12.1	7.31	7.27	0.04	7.30	7.08	0.22
51.3	12.5	7.32	7.31	0.01	7.34	7.14	0.20
55.0	13.0	7.28	7.24	0.04	7.30	7.05	0.25

C.   Normal Blood + 1 μM Uncoupler   Normal Blood

11.1	2.3	7.30	7.29	0.01	7.30	7.09	0.21
14.3	3.5	7.29	7.26	0.03	7.28	7.09	0.19
18.7	4.8	7.33	7.30	0.03	7.34	7.12	0.22
22.4	6.2	7.32	7.30	0.02	7.35	7.13	0.22
26.2	6.7	7.33	7.31	0.02	7.32	7.13	0.19
27.3	7.5	7.35	7.34	0.01	7.35	7.15	0.20
28.0	7.9	7.34	7.31	0.03	7.37	7.19	0.18
30.3	8.8	7.32	7.30	0.02	7.31	7.09	0.22
31.8	9.6	7.31	7.29	0.02	7.31	7.07	0.24
36.3	10.6	7.33	7.30	0.03	7.32	7.11	0.21
38.1	11.2	7.34	7.30	0.04	7.37	7.17	0.20
39.4	11.4	7.30	7.27	0.03	7.34	7.15	0.19
44.0	11.9	7.37	7.36	0.01	7.36	7.16	0.20
48.3	12.2	7.35	7.31	0.04	7.36	7.17	0.19
50.4	12.9	7.36	7.34	0.02	7.38	7.16	0.22

DISCUSSION

Hyperchylomicronemia appears to increase the affinity of hemoglobin for oxygen by slowly abolishing the pH gradient across the red cell membrane. The increased intracellular pH produced the observed fall in P-50.

Evidence for this stems from the observation that an oxidative phosphorylation uncoupling agent immediately produced effects similar to HC. The uncoupling agent inhibits oxidative phosphorylation by abolishing the proton gradient across the inner mitochondrial membrane. It appears to affect the red cell membrane in the same way.

The addition of both chylomicra and uncoupling agent affected the red cell membrane so that the intracellular pH rose to approximate the extracellular pH. As a result a Bohr shift was produced and the affinity of hemoglobin for oxygen was increased. A literature search indicated that no similar observations have been reported.

In order for the proton gradient to be abolished, some physical modification of the red cell membrane must occur. Ditzel and Dyerburg suggested that a loss of membrane cholesterol might contribute to increased membrane fluidity and consequently an increased permeability. This seemed a plausible suggestion considering the agreement between our findings and those in the literature.

Hagerman and Gould (1951) and Murphy (1962) showed that non-esterified membrane cholesterol freely exchanged with plasma and came to equilibrium after 4 hrs. The time for equilibration was approximately equal to that in which we observed an increase in the affinity of red cells for oxygen. After equilibration in HC plasma, red cell ghosts were found to have a decreased cholesterol content (Bagdade and Ways, 1970). Furthermore, Kroes and Osterwald (1971) noted that cholesterol deficient red cells exhibit an increased permeability to several non-electrolytes and Na^+.

We hoped to confirm the theory of Ditzel and Dyerburg by showing that red cells exposed to 6% lipid emulsions exhibit a slow decrease in membrane cholesterol content which follows the rise in intracellular pH and the consequent increase in oxygen affinity. However we found no such decrease in membrane cholesterol content. Therefore the proton gradient across the red cell membrane must be altered by some other mechanism.

One possibility is that a constituent of the lipid droplets could act as an ionophore. The H^+ might then be exchanged across the membrane and the proton gradient gradually abolished. Fukuzaki (1975) reported that red cells exposed to HC plasma adsorbed large amounts of fat on the membrane.

Another suggestion is that fatty acids with a carboxyl group pK_a around 7.0 could exchange H^+ in a way similar to oxidative phosphorylation uncouplers. If the long lipophilic moiety diffused across the phospholipid bilayer, the carboxyl group would become protonated in the red cell cytoplasm. It could then shuttle H^+ across the red cell membrane until the proton gradient was abolished.

In vitro analysis of the molecular events which contribute to the relationship between HC and increased red cell affinity for oxygen were performed. The actual molecular mechanism remains unclear. Nevertheless the significance of such an increase in the affinity of the red cell for oxygen can be appreciated. If the red cell affinity for oxygen were increased, less oxygen would be available to the myocardium. This event suggests a causal relationship between molecular events associated with oxygen transport and the pathology of cardiovascular disease.

177

SUMMARY

Previously we have shown that HC alters significantly the affinity of hemoglobin for oxygen. This evidence stands in support of earlier investigations presented by Ditzel and Dyerburg (1977).

Exposure to 6% chylomicra, slowly produced a decrease in the pO_2 of human blood (37° C, at in vivo pH) amounting to a 6.0 mm Hg leftward shift in the P-50.

This increase in oxygen affinity may be due to the Bohr shift since exposure to HC produced an increase in intracellular pH by abolishing the proton gradient across the red cell membrane.

We postulate that exposure to HC might physically modify the composition of the red cell membrane so that proton passage is permitted and the intracellular pH rises to approximate that of the extracellular pH.

An oxidative phosphorylation uncoupling agent (CCCP), abolishes the proton gradient and produces an immediate decrease in the pO_2 similar to that produced by HC.

We explored the suggestion that proton passage might be facilitated by an increase in the membrane fluidity produced by a decrease in membrane cholesterol. However no decrease in red cell membrane cholesterol was found on exposure to 6% chylomicra.

Although the molecular mechanism remains unclear, this event would dramatically reduce oxygen delivery to the myocardium and might prove to be an important factor in the incidence of myocardial hypoxia.

ACKNOWLEDGEMENTS

We greatly appreciate the generosity of Lynda Hart and her colleagues at Yellow Springs Instruments Incorporated. By loaning our laboratory the use of a Biological Oxygen Monitor 5300, they enabled us to complete our study efficiently.

REFERENCES

Bagdade, J. D., and Ways, P. O., 1970, Erythrocyte membrane lipid composition in endogenous hypertriglyceridemia, J. Lab. Clin.. Med., 75:53.

Ditzel, J., and Dyerburg, J., 1977, The oxyhemoglobin dissociation curve in patients with familial hyperchylomicronemia, J. Lab. Clin. Med., 89:573.

Fukuzaki, H., Okamoto, R., and Matsuo, T., 1975, Studies on pathophysiological effects of postalimentary lipemia in patients with ischemic heart disease, Jpn. Circ. J., 39:31.

Hagerman, J. S., and Gould, R. G., 1951, The in vitro interchange of cholesterol between plasma and red cells, Proc. Soc. Exp. Biol. Med., 78:329.

Hilpert, P., Fleischmann, R. C., Kempe, and Bartels, H., 1963, The Bohr effect related to blood and erythrocyte pH, Am. J. Physiol., 205:331.

Kroes, J., and Ostwald, R., 1971, Erythrocyte membranes--effect of increased cholesterol on permeability, Biochim. Biophys. Acta, 249:647.

Longmuir, I.S., and Chow, J., 1970, Rapid method for determining effects of agents on oxyhemoglobin dissociation curves, J. Appl. Physiol., 28:343.

Lubin, H. L., Kuypers, F. A., Chiu, D. T.-Y., and Shohet, S. B., 1988, Analysis of red cell membrane lipids, in: "Red Cell Membranes," S. B. Shohet and N. Mohandes, eds., Churchhill Livingstone, New York.

MacFarlane, J. W., Trevan, J. W., and Attwood, A. W., 1941, Participation of fat soluble substances in coagulation of blood, J. Physiol., 99:7P.

MacLagen, N. F., and Billimoria, J. D., 1956, Food lipids and blood coagulation. Lancet, i:235.

Murphy, J. R., 1962, Erythrocyte metabolism. IV. Equilibrium of cholesterol-4-C^{14} between erythrocytes and various treated sera, J. Lab. Clin. Med., 60:571.

Neville, J. R., and Clemmer T., 1977, Hemoglobin-Oxygen Affinity in Organic Heart Disease, in: "Advances in Experimental Medicine and Biology," Vol. 94, "Oxygen Transport to Tissue III," I. A. Silver, M. Erecinska, and H. I. Bicher, eds., Plenum Press, New York.

Poss, M. J. and Longmuir, I. S., in press, The effect of hyperchylo-micronemia on oxygen affinity in human blood, in: "Advances in Experimental Medicine and Biology," Vol. undetermined, "Oxygen Transport to Tissue XI," K. Rakusan and T. K. Goldstick, eds., Plenum Press, New York.

Selwyn, A. p. and Ganz, P., 1988, Myocardial ischemia in coronary disease, New Eng. J. Med., 318:1058.

Shafiroff, G. P., and Frank, C., 1947, a homologous emulsion of fat, protein and glucose for intravenous administration, Science, 106:474.

FACILITATED TRANSPORT OF OXYGEN THROUGH

HEMOGLOBIN SOLUTIONS

Masafumi Hashimoto, Ryuji Hata, Takeshi Shiga,
Akio Isomoto*, and Mitsuro Uozumi**

Department of Physico-chemical Physiology, Osaka University
Medical School, Osaka 530, *Department of General Education,
Kinki University, Higashiosaka 577, and **Osaka Prefectural
Institute of Public Health, Osaka 537, Japan

INTRODUCTION

In a previous study on the facilitated diffusion of oxygen by hemoglobin, we developed a new method using an image-input and -processing
system composed of a 3-tube video camera and a digital image analyzer.
With this system, diffusion of oxygen in hemoglobin solutions was observed
through a microscope, i.e., a deoxyhemoglobin solution was put in a glass
capillary tube, and the change in hemoglobin color as oxygen diffused from
the open end of the tube was observed and analyzed quantitatively by using
our computer algorithm to estimate the oxygen saturation of hemoglobin.
Thus, we were able to identify the diffusion profile as resolvable into
intervals of 5.5 μm for position and 2.9 sec for time. The experimental
results provided new information about the oxygen transport (Hashimoto, et
al., 1988).

The fact that the diffusion of oxygen in aqueous solution is facilitated in the presence of hemoglobin was first found by Wittenberg (1959)
and Scholander (1960). Several models have been proposed with the following one being generally accepted. While association and dissociation are
taking place, oxygen and oxyhemoglobin diffuse independently at rates of
transport which are in proportion to their concentration gradients with
their specific diffusion coefficients; thus the total oxygen flux will be
the sum of those of oxyhemoglobin and of unbound oxygen (Snell, 1965;
Wyman, 1966; Moll, 1968; Kutchai, Jacquez, and Mather, 1970; Kreuzer and
Hoofd, 1970; Kreuzer and Hoofd, 1972; Kreuzer and Hoofd, 1976). However,
in contradiction to this model, our experimental results indicated that
the oxygen flux estimated from the time course of the oxygen saturation
(oxyhemoglobin concentration) profile, was in proportion to the oxyhemoglobin concentration itself rather than its gradient, giving the empirical
formula, $J = kY$, where J is the oxygen flux, Y is the oxygen saturation of
hemoglobin, and k is a constant. The equation suggested a unidirectional
transport model such as found in convection or electrophoresis. Nevertheless, these are unlikely to occur in the absence of a disorder of the
diffusion front and electric field.

Another model, proposed by Scholander and Hemmingsen (1960) and Enns
(1964), is the "bucket-brigade" model, i.e., oxygen exchange takes place

unidirectionally when hemoglobin molecules collide. This model can explain our experimental results, but the question of why it occurs unidirectionally still remains.

In the present study, we investigated the phenomenon further, examining the effects of oxygen affinity to hemoglobin (Experiment I) and hemoglobin concentration (Experiment II) on the transport. Our results, together with consideration of earlier findings, led to the development of new mathematical expression to express a new transport model of oxygen by hemoglobin.

MATERIALS AND METHODS

Hemoglobin Sample

Hemoglobin, obtained from human erythrocytes using Drabkin's method (1946), was dissolved in 0.1 M potassium phosphate buffer. Oxygen affinity to hemoglobin was varied using different buffer pH as well as by adding inositol hexaphosphate (IHP). The oxygen binding property in each experiment was examined using a tonometric procedure to obtain the oxygen equilibrium curve. Oxygen saturation and hemoglobin concentration were determined with a spectrophotometer (Hitachi 320L). All chemicals were of reagent grade.

Experimental System

Instrumentation. As reported in the previous study, our method uses an image-input and -processing system composed of a 3-tube video camera (Ikegami Tsushinki Co, Tokyo, Model ITC-350M) and a digital image analyzer (Kashiwagi Res. Inc., Tokyo, Nexus 6400). The Nexus 6400 resolves and quantizes an image obtained via the video camera into 512 × 480 pixels and the color of each pixel into 256 (= 2^8) stages of red (R), green (G), and blue (B) brightness, then stores the digital information of 512 × 480 × 3 × 8 bits in R-, G-, and B-frame memories addressed for each pixel and, if necessary, in a floppy disk of 1Mbyte. By connecting this system to a microscope (Olympus Inc., Tokyo, Model BH-2), we were able to observe the phenomena microscopically. The Nexus 6400 was driven according to programmed commands that had been entered into a host computer (NEC, Tokyo, Model PC-9801-F2) beforehand. The gamma-correction channel of the video

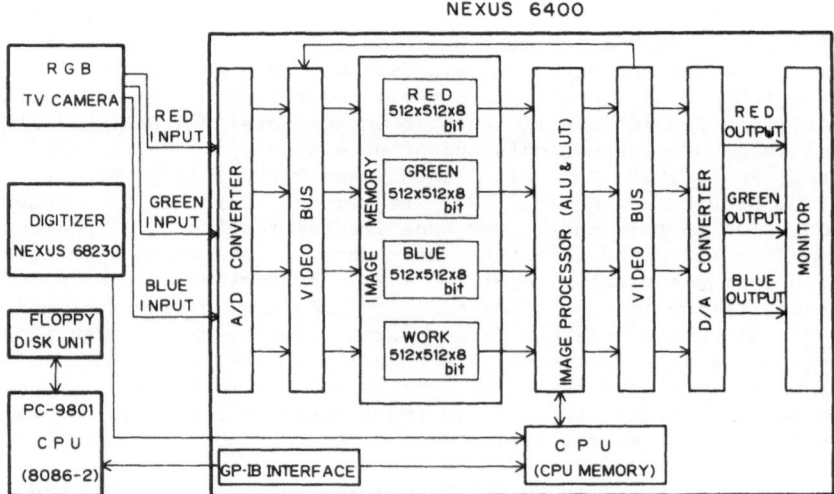

Fig. 1. Schematic presentation of the image-input and -processing system.

182

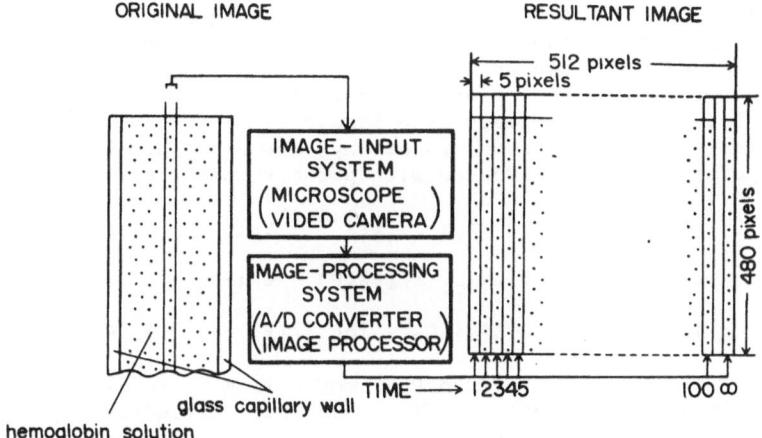

ORIGINAL IMAGE   RESULTANT IMAGE

IMAGE-INPUT SYSTEM (MICROSCOPE VIDED CAMERA)

IMAGE-PROCESSING SYSTEM (A/D CONVERTER IMAGE PROCESSOR)

glass capillary wall

hemoglobin solution

Fig. 2.   Schematic presentation of the experimental procedure.

camera was switched off, to have the input values of R, G, and B brightness proportional to the corresponding light energies (Fig. 1).

Experimental Procedure. A deoxyhemoglobin solution prepared by flushing with pure nitrogen gas (99.99 %) was sent into a glass capillary tube with a 1 ml syringe connected anaerobically to one end of the tube. Oxygen began to diffuse into the hemoglobin solution from the other end of the tube that was open to air. The experiment was started when a portion of the oxygenated hemoglobin was discarded by operating the syringe and a new surface of the deoxyhemoglobin solution appeared at the open end of the tube. The 5-pixel width and 480-pixel length images along the center axis of the tube were entered into the R-, G-, and B-frame memories at intervals of 2.9 sec (Experiment II) or 3.8 sec (Experiment I), thus arranging 100 images of 5-pixel width from left to right with time. The 1-pixel length corresponded to the actual 5.5-μm diffusion path length (Fig. 2). The resultant image was transformed as the R, G, and B values of any pixel were replaced with the median of the neighboring 5×5 pixels, which allowed us to reduce noise in the image. Thus, using an algorithm for estimating the oxygen saturation of hemoglobin from the R, G, and B brightness values, we were able to find the oxygen saturation at any desired time and position of the diffusion process.

Computer Algorithm for Estimating Oxygen Saturation and Oxygen Flux

Oxygen Saturation. Since the R, G, and B brightness values indicate the intensities $[I(i), i = R,G,B]$ of the light transmitted through an object in the corresponding wave bands, we can employ the Beer-Lambert law:

$$- \log I(i)/I_0(i) = \varepsilon(i)D, \quad i = R,G,B, \tag{1}$$

where $I_0(i)$ is the intensity of the incident light; $\varepsilon(i)$ is the absorption coefficient; D is the product of the object's concentration and the effective light path length. Equation [1] gives the following expression for the optical densities of the two different color components:

$$- \log I(i)/I(j) + \log I_0(i)/I_0(j) = D[\varepsilon(i) - \varepsilon(j)], \tag{2}$$

$$i = R,G,B, \quad j = R,G,B, \quad i \neq j.$$

Since $\log I_0(i)/I_0(j)$ has a constant value, this equation indicates that $\log I(i) - \log I(j)$ is linear to D regardless of the luminosity. In the xy-coordinate space defined as $x = \log I(G) - \log I(B)$ and $y = \log I(R) - \log I(G)$, a point S for the color of a hemoglobin solution at the oxygen saturation Y ($0 \leq Y \leq 1$) appears as a linear combination of the data points S_1 and S_0 for the respective colors of oxy- and deoxy-hemoglobin, so the xy values of point S can be expressed with parameters D and Y:

$$x = DY[\varepsilon_1(B) - \varepsilon_1(G)] + D(1 - Y)[\varepsilon_0(B) - \varepsilon_0(G)] - C_1, \qquad [3]$$

$$y = DY[\varepsilon_1(G) - \varepsilon_1(R)] + D(1 - Y)[\varepsilon_0(G) - \varepsilon_0(R)] - C_2, \qquad [4]$$

$$C_1 = \log I_0(G) - \log I_0(B), \quad C_2 = \log I_0(R) - \log I_0(G).$$

When Y varies from 0 to 1 at fixed D, S moves from S_0 to S_1 along the line $S_0 S_1$ in proportion to the value of Y. Thus, we can readily find the Y value from the actual locations of the points $S_1(x_1, y_1)$, $S_0(x_0, y_0)$, and $S(x,y)$ in the space. These relationships among (x,y), D, Y were valid in our experiments as previously reported. In the present study, the xy-coordinates of the points S_0 and S were easily obtained from the R-, G-, and B-data of hemoglobin solution at the beginning and the middle of an experiment, respectively. However, the final xy-coordinates of the point S did not always correspond to full saturation. Therefore, we introduced a $(\partial x/\partial t)$ vs. x plot, where $(\partial x/\partial t)$ is the partial derivative of x with respect to time (t) at fixed position (p), so that we were able to obtain the x_1 value by extrapolating x at $(\partial x/\partial t) = 0$ (at infinite t), and similarly to obtain the y_1 value. From Eqs. [3] and [4], we had the value of Y:

$$Y = (x - x_0)/(x_1 - x_0) = (y - y_0)/(y_1 - y_0), \qquad [5]$$

(Hashimoto et al., 1987).

Oxygen Flux. As mentioned in the preceding section, we could obtain the diffusion profile (time course of the oxygen saturation profile) which was resolvable into intervals of 5.5 μm for position and 2.9 (3.8) sec for time. Therefore, we can estimate the approximate oxygen flux (J) per unit cross-sectional area at a certain position (p) and time (t) in the diffusion process under the condition of: the oxyhemoglobin concentration >> the concentration of unbound oxygen:

$$J = C\{[\textstyle\sum_p^\infty Y]_{t+\Delta t} - [\textstyle\sum_p^\infty Y]_t\}(\Delta p/\Delta t), \qquad [6]$$

where C is the hemoglobin concentration, and Δp and Δt correspond to 5.5 μm and 2.9 (3.8) sec, respectively.

RESULTS

Figure 3a shows the profiles of oxygen saturation (Y) (open circles) and oxygen flux (J) (closed circles) against their position (p, p = 0 corresponds to the entrance of oxygen from air to aqueous solution) at a specific time in our experiment, and Fig. 3b shows the relationship between Y and J at each position. Figure 4a shows the profiles of Y (open circles) and J (closed circles) against t at a specific position, and Fig. 4b shows the relationship between Y and J at each time. The relationships shown in Figs. 3b and 4b indicate that J is in proportion to Y at a fixed time but

184

the coefficient value varies with time. Then, we tried plotting $J \cdot d$ vs. Y for the data in Fig. 4, where d is the p value of the diffusion front which increases as oxygen diffuses (Fig. 4c). The plot indicates that $J \cdot d$ is in proportion to Y, which is also cosistent with the results shown in Fig. 3. We found the same relationships for all experiments we coducted. These relationships led to an empirical formula $J(p,t) = k'CY(p,t)/d(t)$, where C is the hemoglobin concentration and k' is a constant.

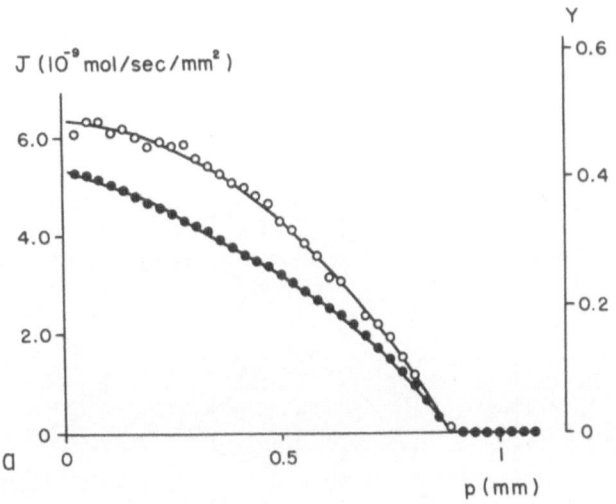

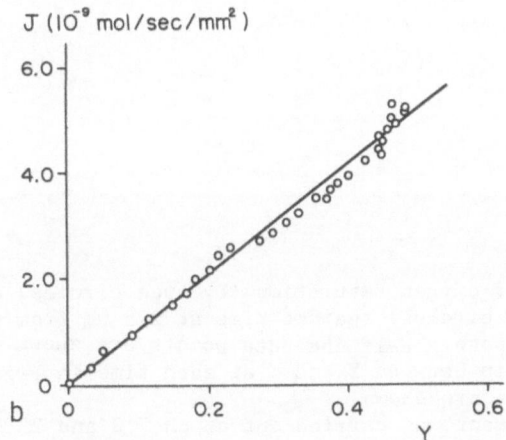

Fig. 3a. Profiles of oxygen saturation (Y) and oxygen flux (J) at 3 min after the beginning of the experiment. Twenty per cent of the data points are shown.

3b. Relationship between Y and J at each position in Fig. 3a. The experiment was carried out at pH 7.0 and $22^\circ C$ (room temperature). The hemoglobin concentration was 2.5 mM.

185

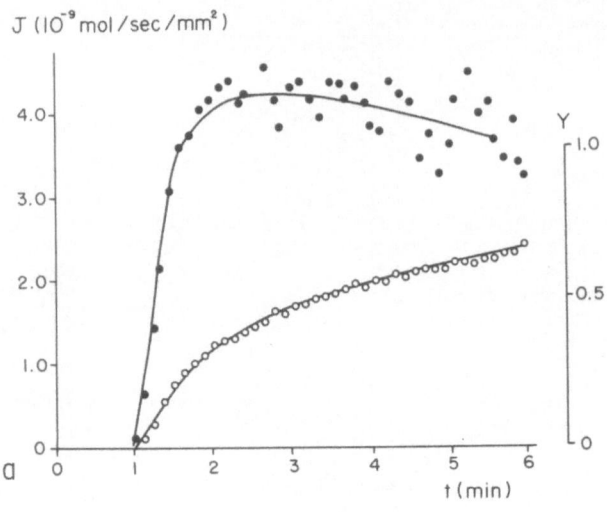

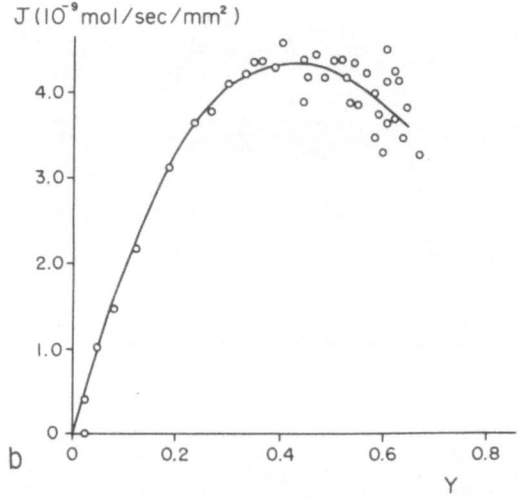

Fig. 4a. Profiles of oxygen saturation (Y, open circles) and oxygen flux (J, closed circles) against time at 275 μm from the entrance of diffusion path. Half the data points are shown.

4b. Relationship between Y and J at each time in Fig. 4a. Half the data points are shown.

The experiment was carried out at pH 7.0 and 22°C (room temperature). The hemoglobin concentration was 2.5 mM.

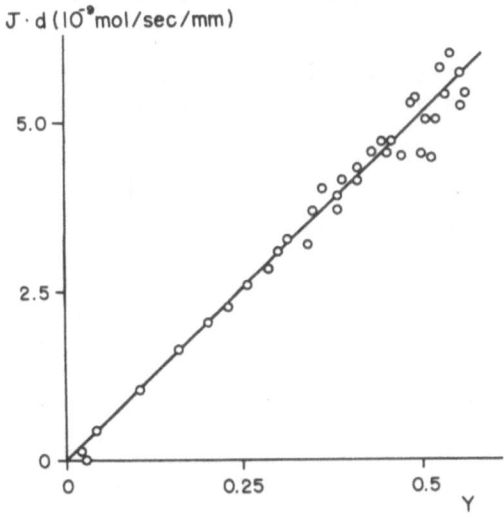

Fig. 4c. Relationship between Y and J•d at fixed p (p = 275 μm).

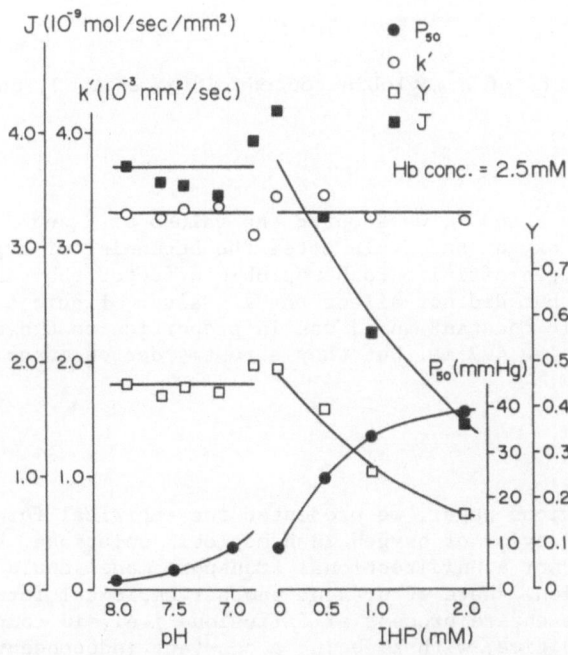

Fig. 5. Effects of oxygen affinity to hemoglobin on Y, J, and k' at 22°C.

Based on the equation J = k'CY/d, we evaluated k' by obtaining the coefficient of linear regression in the J•d vs. Y plot. The values of J, Y, and k' obtained when we varied oxygen affinity to hemoglobin (Experiment I) and hemoglobin concentration (Experiment II) are shown in Figs. 5 and 6, respectively. Oxygen affinity to hemoglobin is expressed by P_{50}, i.e., the oxygen pressure to give 50 per cent saturation of the hemoglobin. For com-

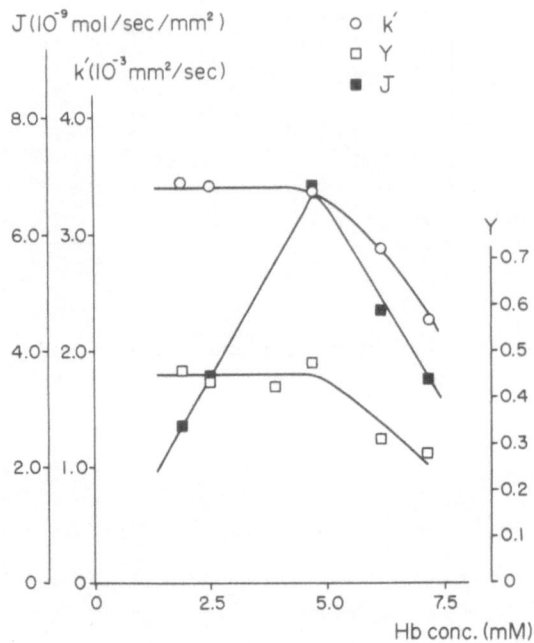

Fig. 6. Effects of hemoglobin concentration on Y, J, and k' at 20°C.

parison based on Y and J, we adopted the values of Y and J at 275 μm from the entrance of oxygen and 3 min after the beginning of experiment. Figure 5 shows that oxygen affinity to hemoglobin affected the value of J through the change of Y but did not affect the k' value. Figure 6 shows that k' and Y were nearly constant and J was in proportion to C over the hemoglobin concentration below 4.7 mM, but they abruptly decreased as C increased to more than 4.7 mM.

DISCUSSION

In our previous paper, we presented the empirical formula J = kY for the diffusion behavior of oxygen in hemoglobin solutions, but neither Fick's diffusion model nor a unidirectional transport model could be used to explain the equation. Here we present another empirical formula, J = k'CY/d, which covers the entire process of diffusion, i.e., in comparison with k which depends on time, with k' being a constant independent of time. The division by d in the equation suggests the relation of a concentration gradient to the diffusion. However, since Y/d does not correspond to ∂Y/∂p in the absence of the linear profile of Y against p, this equation J = k'CY/d does not directly lead to Fick's diffusion model. We thus propose the following transport model in which $k'CY/d$ is divided into two parts as $k_1'CY$ and k_2'/d ($k' = k_1'k_2'$). We compared it with Fick's diffusion equation, $J = D \cdot [grad]$, where D is the diffusion coefficient and $[grad]$ is the concentration gradient, assuming the correspondences between $k_1'CY$ and D and between k_2'/d and $[grad]$. This assumption indicates that oxygen affinity to hemoglobin affects the oxygen diffusibility through the change of Y, and there exists a linear concentration gradient of oxygen from the entrance of oxygen to the diffusion front, which causes the diffusion.

This model is consistent with our experimental results. The decrease in oxygen affinity to hemoglobin caused the decrease of J through the decrease of Y. k_2' may be the concentration of unbound oxygen at the entrance, which we think does not always equilibrate with atmospheric oxygen, but has nearly a constant value under our experimental conditions. Therefore, the k' value was almost constant in our experiments except when an extremely high concentration of hemoglobin affected the k_1' value by raising solution viscosity. Questions still remain about this transport model, and studies are under way to solve them, including the estimation of the concentration profile of unbound oxygen.

SUMMARY

A method for quantitative analysis of hemoglobin color to estimate the oxygen saturation was developed. The method uses an image-input and -processing system composed of a 3-tube video camera and a digital image analyzer. When the system is connected to a microscope, the facilitated diffusion of oxygen in hemoglobin solutions can be observed and analyzed in a position-sensitive manner. The results led to a new transport model expressed as: $J = k'CY/d$, where J is the oxygen flux, C is the hemoglobin concentration, Y is the oxygen saturation of hemoglobin, d is the distance between the entrance of oxygen and the diffusion front, and k' is a constant.

REFERENCES

Drabkin, D. L., 1946, The crystallographic and optical properties of the hemoglobin of man in comparison with those of other species, J. Biol. Chem., 164:703-727.

Enns, T., 1964, Molecular collision-exchange transport of oxygen by hemoglobin, Proc. Nat. Acad. Sci., 51:247-252.

Hashimoto, M., Hata, R., Isomoto, A., Tyuma, I., and Fukuda, M., 1987, Color analysis method for estimating the oxygen saturation of hemoglobin using an image-input and processing system, Anal. Biochem., 162:178-184.

Hashimoto, M., Hata, R., Tyuma, I., Isomoto, A., and Uozumi, M., 1988, Color analysis method for studying oxygen transport in hemoglobin solutions using an image-input and -processing system, Adv. Exp. Med. Biol., 222:231-238.

Hemmingsen, E., and Scholander, P. F., 1960, Specific transport of oxygen through hemoglobin solutions, Science, 132:1379-1381.

Kreuzer, F., and Hoofd, L., 1970, Facilitated diffusion of oxygen in the presence of hemoglobin, Respir. Physiol., 8:280-302.

Kreuzer, F., and Hoofd, L., 1972, Factors influencing facilitated diffusion of oxygen in the presence of hemoglobin and myoglobin, Respir. Physiol., 15:104-124.

Kreuzer, F., and Hoofd, L., 1976, Facilitated diffusion of CO and oxygen in the presence of hemoglobin and myoglobin, Adv. Exp. Med. Biol., 75:207-215.

Kutchai, H., Jacquez, J., and Mather, F., 1970, Nonequilibrium facilitated oxygen transport in hemoglobin solution, Biophys. J., 10:38-54.

Moll, W., 1968/1969, The influence of hemoglobin diffusion on oxygen uptake and release by red cells, Respir. Physiol., 6:1-15.

Scholander, P. F., 1960, Oxygen transport through hemoglobin solutions, Science, 131:585-590.

Snell, F. M., 1965, Facilitated transport of oxygen through solutions of hemoglobin, J. Theoret. Biol., 8:469-479.

Wittenberg, J. B., 1959, Oxygen transport — a new function proposed for myoglobin, Biol. Bull., 117:402-403.

Wyman, J., 1966, Facilitated diffusion and the possible role of myoglobin as a transport mechanism, J. Biol. Chem., 241:115-121.

INTERACTION BETWEEN ORGANIC PHOSPHATES

AND SHEEP HEMOGLOBINS

Robert A.B. Holland, Elisabeth A. Tibben, and Jillian F. Hallam

School of Physiology and Pharmacology,
University of New South Wales, Kensington
NSW, 2033 Australia

INTRODUCTION

It has been widely accepted that 2,3-diphosphoglycerate, which lowers the oxygen affinity (raises P_{50}) of most mammalian blood, has little, if any, effect on sheep hemoglobins. Its small effect in sheep, relative to the effect in most mammals, was reported by Bunn (1971) who worked with dilute Hb solutions. Also, using Hb solution where the Hb was about 1/30 of the concentration found in red cells, Perutz and Imai (1980) showed that bovine Hb has little interaction with 2,3-DPG when compared with human Hb. Both sheep and cattle are Bovidae. This group, which also includes the goat, all have low affinity haemoglobins (relatively high P_{50} in the absence of organic phosphates) and very low or zero 2,3-DPG levels in adult red cells.

In the Bovidae, the first amino acid in the non-α chains is methionine, not the more usual valine, and the second residue, normally histidine, is not present at all. The low interaction with 2,3-DPG has been attributed to the deletion causing a shortening of the non-helical segment of the amino end of the chain and also removing one of the four positively charged binding sites (side chain of HIS β_2) that are found in each β chain. Perutz and Imai attributed the high affinity of these hemoglobins to the N-terminal methionine in the β chains, suggesting that its long hydrophobic side chain pointing into the interior of the molecule, abolishes the interaction with organic phosphates, and locks the A helix into place. This would stabilize the T structure and give the Hb a low O_2 affinity just as organic phosphates do in other hemoglobins (such as human) with an intrinsically high affinity.

Both Bunn (1971) and Perutz and Imai (1980) worked with dilute Hb solutions in the absence of CO_2. Bauer and Jung (1975) working on sheep Hb A and B also used dilute solutions but added higher concentrations of 2,3-DPG. They found a clear interaction between DPG and both Hb types at high DPG concentrations in the absence of CO_2, but the effect was less than is found with human Hb. In the presence of CO_2 they found no interaction at all. Also Baumann et al (1972) found that in haemolysates of adult sheep cells with Hb A or AB, and of new born sheep blood (all at about one quarter of the normal intracellular Hb concentration) the O_2 affinity was not altered by the addition of 2,3-DPG to give 3 mM concentration. These experiments were performed in the presence of 40

Torr CO_2

There were several reasons further studies of these interactions. It is known that the concentration of 2,3-DPG in red cells of new born lambs is up to 10 mM (Baumann et al, 1972), and therefore an effect of 2-3 DPG directly on the haemoglobin molecule appeared possible.

Also the availability of thin film methods for determining the oxygen-hemoglobin equilibrium curve (OEC) has made it possible to measure the curves rapidly with a more concentrated Hb solution. Therefore it was decided to measure the effects of 2,3-DPG on the OEC of sheep Hb A and Hb B using concentrations as close as possible to those found in red cells.

Further there have been no reports of interaction between inositol hexaphosphate (IP6 or phytic acid) with sheep hemoglobin. As this is the most powerful organic phosphate in its effect on the OEC (Benesch and Benesch, 1974) measurements of this action were also included.

The results reported here are preliminary as they do not cover a range of pH or chloride concentrations; nor have measurements yet been done in the presence of CO_2. Nevertheless they do show clearly that there are greater interactions between organic phosphates and sheep Hb than has been previously realized.

METHODS

Blood containing Hb A or Hb B was obtained by jugular venepuncture of lightly restrained adult sheep. Only one sheep with Hb A was available, and two sheep with Hb B were studied. Blood containing Hb F was obtained from indwelling catheters in fetuses in utero being prepared for other experiments in the laboratory of Professor Eugenie Lumbers of this School. The age from conception of the fetuses was: 1F, 140 days; 2F, 135 days; 3F, 118 days. The mothers of the first two fetuses had been treated with glucagon. The haemoglobins present were identified by isoelectric focussing on polyacrylamide ampholine gels and in the case of fetuses 1 and 2, where more than one Hb type was present, the proportions were determined by cutting out the relevant areas from the gel, eluting the Hb, and measuring of the total amount from each area by spectrophotometry at 415 nm.

The red cells were washed three times in 10-20 times their volume of physiological NaCl and then hemolyzed by 20 minutes in liquid nitrogen followed by thawing. To the hemolysates we added twice the volume of 50 mM bis tris in 100 mM NaCl containing no phosphate, 2,3-DPG, or inositol hexaphosphate (IP6 or phytic acid) to give the required additional phosphate concentration. Titration to the relevant pH was with 1.0 M or 0.1 M HCl or NaOH. The 2,3-DPG was the tris salt, and the IP6 was the sodium salt. Both were obtained from Sigma Chemical Co, St. Louis MO, U.S.A.

The oxygen equilibrium curves (OECs) were determined on a modified HEM-O-SCAN, in which a very thin film of blood or Hb solution is exposed to a changing PO_2 and the relation between PO_2 and O_2 saturation is obtained on an X-Y plotter. The main modifications were: (I) that the Hb film was between two layer of gas-permeable membrane, both sides being in contact with gas; and (II) the gas was admitted in intermittent pulses rather than continuously. The methods were more fully described by Holland et al (1988).

In the present experiments only Hb solutions were studied and no CO_2 was present in the gases used. The temperature was $39^{\circ}C$, the normal body temperature of sheep. The OECs were also plotted as log_{10} {(% saturation)/(100 - % saturation)} against log_{10} PO_2, this being the logarithmic form of the standard Hill equation. The Hill coefficient (n_H) was determined by the method of least squares over the middle linear part of the plot.

RESULTS

The results are shown in table 1. The effects of adding a high concentration of 2,3-DPG and a relatively low concentration of IP6 to sheep Hb B are also shown in figure 1, the Hill plots being shown in figure 2.

The P_{50} values without added phosphates show the expected values for Hb A and Hb B, the latter having a considerably lower affinity (higher P_{50}). Also the pure fetal hemoglobin (3,F) showed a high affinity (p_{50} = 11.0 Torr), while the affinity of the fetal lysates containing HbB (1,F, and 2,F) was somewhat lower.

The n_H values without added phosphates were in the normal range except for 2F (n_H = 2.20) and 3F (n_H = 2.30). The low n_H in 2F was to be expected because of the mixture of Hb types (F and B) with different P_{50}s but the low value of 2.30 (95% confidence limits: 2.23 to 2.36) in 3F, which had only fetal Hb, was surprising.

Addition of 2,3-DPG to give a 2 mM increase in its concentration had very little effect on the P_{50}. However when the 2,3-DPG addition gave an increase of 6 mM, the P_{50} was considerably raised. For Hb A at pH 7.21 - 7.22 the increase was by a factor of 1.59; and for Hb B at pH 7.08-7.10, the increase was by a factor of 1.41. The increase was less in the two fetal samples containing Hb B but in the pure Hb F sample the increase was by a factor of 1.51. It should be noted that the pH values were not the same in the different samples but the effect of 6mM additional 2,3-DPG in raising the P_{50} is clear. Addition of further 2,3-DPG (10 mM final increase) was done in the fetal samples and gave little change in P_{50}.

Inositol hexaphosphate (IP6) when added to give 2 mM total IP6 in the samples caused a considerable right shifting of the OEC of Hb A, Hb B, and Hb F (factors of 1.58, 1.76, and 1.84 respectively). For this calculation the control P_{50} values were corrected assuming the Bohr factor of -0.30 calculated for sample 1B. Although this is not necessarily applicable to Hb A and Hb F, the pH differences were small and the assumption could give very little error. Addition of further IP6 had little effect on P_{50} in the fetal samples. The fall in 3F may have been caused by the higher pH when 4.0 mM IP6 was added.

In five of the six blood samples, addition of 2,3-DPG had negligible effect on n_H. In the same five samples IP_6 caused a considerable decrease in n_H. The exception was the pure fetal Hb sample (3F) where both 2,3-DPG and IP6 caused a clear increase in n_H.

DISCUSSION

The results clearly show that both 2,3-DPG and IP6 in adequate concentrations give a marked right shift to the OEC of sheep Hbs A, B, and F. It is likely that the earlier reports of there being little or no

TABLE 1   OEC Data for Sheep Hbs A, B, and F at 39°C.
All in 50 mM bis tris, 100 mM NaCl, no CO_2.
$[Hb_4]$ = 1.0 – 1.5 mM solution. DPG is 2,3-DPG,
IP6 is inositol hexaphosphate.

Hb Type	pH	Additional phosphate	P_{50} (Torr)	n_H
A	7.22	0	26.9	2.83
A	7.23	0	27.3	2.90
A	7.29	DPG, 2 mM	27.7	2.75
A	7.21	DPG, 6 mM	42.7	2.80
A	7.19	IP6, 2 mM	44.2	2.30
1,B	7.10	0	41.4	2.75
1,B	7.28	0	36.5	2.67
1,B	7.08	DPG, 6 mM	58.4	2.61
2,B	7.22	0	37.4	2.72
2,B	7.29	DPG, 2 mM	40.3	2.61
2,B	7.06	DPG, 6 mM	65.7	–
2,B	7.19	IP6, 2 mM	68.2	2.14
1,F†	7.23	0	16.2	2.64
1,F	7.23	DPG, 6 mM	20.6	2.47 *
1,F	7.23	DPG, 10 mM	23.6	2.63
1,F	7.27	IP6, 2 mM	17.9	2.18
2,F†	7.28	0	19.4	2.26
2,F	7.24	DPG, 6.3 mM	27.7	2.32
2,F	7.29	DPG, 10.0 mM	26.4	2.39
2,F	7.23	IP6, 2.0 mM	28.9	2.08
2,F	7.28	IP6, 4.0 mM	25.6	2.07
3,F†	7.23	0	11.0	2.30
3,F	7.32	DPG, 6.3 mM	16.6	2.96
3,F	7.32	DPG, 10.0 mM	17.4	2.66
3,F	7.20	IP6, 2.0 mM	19.9	2.56
3,F	7.40	IP6, 4.0 mM	14.4	2.59

---

* $[Hb_4]$ = 0.95 mM

†   Fetal sample 1 contained 19% Hb B; 2,F contained 36% Hb B; 3,F
contained only Hb F.

n_H determined over middle range of saturation -- no significant
bend in Hill plot in this range

194

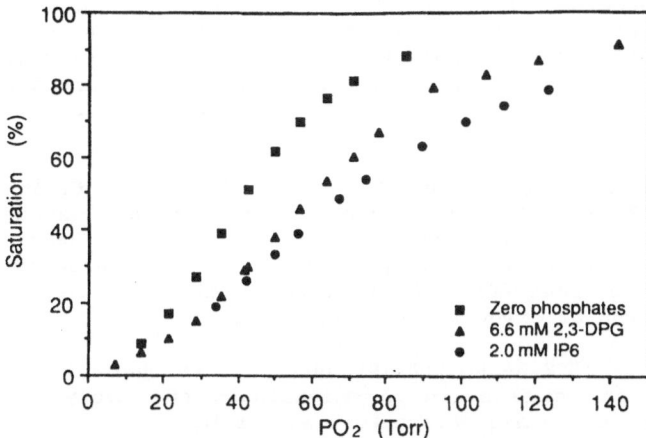

Figure 1   The oxygen-hemoglobin equilibrium curves of sheep Hb B without organic phosphate and with phosphates added to give concentrations as shown. [Hb$_4$] = 1.25 - 1.50 mM; buffer 50 mM bis tris, 100 mM NaCl; temperature 39°C.

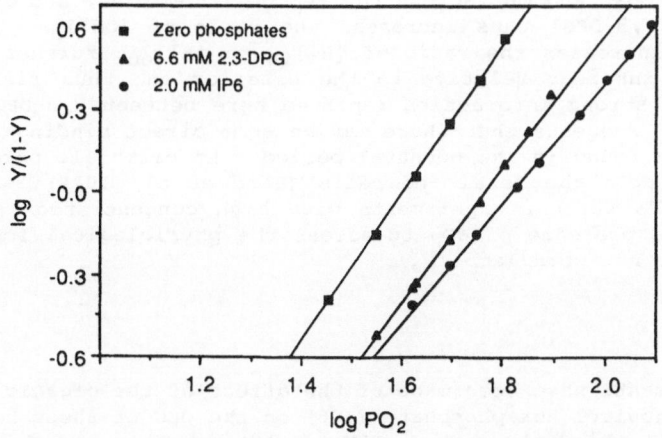

Figure 2   Hill plots of the OECs of sheep Hb B with and without added phosphates. The OECs are from Figure 1. Temperature = 39°C. The n_H value with added IP6 was significantly different from the control (p < 0.001)

interaction between sheep Hb and 2,3-DPG were due to the Hb being in too dilute a solution where the binding of 2,3-DPG and the Hb tetramer would, by the Law of Mass Action, be less than at the normal red cell concentration. [Hb_4] is normally about 5 mM in the red cell; [2,3-DPG] varies among species but total red cell [2,3-DPG] is 4-5 mM in man. In the work of Bauer and Jung (1975) the factor by which 6 mM 2,3-DPG increased P_{50} of dilute Hb solution (0.03 mM Hb_4) can be calculated approximately from their graph. It was 1.26 for Hb A and 1.23 for Hb B, each factor being less than half of the corresponding factor reported here.

Our finding of the interaction with IP6 is the first reported for sheep haemoglobins with this substance. Its decrease in the affinity of all three haemoglobins was expected but its effect in causing a lowering of n_H, an index of co-operativity was not. This decrease in n_H, suggests that IP6 has a powerful action to keep the tetramer in the deoxy conformation even as oxygenation is proceeding.

The effect of 2,3-DPG and IP6 on P_{50} shows that, despite any effect of the methionine side chain in maintaining the deoxy conformation, a further effect of organic phosphates is possible.

The physiological action that has been attributed to 2,3-DPG in sheep blood is to lower the oxygen affinity in the neonatal period. Within the first ten days of life the concentration of 2,3-DPG in sheep red cells has been reported as 10.5 mmol/litre of red cells by Baumann et al (1972), and as 32 µmol/g Hb and 28 µmol/g Hb by Bard et al (1976) and Noble et al (1983) respectively. These latter figures are equivalent to a 2.1 and 1.8 molar ratio of 2,3-DPG to Hb_4 and so give a 2,3-DPG level of about 10 mM, in agreement with the results of Baumann et al (1972). The action of 2,3-DPG in right-sihifting the OEC at this time is of value in adapting the lamb to life outside the uterus. It has been attributed to the effect of DPG on the Donnan ratio. As a non-diffusible anion, 2,3-DPG increases the Donnan ratio for distribution of diffusible anions such as Cl^-, HCO_3^- and OH^- across the red cell membrane, and any increase in red cell [2,3-DPG] thus increases the ratio of [OH^-]$_{out}$ to [OH^-]$_{in}$. This in turn increases the ratio of [H^+]$_{in}$ to [H^+]$_{out}$ further acidifying the red cell interior relative to the exterior and thus right-shifting the OEC. The strong interaction reported here between 2,3-DPG and sheep Hb A, B, and F suggests that there may be some direct binding at the high concentrations found in the neonatal period. At birth all three Hb types may be present in the fetal red cells (Bard et al, 1976). Thus it is important to do further experiments with high concentrations of 2,3-DPG and Hb_4 in the presence of CO_2 to assess the physiological importance of 2,3-DPG in the new-born lamb.

SUMMARY

Measurements have been made of the effect of the organic phosphate 2,3-DPG and inositol hexaphosphate (IP6) on the OEC of sheep hemoglobins A, B, and F. A thin film method (HEM-O-SCAN) was used and the hemoglobin concentration was about one quarter to one third of that normally found in red cells. All experiments were in bis-tris-saline buffer and in the absence of CO_2. At low concentrations of IP6 and at concentrations of 2,3-DPG similar to red cell concentrations, the OEC was right-shifted in all cases, P_{50} being increased by a factor of 1.58-1.84 by IP6 and by a factor of 1.41-1.59 by 2,3-DPG. The value of Hill n, the index of co-operativity was not generally changed by 2,3-DPG but was decreased by IP6.

196

ACKNOWLEDGEMENTS

This work was supported by a grant from the Australian Research Council (A.R.C.).

REFERENCES

Bard, H, Fouron, J.-C., Robillard, J.E., Cornet, A., and Soukini, M.A., 1978, Red cell oxygen affinity in fetal sheep: role of 2,3-DPG and adult hemoglobin, J. Appl. Physiol., 45:7-10.

Bard, H, Fouron, J.-C., Grathic, A.M., Soukini, M.A. and Cornet, A., 1976, The adaptation of the fetal red cells of newborn lambs to extrauterine life: the role of 2,3-diphosphoglycerate and adult hemoglobin, Pediat. Res., 10:823-825.

Baumann, R., Bower, C., and Rathschlag-Schaeffer, A.M., 1972, Causes of the postnatal decrease in oxygen affinity in lambs, Respir. Physiol., 15, 151-158.

Bauer, C., and Jung, H.D., 1975, A comparison of respiratory properties of sheep haemoglobin A and B, J. Comp. Physiol., 102:167-172.

Benesch, R.E. and Benesch, R., 1974, The mechanism of interaction of red cell organic phosphates with hemoglobin, Adv. Prot. Chem., 28:211-235.

Bunn, H.F., 1971, Differences in the interaction of 2,3-diphosphoglycerate with certain mammalian hemoglobins, Science, 172, 1049-1050.

Dayhoff, M., 1972, Atlas of Protein Sequence and Structure. Volume 5. Nat. Biomed. Res. Found., Silver Spring, MD, USA.

Holland, R.A.B., Rimes, A.F., Comis, A., and Tyndale-Biscoe, C.H., 1988, Oxygen carriage and carbonic anhydrase activity in the blood of a marsupial, the Tammar Wallaby (Macropus eugenii) during early development, Respir. Physiol., 73:69-86.

Noble, N.A., Jansen, C.A.M., Nathanielsz, P., and Tanaka, K.R., 1983, Mechanism of red cell 2,3-diphosphoglycerate increase in neonatal lambs, Blood, 61:920-924.

Perutz, M.F., and Imai, K., 1981, Regulation of oxygen affinity of mammalian haemoglobins. J. Mol. Biol., 136:183-191.

CARBON MONOXIDE BINDING IN A MODEL OF HEMOGLOBIN DIFFERS BETWEEN THE T AND

THE R CONFORMATION

Jacob P. Zock

Dept. of Physiology, School of Medicine
University of Groningen
Bloemsingel 10, 9712 KZ Groningen, The Netherlands

INTRODUCTION

Carbon monoxide impedes the transport of oxygen in blood by competitive binding to the oxygen binding sites on hemoglobin. The affinity of these sites for carbon monoxide is much greater than that for oxygen. The ratio between carbon monoxide affinity and oxygen affinity (M) is usually assumed to have a fixed value (Haldane's law), which is about 200. Reported values differ between authors(Douglas et al., 1912, Killick, 1936, Sendroy et al., 1929). However, this ratio is not a constant but depends on the level of saturation (Roughton, 1970) as well as on pH (Joels and Pugh, 1958). This means that simple procedures cannot be used to calculate the combined effects of simultaneous oxygen and carbon monoxide binding. Application of our mathematical model of hemoglobin (Zock, 1987) shows that this model can account for the above-mentioned phenomena in a straightforward way.

In 1912, Douglas, Haldane, and Haldane published quantitative data on the competitive binding of oxygen and carbon monoxide to hemoglobin under equilibrium conditions (Douglas et al, 1912). Their finding that the oxygen equilibrium curve is altered by the presence of carbon monoxide could be explained by the high affinity of carbon monoxide to hemoglobin, as was shown by J.B.S. Haldane (1912). This provided an explanation for the high toxicity of carbon monoxide. His analysis made clear that the presence of carboxyhemoglobin (HbCO) results in a shift of the overall oxygen equilibrium curve of hemoglobin to the left. The relation between oxygen affinity and carbon monoxide affinity is characterized by a more or less fixed ratio between oxygen pressure and carbon monoxide pressure at a certain value of saturation. This is known as Haldane's first law. Values of 224 and 290 were found for this ratio in two samples of human blood. Haldane's law implies that by multiplying either the oxygen or the carbon monoxide pressure with a suitable, but fixed factor, the equilibrium curves for oxygen and carbon monoxide can be made to exactly superpose on each other. Thus, taking M as a constant, independent of the state of oxygenation of the molecule, would result, theoretically, in equilibrium curves for oxygen and carbon monoxide which could be mapped onto each other by suitable scaling of the gas tension axis. Experimental curves are at variance with this conclusion as is shown in figure 1 taken from Roughton (1970).

Identical shapes of the equilibrium curves would furthermore indicate that binding of carbon monoxide and of oxygen have the same effect and that proton Bohr effects are equal after appropriate transformation of the carbon monoxide pressure. In this respect experimental results leave room to diffe-

Oxygen Transport to Tissue XII. Edited by J. Piiper *et al.*
Plenum Press, New York, 1990

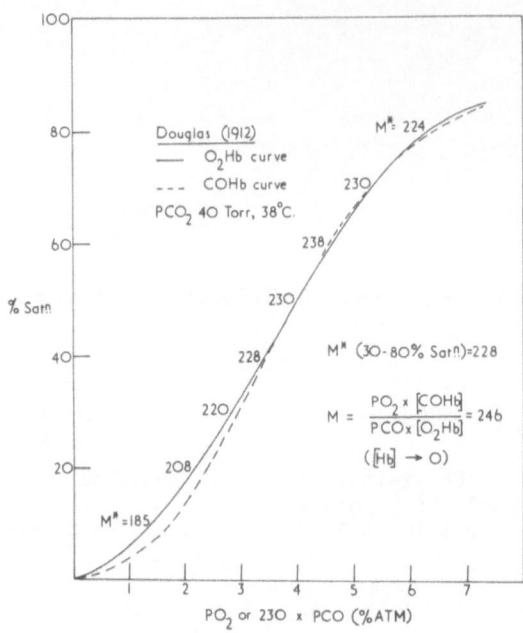

Fig. 1   Comparison of oxygen and carbon monoxide hemog-
          lobin equilibrium curves of human blood. (Repro-
          duced from Roughton (1970) with permission).

rent interpretations (Hlastala et al., 1976, Okada et al., 1976, Zwart et
al., 1984)

Recently, we developed a model which describes allosteric interactions in hemoglobin between oxygen, protons, carbon dioxide, and 2,3-DPG (Zock,
1987). This model could easily be extended as to become suitable for calculation of the effects of the competition between the binding of CO and O2 to
hemoglobin. Then the model enables one to study the effects of the presence
of CO on the relative amounts of oxyhemoglobin and carboxyhemoglobin under
a variety of conditions. It is possible to give the alpha and the beta chains
different affinities for oxygen and carbon monoxide or to assume that the
ratio between oxygen affinity and carbon monoxide affinity is different in
the R conformation from that in the T conformation. The first possibility
turned out to be immaterial for the explanation of the difference between
oxygen and carbon monoxide equilibrium curves. Results of the latter assumption are presented in this paper. They show that the model gives a good representation of the properties of hemoglobin with respect to the competitive
binding between oxygen and carbon monoxide. The model is based on subsequent
principles:
- The hemoglobin molecule is one of the two conformations T or R
- The four monomers in a molecule and their binding sites do not directly
  interact
- Interaction occurs through the equilibrium between the two conformations
- The conformations differ in their equilibrium constants for the groups
  formerly called 'oxygen labile' as well as in those for the oxygen binding
  sites themselves
- The equilibrium between the two conformations is determined by the difference in free energy between them
- Every binding of a substance influences the free energy according to the
  binding potential associated with its site
- Substances accounted for are oxygen, carbon monoxide, H+, carbon dioxide,
  and 2,3-DPG.

The effects of pH on oxygen and carbon monoxide affinity have already

200

been introduced in the model. The fact that both saturation and pH influence the value of M can be easily implemented by assuming that the ratio between the carbon monoxide affinity and the oxygen affinity differs between the two conformations, so that in the T conformation M is lower than in the R conformation. When used to simulate equilibration with gas mixtures containing various proportions of oxygen and carbon monoxide, the results obtained were comparable with experimental outcomes. Furthermore, insight can be gained as to how variation of the gas mixture influences the equilibrium between the conformations. Apart from the importance of a detailed understanding of the interactions in hemoglobin, a functional description of carbon monoxide binding can be very useful in physiological applications such as calculations concerning the diffusion capacity of the lung.

THEORY

The binding of the ligands O_2, CO, H^+, CO_2 and 2,3-DPG to hemoglobin are not independent: each one of these substances influences all the other equilibria and these in turn exert a reciprocal influence on the binding of that substance. Various mechanisms underlie this mutual influence. The first one is, evidently, the competition for one binding site between two or more substances; another one results from the property that hemoglobin molecules take one of two quaternary conformations. The partition over the two conformations itself forms a chemical equilibrium; its actual value is determined by the difference in Gibb's free energy between the two conformations. The binding sites of the above-mentioned substances have conformation-dependent affinities for their ligands, i.e. binding of a substance has, in either state, different effects on the free energy. On the one hand, this makes that ligand binding may result in a new partition between the two conformations; on the other hand, such a new partition means a different overall affinity for the ligands.

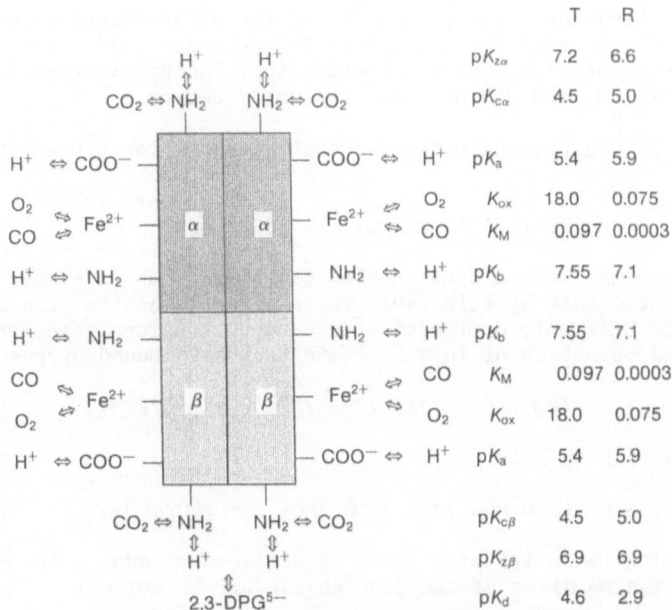

Fig. 2. Model representation of human hemoglobin on which the mathematical model is based. The conformation-dependent pK values used in the calculations are given on the right hand side.

201

There are indications that these two mechanisms do not give a complete description of the interactions. However, it leads to a consistent model which provides a fairly adequate description, useful for physiological calculations. The mathematical model is based on the molecule as given in figure 2. This molecule has one of two conformations or states: T or R. There are a number of binding sites each with two, conformation-dependent, pK values and associated with it one binding potential function (Wyman,1965). For each conformation the binding potentials of all the sites on the molecule add up to the total binding potential. The equilibrium between the two conformations is determined by the difference in free energy. There is no direct interaction between the binding sites: mutual influences result from the way each site contributes to the partition over the two conformations.

The model is clearly a strong simplification. The many binding sites at which H^+ binds are replaced by three Bohr groups per monomer. Two of them only bind H^+, the other one, representing the terminal valine, binds CO2 as well. Between these two valines on the beta chains one molecule of 2,3-DPG can be accommodated. Each chain has one binding site for oxygen. Various aspects of this model have been discussed elsewhere (Zock, 1987). As far as the interaction between the binding of carbon monoxide and of oxygen is concerned, this is introduced into the model by assuming competitive binding at the iron atoms of the four chains. The introduction of competition between carbon monoxide and oxygen into the equations of the model is straightforward. It means that in the expression which gives the contribution of the oxygen binding site to the binding potential, carbon monoxide binding is accounted for as well. The equations concerning oxygen and carbon monoxide binding are given below.

The only distinction made between these sites is the influence of the state on the affinities for oxygen and carbon monoxide and their ratio M; possible differences between oxygen binding to the α and the β chains are not implemented. Although there are strong indications that the two chain types differ in oxygen affinity these differences are immaterial to our argument that M must be different between the two states to explain differences in the shape between the oxygen and carbon monoxide equilibrium curves (Bishop and Gill, 1986). However, if needed and values available, this addition would only mean the change of eight parameters of the mathematical model.

Equations concerning the binding of H^+ are essentially similar to those for oxygen and carbon monoxide but because they include interactions and competition with CO_2 and 2,3-DPG they are more complex.

The part of the binding potential that accounts for oxygen and carbon monoxide is

$$P_{Fe,s} = RT \cdot \ln(1 + pO_2/kO_{2,s} + pCO/kCO_s) \tag{1}$$

In the equation the index S denotes the state which can be either T or R. It should be noted that at this level the binding to a site of a molecule in a certain state S is only dependent on O_2 and CO tensions. The fraction of sites belonging to molecules in this state that have bound oxygen is

$$fHbO_{2,s} = (pO_2/kO_{2,s})/(1 + pO_2/kO_{2,s} + pCO/kCO_s) \tag{2}$$

For carbon monoxide this is

$$fHbCO_s = (pCO/kCO_s)/(1 + pO_2/kO_{2,s} + pCO/kCO_s) \tag{3}$$

The total binding potential of a state is obtained by adding the contributions of all binding sites of the four chains of the molecule

$$P_s = \sum_{\alpha_1}^{\beta_2} P_{Fe,s} + P_{OTHER,s} \tag{4}$$

202

The partition depends on the difference in binding potential between the two states

$$F_R \quad = \quad 1/(1 + L_o \cdot \exp(P_T - P_R)/RT) \qquad (5)$$

and

$$F_T \quad = \quad 1 - F_R \qquad (6)$$

The fraction of all oxygen/carbon monoxide binding sites occupied by oxygen is

$$fHbO_2 \quad = \quad fHbO_{2,T} \cdot F_T + fHbO_{2,R} \cdot F_R \qquad (7)$$

and by carbon monoxide

$$fHbCO \quad = \quad fHbCO_T \cdot F_T + fHbCO_R \cdot F_R \qquad (8)$$

The total fractional occupation of these sites is the sum of (7) and (8)

$$s(O_2 + CO) \quad = \quad fHbO_2 + fHbCO \qquad (9)$$

The oxygen saturation is according to its definition

$$sO_2 \quad = \quad fHbO_2/(fHbO_2 + fHb) = fHbO_2/(1 - fHbCO) \qquad (10)$$

CALCULATIONS

Equilibration of a solution as contained in the erythrocyte, henceforth called erythrocyte fluid, was simulated by using the mathematical model under various assumptions concerning the conditions of this solution. First, it was assumed that pH was kept constant at either 7.2 or 6.8. Carbon dioxide tension was taken constant at 5.33 kPa. The hemoglobin (Hb_4) concentration was 5.0 mmol/L and total 2,3-DPG concentration 4.5 mmol/L.

Calculations were done for various values of pO_2 and pCO. Every combination leads to a new partition over the two states of the molecule and thus to a new distribution of 2,3-DPG between bound and free 2,3-DPG. This distri-

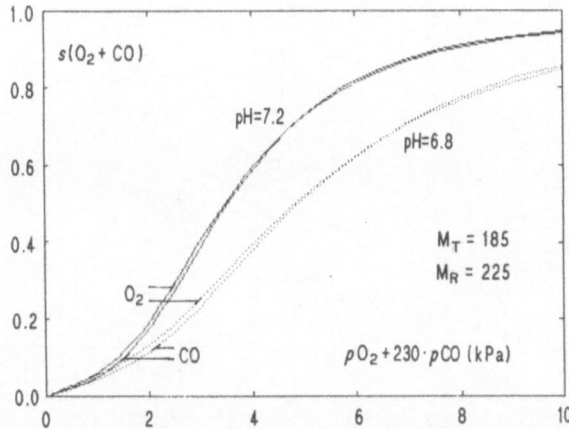

Fig. 3. Oxygen and carbon monoxide equilibrium curves of hemoglobin as calculated for two values of pH. The two curves are those of the pure gases. They form the limiting curves for mixtures of O_2 and CO. The value 230 was chosen to get results comparable with those in figure 1.

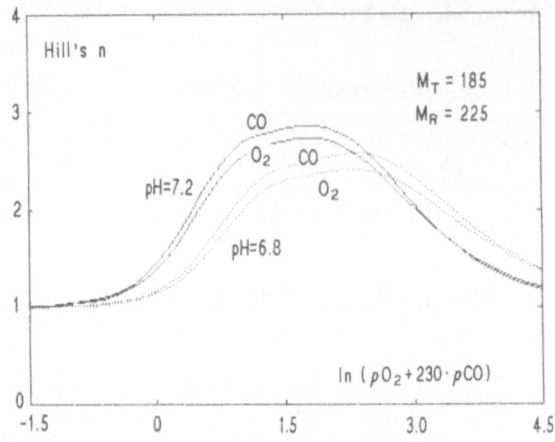

Fig. 4.   Hill's n calculated for the curves of figure 3.

bution was calculated by numerical approximation for solving the set of non-linear equations.

The calculations were all done with the ratios between oxygen and carbon dioxide affinity given in figure 1, i.e. M=185 for the T state and M=225 for the R state

RESULTS AND DISCUSSION

The calculations give the results of a simulated equilibration of erythrocyte fluid with O_2 and CO. Erythrocyte fluid was assumed to be kept at a fixed value of pH, either 7.2 or 6.8, as indicated in the figures.

Figure 3 shows the saturation curves of hemoglobin with O_2 and CO at pH values of 7.2 and 6.8. The calculated equilibrium of carbon monoxide is steeper than that of oxygen which is in agreement with Roughton (1970).

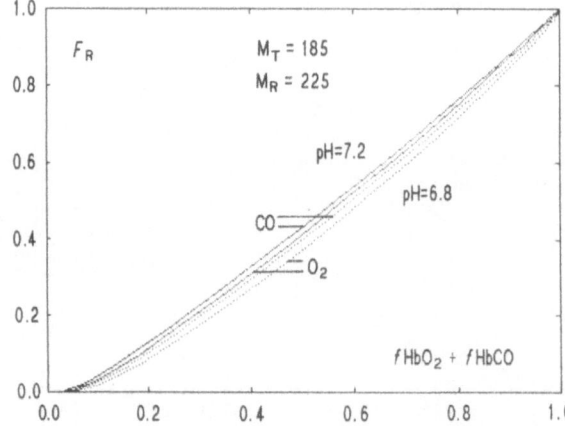

Fig. 5.   Fraction of molecules in the R state dependent
          on occupation with oxygen and carbon monoxide.
          The figure shows that apart from total occupati-
          on the partition depends on whether O_2 or CO
          is bound and on the pH of the solution.

204

The difference between the carbon monoxide curve and that of oxygen is less pronounced than found by Roughton. The difference between the calculated curves could be made larger by taking larger differences between the values of M. Calculations with $M_T=125$ and $M_R=250$ (not shown) gave larger differences indeed and curves closer resembling those of figure 1. These values of M_T and M_R are in line with values found by Di Cera et al. (1987). They ascribe these differences to differences in affinity ratio's between the α and the β chains. However, at least for one of the chain types this ratio must also differ between the two conformations in order to get different shaped oxygen and carbon monoxide equilibrium curves. Otherwise, these curves can always be made to superpose on each other by multiplication of one of the two tension by a fixed factor.

The form of the curves can be quantified by calculation of Hill's n. This was done for the saturation curves of figure 3. The results are shown in figure 4. The CO saturation curves are steeper than the O_2 curves because the transition from the T state to the R state in the presence of CO alone is different from that in the presence of O_2 alone (fig. 5).

The fact that the transition from molecules being in the T conformation to molecules in the R conformation depends on whether oxygen or carbon monoxide is bound, makes that the quantity

$$M = ([HbCO] \cdot pO_2)/([HbO_2] \cdot pCO) \tag{11}$$

used to quantify the affinity ratio is not uniquely defined but depends on the composition of the mixture of oxygen and carbon monoxide (Haldane, 1912).

To study the course of this quantity under the assumptions of the model and equilibration taking place with mixtures of carbon monoxide and oxygen, two types of experiments were simulated. In the first type a fixed ratio was assumed to exist between the pressure of oxygen and that of carbon monoxide, while varying the oxygen pressure. The curves were calculated for several ratios between oxygen and carbon monoxide pressure. The two curves for each pH in fig. 6 belong to the extreme mixtures simulated: one mixtures almost completely consisting of carbon monoxide, with only 0.1% of O_2, and one almost completely consisting of oxygen, with only 0.1% of CO. Less extreme mixtures gave curves lying in between these two curves. The figure shows an increase of M with saturation. Apart from saturation, in the model this value is also influenced by pH (Joels & Pugh, 1958).

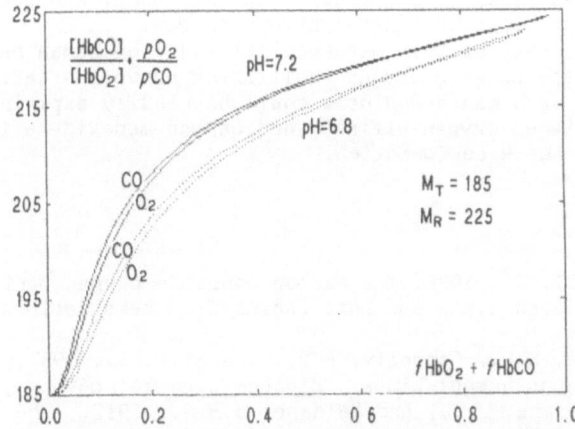

Fig. 6. Influence of the composition of the gas mixture with which hemoglobin equilibrates and pH on the value of M as dependent on the fractional occupation of hemoglobin with O_2 and CO.

205

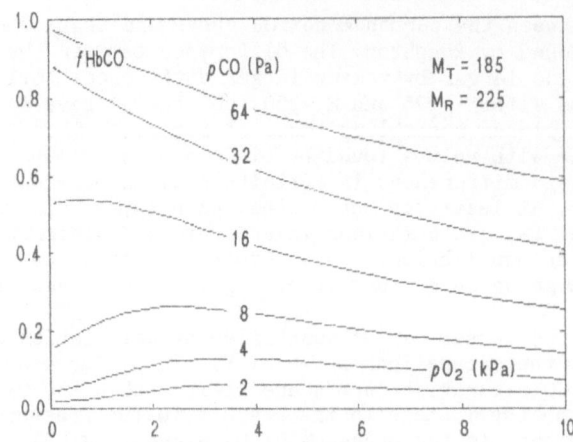

Fig. 7.   The fractional amount of HbCO as a function
           of oxygen tension at various fixed tensions
           of CO.

In the second type of simulation a constant partial pressure of carbon
monoxide was assumed to exist in combination with a varying pressure of
oxygen. In fig. 7 the calculated fractional occupation of the oxygen binding
sites with carbon monoxide is given as dependent on the oxygen pressure.
These curves show the enhanced carbon monoxide binding on an increased oxygen
tension at a pCO of about 10 Pa and pO_2 of about 2 kPa, a phenomenon already
found in the experiments by Douglas et al. (1912). No special effects of
different values of M_T and M_R were found.

SUMMARY

The ratio between carbon monoxide affinity and oxygen affinity (M) is
usually assumed to have a fixed value (Haldane's law), which is about 200.
However, this ratio is not a constant but depends on the level of saturation
(Roughton, 1970) as well as on pH (Joels and Pugh, 1958). This means that
simple procedures cannot be used to calculate the combined effects of simultaneous oxygen and carbon monoxide binding. Application of our mathematical
model of hemoglobin (Zock, 1987) shows that this model can account for the
above-mentioned phenomena in a straightforward way.
Within the context of this mathematical model of human hemoglobin the
differences in shape between oxygen equilibrium curve and carbon monoxide
equilibrium curve of human hemoglobin could be readily explained by assuming
that the ratio between oxygen affinity and carbon monoxide affinity differs
between the T and the R conformation.

REFERENCES

Bishop, G. and Gill, S., 1986, The carbon monoxide-oxygen partition coeffi-
      cient of isolated alpha and beta chains from hemoglobin A$_o$, <u>Biopolymers</u>
      25: 1381.
Di Cera, E., Doyle, M.L., Connelly, P.R., and Gill, S., 1987, Carbon monoxide
      binding to human hemoglobin A$_o$, <u>Biochemistry</u> 26: 6494.
Douglas, C.B., Haldane, J.S., and Haldane, J.B.S., 1912, The laws of combina-
      tion of hemoglobin with carbon monoxide and oxygen, <u>J. Physiol. (London)</u>
      44: 275.

Haldane, J.B.S., 1912, The dissociation of oxyhemoglobin in human blood during partial CO poisoning, J.Physiol. (London) 45: XXII.

Hlastala, M.P., H.P. McKenna, R.L. Franada, and Detter, J.C., 1976, Influence of carbon monoxide on hemoglobin-oxygen binding, J.Appl. Physiol. 41: 893, 1976.

Joels, N., and Pugh, L.G.C.E., 1958, The carbon monoxide dissociation curve of human blood. J. Physiol. (London) 142: 63.

Killick, E.M., 1936, The acclimatization of the human subject to atmospheres containig low concentrations of carbon monoxide. J.Physiol. 87: 41.

Okada, Y., Tyuma, I., Ueda, Y., and Sugimoto, T., 1976, Effect of carbon monoxide on the equilibrium between oxygen and hemoglobin. Am.J. Physiol. 230: 471.

Roughton, F.J.W., 1970, The equilibrium of carbon monoxide with human hemoglobin in whole blood. Ann. NY Acad. Sci. 174: 177.

Sendroy, J., Liu, S.H., and Van Slyke, D.D., 1929, The gasometric estimation of the relative affinity constant for carbon monoxide and oxygen in whole blood at 38 °C. Am. J. Physiol. 90: 511.

Wyman J., 1965, The binding potential, a neglected linkage concept. J.Molec. Biol. 11: 631.

Zock, J.P., 1987, Mathematical model of the physiological properties of human hemoglobin. Proc. Kon. Ned. Acad. Wet. C90: 493.

Zwart, A., Kwant, G., Oeseburg, B., Zijlstra, W.G., 1984, Human whole blood oxygen affinity: effect of carbon monoxide, J. Appl. Physiol. 57: 14.

CARBON MONOXIDE EQUILIBRIUM CURVE OF HUMAN UMBILICAL CORD BLOOD

D. Schuwey, A. Tempini and P. Haab

Dept. of Physiology, University of Fribourg
CH-1700 Fribourg, Switzerland

INTRODUCTION

The presence of carbon monoxide (CO) in blood may be an important factor impairing tissue oxygenation because a) it decreases the concentration of functional hemoglobin and b) it increases the O2 affinity of the functional hemoglobin, both effects depending critically upon the amount of CO bound to hemoglobin (HbCO) and consequently upon the CO partial pressure (PCO) at which blood has been equilibrated. Because CO is suspected to have deleterious effects on fetal growth (Longo, 1987), we have investigated the CO affinity of the fetal blood by determining the CO dissociation curve of umbilical cord blood with the aim of comparing its position and shape with that of maternal blood. Our interest was prompted by the problem of the foeto-maternal CO equilibration.

A large survey at the beginning of this decade (Bureau et al., 1983) had shown that HbCO of umbilical cord blood (UCB), measured spectro-photometrically, was 4 to 5% higher than that of the maternal blood, both being sampled immediately after birth. The HbCO difference was found to be the same whether the mother smoked or not, i.e. the difference was independent of the absolute value of HbCO. Such results suggested that CO affinity of the UCB was larger than that of adult blood. However, it was shown (Cornelissen et al., 1983) that spectrophotometric methods may yield artificially high values of HbCO when applied to UCB samples and not corrected for the presence of fetal hemoglobin. Thus, the divergence with previous data (Longo, 1977), showing that HbCO is practically the same in UCB and in maternal blood, may not exist. Nonetheless, identity between maternal and UCB HbCO raises the question of knowing whether the CO affinity of UCB is indeed the same as that of adult maternal blood. If so, UCB would have binding properties for CO which

Oxygen Transport to Tissue XII, Edited by J. Piiper *et al.*
Plenum Press, New York, 1990

should be different from those for oxygen, since it is well documented that the O2 affinity of UCB is much higher than that of adult blood.

METHODS AND MATERIAL

UBC and maternal blood samples were obtained via umbilical artery by placenta squeezing and by venipuncture respectively. They were taken shortly after delivery and stored in CPDA (citrate, phosphate dextrose adenine) with adjunction of garamycine to prevent germ proliferation. After correction for pH 7.4 the samples were tonometered at 37°C for 4 hours in a bubble tono-meter capable of equilibrating CO in blood within 3 hours with minimal hemolysis and methemoglobin formation. The efficiency of the tonometer was checked. Fig. 1 compares the efficiency of the bubble tonometer we used to that of usual rotating tonometers. To this end both tonometer types were initially filled with 5 ml of blood fully saturated with CO and equilibrated with pure nitrogen. It is seen that in the bubble tonometer 99.5% of decarboxygenation is obtained after 180 min whereas rotating tonometers would be useless because of being much too slow (Spahr et al., 1986).

The tonometry gas mixtures were moistened at 37°C and contained various amounts of CO with 6% CO_2 (PCO_2=40Torr) in nitrogen; absence of oxygen was carefully controlled and after 3 hours HbO_2 was always smaller than 1%. pH was adjusted every hour at 7.4 with $NaHCO_3$.

CO contents of the equilibrated blood samples were determined by gas chromatographic analysis of the extracted gases. Hbtot was measured by the cyanmethemoglobin method and the concentration of HbF by the procedure of Betcke (1977). 2-3 DPG was measured according to Kiesov and Bless (1973).

Saturations were calculated assuming a Hüffner factor of 1.39 for the functional hemoglobin, Hbtot minus methemoglobin.

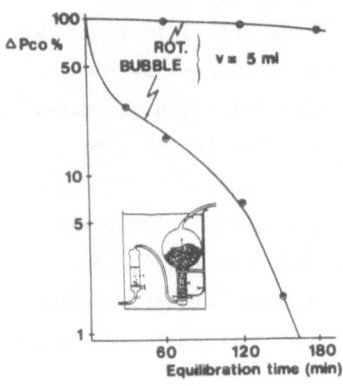

Fig. 1   Efficiency of the bubble and of the rotating tonometers for blood CO equilibration.

210

RESULTS

The results are shown on Fig. 2 on which the data obtained on UCB and on adult blood have been plotted on the same graph. Visual examination shows that both curves coincide exactly. The insert give some basis for quanti-tative comparison : the $P_{50}CO$ are identical as are the Hill coefficients n. These two values indicate that both the position and the shape are the same for the two curves. It should be noted that the 2-3 DPG/Hb_4 ratios are larger than unity, a feature that can result from the long duration tono-metry. However a difference between adult and UCB 2-3 DPG concentrations was observed already before the start of tonometry.

DISCUSSION

Our data raise two important questions : first, why is the CO affinity in the UCB not different from that of adult blood like the one for O_2 is ? Second, should not the HbCO concentrations in maternal and UCB be the same at the moment of delivery ?

As far as the first question is concerned, our findings suggest that the factors explaining the feto-maternal affinity differences for O_2 do not apply to carbon monoxide. For O_2, the difference has been clearly attributed to the absence of sensitivity of fetal hemoglobin toward the effect of 2-3 diphosphoglycerate. In whole blood this has been shown by Bauer et al. (1969), Duc and Engel (1969) and later by Bursaux et al. (1979). On hemoglobin solutions it was observed that the affinity of stripped fetal hemoglobin was much less sensitive to these phosphates than human hemoglobin A (Tyuma and Shimizu, 1970), a feature related to the fact that 2-3 DPG has a binding site on the β chains of HbA and that in HbF the β chains are

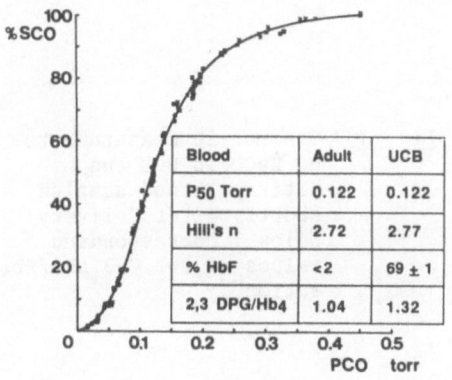

Fig. 2 Comparison of UCB (squares) and adult blood (crosses) CO equilibrium curves at pH = 7.40 and PCO_2 = 40 Torr. 2-3 DPG/Hb_4 ratios obtained at the end of tonometry.

Blood	Adult	UCB
P50 Torr	0.122	0.122
Hill's n	2.72	2.77
% HbF	<2	69 ± 1
2,3 DPG/Hb4	1.04	1.32

replaced by γ chains which have only a small affinity for the phosphates. If the same situation would prevail for CO one should expect the $P_{50}CO$ to be smaller in UCB than in adult blood to an extent which should be in relation with the concentration of HbF in UCB. Thus, our data indicate that the affinity for CO is differently influenced by 2-3 DPG. Unfortunately there are no studies dealing specifically with the effects that heterotropic agents have on UCB CO equilibrium curves, and our conclusion is only indirect.

COEC curves are more difficult to obtain than OEC ones because of the long durations of tonometry required ; with the method of bubble tonometry it was controlled in many instances that equilibrium was reached after four hours. Our curves for adult and maternal blood are essentially the same as those of Joels and Pugh (1958) ; the $P_{50}CO$ of these authors was 0.125 Torr, a figure which does not differ from our value ; the same appears to hold concerning the cooperativity since the value of Hill's coefficient n are also identical. We observed that the 2-3 DPG/Hb_4 ratio slowly increased from 1.17 to 1.32 during tonometry, but the effect of this factor cannot be compared with Joels and Pugh's data since those were obtained before the effects of glycerophosphate were acknowledged in 1967.

We are not aware of any carbon monoxide equilibration curve established on umbilical cord blood, but Engel et al. (1969) measured on this blood the $CO-O_2$ relative affinity factor, M, as defined by Haldane :

$M = HbCO/HbO_2 \cdot PO_2/PCO$

They found that this factor was about 20% lower than in the adult blood, i.e. 180 versus 220 for adult blood. Since we found that the CO affinity was the same in UCB and adult blood, our data also suggest that the 20% decrease in M could be fully accounted for by the difference in O_2 affinity. Hence the $P_{50}O_2$ should be 20% lower in the umbilical cord than in the adult blood. Taking 27 Torr for adult blood $P_{50}O_2$, a value of 21.5 should be expected in the UCB. Such a value is 1-2 Torr lower than that found experimentally (Bauer et al., 1969 ; Duc and Engel, 1969 ; Bursaux et al. (1979).

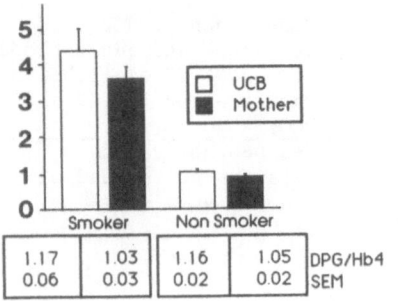

Fig. 3 Gasometric measurements of HbCO in UCB and maternal blood sampled shortly after delivery; below : corresponding values of the 2-3 DPG/Hb_4 ratios.

212

As far as the question of the feto-maternal HbCO concentration ratio is concerned, our data on CO equilibrium curves as well as those of Engel et et al. (1969) do not allow us to predict that these differences should be large. In small series of subjects, we measured HbCO in UCB and in maternal blood using the gas chromatographic technique. We compared data obtained in 8 smoking mothers to those of non smokers together with those of the UCB of their children. The results are presented in Fig. 3 which shows that in non smokers HbCO amounts to about 1% in both maternal and umbilical blood with no significant difference between the two. In the smokers group the absolute value of HbCO are about 4% higher in both mothers and children ; however HbCO is significantly higher in UCB, the UCB/maternal ratio being 1.1.

These data are in full accordance with the earlier data of Longo (1977). They also indicate the necessity of correcting previous data obtained by spectrophotometry as suggested by Cornelissen (1983). In the other hand the fact that the UCB HbCO values are 10% higher than those of the smoking mothers can easily result from a lack of equilibrium between mother and child, the CO elimination being necessarily slower in the fetus (Hill et al., 1977).

CONCLUSION

The shape and the position of the carbon monoxide equilibrium curve of human umbilical cord blood have been found identical to those of adult blood. From this it can be concluded that :

a)   the affinity of umbilical blood for CO is not influenced by heterotropic ligands, mainly 2-3 DPG, in the same manner as O_2 affinity ;
b)   the relative CO-O_2 affinity of umbilical cord blood is larger than that of adult blood only because its O_2 affinity is lower ;
c)   at equilibrium no difference in CO concentration should be expected between umbilical cord and maternal blood.

ACKNOWLEDGEMENT
Study supported by a grant of the Swiss association of cigarettes manufacturers.

REFERENCES

Bauer, C., Ludwig, M., Ludwig, I., Bartels, H., 1969, Factors governing the oxygen affinity of human adult and foetal blood, Respir. Physiol., 7:271-277.
Betke, K., Marti, H. R., Schlicht, L., 1959, Estimation of small percentages of foetal haemoglobin, Nature, 184:1877.
Bureau, M. A., Shapcott, D., Berthiaume, Y., Monette, J., Blouin, D., Blanchard, P., Begin, R., 1983, Maternal cigarette smoking and fetal oxygen transport: a study of P50,2,3-diphosphoglycerate, total hemoglobin, hematocrit, and type F hemoglobin in fetal blood, Pediatrics, 72:22-26.

Bursaux, E., Poyart, C., Guesnon, P., Teisseire, B., 1979, Comparative effects of CO_2 on the affinity for O_2 of fetal and adult erythrocytes, Pflügers Arch., 378:197-203.

Cornelissen, P. J. H., Van Woensel, C. L. M., Van Oel, W.C., De Jong, P. A., 1983, Correction-factors for hemoglobin derivatives in fetal blood as measured with the "IL 282" CO-oximeter, Clinical Chemistry, 29(8):1555-1556.

Duc, G., Engel, K., 1969, Effect of 2,3-DPG concentration on hemoglobin-oxygen affinity of whole blood, Scand. J. Clin. Labor. Invest., 24:405-412.

Engel, R. R., Rodkey, F.L., O'Neal, J. D., Collison, H. A., 1969, Relative affinity of human fetal hemoglobin for carbon monoxide and oxygen, Blood, 33(1):37-45.

Hill, E. P., Hill, J. R., Power, G. G., Longo, L. D., 1977, Carbon monoxide exchanges between the human fetus and mother : a mathematical model, Am. J. Physiol., 232(3):H311-H323.

Joels, N., Pugh, L. G. C. E., 1958, The carbon monoxide dissociation curve of human blood, J. Physiol. Lond., 142:63-77.

Kiesow, L. A., Bless, J. W., 1973, The enzymic determination of 2,3-diphosphoglycerate in a crude cell extract (UV method), Anal. Biochem., 51:91-96.

Longo, L. D., 1977, The biological effects of carbon monoxide on the pregnant woman, fetus, and newborn infant, Am. J. Obstet. Gynecol., 129:69-103.

Longo, L. D., 1987, Respiratory gas exchange in the placenta, in Handbook of Physiology The Respir. system, sect 3:351.

Spahr, I., Tempini, A., Haab, P., 1986, Comparison of rotating and bubble tonometers for their efficienty, Funktionsanalyse biologischer Systeme, 16:23-32.

Tyuma, I., Shimizu, K., 1970, Effect of organic phosphates on the difference in oxygen affinity between fetal and adult human hemoglobin, Federation proceedings, 29(3):1112-1114.

214

EFFECTS OF SO_2 AND pH ON BLOOD-GAS PARTITION COEFFICIENTS OF INERT GASES

Kazuhiro Yamaguchi, Masaaki Mori, Akira Kawai, Kohichiro Asano, Tomoaki Takasugi, Akira Umeda and Tetsuro Yokoyama

Department of Medicine, School of Medicine, Keio University Tokyo, 160 Japan

INTRODUCTION

A potential effect of saturation of hemoglobin with O_2 (SO_2) as well as of pH in the blood on blood-gas partition coefficients (λ = blood/gas concentration ratio in equilibrium) for certain inert gases has been postulated by several authors (Steen, 1963; Robertson et al., 1986). Since such variation in λ might affect inert gas exchange in the lung, the present study systematically reinvestigated SO_2 and pH dependence of λ using human blood.

Furthermore, incorporating the experimental findings on λ, the procedure of multiple inert gas elimination (Yokoyama and Farhi, 1967; Wagner et al., 1974) while breathing room air was performed in order to estimate the effects of variations of λ due to either SO_2 or pH on gas exchange of the inert gases in the lung. The data obtained by the procedure of multiple inert gas elimination were quantitatively analyzed with a numerical inversion technique of Evans and Wagner (1977) but allowance was made for λ of the particular inert gas to vary as blood SO_2 and pH varied with the exchange of O_2 and of CO_2 in the pulmonary capillaries.

METHODS

Measurements of inert gas solubilities

The extraction method of Wagner et al. (1974) was used to determine the dependence of blood-gas partition coefficients for inert gases, λ, on SO_2 as well as on pH in the blood. Blood samples were repeatedly drawn from three healthy adults and were equilibrated for 30 min. in a tonometer with a gas mixture containing SF_6, ethane, cyclopropane, halothane, diethyl ether and acetone, each at a fractional concentration in the gas phase of about 200 ppm. SO_2 values in the samples were kept constant at either zero or 1.0 by altering the O_2 concentration in the equilibrating gas mixture. To achieve 100% SO_2, 30% O_2 was used. pH in the sample was adjusted to 7.2, 7.4 or 7.8 by addition of CO_2 ranging from 1 to 10%. After equilibration, a 5 ml sample was anaerobically transferred into a 50-ml ungreased, gas-tight glass syringe to which was introduced a gas mixture with O_2 and CO_2 at the same concentration as the equilibrating mixture but free of inert gases as described above. Thereafter repeated equilibration up to three times was performed by rotating the syringe in a water bath at $37^{\circ}C$ for 30 min.

Oxygen Transport to Tissue XII, Edited by J. Piiper *et al.*
Plenum Press, New York, 1990

Quantitative analysis of gases released from the blood was made by a gas chromatograph equipped with a detector of flame ionization. Only SF_6 was measured by an electron capture detector. In addition to the measurements on six foreign inert gases, influence of SO_2 and of pH on λ of the physiological inert gas, N_2 was examined. N_2 was measured by the gas chromatograph having a thermal conductivity detector.

Multiple inert gas elimination procedure

Incorporating the experimental data on the dependence of λ upon SO_2 and pH, the multiple inert gas elimination method was applied to patients either with interstitial lung disease of various etiologies or with varied clinical manifestations of chronic obstructive pulmonary disease. A Swan-Ganz catheter was inserted in supine position and was advanced into the pulmonary artery. The patient was allowed to breathe room air during the infusion of solution containing the six foreign gases (SF_6, ethane, cyclopropane, halothane, diethyl ether and acetone). In steady state conditions, 25 ml samples of both arterial and mixed venous blood were taken slowly at a uniform rate over approximately 1 min. 20 ml sample of mixed expired gas was collected simultaneously. Because of high solubility of halothane, ether or acetone in plastics and water, a special heated flow-through system (Wagner et al., 1974) was constructed for collecting the mixed expired gas. PO_2, PCO_2 and pH in blood were measured with electrodes and concentrations of O_2 and CO_2 in gas samples with the Scholander apparatus, while the concentrations of the six foreign gases and N_2 were determined by gas chromatography as described above. Additionally, contents of total hemoglobin, methemoglobin and 2,3-diphosphoglycerate (DPG) were determined using a spectrophotometric method in order to establish the accurate dissociation curves for O_2 and CO_2. Total ventilation, $\dot{V}_E$ was recorded using a calibrated respirometer and cardiac output, $\dot{Q}_T$ was determined by thermodilution.

Fundamental equations for inert gas exchange

Based both on the conservation of mass and on the assumption that alveolar-capillary equilibration is complete for the inert gases, the relationship among alveolar (P_A), end-capillary (P_c') and mixed venous ($P_{\bar{v}}$) partial pressures of the gas in a homogeneous lung unit is expressed by:

$$P_A/P_{\bar{v}} = P_c'/P_{\bar{v}} = \lambda_{\bar{v}}/(\lambda_c' + \dot{V}_A/\dot{Q}) \tag{1}$$

where $\dot{V}_A/\dot{Q}$ is ventilation-perfusion ratio in the unit. λ_c' and $\lambda_{\bar{v}}$ are blood-gas partition coefficients of the inert gas for end-capillary and mixed venous conditions, respectively. In case that λ_c' is equal to $\lambda_{\bar{v}}$, eq. (1) is simplified as:

$$P_A/P_{\bar{v}} = P_c'/P_{\bar{v}} = \lambda/(\lambda + \dot{V}_A/\dot{Q}) \tag{2}$$

which is the original equation derived by Farhi (1967) for steady-state inert gas elimination in the homogeneous lung with complete diffusional equilibrium. Since λ_c' may not be identical with $\lambda_{\bar{v}}$ due to its SO_2 and pH dependence (see results), eq. (1) instead of eq. (2) was taken to be the fundamental formula in the present analysis.

In a real lung which represents a nonhomogeneous distribution of $\dot{V}_A/\dot{Q}$, overall retention (R) and excretion (E) are expressed as:

$$R = P_a/P_{\bar{v}} = \sum_{J=1}^{N} \lambda_{\bar{v}}/(\lambda_c' + \dot{V}_A/\dot{Q})_j * q_j \tag{3}$$

$$E = P_E/P_{\bar{v}} = \sum_{J=1}^{N} \lambda_{\bar{v}}/(\lambda_c' + \dot{V}_A/\dot{Q})_j * v_j \tag{4}$$

216

where N is the total number of gas exchange units in the lung, while q_j and v_j represent fractional perfusion and ventilation of each unit, respectively.

Representative distribution of $\dot{V}_A/\dot{Q}$

In order to quantitatively analyze the potential effects of SO_2 and pH on the estimation of $\dot{V}_A/\dot{Q}$ distribution based on inert gas exchange, distribution of perfusion ($\dot{Q}$) as a function of $\dot{V}_A/\dot{Q}$ was determined in a lung model consisting of 50 compartments with the numerical inversion technique described by Evans and Wagner (1977). Firstly, $\dot{Q}$ distribution was studied on the assumption that λ value of the indicator gas was kept constant throughout the lung, the procedure being the standardized approach to assessing $\dot{V}_A/\dot{Q}$ inequality in the lung. Secondly, taking eqs. (1) and (3) as kernel formulae, thus allowing λ to differ among the lung units, $\dot{Q}$ distribution was again determined and compared to that obtained with the first procedure. In this case, O_2 and CO_2 were assumed to reach full equilibration between alveolar gas and pulmonary capillary blood. This procedure permits one to estimate the λ values of inert gases at the end-capillary point of each lung unit, which may appreciably differ from those at the mixed-venous point. For this computation, use was made of the O_2 dissociation curve data provided by Severinghous (1979) including the CO_2 Bohr and DPG effects, both of which might be influenced by SO_2 (Yamaguchi et al., 1988). The algorithms proposed by Kelman (1967), combined the relationship between pH and PCO_2 given by Von Mengden et al. (1969), were used for obtaining the effective CO_2 dissociation curve.

RESULTS

Blood-gas partition coefficients for inert gases

Among the gases studied, the λ of ethane, cyclopropane, halothane and diethyl ether was 4-10% smaller in the oxygenated blood than in the deoxygenated blood, indicating a potential effect of Hb saturated with heme ligands on λ. Other gases, including SF_6, acetone and N_2, were not influenced by SO_2 in the blood (Figure 1 and Table 1).

With increasing pH in the blood, an increase in λ was found for ethane and a decrease for halothane. The other five gases showed no statistically significant dependence of λ on blood pH (Figure 2 and Table 1).

Effects of SO_2 and pH on inert gas exchange

In Figure 3 the predicted λ values of end-capillary blood in the lung units with $\dot{V}_A/\dot{Q}$ ranging from zero to 100 are shown. The calculation was made assuming that change of λ is related in a linear fashion to that of either SO_2 or pH. In case of the gases influenced by both SO_2 and pH, the effects derived from these factors were taken to be additive. Ethane represented a considerable variation in its λ among different $\dot{V}_A/\dot{Q}$ units, the value at the lung unit with $\dot{V}_A/\dot{Q}$ of 100 being 10% larger than that of mixed venous blood. On the other hand, λ of halothane decreased as the $\dot{V}_A/\dot{Q}$ ratio increased, maximally falling by 20%, as compared to the estimate in mixed venous blood. Cyclopropane and ether diminished their λ in the units with $\dot{V}_A/\dot{Q}$ more than 0.1 but the changes were fairly small, i.e. 1% (cyclopropane) or 3% (ether) lower than that of mixed venous blood. Neither SF_6 nor acetone showed any significant difference of the λ value among varied $\dot{V}_A/\dot{Q}$ units because of little dependence of their λ on SO_2 and pH.

The effects of SO_2 and of pH on the efficiency of inert gas exchange in a homogeneous lung unit were appraised in terms of retentions defined in eqs. (1) and (2). Although four gases, ethane, cyclopropane, halothane and ether, were characterized by an appreciable variation in λ between end-

217

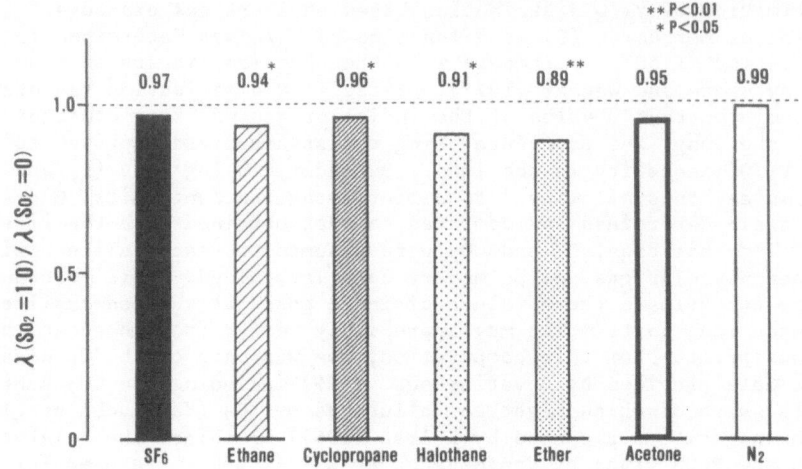

Fig. 1. Dependence of λ on hemoglobin saturation with O_2 (SO_2).

Table 1. Effects of SO_2 and pH on λ.

Gases	$\Delta\lambda/\Delta SO_2$	$\Delta\lambda/\Delta pH$
SF_6	-0.0003	-0.002
Ethane	-0.006^*	0.01^*
Cyclopropane	-0.02^*	0.05
Halothane	-0.25^*	-0.48^{**}
Ether	-1.55^{**}	-1.89
Acetone	-14.38	42.97
N_2	-0.0002	0.0005

**: $P<0.01$,   *: $P<0.05$

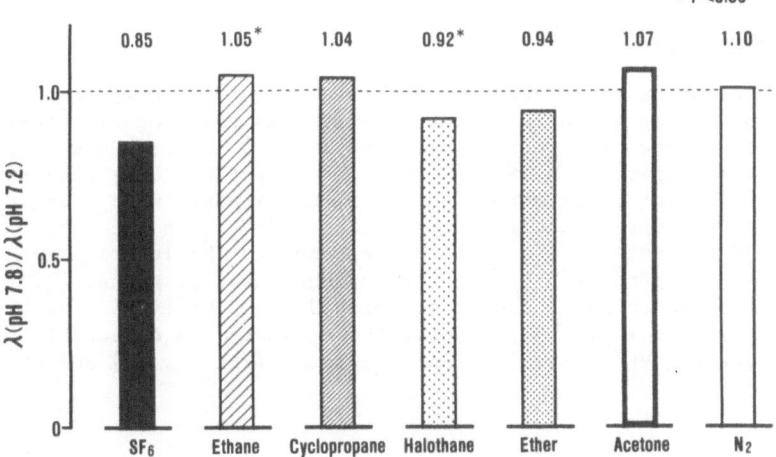

Fig. 2. Dependence of λ on blood pH.

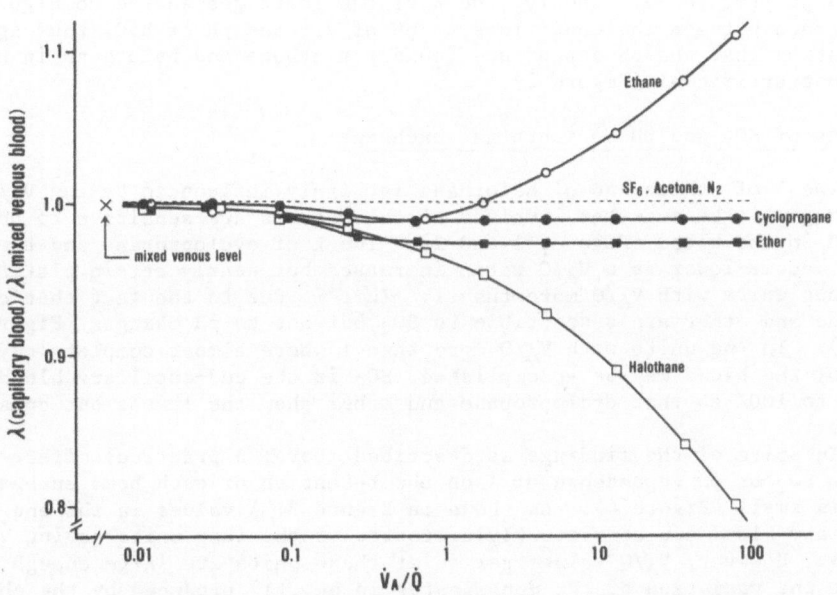

Fig. 3. Variation of λ at varied $\dot{V}_A/\dot{Q}$ units.

capillary and mixed venous blood, the retentions of these gases at any $\dot{V}_A/\dot{Q}$ unit estimated based on eq. (1) differed only little from those calculated with eq. (2) (Figure 4).

Distribution of $\dot{V}_A/\dot{Q}$

$\dot{Q}$ distribution predicted from the present method in which dependence of λ on SO_2 and pH was taken into consideration was compared with that from the classical procedure of Wagner et al. (1974; 1977) in which λ was assumed to be constant irrespective of SO_2 and pH. The inert gas data obtained from a total of 18 patients with varied chronic lung diseases were used for the construction of representative distributions of $\dot{V}_A/\dot{Q}$ in the lung. Although 14 out of 18 patients showed small differences in the location and in the magnitude of the predicted $\dot{V}_A/\dot{Q}$ distributions based on either eq. (1) or eq. (2), essential features of the modes coincided generally with each other (Figure 5). In remaining 4 patients, there existed no detectable differences between the $\dot{V}_A/\dot{Q}$ distributions predicted from eq. (1) and those from eq. (2).

DISCUSSION

Critique of methods

To avoid the artifacts caused by experimental errors, measurements on λ were repeated at least 30 times in the same conditions. Further, in order to ensure the dependence in λ of the inert gas on SO_2 or on pH in the blood, solubilities of the seven inert gases in an isotonic buffer solution with electrolyte composition similar to that of human blood (mmol/L: NaCl 115, KCl 3.5, KH_2PO_4 0.5, $NaHPO_4$ 2.5, $NaHCO_3$ 25.7) were examined. As shown in Figure 6, values observed at PO_2 of 150 Torr did not differ statistically from those without any O_2 in the solution, suggesting that the variation of λ in the blood depending on SO_2 was not artificial (Figure 1). Similar findings were obtained for the λ values in the isotonic buffer solution at varied pH (Figure 7). Namely, the λ of the inert gas showed no significant difference between the conditions at pH of 7.2 and pH of 8.0, thus again indicating that the pH dependence found for ethane and halothane in blood were not artificial (Figure 2).

Effects of SO_2 and pH on inert gas exchange

The λ of ethane and of halothane is highly influenced by the $\dot{V}_A/\dot{Q}$ of the lung unit because the λ values of these gases are sensitive to both SO_2 and pH in the blood (Figures 1 and 2). The λ of cyclopropane and that of ether become lower as a $\dot{V}_A/\dot{Q}$ ratio increases but nearly attain plateaus in the lung units with $\dot{V}_A/\dot{Q}$ more than 1. This is due to the fact that cyclopropane and ether are susceptible to SO_2 but not to pH changes (Figures 1 and 2). In the units with $\dot{V}_A/\dot{Q}$ more than 1 where almost complete oxygenation of the blood may be accomplished, SO_2 in the end-capillary blood comes close to 100% so that cyclopropane and ether show the lowest but constant values.

In spite of the findings as described above, a practical effect of either SO_2 or pH dependence of λ on the retention of each homogeneous $\dot{V}_A/\dot{Q}$ unit is small (Figure 4). As shown in Figure 3, λ values in the end-capillary blood are consistently distorted at the lung units having $\dot{V}_A/\dot{Q}$ over 1. However, $\dot{V}_A/\dot{Q}$ values per se at these units are large enough to offset the variation of the denominator in eq. (1) produced by the change of λ (see eq. (1)).

220

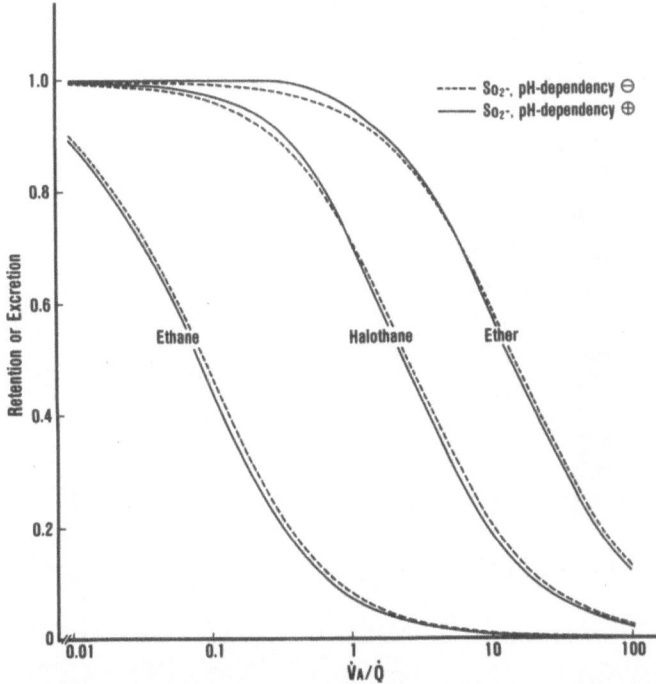

Fig. 4. Effects of SO_2 and pH on retention (or excretion) of a homogeneous $\dot{V}_A/\dot{Q}$ unit.

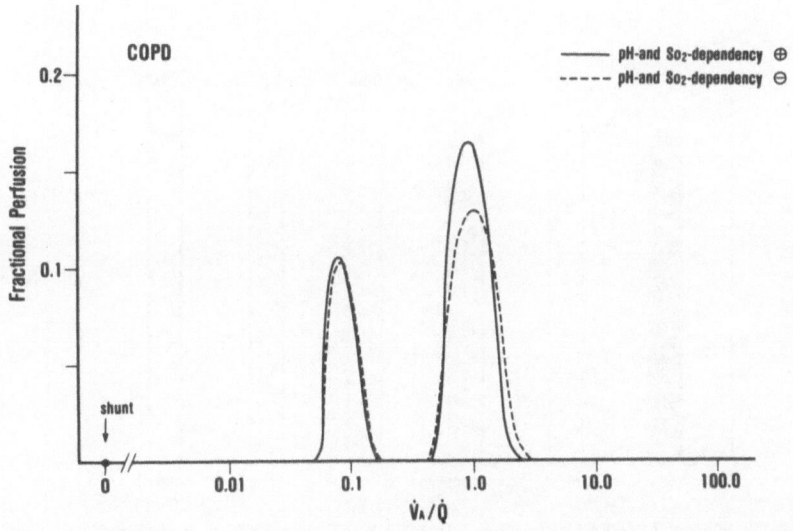

Fig. 5. Effects of SO_2 and pH on recovery of $\dot{V}_A/\dot{Q}$ distribution

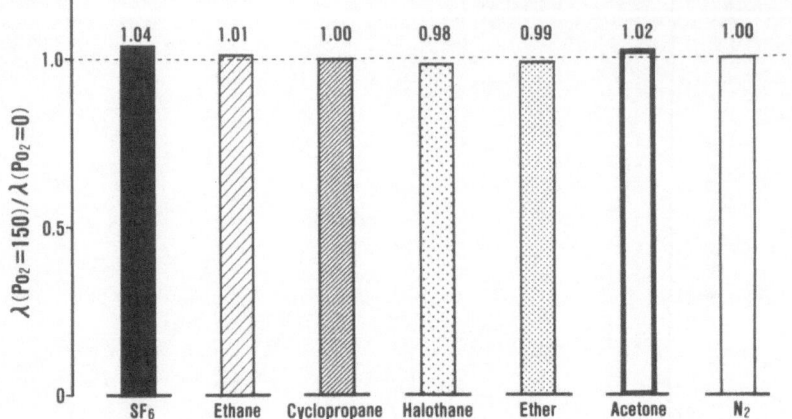

Fig. 6. Effects of PO_2 on λ of isotoxic buffer.

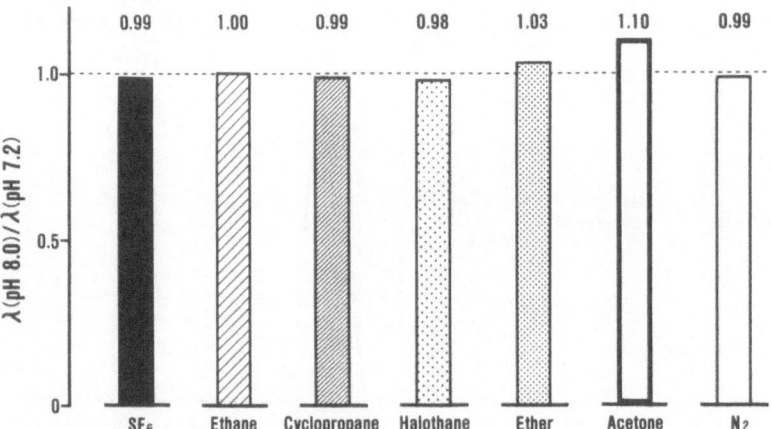

Fig. 7. Effects of pH on λ of isotonic buffer.

Effects of SO_2 and pH on recovery of $\dot{V}_A/\dot{Q}$ distribution

As expected from Figure 4, essential agreement between the $\dot{V}_A/\dot{Q}$ distribution predicted from eq. (1) and that from eq. (2) was observed irrespective of the underlying disease (Figure 5). The findings suggest that the possible effects of SO_2 and of pH on inert gas exchange in the lung are reasonably small. Thereby, fundamental features of impaired gas exchange in diseased lungs caused by $\dot{V}_A/\dot{Q}$ inequality may well be described with the classical procedures of multiple inert gas elimination (Yokoyama and Farhi, 1967; Wagner et al., 1974, 1977), i.e. neglecting the effects of SO_2 and pH on λ.

SUMMARY

Potential effects of SO_2 and of pH on blood-gas partition coefficients, λ, for inert gases, including SF_6, ethane, cyclopropane, halothane, diethyl ether, acetone and N_2, were systematically investigated using human blood. Measurements on λ were performed at $37^{\circ}C$ in conditions of varied SO_2 and pH using gas chromatography. Incorporating the experimental data on λ, multiple inert gas elimination was applied to 18 patients with varied chronic lung diseases, in order to estimate the effects of SO_2 and of pH on both inert gas exchange and resultant recovery of $\dot{V}_A/\dot{Q}$ distribution in the lung. For this purpose, the data obtained by the procedure of multiple inert gas elimination were analyzed with the classical approach but allowance was made for λ of the indicator gas to vary according to exchange of O_2 and of CO_2 in the pulmonary capillary.

Among the gases studied, ethane, cyclopropane, halothane and diethyl ether showed significantly smaller λ values in the oxygenated blood than in deoxygenated blood, whereas SF_6, acetone and N_2 were little dependent on SO_2. An increase in λ was found for ethane and a decrease for halothane with increasing pH in the blood. The other gases were not significantly influenced by pH. In spite of these experimental findings, regional difference of either SO_2 or pH in the lung did not exert important influence on the inert gas exchange or on the predicted $\dot{V}_A/\dot{Q}$ distribution. In conclusion, blood-gas partition coefficients of some inert gases are consistently altered by SO_2 and pH, but their possible effects on inert gas exchange seem to be negligible.

REFERENCES

Evans, J.W., and Wagner, P.D., 1977, Limits on $\dot{V}_A/\dot{Q}$ distributions from analysis of experimental inert gas elimination, J. Appl. Physiol., 42:889.

Farhi, L.E., 1967, Elimination of inert gas by the lung, Respir. Physiol., 3:11.

Kelman, G.R. 1967, Digital computer procedure for the conversion of PCO_2 into blood CO_2 content, Respir. Physiol., 3:111.

Robertson, H.T., Whitehead, J., and Hlastala, M.P., 1986, Diffusion-related differences in elimination of inert gases from the lungs, J. Appl. Physiol., 61:1162.

Severinghaus, J.W., 1979, Simple, accurate equations for human blood O_2 dissociation computations, J. Appl. Physiol., 46:599.

Steen, J.B., 1963, The physiology of the swimbladder of the eel Anguilla Vulgaris. I. The solubility of gases and the buffer capacity of the blood, Acta Physiol. Scand. 58:124.

von Mengden, H.J., Schultehinrichs, D., and Thews, G., 1969, Dependence of plasma pH on oxygen saturation, Respir. Physiol. 6:151.

Wagner, P.D., Saltzman, H.A., and West, J.B., 1974, Measurement of continuous distributions of ventilation-perfusion ratios: theory, J. Appl. Physiol. 36:585.

Yamaguchi, K., Mori, M., Kawai, A., and Yokoyama, T., 1988, Influences of carbon monoxide on the binding of oxygen, carbon dioxide, proton and 2,3-diphosphoglycerate to human hemoglobin, Adv. Exp. Med. Biol. 222:299.

Yokoyama, T., and Farhi, L.E., 1967, The study of ventilation-perfusion ratio distribution in the anesthetized dog by multiple inert gas washout, Respir. Physiol., 3:166.

STROMA-FREE HEMOGLOBIN SOLUTIONS PREPARED BY CRYSTALLIZATION AND

ULTRAFILTRATION METHODS; COMPARISON OF COMPOSITION AND CORONARY

VASOCONSTRICTOR POTENCY

Mia E. Lang, B. Korecky, P.J. Anderson, G.P. Biro

Department of Physiology, Faculty of Medicine
University of Ottawa, Ottawa, Canada, K1H 8M5

INTRODUCTION

Stromafree hemoglobin solution (SFHS), obtained by osmotic lysis of human erythrocytes, has been proposed as a 'blood substitute' (Rabiner, 1975). Its long-term usefulness is limited by the high oxygen affinity and short intravascular retention of the native hemoglobin (Kaplan and Murthy, 1975; Moss et al., 1976; DeVenuto et al., 1977). A further limiting factor was found to be the coronary vasoconstrictor activity found to be present in various SFHS preparations, even after substantial dilution (Vogel et al., 1986; Biro et al., 1988). This vasoconstrictor activity should have been anticipated from the findings in the neurosurgical literature (Toda et al., 1980; Okwuasaba, et al., 1981; Boullin et al., 1983), showing that hemolysate is a powerful constrictor of cerebral blood vessels.

Although hemoglobin has been shown to be a powerful inhibitor of Endothelium-Derived Relaxing Factor (EDRF) (Furchgott et al., 1985; Martin et al., 1985; Fujiwara et al., 1986), this mechanism has not been shown to be responsible for the vasoconstrictor effects of SFHS demonstrated in vitro. Furthermore, SFHS-preparations, produced or purified by various methods, have been shown to exhibit quantitatively different degrees of vasoconstrictor potency (Vogel et al., 1987; Biro et al., 1988). In spite of this, no comprehensive testing program has been undertaken to identify the vasoconstrictor principle and to compare the effects of SFHS's produced by the two principal preparative methods based on crystallization (DeVenuto et al., 1977) and on ultrafiltration (Sehgal et al., 1983). The present experiments were undertaken as a systematic comparison of the two preparative methods as a first step toward identifying the vasoconstrictor principle present in SFHS.

METHODS

Preparation of SFHS

Red cells were obtained from the Red Cross Blood Transfusion Service, no later than five days after their collection. The preparative schema is summarized in Fig. 1. All procedures were conducted in a cold room at 4°C. The red cells were concentrated after the removal of the plasma and buffy coat, and washed three times with sterile physiological saline, in the

Oxygen Transport to Tissue XII, Edited by J. Piiper *et al.*
Plenum Press, New York, 1990

225

original collection bags. They were then lysed with one volume of steril distilled water containing antibiotics. Stromal and other particulate fragments were removed by centrifugation at 2000g and filtration through sterile gauze, under nitrogen. This comprises the initial hemolysate, to which Penicillin (7,500 IU/1) and Streptomycin (75 mg/1) had been added.

At this point, the solution was divided into two aliquots; these were subjected to the crystallization or ultrafiltration methodologies. All subsequent steps were conducted under a nitrogen atmosphere to minimize methemoglobin formation. It should be noted that the two preparative methodologies were used on a common hemolysate, thus eliminating a potential source of variability when preparations are made from different source-material.

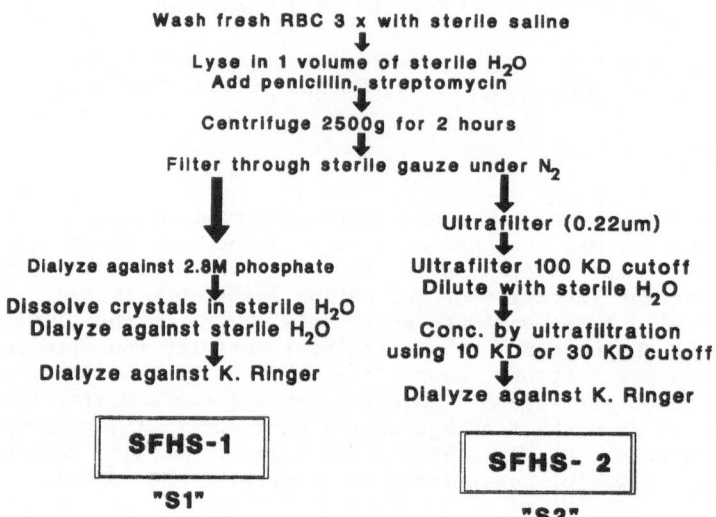

Figure 1. Schematic summary of the steps involved in the preparation of SFHS by the crystallization and ultrafiltration methods (for details, see text).

The crystallization method

This methodology (Fig. 1) is based on that initially described by De Venuto et al. (1977) and involves the precipitation of hemoglobin by a high salt concentration (2.8 M potassium-phosphate at pH = 6.8). The resulting slurry was then dissolved in distilled water and the solution was exhaustively dialyzed, first against water and subsequently against Krebs-Ringer bicarbonate solution (KRB)[1], to remove the large quantity of potassium phosphate. The final dialysate was passed through sterile 0.22 μm Millipore filters (Sterivex-GS, model 11801).

---

[1]The composition of the KRB was: NaCl 7 g/1; KCl 357.5 mg/1; $CaCl_2.2H_2O$ 440 mg/1, $MgSO_4.7H_2O$ 300 mg/1; KH_2PO_4 163 mg/1; $NaHCO_3$ 2.12 g/1; D-glucose 2.97 g/1; EDTA 185 mg/1.

226

The ultrafiltration method

This method (Fig. 1) is based on that described by Sehgal et al. (1983) and is based on a recirculating ultrafiltration of the hemolysate. For this, a Minitan Ultrafiltration System (Model XX42 ASY MT) was used. The filtrate was sterilized by passing through a 0.22 μm Millipore membrane (Durapore, Model GVLPOM04) followed by ultrafiltration using a 100 KD-cutoff Millipore membrane (Durapore Model PTHKDM04). At this stage the solution was highly viscous and was therefore diluted with distilled water. After dilution, the solution was concentrated using either a 10 KD- or 30 KD-cutoff Millipore Polysulfonate Ultrafiltration Plate (Model PTTKOM04 or PTGCOM04) in order to achieve the desired hemoglobin concentration of 5.5-6.5 g/dl. The retentate from the above filtrations thus contains substances in the 10-100 and 30-100 KD-range. These solutions were finally dialyzed against KRB and passed through 0.22 μm Millipore filters for final sterilization.

The crystallized (S1) and ultrafiltered (S2) solutions were stored in Vacutainer tubes, under nitrogen, in the refrigerator at 4°C or in the freezer, at -85°C.

Characteristics of the SFHS's

Hemoglobin concentration of SFHS prepared by both methods was 5.2-6.5 g/dl. Methemoglobin concentration in crystallization-prepared SFHS averaged 5.6%; those having methemoglobin concentration > 10% were rejected. Methemoglobin concentration of all the ultrafiltered SFHS's was less than 5% and was often < 2%. We attempted to measure marker enzyme activities (e.g. lactate dehydrogenase), but these proved to be impractical because of colour interference from hemoglobin.

Phospholipid content was determined by measuring organic phosphate concentration according to the method of Ames and Dubin (1960), in chloroform-methanol-saline extracts (Bowyer and King, 1977). Organic phosphate concentration in the crystallization- and ultrafiltration-produced SFHS's respectively, was 2.70±1.11 and 0.45±0.23 nmole/ml. Thus, ultrafiltration yielded an approximately sixfold purification of phospholipid. These values are at least an order of magnitude less than those reported by Feola et al. (1988) in unextracted bovine SFHS which was associated with deleterious effects.

Testing of cardiac effects in the Langendorff-perfused heart

The method used was similar to that of Vogel et al. (1986); it involves the addition of the SFHS to be tested to the perfusate in isolated rat hearts perfused at constant flow retrogradely in the Langendorff-manner.

Male Sprague-Dawley rats (250-350 g) were anesthetized with sodium pentobarbital (1 mg/Kg) intraperitoneally and were subsequently anticoagulated with sodium heparin (1 mg/Kg) intravenously. After opening the chest, the aorta was cannulated, the atria were removed and the heart was mounted on the perfusion apparatus (Fig. 2). Perfusion, at constant flow, was maintained by a pump, using KRB bubbled with 95% O_2/5% CO_2 at 37°C. The heart was stimulated at 4.5 Hz using a Grass stimulator. A balloon was placed in the left ventricular cavity; its volume was adjusted such that end-diastolic pressure was 0 mmHg. Thus, the non-ejecting hearts worked against a constant preload developing pressure isovolumically. Balloon pressure and its derivative were measured by a transducer and recorded on a Grass polygraph.

227

Coronary perfusion pressure was measured through a sidearm in the cannula, by a transducer and was recorded on the polygraph. The pressure was also monitored using a mercury manometer.

SFHS to be tested was added to the perfusate stream running at 10-14 ml/min. The SFHS was added by a syringe pump for a two-minute period and was delivered directly into the bubble-trap at the rate of 1 ml/min. Since this represented an additional volume, the deadspace of the tubing was loaded with KRB; this was delivered for 40 seconds, resulting in a 2-5 mmHg rise in perfusion pressure. The effects of the SFHS arriving after this "infusion" were determined from this baseline level. Any change in perfusion pressure during the administration of SFHS was thus due to changes in coronary vascular resistance. In order to minimize edema-formation, coronary flow was initially adjusted to maintain perfusion pressure at 50-54 mmHg. Once so adjusted, flow was kept constant.

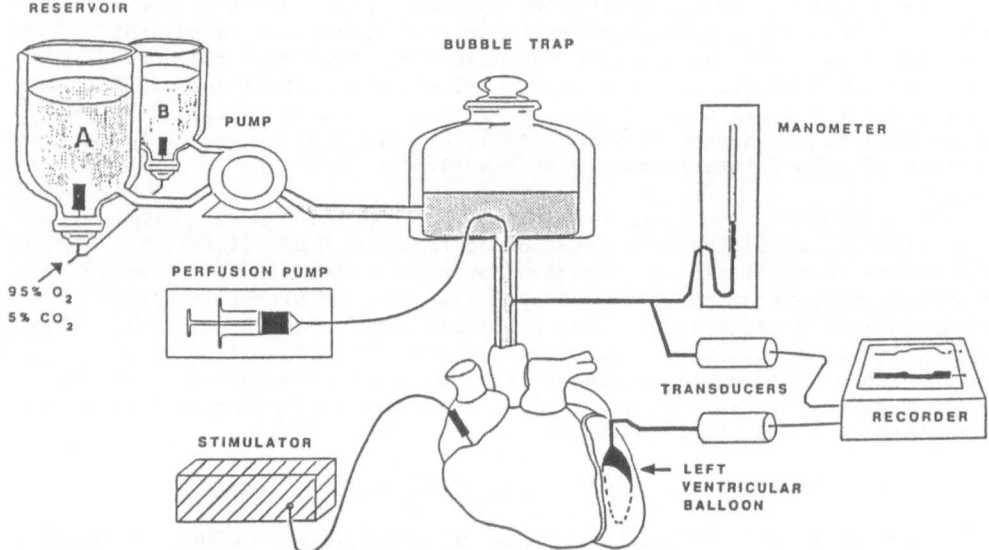

Langendorff Perfusion Apparatus

Figure 2. Schematic diagram of the apparatus used for the perfusion of rat hearts. The perfusate (Krebs-Ringer bicarbonate buffer, oxygenated in temperature-controlled reservoirs) is pumped at constant flow. Into this perfusate stream, the material to be assayed, is mixed at 1 ml/min, by the syringe pump. For further details, see text.

Hearts would slowly accumulate edema during perfusion. We used hearts no longer than 70 minutes, during which time the average weight-gain was <15%; during this period, perfusion pressure rose by <5 mmHg and the fall in left ventricular systolic pressure was <12 mmHg. Since the KRB perfusate contains no protein we used 5-6% fatty acid-free albumin as control, in a manner identical to that used to test SFHS's. 5-6% solutions of albumin caused modest and statistically not significant changes in coronary perfusion pressure (range: 6.6 ± 2.5 (mean $\pm$ SEM) to -7.2 ± 3.2%) and moderate changes in left ventricular systolic pressure (range: $+6.2\pm2.11$ to -7.2 ± 2.3%). Thus, physiologically significant and specific effects of SFHS are considered those which fall outside of the effects of albumin in the same run of hearts. All tests were performed on 6 to 8 hearts.

The results are expressed as the mean induced pressure change $\pm$ SEM. Statistical significance of the induced changes was assessed by one-way analysis of variance and the application of Student's t-test and Tukey's test (Zar, 1984). Significances were established at the $p<.05$ level.

RESULTS

Comparison of crystallization - and ultrafiltration-prepared SFHS's

A large production-run of SFHS produced simultaneously by the two methods was tested at various times during an eight-week period. The purpose of this experiment was to ascertain whether the mode of storage (refrigeration vs. freezing) and the duration of storage had a significant impact on the cardiac actions of SFHS. While the two preparations exhibited various effects upon coronary perfusion pressure and left ventricular developed pressure in the assay hearts, no temporal trend was evident suggesting the absence of a consistent appearance or disappearance of active principle(s) during storage. While there were some small quantitative differences with the ultrafiltration-produced SFHS, overall the cardiac effects were minor (<8% change in perfusion or systolic pressure) and lacking in statistical significance. In contrast, the crystallization-prepared SFHS induced more marked changes in both perfusion and systolic pressures but without a significant temporal trend.

Since duration of storage did not appear to have a major impact, the testing results were pooled over the eight-week period. The averaged pooled results are summarized in Fig. 3. It is evident that the ultrafiltration-produced SFHS caused no significant coronary vasoconstriction or depression of left ventricular contractility, whether stored at 4°C or -85°C. In contrast, the crystallization-produced SFHS induced significant coronary vasoconstriction and depression of contractility. Quantitatively, greater degree of vasoconstriction was seen in frozen SFHS, whereas greater contractility-depressant effect was evident in the refrigerated SFHS.

Batch-to-batch variability

Three batches prepared at different times, by the same methods, were compared with respect to their effects upon coronary perfusion pressure and left ventricular systolic pressure (Fig. 4a and 4b). The three batches of crystallization-produced SFHS showed different degrees of vasoconstrictor and contractility-depressant potencies. These differences are seen, in spite of similar hemoglobin concentration in the three bathches. Moreover, there was little concordance between the magnitude of the coronary vasoconstrictor and contractility-depressant effects in each of the three batches. Differences between batches of the ultrafiltration-produced SFHS were small and lacking in statistical significance.

In general, there was considerable variability in the responses of different hearts, to a given SFHS. The coefficient of variation (S.D./x) ranged from 19 to 41% in different batches, accounting for some of the large variability evident in the figures.

Fractionation to isolate the vasoconstrictor principle

Fractionation on DEAE phosphocellulose column yielded five fractions from crystallization-produced SFHS (Fig. 5). 95% of the hemoglobin present

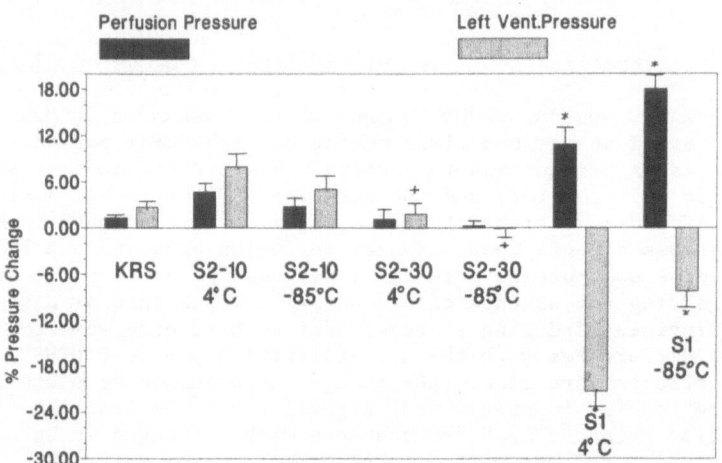

Figure 3. Summary of the changes in perfusion and left ventricular pressures (mean ± SEM) induced by various preparations. These are pooled results of the measurements made over an eight-week period described above. S1 and S2 refer to crystallization and ultrafiltration-prepared SFHS, respectively. The temperatures refer to those at which the solutions were stored. 10 and 30 refer to the size of the molecular weight-cutoff filters used. For details, see text. * indicates significant change from zero, by paired t-test. + indicates significant difference from the effect of S1-induced change, by Tukey's test.

in the starting material was recovered in the five fractions; fraction 4 contained 85% of the hemoglobin and was also found to contain the majority of the vasoconstrictor activity. Because there was relatively little contractility-depressant activity present in the initial SFHS, it was difficult to identify the fraction which contained this activity.

230

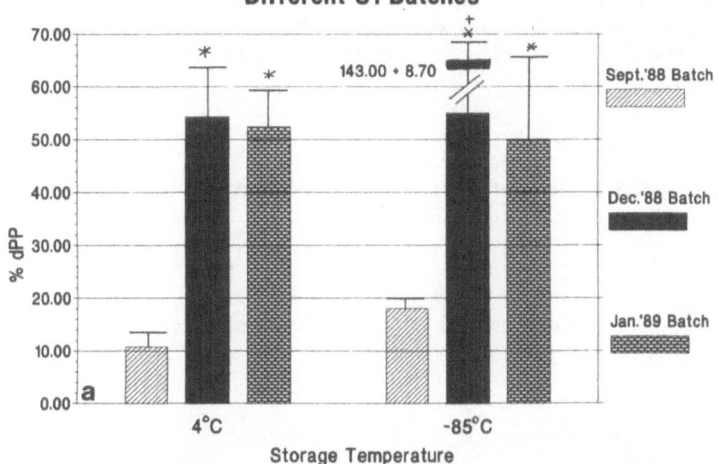

Mean Perfusion Pressure Changes of Different S1 Batches

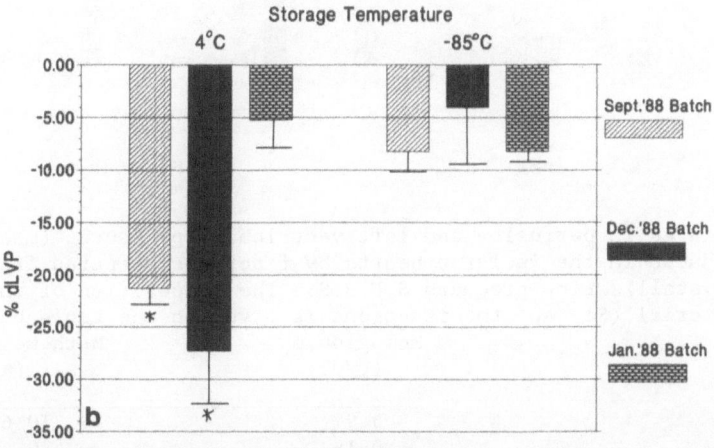

Mean Left Ventricular Pressure Changes of Different S1 Batches

Figure 4 a and b: Summary of the differences in the changes in perfusion (top) and left ventricular (bottom) pressures induced by three different batches of SFHS produced by crystallization. * indicates statistically significant change by paired t-test. + indicates statistically significant difference between batches, by Tukey's test.

231

Particulate matter

Crystallization-produced SFHS on storage appeared to produce a visible particulate material which could be removed by centrifugation or filtration. The pellet, after centrifugation, was difficult to redissolve, but was found to consist largely of protein and, on sodium dodecyl-sulphate polyacrylamide-gel slab electrophoresis, was shown to separate into several bands suggesting hemoglobin and spectrin. When the particulate matter was filtered out, the vasoconstrictor activity seemed to be diminished but was not eliminated (Fig. 6).

DISCUSSION

We compared SFHS produced by the crystallization and ultrafiltration methods. From theoretical considerations it would be expected that a

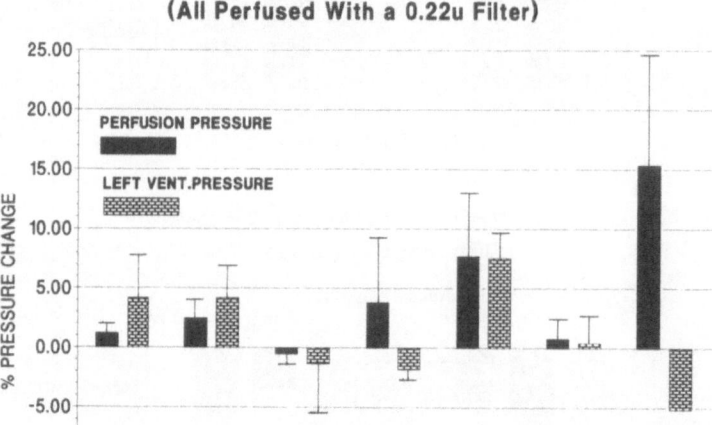

Figure 5. Changes in perfusion and left ventricular pressures (mean ± SEM) induced in the isolated hearts by fractions prepared from crystallization-produced S.F.H.S. The composition of the starting material (S1) and the fractions is given in the table below:

	Hemoglobin (g/dl)	Methemoglobin (%)
S1	5.9	10.6
F1	0.18	0
F2	0	0
F3	0.24	0
F4	4.8	5.1
F5	0.41	0

solution produced by crystallization may be more homogeneous. Our comparison showed that the ultrafiltered product contains less phospholipid, causes little if any significant coronary vasoconstriction and is characterized by far less batch-to-batch variability than that produced by crystallization. These findings suggest that ultra-filtration, particularly when combined with further restriction of molecular size in the retentate (30-100 KD), yields an SFHS which is substantially more suitable for potential clinical use than that produced by crystallization.

One of the vexatious problems facing research on SFHS has been the coronary constrictor potency seen in Langendorff-perfused rat and rabbit hearts (Vogel et al., 1986) as well as in in vivo experiments (Biro, 1982). Nevertheless, some other in vivo experiments failed to find convincing evidence of coronary vasoconstriction which would be sufficient to warrant clinical concern (Moores et al., 1988). This controversy is not yet resolved and the true significance of ex vivo demonstrable coronary vasoconstrictor potency is not yet clear.

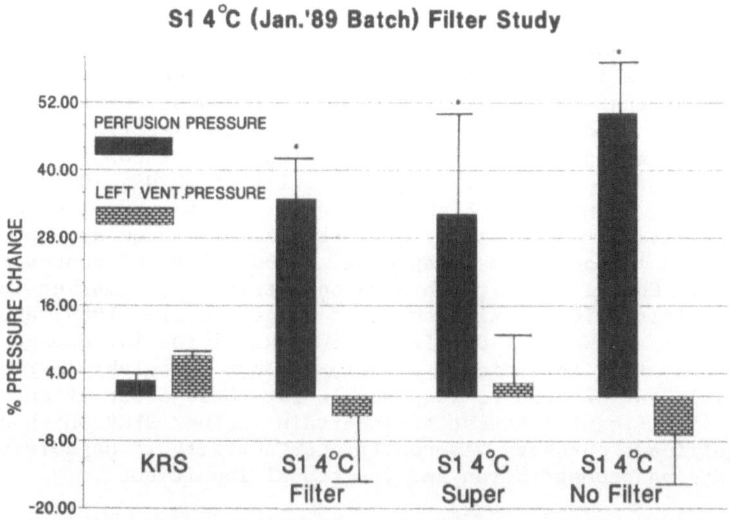

S1 4°C (Jan.'89 Batch) Filter Study

Figure 6. The effect of the removal of particulate matter by filtration ("filter") or centrifugation ("super") from crystallization-prepared SFHS ("S1, no filter"). Removal of the particulate matter reduces the effect quantitatively, but not statistically (mean ± SEM). * indicates a significant change, by paired t-test.

Although hemoglobin is the most abundant of the erythrocytes' contents, it is by no means the only constituent of SFHS. Hemoglobin is also a convenient, easily measurable and abundant marker and some effects may be easily attributable to it. Several cytosolic enzymes and some constituents of the cytoskeleton have molecular weights similar to that of hemoglobin and crude purification, represented by crystallization or ultrafiltration, may not effectively remove these "contaminants". Some observed effects may be due to such "contaminants". In a preparation which exhibited coronary vasoconstrictor potency, the majority of activity was found to be in the fraction which also contained the majority of hemoglobin. Nevertheless, our other findings suggest that tetrameric "native" hemoglobin is not likely to be the constituent responsible for this activity. This conclusion is principally based on our findings that a variety of preparations having essentially identical hemoglobin concentration exhibit widely varying or absent coronary vasoconstrictor potency. This conclusion is not contradicted by the demonstration of a linear dose-response relationship between hemoglobin concentration and constrictor potency demonstrated by Vogel et al. (1986 and 1988), since in these experiments, serial dilutions were made of the SFHS, resulting in dilution of all constituents including the hemoglobin. Furthermore, the demonstration that oxidation of the hemoglobin to methemoglobin (Vogel et al., 1986) or other modification of the molecule quantitatively modifies the vasoconstrictor activity, is not necessarily contradictory to the above conclusion, since the oxidation induced may also affect other protein constituents, including hemoglobin breakdown products, as well as the enzymes and structural constituents of the erythrocyte which are also present in SFHS. Furthermore in spite of extensive washing during the preparative process, contamination with leukocyte- and platelet-contents is a distinct possibility. Such contamination may be significantly variable and may be the source of some of the batch-to-batch variability seen with SFHS prepared by crystallization.

Our experiments do not restrict the field of the possible candidates responsible for the in vitro observed coronary vasoconstriction. They do, however, suggest that adequate purification procedures, based on ultrafiltration methodologies, yield SFHS which is essentially free of this coronary constrictor potency. This removes one of the concerns regarding the feasibility of the future clinical usability of SFHS. Unfortunately, SFHS has recently been shown (Lieberthal et al., 1987; Vogel et al., 1988) to cause significant vasoconstriction in the kidney which appears reversible by the endothelium-independent vasodilator, nitroprusside (Vogel et al., 1989). Coincident with this vasoconstriction, a marked reduction in glomerular filtration is also observed (Vogel et al., 1989) and other experiments (Tam and Wong, 1988) indicate proximal tubular damage. These experiments also fail to indicate clearly whether hemoglobin per se, or another constituent of SFHS is responsible for these observations. It would be of significant interest to ascertain whether SFHS which appears free of significant coronary vasoconstrictor activity is capable of inducing renal vasoconstriction and functional impairment.

We have also observed a variable degree of contractility-depression in the perfused hearts when SFHS was added to the perfusate (Fig. 4b) as indicated by a fall in systolic pressure in the isovolumically contracting left ventricle. This could be due to a non-specific effect, or to a mechanism involving more specific agent(s). The simplest non-specific mechanism may involve calcium-complexing effects due to the presence of protein. This is suggested by the fact that fatty acid-free albumin also caused some depression in left ventricular pressure. The magnitude of contractility-depression observed during perfusion with ultrafiltered hemoglobin of similar concentration, was of the same order as that occurring during perfusion with albumin. While this would suggest that the contractility-depression may be related to calcium-chelation, quantitative considerations negate this interpretation. The concentration of albumin and of hemoglobin in the perfusate reaching the heart was 350-700 mg/l or $55\text{-}110\times10^{-6}M$. Assuming that the calcium-binding capacity of hemoglobin is similar to that of albumin, the amount of calcium complexed by these proteins would be expected to be nearly two orders of magnitude less than the total calcium concentration $(3\times10^{-3}M)$ in the perfusate. Thus, the reduction in free calcium-concentration reaching the myocytes would be of a negligible degree and would not account for the magnitude of contractility-depression observed.

Even if the above considerations seem to exclude the likelihood of a nonspecific origin of the contractile depression based on calcium-availability, our experiments fail to positively identify the mechanism(s) underlying the significant depression observed with the use of crystallization-produced SFHS. In view of the substantial variability of the depression observed, it also seems unlikely that it is related to a nonspecific effect. Regardless of the specific mechanisms responsible, our finding of the presence of significant contractility-depression in crystallization-produced SFHS and its absence in ultrafiltration-produced SFHS (Fig. 3), underlines the latter's greater potential suitability for ultimate use.

SUMMARY

Stroma-free hemoglobin solutions (SFHS) were prepared by the crystallization and ultrafiltration methodologies. The preparations were partially characterized with respect to their effects upon the isolated perfused rat heart. SFHS prepared by ultrafiltration is characterized by a substantially lower content of residual membrane phospholipid and a more restricted protein composition. This preparation is also essentially free of vasoconstrictor and contractility-depressant actions on the ex vivo perfused heart. In contrast, crystallization-produced SFHS is less well purified of both phospholipid and protein constituents, is likely to generate denatured protein aggregates during storage and exhibits vasoconstrictor and contractility-depressant activity which may vary significantly, from batch to batch. These findings indicate that preparative methodology based on ultrafiltration and size-exclusion, yields SFHS which is superior in these respect to that produced by a crystallization method.

ACKNOWLEDGEMENTS

Financial support for these experiments was provided by a contract from the Defence and Civil Institute of Environmental Medicine of Canada. We are grateful to Maryse Ramage, Mary Masika and Diane Mauldin, for technical assistance and to Denyse Longpré for typing this manuscript. We wish to express our thanks to the Medical Directors and technical staff of the Ottawa Center, Canadian Red Cross Blood Transfusion Service, for generously providing us with blood from which the SFHS was prepared.

REFERENCES

Ames, B.N. and Dubin, D.T., 1960, The role of organic polyamines in the neutralization of bacteriophage deoxyribonucleic acid, J. Biol. Chem. 235:769.

Biro, G.P., 1982, Comparison of acute cardiovascular effects and oxygen-supply following haemodilution with dextran, stroma-free hemoglobin solution and fluorocarbon suspension, Cardiovasc. Res. 16:194.

Biro, G.P., Taichman, G.C., Lada, B., Keon, W.J., Rosen, A.L., Sehgal, L.R., 1988, Coronary vascular actions of stroma-free hemoglobin preparations, Artif. Orgs. 12:40.

Boullin, D.J., Tagari, P., du Boulay, G., Aitken, V., Hughes, J.T., 1983, The role of hemoglobin in the etiology of cerebral vasospasm, J. Neurosurg. 59:231.

Bowyer, D.E. and King, J.P., 1977, Methods for the rapid separation and estimation of the major lipids of arteries and other tissues by thin-layer chromatography on small plates followed by microchemical analysis, J. Chromatography 143:473.

DeVenuto, F., Zuck, T.F., Zegna, A.I., Moores, W.Y., 1977, Characteristics of stroma-free hemoglobin prepared by crystallization, J. Lab. Clin. Med. 89:509.

Feola, M., Simoni, J., Dobke, M., Canizaro, P.C., 1988, Complement activation and the toxicity of stroma-free hemoglobin solutions in primates, Circ. Shock 25:275.

Fujiwara, S., Kassell, N.F., Sasaki, T., Nakagomi, T., Lehman, R.M., 1986, Selective hemoglobin-inhibition of endothelium-dependent vasodilation of rabbit basilar artery, J. Neurosurg. 64:445.

Furchgott, R.F., Martin, W., Cherry, P.D., 1985, Blockade of endothelium-dependent vasodilation by hemoglobin: a possible factor in vasospasm associated with hemorrhage, Adv. Prostagl. Thromboxane Leukotr. Res. 15:499.

Kaplan, H.R. and Murthy, V.S., 1975, Hemoglobin solution, a potential oxygen transporting plasma volume expander, Fed. Proc. 34:1461.

Lieberthal, W., Wolf, E.F., Merrill, E.W., Levinsky, N.G., Valeri, C.R., 1987, Hemodynamic effects of different preparations of stroma-free hemolysates in the isolated perfused rat kidney, Life Sci. 41:2525.

Martin, W., Villani, G.M., Jothianandian, D., Furchgott, R.F., 1985, Selective blockade of endothelium-dependent and glyceryl-trinitrate-induced relaxation by hemoglobin and by methylene-blue in the rabbit aorta, J. Pharmacol. Exp. Therap. 232:708.

Moores, W.Y., Mack, R.E., White, F.C., Bloor, C.M., 1988, Coronary flow dynamics in swine following partial exchange transfusions with hemoglobin and albumin solutions, Biomat. Artif. Cells Artif. Orgs. 16:355.

Moss, G.S., DeWoskin, R., Rosen, A.L., Levine, H., Palani, C., 1976, Transport of oxygen and carbon dioxide by hemoglobin-saline solution in the red cell-free primate, Surg. Gynecol. Obstet. 142:356.

Okwuasaba, I., Cook, D., Weir, B., 1981, Changes in vasoactive properties of blood products with time and attempted identification of the spasmogens, Stoke 12:775.

Rabiner, S.F., 1975, Hemoglobin solution as a plasma expander, Fed. Proc. 34:P454.

Sehgal, L.R., Rosen, A.L., Gould, S.A., Sehgal, H.L., Moss, G.S., 1983, Preparation and in vitro characteristics of polymerized pyridoxalated hemoglobin, Transfusion 23:158.

Tam, S.C., Wong, J.T., 1988, Impairment of renal function by stroma-free hemoglobin solution in rats, J. Lab. Clin. Med. 111:189.

Toda, N., Shimizu, K., Ohta, T., 1980, Mechanism of cerebral arterial contraction induced by blood constituents, J. Neurosurg. 53:312.

Vogel, W.M., Dennis, R.C., Cassidy, G., Apstein, C.S., Valeri, C.R., 1986, Coronary constrictor effect of stroma-free hemoglobin solutions, Amer. J. Physiol. 251:H413.

Vogel, W.M., Hsia, J.C., Briggs, L.L., Er, S.S., Cassidy, G., Apstein, C.S., Valeri, C.R., 1987, Reduced coronary vasoconstrictor activity of hemoglobin solutions purified by ATP-agarose affinity chromatography, Life Sci. 41:89.

Vogel, W.M., Lieberthal, W., Apstein, C.S., Levinsky, N., Valeri, C.R., 1988, Effects of stroma-free hemoglobin solution on isolated perfused rabbit hearts and isolated perfused rat kidneys, Biomat. Artif. Cells Artif. Orgs. 16:227.

Vogel, W.M., Lieberthal, W., Valeri, C.R., 1989, Effects of unmodified and DBBF-crosslinked hemoglobin endothelial-dependent vasodilation in isolated rabbit hearts and rat kidneys, Proc. Internat. Symp. Red Cell Substitute, San Francisco, CA, 16-19 May, 1989, p. 56.

Zar, J.H., Biostatistical Analysis, 2nd Ed. Prentice-Hall Inc., New York: 1984, p. 718.

OXYGEN TRANSPORT BY PYRIDOXYLATED POLYHEMOGLOBIN SOLUTION

G. Lenz[1], U. Bissinger[1], H. Benzing[2]

Depts. of Anesthesiology[1] and Physiology[2]
University of Tübingen
Federal Republic of Germany

INTRODUCTION

The principal requirement of any oxygen carrier is its ability to load and unload oxygen effectively. Oxygen-carrying solutions, however, do not necessarily contribute to peripheral oxygen delivery under all conditions. The efficacy of the perfluorochemical emulsion Fluosol-DA as a red cell substitute in acute anemia has been questioned (Gould et al., 1986). Even though stroma-free hemoglobin solutions (SFH) with elevated oxygen affinity (P_{50} < 20 mm Hg) have been shown to maintain effective oxygen delivery and physiological PO_2 after total blood exchange (DeVenuto et al., 1979; Moss et al., 1984), doubt remains about the effectiveness of oxygen unloading at the tissue level after partial exchange transfusion with SFH and polyhemoglobin (PolyHb) preparations with increased oxygen affinity (Rosen et al., 1983). The physiological oxygen affinity (P_{50} 27 mm Hg) of a pyridoxylated PolyHb (PolyHb-PP$_a$) solution (Kothe et al., 1985) therefore takes on added significance.

METHODS

In Vitro Study

Oxygen contents were determined with a Lex-o-con TL (Lexington Instruments). After paraffin coating, plasma was separated by 10-minute centrifugation in a cooling centrifuge and subsequent aerobic removal. Oxygen equilibrium curves were determined with a Hem-O-Scan^{TM} (Aminco).

In Vivo Study

Hemorrhagic shock was induced in 14 minipigs by rapid exsanguination until cardiac output dropped to half its initial level. After a 45 minute shock phase, PolyHb-PP$_a$ was infused to restore cardiac output to its initial level.

The treatment was approximately isovolemic (mean exsanguination volumes were 26 ml/kg BW, mean infusion volumes 28 ml PolyHb-PPa/kg BW, dilution rate 70%, mean PolyHb-PP$_a$ dose 2.4 g/kg BW). All experiments took place under general anesthesia with ketamine and flunitrazepam.

RESULTS

In Vitro Results

The composition and physichemical data of PolyHb-PP$_a$ are given in Table 1.

Table 1. Physichemical data of PolyHb-PP$_a$

Hemoglobin	8.5 g/dl
Mean molecular weight	200,000 d
Non-cross-linked hemoglobin	15%
Albumin	2.5 g/dl
Oncotic pressure	22 mm Hg
Sodium ascorbate	4.5 mmol/l
pH	7.40
Methemoglobin	1.5%
P$_{50}$	27 mm Hg
Hill coefficient	1.7
Oxygen binding coefficient	1.1 ml O$_2$/dl/g Hb
Oxygen capacity	9.3 ml O$_2$/dl

Figure 1 shows the oxygen equilibrium curves of PolyHb-PP$_a$, SFH and human whole blood. Figure 2 gives the O$_2$-content of 8.5% PolyHb-PP$_a$, whole blood (Hb 15g/dl; Hb 8.5g/dl) and 20% Fluosol-DA in relation to PO$_2$.

Isovolemic exchange transfusion of whole blood with PolyHb-PP$_a$ in vitro results in an impressive increase in O$_2$-capacity over that of acute replacement with an asanguineous fluid (hydroxyethyl starch), particularly at high

238

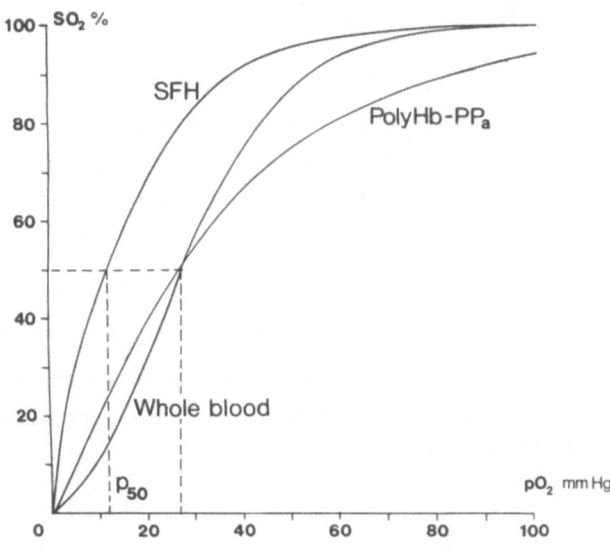

Fig. 1. Oxygen equilibrium curves of PolyHb-PP$_a$ (P$_{50}$ = 27 mm Hg) in comparison to stroma-free hemoglobin solution (SFH) (P$_{50}$ = 12 mm Hg) and human whole blood (P$_{50}$ = 27 mm Hg) (37° C, PCO$_2$ 40 mm Hg).

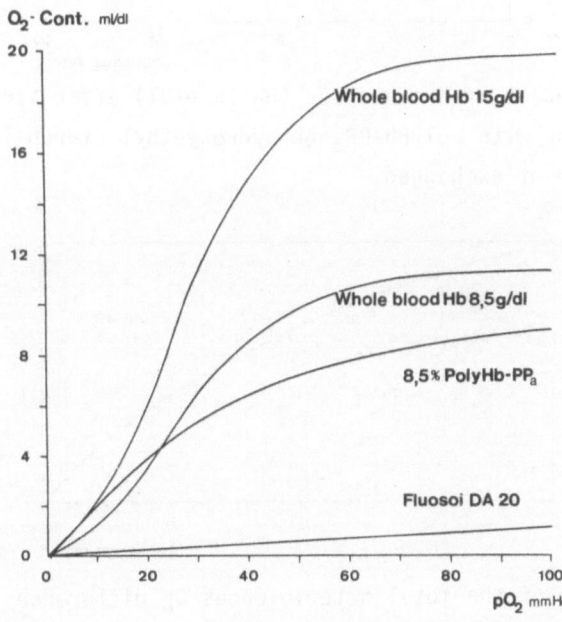

Fig. 2. Oxygen content of 8.5% PolyHb-PP$_a$, human blood (Hb 15 g/dl or 8.5 g/dl) and Fluosol DA 20% in relation to PO$_2$.

replacement volumes (Fig. 3). The hypothetical contribution of plasmatic PolyHb-PP$_a$ to total arteriovenous O$_2$ difference (AVDO$_2$) remains only slight in moderate exchange transfusion (dilution rate up to 50%), assuming constant cardiac output and constant AVDO$_2$ (5 ml O$_2$/dl). Only at an exchange rate of 75% is about 50% of the AVDO$_2$ made available by PolyHb-PP$_a$ (Fig. 4).

Assuming constant cardiac output, calculation of AVDO$_2$ in relation to $P\bar{v}O_2$ for different exchange transfusion rates of 25%, 50%, 75% and 100% (Fig. 5) clearly shows that the absolute and relative contribution of PolyHb-PP$_a$ to total AVDO$_2$ becomes significant only when more than 25% of total blood volume is exchanged.

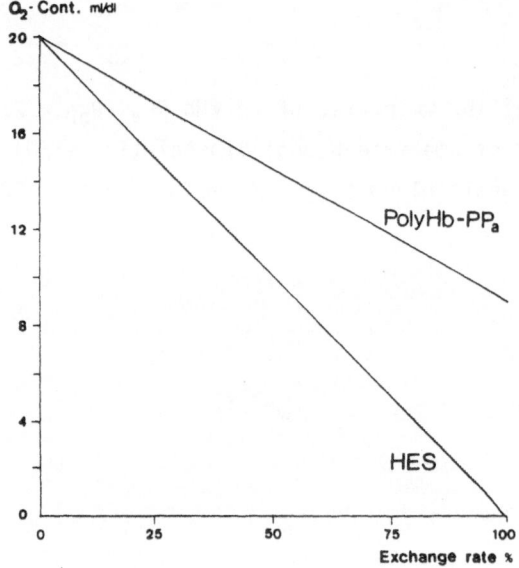

Fig. 3. Oxygen content of human blood (Hb 15 g/dl) after isovolemic exchange transfusion with PolyHb-PP$_a$ or hydroxyethyl starch (HES) in relation to the amount exchanged.

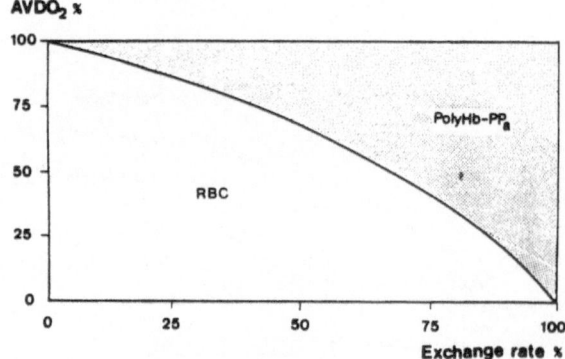

Fig. 4. Percentage of the total arteriovenous O$_2$ difference (AVDO$_2$ = 5 ml/dl) attributable to either blood or PolyHb-PP$_a$ solution. Isovolemic exchange transfusion of human blood (Hb 15 g/dl) with 8.5% PolyHb-PP$_a$ at constant cardiac output.

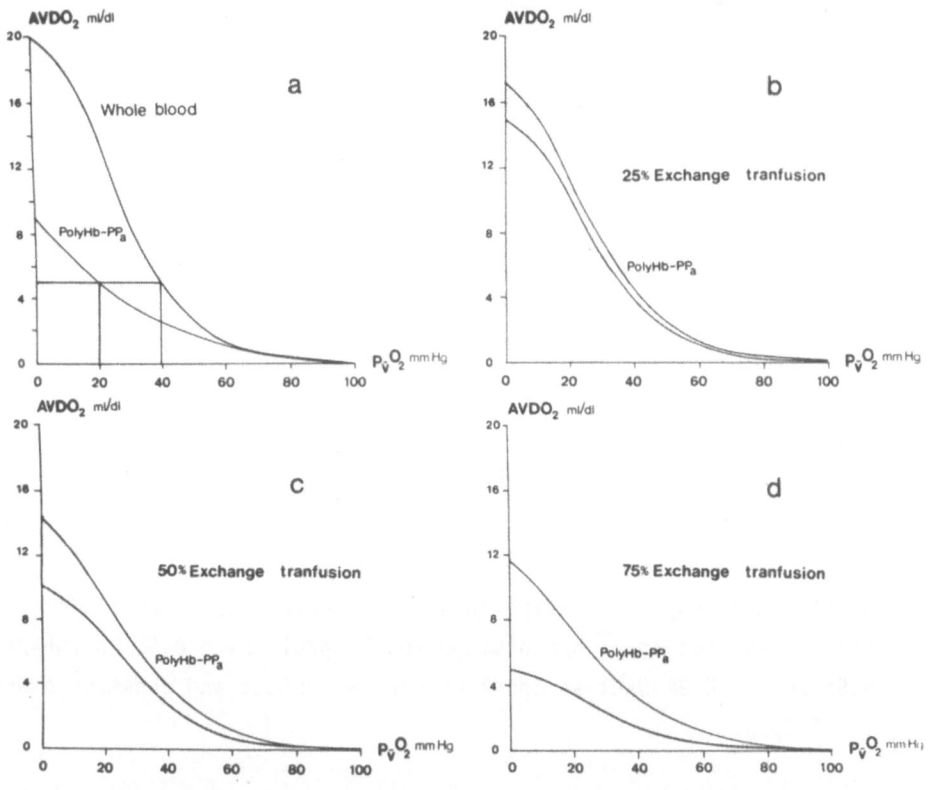

Fig. 5. Arteriovenous O_2 difference in relation to mixed venous PO_2:
(a) Human whole blood, 8.5% PolyHb-PP$_a$ (0% and 100% exchange transfusion) at constant cardiac output; (b), (c), (d) 25%, 50%, 75% exchange transfusion of whole human blood (Hb 15g/dl) with 8.5% PolyHb-PP$_a$ at constant cardiac output. In panels b, c and d the lower curve indicates the contribution of the remaining whole blood by itself, while the upper curve is the total AVDO$_2$, including the contribution by the PolyHb-PP$_a$.

In Vivo Results

Shock-induced hemodynamic changes in minipigs were for the most part reversed by PolyHb-PP$_a$. The mean intravascular half-life of PolyHb-PP$_a$ was 19 h. Following therapy of hemorrhagic shock with PolyHb-PP$_a$, O_2-capacity of whole blood increased considerably; mean PolyHb-PP$_a$ O_2-capacity (4.8 ml O_2/dl) was approximately double that of erythrocytic O_2-capacity, which had dropped to 1/3 of initial levels (2.4 ml O_2/dl). The utilisation of the O_2 additionally delivered by PolyHb-PP$_a$ was demonstrated by an O_2 extraction ratio of about 0.40 directly after shock therapy and 0.25 after 24 hours. In contrast, the erythrocytic extraction ratio was nearly 0.90 after therapy. Mixed venous PO_2 remained lowered to 30 mm Hg after PolyHb-PP$_a$ infusion.

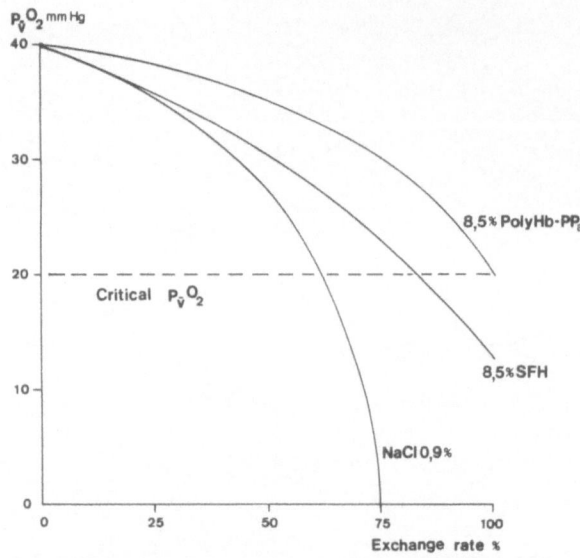

Fig. 6. Mixed venous PO_2 in relation to exchange rate after isovolemic exchange transfusion of human blood (Hb 15 g/dl) with 8.5% PolyHb-PP$_a$, 8.5% SFH or 0.9% NaCl at constant cardiac output and constant AVDO$_2$ (5 ml O_2/dl).

DISCUSSION

Position and shape of the equilibrium curves and oxygen capacity for PolyHb-PP$_a$ and erythrocytic hemoglobin differ clearly.

PolyHb-PP$_a$ exhibits a hyperbolic oxygen equilibrium curve, probably due to hem/hem interactions which have been altered by the cross-linking process. The oxygen affinity of PolyHb-PP$_a$, decisive for oxygen loading and unloading, is the same as that of whole blood (P_{50} 27 mm Hg), whereas PolyHb-PP solutions described by other researchers had lower P_{50} values (< 16-22 mm Hg) (DeVenuto and Zegna, 1983; Keipert et al, 1984; Moss et al., 1984; Keipert and Chang, 1987; Gould et al., 1987). The cardinal significance of this is already evident in vitro when isovolemic exchange transfusion of human blood with PolyHb-PP$_a$ is compared to that with an 8.5 SFH solution (P_{50} 12 mm Hg) and an asanguineous solution (NaCl 0.9%) (Fig. 6). Mixed venous PO_2, as an indicator of tissue oxygenation, decreases with PolyHb-PP$_a$ to a critical value ($P\bar{v}O_2$ 20 mm Hg, i.e., the venous PO_2 at which QO_2 levels cannot be maintained) only after 100% blood exchange, assuming constant cardiac output and constant AVDO$_2$.

242

The oxygen-binding coefficient of PolyHb-PP$_a$ (1.1 ml O$_2$/dl/g Hb), which is nearly 20% lower than that of erythrocytic hemoglobin reflects a corresponding reduction of functional Hb. Since the Met-Hb content of PolyHb-PP$_a$ in oxygenized form is between 3% and 5%, this low oxygen-binding coefficient is apparently due to pyridoxylation and/or cross-linkage of the hemoglobin molecules.

Since the oxygen content of the 8.5% PolyHb-PP$_a$ solution (9.3 ml O$_2$/dl) is only half that of whole blood (Hb 15 g/dl), only a "semianemic" blood exchange is possible. The oxygen capacity of 8.5% PolyHb-PP$_a$ solution is, however, about 10 times greater than that 20% Fluosol-DA at physiological PO$_2$, and the risk of oxygen toxicity is eliminated, since respiratory gas need not be enriched with oxygen.

In vitro analysis, however, shows that the oxygen transported additionally by plasmatic PolyHb-PP$_a$ is markedly utilized only at higher exchange rates. After total exchange with PolyHb-PP$_a$, 2.7 ml O$_2$ at the most are released at a P$\bar{v}$O$_2$ of 40 mm Hg assuming constant cardiac output. A "normal" AVDO$_2$ (5 ml O$_2$/dl) can only be maintained when P$\bar{v}$O$_2$ drops to 20 mm Hg. Since cardiac output, as previous studies have also shown (Lenz, 1988) does not change significantly after isovolemic exchange transfusion with PolyHb-PP$_a$, these considerations are for the most part also valid for conditions in vivo.

After treatment of hemorrhagic shock with PolyHb-PP$_a$ in minipigs, total oxygen capacity of whole blood increased impressively: plasmatic PolyHb-PP$_a$ delivered twice as much oxygen as erythrocytic hemoglobin. The utilisation of this additional oxygen capacity is demonstrated by an O$_2$ extraction ratio of 0.4; the O$_2$ extraction ratio of the erythrocytic hemoglobin, however, was 2.3 times higher. This finding corresponds approximately to the 2.0 of Hobbhahn et al. (1985) after total blood exchange with PolyHb-PP$_a$ in dogs. The oxygen utilisation of PolyHb-PP$_a$ in vivo is therefore about half that of erythrocytic hemoglobin.

Oxygen transport characteristics of PolyHb-PP$_a$ are, however, decidedly better than those reported in studies with hemoglobin solutions with left-shifted O$_2$ equilibrium curves (P$_{50}$ < 20 mm Hg) in which the O$_2$ extraction rate of plasmatic Hb was below 0.15 (Messmer et al., 1978).

P$\bar{v}$O$_2$ values, which declined to 30 mm Hg in the shock phase, remained decreased after PolyHb-PP$_a$ treatment. This finding, however, is in no way comparable to the drastic decrease in P$\bar{v}$O$_2$ which has been observed after exchange transfusion with hemoglobin solutions with increased O_2 affinity (Moss, 1976; Tam et al., 1978).

In view of O_2 unloading after use of PolyHb-PP$_a$, two further aspects are significant. First, the P$_{50}$ of the PolyHb-PP$_a$ solution represents a mean value derived from the P$_{50}$ levels of the individual hemoglobin fractions in the solution. Non-crosslinked pyridoxylated hemoglobin has a P$_{50}$ of 26 to 32 mm Hg (Lenz et al. 1983), while higher molecular weight components have a higher oxygen affinity than those with lower molecular weight; the non-crosslinked hemoglobin fraction, however, is eliminated much more rapidly from the circulation. Secondly, plasmatic hemoglobin autooxidizes to Met-Hb which, depending on concentration, both reduces oxygen content and increases oxygen affinity of the total whole blood hemoglobin (Daniels, 1984).

SUMMARY

Pyridoxylated Polyhemoglobin (PolyHb-PP$_a$) with physiological oxygen affinity (P$_{50}$ 27 mm Hg) was used in vitro and in vivo. The oxygen binding coefficient was 1.1 ml O_2/dl/g Hb, the oxygen capacity of 8.5% PolyHb-PP$_a$ solution 9.3 ml O_2/dl, Hill coefficient 1.7, Met-Hb content below 2%. Exchange transfusion of whole blood with PolyHb-PP$_a$ results in a significant, linear increase in total oxygen capacity, in contrast to exchange transfusion with asanguineous solutions. In vitro analysis of oxygen equilibrium curves of mixtures of whole blood with PolyHb-PP$_a$, however, shows a nonlinear release of the additional blood oxygen capacity. Significant amounts of PolyHb-PP$_a$-bound oxygen are made available only when hemodilution exceeds 25%; only at 75% hemodilution does oxygen delivery by PolyHb-PP$_a$ account for 50% of the unloaded oxygen. In vivo results after therapy for hemorrhagic shock in minipigs clearly shows the efficacy of oxygen transport by PolyHb-PP$_a$. At 70% hemodilution, PolyHb-PP$_a$ bound oxygen was twice that of erythrocytic oxygen. Availability of this additional oxygen capacity was demonstrated by an oxygen extraction ratio of 0.40%. The utility of PolyHb-PP$_a$ solution with "normal" P$_{50}$ in the therapy of moderate anemia, however, remains to be proved.

REFERENCES

Daniels, F. H., McCabe, R. E., Leonard, E. F., 1984, The use of hemoglobin solutions, CRC Crit. Rev. Biomed. Engin., 9:315.

DeVenuto, F., Zegna, A. I., Busse, K. R., 1979, Lyophilisation of crystalline hemoglobin solution and exchange transfusions with lyophilized, reconstituted hemoglobin, Surg. Gynecol. Obstet., 148:69.

DeVenuto, F., Zuck, T. F., Zegna, A. I., Moores, W. Y., 1977, Characteristics of stroma-free hemoglobin prepared by crystallisation, J. Lab. Clin. Med., 89:509.

Gould, S. A., Rosen, A. L., Seghal, L. R., Seghal, H. L., Langdale, L. A., Krause, L. M., Rice, C. L., Chamberlin, W. H., Moss, G. S., 1986, Fluosol-DA as a red-cell substitute in acute anemia, N. Engl. J. Med., 314:1653.

Gould, S. A., Rosen, A. L., Seghal, L. R., Seghal, H. L., Moss, G. S., 1987, Efficacy of polymerized pyridoxylated hemoglobin solution as an O_2 carrier, Biomat. Art. Cells Art. Org., 15:359.

Hobbhahn, J., Vogel, H., Kothe, N., Brendel, W., Peter, K., Jesch, F., 1985, Hemodynamics and oxygen transport after partial and total blood exchange with pyridoxylated polyhemoglobin in dogs, Acta Anaesthesiol. Scand., 29:537.

Keipert, P. E., Chang, T. M. S., 1984, Preparation and in vitro characteristics of a blood substitue based on pyridoxylated polyhemoglobin, Appl. Biochem. Biothechnol. 10:133.

Keipert, P. E., Chang, T. M. S., 1987, Oxygen affinity, viscosity, oncotic pressure, and long-term storage feasibility of pyridoxylated hemoglobin, Biomat. Art. Cells Art. Org., 15:369.

Kothe, N., Eichentopf, B., Bonhard, K., 1985, Characterization of a modified, stroma-free hemoglobin solution as an oxygen-carrying plasma substitute, Surg. Gynecol. Obstet., 161:563.

Lenz, G., Benzing, H., Junger, H., Bissinger, U., Feulner, R., Posininsky, H., Schlerf, S., 1983, Sauerstofftransportcharakterisika polymerisierter pyridoxalierter Hämoglobinlösungen in vitro und in vivo, Anaesthesist (Suppl.), 32:298.

Lenz, G., 1988, Pryidoxalierte Polyhämoglobinlösung als sauerstofftransportierendes Volumenersatzmittel, Enke, Stuttgart.

Messmer, K., Jesch, F., Schaff, J., Schoenberg, M., Pielsticker, K., Bonhard, K., 1978, Oxygen supply by stroma-free hemoglobin, in: "Blood Substitutes and Plasma Expanders," G. A. Jamieson, T. J. Greenwalt, eds., Alan R. Liss, New York.

Moss, G. S., DeWoskin, R., Rosen, A. L., Levine, H., Palani, C. K. (1976). Transport of oxygen and carbon dioxide by hemoglobin-saline solution in the red cell-free primate, Surg. Gynecol. Obstet., 142:357.

Moss, G. S., Gould, S. A., Seghal, L. R., Seghal, H. L., Rosen, A. L., 1984, Hemoglobin solution - from tetramer to polymer, Surgery, 95:249.

245

Rosen, A. L., Gould, S. A., Seghal, L. R., Seghal, H. L., Moss, G. S., 1983, Evaluation of efficacy of stroma free hemoglobin solutions, Prog. Clin. Biol. Res., 122:79.

Tam, S. C., Blumenstein, J., Wong, J. T. F., 1978, Blood replacement in dogs by dextran-hemoglobin, Can. Biochem. J., 56:981.

CORONARY CAPILLARY DEVELOPMENT FOLLOWING TREATED AND UNTREATED

FETAL HYPOXIA IN THE RAT

S.E. Campbell, K. Rakusan, and N.S. Faithfull

Department of Physiology, University of Ottawa
Ottawa, Canada and Department of Anesthesia
University of Manchester, Manchester, U.K.

Placental insufficiency leading to fetal growth retardation is a significant obstetrical problem. Fetal hypoxia resulting from placental insufficiency may have detrimental effects on the fetus as a whole and, depending on the time during organogenesis when the hypoxic insult occurs, may alter normal development of specific organ systems. Hypoxia has been shown to effect normal cardiac growth in neonatal rats (Hollenberg et al., 1976). Within the heart, changes in the growth and maturation of cardiac myocytes and the coronary vasculature may result.

Therapeutic interventions that increase the oxygen-carrying capacity of the maternal blood may enhance oxygen delivery across the placenta and alleviate the hypoxic fetal environment. Vileisis (1985) reported improved survival and growth of fetuses exposed to experimental placental insufficiency, resulting from severe restriction of normal uterine blood flow, after maternal supplemental oxygen inhalation. Other interventions that increase maternal PaO_2 and augment transplacental diffusion of oxygen may prove useful in treatment of the hypoxic fetus. One possible treatment that may prove beneficial is administration of oxygen-carrying perflurocarbon (PFC) emulsions.

PFC emulsions have a high affinity for oxygen and can enhance the microcirculatory distribution of oxygen because of their low viscosity and extremely small size, resulting in good penetrability of the emulsion particles (Faithfull et al., 1985). The potential use of PFC emulsions to enhance transplacental delivery of oxygen has been reported by Cefalo et al. (1984). Under conditions of isovolemic exchange and maternal hyperoxia using Fluosol-DA (20%) in pregnant ewes, fetal blood oxygen tension and oxygen content increased. Faithfull and Marshall (1989) have administered FC-43 and its oncotic component hydroxyethyl starch to pregnant rats with unilateral uterine artery ligation. They reported significant maternal and fetal (in both ligated and unligated uterine horns) weight losses in FC-43 treated animals. The present study was undertaken to determine the effect on the development and maturation of the coronary vasculature in both treated and untreated fetal hypoxia resulting from experimentally-induced placental insufficiency.

Oxygen Transport to Tissue XII, Edited by J. Piiper *et al.*
Plenum Press, New York, 1990

METHODS

Pregnant Sprague Dawley rats were used. On day 17 of gestation, maternal weights were recorded and experimental placental insufficiency was unilaterally produced under ketamine anesthesia (20-30 mg ip) by ligation of the uterine artery (Wigglesworth, 1964). Fetuses from the unligated horns served as internal controls. The number of fetuses in each horn were noted.

One group of pregnant animals received no further treatment. The remaining animals received further treatment consisting of daily iv injections via the tail vein starting on the day of uterine artery ligation. Some animals received 10 ml/kg of the emulsified perfluorocarbon FC-43 with and without inhalation of air enriched with oxygen (in order to achieve an inspired oxygen fraction [FiO_2] of 0.4). Other animals were treated with a 6% hydroxyethyl starch solution (HES), the oncotic component of the FC-43, with and without O_2-enriched air. The final treatment protocol consisted of injection of a stroma-free hemoglobin (non-heat treated bovine hemoglobin) solution (SFH) in association with O_2-enriched air.

On day 21 of gestation, maternal weights were recorded and the animals were sacrificed. Surviving fetuses were recovered and weighed, and their hearts were removed and fixed in 2% buffered glutaraldehyde. Fixed fetal heart weights were determined and the hearts were processed for embedding in Araldite. Transverse sections were cut at 1 μm thickness proceeding from the base to apex of the heart. Sections were collected at 80 μm intervals and stained with toluidine blue. Quantitative analysis of stained sections was done at the light microscopic level using a Bioquant Image Analysis system. Three sections (one toward the base, one approximately half-way between base and apex, and one toward the apex) were analyzed for each heart. Within each section, analysis was done on ten different fields (area 117 μm^2 per field) located in the endomyocardium of the left ventricular portion of the interventricular septum. The percentage of total area occupied by endothelialized and non-endothelialized sinusoids was determined. A comparison of these two populations of sinusoids was used as a measure of maturation of the terminal vascular bed.

Maternal weights were compared using a paired Student's t-test and fetal weights and resorptions were compared using the unpaired Student's t-test. In order to examine the differences in fetal heart weight, total number of sinusoids and the maturing of the terminal vascular bed as a function of group (fetuses from ligated vs. unligated horns) and treatment, a two-way analysis of variance was done. If significant differences were indicated with the ANOVA, multiple comparisons were made using the Bonferroni t-test.

RESULTS

Maternal weight changes between days 17 and 21 of gestation are presented in Table 1. A significant increase in maternal weight was seen in untreated and HES treated animals. FC-43 treatment resulted in a significant decrease (p<0.05) in maternal weight. Although maternal weights were greater at day 21 compared to day 17 of gestation in SFH treated animals, the increase was not significant.

Body weights for fetuses recovered on day 21 of gestation are shown in Figure 1. Fetal weights were significantly greater in unligated horns compared to ligated horns in all groups except the SFH treated

248

Table 1. Maternal weights at days 17 and 21 of gestation

Group	N	Day 17 Wt (g)	Day 21 Wt (g)
Nil	15	304 ± 5	313 ± 6*
FC-43	14	315 ± 10	303 ± 8*
FC-43+O_2	13	315 ± 8	303 ± 8*
HES	16	303 ± 6	312 ± 6***
HES+O_2	12	291 ± 7	302 ± 8*
SFH+O_2	7	301 ± 8	308 ± 11

Values are means ± SEM
Significantly different from day 17 at: *-p<0.05;
***-p<0.001

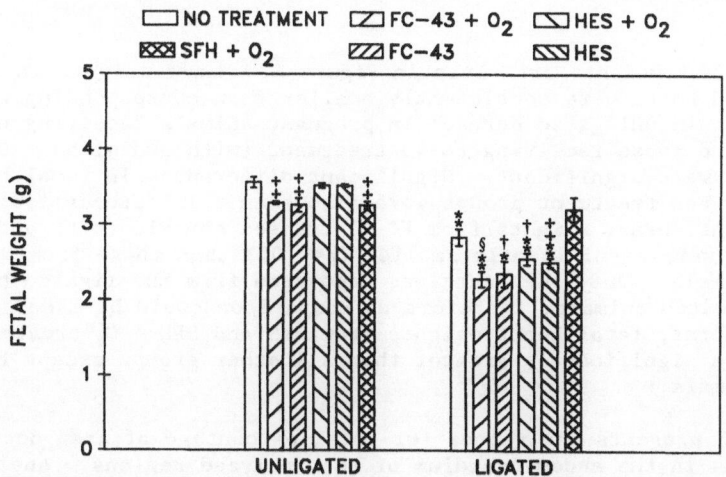

Figure 1. Fetal body weights at day 21 of gestation. Values are means ± SEM. Significantly < corresponding value in unligated horn: * −p<0.01; ** −p<0.001. Significantly < untreated value in same horn: † −p<0.05; ‡ −p<0.01; § −p<0.001.

animals. Within the unligated horns, fetal weights from treated animals were significantly less (p<0.01) than untreated animals in all groups except the HES treated groups. Body weight of fetuses from untreated animals in ligated horns was significantly greater than fetal weights from treated animals in all groups except the SFH treated animals.

The number of fetal resorptions present at day 21 of gestation are given in Table 2. As would be expected, the number of fetal resorptions in the ligated horns of all groups are significantly greater (p<0.001) than the number in unligated horns. Although the incidence of resorptions was slightly greater in the unligated horns of SFH and HES treated animals and the ligated horns of the FC-43 treated animals compared to their respective untreated groups, no significant differences were seen within either horn between treatment groups.

Table 2. Number of fetal resorptions per horn at day 21 of gestation

Group	N	Unligated	Ligated
Nil	18	0.11 ± 0.08	5.94 ± 0.63***
FC-43	14	0.14 ± 0.10	6.57 ± 0.44***
FC-43+O_2	13	0.08 ± 0.08	6.23 ± 0.74***
HES	16	0.31 ± 0.18	4.75 ± 0.69***
HES+O_2	12	0.08 ± 0.08	4.83 ± 0.88***
SFH+O_2	7	0.43 ± 0.20	5.29 ± 0.95***

Values are means $\pm$ SEM
***-Significantly different from corresponding unligated value at p<0.001

Fetal heart weights are given in Figure 2. Heart weights in fetuses from ligated horns were consistently smaller than corresponding weights from fetuses in unligated horns. In pregnant animals receiving no treatment and those receiving FC-43 treatment (with and without O_2), the differences were significant. Significant differences in fetal heart weights between treatment groups were also seen. In fetuses from ligated horns, heart weights from FC-43 treated animals (with and without O_2) were significantly smaller (p<0.05) than those from HES treated animals. Only one heart was recovered from the ligated horns in SFH + O_2 treated animals, therefore no comparison could be made. In unligated horns, fetal heart weights from HES and SFH + O_2 treated animals were significantly greater than all other groups except FC-43 treated animals.

Figure 3 presents the values for total percentage of area occupied by sinusoids in the endomyocardium of the analyzed regions. Analysis of variance indicated a significantly higher (p<0.01) percentage of sinusoids in fetal hearts from ligated compared to unligated horns. These differences could not be pinpointed to any one treatment group comparison. Generally, hearts from ligated horns of all treatment groups had a higher percentage of total sinusoids compared to hearts from unligated horns. Values determined for the SFH + O_2 and FC-43 treated ligated horns did not follow this trend, but no conclusions could be drawn from these groups due to low numbers of recovered hearts. Comparison of hearts from different treatment groups recovered from ligated horns showed a significant decrease (p<0.01) in the number of sinusoids in HES and HES + O_2 treated animals compared to animals receiving no treatment. In unligated horns, only fetal hearts from the

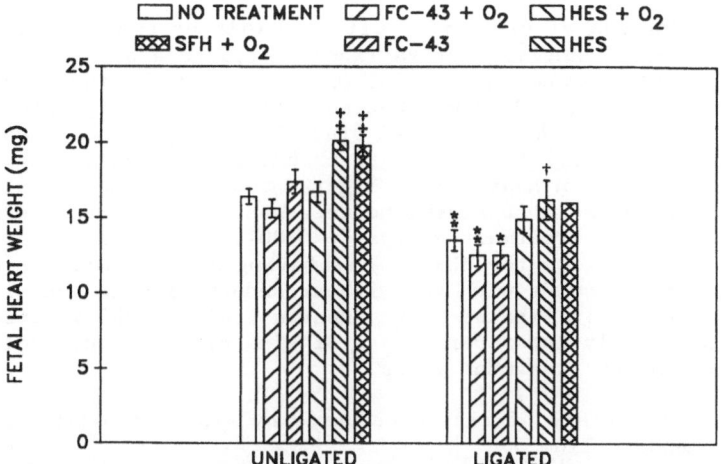

Figure 2. Fetal heart weights at day 21 of gestation. Values are means ± SEM. Significantly < corresponding value in unligated horn at:*–p<0.05; ‡–p<0.01. Significantly > FC–43 values in same horn: †–p<0.05. Significantly > untreated, FC–43 + O$_2$ and HES + O$_2$ values in same horn: ‡– p<0.01.

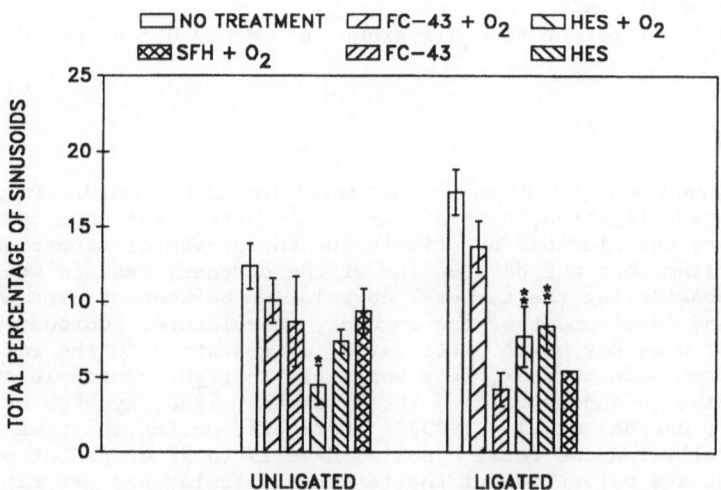

Figure 3. Total percentage of area occupied by sinusoids. Values are means ± SEM. Significantly different from untreated value of same horn at:*–p<0.05; ‡–p<0.01.

HES + O_2 treated group exhibited a significant decrease (p<0.05) in number. The trend in all treatment groups, from both ligated and unligated horns, was a lower percentage of total sinusoids compared to the no treatment group.

The percentages of sinusoids, with and without endothelium, are shown in Table 3. Analysis of variance indicated a significantly greater (p<0.01) percentage of sinusoids with endothelium in hearts from ligated versus unligated horns, but no statistically significant differences in non-endothelialized sinusoids were seen between the two horns. Generally, all treatment groups had a lower percentage of both types of sinusoids in hearts from ligated and unligated horns compared to hearts from animals receiving no treatment. There was a significant difference between treatment groups (p<0.001) and multiple comparisons identified hearts from the HES + O_2 as having a significantly lower percentage of sinusoids with and without endothelium in both ligated and unligated horns. In ligated horns, hearts from the HES treated animals had a significantly lower percentage of sinusoids without endothelium compared to the no treatment group.

Table 3. Percentage of tissue occupied by sinusoids with and without endothelium

Group	N	With endothelium Ligated	N	Unligated	N	Without endothelium Ligated	N	Unligated
Nil	7	11.3 ± 1.2	7	8.5 ± 1.1	7	6.0 ± 0.7	7	3.9 ± 0.6
FC-43	3	2.8 ± 0.6	5	6.1 ± 2.3	3	1.4 ± 0.5	5	2.7 ± 0.7
FC-43+O_2	10	9.6 ± 1.3	11	6.5 ± 1.1	10	4.1 ± 0.5	11	3.6 ± 0.5
HES	6	6.5 ± 1.3	5	4.1 ± 0.6	6	1.9 ± 0.5**	5	3.4 ± 0.8
HES+O_2	6	5.3 ± 1.8*	6	3.0 ± 0.9*	6	2.4 ± 0.7**	6	1.5 ± 0.4*
SFH+O_2	1	3.5	5	6.5 ± 0.9	1	1.9	5	2.9 ± 0.7

Values are means $\pm$ SEM
Significantly different from Nil group at: *-p<0.05; **-p<0.01

DISCUSSION

The present study indicates that fetal hypoxia, brought about by uterine artery ligation, does significantly effect the development of the coronary vascular bed by stimulating the growth of sinusoids. It is not surprising that the development of the coronary vessels would be affected considering the temporal correlation between uterine artery ligation and development of the coronary vasculature. Coronary artery development does not begin until day 17 of gestation in the rat and proceeds from base to apex, left ventricle to right ventricle and epimyocardium to endomyocardium (Dbaly et al., 1968; Rychter et al., 1971, 1972; Ostadal et al., 1975). Therefore, during the time of experimentally induced fetal hypoxia, days 17 to 21 of gestation, the development and maturation of the terminal vascular bed was taking place in all regions of the ventricles of the heart. In fact, augmented proliferation of vascular endothelial cells is still possible in hearts of neonatal rats exposed to prolonged hypoxia (Hollenberg et al., 1976).

FC-43 was administered in an effort to improve oxygen delivery to fetuses in the hypoxic environment of the ligated uterine horns. If hypoxia is a stimulating factor for growth of coronary vessels (sinusoids), then a decrease in the percentage of area occupied by the terminal vascular bed would be expected if oxygen delivery was improved. This was indeed the case in hearts from fetuses in ligated horns. However, considering the significant maternal and fetal weight losses and increased number of resorptions accompanying FC-43 treatment, it is not known whether the decreased vascular response was due to better oxygenation or a toxic effect on vascular growth brought about by the treatment. Additional evidence in support of potential toxicity is that fetal weights and the number of total sinusoids were also decreased in fetuses from unligated horns of FC-43 treated animals.

Altered hemostasis has been reported with administration of Fluosol-DA resulting from activation of the plasma complement system (Vercellotti and Hammerschmidt, 1985). Activation of the alternate pathway results in the aggregation and attachment of polymorphonuclear leukocytes and is related to the emulsifier Pluronic F-68 in the Fluosol-DA (Vercelotti et al., 1982). FC-43 also contains Pluronic F-68 and the toxicity of this component could be associated with known impurities and adjuncts associated with the emulsifier (Biro and Blais, 1987; Faithfull and Marshall, 1989). Cefalo et al. (1984) reported that the PFC particles did not cross the placenta. However, the Pluronic is of small enough size to potentially cross the placenta and effect the fetus directly.

Treatment with HES appears to have the most detrimental effect on the coronary vasculature. Significant decreases in the percentage of sinusoids with and without endothelium in both ligated and unligated horns were seen. HES functions as a plasma expander and also may contribute to stabilizing the PFC micelles in conjunction with surfactants (Pluronic F-68) (Biro and Blais, 1987). Considering a half-life of approximately 40 hours for HES (Doenicke et al., 1977), the expanded blood volume in the maternal vasculature resulting from daily treatments of HES may have resulted in the development of a mild to moderate anemia. This may have had an effect on fetal heart development. It is not known if the relatively large HES molecules can pass the placenta and directly affect the fetus.

Administration of SFH leads to an increase in fetal weight compared to treated and non-treated animals in ligated horns, and hearts from SFH treated animals are greater in mass than those from FC-43 treated animals. Furthermore, as was the case with FC-43 treatment, the vascular response appears to be decreased. SFH was given as an alternative "blood substitute" for comparison to the response elicited by FC-43. The response brought about by SFH treatment varies depending on the procedures for purification of and the individual components of the SFH itself. Although a very mild anemia could result from daily SFH treatments, it is unlikely that the response would be significant. At least a portion of the SFH probably crosses the placenta, resulting in a small increase in oxygenation of the plasma phase of the fetal circulation. Therefore, the decreased stimulation in the terminal vascular bed with SFH treatment is probably due to increased delivery of oxygen to the fetus.

Inhalation of supplemental oxygen does not appear to have a significant effect in the present study. Comparison of values in treated groups with oxygen supplementation to those without supplemental oxygen resulted in no significant differences. Oxygen loading is linearly related to the pO_2 in PFC emulsions (Biro and Blais, 1987). Maintenance of a high alveolar pO_2 is necessary to take full advantage of the PFC oxygen affinity. Cefalo et al. (1984) used 100% oxygen to load Fluosol-DA in their study of enhanced transplacental diffusion of oxygen. The 40% oxygen level used in the present study appears to be inadequate to appreciably alter delivery of oxygen from that in non-oxygen treated animals.

Considering the temporal events of coronary vascular development, a more mature vascular bed should contain a greater percentage of endothelialized sinusoids compared to non-endothelialized sinusoids. Although the vascular response was decreased overall in FC-43 and SFH treated animals, the maturation of the bed was not significantly altered compared to non-treated animals. HES treatment resulted in significant decreases in the percentage of both types of sinusoids in hearts from ligated and unligated horns, but no significant effect on the maturation of the terminal vascular bed was noted.

In summary, FC-43 appears to be toxic to both mother and fetus as indicated by significantly lower birth weights and weight losses in mothers compared to controls. Heart weights were also significantly lower in fetuses from both untreated and FC-43 treated ligated horns. In contrast, SFH treatment appears to augment heart mass. Endomyocardial vascular capacity, represented by endothelialized and non-endothelialized sinusoids, is increased in hearts from fetuses in ligated horns compared to unligated horns. FC-43 and SFH tend to decrease the vascular growth response. HES treatment, in fetuses from both ligated and unligated horns, significantly decreased the number of fetal heart sinusoids. It is possible that components in FC-43 other than HES (e.g. Pluronic F-68) may contribute to the fetal and maternal toxicity. Introduction of more purified fluorocarbons, accomplished by running the pluronic emulsifiers through silica gel columns, may reduce the toxic effects. The methods developed to quantitate the growth and development of the fetal terminal vascular bed in the heart in the present investigation may be used to make that determination.

Supported by the Ontario Heart and Stroke Foundation and the Medical Research Council of Canada

REFERENCES

Biro, G. P. and Blais, P., 1987, Perfluorocarbon blood substitutes. CRC Crit. Rev. Oncol./Hematol., 6:311.
Cefalo, R. C., Seeds, J. W., Proctor, H. J. and Baker, V. V., 1984, Maternal and fetal effects of exchange transfusion with a red blood cell substitute. Am. J. Obstet. Gynecol., 148:859.
Dbaly, J., Ostadal, B. and Rychter, Z., 1968, Development of the coronary arteries in rat embryos. Acta Anat., 71:209.
Doenicke, A., Grote, B. and Lorenz, W., 1977, Blood and blood substitutes. Br. J. Anaesthesiol., 49:681.
Faithfull, N. S. and Marshall, H. W., 1989, The effect of fluorocarbon emulsion on placental insufficiency. In: Oxygen Transport to Tissue-XI, K. Rakusan, G. Biro, T. K. Goldstick and Z. Turek, eds., Plenum Press, New York, p. 357.
Faithfull, N. S., Fennema, M., Erdmann, W., Lapin, R., Smith, A. R., van Alphen, W., Essed, C. E. and Trouwborst, A., 1985, Tissue oxygenation by fluorocarbons. Adv. Exp. Med. Biol., 180:569.
Hollenberg, M., Honbo, N. and Samorodin, A. J., 1976, Effects of hypoxia on cardiac growth in neonatal rat. Am. J. Physiol., 231:1445.
Ostadal, B., Schiebler, T. H. and Rychter, Z., 1975, Relations between development of the capillary wall and myoarchitecture of the rat heart. Adv. Exp. Med. Biol., 53:375.
Rychter, Z., Jelinek, R. and Marhan, O., 1971, Progress of vascularization of the ventricular myocardium in the rat embryo. Physiol. Bohemoslov., 20:527.
Rychter, Z., Jelinek, R. and Marhan, O., 1972, Shape and location of

non-vascularized area of ventricular myocardium in rat embryo during terminal phase of heart vascularization. Folia Morphol., 20:21.

Vercellotti, G. M. and Hammerschmidt, D. E., 1985, Immunological biocompatibility in blood substitutes. Int. Anesth. Clin., 23:47.

Vercellotti, G. M., Hammerschmidt, D. E., Craddock, P. R. and Jacob, H. S., 1982, Activation of plasma complement by perfluorocarbon artificial blood: probable mechanism of adverse pulmonary reactions in treated patients and rationale for corticosteroid prophylaxis. Blood, 59:1299.

Vileisis, R. A., 1985, Effect of maternal oxygen inhalation on the fetus with growth retardation. Ped. Res., 19:324.

Wigglesworth, J. S., 1964, Experimental growth retardation in the foetal rat. Pathol. Bacteriol., 88:1.

PERFLUOROCHEMICAL OXYGEN CARRIERS AND ISCHAEMIC TISSUES

Kenneth C. Lowe

Mammalian Physiology Unit, Dept. Zoology, University of Nottingham, University Park, Nottingham NG7 2RD, UK

INTRODUCTION

There is growing interest in the potential use of synthetic, oxygen-carrying perfusates for the management of hypoxically ischaemic tissues. One such approach is to use perfusates based on emulsified perfluorochemicals (PFCs) and this has received much attention in recent years. This paper considers progress in this area and evaluates the potential beneficial therapeutic value of emulsified PFCs in myocardial, cerebral and intestinal ischaemia. Recent advances in the development and assessment of improved PFC emulsions have also been highlighted.

PERFLUOROCHEMICALS

PFCs are inert, highly fluorinated organic compounds which can dissolve large volumes of respiratory gases. For example, the solubility of oxygen in PFC liquids is over 40 vol. % whereas that of carbon dioxide can exceed 160 vol. % (Riess and Le Blanc, 1988). Gas solubility in PFCs varies linearly with partial pressure according to Henry's Law and this contrasts sharply with the characteristic sigmoid binding curve of oxygen to haemoglobin (Riess and Le Blanc, 1988).

PFC oils are immiscible with blood and other aqueous systems but they can be used intravascularly in an emulsified form. One such emulsion, Fluosol-DA 20% (F-DA; Green Cross, Japan), is produced commercially and has been tested both experimentally and clinically as an oxygen-carrying resuscitation fluid (Faithfull, 1987; Lowe, 1987, 1988). F-DA consists of a 20% (w/v) emulsion of two PFCs, perfluorodecalin and perfluorotripropylamine, emulsified with Pluronic F-68 and egg yolk phospholipids in a balanced electrolyte solution (Table 1). A second commercial emulsion, Oxypherol (FC-43), which contains perfluorotributylamine (Table 1), is also available for animal experiments. While the gas solubilities of F-DA and FC-43 are correspondingly lower than those of their respective PFC components, they are still several times greater than those of plasma (Faithfull, 1987).

Microcirculatory support and ischaemic rescue

Attention has focussed on the potential therapeutic value of emulsified PFCs for providing microcirculatory support in hypoxically ischaemic tissues. In addition to their oxygen-transporting properties, PFCs and their emulsions have very low viscosity which is almost independent of shear rate. This contrasts with the viscosity of blood which increases markedly under the low flow rates normally present in the microcirculation (Fig. 1). For example, F-DA has only 34% of the viscosity of blood under the shear rate conditions found in the microcirculation. This property makes PFC emulsions attractive as haemodiluents in ischaemic disease to reduce blood viscosity and improve perfusion through available collaterals. An additional advantageous feature of PFC emulsions is their small particle size relative to that of erythrocytes: the average particle size of commercially available emulsions is generally $< 0.25 \ \mu m$ - only about 3% of the diameter of a red cell. It has been proposed that PFC emulsion particles can penetrate into hypoxic tissue beds and by-pass circulatory obstructions caused by endothelial surface blisters ("blebs") and sludged red cells (Faithfull et al., 1984). PFCs may also re-oxygenate sludged erythrocytes and hence reverse the membrane stiffening which normally occurs in hypoxic and acidotic tissues (Faithfull, 1987). Thus, the vicious circle of sludged standstill can be arrested.

There is evidence that PFC emulsions may also affect the outcome of ischaemic hypoxia by direct actions on leucocytes. Thus, commercial PFC emulsions can inhibit neutrophil chemotaxis, adherence and superoxide release (Virmani et al., 1983, 1984) and this appears due, at least in part, to the effects of the Pluronic F-68 surfactant component (Lane and Lamkin, 1984; Janco et al., 1985)(Table 1). This mechanism would help to explain the decreased inflammatory response and increased myocardial salvage which follows early post-infarction haemodilution with F-DA (Kolodgie et al., 1985).

Myocardial ischaemia

F-DA and FC-43 have been studied as oxygen-carrying perfusates for myocardial ischaemic rescue in dogs (Nunn et al., 1983; Forman et al., 1985, 1987; Kolodgie et al., 1985, 1986), pigs (Faithfull et al., 1986, 1988) and rabbits (Mushlin et al., 1985) (Table 2). Initial studies showed that, following vascular occlusion, infusion of PFC emulsion improved flow and, in some cases, oxygen supply to the ischaemic region and reduced the area of infarcted tissue. Menasche and colleagues showed that emulsified perfluorotributylamine alone was unable to prevent ischaemic damage in the dog heart but nevertheless offered protection against reperfusion injury, probably due to altered leucocyte functions as discussed above (Menasche et al., 1985). Care must, of course, be taken in interpreting results obtained using different emulsions in different species.

While certain questions concerning the effectiveness of emulsified PFCs to improve myocardial oxygen supply over and above that provided by an oxygenated crystalloid solution (Tabayashi et al., 1988) have not been resolved, there are still very strong indications that an important therapeutic role exists for such compounds in the management of ischaemic heart disease. However, further studies on the cardioplegic properties of PFC emulsions are needed, particularly in relation to their effectiveness both alone and in combination with other therapeutic agents.

Coronary artery angioplasty

One area attracting increasing interest is the use of PFC emulsions as distal oxygen-carrying perfusates during percutaneous transluminal coronary angioplasty (PTCA). This is a well established therapeutic procedure for the treatment of obstructive artery disease. One complication, however, is that balloon inflation during PTCA invariably

TABLE 1. Composition of Fluosol-DA 20% and Oxypherol (FC-43) (all values are w/v %)

	Fluosol-DA	FC-43
Perfluorodecalin	14.0	–
Perfluorotripropylamine	6.0	–
Perfluorotributylamine	–	20.0
Pluronic F-68	2.7	2.56
Yolk phospholipids	0.4	–
Glycerol	0.8	–
NaCl	0.600	0.600
KCl	0.034	0.034
$MgCl_2$	0.020	0.020
$CaCl_2$	0.028	0.028
$NaHCO_3$	0.210	0.180
Glucose	0.180	0.180
Hydroxyethyl starch	3.0	3.0

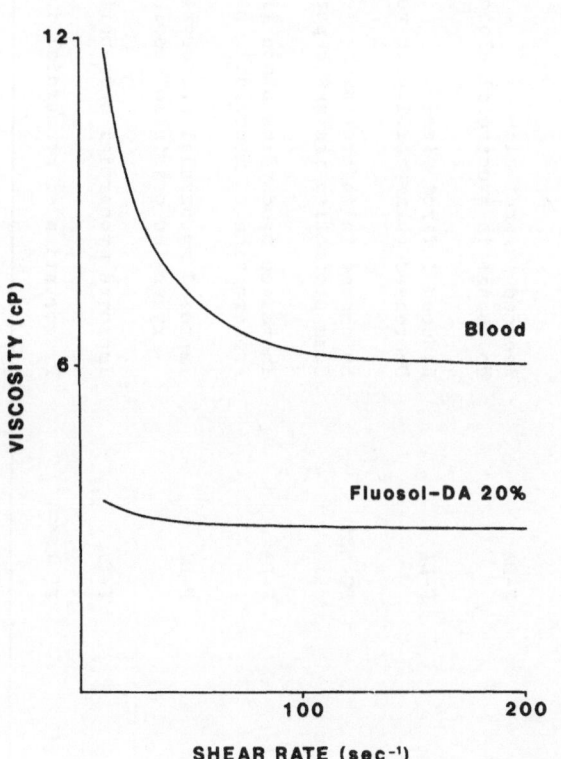

Fig. 1. The viscosity of normal blood vs F-DA.

Table 2. Use of PFC emulsions in experimental myocardial ischaemia

Species	Emulsion	Comments	References
Dog	F-DA*	Reduced infarct size	Nunn et al., 1983; Kolodgie et al., 1985
Dog	F-DA	Reduced infarct size No change in myocardial blood flow	Forman et al., 1985
Dog	F-DA	Reduced infarct size Decreased vulnerability to ventricular arrhythmia	Kolodgie et al., 1986
Dog	FC-43†	No change in infarct size Some protection against reperfusion injury	Menasche et al., 1985
Dog	F-DA	Increased myocardial blood flow Preservation of myocardial structural integrity	Forman et al., 1987
Dog	F-DA	Improved myocardial preservation, similar to oxygenated crystalloid solution	Tabayashi et al., 1988
Pig	F-DA	Improved myocardial oxygenation	Faithfull et al., 1986, 1988
Rabbit	FC-43**	Preservation of myocardial function	Mushlin et al., 1985

* F-DA = Fluosol-DA 20% (Green Cross, Japan)

** FC-43 = Oxypherol/Fluosol-43 (Green Cross, Japan)

† FC-43 = Perfluorotributylamine (20% w/v) emulsified with 2.56% (w/v) Pluronic F-68

Table 3. Use of PFC emulsions in experimental cerebral ischaemia

Species	Emulsion	Comments	Reference(s)
Cat	F-DA*	Maintenance of cerebral integrity	Peerless et al., 1981
Cat	FC-80†	Maintenance of cerebral function	Osterholm et al., 1983, 1984
Cat	F-DA††	Increased cerebral blood flow Improved cerebral oxygenation	Sutherland et al., 1984
Cat	F-DA	Partial protection against ischaemic injury	Peerless et al., 1985
Cat	FC-80	Reduced infarct size	Bose et al., 1984
Cat	F-DA	No change in cerebral blood flow No change in cerebral oxygenation No change in infarct size	Kolluri et al., 1986
Cat	FC-80	Reduced infarct size Protection against ischaemic injury	Bose et al., 1986
Cat	F-DA	No change in infarct size Decreased protection against ischaemic injury	Pereira et al., 1988
Dog	F-DA+	Protection against ischaemic injury	Suzuki et al., 1984
Rabbit	FC-43**	Increased brain surface oxygenation	Laycock et al., 1986
Rat	F-DA	Protection against ischaemic injury	Naruse et al., 1984

*   F-DA = Fluosol-DA 20% (Green Cross, Japan)
**   FC-43 = Oxypherol/Fluosol-43 (Green Cross, Japan)
†   FC-80 = Perfluorobutyltetrahydrofurane (20% w/v) emulsified with 5% (w/v) Pluronic F-68
††   F-DA = Fluosol-DA 35% (Green Cross, Japan; now discontinued)
+   Emulsion administered with various anti-oxidant drugs

261

leads to temporary ventricular ischaemia in those regions supplied by the occluded coronary artery. There have been various attempts to overcome this problem by infusing an oxygenated fluid through the balloon catheter. Blood is an obvious choice but it is technically difficult to achieve perfusion rates high enough to ensure adequate tissue oxygenation (Meier et al., 1984). Studies in both dogs (Spears et al., 1983; Roberts et al., 1986; Tokioka et al., 1987) and humans (Anderson et al., 1985, 1986; Cleman et al., 1986; Jaffe et al., 1986) have shown that the use of oxygenated F-DA as perfusate during PTCA resulted in preservation of both global and regional ventricular function together with decreased ST segment changes. An important finding was that the mean balloon occlusion time was 20% greater in those patients receiving F-DA compared to oxygenated Ringers-lactate solution (Anderson et al., 1985, 1986). Further studies are needed, however, since recent experiments in dogs showed that infusion of F-DA at 30 ml/min could not prevent changes in myocardial ultrastructure characteristic of mild ischaemia (Virmani et al., 1988).

Cerebral ischaemia

The possibility that emulsified PFCs can be used to ameliorate the effects of cerebral ischaemia has been studied in several species (Table 3). For example, Peerless and colleagues showed that infusion of F-DA could reduce, but not completely abolish, the effects of middle cerebral artery occlusion in cats (Peerless et al., 1985). Subsequent studies in dogs and rats showed that the protective effects of F-DA could be improved when used in conjunction with other therapeutic agents, notably anti-oxidants such as mannitol (Suzuki et al., 1984) or glycerol (Naruse et al., 1984).

Perfusion of oxygenated perfluorobutyltetrahydrofurane (FC-80) into the subarachnoid space in cats has also been shown to improve tissue damage following severe global cerebral ischaemia (Osterholm et al., 1983) or spinal cord injury (Osterholm et al., 1984). In addition, Bose and co-workers (Bose et al., 1984, 1986) reported that ventriculocisternal perfusion with oxygenated FC-80 emulsion was effective in preserving neuronal function following prolonged focal cerebral ischaemia (Table 3). One controversial area arising from this work, however, is the recent finding of Pereira and colleagues that administration of F-DA and 100% oxygen after the onset of cerebral ischaemia augmented the effects of reperfusion injury in cats (Pereira at al., 1988). This result, which contradicted many earlier observations (Table 3), draws attention to possible limitations of PFC therapy for brain protection in acute ischaemia.

Handa and others described the use of F-DA in 107 patients suffering from cerebral ischaemia of various origins (Handa et al., 1983). An encouraging finding was that temporary improvement for up to 24 hours was noted in 62% of cases studied. Unfortunately, it is not possible to draw any firm conclusions from this work since adequate controls were not included.

Intestinal ischaemia

In a preliminary report, Baba and Mizutani (1981) noted that intraluminal injection of oxygenated F-DA or FC-43 was effective in reducing necrosis in the ischaemic intestine of rats, dogs or guinea pigs. Subsequent experiments in rats showed that intraluminal injection of oxygenated FC-47 emulsion could minimize tissue damage caused by experimental intestinal ischaemia (Ricci et al., 1985). Further studies are needed to determine how far emulsified PFCs can be used to facilitate

262

rescue and preservation of the ischaemic gut and associated tissues and determine the mechanism(s) involved.

IMPROVED EMULSIONS

It is now generally accepted that the commercial emulsions produced by Green Cross have provided very useful "first generation" oxygen transport fluids for assessment in both clinical medicine and basic biomedical research. More recent attention has, however, focussed on the development of improved formulations containing highly purified components and showing enhanced stability characteristics (Sharma et al., 1987; Riess and Le Blanc, 1988; Bentley et al., 1989). Preliminary biocompatibility studies with one such preparation have been encouraging (Sharma et al., 1987; Lowe et al., 1989) and further studies are in progress.

SUMMARY

This paper has outlined the potential beneficial effects of emulsified PFCs for ischaemic tissue rescue. There are very strong indications that PFC emulsions will come to play an important part in the management of the ischaemic tissues. However, the immediate goal of future studies should be to assess the value of improved formulations in this context together with analysis of their mechanism(s) of action, both alone and in combination with other therapeutic agents.

REFERENCES

Anderson, H.V., Leimgruber, P.P., Roubin, G.S., Nelson, D.L. and Greuntzig, A.R., 1985, Distal coronary artery perfusion during percutaneous transluminal coronary angioplasty. Am. Heart J., 110: 720.

Anderson, H.V., Nelson, D.L., Leimgruber, P.P., Roubin, G.S. and Greuntzig, A.R., 1986, Coronary artery perfusion during percutaneous transluminal coronary angioplasty, in: "Transfusion Medicine: Recent Technological Advances", Murawski, K. and Peetoom, F., eds., p. 3, Liss, New York.

Baba, S. and Mizutani, K., 1981, The intraluminal administration of perfluorochemicals to the ischaemic gastrointestinal tract. Aust. N.Z. J. Surg., 51: 468.

Bentley, P.K., Davis, S.S., Johnson, O.L., Lowe, K.C. and Washington, C., 1989, Purification of Pluronic F-68 for perfluorochemical emulsification. J. Pharm. Pharmac., in press.

Bose, B., Osterholm, J.L. and Triolo, A., 1984, Focal cerebral ischemia: reduction in size of infarcts by ventriculo-subarachnoid perfusion with fluorocarbon emulsion. Brain Res., 328: 223.

Bose, B., Osterholm, J.L., Payne, J.B. and Chambers, K., 1986, Preservation of neuronal function during prolonged focal cerebral ischemia by ventriculocisternal perfusion with oxygenated fluorocarbon emulsion. Neurosurgery, 18: 270.

Cleman, M., Jaffe, C.C. and Wohlgelernter, D., 1986, Prevention of ischemia during percutaneous coronary angioplasty by transcatheter infusion of oxygenated Fluosol-DA 20%. Circulation, 74: 555.

Faithfull, N.S., 1987, Fluorocarbons. Current status and future applications. Anaesthesia, 42: 234.

Faithfull, N.S., Fennema, M., Erdmann, W., Lapin, R., Smith, A.R., van Alphen, W., Essed, C.R. and Trouborst, A., 1984, Tissue oxygenation by fluorocarbons. Adv. Exp. Med. Biol., 180: 569.

Faithfull, N.S., Erdmann, W., Fennema, M. and Kok, A., 1986, Effects of haemodilution with fluorocarbons or dextran on oxygen tensions in the acutely ischaemic myocardium. Br. J. Anaesth., 58: 1031.

Faithfull, N.S., Fennema, M. and Erdmann, W., 1988, Protection against myocardial ischaemia by prior haemodilution with fluorocarbon emulsions. Br. J. Anaesth., 60: 773.

Forman, M.B., Bingham, S., Kopelman, H.A., Wehr, C., Sandler, M.P., Kolodgie, F.D., Vaughn, W.K., Friesinger, G.C. and Virmani, R., 1985, Reduction of infarct size with intracoronary perfluorochemical in a canine preparation of reperfusion. Circulation, 71: 1060.

Forman, M.B., Puett, D.W., Bingham, S.E., Virmani, R., Tantengo, M.V., Light, R.T., Bajaj, A., Price, R. and Friesinger, G., 1987, Preservation of endothelial cell structure and function by intracoronary perfluorochemical in a canine preparation of reperfusion. Circulation, 76: 469.

Handa, H., Nagasawa, S., Yonekawa, Y., Naruo, Y. and Oda, Y., 1983, New treatment of cerebral vasospasm with Fluosol-DA 20%: protective effect on cerebral ischemia and change of cerebral blood flow, in: "Advances in Blood Substitute Research", Bolin, R.B., Geyer, R.P. and Nemo, G.J., eds., p. 299, Liss, New York.

Jaffe, C.C., Wohlgelernter, D., Highman, H.A. and Cleman, M., 1986, Oxygenated Fluosol-DA 20% distal infusion during coronary angioplasty protects myocardial function, in: "Transfusion Medicine: Recent Technological Advances", Murawski, K. and Peetoom, F., eds., p.21, Liss, New York.

Janco, R.L., Virmani, R., Morris, P.J. and Gunter, K., 1985, Perfluorochemical blood substitutes differentially alter human monocyte procoagulant generation and oxidative metabolism. Transfusion, 25: 578.

Kolluri, S., Heros, R.C., Hedley-Whyte, E.T., Vonsattel, J.P., Miller, D. and Zervas, N.T., 1986, Effect of Fluosol on oxygen availability, regional cerebral blood flow, and infarct size in a model of temporary focal cerebral ischemia. Stroke, 17: 976.

Kolodgie, F.D., Dawson, A.K., Forman, M.B. and Virmani, R., 1985, Effect of perfluorochemical (Fluosol-DA) on infarct morphology in dogs. Virchows Arch., 50: 119.

Kolodgie, F.D., Dawson, A.K., Roden, D.M., Forman, M.B. and Virmani, R., 1986, Effect of Fluosol-DA on infarct morphology and vunerability to ventricular arrhythmia. Am. Heart J., 112: 1192.

Lane, T.A. and Lamkin, G.E., 1984, Paralysis of phagocyte migration due to an artifical blood substitute. Blood, 64: 400.

Laycock, J.R.D., Coakham, H.B., Silver, I.A. and Walters, F.J.M., 1986, Effect of carotid artery ligation and infusion of Fluosol FC-43 on brain surface oxygen tensions. Stroke, 17: 1242.

Lowe, K.C., 1987, Perfluorocarbons as oxygen-transport fluids. Comp. Biochem. Physiol., 87A: 825.

Lowe, K.C. (ed.), 1988, "Blood Substitutes: Preparation, Physiology and Medical Applications", Ellis Horwood, Chichester.

Lowe, K.C., Bollands, A.D. and Raven, P.D., 1989, Effects of a novel perfluorochemical emulsion on lymphoid tissues and immunocompetence in rats: time course effects relative to immunological challenge. Comp. Biochem. Physiol., in press.

Meier, B., Greuntzig, A.R. and Brown, J.E., 1984, Percutaneous arterial perfusion of acutely occluded coronary arteries in dogs. J. Am. Coll. Cardiol., 3: 505.

Menasche, P., Escorsin, M., Birkui, P., Lavergne, A., Fauchet, M., Commin, P., Lorente, P., Geyer, R.P. and Piwnica, A., 1985, Limitations of fluorocarbons in reducing myocardial infarct size. Am. J. Cardiol., 55: 830.

Mushlin, P.S., Boucek, R.J., Parrish, M.D., Graham, T.P. and Olson, R.D., 1985, Beneficial effects of perfluorochemical artificial blood on cardiac function following coronary occlusion. Life Sci., 36: 2093.

Naruse, S., Horikawa, Y., Tanaka, C., Hirakawa. K., Nishikawa, H. and Watari, H., 1984, Measurements of in vivo energy metabolism in experimental cerebral ischaemia using 31P-NMR for the evaluation of protective effects of perfluorochemicals and glycerol. Neurol. Res., 6: 169.

Nunn, G.R., Dance, G., Peters, J. and Cohn, L.H., 1983, Effect of fluorocarbon exchange transfusion on myocardial infarction size in dogs. Am. J. Cardiol., 52: 203.

Osterholm, J.L., Alderman, J.B., Triolo, A.J., D'Amore, B.R. and Williams, H.D., 1983, Severe cerebral ischaemia treatment by ventriculosubarachnoid perfusion with an oxygenated fluorocarbon emulsion. Neurosurgery, 13: 381.

Osterholm, J.L., Alderman, J.B., Triolo, A.J., D'Amore, B.R. and Williams, H.D., 1984, Oxygenated fluorocarbon nutrient solution in the treatment of experimental spinal cord injury. Neurosurgery, 15: 373.

Peerless, S.J., Ishikawa, R., Hunter, I.G. and Peerless, M.J., 1981, Protective effect of Fluosol-DA in acute cerebral ischaemia. Stroke, 12: 558.

Peerless, S.J., Nakamura, R., Rodriguez-Salazar, A. and Hunter, I.G., 1985, Modification of cerebral ischaemia with Fluosol. Stroke, 16: 38.

Pereira, B.M., Weinstein, P.R. and Rodriquez y Baena, R., 1988, Effect of treatment with Fluosol and mannitol during temporary middle cerebral artery occlusion in cats. Neurosurgery, 23: 139.

Ricci, J.L., Sloviter, H.A. and Ziegler, M.M., 1985, Intestinal ischemia: reduction of mortality utilizing intraluminal perfluorochemical. Am. J. Surg., 149: 84.

265

Riess, J.G. and Le Blanc, M., 1988, Preparation of perfluorochemical emulsions for biomedical use: principles, materials and methods. in: "Blood Substitutes: Preparation, Physiology and Medical Applications", Lowe, K.C., ed., p. 94, Ellis Horwood, Chichester.

Roberts, C.S., Anderson, H.V., Carboni, A.A., Justicz, A.G.S., Leimgruber, P.P., Klonar, R.A. and Greuntzig, A.R., 1986, Usefulness of intracoronary infusion of fluorocarbon distal to prolonged coronary occlusion by angioplasty balloon in dogs. Am. J. Cardiol., 57: 1202.

Sharma, S.K., Bollands, A.D., Davis, S.S. and Lowe, K.C., 1987, Emulsified perfluorochemicals as physiological oxygen-transport fluids: assessment of a novel formulation, in: "Oxygen Transport to Tissue", Vol. IX, Silver, I.A. and Silver, A., eds., p. 97, Plenum, London.

Spears, J.R., Serur, J., Baim, D.S., Grossman, W. and Paulin, B., 1983, Myocardial protection with Fluosol-DA during prolonged balloon occlusion in the dog. Circulation, 68 (Suppl.III): III-80.

Sutherland, G.R., Farrar, J.K. and Peerless, S.J., 1984, The effect of Fluosol-DA on oxygen availability in focal cerebral ischaemia. Stroke, 15: 829.

Suzuki, J., Fujimoto, S., Mizoi, K. and Oba, M., 1984, The protective effect of combined administration of anti-oxidants and perfluorochemicals on cerebral ischemia. Stroke, 15: 672.

Tabayashi, K., McKeown, R.P., Miyamoto, M., Luedtke, A.E., Thomas, R., Allen, M.D., Misbach, G.A. and Ivey, T.D., 1988, Ischaemic myocardial protection. J. Thoracic Cardiovasc. Surg., 95: 239.

Tokioka, H., Miyazaki, A., Fung, P., Rajagopalan, R.E., Kar, S, Meerbaum, S., Corday, E., Drury, J.K., 1987, Effects of intracoronary infusion of arterial blood or Fluosol-DA 20% on regional myocardial metabolism and function during brief coronary artery occlusions. Circulation, 75: 473.

Virmani, R., Warren, D., Rees, D., Fink, L.M. and English, D., 1983, Effects of perfluorochemical on phagocytic function of leukocytes. Transfusion, 23: 512.

Virmani, R., Fink, L.M., Gunter, K. and English, D., 1984, Effect of perfluorochemical blood substitutes on human neutrophil function. Transfusion, 24: 343.

Virmani, R., Kolodgie, F.D., Osmialowski, A., Zimmerman, P., Mergner, W. and Forman, M.B., 1988, Myocardial protection by perfluorochemical infusion during transient ischaemia produced by balloon coronary occlusion. Am. Heart J., 116: 421.

OXYGEN-TRANSPORT FLUIDS BASED ON PERFLUOROCHEMICALS: EFFECTS ON LIVER
BIOCHEMISTRY

Kenneth C. Lowe and Fiona H. Armstrong

Mammalian Physiology Unit, Department of Zoology

University of Nottingham, Nottingham NG7 2RD, UK

INTRODUCTION

Emulsified perfluorochemicals (PFCs) have been tested as oxygen transport fluids in several species, including man (Lowe, 1987, 1988a,b). The most widely used preparation is the "first-generation" commercial emulsion, Fluosol-DA 20% (F-DA) which contains perfluorodecalin (FDC) and perfluorotripropylamine (FTPA) emulsified with the poloxamer surfactant, Pluronic F-68 (Naito and Yokoyama, 1978). An additional commercial formulation, Oxypherol (FC-43), which consists of emulsified perfluorotributylamine (FTBA), is also available for animal experiments (Lowe and Bollands, 1985).

PFC emulsion particles can accumulate in lymphoid tissues and this depends on emulsion composition, dose and route of injection together with both tissue and species studied (Bollands and Lowe, 1986, 1987; Lowe, 1987; Lowe and Bollands, 1988). However, little attention has focussed on changes in tissue biochemical function following injection of PFCs and their emulsions. Previous work has shown that FDC oil is an effective inducer of hepatic microsomal cytochromes P-450 in both rats (Obraztsov et al., 1985) and mice (Khlopushina et al., 1986) but the effects of different PFC emulsions or their constituents on this and related enzyme systems have not been studied systematically.

In the present experiments we have studied the effects of injecting different PFCs, their emulsions or the Pluronic F-68 surfactant on liver cytochromes P-450 in rats. The effects of one such preparation, a novel emulsion of FDC stabilized by the addition of a small quantity of a perfluorinated, polycyclic, C-16 higher boiling point oil (HBPO) additive (Sharma et al., 1987) have hitherto been untested. Additional objectives were to study the time course of response over 7 days and identify any differences between male and female animals. Some of these results have already been published in a preliminary form (Armstrong and Lowe, 1988).

MATERIALS AND METHODS

Care of animals and experimental procedures

Mature male or female Wistar rats (body weight 180 - 310g; N = 186) were used. They were maintained in the laboratory animal house under

Oxygen Transport to Tissue XII, Edited by J. Piiper *et al.*
Plenum Press, New York, 1990

controlled conditions (13 hr light, 11 hr dark; temperature 24 ± 1 °C) and fed on a standard diet (Rat and Mouse Breeding Diet, Haygates, Birmingham) ad libitum. Prior to any experimental treatment, male or female animals were allocated randomly into 1 of 12 experimental groups as follows: Group I ($N = 24$) saline controls, intravenous (i.v.) injection; Group II ($N = 24$) F-DA i.v.; Group III ($N = 23$) FC-43 i.v.; Group IV ($N = 23$) novel emulsion i.v.; Group V ($N = 23$) commercial grade Pluronic F-68 i.v.; Group VI ($N = 24$) purified Pluronic i.v.; Group VII ($N = 7$) saline controls, intraperitoneal (i.p.) injection; Group VIII ($N = 7$) corn oil i.p. injection; Group IX ($N = 7$) FDC oil i.p.; Group X ($N = 8$) FTPA oil i.p.; Group XI ($N = 8$) FTBA oil i.p.; Group XII ($N = 8$) C-16 HBPO i.p.

I.v. injections were made via a tail vein and i.p. injections were made in the inguinal region under light ether anaesthesia. FDC (Flutec PP5), FTPA, FTBA and the C-16 HBPO, perfluoroperhydrofluoranthrene, were obtained from RTZ Chemicals (ISC Division), Avonmouth. F-DA 20% (Green Cross, Osaka, Japan) was freshly prepared according to the manufacturer's instructions while FC-43 was used as supplied (Naito and Yokoyama, 1978; Table 1). The novel emulsion consisted of 20% (w/v) FDC and 1% (w/v) of the C-16 oil emulsified with 4% (w/v) Pluronic F-68 (Atochem/ICI Petrochemicals, Runcorn) in an aqueous phase containing 0.9% (w/v) NaCl (Sharma et al., 1987; Table 1). Animals in Group V were injected with a 4% (w/v) solution of commercial grade Pluronic F-68 whereas those in Group VI received an identical dose of Pluronic solution after it had been purified using a silica-Amberlite resin column (Bentley et al., 1989). The dose of F-DA, FC-43, novel emulsion or Pluronic solution injected was 10 ml/kg body weight throughout; control animals in Group I received an identical dose of 0.9% (w/v) sterile NaCl solution (Viaflex; Travenol, Thetford).

Animals in Groups IX-XII were injected with 20 ml/kg body weight of PFC oil corrected for a volume equivalent to that in the approriate emulsion studied. Control animals in Group VII and VIII received equivalent doses of either sterile saline or corn oil (Sigma, Poole) respectively.

At either 24 hr, 72 hr or 7 days after injection, animals in Groups I-VI were killed by cervical dislocation and their livers removed and weighed; animals in Groups VII-XII were sacrificed after 72 hr.

TABLE 1. Composition of Fluosol-DA 20%, Oxypherol (FC-43) and the Novel Emulsion (all values are w/v %)

	Fluosol-DA	FC-43	Novel Emulsion
Perfluorodecalin	14.0	–	20.0
Perfluorotripropylamine	6.0	–	–
Perfluorotributylamine	–	20.0	–
Perfluoroperhydrofluoroanthrene	–	–	1.0
Pluronic F-68	2.7	2.56	4.0
Yolk phospholipids	0.4	–	–
Glycerol	0.8	–	–
NaCl	0.600	0.600	0.900
KCl	0.034	0.034	–
$MgCl_2$	0.020	0.020	–
$CaCl_2$	0.028	0.028	–
$NaHCO_3$	0.210	0.180	–
Glucose	0.180	0.180	–
Hydroxyethyl starch	3.0	3.0	–

268

Treatment of tissue samples and analytical procedures

Liver samples (ca. 7 g) were homogenized in 21 ml 1.15% (w/v) KCl solution buffered with 20 mM Tris/HCl (pH 7.4). Homogenates were centrifuged at 10,000g for 30 min. at 4 °C. The supernatants were decanted off and re-centrifuged at 4 °C for 1 hr at 104,000g. Supernatants were discarded and the cell pellet re-suspended in 21 ml ice-cold buffered KCl solution. Samples were then stored at -20 °C until required for analysis.

Hepatic microsomal total protein concentrations and cytochromes P-450 concentrations were measured by conventional spectrophotometric assays (Lowry et al., 1951; Omura and Sato, 1964a,b) using a Kontron Uvikon 860 double-beam spectrophotometer.

Statistical analyses

Statistical analyses were performed according to the methods of Snedecor and Cochran (1980). Means and standard errors (SEM) have been used throughout and statistical significance between mean values was assessed using a conventional Student's t test. A probability of $P < 0.05$ was considered significant.

RESULTS

Changes in liver weights

Mean liver weights in male rats were increased by 27% and 32% ($P<0.01$) at 72 hr and 7 days respectively following injection of the novel emulsion (Table 2). Liver weight in male rats was similarly increased by 17% at 7 days after injection of F-DA ($P<0.05$); no comparable changes occurred following injection of FC-43. In female rats, liver weight was increased by 10% at 72 hr after injection of the novel emulsion ($P<0.05$) but was unaffected by injection of either F-DA or FC-43 (Table 2). Liver weight increased by 12% ($P<0.05$) in female rats injected with FDC but was otherwise unchanged in both male and female rats injected with PFC oils.

Changes in liver cytochromes P-450

Mean liver cytochromes P-450 concentrations in male rats were increased 2-3 fold at 72 hr following injection of the novel emulsion ($P<0.05$; Fig. 1) or FDC oil alone. A similar but much less pronounced increase also seen in response to F-DA injection ($P<0.05$; Fig. 1). Cytochromes P-450 were also up to 65% greater than control in livers from male rats at 7 days after injection of the novel emulsion but were similar to control in F-DA-injected animals (Fig. 1). Liver cytochromes P-450 in male rats were unchanged in response to injection of FC-43 (Fig. 1) or Pluronic solution, irrespective of purity (data not shown). None of the experimental treatments had any significant effect on cytochromes P-450 in female rats at any time: the overall mean enzyme concentration in livers from the 3 groups of i.v.-injected control females was 0.64 ± 0.09 nmol/mg protein; n = 12).

DISCUSSION

These results show that injection of a single low dose of either FDC oil or its emulsions into rats can increase synthesis of the cytochromes P-450 complex in the liver. However, the response was highly variable, depending on composition of emulsion, route of injection and sex of the recipient.

269

Table 2. Liver weights in male and female rats for up to 7 days following injection of saline (Group I), Fluosol-DA (Group II), FC-43 (Group III) or novel emulsion (Group IV)

Experimental Group	Liver weight (% body weight)		
	24 hr	72 hr	7 days
A. Male rats			
Group I	5.25 ± 0.20 (4)	5.27 ± 0.12 (4)	4.56 ± 0.10 (4)
Group II	5.63 ± 0.36 (4)	5.49 ± 0.25 (4)	5.34 ± 0.20 (4)*
Group III	5.26 ± 0.22 (4)	5.40 ± 0.10 (4)	4.72 ± 0.08 (4)
Group IV	5.42 ± 0.26 (4)	6.68 ± 0.25 (4)**	6.03 ± 0.21 (4)**
B. Female rats			
Group I	5.10 ± 0.26 (4)	5.30 ± 0.12 (4)	3.85 ± 0.28 (4)
Group II	5.14 ± 0.36 (4)	5.27 ± 0.29 (4)	4.05 ± 0.23 (4)
Group III	4.96 ± 0.05 (4)	4.84 ± 0.20 (3)	4.07 ± 0.11 (4)
Group IV	5.59 ± 0.24 (4)	5.82 ± 0.10 (4)*	4.52 ± 0.11 (3)

Values are given as mean $\pm$ s.e. of mean (no. observations in parenthesis).

* $P < 0.05$; ** $P < 0.01$ compared to corresponding control (Group I) mean value.

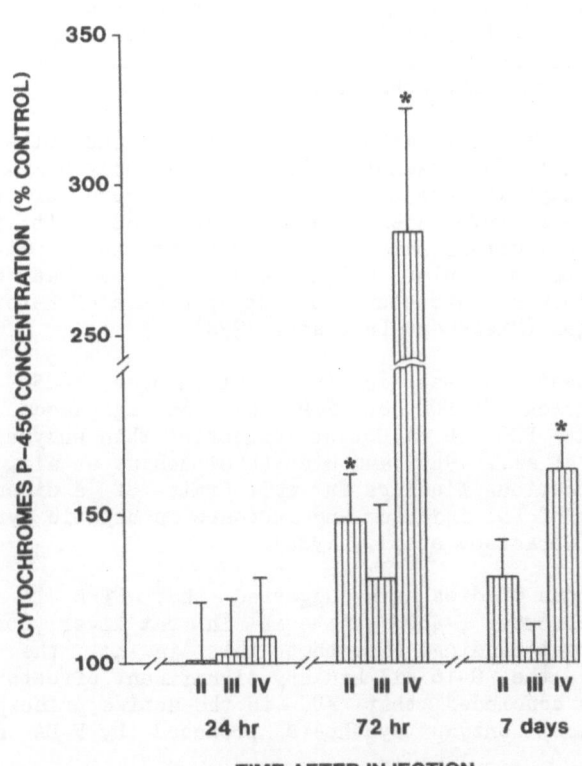

Fig. 1. Hepatic cytochromes P-450 concentrations in male rats at 24 hr, 72 hr and 7 days following injection of either F-DA (Group II), FC-43 (Group III) or novel emulsion (Group IV). Values are expressed as mean percentages of corresponding control (Group I) mean value; vertical bars represent s.e.m. (n = 4 throughout); *P < 0.05.

The increases in liver weight at 7 days after injection of the novel emulsion were in accord with previous observations in rats (Bollands et al., 1987). However, the present experiments also show that liver weight had increased significantly at 72 hr after injection and, in male rats, persisted for at least 7 days. These results also show that FDC is the emulsion component responsible for such changes. The finding that liver weight in female rats was unchanged for up to 7 days after injection of either F-DA or FC-43 was consistent with previous studies using comparable doses of these emulsions (Caiazza et al., 1984; Bollands and Lowe, 1986).

Earlier studies showed that PFCs can accumulate in the liver in both Kupffer cells and hepatocytes where they are retained within membrane-bound vesicles giving the cells a "foamy" appearance (Caiazza et al., 1984). This retention of PFCs leads to cellular hypertrophy although an increase in cell number can also occur following repeated injection of large doses (Pfannkuch and Schnoy, 1983).

Individual PFCs are retained in the liver and other tissues for differing periods: for example, FDC has a body half-life in rats of approximately 7 days whereas the corresponding figure for FTBA is 65 days (Naito and Yokoyama, 1978; Riess and Le Blanc, 1982). It is now accepted that the principal determining factor governing tissue retention times of different PFCs is molecular weight and not, as has been suggested previously, molecular structure or the presence of heteroatoms such as oxygen or nitrogen (Riess and Le Blanc, 1988).

The transient increase in liver cytochromes P-450 in male rats following injection of FDC or F-DA was in agreement with previous observations that FDC is a potent inducer of this enzyme system in both rats (Obraztsov et al., 1985) and mice (Kholpushina et al., 1986). We have also confirmed previous findings in male rats of a direct relationship between quantity of FDC injected and increase in hepatic cytochromes P-450 concentrations (Obraztsov et al., 1988).

While previous studies have suggested that FTPA is also able to stimulate cytochromes P-450 synthesis in rat liver (Obraztsov et al., 1988), the present findings show that this is not the case. Because neither FTPA or the C-16 oil had any significant effects on cytochromes P-450, it may be concluded that FDC is the active principle responsible for the increased enzyme synthesis produced by F-DA and the novel emulsion.

The finding that cytochromes P-450 concentrations were unchanged in both male and female animals injected with FTBA or the FC-43 emulsion was consistent with previous studies in male rats (Khlopushina et al., 1986; Huang et al., 1987; Obraztsov et al., 1988). This emphasises the existence of marked variability in the ability of different PFC oils to stimulate microsomal cytochromes P-450 synthesis in rat liver. Further studies are nevertheless needed to assess the effects of other variables, including age and strain of animals, which may also affect enzyme synthesis.

There have been several attempts to explain the differing effects of PFCs on rat liver cytochromes P-450: for example, Ullrich and Diehl (1971) showed that perfluorohexane formed an enzyme-substrate complex with cytochromes P-450. They concluded that this compound acted as a dead end inhibitor which uncoupled electron transport of the microsomal oxidases. Other PFCs are similarly able to form complexes with the cytochromes P-450 enzyme systems (Obraztsov et al., 1988) and are therefore believed to act in a manner comparable to other xenobiotics (Guengerich, 1988). In this regard, FDC emulsion containing 7.6% (w/v) perfluoromethylcyclo-hexylpiperidine can promote synthesis of cytochromes P-450 isomers identical with those which appear following injection of phenobarbital, a known enzyme inducer (Grishanova et al., 1987). However, PFCs are not metabolized either in vivo or in vitro, neither do they appear to affect other enzymes involved in xenobiotic metabolism in the short term (Guaitani et al., 1983). Nevertheless, the effects of PFCs on microsomal enzymes can persist for a considerable time after the compound(s) have been cleared from the liver (Huang et al., 1987).

The observed differences in liver response to individual PFCs are more difficult to explain: Huang and colleagues (Huang et al., 1987) proposed that the rate of liver clearance may be an important determinant of the differences between FDC and FTBA in their ability to induce cytochromes P-450. More recently, Obraztsov and others (Obraztsov et al., 1988) reported that the ability of different PFC compounds to act as inducers of cytochromes P-450 was related to their solubility in lipids and aqueous systems together with their saturated vapour pressure.

It is noteworthy that, in addition to any direct effects of PFCs on the cytochromes P-450 complex, increased enzyme synthesis may also be a consequence of PFC-induced liver cell hypertrophy. This is supported by the present finding that liver cytochromes P-450 concentrations in male rats tended to be highest in animals also having the greatest increases in liver weight.

The absence of changes in liver weight in animals injected with the unpurified pluronic solution contrasted with previous observations (Bentley et al., 1989) and show that this surfactant can have variable effects on this tissue. Neither commercial grade pluronic or the purified fraction had any significant effect on cytochromes P-450 concentrations in male or female rats and it may therefore be concluded that this compound does not contribute to the increased enzyme biosynthesis which occurred following injection of either F-DA or the novel emulsion.

The present results show a marked variation in response to FDC and its emulsions between male and female animals. Previous work indicates that this is due, at least in part, to differences in circulating androgens which are known to affect liver cytochromes P-450 in this species (Kato, 1977). Castration of adult male rats decreases the activities of sex-dependent hepatic microsomal drug-metabolizing enzymes, including cytochromes P-450, and this can be reversed by administration of androgens (Kato and Onoda, 1970). Moreover, androgens are able to increase the binding affinity of drugs to cytochromes P-450 while oestrogens have the opposite effect (Kato and Onoda, 1970). It is thus reasonable to conclude that the clear sex-dependent response seen in the present experiments was steroid-related, but this is under further investigation.

The principal finding from this study, namely that certain PFCs are able either directly or indirectly to increase cytochromes P-450 concentrations in male rats, could have important consequences in subsequent metabolism of hepatically-eliminated drugs. For example, partial exchange-transfusion of rats with F-DA has been shown to produce a marked increase in the clearance of antipyrine, a drug commonly used to assess hepatic metabolism (Shrewsbury et al., 1986; Shrewsbury, 1987). Such findings are particularly relevant in view of the proposed clinical use of emulsified PFCs since alterations in liver metabolism could impair drug efficacy and perhaps lead to toxicity problems.

SUMMARY

The effects of (1) i.v. injection of various perfluorochemical (PFC) emulsions, (2) different fractions of the non-ionic poloxamer surfactant, Pluronic F-68, or (3) i.p. injection of component PFC oils have been studied separately in male and female rats. Injection of 10 ml/kg body wt of either Fluosol-DA 20% (F-DA) or a novel perfluorodecalin emulsion containing a C-16 oil additive in male rats increased liver weight up to 7 days later; no corresponding effect occurred in response to injection of Oxypherol (FC-43). Liver weight was also increased in female rats at 72 hr after injection of the novel emulsion but this was less pronounced than in males; liver weight in female rats was unchanged in response to injection of either F-DA or FC-43 but was 12% greater at 72 hr after injection of FDC oil. Mean liver microsomal cytochromes P-450 concentrations in male rats were increased 2-3 fold at 72 hr after injection of either FDC, F-DA or the novel emulsion; a less pronounced increase was also seen at 7 days in animals receiving the novel emulsion. No significant alterations in cytochromes activity occurred in response to injection of FTPA, FTBA, the C-16 oil, FC-43 or either commercial grade or purified pluronic solution. Liver cytochromes P-450 concentrations in female rats were unaffected by any of the experimental treatments.

Acknowledgements

This work was supported by research grants from RTZ Chemicals Ltd (ISC Division), Avonmouth. F.H.A. was the recipient of a research studentship from the Science and Engineering Research Council. We are grateful to Dr T. Suyama of the Green Cross Corporation, Japan, for the generous gifts of Fluosol-DA and Oxypherol (FC-43). We are grateful to Dr R. Berry for helpful comments.

REFERENCES

Armstrong F.H. and Lowe K.C., 1988, Effects of perfluorochemical emulsion components on rat liver cytochromes P-450. Br. J. Pharmac., 95: 610P.

Bentley P.K., Davis S.S., Johnson O.L., Lowe K.C. and Washington C., 1989, Purification of Pluronic F-68 for perfluorochemical emulsification. J. Pharm. Pharmac. (in press).

Bollands A.D. and Lowe K.C., 1986, Effects of a perfluorocarbon emulsion, Fluosol-DA on rat lymphoid tissue and immunological competence. Comp. Biochem. Physiol., 85C: 309.

Bollands A.D. and Lowe K.C., 1987, Lymphoid tissue responses to perfluorocarbon emulsion in mice. Comp. Biochem. Physiol., 86C: 431.

Bollands, A.D., Lowe, K.C., Sharma, S.K. and Davis, S.S., 1987, Lymphoid tissue responses to a novel perfluorochemical emulsion in rats. J. Pharm. Pharmac., 39: 1021.

Caiazza S., Fanizza M. and Ferrari M., 1984, Fluosol 43 particle size localization pattern in target organs of rats. An electron microscopical study. Virchows Arch. path. Anat. Physiol. A., 404: 127.

Grishanova A.Y., Obraztsov V.V., Shekhtman D.G. and Lyakhobich, V.V., 1987, Phenobarbital type of induction of cytochrome P-450 of liver microsomes by perfluorodecalin. Biokhimiya, 52: 981.

Guaitani A., Villa P. and Bartosek I., 1983, Effect of perfluorodecalin on the microsomal mono-oxygenase system in perfused rat livers. Xenobiotica, 13: 39.

Guengerich F.P., 1988, Cytochromes P-450. Comp. Biochem. Physiol. 89C: 1.

Huang R., Cooper D.Y. and Sloviter H.A., 1987, Effects of intravenous emulsified perfluorochemicals on hepatic cytochromes P-450. Biochem. Pharmacol., 36: 4331.

Kato R., 1977, Drug metabolism under pathological and abnormal physiological states in animals and man. Xenobiotica, 7: 25.

Kato R. and Onoda K., 1970, Studies on the regulation of the activity of of drug oxidation in rat liver microsomes by androgen and estrogen. Biochem. Pharmacol., 19: 1649.

Khlopushina T.G., Kovalev I.E. and Lysenkova E.M., 1986, Effect of perfluorodecalin and perfluorotributylamine on the cytochrome P-450 system of the liver. Biokhimiya, 51: 664.

Lowe K.C., 1987, Perfluorocarbons as oxygen-transport fluids. Comp. Biochem. Physiol., 87A: 825.

Lowe K.C., 1988a, Emulsified perfluorochemicals for oxygen transport to tissues: effects on lymphoid system and immunological competence, in: "Oxygen Transport to Tissue", Vol. X, M. Mochizuki, C.R. Honig, T. Koyama, T.K. Goldstick and D.F. Bruley, eds., p. 655, Plenum, New York.

Lowe K.C., 1988b, Biological assessment of perfluorochemical emulsions. In in: "Blood Substitutes: Preparation, Physiology and Medical Applications", K.C. Lowe, ed., p. 149, Ellis Horwood, Chichester.

Lowe K.C. and Bollands A.D., 1985, Physiological effects of perfluorocarbon blood substitutes. Med. Lab. Sci., 42: 367.

Lowe K.C. and Bollands A.D., 1988, Lymphoid tissue responses to emulsified perfluorochemicals: comparative aspects. Biomat. Art. Cells Art. Org., 16: 495.

Lowry O.H., Rosebrough N.J., Farr A.L. and Randall R.J., 1951, Protein measurement with the Folin Phenol reagent. J.biol. Chem., 193: 265.

Naito R. and Yokoyama K., 1978, "Perfluorochemical Blood Substitutes", Green Cross Tech. Inform. Ser. 5, Green Cross, Osaka.

Obraztsov V.V., Shekhtman D.G., Sologub G.R. and Beloyartsev F.F., 1985, Induction of microsomal cytochromes in the rat liver after intravenous injection of an emulsion of perfluoroorganic compounds. Biokhimiya, 50: 1220.

Obraztsov V.V., Shekhtman D.G., Sklifas A.N. and Makarov K.N., 1988, Analysis of physico-chemical properties of fluorocarbon inducers of cytochrome P-450 in membranes of endoplasmic reticulum of liver. Biokhimiya, 53: 535.

Omura T. and Sato R., 1964a, The carbon monoxide-binding pigment of liver microsomes. I. Evidence for its hemoprotein nature. J. biol. Chem., 239: 2370.

Omura T. and Sato R., 1964b, The carbon monoxide-binding pigment of liver microsomes. II. Solubilization, purification, and properties. J. biol. Chem, 239: 2379.

Pfannkuch F. and Schnoy N., 1983, Long term observation of PFC storage in organs of rats after various doses, in "Advances in Blood Substitute Research", R.B. Bolin, R.P. Geyer and G.J. Nemo, eds., p. 209, Liss, New York.

Riess J.G. and Le Blanc M., 1982, Solubility and transport phenomena in perfluorochemicals relevant to blood substitution and other biomedical applications. Pure Appl. Chem., 54: 2383.

Riess J.G. and Le Blanc M., 1988, Preparation of perfluorochemical emulsions for biomedical use: principles, materials and methods, in "Blood Substitutes: Preparation, Physiology and Medical Applications", K.C. Lowe, ed., p. 94, Ellis Horwood, Chichester.

Sharma S.K., Bollands A.D., Davis S.S. and Lowe K.C., 1987, Emulsified perfluorochemicals as physiological oxygen-transport fluids: assessment of a novel formulation, in: "Oxygen Transport to Tissue", Vol. IX, I.A. Silver and A. Silver, eds., p. 97, Plenum, London.

Shrewsbury R.P., 1987, Effect of Fluosol-DA hemodilution on the kinetics of hepatically eliminated drugs. Res. Comm. Chem. Path. Pharmac., 55: 375.

Shrewsbury R.P., White S.G., Pollack G.M. and Wargin W.A., 1986, Antipyrine kinetics following partial blood exchange with Fluosol-DA in the rat. J. Pharm. Pharmac., 38: 883.

Snedecor G.W. and Cochran W.G., 1980, "Statistical Methods", 7th Edn., Iowa State College Press, Ames.

Ullrich V. and Diehl H., 1971, Uncoupling of monooxygenation and electron transport by fluorocarbons in liver microsomes. Eur. J. Biochem., 20: 509.

PERFLUOROCHEMICALS AND PHOTODYNAMIC THERAPY IN MICE

M.C. Berenbaum, S.L. Akande, F.H. Armstrong, P.K. Bentley, R. Bonnett, R.D. White and K.C. Lowe

Dept. Pathology, St. Mary's Hospital Medical School, London
Dept. Zoology, University of Nottingham, Nottingham; Dept.
Chemistry, Queen Mary College, London, UK

INTRODUCTION

Photodynamic therapy (PDT) is currently being evaluated for the treatment of tumours in man. It depends on the use of tumour-localising porphyrins that are activated by visible light to produce tumour destruction (Fingar et al., 1988). PDT is oxygen dependent and recent attention has focussed on strategies for improving tumour oxygenation and thereby increasing the effectiveness of PDT.

One approach for improving oxygen supply to tumours is to use emulsions of inert organic compounds called perfluorochemicals (PFCs). PFCs can dissolve substantial quantitites of respiratory gases (Riess and Le Blanc, 1982) and this makes them attractive as oxygen transport fluids. Although PFC oils are immiscible with aqueous systems, they can be used in vivo in an emulsified form.

One "first-generation" commercial PFC emulsion, Fluosol-DA 20% (F-DA; Green Cross, Japan), has been tested as an adjunct to cancer therapy to enhance tumour oxygenation and thus increase sensitivity to ionizing radiation (Lee et al., 1987; Martin et al., 1987), chemotherapy (Teicher and Holden, 1987) and PDT (Fingar et al., 1987). However, problems with F-DA include inadequate emulsion stability and adverse effects (Lowe and Bollands, 1985; Lowe, 1987, 1988). In the present experiments we have therefore studied the effects of PDT on skin and tumours in mice previously injected with a novel perfluorodecalin (FDC) emulsion having improved stability and biocompatibility characteristics (Sharma et al., 1987, 1988).

MATERIALS AND METHODS

Animals; experimental and analytical procedures

Inbred BALB/c female mice with subcutaneous implants of a plasma cell tumour (Berenbaum et al., 1986) were given various doses of meta-tetra(hydroxyphenyl) porphyrin (m-THPP; Fig. 1), a new, potent, tumour photosensitizer (Berenbaum et al., 1986). 24 hr later, tumours and depilated dorsal skin were exposed to 10 J cm^{-2} of light at the activating wavelength (648 nm) from a dye laser pumped by a copper vapour laser.

Oxygen Transport to Tissue XII, Edited by J. Piiper *et al.*
Plenum Press, New York, 1990

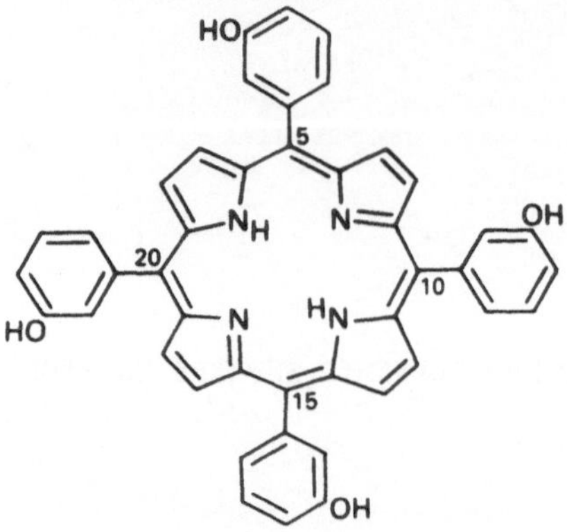

Fig. 1. Structure of m-THPP

Some mice were given, 2.5 hr before PDT, an intravenous (i.v.) injection of 0.5 ml/10g body weight of a novel 20% (w/v) emulsion of FDC containing 4% (w/v) Pluronic F-68 and stabilized with 1% (w/v) of a C-16 oil additive, perfluoroperhydrofluoranthrene (Sharma et al., 1987, 1988). The next day, 0.2 ml of 1% (w/v) Evans Blue (Sigma, Poole) in saline (0.9% w/v NaCl) was injected i.v.; 1 hr later, animals were killed and tumours removed into formol saline. The depth of necrosis was measured on slices of the fixed tumour using a dissecting microscope fitted with an eyepiece graticule (Berenbaum et al., 1986). Additional groups of animals were injected with m-THPP either 24 hr or 48 hr before light exposure and with emulsion either 2.5 hr or 24 hr before light exposure.

4 hr after illumination, animals were killed and a 1 cm diameter disc of depilated dorsal skin was punched out, rapidly weighed and placed in sodium suphate-acetone for extraction of Evans Blue; dye concentration was determined by absorbance at 620 nm (Berenbaum et al., 1986).

Statistical analyses

Statistical analyses were performed according to the methods of Snedecor and Cochran (1980). Means and standard errors (SEM) have been used throughout and statistical significance between mean values was assessed using a conventional Student's $\underline{t}$ test. A probability of $P < 0.05$ was considered significant.

RESULTS

The mean depth of tumour necrosis 24 hr after PDT was 2.5 ± 0.1 mm (n = 7) in mice given FDC emulsion and 2.14 ± 0.6 mm (n = 7) in controls; this difference was not significant ($P > 0.05$). However, PDT-induced skin damage was significantly less in mice injected with the FDC emulsion, as shown by shifts in the dose-response curves for skin disc weight (oedema) and uptake of Evans blue dye (vascular permeability). For example, in mice injected with 12.5 µm/kg m-THPP 24 hr before light and emulsion 2.5 hr before light, mean skin disc weight was up to 60% lower than control (Fig. 2). In addition, in animals injected with 12.5 - 50.0 µm/kg m-THPP 48 hr before light and emulsion 24 hr before light, skin Evan's blue concentration was up to 44% lower than in controls (Fig. 3).

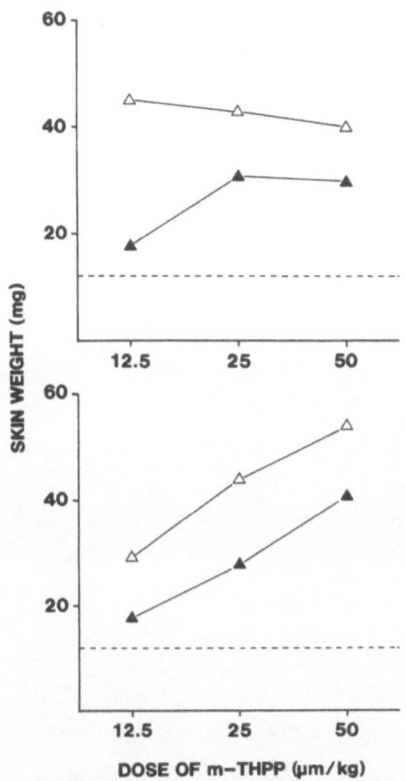

Fig. 2. Mean weights of skin discs in mice injected with different doses of m-THPP and either saline (△)(controls) or the novel PFC emulsion (▲) as follows:

Upper panel: m-THPP injected 24 hr and emulsion 2.5 hr before exposure to light.

Lower panel: m-THPP injected 48 hr and emulsion 24 hr before exposure to light.

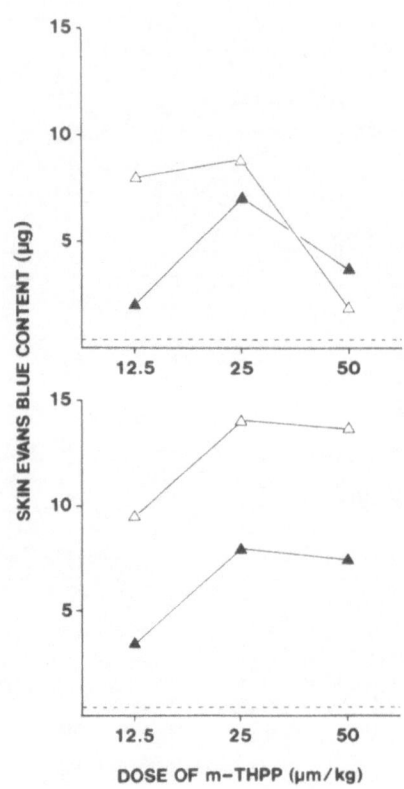

Fig. 3. Mean skin Evans Blue content in mice injected with different doses of m-THPP and either saline (△)(controls) or the novel PFC emulsion (▲) as follows:

Upper panel: m-THPP injected 24 hr and emulsion 2.5 hr before exposure to light.

Lower panel: m-THPP injected 48 hr and emulsion 24 hr before exposure to light.

DISCUSSION

There have been many attempts to use emulsified PFCs for improving oxygenation of solid tumours and thereby increasing their sensitivity to both ionizing radiation and cytotoxic drugs: for example, studies in rats (Martin et al., 1987), mice (Song et al., 1985; Lee et al., 1987; Rockwell et al., 1988) and humans (Rose et al., 1986; Lustig et al., 1988) have shown that injection of F-DA together with breathing either oxygen or carbogen can increase the efficacy of radiation in killing malignant cells and perturbing tumour growth. The effectiveness of PFCs as adjuncts to cancer therapy depends upon the ability of the gas-tansporting emulsion particles to penetrate regions otherwise inaccessible to red cells.

The present results show that, while pre-treatment with emulsified PFCs did not alter PDT-induced tumour necrosis in mice, it nevertheless provided protection against photodynamic skin damage. It is possible that the ineffectiveness of the emulsion to enhance tumour destruction was related to oxygen supply: animals in the present studies breathed only ambient air, and therefore, any increase in tumour oxygenation due to emulsion injection may have been insufficient to enhance PDT-induced necrosis. However, in previous studies with mice bearing RIF tumours and injected with F-DA while breathing carbogen prior to PDT, there was no difference in response compared to saline-injected controls breathing air (Fingar et al., 1988). Further studies are needed to investigate the effects of other variables such as type and dose of photosensitizer used together with emulsion composition and dose administered.

The apparent protective effects provided by the novel emulsion against photodynamic damage suggest that, despite the absence of enhanced tumour destruction, there may nevertheless be specific advantages in using PFCs in conjunction with PDT. We speculate that such protective effects of PFCs may be due to alterations in skin blood flow, inhibition of leucocyte-mediated inflammatory responses by emulsion components in accord with previous observations (Lowe, 1987), haemodilution effects, or a combination of any of these. Further studies are needed to analyse the mechanism of protection.

SUMMARY

The effects of pre-treatment with a novel PFC emulsion on PDT-induced tumour necrosis have been studied in mice. Injection of emulsion either 2.5 hr or 24 hr before PDT did not affect the depth of tumour necrosis. However, pre-treatment with the emulsion appeared to protect skin against photodynamic damage although the mechanism(s) and active principle(s) involved were not identified. These results suggest that there may be specific advantages in using emulsified PFCs in conjunction with PDT which may be independent of changes in tumour oxygenation. .

ACKNOWLEDGEMENTS

This work was supported by grants from the Cancer Research Campaign, Scotia Pharmaceuticals and RTZ Chemicals (ISC Division). F.H.A. and P.K.B. are recipients of research studentships from the SERC.

REFERENCES

Berenbaum, M.C., Akande, S.L., Bonnett, R., Kaur, H., Ioannou, S., White, R.D and Winfield, U.-J., 1986, Meso-tetra(hydroxyphenyl)-porphyrins, a new class of potent tumour photosensitizers with favourable selectivity. Br. J. Cancer, 54: 717.

Fingar, V.H., Mang, T.S. and Henderson, B.W., 1988, Modification of photodynamic therapy-induced hypoxia by Fluosol-DA (20%) and carbogen breathing in mice. Cancer Res., 48: 3350.

Lee, I., Levitt, S.H. and Song, C.W., 1987, Effects of Fluosol-DA 20% and carbogen on the radioresponse of SCK tumors and skin of A/J mice. Radiat. Res., 112: 173.

Lowe, K.C., 1987, Perfluorocarbons as oxygen-transport fluids. Comp. Biochem. Physiol., 87A: 825.

Lowe, K.C., 1988, Biological assessment of perfluorochemical emulsions, in: "Blood Substitutes: Preparation, Physiology and Medical Applications", K.C. Lowe, ed., p. 149, Ellis Horwood, Chichester.

Lowe, K.C. and Bollands, A.D., 1985, Physiological effects of perfluorocarbon blood substitutes. Med. Lab. Sci., 42: 367.

Lustig, R.A., Rose, C.M. and McIntosh-Lowe, N.L., 1988, Fluosol-DA and oxygen as an adjuvant to the radiotherapeutic management of advanced head/neck carcinoma. Biomat. Art. Cells Art. Org., 16: 511.

Martin, D.F., Porter, E.A., Fischer, J.J. and Rockwell, S., 1987, Effect of a perfluorochemical emulsion on the radiation response of BA1112 rhabdomyosarcomas. Radiat. Res., 112: 45.

Riess, J.G. and Le Blanc, M., 1982, Solubility and transport phenomena in perfluorochemicals relevant to blood substitution and other biomedical applications. Pure Appl. Chem., 54: 2383.

Rockwell, S., Irvin, C.G. and Kelley, M., 1988, Preclinical studies of a perfluorochemical emulsion as an adjunct to radiotherapy. Int. J. Radiat. Oncol. Biol. Phys., 15: 913.

Rose, C., Lustig, R., McIntosh, N. and Teicher, B., 1986, A clinical trial of Fluosol-DA 20% in advanced squamous cell carcinoma of the head and neck. Int. J. Radiat. Oncol. Biol. Phys., 12: 1325.

Sharma, S.K., Bollands, A.D., Davis, S.S. and Lowe, K.C., 1987, Emulsified perfluorochemicals as physiological oxygen transport fluids: assessment of a novel formulation, in: "Oxygen Transport to Tissue", Vol. IX, I.A. Silver and A. Silver, eds., p. 97, Plenum, London.

Sharma, S.K., Lowe, K.C. and Davis, S.S., 1988, Novel compositions of emulsified perfluorochemicals for biological uses. Biomat. Art. Cells Art. Org., 16: 447.

Snedecor, G.W. and Cochran, W.G., 1980, "Statistical Methods", 7th Edn., Iowa State College Press, Ames.

Song, C.W., Zhang, W.L., Pence, D.M., Lee, I. and Levitt, S.H., 1985, Increased radiosensitivity of tumors by perfluorochemicals and carbogen. Int. J. Radiat. Oncol. Biol. Phys., 11: 1833.

Teicher, B.A. and Holden, S.A., 1987, Survey of the effect of adding Fluosol-DA 20%/oxygen to treatment with various chemotherapeutic agents. Cancer Treatment Rep., 71: 173.

PERFLUOROCHEMICALS FOR GAS TRANSPORT AND IMPROVEMENT OF CELL CULTURES

Alastair T. King, Bernard J. Mulligan* and Kenneth C. Lowe

Departments of Zoology and *Botany, University of

Nottingham, Nottingham, NG7 2RD, UK

INTRODUCTION

Much attention has focussed on the use of perfluorochemicals (PFCs) as organic oxygen carriers in biological systems (Sharma et al., 1987; Lowe, 1988a). Previous work has investigated the use of PFC emulsions as so-called "blood substitutes" and vehicles for tissue oxygenation (Lowe, 1988a). However, it is clear that great potential for such compounds also exists in other areas, especially in biotechnologically-important cell cultures. An additional attraction of using PFCs in cell culture systems is their immiscibility with aqueous media; in principle, they are therefore re-cycleable and recoverable, an obvious economic advantage. This paper considers the effects of PFCs and their emulsion components on both prokaryotic and eukaryotic cells and assesses their use for improving gas supply and regulating cellular growth.

MICROBIAL CELL CULTURES

In an early U.S. patent (Chibata et al., 1974) it was suggested that growth, propagation and productivity of various microorganisms could be increased by incubation with the PFC oil, perfluorotributylamine (FTBA). Another patent demonstrated that microbial growth in solid or liquid media could be stimulated in the presence of an admixture of hydrogen peroxide and the PFC oil, FC-75 (Hertl and Ramsey, 1979). Subsequently, there have been several attempts to use PFCs and their emulsions to stimulate microbial growth and productivity, with varying degrees of success (Table 1). For example, FC-72 emulsified with the polyol surfactant, Pluronic F-68, and mixed with a substrate medium was effective in increasing oxygen transfer to immobilized Gluconobacter oxydans provided the emulsion was pre-oxygenated (Adlercreutz and Mattiasson, 1982; Mattiasson and Adlercreutz, 1984). An important finding was that in such oxygen-enriched cultures, the conversion of glycerol to dihydroxyacetone was almost 6 times greater than in control cultures. This showed that the additional oxygen supplied could be readily utilised by cells. In similar studies using emulsified perfluorodecalin (FDC) (Leonhardt et al., 1985) or Forane F-66E (F-66E) (Rols et al., 1988) oxygen transfer was enhanced in cultures of the Providencia sp. PCM 1298 and Klebsiella oxytoca respectively.

The PFC oil, perfluoromethyldecalin, has been used to supply oxygen to Escherichia coli growing in a spray column fermenter by liquid-liquid

Oxygen Transport to Tissue XII, Edited by J. Piiper et al.
Plenum Press, New York, 1990

TABLE 1

MICROBIAL CELL RESPONSES TO CULTURE WITH PERFLUOROCHEMICALS

SPECIES	PFC*	RESPONSE(S)	REF.
Escherichia coli	FC-75**	Enhancement of culture growth	[12]
Escherichia coli	FDC	Increased oxygen transfer	[7]
Providencia sp.+	FDC++	Increased oxygen transfer	[21]
Klebsiella oxytoca	F-66E++	Increased oxygen transfer	[29]
Acetobacter suboxydans	FTBA	Increased growth and productivity	[5]
Gluconobacter oxydans+	FC-72++	Increased oxygen transfer Enhanced production of dihydroxyacetone	[1,24]
Clostridium perfringens	F-66E	Regulation of carbon dioxide supply	[3]
Staphylococcus aureus	FTBA	Enhancement of culture growth	[5]
Streptomyces sp.	FC-77	Growth at increased hydrostatic pressure	[11]
Streptomyces fradiae	FTBA	Increased growth and productivity	[5]
Streptomyces humidas	FTBA	Increased growth and productivity	[5]
Mycobacterium bovis	FDC	Enhancement of culture growth	[28]
Mycobacterium lufu	FDC	Enhancement of culture growth Increased superoxide dismutase activity	[28]
Saccharomyces cerevisiae	FTBA	Enhancement of culture growth	[5]
Saccharomyces cerevisiae	FDC	Increased oxygen supply	[19]

* See reference 1 for full chemical names and structures

** Hydrogen peroxide added to culture system

+ Immobilized cells

++ Emulsified with Pluronic F-68

contacting (Damiano and Wang, 1985). Similarly, FDC has also been used to supply oxygen to suspensions of the yeast, Saccharomyces cerevisiae (King et al., 1989). In these experiments, the specific growth rates of cells were increased in the presence of pre-oxygenated PFC oil and no adverse effects on cell viability or morphology were reported. Culture under ambient conditions with non-oxygenated PFC oil or emulsion produces no stimulation of growth of E. coli or yeast (King et al., 1988a); pre-oxygenation of the PFC component is therefore required to achieve increased gas supply. PFCs have also been used as dispersion media for immobilization of Lactobacillus sp., E. coli and S. cerevisiae (Keller and Siegemund, 1984).

Gases other than oxygen

PFCs can be used to enhance the supply of other gases, including carbon dioxide, to microbial systems: experiments have shown that the latent phase of Clostridium perfringens cultures supplemented with F-66E oil and bubbled with carbon dioxide was greatly reduced compared with controls (Ceschin et al., 1985). These results demonstrate that the potential use for PFCs in regulating gas supply in cell cultures can be extended beyond aerobic systems. A wide range of gases are soluble in liquid PFCs (Sharma et al., 1987; Lowe, 1988a) and an interesting but as yet untested application for these compounds in biological systems would be as scavengers of toxic gaseous by-products.

Microbial structure and function

Relatively few studies have examined the effects of PFCs and their emulsions on microbial structure and function. In a recent paper (Popkova et al., 1988), it was reported that, although FDC increased oxygen transfer to cultures of mycobacteria (Table 1), it produced a species-specific differential effect on the oxygen-dependent superoxide dismutase (SOD) enzyme: SOD activity in Mycobacterium lufu was increased in response to culture with FDC whilst no corresponding change occurred in M. bovis. Electron microscopical studies of yeast cells have revealed no obvious changes in structure following culture with FDC, FTBA or perfluorotripropylamine (FTPA); similarly, no ultrastructural changes were seen when yeast cells were cultured with FDC emulsion (King et al., unpublished observations). This contrasted with earlier preliminary observations following incubation of yeast with a similar emulsion (Chandler et al., 1987). One explanation for this variation may be the use of different methods for PFC emulsification in these studies. In the former case, a microfluidizer was used whilst in the latter study, sonication was employed (Chandler et al., 1987) and this is known to generate potentially toxic fluoride ions (Lowe, 1988a).

Effects of Pluronic F-68

An interesting feature has emerged in work with PFC emulsions, namely that the Pluronic F-68 surfactant component of proprietary PFC emulsions (Lowe, 1988a) can induce physical changes in bacterial or yeast membranes/cell walls producing altered culture optical density characteristics (King et al., 1988b). However, these effects are not of sufficient magnitude to interfere with cell viability (King et al., 1988b). Pluronic F-68 can also alter the rate of fluorescein diacetate uptake in yeast and the magnitude of this response depends on both the source and degree of purity of the pluronic used (King et al., 1988c). In contrast, no effects of the surfactant on the intracellular enzymes responsible for dye hydrolysis were observed (King et al., 1988c), neither were there any marked changes in cell structure or polypeptide profiles of crude membrane extracts. It is clear that pluronic can influence microbial cell function and thus, any studies using PFC emulsions prepared with this surfactant must take this property into account.

ANIMAL CELL CULTURES

Experiments using high density suspension cultures of mouse hybridoma cells showed that dropwise addition of an oxygenated PFC oil, FC-40, was able to supply sufficient oxygen to allow them to grow normally (Hamamoto et al., 1987). Similarly, Cho and Wang showed that oxygenated FDC was effective in increasing oxygen supply to such cells (Cho and Wang, 1988). This work demonstrated that PFCs may have value in fragile animal and tumour cell cultures for supplying oxygen and also reducing or eliminating mechanical damage produced by conventional aeration methods (Lee et al., 1988). Indeed, because of the growing interest in the potential use of PFC emulsions as adjuncts to cancer therapy (Lowe, 1988a), further detailed studies of the effects PFCs and emulsion components on tumour cells are needed.

Growth at oil-medium interfaces

A novel approach for improving growth of animal cells under anchorage-independent conditions is to grow cells at the interface between PFC oil and aqueous culture media. This method was pioneered by Rosenburg in 1964 (Rosenburg, 1964) and subsequently employed for in vitro growth of human newborn foreskin fibroblasts (Keese and Giaever, 1983a), placental cells (Keese and Giaever, 1983b) and various cell lines from mice (Giaever and Keese, 1983). More recently, Sanfilippo and co-workers showed that both normal and oncogene-transformed mammalian cell lines would proliferate on a PFC-medium interface provided that appropriate growth factors were added (Sanfilippo et al., 1988). It has been suggested (Rosenburg, 1964) that binding of trace impurities normally present in PFC oils with denatured serum proteins in the culture media promoted the formation of a rigid surface on which cells could spread and grow as a monolayer. In this regard, purified PFCs are poor substrates for normal anchorage-independent growth because cells cannot spread on the oil surface. This problem has been overcome by increasing the strength of the interfacial protein film using a PFC emulsion stabilized by polylysine (Keese and Giaever, 1983a,b).

Effects of Pluronic F-68

In addition to testing the potential value of PFC oils in animal cell cultures, related work has also assessed the effects of commercial PFC emulsions and their components in such systems. Much of this work has focussed on the effects of emulsified PFCs on in vitro function of cells from different species and this aspect has been reviewed in detail elsewhere (Lowe, 1988a,b). Because many of the PFC emulsions prepared as oxygen-transport fluids contain the Pluronic F-68 surfactant, it is inevitable that cellular effects of this compound have been studied in detail (Lowe, 1988a,b). While there has been some concern about possible adverse effects of Pluronic F-68 on immune cells (Lowe, 1988a,b) it is somewhat paradoxical that a substantial literature is available describing its beneficial effects in high density animal cell cultures. For example, Pluronic F-68 can protect cultured human lymphoblastoid cells (Mizrahi, 1984) and insect cells (Murhammer and Goochee, 1988) against damage caused by sparging. Pluronic is also able to alter growth of animal cells in culture but the extent of this effect depends upon source, purity and concentration of compound used (Mizrahi, 1975; Bentley et al., 1989).

286

PLANT CELL CULTURES

Plant cell cultures are widely used to produce commercially-important products including pharmaceuticals, food additives, cosmetics and agrochemicals (Fujita and Tabata, 1987). Thus, any improvements in plant cell culture methods which can be readily applied to commercially-important species would be a major technological advance. Recent experiments have shown that incubation of Solanum dulcamara cells, which produce medicinal alkaloids, with FDC, FTPA or FTBA had no detrimental effects on growth or structure (King et al., 1989). In addition, preliminary experiments with FDC suggested a possible stimulatory effect on cell growth (Fig. 1) but this needs to be confirmed. Further studies are also needed to determine the biocompatibility of different PFCs in plant cell cultures and identify any useful role in regulating oxygen delivery. Furthermore, such studies should establish the extent to which PFCs may protect more fragile plant cells against mechanical damage caused by physical stress (Tanaka et al., 1988). Obviously, the effects of PFCs on yields of secondary and other products is an important consideration in work with plant cells.

SUMMARY

This paper has considered the effects and potential applications of PFCs, their emulsions and emulsion components for regulating growth and metabolic functions of microbial, animal and plant cells in culture. PFCs will help to overcome problems encountered in conventional culture systems (e.g. limited gas supply, mechanical damage), especially where cells are grown to high density. While the commercial potential of PFCs for in vitro systems has not yet been fully exploited, the most exciting areas for future developments are in the culture of animal and plant cell lines of importance in biotechnology and medicine.

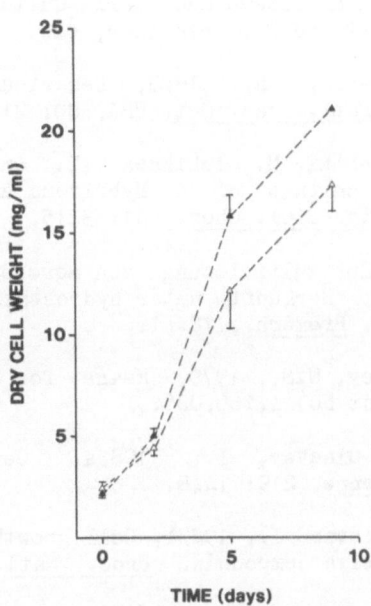

Fig. 1. Changes in mean dry cell weights (mg/mL) in S. dulcamara cell suspensions cultured for up to 9 days either in medium alone (controls △) or with 20% (v/v) FDC oil (▲). Vertical bars represent s.э.m. (n = 3–5).

287

REFERENCES

1. Adlercreutz, P. and Mattiasson, B., 1982, Oxygen supply to immobilized cells. 3. Oxygen supply by hemoglobin or emulsions of perfluorochemicals. Eur. J. Appl. Microbiol. Biotechnol., 16: 165.

2. Bentley, P.K., Gates, R.M.C., Lowe, K.C., de Pomerai, D.I. and Walker, J.A.L., 1989, In vitro cellular responses to a non-ionic surfactant, Pluronic F-68. Biotechnol. Lett., 11: 265.

3. Ceschin, C., Malet-Martino, M.C., Michel, G. and Lattes, A., 1985, Optimization of anaerobes growth with a perfluorinated oil. C.R. Acad. Sc. Paris., Serie III, 18: 669.

4. Chandler, D., Davey, M.R., Lowe, K.C. and Mulligan, B.J., 1987, Effects of emulsified perfluorochemicals on growth and ultrastructure of microbial cells in culture. Biotechnol. Lett., 9: 195.

5. Chibata, I.S., Yamada, S.T., Wada, M.N., Izuo, N.Y. and Yamaguchi, T.Y., 1974, Cultivation of aerobic microorganisms. U.S. Patent No. 3,850,753.

6. Cho, M.H. and Wang, S.S., 1988, Enhancement of oxygen transfer in Hybridoma cell culture by using a perfluorocarbon as an oxygen carrier. Biotechnol. Lett., 10: 855.

7. Damiano, D. and Wang, S.S., 1985, Novel use of a perfluorocarbon for supplying oxygen to aerobic submerged cultures. Biotechnol. Lett., 7: 81.

8. Fujita, Y. and Tabata, M., 1987, Secondary metabolites from plant cells – pharmaceutical applications and progress in commercial production, in: "Plant Tissue and Cell Culture", C.E. Green, D.A. Somers, W.P. Hackett and D.D. Biesboer, eds., p. 169, Liss, New York.

9. Giaever, I. and Keese, C.R., 1983, Behaviour of cells at fluid interfaces. Proc. Natl. Acad. Sci. USA, 80: 219.

10. Hamamoto, K., Tokashiki, M., Ichikawa, Y. and Murakami, H., 1987, High cell density culture of a Hybridoma using perfluorocarbon to supply oxygen. Agric. Biol. Chem., 51: 3415.

11. Helmke, E., 1979, Zur kultivierung von aeroben Actinomyceten mariner und terrestrischer Herkunft unter hydrostatischem druck. Veroff. Inst. Meeresforsch. Bremerh., 18: 1.

12. Hertl, W. and Ramsey, W.S., 1979, Means for stimulating microbial growth. U.S. Patent No. 4,166,006.

13. Keese, C.R. and Giaever, I., 1983a, Cell growth on liquid microcarriers. Science, 219: 1448.

14. Keese, C.R. and Giaever, I., 1983b, Cell growth on liquid interfaces: role of surface active compounds. Proc. Natl. Acad. Sci. USA, 80: 5622.

15. Keller, R. and Siegemund, G., 1984, German Patent No. 3 237 341.

16. King, A.T., Lowe, K.C. and Mulligan, B.J., 1988a, Emulsified perfluorochemicals for oxygenation of microbial cell cultures? in: "Oxygen Transport to Tissue", Vol. X. M. Mochizuki, C.R. Honig, T. Koyama, T.K. Goldstick, and D.F. Bruley, eds., p. 579, Plenum, New York.

17. King, A.T., Lowe, K.C. and Mulligan, B.J., 1988b, Microbial cell responses to a non-ionic surfactant. Biotechnol. Lett., 10: 177.

18. King, A.T., Lowe, K.C. and Mulligan, B.J., 1988c, Microbial cell responses to a non-ionic surfactant. II. Effects as assessed by fluorescein diacetate uptake. Biotechnol. Lett., 10: 873.

19. King, A.T., Bray, J., Mulligan, B.J. and Lowe, K.C., 1989, Biocompatibility assessment of perfluorochemical oils in microbial and plant cell cultures. Biotechnol. Lett., submitted for publication.

20. Lee, G.M., Huard, T.K., Kaminski, M.S. and Palsson, B.O., 1988, Effect of mechanical agitation on hybridoma cell growth. Biotechnol. Lett., 10: 625.

21. Leonhardt, A., Szwajcer, E. and Mosbach, K., 1985, The potential use of silicon compounds as oxygen carriers for free and immobilized cells containing L-amino acid oxidase. Appl. Microbiol. Biotechnol., 21: 162.

22. Lowe, K.C. (ed.), 1988a, "Blood Substitutes: Preparation, Physiology and Medical Applications", Ellis Horwood, Chichester.

23. Lowe, K.C., 1988b, Emulsified perfluorochemicals for oxygen-transport to tissues: effects on lymphoid system and immunological competence, in: "Oxygen Transport to Tissue, Vol. X, M. Mochizuki, C.R. Honig, T. Koyama, T.K. Goldstick, T.K. and D.F. Bruley, eds., p. 655, Plenum, New York.

24. Mattiasson, B. and Adlercreutz, P., 1984, Use of perfluorochemicals for oxygen supply to immobilized cells. Ann. N.Y. Acad. Sci., 413: 545.

25. Mizrahi, A, 1975, Pluronic polyols in human lymphocyte cell line cultures. J. Clin. Microbiol., 2: 11.

26. Mizrahi, A., 1984, Oxygen in human lymphoblastoid cell line cultures and effect of polymers in agitated and aerated cultures. Develop. Biol. Stand., 55: 93.

27. Murhammer, D.W. and Goochee, C.F., 1989, Scaleup of insect cell cultures: protective effects of Pluronic F-68. Biotechnology, 6: 1411.

28. Popkova, N.I., Yushchenko, A.A., Yurkiv, V.A. and Irtuganova, O.A., 1988, Functional characteristics of the antioxidative system of Mycobacteria grown on media modified by perfluorodecalin. Bull. Exp. Biol. Med., 105: 215.

29. Rols, J.L., Condoret, J.S., Fonade, C. and Goma, G., 1988, Mechanism of enhanced oxygen transfer in fermentation using emulsified oxygen-vectors. Proc. 8th Internat. Biotechnol. Symp. Paris., Abst. B94.

30. Rosenburg, M.D., 1964, Cell surface interactions and interfacial dynamics, in: "Cellular Control Mechanisms and Cancer", P. Emmelot and O. Muhlbock, eds., p. 146, Elsevier, Amsterdam.

31. Sanfilippo, B., Ciardiella, F., Salomon, D.S. and Kidwell, W.R., 1988, Growth of cells on a perfluorocarbon-medium interface: a quantitative assay for anchorage-independent cell growth. In vitro Cell Dev. Biol., 24: 71.

32. Sharma, S.K., Bollands, A.D., Davis, S.S. and Lowe, K.C., 1987, Emulsified perfluorochemicals as physiological oxygen-transport fluids: assessment of a novel formulation, in: "Oxygen Transport to Tissue," Vol. IX, I.A. Silver and A. Silver, eds, p. 97, Plenum, London.

33. Tanaka, H., Semba, H., Jitsufuchi, T. and Harada, H., 1988, The effect of physical stress on plant cells in suspension cultures. Biotechnol. Lett., 10: 485.

THE CARDIOVASCULAR EFFECTS OF THE SURFACTANT PLURONIC F68 IN ANESTHETIZED DOGS

Anne-Marie Gosselin and G.P. Biro

Department of Physiology, Faculty of Medicine
University of Ottawa, Ottawa, Ont. K1H 8M5
Canada

INTRODUCTION

Perfluorocarbons (PFC's) have a high solubility for oxygen and carbon dioxide. This property suggests their potential usefulness as oxygen transporting 'blood substitutes'. Unfortunately, PFC's are neither miscible, nor soluble in water and therefore, they must be prepared as emulsions for use as 'blood substitutes' (Biro and Blais, 1987). The preparation most extensively investigated is Fluosol-DA^R. When it was administered to animals or humans, a variety of "beneficial" actions have been reported, including successful resuscitation from hemorrhagic shock, salvage of ischemic organs and a general improvement in oxygen supply to the tissues (Biro, 1985; Biro and Blais, 1987). In many cases, most, if not all, "beneficial" effects have been attributed to the PFC component and the extra oxygen carried in solution in this phase.

Critical analysis of these results suggests (Biro and Blais, 1987) that some of the observed effects may be attributed to other factors. Fluosol-DA consists of only about 11% by volume of PFC; the remaining 88-89% aqueous phase results in a large degree of hemodilution when mixed with the residual blood volume and a large dilution of the PFC phase also occurs (Biro, 1988). The hemodilution, together with a significant rheological improvement of the blood (Biro, 1985), may have an important impact on tissue oxygen supply, independently of the relatively modest amounts of oxygen present in the PFC-phase.

The preparation of Fluosol-DA and related PFC-based 'blood substitutes' requires emulsifiers and surfactants. The principal surfactant used in Fluosol-DA is a linear polyethoxyglycol, Pluronic F68, whose average molecular weight is about 7000D:

$$HO-[CH_2-CH_2-O]_n-[CH_2-\underset{\underset{CH_3}{|}}{C}HO]_m-[CH_2-CH_2-O]_n-H$$

where n=30, m=75

The potential physiological actions of Pluronic-type surfactants have been largely neglected in recent reports on the PFC-based 'blood substitutes', although earlier reports have shown interesting properties. The latter included the ability to reduce blood viscosity and platelet adhesiveness during cardiopulmonary bypass (Grover et al., 1969), inhibition of blood "sludging" during cold injury (Knize et al., 1969), dissipation of the microcirculatory obstruction induced by amniotic fluid embolism (Hymes et al., 1970) and a general improvement of hemorrhagic shock (Hymes et al., 1971). These observations were made when Pluronic F68 was administered in amounts substantially less than the doses of Pluronic received by experimental animals and human subjects given Fluosol-DA for resuscitation of hemorrhagic shock. The observations therefore suggest that Pluronic F68 may have significant microcirculatory and/or hemorheological effects and that some of these effects may contribute to those observed after the administration of Fluosol-DA. For this reason, we studied the cardiovascular effects induced by the intravenous administration of Pluronic F68 in anesthetized dogs, in doses calculated to be equivalent to that achieved by the replacement of 30 and 80% of the blood volume by Fluosol-DA.

METHODS

The preparation of Pluronic F68

We obtained Pluronic F68, manufactured by BASF Wyandotte Chemicals Corp. The dry powder was kept in sealed containers in a refrigerator. When needed, it was dissolved in cold 0.9% NaCl solution. As the solution tends to expand, the final volume was made up, after allowing slow dissolving overnight in the refrigerator, without stirring. In order to prevent excessive hemodilution, the solution to be administered was made up as 20% w/v.

We cleaned the Pluronic solution with activated coconut charcoal (Fisher Scientific Co.) and filtered it through 0.2 μm disposable filters (Nalgene Products Co.), prior to its use.

Animal preparation

Adult mongrel dogs of both sexes (14-27 Kg) were used. Food was removed the night before the experiment but water was provided ad libitum. Pre-anesthetic medication consisted of Demerol (6.7 mg/Kg) and Acepromazine (0.33 mg/Kg). Fifteen-to-twenty minutes thereafter, anesthesia was induced by intravenous sodium pentobarbital (26 mg/Kg). The dogs were intubated and ventilated with oxygen (80%):nitrous oxide (20%):methoxyflurane (0.2-0.5%) gas mixture. Arterial blood gas samples were obtained periodically and ventilation was adjusted as required to maintain pCO_2 in the 35-42 mmHg-range. Sodium bicarbonate was administered as required to maintain pH between 7.35 and 7.45. Arteries and veins were cannulated and a thoracotomy was performed in the left fourth intercostal space. A catheter was advanced in the left ventricle and a Swan-Ganz catheter was placed in the pulmonary artery. The left carotid artery was exposed and wrapped with a moist swab; before the start of the experiment, the presence of the baroreflex was tested by a twenty-second carotid occlusion: a normal response consisted of a 12-15-beat acceleration of the heart rate and a 5-10 mmHg rise in mean blood pressure. The pericardium was opened and a catheter was placed in the left atrium for the administration of microspheres. The catheters were connected to transducers for pressure-measurements and the pressures were recorded on a Grass four-channel recorder. Cardiac output was measured in duplicate, at intervals, using the Swan-Ganz catheter and an Edwards Laboratories thermodilution computer.

292

Experimental protocol

After stabilization, measurements were made in the control period; these included pressures, cardiac output and the injection of radionuclide-labelled microspheres for the determination of regional blood flows. This was followed by the administration of hydrocortisone sodium succinate (Solu-Cortef, Upjohn) by a ten-minute intravenous infusion (5 mg/Kg). The hemodynamic measurements were repeated after this infusion was completed. Pluronic F68 (20% solution) was administered intravenously over a fifteen-minute period (0.66 g/Kg). At the end of this infusion, hemodynamic and blood flow-measurements were repeated. After a fifteen-minute pause, a second infusion of Pluronic F68 (1.11 g/Kg) was given, over 15 minutes. At the end of this infusion, hemodynamic and blood flow-measurements were made.

Blood flow-measurements

Carbonized plastic microspheres (NEN-TRAC, 15±2 μm) were obtained from Dupont-NEN; they had been labelled with ^{57}Co, ^{85}Sr, ^{113}Sn or ^{43}Sc. Blood flow was determined by the use of an arterial reference sample (Bartrum et al., 1974) and the customary precautions of the handling of microspheres were observed (Heymann et al., 1977; Baer et al., 1984). Spillover-correction during counting of the activity of mixtures of nuclides was made by a resident program in the Compu-gamma (LKB Instruments) five-channel gamma-spectrometer, using a 256-channel pulse-height analyzer.

At the end of the experiment, the heart was arrested with a saturated solution of KCl and tissue samples were taken, weighed, dried and radioactivity was determined for the subsequent calculation of blood flow, using the formula:

$$\text{blood flow (ml/g/min)} = \frac{\text{(reference blood flow) x (organ microsphere content)}}{\text{microsphere content of reference blood sample}}$$

All data are reported as mean ± SEM. Statistical analysis was performed using analysis of variance and by Tukey's and Scheffe's post-hoc tests and paired t-tests (Zar, 1989). Dose-dependence of an effect was tested by determining of the regression of the variable against the dose of Pluronic.

RESULTS

The study used eleven dogs of which eight experiments were considered to be technically completely satisfactory. One experiment was excluded because of a severe reaction to Pluronic, involving hypotension, mottling of the skin and subsequent edema.

Table I presents a summary of the hemodynamic variables. The prophylactic administration of hydrocortisone caused no significant change in any of the hemodynamic variables. The administration of Pluronic in the lower dose was followed by a rise in cardiac index, systemic and pulmonary arterial blood pressures, as well as in the right and left atrial and left ventricular end-diastolic pressures; of these, only the last four were of a statistically significant magnitude. Heart rate and left ventricular dP/dt remained unchanged. The subsequent infusion of the higher dose of Pluronic induced further significant increments in systemic and pulmonary arterial blood pressures and in the right and left heart filling pressures, but no further increments were observed in cardiac output. The heart rate and left ventricular dP/dt remained at the control levels.

It must be noted that the magnitude of the rises in systemic and pulmonary arterial pressures were quite different: mean systemic arterial pressure rose by 9 and 15%, respectively, whereas in the pulmonary artery, the corresponding pressure increments were 48 and 100% ($p<0.02$). In the systemic circulation, vascular resistance fell by nearly the same proportion as the rise in cardiac output, resulting in no significant change in arterial blood pressure. In the pulmonary circulation, vascular resistance fell by $41\pm22\%$ ($p>0.05$) after the lower dose of Pluronic and returned to the control value after the higher dose. These would suggest that principally the increment in cardiac output and left atrial pressure contributed to the marked rise in pulmonary arterial blood pressure.

Table I. Absolute values of various hemodynamic parameters in the control period and following administration of 0.66g/Kg and 1.11g/Kg of Pluronic F-68. The parameters are expressed in the following units: cardiac index: ml/min/Kg; heart rate: beats/min; pressures: mmHg; vascular resistances: mmHg/l/min/Kg; dP/dt of left ventricle: mmHg/sec; hematocrit: %.

	Control	Cortisone	Pluronic F-68	
			0.66g/Kg	1.11g/Kg
Cardiac Index :	140 ± 71	137 ± 70	283 ± 253	195 ± 93
Heart Rate :	127 ± 16	121 ± 17	126 ± 12	126 ± 14
Pressures :				
arterial systolic:	130 ± 18	126 ± 18	144 ± 20	151 ± 23
arterial diastolic:	94 ± 22	89 ± 17	101 ± 18	106 ± 18
L.V. diastolic :	3.0 ± 2.1	2.8 ± 1.6	6.2 ± 3.4	$10.0\pm3.5*$
pulmonary art. mean :	10.5 ± 2.5	9.8 ± 1.9	$15.5\pm1.8*$	$21.0\pm3.5@*$
right atrial :	1.0 ± 2.4	1.3 ± 2.3	3.8 ± 3.1	$7.0\pm3.2*$
left atrial :	3.8 ± 2.2	4.3 ± 2.4	$9.3\pm2.7*$	$14.8\pm3.1@*$
L.V. dP/dt :	1930 ± 575	3003 ± 486	2003 ± 515	2106 ± 486
Vascular resistance:				
systemic	731 ± 32	711 ± 41	$372\pm62*$	541 ± 69
pulmonary	72 ± 8	60 ± 6	$41\pm12*$	72 ± 11
Hematocrit :	36.0 ± 2.1	-	33.5 ± 2.2	30.3 ± 5.0

*: significantly different from control values as determined using a One-Way ANOVA and a Tukey test ($p < 0.05$)
@: significantly different from the lower Pluronic dose values, determined as above. ($p < 0.05$)

The hematocrit exhibited a modest and dose-dependant fall; it decreased from 36 to 33.5 and 30.3%, respectively, with the two doses of Pluronic.

The changes in blood flow-distribution are summarized in Table II. Myocardial blood flow increased after Pluronic but we were unable to document dose-dependence of the increments by finding statistically significant regression. Blood flow to all parts of the heart increased. It is noteworthy that right and left ventricular blood flows increased in nearly identical proportion, even though the increment in their respective afterloads differed substantially. Bronchial arterial and renal cortical blood flow also increased in an apparently dose-independent manner. In contrast, blood flow to skeletal muscle, brain, various parts of the gastrointestinal tract, pancreas and liver (hepatic arterial flow only) did not change significantly although a modest trend of increasing flows was evident in nearly all organs sampled.

Table II. The effect of Pluronic F68 on blood flow to various organs in six dogs. The flow is expressed as the percentage of the flow determined in the control period. (i.e. = 100%).

		Pluronic F-68	
		0.66g/Kg	1.11g/Kg
Atria	:	190.0 ± 74.9	$250.3\pm99.4*$
Right Ventricle	:	138.8 ± 23.4	$175.5\pm46.6*$
Left Ventricle	:	141.2 ± 32.5	$181.0\pm64.0*$
Lung	:	253.8 ± 90.1	$251.4\pm128.0*$
Muscle	:	141.8 ± 28.7	164.4 ± 158.7
Stomach	:	130.0 ± 26.7	169.4 ± 116.4
Jejunum	:	119.8 ± 26.9	115.8 ± 62.5
Ileum	:	139.6 ± 51.8	148.0 ± 105.0
Colon	:	148.6 ± 43.2	137.8 ± 61.8
Kidney Cortex	:	$165.8\pm39.1*$	$183.4\pm41.0*$
Kidney Medulla	:	144.2 ± 54.4	128.0 ± 95.1
Liver	:	60.0 ± 10.3	86.6 ± 40.9
Spleen	:	121.0 ± 15.5	120.8 ± 44.7
Brain Cortex	:	126.4 ± 14.4	192.2 ± 145.4
Brain White Matter	:	115.4 ± 19.9	135.0 ± 68.2

*: significantly different from control values as determined using a One-Way ANOVA and Tukey test. $(p < 0.05)$

DISCUSSION

In this experiment we sought to determine the effects of the surfactant Pluronic F68 upon gross hemodynamic variables and regional blood flow-distribution in anesthetized dogs. This was done, because the surfactant is a constituent of a complex 'blood substitute', Fluosol-DA, the administration of which has been documented to result in substantial hemodilution and other cardiovascular changes. Our experiments attempted to define the effects due to Pluronic alone, in the absence of major hemodilution which always accompanies the administration of Fluosol-DA and is known to have major hemorheologic effects (Biro and Blais, 1987; Biro, 1988).

Our experiments have shown that under these conditions, Pluronic F68 alone has significant hemodynamic effects. These consist of increments in the filling pressures on both sides of the heart, as well as in systemic and pulmonary arterial pressures and in the cardiac output, without significant changes in heart rate or left ventricular contractility. These changes were accompanied by increments in blood flow to various organs, but the distribution of the flow increments was not uniform. The largest and statistically significant flow increments were noted in the myocardium, renal cortex and bronchial circulation; the relative magnitude of these increments was greater than that to other organs with modest or negligible (and statistically not significant) increments of blood flow (skeletal muscle, brain and in the abdominal viscera).

Moderate to severe reactions have been reported to occur rapidly after the administration of Fluosol-DA (Police et al., 1985); those reactions principally involve the activation of the complement system (Vercelotti et al., 1982) and have been traced to the presence of Pluronic F68 and contaminants (Hammerschmidt and Vercelotti, 1988). In vivo, the reactions could be ameliorated or eliminated by the prophylactic administration of glucocorticoids (Vercelotti et al., 1982). The in vitro cytotoxic effects of unpurified Pluronic F68 could be removed by charcoal treatment (Chubb, 1985). The activation of polymorphonuclear leukocytes and the subsequent trapping of activated leukocyte-aggregates in the pulmonary circulation are the likely cause of the soaring pressures observed in the right heart and pulmonary artery in dogs and pigs (Biro et al, 1986) exhibiting reactions and of the subsequent systemic hypotension. Although corticosteroid pretreatment may alleviate these reactions in vitro and in vivo (Vercelotti et al., 1982), the reactions are not completely eliminated (Kingma et al., 1988). While it is possible that some residual complement activation may have occurred in our animals, the absence of significant increments in pulmonary vascular resistance negates this possibility. Our experiments combined charcoal purification and the prophylactic administration of glucocorticoid and this combined treatment may have succeeded in reducing the reactions which occurred without such treatment in our preliminary experiments. A report (Lane and Krukonis, 1988) has appeared since these experiments had been completed, describing the supercritical purification of Pluronic F68. It is clearly desirable to repeat the present experiments using supercritically purified Pluronic, to assess the extent to which the effects herein observed may be attributable to contaminants which may be removed by such a purification process.

Marked increases were observed in the filling pressures of both ventricles; these increases were statistically and hemodynamically highly significant. The mechanisms responsible for these effects are not clear. Since cardiac output failed to rise in proportion to the increments in filling pressure, there may have been a substantial degree of cardiac failure involved. The reason for this is unclear and requires further investigation. Nevertheless, a significant or large proportion of the rise in filling pressures may well have a more peripheral cause. Right atrial pressure represents the balance of venous return and cardiac output. The former is determined by mean systemic pressure, venous capacitance and

296

resistance to venous return (Guyton et al., 1973). It is conceivable that a direct micro-hemorheologic effect of Pluronic (Hymes et al., 1971) may have had an impact on the aggregation-tendency of erythrocytes in venules and on the hydraulic hindrance to blood flow in veins, to reduce venous resistance and thereby reduce the pressure-dissipation in the veins. Larger linear polymeric macromolecules are known to improve flow conditions by reducing drag (Stein et al., 1972). The possibility of a direct hemorheologic effect of Pluronic F68 is currently under investigation and other hemodynamic causes are also explored.

Even though the above considerations suggest the possibility of a direct "physical" effect of Pluronic F68, the apparently differential effects upon the distribution of the cardiac output to different organs also suggests the possibility of an action on regulatory mechanisms. At least one report (Saeed et al., 1987) vaguely suggests the possibility that some constituents of Fluosol-DA, including Pluronic, may affect the disposal and/or action of vasoactive substances. Although nothing definitive is known about the pharmacology of this substance, such a possibility may also be worth exploration.

Lastly, commercial but scientifically unsubstantiated reports (Raymond, 1989) circulate suggesting that Pluronic may enhance the thrombolytic action of tissue plasminogen activator and reperfusion of ischemic myocardium. This suggestion and the above-described cardiovascular effects of Pluronic indicate the necessity to evaluate critically this and related compounds after appropriate purification.

SUMMARY

We tested the cardiovascular actions of the surfactant Pluronic F68 by infusing it into anesthetized dogs, in doses approximating those which would be received when Fluosol-DA were used to resuscitate moderate or severe hemorrhage (0.66 and 1.11 g/Kg). In order to alleviate the reactions to Pluronic, the surfactant was purified by treatment with activated charcoal and the dogs were pretreated with corticosterone. Pluronic F68 caused dose-dependant increments in cardiac filling pressures and in systemic and pulmonary arterial blood pressure. Heart rate and contractility remained unchanged. There was an increase in the cardiac output which was dose-independant and unrelated to the filling pressures. Regional blood flow, as determined by radionuclide-labelled microspheres, tended to increase, but only that to the heart, kidney cortex and lung (bronchial arterial) increased to a statistically significant extent. It would appear that Pluronic F68, when purified with charcoal and after steroid prophylaxis, possesses significant hemorheologic and cardiovascular effects, indicating the need for further investigation of various purification methods and the effects of purified preparations.

ACKNOWLEDGEMENTS

We are grateful to Dr. Leland C. Clark, Children's Hospital Research Foundation, Cleveland, OH., for supplying us with Pluronic F68. Our gratitude is extended to Mrs. M. Bosc-Davy, Mr. R. Seymour, Mrs. J. Sistek and Mrs. D. Mauldin, for their excellent technical assistance and to Mrs. D. Longpré and Mrs. S. Dunn for their secretarial assistance. A grant to George P. Biro by the Heart and Stroke Foundation of Ontario (AN-1214) enabled us to do these experiments; we are grateful to the volunteers and donors whose contribution of time and money enabled the prosecution of this and all other research supported by the Foundation.

REFERENCES

Baer, R.W., Payne, B.D., Verrier, E.D., Vlahakes, G.J., Molodowitch, D., Uhlig, P.N., and Hoffman, J.I.E., 1984, Increased number of myocardial blood flow measurements with radionuclide-labelled microspheres, Am. J. Physiol., 246:H418.

Bartrum, R.J., Berkowitz, D.M., Hollenberg, N.K., 1974, A simple radioactive microsphere method for measuring regional flow and cardiac output, Invest. Radiol., 9:126.

Biro, G.P., 1985, Fluorocarbons in the resuscitation of hemorrhage, Int. Anesthesiol. Clin., 23:143.

Biro, G.P., 1988, Blood substitutes and the cardiovascular system, Biomater. Art. Cells Art. Organs, 16:595.

Biro, G.P. and Blais, P., 1987, Perfluorocarbon blood substitutes, CRC Crit. Rev. Oncol. Hematol., 6:311.

Biro, G.P., White, F.C., and Guth, B.G., Breisch, E.A., Bloor, C.M., 1986, The effect of hemodilution with fluorocarbon or dextran on regional myocardial flow and function during acute coronary stenosis in the pig, Amer. J. Cardiovasc. Pathol., 1:99.

Chubb, C., 1985, Reversal of the endocrine toxicity of commercially produced perfluorochemical emulsions, Biol. Reprod., 33:854.

Grover, F.L., Heron, W.H., Newman, M.M., and Paton, B.C., 1969, Effect of a nonionic surface-active agent on blood viscosity and platelet adhesiveness, Circulation, 39(Suppl 1):I-249.

Guyton, A.C., Jones, Coleman, T.G., 1973, "Circulatory Physiology: Cardiac Output and its Regulation", W.B. Saunders Company, Philadelphia.

Hammerschmidt, D.E., and Vercellotti, G.M., 1988, Limitation of complement activation by perfluorocarbon emulsions: superiority of lecithin-emulsified preparations, Biomater. Art. Cells Art. Organs, 16:431.

Heyman, M.A., Payne, B.D., Hoffman, J.I.E., and Rudolf, A.M., 1977, Blood flow measurements with radionuclide-labelled particles, Prog. Cardiovasc. Dis., 20:55.

Hymes, A.C., Robb, H.J., and Margulis, R.R., 1970, Influence of an industrial surfactant (Pluronic F-68) on human amniotic fluid embolism, Am. J. Obstet. Gynecol., 107:1217.

Hymes, A.C., Safavian, M.H., and Gunther, T., 1971, The influence of an industrial surfactant Pluronic F-68, in the treatment of hemorrhagic shock, J. Surg. Res., 11:191.

Kingma, J.G., Rouleau, J.R., Magrina, J., Dagenais, G.R., 1988, Effects of perfluorochemical hemodilution on coronary blood flow distribution in dogs, Circulation, 78:746.

Knize, D.M., White, R.C., and Paton, B.C., 1969, Use of antisludging agents in experimental cold injuries, Surgery Gynecol. and Obstet., 129:1019.

Lane, T.A., and Krukonis, 1988, Reduction in the toxicity of a component of an artificial blood substitute by supercritical fluid fractionation, Transfusion, 28:375.

Police, A.M., Waxman, K., and Tominaga, G., 1985, Pulmonary complications after Fluosol administration to patients with life-threatening blood loss, Crit. Care Med., 13:96.

Raymond, C., 1989, Copolymer, undergoing trials, could improve fibrinolytics' effectiveness, JAMA, 261:2475.

Saeed, M., Hartmann, A., and Bing, R.J., 1987, Inhibition of vasoactive agents by perfluorochemical emulsion, Life Sci., 40:1971.

Stein, P.D., Parsons, E.D., and Blick, E.F., 1972, Modifications of dynamic flow properties of turbulent flowing human blood by long chain polymers, Med. Res. Eng., 11:6.

Vercellotti, G.M., Hammerschmidt, D.E., Craddock, P.R., Jacob, H.S., 1982, Activation of plasma complement by perfluorocarbon artificial blood: probable mechanism of adverse pulmonary reactions in treated patients and rationale for corticosteroid prophylaxis, Blood, 59:1299.

Zar, J.H., Biostatistical Analysis, 2nd Ed. Prentice-Hall, New York: 1984, p. 718.

CENTRAL NERVOUS SYSTEM

OXYGEN SUPPLY AND BRAIN FUNCTION IN VIVO: A MULTIPARAMETRIC MONITORING APPROACH IN THE MONGOLIAN GERBIL

Mayevsky A.*, Frank K. H.**, Nioka S.*, Kessler M.** and Chance B.*

* Dept. of Biochemistry and Biophysics, Medical School University of Pennsylvania, Philadelphia PA 19104-6089, USA

** Institut für Physiologie und Kardiologie, Universität Erlangen-Nürnberg, D-8520 Erlangen, West Germany

INTRODUCTION

The on-line evaluation of brain physiological and biochemical functions in vivo is expected to contribute to a better understanding of the development of a pathophysiological state. The development and usage of a multiprobe assembly (MPA) for the monitoring of the metabolic, ionic and electrical activities from the cerebral cortex of experimental animals was the subject of our previous publications (Friedli et al 1982, Mayevsky et al 1980, 1985, Mayevsky and Zarchin 1987). In the present study we further extended the MPA to include a second fiber optic bundle for the monitoring of intracapillary hemoglobin oxygenation (oxy-deoxy Hb) using the EMPHO I device (Frank et al 1984a,b, 1989).

As an animal model for the studies we adopted the Mongolian gerbil (Meriones Unguiculatus) which provides an easy way to produce reversible partial or complete ischemia (Levine and Payan 1966, Levy and Brierley 1974, Mayevsky 1978, Donadio et al 1984, Mayevsky and Breuer 1989). In the same animal we tested also the effects of graded hypoxia as well as cortical spreading depression (SD) (Leao 1944a,b, 1947, Marshall 1957) which recently was suggested as a possible model for studying migraine (Lauritzen 1987). The effect of SD wave on brain oxygenation as evaluated by oxy-deoxy Hb spectral analysis together with NADH redox state is described here for the first time.

The newly developed MPA provides simultaneous readout of tissue energy state, ionic and electrical activities. Tissue oxygen delivery was monitored by the oxy-deoxy Hb ratio and, intracellular O_2 balance was evaluated by the intramitochondrial NADH redox state. The ionic homeostasis was evaluated by monitoring the extracellular K^+ and Ca^{2+} activities reflecting the permeability changes of cation channels as well as the activities of Na^+-K^+-ATPase and other ion-linked transport processes. The electrical activities were monitored by a bipolar electrocortical surface electrode (ECoG) and the changes in the DC steady potential.

Oxygen Transport to Tissue XII, Edited by J. Piiper *et al.*
Plenum Press, New York, 1990

METHODS

Multiprobe assembly (MPA)

The basic concept and features of the MPA were described previously (Friedli et al 1982, Mayevsky et al 1985, Mayevsky and Zarchin 1987). The conical plexiglass probes holder had a diameter of 5.6 mm (located on brain surface) and includes 7 holes to accomodate the following:

1.   Y-shape light guide (quartz fibers) for NADH monitoring.
2.   Y-shape light guide for HbO_2 monitoring.
3.+4. Surface mini K^+ and Ca^{2+} electrodes .
5.+6. Stainless steel wires (0.5 mm diameter) for ECoG recording.
7.   Needle type temperature probe (YSI).

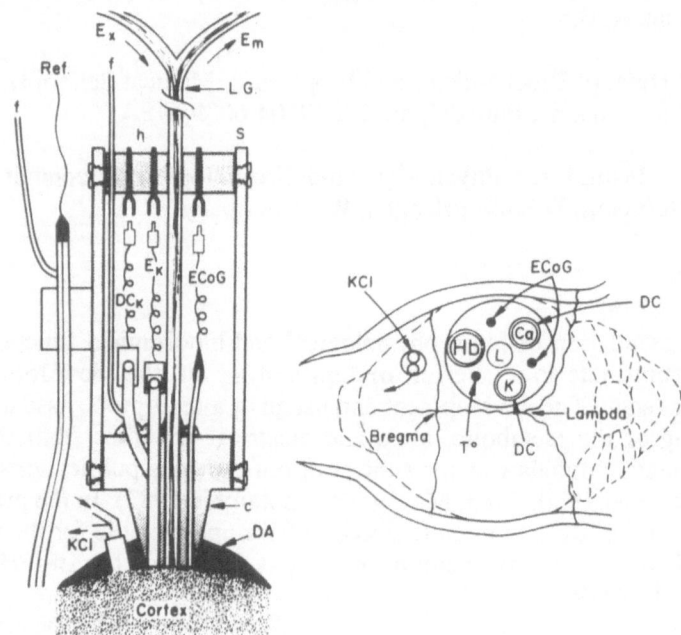

Fig. 1. Longitudinal section of the MPA (multiprobe assembly) used (left side) and its topical view when located on the gerbil brain (right side).

ECoG   - Electrocorticography electrodes.
Ek, DCk - Extracellular K+ and the surrounding DC steady potential electrodes.
Ex, Em   - Excitation and emission fiber bundles used for the NADH or HbO_2 monitoring.
L.G.   - Light guide.
h   - Connector holder.
s   - Plexiglass sleeve.
Ref   - Reference electrode.
f   - Filling tube for the DC or the Ref electrodes.
c   - Plexiglass conical probe holde.
DA   - Dental acrylic cement.
KCl   - Push pull cannula for topical application of KCl solution.
L   - Light guide.
T   - Needle probe for temperature measurements.

The DC steady potential was measured by two Ag/AgCl electrodes connected to a salt bridge concentric to the K^+ and Ca^{2+} electrodes.

Figure 1 shows in a schematic way the construction of the MPA and its location on the gerbil brain.

All other details regarding the electrodes, connectors, amplifiers as well as light guides, spectrophotometer and fluorometer were published (Mayevsky and Chance 1982, 1983, Mayevsky 1984, Frank et al 1984a,b, 1989).

Animal preparation

Adult female gerbils (Tumblebrook Farms, Mass., USA) were anesthetized by IP injection (0.3 ml/100 gr) of Equithesin (each ml contains: pentobarbital, 9.72 mg; chloral hydrate, 42.51 mg; magnesium sulfate, 21.25 mg; propylene glycol, 44.34%w/v; alcohol, 11.5%; and water). In order to keep the gerbil in the anesthetized state during the experiment, additional injection of Equithesin (0.03-0.05 ml) was made every 30 minutes. After making midline incision of the skin, the skull was exposed and a 6 mm hole was drilled in the parietal bone. Anterior to this hole, a 2 mm hole was drilled (frontal bone) to accomodate a push-pull washing device (cannula) in order to elicit cortical spreading depression by topical application of KCl solution. Two stainless steel screws were used to connect the plexiglass holder and the other cannula to the skull by dental acrylic cement. The two common carotid arteries were isolated and ligatures of 3-0 silk thread were placed around them.

RESULTS

Figure 2 shows a typical response to graded hypoxia covering the range from 100% to 0% oxygen in the breathing mixture and then a recovery to 100% O_2 for another 5 minutes. As seen, a gradual decrease in hemoglobin saturation (HbO_2) was recorded simultaneously with the increase in the intramitochondrial redox state (CF). In parallel, changes were recorded in the reflected light intensity (R) and the Hb concentration representing the autoregulation response due to the hypoxia. The first change in the ECoG was noted at FiO_2 level below 9% and was severely depressed at 4.8% and became isoelectric under N_2 breathing.

The resting level of the extracellular K^+ level was 2.5 mM and increased gradually to a maximum of 3.9 mM during the short complete anoxic period. The resting level of Ca^{2+} was quite low as compared to other brains monitored (due to a film of blood below the electrode) and was decreased from 0.6 mM (normoxia) to 0.5 mM during the anoxia. The DC steady potential around the K^+ and Ca^{2+} electrodes showed a 1-2 mV positive change during the hypoxia but we suspect it is because of a small instability of the reference electrode since it did not recover during the normoxic recovery period.

The recovery period presents a short period of reoxygenation followed by a wave of spreading depression (SD) developed probably by the increased K^+ level during the anoxic period. As seen, the brain became hyperoxygenated due to the vasodilated state (at the end of the anoxia) and the breathing of 100% oxygen. In order to describe the exact relationship between the various parameters we analyzed the analog signals. These normalized results are presented in Figure 3. The initial value (for each parameter), under 100% O_2 was taken as a 100% and all the rest of the numbers were related to this control value. As seen, during the first 25 seconds of recovery oxygen

305

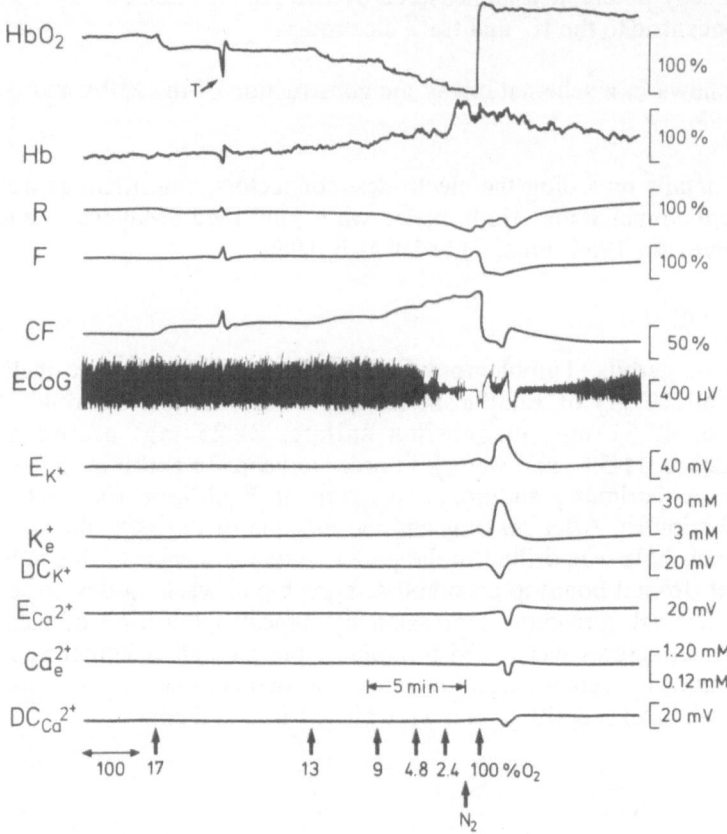

Fig. 2. Metabolic, ionic and electrical responses to graded hypoxia induced in the gerbil brain

R, F, and CF — Reflectance, fluorescence and corrected fluorescence.
ECa^{2+}, EK^+ — Uncorrected calcium and potassium potentials.
Cae^{2+}, Ke^+ — Corrected ion activities.
$DCCa^{2+}$, DCK^+ — DC steady potential measured around the calcium and potassium electrodes.
ECoG — Electrocorticogram.
HbO_2 — Hemoglobin oxygenation.
Hb — Hemoglobin volume or content.

availability and intramitochondrial redox state completely recovered. The time for the complete recovery of the vascular changes was longer, especially the Hb concentration signal.

The SD episode occured immediately after the reoxygenation of the brain showing the typical ionic homeostasis disturbances, namely an increase in extracellular K^+ (26.3 mM) and a decrease in Ca^{2+} (0.22 mM). During the recovery from the SD wave the K^+ showed an undershoot due to the activated Na^+-K^+-ATPase or the higher permeability of K^+ through the plasma membrane. The NADH became more oxidized during the SD cycle while the blood HbO_2 showed a very small, if at all, transient deoxygenation.

The transient change in blood oxygenation and NADH redox state seen during the first step of the hypoxia is due to a short bilateral carotid artery occlusion. This procedure enabled us to determine indirectly the O_2 uptake rate and will be discussed separately.

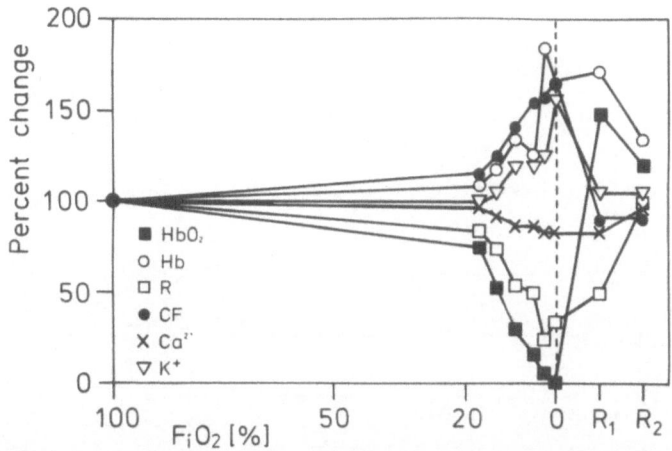

Fig. 3. The effect of graded hypoxia on the various parameters monitored from the gerbil brain as shown in Fig. 2. Each data point represents the value calculated at the end of the period of lower O_2 breathing. R_1 and R_2 values represent 25 seconds and 5 minutes recovery at 100% O_2.

The effects of ischemia on the brain are shown in Figure 4. The ischemia was induced in two steps. After right artery occlusion (the MPA was located on the right hemisphere) the NADH increased by 10-12% as compared to the normoxic level which represents about 20% of the total signal increase (ischemia + anoxia). The HbO_2 decreased also by 20-22% from the normoxic value. This small change in the O_2 supply led only to small changes in ions level and ECoG activity. The occlusion of the left carotid artery led to a more severe low flow state but not to a complete one as seen by the further increase of NADH upon addition of N_2 to the bilateral carotid occluded brain. This specific gerbil had probably a well developed posterior communication system in the circle of Willis (Mayevsky and Breuer 1989). The HbO_2 shows the same pattern of changes during the occlusionanoxia combined episode. The K^+ response to the ischemia was significant only when the two carotides were occluded. The addition of O_2 while the two carotides were occluded led to a 30-40% increase of HbO_2 and then the ID was recorded as seen by the ions and DC potential changes. Only when the blood circulation was restored (L+R open) did the HbO_2 and NADH become fully recovered to the normoxic level after a hyperoxygenated state. During the ID a massive vasoconstriction was noted as seen by the large increase in reflectance (SRI), and a

307

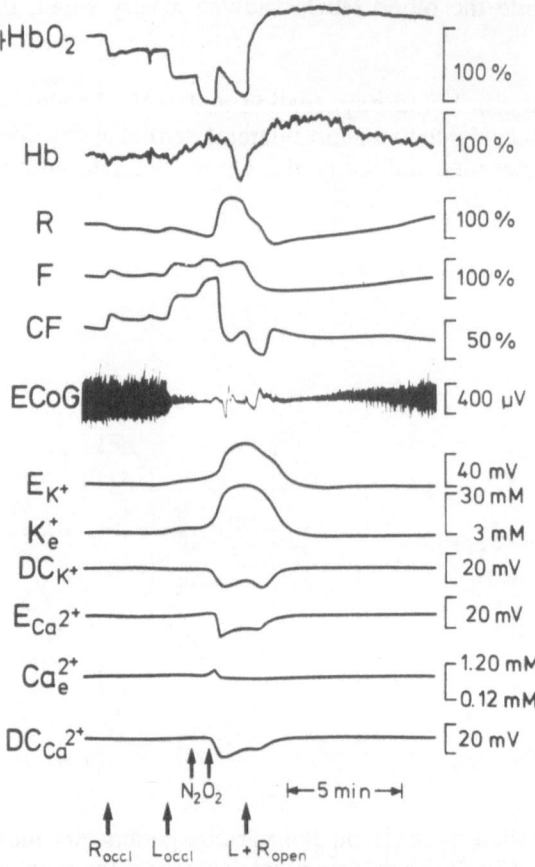

Fig. 4. The effects of unilateral and bilateral carotid occlusion as well as N2 on the various parameters monitored from a gerbil brain. All abbreviations are as in Fig. 2.

large decrease in Hb volume. The recovery from the ID episode includes the typical hyperaemic response as indicated by the decrease in the R signal and overshoot of the Hb volume. The development of the ID event started in the Ca^{2+} electrode area and appeared later below the other probes. Due to the coincidence in the ID appearance and the reoxygenation of the brain, the ID response was very similar to a SD wave type of responses which are shown in Fig. 5.

The SD wave was induced by insertion of a needle into the cortical area next to the MPA and the propagated wave was monitored. The Ca^{2+} electrode was the first to detect the wave due to the MPA positioning on the brain in relation to the SD initiation spot. The ionic homeostasis disturbances due to the SD waves, including the increase in K^+, and NADH response, had started about the same time as can be seen also from the probe location scheme. The main new results are the HbO2 responses to the SD wave and its correlation to the NADH kinetic changes. The main change seen in the HbO2 trace is the increase in oxygenation although a small short initial deoxygenation

308

could be detected. In parallel, the Hb volume was increased and it correlated well to the decrease in the R signal. Those changes in the HbO_2 oxygenation and volume are secondary to the primary oxidation of the NADH as seen by the decrease in the CF sig-

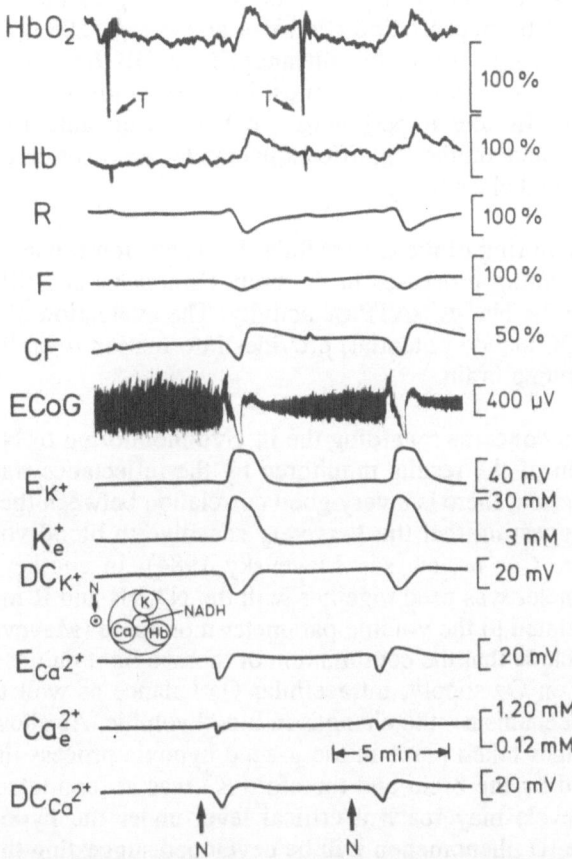

Fig. 5. Metabolic ionic and electrical responses to cortical spreading depression induced by needle insertion two milimeters anterior to the MPA. Two short R+L occlusion episode can be seen and were used for the determination of HbO_2 unloading.

nal. Those metabolic changes recorded in tissue oxygenation as well as the oxidation of the mitochondrial redox state correlated very well to the increase in energy demand by the Na^+-K^+-ATPase and other active transport systems. The short delay in the HbO_2 recorded responses is due to location of the HbO_2 probe in relation to the needle insertion site.

DISCUSSION

In vivo monitoring of the functioning brain provides information about physiological and pathological processes which have fast kinetic changes. The direct coupling between energy supplying and producing processes and the energyconsuming mechanisms (Na^+-K^+-ATPase, Ca^{2+}-ATPase) is part of the normal autoregulated functioning brain. In order to investigate those relationships it is important and unavoidable to monitor all possible parameters representing the various steps in this most complicated organ - the brain. To do so we developed the multiprobe assembly (MPA) few years ago (Friedli et al 1982, Mayevsky et al 1980, 1985), and in the present study we added to the MPA an optical probe for the monitoring of the HbO_2 oxygenation level as well as total Hb concentration (Frank et al 1984a,b, 1989). This combination of monitoring the range from the intracapillary Hb oxygenation to the intramitochondrial redox state provides the most important information about O_2 supply to the brain in vivo. In few experiments we have been able to combine a pO_2 microelectrode probe, and another significant parameter was monitored simultanuously (results will be presented elsewhere).

The on line monitoring of the extracellular K^+ levels represents, indirectly, one of the main energyconsuming processes in the brain (Erecinska and Silver 1989) due to its direct coupling to the Na^+-K^+-ATPase activity. The evaluation of the electrical activities (ECoG and DC steady potential) provides information regarding the integrated activity of the functioning brain.

One of the main concerns regarding the in vivo monitoring of NADH redox state was the interpretation of the results monitored by the reflectance trace (366 nm). As seen in the results section, there is a very good correlation between the R signal and the Hb concentration suggesting that the R is very sensitive to blood volume changes as also suggested before (for review see Mayevsky 1984). In another set of studies a Laser-Doppler flowmeter was used together with the NADH and R monitoring and indeed the R was correlated to the volume parameter monitored (Mayevsky, unpublished results). We can conclude that the combination of the two light guides provides the significant information on O_2 supply, intracellular O_2 balance as well as the product of the autoregulation mechanism - the changes in blood volume. As shown in Figs. 2 and 3 those volume changes taken place in the graded hypoxia process did not supply the amount of O_2 needed by the brain and therefore K^+ was accumulated in the extracellular space. If K^+ levels may reach a critical level under the hypoxic, ischemic or anoxic state then, the ID phenomenon will be developed suggesting that ionic homeostasis failed (Vyskocil et al 1972). In the episode shown in Fig. 2, the oxygen was returned to the brain at the same time of reaching the critical accumulation of K^+, and the SD wave was developed. The difference between the ID and the SD is in the kinetics of the recovery processes under those two situations. Under the ID, the massive vasoconstriction occurrence led to a decrease in blood volume which even when O_2 is available will take longer to supply enough O_2 to the entire tissue. Under SD we recorded mainly vasodilationincrease in blood volume as well as increase in HbO_2 oxygenation. Therefore the ability of the brain to restore the ionic homeostasis is greater and could be completed within a shorter period of time.

The possibility that a short vasoconstriction event preceeded the vasodilation cannot be excluded and more experiments are needed in order to clearify this point.

It is very clear from the data presented in Fig. 5 that during SD wave passage, the large increase in blood volume is able to provide enough O_2 to keep the Hb in hyperoxygenated state although O_2 consumption was massively stimulated as indicated by the oxidation of the intramitochondrial NADH. This change in O_2 consumption is associated mainly with the increase in Na^+-K^+-ATPase activity, which led to a state 4 to state 3 transition (Chance and Williams 1955) and thus NADH became more oxidized. The same type of results were described by other groups (for example Somjen et al 1976, Rosenthal and Sick 1988). This oxidation response of the NADH to SD wave is typical to the normoxic brain and may change toward an increase of NADH according to the O_2 balance created in the brain (Mayevsky et al 1982).

In summary, the new approach used in this study provided very significant information regarding the kinetic changes taken place in the gerbil brain exposed to various pathological situations.

A clear correlation was recorded between the changes in Hb oxygenation, tissue oxgenation and intramitochondrial redox state. This relationship was kept in oxygen deprived situation such as hypoxia and ischemia as well as during cortical spreading depression.

Blood volume changes monitored by the 366 nm reflectance and the Hb concentratin of the EMPHO I unit were correlated very well.

During oxygen deprived situation the degree of the Na^+-K^+-ATPase (evaluated by extracellular level of K^+) was correlated to the energy state.

Stimulation of the brain by a wave of spreading depression induced metabolic activation in response to the increased Ke^+ levels.

We believe that the usage of the multiparametric monitoring approach will provide significant information for basic brain research as well as for clinical situations.

ACKNOWLEDGEMENTS

This work was supported by the NIH grant - NS 22881.

REFERENCES

Chance, B. and Williams, G.R. (1955) Respiratory enzymes in oxidative phosphorylation, J. Bio. Chem.,217,383-393.

Donadio, M.F., Kozlowski, P.B., Kaplan, H., Wisniewski, H.M. and Majkowski, J. (1982) Brain vasculature and induced ischemia in seizureprone and non-seizureprone gerbils, Brain Res., 234,263-275.

Erecinska M. and Silver I.A. (1989) ATP and brain function. J. CBF and Metabol. 9 : 2-19

Frank K.H., Schabert A., Friedl A., Brunner M., Höper J., Kerl G., Kessler M. (1984a) Correlation between tissue pO2 and intracapillary Hb spectra, Adv. Exp. Med. Biol. pp 811-818

311

Frank KH., Rettig V., Friedl A., Brunner M., Ellermann R. Kerl G., Höper J. (1984b) Measurements of intracapillary haemoglobin spectra in the beating heart, skeletal muscle and the liver, using the Erlangen micro-lightguide spectrophotometer, Pflüger Archiv 400:Suppl pp. 219

Frank K.H., Kessler M., Appelbaum K., Dümmler W. (1989) The Erlangen micro-lightguide spectrophotometer EMPHO I, Phys. Med. Biol. in press

Friedli, C.M., Sclarsky, D:L. and Mayevsky, A. (1982) A New Multiprobe Assembly for Surface Monitoring of Ionic Metabolic and Electrical Activities in the Awake Brain. Am. J. Physiol 243:R462-R469

Lauritzen M. (1987) Cerebral blood flow in migraine and cortical spreading depression, Acta Neurol. Scan. Vol 76, Suppl. 113 (40 pages)

Leao A.A.P. (1944a) Spreading depression of activity in the cerebral cortex, J. Neurophysiol. 7:359-390

Leao A.A.P. (1944b) Pial circulation and spreading depression of activity in cerebral cortex, J. Neurophysiol 7:391-396

Leao A.A:P. (1947) Further observations on the spreading depression of activity in the cerebral cortex. J. Neurophysiol. 10:409-419

Levine, S., and Payan, H. (1966) Effects of ischemia and other procedures on the brain and retina of the gerbil (Meiones unguiculatus). Exp. Neurol.16:255-262.

Levy, D.E. and Brierley, J.B. (1974) Communications between vertebrobasilar and carotid arterial circulations in the gerbil. Exp. Neurol.45:503-508

Marshall W.H. (1959) Spreading cortical depression of Leao. Physiol. Rev.39:239-279

Mayevsky, A. (1978) Pyridine nucleotide oxidation-reduction state of the cerebral cortex in the awake gerbil. J. Neurosci. Res. 3:369-374

Mayevsky, A. (1984) Brain NADH redox state monitored in vivo by fiber optic surface fluorometry. Brain Res. Rev. 7:49-68

Mayevsky, A. and Chance, B. (1982) Intracellular oxidation reduction state measured in situ by multichannel fiberoptic-surface fluorometer, Science. 217:527-540

Mayevsky, A. and Chance, B. (1983) Multisite measurements of NADH redox state from cerebral cortex of the awake animal, Oxygen Transport in Tissue IV, Bicher, H:I: and Bruley, J.F. eds., Plenum Press, New York, pp 143-155.

Mayevsky, A. and Zarchin, N. (1987) Metabolic Ionic and electrical activities during and after incomplete or complete cerebral ischemia in the Mongolian gerbil. In: Oxgen Transport to Tissue IX, Silver, I.A. and Silver, A. (eds), Plenum Publishing Corp. pp. 265-273

Mayevsky, A. and Breuer. (1989) The Mongolian gerbil as a model for cerebral Ischemia. In: Cerebral Ischemia and Cerebral Resuscitation, Schurr, A. and Rigor B.M. (eds), CRC (in press)

Mayevsky, A. Lebourdais, S. and Chance, B. (1980) The interrelation between brain pO2 and NADH oxidation-reduction state in the gerbil, J. Neurosci. Res., 5:173-182

Mayevsky, A., Zarchin, N. and Friedli, C.M.(1982) Factors affecting the oxygen balance in the awake cerebral cortex exposed to spreading depression, Brain Res., 236:93-105

Mayevsky, A., Friedli C.M. and Reivich, M. (1985) Metabolic, Ionic and Electrical Responses of Gerbil Brain to Ischemia. Am. J. Physiol. 17:R99-R107

Rosenthal, M. and Sick, T.J. (1988) Measurement of Metabolic Activity Associated with Ion Shifts In: Neuromethods; the Neuronal Microenvironment. A.A. Boulton, G.B. Baker and W.Walz (eds), The Humana Press pp. 187-245

Somjen, G.G., Rosenthal, M., Cordingley, G. LaManna, J. and Lothman, E. (1976) Potassium, Neuralgia and Oxidative Metabolism in Central Gray Matter. Fed. Proc. 35:1266-1271

Vyskocil, F., Kriz, N. and Bures, J. (1972) Potassium selective Microelectrodes used for Measuring Brain Potassium during Spreading Depression and Anoxic Depolarization in Rats. Brain Res. 39:255-259

BRAIN ISCHEMIC DEPOLARIZATION AND VASOSPASM IN THE MONGOLIAN GERBIL:

THE DEPENDENCE ON ENERGY DEPLETION LEVELS

Avraham Mayevsky* and Shlomo Cohen

Department of Life Sciences, Bar Ilan University, Ramat Gan, 52900, Israel, and *Department of Biochem. and Biophys. Univ. of Penna., Philadelphia, PA 19104-6089 USA

INTRODUCTION

The understanding of the various pathological events occuring in the ischemic brain of an animal model has significant implications to the treatment of patients undergoing stroke. One of the events that occurs under ischemia is the ischemic depolarization (ID) during which various pathological changes take place in the brain (Harris et al 1981, Hansen 1985, Siesjo and Bengtson 1989). The mechanism behind the development of ID is not clear as yet, although ion homeostasis disturbances are involved (Mori et al 1987). The inhibition of the Na^+-K^+-ATPase is energy depletion dependent, and an immediate response of the brain to ischemia (Erecinska & Silver 1989) will result in extracellular K^+ accumulation. In the Mongolian gerbil (Meriones Unguiculatus), unilateral carotid artery occlusion creates a variable level of ischemia in the ipsilateral hemisphere (in different animals) while the contralateral side is not affected significantly (Levine and Payan 1966, Levy and Brierly 1974, Mayevsky 1978, Mayevsky and Breuer 1989).

In the Mongolian gerbil brain we tested the correlation between energy depletion level (evaluated by intramitochondrial NADH redox state) and the development of ischemic depolarization (ID) developed under ischemia, using the multiparametric monitoring assembly--MPA (Friedli et al 1982, Mayevsky et al 1985).

Surface fluorometry/reflectometry was used in order to evaluate quantitatively the level of ischemia created under unilateral carotid artery occlusion and to correlate it to the development of the ischemic depolarization (ID) measured by the appearance of the secondary reflectance increase (SRI), a typical optical correlate of the ID (Mayevsky and Chance 1976, Mayevsky et al 1985, Mayevsky and Zarchin 1987).

METHODS

Fluorometer/reflectometer

In the present study, a two-channel DC fluorometer/reflectometer was used (Mayevsky 1978, Mayevsky and Chance 1982, Mayevsky 1984). The source for the 366 nm excitation light was a 100 W air-cooled mercury lamp passing through the two excitation fiber optic bundles to the brain. The emitted light from the brain that was transmitted through another fiber optic bundle was split into a 90:10 ratio in order to measure the fluorescence light (F: 90%) and the reflected light (R: 10%). The two fluorescence signals and the two reflectance signals, as well as the two corrected fluorescence signals, were recorded on a multichannel Grass instrument. The correction of the fluorescence signals for hemodynamic artifacts was done by subtracting the R from the F signal in a 1:1 ratio (Mayevsky, 1984).

Multiprobe Assembly

The multiprobe assembly (MPA) used in the present study was described in detail in our previous publications (Friedli et al 1982, Mayevsky and Zarchin 1987). Hence, only major points will be described herein. The MPA has the following main features: 1) all components have the same type of non-invasive surface contact with the cortex; 2) it can very easily be removed from the skull at the end of the experiment without any damage, so that repetitive experiments can be performed in a short period of time; 3) all signals monitored by the MPA have a very low sensitivity to movements of the animal, even while exposed to convulsions. In the present study, the pH electrode was substituted by a Ca^{2+} electrode. All the details regarding this new electrode will be published elsewhere (Mayevsky et al, in preparation).

Animal Preparation

Adult Male Gerbils (Tumblebrook Farms, Mass, USA) were anesthesized by IP injection (0.3 ml/100 gr) of Equithesin (each ml contains: pentobarbital, 9.72 mg; chloral hydrate, 42.51 mg; magnesium sulfate, 21.25 mg; propylene glycol, 44.34%w/v; alcohol, 11.5% and water). In order to keep the gerbil in the anesthetized state during the experiment and the ischemic period, an additional injection of Equithesin (0.03-0.05 ml) was made every 30 minutes. After making a midline incision of the skin, the skull was exposed and two holes of 3.5 mm were drilled in the parietal bone area to accomodate the light guide holders (located epidurally). Two pairs of stainless-steel screws (used for ECoG recording) were also placed in the skull and dental acrylic cement was applied.

In experiments where the MPA was used, a 6 mm hole was drilled in one hemisphere and after gentle removal of the dura mater , the MPA was placed on the brain using a micromanipulator. Two stainless steel screws were used to connect the MPA and the dental acrylic cement to the skull. The two common carotid arteries were isolated and ligatures of 3-0 silk thread were placed around them.

RESULTS

In the first part of the results a description of the SRI phenomenon is given. Figure 1 shows the responses of the gerbil brain to unilateral carotid artery occlusion repeated twice in the same animal (Roccl). During the five minutes of ischemia, the NADH became more reduced (increased CF signal) while the reflectance trace showed a small decrease. The percent change of the NADH was about 20% as compared to the preischemic baseline, and as shown in Figure 2, under complete ischemia, the change was about 50%. Thus we can calculate the level of ischemia under unilateral occlusion to be 40% of the maximum (20/50). A small increase in extracellular K^+ was recorded (K_e^+) as well as a depression in the intensity of the ECoG signal especially in part B. The extracellular Ca^{2+} (Ca_e^{2+}) as well as the DC steady potential around the two ion sensitive electrodes were not changed. As can be seen, the ID was not developed during the five minutes of unilateral occlusion.

316

When in the same animal the two carotid arteries were occluded (Loccl+Roccl) as seen in Figure 2, the brain was exposed to complete ischemia as seen in the CF trace as well as the isoelectric ECoG trace. As expected, the extracellular K^+ started to increase immediately after the right carotid occlusion (the MPA was implanted in the right brain hemisphere). After two to five minutes of complete ischemia the ischemic depolarization was recorded (ID) as seen in all traces recorded. The K_e^+ showed a faster increase together with the negativation of the DC potential (DC_{K+}). Also, the extracellular Ca^{2+} (Ca_e^{2+}) showed a large decrease together with the negative shift in the DC potential (DC_{Ca}^{2+}). During the ID period the reflectance trace showed a large increase (SRI) but the corrected fluorescence (CF) only slightly increased. About 15 minutes after the reperfusion, all signals (except ECoG) reached the preischemic level suggesting a complete recovery of the parameters measured.

In the second phase of the study we used the change in the reflectance trace (SRI) as an indicator of the ID phenomenon. We used 18 gerbils, (NADH was monitored bilaterally together with the 366 nm reflectance) in which 65 unilateral carotid occlusions were performed. In 32 cases, the SRI ($^+$SRI) was developed, while in 33 no secondary change in the reflectance trace was recorded (-SRI). Figure 3 shows four typical responses measured from 4 different gerbils. The two responses shown in A and B were taken from gerbils

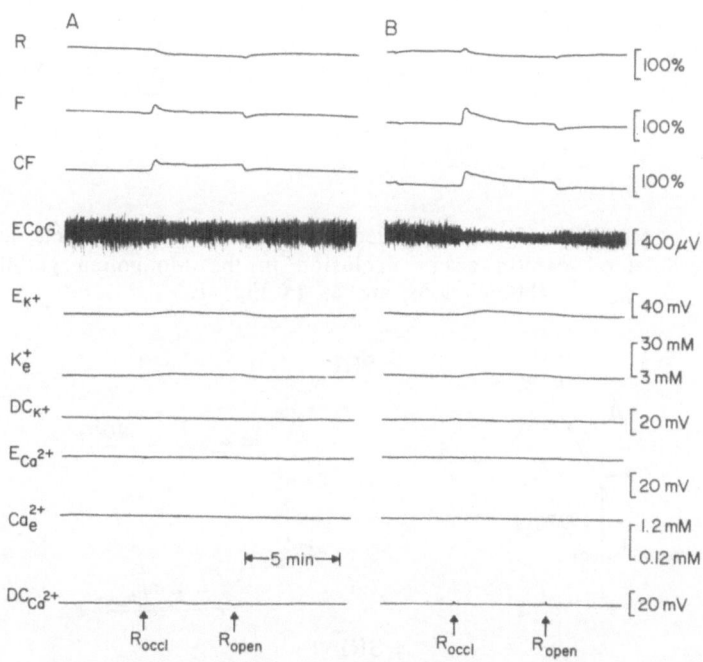

Figure 1   The effects of two sessions of 5 minute unilateral carotid artery occlusion (A,B) on the metabolic, ionic and electrical activities in the gerbil brain. The MPA was implanted above the right brain hemisphere.

R, F and CF - reflectance, fluorescence and corrected fluorescence.
E_{Ca}^{2+}, E_{K^+} - uncorrected calcium and potassium potentials
Ca_e^{2+}, K_e^+ - corrected ion activities
DC_{Ca}^{2+}, DC_{K^+} -DC steady potential measured around the calcium and potassium electrodes.
ECoG - Electrocorticogram

317

which developed a mild ischemia (18%-20%) under unilateral carotid occlusion and therefore SRI was not recorded. The responses shown in part C and D were obtained from gerbils which developed the SRI; the level of ischemia for C and D was twice that of A and B (32%-38%). As seen, there are two different kinetics in

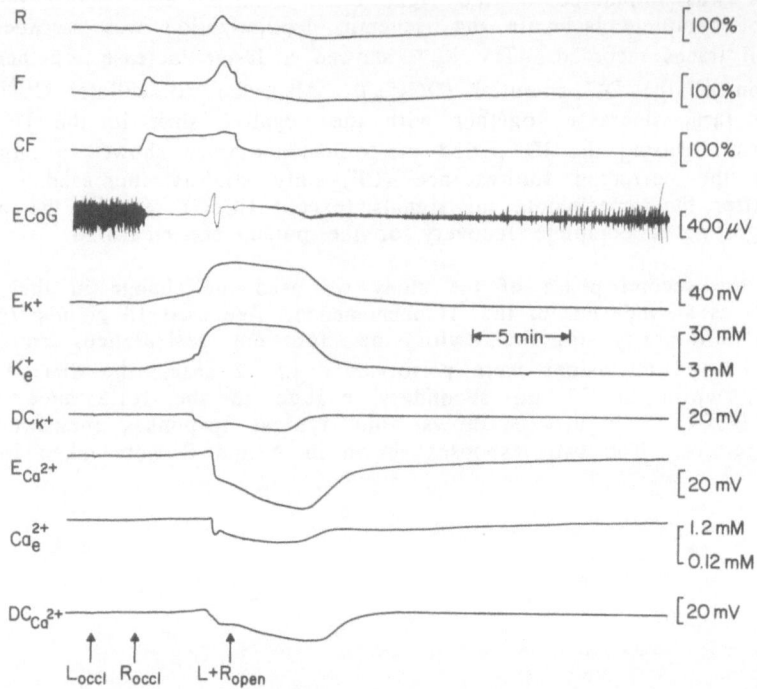

Figure 2   Metabolic, ionic, and electrical responses to complete ischemia induced by bilateral carotid artery occlusion in the Mongolian gerbil. All abbreviations are as in Fig. 1.

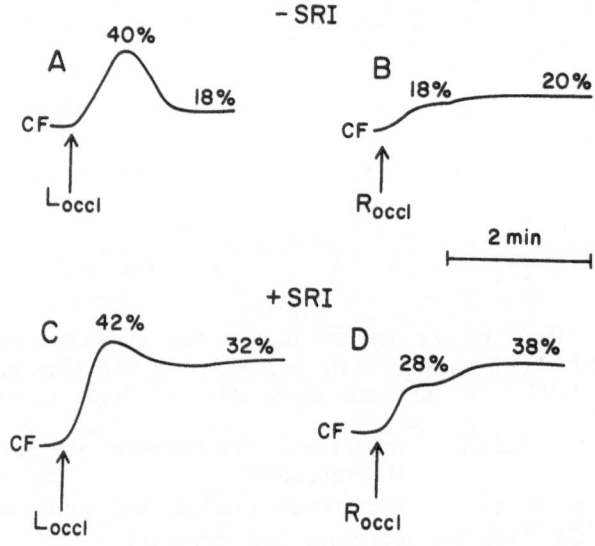

Figure 3   Typical metabolic responses to unilateral carotid artery occlusion in the gerbil brain. The numbers represent the percent increase of NADH level (CF) during the initial step as well as during the steady state phase. $^+$SRI, $^-$SRI represent two groups of gerbils as explained in the text.

318

<u>Table 1</u>. The effects of unilateral carotid artery occlusion on various NADH levels measured from the gerbil brain (Means± S.E.M.).

Group	%CF Initial increase	%CF Steady State increase	%CF St. St. max Inc.	Δ% CF St. St.--Init.
$^+$SRI	41.6 ± 2.6	42.6 ± 3.1	85.5 ± 4.6	1.03 ± 2.0
$^-$SRI	28.2 ± 2.1	17.4 ± 1.8	34.4 ± 3.8	$^-$10.9 ± 1.0
p< (Level of significance)	0.005	0.005	0.005	0.005

the reponses of the NADH (CF). In parts A and C, the initial increase was followed by a decrease which later stabilized at a steady state level. In parts B and D, the initial increase was followed by an additional increase up to the steady state level. In each gerbil the NADH was monitored from the two hemispheres but only the ipsilateral hemisphere of the occlusion side is shown. The contralateral NADH level remained unchanged in all gerbils tested.

In order to quantitate the results, the 65 responses were analyzed for changes in NADH level during the initial phase of the occlusion as well as during the steady state (st. st.). We also calculated the steady state level as compared to the maximal increase of CF during anoxia (100% N_2). The results are shown in Table 1 and Figure 4, and as seen, a large difference was found between the $^+$SRI and the $^-$SRI groups of gerbils. The difference was statistically significant in all

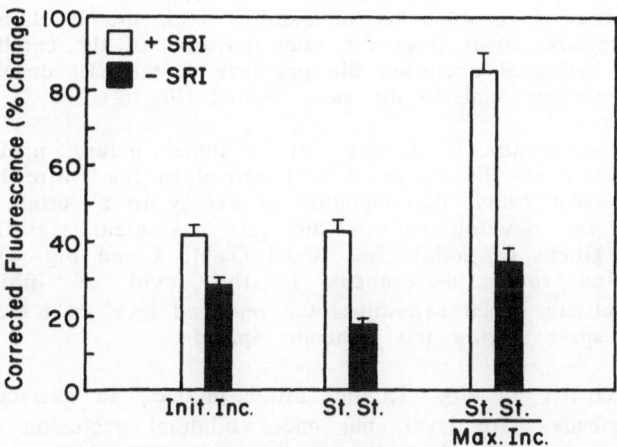

<u>Figure 4</u> The effects of unilateral carotid artery occlusion on the NADH level (corrected fluorescence) in two groups of gerbils (Means ± S.E.M).

Init Inc -   Initial increase after occlusion
st. st.-   Steady State level during the occlusion
st.st./Max Inc - Steady State level as compared to the maximum increase during anoxia.
-SRI, $^+$SRI -   Gerbils among which the Secondary Reflectance Increase during the ischemia developed (+) or did not develop (-)

319

three comparisons (p<0.005). Also when we calculated the ΔCF between the st.st. and the init. levels, the difference between the two groups was significant (p<0.005).

In the $^+$SRI group the average level of NADH was stable during the occlusion period (41.6 vs. 42.6) while in $^-$SRI group a large decrease was recorded (28.2 vs. 17.4), suggesting that the anterior circulation compensated for part of the ischemia.

DISCUSSION

One of the main results of brain ischemic event is the development of the so called ID-ischemic depolarization. This depolarization is very similar to the one developed under severe anoxia - anoxic depolarization (Leão 1947, Marshall 1959) and includes similar ion homeostesis disturbances (Vyscocil et al 1972, Dora and Zeuthen 1976) as well as vascular spasm (Mayevsky and Chance 1976).

The main questions are: what is the triggering mechanism that initiates such a massive pathological state?; and whether or not the development of the ID is energy depletion level dependent. To date, there is no published explanation for the ID development although few correlative events were described (Harris et al 1981). In our previous work we showed the difference between unilateral and bilateral carotid artery ligation in the Mongolian gerbil while monitoring the metabolic ionic and electrical activities (Mayevsky and Zarchin 1987). We presented the results only from gerbils who developed the SRI under unilateral as well as bilateral carotid artery occlusion. In the present study, we are presenting the differences between the responses of two groups of gerbils to unilateral carotid occlusion in the development of the ID as evaluated by the appearance of the SRI response.

In order to determine the level of ischemia during the unilateral occlusion we monitored the NADH redox state which represent the intracellular oxygen availability under such conditions. The usage of the fiber optic surface fluorometry/reflectometry enabled us to monitor both the level of ischemia and the vasospasm response to it from the same territory of the cerebral cortex. As shown in Fig. 1 unilateral occlusion did not lead to the SRI development as seen under bilateral occlusion done in the same animal (Fig. 2).

The ID phenomenon is of the "all or none" nature, namely, once it is triggered in one spot of the cortex it will spread to the entire hemisphere. As seen from the present study, the depletion of energy to a certain low level is a pre-condition to the development of the ID. A clear significant difference between the two groups of gerbils was found (Table 1 and Fig. 4). The possible mechanism behind these differences is the level of inhibition of the $Na^+K^+ATPase$ activity which determines the rate and level of K^+ accumulation in the extracellular space during the ischemic episode.

As seen clearly in Fig. 1, the initial increase in extracellular K^+ was followed by a steady state level, but under bilateral occlusion (Fig.2) the K^+ increased continuously until the appearance of the ID. In our previous work (Mayevsky and Zarchin 1981), we showed that the ischemic hemisphere had developed the SRI much faster than the control. This also indirectly supports the finding that a different K^+ leakage rate was found in the mild ischemic side as compared to the more severe level of ischemia under unilateral occlusion in other animals.

The factor which determined the level of ischemia under unilateral occlusion is the structure of the anterior communication system in the Circle of Willis (Mayevsky and Breuer 1989). Since the SRI is an "all or none" phenomenon it is not possible to quantitate this parameter and correlate it with

320

the level of ischemia. We assume that determination of the K^+ level in parallel will contribute also to the understanding of the ID development.

In summary, the results presented here show very clearly that the development of the ID state under partial ischemia is dependent upon the degree of energy depletion. The nature of the development of the ID is of the "all or none" nature, namely that the depletion of energy has to reach a critical low level in order to initiate the process of ID. It would seem to us that the accumulation of extracellular K^+ and its relationship to the changes in Ca^{2+} are involved in it, but further studies are needed to clarify this mechanism.

ACKNOWLEDGEMENTS

This work was supported by NIH grant NS-22881, by the Health Sciences Res. Center, Dept. of Life Sciences Bar Ilan Univ. Israel and by the Chief Scientist's Office, Ministry of Health, Israel.

REFERENCES

Dora, E. and Zeuthen, T. (1976) Brain Metabolism and Ion Movements in the Brain Cortex of the Rat during Anoxia. In: Ion and Enzyme Electrodes in Biology and Medicine. M. Kessler, L.C. Clark, D.W. Lubbers, I.A. Silver and W. Simon, eds. University Park Press, Baltimore. pp. 294-298.

Erecinska M. and Silver I.A. (1989) ATP and brain function. J. CBF and Metabol. 9:2-19.

Friedli, C.M., Sclarsky, D.L. and Mayevsky, A. (1982) A New Multiprobe Assembly for Surface Monitoring of Ionic Metabolic and Electrical Activities in the Awake Brain. Am. J. Physiol. 243:R462-R469.

Hansen, A.J. (1985) The Effect of Anoxia on Ion Distribution in the Brain. Physiol. Rev. 65:101-148.

Harris, R.J., Symon, L., Branston, N. M. and Bayhan, M. (1981) Changes in extracellular calcium activity in cerebral ischemia. J. CBF and Metabol. 1:203-209.

Leão A.A.P. (1947) Further ovservations on the spreading depression of activity in the cerebral cortex. J. Neurophysiol. 10: 409-419.

Levine, S., and Payan, H. (1966) Effects of ischemia and other procedures on the brain and retina of the gerbil (Meiones unguiculatus). Exp. Neurol., 16:255-262.

Levy, D.E. and Brierley, J.B. (1974) Communications between verte- brobasilar and carotid arterial circulations in the gerbil. Exp. Neurol.45:503-508.

Marshall W.H. (1959) Spreading cortical depression of Leão. Physiol. Rev. 39:239-279.

Mayevsky, A. (1978) Pyridine nucleotide oxidation-reduction state of the cerebral cortex in the awake gerbil. J. Neurosci. Res. 3:369-374.

Mayevsky, A. (1984) Brain NADH redox state monitored in vivo by fiber optic surface fluorometry. Brain. Res. Rev. 7:49-68.

Mayevsky, A. and Chance, B. (1976) The Effect of Decapitation on the Oxidation Reduction State of NADH and ECoG in the Brain of the Awake Rat, In:

Oxygen Transport to Tissue II. J. Grote, D. Reneau and G. Thews, eds. Plenum Pub. Corp. pp. 307-312.

Mayevsky, A. and Zarchin, N. (1981) The effects of unilateral carotid Occlusion on the responses to decapitation in the gerbil brain. Brain Res. 206:155-160.

Mayevsky, A. and Chance, B. (1982) Intracellular oxidation reduction state measured in situ by multichannel fiber-optic surface fluorometer. Science. 217:527-540.

Mayevsky, A. and Zarchin, N. (1987) Metabolic ionic and electrical activities during and after incomplete or complete cerebral ischemia in the Mongolian gerbil. In: Oxygen Transport to Tissue IX, Silver, I.A. and Silver, A., eds). Plenum Publishing Corp. pp. 265-273.

Mayevsky, A. and Breuer. (1989) The Mongolian gerbil as a model for cerebral Ischemia. In: Cerebral Ischemia and Cerebral Resuscitation. Schurr, A. and Rigor B.M. (Eds.). CRC (In press).

Mayevsky, A., Friedli C.M. and Reivich, M. (1985) Metabolic, Ionic and Electrical Responses of Gerbil Brain to Ischemia. Am. J. Physiol. 17: R99-R107.

Mori, K. Iwayama, K., Kawano, T. and Kaminogo, M. (1987) DC potential and extracellular K^+ and Ca^{2+} at critical levels of brain ischemia in cats. J. CBF and Metabol. 7, Suppl. 1, S112

Rosenthal, M. and Sick, T.J. (1988) Measurement of Metabolic Activity Associated with Ion Shifts In: Neuromethods; The Neuronal Microenvironment. A. A. Boulton, G.B. Baker and W. Walz, eds. The Humana Press pp. 187-245.

Siesjo, B.K. and Bengtson, F. (1989) Calcium fluxes, calcium anagonists and calcium related pathology in brain ischemia. J. CBF and Metabol. 9:127-140.

Somjen, G.G., Rosenthal, M., Cordingley, G., LaManna, J. and Lothman, E. (1976) Potassium, Neuralgia and Oxidative Metabolism in Central Gray Matter. Fed. Proc. 35:1266-1271.

Vyskocil, F., Kriz, N. and Bures, J (1972) Potassium selective Microelectrodes used for Measuring Brain Potassium during Spreading Depression and Anoxic Depolarization in Rats. Brain Res. 39:255-259.

THE NON-INVASIVE MONITORING OF CEREBRAL TISSUE OXYGENATION

M.S. Thorniley, L.N. Livera*, Y.A.B.D. Wickramasinghe,
S.A. Spencer*, and P. Rolfe

School of Postgraduate Medicine and Biological
Sciences, Department of Biomedical Engineering
and Medical Physics, University of Keele,
Thornburrow Drive, U.K., and the *Neonatal
Department, North Staffordshire Maternity
Hospital, Stoke-on-Trent, U.K.

INTRODUCTION

In adult and neonatal intensive care there is an obvious
need for reliable, non-invasive methods of monitoring
cerebral oxygenation and circulation. The technique of near
infra-red spectroscopy (NIRS) may have the potential to
fulfill this need.

The technique of NIRS utilises the known absorption
properties of haemoglobin and of cyt aa3, the terminal member
of the respiratory chain, and has been well described
(Jobsis,1977, Rea et al.,1985,Wyatt et al.,1986). However
some of the major areas of concern with this method relate to
the sensitivity and specificity of the technique for the
measurement of changes in the oxygenation level of
haemoglobin and , more particularly, in the redox state of
cyt aa3 (Wray et al.,1988; Carta et al.,1987). The use of
difference measurements (Jobsis,1977, Ferrari et al.,1983,
Thorniley et al.,1988a,b) has played an important role in the
assessment of cerebral oxygenation but this is not the most
accurate method of determining changes in concentration of
the species of interest. An additional problem in
quantitation is in the determination of the optical path
length; this is being addressed by several groups.

INSTRUMENTATION

Instrumentation that was originally developed by the
Dept of Biomedical Engineering and Medical Physics,
University of Keele, has since been made available to several
groups by Radiometer, Copenhagen (Rea et al.,1985;
Wickramasinghe et al.,1986). This instrument uses 4 single
heterojunction laser diodes with nominal wavelengths of 775,
805,845 and 904nm. The laser firing circuit is controlled by

Oxygen Transport to Tissue XII, Edited by J. Piiper *et al.*
Plenum Press, New York, 1990

a microprocessor and provides light pulses of 200ns duration. The output power of laser diodes generally falls with temperature, and in order to combat this the laser diodes are temperature stabilised using Peltier elements. The transmitted light after absorption by the biological organ is detected by a cooled avalanche photodiode module, which converts the light to a current and then to a voltage for amplification. The signal is
then digitised and processed, using signal averaging techniques to reduce the noise level.

A computer then analyses the absorption measurements and using NIR multiplier coefficients, is able to display the calculated change in concentrations of oxygenated/ deoxygenated haemoglobin, redox state of cyt aa3, and the change in total haemoglobin during monitoring. Thus for in-vivo monitoring of infants, these variables can be observed in "real time", to enable the clinician to assess the physiological and biochemical status of the patient. The system achieves a coefficient of variation of 0.01% over a 30 second period whilst monitoring a basal level of 8 optical density units. System drift was found to be less than 0.004 OD/hr for all 4 laser diodes.
It is the oxidised form of cytochrome oxidase (hereafter described as cyt aa3) that is considered and the results are presented as in changes in the oxidation level (or redox state) of cyt aa3.

ALGORITHM DEVELOPMENT

The basic assumption made is that it is possible to apply the Beer - Lambert Law to biological organs. If a substance of concentration C and path length d is transilluminated by light of intensity Io, and results in a transmitted light of I, then the absorbance
A= log10 Io - log10 I=eCd,

where e = absorption coefficient of the substance.

This method can be applied to a mixture of four substances, having individual concentrations of C_1, C_2, C_3, & C_4, with absorption coefficients e_1, e_2, e_3, & e_4 respectively. If the total path length of the mixture is d then total absorbance =

$(e_1C_1 + e_2C_2 + e_3C_3 + e_4C_4)d$.

We consider the biological organ as a multicomponent mixture comprising those chromophores which we have observed to change their absorbance according to the availability of oxygen; these are assumed to be HbO2, HbR and cyt aa3. For such a situation the total absorbance can be represented as:

$A = (e_1C_1 + e_2C_2 + e_3C_3)d + T$,

where T corresponds to absorbers that do not depend on oxygenation. It is not possible to evaluate T accurately for biological organs. If we consider changes in concentration with time , then the above equation can be rewritten to represent change in concentration instead of absolute

324

concentrations:

Change in absorbance = $(e_1 C_1 + e_2 C_2 + e_3 C_3)d$, If absorbance is monitored at three wavelengths, 775nm, 845nm and 904nm, then the following three equations can be written for <u>change</u> <u>in</u> <u>absorbance</u>(A_1 A_2 & A_3):

$$A_1 = (e_{1.1}C_1 + e_{2.1}C_2 + e_{3.1}C_3)d$$

$$A_2 = (e_{1.2}.C_1 + e_{2.2}C_3 + e_{3.2}C_3)d$$

$$A_3 = (e_{1.3}C_1 + e_{2.3}C_3 + e_{3.3}C_3)d$$

(With this notation e_{ij}, for example, is the absorbtion coefficent of component i at wavelength j).

These three equations can be solved for C1,C2, & C3 (changes in concentration) using matrix inversion. The concentration changes can then be represented as:

$$C_1 = P_{11}A_1 + Q_{12}A_2 + R_{13}A_3$$
$$C_2 = P_{21}A_1 + Q_{22}A_2 + R_{23}A_3$$
$$C_3 = P_{31}A_1 + Q_{32}A_2 + R_{33}A_3$$

The nine values for P,Q & R are purely functions of the absorbtion coefficients of HbO_2, HbR and oxidised cyt aa3.

<u>In-vitro</u> <u>and</u> <u>in-vivo</u> <u>absorption</u> <u>coefficients</u>

In vitro absorbtion data for oxygenated and deoxygenated haemoglobin have been published (Wray et al.,1988). We have performed in - vivo experiments on rats following fluorocarbon transfusion to remove blood and have obtained in-vivo absorbtion data for cyt aa3 (oxidised minus reduced). Using these values, we have obtained the following NIR multiplier coefficients:

	775nm	845nm	904nm
for HbO2 --------	-1.156	0.074	1.428
for HbR --------	1.641	-0.935	-0.178
for cyt aa3 (ox-red)	-0.014	0.582	-0.491
for total haemoglobin	0.485	-0.861	1.249

Thus, the change in concentration of HbR, for example, is given as:

A1(1.641)-A2(.935)-A3(.178)

where A_1,A_2,A_3 are the changes in absorbance at 775nm,845nm and 904nm respectively.

<u>Sensitivity</u>

In the development of the instrumentation for in -vivo studies it is necessary for the instrument to be sufficiently sensitive to detect changes in the redox state of cyt aa3 and in the oxygenation level of haemoglobin.

There is no doubt that the instrument is sufficiently sensitive to detect the Hb/HbO2 changes which occur physiologically. However its ability to detect physiological changes in the redox level of cyt aa3 is more questionable and may be a reflection of the kinetics and properties of cyt.aa3.

The maximum absorbance expected for the complete oxidation of cyt aa3 (22 micromolar) reported to be present in human brain was calculated to be 0.05 absorbance units per cm of path length (using information derived from studies by Purves et al., 1972) Cope and Delpy 1988. This is dependent upon a difference value for the extinction coefficient (ox-red) of cyt aa3 (Brunori etal.,1981) and assuming that the Beer-Lambert Law holds, for a 1cm path length, and with negligible scattering. Our present instrument is capable of resolving changes of better than $\pm$ 0.01 absorbance units with an overall absorbance of 8 absorbance units. There have been no better estimates of the sensitivity of any NIR instrument so far described.

Specificity

The specificity of the determination of changes in the oxygenation level of haemoglobin are relatively well established. To date there have been no reported significant measurements of other chromophores in the NIR region whose spectral properties are oxygen dependent. (Brunori et al.,1981;Griffiths and Wharton, 1961; Porter et al.,1966).

RESULTS

Animal Experiments

These were devised to determine if NIRS can be used to detect variations in the oxygenation level of haemoglobin and in the oxidation state of cyt aa3, in one series of studies cerebral transmission measurements were made in which oxygenation changes were induced without hypoxia and in the second series studies were undertaken in which extreme hypoxia is induced followed by recovery.

In Fig 1(a and b). is shown a transmission study on a rat head in which NIRS measurements were made in response to variations in the inspired oxygen levels. Initially the FiO2 was reduced and a lag phase was observed prior to an increase in the level of HbR and a decrease in the level of HbO2. The animal was not hypoxic, the pH and pCO2 were unaltered and there was no effect on cardiovascular parameters. The inspired oxygen was then returned to 100% leading to an arterial pO2 of 402mmHg, and the NIRS recordings showed an increase in the level of HbO2 and a drop in the level of HbR. The cerebral blood volume trace (not shown, average total Hb) showed an increase on re-administration of 100% oxygen consistent with a reactive hyperaemia in which there is a compensatory arterial dilatation.

There was a slight drop in the level of cyt aa3 (and can be seen in fig.1b ,enlarged version of Fig.1a).

326

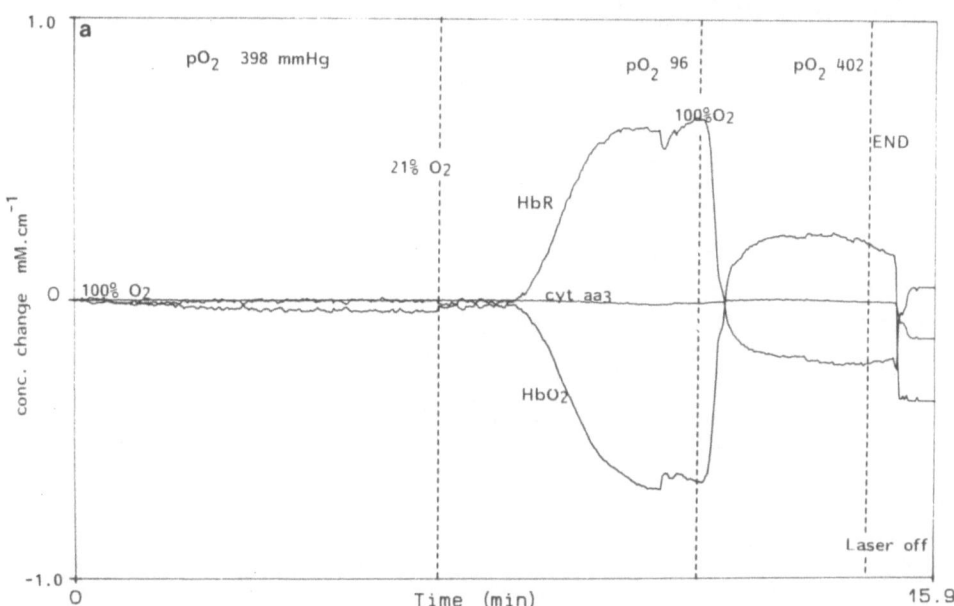

Figure 1a. A rat cerebral transmission study showing the effects of a change in inspired oxygen without hypoxia on the concentrations of Hbr,HbO2 and cyt.aa3 as detected by NIRS.

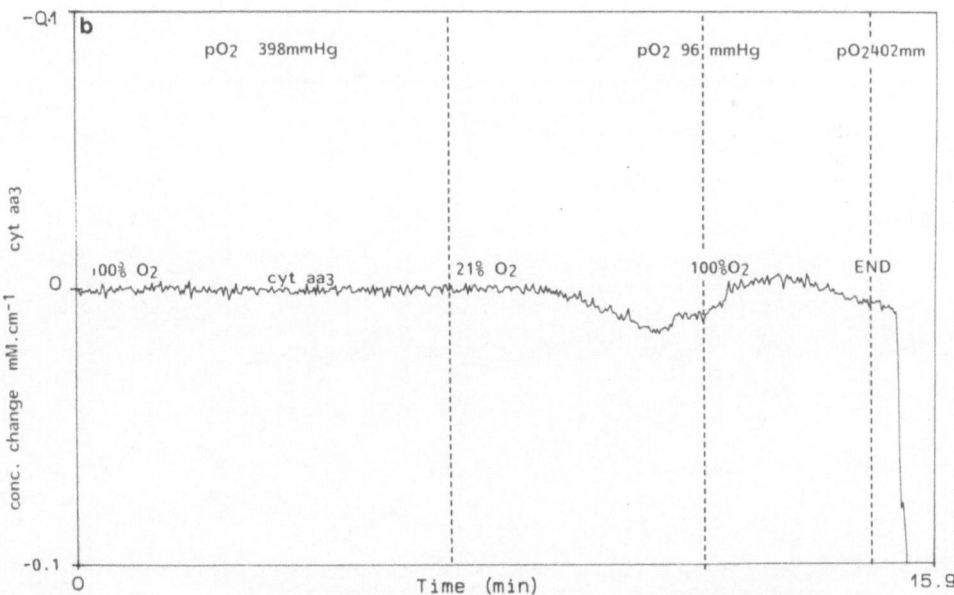

Figure 1b. An enlarged section of Figure 1a showing the effect of changing the inspired oxygen on the concentrations of cyt.aa3.

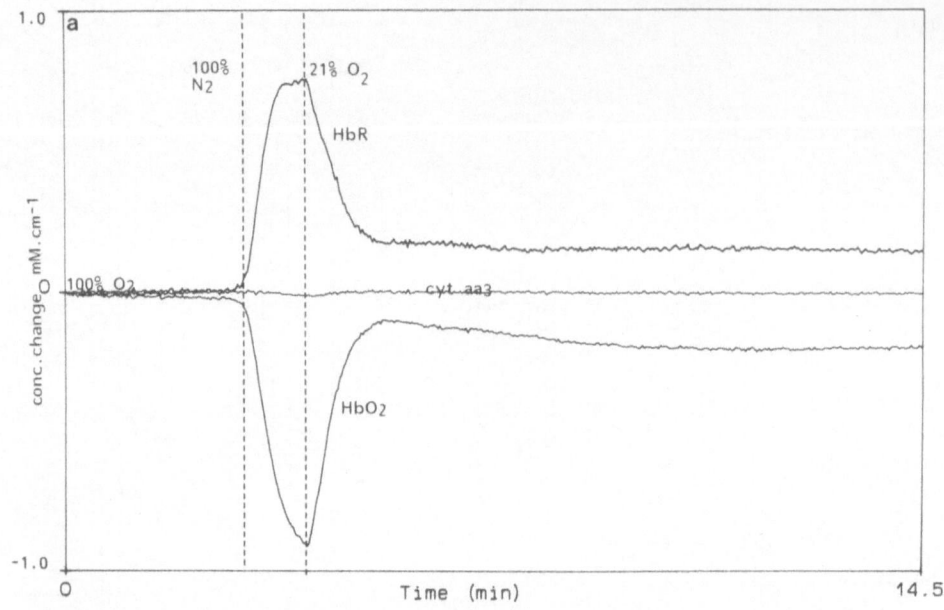

Figure 2a. A rat cerebral transmission study showing the effects of profound hypoxia and recovery on NIRS measurements of HbR, HbO2 and cyt.aa3.

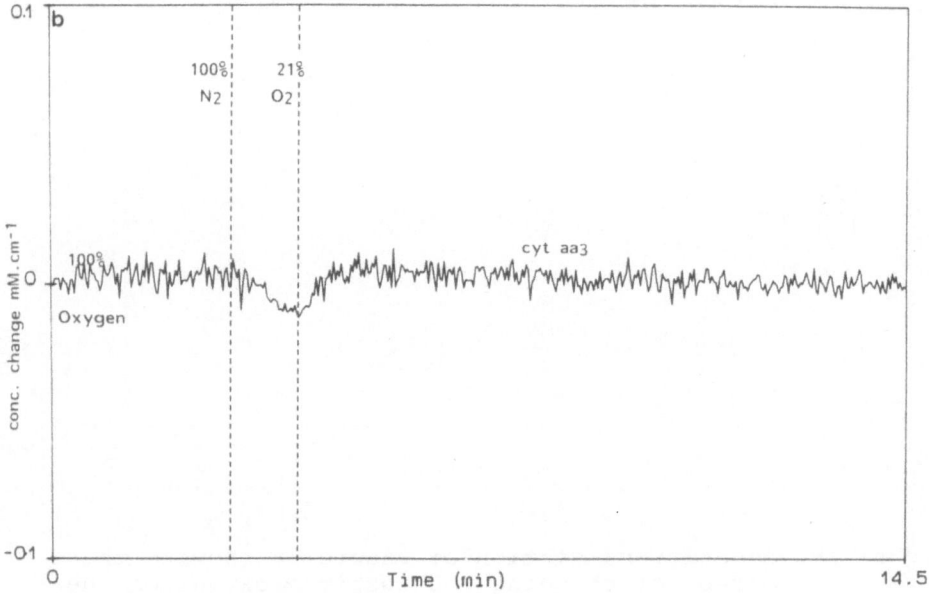

Figure 2b. The cytaa3 section of Figure 2a is enlarged.

328

In the second series of studies hypoxia was induced followed by recovery. In Fig.2 is shown a rat cerebral transmission study. NIRS recording were made in response to changing the inspired oxygen from 100% (pO2 446mmHg) to 100% nitrogen. There was a rapid increase in the level of de-oxyhaemoglobin and this was accompanied by a fall in the oxyhaemoglobin level. There was also a fall in the oxidation level of cyt aa3(as oberved in Fig 2b, enlarged cyt.aa3 trace from Fig2a). Administration of 21% oxygen resulted in the cyt aa3 becoming reoxidised, and an increase in the oxygenation level of haemoglobin and a fall in the level of HbR.

The change in the redox state of cyt aa3 is significant, and is greater than the noise level.

CLINICAL STUDIES

These were carried out on babies of varying gestation and postnatal age. None of the babies had ultrasound evidence of periventricular leukomalacia or periventricular haemorrhage.

Oxygen dependent babies were studied during changes in oxygenation, either induced by small alterations in inspired oxygenation or occurring spontaneously with rapidly fluctuating changes in the haemoglobin saturation level, or spontaneous bradycardias.

In Fig 3, a cerebral transmission study is presented on a female infant of 27 week gestation with a birth weight of 985 g studied on day 9 at a weight of 870 g. At this stage she had resolving hyaline membrane disease with recurrent apnoeas/ bradycardias being treated with caffeine, and requiring an increase in inspired oxygen. Simultaneous pulse oximetry measurements were carried out.

The NIRS recordings show an episode in which a spontaneous fall in oxygen saturation occurs from point 1 (97%) to a lowest value of 84%, resulting in a fall in concentration of oxyhaemoglobin and a simultaneous rise in the de-oxyhaemoglobin concentration. At point 2 the infant had a brief, spontaneous bradycardia down to 80bpm, without apnoea, lasting 20sec. There is a marked fall in HbO2 at this point with a smaller rise in HbR which was not seen on the pulse oximeter. Following recovery of this bradycardia there is a steady rise in haemoglobin saturation to 95% at point 3.

The oxy and de-oxyhaemoglobin levels then returned to resting levels.

In Figures 4a and b are shown NIRS recordings on an infant who was having episodes of bradycardia; the changes in cerebral blood volume are shown in Fig4b. In this study cerebral transmission NIRS measurements were carried out (Fig4a) and a brief, spontaneous bradycardia episode occured between points 1 and 2 (Fig.4a) with a subsequent fall in Hb saturation to 82% measured by pulse oximetry. The oxygen saturation level then increased to the resting value of 95%

saturation. These changes in the oximetry measurements can be correlated with the NIRS recordings, in which it can be clearly seen that there is a marked fall in HbO2 and rise in HbR during and after the bradycardia which then returns to previous levels. In Fig.4b is shown the effects of the spontaneous bradycardia on total cerebral haemoglobin which is related to cerebral blood volume. It can be readily observed that at the same time as the bradycardia there is a fall in total haemoglobin, which recovers more rapidly than the oxyhaemoglobin level.

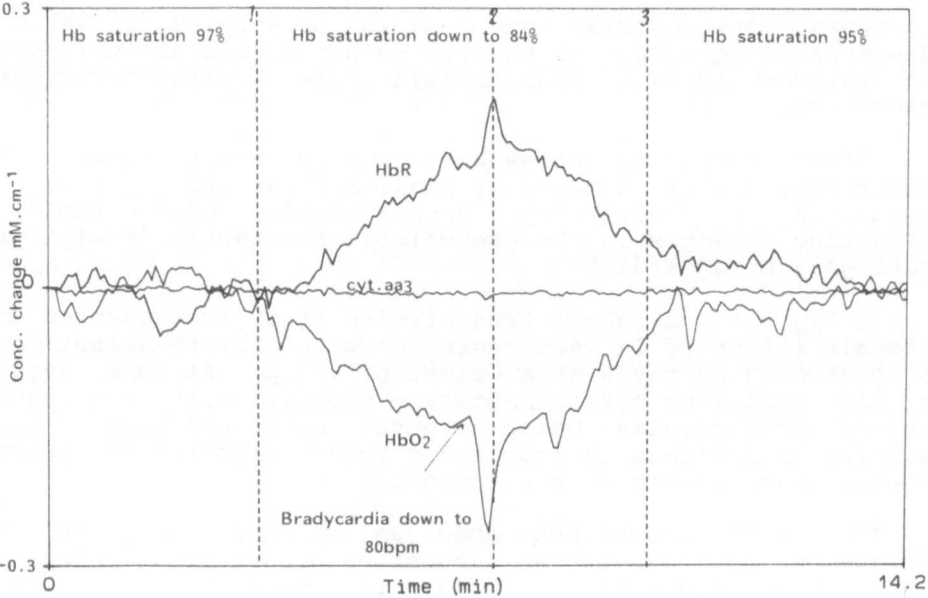

Figure 3. A neonatal cerebral transmission study showing the effect of a spontaneous bradycardia on the NIRS measurements of HbR, HbO2 and cyt aa3.

As also observed in Fig.3. there is an insignificant fall in the oxidation state of cyt aa3 under these conditions, it could be that if the desaturation episodes were of longer duration then changes in the redox state of cyt.aa3 would be observed.

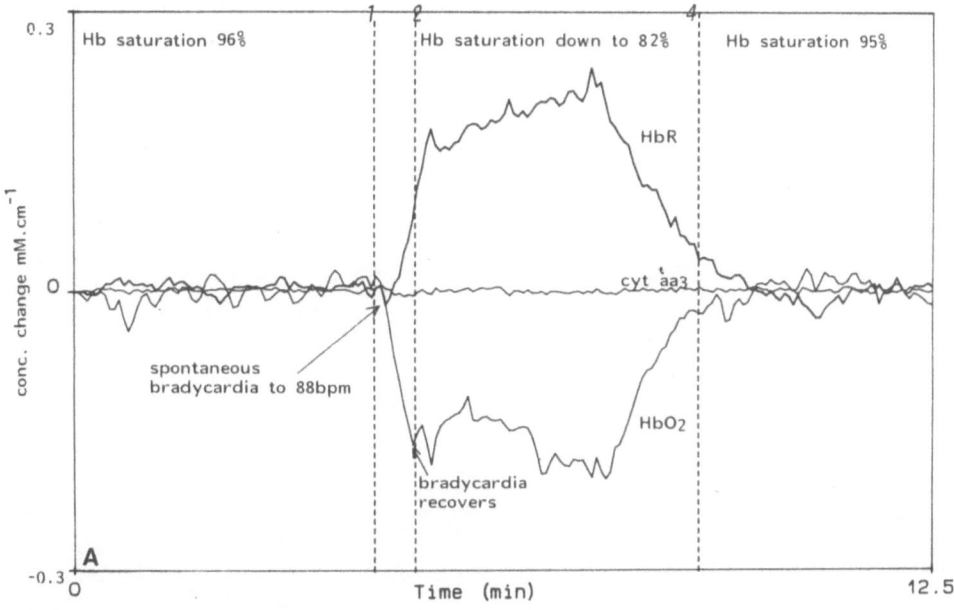

Graph A — y-axis: conc. change $mM.cm^{-1}$, from -0.3 to 0.3; x-axis: Time (min), from 0 to 12.5.

Hb saturation 96%   Hb saturation down to 82%   Hb saturation 95%

HbR

cyt aa3

spontaneous bradycardia to 88bpm

bradycardia recovers

HbO2

A

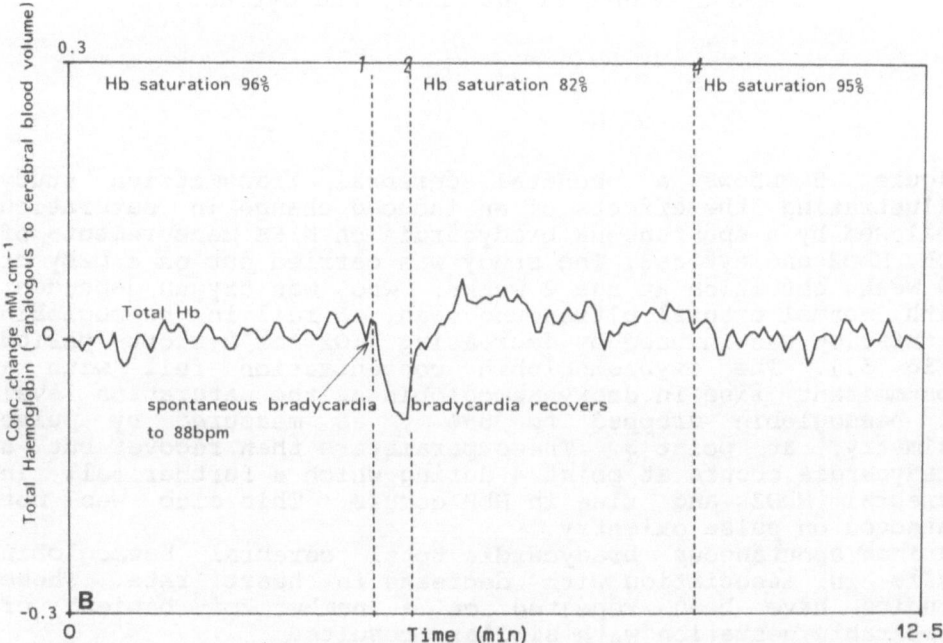

Graph B — y-axis: Total Haemoglobin (analogous to cerebral blood volume) / Conc.change $mM.cm^{-1}$ (analogous to cerebral blood volume), from -0.3 to 0.3; x-axis: Time (min), from 0 to 12.5.

Hb saturation 96%   Hb saturation 82%   Hb saturation 95%

Total Hb

spontaneous bradycardia to 88bpm   bradycardia recovers

B

Figure 4. A neonatal cerebral transmission study showing the effect of a spontaneous bradycardia on NIRS measurements of A: HbR,HbO2 and cyt.aa3 and B: total haemoglobin(analogous to cerebral blood volume).

331

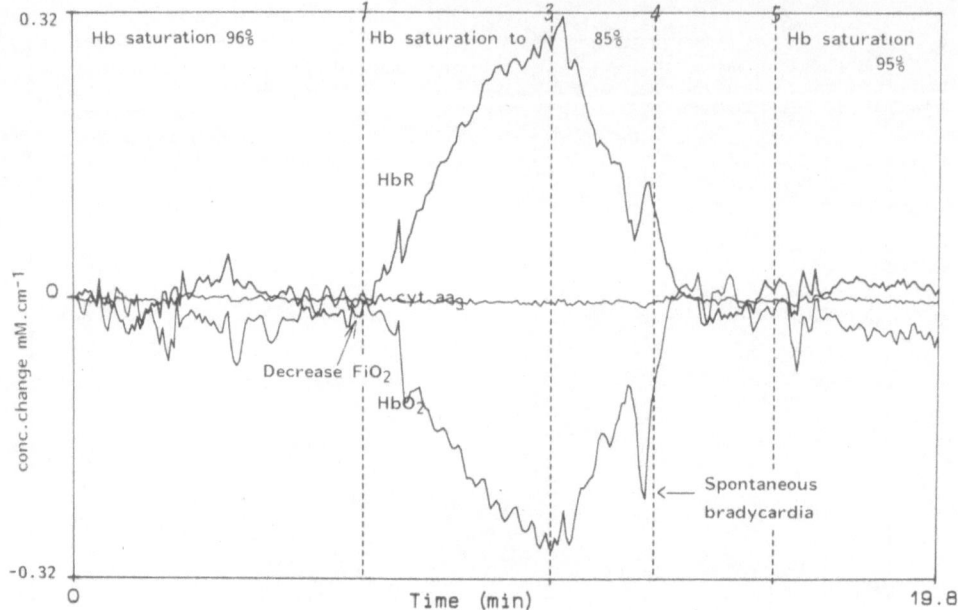

Figure 5. A neonatal cerebral transmission study showing
the effects of an induced change in saturation
followed by a spontaneous bradycardia on NIRS
measurements of HbR,HbO2, and cyt.aa3.

Figure 5 shows a neonatal cerebral transmission study
illustrating the effects of an induced change in saturation
followed by a spontaneous bradycardia on NIRS measurements of
HbR, HbO2 and cyt aa3. The study was carried out on a baby of
28 weeks gestation at age 2 weeks, who was oxygen dependent
with normal cranial ultrasound scan. A fall in haemoglobin
saturation was induced by decreasing FiO2 for a short period
(Fig 5.). The oxyhaemoglobin concentration fell with a
concomitant rise in deoxyhaemoglobin as the saturation level
of haemoglobin dropped to 85% , as measured by pulse
oximetry, at point 3. These parameters then recover but a
bradycardia occurs at point 4 during which a further fall in
cerebral HbO2 and rise in HbR occurs. This also was not
detected on pulse oximetry.
During spontaneous bradycardia total cerebral haemoglobin
falls in association with decrease in heart rate. These
studies have been repeated on a number of babies of
comparable gestation with similar results.

SUMMARY

The instrument drift was found to be less than 0.004
OD/hour and from measurements on glass filters of 8 optical
density units, a coefficient of variation of 0.01 over the 30
second averaging time was observed. The instrument is

332

sufficiently sensitive to enable monitoring of changes in the cerebral oxygen saturation level of haemoglobin and to enable changes in the concentration of cyt aa3(oxidised form), to be measured with reasonable confidence. It is of the utmost importance in NIRS investigations to be certain of the specificity of the technique, and it is vital that reliable determinations of the amount of cytochrome aa3 in brain are made in addition to measurements of the extinction coefficients. This is still a matter of considerable debate, not only in the isolation and properties of the multisubunit structured membrane bound protein, which many enzymologists have been investigating for the last 50 years, (Keilin and Hartree,1939; Brunori et al, 1981) but also in relating to in-vivo versus in vitro comparisons.

An additional point for consideration is the validity of employing multiplier coefficients derived from the rat brain, as several groups have done (Wyatt et al 1986, Ferrari et al 1985,) and applying them to the human brain and to the infant brain. There may be significant differences in the activity of the enzyme, and the profound physiological effects which will arise during the end point fluorocarbon studies: the presence of fetal haemoglobin must also be considered(Carta et al.,1987). The clinical determination of the optical path length is a considerable problem (Cope etal.,1988,) and once again estimates of the pathlength correction factor made in animals and dead fetuses may not be valid for living human tissue.

Results from our animal studies indicate that NIRS can be used to monitor changes in the oxygenation level of Hb and, in extreme hypoxia, changes in the level of the redox state of cyt aa3 can be reliably measured and are well within the sensitivity of the instrument. The results indicate that under small change in saturation the redox state of cyt aa3 appears to be unaltered. It may be that under normal physiological conditions the redox state of aa3 appears to be apparently unchanged under episodes of mild hypoxia of short duration.

AKNOWLEDGEMENTS
We gratefully thank the SERC, MRC, Welcome Trust, Wolfson Foundation, North Staffordshire Medical Institute, Regional Health Authority and Birthright for financial support.

REFERENCES

1. Brunori M., Antonini E., & Wilson M. T., 1981, In: Metal Ions in Biological Systems (Seigel, H. ed), vol. 23. pp 187 - 228, Marcel Dekkers, New York.
2. Carta F., Ferrari M., and Giannini I., 1987 Adv.Exp.Med.Biol., 215:275-282.
3. Cope M., Delpy D.T., 1988, Med. Biol. Eng. Comput., 26: 289 - 94.
4. Cope M., Delpy D.T., Reynolds E.O.R., Wray S., Wyatt J., Van der Zee P., 1988, Adv. Exp. Med. Biol., 222: 183 - 189.
5. Ferrari M., Giannini L., Capri A., Fasella P., 1983 Physiol. Chem. Phys. Med NMR. 15: 109 - 113.

6. Ferrari M., Giannini T., Sideri G., Zanette E.,1985, *Adv. Exp. Med. Biol.*, *191*: 873 - 82.

7. Griffiths D.E. & Wharton D.C., 1961, *J.Biol. Chem.* *236*: 1857 - 1860.

8. Jobsis F. F., 1977, *Science* *198*: 1264 - 1266.

9. Jobsis - Vander Vliet F.F., *Adv. Exp. Med. Biol.*, *191*: 833 - 841.

10. Keilin D. & Hartree E.F., 1939, *Proc.Roy.Soc.B.*, *127*: 167-170.

11. Piantadosi C. A. & Sylvia A.L., 1985, *Adv. Exp. Med. Biol.*, *191*: 823 - 832.

12. Porter H., 1966, The Biochemistry of Copper, Peisach J. et al., (eds.), Academic Press, New York, 159-172.

13. Purves M.J., 1972, The Physiology of the Cerebral Circulation, Cambridge University Press.

14. Rea P.A., Crowe J., Wickramasinghe Y. and Rolfe P., 1985 *J. Med. Bio. Eng. Tech.*, 9: 160

15. Sylvia A.L., Piantadosi C.A., and Jobsis-Vander Vleit F.F., 1985, *Neurol. Res.*, 7: 81-88.

16. Thorniley M. S., Wickramasinghe Y. & Rolfe P. 1988, *Biochem. Soc. Trans.*, *16*: 978 - 982.

17. Thorniley M. S., Wickramasinghe Y. & Rolfe P., 1988, *Biochem. Soc. Trans.*, *16*: 980 -982.

18. Tamura M., Seiyama A. & Hazeki., 1987, *Adv. Exp. Med. Biol.*, *215*: 292 - 300.

19. Wickramasinghe Y., Crowe J. & Rolfe P., 1986, *Proc. 8th Annual Conference IEEE Eng. in Med. and Biol. Sci.*, pp 1172 - 1174, Dallas, Fortworth, U.S.A.

20. Wray S., Cope M., Delpy D.T., Wyatt J.S., Reynolds E.O.R., 1988 *Biochem. Biophys. Acta.*, 183 - 92.

21. Wilson B.,Lowe D.M., 1985, *Photochem.Photbiol.*, *42*:153 - 62.

22. Wyatt J.S., Delpy D.T., Cope M., Reynolds E.O.R., 1986, *Lancet* *ii*: 1063 - 1066.

CARBONIC ANHYDRASE INHIBITION AND CEREBRAL CORTICAL OXYGENATION IN THE RAT

Joseph C. LaManna and Kimberly A. McCracken

Departments of Neurology and Physiology/Biophysics, Case
Western Reserve University School of Medicine and University Hospitals
Cleveland, Ohio 44106, USA

INTRODUCTION

The carbonic anhydrase inhibitor, acetazolamide (Maren, 1967), has been used clinically for many years as a treatment for the prevention of the symptoms, primarily headaches, of acute exposure to altitude hypoxia, as originally demonstrated by Cain and Dunn (1966). This effect was thought to be due to a drug-induced metabolic acidosis that compensated for an altitude-induced respiratory alkalosis. Furthermore, it had already been noticed that acetazolamide administration was followed by increased cerebral blood flow (Cotev et al., 1968; Ehrenreich et al., 1961; Posner and Plum, 1960), an observation that has often been confirmed since. Nevertheless, the effects of acetazolamide on metabolism and blood flow in the brain and their relationships to the physiological mechanisms of action of acetazolamide have not been well studied.

Most of the reported consequences for brain function of acetazolamide administration can be explained through the effects of increased CO_2 retention in the tissue (Kjällquist et al., 1969; Meyer et al., 1961). However, we have reported an effect of acetazolamide on the dynamic metabolic response to electrical stimulation that is not explained by increased CO_2 (Lockwood et al., 1984). That previous study involved a single dose of acetazolamide at a single time point, and only considered the metabolic and vascular response to direct electrical stimulation in the anesthetized rat preparation in which respiration was artificially controlled. The purpose of the studies we report here in this paper thus was twofold: 1) to confirm those previous findings, extending the investigations to include the time course of the effect and the dose dependency; and 2) to determine the time course of the vascular consequences of acetazolamide administration in awake, normoxic rats. These studies were to serve as a prelude to further investigations concerning the physiological mechanisms of adaptation to both acute and chronic hypoxia in a rat model (LaManna et al., 1989).

METHODS

Two sets of experimental protocols were used in these studies. For the first, dynamic metabolic and vascular responses to direct cortical electrical stimulation were recorded from anesthetized rats. These rats were prepared and studied under maintained chloral hydrate anesthesia (400 mg/kg, i.p.; supplemented as needed). Cannulae were placed in the femoral artery and vein for blood gas and pressure monitoring and drug administration. These rats were placed in a head holder and the skin and muscle overlying the parietal skull was reflected and a portion of the bone removed to expose the dura-covered cortical surface. A small reservoir was constructed around this opening by sealing a 5 mm high section of 10 mm diameter polyethylene tubing to the remaining bone using glue and vacuum grease. This reservoir was filled with mineral oil. A forked stimulating electrode and an Ag-AgCl recording electrode were placed on the dural surface and a reference electrode was sutured under the retracted scalp. A dual wavelength reflectance spectrophotometer was used to monitor hemoglobin and cytochrome oxidase redox state changes in response to electrical

Oxygen Transport to Tissue XII, Edited by J. Piiper *et al.*
Plenum Press, New York, 1990

stimulation (LaManna et al., 1987). In an additional set of experiments in rats prepared in an identical fashion, brain tissue oxygen tension was monitored by oxygen-sensitive platinum microelectrodes inserted to a depth of 800 microns below the parieto-temporal cortical surface, as previously described (Kreisman et al., 1981).

For the second series of experiments, rats were initially anesthetized with chloral hydrate for the placement of a tail arterial cannula and a cannula placed into the right atrium via the external jugular vein. These rats were allowed to recover from anesthesia and restrained in plaster casts. The tail artery cannula was connected to a syringe fitted in a

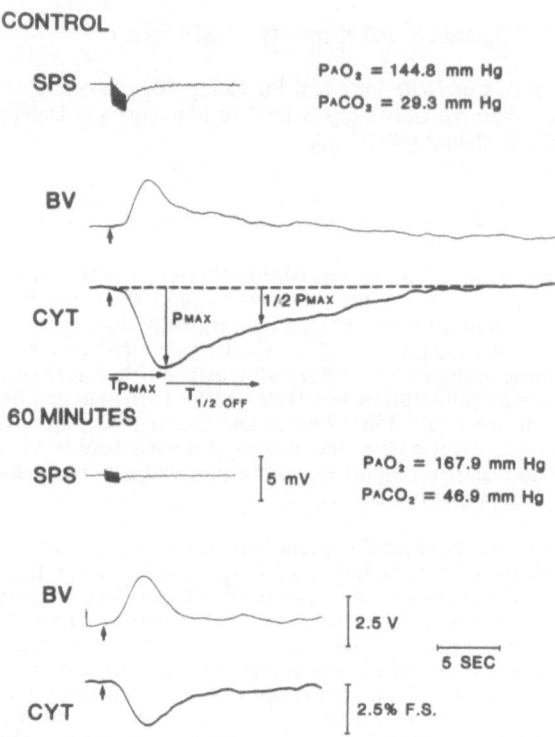

Figure 1. Metabolic and vascular responses to direct cortical electrical stimulation (20 V, 20 Hz, 0.5 msec duration, 2 sec train) before, and 1 hour after 50 mg/kg (i.v.) acetazolamide in chloral hydrate anesthetized rat brain in situ.

pump which was calibrated to withdraw blood at a constant rate (1.6 ml/min). The withdrawal pump was started just before rapid injection of a 150 μl bolus of a buffer solution into the right atrium. The bolus contained 10 mM HEPES buffer, pH 7.4, and a mixture of 10 μCi [^{14}C]n-butanol and 20-50 μCi of [^{3}H]D- or L-glucose. Ten seconds after the intra-atrial bolus injection the rat was decapitated and the withdrawal pump simultaneously stopped. The withdrawn arterial blood was sampled for radioactive content. The brain was rapidly removed and bilateral samples from the frontal cortex, parietal cortex, hippocampus, striatum, and cerebellum were taken. A sample of blood oozing from the foramen magnum was collected in heparinized tubes and an aliquot of the plasma obtained was used to estimate the radioactive content of the cerebral intravascular

336

compartment at decapitation (Sage et al., 1981). Regional cerebral blood flow, glucose extraction fraction, unidirectional blood to brain glucose influx, and permeability-surface area (PS) product were calculated as previously described (Harik and LaManna, 1988).

The effects of acetazolamide were studied at two intravenous dose levels: high dose, 50 mg/kg; and low dose, 10 mg/kg. The effects of acetazolamide on systemic arterial blood gases and pH were measured from blood samples removed from the femoral or tail arterial cannulae with a Radiometer ABL-2 blood gas analyzer. The blood gas and pH data was reported after core temperature compensation; bicarbonate ion concentration was calculated, but not measured directly.

RESULTS AND DISCUSSION

The first part of the study involved confirming and extending our original observations of the effects of acetazolamide in anesthetized rats (Lockwood et al., 1984). Figure 1 demonstrates the effect of acetazolamide on the transient response of the cerebral cortex to direct cortical electrical stimulation. The characteristic response of the cerebral cortex to stimulation of this kind is shown in the upper half of the figure. Direct electrical stimulation for 2 seconds at 20 Hz produced a negative shift of a few millivolts in the local

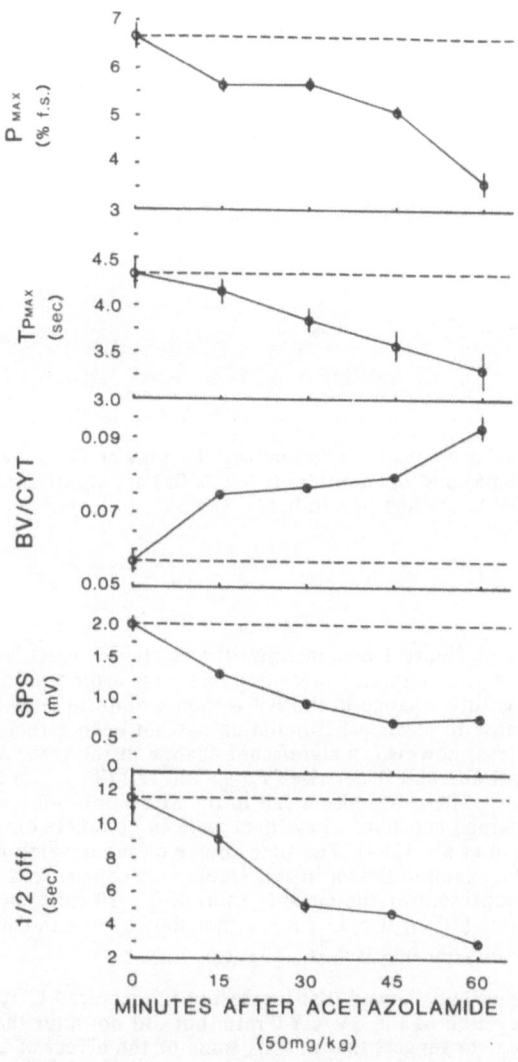

Figure 2. Metabolic, vascular and electrical responses to direct cortical stimulation as a function of time after 50 mg/kg (i.v.) acetazolamide. Dashed lines indicate control values.

337

cortical steady potential (SPS). This shift was accompanied by a change in the reflected light spectrum which was manifested as 1) a transient decrease in intensity of light reflected at 590 nm which could be interpreted as an increase in local blood volume (BV); and a transient increase in the light reflected at 605 nm compared to that at 590 nm which could be interpreted as indicating either, or both, an oxygenation of hemoglobin or an oxidation of cytochrome a. This latter signal, labeled CYT in the figure is plotted so that oxidation of cytochrome and oxygenation of hemoglobin is in the downward direction. Although the response to electrical stimulation detected by dual-wavelength spectrophotometry in the visible spectrum has been shown to be only qualitative at best (LaManna et al., 1987), it is apparent that, regardless of interpretation, a downward going trace produced by either an oxidation of cytochrome oxidase or an increase in oxygenated hemoglobin would be indicative of increased capillary oxygen availability. This trace also includes a graphical representation of the variables that characterize the transient response for analysis. These include the amplitude (Pmax); the time from the onset of the stimulus to the peak of the response ($T_{P_{max}}$); and the time required to return halfway to baseline from the peak ($T_{1/2\ off}$).

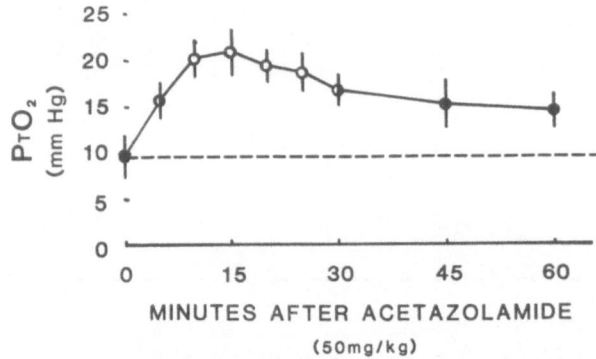

Figure 3. Effect of acetazolamide (50 mg/kg, i.v.) on brain tissue oxygen tension Half-filled (p < 0.05) and open circles (p < 0.0005) are significantly different from controls by ANOVA. Dashed line indicates the control level.

The lower portion of figure 1 demonstrates the electrical, vascular and metabolic responses to exactly the same stimulus presented one hour after administration of 50 mg/kg acetazolamide. There is little change in the BV response amplitude or time course suggesting that the vascular response to electrical stimulation has not been affected by even high dose acetazolamide. There was, however, a significant change in the trace marked CYT which displays a smaller amplitude and faster rise ($T_{P_{max}}$) and fall ($T_{1/2\ off}$) times. Thus, the ratio, BV/CYT was increased. There was also a fall in the SPS amplitude even though the stimulus intensity remained constant. These data were in complete agreement with previous observations (Lockwood et al., 1984). The time course of the development of these effects is shown in Figure 2. The example shown in this figure was constructed from the data from a single rat, each point representing the mean ($\pm$ sem) of 7 - 13 transients. The effect of lower dose acetazolamide (10 mg/kg, i.v.; n = 2, not shown) was indistinguishable from that of the higher dose (50 mg/kg, i.v.; n = 3).

Our previous studies had showed that increasing the inspired CO_2 content to 10% produced the same elevation of the BV/CYT ratio but did not alter the rise or fall times of the CYT signal. These data suggest that at least some of the effect of acetazolamide can be attributed to a central rather than systemic mechanism. This effect could probably be best explained by an interaction between the acetazolamide which has crossed the blood-barrier, a small but significant amount after 1 hour in the rat (Kjällquist et al., 1969), and one or more of the tissue carbonic anhydrase pools (Giacobini, 1962). Also, the effect on tissue

338

acid-base status due to acetazolamide inhibition of cerebrospinal fluid production by the choroid plexus must be considered (Johanson, 1984).

Although it had been suggested at one time that acetazolamide might produce tissue hypoxia in the brain (Laux and Raichle, 1978), our findings rather were more compatible with increased brain tissue oxygen content. For example, the effect of tissue hypoxia on the metabolic and vascular responses to direct cortical electrical stimulation (LaManna et al., 1984) were much different than those shown above after acetazolamide. Nevertheless, direct evidence of tissue oxygenation would be more definitive, and indeed, increased brain tissue oxygenation after acetazolamide appears to be a now well documented finding (Cotev et al., 1968; Grieb and Forster, 1981; Meyer et al., 1961). Figure 3 summarizes our results recorded from chloral hydrate anesthetized rats in which tissue PO_2 was monitored for 1 hour following high dose acetazolamide. Each data point represents the mean ($\pm$ sem) fro' 3 - 8 rats. Elevated tissue oxygen tension was observed at each site.

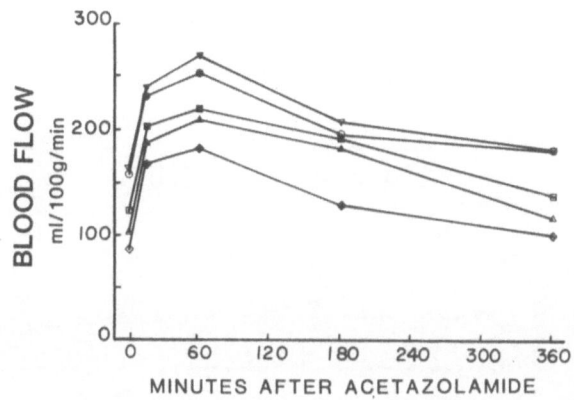

Figure 4. Effect of acetazolamide on regional cerebral blood flow. From top to bottom: frontal and parietal cerebral cortex, striatum, hippocampus and cerebellum. Filled circles are significantly different from control (p < 0.005, ANOVA).

These previous studies were done in the anesthetized preparation where respiration was artificially controlled. To determine the effects of acetazolamide in adaptation to hypoxia, we continued our studies in the awake, restrained rat model. We found, as expected, a rather large increase in regional cerebral blood flow, the effect reaching a maximum at 1 hour, and returning to baseline slowly over 6 hours (figure 4). This effect was observed in all regions studied. Hippocampal, striatal and cerebellar blood flows were significantly higher than control even at three hours after acetazolamide while the cerebral cortical blood flows had returned almost to control levels. One hour after high dose acetazolamide, blood flow to the frontal and parietal cerebral cortex was increased by over 60%, that to the striatum by 80%, and that to the hippocampus and cerebellum by more than double. Low dose acetazolamide administration (10 mg/kg, i.v.) also increased cerebral blood flow to these regions but to a lesser extent in cerebral cortical structures where blood flow was increased by 20% and in cerebellum where the increase was 50%. However, even the lower dose of acetazolamide was effective in doubling blood flow to the hippocampus and striatum (n = 5). Cerebral blood flow increases of 30 - 100% after acetazolamide administration is a common observation in humans (Gotoh and Shinohara, 1977; Hauge et al., 1983; Posner, and Plum, 1960; Vorstrup et al., 1984), monkeys (Laux and Raichle, 1978) and dogs (Cotev et al., 1968; Grieb and Forster, 1981). The increased blood flow occurs without an apparent increase or decrease in tissue oxygen consumption (Grieb and Forster, 1981; Vorstrup et al., 1984).

Changes in blood flow rates by themselves might not be indicative of functional alterations. To determine if there were functional alterations which accompanied the cerebral blood flow changes, we measured the unidirectional transport of glucose across the blood-brain barrier (figure 5). Data are shown only for the parietal cerebral cortex, but responses of the other brain regions were qualitatively similar. Each point on the curve represents the mean ($\pm$ sem) of determinations from 4 - 8 rats. We found an early increase in glucose influx which could be explained as due to the measured increase in PS product. The increased PS product would certainly be compatible with an acetazolamide-induced capillary recruitment. Whatever the cause, the increased PS product and increased glucose influx were rapidly reversed by 1 hour after acetazolamide, despite the continued elevation

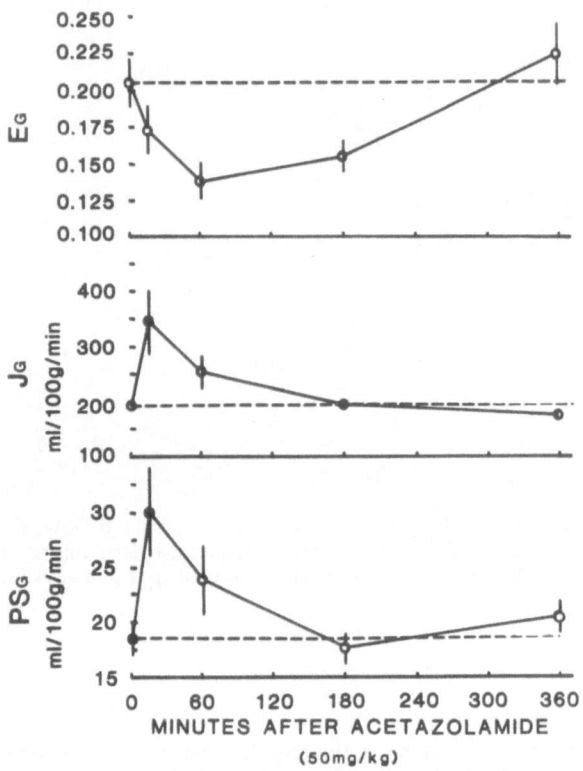

Figure 5. Effect of acetazolamide on regional cerebral glucose delivery in parietal cerebral cortex. Half-filled ($p < 0.05$) and filled ($p < 0.005$) circles are significantly different from control by ANOVA. Dashed lines are control values.

of regional blood flow. This temporal pattern was identical to that for cerebral oxygen content (figure 3), which would also be expected to vary with capillary surface area rather than just blood flow rates. The time course in cerebral blood flow, however, was mirrored by the extraction fraction for glucose, as expected from the known relationship between these two variables (Lund-Andersen, 1979), i.e. that changes in blood flow rate have little net effect on glucose delivery since extraction fraction falls as blood flow rate rises. Low dose acetazolamide also produced a decreased extraction for glucose, but the decrease was exactly proportional to the blood flow increase and there was thus no change in either PS product or influx at any time after the 10 mg/kg, i.v. dose.

340

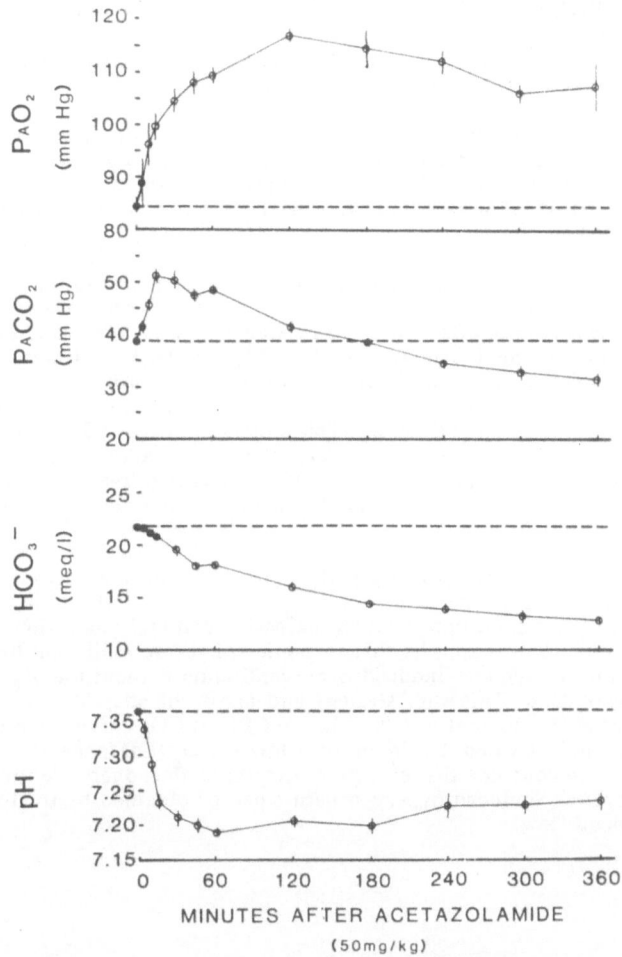

Figure 6. Effect of acetazolamide on blood gas, pH and HCO_3^- in serial arterial samples from awake, restrained rats. Dashed lines indicate control values. Half-filled (p < 0.05) and open (p < 0.001) circles are different from control by ANOVA.

The effects of acetazolamide administration on systemic blood gas and pH values are shown in Figure 6. Acetazolamide treatment induces an increased ventilatory rate (Mithoefer, 1959) which increases arterial oxygen partial pressure. The increased elimination of bicarbonate ion over 6 hours also results in arterial blood acidification. Interestingly, arterial carbon dioxide levels initially rise, but then fall back toward and then below control values. For comparison, the effect of low dose acetazolamide on these same variables was about the same except that the $PaCO_2$ never really rose significantly. In chloral hydrate anesthetized rats, the effects on $PaCO_2$ of each dose of acetazolamide were much larger than for the corresponding dose in awake, restrained rats because of the prevention of hyperventilation in the anesthetized animals. In comparing the time course of the response of the blood variables to acetazolamide it is apparent that it is the change in $PaCO_2$ that corresponds most nearly to the time course of the cerebrovascular and metabolic responses. Thus, the changes in blood flow and glucose transport across the blood-brain barrier occurred at a time when arterial CO_2 was returning toward baseline but while pH was still low and arterial oxygen was still elevated.

SUMMARY and CONCLUSIONS

We report here the results of our study of the carbonic anhydrase inhibitor acetazolamide on cerebral vascular and metabolic function, correlated with the effects of this agent on systemic arterial blood gases and pH. We found that the effects of acetazolamide were to increase PaO_2, decrease bicarbonate ion concentration and decrease pH. While these effects were maintained for many hours after both high and low dose acetazolamide, the cerebral metabolic and vascular effects of the drug were transient. The central effects of carbonic anhydrase inhibition were consistent with increased oxygen delivery and increased tissue oxygenation.

Hypoxia such as encountered at altitude, represents a challenge to the mechanisms which control blood flow in the brain. The decreased arterial oxygen content at altitude is a ventilatory drive which has the effect of 1) increasing somewhat the PaO_2; 2) decreasing the $PaCO_2$; 3) alkalinizing the blood. The decreased $PaCO_2$ then leads to decreased CBF compounding the problem of hypoxemia. In this situation, increasing CBF helps to relieve the tissue hypoxia. This has been done by either increased inhalation of CO_2 (Harvey et al., 1988) or by acetazolamide (Cain and Dunn, 1966; Forwand et al., 1968). A common feature in both treatments might be increased tissue CO_2 retention (Kjällquist et al., 1969; Meyer et al., 1961) and tissue acidification (Heuser et al., 1975). The two treatments are not identical since acetazolamide seems to have additional effects on cerebral metabolism that elevated CO_2 does not.

Thus, we can deduce that the primary pathologic effects of acute hypoxia are due to the decreased cerebral blood flow produced by hyperventilation-induced hypocapnia. Effective treatment of this condition involves increasing cerebral blood flow to increase oxygen delivery to the brain. It appears that carbonic anhydrase inhibition by acetazolamide has the effect of allowing hypoxia-induced hyperventilation without the negative effects of reduced cerebral blood flow. This same strategy appears in the adaptation to chronic hypoxia where there is an apparent left shift in the CBF vs CO_2 curve, such that cerebral blood flow is maintained elevated despite chronic low $PaCO_2$ (LaManna et al., 1989). Based on this evidence we can conclude that effective reversal of the adverse central nervous system effects of hypoxia-induced hyperventilation can be obtained by treatments which increase cerebral blood flow.

REFERENCES

Cain, S.M., and Dunn, J.E., 1966, Low doses of acetazolamide to aid accommodation of men to altitude, J. Appl. Physiol., 21: 1195-1200.

Cotev, S., Lee, J., and Severinghaus, J.W., 1968, The effects of acetazolamide on cerebral blood flow and cerebral tissue P_{O2}, Anesthesiol., 29: 471-477.

Ehrenreich, D.L., Burns, R.A., Alman, R.W., and Fazekas, J.F., 1961, Influence of acetazolamide on cerebral blood flow, Arch. Neurol., 5: 125-130.

Forwand, S.A., Landowne, M., Follansbee, J.N., and Hansen, J.E., 1968, Effect of acetazolamide on acute mountain sickness, New Engl. J. Med., 279: 839-845.

Giacobini, E., 1962, A cytochemical study of the localization of carbonic anhydrase in the nervous system, J. Neurochem., 9: 169-177.

Gotoh, F., and Shinohara, Y., 1977, Role of carbonic anhydrase in chemical control and autoregulation of cerebral circulation, Int. J. Neurol., 11: 219-227.

Grieb, P., and Forster, R.E., 1981, The effect of acetazolamide on brain O_2 metabolism, in: "Oxygen Transport to Tissue (Adv. Physiol. Sci., vol 25)," A.G.B. Kovách, E. Dora, M. Kessler, and I.A. Silver, eds., pp. 271-272, Akademiai Kiado, Budapest.

Harik, S.I., and LaManna, J.C., 1988, Vascular perfusion and blood-brain glucose transport in acute and chronic hyperglycemia, J. Neurochem., 51: 1924-1929.

Harvey, T.C., Raichle, M.E., Winterhorn, M.H., Jensen, J., Lassen, N.A., Richardson, N.V., and Bradwell, A.R., 1988, Effect of carbon dioxide in acute mountain sickness: a rediscovery, Lancet, ii: 639-641.

Hauge, A., Nicolaysen, G., and Thoresen, M., 1983, Acute effects of acetazolamide on cerebral blood flow in man, Acta Physiol. Scand., 117: 233-239.

Heuser, D., Astrup, J., Lassen, N.A., and Betz, E., 1975, Brain carbonic acid acidosis after acetazolamide, Acta Physiol. Scand., 93: 385-390.

Johanson, C.E., 1984, Differential effects of acetazolamide, benzolamide and systemic acidosis on hydrogen and bicarbonate gradients across the apical and basolateral membranes of the choroid plexus, J. Pharmacol. Exp. Ther., 231: 502-511.

Kjällquist, A., Nardini, M., and Siesjö, B.K., 1969, The effect of acetazolamide upon tissue concentrations of bicarbonate, lactate, and pyruvate in the rat brain, Acta Physiol. Scand., 77: 241-251.

Kreisman, N.R., Sick, T.J., LaManna, J.C., and Rosenthal, M., 1981, Local tissue oxygen tension - cytochrome a,a_3 redox relationships in rat cerebral cortex in vivo, Br. Res., 218: 161-174.

LaManna, J.C., Light, A.I., Peretsman, S.J., and Rosenthal, M., 1984, Oxygen insufficiency during hypoxic hypoxia in rat brain cortex, Br. Res., 293: 313-318.

LaManna, J.C., Sick, T.J., Pikarsky, S.M., and Rosenthal, M., 1987, Detection of an oxidizable fraction of cytochrome oxidase in intact rat brain, Am. J. Physiol., 253: C477-C483.

LaManna, J.C., McCracken, K.A., and Strohl, K.P., 1989, Changes in regional cerebral blood flow and sucrose space after 3-4 weeks of hypobaric hypoxia (0.5 ATM), in: "Oxygen Transport to Tissue XI (Advances in Experimental Medicine and Biology, v. 247)," K. Rakusan, G.P. Biro, T.K. Goldstick, and Z. Turek, eds., pp. 471-477, Plenum Publishing Corp, New York.

Laux, B.E., and Raichle, M.E., 1978, The effect of acetazolamide on cerebral blood flow and oxygen utilization in the rhesus monkey, J. Clin. Invest., 62: 585-592.

Lockwood, A.H., LaManna, J.C., Snyder, S., and Rosenthal, M., 1984, Effects of acetazolamide and electrical stimulation on cerebral oxidative metabolism as indicated by the cytochrome oxidase redox state, Br. Res., 308: 9-14.

Lund-Andersen, H., 1979, Transport of glucose from blood to brain, Physiol. Rev., 59: 305-352.

Maren, T.H., 1967, Carbonic anhydrase: chemistry, physiology, and inhibition, Physiol. Rev., 47: 595-781.

Meyer, J.S., Gotoh, F., and Tazaki, Y., 1961, Inhibitory action of carbon dioxide and acetazolamide in seizure activity, Electroenceph. clin. Neurophysiol., 13: 762-775.

Mithoefer, J.C., 1959, Inhibition of carbonic anhydrase: Its effect on carbon dioxide elimination by the lungs, J. Appl. Physiol., 14: 109-115.

Posner, J.B., and Plum, F., 1960, The toxic effects of carbon dioxide and acetazolamide in hepatic encephalopathy, J. Clin. Invest., 39: 1246-1258.

Sage, J.I., Van Uitert, R.L., and Duffy, T.E., 1981, Simultaneous measurement of cerebral blood flow and unidirectional movement of substances across the blood-brain barrier: theory, method, and application to leucine, J. Neurochem., 36: 1731-1738.

Vorstrup, S., Henriksen, L., and Paulson, O.B., 1984, Effect of acetazolamide on cerebral blood flow and cerebral metabolic rate for oxygen, J. Clin. Invest., 74: 1634-1639.

PLATELET ACTIVATING FACTOR ANTAGONISTS DO NOT ALTER NORMAL CEREBRAL BLOOD FLOW OR CEREBRAL OXYGEN CONSUMPTION

Patrick M. Kochanek, John A. Melick, Rebecca J. Schoettle, Mary Jo Magargee, Rhobert W. Evans, and Edwin M. Nemoto

University of Pittsburgh, School of Medicine, Department of Anesthesiology and Critical Care, Medicine and Pediatrics
Room 1081 Scaife Hall, Pittsburgh, PA 15261, USA

INTRODUCTION

Platelet activating factor (PAF), (1-o-alkyl-2-acetyl-sn-glyceryl-3-phosphorylcholine) an endogenous lipid, is involved in the cerebrovascular response to ischemia (1-5). It potently vasoconstricts (6,7), increases microvascular permeability (8), activates granulocytes (9) and stimulates arachidonate-independent platelet aggregation (10). Infused IV, PAF causes cerebral hypoperfusion and hypermetabolism similar to that observed during reperfusion following cerebral ischemia (11,12). PAF-receptor blockade attenuates the development of postinsult hypoperfusion after cerebral embolism in rats (1) and dogs (3) and after carotid occlusion in gerbils (4,5). Its effects on cerebrometabolic rate for oxygen ($CMRO_2$) have not been investigated. Thus, PAF plays a role in the pathophysiological response of the cerebral circulation and perhaps metabolism, but whether it plays a similar role under normal physiological circumstances is unknown.

Two values of brain tissue PAF levels have been reported. Tokumura et al. (13), reported bovine brain tissue PAF levels of 50 pmol/g of brain or 50 nM. Kumar et al. (14) using a bioassay technique measured PAF levels of about 0.25 nM in the normal rat brain. Since 1 nM PAF activates granulocytes and other cell types and PAF induced oial artery constriction occurs at 20 nM, if PAF levels are hormally in the range of 50 nM, it should be involved in regulating cerebrovascular tone under normal physiological circumstances. To determine whether PAF does play a role in regulating cerebrovascular tone under normal physiological circumstances, the first objective of the study is to evaluate the effect of PAF receptor blockade with BN 52021 and WEB 2086 infused intravenously on CBF and $CMRO_2$ in the normal rat brain. Comparisons are also made with effects of indomethacin which has been used to show that cyclooxygenase products do regulate CBF under normal physiological circumstances without affecting $CMRO_2$. Endogenous eicosanoids which can be derived from a common precursor phospholipid (15) modulate both normal (16) and postischemic CBF (17,18).

MATERIAL AND METHODS

This protocol was approved by the Animal Care and Use Committee of the University of Pittsburgh. Barrier bred laboratory acclimated male Wistar rats had free access to food and water up to the time the study. Anesthesia was induced in plastic jars insufflated with 5% halothane in O_2. Their tracheas were intubated and the rats mechanically ventilated on 1% halothane/66% N_2O/33% O_2 with pancuronium bromide 0.1 mg/kg/h, immobilization. Rectal temperature and arterial blood pressures were continuously monitored.

Under 20X magnification, a burr hole was made over the superior saggital sinus and a platinum microelectrode (50 microns) was advanced into saggital sinus for monitoring hydrogen (H_2) washout after equilibration on 5-6% (H_2) in the inspired gas. CBF (ml/100 g/min) was calculated by the T 1/2 method from the H_2 clearance curves (19). Cerebral venous blood samples were obtained for measurement of O_2 content and $CMRO_2$ calculated as the product of CBF and the arterial-venous O_2 gradient across the brain.

Following surgical preparations, the animals were allowed to stabilize for 60 min on 0.4% halothane in N_2O/O_2 (2/1). Baseline measurements of MAP, CBF and $CMRO_2$ were made. Blood gases, pH and hemoglobin were maintained within the following limits: 7.40 + 0.05 for pH; 125 $\pm$ 25 torr for PaO_2; 35 to $\pm$ 5 torr for $PaCO_2$; and 0 $\pm$ 5.0 mM/l for base excess and 13 $\pm$ 1.5% for hemoglobin. After baseline measurements, BN 52021 (10 mg/kg, n=4, 30 mg/kg, (n=2), WEB 2086 (5 mg/kg, n=6), and indomethacin (10 mg/kg, n=4) were administered IV over 1 min (20, 21). Monitored physiological variables were again measured in each animal at 15 and 60 min after drug administration. Subsequently, the rats were killed with an IV injection of saturated KCl.

All data presented are mean $\pm$ SEM. Statistical comparisons within groups were made using the one-way ANOVA for repeated measures followed by Student-Neumann Keul's test for multiple comparisons. Comparisons between groups were made using the one- way ANOVA followed by Student-Neumann Keul's test. Differences of P < 0.05 were considered to be statistically significant.

RESULTS

Physiological variables were maintained within similar ranges in all three groups without any significant differences between the groups (P > 0.05).

Baseline CBF and $CMRO_2$ values did not differ among the three experimental groups (figures 1 and 2). Compared to baseline neither of the PAF antagonists altered CBF at 15 or 60 min after drug administration. CBF and $CMRO_2$ data from rats treated with BN 52021 at 10 and 30 mg/kg were combined since neither had an effect on CBF or $CMRO_2$. In contrast, indomethacin significantly ($P < 0.05$) decreased CBF by 35% after 15 min and by 29% after 60 min compared to baseline (figure 1). Neither BN 52021, WEB 2086, nor indomethacin affected $CMRO_2$ (figure 2).

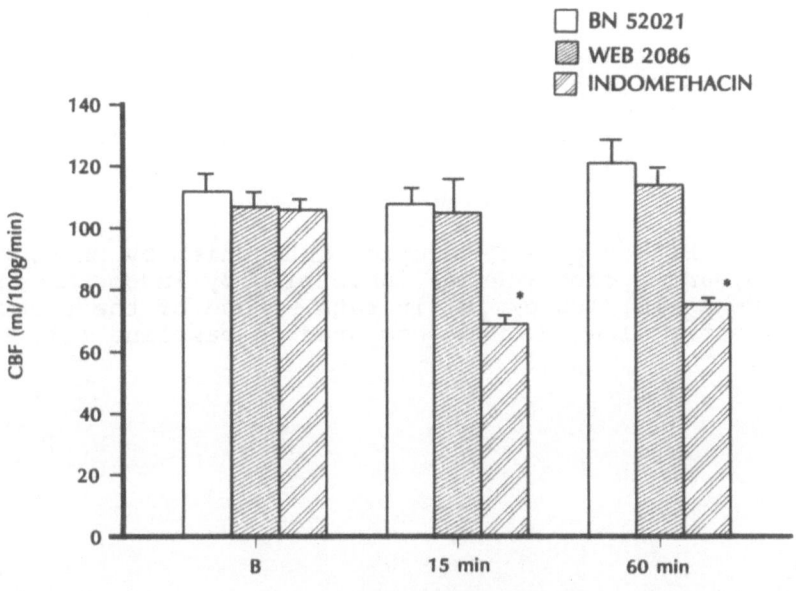

Figure 1. Effect of PAF-receptor antagonism by BN 52021 or WEB 2086, and cyclooxygenase inhibition by indomethacin, on global CBF (ml/100 g/min in rat) (* significantly different from baseline [B] value at $P < 0.05$).

DISCUSSION

Our findings are consistent with the observation of Armstead, et al. (7) that resting pial artery diameter in piglets was unaffected by PAF receptor blockade with U 69985. The threshold concentration at which PAF causes vasoconstriction in the pial artery is appoximately 20 nM (7). Thus, our data suggest that PAF in the normal rat brain is less than this value and more consistent with the levels of 0.25 nM reported by Kumar, et al. (14) than with those reported by Tokumura, et al. (13).

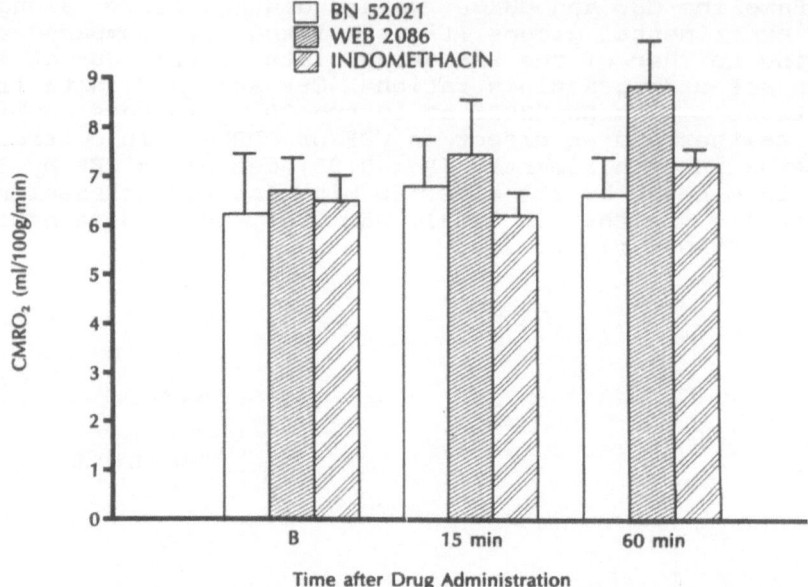

Figure 2. Effect of PAF-receptor antagonism by BN 52021 or WEB 2086, and cyclooxygenase inhibition by indomethacin, on global $CMRO_2$ (ml/100 g/min) in rats. None of the treatments significantly altered $CMRO_2$ compared to baseline (B).

We also confirmed a report by Dahlgren (16) that cyclooxgenase inhibitor indomethacin, consistently and significantly increases CBF in normal brain without affecting $CMRO_2$ suggesting that in normal brain, PAF, unlike the eicosanoids, are subthreshold and exert no influence on global CBF.

The doses of BN 52021 and WEB 2086 used in this study are effective in reversing or attenuating PAF related effects in rats including vasospasm, hypotension and increased vascular permeability (20,21,24-26). Consistent with complete inhibition of any PAF effects, unpublished results from our laboratory using Bligh-Dyer extraction in vapor phase chromatography indicate that a 10 mg/kg dose of WEB 2086 in Wistar rats results in a plasma concentration of 42 ± 17 µM after 15 min. This is well above the IC50 values of 0.17 and 0.36 µM IC_{50} reported for WEB 2086 inhibition of PAF-induced (50 nM) platelet and neutrophil aggregation, respectively (21). Thus, in conclusion, PAF receptor antagonism with BN 52021 or WEB 2086 does not alter normal global CBF and $CMRO_2$ in rats indicating that in the normal rat brain PAF levels are subthreshold and do not modulate normal CBF and $CMRO_2$.

ACKNOWLEDGEMENTS

We thank Dr. P. Braquet at Institut Henri Beaufor and Dr. H. Heuer at Boehringer Ingelheim for their generous supplies of BN 52021 and WEB 2086, respectively. This study was supported in part by a grant from the American Heart Association, Pennsylvania Affiliate.

REFERENCES

1. L. M. Le Poncin, J. Rapin, and J. R. Rapin: Effects of Ginkgo biloba on changes induced by quantitative cerebral microembolization in rats. Arch Int Pharmacodyn 243:236-244,(1980)

2. L. Karcher , P. Zagermann, and J. Krieglstein: Effect of an extract of Ginkgo biloba on rat brain energy metabolism in hypoxia. Naunyn-Schmiedeberg's Arch Pharmacol 327:31-35 (1984)

3. P. M. Kochanek, A. J. Dutka, K. K. Kumaroo, and J. M. Hallenbeck: Platelet-activating factor blockade enhances early neuronal recovery after multifocal brain ischemia in dogs. Life Sci 41:2639-2644 (1987)

4. T. Panetta, VL Marcheselli, P. Braquet, B. Spinnewyn, and N. G. Bazan: Effects of a platelet activating factor antagonist (BN 52021) on free fatty acids, diacylglycerols, polyphosphoinoitides and blood flow in the gerbil brain: Inhibition of ischemia-reperfusion induced cerebral injury. Biochem Biophys Res Commun 149:580-587 (1987)

5. B. Spinnewyn, N. Blavet, B. F. Clostre, N. Bazan, and P. Braquet: Involvement of platelet-activating factor (PAF) in cerebral postischemic phase in mongolian gerbils. Prostaglandins 34:337-350 (1987)

6. G. Feuerstein, L. M. Boyd, D. Ezra, and R. E. Goldstein: Effect of platelet activating factor on coronary circulation of the domestic pig. Am J Physiol 246:H466 (1984)

7. W. M. Armstead, M. Pourcyrous, R. Mirro, C. W. Leffler, and D. W.Busija: Platelet activating factor: A potent constrictor of cerebral arterioles in newborn pigs. Circ Res 62:1-7 (1988)

8. D. M. Humphrey, L. M. McManus, K. Satouchi, and D. J. Hanahan, and N. Pinckard: Vasoactive properties of acetyl glyceryl ether phosphorylcholine and analogues. Lab Invest 46:422-427 (1982)

9. J. O. Shaw, N. Pinckard, K. S. Ferrigni, L. M. McManus, and D. J. Hanahan: Activation of human neutrophils with 1-O-hexadecyl/hexadecyl/octadecyl-2-acetyl-sn-glyceryl-2-phosphorylcholine (platelet-activating factor). J Immunol 127:1250-1255 (1981)

10. D. H. Namm, A. S. Tadephalli, and J. A. High: Species specificity of the platelet response to 1-O-alkyl-2-acetyl-sn-glycero-3-phosphocholine. Thromb Res 25:341-350 (1982)

11. P. M. Kochanek, E. Nemoto, J. Melick, R. Evans, and D. Burke: Cerebrovascular and cerebrometabolic effects of intracarotid infused platelet-activating factor in rats. J Cereb Blood Flow Metab 8:546-551 (1988)

12. K. A. Hossmann: Treatment of experimental cerebral ischemia. J Cereb Blood Flow Metab 2:1275-297 (1982)

13. A. Tokumura, K. Kamiyasu, K. Takauchi, and H. Tsukatani: Evidence for existence of various homologues and analogues of platelet-activating factor in a lipid extract of bovine brain. Biochem Biophys Res Commun 145:415-425 (1987)

14. R. Kumar, S. Harvey, M. Kester, D. Hanahan, and M. Olson: Production and effects of platelet-activating factor in the rat brain. Biochim Biophys Acta 963:375-383 (1988)

15. C. L. Swendsen, J. M. Ellis, F. H. Chilton, J. T. O'Flaherty, and R. L. Wykle: 1-Q-alkyl- 2-acyl-sn-glycero-2-phosphocholine: A novel source of arachidonic acid in neutrophils stimualted by the calcium ionophore A23187. Biochem Biophys Res Commun 113:72-79 (1983)

16. N. Dahlgren, B. Nilsson, T. Kakabe, and B. K. Siesjo: The effect of indomethacin on cerebral blood flow and oxygen consumption in rat at normal and increased carbon dioxide tension. Acta Physiol Scand 111:475-485 (1981)

17. T. W. Furlow and J. M. Hallenbeck: Indomethacin prevents impaired perfusion of dog's brain after global ischemia. Stroke 9:591-594 (1948)

18. S. Shigeno, E. Fritschka, T. Shigeno, and M. Brock: Effects of indomethacin on rCBF during and after focal cerebral ischemia in the cat. Stroke 16:235-240 (1985)

19. W. Yong: H_2 clearance measurement of blood flow: A review of technique and polarographic principles. Stroke 11:552-564 (1930)

20. J. Casals-Stenzel: Protective effect of WEB 2086, a novel antagonist of platelet activating factor, in endotoxin shock. Euro J Pharmacol 135:117-122 (1987)

21. J. Casals-Stenzel, G. Muacevic, and K. Heinz-Weber: Pharmacological actions of WEB 2086, a new specific antagonist of platelet activating factor. J Pharmacol Exper Ther 241:974-981 (1987)

22. S. A. Glantz: Primer of Biostatistics, 2d ed, McGraw-Hill, Inc., New York, 1981, pp 156-157

23. E. Kornecki and Y. H. Ehrlick: Neuroregulatory and neuropathological actions of the ether-phospholipid platelet-activating factor. Science 240:1792-1794 (1988)

24. S. Adnot, J. Lefort, V. Lagente, P. Braquet, B. B. Vargaftig: Interference of BN 52021, a PAF-acether antagonist, with endotoxin-induced hypotension in the guinea-pig. Pharmacol Res Commun 18:197-200 (1986)

25. Braquet P: The ginkolides: Potent platelet-activating factor antagonists isolated from ginkgo biloba L: Chemistry, pharmacology and clinical applications. Drugs of the Future 12:643-699 (1987)

26. Pretolani M, Lefort J, Malanchere E, Vargaftig BB: Interference by the novel PAF-acether antagonist WEB 2086 with the bronchopulmonary responses to PAF-acether and to active and passive anaphylactic shock in guinea-pigs. Euro J Pharmacol 140:311-321 (1987)

CEREBROCORTICAL OXYGEN SUPPLY OF SCLEROTIC RATS AND

ACUTE DILTIAZEM THERAPY

Hermann P.Metzger, Heidrun Pante and
Sabine Heuber-Metzger

Biosignalprocessing and Cybernetics
Medizinische Hochschule Hannover
P.O.Box 610 180, 3000 Hannover 61 FRG

ABSTRACT

The therapeutic effects of calcium antagonist diltiazem on cerebrovascular smooth muscle of sclerotic and normal rats has been investigated. Arteriosclerosis was caused by treatment of the rats with a water soluable 1,25-dihydro-cholecalciferol in a dose of 97 000 IU. After 2-4 days, serum urea of the treated animals was increased from 56 ± 8 to 78 ± 17 mg/dl. Histological examinations of tissue probes from kidneys, heart, aorta and the major arteries by means of the staining method according to von Kossa clearly showed the picture of the Mönckeberg-type atherosclerosis. The calcified rats (N= 23) and the control group (N= 25) were anesthetized with ketamin/xylazine. Following craniotomy and tracheotomy for artificial ventilation, surface PO_2 of the cerebrocortex and cerebral blood flow via the inhalatory hydrogen clearance technique were measured.

Surface PO_2 frequency distribution of vitamin D_3 treated animals showed a left shift (mean$\pm$SD = 29 ± 8 compared with 34 ± 6 mm Hg for controls) and a CBF decrease (1.36 ± 0.51 compared to 1.50 ± 0.60 ml/g.min for the controls). Diltiazem infusion of 2 ml/h corresponding to 5 ug/kg bw after 5 minutes caused a slight CBF decrease ($1.50+0.60$ to $1.43+ 0.78$ ml/g.min) in controls and an insignificant CBF reduction in sclerotic rats. Mean arterial blood pressure was lowered by diltiazem infusion. The hypertensive sclerotic rats became almost normal ($148+18$ to $91+15$ mm Hg) while the control animals decreased in MAP from 118 to $80+15$ mm Hg. Despite the MAP decrease, the effect of diltiazem on CBF and oxygen supply of the cerebrocortex were kept at a reasonable level in both, sclerotic and control rats. The antihypertensive effect has the potential to be beneficial for the atherosclerotic patient.

INTRODUCTION

Vasospams and high blood pressure are the classical indications for calcium antagonist therapy which improves coronary and peripheral blood flow as well as oxygen supply by lowering the effects of partially constricted blood vessels through vasodilatation. If, however, in the case of already manifested chronic atherosclerosis a beneficial longterm effect might be achieved by the use of calcium antagonists, this has as yet not been investigated.

Fleckenstein et al. (1977,83) have postulated that excessive calcium uptake into vascular smooth muscle has to be prevented in order to reduce the activation of myosin light chain kinase. Following this hypothesis, calcium antagonists have been used which are known to reduce the pathological calcium uptake into vascular smooth muscle and, therefore, might reduce the danger of deleterious vasoconstriction.

The following study attempts to clarify whether the calcium antagonist diltiazem can improve an already manifested vitamin D_3 induced atherosclerosis or not. Cerebrocortical oxygen supply and blood flow have been measured by use of polarographic surface electrodes placed on top of the open brain through a cranial window. Sclerotic and normal rats were treated with diltiazem or kept as controls. The development and degree of atherosclerosis within the walls of major arteries and veins have been successfully demonstrated by use of histological methods, especially according to the method of von Kossa.

MATERIAL AND METHODS

Animals and anesthesia

Female white rats (N= 50, Wistar-Munich-Frömter (WFS) strain) were subdivided into two groups: group I was treated with vitamin D_3, group II served as control. The body weight of the animals ranged from 200 to 250 g. The animals were anesthetized with a combination anesthesia of ketamin/xylazine (Ketavet o.15 mg/kg bw and Rompun o.o15 mg/kg bw), craniotomized, tracheotomized and cannulated. A small piece of the os parietale of 0.5x0.5 cm^2 was carefully removed. Then the animals were placed in the supine position for cannulation of the arteria carotis communis for MAP monitoring and the vena jugularis for ketamin-xylazine and diltiazem infusion.

Construction and calibration of the PO_2 and PH_2 electrodes

Surface oxygen tension (sPO_2) of the cerebral cortex of the rats was measured by use of polarographic oxygen electrodes consisting of six small gold wires (15 um in diameter per cathode). A centrally located Ag/AgCl wire (2oo um in diameter) served as reference. The whole surface of the six cathode-electrode (5 mm in diameter) was covered with a thin polystyrol membrane (6 um thickness) which enabled the registration of sPO_2 including the small vasomotions independent of the extracellular milieu surrounding the individual cathodes. The hydrogen sensitive electrode consisted of four platinum-iridium wires (100 um in diameter)

354

arranged 2o um apart at the corners of a square and soldered on a piece of printed plate. As a reference electrode for hydrogen clearance determination, an external calomel electrode was used which was connected via a KCl-saturated agar bridge to the abdominal cavity. Current amplifiers for the registration of tiny polarographic currents were constructed by use of FET high input impedance electrometer amplifiers. Polarographic currents in the range of 10^{-12} ampere were measured in the case of low sPO_2 values as well as small hydrogen responses. Both types of electrodes were fixed with a two-component glue which is known for its mechanical stability and high electrical resistance as well as its nontoxicity from heart pacemaker techniques (Dexter Corp.,N.Y.).

The sPO_2 values were registered during the animal experiment over 10-20 minute intervals until stable sPO_2 and sPH_2 as well as MAP registrations were obtained. From the hydrogen clearance curves in response to inhalation of a gas mixture of 69% N_2, 21 % O_2 and 1o% H_2, local cerebrocortical blood flow was calculated using Auklands et al. equation (1972) modified by Metzger (1989).

Experimental protocol and induction of atherosclerosis

Three days before an experiment, vitamin D_3 was injected IM in the form of vit. D_3 hydrosol, a water soluable cholecalciferol, in the dosage of 97 000 IU according to Fleckenstein (1977,1983). Both groups of animals, the vit.D_3 treated and the controls had free access to food and water. About 30 minutes before the beginning of the experiments, the animals were anesthetized, craniotomized and thracheotomized. The head was fixed by stereotactic equipment. After sPO_2 or sPH_2 registrations approximated steady-state values, a transient change of the inspiratory gas mixture was made in order to test the reactivity of the whole system, electrode and amplifiers as well as the animal itself. Transient ventilation with a hypoxic gas mixture consisting of 10% O_2 and 90% N_2 was performed for two minutes duration. Then the gas mixture was returned to room air. As a second test, the brain surface was flushed with pure nitrogen or hydrogen in order to detect mechanical and electrical instabilities between the measuring points and its surrounding. Registrations were not started before the external influences had been excluded.

After return to the initial sPO_2 level, diltiazem infusion was started with a speed of 2 ml/h for 15 minutes. Stable sPO_2 and sPH_2 values were reached within 10-20 minutes. Then the animal was sacrificed and tissue samples from the aorta, arteria carotis communis, arteria radialis, arteria renalis and the renal cortex as well as the heart were taken and stained according to the method of von Kossa. Prior to the end of the infusion, blood samples for glood gas and pH as well as serum urea and dehydrated creatine were withdrawn.

MAP and sPO_2 values have been measured immediately at the beginning of the experiment when the preparation was com - pleted. After a steady-state was reached (about 15 minutes later), the continous data collection was started as documented in Fig. 1 and Fig. 2. The MAP values reported in the abstract and discussion are corresponding to the initial values obtained at the beginning while the 15 min.values are the steady-states described in figures 1 and 2.

RESULTS

The sPO_2 and cortical blood flow of sclerotic and control rats

sPO_2 and MAP registrations showed the same temporal variations and ran somewhat parallel to each other. During the initial 2 minutes of 10% O_2 in 90% N_2 breathing (to test the reactivity of the vascular system and the experimental setup), a transient MAP and sPO_2 decrease in response to respiratory hypoxia was registered. The mean sPO_2 reached values between 10 and 20 mm Hg (Fig.1). After returning to room air breathing, MAP and sPO_2 showed slight overshoot due to the hypoxic vasodilatation before returning to the original vascular diameters. Thereafter, they remained at about two-third of the initial level. Somewhat lower sPO_2 values were observed in the slcerotic group compared with the controls (Fig.1), but vitamin D_3 treatment caused slight hypertension.

Interpretation of the data becomes clearer from the sPO_2 frequency distribution (Fig. 2). The sclerotic animals showed slightly lower sPO_2 values at the beginning of the experiment (normoxia) with about 10% of the sPO_2 values in the 0 - 10 mm Hg range compared with none in the controls. Transient hypoxic-hypoxia caused a more pronounced sPO_2 decrease in the treated compared with the control group. Following diltiazem infusion, a pronounced left-shift in sPO_2 of the vit.D_3 treated anmials was observed while the controls showed somewhat better oxygen supply situation with less than 5% hypoxic animals during during diltiazem application. At the end of infusion, a small decrease in supply has been registered in the untreated animals.

Cerebrocortical blood flow of sclerotic and control animals

Vit.D_3 treated rats showed a CBF decrease from 1.50± 0.60 to 1.36+o.51 ml/ g.min(N= 8 rats per group, 4 determinations for each group). Diltiazem infusion of 2 ml/h caused a slight CBF decrease (1.5o to 1.43 ml/g.min)in control, and an insignificant CBF reduction (1.36 to 1.33 ml/g.min) in arteriosclerotic animals.

Urea and dehydrated creatine concentrations

Mean urea concentration values +SD of 56.4+8.8 mg/dl (N= 8 animals) were observed for the controls while the sclerotic animals showed a pronounced increase to 78.4+ 17.2 ml/dl. The long inbreeding times of the WFS-rats may be responsible for the elevated MAP and urea concnetrations values. Under the condition of atherosclerosis the tendency to higher MAP and urea values are typical signs of hypertension. Mean dehydrated creatine of 0.40+0.1 mg/dl for the control and 0.49+ 0.1 mg/dl for the sclerotic rats have been determined. The data are somewhat lower than those reported for Sprague-Dawly rats of 0.66± 0.05 mg/dl. Sprague-dawly rats are characterized by lower urea values of 32± 1.4 mg/dl.

Histological examination of the vascular calcinosis

Tissue slices from the major arteries have been cut and stained according to the method of von Kossa. Sections from heart muscle, aortic arch, a.carotis communis sinister, kidney cortex abd papilla as well as pelvis ureter showed pronounced depositions of calciumphosphate due to vit.D_3 treatment while the controls were free from signs of calcinosis (Fig.3).

356

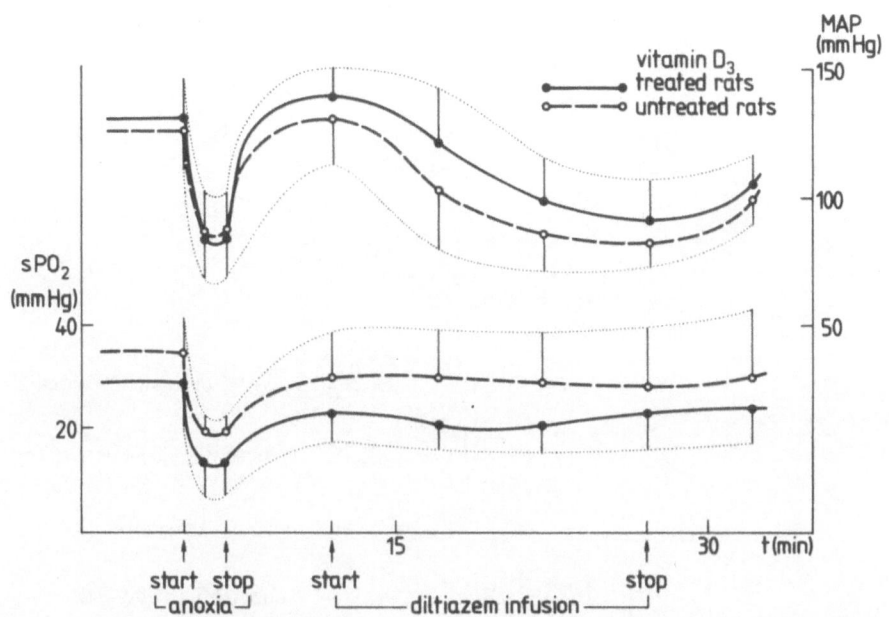

Fig. 1. Mean arterial pressure (MAP) of vitamin D_3 treated rats (continuous line) and controls (dashed line). Lower part of the figure: Mean surface PO_2 (sPO_2) of the cerebral cortex of treated and untreated rats. Each point = mean value of 5 registrations in 6 rats. Verticle lines are $\pm$SD. n= 3o measurements per point.

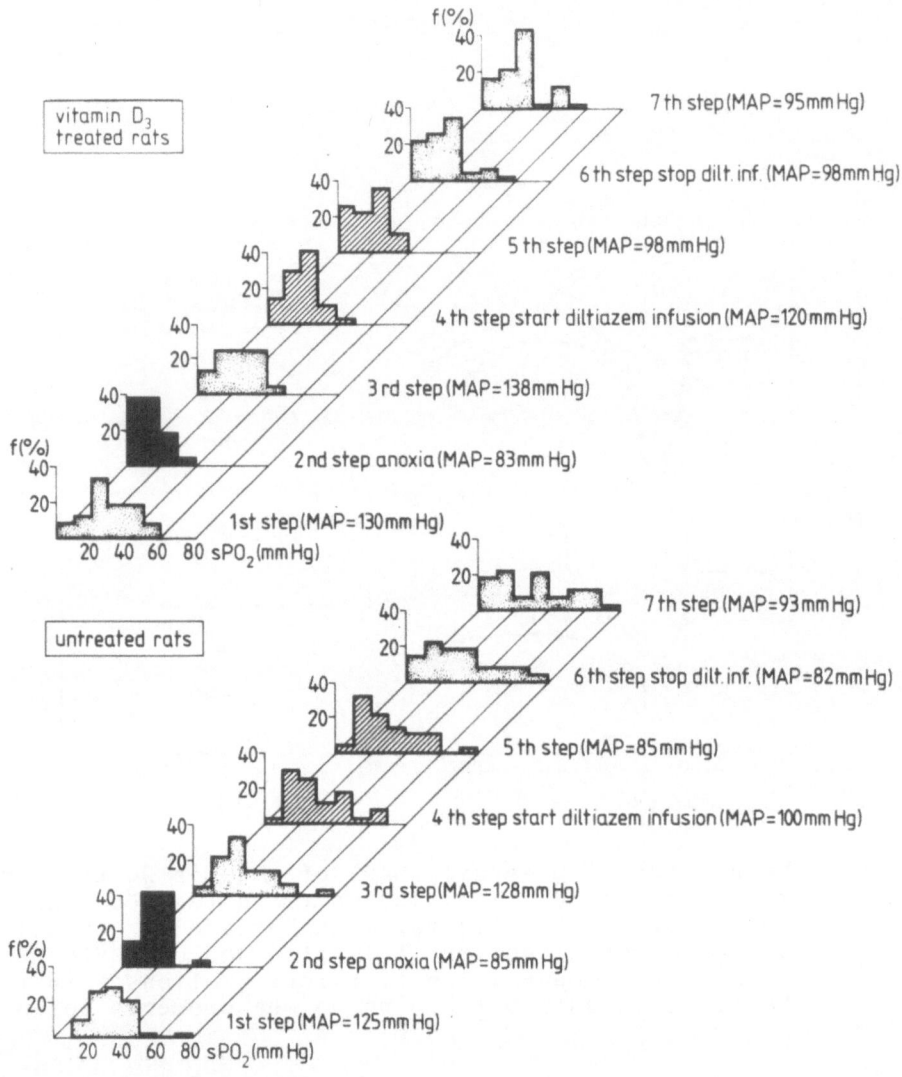

Fig.2. Surface PO_2 frequency distributions of vit.D_3 treated animals and controls. Time of the experiment: 35 minutes. Intervals of 5 minutes have been taken for the figure. Start of diltiazem infusion at step four, stop at step 6.

DISCUSSION

Both, IM or IP administration of the chemically synthesized water soluable 1,25-dihydro-cholecalciferol (vit.D_3) have induced hypercalcemia (Hartenbauer et al.,1977). Within four to five days of treatment with a dosage of about 5 ug/day an extensive calcification and degeneration of kidney cortex tubules and glomeruli as well as of the aorta and the major arteries have been observed. Histological examination by means of the method of von Kossa showed pronounced plaques especially within the media cells. This type of sclerosis is called Mönckeberg-atheroslcerosis (Sleyle, 1958). It is assumed that vit.D_3 enhances the calcium permeability of vascular smooth muscle cells by activating phospholipase A_2; the enzyme A_2 releases free fatty acid (FFA) from cell membranes and, among many effects, triggers oxygen free radicals (Gelmers,1985). Cellular integrity within the arterial and arteriolar wall is damaged and tissue necrosis results.

The decisive role of arterial lipid accumulation, particularly cholesterol, has been discussed as primary source for the pathogenesis ofatherosclerosis whereas the simultaneously observed calcinosis was considered to be a phenomenon of secondary importance. However, the cytoplasmic and mitochondrial calcium overload as well as the vascular damage of the arterial walls might be as important as the cholesterol accumulation. Moreover, the importance of calcium in ischemic cell death has been investigated. During cerebral ischemia, there is a massive calcium shift from the extra- to the intracellular space which is accompanied by a decrease of ATP necessary to maintain the intracellular homeostasis (Schanne et al.,1979). On the other hand, a group of chemically completely different molecules have been detected during the last decade which are capable of protecting the vascular walls and ischemic cells against calcium overload. However, these drugs act very unspecifically and induce a general vasodilatation and hypotension, even within the normal, nonischemic animal. Nevertheless, a preferential cerebrocortical supply has been detected for the ketamin/xylazine anesthetized rat for verapamil and, to a lower extent, for flunarizine (Metzger and Savas,1988).

The beneficial influence of the calcium antagonist diltiazem can be visualized through the MAP. The atherosclerotic rats showed chronic hypertension with a MAP of 148 mm Hg. Following diltiazem infusion a decrease to 91±15 mm Hg was registered. The control animals without atherosclerosis were characterized by an initial MAP of 118±14 mm Hg. Following diltiazem infusion 8o mm Hg was registered. After 5 minutes of infusion, a final concentration of 75 ug/dl bw had been approached but cardiac arrythmas have not been detected.

Vitamin D_3 treated rats showed a CBF decrease from 1.5o to 1.36 ml/g.min while sPO_2 was lowered from 34 to 29 mm Hg and shifted slightly to the left of the sPO_2 histogram. Infusion of 75 ug/dl dilatiazem caused a slight CBF reduction of about 5% in sclerotic and control animals as well. The results of the investigation demonstrate a stable cerebrocortical supply of the sclerotic rats during diltiazem infusion with a slight left shift to the hypoxic sPO range and an in-

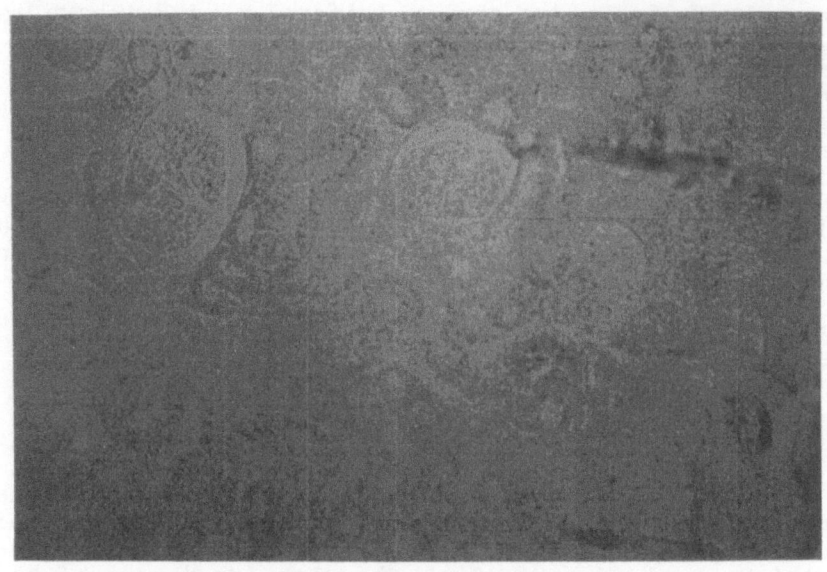

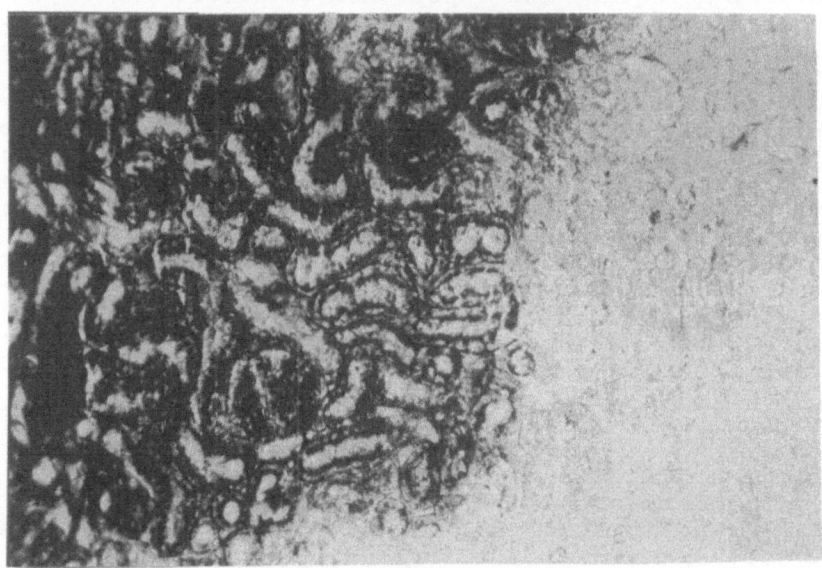

Fig. 3. Comparison of histological sections from controls and vitamin D_3 treated rats obtained from the kidney cortex. The staining procedure according to von Kossa enabled the investigation of calcium depostion in response to vitamin D_3 treatment. Tissue sections stained black indicate the intensity of calcium deposition.
Upper part: control animal, lower part: vit.D_3 treated.

creasing number of low sPO_2 values. Nevertheless, the positive aspect of the diltiazem therapy under atherosclerosis is documented by the fact that CBF as well as mean cerebrocortical sPO_2 are kept almost constant despite the induced lowering of the mean arterial blood pressure.

ACKNOWLEDGEMENT

The authors are grateful to Dr. Thomas Goldstick from the Northwestern University, Evanston Illinois, for his kind help and advice in improving the manuscript.

REFERENCES

Aukland,K., B.F. Bower and R.W. Berliner (1964): Measurement of local blood flow with hydrogen gas. Circ.Res.14:164-187.

Fleckenstein, A.(1977): Specific pharmacology of calcium in myocardium, cardiac pacemakers and vascular smooth muscle. Ann.Rev.Pharmacol.Toxicol. 17: 149-166

Fleckenstein, A.(1983): History of calcium-antagonists. Circ.Res. 52: 3-16

Gelmers, H.J. (1985): Calcium-channel blockers: effects on cerebral blood flow and potential uses for acute stroke. Am.J.Physiol. 55; 144B-148B

Hartenbower, D.L., T.M.Stanley, J.W.Coburn and A.W. Norman (1977): Serum and renal histologic changes in the rat following administration of toxic amounts of 1,25-dihydro-vitamin D_3, in: Vitamin D. Biochemical, chemical and clinical aspects related to calcium metabolism, A.W. Norman, K. Schaefer, J.W. Coburn, H.F.Deluca, D. Fraser, H. G.Grigoleit, D.v.Herrath (eds)

Metzger, H.P. and Y.Savas (1988): The influence of the calcium antagonists flunarizine and verapamil on cerebral blood flow and oxygen tension of anesthetized WFS-rats. Adv.Exp.Med.Biol.222:411-418

Metzger, H.P.(1989): The hydrogen gas clearance method for liver blood flow examination: inhalation or local application of hydrogen?Adv.Exp.Med.Biol.247:141-149

Schanne,F.A., A.B.Kane, E.E.Young, J.L.Farber (1979): Calcium dependence of toxic cell death: a final comman pathway. Science 2o6:7oo-7o2

Selye, H.(1958): Prophylactic treatment of an experimental arteriosclerosis with magnesium and potassium salts. Am.Heart J.55: 8o5-8o9

SPINAL CORD REPAIR: IS TISSUE OXYGENATION AN IMPORTANT VARIABLE?

Bradford T. Stokes and Paul J. Reier*

Ohio State University and University of Florida*

Columbus, Ohio 43210 and Gainesville, FL 32610*

INTRODUCTION

It is now clear that a broad range of grafting techniques can be used to successfully introduce fetal homografts of the vertebrate spinal cord into an adult injury site (Sladek and Gash, 1984; Houle and Reier, 1988). In general, these grafts are capable of long-term survival, and undergo extensive differentiation even to the extent that many exhibit features characteristic of homologous sites of the normal adult CNS. Furthermore, these grafts can form axonal interactions in varying degrees with the recipient spinal cord; host axons often project to the transplants, and dorsal root afferents penetrate the graft neuropil (Tessler et al., 1988). Thus, fetal spinal grafts also provide an excellent experimental condition in which to investigate problems of development, plasticity, and regeneration (Sladek and Gash, 1984). Although important questions remain about the mechanism by which such repair processes take place, various lines of evidence suggest that this could occur either by restoration of neurotransmitter stores, by provision of a cellular bridge which can facilitate axonal elongation across a lesion, or by anatomical and physiological reconstruction of damaged synaptic circuitries. In this regard, transplantation seems to offer great promise as a possible therapeutic approach to a variety of brain disorders related to disease and trauma.

Recent studies have also demonstrated that fetal CNS tissue can be successfully grafted into both acute (Reier, 1985; Reier, 1986; Bregman and Reier, 1986) and chronic (Houle and Reier, 1988) lesions of the spinal cord. Together, these studies have demonstrated that, as in the case of intracranial grafts, intraspinal transplants exhibit prolonged survival, considerable growth, and varying degrees of organotypic differentiation. Neuronal projections between host and graft are apparently facilitated by focal sites of host/graft fusion with an absence of intervening glial or connective tissue scar formation (Reier and Houle, 1988).

Oxygen Transport to Tissue XII, Edited by J. Piiper *et al.*
Plenum Press, New York, 1990

The progress of such work, coupled with the recent interest in transplants used in treating Parkinsonian patients, has led to the evolution of several major hypotheses that attempt to explain the relative success (or lack there of) of the various grafting procedures. One such tenet holds that the "microenvironment" that is created by the fetal graft after placement seems to be the critical factor in transplant survival. Because of the robust regenerative capacity that adult spinal tissue shows outside the CNS (David and Aguayo, 1981), a series of inhibitory growth factors, lack of trophic influences, and the absence of the substrate on which neurons usually grow (e.g. basal lamina), collectively appear to explain the minimal growth that occurs after spinal damage. The role of physiological processes in these events, including the development of the microcirculation, has yet to be explored.

BACKGROUND

As indicated by this general overview, the majority of studies of fetal CNS grafts into the brain and spinal cord have focused on basic anatomical features of the transplants and their functional neuronal relationships with the recipient CNS. Little attention, however, has been given thus far to the physiological microenvironment of these grafts and, in turn, to what characteristics these transplants might impart on the physiological microenvironment of the host CNS. This point especially applies to the important vascular relationships that are established at the host/graft interface. Although various morphological aspects of graft vascularization have been documented (Lawrence et al., 1984; Krum and Rosenstein, 1988; Joyner et al., 1988), only the study of Rosenstein and Brightman (1983), showing a PNS graft-induced local modification in the host blood-brain barrier, has begun to approach this important issue from a more physiological perspective. Clearly, additional information of this kind would not only be useful in providing a more detailed understanding of some of the basic principles upon which successful transplantation is dependent, but would also yield a greater insight into the physiology of developing neural tissue when divorced from its natural embryonic surroundings.

One important consideration in this regard is the degree of tissue oxygenation (i.e., tissue oxygen tension; P_tO_2) both in the transplant and in the surrounding regions of the host CNS. Although little is known about the appropriate oxygen conditions under which neurons will show optimal growth and survival, some clues as to the importance of appropriate P_tO_2 levels have come from developmental studies. We have observed, for example, that the embryonic spinal cord is uniquely sensitive to alterations in its oxygen microenvironment and that tissue oxygen tensions are maintained within a narrow range independently of environmental challenges (Gonya-McGee and Stokes, 1980). Moreover, oxygen tensions in the embryonic chick cord are remarkably stable and low when compared to adult values (Stokes, 1982; Stokes et al., 1981). Erdmann (1977) has also confirmed the occurrence of low oxygen tensions in the fetal rat cortex during the last week of gestation. Other

364

evidence for a critical P_tO_2 in neogenic phenomena has been found in studies of oxygen tensions in premature retinopathies (Ernest and Goldstick, 1984).

In view of these observations, it would be of considerable interest to learn to what extent oxygenation of fetal CNS tissue is affected by transplantation into the radically different tissue environment of the injured adult CNS and how this parameter varies with maturation of the graft. With emphasis on spinal cord injury, it would also be important to establish whether or not fetal transplants can influence oxygen tension in the surrounding regions of the host CNS. Along these lines, it is interesting that studies of wound healing (Knighton et al., 1981) and oxygen treatment of the spinal cord during decompression sickness (Leitch and Hallenbeck, 1985) all point to the ability of adult tissues to repair themselves being dependent upon a narrow range of tissue oxygen tensions after injury. In the injured spinal cord, spontaneous increases (>3 hrs) in tissue P_tO_2 have been associated with tissue necrosis and the absence of functional recovery (Stokes et al., 1985; Stokes and Garwood, 1982). Accordingly, it is quite conceivable, that the control of tissue oxygenation may be important for sparing of gray matter near the site of injury. That fetal grafts may by virtue of angiogenic or other properties may enhance tissue survival has been suggested (Houle and Reier, 1988). Analysis of the influences which fetal spinal cord grafts might exert on the hemodynamic properties of the host spinal cord (reflected in part by P_tO_2 levels) could thus demonstrate another important attribute of such transplants which has thus far gone unnoticed.

The studies described here also permit examination of tissue oxygenation in cellular microenvironments that may not be compatible with axonal outgrowth. Axonal elongation is clearly a function of the interdependencies established between the intrinsic and extrinsic environment of the neuron (Lasek et al., 1981), and recent peripheral nerve grafting experiments in the CNS have especially illustrated the importance of a wide range of cellular components that can affect axonal outgrowth. Extracellular matrix factors, alterations in the glial components (e.g., absence of Schwann cells and/or astrogliosis), absence of appropriate trophic (or tropic) factors, and a large number of receptor complexes which might be used for axonal guidance are among the more prominent candidates. Of these, reactive gliosis seems to be a major determinant of the success or failure of CNS regeneration (Reier and Houle, 1988). Little is known, however, about the physiological environment of astrocytic scars and the role it may play in axonal elongation. It is noteworthy, that reactive glia undergo a number of changes (Nathaniel and Nathaniel, 1981) which seem to be indicative of significant shifts in metabolic substrate utilization. A major factor in inducing such alterations could be the availability of molecular oxygen to carry on oxidative phosphorylation. Damaged gliotic areas of the central nervous system would also be invaded by connective tissue or cells such as macrophages. Oxygen tensions in these macrophage rich microenvironments seem also to control the release of angiogenic factors. It thus appears that vascular neogenesis, a factor critically linked to graft survival (Stenevi et al., 1976), could also be dependent on oxygen transport phenomena. Collectively, these factors augur for a closer examination of oxygen transport variables during the process of graft-mediated repair.

EXPERIMENTAL PROCEDURES

These experiments were conducted on a series 24 adult Sprague-Dawley rats (250-350 gm) that received fetal transplants acutely from 14-day fetal donors. The transplants were surgically reexposed at various times and stages of development. Recordings were made of P_tO_2 in laminectomized controls, in and around the transplanted fetal spinal cord and in adjacent host spinal tissue. The electrodes used were of the Whalen type and of extreme taper to avoid tissue dimpling. Procedures used for data collection were as before (Stokes, 1982; Stokes et al., 1985; Stokes et al., 1981).

All of the data shown here is displayed in the form of frequency histograms. All transplant values were taken from animals that had acute aspiration (2-4 mm hemisections) lesions with immediate placement of 14 day fetal spinal grafts soon after the lesion; days cited are times after transplant procedures. In all cases strip chart recordings were made from the spinal cords (L1-L2 laminectomy site) of anesthetized (Ketamine-70 mg/kg; Xylazine-10 mg/kg) animals. These were then reduced by a combination of a SIGMA-SCAN digitizing system (sample rate=1 pt./2 sec) in conjunction with a Leading Edge D computer system. The data was subsequently plotted (bin size= 3 mm Hg; P_tO_2 torr) in the form displayed here by an interfaced Hewlett Packard 9872B digital plotter. Arterial blood analysis done at the termination of experimental procedures revealed a pH= 7.22 (+/-.04 S.D.), p_aCO_2= 54 (+/-10.4 S.D.), p_aO_2= 123 (+/-11 S.D.) and HCO_3= 24.2 (+/-3.5 S.D.). This slight acidosis was likely produced by the anesthetic regimen that was chosen for its minimal effects on oxygen delivery (Wixson et al., 1987).

OXYGEN TENSIONS IN THE HOST-GRAFT NEUROPIL

The first series of histograms were constructed from a control population of animals who only received a laminectomy in the upper lumbar cord. The general protocol calls for several penetrations to be made at slow rates of electrode travel (< 2.0 uM/sec) to various depths (usually 500 and 1000 uM from the surface) where sampling for longer times (5-10 min) to check P_tO_2 stability is accomplished. Care is taken to avoid tissue dimpling (e.g. vibration techniques for example) and records are checked to see if P_tO_2 profiles are similar on entrance and withdrawal. The first histogram (LAMCON) was constructed from three animals each of which had three electrode penetrations; the second (LAM24) is taken from two electrode penetrations at a reopened laminectomy site in one of these three animals 24 hrs after the original procedure.

366

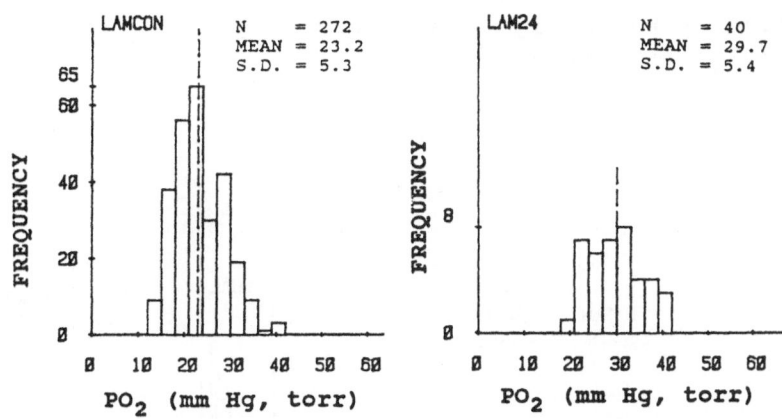

FIGURE 1.   LAMINECTOMY CONTROL LAMINECTOMY + 24 HRS

Absolute P_tO_2 values in these histograms have a range of approximately 12-40 mm Hg with means well above 20 mm Hg. Note also that P_tO_2 values below 10 mm Hg in our sample population rarely occur. Otherwise these control values confirm those reported elsewhere for the rat spinal cord (Hayashi et al., 1983).

In the next series of illustrations, histograms have been constructed from animals that received recent transplants. The first examples are of recordings made in transplanted animals after about a week in situ (Figure 2). At 5 days normal P_tO_2 values (T5DAA) can be found in host tissue near to the transplant site (<1 mm from edge, 2 penetrations). In the same animal, however, areas of low P_tO_2 can be found in the transplant itself (T5DAL, 2 sample sites).

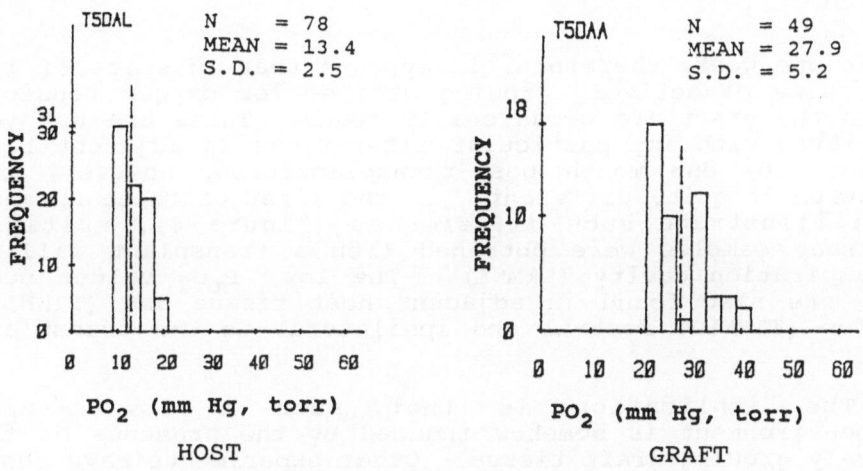

FIGURE 2.   5 DAY TRANSPLANT

In two other animals one week (Figure 3) after transplantation (T7DAH, T8DAL), distinct differences are noted. In one area from the first animal which was later

367

identified as being highly necrotic, high P_tO_2 values are seen during the two sampling periods (T7DAH). In contrast three penetrations into a well developed 8 day transplant reveal fairly low P_tO_2 tensions.

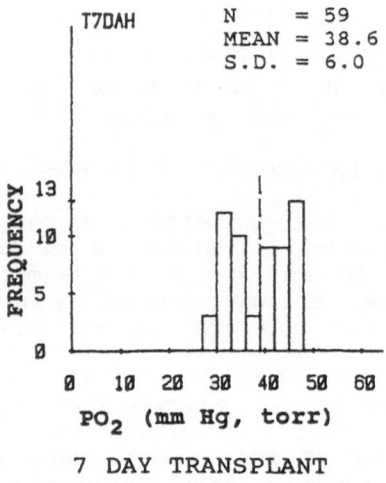

7 DAY TRANSPLANT

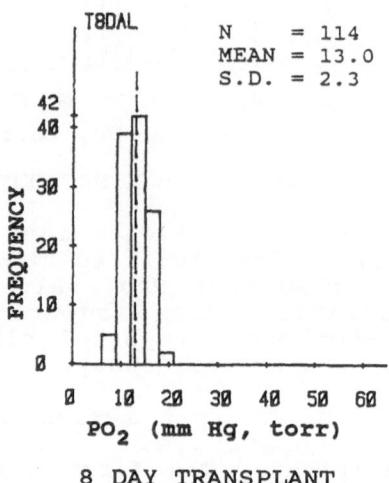

8 DAY TRANSPLANT

FIGURE 3

At one week, therefore, it appears that in spite of the small size of actively growing grafts, low oxygen tensions within the graft are occasionally found. These are not yet associated with any particular alterations in adjacent host tissue. By one month post-transplantation, however, the situation is quite different. In the first of these animals which illustrate such differences (Figure 4), multiple electrode samples were obtained from a transplant filling the aspiration cavity (T1MNL). The low P_tO_2 values seen there are also found in adjacent host tissue 3mm (T1MNL1) and 6mm (T1MNL2) rostral and ipsilateral to the transplant site.

The implication is that such a low P_tO_2 microenvironment is somehow induced by the presence of the actively growing graft tissue. Other experiments have shown that less well developed grafts at one month do not have such P_tO_2 tensions (Figure 5, T1MNH, 2 penetrations). Indeed, in some cases P_tO_2 values are well above those found in normal spinal tissue (T1MNH2).

At two months, similar profiles apply (Figure 6). A poor graft with obvious surface cavitation shows elevated P_tO_2 tensions (T2MNH, 3 penetrations).

368

T1MNL
N = 150
MEAN = 11.8
S.D. = 3.6
FREQUENCY
PO2 (mm Hg, torr)
GRAFT
T1MNLI
N = 83
MEAN = 7.1
S.D. = 1.7
FREQUENCY
PO2 (mm Hg, torr)
HOST + 3 mm
T1MNL2
N = 53
MEAN = 12.0
S.D. = 3.5
FREQUENCY
PO2 (mm Hg, torr)
HOST + 6 mm

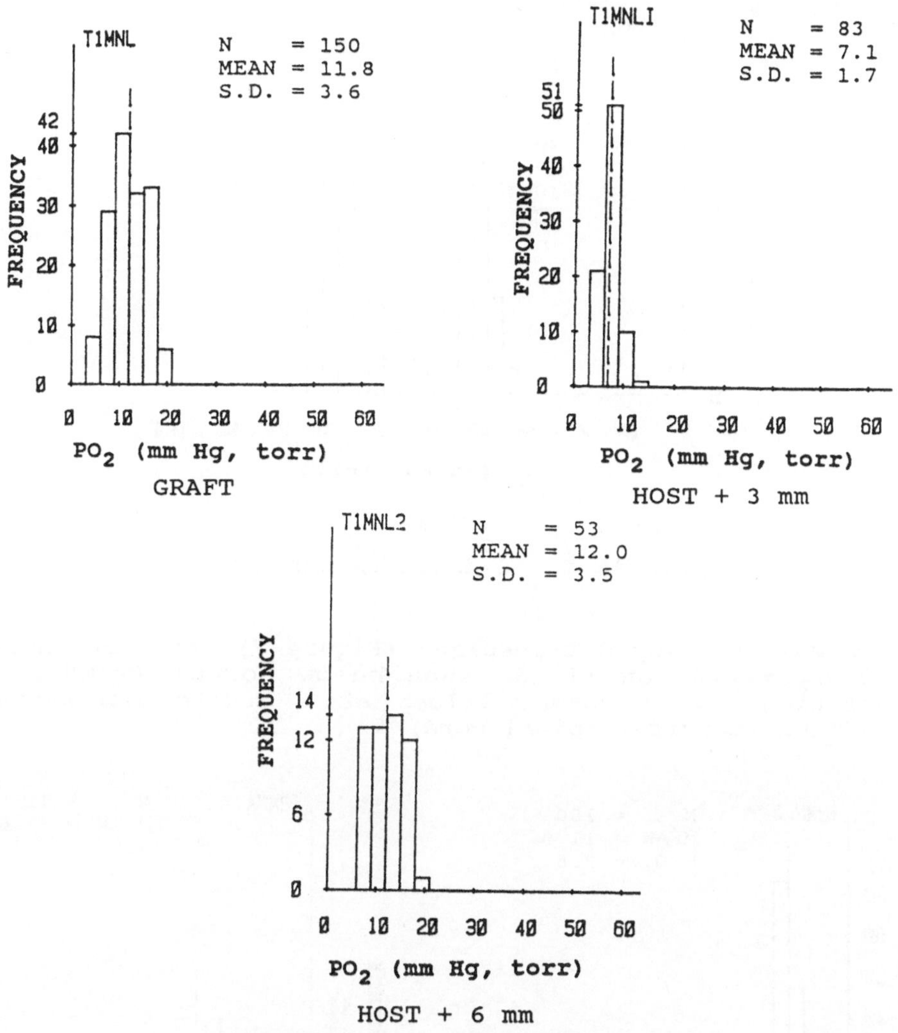

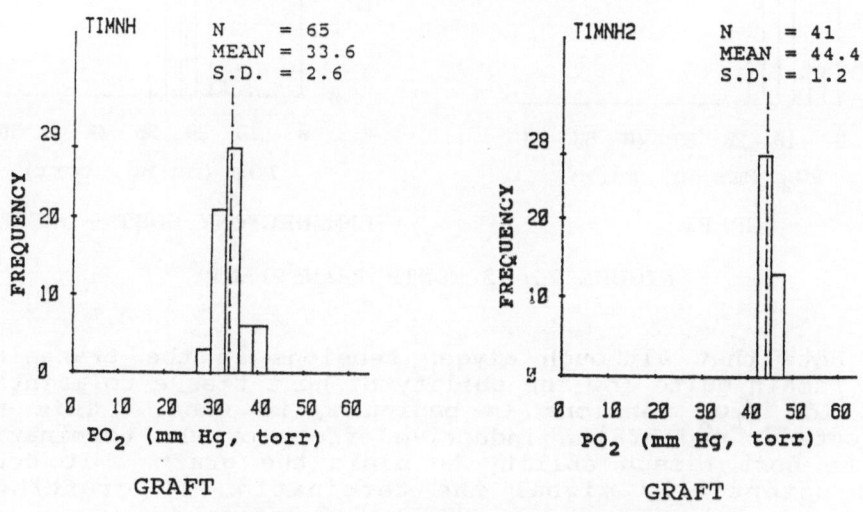

FIGURE 4. 1 MONTH TRANSPLANT

T1MNH
N = 65
MEAN = 33.6
S.D. = 2.6
FREQUENCY
PO2 (mm Hg, torr)
GRAFT
T1MNH2
N = 41
MEAN = 44.4
S.D. = 1.2
FREQUENCY
PO2 (mm Hg, torr)
GRAFT

FIGURE 5. 1 MONTH TRANSPLANT

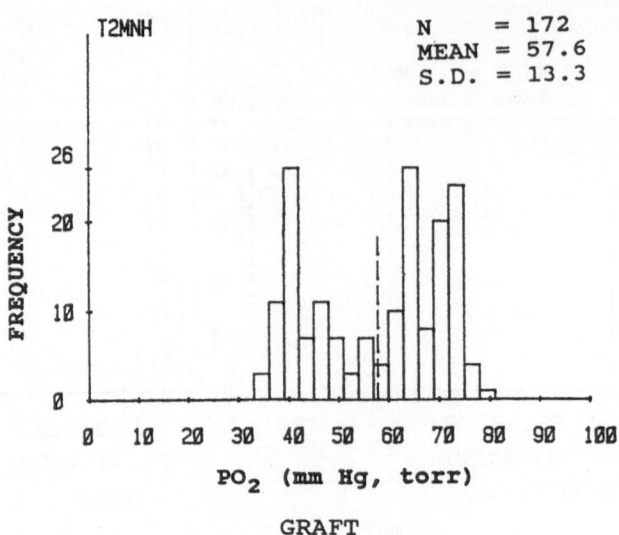

GRAFT

FIGURE 6.   2 MONTH TRANSPLANT

A well developed transplant (Figure 7), however, again has P_tO_2 diffusion fields much below normal (T2MNL2, 5 penetrations) with normal values at a remote laminectomy site (T2MNA-cervical spinal cord)

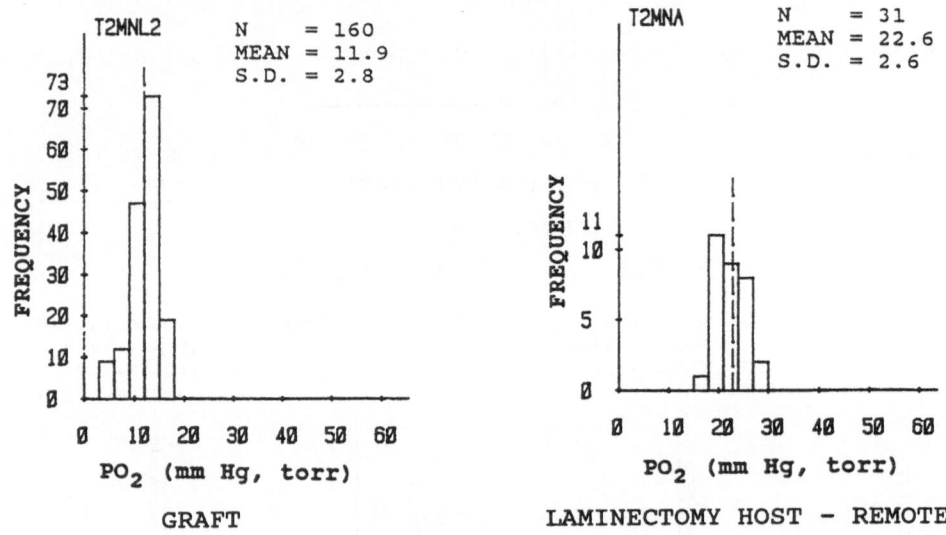

GRAFT   LAMINECTOMY HOST - REMOTE

FIGURE 7.   2 MONTH TRANSPLANT

Note that although oxygen tensions in the transplant site remain quite low the ability of host tissue to maintain such low P_tO_2 tensions is beginning to wane. This may indicate an end of the inductive effect or the termination of the host tissue ability to mimic the graft. It could also potentially signal the termination of graft/host integration.

370

Finally, we have illustrated recordings taken from a three month graft (Figure 8). While P_tO_2 tensions remain low, note the increasing number of samples found above 20 mm Hg (2 penetrations). In this case the graft now seems to have oxygen tensions approaching those normally found in the host cord.

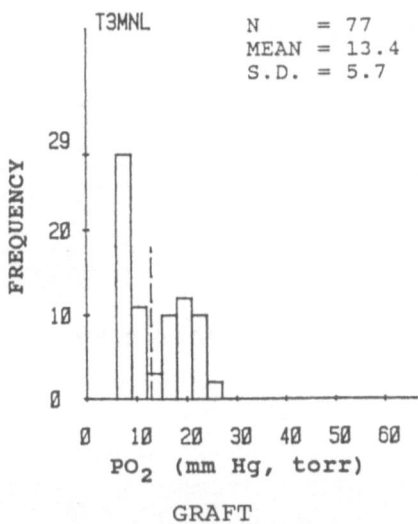

T3MNL   N   = 77
         MEAN = 13.4
         S.D. = 5.7

GRAFT

FIGURE 8.   3 MONTH TRANSPLANT

TRANSPLANT MORPHOLOGY

The following figures illustrate the ability to fetal graft tissue to develop in an aspiration cavity and fuse with the host neuropil. In Fig. 9, a low power micrograph (27X) of a horizontal section reveals the extensive development of the graft which in this case is positioned over a zone of host white matter (WM) due to the incompleteness of the aspiration lesion. Note the partial integration with the host gray matter (white arrows) caudally; rostral to the graft itself is a large cavity (C) often seen in areas of incomplete host/graft anastomosis.

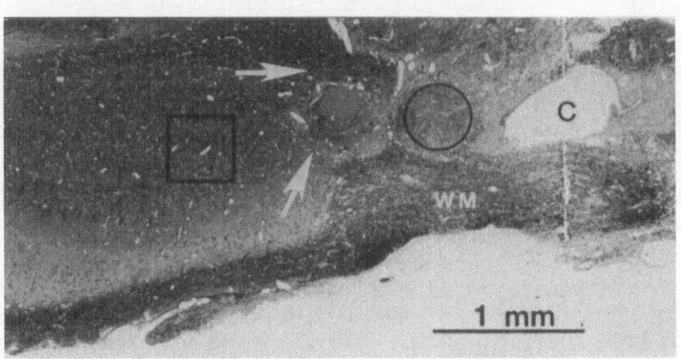

FIGURE 9.   LOW POWER HOST/GRAFT MORPHOLOGY

The circle and box in Fig. 9 represent areas of graft and host respectively that are shown in higher power in thin section micrographs in Fig. 10.

371

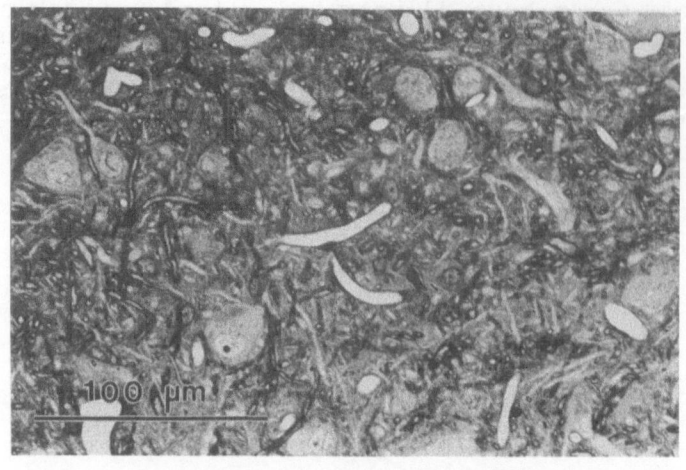

FIGURE 10A.   GRAFT MORPHOLOGY

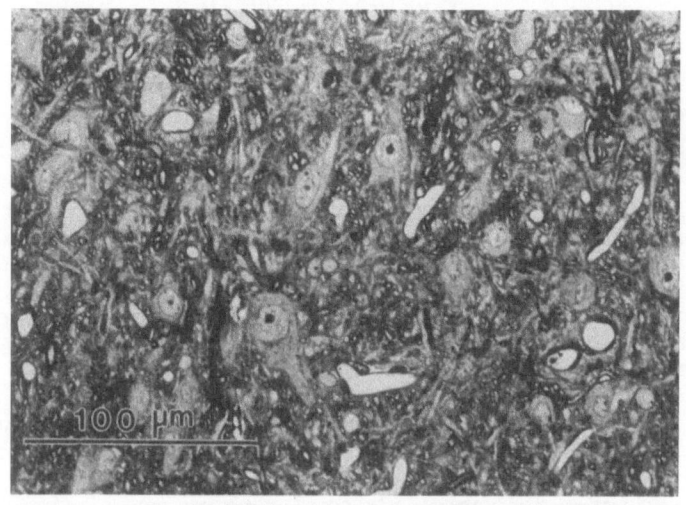

FIGURE 10B.   HOST MORPHOLOGY

It becomes difficult to differentiate between the host and graft areas sampled because of the remarkable cytological differentiation of the transplanted graft tissue.

SUMMARY

We have demonstrated the reliability and feasibility of making P_tO_2 recordings from graft and host tissue in the injured spinal cord. The data suggest that the oxygen microenvironment of developing graft and host spinal tissue is clearly different from that found in normal spinal tissue or in transplants that have not survived or integrated well. These same constraints seem to apply to cavitation in developing grafts and poorly developed graft/host interfaces. The similarity between these findings and those from previous studies in other fetal vertebrates suggests that oxygen tensions in the spinal cord probably reflect the developmental status of the regenerating tissue. Our future studies will seek to define the relationship between anatomical development of transplant tissue and these functional (P_tO_2, microvascular development and tissue metabolism) indicators of graft development. These investigations should also provide a background for those later studies which seek to establish the mechanisms by which these relationships come about i.e. oxygen consumption of host/transplant tissue, blood flow to transplants, studies of glycolytic metabolism (2-DG autoradiography), etc. In this way, we can begin to understand the role of tissue metabolism in graft-mediated repair.

REFERENCES

Bregman, B.S. and Reier, P.J., 1986, Neural tissue transplants rescue axotomized rubrospinal cells from retrograde death. J.Comp.Neurol., 244:86-95.

David, S. and Aguayo, A.J., 1981, Axonal elongation into peripheral nervous system 'bridges' after central nervous system injury in adult rats. Science., 214:932.

Erdmann W., 1977, Microelectrode studies in the brain of fetal and newborn rats. Adv.Exp.Med.Biol., 94:455-461.

Ernest, J.T. and Goldstick, T.K., 1984, Retinal oxygen tension and oxygen reactivity in retinopathy of prematurity in kittens. Invest.Ophthalmol.Vis., 25:1129-1134.

Gonya-Mcgee, T. and Stokes, B.T., 1980, Acute modification of embryonic spinal cord activity induced by hypoxia. Dev.Neurosci., 3:11-18.

Hayashi, N., Green, B., Gonzalez-Carvajal, M., Hora, J. and Vera, R., 1983, Local blood flow, oxygen tension, and oxygen consumption in the rat spinal cord, Part 1: oxygen metabolism and neuronal function. J.Neurosurg., 58:516-525.

Houle, J.D. and Reier, P.J., 1988, Transplantation of fetal spinal cord tissue into the chronically injured adult rat spinal cord. J Comp.Neurol., 269:535-547.

Joyner, W.L., Young, R., Blank, D., Eccleston-Joyner, C.A. and Gilmore, J.P., 1988, In vivo microscopy of the cerebral microcirculation using neonatal allografts in hamsters. Circ.Res, 63:758-766.

Knighton, D.R., Silver, I.A. and Hunt, T.K., 1981, Regulation of wound-healing angiogenesis - Effect of oxygen gradients and inspired oxygen concentration. Surgery., 90:262-270.

Krüm, J.M. and Rosenstein, J.M., 1988, Patterns of angiogenesis in neural transplant models: II. Fetal neocortical transplants. J Comp.Neurol., 271:331-345.

Lasek, R.J., McQuarrie, I.G. and Wujek, J.R., 1981, The central nervous system regenerative problem: neuron and environment, in: "Post-Traumatic Peripheral Nerve Regeneration," H. Millesi, S. Mingrino and A. Gorio, ed., Raven Press, New York.

Lawrence, J.M., Huang, S.K. and Raisman, L.G., 1984, Vascular and astrocytic reactions during establishment of hippocampal transplants in adult host brain. Neuroscience., 12:745-760.

Leitch, D.R. and Hallenbeck, J.M., 1985, Oxygen in the treatment of spinal cord decompression sickness. Undersea Biomed.Res., 12:269-289.

Nathaniel, E.J.H. and Nathaniel, D.R., 1981, The reactive astrocyte. Adv.Cell Neurobio., 2:249-301.

Reier, P.J., 1985, Neural tissue grafts and repair of the injured spinal cord. Neuropathol.Appl.Neurobiol., 11:81-104.

Reier, P.J., Bregman, B.S. and Wujek, J.R., 1986, Intraspinal transplantation of fetal spinal cord tissue: An approach toward functional repair of the injured spinal cord, in: "Proceedings of the International Symposium on Plasticity and Development of the Mammalian Spinal Cord," M. Goldberger, A. Gorio and M. Murray ed.,

Reier, P.J. and Houle, J.D., 1988, The glial scar: its bearing on axonal elongation and transplantation approaches to CNS repair. Adv.Neurol., 47:87-138.

Rosenstein, J.M. and Brightman, M.W., 1983, Circumventing the blood-brain barrier with autonomic ganglion transplants. Science., 221:881-887.

Sladek, J.R. and Gash, D.M., 1984, Morphological and functional properties of transplanted vasopressin neurons, in: "Neural Transplants: Development and Function," J. Sladek and D. Gash, ed., Plenum Press, New York.

Stenevi, U., Bjorklund, A. and Svendgaard, N-NA., 1976, Transplantation of central and peripheral monoamine neurons to adult rat brain: techniques and conditions for survival. Brain Res., 114:1-20.

Stokes, B.T., 1982, O2 tension in the spinal cord of the avian embryo. J.Appl.Physiol., 53:1455-1460.

Stokes, B.T., Fox, P. and Hollinden, G., 1985, Extracellular metabolites: their measurement and role in the acute phase of spinal cord injury, in: "Trauma of the Central Nervous System," R.G. Dacey ed., Raven Press, New York.

Stokes, B.T. and Garwood, M., 1982, Traumatically induced alterations in the oxygen fields in the canine spinal cord. Exp.Neurol., 75:665-677.

Stokes, B.T., Garwood, M. and Walters, P., 1981, Oxygen fields in specific spinal loci of the canine spinal cord. Am.J.Physiol., 240:H761-H766.

Tessler, A., Himes, B.T., Houle, J. and Reier, P.J., 1988, Regeneration of adult dorsal root axons into transplants of embryonic spinal cord. J Comp.Neurol., 270:537-548.

Wixson, S.K., White, W.J., Hughes, H.C., Jr., Lang, C.M. and Marshall, W.K., 1987, The effects of pentobarbital, fentanyl-droperidol, ketamine-xylazine and ketamine-diazepam on arterial blood pH, blood gases, mean arterial blood pressure and heart rate in adult male rats. Lab. Animal Sci., 37:736-742.

CARDIOVASCULAR SYSTEM

MORPHOMETRIC ANALYSIS OF CAPILLARY

NETS IN RAT MYOCARDIUM

S. Batra and K. Rakusan

Department of Physiology
University of Ottawa
Ottawa, Ontario Canada K1H 8M5

INTRODUCTION

The importance of relevant data with respect to capillary geometry cannot be under-estimated in the modelling of oxygen transport to tissue. C. R. Honig stated "knowledge of capillary length and it's frequency distribution is essential in modelling O_2 transport and indicator dilution data" (Honig et al., 1977). To this end, data have been collected for capillary lengths in skeletal muscle (eg. Honig et al., 1977; Skalak and Schmid-Schonbein, 1986; Potter et al., 1988). The situation in cardiac muscle, to date, is certainly more obscure. This is due to the exigency of developing appropriate facilities for viewing surface vessels of a beating heart in situ; or in the development of morphometric techniques that accurately demarcate capillaries from histological sections.

The approach in our lab has been to use quantitative morphology to study microcirculatory architecture. From tissue cross-sections, it has been possible to measure capillary density, and further the heterogeneity of capillary spacing. However, the task of following capillaries in longitudinal sections has proven to be more difficult, which may explain why the data are relatively sparse. Part of the hindrance lies in the procurement of tissue sections that capture a significant part of the capillary network. Further, it has been difficult to establish valid standards for defining capillaries from arteriole to venule. The present study employed a combination of histochemical methods that distinguished, by colour, the arteriolar and venular portion of individual capillaries. With this information, it was possible to follow individual capillaries from arteriole to venule. Our previous paper (Batra et al., 1989) used this staining technique to study the distribution of coronary capillaries on cross-section, as a function of their arteriolar and venular portions. The present study takes advantage of this differential stain sensitivity, to study capillary branching phenomena from terminal arteriole to collecting venule in well defined capillary nets from longitudinal sections.

Oxygen Transport to Tissue XII, Edited by J. Piiper *et al.*
Plenum Press, New York, 1990

METHODS

Eight male rats were used in this study (Sprague-Dawley; 250-300 g). Subsequent to anaesthesia with sodium pentobarbitol, the heart was quickly removed and frozen in liquid nitrogen. Sixteen μm sections were prepared of the mid-myocardium of the anterior left ventricle. The staining protocol has been given elsewhere (Batra et al., 1989). Briefly, tissue sections were prefixed in a mixture of chloroform and acetone (1:1). The sections were then transferred to a solution sensitive to Dipeptidyl Peptidase IV in the capillary endothelium for 100 minutes. This treatment stained the venous portion of capillaries red. Next, the sections were transferred to a solution sensitive to Alkaline Phosphatase for 25 minutes. This treatment stained the arterial portion of capillaries blue (Lojda, 1979).

With the aid of a drawing tube attached to the microscope, scale drawings were made of capillaries on longitudinal section. A capillary net, defined as all of the micro-vessels connected to the same feeding arteriole and collecting venule, served as the principle unit of study. Forty capillary nets, 4-6 per heart were delineated, and subjected to detailed morphometry. From the tracings of each capillary net, the following data were collected:

1. Minimal Capillary Length (MCL): the shortest contiguous pathway from terminal arteriole to collecting venule.

2. Capillary Set Length (CSL): equivalent to the defined capillary net; that is, all the micro-vessels connected to the same arteriole and venule.

3. Mean Capillary Pathlength: mean measure of all possible pathways that could be traced from arteriole to venule.

4. Number of Segments per Path: the number of segments constituting each pathlength.

5. Arteriolar Segment Length: the length of segments (the distance between two ensuing branch points) that stained positive for Alkaline Phosphatase; i.e. blue in colour.

6. Venular Segment Length: the length of segments that stained positive for Dipeptidyl Peptidase IV; i.e. red in colour.

7. Mixed Segment Length: the length of segments that stained positive for both Alkaline Phosphatase and Dipeptidyl Peptidase IV; i.e. blue and red in colour, including the transition zone from blue to red.

8. Arteriolar Segment Number: the number of arteriolar segments in each net.

378

9. Venular Segment Number: the number of venular segments in each net.

10. Mixed Segment Number: the number of mixed segments in each net.

RESULTS

Mean values for the above listed parameters are given in table 1. All data were collected from scale drawings of 40 capillary nets. A net was defined as all of the micro-vessels connected to the same arteriole and venule. As the data for segment length and segment number was poorly characterized by the normal distribution, non-parametric statistical analyses were used. To compare segment length and segment number from arteriole to venule (including mixed) regions of the capillary, the Kruskal-Wallis One-Way Analysis of Variance by Ranks was employed.

Table 1. Capillary length data (mean ± SE) from normal rat myocardium.

	Mean Value
Minimal Capillary Length (μm)	596.75 ± 15.19
Capillary Set Length (μm)	3180.47 ± 115.85
Mean Capillary Pathlength (μm)	638.62 ± 6.80
Number of Segments per path	7.6 ± .15
Arteriolar Segment Length (μm)	93.41 ± 3.10[*]
Venular Segment Length (μm)	75.81 ± 2.83[*]
Mixed Segment Length (μm)	228.88 ± 12.17[*]
Arteriolar Segment Number	19.85 ± 1.08[#]
Venular Segment Number	9.08 ± .67[#]
Mixed Segment Number	3.18 ± .24[#]

[*][#], significantly different at p<0.01; Kruskal-Wallis ANOVA.

Frequency distributions of minimal capillary length and capillary set length are shown in figures 1 and 2. Minimal capillary length ranged from 342 to 765.5 μm, with an interquartile range of 111.5 μm. Capillary set length ranged from 1807.5 to 4820 μm, with an interquartile range of 1032.75 μm. There was a positive correlation between both of these parameters (R=.72) as shown in figure 3. Mean capillary pathlength (Fig. 4) ranged from 466.5 to 805.5 μm, which elucidates the variability that exists in pathlength from terminal arteriole to collecting venule. Figure 5 illustrates frequency distributions for arteriolar and venular segment length. Mixed segment length was not included in this figure in order to highlight the differences between arteriolar and venular segment length.

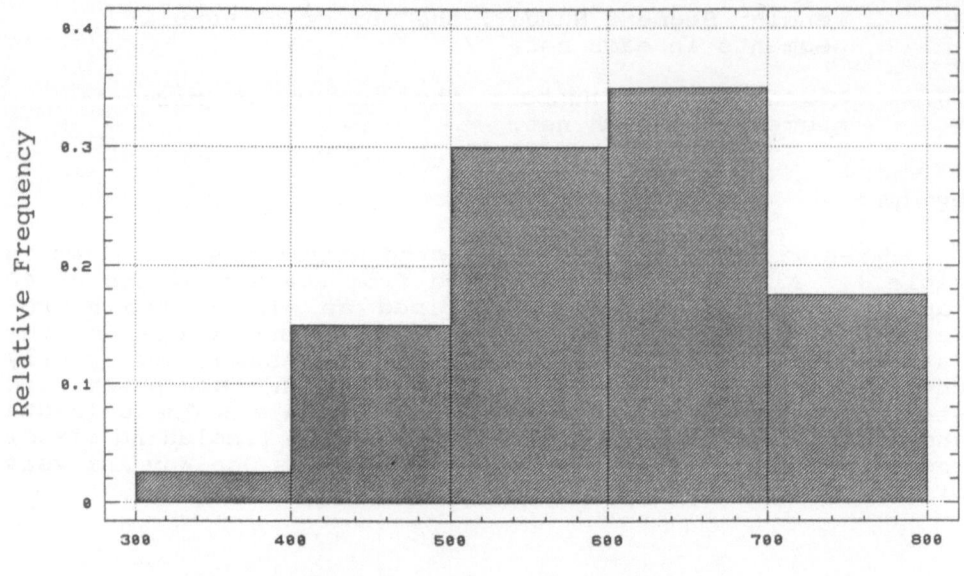

Minimal Capillary Length (μm)

Figure 1. Frequency Histogram of Minimal Capillary Length, the shortest contiguous pathway from terminal arteriole to collecting venule; n=40.

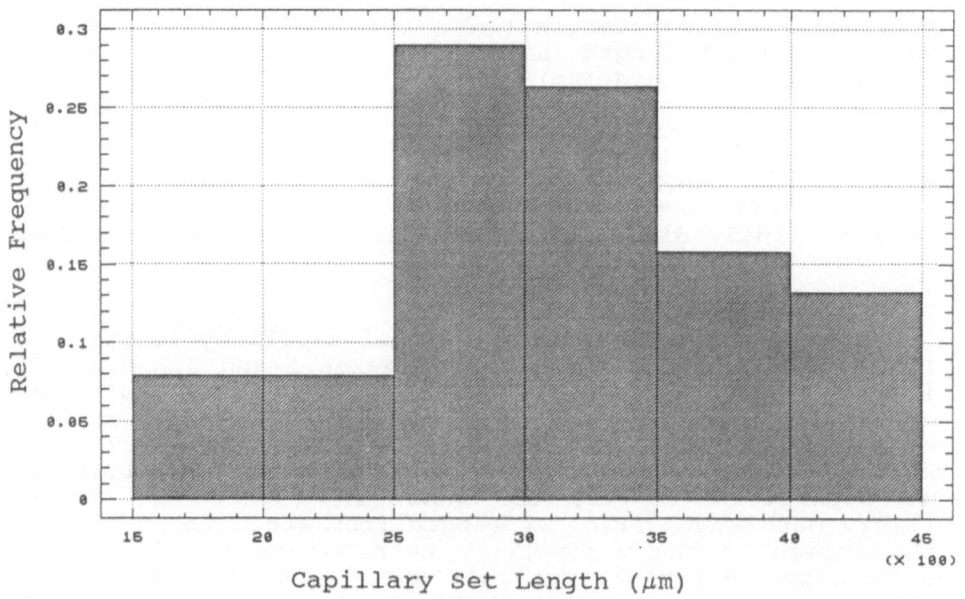

Capillary Set Length (μm)

Figure 2. Frequency Histogram of Capillary Set Length, the sum of all micro-vessels connected to the same terminal arteriole and collecting venule; n=40.

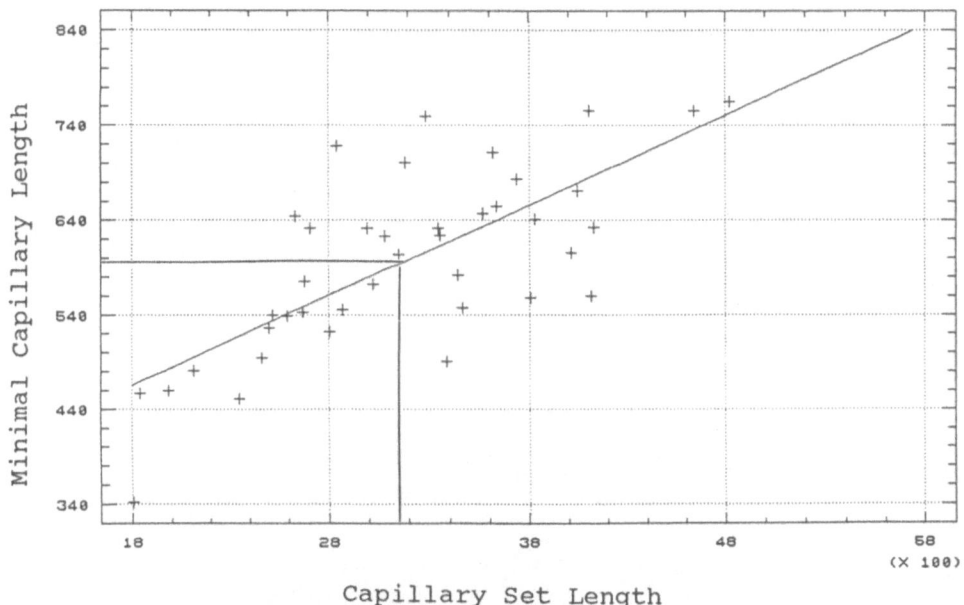

Figure 3. Regression of Minimal Capillary Length on Capillary Set Length. R=.72; Note that the mean values for both parameters fall on regression line.

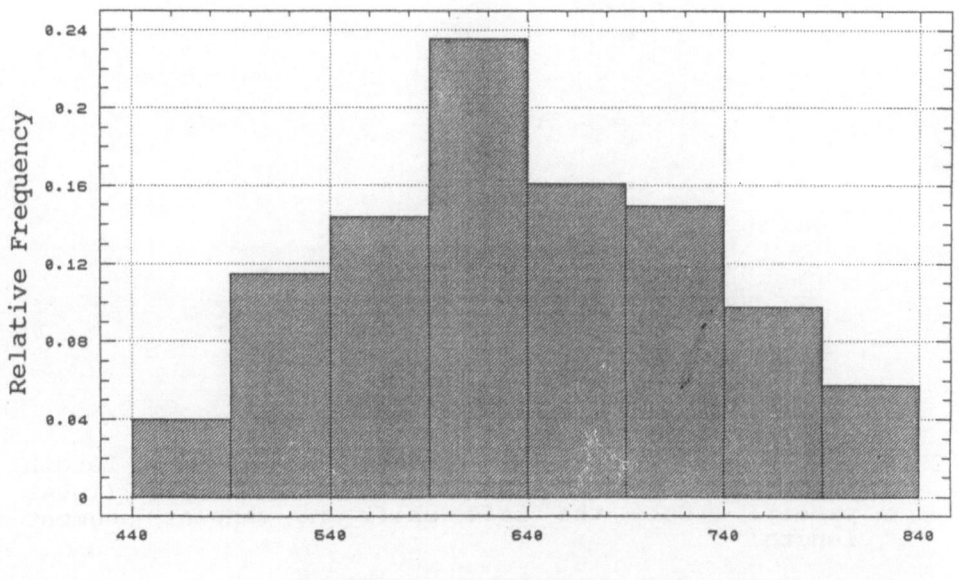

Capillary Pathlength (μm)

Figure 4. Frequency Distribution of Capillary Pathlengths. Pooled data for 40 nets; 362 paths in total were traced, representing an average of 9 paths per net.

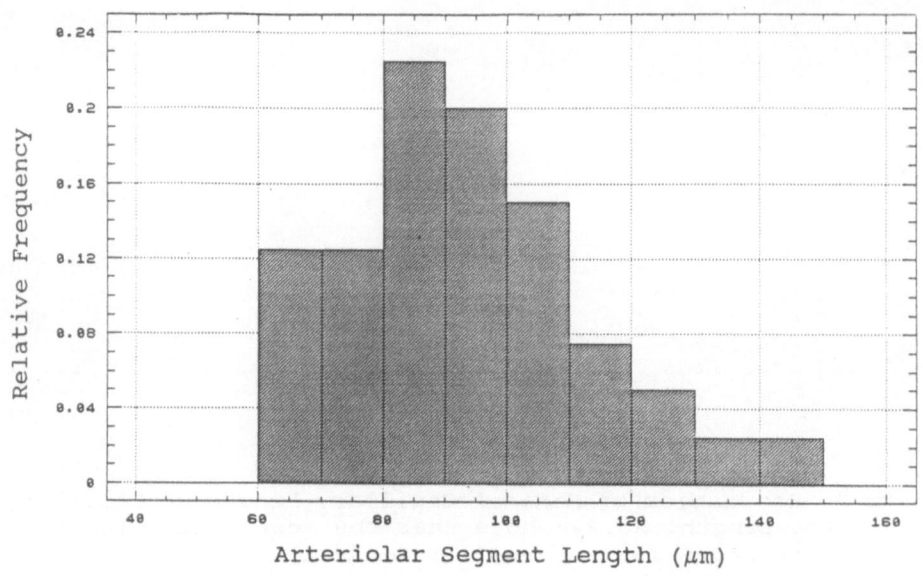

Arteriolar Segment Length (μm)

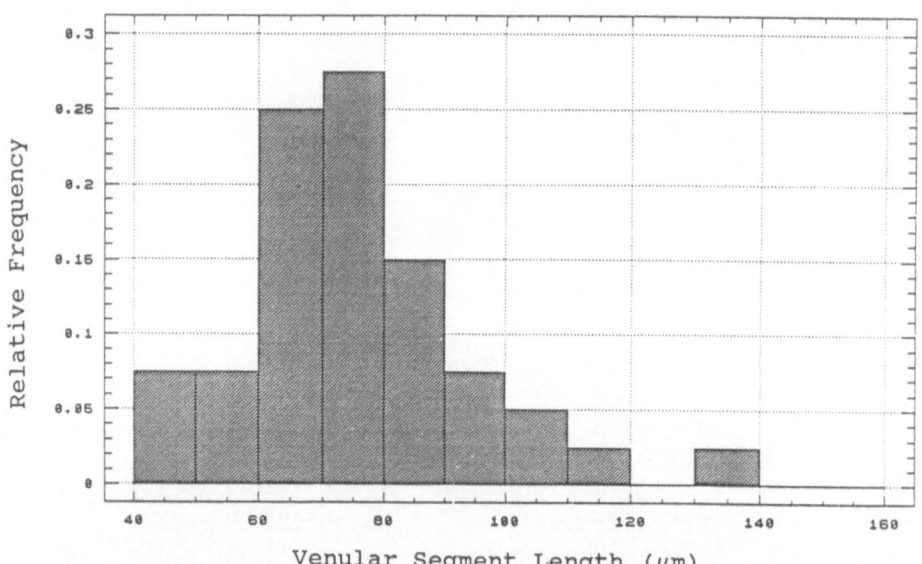

Venular Segment Length (μm)

Figure 5. Frequency Histograms of Arteriolar Segment Length (upper panel) and Venular Segment Length (lower panel). Note the left shift in venular segment length;

DISCUSSION

The results presented in this paper represent, to the best of our knowledge, the first attempt at characterizing longitudinal capillary branching patterns in myocardium using quantitative morphology. However, the data should be considered in light of the several limitations. Firstly, the data are based upon the traced drawings of the previously defined capillary nets. Clearly, some nets may be larger or smaller, and thus render different measurements. The nets used in this study served as a repeatable working construct, and as such provided a frame of reference. It is further accepted that some vessels may be leaving the net, and others entering it, as the coronary microcirculation appears to be an infinite branching network. Therefore, the capillary nets studied do not represent an isolated set of exchange vessels. Indeed they do form contiguous pathways from terminal arteriole to collecting venule, but this circulation is neither exclusive nor exhaustive.

The data presented in this paper show that heterogeneity exists along the length of individual capillaries. Arteriolar segment length was found to be significantly longer than venular segment length (93.41 ± 3.10 μm vs. 75.18 ± 2.83 μm; mean $\pm$ SE, $p<.01$). This configuration would allow for more variability in the path a red blood cell could take on the venular side of capillaries where PO_2 values are generally lower. This observation is even more striking in light of our previous finding (Batra et al., 1989) that the average inter-capillary distance is significantly lower on the venular side of neighbouring capillaries ($25.34 \pm .27$ μm for arteriolar neighbours and $22.93 \pm .22$ μm for venular neighbours; mean $\pm$ SE). Thus not only are the venular segments shorter, but are also closer together on cross-section. This would allow for more favourable geometrical conditions for oxygen diffusion on the venous side of the capillary nets, where PO_2 values are lower.

Previous experiments in coronary capillary architecture from longitudinal sections have generally been descriptive and non-rigorous in nature. Ludwig (1971) measured distances from arteriole to venule in rabbit heart using an India ink suspension. Mean distances were 400 μm for the subepicardial layers, 300 μm for the intramural layers and 500 μm for the subendocardial layers. These values were reported only as approximations and the number of measurements taken were not given. Toborg (1972) perfused the heart with different dye suspensions to illustrate the coronary capillary bed in rat and cat. From measurements in two hearts, he reported mean capillary lengths of 310 μm (rat) and 400 μm (cat). Bassingthwaighte (1974) injected a silicon rubber elastomer in the left anterior descending coronary artery while the circumflex coronary artery remained intact. Functional capillary lengths, the distance from the fringe of the elastomer filled region to the nearest venule, were observed to range from 300-1200 μm.

From the present study, minimal capillary length was found to be 596.75 ± 15.19 μm ($\pm$ SE), which represented approximately 20% of the capillary set length. Mean capillary pathlength was 638.62 ± 6.80 μm ($\pm$ SE), which was only 7-8% greater than the minimal capillary length. This would indicate that the majority of paths closely followed the shortest path from arteriole to venule without much aberration. The number of segments per path was $7.6 \pm .15$, which was 20-25% of the total number of segments per net; this value directly corresponded with the 20% proportion of the minimal capillary length to the capillary set length. Capillary nets stained predominately for alkaline phosphatase: approximately 70-75% of the nets were blue. This would explain the greater number of arteriolar versus venular segments per net (19.85 ± 1.08 vs. $9.08 \pm .67$, respectively). Of particular interest were the mixed segments, where the transition from blue to red was noted. Although relatively few in number ($3.18 \pm .24$ per net), their length was 2-3 fold the length of the strictly arteriolar or venular segments. The reason for these very long mixed segment lengths is not clear at the present time.

Similar experiments have been conducted for hypertrophic rat myocardium. Hypertrophy was produced in 5 day old rats by constriction of the abdominal aorta. Preliminary data indicate that minimal capillary length is not significantly different between sham-operated control and hypertrophic groups. Arteriolar as well as venular segment length increased in hypertrophy. Most interestingly, average inter-capillary distance, which decreased from the arteriolar to the venular side in control hearts, did not show the same trend. Average inter-capillary distance was significantly longer than in sham-operated controls, on the order of 27 μm, however this was the same on the arteriolar and venular side of individual capillaries in hypertrophic myocardium. This would suggest some sort of mal-adaptation in this model of pressure overload cardiac hypertrophy at the capillary level.

Employing a staining technique that distinguished the arteriolar and venular portions of capillaries, it was possible to study capillary geometry in selected longitudinal sections. The results may be interpreted as an adaptation of the capillary architecture to the lower PO2 values present in the capillaries close to venules. This adaptation appears not to be preserved in cardiac hypertrophy.

Acknowledgement: Supported by the Medical Research Council of Canada. The authors wish to thank Jimmy Gao and Ching Kuo for their expert technical assistance.

REFERENCES

Bassingthwaighte, J.B., Yipintsoi, T., and Harvey, R.B., 1974, Microvasculature of the dog left ventricular myocardium. Microvasc. Res. 7:229-249.

Batra, S., Rakusan, K., and Kuo, C., 1989, Spatial distribution of coronary capillaries: A-V segment staggering. in: Oxygen Transport to Tissue-XI. Eds K. Rakusan, G.P. Biro, T.K. Goldstick, Z. Turek, Plenum Press, New York and London, pp 241-247.

Honig, C.R., Feldstein, L., and Frierson, J.L., 1977, Capillary lengths, anastomoses, and estimated capillary transit times in skeletal muscle. Am. J. Physiol. 233(1):H122-129.

Lojda, Z., 1979, Studies on Dipeptidyl(Amino)Peptidase IV (Glycyl-Proline Naphthylamidase. Histochem. 59:153-166.

Ludwig, G., 1971, Capillary pattern of the myocardium. in: Meth. Achievm. exp. Path., vol. 5. Eds E. Bajusz, and G. Jasmin. Karger, Basel, pp 238-271.

Potter, R.F., Houghton, S., and Groom, A.C., 1989, Capillary lengths and anastomoses in rat hindlimb muscles, studied by aquablack perfusion during rest versus exercise. in: Oxygen Transport to Tissue-XI. Eds K. Rakusan, G.P. Biro, T.K. Goldstick, Z. Turek, Plenum Press, New York and London, pp 313-322.

Skalak, T.C., and Schmid-Schönbein, G.W., 1986, The microvasculature in skeletal muscle. IV. A model of the capillary network. Microvasc. Res. 32:333-347.

Toborg, M., 1972, The Microcirculatory bed in the myocardium of the rat and the cat. Z. Zellforsch. 123:369-394.

FINE STRUCTURE OF CAPILLARY PROLIFERATION IN MYOCARDIUM OF

VOLUME OVERLOADED RATS

Koichi Kawamura, Kohei Tohda, Mikio Kobayashi, Hirotake Masuda
and Takeshi Shozawa

2nd Department of Pathology
Akita University School of Medicine
1-1-1 Hondo, Akita 010, Japan

INTRODUCTION

Ultrastructural changes of the capillary in the myocardium of the volume over-loaded hearts are not well understood. Recently we have shown that endothelial cell proliferation is observed in the flow loaded canine and rat carotid arteries (Masuda et al., 1985; Tohda et al., 1988). We suppose that the endothelial cell will respond to the flow change in the capillaries of the myocardium of the volume-loaded hearts, because flow is expected to be increased in them. In this report, we describe ultrastructural changes of the capillaries of the left ventricle myocardium of volume-loaded rat hearts, produced by arterio-venous shunt (A-V shunt) between the common carotid artery and the external jugular vein.

MATERIALS AND METHODS

Thirty male Sprague-Dawley rats (8 weeks olds, 288 ± 10 g) were used. Fifteen were anesthetized with sodium pentobarbital (50 mg/kg) intraperitoneally. After the left common carotid artery and the ipsilateral external jugular vein were exposed, a side-to-side anastomosis was constructed at about 2 cm from the aortic arch. The detailed operative procedures were the same as described in our previous report (Tohda et al., 1988). The animals were kept for 1, 2 and 4 weeks (5 animals each). Five age-matched control animals were used in each group. Blood flow rate of the right and left common carotid artery was measured before and immediately after anastomosis, and at the sacrifice for the blood flow index, which was expressed in ml/min/kg. Normotensive perfusion fixation of the heart was carried out. A cannula was introduced into the abdominal aorta and the heart was washed out with 75 ml of oxygen saturated heparinized lactate Ringer's solution. Then 50 ml of 3% glutaraldehyde solution in 0.1 M cacodylate buffer (pH 7.4) was perfused under 100 mmHg at room temperature.

The weight and the volume of the left ventricle were measured. A transverse slice of left ventricle was cut at the level of the upper one third of the left ventricle. For the electron microscopic study, exact transverse and longitudinal specimens were made from the slice. They were post-fixed with 1% osmic acid, dehydrated through alcohols and embedded in Epon 812 resin.

Two different characteristic areas were trimmed from the middle layer of the myocardium. One was the area where the muscle fibers were precisely cut transversely and the other was the area where the muscle fibers were precisely cut longitudinally. The super-wide

Oxygen Transport to Tissue XII, Edited by J. Piiper et al.
Plenum Press, New York, 1990

387

meshes (grid size: 0.5 x 0.67 mm) were used, because more than 500 capillaries were necessary to be observed under more than x 6,000 magnification. The ultrathin sections were observed with LEM 2000 transmission electron microscope (TEM). A wide field of the grid was continuously exposed with panorama-photo-system (LM-PVP). Then the composite panorama photograph was made with 24 TEM photographs (6 x 4 sheets, x 1600 magnification). The number of the endothelial cells per capillary was directly counted under TEM in the area where the muscle fibers were precisely cut transversely. In each animal, we counted the number of endothelial cells of more than 500 round or oval capillaries which were cut almost transversely. This counting was carried out in 3 shunted animals (one of 1 week, one of 2 weeks and one of 4 weeks) and 3 age-matched controls.

RESULTS

Blood flow rate of the left common carotid artery was significantly increased immediately after operation (22.2 ± 4.9 ml/min) as compared with that of before operation (3.2 ± 1.1 ml/min) ($p < 0.001$). It was slightly increased throughout the experiment (1 week: 19.6 ± 3.9 ml/min, 2 weeks: 22.8 ± 3.8 ml/min, 4 weeks: 26.6 ± 5.5 ml/min) and was significantly larger than the age-matched controls ($p < 0.001$). And blood flow rate of the right common carotid artery was slightly decreased from 3.5 ± 1.2 ml/min to 3.0 ± 0.9 ml/min. Blood flow index (ml/min/kg) of both carotid arteries was increased from 23.6 ± 7.3 ml/min/kg before operation to 90.1 ± 17.2 ml/min/kg. This high level of blood flow index was kept throughout the experiment and was significantly larger than the age-matched controls ($p < 0.001$).

The left ventricular cavity of the animals with A-V shunt was 0.442 ± 0.043 ml in 2 weeks and 0.527 ± 0.052 ml in 4 weeks, which were 1.37 times ($p < 0.001$) and 1.54 times ($p < 0.001$) larger than the age-matched controls, respectively. The left ventricular weight of the animals with A-V shunt was 0.918 ± 0.045 g in 2 weeks and 1.043 ± 0.039 g in 4 weeks, which were 1.18 times ($p < 0.001$) and 1.24 times ($p < 0.001$) larger than the age-matched controls, respectively.

The percentage of capillaries with 1, 2, 3, 4 or 5-7 endothelial cells of each group was summarized in Fig. 1. In 2 weeks shunted animals, incidences of over 3 endothelial cells per capillary were definitely increased. In 4 weeks shunted animals, they were almost same as the age-matched control. They were almost constant in 1 week shunted and control animals. The average number of endothelial cells per capillary (2.436) was definitely increased in 2 weeks shunted animals as compared with those of other groups and the age-matched control.

In 1-week shunted animals, endothelial cells were mostly thin, but their nuclei were relatively enlarged and clearly protruded into the lumen. Roundly bulged cytoplasmic protrusions with increased number of pinocyotic vesicles (Fig. 2) and overlapping of two endothelial cells were observed frequently (Fig. 3). Thin cytoplasmic projections were observed adjacent to the intercellular junctions. Basement membrane was mostly thin. In 2 weeks, endothelial cells were distinctly thick with rich cytoplasmic organelles. Overlapping of two endothelial cells and thin cytoplasmic projections adjacent to the intercellular junctions were still observed (Fig. 4). Microvilli were abundant in some endothelial cells. Roundly bulged cytoplasmic protrusions were observed frequently (Fig. 5). Intercellular junctions were slightly elongated and basement membrane was moderately thick. In 4 weeks, endothelial cells were thick with rich cytoplasmic organelles. Intercellular junctions were clearly elongated with distinct tight junctions (Fig. 6). The basement membrane was more thick than that of 2 weeks. The bulged endothelial cells were not so frequently observed. In a case of 4 weeks, there was an intracellular capillary which runs through intracellular sarcolemmal tunnel (Kawamura et al., 1976) (Fig. 7). In controls, a few capillaries showed same kinds of endothelial changes as shown in 2 weeks, but they were mild.

388

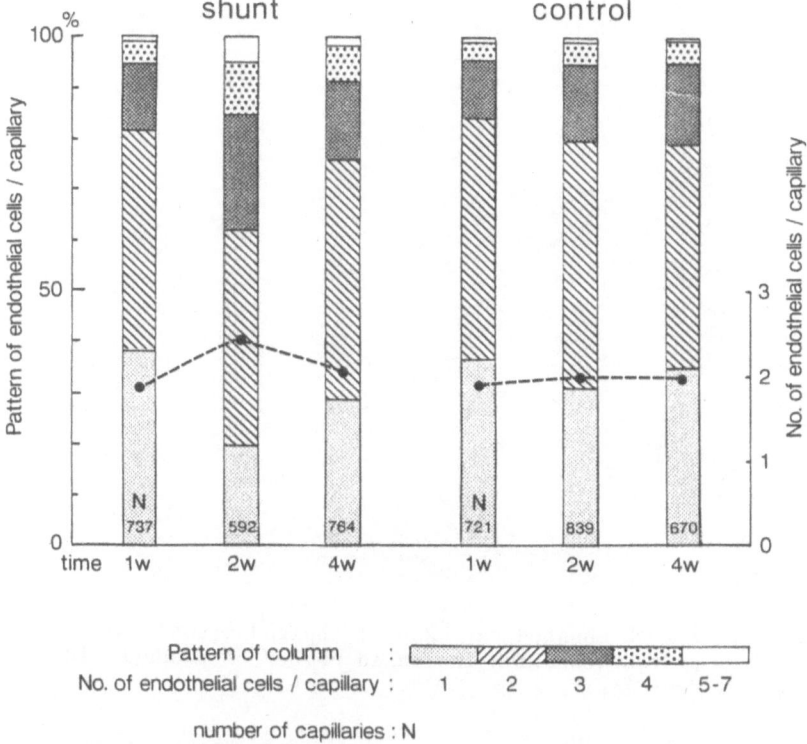

Fig. 1. The percentage of capillaries with 1, 2, 3, 4 or 5-7 endothelial cells and the number of endothelial cells per capillary.

DISCUSSION

Cardiac hypertrophy induced by volume overload (volume-loaded hypertrophic heart) is made by A-V fistula (A-V shunt) (Dart and Holloszy, 1969; Masuda et al., 1985), swimming (Ljungqvist et al., 1976; Carlsson et al., 1978), motor-driven treadmill (Mattfeldt et al., 1986) or dypyridamole administration (Tornling et al., 1978; Tornling, 1982; Mattfeldt and Mall, 1983). Some of these reports showed neoformation of myocardial capillaries (Ljungqvist et al., 1976; Carlsson et al., 1978; Tornling et al., 1978; Tornling, 1982; Mattfeldt and Mall, 1983; Mattfeldt et al; 1986). They observed 3H-thymidine incorporation into the endothelial nuclei (Ljungqvist et al., 1976; Carlsson et al., 1978; Tornling et al., 1978) and counted endothelial nuclei directly (Mattfeldt and Mall, 1983) or the number of capillaries per muscle fiber with light microscopy (Tornling, 1982; Mattfeldt et al., 1986). However, there were no precise ultrastructural studies about intramyocardial proliferation of capillaries. In our present report, we studied the fine structure of intramyocardial capillary changes in volume over-loaded cardiac hypertrophy of rats using A-V shunt method.

In our experiment, volume overload of the heart was constantly induced, because blood flow index was increased throughout the experiment and the volume and weight of the left ventricle were significantly increased in 2 and 4 weeks as compared with the age-matched controls. Marked thickening and bulging of the endothelial cells with increasing of organelles, enlargement of the nucleus and increasing of microvilli on the luminal surface were mostly prominent in 2 weeks. These findings seemed to be characteristic in proliferating phase of the endothelial cells. In 4 weeks, these changes were not so distinct as in 2 weeks. Morphometric

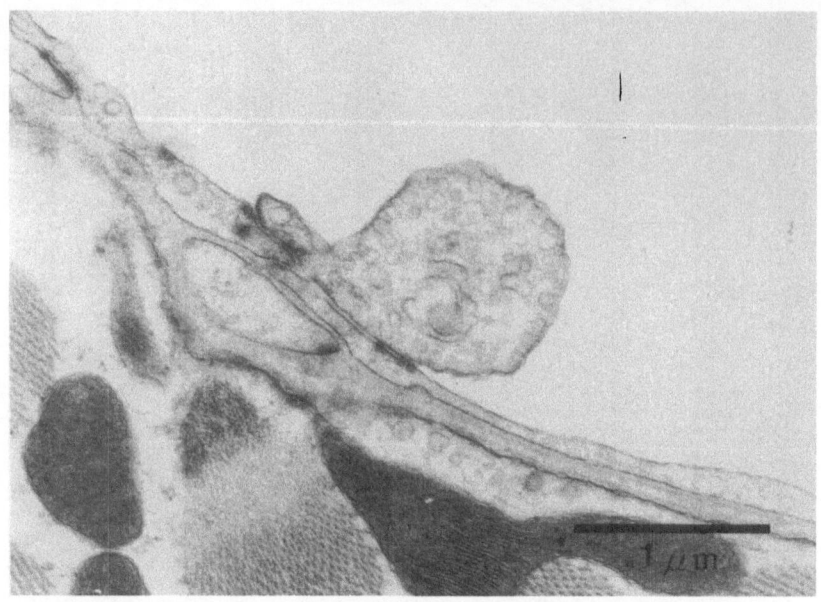

Fig. 2. 1 week shunted rat. Roundly bulged cytoplasmic protrusions with increased number of pinocytotic vesicles.

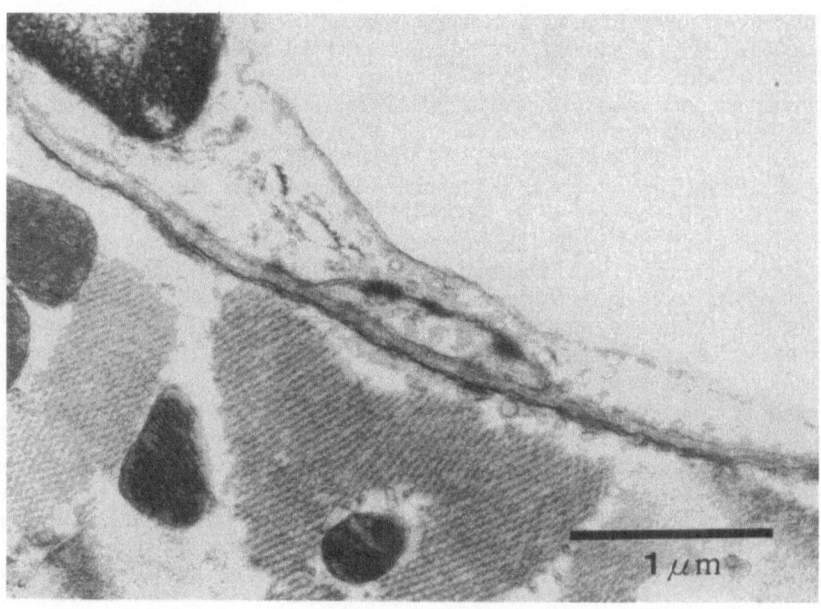

Fig. 3. 1 week shunted rat. Overlapping of two endothelial cells.

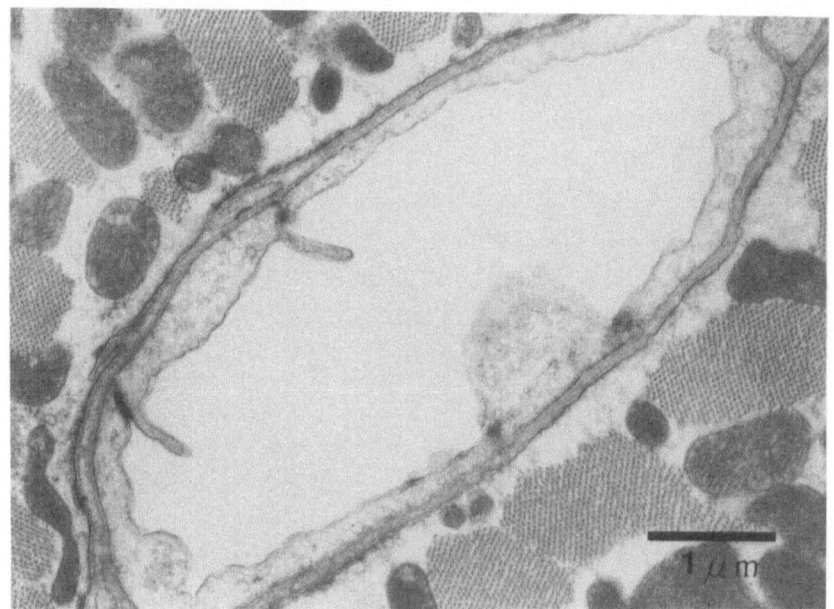

Fig. 4. 2 weeks shunted rat. Thin cytoplasmic projection adjacent to the intercelluar junction and bulged cytoplasmic protrusions.

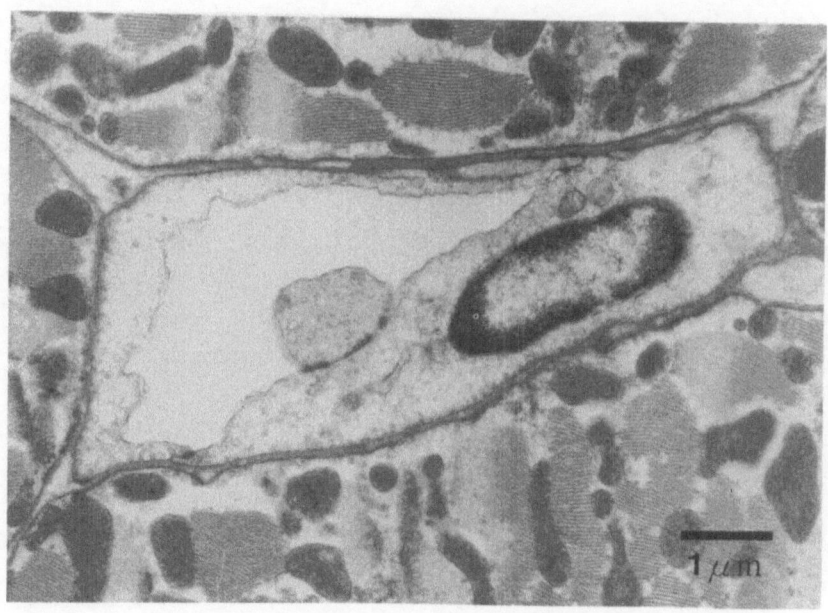

Fig. 5. 2 weeks shunted rat. Roundly bulged cytoplasmic protrusions.

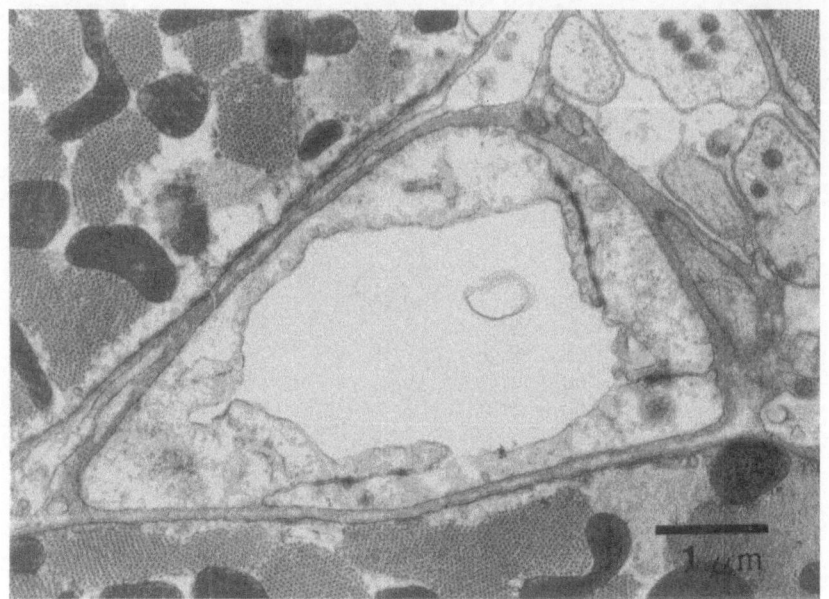

Fig. 6. 4 weeks shunted rat. Elongated intercellular junctions with distinct tight junctions and thickened endothelial cytoplasm.

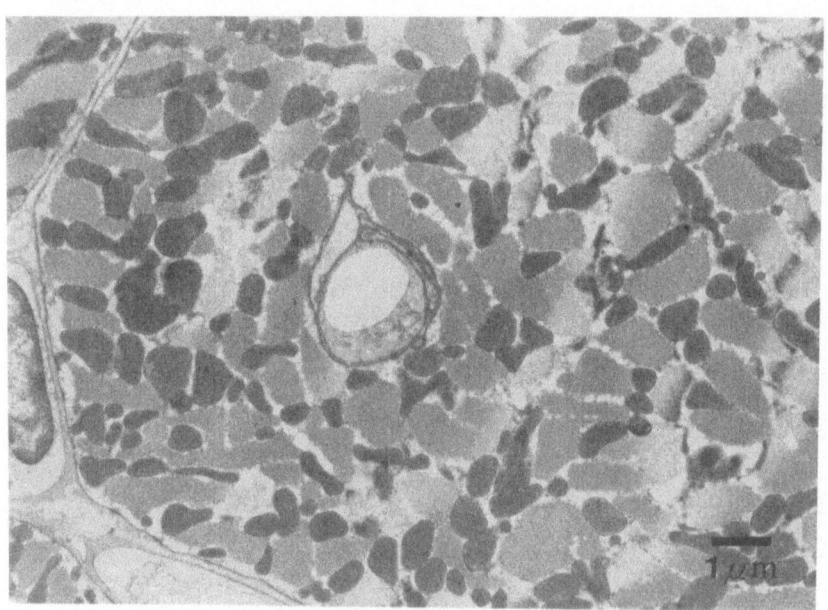

Fig. 7. 4 weeks shunted rat. Intracellular capillary which runs through intracellular sarcolemmal tunnel.

data showed that the number of endothelial cells per capillary were clearly increased in 2 weeks and almost normalized in 4 weeks. In 4 weeks, however, the total number of endothelial cells of the heart might be almost same as in 2 weeks, because the weight of the left ventricle increased.

From these finding, endothelial cells of intramyocardial capillaries are considered to be activated as early as 1 week and proliferated in around 2 weeks after volume overload as reported by other authors (Ljungqvist et al., 1976; Carlsson et al., 1978; Tornling et al., 1978). Kamiya and Togawa (1980) showed adaptive regulation of wall shear stress to flow-volume changes in the canine carotid arteries. Masuda et al. (1985) have shown endothelial proliferation in the arterial wall if the artery is loaded by increased blood flow. They thought that wall shear stress stimulates endothelial cells. Hemodynamic of intramyocardial capillaries are much different from the common carotid artery and coronary blood flow increases with increase of heart work. Thus, in this experiment, an increase of coronary blood flow might be one of the most important causes of the endothelial proliferation and regulated by dilatation of the capillary lumen.

Further studies of total length and luminal volume of intramyocardial capillary, and direct measurement of intramyocardial capillary flow are necessary to clear mechanisms of endothelial proliferation of intramyocardial capillary by volume hypertrophy.

SUMMARY

We studied the fine structure of capillary endothelial cells of volume over-loaded cardiac hypertrophy in rats using A-V shunt between the left common carotid artery and the left external jugular vein. The duration of experiment was 1, 2 and 4 weeks. Volume overload of the heart was constantly induced, because blood flow index (ml/min/kg) of both carotid arteries was increased throughout the experiment and the volume and the weight of the left ventricle were significantly increased in 2 and 4 weeks. After the making of the large panorama photograph composed of 24 TEM photographs, the number of the endothelial cells of each capillary was directly counted under TEM. In 2-week shunted animals, incidences of over 3 endothelial cells per capillary were definitely increased. Also bulging of the endothelial cells, overlapping of two endothelial cells and increasing of microvilli on the luminal surface were mostly prominent in 2 weeks. Endothelial cells of intramyocardial capillaries are considered to be activated as early as 1 week and proliferated in around 2 weeks after volume-overload. An increase of coronary blood flow might be one of the most important causes of the endothelial proliferation and regulated by dilatation of the capillary lumen.

REFERENCES

Carlsson, S., A. Ljungqvist, G. Tornling and G. Unge, 1978, The myocardial capillary vasculature in repeated physical exercise, Acta Path. Microbiol. Scand. Sect. A., 86: 117-119.
Dart, C.H. and J.O. Holloszy, 1969, Hypertrophied non-failing rat heart - partial biochemical characterization, Circ. Res., 25: 245-253.
Kamiya, A. and T. Togawa, 1980, Adaptive regulation of wall shear stress to flow change in the canine carotid artery, Am. J. Physiol., 239: H14-21.
Kawamura, K., C. Kashii and K. Imamura, 1976, Ultrastructual changes in hypertrophied myocardium of spontaneously hypertensive rats, Jap. Circ. J., 40: 1119-1145.
Ljungqvist, A., G. Unge and S. Carlsson, 1976, The myocardial capillary vasculature in exercising animals with increased cardiac pressure load, Acta Path. Microbiol. Scand. Sect. A, 84: 244-246.
Masuda, H., T. Shozawa, S. Hosoda, M. Kanda and A. Kamiya, 1985, Cytoplasmic microfilaments in endothelial cells of flow loaded canine carotid arteries, Heart Vessels, 1: 65-69.
Mattfeldt, T. and G. Mall, 1983, Dipyridamole-induced capillary endothelial cell proliferation in the rat heart - a morphometric investigation, Cardiovasc. Res., 17: 229-237.

Mattfeldt, T., K.L. Kramer, R. Zeitz and G. Mall, 1986, Stereology of myocardial hypertrophy induced by physical exercise, Virchows Arch. (Pathol. Anat.), 409: 473-484.

Tohda, K., H. Masuda, K. Kawamura and T. Shozawa, 1988, Ultrastructural changes in the endothelial surface of rat carotid artery induced by blood flow load - a scanning electron microscopical study, in "Role of blood flow in atherogenesis", Y. Yoshida et al., eds., Springer-Verlag, Tokyo, 171-177.

Tomling, G., G. Unge, L. Skoog, A. Ljungqvist, S. Carlsson and J. Adolfsson, 1978, Proliferative activity of myocardial capillary wall cells in dipyridamole-treated rats, Cardiovasc. Res., 12: 692-695.

Tomling, G., 1982, Capillary neoformation in the heart of dipyridamole-treated rats, Acta Path. Microbiol. Immunol. Scand. Sect. A., 90: 269-271.

EFFECT OF TACHYCARDIA ON INTRACELLULAR PO$_2$ AND RESERVES OF O$_2$ TRANSPORT IN SUBENDOCARDIUM OF MOUSE LEFT VENTRICLE

C.R. Honig and T.E.J. Gayeski

The University of Rochester, School of Medicine and Dentistry
Rochester, NY 14642, USA

The purpose of this study is to evaluate reserves of O$_2$ transport in myocardium using tachycardia as a stimulus. The O$_2$ saturations of hemoglobin (Hb) in microvessels and myoglobin (Mb) in individual myocytes are the measured responses. The latter is of particular interest in that it represents the net balance of all determinants of O$_2$ supply and demand.

Coronary blood flow occurs mainly in diastole. It therefore depends on the ratio of diastolic time to systolic time, as well as on vasomotor tone. Since tachycardia decreases diastolic time more than systolic time, and also increases O$_2$ demand (Honig, 1988), it severely stresses reserves of convective and diffusive transport. The subendocardium is stressed more than subepicardium because dependence on diastolic flow increases progressively with depth beneath the pericardium (Buckberg et al., 1972). Moreover, O$_2$ extraction and myocardial O$_2$ consumption ($\dot{V}O_2$) are greater in subendocardium, even at normal heart rates (Weiss et al., 1978). Nevertheless, tachycardia is an essential, well tolerated response to the exigencies of life. Results to be described indicate that tolerance of tachycardia depends not only on the flow reserve but also on a large reserve of diffusive transport between blood and mitochondria.

METHODS

Choice of species

Cryospectroscopic measurements of Mb saturation depend on trapping the O$_2$ distribution on a subcellular scale. Observations on dog and rat hearts indicate that only the outermost mm of tissue can be frozen rapidly enough to accomplish this (Gayeski and Honig, 1989). To permit measurements in subendocardium we turned to mouse left ventricle, which is less than one mm thick near the apex and weighs ˜ 130 mg. Mean arterial pressure and the ratio of ventricular radius to wall thickness are about the same as in large species. Consequently, systolic wall stress, the principal determinant of O$_2$ demand, should also be about the same. Recent models of cardiac mechanics that take account of residual strains in diastole, and the non-isotropic nature of cardiac tissue, predict that the transmural distribution of wall stress is remarkably

Oxygen Transport to Tissue XII, Edited by J. Piiper *et al.*
Plenum Press, New York, 1990

uniform and scaled to wall thickness (Chuong and Fung, 1986; Omens, 1989; Waldman et al., 1985). The mouse left ventricle should therefore furnish information of general relevance.

Preparative Procedures

Dog hearts were exposed as described previously (Gayeski and Honig, 1989). Cardiac output was determined in duplicate by dilution of indocyanine green. Immediately thereafter hearts were frozen in situ by applying a copper heat sink cooled in liquid N_2 to the left ventricular epicardium.

Mice weighing ~ 40 g were anesthetized with intraperitoneal pentobarbital 6 mg/100g. body wt. The abdominal aorta and inferior vena cava were catheterized. Heart rate, mean and phasic arterial pressure and the electrocardiogram were monitored. Arterial pH was maintained within normal limits with intravenous HCO_3^-. Because of the high glycolytic capacity of mouse muscle, severe lactate acidosis and pump failure ensued if HCO_3^- was not given. Blood lost in sampling and preparative procedures was replaced with blood from donor mice collected the same day and adjusted to 45% hematocrit and pH7.4. Body fluid volume was sustained with 5% glucose solution administered by clysis. The heart was exposed through a midline thoracotomy; VA and FIO_2 were adjusted to insure at least 90% saturation of Hb. The heart was covered with saran^R, and the temperature of the ventricular surface was adjusted to 37°C.

Hearts were frozen in situ with a Wollenberger clamp cooled to -196°C. Specimens were cut under liquid N_2 transverse to the long axis of the ventricle at a point 3-4 mm from the apex. The apical tissue blocks were mounted on a cold stage regulated at -110°C.

Microscopy

The cut surface of the heart was viewed in reflection with a Leitz MPV I microscope system. The specimen was scanned at 250x to locate the lumen, and to measure wall thickness at sites chosen for saturation determinations. A reticle in the microscope oculars (4.3 μm/minor division) was used for this purpose. The left ventricular lumen could be traced even where it contained little or no blood. Measuring sites were chosen at fixed distances from the epicardium (100 μm, 200 μm etc.) until the endocardium was reached. Each cell chosen was at least 50 μm from any other. All venules larger than 16 μm in diameter were sampled.

Positioning the measuring diaphragm was done at 600x with a monocular eyepiece equipped with a second reticle (0.63 μm/minor division). The specimen, the image of the measuring diaphragm of the photometer, and the reticle could all be visualized simultaneously. This allowed exact placement of the measuring diaphragm within the image of an individual vessel or myocyte. Light did not traverse any other structure to reach the one chosen.

Spatial Resolution and Optical Catchment Volume

Optical theory predicts that the catchment volume of the photometer extends 1½ μm beyond the perimeter of the measuring diaphragm and 1½ μm beneath surface of the frozen sample (Gayeski, 1981). A 6 x 6 μm diaphragm was used for vessels; a 1½ x 1½ μm measuring diaphragm was used for myocytes. Mouse cardiac myocytes are 8-10 μm in diameter. A 1½ x 1½ μm measuring diaphragm therefore should collect light from about half the volume of one mouse myocyte. Position of the measuring diaphragm was checked for specimen drift after each scan.

Six empirical tests of spatial resolution in myocardium are described in (Gayeski and Honig, 1989). As a further test, a 30 μm venule in dog gracilis muscle was found with a 45 μm ice crystal directly abutting it. (This ice crystal on the cut surface of the sample was formed from condensed water vapor. The ice crystals intrinsic to the sample average 0.75 μm.) The intensity of light reflected from the 45 μm crystal was 2 orders of magnitude greater than from the venule. Nevertheless, light intensity, saturation, and a ratio of isosbestic wavelengths (RISO) were same when a 4 x 6 μm measuring diaphragm was centered in the cross-sectional profile of the venule, and when the leading edge of the measuring diaphragm was 3 μm from the ice crystal. When the measuring diaphragm straddled the venule and the ice crystal, light intensity and RISO were comparable to values obtained when the measuring diaphragm was positioned entirely on the ice crystal. We again conclude that the volume from which light is collected is smaller than the diameter of a cardiac myocyte.

Determination of Saturations and O_2 Tensions

Mb saturation was determined with a 4-wavelength method based on an adaptation of Beer's law that takes account of light scattering in the intrinsic ice crystals of the sample (Gayeski, 1981). The principal determinant of light scattering in our experiments is the size of these ice crystals. Crystal size depends mainly on the rate of freezing, which decreases with depth beneath the epicardium. Since RISO increases with light scattering, it furnishes a check on the adequacy of freezing, and on the over-all performance of the system. PO_2 in equilibrium with Mb ($PmbO_2$) was calculated from Mb saturation using a Mb P50 of 5.3 torr at 37°C (Gayeski, 1981). $PmbO_2$ can be interpreted at the PO_2 to which the terminal oxidase is exposed because the ΔPO_2 between the cytosol and the surface of a mitochondrion is < 0.05 torr at maximum VO_2 (Clark et al., 1987). Hb saturation was determined with the same method used for Mb but with wavelengths appropriate for Hb (Degner and Gayeski, 1987). The measured pHa and the oxydissociation curve of mouse blood (Gray and Steadman, 1964) were used to approximate PvO_2 from SvO_2.

RESULTS AND DISCUSSION

Characteristics of Animal Model

Heart rate in 5 mice anesthetized but otherwise undisturbed averaged 420/min. After thoracotomy and immediately prior to cardiectomy heart rate in these 5 animals ranged from 600-670/min (43-60% greater than normal). Thus all hearts were stressed by tachycardia incident to the preparative procedures. Mean arterial pressure ranged from 75-118, hematocrit from 37-43%, PaO_2 from 119-242 torr, and pHa from 7.32-7.43.

Myoglobin Saturation and Intracellular PO_2

Mb saturation and $PmbO_2$ are shown as functions of depth beneath the epicardium in Figure 1. Panel A is a summary of data for 5 mice which experienced no stresses other than those attributable to preparative procedures. These will be designated the reference group. Panel B shows results for each myocyte in one of the animals in the reference group. Mean Mb saturation was 55%, not statistically different from values obtained by Mb cryospectroscopy in dog, cat, rabbit, ferret and rat

(Gayeski and Honig, 1989). Almost the same Mb saturation was obtained by blood-tissue partition of CO in dogs (Coburn et al., 1973). Uniformity across species demonstrates that convective and diffusive transport are scaled to hold Mb saturation near its P_{50} over a wide range of VO_2. It is probably no coincidence that the transport functions of Mb are optimal near the P_{50}.

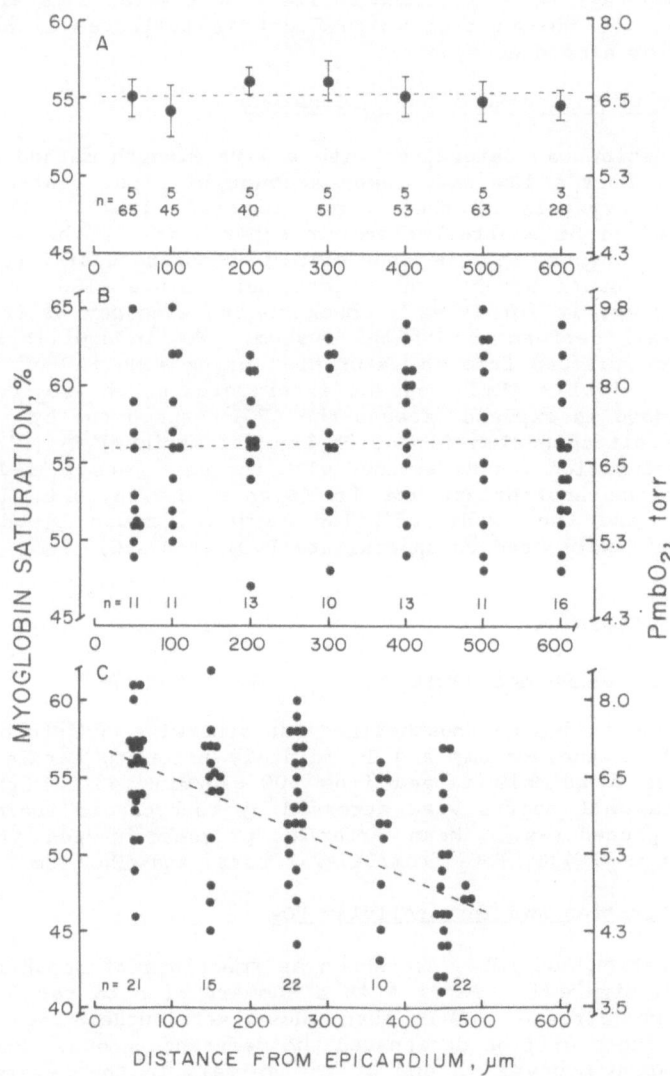

Figure 1A. Summary of data in reference group; n = number of mice and number of myocytes. B: Data for one mouse; each data point represents one myocyte. C: Data for one mouse frozen at maximum heart rate.

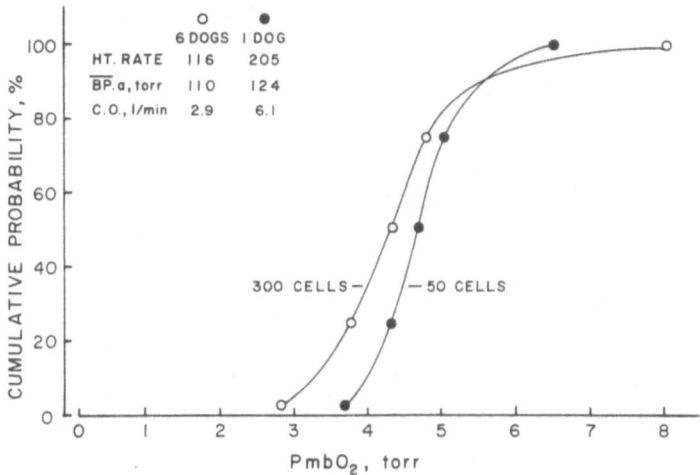

Figure 2. Influence of near-maximum heart rate on O_2 distribution in dog subepicardium.

Mean Mb saturation and its variance were uniform across the mouse left ventricle despite moderately elevated heart rate. The minimum $PmbO_2$ found among the 345 myocytes sampled in 5 hearts was 4.2 torr, or about 10x time minimum PO_2 required to support maximum cytochrome turnover (Gayeski, Connett and Honig, 1987).

Effect of Maximum Heart Rate

Subepicardium. For purposes of this report subepi and subendocardium are defined as tissue less than 25% and more than 75% of wall thickness, respectively. Spontaneous supraventricular tachycardia was observed just prior to freezing in an otherwise representative dog. Figure 2 compares the probability distribution of subepicardial $PmbO_2$ in this animal with data collected in 6 dog hearts frozen in normal sinus rhythm. Near-maximum heart rate did not decrease the median or lower the minimum PO_2. Near-maximum heart rate also had no effect on subepicardial $PmbO_2$ in a cat heart (Gayeski and Honig, 1989).

Transmural $PmbO_2$. To evaluate the effect of tachycardia on subendocardium we explored a mouse heart which developed spontaneous atrial tachycardia at 860/min. This would appear to be the maximum for a mouse, for we could not drive heart rate above 840/min in closed-chest anesthetized mice given isoproterenol intravenously. At 860/min the median Mb saturation in subepicardium was identical to that in the reference group. Mb saturation and $PmbO_2$ decreased linearly with depth as shown in Figure 1 panel C. The slope of the least-squares regression of Mb saturation on depth is significantly greater than zero. Since no transmural gradient was found in the reference group in which RISO was higher in subendocardium and the ventricular wall was thicker, the transmural gradient in tachycardia cannot be a freezing artifact. Mean Mb saturation in subendocardium was 48.1% vs. an average of 53.5% in the subendocardium in the reference group (P<.01). The transmural gradient in Mb saturation corresponds to a transmural ΔPO_2 of only 1.6 torr, because the oxydissociation curve of Mb acts as a PO_2 buffer (Gayeski and Honig, 1986; 1989). Moreover, the minimum subendocardial $PmbO_2$ found was 3.8 torr vs. 4.1 torr in the reference group. Thus the transmural $PmbO_2$

399

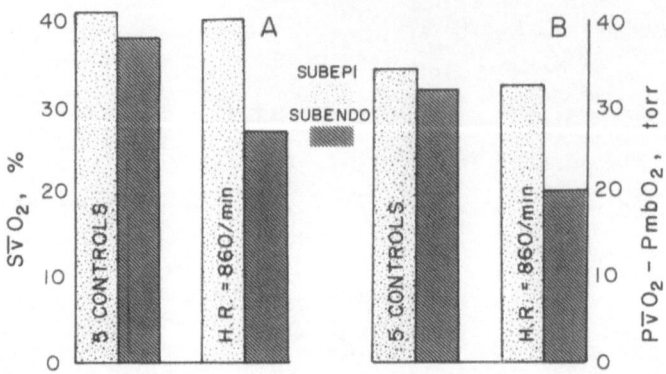

Figure 3. Effect of maximum heart rate on SvO_2 and O_2 extraction (Panel A) and the driving force for blood-tissue O_2 diffusion (Panel B).

gradient at maximal heart rate is of little physiological significance in mouse heart. This might be expected, for systole has little effect on diastolic coronary blood flow (Katz and Feigl, 1988), and a large coronary vasodilator reserve exists in small animals (Vetterlein and Schmidt, 1985). In accord with the above expectation, pacing had no effect on phosphate metabolites when the NMR signal was preferentially weighted by deep myocardium (Balaban and co-workers, 1986). Furthermore, heart rate increases almost linearly with whole-body $\dot{V}O_2$ in normal exercise with no evidence of subendocardial ischemia (Honig, 1988). In contrast, BPa is regulated within close limits. Large, metabolically significant transmural O_2 gradients are found in mouse left ventricle stressed by either hypotension or hypertension (manuscript in preparation).

SvO_2 and O_2 Extraction

Hb saturation in venules (SvO_2) can be interpreted as end-capillary PO_2 because precapillary O_2 losses and arteriovenous diffusive shunting are below limits of detection myocardium (Honig and Gayeski, 1989). Maximum heart rate had no effect on $S\bar{v}O_2$ and O_2 extraction in mouse subepicardium; see stippled bars in Figure 3A. The transmural ΔSvO_2 was not statistically significant in the reference group. At 860/min, however, SvO_2 in subendocardium was below a 95% confidence interval constructed for the reference group. The transmural ΔPvO_2 was 14 torr at maximal heart rate. This ΔPvO_2 is almost 9 times the transmural $\Delta PmbO_2$, suggesting that reserves of diffusive transport in subendocardium contribute to adaptation to tachycardia.

Diffusive Transport from Hb to Mb

The difference between $P\bar{v}O_2$ and $PmbO_2$ is a lower bound on the driving force for O_2 diffusion from blood to myocytes (Federspiel and Popel, 1986; Honig, Gayeski and Frierson, 1989). In the reference group this driving force is about the same throughout the heart. In contrast,

400

the estimated minimum driving force for O_2 diffusion is a third smaller in subendocardium at maximum heart rate; see Figure 3B. Tachycardia should increase transcapillary O_2 flux throughout the heart (Honig, 1988), and but especially in subendocardium (Weiss et al., 1978). Since a larger flux was driven by the same ΔPO_2 in subepicardium and a smaller ΔPO_2 in subendocardium we conclude that <u>tachycardia sharply decreases resistance to diffusive O_2 transport</u>, especially in deeper layers of the wall. Mechanisms that may account for this adaptive behavior include (a) lengthening red cell transit time through recruitment of flow paths, and decreasing O_2 flux density by increasing functional red cell surface area (Federspiel and Popel, 1986; Honig and co-workers, 1989). In addition, greater Mb-facilitated diffusion promotes O_2 extraction as well as diffusive transport (Bailey and Driedzic, 1986; Groebe and Thews, 1986; Wittenberg and Wittenberg, 1989). In the experiment shown in Figure 1C, greater Mb-facilitated transport was limited to subendocardium, since it depends on a fall in $PmbO_2$ (Honig and co-workers, 1989; Kreuzer and Hoofd, 1987; Wittenberg and Wittenberg, 1989). We conclude that reserves of convective and diffusive transport interact to permit uniformly aerobic metabolism over the full range of heart rate and cardiac output.

SUMMARY

Intracellular Po_2 ($PmbO_2$) was determined by cryospectrophotometry in individual cardiac myocytes. The rate of progression of the freezing front was sufficient to trap the O_2 distribution across the wall of the mouse left ventricle. The transmural $PmbO_2$ distribution was uniform despite moderate tachycardia. Maximal heart rate produced a small but statistically significant transmural O_2 gradient but no hypoxic myocytes in subendocardium. Reserves of diffusive as well as convective transport contribute to maintenance of aerobic metabolism during tachycardia.

ACKNOWLEDGEMENTS

The technical skills of J.L. Frierson and W.R. Reaves made development of the mouse heart preparation possible. Our research is supported by Grant HLB03290 from the United States Public Health Service.

REFERENCES

Bailey, J.R., and Driedzic, W.R., 1988, Perfusion-independent oxygen extraction in myoglobin-rich hearts, <u>J. Exp. Biol.</u>, 135:301.

Balaban, R.S., Kantor, H.L., Katz, L.A., and Briggs, R.W., 1986, Relation between work and phosphate metabolites in the in vivo paced mammalian heart, <u>Science</u>, 232:1121.

Buckberg, D.D., Fixler, F.E., Archie, J.P., and Hoffman, J.E., 1972, Experimental subendocardial ischemia in dogs with normal coronary arteries, <u>Circ. Res.</u> 30:67.

Chuong, C.J., and Fung, Y.C., 1986, Residual stress in arteries, <u>in</u>: "<u>Frontiers in Biomechanics</u>," G.W. Schmid-Schoenbein, S.L.-Y. Woo, and B.W. Zweifach, eds., Springer, New York.

Clark, A., Clark, P.A.A., Connett, R.J., Gayeski, T.E.J., and Honig, C.R., How large is the drop in PO_2 between cytosol and mitochondrial? <u>Am. J. Physiol.</u>, 252:C583, 1987.

Coburn, R.F., Ploegmakers, F., and Gondrii, P., and R. Abboud., 1973, Myocardial myoglobin O_2 tension, <u>Am. J. Physiol.</u>, 224:870.

Degner, F., and Gayeski, T. E. J., 1987, A comparison of a four wavelength analysis and multicomponent wavelength analysis applied to determination of hemoglobin saturation, Adv. Exper. Med. Biol., 215:153.

Federspiel, W. J., and Popel, A. S., 1986, A theoretical analysis of the effect of the particulate nature of blood on oxygen release in capillaries, Microvasc. Res., 32:164.

Gayeski, T. E. J., 1981, A Cryogenic Microspectrophotometric Method for Measuring Myoglobin Saturation in Subcellular Volumes; Application to Resting Dog Muscle, (Ph.D. Thesis), Rochester, NY: University of Rochester, Univ. Microfilms, No. DA9224720, Ann Arbor, MI.

Gayeski, T. E. J., and Honig, C.R., 1986, O_2 gradients from sarcolemma to cell interior in a red muscle at maximal VO_2, Am. J. Physiol., 251:789.

Gayeski, T. E. J., and Honig, C. R., 1989, Intracellular PO_2 in individual cardiac myocytes in dog, cat, rabbit, ferret and rat, Am. J. Physiol., in press.

Gayeski, T.E.J., Connett, R.J., and Honig, C.R., 1987, The minimum intracellular PO_2 for maximum cytochrome turnover in red muscle in situ. Am. J. Physiol., 252:H906.

Gray, L.H., and Steadman, J.M. 1964, Determination of the oxyhaemoglobin dissociation curves for mouse and rat blood, 175:161.

Groebe, K., and Thews, G., 1986, Theoretical analysis of oxygen supply to contracted skeletal muscle, Adv. Exper. Med. Biol., 200:495.

Honig, C.R. 1988. Modern Cardiovascular Physiology, 2nd Ed, Little Brown, Boston.

Honig, C.R., and Gayeski, T.E.J., 1989, Precapillary O_2 loss and arteriovenous O_2 diffusion shunt are below limit of detection in myocardium, Adv. Exper. Med. Biol., 247:591

Honig, C.R., Gayeski, T.E.J., and Frierson, J.F., 1989, Anatomic determinant of O_2 flux density at cardiac capillaries, Am. J. Physiol., 256:H375.

Katz, S.A., and Feigl, E.O., 1988, Systole has little effect on diastolic coronary artery blood flow, Circ. Res., 62:443

Kreuzer, F., and Hoofd, 1987, Facilitated diffusion of oxygen and carbon dioxide, in: "Handbook of Physiology, Section 3: The Respiratory System, Volume IV: Gas Exchange, L.E.Farhi, and S.M. Tenney, eds., Am. Physiol. Soc., Bethesda, Maryland.

Omens, J.H., 1988, Left ventricular strain in the no-load state due to the existence of residual stress. (Thesis). University of California, San Diego, LaJolla.

Vetterlein, F., and Schmidt, G., 1985, Dilatory capacity of the coronary system in the anesthetized rat, Basic Res. Cardiol., 80:661.

Waldman, L.K., Fung, Y.C., and Covell, J.W., 1985, Transmural myocardial deformation in the canine left ventricle. Normal in vivo three-dimensional finite strains. Circ. Res., 57:152.

Weiss, H.R., Neubauer, J.A., Lipp, J.A., and Sinha, A.K., 1978, Quantitative determination of regional oxygen consumption in the dog heart, Circ. Res., 42:394.

Wittenberg, B.A., and Wittenberg, J.B., 1989, Transport of oxygen in muscle, Ann. Rev. Physiol., 51:857.

COMPARISON OF TYRODE AND BLOOD PERFUSED WORKING ISOLATED RAT HEARTS

J. Olders, Z. Turek, J. Evers, L. Hoofd, B. Oeseburg, and F. Kreuzer

Dept. Physiology, University of Nijmegen
Nijmegen, The Netherlands

INTRODUCTION

The isolated heart, perfused with an electrolyte solution (usually a modified Krebs-Henseleit or Tyrode buffer), is a widely used experimental model for the study of cardiac performance. The hearts can be perfused in a retrograde (Langendorff heart) or antegrade way (working heart). However, the oxygen supply of these hearts is very unphysiological, with very high P_{O_2} values but a limited oxygen concentration of the perfusion medium. There exists a great deal of disagreement about the adequacy of oxygenation of this preparation, and one may conclude that it is on the brink of hypoxia (see review by Kreuzer, in press; Figulla et al., 1983; Opie, 1984; Robiolio et al., 1989).

Recently, we have presented some results of a study on myocardial function and response to hypoxia of the isolated working rat heart perfused with Tyrode buffer (Olders et al., 1988). We extended and modified the experimental set-up to make it suitable for perfusion with a suspension of erythrocytes in buffer ('blood'). A series of experiments was carried out in order to compare left ventricular function and response to hypoxic hypoxia of the blood perfused vs. the Tyrode perfused hearts.

METHODS

Perfusion Media

In the first series of experiments the hearts were perfused with a modified Tyrode solution consisting of (mmol/l): NaCl (130), KCl (5,6), $CaCl_2$ (2.16), $MgCl_2$ (0.56), $NaHCO_3$ (25), NaH_2PO_4 (1.2) to which glucose (11.1) and pyruvate (5.0) were added as substrates (Snoeckx et al., 1986). For the experiments with blood perfusion the method described by Duvelleroy et al. (1976) was used: The suspending medium was a Krebs-Henseleit buffer, consisting of NaCl (120), KCl (4.7), $CaCl_2$ (2.5), MgSO4 (1.2), KH_2PO_4 (1.2), $NaHCO_3$ (30.0) and glucose (5.5), modified with calcium EDTA (0.5) and pyruvate (2.0). Bovine serum albumin (fraction V) was added to a final concentration of 15 g/l. Pig blood was collected in reservoirs containing heparin (50 mg), filtered (pore-size 45 μ) and centrifuged 3 min at 3,500 rpm. The plasma was discarded and the erythrocytes were washed three times with equal volumes of saline and mixed with the electrolyte solution to hematocrit of appr. 25 %.

Oxygen Transport to Tissue XII, Edited by J. Piiper *et al.*
Plenum Press. New York. 1990

Perfusion System

The perfusion system used for the Tyrode perfusion is based on that described by Snoeckx et al. (1986): The hearts were perfused in a non-recirculating manner and the driving force for the aortic pressure (during both the retrograde and antegrade perfusion) was a peristaltic pump. The experimental set-up for the blood perfusion is a modification of that described by Duvelleroy et al. (1976). It is presented in detail elsewhere (Olders et al., this volume) and only briefly summarized here: The left atrium of the heart was perfused from three identical but separate circuits (atrial line). Rapid changes in the S_{O_2} of the blood could be made by conversion from one circuit to another. The left atrial filling pressure or preload (PL) could be changed by varying the height of an overflow before the heart. The perfusate was pumped back to the reservoir by the left ventricle against a hydrostatic pressure column (aortic line) with a controllable height, determining the aortic pressure (AP) or afterload (AL). Venous coronary blood was collected through a catheter in the pulmonary artery (coronary line).

In both systems the following variables were measured. Aortic flow (AF) and mean aortic pressure (AP) were measured by means of an electromagnetic flow probe (Skalar) and a pressure transducer (Gould) in the aortic line. The P_{O_2} values of the perfusate in the atrial (Pa_{O_2}) and coronary lines (Pv_{O_2}) were monitored with P_{O_2} electrodes. Total coronary flow (CF) was collected and measured by means of a balance. In the experiments with blood the arterial or venous S_{O_2} could also be measured by means of a saturation probe. Before these experiments were started the oxygen dissociation curve (ODC) of the blood was established by calculation of the P50 and Hill's n. During the experiments several blood samples were taken for measurements of P_{O_2}, P_{CO_2}, pH and Hb derivatives (IL 1312 and 482), both in the atrial and the coronary line. The P_{O_2} and S_{O_2} values of the samples were used for the calibration of the P_{O_2} electrodes and S_{O_2} probes.

In both systems, a microcomputer data acquisition and analysis set-up was used for the sampling of the measured variables. It consisted of a data-acquisition device with signal condition units and interface- and AD-charts for analog and digital data sampling. The main sampling and analyzing program was running on an XT-microcomputer (ITT). The sample interval used was 30 s. At each sample point the following variables were calculated: Cardiac output (CO) was calculated as the sum of aortic and coronary flow. The flow ratio (FR) was defined as the ratio of coronary flow to cardiac output. External cardiac work (ECW) was computed as the product of mean aortic pressure, cardiac output and a coefficient (1.33322 x 10^{-4}) which expresses work in J/min. Myocardial oxygen consumption (MV_{O_2}), ml/min, was calculated as the product of coronary flow times the arteriovenous O_2 content difference, ml O_2/ml. Efficiency (EFF) was computed as $ECW/(MV_{O_2}$ x 20.9) %. Calculations could be carried out either on line or off line. For the blood experiments, S_{O_2} values could be calculated from P_{O_2} values (or vice versa) by means of the pH, PCO2 and ODC.

Experimental Protocol

Male rats of Wistar strain were anesthetized with ether. The hearts were quickly removed from the thorax and immersed in ice-chilled buffer solution. Remnant lung tissue and fat were removed. The hearts were attached to the aortic cannula of the perfusion apparatus and retrograde perfusion was started immediately at a pressure of 60 mm Hg, using a well oxygenated hemoglobin-free buffer solution (Tyrode or Krebs-Henseleit).

404

During this period, which lasted for 10 minutes, the left atrial and pulmonary artery were cannulated and two small platinum wires were attached to the right atrium for stimulation. After this period the left atrial cannula was opened and the hearts started to eject at a preload of 15 mm Hg and a mean aortic pressure which was adjusted to 90 mm Hg. Heart rate was kept constant by pacing the hearts with a frequency of 5 Hz.

During the first 30 min of the experiments the left ventricular function was established in 9 steady state situations: the combination of three left atrial filling pressures (preload: 10, 15 and 20 mm Hg) and three aortic pressures (afterload: 60, 90 and 120 mm Hg). During this period the mean oxygen pressure (Pa_{O_2}) of the perfusate was 660 mm Hg for the Tyrode perfused hearts and 100 mm Hg for the blood perfused hearts.

After this period, hypoxic hypoxia was induced at a preload of 15 and an afterload of 90 mm Hg, in the Tyrode perfused (TP) hearts by lowering the P_{O_2} to a mean value of 450 mm Hg, in the blood perfused (BP) hearts by changing the S_{O_2} in a few steps between 95-40 %. At the end of each hypoxic period, which lasted about 10 min, myocardial function was also established at afterload values of 60 and 120 mm Hg, before turning back to 90 mm Hg. After the experiments, the hearts were weighed.

Table 1. Blood parameters. means and standard deviations. P50 corrected for pH 7.4.

	P50	Hill's n	O2 capacity	pHa	pHv	Pa_{CO_2}	Pv_{CO_2}
n=	8	8	8	11	11	11	11
	31.6	2.9	11.7	7.23	7.21	33.2	37.2
	1.6	0.25	1.0	0.07	0.07	3.0	3.0

RESULTS

The weights of the body and of the left ventricle + septum (wet weight) of the rats used for Tyrode perfusion (402 ± 31 g and 0.80 ± 0.08 g, respectively, n = 14) did not differ from those of the rats used for blood perfusion (408 ± 36 g and 0.81 ± 0.11 g, n = 11). Therefore, we did not correct our results for heart weight. We used 8 blood preparations in 11 experiments with blood perfusion. The results of some blood parameters are shown in Table 1. From this Table it appears that the blood can be prepared in a reproducible manner, as is also illustrated by Fig. 1. Mean P50 agreed very well with the value of 30.6 mm Hg reported for pig blood by Bartels and Harms (1959).

Left Ventricular Function

The results of the left ventricular function at the 9 combinations of pre- and afterload levels are shown in Figs. 2, 3 and 4. It appears that the aortic flow in the BP hearts is significantly higher, especially at the higher workload levels (2 a), whereas the coronary flow is always much lower (2 b). Cardiac output of the BP hearts tends to be somewhat higher, but this is only significant in two situations with high preload (2 c). So, in the BP hearts cardiac output is about the same as in the TP hearts with a higher aortic and a lower coronary flow. This is clearly reflected in the flow ratio (coronary flow divided by cardiac output) shown in panel 3 a, where the flow ratios of TP hearts at high afterload levels can reach values up to 50 %. Mean arterial P_{O_2} was 660 mm Hg for the TP hearts and 105 mm Hg for the BP hearts (3 b). Mean venous P_{O_2} was in the range between 120 and 200 mm Hg for the TP hearts and between 40 and 50 mm Hg for the BP hearts (3 c). In panels 4 a and b it can be seen that MV_{O_2} and ECW did not differ in the two series, an exception being the somewhat higher ECW (due to the higher CO) at the highest preload level. Efficiency tends to be somewhat higher in the BP hearts, but this difference is not significant (4 c).

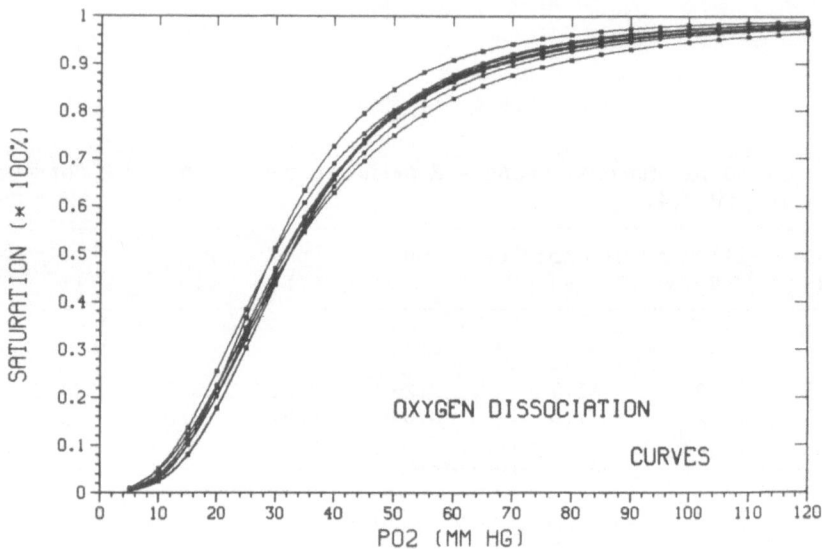

Fig. 1. Computed oxygen dissociation curves (pH = 7.4) of the 8 blood preparations used.

Response to Hypoxia

The response of the hearts to hypoxic hypoxia is shown in Fig. 5. Hypoxia was induced in the TP hearts at a preload of 15 and an afterload of 90 mm Hg by lowering the Pa_{O_2} from a mean value of 660 mm Hg to a mean value of 450 mm Hg. This decreased arterial O_2 content (physically dissolved oxygen) to about 65 % of the initial value of 0.02 ml/ml. It took about 8 min to reach this value and the various heart variables measured and calculated at intervals of 1 min were used (nonsteady-state). In the BP hearts, at the same workload level, the Pa_{O_2} was varied in each experiment in a few steps between 120 and 30 mm Hg (a variation in Sa_{O_2} from 95 to 45 %). Arterial O_2 content therefore decreased from a mean value of 0.14 ml/ml

406

to minimum values of 0.04 ml/ml. Each step lasted about 10 min and variables measured and calculated at the end of the step were used (steady-state). From the figure it can be seen that the TP hearts responded immediately to lowering the Ca_{O_2} by a decrease in aortic flow, while coronary flow remained unchanged, indicating an already maximal dilatation of the vascular bed at the high Pa_{O_2}. Although the Pv_{O_2} decreased (from a mean value of 150 to a mean value of 50 mm Hg), this decrease was not

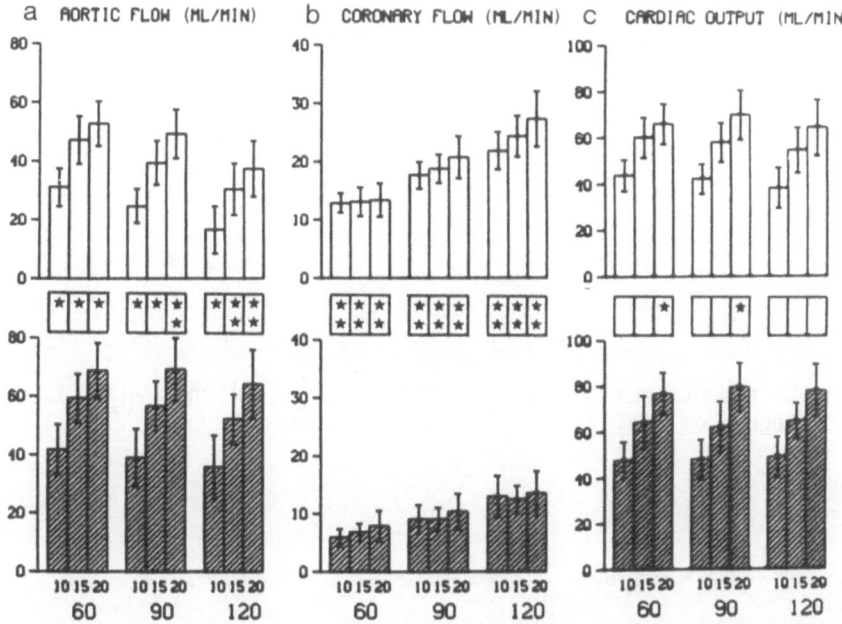

Fig. 2. Comparison of the left ventricular function of the hearts of the two series. The values of the various measured and calculated variables for the 9 pre- and afterload situations are shown in the various panels: AF (left), CF (middle) and CO (right).
PL and AL values are indicated at the bottom of the panel. Values for the TP hearts are represented in the upper part of each panel by the empty columns and error bars (means and standard deviations of 11, 14 and 11 experiments for the preload values of 10, 15 and 20 mm Hg, respectively). The BP hearts are indicated in the lower part of each panel by the hatched bars (preload 10, 15 and 20 mm Hg measured in 7, 10 and 9 experiments respectively). A Wilcoxon 2-sample test was used to compare the results of the two series for each variable in each workload situation. * : p < 0.05; * * : p < 0.01.

sufficient to prevent a decrease in arterio-venous O_2 content difference and thus a decrease in MV_{O_2}. The decrease in cardiac work (not shown in the figure) corresponds with the decrease of aortic flow. The BP hearts responded to a decrease in Ca_{O_2} by both an increase in coronary flow and a decrease in Pv_{O_2} (and therefore Sv_{O_2}). This enabled them to keep MV_{O_2} on a high level, while aortic flow and cardiac work remained stable. Only below Ca_{O_2} values of 60-70 % of the maximum, aortic flow tends to decrease.

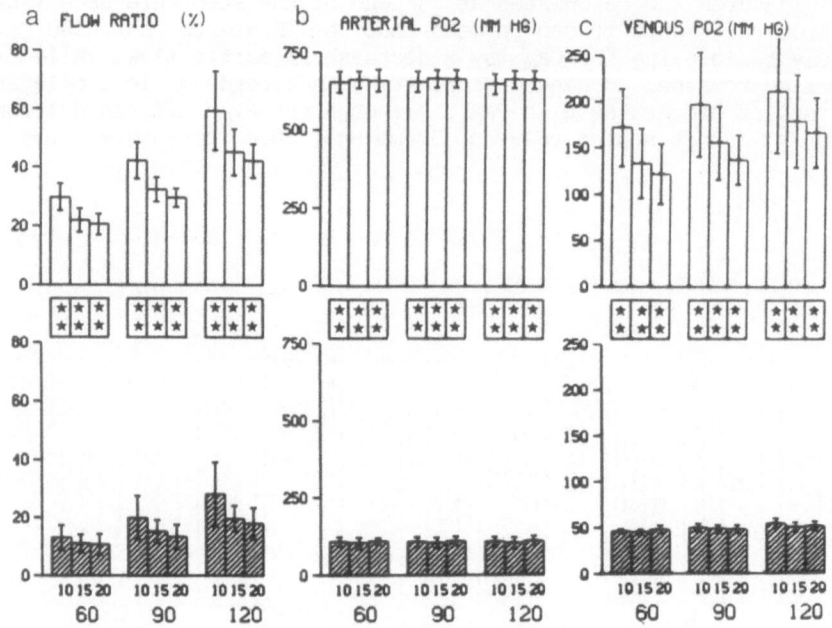

Fig. 3. Left ventricular function: FR (a), Pa_{O_2} (b) and Pv_{O_2} (c). See legend of Fig. 2.

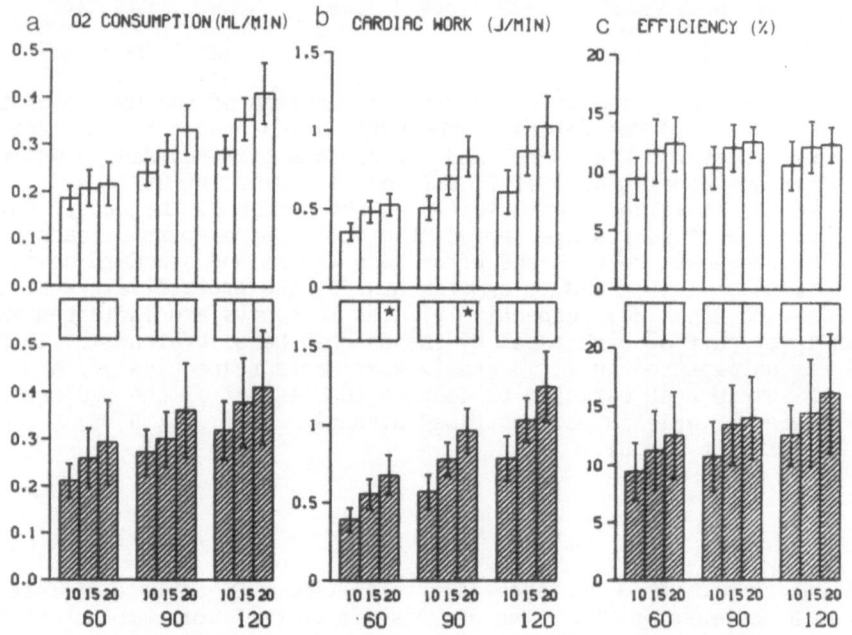

Fig. 4. Left ventricular function: MV_{O_2} (a), ECW (b) and EFF (c). See legend of Fig. 2

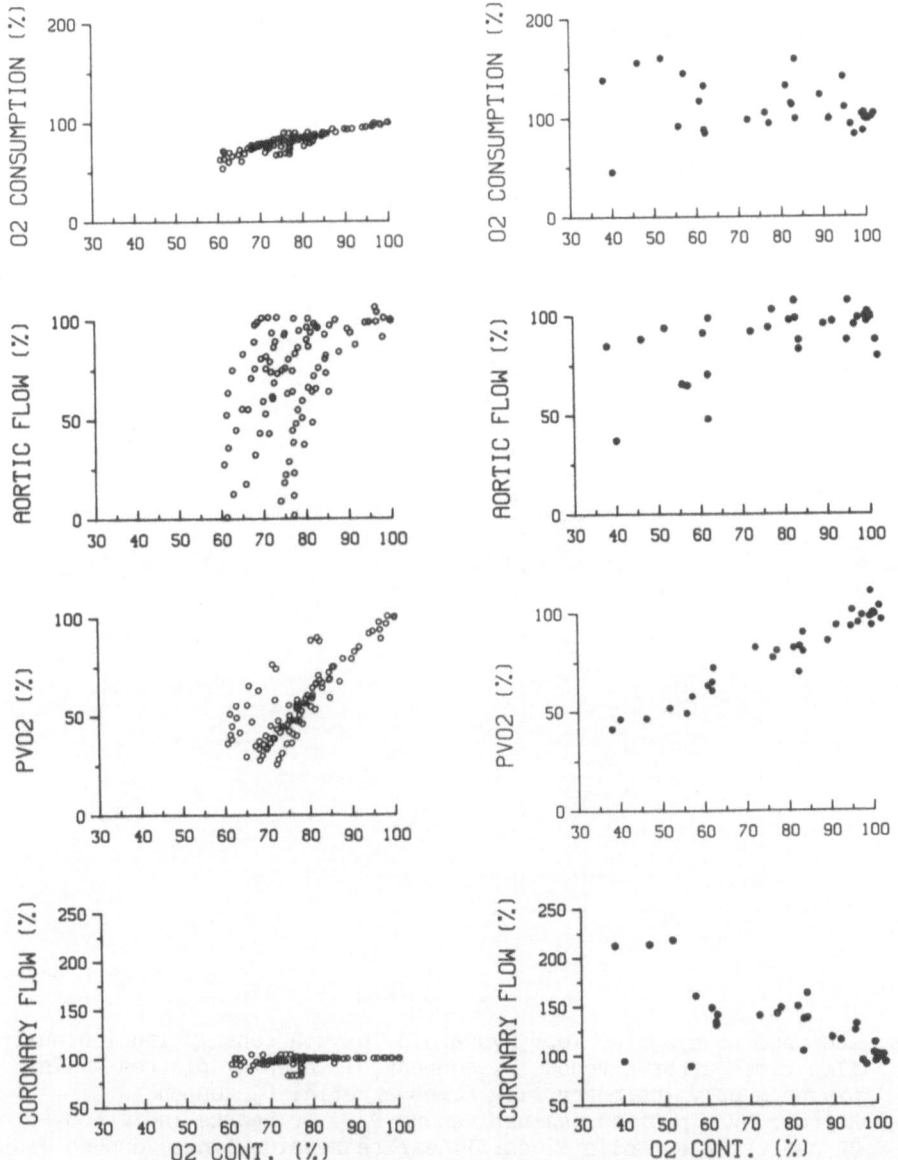

Fig. 5. Comparison of the response to hypoxic hypoxia of the hearts from the two series. The relative changes (% of initial value in the experiment) of the various variables in relation to the relative change in arterial O_2 content (% of initial value, represented at the horizontal axes) are shown in the various panels. The results for the TP hearts (indicated by o, n = 108 in 13 experiments) in the left, the results for the BP hearts (indicated by •, n = 39 in 10 experiments) in the right part of the figure.

The same results are shown in Fig. 6 in a different manner. In panel a it can be seen that the oxygen consumption of the TP hearts decreases linearly with decreasing oxygen supply. This means that oxygen consumption is always limited by oxygen supply in this situation. Whether this is also the case in the BP hearts is not clear. A slightly decreasing regression line with decreasing oxygen supply results when all values are used in the

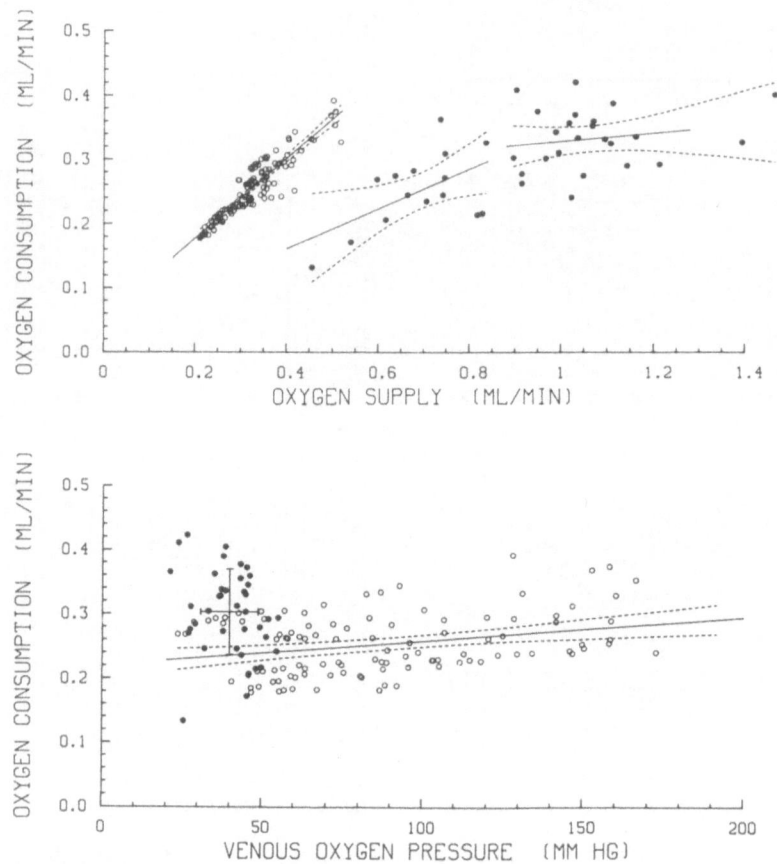

Fig. 6. Response to hypoxia. Top: Myocardial oxygen consumption (coronary flow times arterio-venous O_2 content difference) plotted against oxygen supply (coronary flow times arterial O_2 content). Bottom: MV_{O_2} plotted against venous P_{O_2}. TP hearts indicated by o; BP hearts by ●. Solid lines: linear regression lines; dashed lines: 95 % confidence belts.

regression analysis (not shown in the figure). However, when the results for the BP hearts are divided into two samples, with O_2 supply values either below or above 0.85 ml/min, an increase in oxygen consumption is shown with increasing oxygen supply until a plateau is reached at values above 0.85 ml/min. In panel b a significant difference is seen in oxygen consumption levels at venous P_{O_2} values in the range between 30 and 50 mm Hg. At these values, oxygen consumption of the TP hearts (extrapolated from regression analysis) is much lower.

410

DISCUSSION

Cardiac Performance

Table 2 compares our results concerning the hemodynamics of the hearts with those reported by Duvelleroy et al. (1976). The hematocrit in our experiments was appr. 25 % , so the means of the data of Duvelleroy et al. on perfusion with pig red cell suspension with hematocrit 20 and 30 % were taken. Their measurements were carried out at a PL of 13.5 and an AL of 72 mm Hg. In order to compare the results, we estimated values for PL 15 and AL 72 mm Hg from our data at PL 15 and AL 60 and 90 mm Hg. Also, data from Duvelleroy et al., expressed per g wet weight of the ventricles, were recalculated for whole heart using a mean value of 1.2 g for the weight of the ventricles. From the table it can be seen that AF, ECW and MV_{O_2} in their blood perfused hearts were higher. This may be due to the fact that they used highly oxygenated blood, with P_{O_2} values of 495 mm Hg. However, our data on CO are still in the physiological range reported for rats (70 ml/min, Dowell et al., 1975). The results of the buffer perfused hearts show that cardiac performance reported by Duvelleroy et al. was lower, probably due to the relatively low Pa_{O_2} of 495 mm Hg in their experiments, which is also reflected in their low $P\bar{v}_{O_2}$ values.

Summarizing, we can conclude that our results of left ventricular function agree reasonably well with those reported by Duvelleroy et al. The main difference between the blood and Tyrode perfused hearts in our experiments is the partition of the cardiac output. CO in both hearts is about the same, but in the Tyrode perfused hearts a much greater and substantial part of it is used for myocardial perfusion.

Hypoxic Hypoxia

The results concerning the response to hypoxia reveal that the TP hearts responded to lowering the Pa_{O_2} by an immediate decrease in Pv_{O_2}, MV_{O_2} and AF and ECW, whereas CF remained unchanged. The BP hearts responded to hypoxia by an increase in CF and a decrease in Pv_{O_2} and Sa_{O_2}. MV_{O_2} and AF and ECW remained high, until O_2 supply levels decreased because of a decrease in coronary perfusion (Fig. 5). We made only one step in lowering Pa_{O_2} in the TP hearts because cardiac performance rapidly went down and the hearts recovered only poorly when Pa_{O_2} was subsequently increased again to high levels (Olders et al., 1988).

Table 2. Comparison of the hemodynamic data of the present study (PS) with those of Duvelleroy et al. (1976) (DUV). Blood = blood perfused hearts; Buffer = buffer perfused hearts.

	Blood				Buffer			
	PS (n=10)		DUV (n=5)		PS (n=14)		DUV (n=7)	
	mean	sd	mean	sd	mean	sd	mean	sd
CF (ml/min)	7.3	1.4	7.2	1.0	15.4	2.5	15.1	1.0
AF (ml/min)	58.5	9.0	84.0	10.0	44.1	8.0	38.4	2.5
FR (%)	12.9	3.5	8.6	1.7	26.2	4.0	29.0	1.0
ECW (J/min)	0.69	0.11	0.84	0.10	0.57	0.09	0.44	0.10
MV_{O_2} (ml/min)	0.28	0.06	0.39	0.05	0.24	0.04	0.18	0.01
EFF (%)	12.7	3.6	10.5	1.3	11.9	2.6	12.0	0.5
Pa_{O_2} (mm Hg)	105.0	18.0	467.0	12.0	667.0	28.0	495.0	2.0
Pv_{O_2} (mm Hg)	46.3	4.0	45.5	4.0	142.7	40.0	88.0	8.0

411

In Fig. 6a it can be seen that at the same O_2 supply levels, MV_{O_2} in the TP hearts is significantly higher. Our conclusion is that the TP hearts are better able to extract the oxygen from the perfusate at these low O_2 supply levels, probably due to the high Pa_{O_2} values and therefore a steeper P_{O_2} gradient. In this respect, it is interesting that Chinet and Mejsnar (1989) reported that O_2 consumption of blood-perfused skeletal muscle in vitro is much below its maximum as reached when perfusion with a high P_{O_2} albumine-saline medium is carried out. On the other hand, Fig. 6b reveals that at venous P_{O_2} values below 50 mm Hg, MV_{O_2} is significantly higher in the BP hearts.

The P_{O_2} gradient from capillary to tissue depends on the sum of diffusive and perfusive resistances to O_2 transport. In the literature it is often questioned which of these two resistances is mainly limiting (see review by Kreuzer et al., in press). Mathematical modeling (Krogh and Hill models) contribute a valuable approach to asses the general properties of O_2 transport. The most critical location in the tissue cylinder with respect to O_2 supply is at the periphery of the venous end of the cylinder (lethal corner). Scheid and Piiper (1983) calculated the maximum P_{O_2} gradient from the artery to that point with a formula expressing this gradient as the sum of the diffusive and perfusive (convective) O_2 transfer resistances. The models stipulate a limitation by diffusion. Gutierrez et al. (1988) found in experiments with low vs. high flow in the rabbit hindlimb preparation that O_2 consumption is primarily limited by diffusion during hypoxemia. This was confirmed by Hogan et al. (1989) in isolated canine gastrocnemius muscle in situ, comparing low flow-high Ca_{O_2} with high flow-low Ca_{O_2} at the same O_2 supply. Cain (1977), however, comparing anemic and hypoxic hypoxia in the dog, concluded that consumption was not limited by diffusion but by O_2 supply.

The response to hypoxia of the TP and BP hearts reveals the important difference concerning the oxygenation of the hearts. In TP hearts the supply is limited and although these hearts can extract the O_2 better (at comparable O_2 supply levels) because of the high P_aO_2 values, a decrease in arterial O_2 content results in a concomitant decrease in MVO_2 and cardiac performance because the coronary flow is already maximal and the decrease in P_aO_2 can not be fully compensated for by a decrease in P_VO_2. In contrast, the BP hearts during hypoxia can improve their O_2 supply by increasing the coronary flow. Moreover, because of the high O_2 capacity of the blood compared to an electrolyte solution and the steep oxygen dissociation curve, a decrease in arterial O_2 content can be fully compensated for by only a slight decrease in P_VO_2. These mechanisms prevent the BP hearts from anoxia even when S_aO_2 decreases to 60-70 % and enable them to maintain cardiac performance on a high level.

It is tempting to conclude from our experiments that in myocardial O_2 consumption both oxygen diffusion and oxygen supply play a role and that their relative importance depends on the situation. It is not easy to separate their effects and it is difficult to conclude isolated limitation by diffusion only or supply only. However, one has to keep in mind that another possible explanation for our results could be a different flow heterogeneity in the Tyrode vs. the blood perfused hearts. Further experiments have to be carried out to answer these questions.

ACKNOWLEDGEMENT

This work was supported in part by NWO, the Netherlands Organization for Scientific Research.

REFERENCES

Bartels, H., und Harms, H., 1959, Sauerstoffdissoziationskurven des Blutes von Säugetieren, Pflügers Arch., 268: 334-365.

Cain, S.M., Oxygen delivery and uptake in dogs during anemic and hypoxic hypoxia, 1977, J. Appl. Physiol., 42: 228-234.

Chinet, A.E., and Mejsnar, J., 1989, Is resting muscle oxygen uptake controlled by oxygen availability to cells?, J. Appl. Physiol., 66: 253-260.

Dowell,R.T., Cutilletta, A.F., and Sodt, P.C., 1975, Functional evaluation of the rat heart in situ, J. Appl. Physiol., 39: 1043-1047.

Duvelleroy, M.A., Duruble, M., Martin, J.L., Teisseire, B., Droulez, J., and Cain, M., 1976, Blood-perfused working isolated rat heart, J. Appl. Physiol., 41: 603-607.

Figulla, H.R., Hoffmann, J., and Lübbers, D.W., 1983, Coronary conductivity and tissue oxygenation as measured by the myoglobin O_2 saturation and the cytochrome aa_3 redox state in the nLangendorff guinea pig heart preparation, Adv. Exp. Med. Biol., 159: 579-585.

Gutierrez, G., Pohil, R.J., and Strong, R., 1988, Effect of flow on O_2 consumption during progressive hypoxemia, J. Appl. Physiol., 65, 601- 607.

Hogan, M.C., Roca, J., West, J.B., and Wagner, P.D., 1989, Dissociation of maximal O_2 uptake from O_2 delivery in canine gastrocnemius in situ, J. Appl. Physiol., 66: 1219-1226.

Kreuzer, F., in press, Critical oxygen supply to striated muscle; influence of various factors.

Kreuzer, F., Turek, Z., and Hoofd, L., in press, Oxygen transfer from blood to mitochondria.

Olders, J., Turek, Z., Evers, J., and Hoofd, L., 1988, Myocardial function and response to hypoxia of the isolated working heart of normal and simulated high altitude rats, Abstract, ISOTT Meeting, Ottawa, Canada.

Olders, J., Boumans, T., Evers, J. and Turek, Z., this volume, An experimental set-up for the isolated working rat heart perfused with blood.

Opie, L.H., 1984, Adequacy of oxygenation of the isolated perfused rat heart, Bas. Res. Card., 79: 300-306.

Robiolio, M., Rumsey, W.L., and Wilson D.F., 1989, Oxygen diffusion and mitochondrial respiration in neuroblastoma cells, Am. J. Physiol., 256: C1207-C1213.

Scheid, P., and Piiper, J., 1983, Oxygen exchange at external and internal surfaces. in: "Oxygen: Physiological Adjustment to Changes in its Supply and Demand", Jones, D.R., Satchel, G.H., eds., Proc. Physiol. Soc. New Zealand, 3: 57-65.

Snoeckx, L.H.E.H., Vusse van der, G.J., Coumans, W.A., Willemsen, P.H.M., Nagel van der, T., and Reneman, R.S., 1986, Myocardial function in normal and spontaneously hypertensive rats during reperfusion after a period of global ischaemia, Cardiovasc. Res., 20, 67-75.

RESPONSE TIME OF MITOCHONDRIAL OXYGEN CONSUMPTION

FOLLOWING STEPWISE CHANGES IN CARDIAC ENERGY DEMAND

J.H.G.M. van Beek and N. Westerhof

Laboratory for Physiology, Free University
van der Boechorststraat 7, 1081 BT Amsterdam
The Netherlands

INTRODUCTION

When the metabolic demand of the heart changes, the mitochondria alter their ATP-production which is linked to mitochondrial oxygen consumption. The control mechanisms of mitochondrial respiratory rate are under debate (Chance et al., 1986; Katz et al., 1989; Erecinska and Wilson, 1982). Knowledge of the time course of mitochondrial oxygen consumption following changes in ATP-hydrolysis could give important information on mitochondrial metabolic control mechanisms and adaptation to cellular energy demand.

The uptake of oxygen in the heart following a change in cardiac energy demand can be measured but differs from the true mitochondrial oxygen consumption by the rate of change in the amount of oxygen which is stored in the heart, either in the physically dissolved form or reversibly bound to myoglobin or hemoglobin. In this paper we describe how the time course of oxygen uptake in the isolated rabbit heart was measured during stepwise changes in heart rate and how the true response time of the mitochondrial oxygen consumption was derived from these measurements by taking the change of oxygen stores in tissue into account.

MEASUREMENTS OF OXYGEN UPTAKE IN ISOLATED RABBIT HEARTS DURING HEART RATE STEPS

Methods

Isolated rabbit hearts were perfused according to Langendorff at 37 $^\circ$C with a constant flow of Tyrode solution, containing 11 mM glucose as substrate. The left ventricle was vented, so that the Thebesian venous flow was drained. The left ventricle therefore developed no pressure. The veins entering the right atrium were ligated. Part of the coronary venous effluent was drawn through a cuvette which contained a Clark type oxygen electrode, just after leaving the cannulated pulmonary artery. Details of the experimental methods have been described elsewhere (van Beek and Elzinga, 1987; van Beek et al., 1989). The hearts were paced. Changes in energy demand were accomplished by changes in heart rate.

To obtain information on oxygen transport and oxygen store changes in the myocardium, the arterial oxygen concentration was changed

Oxygen Transport to Tissue XII, Edited by J. Piiper *et al.*
Plenum Press, New York, 1990

stepwise, keeping the perfusion flow constant (van Beek and Elzinga, 1987), and the response of the venous oxygen concentration was measured. In 8 isolated rabbit hearts in which the coronary vessels were maximally dilated with 10 μM adenosine, the oxygen consumption decreased in each case after the arterial O_2 content, was lowered by 10%. The oxygen consumption decreased significantly by 5.8% on average per 10% decrease in oxygen content at constant flow (van Beek et al., 1989). When no adenosine was added to the perfusate, O_2 uptake changed by less than 1% when arterial O_2 content was decreased by 10% in 3 of 8 experiments. We selected these 3 experiments for further analysis because the step response in venous oxygen concentration was negligibly distorted.

Results

The responses of venous P_{O2} to stepwise changes in cardiac energy demand, achieved by a step in heart rate, were measured in the 3 selected hearts (see Fig. 1). Since arterial P_{O2} and coronary flow were kept constant, venous P_{O2} directly reflects oxygen uptake by the heart. The P_{O2} is proportional to oxygen content in this saline perfusate, with oxygen solubility (α) in Tyrode solution $2.987 \cdot 10^{-5}$ $ml.ml^{-1}.mmHg^{-1}$ at 37 ^{O}C. After the stepwise change in heart rate, oxygen uptake decreased but after 50 s a slow increment of oxygen uptake is seen until a new steady state is reached. The time course of oxygen uptake in the heart lags the mitochondrial oxygen consumption due to the change in tissue O_2 stores. In the following sections we will outline the calculation of the true mitochondrial response time from the externally measured venous P_{O2} time course.

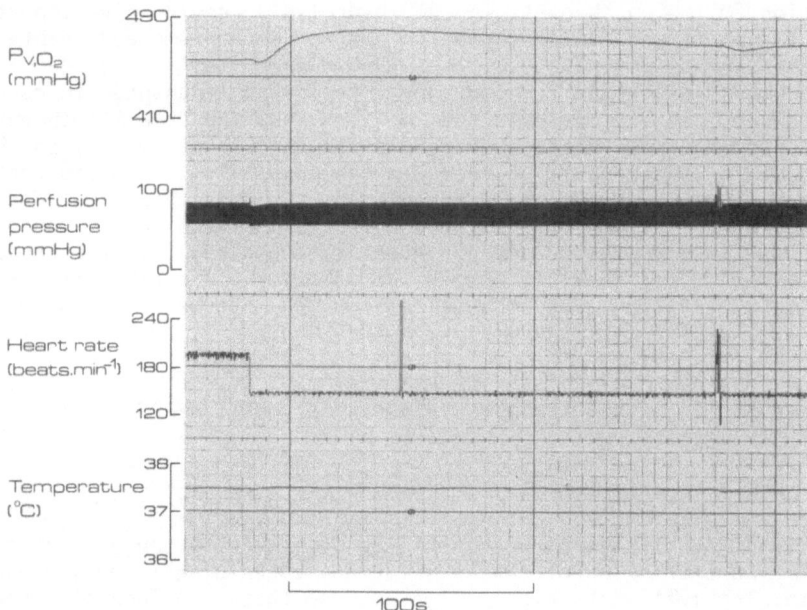

Fig. 1. Reponse of coronary venous oxygen tension to a step in metabolic demand as achieved by a change in heart rate. Arterial P_{O2} and coronary perfusion flow are constant. The left ventricle is vented. Perfusion pressure oscillates due to cardiac contraction. Heart rate is determined with a cardiotachometer triggered by perfusion pressure. An arrythmia at the end of the track leads to a transient change in oxygen uptake, but the $P_{v,O2}$ returns to the same steady-state level. Temperature is measured in coronary venous effluent.

416

RESPONSE TIME DERIVED FROM MASS BALANCE EQUATIONS

The oxygen uptake in the heart differs from the mitochondrial oxygen consumption by the amount of oxygen being stored in tissue:

oxygen uptake = oxygen consumption + rate of change of stored O_2 \hfill (1)

where the terms are instantaneous rates and may vary with time. After integration over time of this equation in a suitably arranged form, starting at t = 0 (i.e. the time at which the change in heart rate is made), we obtain:

$$t_{mitochondria} = t_{measured} - t_{transport} \hfill (2)$$

where $t_{transport} = (-\Delta Q / \Delta m\dot{V}_{O2})$. Here $-\Delta Q$ is the decrease in the total amount of oxygen present in the heart, both physically dissolved and bound to myoglobin, following the step increase in oxygen consumption by $\Delta m\dot{V}_{O2}$. The mitochondrial response time, $t_{mitochondria}$, is precisely defined by:

$$t_{mitochondria} = \int_0^\infty [1-\Delta m\dot{V}_{O2}(t)/\Delta m\dot{V}_{O2}(\infty)].dt \hfill (3)$$

where $\Delta m\dot{V}_{O2}(t)$ is the change in mitochondrial oxygen consumption from the initial steady-state value. The steady-state change, $\Delta m\dot{V}_{O2}(\infty)$ is reached after long times.

The externally measured response time, $t_{measured}$, is defined as:

$$t_{measured} = \int_0^\infty [1-\Delta P_{v,O2}(t)/\Delta P_{v,O2}(\infty)].dt \hfill (4)$$

where $\Delta P_{v,O2}(t)$ is the change in venous oxygen tension from the initial steady-state value at t=0, which is proportional to the change in oxygen content in the saline perfusate. Subtracting the transport time from $t_{measured}$ thus corrects for oxygen diffusion and convective transport in the perfusate. The steady-state change is $\Delta P_{v,O2}(\infty)$. The definition of measured response time is mathematically similar to the definition of mean transit time in indicator dilution theory for a stepwise change in arterial indicator concentration (Zierler, 1961).

The measured response time for the experiments, obtained from 4 downward steps in heart rate (see Fig. 1), and 4 upward steps in heart rate, by 40 beats.min^{-1}, was found to be 17.8 $\pm$ 1.6 s (mean $\pm$ SE, 3 hearts). We see from equation 2 that the mitochondrial response time can be calculated when we can estimate the change in amount of oxygen in tissue due to the increase in oxygen consumption. This stored amount can be calculated from the venous and arterial oxygen contents using the model for P_{O2} gradients in tissue discussed in the following section.

A schematic picture of the experiment with a stepwise increase in heart rate, with the response times of equations 3 and 4 given graphically, is shown in Fig. 2. In those cases where the time course is purely exponential, the response time is exactly equal to the time constant.

417

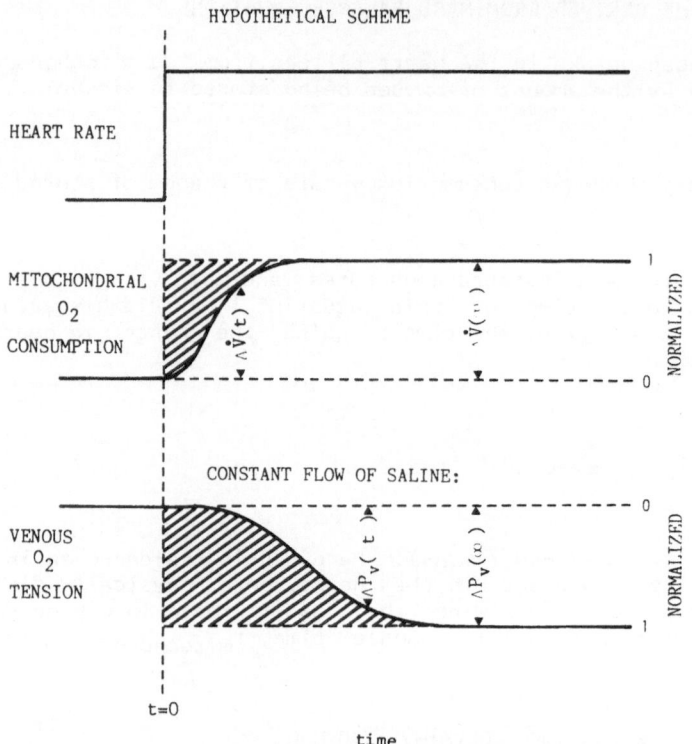

HYPOTHETICAL SCHEME

Fig. 2. Hypothetical scheme for the experiment with a stepwise increase in heart rate. The changes in mitochondrial O_2 consumption, $\Delta\dot{V}(t)$, and in venous oxygen tension, $\Delta P_v(t)$, can be normalized to 0 in the initial steady state, and to 1 in the new steady state. The hatched areas, calculated using the normalized scale, give the mitochondrial response time (middle trace, eq. 3) and the measured response time (lower trace, eq. 4).

Model for amount of oxygen stored in tissue

The amount of oxygen present in all tissue, Q, can be expressed as a volume integral over the whole organ:

$$Q = \int C_{O_2} \cdot dV \qquad (5)$$

For the hypothetical situation that oxygen concentration is constant in tissue this becomes:

$$Q = V \cdot \lambda \cdot C_{O_2} = V \cdot \lambda \cdot \alpha \cdot P_{O_2} \qquad (6)$$

where λ is the tissue-to-perfusate partition coefficient, α is oxygen solubility, V is the organ's volume, and $V \cdot \lambda$ is the volume of distribution.

When there is a gradient of oxygen tension from arterial to venous side (see Fig. 3), we calculate the oxygen stores in various subsections of the vasculature and tissue as shown below, and add them. From the

418

arterial oxygen concentration and the volume of perfusate present in the arterial system the amount of oxygen present in the arterial vessels can be calculated. We assume that the loss of O_2 from arteries and arterioles is negligible, as was found with cryospectrophotometry in blood-perfused hearts (Honig and Gayeski, in press). Diffusional shunting of O_2 directly from the arterial to the venous vessels is indeed negligible in saline-perfused rabbit heart (van Beek and Elzinga, 1987). Similarly, the amount of oxygen in the venous vessels can be calculated from the measured P_{O2} in the coronary venous effluent, and the volume in the venous vessels. In our experimental preparation, where oxygen from the coronary veins leaves the heart via the right ventricle, the volume present in the right heart chambers should be included in the venous volume. In the capillaries the oxygen tension will decrease from the arterial to the venous side. If the diffusive oxygen flux from lumen to tissue is uniform along the length of the capillary, the average capillary P_{O2} is (arterial P_{O2} + venous P_{O2})/2. This average capillary P_{O2} can be used with the capillary volume to estimate the amount of oxygen in the capillary vessels.

To calculate the oxygen tension in the tissue surrounding the capillaries we need to know the diffusion gradient between capillary lumen and tissue. The oxygen tension in the periphery of the Krogh (1919) tissue cylinder is thus calculated to be about 4 mmHg lower than in the capillary (van Beek et al., 1989). This gradient would increase by 10% (order 0.4 mmHg) during a 10% increase in oxygen consumption. Since the change in venous oxygen tension is found to be of the order of 30 mmHg for a 10% increase in O_2 consumption, the steepening of the diffusion gradient between vessel lumen and myocyte can in the saline-perfused heart be neglected relative to the steepening of the longitudinal gradient in the capillary. However, the measured oxygen tension in the cytosol of the cardiac myocyte is around 5 mmHg in the blood perfused heart according to cryospectrophotometric measurements by Honig and Gayeski (in press). About 30% of myoglobin in the saline-perfused rat heart at 37 $^{\circ}$C was deoxygenated (Araki et al., 1983). This might indicate a lower diffusivity in tissue than we assumed in our calculations. A lower bound for the estimate of oxygen diffusivity is obtained as follows. Under normal circumstances the amount of oxygen taken up in blood-perfused rabbit heart is about 0.1 ml.ml^{-1}.min^{-1}, while the average oxygen tension gradient available for diffusion can not be more than 60 mmHg, since the lowest possible value for P_{O2} in the cell is 0 mmHg and the average P_{O2} in the vessels is not higher than 60 mmHg. Since oxygen uptake in our saline-perfused preparation was about 0.05 ml.ml^{-1}.min^{-1}, the oxygen tension gradient from lumen to tissue should be lower than 30 mmHg. This gradient will increase by 3 mmHg for a 10% increase in oxygen consumption. This yields an upper bound for the estimate of the change in lumen-to-tissue gradient which is still an order of magnitude lower than the change in the longitudinal oxygen tension gradient in the vessel lumen. The lumen-to-tissue gradient was therefore ignored, and the change in amount of oxygen in tissue was calculated from the average change in intracapillary P_{O2}, which is obtainable from the arterial and venous P_{O2} (see Fig. 3). However, to calculate the amount of oxygen in tissue, the oxygen solubility in tissue and the possible involvement of myoglobin has to be known.

Oxygen solubility in tissue

We can measure the amount of oxygen washed out of tissue when the arterial oxygen content is lowered stepwise, since we did select experiments in which oxygen consumption did not change. Venous P_{O2} at constant perfusion flow is eventually, in the new steady state, changed by the same amount as arterial P_{O2}. Venous P_{O2} changes more slowly than

419

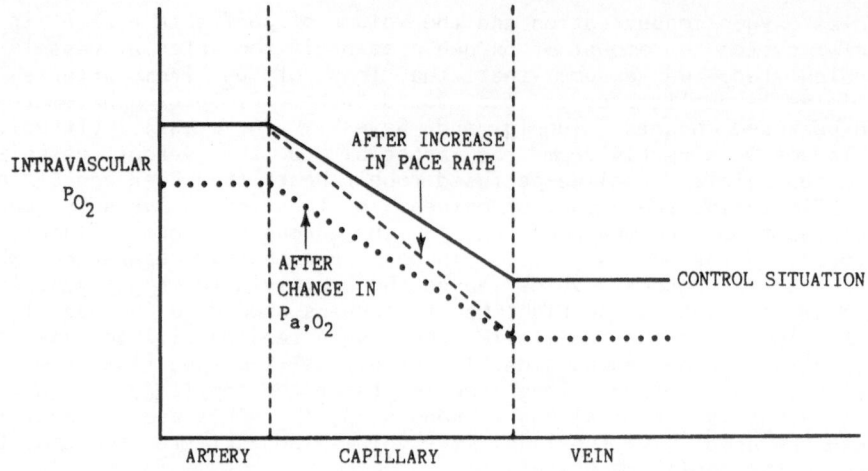

INTRAVASCULAR P_{O_2}

AFTER INCREASE IN PACE RATE

AFTER CHANGE IN P_{a,O_2}

CONTROL SITUATION

ARTERY   CAPILLARY   VEIN

Fig. 3. The profile of intravascular oxygen tension in the blood vessels in the control situation, after an increase in cardiac oxygen consumption, and after a decrease in arterial oxygen content. We assume that the blood vessels are perfused with saline solution.

arterial P_{O_2} since there is washout of oxygen. The total amount of washed out oxygen is equal to $\lambda.V.\alpha.\Delta P_{v,O_2}$, where $\Delta P_{v,O_2}$ stands for the measured change in venous oxygen tension in the new steady state. This is true since the change in P_{O_2} is the same everywhere in tissue (see Fig. 3).

The intravascular volume is determined by measuring the mean transit time of dye-tagged albumin, an indicator which stays in the blood vessels, at a known perfusion flow. Since the partition coefficient is by definition 1 for the Tyrode in the intravascular space, the extravascular volume of distribution, $\lambda.V$, of O_2 is determined by subtracting the intravascular volume from the measured total volume of distribution of O_2. In this way we found 0.72 ± 0.11 ml.g^{-1} for the extravascular volume of O_2 distribution. The anatomical extravascular volume in rabbit heart is 0.79 ml.g^{-1} (Bassingthwaighte et al., 1985). This means that the partition coefficient is 0.91 ± 0.14. Since this shows that the effective oxygen solubility in tissue is 91% of that of Tyrode solution, it is likely that myoglobin O_2 saturation is not appreciably changed during the steps in arterial oxygen content and that the partition coefficient for oxygen reflects a physical oxygen solubility in tissue which is slightly lower than in Tyrode solution, due to the presence of macromolecules in tissue which do not dissolve oxygen.

CALCULATION OF TRUE MITOCHONDRIAL RESPONSE TIME

The measured response time for 4 upward steps and 4 downward steps in heart rate was 17.8 ± 1.6 s (mean $\pm$ SE, 3 hearts). The transport time, calculated from equation 2 according to the previous sections equals 10.3 ± 1.3 s. The true mitochondrial response time is therefore equal to 7.5 ± 1.0 s. One could assume that the time course of mitochondrial oxygen consumption following a stepwise change in metabolic demand is monoexponential, as in skeletal muscle (Meyer, 1988). The time constant of mitochondrial oxygen consumption following a step in heart rate is in that case equal to 7.5 s.

420

DISCUSSION

In the correction of the measured oxygen response for oxygen storage, heterogeneity of perfusion and metabolism was not taken into account. Analysis with a model of heterogeneously perfused tissue shows that uneven perfusion has no effect on the calculations as long as local perfusion and metabolism are strongly correlated (r=1). For the case that metabolism and perfusion are not correlated (r=0), calculations show that the change in O_2 stores in tissue during a metabolic step is increased by about 30%. We assumed for this calculation that the heterogeneity of flow is similar to the one found by Bassingthwaighte et al. (in press) for 0.1 gram sample pieces and that oxygen consumption is homogeneous. It is true that for greater spatial resolution of the flow measurement the dispersion of flows may become larger (Bassingthwaighte et al., in press), but this increase is limited for the fractal network model of distribution of myocardial flow, which fits the dispersion of myocardial flow in the observable range (van Beek et al., in press). We therefore estimate that the transport time can be up to 30% higher due to heterogeneous perfusion. The precise correction can not be calculated, since the relation between local perfusion and O_2 consumption is unknown. We conclude that the cardiac mitochondrial response time is in the range 4-8 s at 37 $^{\circ}$C.

When ATP-hydrolysis is increased and ATP-synthesis is still adapting to the increase, the balance will be taken from the stored creatine phosphate (CrP). The change in CrP following a step increase in heart rate has been measured by us with ^{31}P-NMR in the pressure developing rabbit heart perfused with Tyrode solution at 37 $^{\circ}$C (van Echteld et al., 1988). We saw a decrease in CrP and an increase in P_i with a time constant of 3.5 s. This is slightly faster than the time constant for oxygen consumption, suggesting that there is non-mitochondrial ATP production early in the response. Indeed, it has been found that after a step change in cardiac metabolic demand there is a transient burst of lactate production (Achs et al., 1979 and 1982). The early ATP-production associated with lactate production may cause the change in CrP to be faster than the change in mitochondrial oxygen consumption.

The time constant for adaptation of oxygen consumption was determined for the first phase of the response (up till 50 s). We often saw a secondary change in the opposite direction later on (see Fig. 1). This is not yet accounted for in the analysis. Sufficient separation in time of the two phases was assumed, and the first phase was analyzed. The overshoot in ATP-production suggested by our oxygen measurements was also reflected in the CrP-measurements of Achs et al. (1982), showing a similar time course.

In skeletal muscle a monoexponential change in CrP with a time constant of 1.5 min was found (Meyer, 1988). The time course was explained with a linearized model, and was found to depend on the mitochondrial capacity and the total concentration of CrP and Cr (Meyer, 1988). Since the mitochondrial volume density in heart is larger than that in skeletal muscle, Meyer's model indeed predicts that the mitochondrial response is faster in the heart.

SUMMARY

We determined the speed with which mitochondrial oxygen consumption and therefore the mitochondrial ATP-synthesis adapted to changes in metabolic demand in the rabbit heart. This was done by measuring the oxygen uptake of the whole heart during a stepwise change in heart rate and correcting for the time taken by diffusion and by convective transport in the blood vessels. Data for the correction for transport time were obtained from the response of venous oxygen concentration to a stepwise change of arterial oxygen concentration. The time constant of the response of mitochondrial oxygen consumption to a step change in heart rate was found to be 4-8 s.

REFERENCES

Achs, M.J., Garfinkel, D., and Opie, L.H., 1982, Computer simulation of metabolism of glucose-perfused rat heart in a work-jump, Am. J. Physiol. 243: R389-R399.

Achs, M.J., Kohn, M.C., and Garfinkel, D., 1979, Computer simulation of metabolism in pyruvate-perfused rat heart. IV. Model behavior, Am. J. Physiol. 237: R174-R180.

Araki, R., Tamura, M., and Yamazaki, I., 1983, The effect of intracellular oxygen concentration on lactate release, pyridine nucleotide reduction, and respiration rate in the rat cardiac tissue, Circ. Res. 53: 448-455.

van Beek, J.H.G.M., and Elzinga, G., 1987, Diffusional shunting of oxygen in saline-perfused rabbit heart is negligible, Pflügers Arch. 410: 263-271.

van Beek, J.H.G.M., Bouma, P., and Westerhof, N., 1989, Oxygen uptake in saline-perfused rabbit heart is decreased to a similar extent during reductions in flow and in oxygen concentration, Pflügers Arch. 414: 82-88.

van Beek, J.H.G.M., Bassingthwaighte, J.B., and Roger, S.A., in press, Fractal networks explain regional myocardial flow heterogeneity, in: "Oxygen transport to tissue XI," K. Rakusan, G. Biro, T.K. Goldstick, and Z. Turek, eds., Plenum, New York.

Bassingthwaighte, J.B., Kuikka, J.T., Chan, I.S., Arts, T., and Reneman, R.S., 1985, A comparison of ascorbate and glucose transport in the heart, Am. J. Physiol. 249: H141-H149.

Bassingthwaighte, J.B., King, R.B., and Roger, S.A., in press, Fractal nature of regional myocardial blood flow heterogeneity, Circ. Res.

Chance, B., Leigh, J.S. Jr., Kent, J., McCully, K., Nioka, S., Clark, B.J., Maris, J.M., and Graham, T., 1986, Multiple controls of oxidative metabolism in living tissues as studied by phosphorus magnetic resonance, Proc. Natl. Acad. Sci. USA 83: 9458-9462.

van Echteld, C.J.A., van Beek, J.H.G.M., Kirkels, J.H., van der Meer, P., Ruigrok, T.J.C., and Westerhof, N., 1988, ^{31}P-NMR study of the response of myocardial energy metabolism to heart rate steps, III International Congress on Muscle Energetics, Yufuin, Japan.

Erecinska, M., and Wilson, D.F., 1982, Regulation of cellular energy metabolism, J. Membrane Biol. 70: 1-14.

Honig, C.R., and Gayeski, T.E.J., in press, Precapillary O_2 loss and arteriovenous O_2 diffusion shunt are below limit of detection in myocardium, in: "Oxygen transport to tissue XI," K. Rakusan, G. Biro, T.K. Goldstick, and Z. Turek, eds., Plenum, New York.

Katz, L.A., Swain, J.A., Portman, M.A., and Balaban, R.S., 1989,

Relation between phosphate metabolites and oxygen consumption of heart in vivo, Am. J. Physiol. 256: H265-H274.

Krogh, A., 1919, The number and distribution of capillaries in muscles with calculations of the oxygen pressure head necessary for supplying the tissue, J. Physiol. 52: 409-415.

Meyer, R.A., 1988, A linear model of muscle respiration explains monoexponential phosphocreatine changes, Am. J. Physiol. 254: C548-C554.

Zierler, K.L., 1961, Theory of the use of arteriovenous concentration differences for measuring metabolism in steady and non-steady states, J. Clin. Invest. 40: 2111-2125.

LEFT VENTRICULAR SURFACE TISSUE OXYGEN PRESSURES DETERMINED BY OXYGEN SENSITIVE MULTIWIRE ELECTRODES IN PIGS

Peter F. Conzen*, Helmut Habazettl, Michael Christ, Hans Baier, Jonny Hobbhahn*, Brigitte Vollmar, A. Goetz, Klaus Peter*, and Walter Brendel

Institute of Surgical Research and *Institute of Anesthesiology
Marchioninistr. 15, 8000 Munich 70, FRG

INTRODUCTION

Oxygen sensitive needle electrodes have been used during the past years to describe intramyocardial oxygen tensions (Kirk and Honig, 1964; Koyama et al., 1979; Mendler et al., 1973; Moss, 1968; O'Riordan, 1977; Schubert et al., 1978; Winbury et al., 1971). However, a limitation of the intramyocardially inserted electrodes is that measurements at identical locations of tissue over longer periods of time are limited. This is mainly caused by the relatively large diameters of the electrodes, interfering with oxygen transport to the tissue under investigation (Moss, 1968). The average intramyocardial oxygen pressures obtained by these devices are usually below coronary venous PO_2: Such a gradient is assumed to be necessary to enable diffusion of a sufficient amount of oxygen molecules from red cells into tissue.

Determinations of myocardial surface tissue oxygen pressures by means of Clark type multichannel electrodes (Mehrdraht-Dortmund-Oberflächen Elektrode, MDO) revealed narrow oxygen pressure distribution curves with averages above coronary venous PO_2 (Conzen et al., 1989; Habazettl et al., 1989; Walfridsson et al., 1985; Hobbhahn et al., 1989). These electrodes are non-traumatic for the tissue under investigation and enable measurements during different experimental conditions without a necessity of changing their position. The purpose of this investigation was to gain further insights into the meaning of the surface oxygen pressures for the oxygenation of the myocardium. This paper is mainly an attempt to interpret and to discuss the results of these measurements with respect to current theories for oxygen transfer between red cells and cardiac myocytes.

Oxygen Transport to Tissue XII, Edited by J. Piiper *et al.*
Plenum Press, New York, 1990

METHOD

Experimental Preparation

Experiments were conducted in 17 pigs with an average weight of 34.5 kg. Anesthesia was induced by i.m. injection of ketamine (15 mg/kg), flunitrazepam (0.01 mg/kg) and atropin (0.5 mg). Adequate anesthesia was then maintained by infusion of the narcotic piritramid (1 mg•kg^{-1}•h^{-1}; Janssen, Denmark) and by nitrous-oxide. Mechanical ventilation was used to keep end-expiratory CO_2 at 4.5 vol% and arterial PO_2 at 100-110 mmHg. The animals were placed on a heating pad to maintain core temperature throughout the experiments. Catheters were placed to determine blood pressures and cardiac output, and to inject radioactive microspheres into the left atrium.

The pericardium was opened through a sternotomy and the proximal part of the left anterior descending artery (LAD) dissected free of surrounding connective tissue. A teflon coated copper wire was placed round the vessel and connected to a micrometer screw which served for producing stable and reproducible degrees of stenosis of the artery.

Determination of left ventricular surface oxygen tensions

Two 8-channel oxygen sensitive electrodes (MDO) were placed on the epicardial surface of the left ventricle: One electrode served for detection of PO_2 values on the poststenotic LAD area, the second was placed on the area supplied by the circumflex (CX) artery. Both electrodes were kept in place by flexible silicone rubber disks, fixed at the myocardium as described earlier (Conzen et al., 1989, Habazettl et al., 1989).

The measuring devices consist of Clark type electrodes (Clark, 1956) and have 8 thin platinum cathodes and 1 Ag/AgCl anode incorporated. At a constant polarisation voltage of approximately -700 mV, a single platinum wire registers a reduction current linearly dependent on the partial pressure of oxygen. The wires are randomly distributed on the electrode surface and register oxygen pressures independently of each other. The radius of the oxygen sensitive surface area of the platinum wires depends on the thickness of the membrane used and is 20-25 μm with the membrane of 25 μm used in the experiments (Kessler et al., 1976). Both electrodes were calibrated in physiologic saline solutions equilibrated with pure nitrogen, 5% and 10% oxygen in nitrogen before the first measurement on the heart and then recalibrated in hourly intervals.

Correct position of the electrodes was verified by a short complete occlusion of the LAD: signals of the electrode placed on the LAD territory rapidly fell towards 0 mmHg, whereas the recordings on the CX zone remained unchanged. The average PO_2 values from each electrode were used for further analyses.

Regional left ventricular function

Ultrasonic dimension gauges (5 MHz) were placed in the LAD supplied left ventricular myocardium to measure regional wall function: One pair of crystals was positioned in the LAD supplied subepicardium parallel with the superficial fiber orienta-

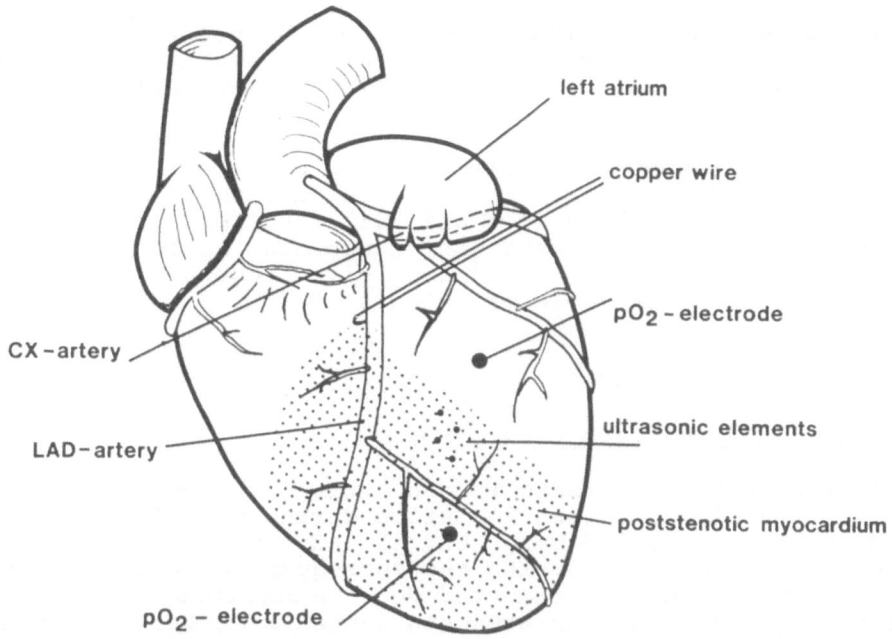

left atrium

copper wire

pO_2 - electrode

ultrasonic elements

poststenotic myocardium

CX - artery

LAD - artery

pO_2 - electrode

Figure 1.   *Schematic drawing of the heart preparation used in the experiments. The approximate locations for stenosis of the LAD and for the measuring devices for epicardial PO_2 (PO_2 electrodes) and segment length shortening (ultrasonic elements) are given.*

tion (Gallagher et al., 1985). A second pair of crystals was positioned in the left ventricular subendocardial layer approximately parallel to the longitudinal heart axis.

The extent of systolic segment length shortening (SLS) as the difference between end-diastolic and end-systolic dimensions was calculated and expressed as percentage change from end-diastolic length: (EDL-ESL)/EDL·100.

Determination of regional myocardial blood flow

Regional blood flow (RBF) to the left ventricular myocardium was determined by radioactive microspheres (Heymann et al. 1977). 6-9 million microspheres (diameter: 9μm) were injected for each measurement. Immediately before killing the animals by i.v. KCl, India ink was injected into the occluded LAD artery next to the constrictor wire to delineate the ischemic myocardium. The left ventricular free wall was subdivided into three layers of almost equal thickness (subendocardium, midwall layer, subepicardium).

Experimental protocol and statistical analyses

Control recordings were obtained after 30 minutes of stable conditions. Stenosis of the LAD was then produced by narrowing the micrometer screw driven copper wire placed round the LAD artery. Varying levels of stenosis as defined by decreases in tissue PO_2 or by deteriorations of regional wall

427

function were induced. The severity of stenoses was varied randomly. Recordings were repeated after stable conditions had been achieved for at least 5 minutes. The stenosis was then opened and the hearts allowed to recover.

Data are presented as mean values ± SEM. Regression analysis was applied to relate changes in surface tissue PO_2 to changes in regional blood flow and wall function. The results of linear regression analyses (which appeared appropriate after visual inspection of the data) are given together with the Spearman rank correlation coefficients.

RESULTS

Left ventricular surface tissue PO_2

Average surface tissue PO_2 was approximately 45 mmHg on both left ventricular measuring sites during the first control recording (LAD: 46.3±1.3 mmHg; CX: 43.1±3.3 mmHg). No significant changes were noted during the subsequent control periods up to the end of the experiments. PO_2 measurements on the LAD supplied myocardium during stenosis ranged from a 5 mmHg decrease down to values not detectably different from 0 mmHg. As an average, surface PO_2 was reduced to 19.4 mmHg when all ischemic periods are considered.

The decrease in surface oxygen pressures coincided with relatively small changes of subepicardial blood flow: Blood supply of the subepicardium was 20-30% lower than its control value in the CX supplied myocardium (101 ± 9 ml•$100g^{-1}$•min^{-1}) when the LAD electrodes registered PO_2 values not detectably different from 0 mmHg (figure 2). At that level of constriction, subendocardial blood flow values were restricted to approximately 50% (figure 2). Thus, the endo:epi blood flow ratio of the ischemic myocardium was decreased significantly from 1.03±0.02 at control to 0.59±0.03. PO_2 and segment shortening registrations returned to normal within 30 seconds to 3 minutes following discontinuation of stenosis.

Segment length changes

Segment length shortening was almost identical in subendo- and subepicardial layers of the LAD supplied area at control and amounted to 14.1±0.6% and 14.2±0.7%, respectively. Stenosis of the LAD mainly depressed function of the subendocardium, whereas virtually no deterioration was observed in the superficial layers (figure 3). When all periods during coronary artery stenoses with surface PO_2 values not detectably different from 0 mmHg are considered, mean values of segment shortening were 9.8±1.2% (subendocardium; p<0.01 vs control) and 15.8±2.4% (subepicardium; n.s.).

Hence, stenosis of the LAD only slightly affected epicardial segment shortening. This was apparently due to the fact, that subepicardial perfusion was reduced considerably less than subendocardial blood flow. Therefore also, no close correlation was obtained between epicardial function and PO_2 in the MDO sensitive range (reduction of subepicardial RBF by 20-30%). However, when LAD constriction was aggravated and blood supply further reduced, a significant correlation for SLS with blood

428

flow (x) was obtained: y = -15.1+1.14x; r=0.65; p<0.001.

<u>Hemodynamic parameters</u>

Control value for heart rate was 91±3 bpm, mean arterial pressure and left atrial pressure were 80±2 mmHg and 6±1 mmHg, respectively. Cardiac output was 3.8±0.2 l/min. Systemic hemo-

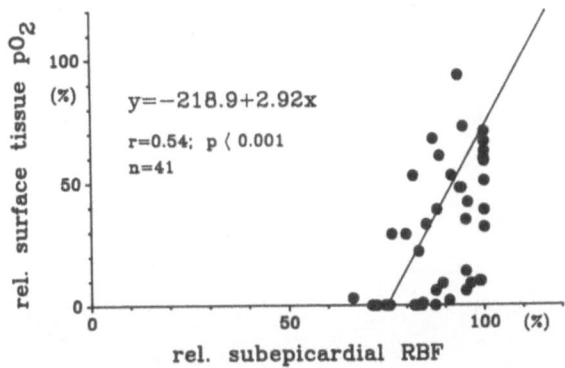

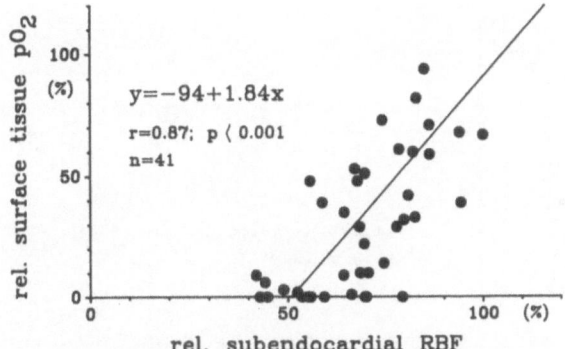

<u>Figure 2.</u>   *The linear relationships between % decreases in left ventricular surface tissue PO_2 (ordinate) and regional blood flow (x). Decreases of subepicardial blood flow are plottet in the upper, decreases of subendocardial flow in the lower part of the figure (n, number of observations; r, Spearman's rank correlation coefficient).*

dynamic variables were not subjected to significant alterations during the experiments. Also, the individual stenoses of the LAD did not affect hemodynamics detectably: Comparisons of mean values obtained during control and during the highest degree of stenosis of the LAD indicated that global left ventricular function was not impaired significantly.

DISCUSSION

A major advantage of oxygen sensitive surface electrodes over the use of needle electrodes is that surface electrodes avoid trauma of the organ produced by insertion of the measuring devices (Moss, 1968) and allow measurements at identical locations of tissue during different experimental variations. The results of these PO_2 measurements are considered to reflect

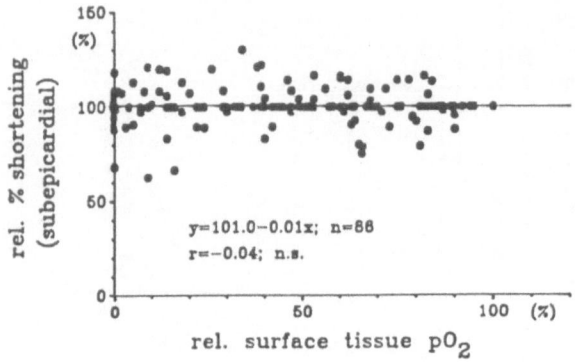

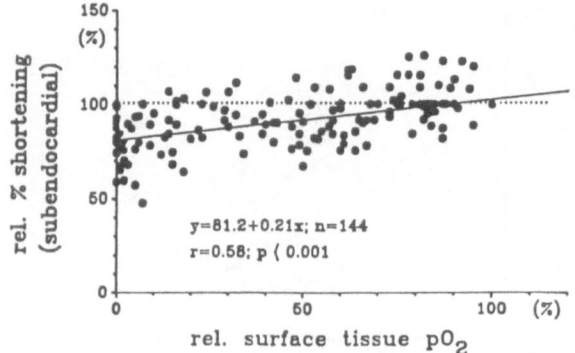

Figure 3.   Linear relationships between % changes in regional segment shortening (SLS, % shortening, ordinate) are plottet against decreases in surface tissue PO_2. Epicardial segment shortening is given in the upper, endocardial shortening in the lower part of the figure. Only the regression coefficient with subendocardial function is significantly different from zero, indicating that function is maintained at low values of surface tissue PO_2.

local tissue oxygenation, giving the net result of oxygen delivery and oxygen demands of the tissue (Kessler et al., 1976; Lübbers, 1977). This was also confirmed for the myocardium in a recent study from this laboratory, when tissue PO_2 distal to a coronary artery constriction decreased from 23 mmHg to almost 0 mmHg as a consequence of atrial pacing (+40 bpm), albeit subepicardial blood flow was increased by 10%

(Hobbhahn et al., 1989). Due to the gradient of oxygen pressures within the left ventricular wall (Kirk and Honig, 1964, Koyama et al., 1979; Winbury et al., 1971; Yokoyama et al., 1978), the MDO recordings are certainly not identical with PO_2 values in inner left ventricular layers. But the determinations might nevertheless provide new insights into physiological principles of the regulation of myocardial oxygenation.

Measurement of genuine myocardial PO_2 includes the avoidance of diffusion of room air into the space between the electrode and the myocardium, as well as a resorption of oxygen by the exposed microvessels of the epicardial surface which might be detected subsequently by the measuring device. To exclude such interferences, the electrodes were suffused before each measurement with pure oxygen and 100% nitrogen for several minutes. The experiments were continued only, if correct contact of the electrodes with the heart was evident from unchanged PO_2 registrations. In some cases, a reversible decrease of the PO_2 signals by 3-6 mmHg was noted which was independent of the application of either oxygen or nitrogen: We assume that this was caused by local cooling of the cardiac surface and a subsequent constriction of epicardial vessels. A second way of testing close contact of the electrode with the myocardium is to occlude the supplying artery completely for a short period. If the electrode is placed correctly, - in accordance with results obtained by others (Forst et al., 1987) - all wires register a PO_2 of 0 mmHg within about one minute. This procedure also helps for deciding if the electrode is placed on a myocardial area perfused by the occluded artery solely, or if there is an admixture of blood supply through collateral vessels (Walfridsson et al., 1985).

In contrast to the subendocardium, segment shortening in the subepicardium was not deteriorated during the experiments, probably because epicardial blood flow was quite well maintained and because oxygen consumption is lower in the subepicardium. SLS in outer layers was even slightly, yet not significantly increased during coronary constriction. Differences in subepicardial segment function were reported during coronary stenosis with respect to fiber orientation despite sustained epicardial blood flow (Gallagher et al., 1982). However, the authors could not exclude a contribution of already ischemic inner left ventricular layers to epicardial dyskinesis when function was measured in the circumferential axis, since this orientation also reflects the motion of the midwall layers of left ventricular muscle, which are primarily of circumferential alignment (Streeter et al., 1969). To avoid such interferences, the ultrasonic dimension gauges were placed parallel to the superficial myocardial fiber orientation. Therefore, epicardial blood flow reductions of approximately 25-30% - corresponding to the results obtained in subendocardium - did not markedly affect intrinsic dimensions.

Significant lactate production in the porcine left ventricle was observed when transmural blood flow decreased by 25% albeit regional function was still almost normal (Stowe et al., 1978). In a previous study of this laboratory, left ventricular lactate extraction commenced to fall when surface PO_2 was decreased to approximately 35 mmHg (Conzen et al., 1987). When a PO_2 of 25-30 mmHg was attained, a value where neither segment shortening nor transmural oxygen supply were markedly worsened

431

in the present study, lactate metabolism changed to lactate production indicating a considerable alteration of myocardial oxidative metabolism already at these minor degrees of coronary artery stenosis (Conzen et al., 1987). Therefore, the MDO detected epicardial oxygenation sensitively at normal and mild ischemic conditions, whereas more pronounced levels of subepicardial flow reduction (>30%) were not differentiated by this device.

The study was performed to further clarify the role of the epicardial tissue PO_2 determinations for oxygenation of the myocardium (Conzen et al., 1989; Habazettl et al., 1989; Walfridsson et al., 1985; Hobbhahn et al., 1989). A problem of the average PO_2 values obtained by the MDO is, that they exceed coronary venous PO_2, which is often also regarded as a measure of average tissue PO_2. However, the contrary should be expected (tissue PO_2 below coronary venous PO_2), since a gradient of partial pressures from capillaries into tissue is required to enable diffusion of a sufficient amount of O_2 molecules into the myocardial tissue.

One possible hypothesis to explain this discrepancy is that there exist gradients within the left ventricular wall for tissue PO_2 as well as for local coronary venous PO_2. O_2 pressures might be high in the outer myocardial layers, but considerably lower near the subendocardium. PO_2 determinations in blood from the coronary sinus would reflect the average of these values. In fact, results obtained by intramyocardially inserted needle electrodes have demonstrated such gradients of O_2 pressures. However, these gradients were shallow and would not allow to explain surface tissue O_2 pressures of approximately 45 mmHg. Therefore, other explanations must be sought for the high O_2 pressures observed in this and in other studies.

Implications for tissue oxygenation

Since there is only a limited number of non-polarographic determinations of cardiac tissue oxygenation available at present, it seems justified to relate the results of the surface electrode primarily to direct measurements of intracellular PO_2 in working skeletal muscle. Due to the relatively large hemispheric catchment area of the single electrode wires, the recordings are averaged values on cardiac muscle cells, capillaries, arteriolar and venular vessels. The PO_2 values then primarily reflect interstitial PO_2 in this tissue volume rather than intracellular, since PO_2 in the space surrounding the myocytes must exceed intracellular PO_2 to enable a net flux of oxygen molecules into the muscle cells. If the Krogh Erlang model for transcapillary oxygen exchange, which does not include diffusional or functional resistances for oxygen flux is applied (Krogh, 1919), the measured average oxygen pressures of 45 mmHg would account for extracellular PO_2 as well as for intracellular PO_2, at least for the layer immediately adjacent to the sarcolemma. The principal diffusion resistance is inside the myocyte, and hence the PO_2 gradient from the red cell to the sarcolemma is assumed to be a small fraction of the PO_2 gradient from the sarcolemma to the cell interior (Gayeski and Honig, 1986). And indeed, if the Krogh Erlang formula is used to calculate myocardial tissue oxygen pressures, mean PO_2 values of the magnitude found in this study are

432

obtained (Lübbers, 1982; Turek et al., 1986).

However, the concept of an unimpeded oxygen diffusion and a curvilinear decrease of the PO_2 values from the capillaries into tissue is not accepted unanimously: In cardiac muscle cells (Honig and Gayeski, 1987) and in heavily working skeletal muscle intracellular PO_2 values obtained by cryospectrophotometry were largely below 4 Torr (Connett et al., 1985; Gayeski and Honig, 1987) and hence considerably below venous PO_2 (Gayeski and Honig, 1987). Intracellular PO_2 values between 4 and 6 mmHg were also obtained for the heart by measurement of carbon monoxide binding to myoglobin (Coburn et al., 1973). Low intracellular PO_2 values (found in the center of the muscle cells as well as adjacent to the sarcolemma) would imply a steep gradient of PO_2 from capillaries into muscle cells (Connett et al., 1985; Gayeski and Honig, 1987). Precise nature and location of such a diffusion limiting resistance are not exactly defined yet: The sarcolemma of muscle cells, tight binding of oxygen to hemoglobin (Rose and Goresky, 1985) as well as the capillary itself (Gayeski and Honig, 1986, 1987) have been proposed. According to the latter theory, a boundary layer develops beneath the red cell membrane, slowing O_2 release from the red cells by chemical processes (Gayeski and Honig, 1986). Furthermore, the oxygen flux density in the oxygen carrier free space surrounding the red cells is high. High flux density and absence of facilitated diffusion require a steep oxygen concentration gradient to support flux (Gayeski and Honig, 1986). Both effects would explain the steep PO_2 gradient from red cells into myocytes especially during near maximal work and oxygen consumption. However, the PO_2 drop would be reduced if oxygen consumption by the myocytes were less (Connett et al., 1985; Gayeski et al., 1987). Although we did not determine left ventricular oxygen consumption in these experiments, systemic hemodynamic variables indicate that left ventricular oxygen consumption was far below the maximal oxygen consumption of the heart. This might explain the unexpectedly high tissue oxygen pressures. Since the sarcolemma is not an important oxygen diffusion impeding barrier (Gayeski and Honig, 1986), intracellular PO_2 beneath the sarcolemma in subepicardial layers of the normally beating left ventricle then would be slightly lower than the average polarographic PO_2. An intracellular PO_2 gradient is expected in this situation because of absence of facilitated oxygen diffusion at a PO_2 well above the p50 of myoglobin.

Another important point is that PO_2 recordings commenced to decrease at coronary stenoses which did not strikingly influence subepicardial blood flow, and that PO_2 values of 0 mmHg were achieved when relative oxygen supply was reduced by only 20-30% and when subepicardial function still was normal. This could be explained by anaerobic cardiac metabolism which effectively compensated the reduction in local oxygen supply. However, this compensation would not account for an equivalent reduction in global left ventricular blood supply (Lekven et al., 1973). More likely, a compensatory (yet not significant) increase in subepicardial function and hence oxygen consumption at a limited oxygen availability explains the pronounced drop in surface tissue PO_2 even at low grades of coronary constriction. This would explain the steep slope of the regression line between PO_2 and subepicardial blood flow.

A further explanation for the steep decrease in surface tissue PO_2 might be that due to hypoxic coronary vasodilation and a concomitant decrease in red cell velocity, diffusional oxygen shunting is increased in the ischemic myocardium during coronary stenosis (Schubert et al., 1978). Finally, a reduced poststenotic driving pressure by itself could induce a redistribution of capillary blood flow and deteriorate tissue PO_2 (Kessler et al., 1976). These phenomenons would also explain the high sensitivity of the MDO when blood flow is only minimally impaired.

We stress that PO_2 recordings of 0 mmHg do not necessarily indicate that cellular oxygen availability or intracellular PO2 are in fact zero. Because of the diffusion resistance of a thin layer of connective tissue covering the left ventricle or due to minor electrode drifts, the MDO may not be very effective for measuring oxygen pressures between 0 and 0.5 torr, albeit adequate myocyte function may be well maintained (Gayeski et al., 1987; Wittenberg and Robinson, 1981). Nevertheless, this technique of determining tissue PO_2 appears to be useful for studying effects of interventions on oxygenation of the normal and of the slightly ischemic myocardium.

ACKNOWLEDGMENT

The authors gratefully acknowledge the technical skill of A. Holzer and A. Schmidbauer as well as the secretarial expertize of C. Chaudry.

The study was supported by Friedrich Baur Foundation.

REFERENCES

Clark, L.C., 1956, Monitor and control of blood oxygen tension, Am. Soc. Artif. Intern. Organs, 2: 41-48.
Coburn, R.F., F. Ploegmakers, P. Gondrie, R. Abboud, 1973, Myocardial myoglobin oxygen tension, Am. J. Physiol, 224: 870-876.
Connett, R.J., T.E.J. Gayeski, C.R. Honig, 1985, An upper bound on the minimum pO2 for O2 consumption in red muscle, Adv. Exp. Med. Biol., 191: 291-300.
Conzen P.F., J. Hobbhahn, A.E. Goetz, P. Gonschior, G. Seidl, K. Peter, W. Brendel,1989, Regional blood flow and tissue oxygen pressures of the collateral-dependent myocardium during isoflurane anesthesia in dogs, Anesthesiology,70: 442-452.
Conzen, P., J. Hobbhahn, A. Goetz, G. Seidl, P. Gonschior, K. Peter, W. Brendel, 1987, Myocardial surface PO_2 correlates with transmural lactate metabolism in a porcine model of acute coronary stenosis, in: "Microcirculation - an update," Tsuchiya, M., M. Asano, Y. Mishima, M. Oda, eds., Excerpta medica, Amsterdam.
Forst, H., J. Racenberg, R. Schosser, K. Messmer: Right ventricular tissue pO2 in dogs, 1987, Effects of hemodilution and acute right coronary artery occlusion, Res. Exp. Med., 187: 159-174.
Gallagher, K.P., G. Osakada, O.M. Hess, J.A. Koziol, W.S. Kem-

per, J. Ross jr, 1982, Subepicardial segmental function during coronary stenosis and the role of myocardial fiber orientation, Circ. Res, 50: 352-359.

Gallagher, K.P., M.C. Stirling, M. Choy, C.A. Szpunar, R.A. Gerren, M.J. Botham, J.H. Lemmer, 1985, Dissociation between epicardial and transmural function during acute myocardial ischemia, Circulation, 71: 1279-1291.

Gayeski, T.E.J., R.J. Connett, C.R. Honig, 1987, Minimum intracellular pO_2 for maximum cytochrome turnover in red muscle in situ, Am. J. Physiol., 252: 906-915.

Gayeski, T.E.J., C.R. Honig,1986, O_2 gradients from sarcolemma to cell interior in red muscle at maximal VO_2, Am. J. Physiol., 251: 789-799.

Gajeski, T.E.J. and C.R. Honig, 1987, Shallow intracellular O_2 gradients and absence of perimitochondrial O_2 "wells" in heavily working red muscle, Adv. Exp. Med. Biol., 200: 487-494.

Habazettl, H., P.F. Conzen, J. Hobbhahn, T. Granetzny, A.E. Goetz, K. Peter, W. Brendel, 1989, Left ventricular oxygen tensions in dogs during coronary vasodilation by enflurane, isoflurane and dipyridamole, Anesth. Analg., 68: 286-294.

Heymann, M.A., B.D. Payne, J.I.E. Hoffman, A.M. Rudolph, 1977, Blood flow measurements with radionuclide-labeled particles, Prog. Cardiovasc. Dis., 20: 55-79.

Hobbhahn, J., P.F. Conzen, E. Hansen, A.E. Goetz, G. Seidl, P. Gonschior, W. Brendel, K. Peter, 1989, Myocardial surface oxygen tension is an indicator of transmural tissue oxygenation of the in vivo beating pig heart, Cardiovasc. Res., 23: 529-540.

Honig, C.R. and T.E.J. Gayeski, 1987, Capillary function and the role of myoglobin in myocardium. Abstracts of the Fourth World Congress for Microcirculation, Tokyo.

Kessler, M., J. Höper, B.A. Krumme, 1976, Monitoring of tissue perfusion and cellular function, Anesthesiology, 45: 184-197.

Kirk, E.S. and C.R. Honig, 1964, Nonuniform distribution of blood flow and gradients of oxygen tension within the heart, Am. J. Physiol., 207: 661-668.

Koyama, T., M. Horimoto, Y. Kikuchi, Y. Kakiuchi, T. Arai, 1979, Non-uniform oxygen supply to the left ventricular myocardium by systolic perfusion of coronary artery, Jap. J. Physiol., 29: 267-274.

Krogh, A., 1919, The number and distribution of capillaries in muscles with calculations of the oxygen pressure head necessary for supplying the tissue, J. Physiol. (Lond), 52: 409-415.

Lekven, J., O.D. Mjos, J.K. Kjekshus, 1973, Compensatory mechanisms during graded myocardial ischemia, Am. J. Cardiol., 31: 467-473.

Lübbers, D.W., 1977, Quantitative measurement and description of oxygen supply to the tissue, in: "Oxygen and Physiological Functions", F.F. Jöbsis, ed., Professional Information Library, Dallas.

Lübbers, D.W., 1982, Oxygen supply to the myocardium, in: "Microcirculation of the Heart", H. Tillmanns, W. Kübler, H. Zebe, eds., Berlin, Springer-Verlag, 1982, p 119

Mendler, N., S. Schuchhardt, F. Sebening, 1973, Measurement of intramyocardial oxygen tension during cardiac surgery in man, Res. Exp. Med., 159: 231-238.

Moss, A.J., 1968, Intramyocardial oxygen tension, Cardiovasc. Res., 3: 314-318.

O'Riordan, J.B., J.T. Flaherty, S.F. Khuri, R.K. Brawley, B. Pitt, V.L. Gott, 1977, Effects of atrial pacing on regional myocardial gas tensions with critical coronary stenosis, Am. J. Physiol., 232: 49-53.

Rose, C.P. and C.A. Goresky, 1985, Limitations of tracer oxygen uptake in the canine coronary circulation, Circ. Res., 56: 57-71.

Schubert, R.W., W.J. Whalen, P. Nair, 1978, Myocardial PO_2 distribution: relationship to coronary autoregulation, Am. J. Physiol., 234: 361-370.

Stowe, D.F., D.G. Mathey, W.Y. Moores, S.A. Glantz, R.M. Townsend, P. Kabra, K. Chatterjee, W.W. Parmley, J.V. Tyberg, 1978, Segment stroke work and metabolism depend on coronary blood flow in the pig, Am. J. Physiol., 234: 597-607.

Streeter, D.D., H.M. Spotnitz, D.P. Patel, J. Ross, E.H. Sonnenblick, 1969, Fiber orientation in the canine left ventricle during diastole and systole, Circ. Res., 24: 339-347.

Turek, Z., L. Hoofd, K. Radusan: Myocardial capillaries and tissue oxygenation, 1986, Can. J. Cardiol., 2: 98-103.

Walfridsson, H., D.H. Lewis, F. Sjöberg, N. Lund, 1985, Acute coronary occlusion: Oxygen pressure in the border zone studied in the pig, Int. J. Microcirc.: Clin. Exp., 4: 109-119.

Winbury, M.M., B.B. Howe, H.R. Weiss, 1971, Effect of nitroglycerin and dipyridamole on epicardial and endocardial oxygen tension - further evidence for redistribution of myocardial blood flow, J. Pharmacol. Exp. Ther., 176: 184-199.

Wittenberg, B.A. and T.F. Robinson, 1981, Oxygen requirements, morphology, cell coat and membrane permeability of calcium tolerant myocytes from hearts of adult rats, Cell Tissue Res., 216: 231-251.

Yokoyama M., K. Maekawa, Y. Katada, Y. Ishikawa, T. Azumi, T. Mizutani, H. Fukuzaki, T. Tomomatsu, 1978, Effects of graded coronary constriction on regional oxygen and carbon dioxide tensions in outer and inner layers of the canine myocardium, Jap. Circ. J., 42: 701-709.

EPICARDIAL OXYGEN TENSIONS DURING CHANGES IN ARTERIAL PO$_2$ IN PIGS

Helmut Habazettl, Peter F. Conzen*, Hans Baier,
Michael Christ, Brigitte Vollmar, Alwin Goetz,
Klaus Peter*, Walter Brendel

Institute of Surgical Research and *Institute of
Anesthesiology, University of Munich
Marchioninistr. 15, 8000 Munich 70, FRG

INTRODUCTION

Myocardial energy production depends mainly on aerobic metabolism and oxygen extraction in the coronary vascular bed is near maximal already during normoxemia. Thus the myocardium may be more susceptible to hypoxemia than other organs.

The first aim of the present study was to determine the net effects of arterial hypoxemia and the compensatory increase in blood flow on oxygenation and metabolism of the left ventricular myocardium. To determine myocardial oxygenation, we applied the technique of measuring surface tissue PO$_2$ by means of a Clark-type multi channel electrode (Kessler et al., 1976) to the in situ beating porcine heart and measured epicardial PO$_2$ during different degrees of hypoxemia.

The second aim of the present study was to gain further insight into the interrelations between epicardial PO$_2$, coronary venous PO$_2$ and myocardial lactate extraction during changes of arterial PO$_2$. This was done by extending measurements to high arterial PO$_2$ and by correlating epicardial tissue PO$_2$ with arterial and coronary venous PO$_2$ and with lactate extraction.

METHODS

Experimental Preparation

Seven pigs with a mean body weight of 31 kg were premedicated with ketamine (10 mg/kg), flunitrazepam (0.01 mg/kg) and atropine (0.5 mg). Following tracheotomy and endotracheal intubation the animals were mechanically ventilated with a mixture of oxygen and N$_2$O. The inspiratory oxygen fraction was adjusted to 0.22. At a respiratory rate of 18/min tidal volume was set to keep end-expiratory CO$_2$ at 4.5 vol%. End-expiratory

Oxygen Transport to Tissue XII, Edited by J. Piiper *et al.*
Plenum Press, New York, 1990

pressure was set +3 cm H_2O to avoid atelectasis. Sufficient anesthetic depth was achieved by 0.4 vol% inspired concentration enflurane and by continuous intravenous infusion of the narcotic piritramid (1 mg/kg/h) throughout the study.

Catheters were placed in the abdominal aorta and in the pulmonary artery. Following a left side thoracotomy the pericardium was opened and a catheter inserted into the left atrium for measuring atrial pressure and for injecting radioactive microspheres. A catheter was placed in the great cardiac vein by direct puncture. Finally two silicon rubber discs were fixed with atraumatic sutures to the surface of the beating left ventricle to serve as electrode holders for the O_2 sensitive multiwire electrodes.

Measurement of Myocardial Blood Flow

The reference withdrawal method was applied to measure blood flow to the myocardium with radioactive microspheres (Heymann et al., 1977). We used standard carbonized microspheres with mean diameters of 9.1 ± 0.5 µm labelled with ^{141}Ce, ^{85}Sr or ^{95}Nb. For each blood flow determination 5-6 million microspheres suspended in 0.9% saline were injected over a period of 25-30 sec through the left atrial catheter which was subsequently flushed with 10 ml 37 °C saline. At the end of the experiment the animal was killed and the heart removed. The left ventricle was dissected into 74 tissue specimens. The free wall of the left ventricle was sliced into 3 transmural layers (subepicardial; mid; subendocardial). The specimens were immediately weighed and put into plastic vials for gamma counting.

Determination of Myocardial Surface Oxygen Tensions

The method used herein to determine oxygen pressure fields has been described in detail (Kessler et al., 1976; Lübbers, 1977). In brief, it consists of a Clark type electrode (Clark, 1956) with 8 thin platinum cathodes and one Ag/AgCl anode. The electrode is covered with a teflon membrane and placed on the surface of the left ventricle. At a constant polarisation voltage of approximately -700 mV a single platinum wire (diameter 15 µm) registers a reduction current linearily dependent on the partial pressure of oxygen. The mean distance between 2 wires is 100 µm. The radius of the oxygen sensitive surface area of each single platinum wire is about 30 µm with the membrane of 12.5 µm used in the experiments. Thus, the hemispheric catchment area of a single wire measures an averaged oxygen tension from a tissue volume including several myocardial muscle cells, small supplying vessels and connective tissue. Care was taken not to place the electrodes on visible epicardial vessels.

Reduction current signals from each of the 8 channels are amplified, digitized and processed, using a PDP 11/23 computer (Digitial Equipment Corporation, Maynard, MS). The results are presented both as mean values and as frequency distribution curves which reflect tissue oxygenation, giving the net result of nutritive blood flow and tissue oxygen consumption (Nylander et al., 1983; Winbury et al., 1971). The electrodes were calibrated before and after each recording using calibration gases with 0%, 5% and 10% oxygen in nitrogen.

438

Blood Gas and Lactate Analyses

Hemoglobin concentrations and blood gas parameters in arterial and coronary venous blood samples were derived from a gas analyzer (ABL 300, Radiometer, Copenhagen, Denmark). Left ventricular oxygen delivery and uptake were calculated by arterial oxygen content times total left ventricular blood flow and by the arterial to coronary venous oxygen content difference times blood flow, respectively. Lactate concentrations were determined enzymatically in non deproteinized plasma (Boehringer Mannheim GmbH, W-Germany).

Experimental Protocol

At the end of the surgical preparation period the animals were allowed to stabilize for 30 minutes at 22% inspired oxygen. Baseline recordings of hemodynamics, myocardial surface PO_2 and myocardial blood flow were then obtained and blood samples withdrawn. Inspiratory O_2 was lowered from 22% to 19%, 16%, 13% and 10% by replacing O_2 for N_2O in the inspired air. At each step tissue PO_2 recordings and arterial and coronary venous blood samples were obtained simultaneously after 15 minutes stabilization. Hemodynamics and blood flow were measured at 22%, 16% and 13% insp. O_2. The inspiratory O_2 concentration was then switched to 22% again and measurements were repeated. The hypoxic challenges did not produce irreversible damage to the myocardium, as was indicated by complete recovery of the hemodynamic and tissue PO_2 values. Insp. O_2 was then raised to 25%, 30%, 40%, 60% and 80% and readjusted to 22% again. Simultaneous measurements of tissue PO_2 and withdrawal of blood samples were performed at each step after a stabilization time of 5 min.

Data Analyses

Summary data of hemodynamic and blood gas variables are presented as means ± standard error of the mean (SEM). Epicardial PO_2 distribution curves are given as summary histograms of all animals with PO_2 classes of 5 mmHg. Friedman's rank analysis of variance was performed, followed by adequate multiple comparisons: $p < 0.05$ was considered statistically significant. Linear regression analyses were performed using the mean values obtained at each O_2 concentration. Correlation coefficients were calculated by Spearman's rank correlation analysis.

RESULTS

Blood gases

Alveolar hypoxia reduced arterial oxygen tension from a baseline value of 103±4 mmHg to 57±3 (16% insp. O_2) and 44±2 mmHg (13% insp O_2). Arterial oxygen content decreased from 12.2±0.5 to 9.9±0.3 and 8.4±0.5 vol%. Coronary venous PO_2 and oxygen content also decreased significantly.

Hemodynamic parameters

Arterial hypoxemia resulted in a significant increase of

439

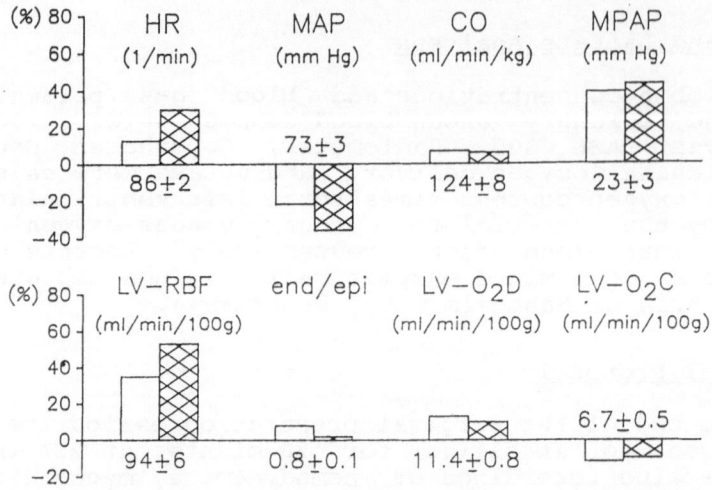

Figure 1.   *Percent changes in hemodynamic and blood flow parameters during arterial hypoxemia at 16 % (blank bars) and 13 % (crosshatched bars) inspired oxygen concentrations. Baseline values ± SEM are given for all parameters. HR, heart rate; MAP, mean arterial pressure; CO, cardiac output; MPAP, mean pulmonary artery pressure; LV-RBF, left ventricular regional blood flow; end/epi, endocardial to epicardial blood flow ratio; LV-O$_2$D and LV-O$_2$C, left ventricular oxygen delivery and consumption.*

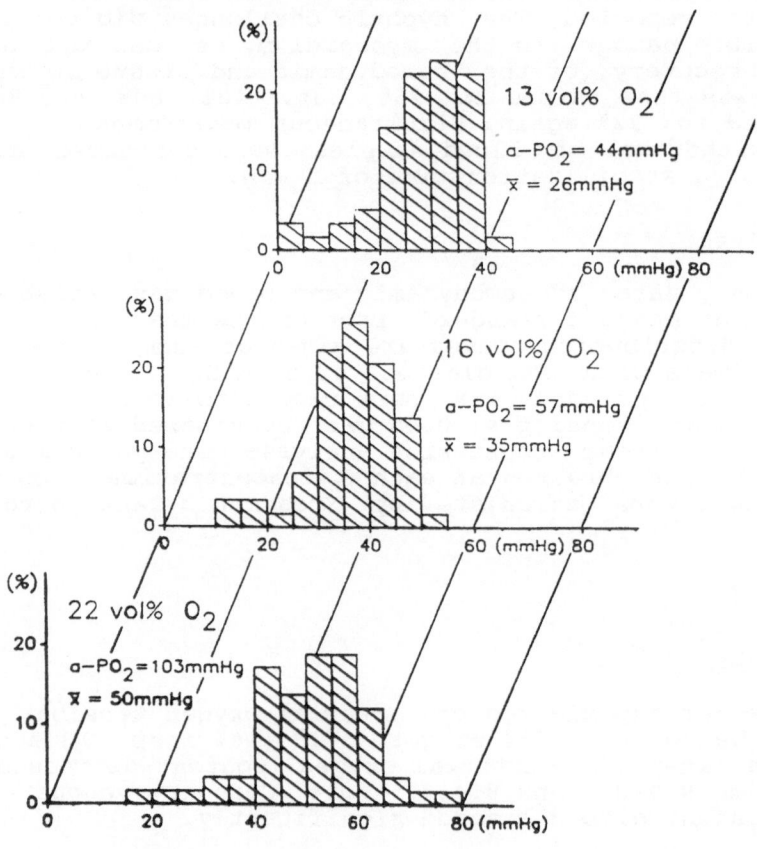

Figure 2.   *Epicardial PO$_2$ histograms during baseline (22 %) and during hypoxemia (16 % and 13 % inspired oxygen concentrations). The corresponding arterial PO$_2$ (a-PO$_2$) and the mean (x) epicardial PO$_2$ are given for each histogram.*

heart rate and, as a result of peripheral vasodilation, a significant decrease of mean arterial pressure. Cardiac output increased slightly but not significantly, mean pulmonary artery pressure increased considerably (Figure 1).

Myocardial oxygen balance and epicardial PO_2

During hypoxia left ventricular blood flow increased, thus maintaining O_2 delivery to the left ventricular myocardium close at its baseline value. Left ventricular O_2 demand, estimated by heart rate times systolic pressure product, as well as O_2 consumption remained essentially unchanged (Figure 1). Epicardial PO_2 decreased during hypoxia as indicated by a leftward shift of the PO_2 distribution curve (Figure 2) and a significant decrease of mean epicardial PO_2 The unchanged distribution patterns of the histograms indicate a uniform reduction of oxygenation in the myocardium.

Myocardial lactate metabolism

At 16% insp. O_2 one and at 13% insp. O_2 five out of seven hearts produced lactate. The mean lactate extraction values decreased significantly from 27.5±4.6% at baseline to 16.5± 3.3% at 16% insp.O_2 and to -3.3±5.3% at 13% insp. O_2

Regression analyses

Regression analyses revealed significant linear correlations of 1/arterial PO_2 with both epicardial PO_2 ($r=0.98$) and coronary venous PO_2 ($r=0.96$) and of epicardial PO_2 with coronary venous PO_2 ($r=0.98$). Significant linear relationships were also calculated for arterial with coronary venous O_2 contents ($r=0.9$), for arterial O_2 content with epicardial PO_2 ($r=0.86$) and for coronary venous O_2 content with epicardial PO_2 ($r=0.76$). Myocardial lactate extraction, an indicator of myocardial metabolism, correlated linearly with the reciprocal value of arterial PO_2 ($r=0.89$;), of coronary venous PO_2 ($r=0.81$) and of epicardial PO_2 ($r=0.88$) (Figures 3 and 4).

DISCUSSION

Methodological Remarks

Blood flow to the heart was determined by radioactive microsphere method using a reference sample technique (Heymann et al., 1977). Particles were injected into the left atrium to guarantee homogenous mixing with blood already at the aortic root (Buckberg et al., 1971). Homogenous mixing was verified by comparing the left and right renal cortical flow values, which did not differ by more than 10 in any animal of this study. Mean diameter of the microspheres used was 9.1 µm. With this size distribution of the streaming microspheres is not significantly different from red blood cells (Phibbs and Dong, 1970). By injecting 5-6 million microspheres 1000 to 2000 spheres were obtained per tissue sample, providing a accuracy of ± 5% at the 95% confidence level (Buckberg et al., 1971).

Our technique of measuring epicardial PO_2 provides the

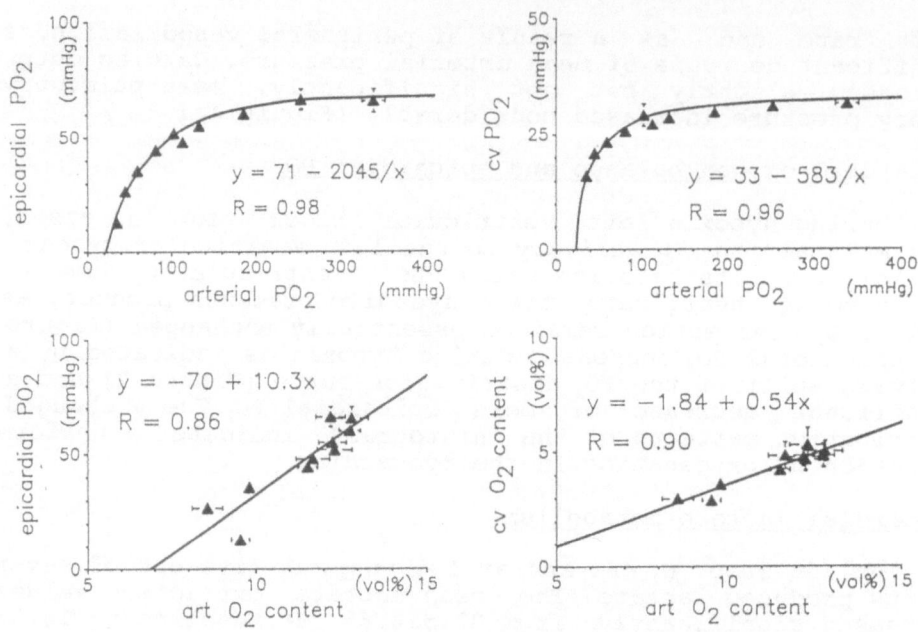

Figure 3. *Regression curves of arterial PO$_2$ with epicardial and coronary venous (cv) PO$_2$ and of arterial O$_2$ content with epicardial PO$_2$ and cv O$_2$ content. The mean values ± SEM at each inspired O$_2$ concentration are given. Error bars are within the symbols when not plotted.*

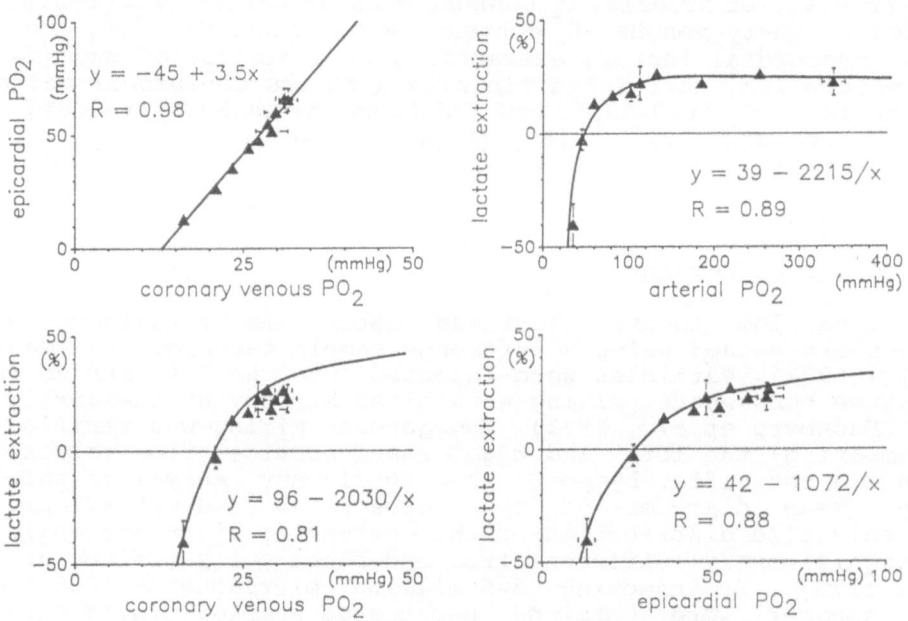

Figure 4. *Regression curves of coronary venous PO$_2$ with epicardial PO$_2$ and of arterial PO$_2$, coronary venous PO$_2$ and epicardial PO$_2$ with myocardial lactate extraction. The mean values ± SEM at each inspired O$_2$ concentration are given. Error bars are within the symbols when not plotted.*

442

advantage of atraumatic measurement of local tissue oxygenation. This is in contrast to needle electrodes which may produce trauma and hemorrhage in the tissue, thereby disturbing microcirculation. Additionally, the teflon covered surface electrode is protected from contamination with biologic material, which would alter PO_2 determinations and lead to a considerable electrode drift (Clark, 1956).

The PO_2 measurements on organ surfaces do not necessarily reflect oxygen tensions in deeper layers of tissue. Measurements of transmural myocardial oxygen tensions with needle electrodes suggest a lower tissue PO_2 in the subendocardium compared to the subepicardium (Kirk and Honig, 1964; Winbury et al., 1971). The data obtained by means of a surface electrode, therefore, do certainly not describe subendocardial tissue oxygenation. However, results obtained by means of needle electrodes (Winbury et al., 1971; Yokoyama et al., 1978) also suggest that transmural changes of myocardial tissue PO_2 can be predicted by measuring epicardial PO_2 only.

Earlier experiments have shown that the electrode can be kept in place without changing its position for more than 1 hour without alterating tissue PO_2 (Habazettl et al., 1989). Pilot studies also demonstrated that there is no influence of oxygen from the ambient air, since suffusion of the electrode with pure oxygen or pure nitrogen had no effect on the resulting PO_2 values. Also in pilot studies, killing of the animals by i.v. potassium-chloride resulted in a decrease of epicardial tissue PO_2 to 0 mmHg within 2 minutes, thereby again excluding an influence of ambient air on the results.

We applied the method of measuring tissue PO_2 by a surface electrode, because it instantly and reproducibly detects tissue oxygenation without major trauma and because it has proved useful in many experimental approaches on different tissues such as skeletal muscle (Lund et al., 1980; Nylander et al., 1983), brain (Seyde and Longnecker, 1986), liver (Conzen et al., 1988) and myocardium (Conzen et al., 1989; Habazettl et al., 1989; Kessler et al., 1984).

Myocardial oxygenation

The increase in left ventricular blood flow during hypoxemia was sufficient to maintain O_2 delivery to the left ventricle at its baseline level. The increase in heart rate and the simultaneous decrease in arterial pressure during hypoxemia resulted in an unchanged left ventricular O_2 demand, as is indicated by unchanged rate pressure product and O_2 consumption. These results may indicate that myocardial oxygenation was still adequate. On the other side, the decreases of coronary venous and epicardial PO_2 and the decrease of myocardial lactate extraction with a net lactate production in 5 out of 7 animals indicate severe tissue hypoxia.

However, coronary venous PO_2 or oxygen content do not necessarily reflect actual tissue oxygenation (Coburnet al., 1973; Habazettl et al., 1989). One possible reason for this might be a large gradient between intracapillary and intracellular PO_2. Estimating tissue oxygen pressures by determining cardiac myoglobin oxygen saturations with tracer carbon monoxide, Coburn and coworkers detected an intracellular PO_2 of

4-6 mmHg which was rather constant over a large variety of arterial oxygen tensions. Only below a critical arterial PO_2 of 30 to 35 mmHg tissue oxygen tensions decreased and net lactate production occurred (Coburn et al.,1973). A low intracellular PO_2 was also obtained spectrophotometrically in the myocardium and in red muscle. Medians of the PO_2 distribution curves between 1 mmHg and 10 mmHg (Connett et al., 1986; Honig and Gayeski, 1987) were reported. In another study on cell free perfused heart preparations intracellular PO_2 was estimated from myoglobin (Mb) reflection spectra. During hypoxia a reduction of O_2 consumption, as an indicator of tissue O_2 deficiency, occurred at about 50% MbO_2 (about 2-3 mmHg), but lactate release commenced to increase already at about 70 % of MbO_2 (Araki et al., 1983). An increase in lactate release from a perfused rabbit heart preparation at mild levels of hypoxia, when O_2 consumption was maintained at baseline levels has been described (Edlund et al., 1983). Thus, it appears that lactate release from the myocardium already increases, when the intracellular O_2 concentration is still sufficient to maintain aerobic energy production and O_2 consumption (Katz et al., 1987).

Summarizing these data from literature with respect to our own results it may be concluded that the decrease of both, coro-nary venous PO_2 and lactate extraction may indicate limited O_2 supply to the myocardium. However, both parameters fail to reflect myocardial tissue oxygenation on the cellular level, since left ventricular O_2 delivery and consumption are unchanged.

Considering the presumably low intracellular PO_2 (Coburn et al., 1973; Honig et Gayeski, 1987), the baseline histograms in this and in other studies (Kessler et al., 1984; Habazettl et al., 1989) with mean epicardial PO2 between 28 and 60 mmHg indicate, that the polarographic epicardial PO_2 values are preferentially determined by the structures with high PO_2, i.e. microvessels and interstitial space. Then the low values on the left side of the PO_2 histograms represent tissue sites around venules or venular capillaries while the high values on the right side reflect the situation near arterioles and the arteriolar ends of the capillaries.

We performed correlation analyses to gain further information about the relevance of epicardial PO_2 compared to coronary venous PO_2 and lactate extraction. The y=a+b/x shape of the correlations of both, coronary venous and epicardial PO_2 with arterial PO_2 may result from the hemoglobin-O_2 dissociation curve, as the correlations of both parameters with arterial O_2 content are linear. The unchanged shapes of the PO_2 histograms during hypoxemia indicate that there were no major disturbances of microvascular blood flow distribution. The close linear relationship between coronary venous PO_2 and epicardial PO_2 shows, that during the conditions of this study both parameters are equally effective in reflecting changes of the oxygenation state of the myocardium.

This, however, is only true for the arithmetic mean of the PO_2 histograms. If the PO_2 histograms are analyzed, especially with respect to the lowest PO_2 values in the distribution curves, the reduction of arterial PO_2 from 103 to 57 mmHg results in a reduction of mean epicardial PO_2 by 15 mmHg,

444

while the lowest values decrease by only 5 mmHg and are still well beyond the lowest PO_2 class of 0 to 5 mmHg, which is supposed to be the critical range for tissue hypoxia (Seyde and Longnecker, 1986; Kessler et al., 1976; Nylander et al., 1983). Even with 13% insp. O_2 at an arterial PO_2 of 44 mmHg and a mean epicardial PO_2 of 26 mmHg only 4 percent of the single PO_2 values were below 5 mmHg. Thus, only few tissue sites appear to be compromised by hypoxia and the decrease in mean epicardial PO_2 may reflect a decrease in O_2 availability rather than tissue hypoxia on a cellular level.

In the present study lactate extraction remained unchanged over the wide variety of arterial PO_2 values but decreased slightly between arterial PO_2 values of about 100 and 60 mmHg and steeply below 60 mmHg. The correlations of lactate extraction with coronary venous PO_2 and epicardial PO_2 are of similar shape: no change in lactate extraction at normoxemia and hyperoxemia, and a steep decrease of lactate extraction below 25 mmHg coronary venous and 35 mmHg epicardial PO_2 (Fig 4). An unchanged lactate extraction during hyperoxemia is expected, since the metabolism of the well oxygenated myocardial myocytes should not be influenced by further increases in O_2 supply. But the decrease of lactate extraction commenced at arterial PO_2 values, where neither left ventricular performance or oxygen balance nor the lowest values of the PO_2 histograms indicated tissue hypoxia. The steep decrease of lactate extraction with decreasing arterial, coronary venous and epicardial PO_2 and the wide interindividual variability of the lactate data prevent prediction of epicardial PO_2 by measuring lactate extraction.

In conclusion, moderate systemic hypoxemia by reducing insp. O_2 to 16% and 13% did not induce left ventricular myocardial hypoxia in our pig model as estimated from hemodynamic performance, myocardial oxygen balance or the lowest values of the epicardial PO_2 histograms. Obviously, the increase in myocardial blood flow by metabolic coronary dilation was sufficient to maintain adequate O_2 metabolism. Decreases in coronary venous PO_2 and mean epicardial PO_2 seem to reflect a decrease in capillary and interstitial PO_2 rather than tissue hypoxia on a cellular level. Mean epicardial PO_2 was equally effective in reflecting myocardial oxygenation as coronary venous PO_2. Analysis of the PO_2 histograms, however, allowed to differentiate for the lowest PO_2 values, which presumably represent the dead ends of the oxygen distribution in the tissue and did not indicate tissue hypoxia. We assume that in experiments with regional changes in myocardial oxygenation, as during coronary stenosis, O_2 sensitive electrodes with their well defined spatial resolution are superior to O_2 or metabolite measurements in coronary venous blood.

SUMMARY

Arterial hypoxemia decreased epicardial tissue PO_2, measured by means of a multiwire surface electrode, as well as coronary venous PO_2 and myocardial lactate extraction. Left ventricular blood flow increased, O_2 delivery, O_2 demand and O_2 consumption of the left ventricle remained unchanged. Thus, epicardial and coronary venous PO_2 indicated decreased capillary and interstitial PO_2 rather than cellular hypoxia.

445

A linear relation between mean epicardial PO_2 and coronary venous PO_2 proves both parameters equally effective in reflecting changes in myocardial tissue oxygenation. However, PO_2 distribution curves provide additional information and epicardial PO_2 is superior in models with regional changes of myocardial oxygenation.

ACKNOWLEDGEMENT

The authors thank Dr. J. Hobbhahn for his valuable advice and A. Holzer, and S. Schneider for their technical assistance.

The study was supported by the Friedrich Baur Foundation.

REFERENCES

Araki, R., Tamura, M., and Yamazaki, I., 1983, The effect of intracellular oxygen concentration on lactate release, pyridine nucleotide reduction, and respiration rate in the rat cardiac tissue. Circ. Res. 53:448-455

Buckberg, G.D., Luck, J.C., Payne, D.B., Hoffmann, J.I.E., Archie, J.P., and Fixler, D.E., 1971, Some sources of error in measuring regional blood flow with radioactive microspheres. J. Appl. Physiol. 31:598-604

Clark, L.C. 1956, Monitor and control of blood oxygen tension. Am. Soc. Artif. Intern. Organs 2:41

Coburn, R.F., Ploegmakers, F., Gondrie, P., and Abboud R., 1973, Myocardial myoglobin oxygen tension. Am. J. Physiol. 224:870-876

Coetzee, A., Foex, P., Holland, D., Ryder, A., and Jones L., 1984, Effect of hypoxia on the normal and ischemic myocardium. Crit. Care Med. 12:1027-1031

Connett, R. J., Gayeski, T.E.J., and Honig, C.R., 1986, Lactate efflux is unrelated to intracellular PO_2 in a working red muscle in situ. J. Appl. Physiol. 61:402-408

Conzen, P.F., Hobbhahn, J., Goetz, A.E., Habazettl, H., Granetzny, T., Peter, K., and Brendel, W., 1988, Splanchnic oxygen consumption and hepatic surface oxygen tensions during isoflurane anesthesia. Anesthesiology 69:643-651

Conzen, P.F., Hobbhahn, J., Goetz, A.E., Gonschior, P., Seidl, G., Peter, K., and Brendel, W., 1989, Regional blood flow and tissue oxygen pressures of the collateral dependent myocardium during isoflurane anesthesia in dogs. Anesthesiology 70:442-452

Edlund, A., Fredholm, B.B., Patrignani, P., Patrono, C., Wennmalm, A., and Wennmalm, M., 1983, Release of two vasodilators, adenosine and prostacyclin, from isolated rabbit hearts during controlled hypoxia. J. Physiol. 340:487-501

Habazettl, H., Conzen, P.F., Hobbhahn, J., Granetzny, T., Goetz, A.E., Peter, K., and Brendel, W., 1989, Left ventricular oxygen tensions in dogs during coronary vasodilation by enflurane, isoflurane and dipyridamole. Anesth. Analg. 68:286-294

Heymann, M.A., Payne, B.D., Hoffmann, J.I.E., and Rudolph, A.M., 1977, Blood flow measurement with radionuclide-labeled particles. Prog. Cardiovasc. Dis. 20:55-79

Hobbhahn, J., Conzen, P.F.M., Goetz, A., Seidl, G., Gonschior, P., Brendel, W., Peter, K., 1989, Myocardial surface PO_2 -an indicator of myocardial oxygenation? Cardiovascular Research 23:529-540

Honig, C.R., and Gayeski T.E.J., 1987, Comparison of intracellular PO_2 and conditions for blood-tissue O_2 transport in heart and working red muscle. Adv. Exp. Med. Biol. 215:309-321

Katz, A., Edlund, A., and Sahlin, K., 1987, NADH content and lactate production in the perfused rabbit heart. Acta Physiol. Scand. 130:193-200

Kessler, M., Hoeper, J., and Krumme, B.A., 1976, Monitoring of tissue perfusion and cellular function. Anesthesiology 45:184-197

Kessler, M., Klövekorn, W.P., and Höper, J., 1984, Local oxygen supply and regional wall motion of the dog's heart during critical stenosis of the LAD. Adv. Exp. Med. Biol. 169:331-340

Kirk, E.S., and Honig, C.R., 1964, Non-uniform distribution of blood flow and gradients of oxygen tension within the heart. Am. J. Physiol. 207:661-668

Lübbers, D.W., 1977, Die Bedeutung des lokalen Gewebesauerstoffdruckes und des pO_2-Histogrammes für die Beurteilung der Sauerstoffversorgung eines Organs. Prakt. Anästh. 12:185-193

Lund, N., Jorfeldt, L., and Lewis, D.H., 1980, Skeletal muscle oxygen pressure fields in healthy human volunteers. Acta Anaesth. Scand. 24:272-278

Nylander, E., Lund, N., and Wranne, B., 1983, Effect of increased blood oxygen affinity on skeletal muscle surface oxygen pressure fields. J. Appl. Physiol. 54:99-104

Phibbs, R.H. and Dong, L., 1970, Nonuniform distribution of microspheres in blood flowing through a medium-size artery. Can. J. Physiol. Pharmacol. 48:415-421

Seyde, W.C., and Longnecker, D.E., 1986, Cerebral oxygen tension in rats during deliberate hypotension with sodium nitroprusside, 2-chloroadenosine, or deep isoflurane anesthesia. Anesthesiology 64:480-485

Winbury, M.M., Howe, B.B., and Weiss, H.R., 1971, Effect of nitroglycerine and dipyridamole on epicardial and endocardial oxygen tension - further evidence for redistribution of myocardial blood flow. J. Pharmacol. Exp. Ther. 176:184-199

Yokoyama, M., Maekawa, K., and Katada, Y., 1978, Effects of graded coronary constriction on regional oxygen and carbon dioxide tensions in outer and inner layers of the canine myocardium. Japan. Circ. J. 42:701-709

OXYGEN DEPENDENCE OF ENERGY STATE AND CARDIAC WORK IN THE PERFUSED RAT HEART

Kinji Ito, Shoko Nioka, and Britton Chance

The Department of Biochemistry and Biophysics, School of Medicine
University of Pennsylvania
Philadelphia, PA, 19104. USA

INTRODUCTION

The effects of hypoxia can be observed by noting changes in the circulatory system and tissue metabolism. Theoretical and experimental studies (Sugano et al., 1974) have shown a dependence of mitochondrial oxidative phosphorylation on oxygen tension. Thus, a reduction in oxygen tension results in a decrease of energy production. To date, there have been few studies concerning the relation between cellular oxygen concentration and high energy phosphates in heart tissue (Fukuda et al., 1989).

In tissue, an oxygen gradient is required for capillaries to supply oxygen to the mitochondria. The amount of tension required across the two sites, however, is a subject of controversy (Tamura et al., 1978; Wittenberg and Wittenberg, 1985; Gayesky and Honig, 1986). The oxygen gradient theoretically increases when oxygen consumption is increased. Thus, the dependence of cellular energy metabolism on oxygen and the oxygen gradient is critical for understanding biochemical and pathophysiological mechanisms which appear in the hypoxic heart.

The relationships between intra- and extra-cellular oxygen concentrations, and between high-energy metabolism and cardiac work with or without high work load were studied on the hypoxic rat myocardium. Two preparations, Langendorff (low work load) and working (high work load), induced two levels of cardiac response. These preparations were used to model differing levels of oxygen demand. The approach is similar to that of Tamura et al. (1978) and Chance et al. (1986).

METHOD

Male rats (250-450 g) were anesthetized with pentobarbital (100 mg/kg). The hearts were excised rapidly and perfused with a blood-free solution using the modified Langendorff method. The perfusate consisted of Krebs-Henseleit bicarbonate solution containing 10 mM glucose without phosphate and bubbled with 95% O_2 + 5% CO_2 at pH 7.4. The temperature of the influent perfusate was maintained at 32-35 °C. The perfusion circuit included an overflow reservoir which was placed 90 cm above the level of the aortic cannula to maintain a constant coronary perfusion pressure. The hearts were beating spontaneously.

For the working heart preparation, the left atrium was cannulated with a vinyl tube. This tube was connected to a reservoir placed 30 cm above the level of the heart to maintain constant filling pressure. This caused a twofold increase in the heart work load. Left ventricular pressure was measured by a pressure transducer (Gould 1 nc), through a 22-gauge teflon catheter placed in the left ventricle. A catheter was placed in the pulmonary artery. Oxygen tensions in the aorta (P_aO_2) and coronary sinus (P_vO_2), sampled constantly by peristaltic pump, were measured by two flow-type miniature Clark electrodes (MI-730, Microelectrodes, Inc.) connected to a dual oxygen monitor (University of Pennsylvania

Oxygen Transport to Tissue XII, Edited by J. Piiper *et al.*
Plenum Press, New York, 1990

Biomedical Instrumentation Group). Coronary flow (CF) was measured by collecting outflow drained from the NMR tube. Myocardial oxygen consumption (M_vO_2, μ mole/min/g wet weight) was calculated from the A-V oxygen difference and CF.

The ^{31}P-NMR spectra were obtained with a Bruker AM-500 spectrometer at 11.7 Tesla, which was operated in the pulsed Fourier transform mode and was interfaced to an Aspect 3000 computer. The NMR tube (diameter 16 mm) containing a heart, was placed in a specially designed probe (Helmholtz coil; diameter 18 mm) tuned to 202.5 MHz to obtain ^{31}P spectra. To achieve a homogeneous magnetic field, shimming was done by using a heart phantom containing a concentrated inorganic phosphate solution. Each scan was acquired for 2 minutes, using the following pulsing parameters; pulse width = 40 μsec (60° pulse), acquisition time = 200 msec, delay time = 3.6 sec ; 32 FIDs (free-induction-decay) were collected for each spectrum. FIDs were Fourier-transformed with 40 Hz line broadening. The saturation factor was determined by measuring control spectra with 10 sec delay time and the same pulse width. The areas of PCr (phosphocreatine), Pi (inorganic phosphate) and ATP peaks were calculated by multiplying peak height times one half the peak width.

Spectrophotometric studies were performed separately from NMR studies, using the same perfusion condition described above. Dual-wavelength reflectance spectrophotometry was used to measure myoglobin and cytochrome *aa3*. A flexible light guide (2 mm diameter), containing glass fiber bundles (70 μm diameter) for mixing incident and emergent light, was placed so as to touch the surface of the left ventricle. The emergent light guide was connected to an Otsuka MCPD 1000 (Otsuka Electronics). Absorbance differences between 580 nm or 605 nm and 620 nm ($OD_{580-620}$, $OD_{605-620}$) were chosen for myoglobin and cytochrome aa3, respectively (Tamura et al., 1978; Araki et al., 1983). After base line correction, % oxygenation of myoglobin (%MbO$_2$) and % oxidation of cytochrome aa3 (%Cyt$_{aa3}$) were calculated from these differences. These values were then normalized using 0% to represent absorbance at death and 100% to represent absorbance during aerobic perfusion (control).

<u>Experimental Protocol</u>

Langendorff and working hearts were perfused for 20-30 minutes to stabilize heart tissue. The P_aO_2 was then reduced from 600 to 50 mm Hg at a rate of 10 mm Hg/min, manipulated by decreasing influent O_2. Six NMR experiments and four optic studies were performed on Langendorff hearts while four NMR experiments and five optic studies were performed on working hearts.

<u>Data Sampling and Analysis</u>

NMR data were acquired every two minutes and were expressed as a percentage of the initial control value (100%). All other measured parameters were averaged over the corresponding two minute intervals. The product of pressure and rate development, an indicator of cardiac work, was calculated by multiplying ventricular pressure and heart rate and expressed as percent of the initial value (%PPR).

RESULTS

Biochemical and physiological measurements for both Langendorff and working hearts were taken during the control period. Significant differences were observed for Pi/PCr, MvO$_2$, and PPR (Table 1). Working hearts show an increase in O$_2$ consumption, cardiac work and a decreased energy state.

Table 1. Working Heart Parameters Expressed as Percentage of
Langendorff Heart Parameter Values During Control.

Parameter	Mean ± S.D.
PCr	90.1 ± 5.3
ATP	98.3 ± 6.8
Pi	119.0 ± 22.2
Pi/PCr	1.32 ± 0.08
M_vO_2	215.8 ± 65.4
PPR	140.0 ± 15.9

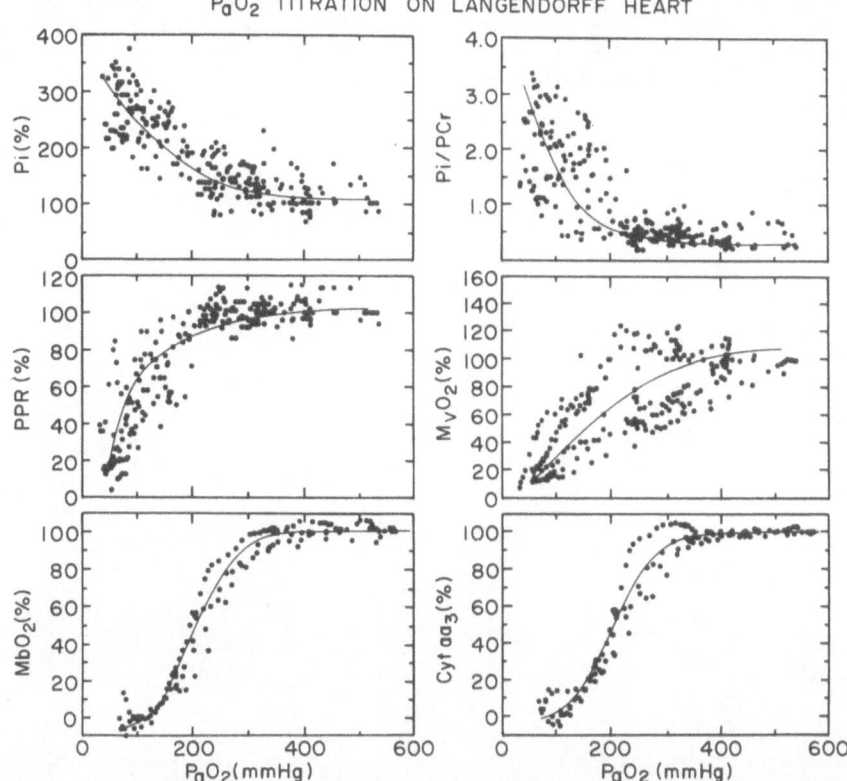

P_aO_2 TITRATION ON LANGENDORFF HEART

Figure 1. The effect of hypoxia on biochemical and physiological parameters in Langendorff hearts. With decreasing PaO_2, %PPR, %MvO_2, %MbO_2 and %Cytaa3 decreased while %Pi and Pi/PCr increased.

Table 2. P_aO_2 Values at Which Biochemical Parameters Changed Twofold During Severe Hypoxia

	Langendorff	Working
Parameter	P_aO_2 mm Hg	P_aO_2 mm Hg
PCr	110	130
Pi	120	120
Pi/PCr	150	180
PPR	90	180
M_vO_2	130	120
MbO_2	210	250
Cyt_{aa3}	210	280

451

During hypoxia (decreasing P_aO_2) Langendorff hearts and working hearts show a decrease in %PPR, $\%M_vO_2$, $\%MbO_2$ and $\%Cyt_{aa3}$ and an increase in %Pi and Pi/PCr (Figures 1 and 2). During severe hypoxia, %ATP decreased to 70% and %PCr decreased to 40%. The changes in working hearts, however, occurred at much higher P_aO_2 values than those in Langendorff hearts (Table 2).

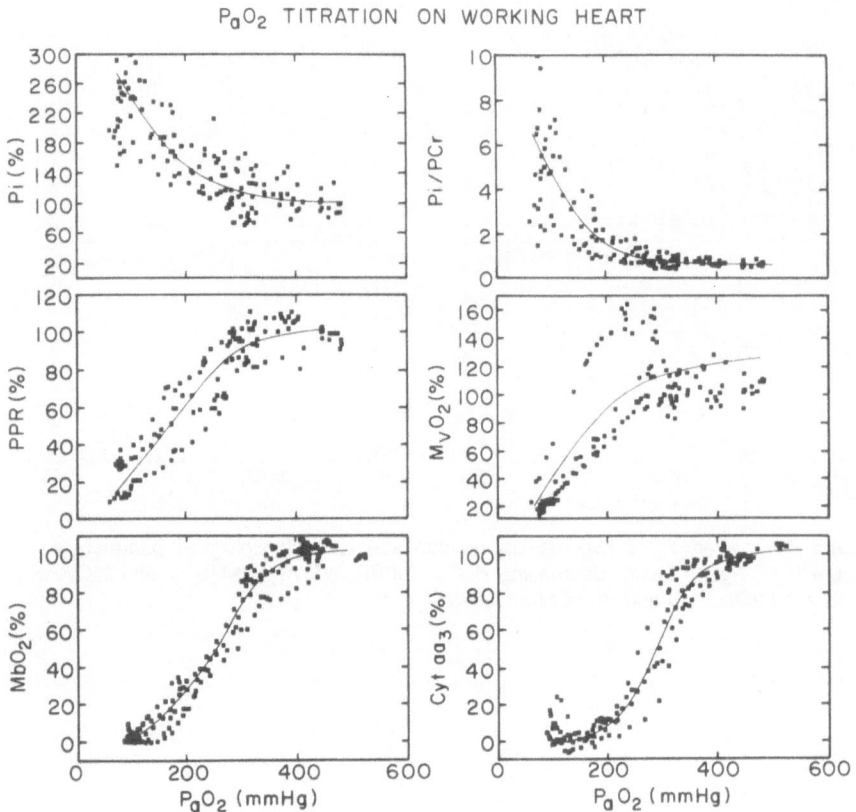

P_aO_2 TITRATION ON WORKING HEART

Figure 2. The effect of hypoxia on biochemical and physiological parameters in working hearts. With decreasing P_aO_2, %ATP, %PPR, $\%M_vO_2$, $\%MbO_2$, and $\%Cyt_{aa3}$ decreased while %Pi and Pi/PCr increased as in Langendorff hearts.

Correlation curves between biochemical parameters and P_aO_2 shown in Figures 1 and 2 were extracted and shown together in Figure 3. Biochemical changes in working hearts occurred at much higher P_aO_2 values than in Langendorff hearts. Figure 4 represents the correlations of cardiac work and oxygen indicators to PaO_2. Changes in cardiac work and cellular oxygen concentration in working hearts occurred at higher PaO_2 values than in Langendorff hearts. Both workload treatments are represented in Figures 3 and 4. Cardiac work (%PPR) was significantly affected at P_aO_2 values of 105 mm Hg in Langendorff hearts and 200 mm Hg working hearts. At these levels of oxygen energy states and redox states have decreased. Cytosolic oxygen decreased to 20% of the initial MbO_2 levels in Langendorff hearts and to 40% of the initial MbO_2 levels working hearts.

452

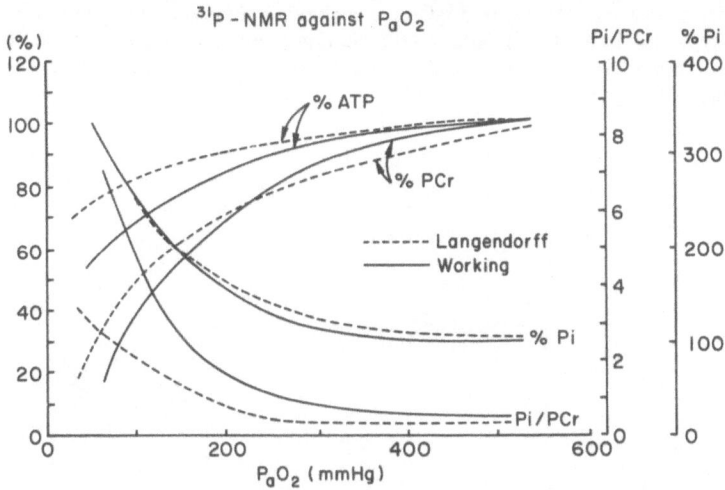

Figure 3. Biochemical changes in working hearts occurred at much higher P_aO_2 values than in Langendorff hearts.

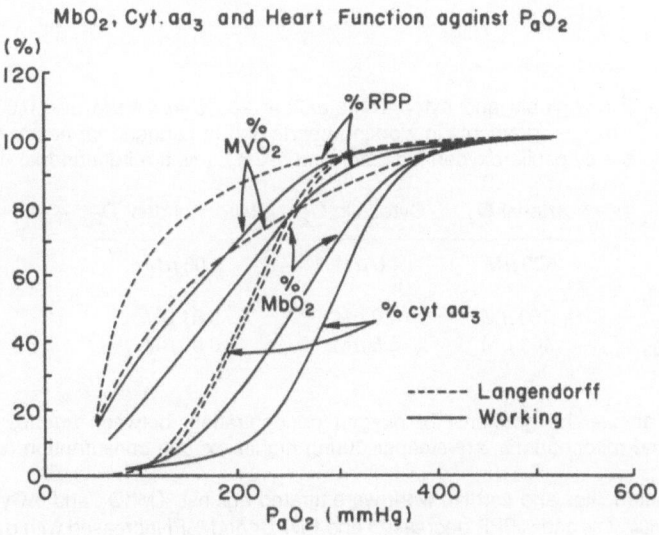

Figure 4. Changes in cardiac work and cellular oxygen concentration in working hearts occurred at higher P_aO_2 values than in Langendorff hearts.

453

The relationship between %MbO$_2$ and %Cyt$_{aa3}$ is shown in Figure 5. Langendorff hearts displayed a 1:1 linear relationship. In working hearts, a 50% change of %MbO$_2$ occurred at 250 mm Hg of P$_a$O$_2$ with %Cyt$_{aa3}$ showing a 75% change. A 50% change in %Cyt$_{aa3}$ occurred at 280 mm Hg of P$_a$O$_2$ with %MbO$_2$ showing a change of 60%.

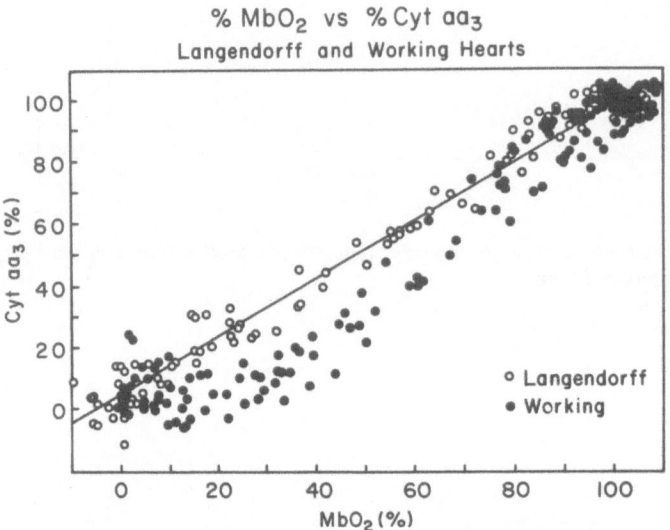

Figure 5. A 1:1 linear relationship is demonstrated in Langendorff hearts. In working hearts, values are found below the linear relationship established by Langendorff hearts. Oxygen gradients are found to be steeper during high oxygen consumption state.

Assuming that P$_{50}$ of myoglobin and cytochrome $aa3$ at 23 $^{\circ}$C are 4 μM and 0.06 μM respectively (Sugano, et al.,1974), oxygen gradients in working hearts and in Langendorff hearts can be expressed as follows (%MbO$_2$ is a cytosolic oxygen indicator and %Cyt$_{aa3}$ is a mitochondrial oxygen indicator):

	Arterial O$_2$	Cytosolic O$_2$	Mitochondrial O$_2$
Langendorff	300 μM	4.0 μM	0.06 μM
Working Hearts			
at 50 %M$_b$O$_2$	360 μM	4.0 μM	0.01 μM
at 50 %Cyt$_{aa3}$	400 μM	5.5 μM	0.06 μM

As demonstrated above, the gradients of oxygen concentration, between arteries and cytosol and between cytosol and mitochondria, are steeper during higher oxygen consumption (working heart).

Phosphorus metabolites and cardiac work were titrated against %MbO$_2$ and %Cyt$_{aa3}$ (Figures 6,7). The %ATP, %PCr, %M$_v$O$_2$, and %PPR decreased and Pi/PCr and %Pi increased with decreasing %MbO$_2$ and %Cyt$_{aa3}$. Thus high energy metabolism and cardiac work are also dependent upon cytosolic and mitochondrial oxygen concentration. When comparing changes in working hearts with those in Langendorff hearts, larger biochemical and cardiac work changes were observed in working hearts.

454

DISCUSSION

The dependence of energy state, redox state and cardiac work on oxygen tension was investigated in perfused rat hearts when subjected to two workload treatments. Using %MbO$_2$ and %Cyt$_{aa3}$ as indicators of intracellular and mitochondrial oxygen, it was found that energy state and cardiac work decreased in response to the decrease in P$_a$O$_2$, MbO$_2$, and Cyt$_{aa3}$ and that these changes were enhanced in working hearts. The O$_2$ gradients between the capillaries and cytosol and between the cytosol and the mitochondria are greater in hearts subjected to higher work loads. The plots %MbO$_2$ versus %Cyt$_{aa3}$ showed a 1:1 straight line relationship in Langendorff hearts and show a more inflected curve in working hearts, suggesting a steeper intracellular oxygen gradient during high oxygen consumption.

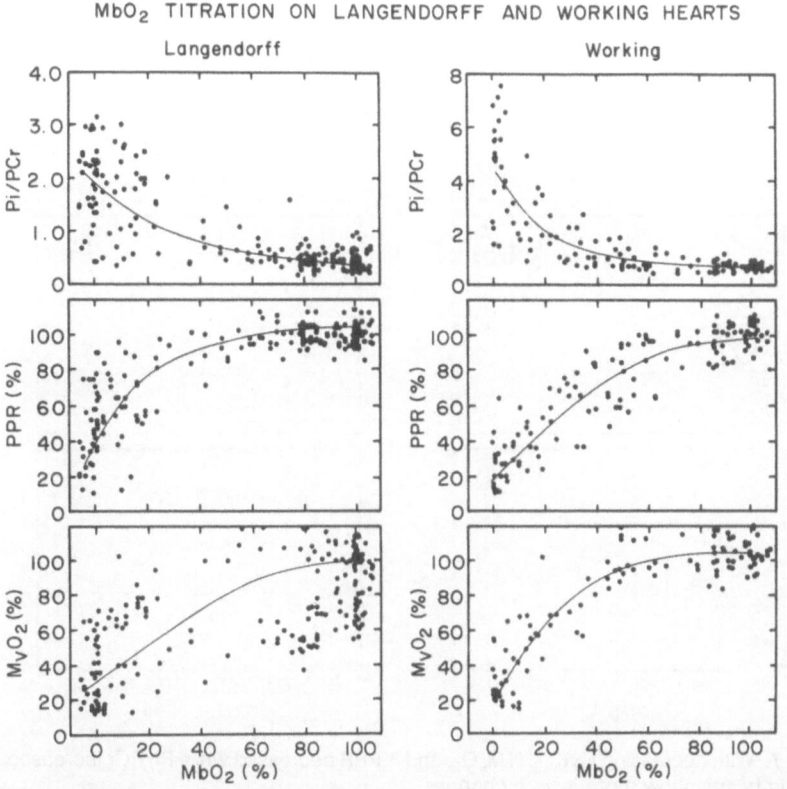

Figure 6. With decreasing MbO$_2$, %M$_v$O$_2$ and %PPR decreased while Pi/PCr increased. Working hearts show more severe changes.

Intracellular oxygen gradient studies have been reported by several investigators. Tamura et al. (1978) reported, using these two oxygen indicators that, during graded hypoxia, the ratio of %MbO$_2$ to %Cyt$_{aa3}$ was a 1:1 relationship in Langendorff perfused rat hearts. Caspary et al. (1985) reported that these relationship between the two oxygen indicators was altered in cooled and arrested perfused hearts, where oxygen consumption is low. In his study he reported relationships similar to the ones stated in this study. These studies clearly showed that the intracellular oxygen gradient is a function of oxygen consumption. These phenomena were also reported in studies of isolated cardiac myocytes (Wittenberg and Wittenberg 1985).

The relationships between high energy metabolites, and cytosolic and mitochondrial oxygen tension were also studied in this experiment. The Pi/PCr ratio is an indicator of phosphorylation potential and a direct measurement of the cellular energy state (Chance et al., 1986). This ratio is affected when mitochondrial oxygen limits ATP synthesis. Cardiac work correlated well with Pi/PCr values. Araki et al. (1983) reported that aerobic metabolism switched to anaerobic at 80% of %MbO$_2$. In this study, at about 10μM of cytosolic oxygen concentration (80% MbO$_2$), energy metabolism and cardiac work started to decline in Langendorff hearts.

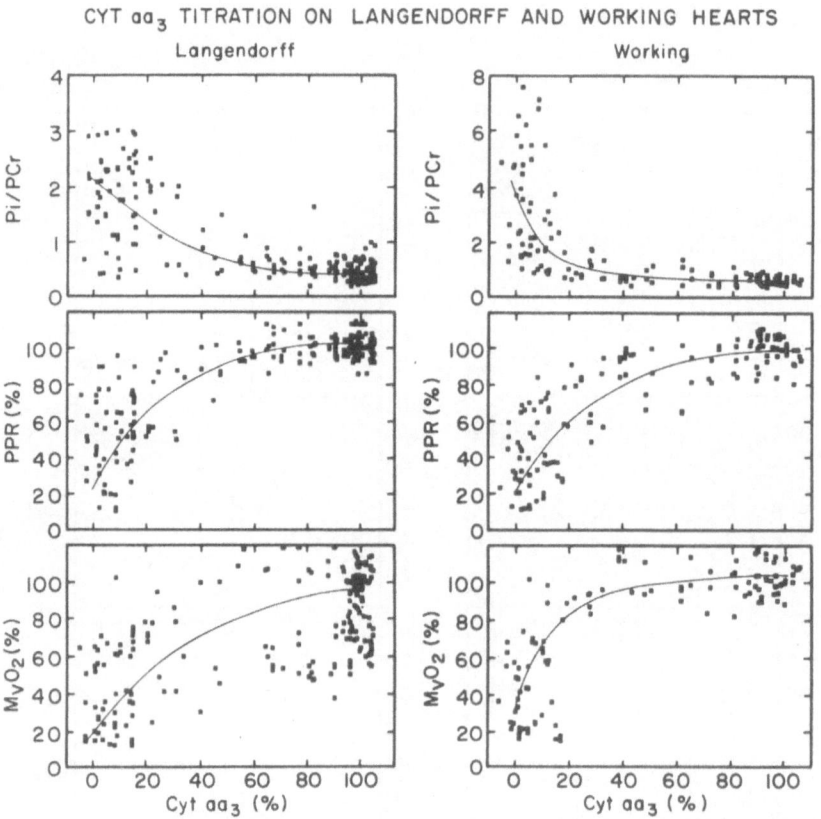

Figure 7. With decreasing Cyt$_{aa3}$, %M$_v$O$_2$, and %PPR decreased while Pi/PCr increased. Working hearts show more severe changes.

SUMMARY

The relationship between oxygen concentrations in arteries, cytosol, and mitochondria and high energy phosphate metabolism was studied in perfused rat hearts subjected to low and high workloads during gradual hypoxia. PCr, ATP, and Pi were measured by ^{31}P-NMR. Myoglobin oxygenation and cytochrome aa3 oxidation were measured by the optical method. When influent oxygen tension was decreased gradually, PCr, ATP, %MbO$_2$, %Cyt$_{aa3}$, cardiac work and M$_v$O$_2$ decreased while Pi and Pi/PCr increased in Langendorff and working hearts. These changes occurred, however, at higher P$_a$O$_2$ values in working hearts. The decrease of %MbO$_2$ and %Cyt$_{aa3}$ in Langendorff hearts was parallel where the ratio of %MbO$_2$ / %Cyt$_{aa3}$ was 1:1. However, this ratio was more than 1:1 in working hearts. It had been demonstrated that oxygen gradients change with changing oxygen consumption. Metabolic and heart work changes occurred simultaneously and significant changes occurred at low levels of $\%MbO_2$ and $\%Cyt_{aa3}$. Some differences were observed between Langendorff and working hearts. Oxidative phosphorylation is a good indicator of ATP synthesis during hypoxia and is regulated by the intracellular oxygen concentration as well as oxygen gradients.

ACKNOWLEDGEMENT

This work was supported by the National Institute of Health Grant HL 18708.

REFERENCES

Araki, R., Tamura, M., and Yamazaki, I., 1983, The effect of intracellular oxygen concentration on lactate release, pyridine nucleotide reduction, and respiration rate in rat cardiac tissue, Circ. Res. 53:448-455.

Caspary, L., Hoffmann, J.,Ahmad, H.R., and Lubbers ,D.W., 1985, Multicomponent analysis of reflection spectra from the guinea pig heart for measuring tissue oxygenation by quantitative determination of oxygen saturation of myoglobin and of the redox state of cytochrome aa_3, c , and b, Adv. Exp. Med. Biol. 191:263-270.

Chance, B., Leigh, J.S. Jr., Kent, J., Mc Cully, K., Nioka, S. Clark, B.J., Maris, J.M., and Graham, T., 1986, multiple controls of oxidative metabolism in living tissues as studied by phosphorous magnetic resonance, Proc. Natl. Acad., 83:9458-9462.

Fukuda, H., Yasuda H., Shimokawa, S. and Tamura T., 1988, The oxygen dependence of the energy state of cardiac tissue. ^{31}P-NMR and optical measurement of myoglobin in perfused rat heart, Adv. Exp. Med. Biol. 248:567-573.

Gayeski, T.E.J. and Honig, C.R., 1986, O_2 gradients from sarcolemma to cell interior in red muscle at maximal $V.O_2$. Am. J. Physiol. 251:H789-H799.

Kennedy, F.G. and Jones, D.P., 1986, Oxygen dependence of mitochondrial function in isolated cardiac myocytes. Am. J. Physiol. 250:C374-C383.

Sugano, T., Oshino, N. and Chance, B., 1974, Mitochondrial function under hypoxic conditions the steady state of cytochrome c reduction and energy metabolism, Biochim. et Biophy. Acta, 347:340-358.

Tamura, M., Oshino, N., Chance, B.,and Silver, I.A., 1978, Optical measurement of intracellular oxygen concentration of rat heart in vitro, Arch. Bioch. Biophy. 191:8-22.

Wittenberg, B.A. and Wittenberg, J.B., 1985, Oxygen pressure gradients in isolated cardiac myocytes. J. Biol. Chem. 260:6548-6554.

EFFECT OF LOCAL ANAEROBIOSIS ON HEART RATE

Franz Thimm

Physiologisches Institut
Universität Freiburg
Freiburg, F.R.G.

INTRODUCTION

Early occlusion experiments of legs showed that anaerobiosis combined with an accumulation of metabolites initiated an increase in the heart rate (Alam and Smirk, 1938; Stegemann, 1963). The authors concluded that this reflex was caused by metabolites which stimulated afferent nerve fibres. The connection between diminished oxygen supply and vegetative reactions has been demonstrated meanwhile by experiments in animals and in man, including exercise and training experiments (Bonde-Petersen et al., 1978; Barnas et al., 1986).

To prove that there exist receptors in skeletal muscle that are stimulated by products of exercise metabolites, one should separate the humoral information pathway from the neural one from skeletal muscle to the central nervous system. Therefore we designed experiments in which the hind leg of a rat was detached from the body except for nerve and bone. The leg was perfused with control Tyrode solution equilibrated with 95% O_2 and 5% CO_2 via the femoral artery and vein. During the test anaerobiosis was produced by diminishing the oxygen supply (equilibration with 80% N_2, 15% O_2 and 5% CO_2). The experimental model allowed us to control the oxygen supply to the preparation and to simultaneously measure the time course of several metabolic parameters in the venous outflow. The aim of our experiments was to correlate changes in these parameters with changes in heart rate.

In the above-mentioned occlusion experiments muscles were also stimulated to contract. Therefore we planned experiments in which the muscles were electrically stimulated during anaerobiosis. Because animals were anaesthetized a central control of movement could be excluded. Thus the infuence of the reflex component could be tested independently from central command effects. However, the electrically stimulated muscle contractions could also affect mechanoreceptors and thermoreceptors (Kao, 1963; Kumazawa and Mizumura, 1977; Kniffki et al., 1978; Foreman et al., 1979). The temporal relationship between cardiorespiratory responses and metabolic parameters could reveal whether the cardiorespiratory response could be sufficiently explained by metabolic changes and to what extent mechanoreceptors and thermoreceptors had to be assumed to participate in cardiorespiratory control from peripheral skeletal muscles.

Oxygen Transport to Tissue XII, Edited by J. Piiper *et al.*
Plenum Press, New York, 1990

METHODS

Experiments were performed on male Wistar rats (180-240 g, n = 12). There were anaesthetized with Inactin intraperitoneally (60 mg · kg⁻¹). The right hind leg was sectioned 0.5 cm proximally to the knee joint. The femur bone and the sciatic nerve were the only connections to the body (Fig. 1). The skin was removed. The femoral artery and vein were proximally ligated and distally cannulated using polyethylene cannulas. The hind leg was perfused with Tyrode solution (109 mmol/l NaCl, 4 mml/l KCl, 30 mmol/l NaHCO₃, 1 mmol/l Na₂HPO₄, 2.5 mmol/l CaCl₂, 0.7 mmol/l MgSO₄, 10 mmol/l glucose, 10 mmol/l sucrose, T = 37°C). For normoxic perfusion the Tyrode solution was equilibrated with 95% O_2 and 5% CO_2. Isolation of the preparation was tested by means of Evans blue solution. No Evans blue could be detected in the blood of the other hind leg.

Heart rate and respiratory rate (in second experimental series, see below) were continuously recorded by means of an ECG and a Fleisch sensor, respectively. The data were stored on magnetic tape and analysed using computer programs for time series analysis and statistical computation. The appropriate method for comparing two time series of synchronous measurements is cross correlation computing. The aim is to determine whether there is a temporal interrelationship between the two time series and, this being true, to determine the delay of optimum correlation. Indication of possible causal relationships are the magnitude of high positive and negative correlation, the slopes of the flanks and the time shifting position. A modified ANOVA with repeated measurements was used. In the perfusion outflow (femoral vein) the following parameters were measured: P_{O_2}, P_{CO_2} and pH, by a blood gas analyser (Corning 168), $[K^+]$ and $[Na^+]$, by ion sensitive electrodes (Corning 902) and lactate concentration ($[Lac]$), using the method described by Hohorst (1970). P_{O_2} and P_{CO_2} in the inflow (femoral artery) were determined at the start and the end of the experiment.

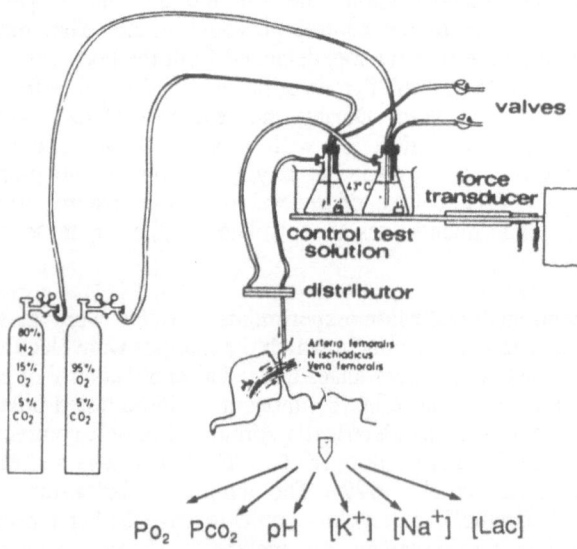

Fig. 1. Schematic diagram of the preparation of rat's hind leg and the experimental design of normoxic and hypoxic perfusion of hind leg isolated from the body except for nerve and bone. In venous outflow P_{O_2}, P_{CO_2}, pH, $[K^+]$, $[Na^+]$ and lactate [Lac] were measured.

460

Two series of experiments were performed: 1.) experiments with perfusion of hypoxic Tyrode solution (n = 6), 2.) experiments with perfusion of hypoxic Tyrode solution and electric stimulation of the hind leg muscles.

In the first series the hind leg was perfused during 17 min with control Tyrode solution and then up to 80-90 min with hypoxic Tyrode solution which was prepared by equilibration with 80% N_2, 15% O_2 and 5% CO_2. Flow rates (25-30 ml $\cdot$ g^{-1} $\cdot$ min^{-1}) were controlled by the pressure of the gas used for equilibration (Fig. 1).

In the second series a 20 min control perfusion (normoxic perfusion without stimulation) was followed by a 40 min test period (hypoxic perfusion and electric stimulation) and a second 20 min control perfusion. The sciatic nerve was stimulated by bipolar platinum electrodes. In order to avoid stimulation of small afferent fibres (group III and IV) the voltage was kept at a low level. Before each experiment the motor threshold was determined. During the test period a voltage twice the threshold voltage (0.4 - 0.8 V) was used for stimulation (Tallarida et al., 1981). Stimulation frequency was 50 s^{-1}, stimulation period 1 s followed by an interval of 2 s.

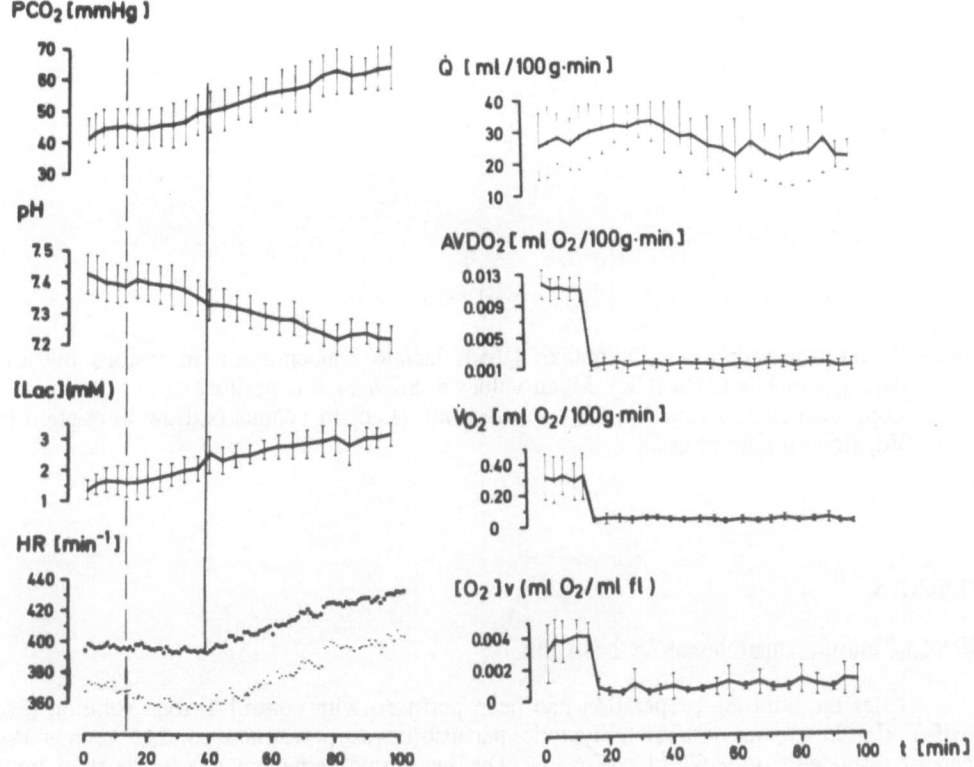

Fig. 2. Average time courses ($\pm$ SD, n = 6).
    Left hand panels: metabolic parameters (Pco_2, pH, and [Lac]) measured in venous outflow and heart rate. After 17 min the inflow was changed from control to hypoxic perfusion (dashed vertical line). The continuous vertical line indicates the time at which heart rate started to increase.
    Right hand panels: flow ($\dot{Q}$), arterio-venous difference of oxygen ($AVDO_2$), oxygen uptake ($\dot{V}o_2$) and O_2 concentration in venous outflow $[O_2]_v$.

To determine whether afferent nerve fibres were stimulated we performed experiments in which the signal flow from nerve to muscle was interrupted. The nerve was cut and proximally stimulated (n = 2). In other experiments the hind leg was perfused with Tyrode solution containing the muscle relaxans Alloferin (Alcuroniumchlorid, 1.6 ml per 1 l Tyrode solution). Thereafter the nerve was stimulated electrically. The muscle did not contract. The heart and respiratory rate did not change.

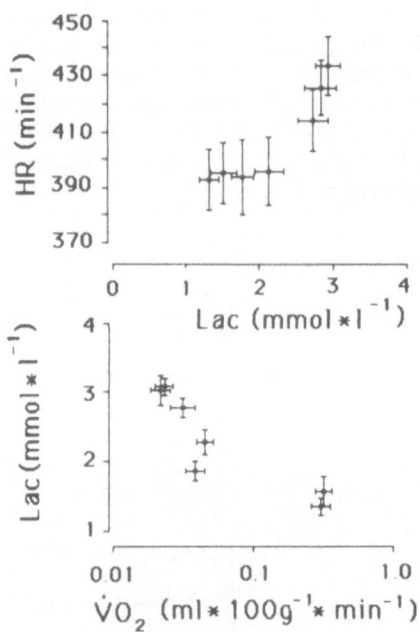

Fig. 3. Relationships between O_2 uptake ($\dot{V}O_2$), lactate concentration in venous outflow ([Lac]) and heart rate (HR). Mean values ± SEM of 6 experiments.
Top: heart rate as function of [Lac]. Bottom: [Lac] in venous outflow in relation to $\dot{V}O_2$ (logarithmic scale).

RESULTS

Effect of muscle anaerobiosis on heart rate

After the hind leg preparation had been perfused with control Tyrode solution (PO_2 = 65.3 kPa in venous outflow), hypoxic perfusion was performed so that venous PO_2 decreased suddenly to 8.67 kPa (Fig. 2). The heart rate remained nearly constant until 20 min after the change to hypoxia, thereafter increasing continuously. At this time the mean values ± SD (n = 6) in the venous outflow were: PCO_2 = 6.68 ± 0.89 kPa, [Lac] = 2.07 ± 0.37 mmol/l and pH = 7.352 ± 0.056. These values represent approximate limit values for the effects of transition from rest to exercise (Tibes, 1981).

Fig. 3 shows the relationship between O_2 uptake, $\dot{V}O_2$, and [Lac] as well as between [Lac] and heart rate. The values were determined at each 15 min interval and then averaged over the experiments. $\dot{V}O_2$ and [Lac] were negatively correlated. The threshold of heart rate increase was at 2 - 2.5 mmol/l [Lac].

462

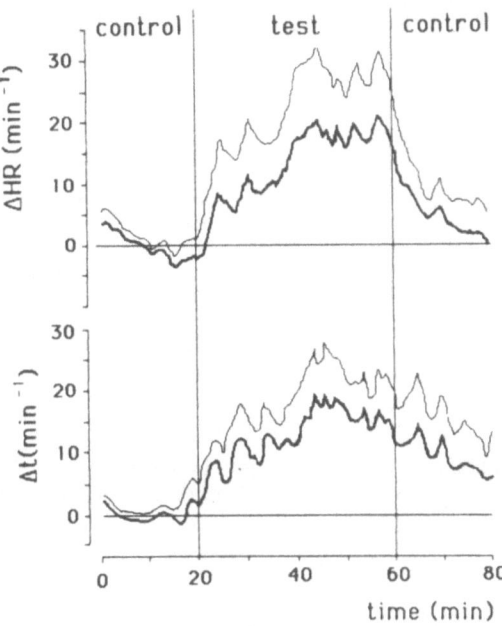

Fig. 4. Changes of heart rate, ΔHR, and changes of respiratory rate (Δt) during control perfusion (control 0-20 min), during hypoxic perfusion and electrical stimulation (test, 20 - 60 min), and during second control perfusion (control, 60 - 80 min). Thick lines: mean values. Thin lines: + SEM (6 experiments).

Cardiorespiratory responses to muscle contraction in hypoxic conditions

A short time (up to 1 min) after the start of electrical stimulation both heart and respiratory rate began to increase (Fig. 4). Maximum values were reached in 20 min. The mean maximal increase ($\pm$ SEM) of heart rate was 20.2 ± 13.2 min^{-1} (n = 6). The relative change of the heart rate was by 5.8%. Respiratory rate was increased by 24.3% from 80.0 min to 99.4 min. ANOVA showed the changes of both heart and respiratory rate to be significant (p < 0.005). The heart rate maintained its value until the end of the test period, and decreased to the control values during the second control period. Respiratory rate began to decrease in the second half of the test period and did not reach control values in 20 min of recovery.

During the first control period the metabolic parameters Pco_2, [Lac] and pH in the venous outflow were close to the values expected in resting conditions (Fig. 5) (Streter and Friedman, 1958; Steinhagen et al., 1974; Tibes, 1977). After the start of both electrical stimulation and hypoxia, Pco_2 and [Lac] rose and pH and Po_2 decreased (p < 0.001, n = 6, AVOVA) (Fig. 5). $[K^+]$ increased only at the beginning of the test and $[Na^+]$ did not change.

The time course of [Lac], Po_2, Pco_2 and pH were very similar to those of heart and respiratory rate. By means of cross correlation analysis it could be shown that the best temporal coincidence was between [Lac] and heart rate, [Lac] and respiratory rate (Fig. 6). The other parameters showed also high positive or negative correlation but maximal correlation of these relations were not found near time = 0 s (Fig. 6).

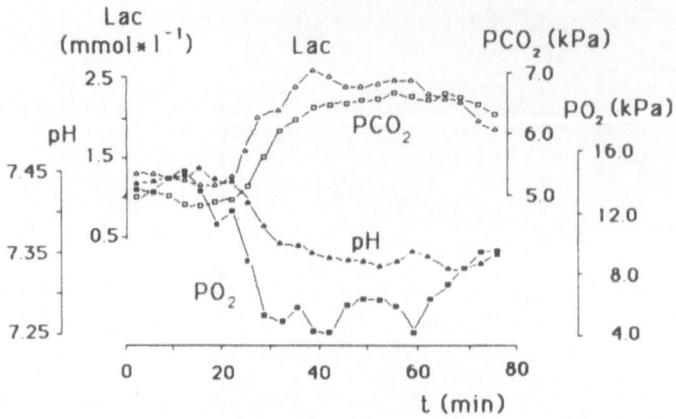

Fig. 5. Mean values of venous outflow parameters [Lac], Pco$_2$, pH, Po$_2$ during first control period (0 - 20 min), test period (20 - 60 min), and second control period (60 - 80 min).

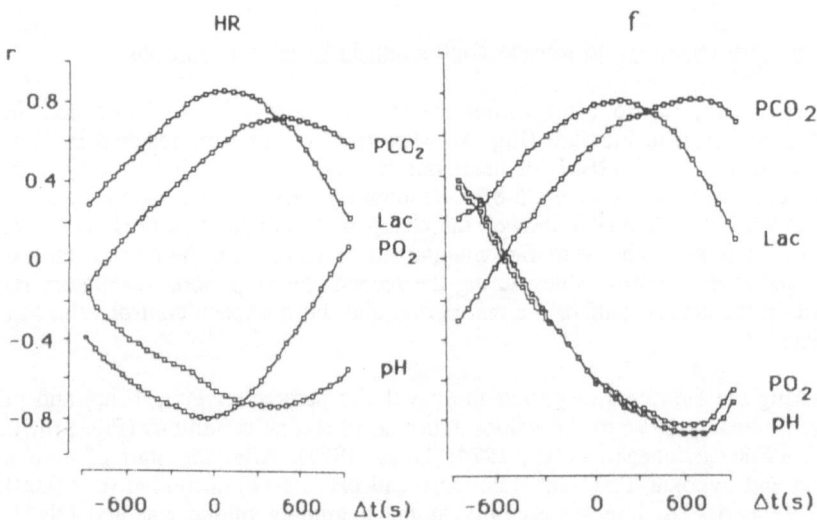

Fig. 6. Cross-correlation between heart rate (HR, left) and respiratory rate (f, right) and the venous outflow parameters: Pco$_2$, Lac, Po$_2$ and pH. The relationship (Lac - HR), and (Lac - f) is characterized by high positive correlation coefficients near 0 and steep flanks to the right and left side of the extreme values of correlation coefficients.

DISCUSSION

The muscles of the hind leg of rats were perfused separately from the body and not connected to the the body except for nerve and bone. If cardiorespiratory responses were caused by metabolic changes in muscles they could be only mediated by afferent nerve fibres. A humoral influence via blood vessels associated with the nerve or through the bone was very unlikely because no Evans blue added to the perfusion fluid of the experimental leg could be found in the other leg.

When oxygen supply to the muscle was diminished heart rate increased. Because the anaerobic muscle in the first experimental series was not stimulated to contract and the temperature did not change, we suggest that mechanoreceptors and thermoreceptors cannot alone be responsible for the heart rate increase (cf. Kniffki et al., 1981). The correlation of the changes of the metabolic parameters Pco_2, pH and [Lac] with increasing heart rate is in agreement with the hypothesis that chemosensitive muscle receptors (metabolic muscle receptors) are responsible for heart rate increase in exercise (Stegemann, 1963; McCloskey and Mitchell, 1972; Tibes, 1977; Thimm and Baum, 1987). In our earlier experiments in which the vascularly isolated hind leg of a rat was perfused with Tyrode solution enriched isotonically with lactic acid, a dose response curve showed a linear relationship between heart rate and venous [Lac] in the range of 5-10 mmol/l (Thimm et al., 1984). When the perfusion fluid was enriched isotonically with HCl the heart rate also increased. At low pH heart rate increased at both high and low [Lac], but at high [Lac] the increase was significantly higher than at low [Lac]. In hypoxia experiments mentioned above in which muscle had to produce lactic acid for energy supply, a venous [Lac] of about 2 mmol/l and pH of 7.35 was found when the heart rate started to rise. These metabolic values lay slightly over the limit values for the effects of transition from rest to exercise. Thus we suppose that these values mark the threshold of the chemosensitive receptors.

When muscles contract in hypoxia (second experimental series) a participation of mechanoreceptors could occur. It is possible that they are involved particularly at the beginning of physical exercise. However, even at the start of our test period of the second experimental series the temporal coincidence between heart rate, respiratory rate and venous [Lac] was very high. Therefore it seem to be not necessary to assume that mechanoreceptors participate in the cardiorespiratory reflexes of the working muscle.

In the recovery (second control period), the heart remained at a high level although [Lac] and pH had decreased almost to normal levels. It could be that yet other kinds of receptors (like thermoreceptors, Kumazawa and Mizumura, 1977) were operative. However, an other interpretation is possible. A prolonged reaction of this kind might be caused by continuing activation of the sympathetic system possibly due to release of catecholamines (Davies et al., 1974). In this respect it is interesting to note that the increase in heart rate was prolonged but that of the respiratory rate was not.

SUMMARY

The isolated leg of a rat was connected to the body only by nerve and bone and was perfused with hypoxic Tyrode solution. Heart rate increased when metabolic parameters (Pco_2, pH and lactate) reached values similar to those observed at the beginning of exercise. When the muscle was additionally stimulated by electric stimuli a significant temporal correlation between lactate and heart or respiratory rate was found. Metabolic changes caused by hypoxia and muscular contraction, in particular lactic acid, appear to act as chemical stimuli for metabolic muscle receptors participating in the generation of circulatory and respiratory responses to physical exercise.

REFERENCES

Alam, M., and Smirk, F.H., 1938, Observation in man on a pulse-accelerating reflex from the voluntary muscles of the legs, J. Physiol., 92: 167.

Barnas, G., Gleeson, M., and Rautenberg, W., 1986, Cardiorespiratory responses to hypoxia in intact and bilaterally vagotomized pigeons, J. Appl. Physiol., 61: 1340.

Bonde-Petersen, F., Rowell, L.B., Murray, R.G., Blomqvist, C.G., White, R., Karlson, E., Campbell, W., and Mitchell, J.H., 1978, Role of cardiac output in the pressor responses to graded muscle ischemia in man, J. Appl. Physiol., 45: 547.

Davies, C.M.T., Few, J., Foster, K.G., and Sargeant, A.J., 1974, Plasma catecholamine concentration during dynamic exercise involving different muscle groups, Eur. J. Appl. Physiol., 32: 195.

Hohorst, H.J., 1970, L(+)-Lactat. Bestimmung mit Lactat-Dehydrogenase und NAD, in: Methoden der enzymatischen Analyse, Vol. II, H.U. Bergmeyer, ed., Verlag Chemie, Weinheim.

Kao, F.F., and Ray, L.H., 1954, Regulation of cardiac output in anesthetized dogs during muscular work, Am. J. Physiol., 179: 255.

Kniffki, K.D., Mense, S., and Schmidt, R.F., 1978, Responses of group IV afferent units from skeletal muscle to stretch contraction and chemical stimulation, Exp. Brain Res., 31: 511.

Kniffki, K.D., Mense, S., and Schmidt, R.F., 1981, Muscle receptors with fine afferent fibers which may evoke circulatory reflexes. Circ. Res., 48: I 25.

Kumazawa, A., and Mizumura, K., 1977, Thin-fibre receptors responding to mechanical, chemical, and thermal stimulation in skeletal muscle of the dog, J. Physiol. (London), 273: 179.

Leiner, B., 1986, "Einführung in die Zeitreihenanalyse", 2nd edition, Oldenbourg-Verlag, München.

McCloskey, D.I., and Mitchell, J.H., 1972, Reflex cardiovascular and respiratory responses originating in exercising muscle, J. Physiol. (London), 224: 173.

Sreter, F.A., and Friedman, S.M., 1958, The effect of muscular exercise on plasma sodium and potassium in the rat, Can. J. Biochem. Physiol., 36: 333.

Stegemann, J., 1963, Zum Mechanismus der Pulsfrequenzeinstellung durch den Stoffwechsel I - IV, Pflügers Arch., 276: 481.

Steinhagen, C., Hirche, H.J., Nestle, H.W., Bovenkamp, U., and Hosselmann, J., 1976, The interstitial pH of working gastrocnemius muscle of the dog, Pflügers Arch., 367: 151.

Thimm, F., Carvalho, M., Babka, M., and Meier zu Verl, E., 1984, Reflex increase in heart rate induced by perfusing the hind leg of the rat with solution containing lactic acid, Pflügers Arch., 400: 288.

Thimm, F., and Baum, K., 1987, Response of chemosensitive nerve fibers of group III and IV to metabolic changes in rat muscles, Pflügers Arch., 410: 143.

Tibes, U., 1981, "Kreislauf und Atmung bei Arbeit und Sport", Richarz, St. Augustin.

Tibes, U., Hemmer, B., and Böning, D., 1977, Heart rate and ventilation in relation to venous $[K^+]$, osmolality, pH, Pco_2, Po_2 [orthophosphate], and [lactate] at transition from rest to exercise in athletes and non-athletes, Eur. J. Appl. Physiol., 36: 127.

HYPOXIA TOLERANCE OF CORONARY ENDOTHELIAL CELLS

T. Noll*, P. Wissemann*, S. Mertens, A. Krützfeldt,
R. Spahr, and H.M. Piper

Institut für Physiologische Chemie I*, and
Institut für Physiologie I,
Heinrich-Heine-Universität Düsseldorf, F.R.G

INTRODUCTION

The coronary endothelium performs a great number of specific metabolic and physiological functions, e.g. the synthesis of autacoids, and the control of the transport of material into and out of the blood stream. For these performances a sufficient supply of metabolic energy is required. Alterations of the energetic state, e.g. due to an insufficient O_2 supply, can lead to serious malfunctions and injury of endothelial cells, which thus can crucially contribute to ischemic myocardial damage.

The rationale of the present experiments was to study the response of the energy metabolism of coronary endothelial cells to hypoxia and reoxygenation. To meet the experimental requirements a special technique was applied, the oxystat system (Noll et al., 1986), which permitted continuous incubations of coronary endothelial cells at defined Po_2 levels and simultaneously monitoring of their energy metabolism.

MATERIALS AND METHODS

Cells. Coronary microvascular endothelial cells from the rat were maintained in monolayer cultures, as previously described (Buderus et al. 1989). Cells from 14 day-old cultures

Oxygen Transport to Tissue XII, Edited by J. Piiper *et al.*
Plenum Press, New York, 1990

were harvested and suspended in an incubation medium (composition in mM: 140.0 NaCl, 5.0 KCl, 1.0 $CaCl_2$, 1.2 $MgSO_4$, 30.0 HEPES (N-2-hydroxyethylpiperazine-N'-2-ethanesulfonic acid); pH 7.4, 37 °C). As indicated, this medium was supplemented with either 10 mM glucose, or 0.1 mM palmitate (complexed 5:1 to bovine serum albumin) plus 0.5 mM glutamine. The density of the suspension was 10^7 cells x ml^{-1}. During incubations the pH was continuously monitored. It remained in the range of 7.3 - 7.4 in all experiments.

Oxystat system. Incubations of suspended coronary endothelial cells at defined Po_2 levels between 0.1 and 100 mm Hg were performed in the oxystat system (Fig. 1). This system was based on a computer-supported feedback control circuit. The components were: (1) a 10 ml incubation chamber with a membrane filter as bottom (12 μm pore size). The chamber was equipped with a magnetic stirrer, a fluid inlet and outlet and an additional port for direct injections of supplements and for the removal of samples of the cell suspension; (2) an O_2-sensor which continuously monitoring the actual Po_2 of the cell suspension; (3) a personal computer; (4) a liquid chromatography pump. The O_2 uptake of the respiring biological material was balanced by the infusion of appropriate amounts of O_2-saturated

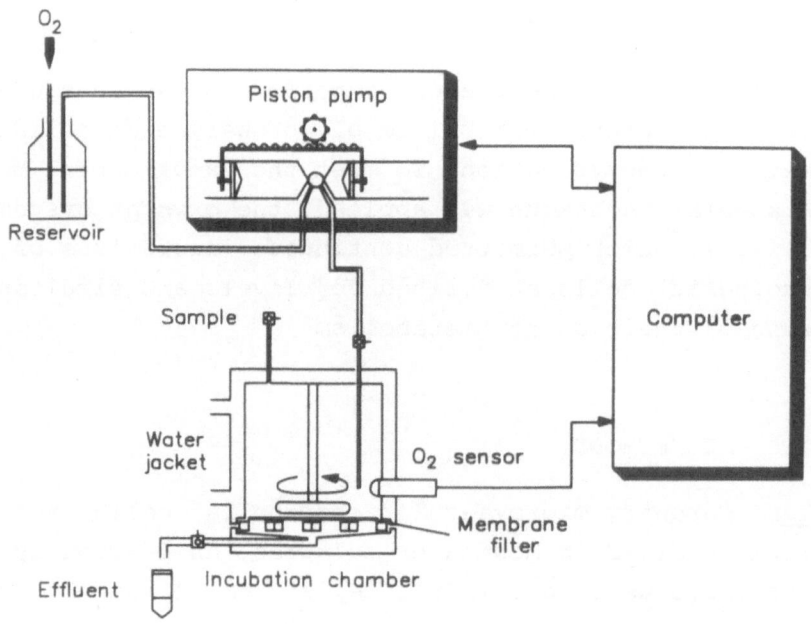

Fig. 1.   Scheme of the oxystat system.

468

medium via the liquid chromatography pump. The rates of oxygen uptake of the respiring cells were calculated from the amounts of O_2-saturated medium added.

Analytical methods. For the analysis of acid soluble metabolites, samples of the cell suspension were taken up directly into 1.2 M $HClO_4$. The neutralised extracts were analyzed for lactate according to Gutman and Wahlefeld (1974), and for ATP, ADP and AMP according to Jüngling and Kammermeier (1980). Effluent samples were also analysed for lactate. In all samples withdrawn from the cell suspension, protein was determined, using bovine serum albumin as standard (Lowry et al., 1951).

If not stated otherwise, all data are given as mean $\pm$ S.E.M. of at least 5 independent culture preparations. Statistical discriminations were performed with the unpaired t-Test; p > 0.05 was regarded as not significant.

RESULTS AND DISCUSSION

Metabolic basis of hypoxia tolerance

To evaluate the importance of exogenous substrates for the maintenance of the energetic state under well-oxygenated conditions, coronary endothelial cells were incubated at a Po_2 of 100 mm Hg. In the absence of any exogenous substrates, the cells consumed O_2 with a rate of about 8.1 $\pm$ 0.2 nmol x min^{-1} x mg $protein^{-1}$ and released lactate with a low rate of about 0.7 $\pm$ 0.1 nmol x min^{-1} x mg $protein^{-1}$, both rates of which were maintained constant for 30 min of observation (Fig. 2).

Compared to the cells without exogenous substrates, O_2 consumption and lactate formation remained unaltered when 0.1 mM palmitate plus 0.5 mM glutamine were added (Fig. 2). Under both conditions, without exogenous substrates and in the presence of palmitate and glutamine, the cellular contents of ATP, ADP and AMP were maintained constant at 20.8 $\pm$ 1.6 , 4.0 $\pm$ 0.3, and 1.6 $\pm$ 0.2 nmol x mg $protein^{-1}$, respectively. The adenylate energy charge (Atkinson, 1971) EC = ([ATP] + 0.5 [ADP]) / ([ATP] + [ADP] + [AMP]) was 0.86 indicating a well-energized metabolic state. When the ATP/O ratio is taken as 2.8, the endothelial oxygen consumption rate is equivalent to an energy production of 44.8 nmol ATP x min^{-1} x mg $protein^{-1}$. For comparison, the energy demand of the arrested heart is about 40 nmol ATP x min^{-1} x mg $protein^{-1}$ and the beating heart at moderate work load requires approximately 100 to 500 nmol ATP x min^{-1} x mg $protein^{-1}$ (Krützfeldt et al., 1989).

These results indicate that the energy demand of coronary endothelial cells is low compared to the working myocardium and is rather of the order of the arrested heart. Under normoxia, endothelial cells supplied with palmitate plus glutamine are capable to maintain their energetic state at high levels. For a limited period of time, these cells dispose of sufficient stores of endogenous substrates to fuel their aerobic energetic need.

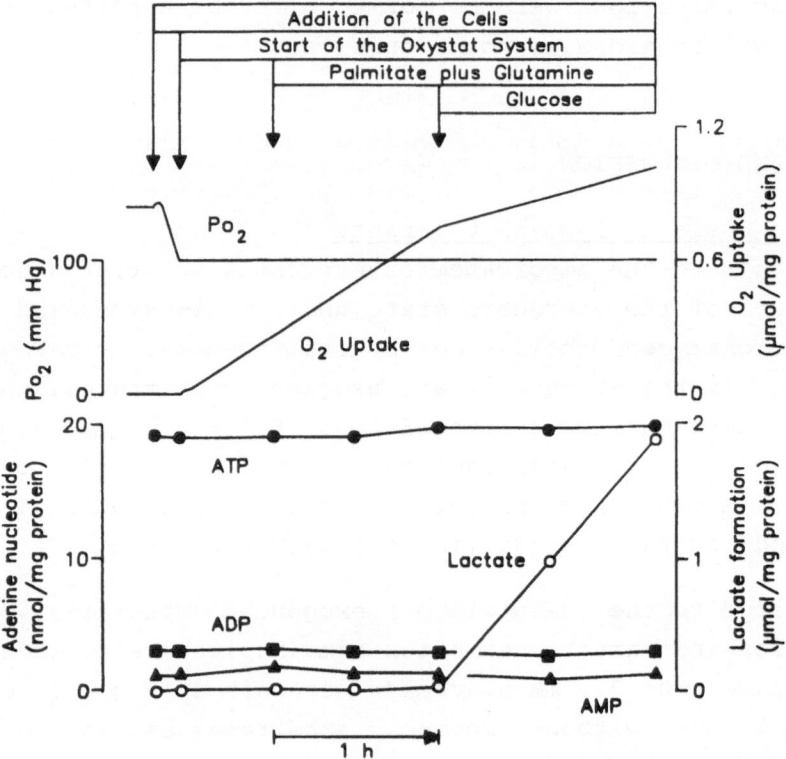

Fig. 2.   Experiment on the effect of different substrates on the endothelial O_2 uptake, lactate formation and adenine nucleotide contents. Coronary endothelial cells were incubated at a steady state Po_2 of 100 mm Hg; 0.1 mM palmitate plus 0.5 mM glutamine, and 10 mM glucose were added as indicated.

470

Upon the further addition of 10 mM glucose (Fig. 2), the endothelial respiration responded instantaneously with a 50 ± 5 % decrease of the oxygen consumption rate and an increase of the rate of lactate production to 25.4 ± 2.3 nmol x min^{-1} x mg $protein^{-1}$. The contents of the adenine nucleotides were not affected. Oxygen consumption, lactate production and adenine nucleotide contents were the same when glucose was supplied without a previous supply of palmitate and glutamine (data not shown).

The pronounced effect of glucose on the endothelial respiration and the aerobic lactate production reflects a considerable shift of the metabolic flux from mitochondrial to glycolytic energy production, while the energetic state remains unaffected. The 50 % drop of the oxygen consumption rate is equivalent to a decrease of the mitochondrial ATP production of 22.4 nmol x min^{-1} x mg $protein^{-1}$. The concomitant increase of the lactate production rate to 25.4 nmol x min^{-1} x mg $protein^{-1}$ corresponds to a rate of glycolytic energy production of 25.4 nmol ATP x min^{-1} x mg $protein^{-1}$ (1 mol ATP/1 mol lactate). The calculated total energy production amounts to 47.8 nmol ATP x min^{-1} x mg $protein^{-1}$. This shows that glucose supplied endothelial cells increase their glycolytic energy production at the expense of an equivalent decrease of their mitochondrial energy production; the total energy turnover and the energetic state is remained constant. Thus, even at normoxia, glucose supplied coronary endothelial cells cover half of their energy demand by the O_2-independent catabolic breakdown of glucose to lactate, via the glycolytic pathway. The oxygen consumption of endothelial cells under these conditions mirrors the oxidation of endogenous substrates.

The response of the energetic state to hypoxia

Since normoxic endothelial cells can cover half of their energy demand by the O_2-independent glycolytic pathway, the question arises if their capacity is sufficient to balance the lack of mitochondrial energy production under hypoxia, i.e. when O_2 becomes limiting at the site of the cytochrome oxydase. To study the role of glycolytic energy production, endothelial cells were exposed to hypoxia either supplied with 10 mM glucose or with 0.1 mM palmitate plus 0.5 mM glutamine. The cells were incubated in approximately 1 hour intervals at different Po_2 levels which were stepwise decreased from 100 to 0.1 mm Hg.

During incubations of glucose supplied coronary endothelial cells, the variation of the Po_2 between 3 and 100 mm Hg left the oxygen consumption unchanged (Fig. 3). Below 3 mm Hg, the rates of oxygen consumption declined. They were half-maximal at a Po_2 of 0.8 mm Hg. Likewise, the lactate production was almost unaffected when the Po_2 level was decreased from 100 to 3 mm Hg (Fig. 3). At Po_2 levels below 3 mm Hg, however, the rates increased, inversely related to the decrease in oxygen consumption. At 0.1 mm Hg, lactate production was 2.2-fold increased as compared to 100 mm Hg (Fig. 3). Under all Po_2 conditions tested, the adenine nucleotide contents of glucose supplied endothelial cells remained unchanged up to 2 hours of observation (Fig. 4), even if the Po_2 was lowered below 1 mm Hg.

From the lactate production at 0.1 mm Hg (Fig. 3), the rate of glycolytic energy production is equivalent to 55.9 nmol ATP x min^{-1} x mg $protein^{-1}$. This corresponds to the energy demand of 47.8 nmol ATP x min^{-1} x mg $protein^{-1}$, calulated for glucose

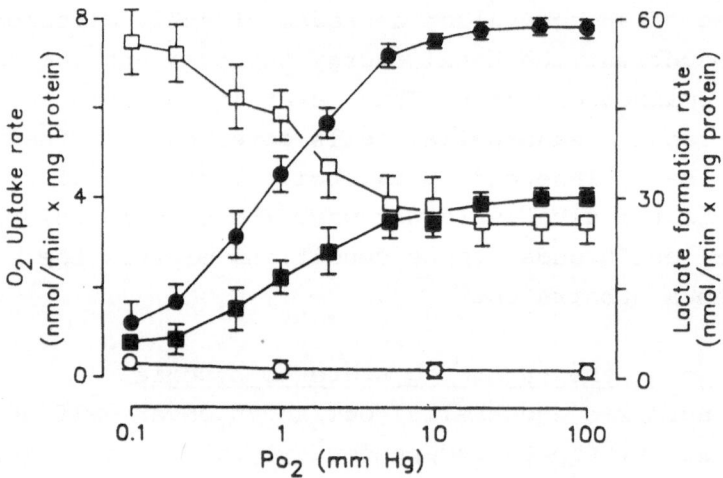

Fig. 3. Rates of O_2 uptake (filled symbols) and lactate formation (open symbols) in the presence of 10 mM glucose (squares) or 0.1 mM palmitate plus 0.5 mM glutamine (circles) as a function of Po_2. The vertical bars denote S.E. of the mean of five separate incubations.

472

supplied endothelial cells at normoxic Po_2 levels (see above). From this it can be concluded that coronary endothelial cells possess a glycolytic capacity sufficient to fully compensate the lack of respiratory ATP supply under hypoxic conditions. Furthermore, these cells are well adapted to maintain their energetic state during transitory hypoxia. The decrease of O_2 consumption at low Po_2 levels was completely reversible. After a 2 hour-incubation below 1 mm Hg, reoxygenation to 10 mm Hg quickly restored the pre-hypoxic rate of oxygen consumption (Fig. 4). Likewise, the energetic state was left unaffected by the Po_2 shift back to 10 mm Hg.

When glucose was replaced by palmitate plus glutamine, the O_2 affinity of the endothelial O_2 consumption was the same as with glucose alone, i.e. the half-maximal rate of oxygen consumption was approached at 0.8 mm Hg (Fig. 3). However, the

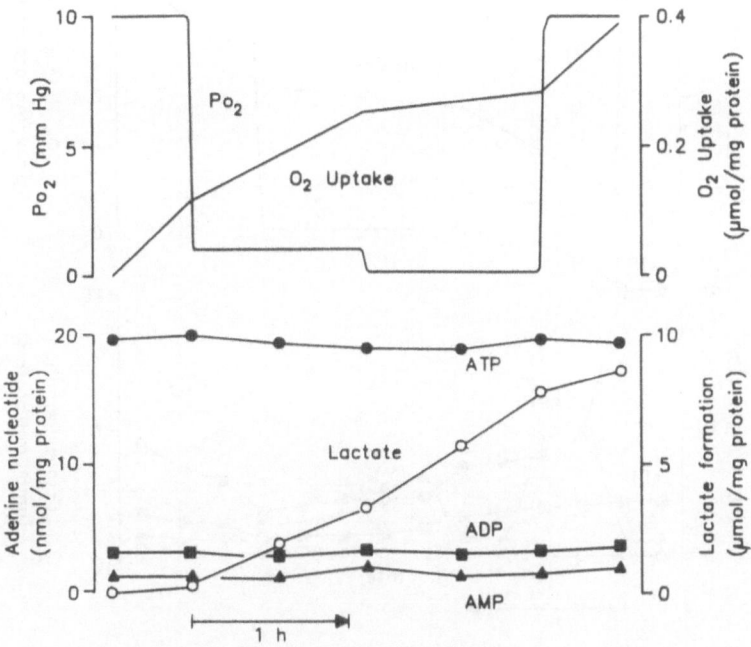

Fig. 4.   Experiment on the effect of hypoxia and reoxygenation on the endothelial O_2 uptake, lactate formation and adenine nucleotide contents in the presence of 10 mM glucose. Coronary endothelial cells were incubated at Po_2 levels of 10, 1 and 0.1 mm Hg. Subsequently the cells were reoxygenated to 10 mm Hg.

maximal rate of O_2 consumption was twice as high as in media containing glucose alone (Fig. 3) or glucose together with palmitate and glutamine (Fig. 2). The endothelial lactate production was negligibly small (below 1 nmol x min^{-1} x mg protein^{-1}).

At Po_2 levels between 10 and 100 mm Hg, the adenine nucleotide contents were identical to those found in incubations with glucose (Fig. 2, 4, 5). But in contrast to the energetic stability in the presence of glucose, the cellular contents of ATP decreased when the Po_2 was lowered to 1 mm Hg (Fig. 5). Ten minutes after changing the exterior Po_2 from 10 to 1 mm Hg, the ATP content of endothelial cells had fallen to half of its initial value. Subsequently, it declined only slowly. The decrease of the ATP content was paralleled by a transitory increase in

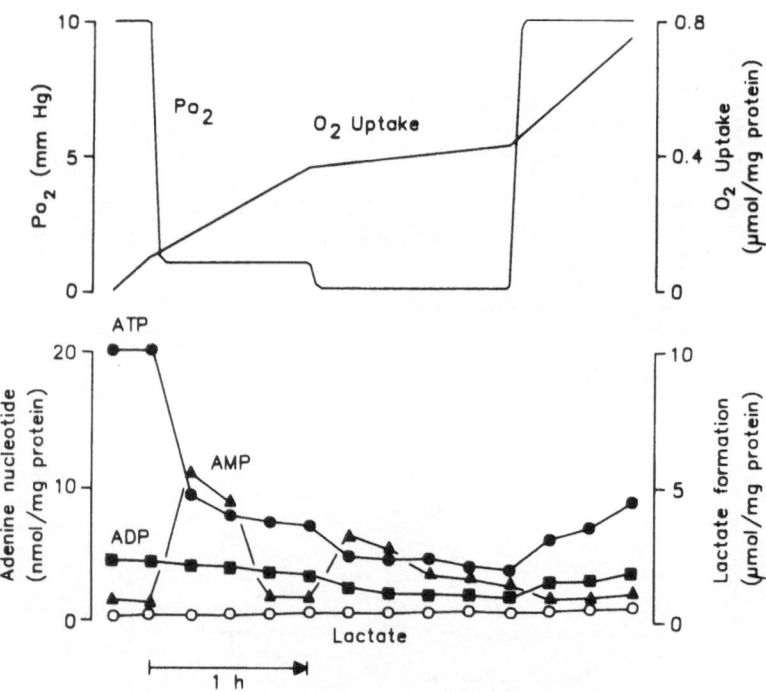

Fig. 5. Experiment on the effect of hypoxia and reoxygenation on the endothelial O_2 uptake, lactate formation and adenine nucleotide contents in the presence of 0.1 mM palmitate plus 0.5 mM glutamine. Coronary endothelial cells were incubated at Po_2 levels of 10, 1 and 0.1 mm Hg. Subsequently the cells were reoxygenated to 10 mm Hg.

the AMP content while the ADP level was changed only a little. The shift of the Po_2 to 0.1 mm Hg led to a further decrease of the ATP content (Fig. 5). After 2 h at Po_2 levels below 1 mm Hg, reoxygenation to 10 mm Hg resulted in the re-establishment of the control rate of oxygen consumption, but led only to a partial restoration of the ATP content within 45 min of observation. The rapid recovery of the initial rates of oxygen uptake after reoxygenation indicates that the mitochondrial integrity is preserved in spite of the hypoxic energy loss. The reason for the incomplete restoration of the cellular ATP contents within 45 min of reoxygention is probably due to a substantial loss of purines from the cells. It is interesting that the loss of ATP proceeds stepwise under hypoxia, i.e. an initial rapid loss is followed by a rather stable new, but lower level of cellular ATP. This indicates that the endothelial cells can establish their energetic states on a lower energy level.

CONCLUSIONS

1. The energy demand of coronary endothelial cells is distinctly lower than that of the working myocardium and rather of the order of the arrested heart.

2. Endothelial respiration has a high affinity to oxygen, closely reflecting the O_2 affinity of the mitochondria (Chance, 1965; de Groot et al., 1985).

3. Even at normoxia, glucose supplied coronary endothelial cells cover half of their energy demand by the O_2-independent catabolic breakdown of glucose to lactate, via the glycolytic pathway. Therefore, these cells release great amounts of lactate already under well-oxygenated conditions. They possess a glycolytic capacity sufficient to fully compensate the lack of respiratory ATP supply under hypoxic conditions; although, the Pasteur effect is small. Moreover, coronary endothelial cells are well adapted to maintain a well-energized metabolic state during transitory hypoxia.

4. Low energy demand and high glycolytic activity may be the cause why the coronary endothelium is less severely injured in

the anoxic-reoxygenated heart than the cardiomyocytes (Kloner et al., 1980; Buderus et al., 1989). Conditions leading to energy depletion of endothelial cells are most likely to occur in complete ischemia where a lack of oxygen comes together with a lack of glucose. But even under such conditions the endo-thelium may be better off than the myocardial cells due to its lower energy demand.

REFERENCES

Atkinson, D. E., 1971, Adenine nucleotides as stoichiometric coupling agents in metabolism and as regulatory modifiers: the adenylate energy charge, in: "Metabolic Pathways", volume V, H. J. Vogel, ed., p. 1, Academic Press, New York.

Buderus, S., Siegmund, B., Spahr, A., Krützfeldt, A., and Piper, H. M., 1989, Resistance of coronary endothelial cells to anoxia-reoxygenation in isolated guinea pig hearts, Am. J. Physiol., in press.

Chance, B., 1965, Reaction of oxygen with the respiratory chain in cells and tissue, J. Gen. Physiol. 49:163.

de Groot, H., Noll, T., and Sies, H., 1985, Oxygen dependence and subcellular partitioning of hepatic menadione-mediated oxygen uptake, Arch. Biochem. Biophys. 243: 556.

Gutman, I., and Wahlefeld, A. W., 1974, L(+)-Lactate, in " Methoden der enzymatischen Analyse, H. U. Bergmeyer, ed., p. 1510, Verlag Chemie, Weinheim.

Jüngling, E., and Kammermeier, H., 1980, Rapid assay of adenine nucleotides or creatine compounds in extracts of cardiac tissue by paired-ion reverse-phase high performance liquid chromatography, Anal. Biochem. 102:358.

Kloner, R. A., Rude, R. E., Carlson, N., Maroko, P. R., DeBoer, L. W. V., and Braunwald, E., 1980, Ultrastructural evidence of microvascular damage and myocardial injury after coronary artery occlusion: which comes first? Circulation 62: 945.

Krützfeldt, A., Spahr, R., Mertens, S., Siegmund, B., and Piper, H. M., 1989, Metabolism of exogenous substrates by coronary microvascular endothelial cells in culture, Am. J. Physiol., in press.

Lowry, O. H., Rosebrough, N. J., Farr, A. L., and Randall, R. L., 1951, Protein measurement with the folin phenol reagent, J. Biol. Chem. 193:265.

Noll, T., de Groot, H., and Wissemann, P., 1986, A computer-supported oxystat system maintaining steady-state oxygen partial pressures and simultaneously monitoring oxygen uptake in biological systems, Biochem. J. 236: 765.

HETEROGENEOUS NADH FLUORESCENCE DURING POST-ANOXIC REACTIVE

HYPEREMIA IN SALINE PERFUSED RAT HEART

C. Ince*, H. Vink, P.A. Wieringa**,
M. Giezeman, J.A.E. Spaan

Department of Medical Physics, University of Amsterdam,
*Department of Surgery, Erasmus University of Rotterdam,
**Department of Mechanical Engineering, Technical University
of Delft, The Netherlands

INTRODUCTION

Measurement of the autofluorescence of reduced pyridine nucleotide (NADH; 360nm excitation, 465nm emission), pioneered by Chance et al. (1962), not only enables the evaluation of cellular metabolic activity but also of cellular oxygen requirement (Chance, 1976). Steenbergen et al. (1977), using epicardial NADH fluorescence photography (Barlow and Chance, 1976), found that under steady-state conditions of low-flow of oxygenated perfusate as well as high-flow of poorly oxygenated perfusate heterogeneous NADH fluorescence patterns could be observed. In this study we investigated whether such patterns are also present during step-wise transitions of tissue oxygen content in the myocard such as occurs during reactive hyperemia.

MATERIALS AND METHODS

Male Wistar rats weighing between 250 and 450 grams were anesthetized with an intraperitoneal injection of sodium pentobarbital (Nembutal, 1.1 ml/kg body weight, Ceva, Paris, France). Artificial ventilation with room air via intubation of the trachea was established by using a mechanical respirator. Before the sternotomy, the rats were heparinized with 2,000 IU/kg heparin (Thromboliquine, Organon Teknika Oss, Holland).

The hearts were perfused in a Langendorff set-up which consisted of two independent circulating circuits each carrying perfusate (one with low and one with high-pO_2). A switch enabled step-wise change from one perfusate to the other. Perfusates were warmed to 37°C in a heat exchanger and equilibrated either with 95% O_2 and 5% CO_2 or 95% N_2 and 5%CO_2. Perfusion pressure was maintained between 70 and 80 mmHg. Perfusate consisted of 115.0 mM NaCl, 5.0 mM KCl, 1.2 mM $MgCl_2.6H_2O$, 2.0 mM $NaH_2PO_4.H_2O$, 1.2 mM Na_2SO_4, 27.0 mM $NaHCO_3$, 1.5 mM $CaCl_2$, 11.0 mM glucose and 10 mM lactate.

For epicardial NADH photography, a glass coverslip was gently pressed against the epicardial surface. A brace holding unit B2-RFA of an Olympus BH2 fluorescence microscope, a 100W Hg lamp attached and a photo camera, was positioned in front of the heart. The B2-RFA unit housed a DMU dichroic mirror with an UG-1 barrier filter providing 360 nm excitation of cellular NADH. A Wratten 2A gelatine barrier filter placed in front of a long-pass filter type L435 Olympus allows the emission of the NADH fluorescence between 460-480nm. Fluorescence photographs were taken with a Nikon FE2 camera fitted with a Micro-Nikkor 105mm macrolens using Kodak Tri-X Pan (400 ASA) film.

Grey level analysis of the NADH fluorescence images was accomplished by reading the photo negatives with a video camera into a PDP 11/23 (DEC) mini computer. Grey levels of the pixels were then determined of the digitized images as a measure of NADH fluorescence. A qualitative measure of the heterogeneity of NADH fluorescence was obtained by determination of the standard deviation divided by the mean grey level of the pixels.

RESULTS AND CONCLUSIONS

With the set-up shown in Fig.1 sharp epicardial NADH fluorescence photographs could be taken with exposure times of 0.5 sec. Blurring of

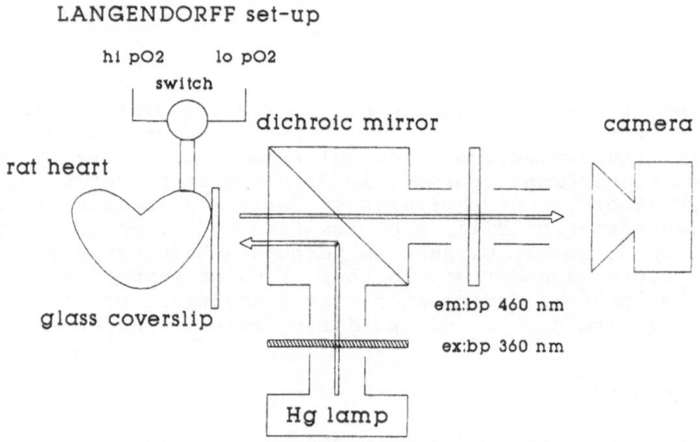

LANGENDORFF set-up

Figure 1. A diagram of the epicardial NADH fluorescence set-up. A rat heart is mounted in a Langendorff set-up with which it is possible to switch between two perfusates under constant pressure conditions. Flow and pressure are measured just above the cannulated heart. A glass coverslip is gently pressed against the epicardium. A camera takes photographs through a light guide fitted with the optical filters needed for NADH fluorescence measurements. Further details can be found in the Materials and Methods section.

images caused by continuous movement of the heart was overcome by resting the heart against a fixed glass coverslip which was taken as the focal plane of the camera. The presence of the coverslip did not cause epicardial ischemia since NADH fluorescence did not change upon removal of the coverslip from the heart surface.

478

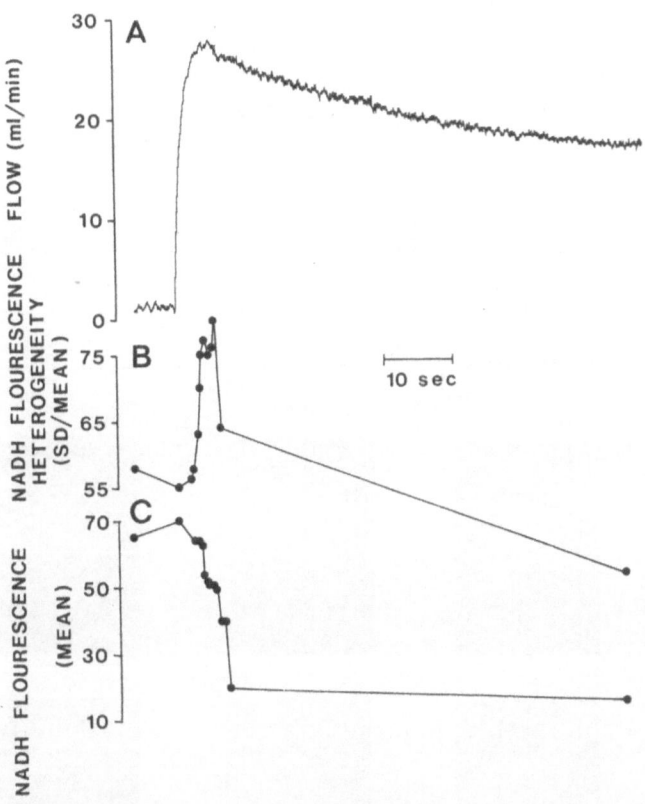

Figure 2. Reperfusion of a heart with a high-pO$_2$ solution following a 60 sec. occlusion results in a peak reactive hyperemic flow being established before control flow level is restored (A). Pixel grey level values of digitized NADH photographs taken during the course of this trace were determined. The standard deviation of the pixel grey levels divided by the mean pixel grey level value of sequential digitized NADH fluorescence images were used as a measure of fluorescence heterogeneity (B). The mean pixel grey level value was used as a measure of the epicardial NADH fluorescence (C). The above results show that reactive hyperemia is accompanied by heterogeneous NADH fluorescence patterns.

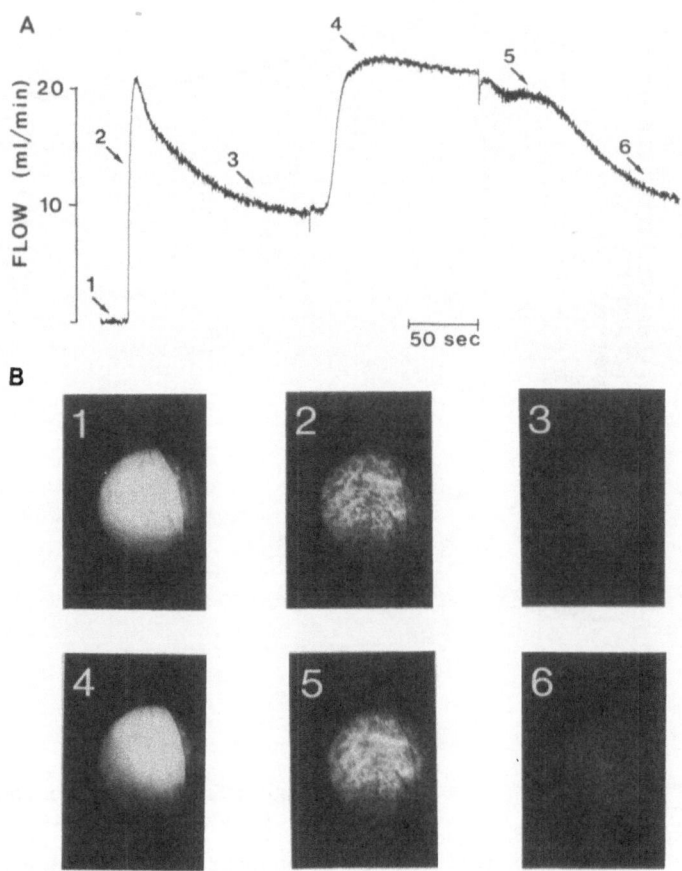

Figure 3. Following restoration from reactive hyperemic flow, perfusate is switched from a high- to a low-pO_2 solution resulting in high-flow hypoxia being established. Restoration from high-flow hypoxia is then achieved by reperfusion of the myocardium with the high-pO_2 solution. The NADH fluorescence images taken at the moments indicated by the arrows in (A) are shown in (B).

To obtain a measure of fluorescence at the epicardium, photo negatives of the NADH fluorescence photographs were digitized and pixel grey level values determined. The mean grey level value was used as a measure of total NADH fluorescence and the standard deviation of the grey level values of the pixels divided by the mean value of the grey level of the image was used as a qualitative measure of NADH fluorescence heterogeneity.

Zero-flow hypoxia was established by closure of the aortic cannula for a period of 60 sec. Restoration of perfusion after such an occlusion resulted in the flow reaching a peak (between 20 and 40 ml/min) after a delay of several seconds before returning to control levels (between 10 and 20 ml/min; Fig.2A). During the 60 sec. occlusion NADH fluorescence reached a steady state enhanced level as can be seen by the mean pixel grey level measurements of the NADH photographs taken during the course of the occlusion (Fig.2C). Reactive hyperemia was accompanied by a diminished fluorescence as oxygen was resupplied to the myocardium and NADH hydrolyzed to NAD+ (Fig.2C). This transition in fluorescence, however, did not occur homogeneously but rather heterogeneously, with some areas of the epicardium returning to control fluorescence level more rapidly than others. Pixel grey level measurements showed that peak heterogeneity occurred prior to or coincided with, the establishment of peak reactive hyperemic flow (Fig.2B).

To establish that the heterogeneous NADH fluorescence observed during reactive hyperemia is due to uneven supply of oxygen to the different parts of the epicardium, the pattern of NADH fluorescence changes was studied when the epicardium recovers from high-flow hypoxia by reperfusion of the myocardium with high-pO_2 perfusate. Such an experiment is illustrated in Fig.3A and was initiated with a 60 second occlusion until full anoxic NADH fluorescence is reached. Restoration of perfusion pressure resulted in a reactive hyperemic flow. Once steady-state flow was achieved, perfusate was switched from the high-pO_2 perfusate to the low-pO_2 perfusate. Once high-flow hypoxia was established the perfusate was switched back again to the high-pO_2 perfusate. As can be seen from the changes in NADH fluorescence during such a protocol (Fig.3B) restoration of tissue oxygen following high-flow hypoxia is also accompanied by a heterogenic fluorescence pattern (Fig.3B-5), supporting the hypothesis that heterogenic NADH fluorescence patterns associated with the reactive hyperemic flow is due to delayed perfusion of small areas in the epicardium. Further support for this view is provided by the observation that high fluorescence areas in the heterogeneous fluorescence pattern seen during reactive hyperemia (Fig.3B-2) correspond to those present during recovery from high-flow hypoxia (Fig.3B-5).

The present study has shown, by use of NADH fluorescence photography, that step-wise transitions in tissue oxygen associated with reactive hyperemia result in heterogeneous flow patterns occurring in the rat epicardium. This observation in combination with the findings by Bellamy et al. (1979) and Downey et al. (1983) that peak reactive hyperemic flow in the sub-endocardium precedes that of the sub-epicardium, indicates that flow patterns in the myocardium can be highly heterogeneous during sudden changes in oxygen tensions.

ACKNOWLEDGEMENTS

This investigation was supported in part by the Netherlands Foundation for Medical Research, MEDIGON, which is subsidized by the Netherlands Organization for the Advancement of Pure Research, NWO.

SUMMARY

In the present study epicardial NADH fluorescence photographs were taken of rat hearts during dynamic transitions of oxygen content of the myocardium. Hearts were perfused in a Langendorff set-up where it was possible to switch between low and high-pO_2 perfusates. NADH fluorescence photographs were taken with a suitable fluorescence set-up and photo negatives digitized and analyzed by use of a computer. Restoration of perfusion with a high-pO_2 solution resulted in a reactive hyperemic flow being established. Prior to the occlusion being lifted high NADH fluorescence was observed. Reactive hyperemic flow was associated with heterogenic NADH fluorescence patterns which diminished as control flow was restored. The patterns observed during reactive hyperemia were identical to those observed when tissue oxygen was restored by high-pO_2 perfusion following high flow hypoxia achieved by low-pO_2 perfusion. This study shows that heterogenic epicardial flow patterns are associated with reactive hyperemia.

REFERENCES

Bellamy, R.F., Lowensohn, H.S. and Olsson, R.A., 1979, Factors determining delayed peak flow in canine myocardial reactive hyperemia, Cardiovasc. Res., 13:147-151.

Barlow, C.H. and Chance, B., 1976, Ischaemic areas in perfused rat hearts: Measurement by NADH fluorescence photography, Science (Wash.D.C.), 193:909-910.

Chance B., 1976, Pyridine nucleotide as an indicator of the oxygen requirements for energy-linked functions of mitochondria, Circ.Res., 38 (suppl I):31-38.

Chance B., Legallais, V. and Schoener, B., 1962, Metabolically linked changes in fluorescence emission spectra of cortex of rat brain, kidney, adrenal gland, Nature (Lond.), 195:1073-1075.

Downey, H.F., Crystal, G.J. and Bashour, F.A., 1983, Asynchonom transmural perfusion during reactive hyperemia. Cardiovas. Res., 17:200-206.

Steenbergen C., Delleuw G., Barlow C., Chance B. and Williamson J.R., 1977, Heterogeneity of the hypoxic state in perfused rat heart, Circ.Res., 41:606-615.

MYOCARDIAL OXYGEN SUPPLY IN CORONARY ARTERY DISEASE

M. Thomas, J. Grote and J. Nitsch*

Department of Physiology and *Department of Cardiology
University of Bonn, D-5300 Bonn, FRG

INTRODUCTION

It is generally accepted that adaptation of coronary blood flow to exercise is limited in coronary artery disease. However, there are still controversial discussions whether coronary blood flow and myocardial metabolism are restricted under resting conditions in these patients (Opie, 1968,1969; Ganz et al., 1971; Klocke et al., 1974; Mymin and Sharma, 1974; Braunwald and Sobel, 1988; Thomassen et al., 1988). To investigate the behavior of myocardial blood flow and metabolism as well as the efficiency of left ventricular work in coronary artery disease, measurements were performed in patients during cardiac catheterization.

METHODS

Fourteen patients with angiographically proven coronary artery disease were compared to 6 patients without signs of coronary artery disease. Myocardial blood flow was determined by using the thermodilution technique. A Webster catheter was placed in the coronary sinus and correct position verified by radiography (Kupper and Bleifeld, 1979). Prior to injection of the indicator solution, blood samples were obtained simultaneously from the aorta and the coronary sinus for measurements of respiratory gas tensions and contents, together with the levels of glucose, lactate and free fatty acids. Left ventricular O_2 consumption and CO_2 production as well as the uptake rates of the metabolites were calculated. ECG was continuously monitored along with the pressure curves of the left ventricle and the aorta. From additional ventriculography stroke volume, ejection fraction and heart index could be determined. A less than 40-60% area change in one or more of 5 areas, defined on a RAO projection of the ventriculogram (computer assisted regional wall motion analysis), was classified as regional contraction abnormality. Mass of left ventricular wall was determined by echocardiography (Devereux and Reichek, 1977). Left ventricular work was calculated according to the standard formula (Just, 1976) and compared to the myocardial energy production as obtained from the uptake rates both of oxygen and of the different metabolites.

Oxygen Transport to Tissue XII, Edited by J. Piiper *et al.*
Plenum Press, New York, 1990

RESULTS AND DISCUSSION

The determined mean values for myocardial blood flow as well as for left ventricular tissue respiration and metabolism of both groups are summarized in Figure 1 and in Table 1.

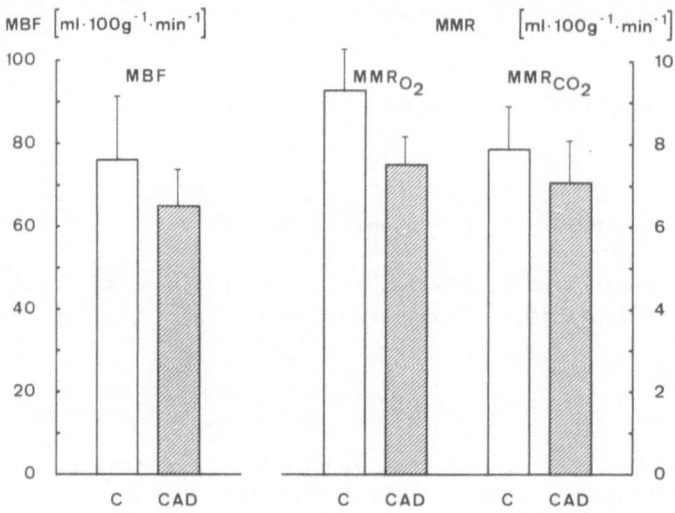

Figure 1. Myocardial blood flow (MBF), O_2 consumption rate (MMR_{O2}) and CO_2 production rate (MMR_{CO2}) in CAD patients (CAD) and in controls (C). Mean values $\pm$ SD

Whereas in the control group a mean flow rate of 76 ml$\cdot$100g$^{-1}\cdot$min^{-1} and a mean O_2 consumption rate of 9.3 mlO$_2\cdot$100g$^{-1}\cdot$min^{-1} were found, patients with coronary artery disease had a reduced myocardial blood flow of 65 ml$\cdot$100g$^{-1}\cdot$min^{-1} and a significantly decreased O_2 consumption of 7.5 mlO$_2\cdot$100g$^{-1}\cdot$min^{-1} (p < 0.01). Despite disagreement with several investigators that did not find a reduction of myocardial blood flow in coronary artery disease under resting conditions (Rowe et al., 1969; Ganz et al., 1971; Kupper and Bleifeld, 1981), more recent findings are in accordance with the present results (Klocke et al., 1974; Mymin and Sharma, 1974; Tauchert, 1975; Braunwald and Sobel, 1988). Along with the O_2 uptake, the CO_2 production of the myocardium was reduced in the coronary artery disease group (s. Fig.1), the resulting respiratory quotient was slightly elevated (0.85 vs. 0.95).

Table 1. Myocardial metabolism in patients with coronary artery disease (CAD) and in controls (C). Mean values $\pm$ SD

	CAD	C
Metabolic rate Glucose μmol$\cdot$100^{-1}g$\cdot$min^{-1}	19 $\pm$ 9	27 $\pm$ 16
Metabolic rate Lactate μmol$\cdot$100^{-1}g$\cdot$min^{-1}	10 $\pm$ 10	7 $\pm$ 7
Metabolic rate FFA μmol$\cdot$100^{-1}g$\cdot$min^{-1}	9 $\pm$ 4	13 $\pm$ 11

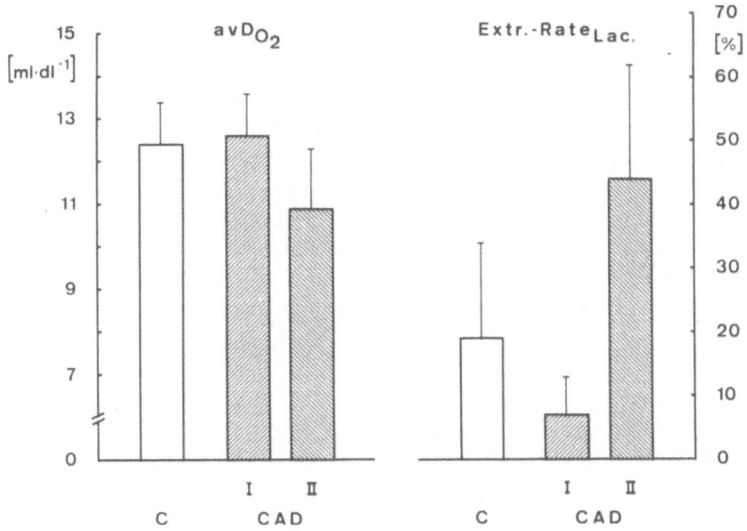

Figure 2. Arterio-coronary sinus differences in O_2 content (avD_{O2}) and lactate extraction ratios for controls (C) and normotensive (I) resp. hypertensive (II) CAD patients (CAD). Mean values $\pm$ SD

As shown in Figure 2, patients with coronary artery disease and concomitant hypertension had a significantly ($p < 0.05$) lower arterio-coronary sinus difference of O_2 content compared to patients with normal arterial blood pressure. To clarify this finding, one should consider the lactate extraction ratio which is accepted as a metabolic indicator of ischemia (Opie et al., 1973; Berland et al., 1984). The normotensive patients within the coronary artery disease group did have a marked depression of lactate extraction ratio which indicates the presence of myocardial ischemia. In the hypertensive patients, however, high lactate extraction ratios were determined ($p < 0.01$). These findings in coronary artery disease may be due to regional blood flow inhomogeneity with luxury perfusion in some myocardial areas of the hypertensives, present in addition to the temporal blood flow inhomogeneity during cardiac cycle (Grote and Thews, 1973; Berne and Rubio, 1979; Braunwald and Sobel,1988). The mean proportions of total myocardial energy production as calculated from the uptake rates of the different metabolites (s. Tab. 1), were 36% for glucose and 57% for free fatty acids in patients with coronary artery disease and 41% and 54%, respectively, in the control group.

In the coronary artery disease group heart function was impaired with a mean heart index of 3.5 $l \cdot min^{-1} \cdot m^{-2}$ compared to 4.2 $l \cdot min^{-1} \cdot m^{-2}$ for controls ($p < 0.05$). In addition, regional contraction abnormalities were present in 77% of these patients (Kreulen et al., 1975) and left ventricular enddiastolic pressure was elevated above 12 mmHg in 70% of them.

Heart function was significantly different between coronary artery disease patients with normal mean arterial blood pressure (MAP) and those with hypertension (MAP > 105 mmHg). The former ones had a mean left ven-

485

tricular enddiastolic pressure (PLV_{ED}) of 25 mmHg, the latter ones of 14 mmHg (s.Tab.2). The decrease in compliance, indicating the presence of fibrosis of left ventricular wall tissue (Hess et al., 1988), was more pronounced in normotensives than in hypertensives. The first derivative of the pressure curve, dp/dt_{max}, revealed abnormal findings only in a minority of patients. Mean left ventricular work was reduced as compared to the controls (101 vs. 132 $J \cdot min^{-1}$). Since the corresponding O_2 consumption was decreased as well, the resulting efficiency of left ventricular work (29%) was comparable to that of the control group. Similar results were obtained using the determined metabolic rates for calculation of left ventricular efficiency (28% vs. 25% for controls).

Table 2. Parameters of heart function in CAD patients (CAD) and in controls (C). Mean values $\pm$ SD

| | | CAD | | C |
		Normotension	Hypertension	
Heart rate	$l \cdot min^{-1}$	74 $\pm$ 11	72 $\pm$ 15	73 $\pm$ 5
MAP	mmHg	96 $\pm$ 10	114 $\pm$ 4	107 $\pm$ 3
PLV_{ED}	mmHg	25 $\pm$ 10	14 $\pm$ 8	11 $\pm$ 4
Compliance	$ml \cdot mmHg^{-1}$	4 $\pm$ 2.7	9 $\pm$ 6.6	10 $\pm$ 3.1
Ejection fraction	%	66 $\pm$ 15	70 $\pm$ 9	75 $\pm$ 11
Heart index	$l \cdot min^{-1} \cdot m^{-2}$	3.4 $\pm$ 0.4	3.7 $\pm$ 0.4	4.2 $\pm$ 0.5

SUMMARY

In patients with coronary artery disease investigated during cardiac catheterisation, myocardial blood flow and myocardial oxygen consumption were significantly decreased as compared to control patients without signs of coronary artery disease. At the same time left ventricular function was impaired. In addition to regional contraction abnormalities observed in 77% of the patients, mean heart index and mean left ventricular work were diminished and mean enddiastolic pressure elevated. Among the coronary artery disease group, patients with normal blood pressure had a significantly higher left ventricular enddiastolic pressure than patients with hypertension.

REFERENCES

Berland, J., Cribier, A., Cazor, J.L., Hecketsweiler, B., and Letac, B., 1984, Angina pectoris with angiograpically normal coronary arteries: A clinical, hemodynamic, and metabolic study, Clin. Cardiol. 7:485-492.

Berne, R.M., and Rubio, R., 1979, Coronary circulation, in: "Handbook of Physiology", Section 2, Vol.I 873-952, The cardiovascular system, R.M. Berne, N. Sperelakis, and S.R. Geiger, eds., American Physiological Society, Bethesda.

Braunwald, E., and Sobel, B.E., 1988, Coronary blood flow and myocardial ischemia, in: "Heart Disease", Vol.II, 1191-1221, E. Braunwald, ed., Saunders, Philadelphia.

Devereux, R.B., and Reichek, N., 1977, Echocardiographic determination of left ventricular mass in man - Anatomic validation of the method, Circulation, 55:613-618.

Ganz, W., Tamura, K., Marcus, H.S., Donoso, R., Yoshida, S., and Swan, H.J.C., 1971, Measurement of coronary sinus blood flow by continuous thermodilution in man, Circulation, 44:181-195.

Grote, J., and Thews, G., 1973, Respiratory gas transport in heart, in: "Oxygen transport to tissue, instrumentation, methods and physiology," H.I. Bicher and D. F. Bruley, eds., Adv. exp. Med. Biol. Vol. 37A., 525-534, Plenum Press, New York.

Hess, O.M., Schneider, J., Nonogi, H., Caroll, J.D., Schneider, K., Turiner, M., and Kreyenbühl, H.P., 1988, Myocardial structure in patients with exercise-induced ischemia, Circulation, 77:967-977.

Just, H., 1976, Herzkatheter-Diagnostik: Methodik, Messungen, Formeln, Nomogramme, Boehringer, Mannheim.

Klocke, F.J., Bunnell, I.L., Greene, D.G., Wittenberg, S.M., and Visco, J.P., 1974, Average coronary blood flow per unit weight of left ventricle in patients with and without coronary artery disease, Circulation, 50:547-559.

Kreulen, T.H., Bove, A.A., McDonough, M.T., Sands, M.J., and Spann, J.F., 1975, The evaluation of left ventricular function in man. A comparison of methods, Circulation, 51:677-688.

Kupper, W., and Bleifeld, W., 1979, Regionale und globale Koronarsinus-flußmessungen mit dem kontinuierlichen Thermodilutionsverfahren I. Methode und experimentelle Untersuchungen, Z.Kardiol., 68:740-747.

Kupper, W., and Bleifeld, W., 1981, Regionale und globale Koronarsinus-flußmessungen mit dem kontinuierlichen Thermodilutionsverfahren II. Klinische Untersuchungen bei Koronarkranken, Z. Kardiol., 70:116-123.

Mymin, D., and Sharma, G.P., 1974, Total and effective coronary blood flow in coronary and noncoronary heart disease, J. Clin. Invest., 53:363-373.

Opie, L.H., 1968, 1969, Metabolism of the heart in health and disease, Am. Heart J., 76:685-698; 77:100-122; 384-410.

Opie, L.H., Owen, P., Thomas, M., and Samson, R., 1973, Coronary sinus lactate measurements in assessment of myocardial ischemia. Comparison with changes in lactate/pyruvate and beta-hydroxybutyrate/acetoacetate ratios and with release of hydrogen, phosphate and potassium ions from the heart, Am. J. Cardiol., 32:295-305.

Rowe, G.C., Thomsen, J.H., Stenlund, R.R., McKenna, D.H., Sialer, S., and Corliss, R.J., 1969, A study of hemodynamics and coronary blood flow in man with coronary artery disease, Circulation, 39:139-148.

Tauchert, M., 1975, Wert und Grenzen klinischer Koronardurchblutungs-messungen, Klin. Wschr., 53:691-707.

Thomassen, A., Bagger, J.P., Nielsen, T.T., and Henningsen, P., 1988, Altered global myocardial substrate preference at rest and during pacing in coronary artery disease with stable angina pectoris, Am. J. Cardiol., 62:686-693.

ARTERIAL O_2-PARTIAL PRESSURE AT POSITVE ENDEXPIRATORY PRESSURE IN HYPEROXIA FOR VERIFICATION OF PATENT FORAMEN OVALE ?

Ludwig Brandt and Friedrich Mertzlufft

Department of Anaesthesiology, University Hospital
Johannes Gutenberg University Mainz
D-6500 Mainz, F.R.G.

INTRODUCTION

As a rule, during anaesthesia a time dependent increase in pulmonary shunting occurs even in states of normal lung function. Therefore positive end-expiratory pressure ventilation (PEEP) is often used to decrease intrapulmonary right/left shunting (RL shunt) and thus to increase arterial O_2 partial pressure (paO_2) without increasing the inspiratory oxygen fraction.

However, due to the accompanying possible decrease in cardiac output with increasing PEEP mixed-venous oxygen saturation may be reduced thus counteracting the effect of pulmonary shunt reduction on paO_2 [Nunn, 1987]. Therefore the effect of PEEP on paO_2 remains only moderate in spite of its sensitivity towards RL shunt.

An additional indication for application of PEEP results from potential risks of certain surgical procedures, especially in neurosurgery [Albin, 1984]: PEEP may decrease venous air embolism (VAE) by increasing venous pressure at the incision level, e.g. in the sitting position (neurosurgical procedures in the posterior cranial fossa). Because PEEP and VAE can both increase pulmonary vascular resistance, it may be possible that PEEP increases right atrial pressure (RAP) relative to left atrial pressure (LAP) during VAE and thereby reverses the normal interatrial pressure gradient, allowing paradoxical air embolism (PAE) in patients with a patent foramen ovale (PFO). Any PAE however may produce devastating neurologic damage [Gronert et al., 1979; Perkins et al., 1984].

The development of contrast M-mode echocardiography introduced the possibility of recognizing PAE immediately. Moreover with 2-D contrast echocardiography, especially transoesophageal echocardiography (TEE) intraoperative verification of VAE plus PAE became possible.

Recently [Guggiari et al., 1988] contrast echocardiography in combination with the Valsalva maneuver has become to

Oxygen Transport to Tissue XII, Edited by J. Piiper *et al.*
Plenum Press, New York, 1990

be recommended as a noninvasive screening method to demonstrate the presence of a PFO preoperatively: during the Valsalva maneuver the normal transatrial pressure gradient is reversed with resultant transient RL shunting across a PFO.

However, contrast echocardiography appears to be a too expensive procedure for being accepted as a routine screening technique for patients at risk for PAE.

It is the aim of this case report to propose a much easier-to-use clinical method for verification of PFO and herewith for potential PAE. During hyperoxic conditions a PEEP-induced decrease in paO_2 may be assumed to result from an increase in extrapulmonary RL shunt. Right-atrial/left-atrial shunt due to a patent foramen ovale then was proved by TEE: during controlled ventilation without PEEP it was functionally closed, applying PEEP it opened thus inducing the intracardial RL shunt with resultant fall in paO_2.

REPORT OF THE CASE

An adult male patient (age 63 years) in a sufficient cardiac state (ejection fraction, cardiac index and left ventricular enddiastolic pressure within normal limits) and with normal lung function was admitted for aorto-coronary bypass surgery (2-vessel disease). He had given written informed consent to participate in a clinical study on arterial and mixed venous blood gas status during hyperoxic anaesthesia using controlled ventilation.

Routine monitoring methods for patients with coronary heart disease in our department include (1.) invasive arterial and pulmonary arterial pressure monitoring and (2.) the Oxyshuttle pulse oxymeter for continuous noninvasive in vivo monitoring of "partial" O_2 saturation. At the end of cardiopulmonary bypass a left atrial catheter for continuous on-line LA pressure monitoring is inserted via the upper right pulmonary vein.

After premedication (2.0 mg p.o Flunitrazepam the evening before operation and another 0.01 mg/kgBW Flunitrazepam i.v. preoperatively) and preoxygenation with pure oxygen (close fitting mask, oxygen flow 10 l/min) induction of anaesthesia is performed with 0.02 - 0.025 mg/kgBW Fentanyl and 0.1 mg/kgBW Pancuronium. With direct laryngoscopy a stomach tube and an oesophageal temperature probe are inserted via the nasal route, followed by endotracheal intubation with a Brandt^{TM} endotracheal tube.

Arterial blood gas samples are taken from the radial artery pressure line, mixed-venous samples from the pulmonary artery catheter, and analyzed immediately using the Nova Biomedical Stat Profile 5 and the Corning CO-Oxymeter 2500.

Blood samples (1.5 ml) are withdrawn (1. immediately before induction of anaesthesia (patient breathing ambient air spontaneously) and (2. five minutes after intubation and controlled ventilation with 100% oxygen to verify the expected hyperoxic situation of paO_2 >400 mmHg.

490

Table 1. Development of paO_2 during different patterns of
          ventilation

TYPE OF VENTILATION	FiO_2	paO_2 [mmHg]
SPONTANEOUS BREATHING	0.21	67
CONTROLLED VENTILATION		
: ZEEP	1.00	362
: + PEEP (5 cmH_2O)	1.00	296
: + PEEP (10 cmH_2O)	1.00	214
: ZEEP	1.00	323

As shown in Table 1 the preanaesthetic paO_2 was measured to be only 67 mmHg (FiO_2 0.21), which is within expected normal limits in this situation (hypoventilation due to the heavy pre-medication; $paCO_2$ was 45.5 mmHg). 5 min after intubation and controlled ventilation (FiO_2 1.0, $paCO_2$ 40 mmHg, pHa 7.44), paO_2 was <400 mmHg (362 mmHg). Assuming an anaesthesia-induced shunt, 5 cmH_2O of PEEP were used to decrease intrapulmonary shunting. However, paO_2 now decreased to 296 mmHg instead of increasing towards 400 mmHg as expected. As mixed venous O_2 content ($c\bar{v}O_2$) remained stable this paO_2 decrease could not be explained by a PEEP-induced decrease in cardiac output. Consequently PEEP was further increased to 10 cmH_2O (a level which normally may maximize O_2 availability) in order to improve paO_2. But, paO_2 continued decreasing to 214 mmHg ($paCO_2$ 38 mmHg, pHa 7.46). Returning to ZEEP (zero endexpiratory pressure) paO_2 increased to 323 mmHg (Table 1).

The observed decrease in paO_2 could only be due to an increase in RL shunt of extrapulmonary origin. Only an intracardial RL shunt between right and left atrium, e.g. PFO, remains as a reasonable explanation. It is an already known fact that PEEP - similar to Valsalva's maneuver - can induce such a RL shunt via a PFO. In order to prove this assumption contrast-TEE was carried out as recommended [Guggiari et al., 1988]. As shown in figs. 1a - b and 2a - b this PFO

   * kept functionally closed during controlled
     ventilation without PEEP, and

   * opened with PEEP thus inducing an atrial RL shunt.

This PEEP-induced opening of the PFO with resultant atrial RL shunt could only be due to a PEEP-induced reversal of left-atrial/right-atrial pressure gradient, i.e. LAP < RAP.

In order to prove this assumption LA and RA pressure measurements were performed applying different levels of PEEP.
As shown in fig. 3a - b RAP may exceed LAP, i.e. LAP < RAP, considerably with increasing PEEP (20 cmH_2O). Thus verification of the PEEP-induced pressure gradient reversal was successful.

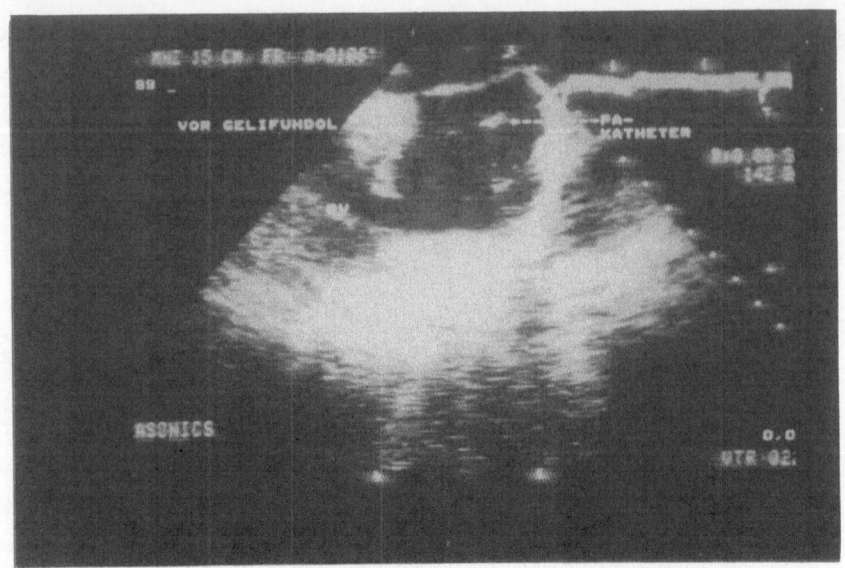

Fig.1a.   2-D TEE showing the situation during ZEEP ventilati-
          on. The dark area in the upper corner of the triangle
          represents the left atrium which is separated from
          the right atrium (dark area below with the shape of
          the PA-catheter in situ) by the interatrial septum
          (bright horizontal structure at the top).

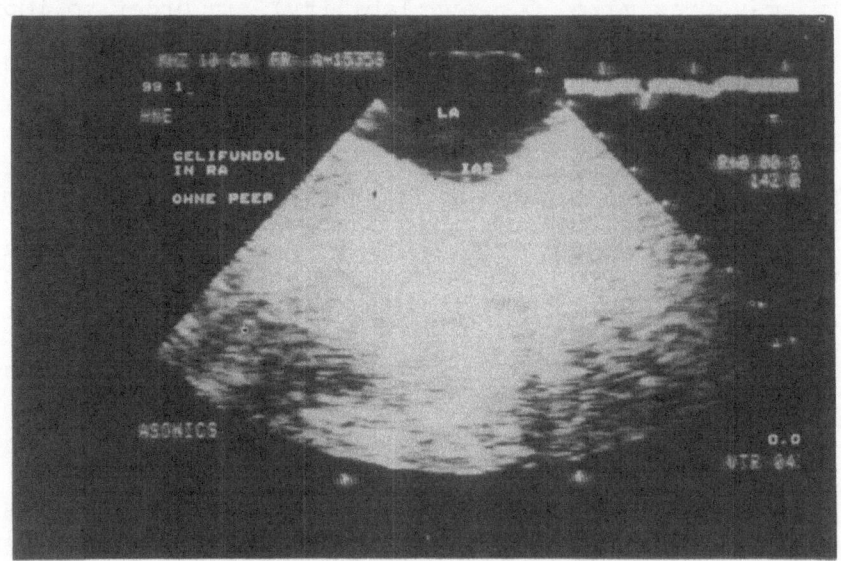

Fig.1b.   2-D contrast-TEE showing the same situation as in 1a.
          The contrast medium (air bubbled hydroxy-aethylic-
          starch [Gelifundol^R]) is injected into the right
          atrium (bright area) through the proximal lumen of
          the PA-catheter. The contrast medium does not pass
          through the interatrial septum (IAS) into the left
          atrium (LA).

492

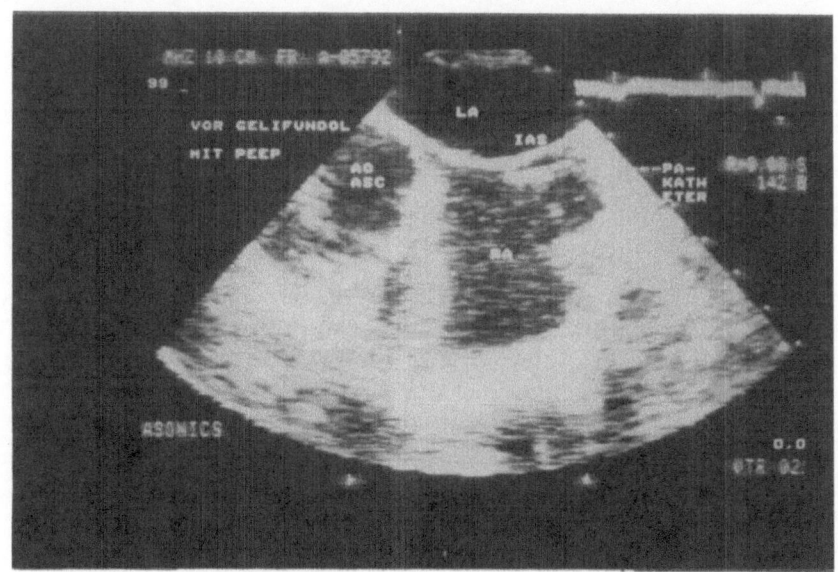

Fig.2a.   2-D TEE showing the situation during PEEP ventilation. LA = left atrium, RA = right atrium, IAS = interatrial septum, PA-KATHETER = pulmonary artery catheter, AOASC = ascending aorta.

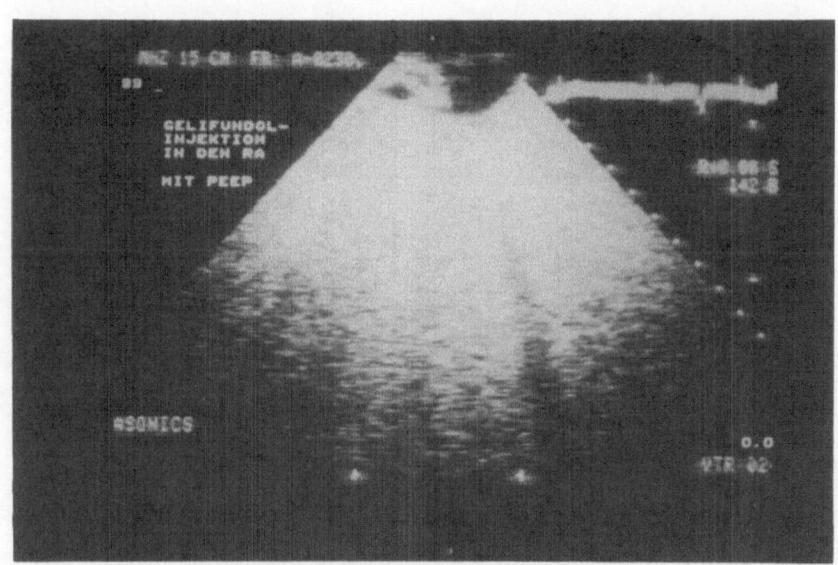

Fig.2b.   2-D contrast-TEE showing the same situation as in 2a. The contrast medium passes the IAS and becomes visible in the LA (bright contrast structure at the top).

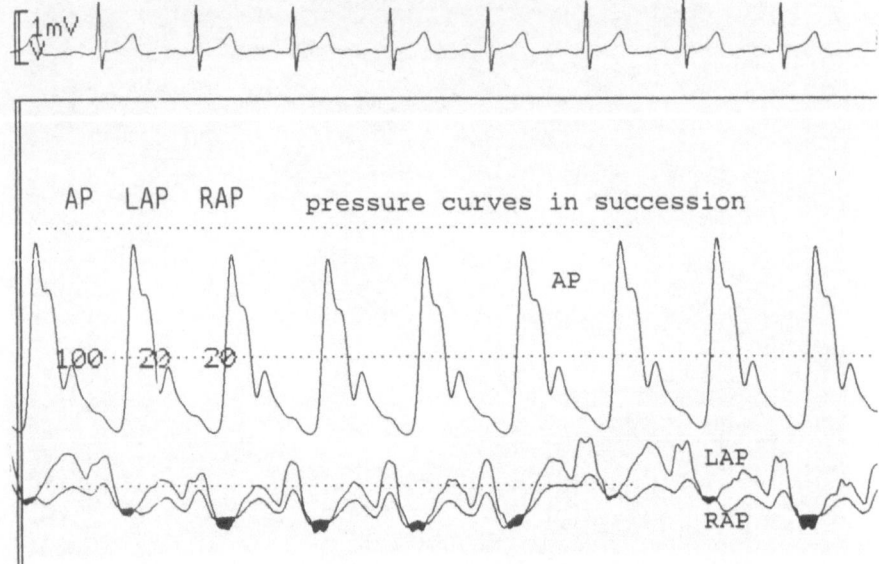

Fig.3a. Simultaneous pressure tracings of arterial (AP),
         right atrial (RAP) and left atrial (LAP) pressures
         (100 20 20 = amplification of gauging AP {100}, LAP
         {20} and RAP {20}); controlled ventilation without
         PEEP. Top tracing: electrocardiogram.
         Black areas: LAP < RAP.

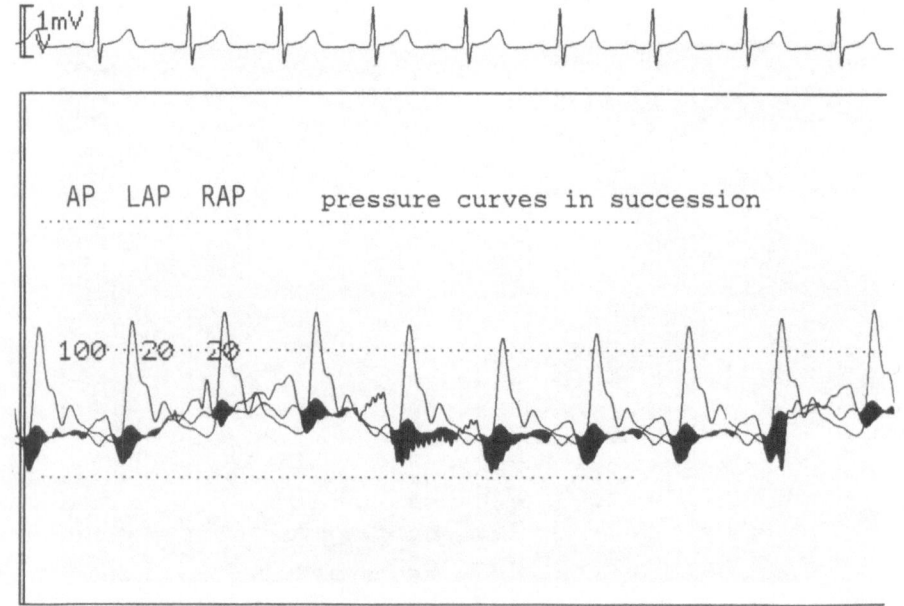

Fig.3b. Cf. fig. 4a; controlled PEEP ventilation (20 cmH_2O).
         Black areas denote LAP < RAP.

CONCLUSIONS

In two-thirds of all persons the tissue of the septum primum lying to the left of the foramen ovale (located at the apogee of the fossa ovalis in line with the axes of the fossa ovalis and the inferior vena cava and is formed by the septum secundum and occupies the right side of the atrial septum) serves as a small flap valve that usually becomes adherent to the left side of the septum (membraneous portion of the atrial septum) within a few months to a few years and forms a permanent closure with the side of the foramen ovale yielding an anatomically closed or sealed foramen [Guyton, 1986; Hagen et al., 1984]. But, even if permanent closure does not occur which is the case in 30 - 35% of the normal population [Hagen et al., 1984; Guggiari et al., 1988; Guyton, 1986], the left atrial pressure throughout life remains 2 to 4 mmHg greater than the right atrial pressure, and the back pressure normally keeps the valve of the foramen ovale (this term implicates that a PFO is present) functionally closed. However, any condition that results in right atrial pressure being greater than left atrial pressure (e.g. use of PEEP; figs. 3a - b) may produce a RL shunt with resultant hypoxaemia and possible PAE [Neidhart et al., 1988; Oliver et al., 1987]. It must be stated that a PFO (a term which should not be confused with the atrial septal defect, i.e. a true deficiency of the atrial septum of the ostium secundum type) is always a potential site of PAE should air enter the venous circulation, and for this reason it is essential to exclude all air from intravenous fluid administration sets. It is of even greater importance in the operating theatre, especially during procedures like posterior cranial fossa surgery in the sitting position [Fischler et al., 1984]. As mentioned above the incidence of PFO is about 30% [Braunwald, 1980; McAlpine, 1975; Schroeckenstein et al., 1972) and VAE per se occurs in 30 - 40% of patients during seated neurological surgery [Gronert et al., 1979; Guggiari et al., 1988]. From this the risk of PAE in the seated position must be estimated close to 6 - 12%. However, the true incidence of PAE still remains unknown. The neurologic or cardiovascular sequelae due to PAE depend on the amount of air passing through the PFO and on the site of PAE.

PEEP ventilation (+10 cmH_2O) is worldwide recommended [e.g. Albin, 1984] and applied for prevention of intraoperative VAE in neurosurgery. However, according to literature data [Fischler et al., 1984] PEEP must be a contraindication in patients with PFO although other investigators [e.g. Zasslow et al., 1986] still contradict or disregard a PEEP-induced reversal of LA/RA pressure gradient. The case report presented here proves undoubtebly (fig. 3) that PEEP obviously causes such a pressure gradient reversal (LA < RA) thus proving that patients with VAE carry the increased actual risk of RL shunt and PAE via PFO.

Therefore, before applying PEEP verification of PFO has vital implications especially in patients at risk and should be performed routinely. For this purpose indicator dilution methods, heart catheterization and contrast-echocardiography have been applied up to date.

All these methods are either too precarious (invasive techniques) or far too cumbersome and expensive (noninvasive techniques) and have therefore not become to be widely used in clinical routine practice. Due to their specific obstacles it

495

would appear to be of great advantage if a more simple and easy-to-use clinical screening method could become available.

The parameter paO_2 has been described as the most sensitive indicator of changes in RL shunting [Nunn, 1987]. In cases of changed LA/RA pressure gradients (LAP < RAP) due to PEEP ventilation an interatrial RL shunt (RAP > LAP) via a PFO may be the cause for decreasing paO_2.
Further investigations will have to give evidence whether the clinical facts observed in this case report are also reproducible in an extended patient population.

REFERENCES

Albin, M.S., 1984, The paradox of paradoxic air embolism - PEEP, Valsalva, and Patent Foramen Ovale. Should the sitting position be abandoned? Anesthesiology, 61:222.

Braunwald, E., 1980, "Heart Disease", W.B. Saunders, Philadelphia; p. 985.

Butler, B.D., Leiman, B.C., Luehr, S., and Katz, J., 1986, Effects of PEEP on the incidence of paradoxical air embolism in the absence of ASD in dogs, Anesthesiology, 65 Suppl:A81.

Fischler, M., Vourc'h, G., Dubourg, O., and Bourdarias, J.P., 1984, Patent Foramen Ovale and sitting position, Anesthesiology, 60:83.

Gronert, G.A., Messick, J.M., Cucchiara, R.F., and Michenfelder, J.D., 1979, Paradoxical air embolism from a patent foramen ovale, Anesthesiology, 50:548.

Guggiari, M., Lechat,P., Garen-Colonne, C., Fusciardi, J., and Viars, P., 1988, Early detection of Patent Foramen Ovale by two-dimensional contrast echocardiography for prevention of paradoxical air embolism during sitting position, Anesth Analg, 67:192.

Gyuton, A.C., 1986, "Textbook of Medical Physiology", W.B. Saunders, Philadelphia; p. 1001.

Hagen, P.T., Scholz, D.G., and Edwards, W.D., 1984, Incidence and size of patent foramen ovale during the first 10 decades of life: An autopsy study of 965 normal hearts, Mayo Clin Proc, 59:17.

McAlpine, W.A., 1975, "Heart and Coronary Arteries", Springer, Berlin; p. 87-100.

Neidhart, P.P. and Suter P.M., 1988, Changes of right ventricular function with positive end-exspiratory pressure (PEEP) in man, Intensive Care Med, 14:471.

Nunn, J.F., 1987, "Applied Respiratory Physiology", Butterworths, London.

Oliver, S., Cucchiara, R., Nishimura, R., and Michenfelder, J., 1987, Parameters affecting the occurrence of paradoxical air embolism, Anesthesiology, 67 Suppl:A435.

Perkins, N.A.K. and Bedford, R.F., 1984, Hemodynamic consequences of PEEP in seated neurological patients - Implications for paradoxical air embolism, Anesth Analg, 63:429.

Schroeckenstein, R.F., Wasenda, G.J., and Edwards, J.E., 1972, Valvular competent patent foramen ovale in adults, Minn Med, 55:11.

Zasslow, M.A., Pearl, R.G., Larson, C.P., and Silverberg, G., 1986, PEEP does not affect left atrial - right atrial pressure gradient in neurosurgical patients, Anesthesiology, 65 Suppl:A304.

SKELETAL MUSCLE

TISSUE OXYGENATION OF THE SKELETAL MUSCLE AND OF THE HEART DURING

HEMODYNAMIC ALTERATIONS IN RATS

Martina Günderoth-Palmowski

Medizinische Universitätsklinik, 7400 Tübingen, FRG

INTRODUCTION

The measurement of local tissue Po_2 provides a unique possibility of directly analysing the actual Po_2 supply to the terminal vascular bed. In recent studies, much data on the tissue Po_2 of rats has been collected. The failure of an in vivo model which allows the parallel measurement of macrocirculatory parameters and of tissue oxygenation made the interpretation of this data more difficult.

The aim of this study was to create an animal model which allows simultaneous measurement of hemodynamic parameters and tissue oxygenation of different organs. To evaluate this model, heart rate changes and subsequently cardiac output changes were induced by electrical stimulation or dissection of the vagal or sympathetic nerve system in the cervical area.

MATERIAL AND METHODS

32 male Sprague-Dawley rats, weighing 312.5 ± 45.9 g were anaesthetized with a combination of Ketanest (Parke Davis) and Dehydrobenzperidol (Janssen) (100 mg and 25 mg/kg body weight).

Following a midline cervical incision, the trachea was intubated, and the animals were ventilated with room air using a constant volume pump (Ugo Basile, Italy) attached to the tracheal cannula. The minute volume was adjusted for the body weight of each animal using the data from a previous study (unpublished).

The chest was opened by a midsternal midclavicular incision. The left carotid artery was cannulated and the blood pressure was monitored with a pressure transducer (Statham P23 ID; Hellige, FRG) and recorded throughout the experiment on a polygraph (Graphtec Corp., Japan). An electromagnetic flow probe (Narco Bio Systems, USA) was placed then around the ascending aorta and blood flow was continuously recorded on the multichannel recorder. A bipolar electrocardiogram (Einthoven) was recorded permanently on the polygraph and heart rate was determined from the interval between successsive P waves. The cardiac output (more correctly: ascending aorta flow) and the total peripheral resistance were calculated.

Oxygen Transport to Tissue XII, Edited by J. Piiper *et al.*
Plenum Press, New York, 1990

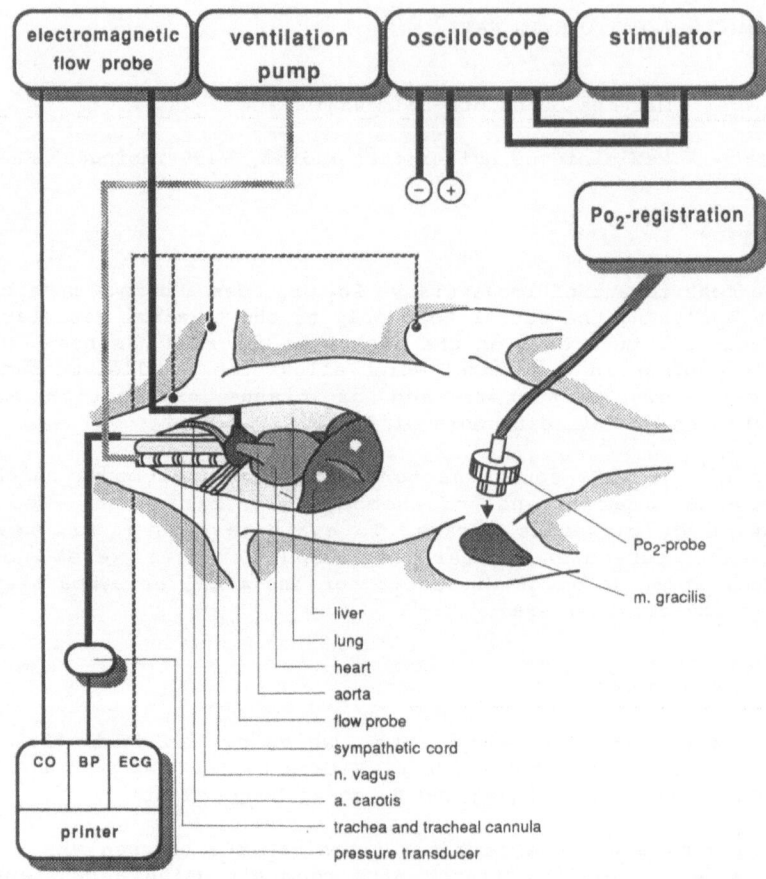

Fig. 1. Experimental design.
CO = cardiac output
BP = blood pressure
ECG = electrocardiogram

Both the left and the right vagus nerves or the sympathetic cord were exposed and suspended over silver bipolar electrodes. The electrodes were connected to a stimulator (self-built) and stimulation with alternating current (2 V, 52 Hz, 60 seconds duration) was chosen to bring the heart rate to a value approximately 30 % below or above the control rate, respectively.

The tissue Po_2 was measured continuously on the right gracilis muscle with a multiwire surface probe according to Kessler and Lübbers (1966) before, during and after electrical stimulation. Additional Po_2 histograms were taken before and 5 minutes after the end of the electrical stimulation. In 5 rats Po_2 histograms of the heart were taken before and 5 minutes after the bilateral dissection of the sympathetic or parasympathetic nerves.

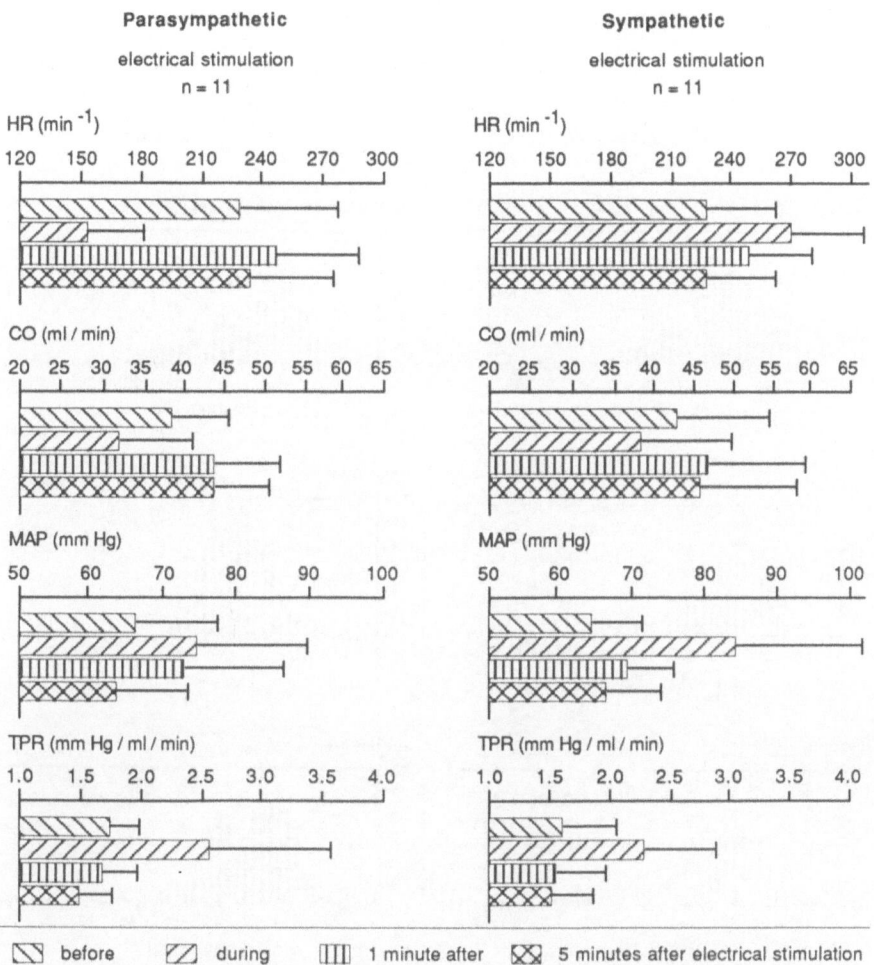

Fig. 2. Mean values and standard deviation of hemodynamic parameters before, during and after electrical stimulation of the parasympathetic and sympathetic nerves.
HR = heart rate   TPR = total peripheral resistance
MAP = mean arterial pressure   n   = number of animals
CO = cardiac output

RESULTS

During the electrical stimulation of the vagus nerves, mean heart rate decreased significantly by approximate 34 % (p<0.005) (Fig. 2, left side). The cardiac output remained almost unchanged whereas the total peripheral resistance increased by approximate 32 %. However, the mean muscle Po_2 did not change (Fig. 3, left side).

After the bilateral dissection of the vagus nerves the heart rate increased (90 %; p>0.0005) (Fig. 4, left side). The cardiac output and the mean arterial pressure were enhanced by approximate 28 and 15 % resp., whereas the peripheral resistance was lowered (by approx. 5.7 %). Mean muscle Po_2 remained unchanged and the Po_2 on the heart increased significantly from 43.7 mm Hg to 66.1 mm Hg (+ 51 %, p < 0.0005) (Fig. 5, left side).

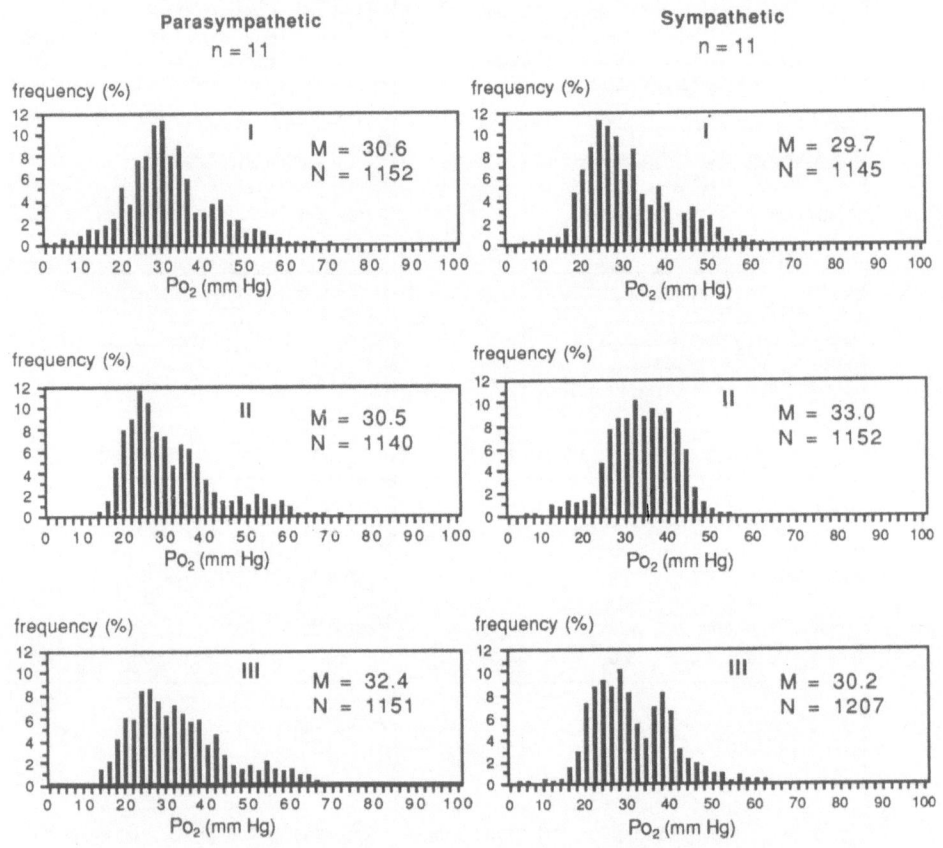

Fig. 3. Pooled histograms of muscle Po_2 of 11 rats before (I) and 5 minutes after the electrical stimulation of the right (II) and the left (III) parasympathetic and sympathetic nerves.
M = mean muscle Po_2
N = number of measuring points
n = number of animals

502

During the electrical stimulation of the sympathetic cord, heart rate was significantly increased (18%, $p < 0.0125$) (Fig. 2, right side). The cardiac output decreased whereas the mean arterial pressure and the total peripheral resistance increased. The mean muscle Po_2 remained unchanged (Fig. 3, right side).

5 min after the bilateral dissection no hemodynamic changes could be observed except an increase in the total peripheral resistance (27 %, $p<0.0005$) (Fig. 4, right side). The mean muscle Po_2 did not change either. However, the Po_2 on the heart increased by approx. 32 % ($p < 0.0025$) (Fig. 5, right side).

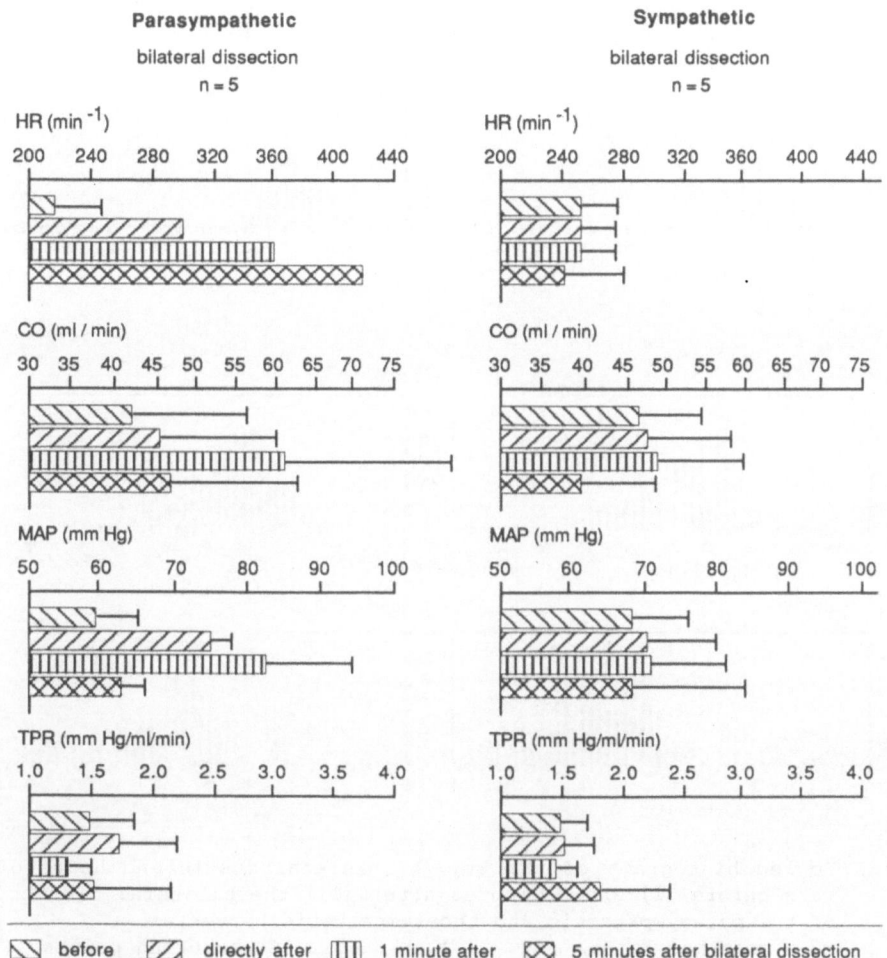

before | directly after | 1 minute after | 5 minutes after bilateral dissection

Fig. 4. Mean values and standard deviation of hemodynamic parameters before and after bilateral dissection of the parasympathetic and sympathetic nerves.
HR = heart rate   TPR = total peripheral resistance
CO = cardiac output   n   = number of animals
MAP = mean arterial pressure

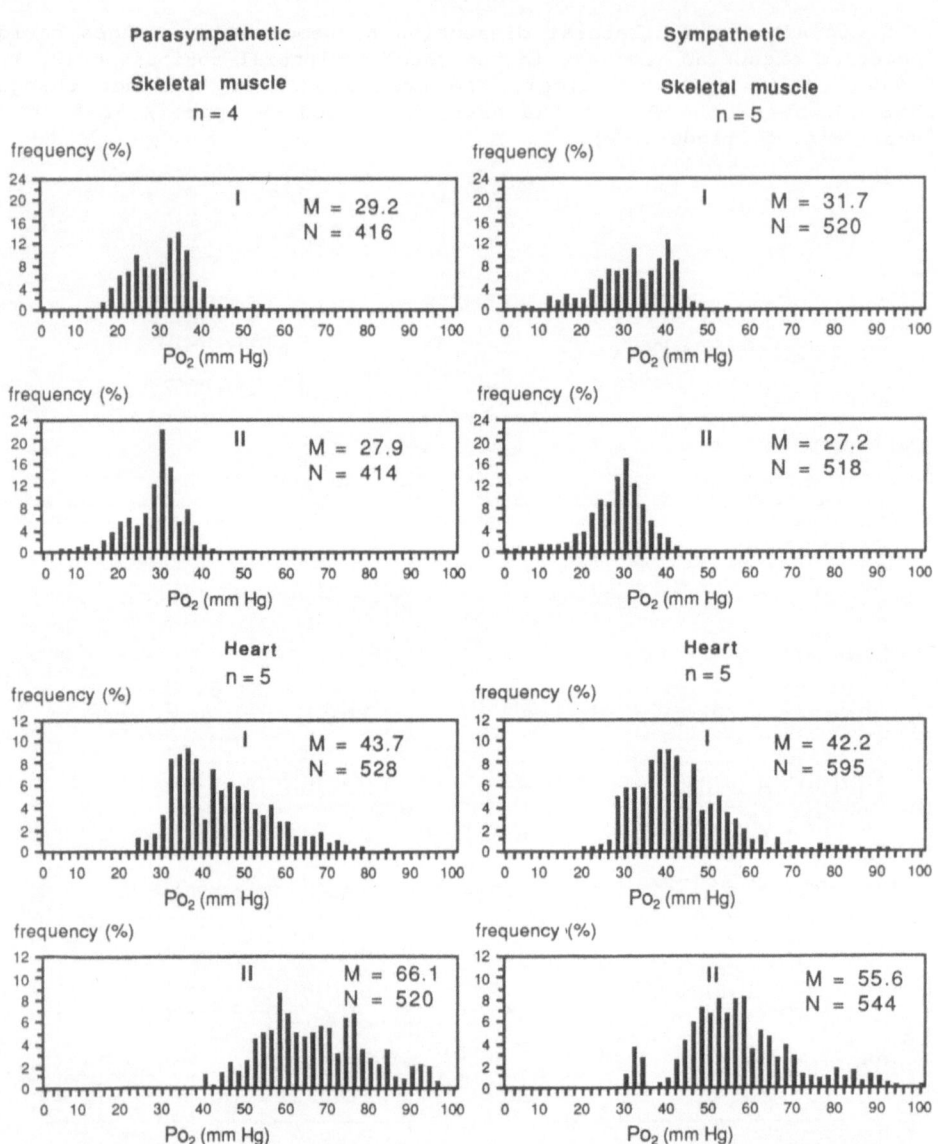

Fig. 5: Pooled histograms of the Po_2 of skeletal muscle and heart of 5 rats before (I) and 5 minutes after (II) the bilateral dissection of the parasympathetic and the sympathetic nerves.
M = mean muscle Po_2 N = number of measuring points
n = number of animals

DISCUSSION

During the electrical stimulation of the parasympathetic nerves, heart rate decreased significantly. However, no decrease in blood pressure or total peripheral resistance occurred. To keep the balance during the electrical stimulation, in spite of the decreased heart rate, the total peripheral resistance had to increase. The increase seemed to be induced by the activation of the sympathetic centers through the baroreceptors. This means that the electrical stimulation of the parasympathetic nerves in the resting rat could decrease the heart rate but there was not enough stimulus to lower blood pressure and total peripheral resistance.

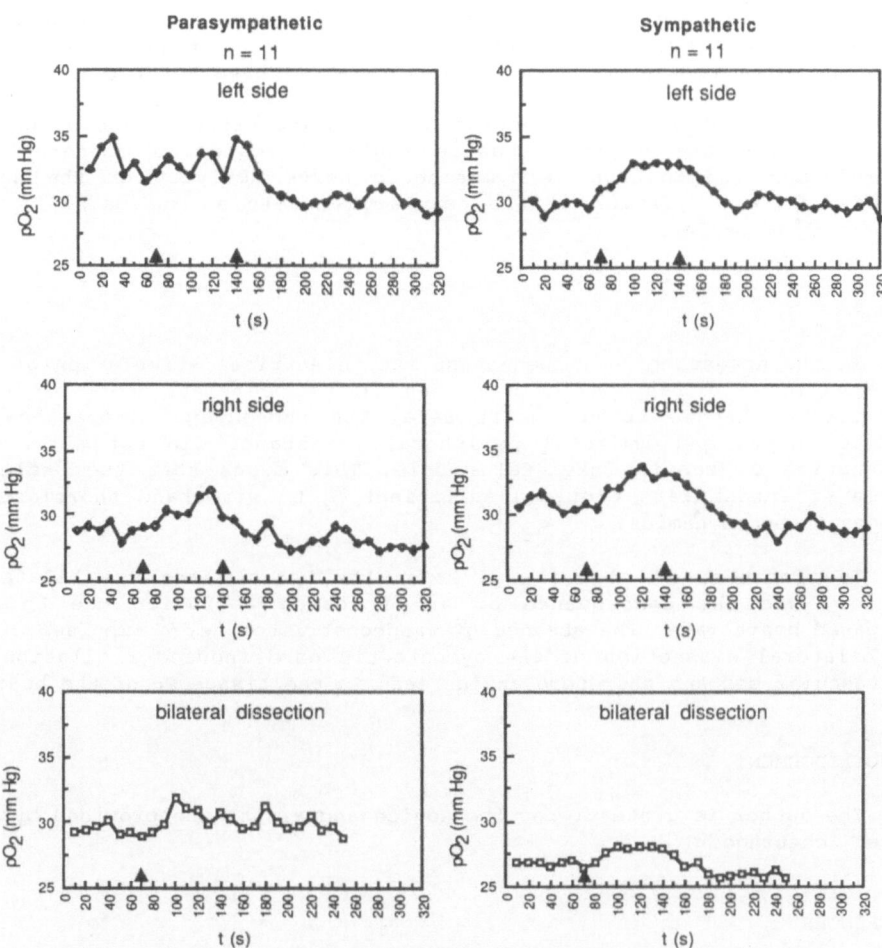

⬆ = Beginning and end of the electrical stimulation or bilateral dissection

Fig. 6: Mean muscle Po_2 of 11 rats continuously recorded before, during, and after electrical stimulation of the parasympathetic and the sympathetic nerves. The arrows indicate the beginning and end of the electrical stimulation.
n = number of animals

The changes in skeletal muscle oxygenation during and after the increased peripheral resistance were only slight and within the normal range of tissue Po_2 oscillations (Fig 6, left side). The skeletal muscle seemed to provide a sufficient O_2-capacity to withstand shortlasting increases in total peripheral resistance.

The elimination of the parasympathetic influence through a bilateral dissection of the vagal nerves induced a significant increase in heart rate. As a consequence of the enhanced heart rate, the tissue Po_2 of the heart muscle increased significantly indicating an augmentation of the coronary circulation.

During the electrical stimulation of the sympathetic nerves, heart rate increased significantly. However, the extent of this increase was not as great as expected. This suggested that the activation of the parasympathetic centers prevented a large increase in the heart rate in the anesthetized rat.

In spite of the enhanced peripheral resistance, tissue oxygenation of the skeletal muscle did not decrease (Fig. 6, right side). This behaviour supports the view that skeletal muscles have sufficient O_2.

After the bilateral dissection of the sympathetic nerves no change in the heart rate occured because higher circulation centers took control. The elimination of sympathetic nerve activity on the heart resulted in the dilation of the coronary vessels as indicated by the increased tissue Po_2.

SUMMARY

In the anaesthetized, open-chest rat, electrical stimulation of the sympathetic or parasympathetic nerves in the cervical area produced significant changes in the heart rate. The subsequent alterations in cardiac output and in total peripheral resistance did not alter the oxygenation of resting skeletal muscle. This means that the skeletal muscle of intact rats provided sufficient O_2 to withstand shortlasting changes in hemodynamics.

The increase in mean tissue Po_2 on the heart after the bilateral vagotomy was the consequence of an increased perfusion due to the increased heart rate. The absence of vasoconstrictor nerve impulses after the bilateral dissection of the sympathetic cord induced a dilation of the vascular bed and therefore an increase in the tissue Po_2 of the heart.

ACKNOWLEDGEMENT

The author is grateful for the advice and assistance provided by Mr. Walter Scheuregger.

REFERENCES

Kessler, M., Lübbers, D. W., 1966, Aufbau und Anwendungsmöglichkeiten verschiedener Po_2-Elektroden, Plügers Arch. ges. Physiol., 291: 82.

506

OXYGEN PARTIAL PRESSURE DISTRIBUTION WITHIN SKELETAL MUSCLE:
INDICATOR OF WHOLE BODY OXYGEN DELIVERY IN PATIENTS?

P. Boekstegers[1], R. Riessen[1], W. Seyde[2]

[1]Department of Physiology, Med. Univ. Lübeck, FRG
[2]Department of Anesthesiology, Univ. Göttingen, FRG

INTRODUCTION

The oxygen supply of body organs is clinically estimated by parameters of whole body oxygen transport and hemodynamics. However, the oxygen delivery to a certain organ is not necessarily indicated by changes in these parameters. Even if cardiac output is measured by Swan Ganz thermodilution technique, thus allowing calculation of the whole body oxygen delivery, its distribution to different organs is unknown and might change, particularly in peripheral organs (e.g. in shock, by administration of vasoactive drugs) (Abrams 1985, Feltes et al. 1987, Slater et al. 1973). Therefore, clinically applicable devices for measurement of the distribution of oxygen partial pressure within peripheral organs have been developed (Kessler and Lübbers 1966, Ehrly 1987, Weiss and Fleckenstein 1986) with the aim to determine the state of oxygen supply more directly.

In the present study in patients undergoing open heart surgery, the distribution of oxygen partial pressure within skeletal muscle was measured simultaneously with parameters of whole body oxygen transport in order to find out whether
1. sustained decrease of local oxygen delivery to skeletal muscle occurs during and after cardiopulmonary bypass,
2. whole body oxygen parameters sufficiently indicate local oxygen delivery to peripheral organs such as the skeletal muscle,
3. the distribution of oxygen partial pressure within the skeletal muscle could be an indicator of whole body oxygen transport.

METHODS

29 patients (aged 41 to 73 years, 24 male, 6 female) were studied up to 6 h after aortocoronary bypass surgery (n = 27) and/or valve replacement surgery (n = 4). Standard procedure of anesthesia included premedication with promethazin (50 mg i.m.) and piritramid (15 mg i.m.), endotracheal intubation, maintenance of anesthesia by fentanyl i.v. (10 µg/kg/h) and nitrous oxide/oxygen (30-50%/70-50%), pancuronium (50 µg/kg/h). Mechanical ventilation was adjusted to maintain arterial Po_2 between 100 and 150 mmHg, arterial Pco_2 between 35 and 45 mmHg. Hypothermic (30.2 ± 1.5°C) extracorporeal circulation (ECC) was performed using nonhemic prime and membrane oxygenator with a pump flow of 2.14 (± 0.25) l/min/m. Hemoglobin concentration content was 8.4 (± 1.2) g/dl during ECC. Patients were rewarmed to at least 35°C core temperature before ECC was terminated. Mechanical ventilation and anesthesia were continued for at least 6 h in the postoperative period.

The distribution of local oxygen partial pressures within biceps muscle of the left arm was measured according to the method described by Weiss and Fleckenstein (1986). Briefly, using a

Po_2-histograph® (Eppendorf, FRG), a hypodermic polarographic needle electrode (tip diameter 350 μm, T90 < 500 ms, probe sensitivity 15-30 pA/mmHg at a polarographic potential of 700 mV, temperature sensitivity 2.44%/°C) was guided into the muscle through the lumen of a previously inserted 20 G Abbocath®. In the muscle the needle probe was stepwise (0.7 mm) moved forward over a total distance of 20 mm. After each step of 0.7 mm, the local Po_2 was measured within less than 1 sec. At the end of each stepwise forward movement of 20 mm the probe was swiftly drawn back to the initial position and its direction of insertion slightly changed. Thereafter, the stepwise forward movement of the needle probe was repeated. Thus, each single Po_2 value was measured at a different randomly situated point within a total muscle volume of about 3 cm³. 200 single local Po_2 values were taken within 8 min. Before and after each measuring procedure of 8 min. the needle probe was two point calibrated (nitrogen/ambient air) in a TRIS-buffered 0.9% saline solution. The temperature of the solution and the temperature within biceps muscle (needle thermocouple, GMS, FRG) was measured. The Po_2-data were corrected for the probe drift and the tissue temperature to calculate a Po_2-histogram and the mean muscle Po_2 (MPo_2m).

Simultaneously with measurement of Po_2-histograms, in 15 patients cardiac output was determined by the thermodilution method (Ganz and Swan 1972) using a cardiac output computer (BN 7206, Fischer, FRG). A 7 F thermodilution catheter (thermokath®, Vygon, FRG) was placed in the wedge position in the pulmonary artery. Three consecutive injections of 10 ml iced saline solution into the right atrium were performed for cardiac output measurement (average taken for further calculation). Arterial blood (radial artery) and mixed venous blood (pulmonary artery) was withdrawn and taken for direct measurement of oxygen saturation, hemoglobin values and oxygen partial pressure. For each measuring period the arterial and mixed venous oxygen concentration, cardiac index, the whole body oxygen delivery (Ox. offer), whole body oxygen consumption ($\dot{V}o_2$), whole body oxygen extraction (Ox. extr.), arterial venous oxygen concentration difference ($CavO_2$) and systemic vascular resistance (SVR) were calculated (Berk and Sampliner 1982). During ECC the above data were determined from blood probes of the arterial and venous line, from blood pressures and from the pump flow (calibrated roller pump). In all patients arterial and central venous blood pressures, arterial blood gases, acid base balances, mixed venous blood lactate (enzyme test combination, Boehringer, FRG), rectal and muscle temperature were measured. These measurements were performed at least hourly up to 6 h after operation.

<u>Statistics</u>

Unless otherwise indicated data are expressed as means (± SD). Pooled Po_2-histograms of the biceps muscle of 14 patients recorded before, during and after ECC, were calculated from all single Po_2 values (n = 2800) of these patients. From each measuring period the paired data of mean muscle Po_2 (MPo_2m) and systemic parameters were taken for linear regression analysis.

RESULTS

The pooled Po_2-histograms recorded from biceps muscle of 14 patients before, during and after ECC are shown in figure 1. 60 min. after beginning of ECC, mean muscle Po_2 (MPo_2m) was decreased from 25 (4) mmHg to 14 (2) mmHg (table 1). The relative frequency of Po_2 values below 5 mmHg was increased from 4% to 20%. In 7 of 14 patients the lowest range of muscle Po_2 (0 - 2.5 mmHg) had the highest incidence. 1 h after ECC mean muscle Po_2 and the shape of the Po_2-histogram were similar to the data before ECC. Except the muscle temperature (r = 0.82), none of the measured or calculated parameters was significantly correlated to the MPo_2m (correlation coefficient -0.4 < r < 0.4) in the three measuring periods during ECC.

In 15 patients cardiac output and Po_2-histograms within biceps muscle were simultaneously determined before and after ECC. Taking into account all paired data the closest correlation existed between whole body oxygen delivery (Ox. offer) and MPo_2m (r = 0.73) and between cardiac index and MPo_2m (r = 0.70) (table 2, figure 2). The correlatian coefficient (Ox. offer/MPo_2m) was higher if only all postoperative data (r = 0.85) (figure 3) or intraoperative data before ECC (r = 0.85) were considered (table 2).

508

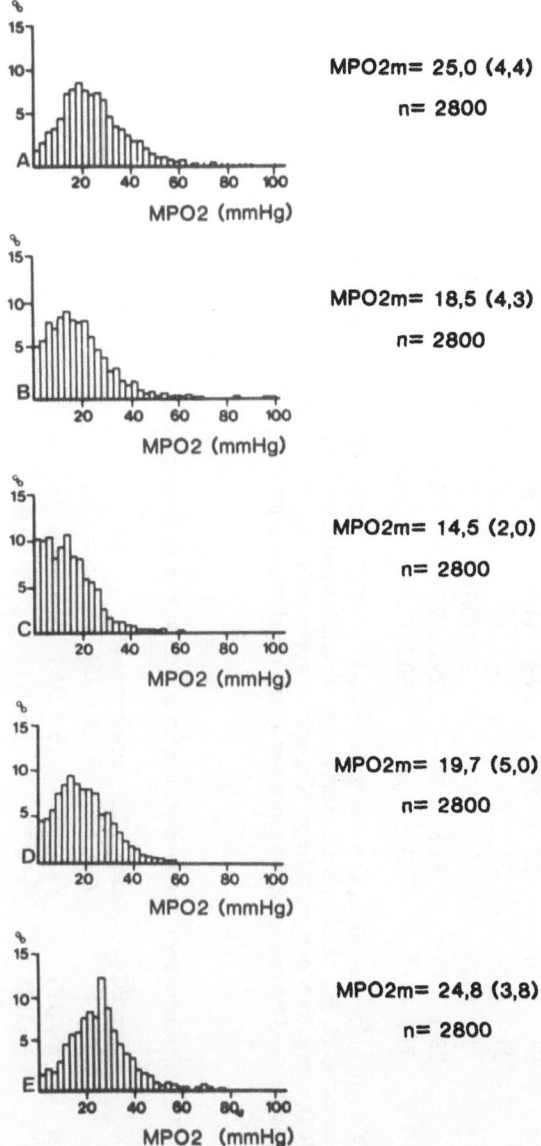

MPO2m= 25,0 (4,4)
n= 2800

MPO2m= 18,5 (4,3)
n= 2800

MPO2m= 14,5 (2,0)
n= 2800

MPO2m= 19,7 (5,0)
n= 2800

MPO2m= 24,8 (3,8)
n= 2800

Fig. 1. Pooled Po_2-histograms of 14 patients. Local Po_2 within biceps muscle (MPo_2), mean local Po_2 within biceps muscle (MPo_2m), number of single Po_2-measurements (n), relative frequency of MPo_2 in ranges of 2.5 mmHg on the ordinate (%). Before ECC (A), 15 min. ECC (B), 60 min. ECC (C), ECC 15 min. after aortic reopening (D), after ECC (E).

Table 1. Summary of data obtained from 14 patients before (A), during (B, C, D) and after ECC (E, F)

	A		B		C		D		E		F	
MPo_{2m} (mmHg)	25.0	(4.4)	18.5	(4.3)	14.5	(2.0)	19.7	(5.0)	24.8	(3.8)	25	(8)
MAP (mmHg)	85	(10)	78	(9)	63	(11)	64	(11)	80	(7)	86	(18)
CI (l/min)	2.2	(0.6)	2.2	(0.2)	2.1	(0.3)	2.2	(0.3)	3.1	(0.9)	3.0	(1)
Ox. offer (ml/min/m²)	368	(114)	255	(40)	246	(42)	265	(41)	375	(120)	436	(126)
$\dot{V}o_2$ (ml/min/m²)	94	(7.3)	55	(16)	69	(13)	96	(13)	130	(19)	150	(50)
Ox. extr. (%)	31	(12)	21	(5)	28	(8)	37	(4)	33	(10)	35	(10)
Lactate (mmol/l)	1.2	(0.2)	2.2	(0.7)	2.1	(0.7)	2.2	(0.8)	1.9	(0.4)	1.4	(0.7)
Tm (°C)	33.0	(1.0)	31.5	(1.1)	29.6	(0.7)	30.5	(1.0)	31.8	(0.9)	31.9	(1.7)
Tr (°C)	35.1	(0.6)	30.0	(1.5)	30.2	(1.5)	33.8	(1.7)	35.1	(0.9)	35.2	(1.7)

A = 60 min. before ECC, B = 15 min. ECC, C = 60 min. ECC, D = ECC, 15 min. after aortic reopening, E = 60 min. after ECC, F = postoperative. Values are means (± SD) of 14 patients. Mean oxygen partial pressure within biceps muscle (MPo_{2m}), mean arterial blood pressure (MAP), cardiac index (CI), whole body oxygen delivery (Ox. offer), whole body oxygen consumption ($\dot{V}o_2$), whole body oxygen extraction ratio (Ox. extr.), mixed venous lactate (Lactate), muscle temperature (Tm), rectal temperature (Tr).

Table 2. Summary of correlation coefficients calculated from paired data of 15 patients before and after ECC

	before ECC (n = 17)	after ECC (n = 39)	postoperative (n = 45)	all (n = 101)
MPo_{2m}/Ox. offer	r = 0.85	r = 0.62	r = 0.85	r = 0.73
MPo_{2m}/CO	r = 0.81	r = 0.72	r = 0.71	r = 0.66
MPo_{2m}/CI	r = 0.79	r = 0.71	r = 0.82	r = 0.70
MPo_{2m}/Ox. extr.	r = -0.63	r = -0.53	r = -0.46	r = -0.55
MPo_{2m}/SvO_2	r = 0.61	r = 0.44	r = 0.45	r = 0.40

Correlation coefficient (r), mean oxygen partial pressure within biceps muscle (MPo_{2m}, mmHg), whole body oxygen delivery (Ox. offer, ml/min/m²), cardiac output (CO, l/min), cardiac index (CI, l/min/m²), whole body oxygen extraction ratio (Ox. extr., %), mixed venous oxygen saturation of hemoglobin (SvO_2).

510

Looking at the relative changes of MPo_2m and whole body oxygen delivery, correlation coefficient was $r = 0.83$ calculated from all paired data (figure 4). Excluding measurements which followed the beginning or end of administration of vasoactive drugs (e.g. dopamine, nitroglycerin), a change of hemoglobin concentration by more than 2 mg/dl or a change of arterial pH by more than 0.04, the correlation coefficient was $r = 0.93$ (figure 5).

DISCUSSION

In order to find out whether sustained decreases of oxygen partial pressure within biceps muscle (MPo_2) occurred during or after extracorporeal circulation (ECC), the distribution of MPo_2 was measured in patients undergoing open heart surgery (figure 1). In hemodynamically stable periods before and after ECC, mean muscle Po_2 (MPo_2m) of 25 mmHg was about 10 mmHg lower than data obtained from healthy human volunteers (Boeksteger 1989). Before and after ECC the relative frequency of low MPo_2 values (< 5 mmHg) in the pooled Po_2-histograms and the shape of the pooled Po_2-histograms of our patients (figure 1) were not different from non anesthetized healthy human subjects (Boeksteger 1989, Ehrly 1987). Therefore, a reduced but sufficient local oxygen delivery within the muscle can be assumed despite decreased mean muscle Po_2.

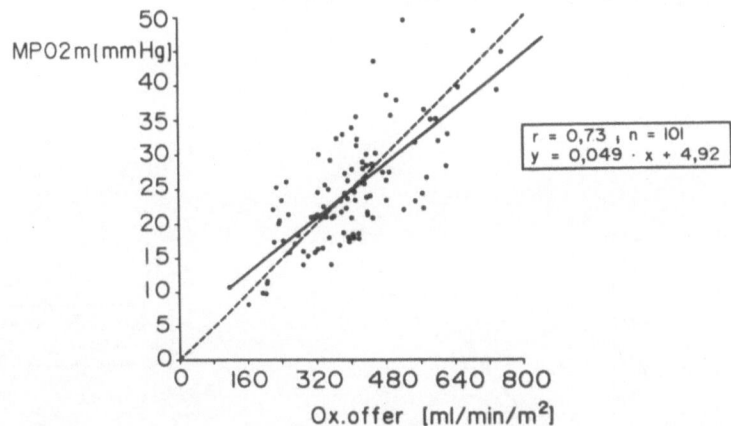

Fig. 2. Paired data of 15 patients, before and after ECC and in the postoperative period. Mean local Po_2 within biceps muscle (MPo_2m), whole body oxygen delivery (Ox. offer), correlation coefficient (r), number of paired data (n), equation of linear regression analysis (y).

During ECC, mean muscle Po_2 (MPo_2m) decreased from 25 mmHg to 14 mmHg and the incidence of Po_2-values below 5 mmHg was increased from 4% to 20% (figure 1). Particularly at the beginning of ECC, there was a strong correlation between MPo_2m and muscle temperature ($r = 0.82$). During ECC the oxygen half saturation point of hemoglobin was calculated with regard to the muscle temperature (table 1) to 19.6 (1.2) mmHg (15 min. ECC) and 17.6 (0.7) mmHg (60 min. ECC). In our opinion, the decrease of MPo_2m during ECC, which was similarly observed by other investigators (Niinikoski et al. 1980), could be sufficiently explained by the left shift of the oxyhemoglobin dissociation curve. Sustained decreases of local oxygen partial pressure within biceps muscle did not occur during and after ECC. Though intermittently in half of the patients the lowest range of muscle Po_2 (0-2.5 mmHg) had the highest incidence during ECC, after ECC the Po_2-histograms of these patients did not differ from the other patients. Postoperatively determined mixed venous blood lactate was not higher and there was no evidence of a higher incidence in postoperative complications in these patients.

511

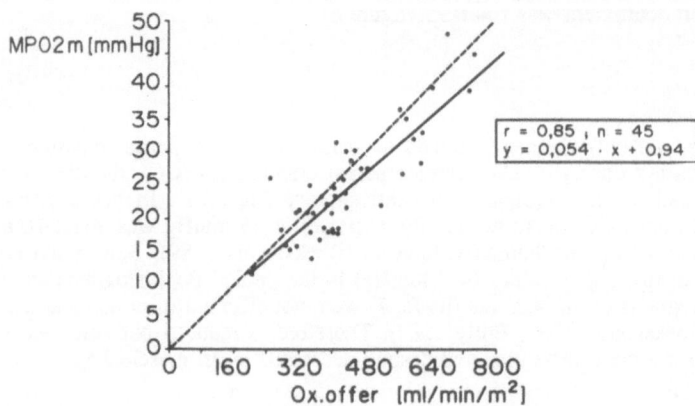

Fig. 3. Paired data of 15 patients, postoperative period. Mean local Po_2 within biceps muscle (MPo_2m), whole body oxygen delivery (Ox. offer), correlation coefficient (r), number of paired data (n), equation of linear regression analysis (y).

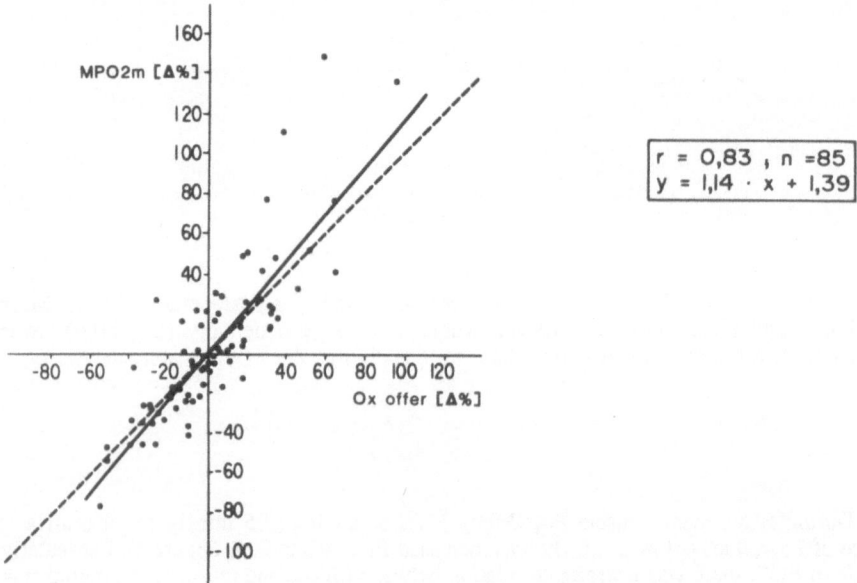

Fig. 4. Paired data of 15 patients, before and after ECC and in the postoperative period. Relative changes of mean local Po_2 (MPo_2m, $\Delta\%$), relative changes of whole body oxygen delivery (Ox. offer, $\Delta\%$), correlation coefficient (r), number of paired data (n), equation of linear regression analysis (y).

Except during ECC, whole body oxygen delivery (Ox. offer) and cardiac index were correlated to mean muscle Po_2 (figures 2 and 3, table 2), particularly if relative changes were considered (r = 0.83) (figure 4). A linear relationship between cardiac output and mean muscle Po_2 was first described in animal shock experiments (Kessler and Lübbers 1966, Furuse et al. 1973). In our patients the highest correlation existed between MPo_2 and whole body oxygen offer before ECC (r = 0.85) and in the postoperative period (r = 0.85) (table 1) indicating that oxygen delivery to biceps muscle depended on cardiac output and arterial oxygen concentration in these periods. It has to be stressed that skeletal muscle relaxation was performed during operation but not in the early postoperative period. Therefore, during operation changes of oxygen consumption within skeletal muscle were not likely to occur in contrast to the postoperative period (e.g. due to muscle shivering). If changes of oxygen consumption within biceps muscle were assumed in the postoperative period, the relationship between MPo_2m and whole body oxygen offer was not influenced hereby.

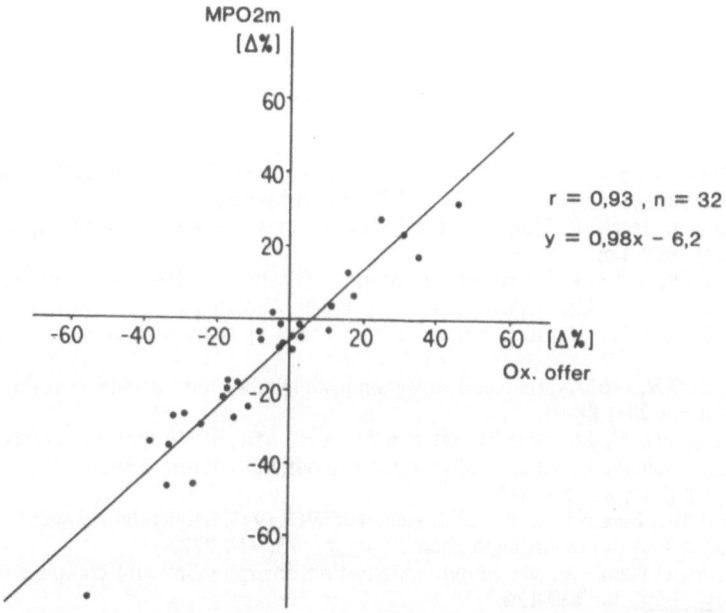

Fig. 5. Paired data of 15 patients, before and after ECC and in the postoperative period. Excluding measurements which followed the beginning or end of administration of vasoactive drugs (dopamine, dobutamine, nitroglycerin), a change of hemoglobin concentration by more than 2 mg/dl or a change of arterial pH by more than 0.04. Relative changes of mean local Po_2 (MPo_2m, $\Delta\%$), relative changes of whole body oxygen delivery (Ox. offer, $\Delta\%$), correlation coefficient (r), number of paired data (n), equation of linear regression analysis (y).

In the initial phase after the beginning or after the end of administration of vasoactive drugs (e.g. nitroglycerin, dopamine and dobutamine) a decrease of the correlation between MPo_2m and whole body oxygen offer occurred suggesting that local oxygen delivery within biceps muscle was changed due to drug induced changes of local muscular blood flow. In addition, only sudden and quantitatively relevant changes of whole body oxygen parameters seemed to intermittenly change the whole body oxygen offer/MPo_2m ratio.

Though some limitations have to be considered, the relatively close correlation between MPo_2m and whole body oxygen offer indicated that measurement of MPo_2m within biceps muscle allows estimation of whole body oxygen offer in patients undergoing open heart surgery. Particularly, if Swan Ganz catheterization of the pulmonary artery is not applied (Tarnow 1982), monitoring of

MPo_2m (with a minimal risk for the patient) might be used for estimation of whole body oxygen offer in other groups of patients as well.

SUMMARY

Simultaneously with determination of cardiac output, the distribution of oxygen partial pressure within biceps muscle was measured during and after open heart surgery in 29 patients. During extracorporeal circulation (ECC) mean muscular oxygen partial pressure (MPo_2m) decreased from 25 mmHg to 14 mmHg with an increase of MPo_2 values below 5 mmHg from 4% to 20%. Sustained decrease of MPo_2m (> 1h) did not occur after ECC. Before ECC and in the postoperative period, MPo_2m was lineary correlated (r = 0.85) to whole body oxygen delivery (Ox. offer) suggesting that local oxygen delivery within biceps muscle was sufficiently indicated only by systemic parameters of oxygen transport which require determination of cardiac output. Particularly with regard to relative changes, MPo_2m might be used for estimation of whole body oxygen offer clinically.

REFERENCES

Abrams J, 1985, Hemodynamic effects of nitroglycerin and long acting nitrates, Am Heart J, 110: 216-224.

Berk JL, Sampliner JE, 1982, Handbook of Critical Care, Little, Brown and Company, Boston.

Boekstegers P, Weiss M, 1989, Tissue oxygen partial pressure distribution within the human skeletal muscle during hypercapnia, Adv Exp Med Biol, (in press).

Ehrly AM, Hauss J, Huch R, 1987, Clinical oxygen pressure measurement, Springer, Berlin Heidelberg New York.

Feltes TF, Hansen TN, Martin CG, Leblanc AL, Smith S, Glesler ME, 1987, The effects of dopamine infusion on regional blood flow in newborn lambs, Pediatr Res, 21: 131-136.

Ganz, W, Swan HJC, 1972, Measurement of of blood flow by thermodiiution, Am J Cardiol, 29: 241-245.

Kessler M, Lübbers DW, 1966, Aufbau und Anwendungsmöglichkeiten verschiedener Po_2-Elektroden, Pflügers Arch, 291: 88-94.

Niinikoski J, Laaksonen V, Meretoja O, Jalonen J, Inberg MK, 1980, Oxygen transport to tissue under normovolemic moderate and extreme hemodilution during coronary bypass operation, Ann Thorac Surg, 31: 134-143.

Slater, GI, Vladeck BC, Bassin R, Kark AE, Shoemaker WC, 1973, Sequential changes in distribution of cardiac output in hemorrhagic shock, Surgery, 73: 714-722.

Tarnow J, 1982, Swan-Ganz catheterization - application, interpretation and complications, Thorac Cardiovasc Surg, 30: 130-136.

Weiss CH, Fleckenstein W, 1986, Local tissue Po_2 measured with "thick" needle probes, Funktionsanalyse biologischer Systeme, 15: 155-166.

DOES ARTERIAL PCO$_2$ INTERFERE WITH HYPOXIA IN MUSCULAR METABOLISM IN MAN ?

J. Raynaud, E. Vargas, M.C. Sant, J. Bordachar, P. Escourrou, O. Bailliart, P. Legros, and Jacques Durand

Instituto Boliviano de Biologia de Altura, La Paz, Bolivia, and CNRS URA-D1159, Centre Marie Lannelongue, Le Plessis 92350 France

INTRODUCTION

The present study examines whether changes in arterial PCO$_2$ (PaCO$_2$), due to exogenous causes, affect muscular metabolism. Few reports in the literature have considered this point of view which is of interest, the degree of arterial oxygenation being usually considered as the main factor affecting muscular metabolism.

It is well documented that acute changes in PaCO$_2$ caused by breathing hypercapnic gas mixtures or by hyperventilating, modify the hemodynamic resistances of some vascular beds, notably cerebral and cutaneous. In contrast, it is generally agreed that they have little or no effect on muscular perfusion (Mellander, 1970) and we are not aware of studies considering the consequences of changes in PaCO$_2$ on muscle metabolism in man.

With regard to chronic changes in PaCO$_2$, little is known because experimental situations providing chronic changes in PaCO$_2$ do not exist, except for some unusual situations of confinement. The only situations of chronic changes in PaCO$_2$, independent of endogenous metabolic production and compatible with a satisfactory maintainance of homeostasis, are high altitude (HA) and chronic alveolar hypoventilation in patients with chronic obstructive lung disease (COLD): PaCO$_2$ decreases at HA whereas it increases in COLD, and in both cases the changes in PaCO$_2$ are accompanied by a decrease in PaO$_2$.

To answer the question posed in the title, we studied the muscular metabolic responses to the same exercise in different conditions of PaCO$_2$:
i) At high altitude (La Paz, 3 800m), in two groups of subjects: one group, native residents of La Paz, fully adapted to HA, and the other, lowlanders during the course of adaptation, after a 3 week stay in La Paz.
ii) At sea level (SL), in two groups of subjects: one made up of COLD patients chosen for their high values of PaCO$_2$, and a second group made up of normal subjects breathing a hypercapnic gas mixture or room air.
Forearm rhythmic exercise provides an adequate experimental device: the muscular mass is so small that exercise induces no measurable alterations in arterial blood composition. Hence, the working muscles continue to be perfused by arterial blood which is unchanged as compared to resting conditions.

Oxygen Transport to Tissue XII, Edited by J. Piiper *et al.*
Plenum Press, New York, 1990

MATERIALS AND METHODS

In all three protocols, which will be described below, the exercise was carried out with the same hand ergometer. The exercise consisted of moving a 6 kg weight upwards, using a pully, over a distance of 4 cm, 30 times per min for 5 min. Power output was 1.5 watt per 100 g of muscular mass, assuming a weight of 180 g for the muscular mass involved in the exercise. The intensity of the exercise corresponds to the maximal aerobic capacity as demonstrated in a previous study (Douguet et al. 1989) using muscular blood flow measured with the pulsed Doppler technique and the muscular arterio-venous O_2 difference.

The experiments were conducted in the morning. The subject sat with the forearm lying on the table, at an angle of 120° to the arm, holding the ergometer handgrip.

A Teflon catheter (Cathlon IV Johnson-Johnson, diam: 16/10, length: 55 mm) was introduced percutaneously upstream into an antecubital vein. Before catheterization, the subcutaneous tissue was infiltrated with 0.5 ml of a 1% lidocaine solution, and a cuff was inflated at about 70 Torr around the arm to obtain suitable filling of the vein and to facilitate the catheter's passage through the valves. The entire length of the catheter was introduced in a retrograde direction into a deep branch of the median cubital vein so that its tip could no longer be palpated. After the cuff had been removed, the vein recovered during 30 min. The insertion of the catheter was always done by the same experimenter for all the protocols.

Two blood samples were drawn with a 5 minutes interval to determine resting values; then, exercise began and muscular venous blood was sampled repeatedly during exercise and recovery as indicated in the figures.

On account of the large number of venous blood samples, arterial blood was drawn from each subject in a separate session. For this purpose a catheter was inserted into the brachial artery. One sample was drawn at rest and another at the end of exercise: this arterial sampling coincided with the routine blood gas checking for the COLD patients.

PO_2, PCO_2 and pH were determined on standard electrodes at 37°C (GM A2 Radiometer and PH M73), lactic acid concentration [LA] using Boehringer test kits, and hemoglobin concentration [Hb] using an OSM2 Radiometer at sea level and by the cyanmethemoglobin technique of Drabkin (Spectrophotometer Jobin et Yvon standard V type) at HA. Samples were analyzed in duplicate.

The concentrations of O_2 [CaO_2], [CvO_2] and of CO_2 [$CaCO_2$], [$CvCO_2$] were calculated with the Gabel's algorithm (1980) using PO_2, PCO_2, pH, [Hb] and blood temperature. The venous muscular blood temperature was taken at 36° at rest and 36.5°C during exercise and the arterial temperature at 36.8°C.

Unfortunately in this study, we could not measure brachial artery blood flow with the pulsed Doppler technique because we could not bring the equipment delivering a 5°C wind, designed to cool the skin of the forearm, to La Paz or to the hospital in Paris. It is evident that reliable measurements of the brachial artery blood flow, particularly when small changes are expected, necessitate the elimination of skin blood flow. To do this, we had previously noted that cooling the skin is a more satisfactory way than inflating a cuff round the wrist on account of the mechanical stimulation of the artery walls due to the sudden stoppage of blood flow.

Different groups of six subjects were examined with the three types of protocols. All subjects were sedentary and were not engaged in arm training. They were right handed and exercised with the right arm.

Protocol at high altitude (La Paz 3 800 m)

Two kinds of HA subjects, all of them laboratory workers, were investigated: six highlanders (age: 35-54 yr) born and living in La Paz, and six lowlanders (40-60 yr) from Paris, after three weeks in La Paz.

Protocol in COLD patients (Paris 45 m)

Six patients (age range: 53-65 yr) with chronic obstructive lung disease, hospitalized following an acute episode, were selected because of their excessively high $PaCO_2$. They were asked whether they agreed to

516

participate in an experiment which was not of clinical interest to them but of physiological interest only. All the subjects asked gave their consent.

Protocol breathing hypercapnic gas mixture and room air (Paris 45 m)

The six subjects who participated in the HA protocols were also studied at SL breathing room air, and also a normoxic hypercapnic gas mixture (5 % CO_2). The subject breathed the hypercapnic gas mixture as soon as the venous catheter was inserted. The exercise begun 30 min later. Both sessions, with the hypercapnic gas mixture or room air were carried out on the same day at a one hour interval.

Three subjects breathed room air first and the hypercapnic gas mixture after, and the three others started with the hypercapnic gas mixture, in order to eliminate possible effects on local vasomotor tone due to the implantation of the catheter in a vein for two hours.

The six normal subjects breathing room air at SL, were taken as the control group (Tables 1 and 2). They are identified by open circles in Figure 2 A, B and C.

RESULTS

Arterial blood

The means ± 1SE of the values observed at the 4th and 5th minutes of exercise were reported in Table 1 and 2. No significant difference appears compared with the corresponding resting values.

Venous blood

The means ± 1SE of the values for the different parameters observed at the 4th and 5th minutes of exercise are reported in Tables 1 and 2.

Figure 2 A, B, and C reported in a following page, illustrates the time courses of the different parameters measured during exercise and recovery in the three protocols.

In addition, results already published (Raynaud et al. 1986) have been included in the Tables. They were obtained with the same methodology, in a study designed to explore the effects of acute hypoxia exposure in a hypobaric chamber. They are therefore comparable to the present results and will help in discussing the HA effects.

Some characteristics in the metabolic responses are common to all groups of subjects: during exercise, PvO_2 decreased within 2 min, then plateaued till the end of exercise. During recovery, the time courses of PO_2 displayed an important rebound overpassing the resting values which had not recovered at the 15th minute.

The time courses of $PvCO_2$ during exercise mirror those of pH: at onset of exercise they changed more slowly than PvO_2 and they reached a steady state value later.

At the beginning of recovery, pHv displayed a transient sharp decrease, probably related to the rebuilding of the creatine phosphate storage as discussed in a previous paper (Raynaud et al. 1986).

Afterwards, $PvCO_2$ and pH returned to the resting values which were reattained within 4 min for $PvCO_2$ and more slowly for pH.

[Hb] rose during the first two minutes of exercise by about 7 %, then plateaued till the end of the muscular activity. This rise in [Hb] during exercise was attributed to a leakage of water from the vessel, but the mechanisms which might account for the drainage of this capillary filtrate remains hypothetical (Douguet et al. 1988).

[LAv] rose continuously during exercise for all groups of subjects, and fell when exercise stopped.

In contrast to the general characteristics relating to the exercise itself, others relate to the experimental conditions. They will be refered in the discussion.

517

DISCUSSION

The discussion of our results is based on a conclusion drawn from a previous study (Douguet et al. 1989) in which it was demonstrated that the 6 kg exercise, used with our subjects corresponds to the maximal aerobic capacity of the muscles involved in the exercise.

Figure 1 illustrates local muscular $\dot{V}O_2$, calculated as the product of the muscular blood flow, $\dot{Q}musc$, (measured in the brachial artery using the pulsed Doppler technique) and the muscular (a-v)O_2 difference, during exercises of increasing intensity carried out in normoxia, and explored with the same methodology as that used here (our six subjects participated to this previous study).

Figure 1 shows that the steady state value of $\dot{V}O_2$ remained unchanged when the exercise passed from 4 kg (17.5 ml.min^{-1}) to 6 kg (18.5 ml. min^{-1}). PvO2 also remained unchanged around 20 Torr for the 4 kg (17.9 $\pm$ 0.7 Torr) and 6 kg exercises (19.4 $\pm$ 1.3 Torr, NS).

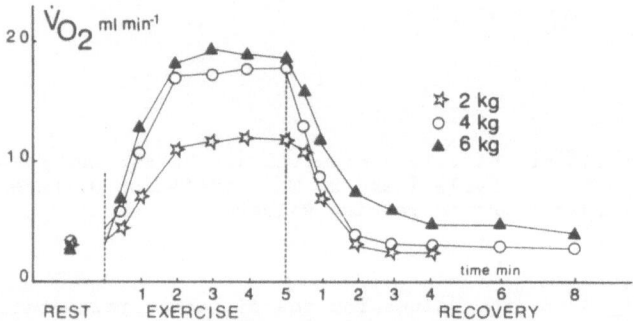

Figure 1. Time course of $\dot{V}O_2$ uptake calculated as the product of brachial artery blood flow (measured using pulsed Doppler technique, a 5°C wind cooling the skin of the forearm in order to eliminate cutaneous blood flow), and of muscular arterio-venous O_2 difference, at rest, during exercise and recovery, for 2, 4 and 6 kg exercises.

We can conclude from this study that the maximal value of the product $\dot{Q}musc$ x (a-v)O_2 difference is accompanied by a maximal decrease of PvO$_2$ to around 20 Torr. This value of PvO$_2$ of about 20 Torr has been underlined by many authors (Doll et al., 1968; Hartley and Saltin, 1969; Hartley et al., 1973; Monod et al., 1961; Raynaud et al., 1986) but we are aware of one study only (Hartley and Saltin, 1969) which firmly established that the muscles were working at $\dot{V}O_{2max}$.

Hogan et al. (1988, 1989) have convincingly demonstrated that O_2 uptake by exercising muscles is limited by O_2 diffusion only. In consequence, the value of 20 Torr observed at $\dot{V}O_{2max}$, at sea level, represents the minimum value of the driving pressure required for O_2 diffusion between blood and mitochondria in which PO$_2$ is presumed to be near 0. If we accept that PvO$_2$ can be taken as being representative of the mean tissular capillary PO$_2$ (Stainby et al. 1988), we can calculate the value of the tissular coefficient for O_2 diffusion (Fick's law) in normoxia, using the mean values of $\dot{V}O_{2max}$ and PvO$_2$ previously found for 4 and 6 kg exercises as: 10 ml.min^{-1}.100^{-1} g of tissue / 18.7 Torr i.e. 0.53 ml. min^{-1}.100^{-1} g.Torr^{-1}, a value close to that found by Hogan et al. (1989) using dog muscle.

Hence, since $\dot{V}O_{2max}$ is the product of maximal muscular blood flow and of maximal muscular arterio-venous O_2 difference, we should deduce that, for a given value of $\dot{V}O_{2max}$, if muscular blood flow varies, necessarily arterio-venous difference varies in the opposite direction.

518

Thus, maximal muscular blood flow might vary on account of changes in the control of the local blood flow, and the maximal (a-v)O_2 difference might vary on account of possible changes in [Hb] associated with a shifting of the Hb/HbO_2 dissociation curve.

In the light of these assumptions and remarks, we can now discuss the results obtained in our protocols.

Hypoxic hypocapnia: Acute and chronic exposure to low barometric pressure

The main findings which result from the comparison of the responses observed in the lowlanders at SL, after one hour in the hypobaric chamber (4 000 m), and after three weeks in La Paz and in the permanent residents of La Paz, are:

i) Permanent residents and translocated lowlanders after a 3 week adaptation period, are quite indistinguishable as regards PvO_2, $PvCO_2$, pHv, [LAv], and arterial and muscular venous O_2 content (Tables 1 and 2 and Figure 2 A), which means that lowlanders after three weeks at 3 800 m reach the same level of adaptation as that of the permanent residents, at least as regards an exercise which does not involve ventilatory and cardiac systems.

ii) PvO_2 in permanent residents and lowlanders at HA display the same value of about 20 Torr as lowlanders at SL, whereas PvO_2 is 14 Torr in acute exposure in the hypobaric chamber (Table 1).

If we accept that the tissular O_2 diffusion coefficient is unchanged at HA according to Sillau et al.1980, we can conclude that $\dot{V}O_{2max}$ declines in acute exposure, and re-attains the normoxic SL value after 3 weeks at 3 800 m, becoming similar to the $\dot{V}O_{2max}$ of the permanent residents.

How does the adaptation to high altitude occur ?

As compared with acute exposure, the adaptation of the lowlanders to HA developed so that the (a-v)O_2 difference increased during adaptation passing from 13.4 ml.100^{-1} ml to 15.7, a larger value than at SL, due to two mechanisms: an increase in [Hb] and a shifting to the right of the O_2 dissociation curve which allows the release of a greater amount of O_2 at the same PvO_2 (Table 2). The shifting of the dissociation curve is most likely to be related to the more pronounced decrease in pH (7.30 in acute exposure versus 7.24 at HA) induced by the higher value of [LA] at HA.

If we accept that $\dot{V}O_{2max}$ is unchanged at HA, the increased (a-v)O_2 difference necessarily argues for a decrease in muscular blood flow in translocated lowlanders after three weeks at 3 800m as well as in highlanders. This conclusion agrees with those of others who measured muscular blood flow directly in translocated subjects (Bidart et al. 1975; Bender et al. 1988).

In contrast to the clear cut conclusion concerning the direction of change in muscular blood flow in response to chronic exposure to HA, it is more questionable to deduce an increase in blood flow from the narrowing of the (a-v)O_2 difference in acute exposure (Table 2), because $\dot{V}O_{2max}$ is probably decreased as suggested by the lower value of the product DO_2 x PvO_2 on account of the lower value of PvO_2 (assuming that DO_2 does not increase in acute hypoxia exposure). Hence, in the absence of accurate measurement of $\dot{V}O_{2max}$ at HA, the conclusion about the direction of change of muscular blood flow cannot be definite. However, it is more than likely that it increased, based on the fact that cardiac output during exercise is higher in acute hypoxia exposure (Cunningham et al. 1965).

Does $PaCO_2$ play a role in muscular metabolism passing from sea level to high altitude ?

TABLE 1

Subjects	PaO_2	$PaCO_2$	[Hba]	PvO_2	$PvCO_2$	pH	[Hbv]
HL	60.0***	30.0***	17.0**	20.8	57.0*	7.24*	17.9*
	1.8	0.9	0.2	2.0	2.4	0.01	0.2
HA							
LL	59.0***	29.0***	16.8**	18.9	59.2*	7.24*	18.0*
	2.5	1.0	0.2	1.7	2.9	0.02	0.2
LL Hypobaric	53.6***	33.5***	14.1	14.3*	59.4**	7.30**	14.5
chamber°	1.6	0.7	0.3	0.6	2.2	0.02	0.3
LL at SL	95.0	39.8	14.0	19.8	64.8	7.28	14.7
control	1.3	0.6	0.3	1.4	2.2	0.02	0.4
COLD patients	47.0***	54.0*	15.7*	25.9*	89.4**	7.20*	16.2*
	2.5	3.2	0.8	2.7	7.1	0.04	0.4
LL 5% CO_2	95.0	42.0	14.1	21.8*	68.7*	7.26	14.8
	1.4	0.8	0.3	1.6	1.9	0.02	0.2

TABLE 2

Subjects	CaO_2	CvO_2	$(a-v)O_2$	$CvCO_2$	$CaCO_2$	$(v-a)CO_2$
HL	20.9	5.7	15.2	56.9	42.2	14.7
HA						
LL	20.5	4.8	15.7	58.8	41.7	17.1
LL Hypobaric chamber	16.3	2.9	13.4	66.9	46.7	20.2
LL at SL control	18.5	4.5	14.0	70.2	56.0	14.2
COLD patients	17.1	6.7	10.4	81.9	74.3	7.6
LL 5% CO_2	18.6	5.2	13.4	71.3	59.1	12.3

The abbreviations have the same meaning as in Table 1.
CaO_2, CvO_2, $(a-v)O_2$ arterio-venous difference in O_2 content, $CaCO_2$,
$CvCO_2$, $(v-a)CO_2$ veno-arterial difference in CO_2 content, are expressed in
$ml.100^{-1}$ ml of blood.
All values of this table were calculated using the mean values of the
data reported in Table 1 (see text for calculations): this is the reason
why SE has not been calculated.

Values are means ± 1SE of the different parameters measured at the 4th
and 5th minutes of exercise.
The lowlanders at sea level have been taken as the references to whom the
subjects in other situations have been compared. The asterisks indicate
the level of statistical significance obtained in such a comparison:
$P < 0.05$: *; $P < 0.01$: **; $P < 0.001$: ***.
HA: high altitude; SL: sea level; HL: highlanders; LL: lowlanders; COLD:
chronic obstructive lung disease.
Arterial and venous partial pressures of O_2 and CO_2 are expressed in
Torr, [Hb] in $g.100^{-1}$ ml of blood,
° The values of this line are reported from a previous study: J. Appl.
Physiol. 60: 1203 - 1208, 1986.

520

Despite the decrease in $PaCO_2$ at HA due to the lowering of barometric pressure, $PvCO_2$ increased during exercise, moving towards the sea level value which however, it did not quite reach.

Hence, the muscles at HA work in conditions of $PvCO_2$, close to those of SL and it would be highly speculative to attribute a role to $PaCO_2$ in muscular metabolism at HA.

In contrast, it seems more important to us to highlight the possible role played by the high values of [LA] in the two groups of subjects at HA, which are probably responsible for the decrease in pH which in turn shifts the O_2 dissociation curve to the right. Thus, O_2 delivery from the O_2 supply is facilitated, the muscular $(a-v)O_2$ difference is enlarged, and muscular blood flow is decreased.

The present protocols do not permit us to discuss whether [LA] in venous blood is higher at HA due to a greater proportion of anaerobic metabolism or to the facilitated passage from the myocytes.

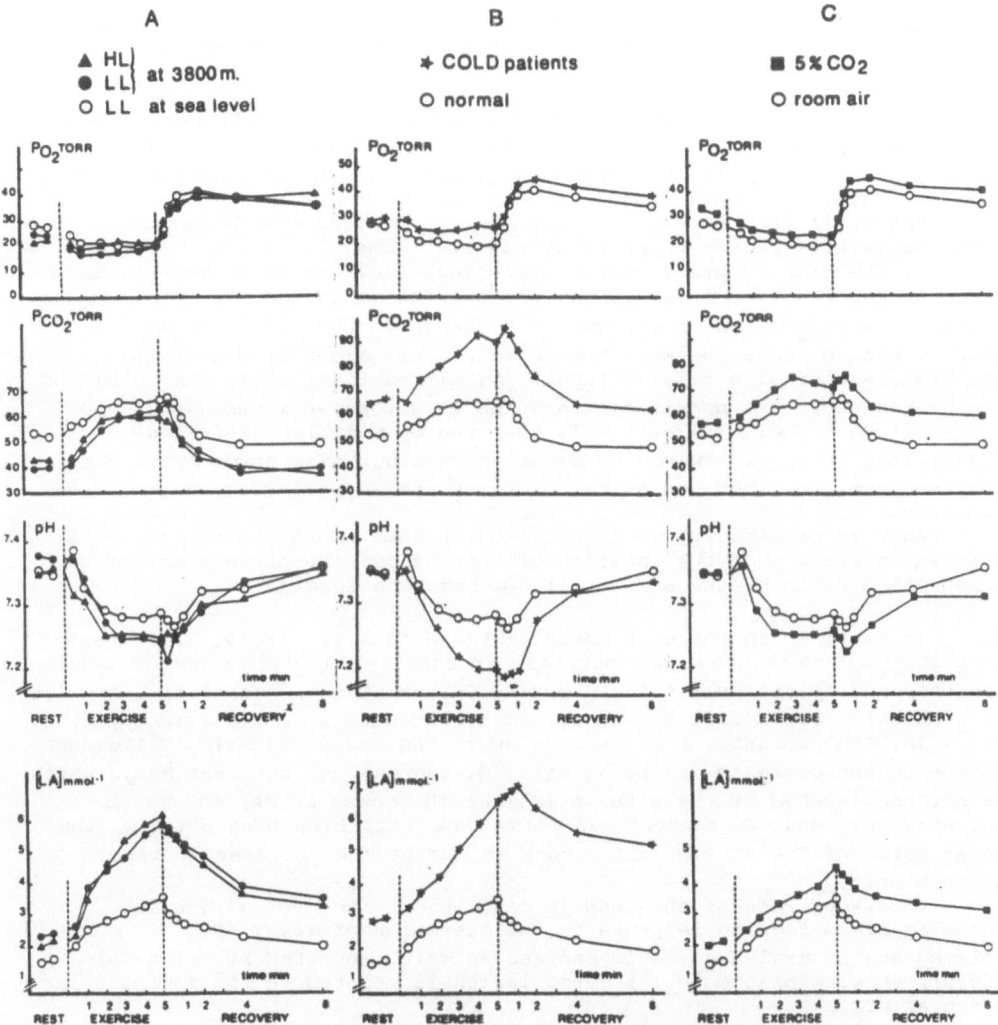

Figure 2. Time course of PO_2, PCO_2, pH and lactic acid concentration [LA] in the muscular venous blood at rest, during exercise corresponding to VO_{2max} of the muscular mass concerned, and recovery. Column A illustrates the results of highlanders (HL) and lowlanders (LL), column B those of chronic obstructive lung disease (COLD) patients and normal subjects, and column C those of lowlanders breathing 5 % CO_2 or room air.

The common characteristic which emerges from the comparison of Figures 2 B and C, is that all the curves obtained in the hypercapnic conditions are displaced in the same direction as compared with those of the normoxic normocapnia subjects i.e. PvO_2, PCO_2 and [LA] are higher and pH lower. However, the difference is much more marked for COLD patients (Fig. 2 B) for whom the levels of PCO_2 in arterial and muscular venous blood are much higher than for normal subjects breathing CO_2 (Fig. 2 C).

It could be deduced from these results that the exercising muscles of COLD patients do not suffer from arterial hypoxia and those of normal subjects breathing CO_2 are in fact more oxygenated.

Surprisingly, despite the fact that the muscular cells in hypercapnia seem to work in a relatively more oxygenated environment than in normocapnia, [LA] in the efferent venous blood is more than double in COLD patients and slightly but significantly higher during CO_2 breathing, inducing in both cases a decrease in pH.

What assumptions can we elicit to explain these apparently contradictory results ?

If we accept that $\dot{V}O_{2max}$ is determined by the product, $DO_2 \times PvO_2$, we should conclude that $\dot{V}O_{2max}$ is higher in hypercapnia since PvO_2 is higher than in normocapnia. Such a conclusion is not credible for either COLD patients, or normal subjects.

Hence, if the theoretical product is too high, the discrepancy might be related to one or other of the two terms.

i) The driving pressure for diffusion, i.e. the difference between the mean capillary O_2 pressure and the intramitochondrial O_2 pressure, might be overestimated on account of a change in the mitochondrial PO_2 which would no longer be near but above 0. This might be due to the metabolic effect of a rise in $PaCO_2$. The marked increase in [LA] could be the consequence of a possible alteration in oxydative metabolism.

ii) The other alternative is that the O_2 tissular diffusion coefficient, DO_2, is lowered as compared with normoxic normocapnia due, for instance, to a reduced number of functional capillaries, which is unlikely.

Another possibility is that diffusion time along the capillary is reduced on account of the increase of blood flow; the consequence of such a mechanism would be the same as if DO_2 had been lowered.

In fact, an increase in muscular blood flow is likely. If we assume that $\dot{V}O_{2max}$ is unchanged, the decrease of the $(a-v)O_2$ difference in COLD patients and to a lesser extent, during CO_2 breathing, necessarily implies that muscular blood flow has increased compared to normoxic normocapnia.

In COLD patients, the lower value of the muscular $(a-v)O_2$ difference is due to the decrease in the arterial O_2 content due to lower PaO_2, which is not compensated by the rise in [Hb] as it occurs at HA, and to the increase in venous O_2 content. In normal subjects breathing 5 % CO_2, the lower value of the $(a-v)O_2$ difference is due to the increase of venous O_2 content only.

Thus, our results obtained in man, which argue for a rise in muscular blood flow in response to the elevation of local PCO_2, corroborate several studies in man and animals, reported by Mellander (1970), showing that muscular vasodilation is related to the increase in muscular PCO_2.

However, apart from the effect of PCO_2 on muscle perfusion during work, can we attribute a muscular metabolic effect to the local change in PCO_2 ?

522

As our results provide no insight into the biochemical steps leading to energy release, we can estimate the possible metabolic consequence of PCO_2 change only from the relative proportion of anaerobic metabolism. We can assume that [LAv] is closely related to anaerobic metabolism.

As regards [LAv], the results show that, compared to normocapnic normoxia, it reaches higher values at the end of exercise, in:

    i)   hypocapnic hypoxia (HA)
    ii) hypercapnic hypoxia (COLD)
    iii) normoxic hypercapnia (breathing 5 % CO_2).

Considering i) and ii), the conclusion would be that the higher increase in [LA] is independent of $PaCO_2$ since $PaCO_2$ decreases at HA and increases in COLD patients, and independent of [CaO_2] since CaO_2 increase at HA and decreases in COLD patients, compared to normal subjects. The increase in [LAv] would be related only to PaO_2 i.e. to the head O_2 pressure at the entrance to the capillary.

Considering ii) and iii), the conclusion would be that a rise in $PaCO_2$, marked or slight, enhances the proportion of anaerobic metabolism which no longer depends on PaO_2 which is decreased in COLD patients, and normal in subjects breathing hypercapnic normoxic gas mixture.

Thus, the relationships of the changes in [LAv] with those in $PvCO_2$ appear so puzzling that our assumption is probably partially incorrect i.e. that [LAv] is not entirely relevant to anaerobic metabolism, but probably also depends on the efflux of lactic acid from the muscular cells as reported recently (Connett et al.). $PvCO_2$ would be one of the factors, which mofifies the permeability of the myocytes to lactic acid.

In conclusion, the present study shows that a rise in $PaCO_2$, which is not directly related to metabolic production of CO_2 (COLD patients and subjects breathing 5% CO_2), is accompanied during exercise by a higher rise in muscular blood flow as compared with normocapnic normoxia, whereas a lowering in $PaCO_2$ related to chronic exposure to low barometric pressure is accompanied by a smaller rise in muscular blood flow. The association of such changes leads us to conclude that local PCO_2 plays a determinant role in muscular vasodilation in man. We cannot advance any conclusion about a possible muscular metabolic effect of PCO_2 except that it may alter myocyte membrane permeability to lactic acid.

SUMMARY

To answer the question whether PCO_2 affects the muscular metabolism, PO_2, PCO_2, pH, lactic acid concentration and hemoglobin were measured in the efferent muscular venous blood from common flexor digitorum, during forearm rhythmic exercise corresponding to VO_{2max}. Exercise was carried out either in hypocapnic hypoxia i.e. in permanent high altitude residents and translocated lowlanders, or in hypercapnic hypoxia i.e. in chronic obstructive lung disease (COLD) patients. The results show that, during exercise: i) PO_2 in muscular venous blood remains around 20 Torr in normoxia and hypocapnic hypoxia and even higher (25 torr) in COLD patients, despite low arterial PaO_2, and ii) arterial and/or local PCO_2 play a role in the control of the muscular blood flow. But we cannot conclude that a change in $PaCO_2$ affects muscular metabolism itself, because lactic acid in the muscular venous blood, that we used to check this effect, is likely dependent on mechanisms other than anaerobic glycolysis, such as a change in lactic acid efflux from the myocytes. The increase in muscular venous PCO_2 may enhance the myocyte permeability to lactic acid during exercise.

REFERENCES

Bender, P.R., Groves, B.M., McCullough, R.E., McCullough, R.G., Huang, S.Y., Hamilton, A.J., Wagner, P.D., Cymerman, A., and Reeves. J.T., 1988, Oxygen transport to exercising leg in chronic hypoxia, J. Appl. Physiol., 65:2592.

Bidart, Y., Drouet, L. et Durand, J., 1975, Débit sanguin dans le muscle squelettique chez les sujets résidant et transplantés en altitude (3 800 m), J. Physiol. (Paris), 70:333.

Connett, R.J., Gayeski, T.E.J., and Honig C.R., 1986, Lactate efflux is unrelated to intracellular PO_2 in a working red muscle in situ, J. Appl. Physiol., 61:402

Cunningham, W.I., Becker, F.J., and Kreuzer. F., 1965, Catecholamines in plasma and urine at high altitude, J. Appl. Physiol., 20:607.

Doll, E., Keul, J., and Maiwald, C., 1968, Oxygen tension and acid-base equilibria in venous blood of working muscle, Am. J. Physiol., 215:23.

Douguet, D., Bordachar, J., Sant, M.C., Gascard, J.P., Legros, P., and Raynaud, J., 1989, Time course of local muscular blood flow and O_2 uptake during forearm rhythmic exercise of various intensities, to be submitted for publication to Clin. Physiol.

Douguet, D., Raynaud, J., Capderou, A., Pannier, C., Reiss, G., and Durand, J., 1988, Muscular venous blood metabolites during rhythmic forearm exercise while breathing air or normoxic helium and argon mixtures. Clin. Physiol., 8:367.

Gabel, R.A., 1980, Algorithms for calculating and correcting blood-gas and acid-base variables, Respiration Physiology, 42:211.

Hartley, L.H., and Saltin, B., 1969, Blood gas tensions and pH in brachial artery, femoral vein, and brachial vein during maximal exercise. in "Biochemistry of exercise. Medicine and sport, Vol. 3", Karger, Basel, New York, p 66.

Hartley, L.H., Vogel, J.A., and Landowne, M., 1973, Central, femoral, and brachial circulation during exercise in hypoxia, J. Appl. Physiol., 34:87.

Hogan, M.C., Roca, J., Wagner, P.D., and West. J.B., 1988, Limitation of maximal O_2 uptake and performance by acute hypoxia in dog muscle in situ, J. Appl. Physiol., 65:815.

Hogan, M.C., Roca, J., West, J.B., and Wagner, P.D., 1989, Dissociation of maximal O_2 uptake from O_2 delivery in canine gastrocnemius in situ. J. Appl. Physiol., 66:1219.

Mellander, S., 1970, Ann. Rev. Physiol., 32:325.

Monod, H., Saint-Saens, M., Scherrer, J., and Soula, C., 1961, Etude du travail musculaire et de la fatigue. III Le sang veineux efferent d'un muscle effectuant un travail dynamique chez l'Homme, J. Physiol. (Paris), 53:697.

Raynaud, J., Douguet, D., Legros, P., Capderou, A., Raffestin, B., and Durand, J., 1986, Time course of muscular blood metabolites during forearm rhythmic exercise in hypoxia, J. Appl. Physiol., 60:1203.

Sillau, A. H., Aquin, L., Bui, M.V., and Banchero, N., 1980, Chronic hypoxia does not affect guinea pig skeletal muscle capillarity, Pflügers Arch., 386:39.

Stainby, W.N., Snyder, B., and Welch, H.G., 1988, A pictographic essay on blood and tissue oxygen transport, Med. Sci. Sports Exercise, 20:213.

TISSUE OXYGEN PARTIAL PRESSURE DISTRIBUTION WITHIN THE HUMAN SKELETAL MUSCLE DURING HYPERCAPNIA

P. Boekstegers and M. Weiss

Department of Physiology
Medical University of Lübeck
Lübeck, F.R.G.

INTRODUCTION

The effects of arterial hypercapnia on the cardiovascular system have been extensively investigated. However, few data exist on the effects of hypercapnia on the peripheral circulation in man. In animal experiments the increase of the systemic sympathetic tone due to hypercapnia (Loeschke 1982, Pelletier 1972, Rose et al. 1983, Soladoye et al. 1985, Suutarinen 1966) elicited an increase in total peripheral resistance (Rose et al. 1983, Rothe et al. 1985). Stimulation of sympathetic nerves resulted in vasoconstriction of vascular beds which were constantly perfused with blood kept at a constant and physiological paO_2, pCO_2, and pH (Lioy et al. 1978, Pelletier 1972, Soladoye et al. 1985). However, vasodilation and an increase of blood flow occurred, if denervated vascular beds were perfused with hypercapnic blood (Daugherty et al. 1967). It is unknown, whether one of these apparently antasonistic effects of hypercapnia on vascular tone changes the peripheral blood flow to an extent that the oxygen delivery to the tissue could be impaired. In order to study the effects of hypercapnia on the oxygen offer to peripheral tissue in man, the distribution of local oxygen partial pressure (pO_2-histogram) within the skeletal muscle was measured in healthy human volunteers during inhalation of a 6.5% carbon dioxide containing air mixture. In order to distinguish between effects on the pO_2-histograms due to hypercapnia and the effects on the pO_2-histograms due to a change of blood pH, the present study was performed with and without buffering the blood pH during carbon dioxide inhalation.

SUBJECTS AND METHODS

The studies were performed on ten healthy nonsmoking male volunteers, aged 21 to 32. All subjects had a resting period of 20 min before measurements in the supine position were started. The inspired gas was a 6.5% carbon dioxide, 20% oxygen, 73.5% nitrogen containing, water vapour saturated gas mixture (CO_2-inhalation). The subjects breathed through a rubber mouthpiece with a nonrebreathing valve. Though CO_2-inhalation caused a huge hyperventilation in our subjects, all subjects remained quite in the supine position during the 30 min. of CO_2-inhalation. Prior to the CO_2-inhalation, blood samples were taken from the arterialized capillary blood of the ear to determine the "arterial" carbon dioxide partial pressure ($PaCO_2$), the "arterial" oxygen partial pressure (PaO_2), the "arterial" oxygen saturation of hemoglobin (SaO_2), the pH (pH), the base excess (BE) and the hemoglobin concentration (Hgb) (ABL 3, Radiometer, Denmark; CO-Oximeter 2500, Corning, USA). 10, 20 and 30 min. after the start of the CO_2-inhalation blood gases and acid-base balance were measured (table 1).

The distribution of local oxygen partial pressures within biceps muscle of the left arm was measured according to the method described by Weiss and Fleckenstein (1986). Briefly,

Oxygen Transport to Tissue XII, Edited by J. Piiper *et al.*
Plenum Press, New York, 1990

using a pO_2-histograph$^{®}$ (Eppendorf, FRG) a hypodermic polarographic needle electrode (tip diameter 350 µm, T90 < 500 ms, probe sensitivity 15-30 pA/mmHg at a polaragraphic potential of 700 mV, temperature sensitivity 2.44%/°C) was guided into the muscle through the lumen of a previously inserted 20 G Abbocath$^{®}$. In the muscle the needle probe was stepwise (0.7 mm) moved forward over a total distance of 20 mm. After each step of 0.7 mm, the local pO_2 was measured within less than 1 sec.. At the end of each stepwise forward movement of 20 mm the probe was swiftly drawn back to the initial position and its direction of insertion slightly changed. Thereafter, the stepwise forward movement of the needle probe was repeated. Thus, each single pO_2 value was measured at a different randomly situated point within a total muscle volume of about 3 cm^3. 200 single local pO_2 values were taken within 8 min.. Before and after the CO_2-inhalation the needle probe was two point calibrated (nitrogen/ambient air) in a TRIS-buffered 0.9% saline solution. The temperature of the solution and the temperature within biceps muscle (needle thermocouple, GMS, FRG) was measured. The pO_2-data were corrected for the probe drift and for the tissue temperature to calculate a pO_2-histogram and the mean muscle pO_2 (MPO_2). PO_2-measurements were performed before the CO_2-inhalation and, subsequently, from the 10th to 20th min. and from the 20th to 30th min. after start of the CO_2-inhalation.

Two weeks after the first series of measurements, five of ten subjects were studied again. During the second measuring session an intravenous infusion of a solution of 8.4%-sodiumbicarbonate was continously administered (Perfusor VI, Braun, FRG) at a constant rate of 300 ml/h. Venous canulation was performed at the right forearm, whereas pO_2-measurements were carried out within left biceps muscle. All other measurements were repeated as in the first series of measurements.

In order to analyze whether differences existed between pO_2-histograms, which were either recorded before or during CO_2-inhalation, the 200 pO_2-values of each of the ten (five) subjects were summarized to yield a "pooled pO_2-histogram" (fig. 1). Following this commonly used procedure (Ehrly et al. 1987), each "pooled pO_2-histogram" was calculated from 2000 (1000) single pO_2-values. The differences in changes of mean muscle pO_2 (each mean was calculated from 200 single pO_2 values) during CO_2-inhalation were analyzed using the paired student's t-test. A difference was considered statistically significant if the p-value was < 0.01.

RESULTS

1. CO_2-inhalation

The effects of inhalation of a 6.5% carbon dioxide, 20% oxygen, 73.5% nitrogen containing gas mixture (CO_2-inhalation) are summarized in table 1. 10 min. after start of the CO_2-inhalation the mean arterial pCO_2 of the ten subjects exceeded 50 mmHg. Mean arterial pCO_2 further increased slightly after 20 min., but was almost stable between 10 and 30 min.. During CO_2-inhalation mean blood pH decreased. The extent of the decrease was similar after 10 min. (pH = 7.33), 20 min. (pH = 7.32) and 30 min. (pH = 7.32). The arterial pO_2 increased from 98 to 125 mmHg during the CO_2-inhalation. Before start of the CO_2-inhalation the calculated mean arterial oxygen content (CaO_2) was 19.30 ml/100ml, during CO_2-inhalation the CaO_2 was 19.39 ml/100ml (10 min.) and 19.38 ml/100ml (20 min.). The mean muscle pO_2 (calculated from 2000 single pO_2-values) of the ten subjects significantly (p < 0.01) increased by more than 35% during CO_2-inhalation (figure 1, tab. 1). Mean muscle pO_2 was not significantly different 10-20 min. and 20-30 min. after start of the CO_2-inhalation. The Gaussian distribution of the relative frequencies in the pooled pO_2-histograms did not markedly change during CO_2-inhalation (figure 1).

2. CO_2-inhalation at constant blood pH

The data of the five subjects who underwent a second series of measurements two weeks after the first series, are summarized in table 2. Buffering of respiratory acidosis by the infusion of a 8.4% sodiumbicarbonate solution kept the mean blood pH constant during the

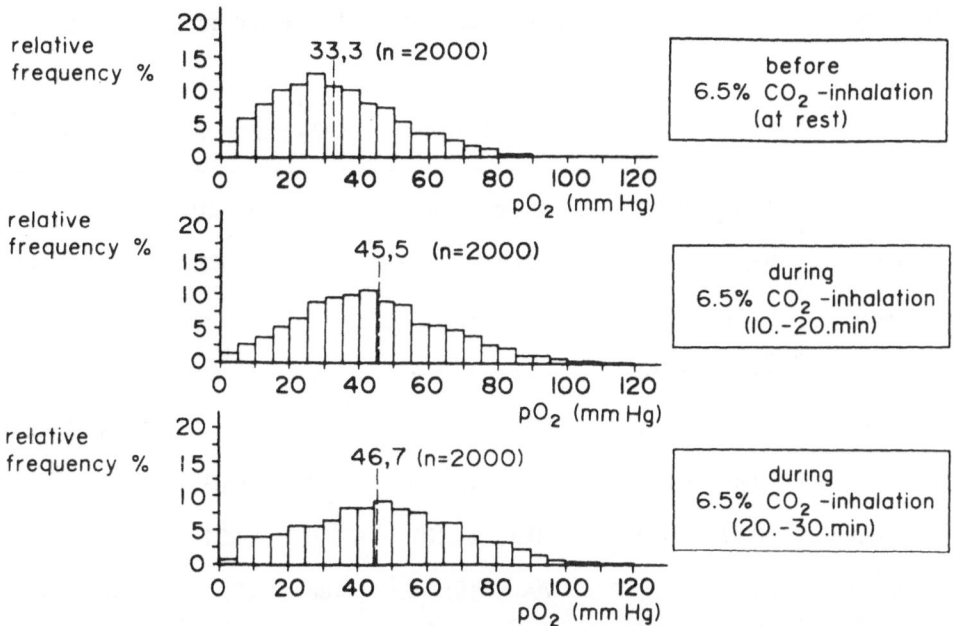

Fig. 1. Pooled pO_2-histograms of ten subjects. Local pO_2 within biceps muscle (pO_2); relative frequency of local pO_2-values in ranges of 5 mmHg (relative frequency). Mean muscle pO_2 indicated by the dashed line.

CO_2-inhalation (table 2). The increase of mean arterial pCO_2 and mean heart rate was similar to the first series of measurements.

The pooled pO_2-histograms of the five subjects are presented in figure 2. 10 to 20 min. after start of the CO_2-inhalation the mean muscle pO_2 (calculated from 1000 single pO_2-values) increased by more than 35%, which was similar to the first series of measurements without buffering the blood pH (figure 1). Though 20 to 30 min. after start of the CO_2-inhalation the mean muscle pO_2 slightly decreased (figure 2), the difference between the mean muscle pO_2 before (32 mmHg) and 20 to 30 min. after start of the CO_2-inhalation (40 mmHg) reached the level of significance ($p < 0.01$). During the CO_2-inhalation at constant blood pH the shape of the pooled pO_2-histogram was not markedly different in comparison to the pooled pO_2-histogram obtained before the CO_2-inhalation.

DISCUSSION

During inhalation of a 6.5% carbon dioxide containing gas mixture (CO_2-inhalation) the pO_2-values within biceps muscle were similarly distributed around the mean compared to the pooled pO_2-histograms obtained before start of CO_2-inhalation (figures 1 and 2). According to published data (Boekstegers et al. 1988, Ehrly et al. 1987, Lund et al. 1980) the Gaussian distribution of the pooled pO_2-histograms suggest that the oxygen delivery was physiologically distributed within the muscle and did not significantly change during hypercapnia. However, the CO_2-inhalation elicited a significant increase in mean muscle pO_2 (table 1, figure 1), which was similar during the first (10th to 20th min.) and second (20th to 30th min.) measuring period. Buffering of the blood pH did not abolish this effect of the CO_2-inhalation (figure 2). Thus, a decrease of peripheral blood pH due to respiratory acidosis cannot explain our results. As

Table 1. Summary of the data obtained from ten subjects before and during hypercapnia

	before CO_2-inhalation	10 min. CO_2-inhalation	20 min. CO_2-inhalation	30 min. CO_2-inhalation
MpO_2 (mmHg)	33.29 (6.71)	45.49 (8.68)	46.87 (7.18)	
PaO_2 (mmHg)	97.98 (12.80)	124.64 (10.28)	120.76 (6.82)	123.48 (4.44)
SaO_2 (%)	97.20 (0.74)	98.26 (1.52)	98.01 (0.38)	98.13 (0.26)
$PaCO_2$ (mmHg)	39.81 (2.16)	50.81 (3.88)	53.07 (3.21)	53.72 (1.47)
Hb (g/dl)	14.62 (1.41)			14.92 (1.04)
pH	7.40 (0.02)	7.33 (0.03)	7.32 (0.03)	7.32 (0.03)
BE (mmol/l)	-0.07 (1.12)	-0.38 (1.10)	0 (1.27)	-0.75 (2.20)
h.r. (/min)	64 (6)	79 (6)	86 (7)	90 (10)

Values are means (± SD) of ten subjects. Mean muscle pO_2 (MpO_2) within biceps muscle; arterial pO_2 (PaO_2); arterial oxygen saturation (SaO_2); arterial pCO_2 ($PaCO_2$); arterial pH (pH); arterial base excess (BE); venous hemoglobin content (Hb); heart rate (h.r.).

Table 2. Summary of the data obtained from five subjects before and during hypercapnia and buffering blood pH

	before CO_2-inhalation	10 min. CO_2-inhalation	20 min. CO_2-inhalation	30 min. CO_2-inhalation
MpO_2 (mmHg)	32.36 (6.08)	44.20 (4.79)	41.62 (4.36)	
PaO_2 (mmHg)	91 (3.95)	116.38 (6.95)	119.66 (6.15)	120.14 (4.68)
SaO_2 (%)	96.73 (0.42)	97.30 (1.48)	98.34 (0.21)	98.42 (0.15)
$PaCO_2$ (mmHg)	39.78 (1.5)	55.08 (5.07)	53.72 (2.88)	55.06 (1.84)
Hb (g/dl)	15.70 (1.23)			14.80 (1.19)
pH	7.40 (0.01)	7.37 (0.01)	7.39 (0.01)	7.41 (0.01)
BE (mmol/l)	0.37 (0.21)	3.78 (0.79)	5.98 (1.97)	7.48 (1.54)
h.r. (/min)	68 (10)	77 (6)	87 (7)	96 (10)

Values are means (± SD) of five subjects. Mean muscle pO_2 (MpO_2) within biceps muscle; arterial pO_2 (PaO_2); arterial oxygen saturation (SaO_2); arterial pCO_2 ($PaCO_2$); arterial pH (pH); arterial base excess (BE); venous hemoglobin content (Hb); heart rate (h.r.).

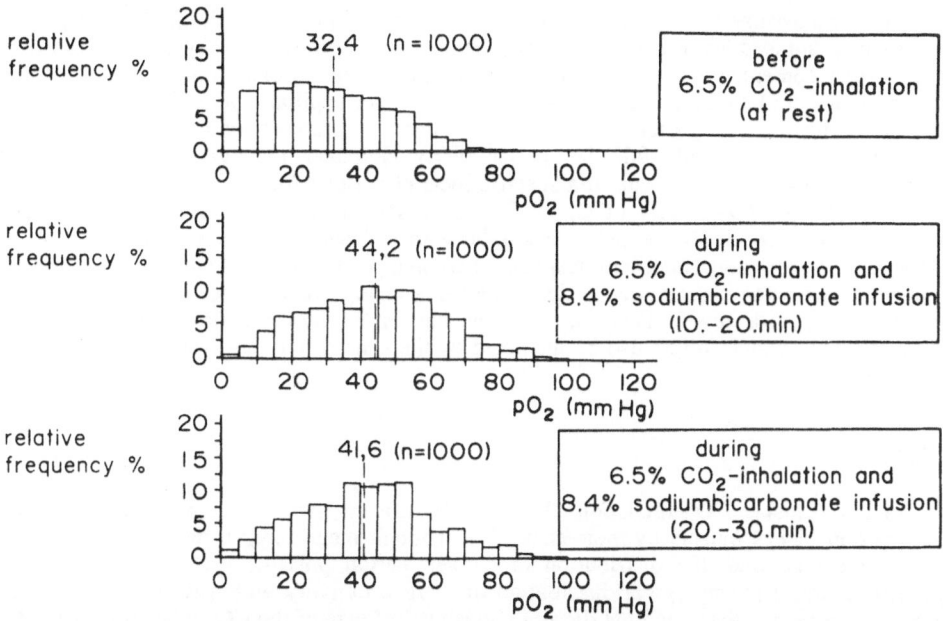

Fig. 2. Pooled pO_2-histograms of five subjects. Local pO_2 within biceps muscle (pO_2); relative frequency of local pO_2-values in ranges of 5 mmHg (relative frequency). Mean muscle pO_2 indicated by the dashed line.

tentative explanation(s) for the increase of mean muscle pO_2 during hypercapnia could be considered:

1. a hypercapnia induced decrease of the oxygen consumption of the muscle,
2. a hypercapnia induced rise of the arterial oxygen content,
3. a hypercapnia induced rise of the mean capillary blood flow within the muscle.

PO_2-measurements were performed in resting skeletal muscle of our subjects after a previous 20 min. resting period in the supine position. Therefore, a change of the oxygen consumption due to changes of muscle activation was not likely to occur during the pO_2-measurements. Data on the oxygen consumption of the human skeletal muscle during hypercapnia could not be found in the literature. However, a decrease of whole body oxygen uptake (VO_2) was observed during hypercapnia (Karetzky and Cain 1970). Since in animal experiments no decrease of VO_2 occurred if blood pH was kept constant during hypercapnia (Cain 1970), the authors inferred that the hypercapnia induced decrease of VO_2 was mostly mediated by a hypercapnia induced decrease of blood pH. The increase of mean muscle pO_2 in our subjects was similar with and without buffering the blood pH (figures 1 and 2). Hence, our data suggest that a decrease of the oxygen consumption in the m. biceps brachii was, if any, only partly responsible for the observed increase of mean muscle pO_2 during hypercapnia.

During the CO_2-inhalation, the arterial pO_2 increased from 99 to 125 mmHg apparently due to hyperventilation of our subjects. The arterial oxygen content only slightly increased by 0.09 ml/100ml (see Results). Though the increase of arterial pO_2 might have contributed to the observed increase of mean muscle pO_2, it seems unlikely that the quantitatively insignificant increase of arterial oxygen content could account for the increase of mean muscle pO_2 by more than 35% during hypercapnia.

529

Since we assume that the observed increase of mean muscle pO_2 during hypercapnia was neither mainly induced by a decrease of the oxygen consumption nor by an increase of the arterial oxygen content, the most plausible explanation of the results would be an increase of mean muscular capillary blood flow. An increase of mean capillary blood flow within biceps muscle could be a consequence of the well documented increase of cardiac output during hypercapnia (Richardson et al. 1962, Suutarinen 1966). Moreover, local vasodilation of muscular vessels was suggested to occur due to a specific local effect of carbon dioxide or a concomitant change of blood pH (Daugherty et al. 1967, Richardson et al. 1962). Since the hypercapnia induced increase of mean muscle pO_2 was similar with and without buffering the blood pH, our results would be more consistent with a specific carbon dioxide mediated vasodilation. However, from our data we cannot definitely decide, whether a change of the arterial blood flow or the local vascular tone occurred. Therefore, the mechanism of hypercapnia induced increase of mean muscle pO_2, which might also be mediated by the central nervous system, requires further investigations.

SUMMARY

In ten subjects CO_2-inhalation elicited a significant increase in mean oxygen partial pressure within biceps muscle by more than 35%. Though mean oxygen partial pressure within biceps muscle increased, the distribution of oxygen partial pressure (pO_2-histogram) did not change suggesting a physiological distribution of oxygen delivery within biceps muscle during hypercapnia. Buffering the blood pH did not abolish the effects of the CO_2-inhalation. Therefore, a decrease of peripheral blood pH could not account for the hypercapnia induced increase of mean oxygen partial pressure within biceps muscle. Our data suggest that oxygen delivery to skeletal muscle was increased during hypercapnia, most probably due to a hypercapnia induced rise of mean capillary blood flow.

REFERENCES

Boekstegers, P., Fleckenstein, W., Rosport, A., Ruschewsky, W. and Braun, U., 1988, Überwachung der Sauerstoffversorgung des Skelettmuskels und der Gesamtsauerstoffaufnahme bei koronarchirurgischen Eingriffen, Anaesthesist, 37: 287-296.

Daugherty, R.M., Scott, .J.B, Dabney, J.M., Haddy, F.J., 1967, Local effects of O_2 and CO_2 on limb, renal, and coronary vascular resistance, Am. J. Physiol., 213: 1102-1110.

Cain, S.M., 1970, Increased oxygen uptake with passive hyperventilation of dogs, J. Appl. Physiol., 28: 4-7.

Ehrly, A.M., Hauss, J. and Huch, R., 1987, Clinical oxygen pressure measurememt, Springer, Berlin Heidelberg New York.

Karetzky, M.S. and Cain, M.S., 1970, Effect of carbon dioxide on oxygen uptake during hyperventilation in normal man, J. Appl. Physiol., 28: 8-12.

Lioy, F., Hanna, B.D. and Polosa, C., 1978, CO_2-dependent component of the neurogenic vascular tone in the cat, Pflügers Arch., 374: 187-191.

Loeschke, H.H., 1982, Central chemosensitivity and the reaction theory, Am. J. Physiol., 332: 1-24.

Lund, N., Jorfeld, L. and Lewis, D.H., 1980, Skeletal muscle oxygen pressure fields in healthy human volunteers, Acta Anaesthesiol. Scand., 24: 272-281.

Pelletier, C.L., 1972, Circulatory responses to graded stimulation of the carotid chemoreceptors in the dog, Circ. Res., 31: 431-443.

Richardson, .D.W, Wassermann, A.J. and Patterson, J.L., 1962, General and regional circulatory responses to change in blood pH and carbon dioxide tension, J. Clin. Invest., 40: 31-43.

Rose, C.E., Althaus, J.A., Kaiser, D.L., Miller, E.D. and Carey, R.M., 1983, Acute hypoxemia and hypercapnia: increase in plasma catecholamines in conscious dogs, Am. J. Physiol., 245: 924-929.

Rothe, C.F., Stein, P.M., MacAnespie, C.L. and Gaddis, M.L., 1985, Vascular capacitance responses to severe systemic hypercapnia and hypoxia in dogs, Am. J. Physiol., 249: 1061-1069.

Soladoye, A.O., Rankin, A.J. and Hainsworth, R., 1985, Influence of carbon dioxide tension in the cephalic circulation on hind-limb vascular resistance in anaesthetized dogs, J. Exp. Physiol., 70: 527-538.

Suutarinen, T., 1966, Cardiovascular response to changes in arterial carbon dioxide tension: An experimental study on thoracotomized dogs, Acta Physiol. Scand., 67: Suppl 266.

Weiss, Ch and Fleckenstein, W., 1986, Local tissue pO_2 measured with "thick" needle probes, Funktionsanalyse biologischer Systeme 15: 155-166.

MUSCLE OXYGENATION AND PERFORMANCE DURING LOW LEVEL CARBON MONOXIDE

EXPOSURE

Cheryl E. King

School of Rehabilitation Therapy, Queen's University

Kingston, Ontario, Canada K7L 3N6

INTRODUCTION

Carbon monoxide (CO) is a toxic, odourless, colourless gas that can cause profound physiological disturbances, even at low concentrations. At low levels of approximately 5% carboxyhemoglobinemia (COHb), selective increases in blood flow to vital organs occur (Stewart, 1975). At approximately 15% COHb, people may develop frontal headaches and experience abnormal visual responses. Further, increases to 20-30% COHb result in a throbbing headache, nausea, and abnormal fine manual dexterity (Stewart, 1975). COHb levels can increase to 10-25% in people who are heavy smokers, when working in underground garages, during exposure to high traffic-congested areas in large cities (especially on a hot, humid day with low cloud cover) and during confinement in military tanks.

Recent findings that transient high levels of CO occur in military tanks following the firing of shells, prompted this investigation (Tikusis et al., 1987). The present study examined the physiological consequences of low level CO exposure on exercise performance. The in situ canine gastrocnemius muscle preparation was used to determine the extent to which low levels of CO affect muscle O_2 uptake (VO_2) and developed tension (TD) at low to moderate levels of work.

METHODS

Male mongrel dogs were anesthetized with pentobarbital sodium (30 mg/kg); additional anesthetic was given as required. The animals were ventilated to maintain an end-tidal PCO_2 of approximately 35 Torr. The left gastrocnemius muscle was exposed and the arterial inflow and venous outflow were isolated. The tendon was freed from its insertion, placed in a metal clamp and attached to a force transducer. The sciatic nerve was sectioned and the distal end was placed in a stimulating electrode. A two-channel cannula was inserted in the popliteal vein draining the muscle. One channel contained an electromagnetic flow probe while the other allowed for an in situ zero flow calibration. Muscle venous outflow was returned to the animal via a reservoir cannula in the right femoral vein. The proximal end of a two-channel cannula was placed in the right femoral artery and the distal end was inserted in the popliteal artery supplying the muscle preparation. One channel allowed for autoperfusion of the muscle while a

Oxygen Transport to Tissue XII, Edited by J. Piiper *et al.*
Plenum Press, New York, 1990

pump was placed in-line in the other channel for pump perfusion. Muscle perfusion pressure was monitored and pump perfusion was used, when necessary, to keep muscle perfusion pressure within a normal range. A catheter was placed in the left carotid artery for measurement of mean arterial pressure and withdrawal of arterial blood samples. A Swan-Ganz catheter was placed in the pulmonary artery via the right external jugular vein in order to determine cardiac output via the thermodilution technique.

Three groups of animals were studied. The first group was a control group (CON, n=5) in which the animals were studied during normoxia. The second and third groups were CO groups in which the animals were ventilated with CO to achieve levels of 10% (10%COHb, n=5) and 25% (25%COHb, n=5) COHb respectively. The animals were ventilated with 1.0% CO in air to elevate the COHb concentration to the desired level and then ventilated with 0.01% or 0.03% CO in air respectively to maintain the target COHb concentration.

The animals were studied at rest during normoxia or CO. Then the muscle was stimulated to contract isometrically (0.5 msec duration, 4-6 volts) for 36 min at 0.5 Hz. Contractions were stopped and a 30 min recovery period ensued. The muscle was then stimulated to contract isometrically at 1.0 Hz for a further 36 min. These contraction frequencies represent approximately 12 and 25% maximal VO_2 in this muscle preparation. The order of the contraction frequency was altered to eliminate any bias of the data with time.

Measurements of mean arterial pressure, muscle perfusion pressure, muscle blood flow, and developed tension were made and arterial and muscle venous blood samples were obtained at rest, 5, 20, and 36 min of contractions. The blood samples were analysed for O_2 concentration, PO_2, PCO_2, pH, and lactate. Muscle VO_2 was calculated using the Fick Principle. During the first 5 min of each contraction frequency, when the physiological responses were not in steady state, arterial and muscle venous blood was sampled continuously at a rate proportional to the blood flow.

All data presented represent the means ± SE. Statistical comparison was performed using a between/within split plot repeated measures design analysis of variance from Human Systems Dynamics. A Duncan's Multiple Range test was used to determine specific differences between means. Significance was accepted at the $p < 0.05$ level.

RESULTS

The values for COHb averaged 0.5 ± 0.2, 12.0 ± 0.7 and 23.4 ± 1.6 percent in the CON, 10%COHb and 25%COHb groups respectively. During CO, arterial O_2 concentration was reduced to 16.7 ± 1.0 and 15.1 ± 0.8 volumes percent in the 10%COHb and 25%COHb groups as compared to 19.8 ± 0.9 volumes percent in the CON group. The P_{50}, the PO_2 at which hemoglobin is fifty percent saturated, was also reduced during CO. The values for P_{50} in the 10%COHb and 25%COHb groups were 20.2 ± 1.1 and 16.8 ± 0.8 mmHg respectively as compared to 26.0 ± 0.6 mmHg in the CON group.

The values for muscle VO_2 are shown in Figure 1. The values for resting VO_2 averaged 4.2 ± 0.9 $\mu l \cdot g^{-1} \cdot min^{-1}$ in all three groups. Oxygen uptake increased to 23.8 ± 3.3 $\mu l \cdot g.^{-1} \cdot min^{-1}$ at 0.5 Hz and 37.9 ± 5.3 $\mu l \cdot g^{-1} \cdot min^{-1}$ at 1.0 Hz in all three groups. There was no significant difference between the three groups at any time during the experiment. These data indicate that muscle VO_2 was not compromised at a level of 25% COHb during continuous work at a level of 25% maximal VO_2.

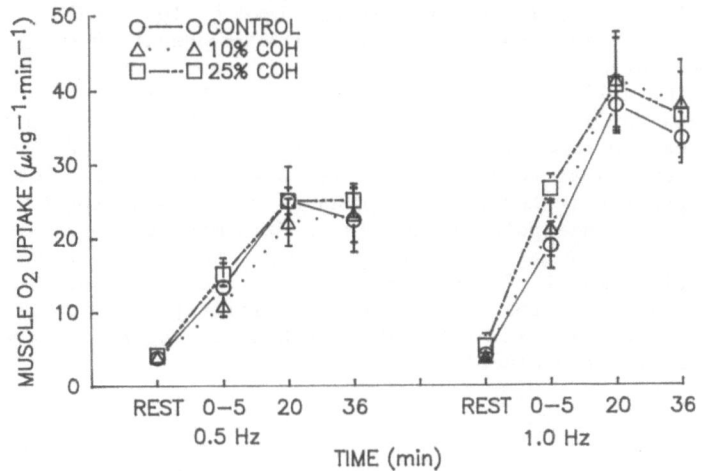

Figure 1. Muscle O_2 uptake in $\mu l \cdot g^{-1} \cdot min^{-1}$ (means ± SE). $\bigcirc$ = control (CON), $\triangle$ 10% COHb, $\square$ 25% COHb, Hz = Hertz. ☆ 25% COHb significantly different from CON. ★ 10% COHb significantly different from CON. ✱ 25% COHb significantly different from 10% COHb.

The values for developed tension (TD) are shown in Figure 2. Developed tension averaged approximately 105.5 ± 17.1 $g \cdot g^{-1}$ and 97.1 ± 14.1 $g \cdot g^{-1}$ at 0.5 and 1.0 Hz respectively. There was no significant difference in TD between the three groups. Further analysis of the muscle twitches is reported in Table 1. The values for rate of rise of the twitch recording, the rate of relaxation, peak twitch height and area (tension-time index) did not differ between the three groups. At 0.5 Hz, however, the peak twitch height was significantly reduced in the 25% COHb group as compared to control.

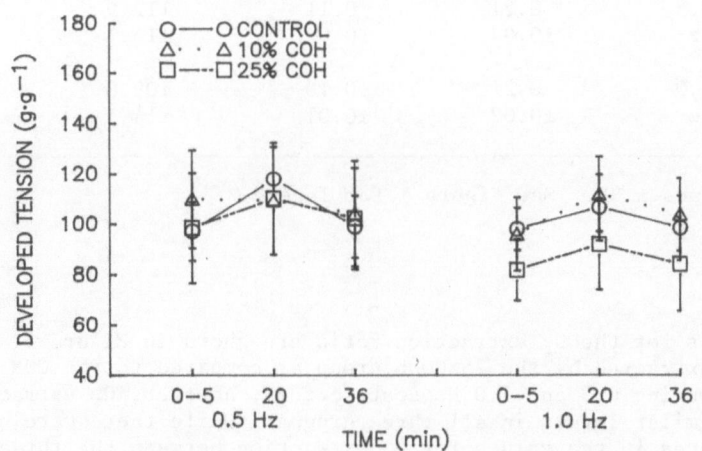

Figure 2. Developed tension in grams of force per gram of wet muscle weight. See Figure 1 for legend.

Both muscle VO_2 and TD were not compromised at low levels of CO exposure and work. The following graphs illustrate the compensatory mechanisms that occurred to prevent any decrement in the above two variables. The muscle blood flow and O_2 delivery responses are shown in Figures 3 and 4 respectively. At rest and during 0.5 Hz stimulation, blood flow did not differ between the three groups. During 1.0 Hz stimulation, however, muscle blood flow tended to be elevated above control levels in the 10%COHb group and was significantly greater in the 25%COHb group as compared to the CON group. At rest, during the 0.5 Hz protocol, muscle O_2 delivery was lower ($p < 0.05$) in the 25%COHb group as compared to the CON group. During the first 5 min of the 0.5 Hz stimulation, muscle O_2 delivery was significantly lower in the 10%COHb group as compared to control. No further significant differences were noted. O_2 delivery was not significantly different between the three groups at any time during the 1.0 Hz protocol.

Table 1. Muscle Force Development Parameters

Group		Rate of Tension Development $(kg.msec^{-1})$	Rate of Relaxation $(kg.msec^{-1})$	Peak Twitch Height $(g \cdot g^{-1})$	Area $(g \cdot msec \cdot g^{-1})$
Control	0.5 Hz	0.28 ±0.02	0.11 ±0.01	131.5 ±10.9	14.8 ±1.4
	1.0 Hz	0.27 ±0.02	0.13 ±0.01	124.1 ±8.3	11.3 ±0.7
10%COHb	0.5 Hz	0.21 ±0.02	0.10 ±0.02	115.8 ±10.6	12.8 ±1.4
	1.0 Hz	0.23 ±0.02	0.14 ±0.02	116.3 ±7.7	10.0 ±0.6
25%COHb	0.5 Hz	0.21 ±0.02	0.11 ±0.02	112.9 ☆ ±15.1	11.3 ±1.7
	1.0 Hz	0.21 ±0.02	0.10 ±0.01	109.8 ±14.1	10.1 ±1.4

Values are means ± SE. See Figure 1 for legend.

The values for the O_2 extraction ratio are shown in Figure 5. At rest, more O_2 was extracted by the 25%COHb group as compared to the CON group ($p < 0.05$). During 0.5 and 1.0 Hz contractions, however, O_2 extraction increased to similar levels in all three groups. While there were no significant differences in the values for O_2 extraction between the three groups, tissue PO_2 was significantly lower in the two CO groups as compared to the CON group at rest and during 0.5 Hz stimulation (Figure 6). At 1.0 Hz, muscle venous PO_2 was significantly less than CON in the 25%COHb group but not the 10%COHb group. These findings can be directly related to the lower P_{50} in the CO groups.

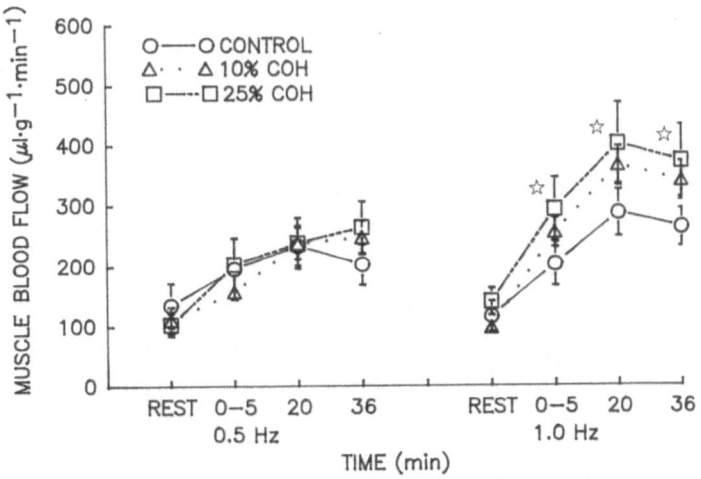

Figure 3. Muscle blood flow in $\mu l \cdot g^{-1} \cdot min^{-1}$. See Figure 1 for legend.

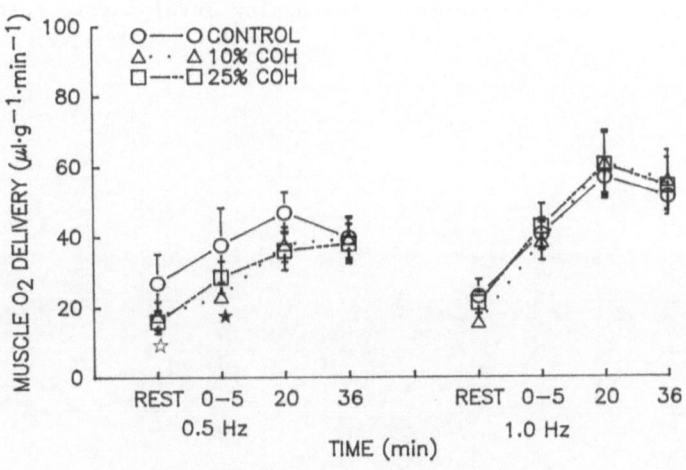

Figure 4. Muscle O_2 delivery in $\mu l \cdot g^{-1} \cdot min^{-1}$. See Figure 1 for legend.

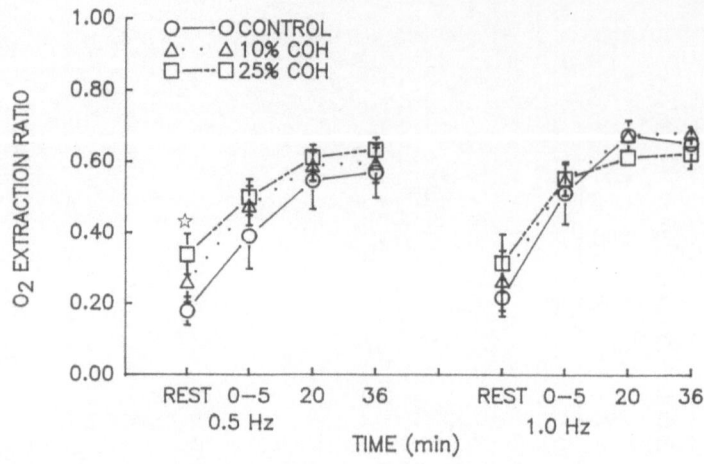

Figure 5. Muscle O_2 extraction ratio. See Figure 1 for legend.

The blood samples were also analysed for lactate concentration. Lactate production was calculated as muscle blood flow times the (v-a) lactate concentration difference. The values for lactate production showed a wide variation, however, at 20 and 36 min of 0.5 Hz stimulation lactate production was greater in the 25%COHb group as compared to the 10%COHb group (p<0.05).

DISCUSSION

The intent of the present study was to determine if muscle VO_2 and/or developed tension were compromised during low level muscle contractions

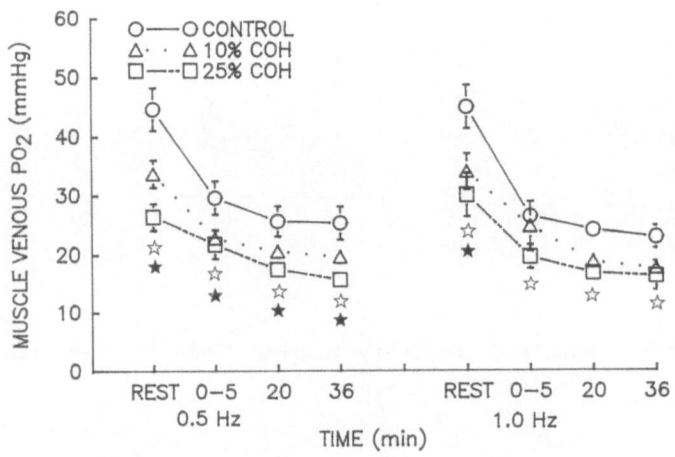

Figure 6. Muscle venous PO_2 in mmHg. See Figure 1 for legend.

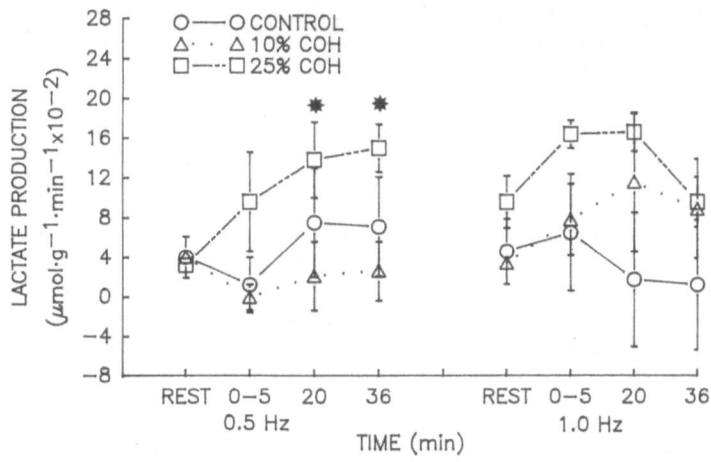

Figure 7. Lactate production in $\mu mol \cdot g^{-1} \cdot min^{-1}$. See Figure 1 for legend.

and low level CO exposure. When the muscle was stimulated to contract at 25% maximal VO_2 (1 Hz) during 25% COHb, for a prolonged period of time, VO_2 and TD were preserved at normoxic levels. Thus, even at these higher than normal pollutant levels of CO, mild exercise was maintained without any decrement in performance. In spite of a decrease in arterial O_2 concentration, compensatory mechanisms of increased blood flow and O_2 extraction provided an adequate flux of O_2 to the muscle cells.

At 0.5 Hz, when VO_2 increased six-fold, both muscle blood flow and O_2 extraction increased in all three groups to accommodate the increased O_2 demand. During 10% and 25% COHb, the degree of reduction in arterial O_2 concentration was such that no further increases in blood flow or O_2 extraction were required to achieve the desired level of O_2 demand. During 1.0 Hz contractions, when O_2 demand was increased approximately ten-fold, compensatory increases in blood flow were required to offset the reduced arterial O_2 concentration during 10% and 25% COHb and, therefore, maintain an adequate O_2 delivery. At both contraction frequencies during CO, O_2 extraction increased to similar levels as observed in the normoxic group. This was achieved at the expense of lowering tissue PO_2, as observed by significantly lower values for muscle venous PO_2. The values for P_{50} were 20.2 and 16.8 mmHg in the 10% and 25% COHb groups respectively. The highest values for O_2 extraction were 0.68, 0.68, and 0.62 in the control, 10% and 25% COHb groups respectively. Earlier data during severe COH (60% COHb) and 1.0 Hz contractions suggested that this is approximately the upper limit of O_2 extraction at these values for P_{50} and that any further increases in O_2 demand might not have been met (King et al., 1987). The values for muscle venous PO_2 decreased to approximately 16 mmHg during COH as compared to 23 mmHg during normoxia. In severe COH during 1.0 Hz contraction, values of 10 mmHg were observed for muscle venous PO_2 (King et al., 1987). It is most likely that further increases in O_2 demand would have resulted in a greater decrease in venous PO_2 without substantial increases in O_2 extraction and, therefore, muscle O_2 uptake would have been compromised.

Muscle lactate production or efflux was highly variable, however, higher lactate production occurred in the 25% COHb group despite no limitation in muscle VO_2. These findings would argue against lactate production being associated with an O_2 lack. Connett et al., (1986) demonstrated a threshold level of lactate efflux of 50% maximal aerobic capacity in the canine gracilis muscle. The lactate data in the present study may be explained by this finding. During normoxia, at 0.5 and 1.0 Hz, muscle VO_2 was less than 50% MVO_2 and no significant increase in lactate efflux was observed. During COH, MVO_2 is limited and has been shown to decrease approximately 1% for every 1% COHb (Ekblom and Huot, 1972). Accordingly at 25% COHb, MVO_2 would be reduced 25% and, therefore, the contractions at both frequencies would represent a greater percentage of the MVO_2. The VO_2 at 25% COHb at 1.0 Hz would then represent approximately 40% MVO_2 which approaches the threshold for lactate efflux reported by Connett et al., (1986).

In conclusion, muscle VO_2 and TD were not compromised at exercise intensities of 12% and 25% MVO_2 and CO concentrations of 10% and 25% COHb. Increases in blood flow offset the decrease in arterial O_2 concentration during COH. The increased lactate production during 25% COHb was most likely associated with the higher relative difficulty of the contraction frequency during COH.

ACKNOWLEDGEMENTS

The author would like to thank Mr. John Taylor for technical assistance and Mrs. Teresa Dwyer and Mrs. Sandy Wymer for assistance with preparation of the manuscript. These experiments were supported by the Canadian Department of National Defense.

REFERENCES

Connett, R.J., Gayeski, T.E., and Honig, C.R., 1986, Lactate efflux is unrelated to intracellular PO_2 in a working red muscle in situ, J. Appl. Physiol., 61:402-408.

Ekblom, B., and Huot, R., 1972, Response to submaximal and maximal exercise at different levels of carboxyhemoglobin, Acta Physiol. Scand., 86:474-482.

King, C.E., Dodd, S.L., and Cain, S.M., O_2 delivery to contracting muscle during hypoxic or CO hypoxia, J. Appl. Physiol., 63:726-732.

Stewart, R.D., 1975, The effect of carbon monoxide on humans, Ann. Rev. Pharmacol., 15:409-423.

Tikusis, R., Madill, H.D., Gill, B.G., Lewis, W.S., Cox, K.M., and Kane, D.M., 1987, A critical analysis of the use of the CFK equation in predicting COHb formation, Am. Ind. Hyg. Assoc. J., 48:208-213.

TEMPERATURE EFFECT ON OXYGENATION AND METABOLISM

OF PERFUSED RAT HINDLIMB MUSCLE

Akitoshi Seiyama, Takeshi Shiga* and Nobuji Maeda

Department of Physiology, School of Medicine, Ehime University
Shigenobu, Onsen-gun, Ehime 791-02, Japan
*Department of Physiology, Shool of Medicine, Osaka University
Nakanoshima, Kita-ku, Osaka 530, Japan

INTRODUCTION

Temperature effect on tissue oxygenation and metabolism is a major concern to activity and/or viability of tissues. In an ischemic state, supplement of oxygen and nutrients is impaired and metabolites are accumulated, thus tissue damage is induced reversibly or irreversibly. This deterioration of tissue accompanied by ischemia is mediated by chemical reactions, and the rates are directly dependent upon temperature. Rat hindlimb is well suited for the optical measurement of biochemicals that participate in oxygen transport (Tamura et al., 1987; Seiyama et al., 1988) as well as the biochemical investigation on the integral function of skeletal muscle (Shiota et al., 1986).

In the present paper, effect of temperature on tissue oxygenation and oxygen metabolism in perfused hindlimb muscle of rat was studied at 35°C and 15°C.

MATERIALS AND METHODS

Preparation of Isolated Perfused Hindlimb

Male Wistar rats (250 to 300 g in body weight) fed on a commercial diet were used. Rats were anesthetized with pentobarbital sodium (3 mg/100 g body weight, i.p.). Surgical procedure was modified from that of Ruderman et al. (1971). The perfusion apparatus was essentially the same as that used for liver (Sugano et al., 1978). Krebs-bicarbonate buffer (35°C and 15°C) containing 4% (w/v) polyvinylpyrrolidone (PVP-4OT; purchased from Sigma Chem. Co., St. Louis, MO) was perfused from the abdominal aorta in a flow-through mode at 14 ml/min/leg, and the effluent after perfusion was collected from the inferior vena cava. The oxygen concentration in the perfusate was varied stepwise by mixing the buffer equilibrated with 95% O_2 + 5% CO_2 and that with 95% N_2 + 5% CO_2. All measurements were started after perfusing buffer equilibrated with 95% O_2 + 5% CO_2 for 30 min.

Spectrophotometric Measurement

Computer-controlled rapid scanning spectrophotometer (USP-501, Unisoku, Osaka, Japan) was used for the analysis of myoglobin (Mb) and cytochromes (Cyts) in thigh (quadriceps) muscle. Difference spectra in visible region (500 - 650 nm) were successively recorded at each steady state of oxygen consumption, in reflectance and/or transmission mode.

Content of Mb and Cyts was determined on the basis of the differences of optical density between 561 and 587 nm for Mb (Fig. 1, curve ① - ②), between 605 and 630 nm for Cyt a (Fig. 1, curve ②), between 564 and 575 nm for Cyt b (Fig. 1, curve ②), and between 550 and 540 nm for Cyt c+c_1 (Fig. 1, curve ②), respectively. Extinction coefficient applied for Mb was 5.6 mM^{-1} cm^{-1} (in the present study), and that for Cyt a was taken from Yonetani (1960) and those for Cyt b and Cyt c+c_1 from Chance and Williams (1955).

Oxygenation of Mb was determined from the difference of optical density between 587 and 630 nm, and oxidation of Cyt aa_3 from that between 605 and 630 nm.

Results obtained by reflectance measurement were in good agreement with those obtained by transmission measurement.

Analytical Methods

Oxygen tension in influent and effluent was monitored with an oxygen electrode (GU-BMS; Iijima Products M.F.G. CO., Ltd., Aichi, Japan). Rate of oxygen uptake was calculated from the flow rate and difference in the oxygen concentration between the influent and effluent. Concentrations of lactate and pyruvate in the effluent were determined enzymatically according to Gutmann and Wahlefeld (1974) and Lamprecht and Heinz (1984).

RESULTS AND DISCUSSION

Temperature effect on metabolism

Changes in O_2 uptake and releases of lactate and pyruvate to the effluent, produced by decreasing of O_2 supply at 35°C and 15°C, are shown in Table 1. Estimating from Table 1 and Fig. 2, O_2 uptake at 35°C was constant above Po_2 = 480 mmHg in the influent (O_2 supply, 10 μmoles/min/leg), and below 480 mmHg O_2 uptake decreased and lactate/pyruvate (L/P) ratio increased. On the other hand, O_2 uptake at 15°C was almost constant above Po_2 = 176 mmHg (O_2 supply 4 μmoles/min/leg), and decreased below that value. The O_2 uptake at 15°C was about one third that 35°C. High L/P ratio at 15°C shows that glycolysis plays a major role in energy metabolism at low temperature even though Cyt aa_3 is maintaining a more oxidized state than that at 35°C (see Fig. 2 and 3). Similar precedence of glycolysis at low temperature has been observed in the kidney (Pegg et al., 1981). Moreover, it should be noted that abrupt decreases in both O_2 uptake and L/P ratio at 15°C were observed at about Po_2 = 550 mmHg in the influent, of unknown mechanism, and that the L/P ratio was decreased with decreasing Po_2. These results suggest that metabolism of skeletal muscle under hypothermia is regulated by the oxygen tension of perfusate in a manner different from that in normothermia.

Optical studies in perfused hindlimb

In order to analyze oxygenation of Mb and oxidation of Cyt aa_3, spectrophotometric separation of Mb and Cyts in the perfused rat hindlimb was performed at 15°C (Fig. 1). The molar ratio of Mb, Cyt b and Cyt c+c_1 to Cyt a obtained from 4 experiments was 3.2, 0.4 and 1.3, respectively. Similar values of molar ratio for Cyts have been obtained in skeletal muscle mitochondria of rat (Makinen and Lee, 1968).

The oxygenation state of Mb in rat hindlimb muscle perfused with buffers of various oxygen tensions at 35°C and 15°C are shown in Fig. 2 (left). In order to estimate the oxygen tension in tissue, the oxygenation curve of Mb was determined by adding 2 mM KCN to the perfusate for the inhibition of respiration. P_{50}'s of Mb were 5.1 mmHg at 35°C, 2.3 mmHg at 25°C and 1.1 mmHg at 15°C. These values were in agreement with those calculated from the P_{50} of purified Mb from rat skeletal muscle at 20°C (Strickland et al., 1959; Antonini and Brunori, 1971). The estimation of mean tissue oxygen tensions at 35°C and 15°C by using these P_{50}'s in perfused hindlimb is shown in Fig. 2, right. When about 30% of Mb was deoxygenated at both 35°C and 15°C, the O_2 uptake started to decrease and lactate release increased (Table 1

542

Table 1. Effect of temperature on oxygen uptake and releases of lactate and pyruvate.

At 35°C (n = 3 except for % Mb-So$_2$ and O$_2$ uptake (n = 6)).

Po$_2$ *	666	527	367	200	60	0
% Mb-So$_2$	87 ± 1	75 ± 2	54 ± 4	24 ± 3	7 ± 1	0
O$_2$ Uptake **	306 ± 4	300 ± 7	234 ± 7	146 ± 9	34 ± 2	0
Lactate Release **	107 ± 6	123 ± 9	129 ± 13	183 ± 24	234 ± 33	308 ± 25
Pyruvate Release **	26 ± 2	25 ± 2	25 ± 3	22 ± 3	15 ± 1	14 ± 2
L/P Ratio	4.3 ± 0.1	4.8 ± 0.1	5.3 ± 0.1	8.5 ± 0.5	15.4 ± 2.2	24.5 ± 2.1

At 15°C (n = 3 except for % Mb-So$_2$ (n = 4))

Po$_2$ *	713	551	357	167	35	0
% Mb-So$_2$	97>	95 ± 1	86 ± 1	67 ± 4	12 ± 3	0
O$_2$ Uptake **	84 ± 16	46 ± 9	67 ± 4	60 ± 7	23 ± 2	0
Lactate Release **	55 ± 11	37 ± 6	53 ± 8	62 ± 23	77 ± 22	104 + 14
Pyruvate Release **	1.7 ± 0.7	2.9 ± 1.5	1.5 ± 0.4	2.2 ± 0.7	2.0 ± 0.5	6.5 ± 1.3
L/P Ratio	31.8 ± 5.9	20.0 ± 6.5	35.3 ± 4.5	28.7 ± 3.9	34.2 ± 5.2	17.9 ± 4.7

* The value of Po$_2$ (mmHg) is that of influent perfusate. Po$_2$ = 0 mmHg was obtained by infusing the buffer containing 2 mM Na$_2$S$_2$O$_4$.

** The values show mean ± S.E. (nmoles/min/g wet weight). The hindlimb was perfused at 14 ml/min/leg in a flow-through mode.

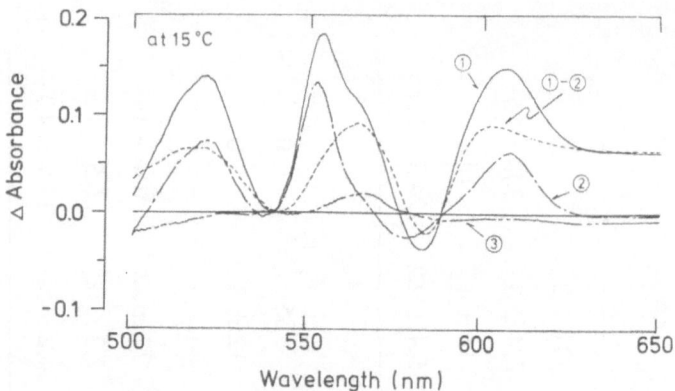

Fig. 1. Spectrophotometric separation of myoglobin and cytochromes in perfused hindlimb muscle of rat. The difference spectra were measured at 15°C: 1, Anaerobic - Aerobic; 2, (Aerobic + 1.7 mM KCN) - Aerobic; 3, (Aerobic + 67 µM Antimycin A) - Aerobic. Aerobic condition was obtained by perfusing the buffer equilibrated with 95% O_2 - 5% CO_2 (Po_2 = 713 mmHg).

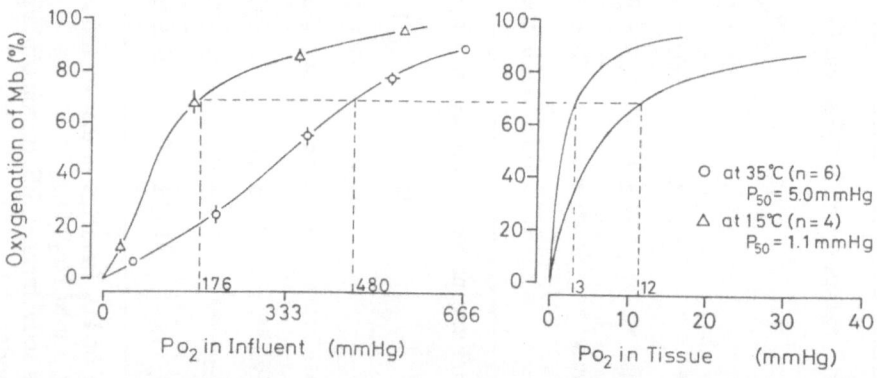

Fig. 2. Temperature effect on myoglobin oxygenation. The isolated hindlimb was perfused with buffers of various oxygen tensions (left) at 35°C ($\bigcirc$) and 15°C ($\triangle$). The oxygenation curves off Mb in the tigh (quadriceps) muscle were drown by using P_{50} = 5.0 mmHg at 35°C and 1.1 mmHg at 15°C which were obtained by separate experiments (right). When about 30% of Mb was deoxygenated, O_2 uptake started to decrease at both 35°C and 15°C (see Table 1).

and Fig. 2). At these points, mean tissue Po_2's were 12 mmHg at 35°C and 3 mmHg at 15°C, where the influent Po_2 was 480 mmHg at 35°C and 176 mmHg at 15°C.

Steady state relationship between oxygenation of Mb and oxidation of Cyt aa_3 in the perfused rat hindlimb is shown in Fig. 3. At 35°C, the reduction of Cyt aa_3 was accompanied by a proportional deoxygenation of Mb. Similar results have been observed in the perfused rat heart (Tamura et al., 1978, Kanaide et al., 1982). This linear relationship between oxygenation of Mb and oxidation of Cyt aa_3 may be due to the following factors: (1) presence of oxygen gradient between cytosol and mitochondrion (Kennedy and Jones, 1986; Tamura et al., 1989), (2) heterogeneity of oxygen tension among cells (Caspary et al, 1985; Clark et al., 1987; Wittenberg and Wittenberg, 1989), (3) difference in oxygen affinity of Cyt aa_3 between in vitro and in vivo (Jobsis, 1974).

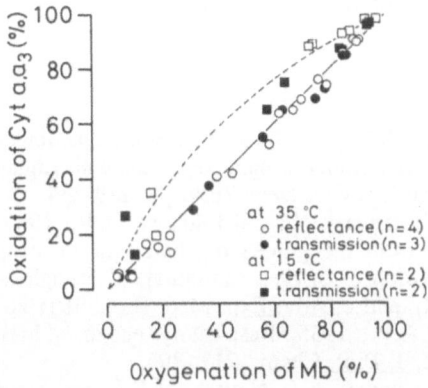

Fig. 3. Steady state relation of oxygenation of myoglobin and oxidation of cytochrome aa_3 in perfused hindlimb. The experiments were performed from the aerobic to the anaerobic conditions by reducing oxygen tension in influent stepwise.

In the present study, the oxidation level of Cyt aa_3 became clearly higher than the oxygenation level of Mb at 15°C. This result indicates that the relationship between the oxygenation level of Mb and the oxidation level of Cyt aa_3 depends on O_2 uptake rate of tissue. Further experiments in hypoxia may be useful in this field.

SUMMARY

The effects of temperature on oxygenation and metabolism in perfused rat hindlimb was studied at 35°C and 15°C. Oxygenation of myoglobin and oxidation of cytochrome aa_3 in the thigh (quadriceps) muscle were estimated from the difference spectra measured with a rapid-scanning spectrophotometer. Simultaneously, oxygen uptake and releases of lactate and pyruvate were measured.

(1) In hypothermia, glycolysis played a major role in energy metabolism even though Cyt aa_3 was maintained in a more oxidized state than in normothermia.

(2) P_{50} of myoglobin in perfused rat hindlimb was 5.0 mmHg at 35°C, 2.3 mmHg at 25°C and 1.1 mmHg at 15°C. The $\Delta H°$ was -13.0 kcal/mol.

(3) When about 30% of myoglobin was deoxygenated at both 35°C and 15°C, the oxygen uptake started to decrease and lactate release increased.

(4) At 35°C, the oxidation level of cytochrome aa_3 was same as the oxygenation level of myoglobin. At 15°C, however, the oxidation level of cytochrome aa_3 was clearly higher than the oxygenation level of myoglobin. The oxygen uptake at 15°C was about one third that at 35°C.

In conclusion, in order to maintain the aerobic condition of cytochrome aa_3 in mitochondria of rat skeletal muscle, a tissue oxygen tension higher than 12 mmHg at 35°C, and higher than 3 mmHg at 15°C is required.

ACKNOWLEDGEMENT

The work was supported in part by grants from the Ministry of Education, Science and Culture of Japan and from the Ehime Health Foundation.

REFERENCES

Antonini, E., and Brunori, M., 1971, Specific aspects of the reactions of myoglobin with ligands, in: "Hemoglobin and Myoglobin in their Reactions with Ligands", A. Neuberger and E.I. Tatum, eds., American Elsevier, New York, pp. 219-234.

Caspary, L., Ahmad, H.R., Hoffmann, Y., and Lübbers, D.W., 1985, Multicomponent analysis of reflection spectra from the guinea pig heart for measuring tissue oxygenation by quantitative determination of oxygen saturation, of myoglobin, and of the redox states of cytochromes aa_3, b, and c, Adv. Exp. Med. Biol., 191: 263.

Chance, B., and Williams, G.R., 1955, Respiratory enzymes in oxidative phosphorylation. II. Difference spectra, J. Biol. Chem., 217: 395.

Clark, A. Jr., Clark, P.P.A., Connett, R.J., Gayeski, T.E.J., and Honig, C.R., 1987, How large is the drop in Po_2 between cytosol and mitochondrion?, Am. J. Physiol., 252: C583.

Gutmann, I., and Wahlefeld, A.W., 1974, L(+) lactate. Determination with lactate dehydrogenase and NAD, in: "Method of Enzymatic Analysis", H.U. Bergmeyer, ed., vol. 3, 2nd edition, Academic Press, New York, pp. 1464-1468.

Jöbsis, F.F., 1974, Intracellular metabolism of oxygen, Am. Rev. Respir. Dis., 110, Suppl: 58.

Kanaide, H., Yoshimura, R., Makino, N., and Nakamura, M., 1982, Regional myocardial function and metabolism during acute coronary occlusion, Am. J. Physiol., 242: H980.

Kennedy, F., and Jones, D., 1986, Oxygen dependence of mitochondrial function in isolated rat cardiac myocytes, Am. J. Physiol., 250: C374.

Lamprecht, W., and Heinz, F., 1984, Pyruvate, in: "Method of Enzymatic Analysis", H.U. Bergmeyer, ed., vol. VI, 3rd edition, Verlag Chemie, Weinheim, Deerfield, Basel, pp. 570-577.

Makinen, M.W., and Lee, C.P., 1968, Biochemical studies of skeletal muscle mitochondria. I. Microanalysis of cytochrome content, oxidative and phosphorylative activities of mammalian skeletal muscle mitochondria, Arch. Biochem. Biophys., 126: 75.

Pegg, D.E., Wusteman, M.C., Foreman, J., 1981, The metabolism of normal and ischaemically injured rabbit kidneys during perfusion for 48 hours at 10°C, Transplantation, 32: 437.

Ruderman, N.B., Houghton, C.R.S., and Hems, R., 1971, Evaluation of the isolated perfused rat hindquarter for the study of muscle metabolism, Biochem. J., 124: 639.

Seiyama, A., Hazeki, O., and Tamura, M., 1988, Noninvasive quantitative analysis of blood oxygenation in rat skeletal muscle, J. Biochem., 103: 419.

Shiota, M., and Sugano, T., 1986, Characteristics of rat hindlimbs perfused with erythrocyte- and albumin-free medium, Am. J. Physiol., 251: C78.

Strickland, E.H., Ackerman, E., and Anthony, A., 1959, Effects of altitude acclimatization on the equilibrium constant of rat oxymyoglobin, Am. J. Physiol., 197: 211.

Sugano, T., Suda, K., Shimada, M., and Oshino, N., 1978, Biochemical and ultrastructural evaluation of isolated rat liver systems perfused with a hemoglobin-free medium, J. Biochem. (Tokyo), 83: 995.

Tamura, M., Oshino, N., Chance, B., and Silver, I., 1978, Optical measurements of intracellular oxygen concentration of rat heart in vitro, Arch. Biochem. Biophys., 191: 8.

Tamura, M., Seiyama, A., and Hazeki, O., 1987, Spectroscopic characteristics of rat skeletal and cardiac tissues in the visible and near-infrared region, Adv. Exp. Med. Biol., 215: 297.

Tamura, M. , Hazeki, O., Nioka, S., and Chance, B., 1989, In vivo study of tissue oxygen metabolism using optical and nuclear magnetic resonance spectroscopies, Ann. Rev. Physiol., 51: 813.

Yonetani, T., 1960, Studies on cytochrome oxidase. I. Absolute and difference absorption spectra, J. Biol. Chem., 235: 845.

Wittenberg, B.A., and Wittenberg, J.B., 1989, Transport of oxygen in muscle, Ann. Rev. Physiol., 51: 857.

SUPERPOSITION OF ARTERIOLAR VASOMOTION WAVES AND REGULATION OF BLOOD FLOW IN SKELETAL MUSCLE MICROCIRCULATION

Antonio Colantuoni, Silvia Bertuglia, Giuseppe Coppini, Luigi Donato

CNR Institute of Clinical Physiology
University of Pisa
56100 Pisa, Italy

INTRODUCTION

The delivery of oxygen from the blood to the tissues is a complex process involving several factors, which ensure adequate oxygen supply to the cell metabolic demands. Tissue oxygenation is deeply affected by the distribution of blood in the arteriolar and capillary network. Many evidences indicate that in the striated muscle the capillary perfusion is inhomogeneous, because velocity of red blood cells, number of red blood cells per unit length of capillary (or capillary hematocrit) and the density of perfused capillaries vary to a considerable extent (Tyml, 1987). It has been shown that these parameters depend on vascular tone, indicating that the arteriolar vessels play a key role in the regulation of muscle perfusion (Tyml, 1987).

Skeletal muscle microcirculation has been studied in many experimental models. Slaaf et al. (1987) and Meyer et al. (1987) showed spontaneous arteriolar vasomotion in rabbit tenuissimus muscle; Bertuglia et al. (1988) described arteriolar luminal changes in hamster skin muscle. Variations in arteriolar diameter influence microvascular hemodynamics and tissue oxygenation. However, no data are available on vasomotion of all arterioles in skeletal muscle microcirculation.

The purpose of the present study was to describe the main features of vasomotion in hamster skin muscle microvasculature, from the largest to the smallest arterioles in the field of observation and to assess the transmission and superposition of waveforms along the vascular tree. Therefore, power spectrum and cross-correlation analyses of sequentially recorded waves were carried out to determine common frequency components.

MATERIALS AND METHODS

10 male Syrian hamsters (Charles River, Italy) weighing 70-80 g

Oxygen Transport to Tissue XII, Edited by J. Piiper *et al.*
Plenum Press, New York, 1990

were used. Two symmetrical teflon-coated aluminum frames were implanted into a dorsal skin fold of hamsters, as previously described (Colantuoni et al., 1985). Briefly, the animal was anesthetized (pentobarbital, 5 mg/100 g body wt., i.p.) (Nembutal, Abbott, USA). A round area of the dorsum skin and underlying skin muscle of 15 mm of diameter was removed from one side of the symmetrical fold, thus exposing the opposite layer of the skin muscle (Musculus Cutaneus maximus) attached to subcutaneous tissue. The tissue was covered by a microcover glass fixed to one of the aluminum frames, while the other part stayed open.

Permanent catheters filled with heparinized saline were implanted in the jugular vein and in the carotid artery. They were passed under the skin to the upper posterior part of the neck and fixed to the upper part of the window. The animals recovered for 48 h in an incubator at 30 ± 0.5 °C with free access to food and water.

Observations were made by placing the animal into a tube that minimized movements. Both the tube and the extending frame of the chamber were fixed to the microscope stage. The window was transilluminated by a 100 W tungsten halogen lamp used in conjunction with a round heat absorbing filter and a 4140 Å band pass filter, and observed under a Leitz Orthoplan microscope fitted with a long working distance objective (x4, n.a. 0.12; x20, n.a. 0.25; x32, n.a. 0.60, and x10 eyepiece). At the end of the period of observation (8-10 days), the microcirculation was studied with fluorescent microscopic technique using a xenon 150 W lamp and a Leitz I2 Ploemopack filter block. The tracer was fluorescein isothiocyanate, bound to dextran Mw 150,000, injected intravenously (50 mg/100 g body wt., as 5% sol) and followed by a COHU 5253 SIT low light level camera. The scenes were monitored and recorded by a Sony U-Matic VO 5800 PS video recorder.

During the observation mean arterial blood pressure (Statham PD 23 transducer, connected to catheter in carotid artery) and ECG were recorded by a Honeywell RM 300 monitor, interfaced to an IBM XT 286 Personal Computer.

The diameter of vessels was measured by a computerized technique. The sequences of interest during play back of recorded experiments were real time A/D converted into 512x512x8 bit matrix by a Tesak VDC 501-A video processor interfaced to a general purpose microcomputer DEC-PDP 11/23. In a window positioned on the vessel of interest, oriented vertically, the edges of vessel were detected automatically and the mean distance between the walls was computed. All the records were processed to assess the variations of diameter; therefore, power spectrum of each record was computed. The digitation rate of time dependent diameter changes was two samples per sec.

Power spectrum analysis was carried out using both Fast Fourier Transform methods and Autoregressive modelling (AR). Despite its computational efficiency FFT spectra are limited by the reciprocal of sampling, finite length of signals and definite bandwidth of truncation window. Higher frequency resolution, with a slight increase in computation time, was obtained by AR method, using maximum entropy (All Poles) algorithm. The spectrum was estimated as an all-poles rational function, that permits the detection of narrow frequency peaks. This procedure is not dependent on the number of time dependent samples and is not affected by windowing process. Moreover, an analytical expression of the spectrum was made available rather than a sequence of spectrum samples as with FFT methods. Therefore, AR spectrum provided the individual spectral components in terms of center frequency and of the corresponding power in absolute, fractional, and normalized values.

Cross-spectral data (amplitude and phase) were computed using the modified periodogram method (FFT). In cross-analysis procedure, FFT algorithms were used to determine the common frequency components in different vasomotion waveforms, recorded sequentially along the same microvascular network. Digitation rate was two samples per sec. Since the spectrum of each record has been previously computed, there was no need of higher frequency resolution as in power spectrum estimation. Therefore, FFT methods allowed us to obtain useful data about wave spreading along the microvasculature.

To map the arteriolar network, photographs of the chamber, at low magnification, were taken directly from microscope with an automatic camera Wild MPS 51S. These low magnification pictures were used as masters to assemble photos taken from monitor, with a final magnification of 300. Terminal arterioles were recorded on video tape with both transillumination and epiillumination. The networks were reconstructed off line during play back of video tape, using frame by frame analysis to take photos for the montage.

To differentiate capillaries from terminal arterioles, it was important to follow the terminal vasculature with the highest magnification objective, scanning systematically capillaries and arterioles filled with the fluorescent tracer. The vessel length was measured with the computer-assisted technique, previously described.

The data were transferred to an IBM XT 286 Personal Computer for statistical analysis. All reported values are mean+SD. Data were analyzed for statistical significance using a statistical package SSP-C.

RESULTS

We studied ten microvascular networks and adopted Strahler ordering scheme to assign order to the arteriolar vessels; order 0 was assigned to the capillaries, order 1 to the capillary supplying arterioles, and so on up to the largest arterioles (order 4) in the preparation. According to Strahler's method, the order is increased by one, when two vessels of the same order join. All microvascular networks showed four orders of arterioles. However, in 6 cases out of 10, we were able to measure order 4 vessel diameter and length. In the remaining we could not measure diameter and length of order 4 arterioles, because they were out of the field of observation. Order 4 vessels can be compared to arcade arterioles described in experimental models.

The average of maximum diameters observed during vasomotion cycles, length, and number of arterioles are reported in Table 1. The main feature of arteriolar microvasculature was the time dependent variations of diameter, whose fundamental frequencies were determined by power spectrum analysis. Order 1 and 2 vessels were characterized by many frequency components; fundamental frequency (i.e., the frequency with the highest amplitude) was in the range 4-15 cycles per min (cpm) in order 1 and 2-11 cpm in order 2 arterioles. Order 3 and 4 vessels presented fundamental frequency in the range 0.5-6 cpm and 0.3-3 cpm, respectively. The percentage amplitude, referred to arteriolar mean diameter, is reported in Table 1. The constriction of order 1 vessels, which closed completely in most cases, caused intermittent blood flow in capillary network, with related changes in red blood cell velocity and numbers.

We recorded rhythmic diameter changes sequentially along the microvascular segments, starting from the largest order 4 arterioles up to order 1 vessels. According to the wave spreading pattern as revealed by cross-correlation analysis, we differentiated 3 groups of networks.

In the first group (n=5), each order of vessels was characterized by peculiar fundamental frequencies of waveforms, that changed at branching points. We observed that low frequency waves of order 3 vessels varied in 8 out of 12 order 2-3 branchings. The active points at order 1-2 branchings were 20 out of 34. Therefore, order 1 vessels were characterized by the highest fundamental frequencies. Indeed, it was possible to note that order 2 vessels branching from the same parent order 3 vessel, did not show the same fundamental frequency. The same trend was observed in active order 1 vessels, originating from the same parent order 2 vessel.

In Fig. 1 we report sequential recordings of waveforms in one vessel of each order and the corresponding power spectrum analysis. Power spectrum of order 4 vessel waves showed a fundamental frequency of 1.5 cpm. The second frequency component was centered around 3 cpm; faster frequency components were also present. In order 3 vessel, power spectrum analysis revealed a fundamental frequency of 4.2 cpm, with other frequency components of lower amplitude, both slower and faster. The main frequency in order 2 vessel was 5.1 cpm, with several other components of lower and higher frequency. In order 1 vessel power spectrum, fundamental frequency was 5.7 cpm, but there were other frequency components of 10 and 15 cpm.

In Fig. 2 we present cross-correlation analysis of the waveforms. Low frequency components, centered aroud 3.5 and 4.5 cpm, were propagated from order 4 and 3 to order 2 and 1 vessels, downstream. Higher frequency groups, centered around 5.5, 10, and 15 cpm, originated in order 1 and 2 vessels and were propagated also upstream, as indicated by negative group delay in cross-spectrum. Moreover, from the data of phase delay, it is possible to suggest a reflection of low and high frequency components; therefore, low frequency components conducted downstream were reflect-

552

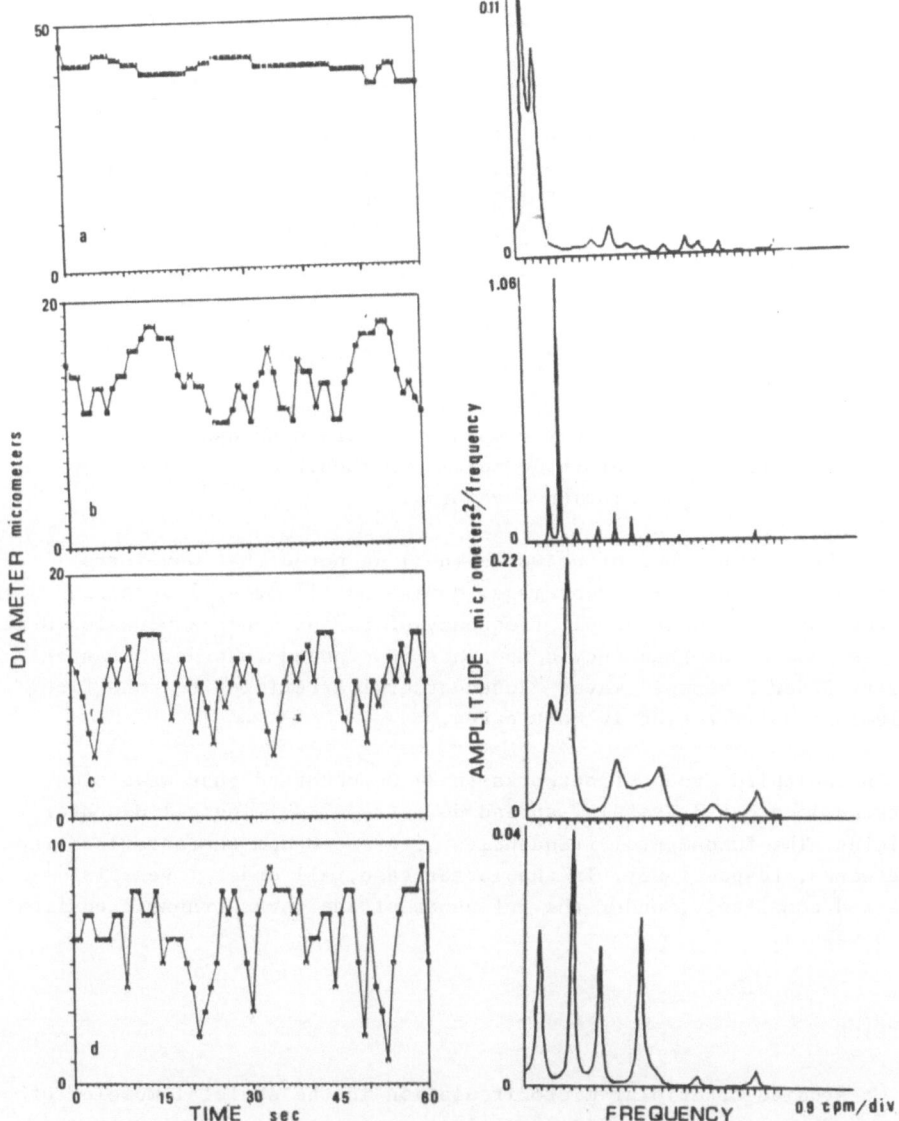

Fig. 1. Rhythmic diameter changes recorded sequentially and the corresponding power spectrum, estimated with Autoregressive Modelling, in one arteriole of each order: order 4 (a), order 3 (b), order 2 (c), order 1 (d). This network showed one order 3 vessel, three order 2 vessels, and eight order 1 arterioles. The fundamental frequencies of order 1 arterioles, originating from the same parent order 2 vessel, were different. The reported "amplitude" is the amplitude of the power spectral density, which is expressed in $micrometers^2/frequency = \mu m^2/cpm$ (cycles per min). See text for details.

Table 1. Number, diameter, length, and percentage amplitude (range) of vasomotion waveforms in arterioles of skeletal muscle microcirculation

Vessel Order	N	Diameter*	Length*	Amplitude**
1	58	7.50 ± 1.16	312 ± 101	60 - 100
2	22	11.62 ± 2.30	533 ± 140	50 - 100
3	10	18.55 ± 3.42	963 ± 361	15 - 50
4	6	28.97 ± 9.55	1212 ± 296	5 - 20

Values: Mean $\pm$ SD
* micrometers
** Amplitude: percentage of mean diameter variations

ed upstream, too. High frequency components, spreading upstream, were reflected downstream. Therefore, complex superposition of frequency components influenced vasomotion waveforms.

In the second group of networks (n=3) we noted that waveforms, originating at order 2-3 branchings, dominated all order 1 vessels, which presented the same fundamental frequency of parent order 2 vessels. Order 3 vessels showed low frequency components superposed to order 2 faster activity. Order 2 vessel waves caused order 1 arterioles to constrict, with luminal obliteration in most cases.

In the third group of networks (n=2) we observed that waveforms, originated in order 3 vessels, spread downstream, driving all daughter arterioles. The fundamental frequencies of 4 and 6 cpm characterized the two networks, respectively. In the latter case, all order 1 vessels contracted completely, under the influence of the waves transmitted from order 3 vessel.

DISCUSSION

We studied arteriolar microcirculation in the skeletal muscle of hamster under control normoxic conditions, at rest. To classify the vessels, we adopted the Strahler's method, that has been widely used by Engelson et al. (1985) in rat spinotrapezius muscle and Koller et al. (1987) in cat sartorius muscle. Moreover, we tried to define the frequency components of arteriolar vasomotion waves in striated muscle microvasculature.

In hamster skin muscle arteriolar networks, the vessels of each order, from the smallest order 1 to the largest order 4, presented specific fundamental frequencies, with low frequency components in order 3 and 4 arterioles, and high frequency groups in order 1 and 2 vessels. Changes in frequencies and in waveforms were observed mainly at or

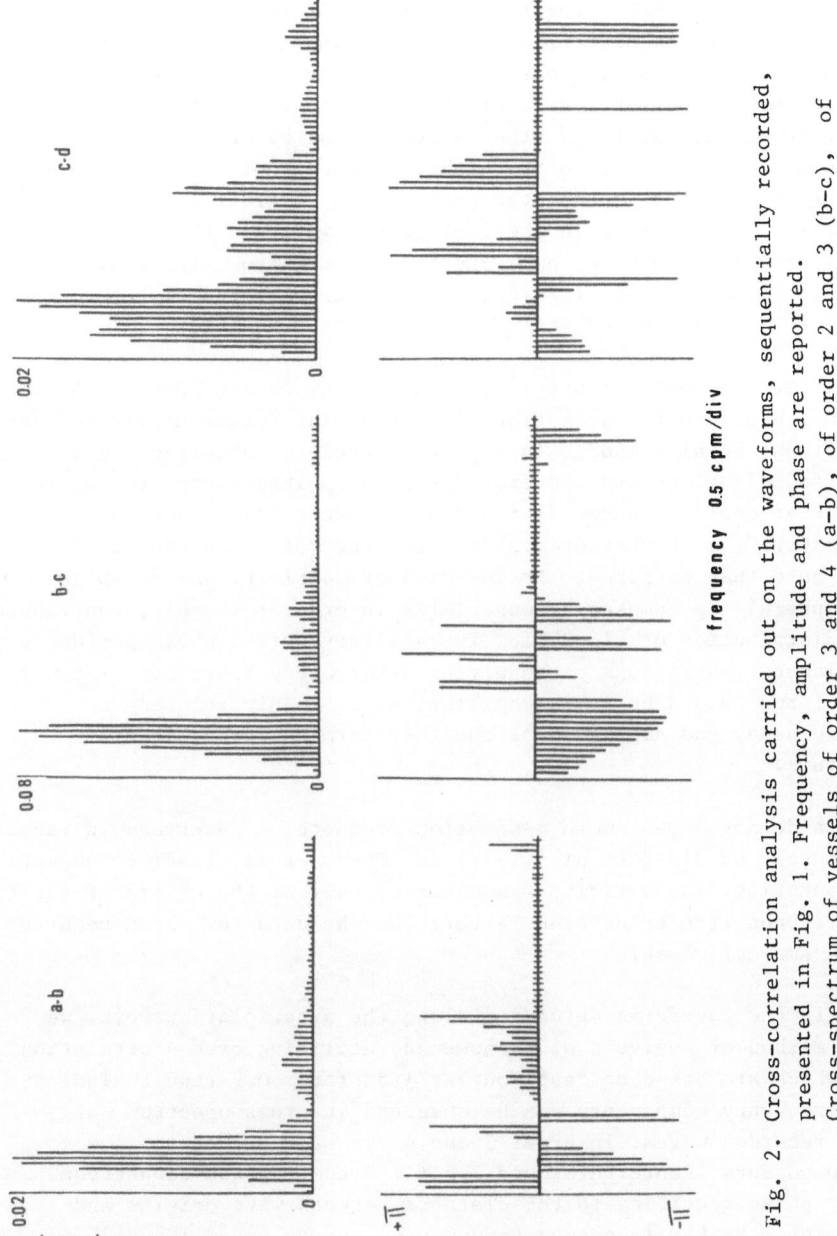

Fig. 2. Cross-correlation analysis carried out on the waveforms, sequentially recorded, presented in Fig. 1. Frequency, amplitude and phase are reported. Cross-spectrum of vessels of order 3 and 4 (a-b), of order 2 and 3 (b-c), of order 1 and 2 (c-d). Two groups of lower and higher frequency components are transmitted along the microvasculature. "Amplitude" is the amplitude of power spectral density. See text for details.

near branchings points, usually at the beginning of daughter vessels or immediately before the branchings, even though it was possible to note active points along the vessels. Usually the waves spread downstream, causing time dependent variations of diameter and affecting blood distribution in the capillary network. According to waveform spreading pattern, we differentiated three groups of networks: the first was characterized by fundamental frequencies typical of vessels of each order; the second was featured by order 2 arterioles dominating all order 1 vessels; the third was characterized by order 3 vessels, that drove all other arterioles. The most networks, however, belonged to the first group. The differences between the first and other groups might be due to impaired activity of order 1 vessels or to the higher frequency of order 2 vessels, that discharged faster and impeded the contraction of order 1 arterioles. It is interesting to note that in the last group, we found that waves of relatively high frequency (6 cpm) were transmitted from the beginning of order 3 vessel, dominating all order 1 and 2 vessels. The high frequency might account for impairment of activity in the other vessels.

Present data support our previous observations on peripheral vascular mechanisms, which originate vasomotion, in arteriolar cutaneous microcirculation (Colantuoni et al., 1985). In skeletal muscle we observed that contractions and dilations started from discrete points along the microvasculature, that could be endowed with smooth muscle cells, acting as "local pacemakers", with characteristic frequency of discharge. It is important to note that different frequencies of activity and shift in phase of peripheral "pacemakers", especially in order 1 vessels, contribute greatly to distribution of blood flow in capillary units, whose perfusion depends on diameter variations of supplying arterioles. Therefore, changes in velocity of red blood cells in capillaries are mainly related to vasomotion frequency and amplitude of the last terminal branchings of microvasculature.

Our data do not agree with vasomotion frequencies described in rabbit tenuissimus muscle by Slaaf et al. (1987) and Meyer et al. (1987). However, in their preparation, the activity was observed only at the origin of first order side branches from transverse arterioles. No data have been reported in larger and smaller vessels.

To clarify the waveforms spreading along the arteriolar network, we improved our method of analysis of phenomenon, utilizing cross-correlation algorithms, which are based on Fast Fourier Transform. Our results indicate that common frequency components can be observed in cross-spectrum of sequentially recorded waves. In order 3 and 4 vessels a group of low frequency components, centered around 2.5 ± 1.0 cpm, spread downstream, with shift in phase according to the distance between wave origins and recording points. Higher frequency components, due to order 1 and 2 vessel waves conducted upstream, were also present in cross-spectrum. Frequency components centered around 4.5 ± 1.5 cpm were observed in cross-spectrum of order 2 and 3 vessels; these waves spread downstream. Higher frequency components, originated in order 1 vessels, were also detected. In some cases, according to negative or positive group delay of frequency components, it is conceivable to suggest that waves could be reflected along the microvasculature. Further studies are required to clarify this phenomenon. The highest frequency components, centered around 5.5 ± 1.5, 10 ± 1.5, 13 ± 2 cpm were found in order 1 and 2 vessel cross-spectrum. The highest frequencies were transmitted upstream, indicating that the origin of waves was located in the smallest arterioles of the microvasculature.

The transmission of waves seems to be decremental, since amplitude of frequency components is reduced in cross-spectrum of vessels far from their origin.

Our data demonstrate in skeletal muscle microcirculation of hamster diameters of arterioles undergo time dependent changes, that are mainly characterized by low and high frequency components, originating at different points along the microvasculature. These variations of vessel lumen contribute to the regulation of blood flow redistribution in arteriolar as well as in capillary networks, causing temporal changes in red blood cell number and velocity in capillaries, parameters that control tissue oxygenation in skeletal muscle.

SUMMARY

In skin muscle microcirculation of Syrian hamsters, rhythmic diameter changes were studied along the arteriolar network, under normoxic conditions, at rest. A teflon coated-aluminum chamber was implanted in the dorsum skin of animals. The microcirculation was investigated using intravital microscopy technique. Vessel diameters were determined by a computer-assisted method. Power spectrum analysis of vasomotion recordings was carried out with Fast Fourier Transform and Autoregressive modelling. To determine vasomotion waveform spreading, cross-spectral data (amplitude and phase) were computed, using the modified periodogram method (FFT).

The arterioles were classified according to Strahler's method. Order 1 vessels (diameter: 7.50 ± 1.16 µm) showed the highest frequency, 4-15 cycles per min, and percentage amplitude in the range 60-100 %. Order 2 and 3 arterioles had intermediate frequencies, and amplitude in the range 50-100 %, and 15-50 %, respectively. The largest order 4 vessels (diameter: 28.97 ± 9.55 µm) had the lowest frequency, 0.3-3 cpm, and amplitude in the range 5-20 %. In most networks, cross-correlation analysis revealed two groups of frequency components. Low frequency group was propagated from order 4 and 3 vessels downstream. High frequency components were transmitted upstream from order 1 and 2 arterioles. Therefore, a complex superposition of waveforms resulted from the activity of discrete points along the microvasculature.

In conclusion, rhythmic diameter changes of arterioles in skeletal muscle microcirculation regulate blood flow distribution in capillary units and control tissue oxygenation.

REFERENCES

Bertuglia, S., Coppini, G., Colantuoni, A., 1988, Systemic hypoxia and
    arteriolar rhytmic diameter changes in hamster skeletal muscle
    microcirculation, Int. J. Microcirc. Clin. Exp., 7:S141.

Colantuoni, A., Bertuglia, S., Intaglietta, M., 1985, Variations of rhythmic
    diameter changes at the arterial microvascular bifurcations, Pflügers
    Arch., 403: 289.

Engelson, E. T., Skalak, T. C., Schmid-Schönbein, G. W., 1985, The micro-
    vasculature in skeletal muscle. I. Arteriolar network in rat
    spinotrapezius muscle, Microvasc. Res., 30:29.

Koller, A., Dawant, B., Liu, A., Popel, A. S., Johnson, P. C., 1987,
    Quantitative analysis of arteriolar network architecture in cat
    sartorius muscle, Am. J. Physiol., 253: H154.

Meyer, J. U., Lindbom, L., Intaglietta, M., 1987, Coordinated diameter
    oscillations at arteriolar bifurcations in skeletal muscle, Am. J.
    Physiol., 253: H568.

Slaaf, D. W., Tangelder, G. J., Teirlinck, H. C., Reneman, R. S., 1987,
    Arteriolar vasomotion and arterial pressure reduction in rabbit
    tenuissimus muscle, Microvasc. Res., 33: 71.

Tyml, K., 1987, Red cell perfusion in skeletal muscle at rest and after
    mild and severe contractions, Am. J. Physiol., 252: H485.

SKELETAL MUSCLE PO$_2$ DURING HYPODYNAMIC SEPSIS

G. Gutierrez, N. Lund*, F. Palizas, and A. Acero

Pulmonary and Critical Care Division. University of Texas
Health Science Center. 6431 Fannin, Houston, Texas 77030
*Department of Anesthesiology. The University of
Rochester Medical Center. 601 Elmwood Ave., Rochester
New York 14642

INTRODUCTION

It has been hypothesized (Nelson et al., 1988, Samsel et al., 1988, Bredle et al. 1989), that sepsis is a disorder of microvascular regulation resulting in decreased O$_2$ transport from the capillaries to the mitochondria. Evidence for this hypothesis was offered by Nelson et al. (1988), who found greater levels of systemic and intestinal critical O$_2$ transport (TO$_2$), defined as the minimum TO$_2$ required to maintain constant O$_2$ consumption (VO$_2$), in hypovolemic septic dogs when compared to a control group. This phenomenon may be related to increases in microcirculatory heterogeneity, resulting in a mismatch of tissue perfusion and cellular O$_2$ needs. This is manifested at the organ level by decreases in O$_2$ extraction capacity.

It is not clear, however, whether skeletal muscle microvascular control is also disrupted during sepsis. To test this hypothesis, we subjected anesthetized, mechanically ventilated rabbits to an intravenous challenge of E. Coli endotoxin and measured hindlimb skeletal muscle PO$_2$ distribution with an array of Clark type PO$_2$ microelectrodes. We characterized increases in microcirculatory heterogeneity by increases in the variance of the tissue PO$_2$ distribution (Lund et al. 1980). We found that endotoxin decreased cardiac output (Q), TO$_2$, and mean tissue PO$_2$, but did not alter tissue PO$_2$ histogram distribution. These findings imply that skeletal muscle microvascular control remains intact in this hypodynamic model of septic shock.

METHODS

The preparation used in the present study has been previously described (Gutierrez et al. 1989). Briefly, New Zealand White rabbits of either sex, weighing 2.3 ± 0.2 kg were used (n = 7). Anesthesia was induced with an intravenous bolus of sodium pentobarbital sodium in a sterile water solution (35 mg/kg). Additional I.V. doses of Na pentobarbital were given at periodic intervals during the experiment (5 to 10 mg/kg) to maintain adequate anesthesia. The rabbit was endotracheally intubated and mechanically ventilated (Siemens Servo Ventilator 900C, Siemens-Elema AB, Solna, Sweden) using airway pressure support of 2 cm H_2O, at a frequency determined by the rabbit's own ventilatory rate. Inspired O_2 fraction (F_IO_2) was maintained at 0.30 throughout the experiment.

Flexible catheters were passed into the right femoral artery and into the internal jugular vein and connected to calibrated pressure transducers (Hewlett-Packard, Waltham, MA). The venous catheter was advanced to the level of the right atrium (RA), as evidenced by the appearance of atrial pressure waves. These catheters were used to obtain arterial and RA venous blood samples and blood pressures. Another catheter was introduced into the right femoral vein and the distal end of the vein was ligated. This catheter was advanced approximately 8 cm, placing the catheter tip in the inferior vena cava (IVC), below the renal veins and above the iliac bifurcation. This catheter was used to obtain venous blood samples from the left hindlimb. Catheter location was visually confirmed at the end of the experiment.

O_2 consumption was measured from the inspired and expired gas O_2 and CO_2 concentrations (Otis 1964), measured with an S-3A O_2 analyzer and CO_2 analyzer CD-3A (Ametek Thermox Instruments, Pittsburgh, PA). Expired gas flow was measured with a mesh flowmeter and electronically integrated, to obtain the expired gas volume with an accuracy of 5% (Siemens-Elema AB, Solna, Sweden). The tubing and connections of the ventilator circuit were tested for leaks with a calibrated air syringe and a water spirometer. The accuracy of the expired volume measurements were also checked with a calibrated syringe prior to the experiments. Normal saline was infused throughout the experiment at a rate of 0.4 ml/kg/min. Esophageal temperature was monitored with a thermistor and it was maintained at 37.5 ± 1.0 °C with a heating pad.

Extracellular tissue PO_2 was measured with an 8 channel, surface Mehrdraht Dortmund Oberflache (MDO) oxygen electrode (Kessler et al. 1976, Lund 1978). This electrode consists of 8 individual Clark-type electrodes (15 um diam.) and an Ag/AgCl reference electrode. A small skin incision was made in the left limb and, following careful dissection of various fascial layers, the electrode was placed over the biceps femoris muscle surface. A radiating heat lamp was used to maintain muscle surface temperature at approximately 34 °C. Details on electrode placing and calibration have been previously described (Gutierrez et al. 1989). The electrode was automatically sampled every 10 seconds during a 140 second sampling period, for a total of 120 measurements per measuring period.

Experimental Protocol

Baseline measurements consisted of 0.5 ml aliquots of arterial (a), right atrial (ra), and limb (l) blood samples drawn simultaneously for measurement of blood gases, pH and hemoglobin O_2 saturation (SO_2) (Corning Blood Gas Analyzer, Corning, N.Y., and IL282 CO-Oximeter, Instrumentation Laboratories, Lexington, MA). Measurements of minute volume inspired O_2, and expired O_2 and CO_2 fractions, and arterial and right

560

atrial blood pressures were made immediately before the drawing of the blood samples.

Following baseline measurements the animals were given 5 ml of an E. Coli endotoxin suspension (Sigma EC-1) 1.5 mg/kg, pH 7.40 infused slowly into the femoral vein catheter over a six minute period (Novotny et al 1988). Measurements similar to those of baseline were taken 15, 30, 60, 90, and 120 minutes after the infusion of endotoxin.

Data Analysis

Cardiac output was calculated using Fick's principle, $Q = VO_2/(C_aO_2 - C_{ra}O_2)$, where the right atrial O_2 content was assumed to be equal to the mixed venous O_2 content (Musch and Larach 1988). TO_2 was calculated as $TO_2 = Q \times C_aO_2$. VO_2, Q, and TO_2 were indexed to the weight of the rabbit.

One-way analysis of variance was used to test for significant changes following the administration of endotoxin. Where differences were detected, the post hoc Neuman-Keuls method was used to identify individual differences (Zar 1984). Tissue PO_2 histogram distributions were compared with the non-parametric two-sample Kolmogorov-Smirnov test as modified by Odman and Lund (1980). Measurements are shown as mean $\pm$ SE.

RESULTS

Table 1 shows the changes in hemodynamic and O_2 transport parameters following the infusion of E. Coli endotoxin in this experimental model. Following the infusion of endotoxin, there were immediate decreases in mean arterial blood pressure (MAP) and Q. There was a rapid decrease in TO_2 to 50% of baseline following the administration of endotoxin, reflecting the sharp decrese in Q. VO_2 decreased to approximately 80% of baseline after endotoxin. Both TO_2 and VO_2 remained lower than baseline throughout the remainder of the experiment.

Table 1. HEMODYNAMIC AND OXYGENATION PARAMETERS DURING SEPSIS

Time	MAP	CVP	Q	TO_2	VO_2
Base	80 ± 4	2 ± 1	297 ± 28	48.8 ± 4.5	14.7 ± 0.6
15	$50 \pm 5+$	3 ± 1	$181 \pm 30*$	$26.7 \pm 5.2+$	$11.4 \pm 0.9*$
30	$54 \pm 5+$	3 ± 1	$185 \pm 32*$	$28.0 \pm 5.4+$	12.1 ± 1.2
60	$53 \pm 3+$	4 ± 2	$186 \pm 31*$	$26.3 \pm 4.5+$	12.8 ± 1.0
90	$54 \pm 3+$	2 ± 1	$174 \pm 28*$	$24.3 \pm 4.3+$	$12.1 \pm 0.7*$
120	$50 \pm 2+$	3 ± 1	$151 \pm 25*$	$21.1 \pm 4.2+$	$10.5 \pm 1.0*$

Time = Time post endotoxin administration (minutes); MAP = Mean arterial pressure (mmHg); CVP = Central venous pressure (mmHg); Q=Cardiac output (ml.min^{-1}.kg^{-1}); TO_2=O_2 transport (ml.min^{-1}.kg^{-1}); VO_2 = O_2 consumption (ml.min^{-1}.kg^{-1}). * P < 0.05; + P < 0.01.

Figure 1 shows changes in ERO_2 during the experiment. The relative maintenace of VO_2, when compared to the decrease in

TO_2, was the result of a rise in the O_2 extraction ratio ($ERO_2 = \dot{V}O_2/TO_2$) from 0.32 ± 0.03 to levels greater than 0.50.

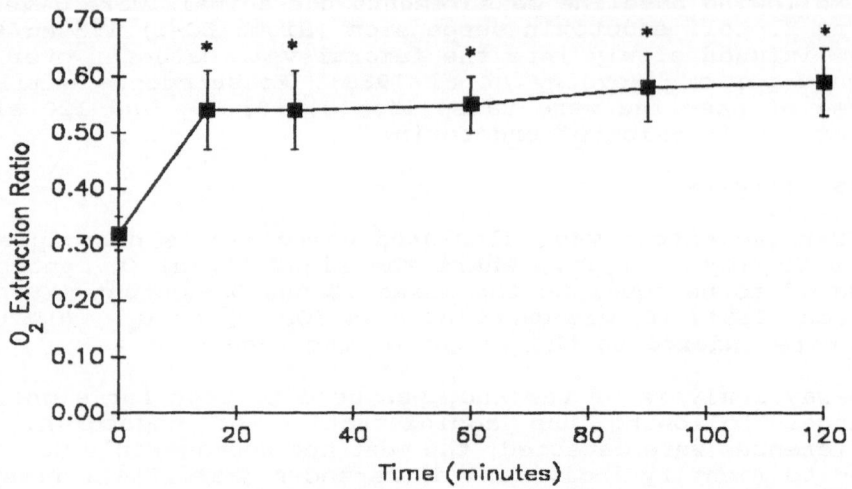

Figure 1 Changes in O_2 extraction ratio after the infusion of endotoxin. * P < 0.05.

The individual TO_2-VO_2 points for all the experiments are shown in figure 2. The unfilled circles represent the baseline data. Taken as a group, these rabbits appeared to maintain baseline VO_2 until TO_2 reached a critical TO_2, shown by a dashed vertical line. A critical TO_2 of 28.7 $ml.min^{-1}.kg^{-1}$ was determined for the grouped data using the polynomial method of Gutierrez et al (1986).

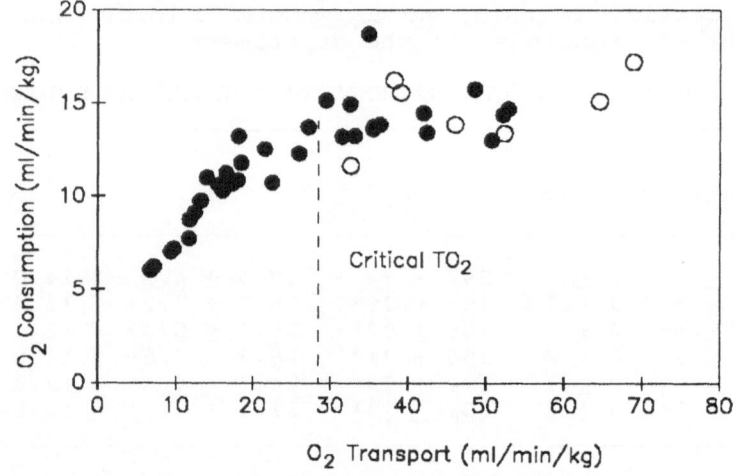

Figure 2 TO_2-VO_2 data pairs obtained during the seven experiments.

Figure 3 shows six PO_2 distributions corresponding to baseline, 15, 30, 60, 90, and 120 minutes after the infusion of endotoxin. The clear vertical bars represent the mean venous blood PO_2 from the hindlimb. With endotoxin the mean of the tissue

562

PO_2 histograms shifted to regions of lower PO_2, but their distributions remained unchanged during the two hours of the experiment, as determined by the Kolmogorov-Smirnov test.

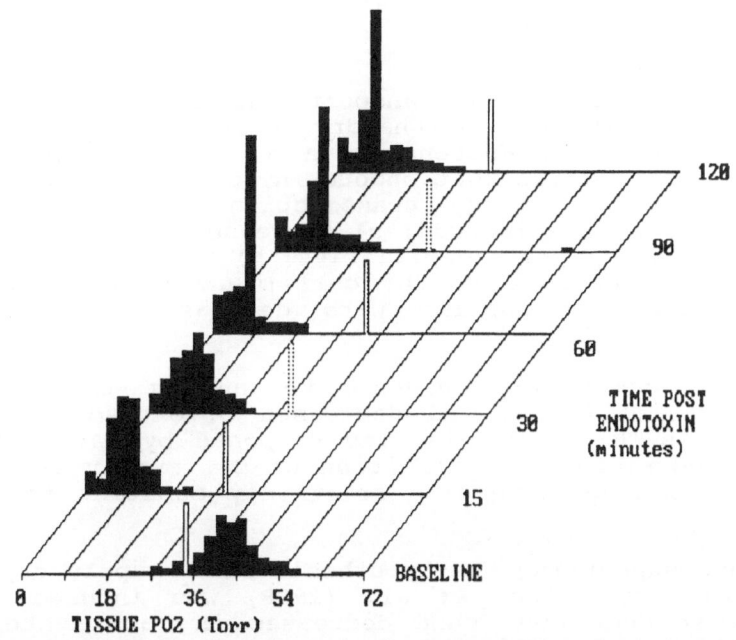

Figure 3 Changes in tissue PO_2 frequency histograms with endotoxin. Shown as a clear vertical line is the mean limb venous PO_2.

Table 2. Changes in arterial, venous, and tissue po$_2$ with endotoxin

Time	P_aO_2	$P_{ra}O_2$	P_lO_2	P_tO_2
Base	133 ± 8	41 ± 3	35 ± 2	43 ± 4
15	100 ± 17	32 ± 4	30 ± 1	8 ± 2 **
30	112 ± 16	33 ± 3	31 ± 1	9 ± 2 **
60	120 ± 14	35 ± 2	33 ± 1	7 ± 2 **
90	123 ± 14	35 ± 2	33 ± 1	9 ± 2 **
120	119 ± 14	37 ± 3	33 ± 2	7 ± 2 **

Time = Time post endotoxin infusion (minutes); P_aO_2 = Arterial PO_2; $P_{ra}O_2$ = Right atrial PO_2; P_lO_2 = Limb venous PO_2; P_tO_2 = Mean tissue PO_2. Values are shown in Torr. ** P< 0.001

Table 2 lists the arterial, central venous, limb venous, and histogram mean tissue PO_2. There were no significant changes in arterial, right atrial, or limb PO_2 levels during the experiment. On the other hand, there was a rapid, and highly significant (P < 0.001) decrease in mean tissue PO_2 immediately following the infusion of endotoxin. Mean tissue PO_2 remained

563

at levels below 10 Torr during the rest of the experiment. Furthermore, mean tissue PO_2 was significantly lower than either limb or right atrial PO_2 during the whole septic period.

DISCUSSION

The hemodynamic, oxygen transport, and metabolic alterations known to occur with sepsis (Chaudry et al. 1979, Clowes et al. 1974, Houtchens and Westenskow 1984, Kilpatrick-Smith and Erecinska 1983) suggest that endotoxin, or a mediator, may act by disrupting the energy transduction process. Possible mechanisms of action include: a) decreased O_2 transport from the lung to the systemic capillaries; b) impaired O_2 diffusion from the capillaries into the cells; and c) alterations in cellular energy metabolism with decreased generation or utilization of ATP.

There is compelling evidence to support the notion of impaired O_2 diffusion and decreased O_2 extracting ability during sepsis. This mechanism was proposed by Cain (1986) in what was described as a condition of O_2 supply dependency, where ERO_2 in sepsis remains relatively constant for a wide range of TO_2.

Evidence supporting the hypothesis of O_2 supply dependency was obtained by Nelson et al. (1988) who infused E. Coli endotoxin in dogs and found decreases in both systemic and intestinal critical ERO_2. Samsel et al. (1988) corroborated the finding of lower systemic critical ERO_2 in sepsis, however, they found that, unlike the intestine, critical ERO_2 in septic skeletal muscle was not decreased. This finding implies that endotoxin elicits a variable response in different tissues, and that skeletal muscle is not affected to the same degree as other tissues. In a recent study, however, Bredle et al. (1989) did find a significant, albeit small, O_2 extraction defect in septic skeletal muscle when compared to control.

O_2 extraction defects may be the result of increased capillary blood flow heterogeneity, creating a mismatch between tissue perfusion and cellular O_2 requirements and increasing diffusion distances. The infusion of endotoxin has been shown to increase the lymphocyte and neutrophil populations in the microvasculature, leading to the release of superoxide anion radicals and endothelial injury (Meyrick and Brigham 1983, Grisham et al. 1988). These changes induce platelet aggregation resulting in microembolization and increased microcirculatory heterogeneity.

In the present study we studied the effect of E. Coli endotoxin infusion on rabbit skeletal muscle. We used the MDO electrode to assess changes in the spatial and temporal distribution of extracellular PO_2. This is an established technique, (Kessler et al. 1976), and the resulting PO_2 frequency distribution has been shown to reflect changes in microcirculatory heterogeneity (Lund et al. 1980). The highest readings are most likely to represent peri-arteriolar PO_2, while the lowest PO_2 values probably correspond to those regions of tissue with the lowest levels of O_2 supply.

564

We found that endotoxin infusion resulted in rapid and substantial decreases in MAP and Q. This appears to be the result of depressed myocardial contractility (Archer 1985, Parrillo 1985), given that ventricular preload, as reflected by the right atrial pressure, did not change after the administration of endotoxin. Decreases in Q resulted in proportional decreases in TO_2 to a level below that required to sustain aerobic metabolism. The fall in TO_2 was partially compensated by a rise in ERO_2 to levels greater than 0.50. Although substantial, it remains to be determined whether this was an appropriate increase in ERO_2, or if O_2 extraction was impaired by sepsis.

In this study we did not have a comparable control group, for example, one with similar reductions in Q produced by mechanical techniques. However, some insight into the adequacy of the critical ERO_2 may be gained by comparing the results of the present study with those obtained with a similar preparation exposed to progressive hypoxemia (Gutierrez et al 1989). The critical TO_2 of 28.7 $ml.min^{-1}.kg^{-1}$ determined in the present study corresponds to an ERO_2 of 0.54 ± 0.7. This value is similar to the ERO_2 of 0.57 ± 4.5 measured in the hypoxemic group at TO_2 and VO_2 levels of 26.1 ± 5.5 and 13.4 ± 1.5 $ml.min^{-1}.kg^{-1}$, respectively. While this comparison does not constitute proof that skeletal muscle ERO_2 was unaffected during sepsis, it suggests that, if present, the impairment in O_2 extracting ability was relatively minor. Furthermore, we did not find increases in histogram distribution following the infusion of endotoxin, implying that sepsis does not increase skeletal muscle microcirculatory heterogeneity.

Decreases in TO_2 produced by the administration of endotoxin resulted in tissue hypoxia, as shown by the decline in mean tissue PO_2 from 43 Torr to 7-9 Torr. On the other hand, as shown in Table 1, limb and central venous PO_2 remained at levels similar to those for baseline. Given that we found no evidence of increased microvascular heterogeneity, a condition favoring the development of functional peripheral shunting, this finding suggests that capillary PO_2 was significantly lower than venous PO_2 during sepsis.

A possible explanation for this condition is the development of a kinetic disequilibrium between RBC and plasma O_2 concentrations produced by increases in capillary transit time or in red cell O_2 affinity (Gutierrez 1986). However, cardiac output decreased following the infusion of endotoxin, which should have produced a slower, not a faster, capillary transit time. Furthermore, to our knowledge, there is no evidence of increases in RBC oxygen affinity during sepsis. Another, and perhaps a more plausible, mechanism to explain the difference between venous and tissue PO_2 is an increase in the rate of O_2 diffusion from the arterioles to the venules during sepsis, akin to the model proposed by Piiper et al. (1984).

In summary, we found that the intravenous infusion of E. Coli endotoxin in rabbits results in decreases in O_2 transport and in tissue hypoxia as the direct result of myocardial depression. The variance of the tissue PO_2 histograms remained unchanged, suggesting that skeletal muscle microcirculatory heterogeneity does not increase in this model of sepsis.

565

ACKNOWLEDGEMENTS

This study was funded in part by a National Institutes of Health Grant HL 41415-01

REFERENCES

Archer, L.T., 1985, Myocardial dysfunction in endotoxin and E.Coli induced shock: pathophysiological mechanisms. Circ. Shock 15:261.

Bredle, D.L., Samsel, R.W., Schumaker, P.T., and Cain, S.M., 1989, Critical O_2 delivery to skeletal muscle at high and low PO_2 in endotoxemic dogs. J. Appl. Physiol. 66:2553.

Cain S.M., 1986, Assessment of tissue oxygenation. Crit. Care Clin. 2:537.

Chaudry, I.H., Wichterman, K.A., and Baue, A.E., 1979, Effect of sepsis on tissue adenine nucleotide levels. Surgery 85:205.

Clowes, G.H.A., O'Donnell, T.F., Ryan, N.T., and Blackburn, G.L., 1974, Energy metabolism in sepsis: Treatment based on different patterns in shock and high output stage. Ann. Surg. 179:684.

Grisham, M.B., Everse, J., and Janssen H.F., 1988, Endotoxemia and neutrophil activation in vivo. Am. J. Physiol. 254:H1017.

Gutierrez, G., 1986, The rate of oxygen release and its effect on capillary O_2 tension: A mathematical analysis. Respir. Physiol. 63:79.

Gutierrez, G., Warley, A.R., and Dantzker, D.R., 1986, Oxygen delivery and utilization in hypothermic dogs. J. Appl. Physiol. 60:751.

Gutierrez, G., Lund, N., Acero, A.L., and Marini C., 1989, The relationship of venous PO_2 to muscle PO_2 during hypoxemia. J. Appl. Physiol. (In Press).

Houtchens, B.A. and Westenskow, D.R., 1984, Oxygen Consumption in Septic Shock: Collective Review. Circ. Shock. 13:361.

Kessler M, Hoper J, Krumme BA., 1976, Monitoring of tissue perfusion and cellular function. Anesthesiology 45:184.

Kilpatrick-Smith, L., and Erecinska, M., 1983, Cellular effects of endotoxin in vitro. I.- Effect of endotoxin on mitochondrial substrate metabolism and intracellular calcium. Circ. Shock. 11:85.

Lund, N., Odman S., and Lewis D.H., 1980, Skeletal muscle pressure fields in rats. A study of the normal state and the effects of local anesthetics, local trauma, and hemorrhage. Acta Anaesth. Scand. 24:155.

Lund, N., 1978, Methods for measuring tissue oxygen tension. Acta Anaesth. Scand. suppl. 70:183.

Meyrick, B. and Brigham, K.L., 1983, Acute effects of **Escherichia coli** endotoxin in the pulmonary microcirculation of anesthetized sheep: structure-function relationships. Lab. Invest. 48:458.

Musch, T.I. and Larach, D.R., 1988, O_2 contents of blood sampled from different venous compartments of the rat. J. Appl. Physiol. 65:988.

Nelson D.P., Samsel, R.W., Wood, L.D.H., and Schumacker, P.T., 1988, Pathological supply dependence of systemic and intestinal O_2 uptake during endotoxemia. J. Appl. Physiol. 64: 2410.

Novotny, M.J., Laughlin, M.H., and Adams, H.R., 1988, Evidence for lack of importance of oxygen free radicals in **Escherichia coli** endotoxemia in dogs. Am. J. Physiol. 254:H954.

Odman, S. and Lund, N., 1980, Data acquisition and information processing in MDO oxygen electrode measurements of tissue oxygen pressure. Acta Anaesth. Scand. 24:161.

Otis, A.B., 1964, Quantitative relationships in steady-state gas exchange, in: "Handbook of Physiology, Vol 1: Respiration." Am Physiol Soc, Washington D.C.

Parrillo, J.E., 1985, Cardiovascular disfunction in septic shock: new insights into a deadly desease. Intl. J. Cardiol. 7:314.

Piiper, J., Meyer, M., and Scheid, P., 1984, Dual role of diffusion in tissue gas exchange: Blood tissue equilibration and diffusion shunt. Respir. Physiol. 56:131.

Samsel, R.W., Nelson, D.P., Sanders, W.M., Wood, L.D.H., and Schumacker, P.T., 1988, Effect of endotoxin on systemic and skeletal muscle O_2 extraction. J. Appl. Physiol. 65:1377.

Zar, J.H., 1984, "Biostatistical Analysis. (2nd Ed.)", Prentice Hall, Englewood Cliffs.

ACTIONS OF A DOPAMINERGIC AND β_2-ADRENERGIC AGONIST ON O_2 EXTRACTION BY CANINE SKELETAL MUSCLE

S. M. Cain and D. L. Bredle

Department of Physiology and Biophysics
University of Alabama at Birmingham
Birmingham, Alabama 35294, U.S.A.

INTRODUCTION

In order for a tissue to extract O_2 from a diminished supply as efficiently as possible, blood flow must be directed within the tissue proportionately to regional O_2 demand. This is accomplished by a balance of vasoconstrictor and vasodilator forces. If O_2 supply is lowered in the whole body, vasoconstrictor tone is increased by increased activity of sensors such as baroreceptors or peripheral chemoreceptors. At the local tissue level, metabolic signals proportional to the imbalance between O_2 supply and demand cause local dilatation to direct more blood flow to those areas and protect them from hypoxic injury. According to this scheme, any intervention that overrides either regulatory factor should be reflected in a diminished ability of the local tissue to extract O_2 under hypoxic conditions.

We have tested this hypothesis in canine skeletal muscle using a structural analog of dopamine, dopexamine, which, in addition to retaining agonist activity at peripheral dopamine receptors, is also a potent agonist at β_2-adrenoceptors (Brown, et al., 1985). This has been found advantageous in cardiac failure because it lowers peripheral vascular resistance by its β-vasodilator activity while preserving the preferential effects of dopaminergic activity on renal and mesenteric beds. The specific question we have asked is whether such vasodilation interferes to any significant degree with the ability of a peripheral tissue, such as resting skeletal muscle, to extract O_2 under ischemic conditions.

METHODS

Dogs were anesthetized (30 mg/kg pentobarbital sodium iv), paralyzed (30 mg succinylcholine chloride im + 0.1 mg/min iv), and pump-ventilated to maintain arterial PCO_2 ~ 35 Torr. Catheters were placed in carotid and pulmonary arteries and in the right femoral vein. Arterial inflow to the left hindlimb was isolated to the femoral artery by ligating the internal and external iliac and deep circumflex arteries at their origin on the abdominal aorta. Perfusion was maintained from the contralateral femoral artery. To control flow, an occlusive roller pump was interposed in the circuit. Perfusion pressure was measured by a pressure transducer at a t-connector placed in the femoral artery catheter. Initial flow was set approximately 90 ml/min per kg estimated muscle weight, which was close to the autoperfused level that was measured prior to pump-perfusion. Venous outflow was isolated to the left femoral vein by two tourniquets placed at the level of the groin and another at the ankle. Blood flow was measured by a cannulating flow probe and by direct timed collections at the time of sampling. Measurements made with this preparation have been shown to be reliably representative of skeletal muscle (Bredle et al., 1988). Completeness of vascular isolation was assured by observation of no flow if the perfusion pump was stopped. Signs of normally reactive vasculature such as reactive hyperemia following brief occlusion of arterial inflow were always seen both at the beginning and end of each experiment.

After all preparations were complete and monitored variables such as whole body gas exchange, arterial blood pressure, heart rate and arterial blood gases were stable, the experimental protocol was begun. In the first group (CONTROL, n = 6) arterial inflow was raised to approximately 120 ml/min per kg estimated leg muscle weight and held there for 15 min. Arterial and venous samples were slowly drawn during the 12th min and the venous outflow measured immediately thereafter. Flow was then progressively lowered in 8 to 10 steps of 15 min each. In a second group (DOPEXAMINE, n = 6), a continuous infusion of dopexamine hydrochloride (12 μg/min per kg body weight iv) was begun and maintained throughout the experimental protocol.

Total O_2 delivery to the limb was calculated as the product of limb blood flow and arterial O_2 content. Limb O_2 uptake was obtained as the product of blood flow and the arteriovenous difference in blood O_2 content. O_2 extraction ratio was obtained by dividing uptake by delivery and was expressed as a percent. Critical O_2 delivery was found by determining the intersecting point of two straight lines fitted by linear regression to the data describing the "plateau" or supply independent zone of O_2 uptake and to the supply dependent zone in which O_2 uptake was directly dependent on the O_2 delivery. The critical O_2 extraction ratio was the ratio of O_2 delivery and uptake at the critical delivery point. Differences between the two groups were tested by Student's t-test with p<0.05 accepted as significant.

RESULTS

The dopexamine infusion had the expected results of lowering arterial blood pressure and whole body vascular resistance. These effects were maintained in the whole body without change during the experimental period.

In the limb, the effect of dopexamine upon vascular resistance can be readily seen in the pressure-flow relationship plotted in Figure 1. At all levels of blood flow above critical (between 30 and 40 ml/kg-min), pressure was significantly lower in DOPEXAMINE. Below that level, metabolic vasodilation could not be distinguished from any vasodilatory effect of dopexamine.

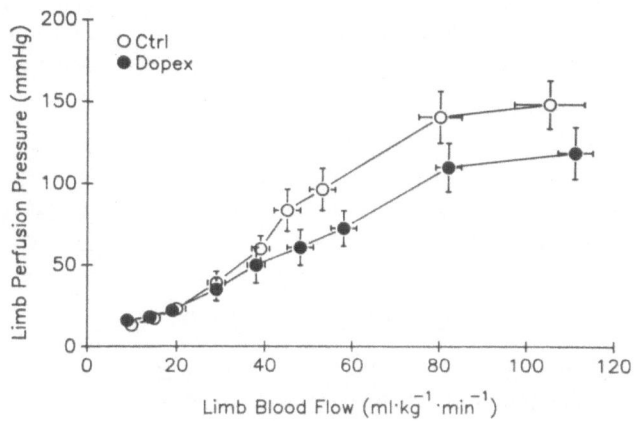

Figure 1. Average values (± SE) for limb perfusion pressure are shown as a function of average blood flow. Dopex and filled symbols represent DOPEXAMINE and Ctrl and unfilled circles represent CONTROL.

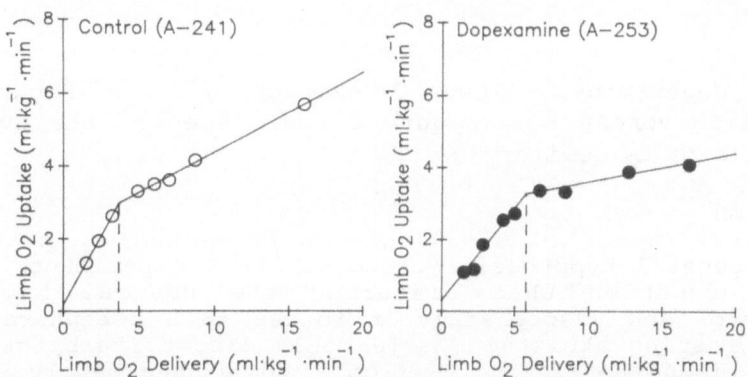

Figure 2. Examples of the 2-line method to obtain critical O_2 delivery. The critical delivery (intercept of dashed line on abscissa) was 3.65 in CONTROL and 5.74 in DOPEXAMINE.

Two examples of the determination of critical O_2 delivery, one from each group, are presented in Figure 2. A positive slope in the "plateau" region was seen in 5 of the 6 CONTROL and

4 of the 6 DOPEXAMINE dogs. In all cases, however, there was a clear demarcation by the more precipitous fall in O_2 uptake with any further lowering of O_2 delivery below the critical point. A similar means was used to arrive at a critical value of PO_2 measured in femoral venous blood at the point that O_2 uptake began to decline more sharply. The critical values are shown in Table 1. Dopexamine raised the critical levels of O_2 delivery and venous PO_2 and lowered the critical O_2 extraction ratio.

TABLE 1. Average values (±SD) for critical O_2 delivery (QO_2), critical O_2 extraction ratio (O_2ER), and critical venous PO_2 (PvO_2).

	CONTROL	DOPEXAMINE
Critical QO_2 (ml/kg-min)	4.01 ±0.82	6.27* ±2.04
Critical O_2ER (%)	81.6 ±4.0	65.4* ±13.5
Critical PvO_2 (Torr)	21.5 ±3.6	32.0* ±5.1

(* denotes a significant difference between groups, $p<0.05$)

With dopexamine, a lower O_2 extraction ratio (Figure 3) and higher limb venous PO_2 (Figure 4) were seen on the average at all levels of O_2 delivery.

DISCUSSION

The general hypothesis tested in these experiments was that any intervention that disturbed the natural balance of vasodilator and vasoconstrictor forces in a peripheral tissue such as resting skeletal muscle would also disturb the ability of that tissue to extract O_2 from a diminished supply. Earlier results had indicated that blocking α-adrenergic tone with phenoxybenzamine caused less complete O_2 extraction in the whole body of anesthetized dogs made hypoxic by ventilation with low O_2 gas mixtures (Cain, 1978). With hypoxia prolonged until the animals began to fail, O_2 extraction in the blocked group was only half that in the unblocked group. On the basis of these results the suggestion was made that efficient O_2 extraction was dependent upon a vigorous and generalized increase in vasoconstrictor tone which was modulated by local need for O_2. When the increase in constrictor tone was prevented, the diminished O_2 supply could not be apportioned according to local need so that organ systems with lower O_2 demand were relatively overperfused. This was tantamount to a functional peripheral shunt and less of the O_2 reserve in venous blood could be utilized after α-adrenergic blockade in hypoxic dogs.

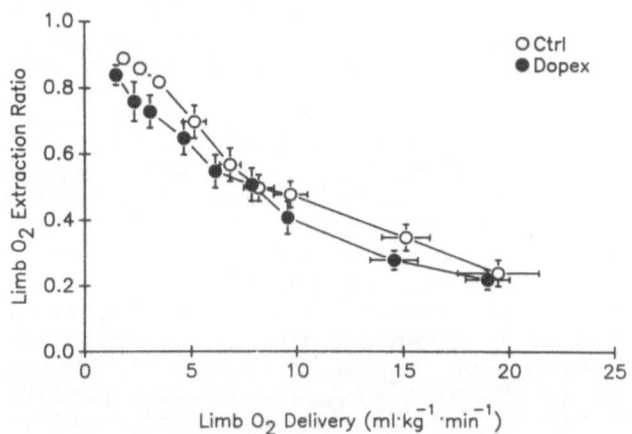

Figure 3. Average values (±SE) of O_2 extraction ratios plotted against average values (±SE) for O_2 delivery for CONTROL (Ctrl, unfilled circles) and DOPEXAMINE (Dopex, filled circles).

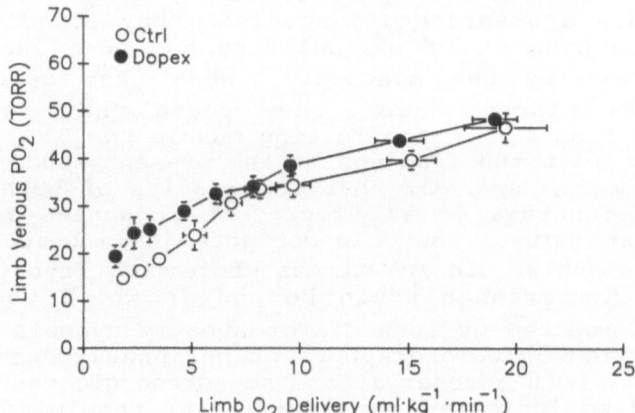

Figure 4. Average values (±SE) of limb venous PO_2 as limb O_2 delivery was decreased in CONTROL (Ctrl, unfilled circles) and DOPEXAMINE (Dopex, filled circles).

In other experiments, the effect of phenoxybenzamine on the specific ability of resting skeletal muscle to extract O_2 was examined. Contrary to the results on whole body O_2 extraction, α-adrenergic blockade did not prevent hindlimb muscles from reaching the same level of extraction as in unblocked hypoxic dogs even though the whole body effect was the same as in the earlier results (Cain and Chapler, 1980). These results did not negate the hypothesis, however, because of the possibility of non-adrenergic vasoconstriction in hypoxia (Rose et al., 1984) and current knowledge that α_2-adrenoceptors contribute to vasoconstrictor tone in skeletal muscle (Chen et al., 1988) and that phenoxybenzamine is a relatively weak blocker of such receptors.

An alternative way to test the hypothesis was employed in the current experiments. ß-vasodilatation is known to occur in skeletal muscle at both resistance vessels (Cain and Chapler, 1979) and at distribution control points (Lundvall and Järhult, 1976; Lundvall and Hillman, 1978). Because of our interest in dopexamine and its possible application to improve peripheral O_2 delivery in various forms of hypoxia, its strong ß-agonist activity was a natural selection. Furthermore, its dopaminergic properties posed no disadvantage for this purpose because skeletal muscle had been shown to be uninfluenced by dopamine during hypoxic hypoxia (Jackson et al., 1982). Our prediction, therefore, was that ß-vasodilation as a result of continuous infusion of dopexamine would cause functional peripheral shunting in resting skeletal muscle. We have demonstrated this by the increased critical O_2 delivery in dopexamine treated animals when O_2 delivery was progressively lowered to the hindlimb. The higher values of femoral venous PO_2 and lower O_2 extraction ratios at the critical point were further evidence that this was the case.

A similar conclusion was reached by Yonekawa et al.(1981) from the results of quite different experiments. They measured tissue O_2 tension in gracilis muscles of anesthetized dogs with a multicathode surface electrode. When they infused isoproterenol, a ß-adrenergic agonist, they found that tissue PO_2 decreased even though O_2 delivery had been increased more than O_2 uptake by the muscle. When they calculated the physiological tissue shunt, they found that isoproterenol increased it from 41% in the resting muscle to 73%. This result is consistent with the findings of the present study. The point needs to be made, however, that the results of both this study and that of Yonekawa et al. represent unopposed increases in vasodilator activity. They did not attempt to limit O_2 delivery to the muscle and we did so without whole body involvement. The more usual circumstance would be one in which whole body O_2 delivery was reduced by ischemia or hypoxic hypoxia with consequent activation of baroreceptors or peripheral chemoreceptors. In that event, both adrenergic and nonadrenergic vasoconstrictor forces would be increased and a different result might thereby be anticipated.

SUMMARY

We have shown that an unopposed stimulation of ß-vasodilator adrenoceptors can cause a decreased ability of skeletal muscle to extract O_2 when O_2 delivery was progressively lowered. The significance of this finding with respect to the application of dopexamine to cardiac failure or to any other condition that causes a decrease in peripheral O_2 delivery, however, needs to be assessed when an appropriate increase in vasoconstrictor tone is allowed to occur.

ACKNOWLEDGEMENT

The authors are grateful for the technical assistance of W. E. Bradley. Funds to support this work were obtained from National Institutes of Health Grant Nrs. HL 26927 and HL 07790 and by a grant from Fisons plc who also supplied the dopexamine.

REFERENCES

Bredle, D. L., C. K. Chapler, and S. M. Cain, 1988, Metabolic and circulatory responses of normoxic skeletal muscle to whole-body hypoxia. J. Appl. Physiol. 65:2063-2068.

Brown, R.A., J. B. Farmer, J. C. Hall, R. G. Humphries, S. E. O'Connor, and G. W. Smith, 1985, The effects of dopexamine on the cardiovascular system of the dog. Br. J. Pharmac. 85:609-619.

Cain, S. M., 1978, Effects of time and vasoconstrictor tone on O_2 extraction during hypoxic hypoxia. J. Appl. Physiol. 45:219-224.

Cain, S. M. and C. K. Chapler, 1979, Oxygen extraction by canine hindlimb during hypoxic hypoxia. J. Appl. Physiol. 46:1023-1028.

Cain, S. M. and C. K. Chapler, 1980, O_2 extraction by canine hindlimb during α-adrenergic blockade and hypoxic hypoxia. J. Appl. Physiol. 48:630-635.

Chen, D. G., X-Z Dai, and R. J. Bache, 1988, Postsynaptic adrenoceptor-mediated vasoconstriction in coronary and femoral vascular beds. Am. J. Physiol. 254 (Heart Circ. Physiol. 23):H984-H992.

Jackson, L. K., B. M. Key, and S. M. Cain, 1982, Total and hindlimb O_2 uptake and blood flow in hypoxic dogs given dopamine. Crit. Care Med. 10:327-331.

Lundvall, J. and J. Hillman, 1978, Noradrenaline evoked beta adrenergic dilatation of precapillary sphincters in skeletal muscle. Acta Physiol. Scand. 102:126-128.

Lundvall, J. and J. Järhult, 1976, Beta adrenergic dilator component of the sympathetic vascular response in skeletal muscle. Acta Physiol. Scand. 96:180-192.

Rose, C. E., Jr., R. L. Godine, Jr., K. Y. Rose, R. J. Anderson, and R. M. Carey, 1984, Role of arginine vasopressin and angiotensin II in cardiovascular responses to combined acute hypoxemia and hypercapnic acidosis in conscious dogs. J. Clin. Invest. 74:321-331.

Yonekawa, H., J. L. Berk, M. R. Neuman, and C. C. Liu, 1981, Tissue hypoxia and increased physiological tissue shunt caused by beta-adrenergic stimulation. Eur. Surg. Res. 13:325-338.

575

RESPIRATORY SYSTEM

PULMONARY CIRCULATION AND SYSTEMIC CIRCULATION:

SIMILAR PROBLEMS, DIFFERENT SOLUTIONS

Leon E. Farhi and Daniel W. Sheehan

Hermann Rahn Laboratory, Department of Physiology

University at Buffalo, Buffalo NY 14214, U.S.A.

The number of sophisticated physical and mathematical systems that have been proposed to simulate the movement of oxygen from the environment to the cell probably equals (and may even exceed!) the number of investigators in this field. In terms of the problem with which we will deal in the next few pages, a much simpler model comes to mind. In this analogy, the integrated O_2 transport to tissue is compared to a tree, in which nutrients are absorbed by finely ramified roots and transported by the sap to a similarly subdivided set of branches through which they reach the delivery site. The two ends of the chain, dissimilar as they may look, share an important requirement, namely the need for proper *functional* distribution: clearly the roots should not be evenly spaced, like the spokes of a wheel, but rather must be more heavily concentrated in the area where more food or water is available; likewise the sap flow must go preferentially to parts that require more nourishment.

The oxygen delivery system mimics the tree in two respects. First of all, it too relies on a pair of fine networks, one at each end, to absorb and release O_2 respectively. More important, the blood flow must also be directed to the terminal branches of each systems not in a constant, uniform fashion, but in a pattern that is responsive to the ever changing local conditions. Since the ability of tissues to maintain aerobic metabolism on the basis of local O_2 reserves is extremely limited -- a matter of seconds for some important organs (Carlisle *et al.*, 1964) -- it is not surprising that oxygen lack, sensed either directly or indirectly, should be one of the crucial factors in determining the distribution of cardiac output. In the pulmonary circulation, blood flow is redirected away from areas that are a poor source of O_2, a diversion that improves both the efficiency of oxygenation and the quality of arterial blood.

Thus, the two circulations share the functional prerequisite of having to meter perfusion to the various terminal branches on the basis of regional oxygen levels, the distribution pattern varying with time as local conditions dictate at that moment. The process of reallocating blood flow creates a number of second order problems, but because of specific constraints, each of the two systems confronts these difficulties in a different way.

CONTROL OF LOCAL P_{O_2} IN THE SYSTEMIC CIRCULATION

In the final analysis, local oxygen tension is determined by the balance between perfusion and O_2 demand. When the latter is high in relation to the former, hypoxia develops and through a number of reflex actions (too numerous and complex to review here) the resistance to blood flow decreases. In the absence of any other

Oxygen Transport to Tissue XII, Edited by J. Piiper *et al.*
Plenum Press, New York, 1990

change, this process increases perfusion, thereby tending to restore the O_2 level. Unfortunately, the efficiency of this set of *local* events -- which constitutes the primary regulatory feedback loop of the system -- is greatly reduced by some *global* effects it produces.

Before we illustrate the importance of these secondary interactions by a numerical example, let us pause briefly to indicate that throughout this article we shall use the word "resistance" only rarely, preferring its inverse, *i.e.* "conductance" to express the relationship between pressure and flow. There are two main advantages to this approach, namely that 1) conductance is directionally related to vessel diameter, and 2) total conductance is the sum of the conductances of the parallel circuits.

Table 1 deals with a hypothetical subject with a cardiac output of 6000 mls per minute, and a total conductance of 60 ml.min-1.mm Hg-1, equally divided between a segment X and the remainder of the circulation. To provide the total blood flow, the individual has to develop a mean arterial pressure of 100 mm Hg. As shown in column A, the perfusion of segment X, given by the product of driving pressure and conductance, will be 100 x 30, that is 3000 ml.min-1. Assume now that a regional mechanism causes extensive vasodilation in X, increasing the conductance threefold (column B). In the absence of any other change, total conductance will rise to 120 ml.min-1.mm Hg-1, and the arterial pressure will decrease to 50 mm Hg, providing a blood flow of 4500 ml.min-1 to our test zone. In this example, an increase in conductance up to 300% of the initial value has boosted the regional perfusion, but only to 150% of control.

TABLE 1. VASODILATION AND INCREASE IN BLOOD FLOW

VARIABLE	A: CONTROL	B: VASODILATION
Cardiac output, ml. min-1	6,000	6,000
Conductance of area X, ml. min-1. mm Hg-1	30	90
Conductance, other beds, ml. min-1. mm Hg-1	30	30
Total conductance, ml. min-1. mm Hg-1	60	120
Mean arterial pressure, mm Hg	100	50
Blood flow of area X, ml. min-1	3,000	4,500

Table 1 emphasizes that local perfusion is dictated not only by regional conductance but also, to an equal extent, by arterial pressure. Since this variable is governed by the combination of cardiac output and total conductance, if total blood flow and the caliber of vessels in the other beds are maintained, local vasodilation cannot produce an equivalent increase in perfusion. But just as the half-empty glass may also be described as half-full, the preceding observation can be viewed as stating that arterial pressure can indeed be safeguarded (thereby restoring the effectiveness of regional arteriolar relaxation) if either conductance of the rest of the circuit or cardiac output is reset appropriately.

In nature, both readjustments occur. The circulatory resistance of the vascular beds of less active organs can change to the point where the blood flow is severely reduced. As is well known, in strenuous physical exercise, perfusion of the renal and splanchnic beds is curtailed, and blood flow is allocated practically *in toto* to the active muscles, myocardium, brain and skin. Although the drop in conductance elsewhere does not make up for its increase in these beds, it does limit the rise in total conductance.

Far more important is the rise in cardiac output, which is such that it can maintain arterial pressure in physiological situations, or even increase it, as occurs in maximum exercise. Indeed, the cardiovascular response demonstrates unequivocally that defense of arterial pressure takes precedence over maintenance of cardiac output.

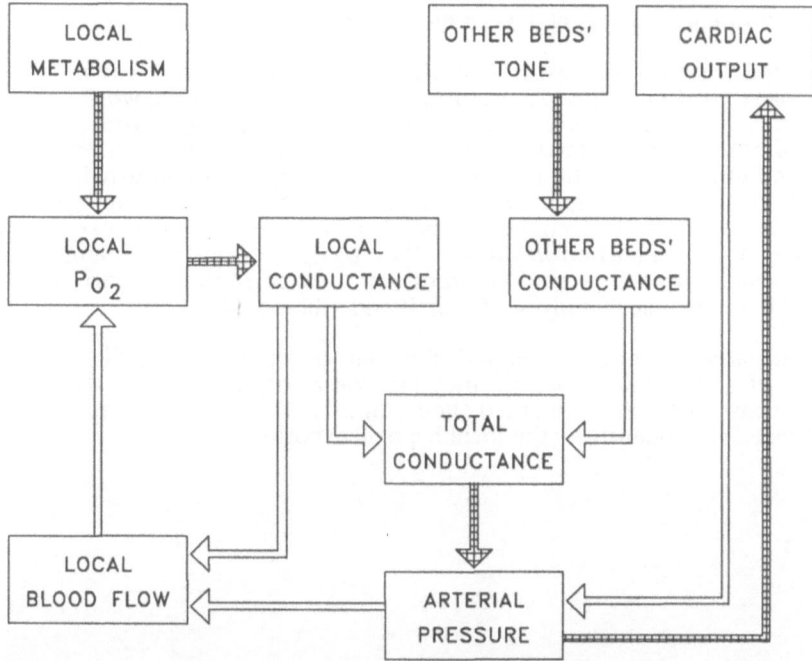

FIG. 1. Interactions in hypoxic control of peripheral circulation. Clear arrows indicate positive relationship, hatched ones denote negative interaction. For details, see text.

The various relationships we have sketched so far and the way in which they interact are shown in Figure 1, on the left edge of which one can see the local P_{O_2} and the two factors that determine it. A first control loop, P_{O_2} to "local conductance" to "local blood flow" and back to the local P_{O_2}, constitutes the primary control negative feedback loop. Outside this loop, there is another one, which takes into account the total conductance and its effects on arterial pressure. As discussed earlier, this second series of interactions limits the efficiency of the primary control system. Finally, at the top right are the two factors which are brought into play to minimize or cancel changes in conductance and blood pressure.

In summary, local vasodilation can increase blood flow substantially only if a concomitant change in arterial pressure can be avoided. This is usually achieved through the combination of vasoconstriction in other areas and rise in cardiac output, which is usually so well controlled that it allows a sizable portion of the vascular bed to increase its perfusion by more than an order of magnitude.

CONTROL OF LOCAL P_{O_2} IN THE PULMONARY CIRCULATION

The four decades that have elapsed since the first report of hypoxic vasoconstriction by von Euler and Liljestrand (1946), have seen publication of such an outpouring of excellent papers (as a result of which we have learnt much about the modalities of this response) that even a cursory review of the literature exceeds the scope of this article. It is however only fitting to mention that our own contribution --

to which we shall turn in a moment -- owes a great debt to the laboratories of Marshall in Philadelphia, Robinson in Auckland and of Sylvester at Hopkins, on whose approach and results we have based some of our work. Because our conclusions indicate that some of the literature data must be used only with great caution, we shall take a few moments to describe those results.

Pulmonary vasomotor effects of oxygen and carbon dioxide

In comparing the laboratory conditions with those that occur in real life, it became clear to us that either the equipment available when the work was done, or the experimental design prevented most investigators from performing studies that closely replicate what happens in nature. In particular, many data were obtained from isolated lungs, an excellent preparation, but one that obviously lacks many of the features of the *in situ* organ. Other authors reported experiments performed on anesthetized and surgically prepared animals, or in models in which a sizable portion of the pulmonary vascular bed was tested, thereby decreasing the amount of normoxic tissue to which blood flow could be redirected, an extremely important factor, as pointed out in the classical study of Marshall and others (1981).

The technique we have developed (Sheehan *et al.*, 1988), like that described by Robinson *et al.* (1978) allows one to study hypoxic vasoconstriction in the right apical lobe of the conscious uninstrumented sheep. In this preparation, arterial blood gases are not affected significantly by the local hypoxic exposure.

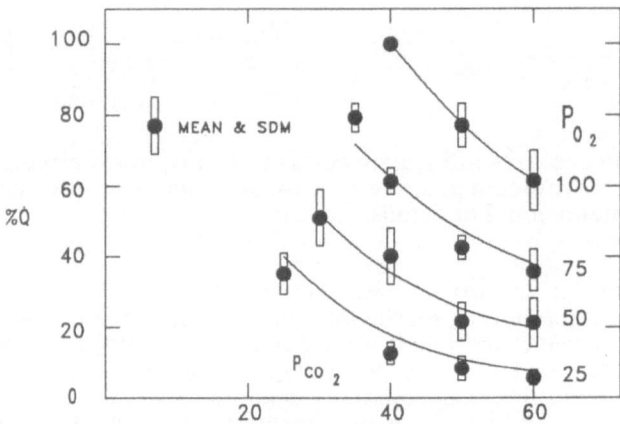

FIG.2. Effect of regional carbon dioxide level (in torr) on regional blood flow in the right apical lobe of the sheep (as % of control).

In some respects, our results only confirm what was already known, namely that a drop in regional alveolar P_{O_2} causes a decrease in local blood flow. More important is the fact that, contrary to the belief held by most, carbon dioxide affects significantly the response to hypoxia (Sheehan and Farhi, 1989). This is shown in Figure 2, where the perfusion of the right apical lobe (expressed as a percentage of the total pulmonary blood flow and normalized to 100% for $P_{A O_2} = 100$ and $P_{A CO_2} = 40$), is plotted against $P_{A CO_2}$. Each of the four lines corresponds to the $P_{A O_2}$ indicated to its right. The top curve, obtained at a normal oxygen level, shows that hypercapnia *per se* has a substantial vasoconstrictor effect: local blood flow drops by about 40% when $P_{A CO_2}$ increases to 60 Torr. At hypoxic levels, CO_2 causes similar changes.

The information from Figure 2 is perhaps more striking when it is presented on a three-dimensional plot, as done in Figure 3. In this representation, the two horizontal axes are P_{O_2} and P_{CO_2} respectively, and the right apical lobe perfusion is on the vertical axis. It is quite obvious that the decrease in this variable is as marked when carbon dioxide rises (left to right) as when oxygen falls (back to front). Of major importance is the fact that whereas P_{O_2} plays a role only when it drops below control value (there is no evidence of a significant increase in flow in local hyperoxia), both hypercapnia and hypocapnia play a role in setting vascular tone, such that a P_{CO_2} change in either direction can alter considerably the hypoxic response.

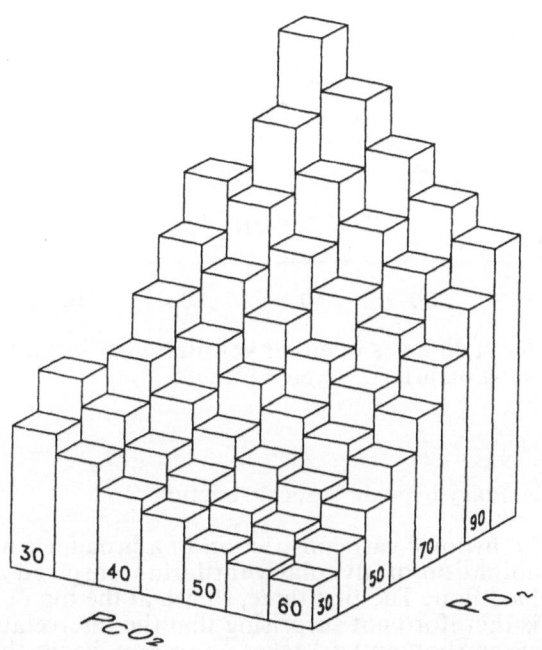

FIG. 3. Effects of local oxygen and carbon dioxide on regional lung perfusion.

The role of CO_2 must be taken into account when applying experimental findings to patients. In their case, regional alveolar hypoxia is caused by local hypoventilation and is, therefore, accompanied by a variable degree of hypercapnia. In the laboratory, where it is impossible to obtain a quantitated, reproducible decrement in ventilation-perfusion ratio, the traditional method for producing regional hypoxia has been to lower the inspired oxygen fraction in a certain area of the lung. The resulting vasoconstriction decreases the CO_2 supply and leads to hypocapnia which partly offsets the vasoconstrictor stimulus. Thus, regional hypoxic vasoconstriction may be expected to be much more pronounced in real life than that reported by several authors.

A powerful vasomotor response will affect arterial blood P_{O_2} in two different ways. First of all, the fraction of the total flow returning from hypoxic areas will decrease. In addition, reduction of perfusion to an underventilated zone will raise its ventilation-perfusion ratio and oxygenation of the blood returning from this region will be more adequate. In the final analysis, arterial P_{O_2} can be expected to be maintained better in real life than in most experimental models. This difference is shown in Figure 4, where PaO_2 is plotted against the reduction of either ventilation or inspired O_2, both expressed as percent of control. The figure is calculated for a hypoxic area covering 15% of the total lung, and shows that a drop in ventilation has much less pronounced effects than a decrease in inspired O_2.

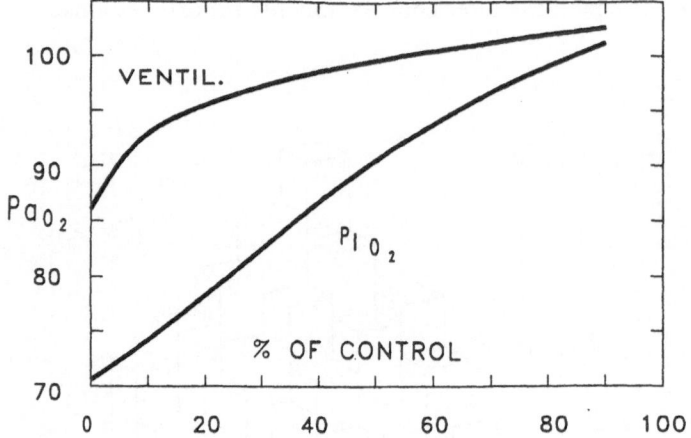

FIG.4. Effect of changes in either ventilation or inspired oxygen in 15% of the lung on arterial oxygen tension.

Interrelations in pulmonary hypoxic vasoconstriction

Figure 5 puts the hypoxic vasoconstriction in a broader context. Local PO_2 is dictated by the combination of alveolar ventilation, inspired and mixed venous oxygen levels, and blood flow. The first three, shown at the top of the figure, provide O_2 to the region; it is therefore not surprising that the interrelation is positive. On the other hand, perfusion (bottom left) takes up oxygen, hence the negative interaction. The next two steps are very close to those we have seen when dealing with the systemic circulation. A primary control loop, PO_2 to conductance, to blood flow and back to PO_2, differs from the one that governs peripheral vessel tone only in the sign of the relationship between alveolar oxygen and conductance. Here also, we must consider changes in local conductance which will affect total conductance and hence pulmonary arterial pressure, creating a second loop that will minimize the effectiveness of the main control path.

The similarity between the two circulations stops here. Whereas the reduction in systemic arterial pressure is normally prevented very efficiently by a combination of readjustment in cardiac output and active vasoconstriction in other areas, the increase in pulmonary arterial pressure caused by the conductance drop in hypoxic lung areas can avail itself of neither of these two mechanisms: total lung perfusion is dictated by systemic needs, and the extremely low resting tone of the pulmonary vasculature leaves little room for effects of relaxation. The only relief from an increase in pressure is provided by a combination of distension and recruitment in the pulmonary vascular bed. That is not an altogether satisfactory solution, since the passive nature of the readjustment implies that pulmonary arterial pressure must increase in order to produce the expansion. In spite of the high distensibility of the vasculature involved, the pressure rise can be significant if both the area affected and the vasoconstriction are of major proportions. Thus, the need to protect the lung against the potentially deleterious effects of hypertension limits -- at times, very severely -- the ability of the lung to redistribute blood flow away from hypoxic areas.

584

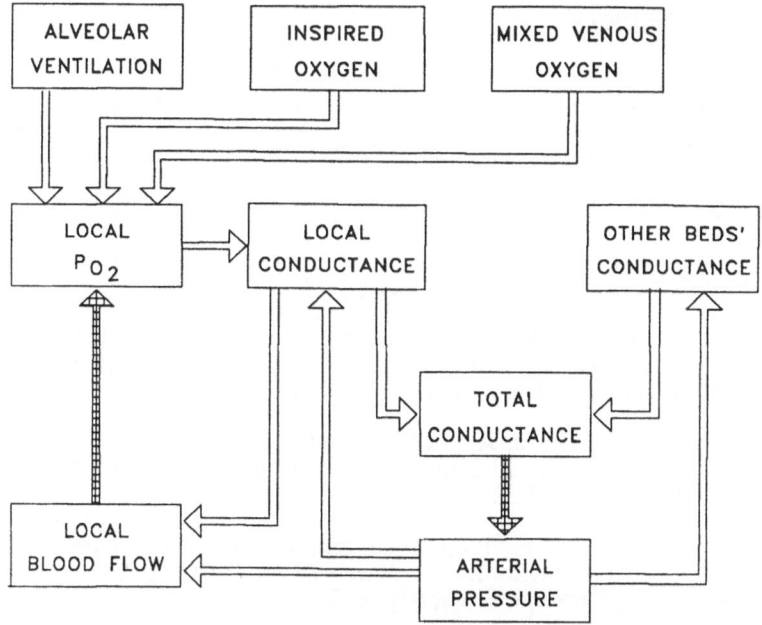

FIG. 5. Interactions in hypoxic control of the pulmonary circulation. Clear arrows indicate positive relationship, hatched arrows denote negative interaction.

At this point, it is interesting to recall that several papers from Sylvester's laboratory have described a paradoxic response to hypoxia (Sylvester *et al.*,1980), in which, at P_{AO_2} levels lower than mixed venous P_{O_2}, a further drop in alveolar oxygen tension actually increases blood flow. Returning to Figure 5, we must note that the sign of some relationships is correct only when P_{AO_2} exceeds P_{vO_2}. When the opposite is true, the circulation provides oxygen to the region, and ventilation removes it. Under these conditions, the main regulatory loop would turn into a positive feedback mechanism: a drop in alveolar oxygen would lower blood flow, further reducing P_{O_2}. The paradoxic response, by changing the direction of the relationship between alveolar oxygen and conductance becomes a stabilizing factor that protects pulmonary arterial pressure at the expense of vasoconstriction and arterial blood gas levels.

SUMMARY

Both the systemic and the pulmonary circulations respond to local hypoxia in the appropriate manner, the former by vasodilating, thereby providing more oxygen, and the latter by constricting and rerouting blood flow to areas where more O_2 is available. In either case, changes in local conductance affect total conductance, and through that variable, the perfusing pressure; as a result, the effects of local vasomotion should be reduced. In the systemic circulation, arterial pressure can be prevented from falling by two important mechanisms: vasoconstriction of other vascular beds, and an increase in cardiac output. There are no similar means for protecting pulmonary arterial pressure against a rise when vessels in hypoxic areas contract; the only defense is provided by passive expansion of the vascular bed. Thus, in the lung regional circulatory readjustments conflict with the need to maintain a reasonably low pulmonary arterial pressure and local regulation (and maintenance of arterial oxygenation) may be subordinate to prevention of pulmonary hypertension.

ACKNOWLEDGEMENTS

The authors are indebted to Misses Margaret Brick and Lisa Brownschidle and Mr. Christopher Eisenhardt for their invaluable technical assistance.

Supported in part by NIH Grant # 5 PO1 HL 34323.

REFERENCES

Carlisle, R., Lanphier, E. H., and Rahn, H., 1964, Hyperbaric oxygen and persistence of vision in retinal ischemia, *J. Appl. Physiol.,* 19:914-918

Marshall, B. E., Marshall, C., Benumof, J., and Saidman, L. J., 1981, Hypoxic pulmonary vasoconstriction in dogs: effects of lung segment size and oxygen tension, *J. Appl. Physiol.: Resp., Environ. Exercise Physiol.,* 51:1543-1551

Robinson, S. M., Cadwallader, J. A., and McN. Hill, P. (1978), An animal model for the study of regional lung function, *J. Appl. Physiol.: Resp., Environ. Exercise Physiol.,* 42:32-324

Sheehan, D. W., and Farhi, L. E., 1989, Local pulmonary vasoconstrictor response at different oxygenation levels. *Proceedings of the XXXI Internatl. Cong. Physiol. Sci.,* Helsinki, Finland, p. 443

Sheehan, D. W., Klocke, R. A., and Farhi, L. E., 1988, Non-invasive on-line measurement of regional pulmonary hypoxic vasoconstriction in the conscious animal, *Physiologist,* 31:A195

Sylvester, J. T., Harabin, A. L., Peake, M. D., and Frank, R. S., 1980, Vasodilator and vasoconstrictor responses to hypoxia in isolated pig lungs, *J. Appl. Physiol.: Respir., Environ., Exercise Physiol.*, 49:820-825

von Euler, U. S., and Liljestrand, G., 1946, Observations on the pulmonary arterial pressure in the cat, *Acta Physiol. Scan.,* 12:301-320

BLOOD-GAS TRANSFER OF O₂ AND CO₂ IN THE LUNGS: NEW MODELS

MEASUREMENTS AND CONCLUSIONS

Masaji Mochizuki

Geriatric Respiratory Research Center
Nishimaruyama Hospital
064 Sapporo/Chuo-Ku, Japan

INTRODUCTION

Three simultaneous differential equations for O_2, CO_2 and HCO_3^- diffusion in the red blood cell (RBC)were solved numerically, taking the Bohr and Haldane effects into account (Mochizuki and Kagawa, 1986). Then, from the numerical solution the relationship of the gas exchange ratio (R) to alveolar PCO_2 ($PACO_2$) during rebreathing was derived (Kagawa and Mochizuki, 1987). R in rebreathing air is linearly related to $PACO_2$. Since CO_2 diffusion in the RBC accompanies HCO_3^- shift and partly results from the Haldane effect, the CO_2 reactions are generally slower than the O_2 reactions. Therefore, the slope of R–CO_2 line (θ) depends not only on the true venous PCO_2 ($trPvCO_2$) and arterial-venous O_2 content difference ($C(a-v)O_2$), but also on the contact time (tc). Using the theoretical equation for the R–PCO_2 line as a gas exchange model, the relationship between tc and $C(a-v)O_2$ is obtained from the experimental data of rebreathing in normal subject (Shibuya et al, 1987).

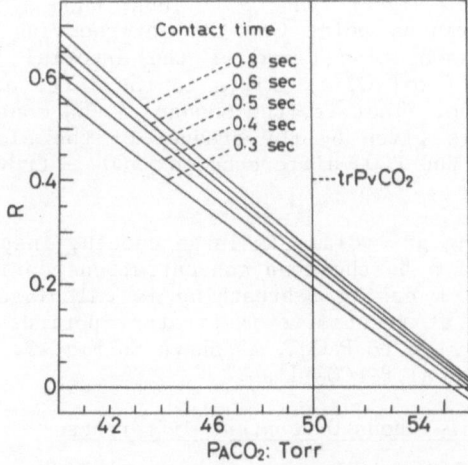

Fig. 1. R–PCO_2 lines at five different contact times ranging from 0.3 to 0.8 sec. Dividing the computed change in CO_2 content by that in O_2 content, R was obtained over the contact time.

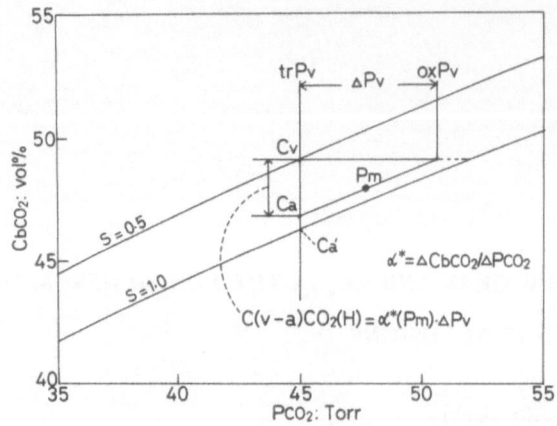

Fig. 2. Schematic illustration of the CO_2 dissociation curves showing the relationship between the Haldane effect component of venoarterial CO_2 difference ($C(v-a)CO_2(H)$), $trPvCO_2$ and $oxPvCO_2$. Pm is the middle between tr- and $oxPvCO_2$. Venous and arterial O_2 saturations are taken to be 0.5 and 1.0, and $trPvCO_2$ is 45 Torr.

RELATIONSHIP BETWEEN GAS EXCHANGE RATIO AND ALVEOLAR PCO_2

1) Linear relationship between R and $PACO_2$

The $R-PCO_2$ line obtained by dividing the computed change in CO_2 content by that in O_2 content is shown in Fig. 1, where the parameter is the contact time (Kagawa and Mochizuki, 1987). The $R-PCO_2$ line shifts downwards as tc is shortened, showing a decrease in R at $trPvCO_2$, i.e. Haldane effect coefficient (HEC). Figure 2 shows the CO_2 dissociation curves at two O_2 saturation (SO_2) levels, 0.5 and 1.0. If venous SO_2 rises from 0.5 to 1.0 in the lung capillary at $trPvCO_2$ of 45 Torr, the CO_2 content has hitherto been expected to decrease from the point, Cv, to the point, Ca', which is located on the CO_2 dissociation curve of SO_2 = 1.0. However, the numerical solution indicates that the arterial CO_2 content, Ca, becomes higher than Ca'. Thus, when a horizontal line is drawn through the venous point Cv, the intersecting point of this line with the CO_2 dissociation curve of the arterial blood, namely, oxygenated venous PCO_2 ($oxPvCO_2$) shifts to the left, as shown by the extrapolated dotted line. That is, the change in CO_2 content due to the Haldane effect which is given by the product of the slope of the CO_2 dissociation curve and the PCO_2 difference, $oxPvCO_2$ - $trPvCO_2$, decreases as tc is shortened.

When a rebreathing gas volume is large enough, inspiratory O_2 and CO_2 levels are assumed to be the mean concentrations during the foregoing expiration. Thus, R during rebreathing is calculated from the O_2 and CO_2 concentrations at successive inspiratory periods. The R value is usually linearly related to $PACO_2$, as shown in Fig. 3, supporting the validity of the theoretical $R-PCO_2$ line.

2) Equation for arterial-venous O_2 content difference

Let α^* be the slope of the CO_2 dissociation curve at the middle (Pm) between tr- and $oxPvCO_2$, and ΔPv be ($oxPvCO_2$ - $trPvCO_2$). Then, the following equation is derived for the Haldane effect component of the venous-arterial CO_2 content difference ($C(v-a)CO_2(H)$):

588

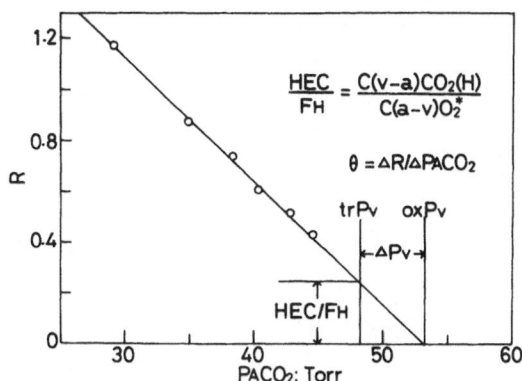

$$\frac{HEC}{FH} = \frac{C(v-a)CO_2(H)}{C(a-v)O_2^*}$$

$$\theta = \Delta R/\Delta PACO_2$$

Fig. 3. Relationship between the slope of a R–PCO$_2$ line and the Haldane effect coefficient (HEC). The product of θ and (oxPv – trPv) gives HEC/FH, where FH is a factor correcting the experimental θ value.

$$C(v-a)CO_2(H) = \alpha^*(Pm)\cdot\Delta Pv. \tag{1}$$

As shown in Fig. 3, the ratio of $C(v-a)CO_2(H)$ to $C(a-v)O_2$, i.e. HEC is approximately given by the product, ΔPv and θ. To be exact, however, HEC receives the influences of tc and $C(a-v)O_2$, and further, $C(a-v)O_2$ is influenced by the Bohr effect. Let $C(a-v)O_2^*$ be the standard $C(a-v)O_2$, where intracellular pH in venous blood is equal to that in arterial blood. Then, from the numerical solution, the ratio, $C(a-v)O_2/C(a-v)O_2^*$ (FavO$_2$), is expressed by a function of PACO$_2$ and $C(a-v)O_2^*$ (Kagawa and Mochizuki, 1987). Furthermore, the relationship between HEC, θ, tc, and $C(a-v)O_2^*$ is derived using a correcting factor, FH, so that HEC can be expressed as follows:

$$HEC = C(v-a)CO_2(H)/C(a-v)O_2^* = \theta\cdot\Delta Pv/FH. \tag{2}$$

Eliminating $C(a-v)CO_2$ and ΔPv from Eqs. (1) and (2), $C(a-v)O_2^*$ is given by

$$C(a-v)O_2^* = FH\cdot\alpha^*(Pm)/\theta. \tag{3a}$$

FH is a function of $C(a-v)O_2^*$ and tc, but since it is close to unity, and the influence of an error in FH to tc is permissible, it can be given by a function of $C(a-v)O_2^*$ alone, using the experimental relation between tc and $C(a-v)O_2^*$ in normal subject. As a result, an equation for $C(a-v)O_2^*$ is derived as follows:

$$C(a-v)O_2^* = 1.09\cdot\{\alpha^*(Pm)/\theta\}^{0.9488}, \text{ (vol\%)}. \tag{3b}$$

Setting θ and $\alpha^*(Pm)$ in Eq. (3b), $C(a-v)O_2^*$ is estimated.

3) Contact time equation

When $C(a-v)O_2^*$ is known, the PCO$_2$-dependent component of the veno-arterial CO$_2$ content difference ($C(v-a)CO_2(P)$) is obtained from the difference in R, Ra – HEC, where Ra is the gas exchange ratio at any PACO$_2$. $C(a-v)CO_2(P)$ equals (Ra – HEC)$\cdot C(a-v)O_2$, and further, $C(a-v)O_2$ is given by FavO$_2\cdot C(a-v)O_2^*$. Since trPvCO$_2$ – PACO$_2$ = (Ra – HEC)/θ, the quotient, $C(v-a)CO_2(P)/(trPvCO_2 - PACO_2)$ is given by $\theta\cdot FavO_2\cdot C(a-v)O_2^*$. From the numerical solution, on the other hand, this quotient is expressed by a hyperbolic function of tc as follows:

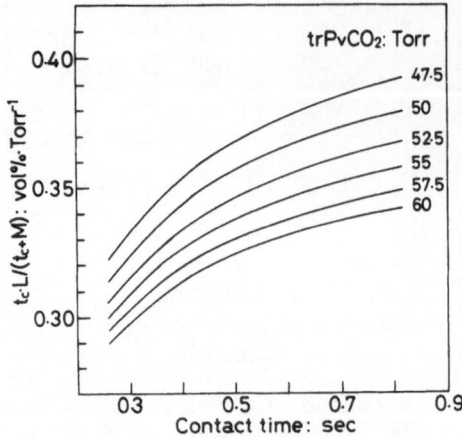

Fig. 4. The quotient, $tc \cdot L/(tc + M)$ shown in Eq. (4) plotted against tc at various $trPvCO_2$ in a range of 47.5 to 60 Torr.

$$\theta \cdot FavO_2 \cdot C(a-v)O_2{}^* - \alpha p' = tc \cdot L/\{Fp \cdot (tc + M)\}, \qquad (4)$$

where $\alpha p'$ is CO_2 solubility in plasma fraction, 0.0377 vol%/Torr, and Fp is a factor correcting linearity of the measured R-PCO_2 line according to curvature of the theoretical R-PCO_2 line. L (vol%/Torr) and M (sec) in Eq.(4) are functions of $trPvCO_2$, as shown in Fig. 4. Fp is approximated to 1.04, when $trPvCO_2$ - $PACO_2$ is about 10 Torr. Ultimately, from the numerical solutions, the relationship between θ, $trPvCO_2$, tc and $C(a-v)O_2{}^*$ is derived as follows:

$$\theta = \{tc/(A + B \cdot tc)\}\{0.405/(0.257 + C(a-v)O_2{}^*)\}, \qquad (5a)$$

where $\quad A = 0.0694 - 0.32 \cdot 10^{-3} \cdot (trPvCO_2 - 60), \qquad (5b)$

and $\quad B = 1 + 0.854 \cdot 10^{-2} \cdot (trPvCO_2 - 60) - 0.18 \cdot 10^{-3} \cdot (trPvCO_2 - 60)^2. \quad (5c)$

RELATIONSHIP BETWEEN tc AND $C(a-v)O_2{}^*$ IN NORMAL SUBJECT

$\alpha^*(Pm)$ in Eq. (3b) is estimated from the mean CO_2 dissociation curve in normal subject (Tazawa et al., 1983) as follows:

$$\alpha^*(Pm) = 3.962 \cdot Pm^{-0.5857}, \text{ (vol%/Torr)}. \qquad (6)$$

Further, since the influence of the error in $\alpha^*(Pm)$ to $C(a-v)O_2{}^*$ is negligible compared with that of θ and $oxPvCO_2$, ΔPv can be approximated to the following equation, using the relationship between $C(a-v)O_2{}^*$ and tc in normal subject as follows:

$$\Delta Pv = 0.433 \cdot \theta^{-0.8}, \text{ (Torr)}. \qquad (7)$$

Since $Pm = oxPv - \Delta Pv/2$, $trPvCO_2$, $\alpha^*(Pm)$ and $C(a-v)O^*$ are obtained from the measured θ and $oxPvCO_2$ values through Eqs. (7), (6) and (3b), respectively. Thus, the relation between tc and $C(a-v)O_2{}^*$ is easily calculated from Eqs. (5a, b and c). tc is almost independent of $trPvCO_2$ in a range of 45 to 60 Torr. Figure 5 shows the relationship between tc and $C(a-v)O_2{}^*$ in normal subject, where the regression line is given by

$$1/tc = -0.384 + 0.352 \cdot C(a-v)O_2{}^* - 0.009 \cdot C(a-v)O_2{}^{*2}. \qquad (8)$$

590

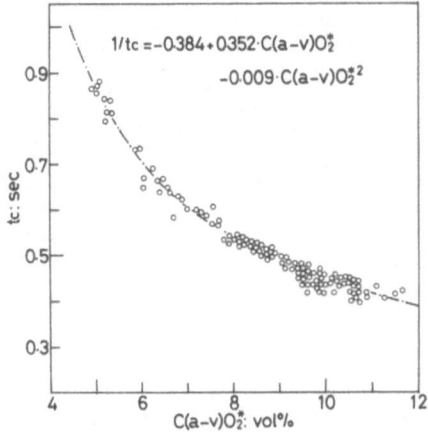

Fig. 5. tc plotted against $C(a-v)O_2^*$ obtained from the $R-PCO_2$ line in normal subject at rest and during exercise.

When the O_2 uptake rate ($\dot{V}O_2$: 1/min) is measured together with the $R-PCO_2$ line, the cardiac output ($\dot{Q}$: 1/min) is estimated by dividing $\dot{V}O_2$ by $C(a-v)O_2^*$. In the measured subjects, $\dot{Q}$ was linearly related to $\dot{V}O_2$ as shown by

$$\dot{Q} = 3.5 + 7 \cdot \dot{V}O_2. \tag{9}$$

Figure 6 shows the relationship between tc and $\dot{Q}$ obtained in 3 normal subjects. The plotted points and arrows are the mean and SD of 5 to 6 measurements at rest and at 5 different exercise levels, respectively. At rest, $\dot{Q}$ was about 6.3 1/min and tc was in a range of 0.6 to 0.9 sec. With a light load of exercise, tc decreased below 0.6 sec. As the exercise level is increased, tc is shortened. Over against the increase in $\dot{Q}$ from 8 to 16 1/min, tc decreases from 0.6 to 0.4 sec. Since dependency of tc on $C(a-v)O_2^*$ attenuates with increasing $C(a-v)O_2^*$ (Fig. 5) and further, $C(a-v)O_2^*$ increases with an increase in $\dot{Q}$, the error in tc diminishes as $\dot{Q}$ is increased by exercise. The SD value of Fig. 6 apparently indicates the above relationship between the extent of error in tc and the $\dot{Q}$ value.

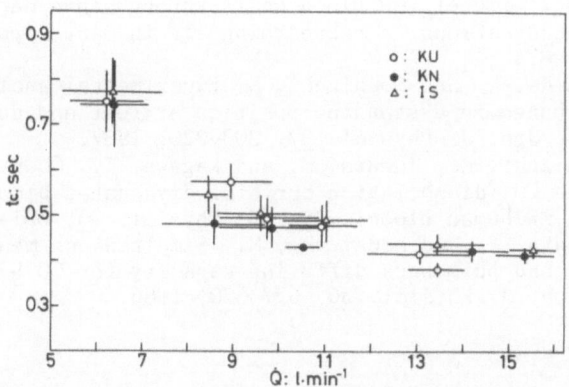

Fig. 6. Relationships between $\dot{Q}$ and tc obtained from the $R-PCO_2$ line during rebreathing. The plotted points and arrows are the mean and SD of 5 to 6 measured values at rest and at 5 different exercise levels in normal subjects. $trPvCO_2$ is within a range between 45 to 65 Torr.

When CO concentration in rebreathing air is measured together with O_2 and CO_2, the pulmonary diffusing capacity (DLCO: ml/(min Torr)) and $\dot{Q}$ are calculated from the same rebreathing experiment (Uchida et al. 1986). DLCO is expressed by the product of the flow rate of RBC through the lung per min, the reaction rate factor of CO with the RBC (FcCO: ml/(sec Torr)) and tc. Since FcCO is already known in the human RBC (Fukui and Mochizuki, 1972), tc can be estimated from DLCO and $\dot{Q}$. Further, since tc is obtained from θ and $trPvCO_2$ using Eqs. (5a, b and c), two tc values are compared to each other in the same subject. Both the tc values obtained in 5 normal subjects agree well with each other, verifying the validity of using the numerical solution as the gas exchange model (Mochizuki et al., 1987).

CONCLUSIONS

1) From the numertical solution of overall O_2 and CO_2 diffusion in the RBC, the relationship between θ, $trPvCO_2$, $C(a-v)O_2{}^*$ and tc was derived. Using the relation as a mathematical model, the relationship between tc and $C(a-v)O_2{}^*$ was estimated from the R-PCO_2 line during rebreathing in 5 normal subjects.

2) The tc value was almost inversely related to $C(a-v)O_2{}^*$.

3) The tc value at rest was distributed in a range of 0.6 to 0.9 sec, and decreased below 0.6 sec, when a load of exercise was given even though slightly. Over against the increase in $\dot{Q}$ from 8 to 16 1/min, tc decreased from about 0.6 to 0.4 sec.

4) Above all, CO_2 diffusion in the RBC is so slow that tc is estimated.

REFERENCES

Fukui, K. and Mochizuki, M. Some basic problems on the pulmonary diffusing capacity for carbon monoxide. I. The reaction rate of CO with oxygenated hemoglobin in the red cell.. Monogr. Ser. Res. Inst. Appl. Electr. Hokkaido Univ. 20, 69-78, 1972.

Kagawa, T. and Mochizuki, M. Theoretical analyses for arterial-venous O_2 content difference and Haldane effect during rebreathing. Jpn. J. Physiol., 37, 267-282, 1987.

Mochizuki, M., Shibuya, I., Uchida, K. and Kagawa, T. A method for estimating contact time of red blood cells through lung capillary from O_2 and CO_2 concentrations in rebreathing air in man. Jpn. J. Physiol. 37, 283-301, 1987.

Shibuya, I., Uchida, K. and Mochizuki, M. Experimental analyses of pulmonary gas exchange in a standing position at rest and during treadmill exercise. Jpn. J. Physiol. 37, 303-320, 1987.

Tazawa, H., Mochizuki, M., Tamura, M. and Kagawa, T. Quantitative analyses of the CO_2 dissociation curve of oxygenated blood and the Haldane effect in human blood. Jpn. J. Physiol. 33, 601-618, 1983.

Uchida, K., Shibuya, I. and Mochizuki, M. Simultaneous measurement of cardiac output and pulmonary diffusing capacity for CO by a rebreathing method. Jpn. J. Physiol. 36, 657-670, 1986.

BLOOD CO_2 AND pH TRANSIENTS DURING APNOEA AFTER O_2 BREATHING IN PATIENTS

Friedrich Mertzlufft and Ludwig Brandt

Department of Anaesthesiology, University Hospital
Johannes Gutenberg University Mainz
D-6500 Mainz, F.R.G.

INTRODUCTION

Endotracheal intubation always is combined with an apnoea the duration of which is dependent on the technique used, the skills of the intubator, and the anatomical situation. The resulting typical potential risks may be (among others) both hypoxaemia and hypercapnia. Therefore the tolerable apnoea time for an intubation procedure is limited in clinical practice to 1 - 2 minutes. It must be noted that the developing hypercapnia is inevitable, whereas hypoxaemia may be avoided even in prolonged apnoea (e.g. >2 min) with "adaequate" preoxygenation [Duda et al., 1988]. Using de-nitrogenation techniques (breathing pure oxygen for 30 to 60 min) "anaesthetized and curarized normal subjects tolerate total apnoea for up to 55 min" [Siggaard-Andersen, 1974]. According to Mertzlufft et al. (1987), $paCO_2$ does not increase linearly during a short-termed (3 min) apnoea. A fast initial increase in the first minute is followed by a slower increase during the ensuing minutes. This fact inspired previous investigations on arterial and mixed-venous acid-base and oxygen status [Brandt et al., 1987]. It could be demonstrated that, in contrast to $paCO_2$, the development of $p\bar{v}CO_2$ was almost linear from the onset of apnoea. Furthermore, in 1987 Mertzlufft et al. and in 1988 Brandt et al. showed that after the first minute of hyperoxic apnoea $paCO_2$ exceeds $p\bar{v}CO_2$. The Haldane effect (for the first time described by Christiansen, Douglas and Haldane in 1914) was found to explain the "pCO_2 reversal" and the accompanying "pH reversal". The Haldane effect describes the reduced CO_2 binding capacity of oxygenated haemoglobin (O_2Hb) compared to deoxygenated haemoglobin (Hb) on the basis of its increased acidity. Because hyperoxic apnoea can be looked at as a closed system for CO_2 (lack of CO_2 elimi- nation by respiration) with ongoing uptake of oxygen (high alveolar pO_2 due to preoxygenation) it seems to be a well suited physiological model for clinical studies on the Haldane effect [Brandt et al., 1989].
The aim of this study was to describe the time course of O_2 and CO_2 transport and acid-base balance during the first two minutes of a hyperoxic intubation apnoea.

METHODS

After institutional approval 12 patients scheduled for cardiac surgery were studied with written informed consent. The mean age of the ten male and two female patients was 60.5 ± 7.4 years (here and in the following text: mean $\pm$ SD).

Monitoring

The monitoring methods include invasive arterial and pulmonary arterial pressure monitoring. Additionally pulse oxymetry (Oxyshuttle, Critikon; at present the most accurate device [Hohmann and Zander, 1988]) for continuous noninvasive monitoring of the "partial" O_2 saturation [Mertzlufft, 1988] was used during intubation apnoea.

Anaesthesia

Premedication consisted of Flunitrazepam 2.0 mg p.o. the evening before operation and another 2.0 mg p.o. 90-120 min before induction of anaesthesia. After insertion of a peripheral venous line 0.01 mg/kgBW i.v. Flunitrazepam were given.

After "adaequate" preoxygenation [Duda et al., 1988] induction of anaesthesia was performed with 0.02-0.025 mg/kgBW Fentanyl and 0.1 mg/kgBW Pancuronium. After cessation of spontaneous respiration assisted/controlled ventilation ($paCO_2$ 38-42 mmHg) was continued with 100% oxygen until intubation. The intubation procedure started three minutes after injection of Pancuronium. With direct laryngoscopy a stomach tube and an oesophageal temperature probe were inserted via the nasal route, followed by endotracheal intubation using a BRANDT^{TM} pressure controlled tube.

Depending on the intubator's experience and on the anatomical situation, the total intubation procedure took up to 4 min only the first two minutes of which were studied.

Blood gas samples

Arterial and mixed-venous blood samples (1.5 ml each) were taken sequentially immediately before (1 sample) and during the first two minutes of apnoea (12 samples, each withdrawn during 10 sec). The samples were stored in an ice-bath [Müller-Plathe and Schlebusch, 1988] and analyzed in a randomized order using a Nova Biomedical Stat Profile 6^{TM} and a Corning 2500 CO-Oxymeter at a temperature of 36.5°C.

RESULTS

As shown in Fig. 1a paO_2 decreased from 485 ± 100 mmHg before apnoea to 376 ± 68 mmHg after two min of apnoea. The $p\bar{v}O_2$ remained almost constant. There was a slightly increasing tendency from 47 ± 5 mmHg before to 50 ± 5 mmHg at the end of apnoea.

"Partial" O_2 saturation (pulse oxymeter) showed stable values of 99%, arterial O_2 saturation (CO-Oxymeter) of about 97%, while $s\bar{v}O_2$ slightly increased from 81.9% to 82.4% until the end of apnoea (Fig. 1b).

594

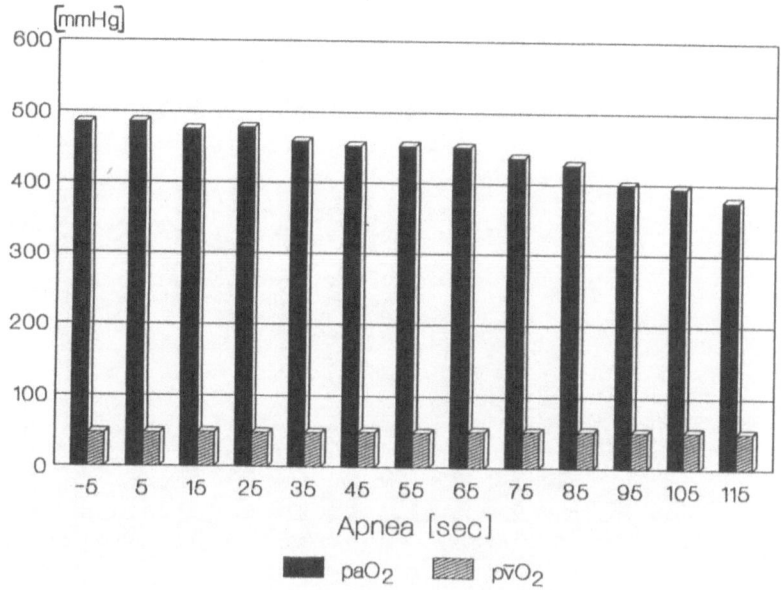

a. paO_2 and $p\bar{v}O_2$

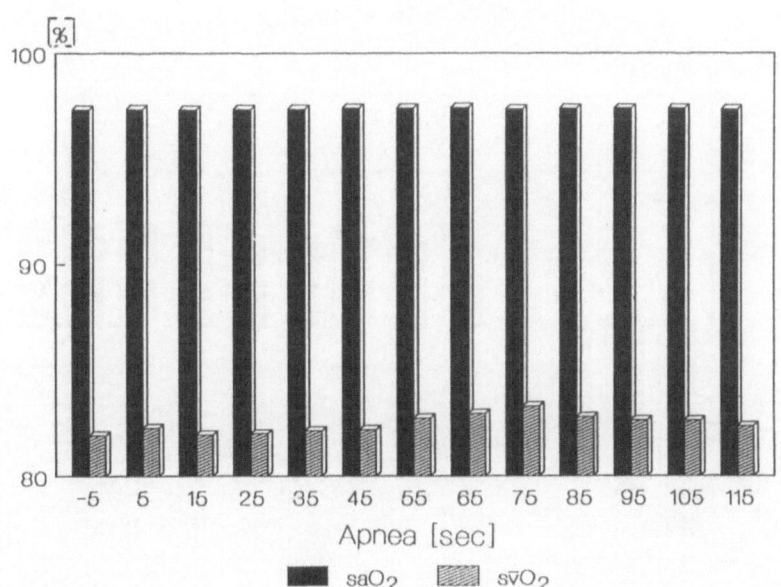

b. saO_2 and $s\bar{v}O_2$

Fig. 1. Changes of paO_2 and $p\bar{v}O_2$ (a.), saO_2 and $s\bar{v}O_2$ (b.) during the first two minutes of hyperoxic apnoea.

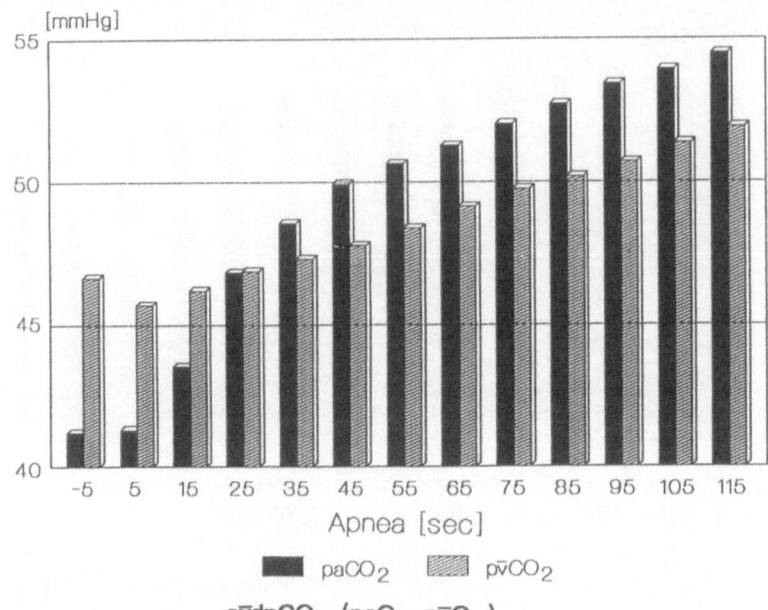

a. $paCO_2$ and $p\bar{v}CO_2$

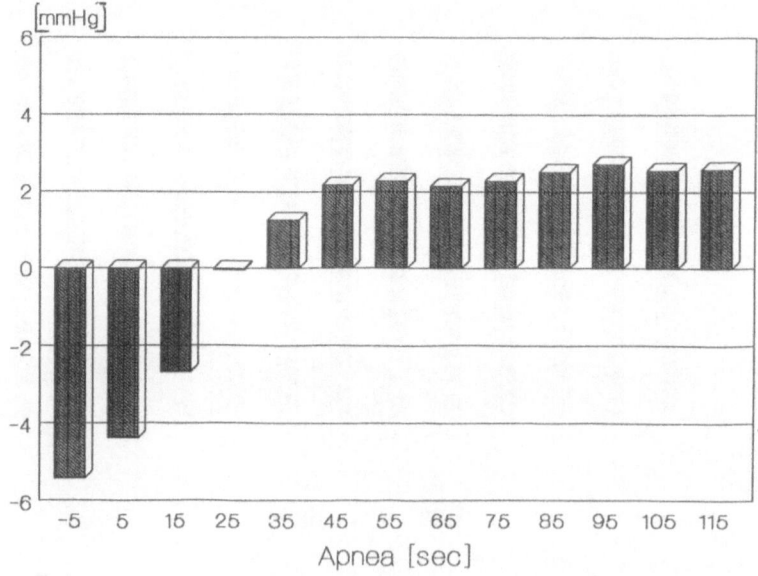

b. $a\bar{v}dpCO_2$ $(paO_2-p\bar{v}O_2)$

Fig. 2. Changes of $paCO_2$ and $p\bar{v}CO_2$ (a.) and $a\bar{v}DpCO_2$ (b.) during the first two minutes of hyperoxic apnoea.

a.

pHa and $pH\bar{v}$

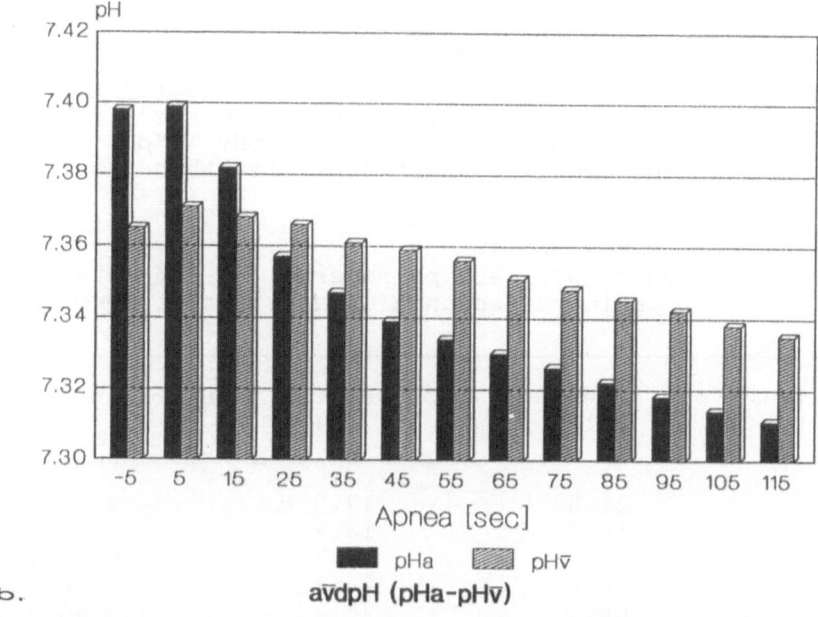

$a\bar{v}dpH$ ($pHa-pH\bar{v}$)

b.

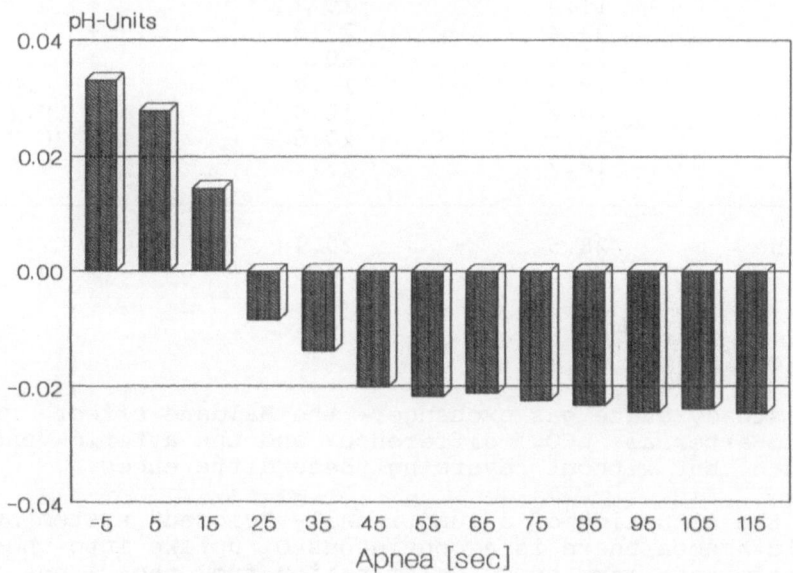

Fig. 3. Changes of pHa and $pH\bar{v}$ (a.) and $a\bar{v}DpH$ (b.) during the first two minutes of hyperoxic apnoea.

A biphasic increase was observed in $paCO_2$ from 41.2 ± 3.4 mmHg before to 54.5 ± 3.9 mmHg at the end of apnoea (Fig. 2a). The increase of $p\bar{v}O_2$ during apnoea was linear, from 45.7 ± 3.9 mmHg to 51.9 ± 4.0 mmHg.
After about 25 sec of apnoea $paCO_2$ exceeded $p\bar{v}CO_2$ (Fig. 2b) ("pCO_2 reversal") and the arterial/mixed-venous pCO_2-difference ($a\bar{v}DpCO_2$) became positive.

As observed for the $paCO_2$ increase, pHa decreased biphasically from 7.40 ± 0.03 to 7.31 ± 0.02 (Fig. 3a). Mixed-venous pH ($pH\bar{v}$) decreased almost linearly from 7.37 ± 0.03 mmHg ("5 sec") to 7.33 ± 0.02 mmHg ("115 sec") (Fig. 3a). After about 20 sec of apnoea pHa exceeded $pH\bar{v}$ ("pH reversal"; Fig. 3b). The "pH reversal" started earlier than the "pCO_2 reversal" ($p < 0,05$; Table 1).

Table 1. Time of pCO_2 and pH reversal in the total of 12 patients (seconds after start of apnoea)

Patient Nr.	"pCO_2 reversal" (a)	"pH reversal" (b)	Difference (a)-(b)
1	21.0	17.0	+ 4.0
2	16.1	18.2	− 2.2
3	31.8	28.7	+ 3.1
4	34.6	28.3	+ 6.3
5	66.7	28.3	+ 38.4
6	11.4	10.7	+ 0.7
7	33.4	27.2	+ 6.2
8	30.4	20.7	+ 9.7
9	18.2	14.2	+ 4.0
10	20.2	12.0	+ 8.2
11	40.5	20.5	+ 20.0
12	18.2	22.2	− 4.0
Mean value	28.5	20.7	+ 7.9
$\pm$ SD	15.0	6.5	0.8

DISCUSSION

In steady-state gas exchange, the Haldane effect reduces the veno-arterial pCO_2 difference and the arterio-venous pH difference, but without reversing these differences.

In the situation of a functionally "closed" system such as hyperoxic apnoea there is a continuous O_2 uptake into the blood with concomitant lack of CO_2 elimination from the lungs which will lead to an increase of $paCO_2$ above $p\bar{v}CO_2$ and to a drop of pHa below $pH\bar{v}$, because CO_2 elimination is blocked (i.e. veno-arterial CO_2 content difference is close to zero) and the Haldane effect comes to full development. Under these conditions (continuing O_2 uptake with lacking CO_2 elimination), it is therefore possible to substantiate the veno-arterial pCO_2 and pH reversal due to Haldane effect in vivo.

The time course of O_2, CO_2 and pH changes during apnoea in man has not been studied to date. In contrast, concerning pCO_2 alone, linear pCO_2 increases of 3 - 6 mmHg/min [Nunn, 1987] or 4 - 5 mmHg/min [Siggaard-Andersen, 1974] have been reported. Obviously the potential risks of hypercapnia have been considered with view to concomitant hypoxia rather than with normoxia or hyperoxia. In clinical practice this lack of knowledge still is responsible for (1.) overestimation of the presumed potential hazards of hypoxaemia plus hypercarbia during hyperoxic apnoea and (2.) misinterpretations concerning hypoxia as well as hypercarbia.

Starting at 485 mmHg paO_2 before apnoea, an average decrease of 55 mmHg/min was observed during the 2 min of apnoea investigated whereas O_2 saturation remained stable. Consequently (and provided no acute anaemia and/or toxaemia being present) the danger of hypoxaemia (decrease in O_2 content) can be excluded on the basis of these findings for at least 6 minutes, adaequate preoxygenation (high initial pAO_2) provided: sufficient O_2 diffusion from the alveoli to the blood is maintained as long as the pulmonary O_2 stores guarantee normal oxygen saturation (96% saO_2) of haemoglobin.

In contrast to literature data [Nunn, 1987] $paCO_2$ development was found to be nonlinear. Its biphasic course (14.5 mmHg/min initially vs. 3.9 mmHg/min during prolonged apnoea) can be explained as follows: The initial steep increase results from the additive effects of ongoing CO_2 production in the tissues, venoarterial aequilibration of CO_2 partial pressures due to lack of alveolar elimination, and the appearance of the Haldane effect. Already after about 28 seconds of apnoea $paCO_2$ reaches $p\bar{v}CO_2$. The steep increase continues until after about 45 sec the Haldane effect has reached its maximum with a nearly constant arterial/mixed-venous pCO_2-difference of 2.4 ± 0.21 mmHg. This value approximates the theoretically expected value of 2.1 mmHg as calculated for an arterial/mixed-venous sO_2-difference of 15% [Brandt et al., 1988]. With continuing apnoea $paCO_2$ and $p\bar{v}CO_2$ show an almost linear increase of 3.9 mmHg/min.

This latter fact seems to be of clinical importance. Provided a sufficient continuing O_2 supply and no impaired lung function are present, it appears meaningful to rediscuss the apnoea time tolerable for the patients. A pCO_2 increase with concomitant hyperoxia during prolonged apnoea appearently is quite moderate (even 250 mmHg pCO_2 may be tolerated without ill effects; [Siggaard-Andersen, 1974]). Under these circumstances only a moderate respiratory acidosis will occur (without concomitant increase of non-carbonic acid). Therefore, monitoring and interpretation of apnoea in clinical practice (e.g. endotracheal intubation, bronchoscopy etc.) should mainly concentrate on oxygen availability rather than hypercapnia.

From the theoretical standpoint it is both noteworthy and surprising that pCO_2 increase as found here during prolonged hyperoxic apnoea seems to be linear after the initial steep increase. With view to the shape of the CO_2 binding curve no almost linear increase in pCO_2 should be expected with ongoing apnoea time. Instead, the rate of increase should diminish with time. Further studies are expected to show whether this dissociation between clinical observation and theoretical prediction might be explained.

REFERENCES

Brandt, L., 1988, Bedeutung des gemischtvenösen O_2-Status als Ergänzung zum arteriellen O_2-Status, in: "Der Sauerstoff-Status des arteriellen Blutes", R. Zander and F. Mertzlufft, eds., Karger, Basel; p. 238-255.

Brandt, L., Mertzlufft, F., Rudlof, B., and Dick, W., 1988, In vivo-Nachweis des Christiansen-Douglas-Haldane-Effektes unter klinischen Bedingungen, Anaesthesist, 37:529.

Brandt, L., Mertzlufft, F., and Dick, W., 1989, Verhalten des arteriellen und gemischtvenösen Blutgasstatus in der Initialphase der Intubationsapnoe, Anaesthesist, 38:167.

Christiansen J., Douglas, C. G., and Haldane J. S., 1914, The absorption and dissociation of carbon dioxide by human blood, J Physiol, 48:244.

Duda, D., Brandt, L., Rudlof, B., Mertzlufft, F., and Dick, W., 1988, Der Einfluß unterschiedlicher Präoxygenationsverfahren auf den arteriellen Sauerstoffstatus, Anaesthesist, 37:408.

Hohmann, C., Zander, R., 1988, Vergleich verschiedener Pulsoxymeter unter Hypoxie bei Rauchern und Nichtrauchern, Anaesthesist (Suppl.), 37:93.

Mertzlufft, F. O., 1988, Nichtinvasive kontinuierliche Messung der arteriellen O_2-Sättigung, in: "Der Sauerstoff-Status des arteriellen Blutes", R. Zander and F.O. Mertzlufft, eds., Karger, Basel; p. 109-119.

Mertzlufft, F. O., Rudlof, B., Jantzen A. H., Brandt, L., and Dick, W., 1987, Christiansen-Douglas-Haldane-Effekt: Auch in vivo? Anaesthesist (Suppl.), 36:374.

Müller-Plathe, O., Schlehbusch, H., 1988, Gewinnung und Aufbewahrung von arteriellem Blut, in: "Der Sauerstoff-Status des arteriellen Blutes", R. Zander and F.O. Mertzlufft, eds., Karger, Basel; p. 12-18.

Nunn, J. F., 1987, "Applied Respiratory Physiology", Butterworth & Co Ltd, London; pp. 295.

Siggaard-Andersen, O., 1974, "The acid-base status of the blood", Munksgaard, Copenhagen (4th ed.), p. 115-117.

CARDIOGENIC OSCILLATIONS OF He AND SF$_6$ IN EXPIRED GAS IN DOGS

M. Meyer, S.M. Lewis, M. Mohr, H.Schulz, K.-D. Schuster and J. Piiper

Department of Physiology, Max Planck Institute for Experimental
Medicine, Göttingen, Federal Republic of Germany

INTRODUCTION

Cardiogenic oscillations synchronous with the heart beat modulating the partial pressure profile of the alveolar plateau (phase III) are generally attributed to the action of the heart altering the distribution of expired flow from lung units with differing partial pressures (Arieli, 1983; Bradley et al., 1975; Fowler and Read, 1961; Langer et al., 1960; West and Hugh-Jones, 1961).The amplitude of cardiogenic oscillations of inert gases with differing diffusivity has been demonstrated to exhibit only minor differences (Arieli, 1983; Arieli et al., 1981; Bradley et al., 1975) but no quantitative analysis has been presented as yet.

In a recent study in anaesthetized, mechanically ventilated supine dogs the mechanisms underlying intrapulmonary gas mixing were studied by single-breath washout of two inert gases with differing diffusivity (He and SF$_6$) in two experimental conditions: during intravenous infusion (*venous loading*) and after equilibration of lung gas with the inert tracer gases (*airway loading*). The technique is of particular interest for the analysis of gas mixing in lungs because in the first instance the test gases undergo continuous gas exchange across the alveolar-capillary membrane while in the latter case their behaviour conforms to that of virtually insoluble gases that remain confined to the gas phase of the lung. As a side aspect of that study the expirograms of He and SF$_6$ displayed distinct and regular cardiogenic oscillations which, for the present communication, were analyzed quantitatively in terms of relative oscillation amplitude and phase relationships.

METHODS

Single-breath expirograms of He and SF$_6$ were recorded from 7 anaesthetized, mechanically ventilated dogs (mean body wt 18 kg). The single-breath maneuver consisted of a passive expiration, an inspiration at constant rate (0.5 L/s) of a volume equal to 50% of the animal's FRC (as determined by helium dilution) followed by a constant-flow expiration of 75% of the individual's FRC at a rate of 100 ml/s. Partial pressures of He and SF$_6$ in the endotracheal tube were continuously analyzed by a modified Varian M3 mass spectrometer.

A femoral artery was cannulated and blood was routed by a roller pump to a Sci-Med 200-2A membrane oxygenator and returned to the ipsilateral femoral vein. Blood flow in the extracorporeal circuit was maintained at 60 ml/min. The oxygenator was flushed at a rate of 1 L/min with N$_2$ during airway loading and a mixture of 50% He and 50% SF$_6$ during venous loading (see below).

Two methods of inert gas loading were studied. (1) In *airway loading* (AL), the lungs were equilibrated with a mixture of 100 ppm He and 100 ppm SF_6. The partial pressures of He and SF_6 were adjusted to approximate the end-tidal partial pressures (about 0.07 Torr) achieved during venous loading (see below). The single-breath maneuver was performed breathing a tracer-free gas mixture (compressed air). (2) In *venous loading* (VL), the test gases were administered by replacing the N_2 flow through the oxygenator by the 50% He - 50% SF_6 mixture. The single-breath maneuver was performed while venous loading continued. A schematic representation of set-up and procedures is presented in Fig. 1.

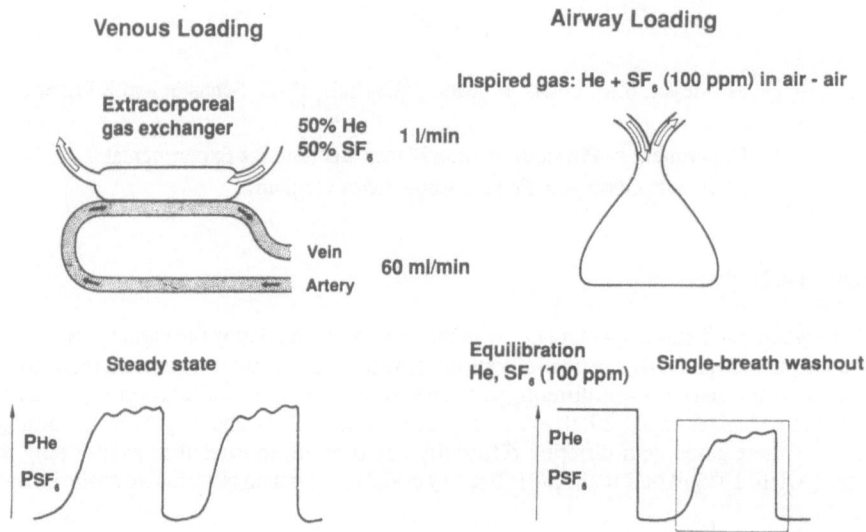

Figure 1. Set-up for single-breath washout of He and SF_6 during venous loading (*left*) and airway loading (*right*)

The expired partial pressure profiles of He and SF_6 on the alveolar plateau were processed for local maxima and minima. The oscillation amplitude, $\triangle P$, was taken as the difference between any regional maximum (minimum) and the mean of the two adjacent minima (maxima). The oscillation amplitudes were normalized with reference to the mixed expired-inspired partial pressure difference and expressed in terms of the relative oscillation amplitude [ROA = $\triangle P/(P_{\bar{E}} - P_I)$].

RESULTS

A direct comparison of the shape of the expirograms (displayed over 40 - 100% of the expired volume) and of the relative amplitude of cardiogenic oscillations of He and SF_6 in venous loading (VL) and in airway loading (AL) is presented in Figs. 2 and 3. The He/SF_6 ratios of the relative oscillation amplitudes for any cardiogenic wave were also included.

Relative oscillation amplitude. The numerical values of the standardized relative He and SF_6 oscillation amplitudes and the He/SF_6 amplitude ratio are compiled in Table 1. The following features are evident. (1) The amplitudes are much higher for venous loading as compared to airway loading, the overall average ratio being 3.0 for both He and SF_6. (2) The oscillations were always larger for He than for SF_6, the He/SF_6 ratio averaging 1.65 for airway loading and 1.88 for venous loading.

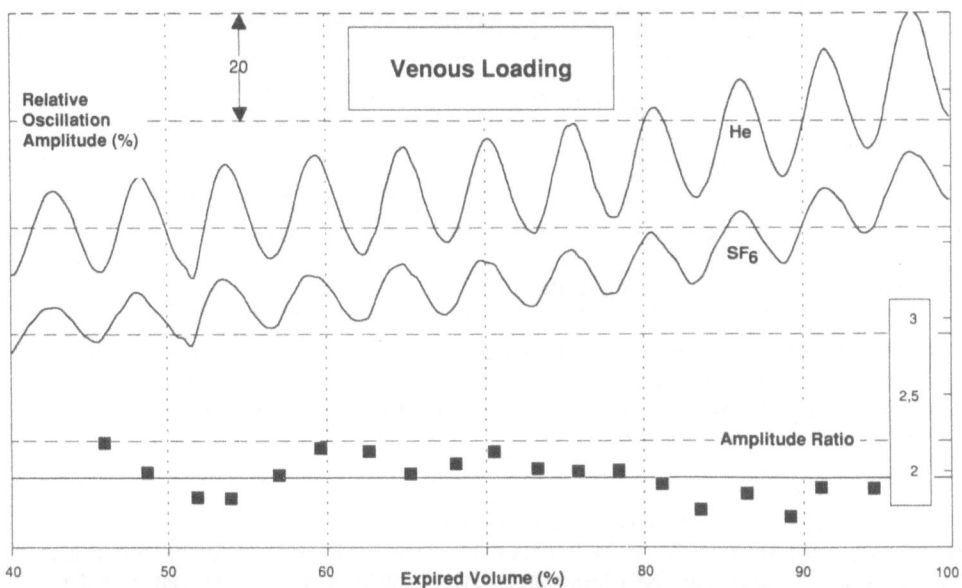

Figure 2. Alveolar plateau of simultaneously recorded He and SF$_6$ expirograms during venous loading. The expirograms were normalized with reference to mixed expired-inspired partial pressure difference and are displayed over 40 - 100% of the expired volume. The ordinate is scaled in terms of relative oscillation amplitude (in per cent). In order to eliminate superposition of curves the expirogram of SF$_6$ is displayed with offset with respect to that of He. Bottom overlay markers and right ordinate axis with box designate the He/SF$_6$ ratio of the relative oscillation amplitudes. The mean value of the amplitude ratio (1.96, SD 0.14) is shown by the continuous line.

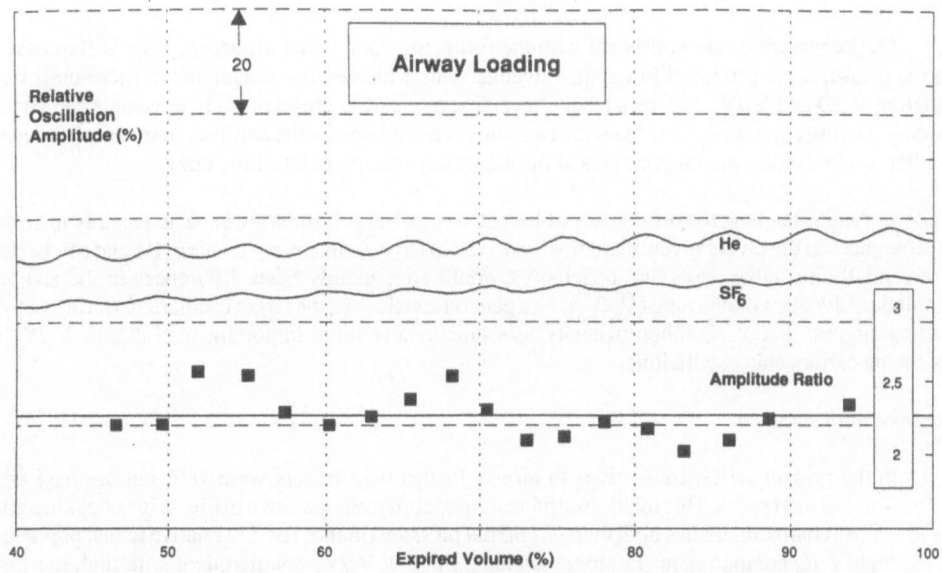

Figure 3. Alveolar plateau of simultaneously recorded He and SF$_6$ expirograms during airway loading. See legend to Fig. 2 for further details. The mean value of the He/SF$_6$ ratio of relative oscillation amplitudes is 2.27 (SD 0.15).

TABLE 1. *Cardiogenic oscillations in expired gas in dogs*

	Venous loading (VL)		Airway loading (AL)	
	He	SF_6	He	SF_6
ROA	6.1 ± 0.4	3.9 ± 0.3	2.1 ± 0.2	1.3 ± 0.1
He/SF_6 ratio	1.88 ± 0.10		1.65 ± 0.06	

Relative amplitude of oscillations (ROA) in per cent of mixed expired-inspired partial pressure difference and He/SF_6 amplitude ratio of cardiogenic oscillations. Results (mean values ± SE from 7 experiments) are displayed for venous loading (VL) and airway loading (AL).

Phase relationships. The phase relationship between the oscillations in airway and venous loading was studied with reference to the R-wave of the electrocardiogram. During venous loading, the oscillation peak was delayed by about 0.4 s compared to airway loading. Since the average cardiac cycle length was about 0.7 s, the oscillations were close to 180° out-of-phase.

DISCUSSION

Mechanisms of Cardiogenic Oscillations

Simple reasoning demonstrates that two essential factors are required to produce changes in the expired partial pressure profiles of respiratory and inert gases in the form of cardiogenic oscillations. First, there must exist lung areas with differing partial pressures. Second, the relative gas flows from these areas with differing partial pressures must be changed by the action of the heart.

Our observations are consistent with the following operational model (Fig. 4). Differences in partial pressures are produced in parallel alveolar units which empty sequentially. These units have differing $\dot{V}_A/\dot{Q}$ and $\dot{V}_A/V_A$ and, as a result, have differing partial pressures in both venous and airway loading. During expiration, flow from the two units is modulated by the cardiac contraction producing oscillations in expired partial pressures at the expiratory mixing point of the units.

In venous loading, the differences of He and SF_6 partial pressures are expected mainly to be due to differences in the alveolar ventilation/perfusion ratio ($\dot{V}_A/\dot{Q}$). In airway loading, He and SF_6 behave as essentially insoluble gases and differences would arise mainly from differences in the alveolar ventilation/alveolar volume ratio ($\dot{V}_A/V_A$). As a general conclusion, the larger oscillations during venous loading suggest that $\dot{V}_A/\dot{Q}$ inhomogeneity was functionally more important than that of $\dot{V}_A/V_A$ in producing cardiogenic oscillations.

Phase relationships

In the present series, oscillations in airway-loaded inert tracers were 180° out-of-phase with venous-loaded inert gases. This relationship and its model analysis is shown in Fig. 4. In venous loading, the low $\dot{V}_A/\dot{Q}$ compartment has high inert gas partial pressure (marked by **) relative to that prevailing in the high $\dot{V}_A/\dot{Q}$ compartment. In airway loading, the low $\dot{V}_A/V_A$ compartment with high inert gas pressure (marked by circled **) relative to its fellow empties in approximate synchrony with the high $\dot{V}_A/\dot{Q}$ compartment while the high $\dot{V}_A/V_A$ unit empties with the low $\dot{V}_A/\dot{Q}$ compartment.

These results suggest that units with high $\dot{V}_A/V_A$ were units with low $\dot{V}_A/\dot{Q}$ and, consequently, with very high $\dot{Q}/V_A$ (Fig. 4, bottom panel). Similarly, earlier studies of intra-regional distribution of ventilation and perfusion by nitrogen-13 clearance in man have demonstrated that the well-ventilated units were also better perfused, hence intra-regional units with high $\dot{Q}/V_A$ also have high $\dot{V}_A/V_A$ (Ewan et al., 1978).

Difference Between He and SF_6 Oscillations

We hypothesize that the differences in amplitude between He and SF_6 oscillations were due to differential dispersion in the airways. Dispersion may be due to four mechanisms: diffusion, laminar Taylor dispersion, turbulent Taylor dispersion, and dispersion due to differences in airway path lengths (Pedley et al, 1977; Piiper and Scheid, 1987; Wilson and Lin, 1970). The latter two mechanisms are essentially independent of diffusivity and would be expected to modulate the He and SF_6 partial

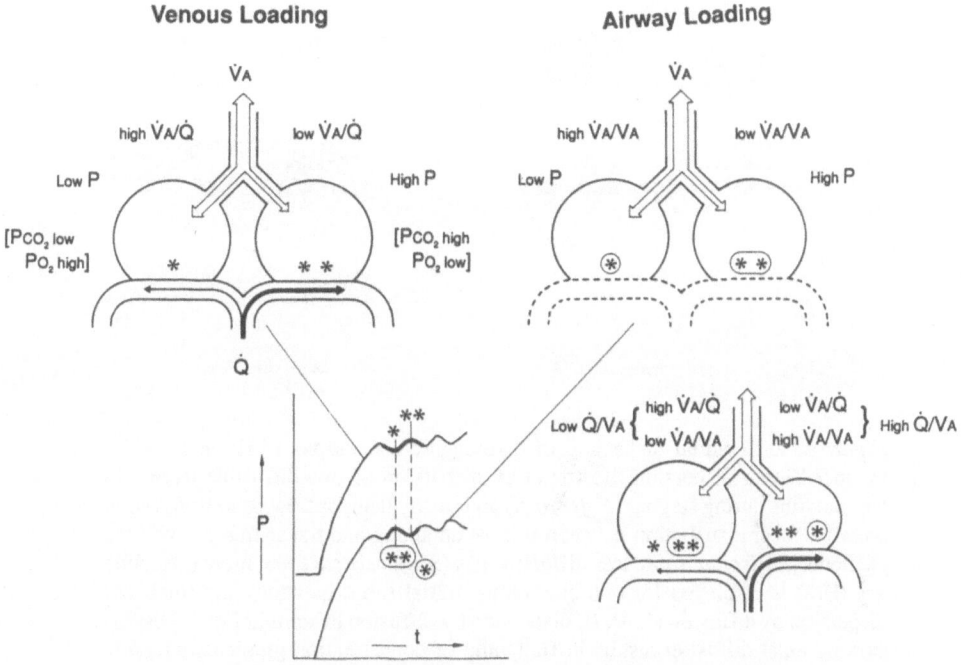

Figure 4. Parallel two-compartment model for analysis of cardiogenic oscillations in venous loading (*left*) and airway loading (*right*). See text for details.

pressures similarly. Diffusion is directly proportional to the diffusion coefficient, whereas laminar Taylor dispersion at moderate flow rates is approximately inversely proportional to the diffusion coefficient (Taylor, 1953). When airflow is smoothly controlled and slow, laminar Taylor dispersion is the dominant dispersive mechanism in the larger airways. Under these conditions, an oscillation in the partial pressure pressure profile of SF_6 would be flattened during passage through the airways to a significantly greater extent than a He oscillation. If this attenuation is significant, differences in the amplitude of oscillations in the two tracers in the sense of the He/SF_6 amplitude ratio exceeding unity will be produced (Fig. 5). An important observation in support of this model is the similarity in the ratio of He and SF_6 amplitudes between venous and airway loading under a wide variety of conditions even though the absolute amplitudes were quite different.

SUMMARY

Quantitative analysis of cardiogenic oscillations of He and SF_6 during airway and venous loading demonstrated that both $\dot{V}_A/\dot{Q}$ and $\dot{V}_A/V_A$ inequalities were involved in the lung gas inhomogeneity producing cardiac oscillations in the expirogram. Both inequalities were coupled in such a manner that low $\dot{V}_A/\dot{Q}$ units had high $\dot{V}_A/V_A$. The oscillations were modified in conducting airways where SF_6 oscillations were attenuated more than He oscillations, probably by laminar Taylor dispersion.

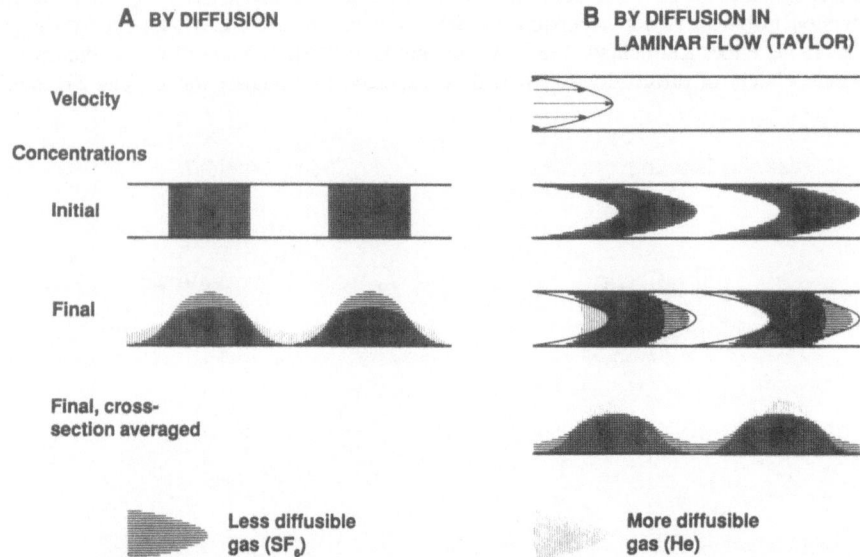

Figure 5. Differential dispersion of cardiogenic oscillations of He and SF_6 by laminar Taylor dispersion illustrated by dispersion of two diffusible tracers in circular tube during stagnant (*left panel*) and steady laminar flow conditions (*right panel*). Tracer distribution between tracer-containing and tracer-free gas volume elements [*horizontal lines*, less diffusible gas (SF_6); *vertical lines*, more diffusible gas (He)] is displayed for two conditions: initial (no dispersion) and final: (*A:* dispersion by diffusion alone, *B:* dispersion by diffusion in laminar flow). Dispersion by axial diffusion results in flattening of concentration gradients which is enhanced for the more diffusible gas. In laminar flow with parabolic radial velocity profile across the tube the initially planar front is distorted by the velocity profile. Tracer molecules of the less diffusible gas species are more dispersed to the velocity profile while radial diffusion of the more diffusible gas counteracts its convective axial dispersion producing a blurred front. The axial dispersion of final mass concentration (averaged over the tube cross section) shows that, by virtue of the Taylor mechanism, axial dispersion of the more diffusible gas is diminished relative to that of the less diffusible gas.

REFERENCES

Arieli, R., 1983, Cardiogenic oscillations in expired gas: origin and mechanism. *Respir. Physiol.* 52: 191-204.

Arieli, R., Olsowska, A.J., and Van Liew, H.D., 1981, Post-inspiratory mixing in the lung and cardiogenic oscillations. *J. Appl. Physiol.* 51: 992-928.

Bradley, G.W., Henderson, A.H., and Mills, R.J., 1975, Cardiogenic oscillations of nitrogen and argon concentrations in expired gas in man. *Clin. Sci. Mol. Med.* 48: 39-45.

Ewan, P.W., Jones, H.A., Nosil, J., Obdrzalek, J., and Hughes, J.M.B., 1978, Uneven perfusion and ventilation within lung regions studied with nitrogen-13. *Respir. Physiol.* 34: 45-59.

Fowler, K.T., and Read, J.,1961, Cardiac oscillations in expired gas tensions, and regional pulmonary blood flow. *J. Appl. Physiol.* 16: 863-868.

Langer, G., Bornstein, D.L., and Sudlow, A.P., 1960, Cardiogenic oscillations in expired nitrogen and regional alveolar hypoventilation. *J. Appl. Physiol.* 15: 855-862.

Pedley, T.J., Schroter, R.C., and Sudlow, M.F., 1977, Gas flow and mixing in the airways. In:*Bioengineering Aspects of the Lung*, edited by J.B. West. New York: Dekker, vol 3, p. 163-265.

Piiper, J., and Scheid, P., 1987, Diffusion and convection in intrapulmonary gas mixing. In:*Handbook of Physiology. The Respiratory System. Gas Exchange.* Bethesda, MD, Am.Physiol. Soc., sect. 3, vol. IV, chap. 4, p. 51-69.

Taylor, G., 1953, Dispersion of soluble matter in solvent flowing slowly through a tube. *Proc. R. Soc. London* Ser. A 219: 186-203.

West, J.B., and Hugh-Jones, P., 1961, Pulsatile gas flow in bronchi caused by the heart. *J. Appl. Physiol.* 16: 697-702.

Wilson, T.A., and K.H. Lin, K.H., 1970, Convection and diffusion in the airways and the design of the bronchial tree. In: *Airway Dynamics*, edited by A. Bouhys. Springerfield, IL,Thomas.

SIGNIFICANCE OF CARDIOGENIC MIXING IN DOG LUNGS

J.M. Schell, A. Rahmel, A. Schwalen, E. Calzia, M. Meyer and J. Piiper

Department of Physiology, Max Planck Institute for Experimental
Medicine, Göttingen, Federal Republic of Germany

INTRODUCTION

Since the early observations of DuBois et al. (1952) and of West and Hugh-Jones (1961) it has been known that the heart beat causes fluctuations of airway pressure. Variations of intrapulmonary pressure that occur synchronously with the heart beat were suggested to result from the cardiac motion itself or from cyclic changes of intrathoracic blood volume. Airflow variations caused by the beating heart in lobar and segmental bronchi have been determined in patients at bronchoscopy and found to occur asynchronously, flow increasing in one airway while decreasing in another (West and Hugh-Jones, 1961). The mechanisms underlying cardiogenic intrapulmonary gas flow, changes in heart volume or alterations in shape and position of the heart, are still controversial. Recent evidence suggests that the total heart volume remains essentially unchanged during the cardiac cycle (Hoffman and Ritman, 1985) and would therefore favour cyclic changes of shape and/or position of the heart causing airway pressure swings.

Apart from the precise mechanisms of heart-lung interactions, the significance of cardiac performance for intrapulmonary gas mixing has been addressed by several author groups. While pulsatile airflows resulting from the cardiac action have been shown to promote mixing at the boundary between inspired and resident gas, mixing distal to this interface, i.e., in the alveolar region, appears not to be enhanced by the heartbeat (Engel et al., 1973a, b; Engel, 1983; Fukuchi, 1976; Horsfield et al., 1982; Jones et al., 1982). Experiments in humans with heart rate varied by cycloergometric exercise suggested that cardiogenic mixing as inferred from a test gas bolus response technique was unresponsive to normal changes in heart rate (Drechsler and Ultman, 1984).

Among the more direct methodological approaches pursued to evaluate quantitatively the mixing effect of the cardiac action have been in vivo and postmortem comparisons with intrapulmonary gas sampling and experiments using cardiopulmonary bypass in open-chest dogs by which the heart could be stopped or stimulated as necessary. These procedures may be critizised for unpredictable side effects by loss of bronchomotor tone after death and changes in airway pressure and lung mechanics resulting from an opened chest.

In the present study, the role of gas mixing by heart-lung interaction was investigated using a less invasive experimental approach. A reversible myocardial arrest was induced in anesthetized, mechanically ventilated dogs by intracoronary injection of acetylcholine. Thus, the contribution of cardiogenic flow to intrapulmonary gas mixing could be studied under in vivo conditions in closed-chest animals.

METHODS

Seven mongrel dogs (mean body weight 22 ± 3 kg) were studied. Anesthesia was induced with sodium pentobarbital, 25mg/kg iv, followed by a continuous intravenous infusion of 3mg/kg/h. They were intubated via the oro-tracheal route and mechanically ventilated, utilizing a computerized piston-type servo-ventilator which also performed the single-breath measurements described below. A detailed description of the respiratory servo system particularly designed for mechanical ventilation and lung function testing in animals is given elsewhere (Meyer and Slama, 1983).

Upon placement of the tip of a 7F non-occluding angiographic catheter (Amplatz) in the left coronary artery (aided by radiographic control of catheter position) a temporary myocardial arrest was repeatedly induced by injecting acetylcholine (35 mg/ml saline) through the catheter into the coronary circulation. Each intervention resulted in a heart arrest of approximately 20s duration and was followed by normal myocardial function. In each of the animals 8 to 15 interventions were performed while the coronary catheter was maintained in place for the duration of the experiment.

Simultaneous single-breath washout measurements of two inert test gases with widely differing diffusivities, He and SF_6, were performed before, during and after acetylcholine-induced cardiac arrest. Prior to the test, lung gas was equilibrated by open-circuit washin of 1% He and 1% SF_6 in air. The single-breath maneuver consisted of an inspiration of air at constant rate ($\dot{V}I$), followed by an expiration at constant rate ($\dot{V}E$). Inspired and expired volumes were adjusted to fixed fractions of the functional residual capacity (FRC). Standard ventilatory parameters were as follows: $VI = 0.5*VFRC$, $VE = 0.75*VFRC$, $\dot{V}I = 0.5$ L/s, $\dot{V}E = 0.1$ L/s. The single-breath maneuvers were automatically performed by the servo-ventilator.

Continuous records of the partial pressures of He and SF_6 in the endotracheal tube were obtained by mass spectrometry (modified Varian M3). Monitoring included arterial blood pressure and continuous non-invasive assessment of cardiovascular function (stroke volume, cardiac output and systolic time intervals) by transthoracic impedance cardiography. Special attention was directed towards the measurement of airway opening pressure, an increase following injection of acetylcholine indicating systemic effects of acetylcholine secondary to accidental catheter displacement. In these cases the single-breath measurement was discarded and repeated.

The He and SF_6 expirograms were analyzed in terms of slope of the relative alveolar plateau (S) and series dead space (VD). While the slope of the alveolar plateau can be considered as an index for alveolar gas inhomogeneity, the series dead space would yield an index for mixing at the interface between inspired and residual lung gas. The relative alveolar slope was defined as the change of expired partial pressure, normalized to mixed expired partial pressure, per unit expired volume $(\Delta PE/\Delta VE)/\Delta\bar{PE}$. Linear least-squares regression analysis was applied to the calculation of the alveolar slope (phase III) in the interval between 40% and 80% of the expired volume. Series dead space was calculated from phase II of the expirogram (Fowler-method).

RESULTS

Comparison of the shape of the expirograms in control conditions (i.e., with the heart beating) and during experimentally induced temporary myocardial arrest revealed no major differences attributable to the mechanical action of the heart, apart from the absence of cardiogenic oscillations in the latter condition.

Series dead space

Average values for series dead space of He and SF_6 in control conditions and during myocardial arrest are summarized in Fig. 1 for the 7 experimental animals studied. The heart arrest/control ratio for the series dead space, 0.95 ± 0.07 for He and 0.94 ± 0.07 for SF_6 (means $\pm$ SD) was not statistically different from unity (p > 0.05) indicating that the myocardial contraction had no measurable effect on

VD. The He/SF$_6$ ratio of the series dead space, 0.904 ± 0.022 and 0.897 ± 0.028 for control and heart arrest respectively, was not affected by the cardiac action suggesting that the diffusion-dependent difference of He and SF$_6$ dead spaces was not modified or eliminated by convective (cardiogenic) mixing.

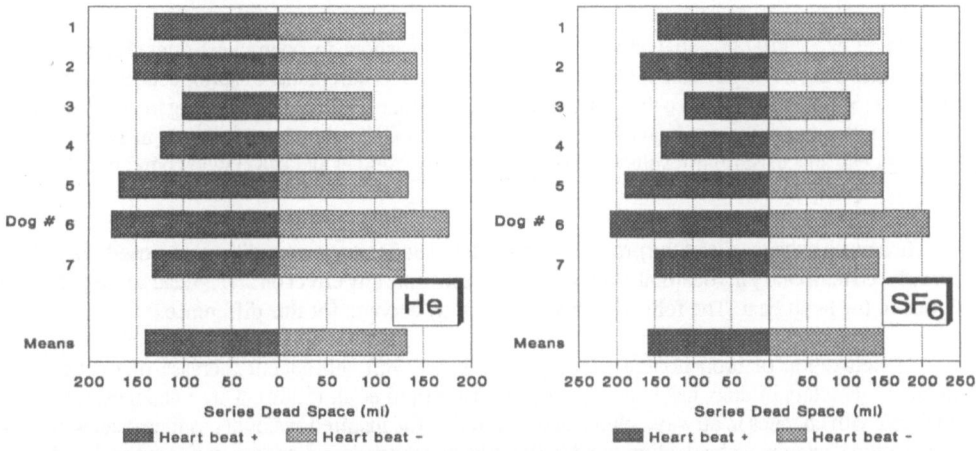

Figure 1. Series dead space of He and SF$_6$ in intact dogs in normal conditions (Heart beat +) and during acetylcholine-induced reversible cardiac arrest (Heart beat -).

Relative alveolar slope

Figure 2 summarizes the results for the relative alveolar slope. Differences from unity between the heart arrest/control ratio of the relative slope, 1.03 ± 0.10 for He and 1.05 ± 0.14 for SF$_6$, were statistically not significant ($p > 0.05$). The He/SF$_6$ ratio of the relative alveolar slope remained unaltered comparing control and heart arrest conditions (0.647 ± 0.093 vs 0.637 ± 0.093, $p > 0.05$). This means that the diffusion-dependent separation of He and SF$_6$ was unaffected by the mechanical action of the heart.

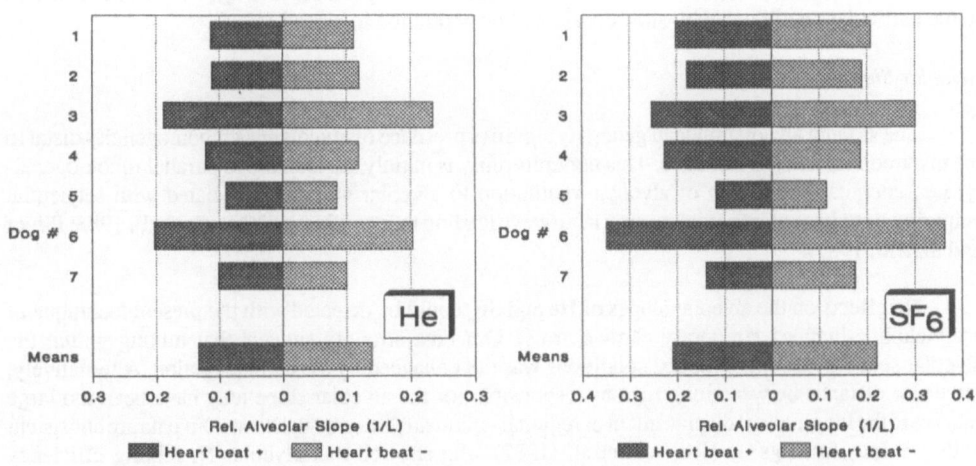

Figure 2. Relative alveolar slope of He and SF$_6$ in intact dogs in normal conditions and during experimental cardiac arrest.

DISCUSSION

In the present study, series dead space and slope of the alveolar plateau were used as indices of intrapulmonary gas mixing. The results revealed no detectable contribution of the beating heart to gas mixing within the lung.

Series Dead Space

Engel et al. (1973a), analyzing single-breath N_2 washout in open-chest dogs during breath-holding following an inspiration of O_2, demonstrated from measurements of series dead space that gas mixing was 5 times faster in vivo than post mortem. This enhancement of gas mixing in vivo suggested a central movement of the stationary front between inspired and residual gas resulting in a reduction of the series dead space. Similar results were obtained by Horsfield et al. (1982) who compared VD of N_2 in vivo and post mortem.

In contrast, the results of the present study obtained under in vivo conditions in closed-chest dogs with induced temporary myocardial arrest did not demonstrate any effect on series dead space of He and SF_6 due to the heart beat. The following two factors may account for this difference.

1. Relaxation of bronchomotor tone post mortem with subsequent increase of tracheal and bronchial diameters in dogs has been observed by Horsfield et al. (1982). These changes, however, associated with changes in airway volume and position of the inspired/residual gas interface, which by experimental design have been eliminated in the present study, do not appear to be responsible for all of the differences in VD observed upon cessation of the heart beat in the post mortem dog studies of Engel et al. (1973a) and Horsfield et al. (1982).

2. A major difference between previous work and the present study resides in the experimental design. In order to achieve reliable comparisons of mixing in two different conditions, flow rates and volume of the single-breath maneuver were carefully controlled. The expiratory flow rate, 100 ml/s, was relatively low compared with normal breathing. Moreover, modulation of total expiratory flow by the heart beat was expected to occur only when expiration was unrestrained and the airway opening was open to the atmosphere. Because the single-breath procedures were performed with linearly-controlled expiration, i.e., in closed system, the mechanical interaction of heart and lungs could only modulate pressures and intrapulmonary airflow between lung regions while the total expiratory flow must have been unaffected by the cardiac action. It is therefore reasonable to assume that the position of the inspired/residual gas interface was stationary and not subject to axial displacement by the cardiac action. Thus, the present experimental findings of absence of any effect on the series dead space by the myocardial contraction should not be considered to be in disagreement with previous findings but may be attributed to the particular conditions with slow constant-flow expiration in closed-circuit.

Alveolar Slope

The sloping alveolar plateau generally signifies presence of alveolar gas inhomogeneity distal to the inspired/residual gas interface. This nonuniformity is mainly attributable to parallel inhomogeneity, i.e., unequal distribution of alveolar ventilation to alveolar volume, associated with sequential emptying or to incomplete axial mixing in airways leading to 'stratification' (Meyer et al., 1983; Piiper and Scheid, 1987).

No effects on the alveolar slopes of He and SF_6 could be detected with the present technique of acetylcholine-induced temporary cardiac arrest. Our data strongly suggest that mixing within the alveolar space in our experimental conditions was not enhanced by the cardiac motion. Alternatively, the linear distance between lung regions responsible for the alveolar slope may have been too large relative to the longitudinal displacement of regional volumes evoked by the heart. Our data are consistent with previous findings by Horsfield et al. (1982) who showed that alveolar N_2 mixing efficiency calculated from single-breath N_2 washout was not significantly altered upon cessation of the heart beat. Absence of a significant role of cardiogenic mixing has also been demonstrated by the observation of similar rebreathing equilibration kinetics of He and SF_6 in dogs maintained on cardio-pulmonary bypass with or without beating heart (Jones et al., 1982).

It is important to emphasize that the measurements of the present series were performed in anesthetized, paralyzed and artificially ventilated supine dogs. The single-breath test expiration was a prolonged piston-controlled constant-flow expiration. These conditions were chosen because they provided a highly accurate and reproducible measurement of the expirogram. Thus, extrapolation of the present findings to different conditions, e.g. freely uncontrolled passive expiration or spontaneous breathing, or different animal species including man, requires experimental confirmation.

SUMMARY

Single-breath washout of two inert gases (He and SF_6) in anesthetized mechanically ventilated dogs in normal conditions with the heart beating and during reversible heart arrest revealed no effects attributable to the action of the beating heart. It is concluded that in the conditions of the experiments convective mixing by the cardiac action played an insignificant role in promoting intrapulmonary mixing and transport.

ACKNOWLEDGEMENTS

Supported by the Deutsche Forschungsgemeinschaft, SFB 330, Göttingen.

REFERENCES

Drechsler, D.M., Ultman, J.S., 1984, Cardiogenic mixing in the pulmonary conducting airways of man?. Respir. Physiol. 56: 37-44.

DuBois, A.B., Fowler, R.C., Soffer, A., Fenn, W.O., 1952, Alveolar CO_2 measured by expiration into the rapid infrared gas analyzer. J. Appl. Physiol. 4: 526-534.

Engel, L.A., 1983, Gas mixing within the acinus of the lung. J. Appl. Physiol. 54: 609-618.

Engel, L.A., Menkes, H., Wood, L.D.H., Utz, G., Joubert, J., Macklem, P.T., 1973a, Gas mixing during breath holding studied by intrapulmonary gas sampling. J. Appl. Physiol. 35: 9-17.

Engel, L.A., Wood, L.D.H., Utz, G., Macklem, P.T., 1973b, Gas mixing during inspiration. J. Appl. Physiol. 35: 18-24.

Fukuchi, Y., Roussos, C.S., Macklem, P.T., Engel, L.A., 1976, Convection, diffusion and cardiogenic mixing of inspired gas in the lung:an experimental approach. Respir. Physiol. 26: 77-90.

Horsfield, K., Gabe, J., Mills, C., Buckman, M., Cumming, G., 1982, Effect of heart rate and stroke volume on gas mixing in dog lung. J. Appl. Physiol. 53: 1603-1607.

Hoffman, E.A., Ritman, E.L., 1985, Invariant total heart volume in the intact thorax. Am. J. Physiol. 249: H883-H890.

Jones, H.A., Chakrabarti, M.K., Davies, E.E., Hughes, J.M.B., Sykes, M.K., 1982, The contribution of heart beat to gas mixing in the lungs of dogs. Respir. Physiol. 50: 177-185.

Meyer, M., Hook, C., Rieke, H., Piiper, J., 1983, Gas mixing in dog lungs studied by single-breath washout of He and SF_6. J. Appl. Physiol. 55: 1795-1802.

Meyer, M., Slama, H., 1983, A versatile hydraulically operated respiratory servo system for ventilation and lung function testing. J. Appl. Physiol. 55: 1023-1030.

Piiper, J., Scheid, P., 1987, Diffusion and convection in intrapulmonary gas mixing. In: Handbook of Physiology. The Respiratory System. Gas Exchange. Bethesda, MD, Am. Physiol. Soc., sect. 3, vol IV, p. 51-69.

West, J.B., Hugh-Jones, P., 1961, Pulsatile gas flow in bronchi caused by the heart beat. J. Appl. Physiol. 16: 697-702.

THEORETICAL ANALYSIS OF FACTORS INFLUENCING RECOVERY OF VENTILATION DISTRIBUTIONS FROM INERT GAS WASHOUT DATA

D. Meyer, K. Groebe, and G. Thews

Physiologisches Institut der Universität Mainz
Saarstr. 21, D-6500 Mainz, West Germany

INTRODUCTION

For the quantitative analysis of intraregional ventilation inhomogeneities, one classically applies the inert gas washout method [7,24] in which an inert gas of negligible solubility in blood and tissue is washed into the lungs. After washin is complete, the inspiratory inert gas fraction is set to a smaller value and the time course of the mixed endexpiratory inert gas fraction during the subsequent inert gas washout is recorded. The determinants of this time course are: The endexpiratory alveolar volume at the instant of change in inspiratory inert gas fraction, the anatomical dead space, the gas exchange ratio, the respiratory frequency, the in- and expiratory tidal volumes, and the inhomogeneous distribution of the tidal volume among the alveolar space. For determining this distribution from the time course of mixed endexpiratory inert gas fractions, several methods have been described which may be grouped into two classes: The depeeling methods and mathematically more advanced methods for assessing continuous distributions.

In the graphical evaluation using the *depeeling method* (see e.g. [7]), the alveolar space generally is assumed to consist of two ore more parallel compartments which are ventilated continuously. The measured time course is graphed in a semi-logarithmic coordinate system and decomposed into several straight lines. From the slopes of these lines one obtains the specific ventilations i.e. the ratios of compartmental ventilations and volumes. Their intersections with the ordinate axis specify the share of compartmental ventilation in total alveolar ventilation. The validity of the depeeling method has widely been questioned (e.g. [7,18]), so it is not used any more in current research.

In order to better take account of the physiological situation in which ventilation is distributed continuously, more complex algorithms for the evaluation of time courses of inert gas washout have been developed. Gomez et al. [11,12], for example, used parametric ansatz functions, and Nakamura et al. [20,21] as well as Okubo and Lenfant [23] applied Laplace transforms to directly invert their time courses. The studies by Evans et al. [8] and by Lewis et al. [16,17] are based on a multicompartment model. Wagner [27] has been using linear programming and Butler and Mohler [4] determined the central moments of the ventilation distribution via its Taylor expansion.

Methodological objections have been raised to most of these approaches [4,13,16,25]. Furthermore, all of these methods make use of a number of assumptions which simplify the real physiological situation e.g. by neglecting the cyclical character of ventilation, physiological variations in tidal volumes and in endexpiratory alveolar volumes, and/or dead space inspiration. As will be shown in this study, these simplifying assumptions may give rise to considerable errors in the resulting ventilation distributions, particularly if measurement errors are involved also.

A further disadvantage of all of these methods lies in the fact that either the specific ventilation [13,16,17,27], its inverse [20,21,23], or the specific tidal volume [4,11,12], i.e. the ratio of local tidal volume and compartmental volume have been used for abscissa of the ventilation distribution. At a given location within the alveolar space, none of these quantities is invariant to variations in tidal volumes or respiratory frequency, even if the distribution pattern of the inspiratory gas stream remains the same. Rather than any of the above quantities, however, it is this distribution pattern which one is interested in when investigating ventilation distribution. Furthermore, any approach which uses the above quantities for abscissa of the ventilation distribution principally does not allow for a comparison of distributions which have been determined at different total alveolar ventilations (e.g. during exercise at different performances) even though the distribution patterns of the inspiratory gas stream are very well comparable. This is because changes in total alveolar ventilation go along with changes in specific ventilation at each site of the lung, by this giving rise to shifts of the entire distribution, the impacts of which upon its momentums are not quantifiable in a straightforward manner.

It may be objected that the errors introduced in former models by the above simplifications are of the order of a few percent only. Still their avoidance is crucial to yield correct results: Even though the time course of an inert gas washout implicitly contains all the information on intraregional ventilation inhomogeneities, the mathematical problem of recovering the underlying ventilation distribution from such a time course is very ill conditioned. This implies that small deviations in the mathematical model from the physiological situation or small errors in the input data may lead to enormous errors in the resulting distribution. Therefore, the mathematical model used has to reflect the physiological situation as closely as possible and measurements have to be performed with the highest possible accuracy. On the other hand, the number of mutually independent values describing the resulting distribution has to be kept well below the number of input data.

Along these guidelines an algorithm for determining ventilation distributions has been developed that avoids the above simplifying assumptions introduced in former models, which implies that the distribution pattern of the inspiratory gas stream is employed rather than the specific ventilation. As a consequence, this algorithm allows for direct comparison of ventilation distributions, independently of respective total alveolar ventilation. The algorithm has been used to perform computer simulations in which the impacts of former simplifying assumptions and of measurement errors upon the recovered ventilation distributions have been studied systematically.

METHODS

The mathematical model. The functionally inhomogeneous lung model on which the new algorithm is based consists of infinitely many parallel homogeneously ventilated alveolar compartments each of which is connected in series to a local dead space. Before entering the parallel arrangement of local dead spaces and alveolar compartments the inspired gas passes through a common dead space. This setup of the conducting airways takes account of the fact that dead space inspiration during each respiratory cycle at first occurs out of the peripheral airways next to the respective alveolar compartment and later on out of a common mixing chamber which corresponds to the upper bronchial tree, trachea, pharynx, oral passage, and additional external breathing devices [2,9,22]. In formulating the model

616

equations, several assumptions have been made some of which are necessary prerequisites to allow for a meaningful definition of a ventilation distribution:

1. It is possible to approximate the dead space gas entering each alveolar compartment by two discrete portions, one originating from the respective compartment and one from a common mixing chamber.

2. There is complete mixing in the common dead space, and the gas fractions there reflect the mixed expired gas fractions of all alveolar compartments weighted according to their expired ventilation. Implicit in this assumption is the absence of serial emptying of alveoli.

3. The ratio of local dead space volume and alveolar compartment volume is spatially constant.

4. The distribution pattern of the inspiratory gas stream is temporally constant during the experiment.

5. Compartmental inert gas exchange takes place through the conducting airways only. There is no inter-compartmental gas transport.

6. In considering the effects of the gas exchange ratio on compartmental volumes, a spatially constant gas exchange ratio is assumed. Thus, inert gas concentrations at the begining and at the end of the washout process are assumed to be spatially constant also.

In order to avoid the above poblems which arise from using the specific ventilation, its inverse, or the specific tidal volume for abscissa of the ventilation distribution, the "normalized specific ventilation" has been introduced into the analysis which is defined as the ratio of specific ventilation over total alveolar ventilation. This new quantity specifies the fraction of the total alveolar ventilation alloted to one unit of volume of the respective alveolar compartment and thus directly reflects the distribution pattern of the inspiratory gas stream. At any location, the normalized specific ventilation is temporally constant, independently of natural variations in tidal volumes and respiratory frequency. Furthermore, it allows for a direct comparison of ventilation distributions which have been determined at different total alveolar ventilations.

The model equations describing the *discontinuous process* of inert gas washout out of a homogeneous alveolar compartment following a change in inspiratory inert gas fraction has been derived from an inert gas mass balance. For an alveolar compartment of the normalized specific ventilation $\mathring{V}_A$ one obtains for the fractional change in endexpiratory inert gas fraction after n inspirations following the change in inspiratory gas

$$F_{r_n}(\mathring{V}_A) = \frac{F_n(\mathring{V}_A) - F_\infty}{F_0 - F_\infty} \tag{1}$$

$$= \prod_{i=1}^{n} \frac{\mathring{V}_A^{-1}\left(1 + V_{DS_{P_i}}/V_{A_i}\right) + V_{A_{i-1}} - V_{A_i} + V_{DS_{c_i}}\frac{F_{i-1}-F_\infty}{F_{i-1}-F_\infty}}{\mathring{V}_A^{-1} + V_{T_{E_i}}},$$

where F_n (F_{i-1}) are the local alveolar inert gas fractions after the n-th ($i-1$-th) breathing cycle, F_0 and F_∞ are the alveolar inert gas fractions before and after the washout process, F_{i-1} is the mixed endexpiratory inert gas fraction after the $i-1$-th expiration, V_{A_i} ($V_{A_{i-1}}$) are the endexpiratory alveolar volumes after the i-th ($i-1$-th) expiration, $V_{T_{E_i}}$ is the i-th expiratory tidal volume, $V_{DS_{c_i}}$ is the volume of the common dead space, and $V_{DS_{P_i}}$ is the sum of all local dead space volumes.

For the fractional change of the mixed endexpiratory inert gas fraction F_{r_n} (which is defined in analogy to F_{r_n} in Eq. (1)) of the functionally inhomogeneous lung in which the normalized specific ventilation is distributed according to an unknown probability density function $\varphi(\mathring{V}_A)$ the following relation holds:

$$\mathbf{F}_{r_n} = \int_{\mathring{V}_{A_{min}}}^{\mathring{V}_{A_{max}}} \prod_{i=1}^{n} \frac{\mathring{V}_A^{-1}\left(1 + \mathbf{V}_{DS_{P_i}}/\mathbf{V}_{A_i}\right) + \mathbf{V}_{A_{i-1}} - \mathbf{V}_{A_i} + \mathbf{V}_{DS_{c_i}} \frac{F_{i-1}-F_\infty}{F_{i-1}-F_\infty}}{\mathring{V}_A^{-1} + \mathbf{V}_{TE_i}} \varphi(\mathring{V}_A)\, d\mathring{V}_A \qquad (2)$$

In the past, the analysis of ventilation distributions has frequently been performed using algorithms which model ventilation as a continuous process during which the alveolar volume is assumed to be constant [13,20,21,23]. In order to assess the errors introduced by specifically this simplifying assumption, a *continuous algorithm* had to be developed in which – other than in former studies – variations in tidal volumes are considered and the normalized specific ventilation and its continuous distribution are utilized. In analogy to Eq. (2) one obtains for the functionally inhomogeneous lung

$$\mathbf{F}_r(\mathbf{V}_{TA_c}) = \int_{\mathring{V}_{A_{min}}}^{\mathring{V}_{A_{max}}} \exp\left(-\mathring{V}_A \mathbf{V}_{TA_c}\right) \varphi(\mathring{V}_A)\, d\mathring{V}_A, \qquad (3)$$

where $\mathbf{V}_{TA_c}$ is the accumulated alveolar expiratory volume which replaces the time as independent variable of the washout process.

From a number of investigations performed with radioactive gases [6,19,29], there is evidence that in healthy subjects the distribution of the alveolar ventilation with respect to the normalized specific ventilation is continuous and unimodal. Therefore, it is an obvious choice to approximate the probability density $\varphi(\mathring{V}_A)$ in Eqs. (2) and (3) by means of a transformed beta distribution. Its parameters mean m_1, variance σ^2, and skewness S are determined by a non-linear least squares algorithm that approximates the measured and the calculated mixed endexpiratory inert gas fractions.

The computer simulations have been performed according to the following scheme: Based on a given ventilation distribution, on given physiological variations in the various volumes, and – in some runs – on a realistic distribution of measurement errors, the discontinuous algorithm (2) was employed to generate time courses of inert gas washout which, in turn, were analyzed using modelling approaches based on various simplifying assumptions. By comparing the original and the recovered distributions the adequacy of each approach may be judged.

After it had been made sure that given ventilation distributions could be reproduced correctly by the discontinuous algorithm even if the starting values of the parameters (mean, variance, and skewness of the distribution) for the non-linear least squares fit were varied systematically to cover a wide physiological range, the following questions were studied:

1. Which errors are to be expected if evaluation is performed with the continuous algorithm?

2. Wich errors are to be expected if physiological variations of the various volumes are neglected in the analysis, and mean values are used instead?

3. Which is the role dead space inspiration plays in the analysis? In particular, which consequences are to be expected if
   (a) only a common dead space is considered?
   (b) only the local dead spaces connected in series to the alveolar compartments are considered?

618

(c) the distribution of total dead space volume among common dead space and local dead spaces is varied?

(d) dead space inspiration is neglected at all?

4. How are random or systematic errors in the volume measurements propagated into the resulting distributions?

The simulated inert gas time courses in this paper are all – unless explicitly stated – based on the following physiological data:

- Expiratory tidal volumes $\mathbf{V}_{T_{E_i}}$ are $0.55\ l \pm 0.05\ l$. These values are typical for healthy subjects at rest (see e.g. [5,14,15]).

- Endexpiratory alveolar volumes $\mathbf{V}_{A_i}$ are $3.075\ l \pm 0.05\ l$ (see e.g. [3,14,28]). Actual expiration and alveolar volumes were generated using a random number generator for normal distributions.

- The total dead space is $0.175\ l$. Its volume depends on alveolar volume according to $\Delta\mathbf{V}_{DS_i}/\Delta\mathbf{V}_{A_i} = 0.03$ (see [1,10,26]). The fraction of common dead space in total dead space was chosen 0.75 in correspondence to morphometric data by NUNN et al. [22] (cf. also [2,9]).

The parameters of the generating ventilation distributions in the below examples have been chosen as follows: mean $m_1 = 0.35$, variance $\sigma^2 = 0.01$ and skewness $S = -0.40$. These values are in the range found in healthy subjects at rest.

RESULTS AND DISCUSSION

In the following, the results of the computer simulations regarding the questions raised above are presented:

1. In the analysis of ventilation distributions considerable errors have to be expected if the cyclical process of ventilation is described as a continuous process (Fig. 1). The resulting ventilation distribution is left-shifted as compared to the generating distribution, and its variance and skewness are about 50 % and 200 % of the original ones, respectively. Dead space inspiration which is neglegted in the continuous algorithm tends to slow down the washout process and – due to gas mixing in the dead space – to level out differences in the time constants of different alveolar compartments (see below). These effects are so to speak compensated for by a left-shifted and narrower ventilation distribution resulting from evaluation using the continuous algorithm.

2. Significant errors are found as well if the discontinuous algorithm is applied, but variations in expiratory tidal volumes are neglected and mean values are used instead (Fig. 3a). Neglecting variations in endexpiratory alveolar volumes is not as critical: Variations of $\pm 0.05\ l$ do not show any effect upon the recovered distribution. Greater variations of e.g. $\pm 0.1\ l$ as they may be found under physiological conditions [3,14,28] lead to errors if evaluation is performed under the assumption of a constant endexpiratory alveolar volume (Fig. 3b). Physiological variations in dead space volumes may be neglected without any consequences for the resulting distribution.

3. The quality of the evaluation largely depends on the choice of the dead space model applied. Dead space inspiration from the peripheral conducting airways (which are modelled as local dead spaces connected in series to the respective alveolar compartments) results in a protraction of inert gas washout which is relatively constant throughout the lung. In addition to a further protraction, inspiration from the central conducting airways (which are modelled as a single dead space common to all alveolar compartments) entails a homogenization of inert gas fractions in compartments of differing ventilation. As a consequence,

619

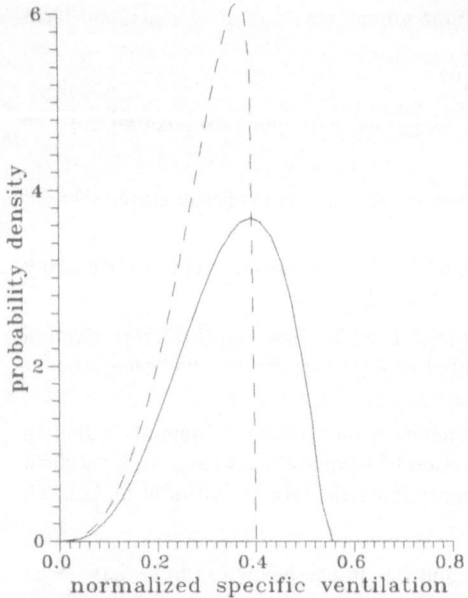

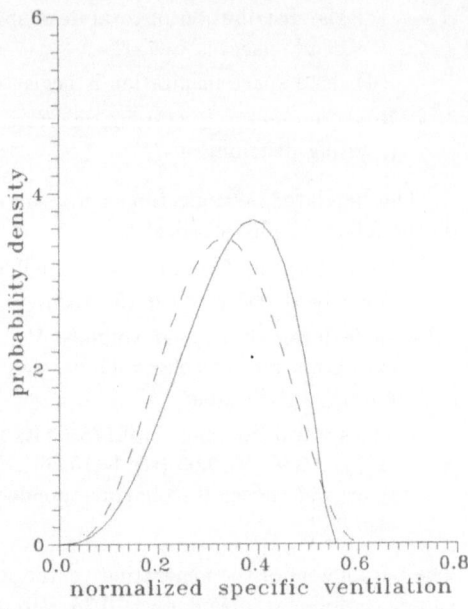

Figure 1. Effects of neglecting the cyclical character of the ventilation process. Generating ventilation distribution (solid, $m_1 = 0.35$, $\sigma^2 = 0.01$, $S = -0.40$) and distribution recovered by means of the continuous algorithm (dashed, $m_1 = 0.295$, $\sigma^2 = 0.005$, $S = -0.78$).

Figure 2. Effects of systematic measurement errors (expiratory tidal volumes by 2 % too large). Generating ventilation distribution (solid, $m_1 = 0.35$, $\sigma^2 = 0.01$, $S = -0.40$) and recovered distribution (dashed, $m_1 = 0.325$, $\sigma^2 = 0.011$, $S = -0.08$).

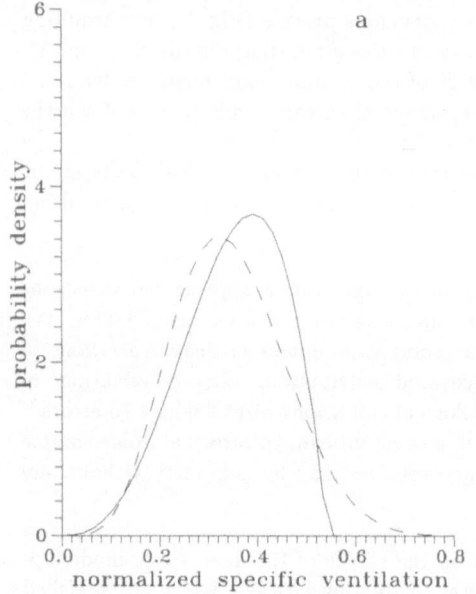

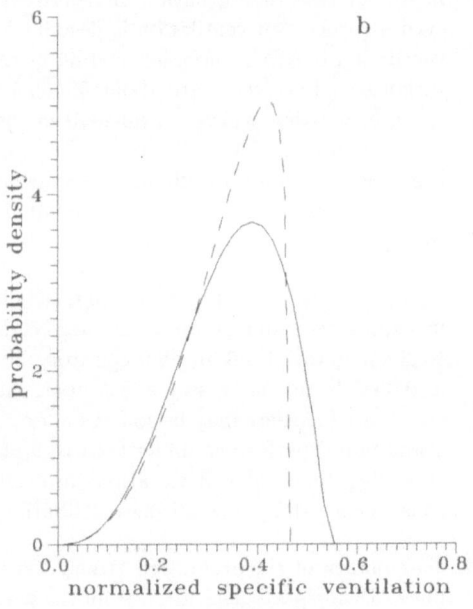

Figure 3. Effects of neglecting variations of $\pm\,0.05\,l$ in expiratory tidal volumes (3a) or of $\pm\,0.1\,l$ in endexpiratory alveolar volumes (3b). Solid: Generating ventilation distribution ($m_1 = 0.35$, $\sigma^2 = 0.01$, $S = -0.40$). Dashed: Recovered distributions (3a: $m_1 = 0.345$, $\sigma^2 = 0.013$, $S = 0.30$. 3b: $m_1 = 0.340$, $\sigma^2 = 0.007$, $S = -0.75$).

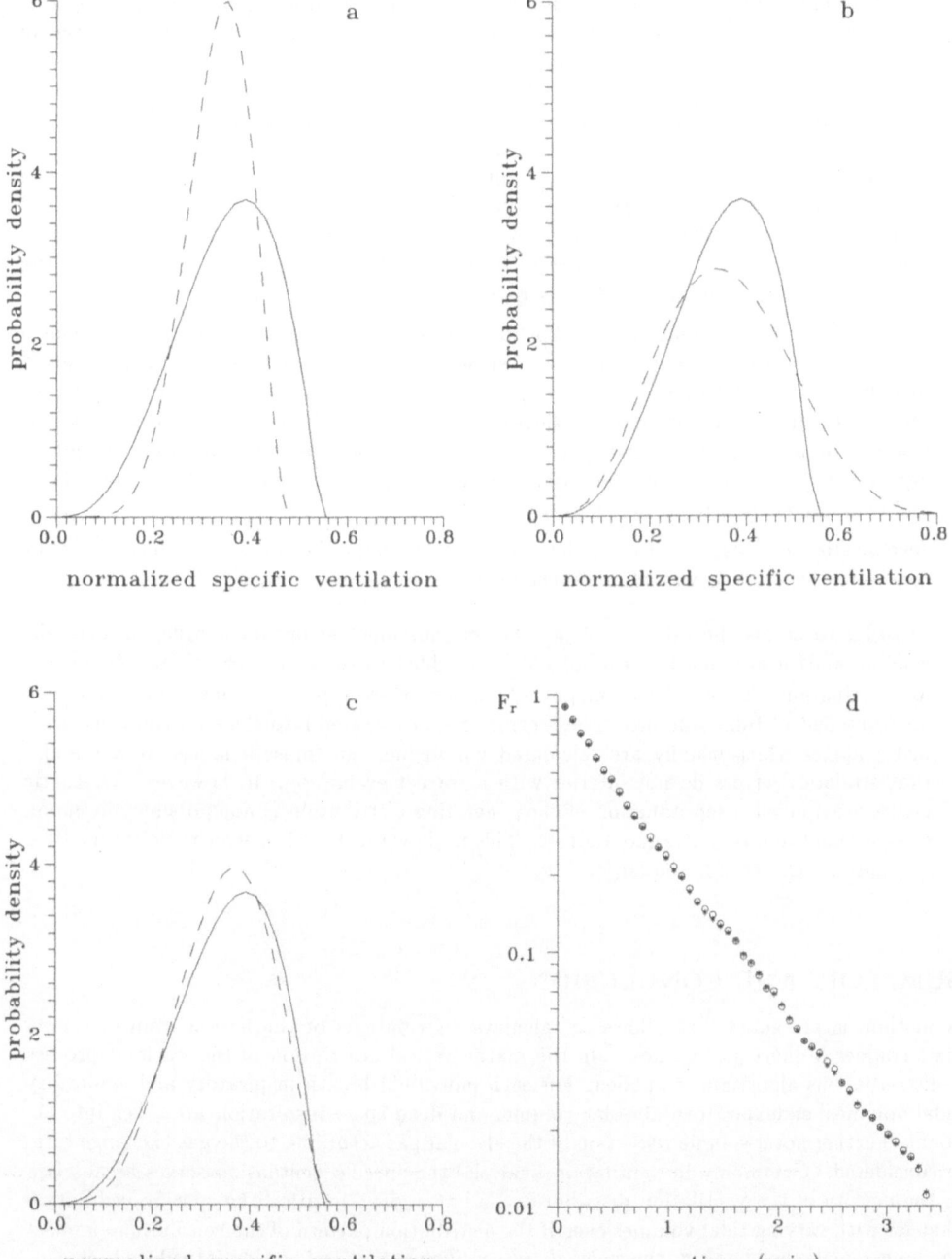

Figure 4. Effects of various dead space models (4a: Total dead space assumed to be local dead space. 4b: Total dead space assumed to be common dead space. 4c: Common and local dead space of same magnitude). Solid: Generating ventilation distribution ($m_1 = 0.35$, $\sigma^2 = 0.01$, $S = -0.40$, common dead space / local dead space = 3). Dashed: Recovered distributions (4a: $m_1 = 0.332$, $\sigma^2 = 0.004$, $S = -0.42$. 4b: $m_1 = 0.360$, $\sigma^2 = 0.017$, $S = 0.20$. 4c: $m_1 = 0.348$, $\sigma^2 = 0.009$, $S = -0.27$). 4d: Generated ($*$) and approximating (o) time course for the case 4a. Even though the original time course is approximated very well by the local dead space algorithm a totally distorted ventilation distribution results.

protraction is more pronounced in the better ventilated compartments. If the analysis is performed assuming the dead space to consist of local dead spaces connected in series to the individual alveolar compartments only (see e.g. BUTLER and MOHLER [4]) the resulting ventilation distribution consequently is much more homogeneous than the generating distribution and shifted to smaller normalized specific ventilations. As Fig. 4a shows, the variance of the resulting distribution is about 60 % smaller than the original one. On the other hand, the variances of the resulting distributions are by 70 % larger than the original ones if dead space is assumed to consist of the common dead space only (Fig. 4b) as has been suggested by EVANS et al. [8]. Fig. 4d demonstrates that in spite of the large differences between generating and resulting distributions in Fig. 4a approximation of the original time course by means of an incorrect model may be excellent.

If the portion of common dead space in total dead space assumed in the analysis deviates from its value used in generating the washout by $\pm$ 0.1, reproduction of the generating distribution is close to perfect. For larger deviations the resulting distributions resemble more and more the ones from one of the extreme cases Figs. 4a,b. Fig. 4c shows a distribution for common and local dead spaces of the same magnitude, indicating that reasonable agreement between recovered and original distribution is achieved even if the size of local and common dead space is not known precisely.

As expected the greatest errors in the evaluation are found if dead space inspiration is not considered at all, as it was common use in the older literature.

4. In order to assess the impacts of possible measurement errors for ventilation analysis, random and/or systematic errors of 2 % were added to the expiratory tidal volumes before evaluation. Errors of this magnitude or larger are typical for pneumotachographic measurement of tidal volumes. These errors are propagated into the endexpiratory alveolar volumes which usually are calculated via an inert gas mass balance. It was found that statistical errors do not interfer with a correct evaluation. If, however, systematic errors are applied a reproduction of the generating distribution is not possible any more, even without additional random errors. This is shown in Fig. 2 in which expiratory tidal volumes are chosen 2 % too large.

SUMMARY AND CONCLUSION

A method is presented that allows to calculate distributions of ventilation from measured time courses of inert gas washout. In the mathematical description of the washout process a discontinuous algorithm is applied: For each individual breath inspiratory and expiratory tidal volumes, endexpiratory alveolar volume, and dead space inspiration are taken into account. Furthermore, volume reduction of the alveolar gas according to the gas exchange ratio is considered. Commonly in ventilation analysis, the specific ventilation serves as abscissa of the density of the ventilation distribution. As at a given location the specific ventilation changes with varying tidal volumes even if the distribution pattern of the ventilation amongst the lung remains unchanged, the *normalized specific ventilation* is newly introduced instead. This quantity is defined to be the ratio of regional alveolar ventilation and regional endexpiratory alveolar volume divided by the total alveolar ventilation. The normalized specific ventilation reflects the distribution of the ventilation independently of variations in tidal volume and respiratory frequency. **Furthermore, it allows direct** comparison of ventilation distributions that are determined at varying alveolar ventilations. Ventilation distributions are approximated by the transformed beta distribution which is parameterized by its mean, variance, and skewness.

In order to evaluate simplifications introduced in former studies and to quantify their effects on the resulting ventilation distributions, washout time courses are generated in a computer simulation from the comprehensive discontinuous algorithm and are used to recover ventilation distributions by means of accordingly simplified algorithms. Furthermore, the influence of errors that may occur in the measurement of tidal volumes are assessed. The results of these studies are summarized as follows: Serious errors are introduced in the recovered distributions if ventilation is modelled as a continuous process and if physiological variations in tidal volumes or endexpiratory alveolar volumes or dead space inspiration are neglected. Modelling the entire dead space as common dead space or as local dead space only, entails significant errors as well. Statistical errors of 2 % in the measured volumes practically do not have any impacts on the recovered distributions whereas systematic errors significantly deteriorate the results.

In conclusion, in ventilation analysis it is essential to apply a discontinuous description of the inert gas washout process that accounts for dead space inspiration and variations in the above mentioned quantities. In addition it is important to obtain all measured values with the highest achievable precision.

REFERENCES

[1] BIRATH, G.: Respiratory dead space measurements in a model lung and healthy human subjects according to the single breath method. *J.Appl.Physiol.* 14: 517–520 (1959)

[2] BOUHUYS, A.: Respiratory dead space. In: *Handbook of Physiology*, W.O. Fenn and H. Rahn (eds.), sect. 3, vol. 1: Respiration. Am. Physiol. Soc., Washington, D.C., 1964, p. 699–714.

[3] BEAVER, W.L., N. LAMARRA, K. WASSERMAN: Breath-by-breath measurement of true alveolar gas exchange. *J.Appl.Physiol.* 51: 1662–1675 (1981)

[4] BUTLER, J.P., J.G. MOHLER: Estimating a distribution's central moments: a specific tidal ventilation application. *J.Appl.Physiol.* 46: 47–52 (1979)

[5] DEJOURS, P., R. PUCINELLI, J. ARMAND, M. DICHARRY: Breath-to-breath variations of pulmonary gas exchange in resting man. *Respir.Physiol.* 1: 265–280 (1966)

[6] DEMEDTS, M., I. CLARYSSE, M. DE ROO: Scintigraphic evaluation of shape of lung and chest in upright and head-down posture. *J.Appl.Physiol.* 60: 427–432 (1986)

[7] ENGEL, L.H.: Intraregional gas mixing and distribution. In: Gas mixing and distribution in the lung, L.A. Engel and M. Paiva (eds.), Dekker, New York, 1985, p. 287–358.

[8] EVANS, J.W., D.G. CANTOR, J.R. NORMAN: The dead space in a compartmental lung model. *Bull.Math.Biophys.* 29: 711–718 (1967)

[9] FORTUNE, J.B., P.D. WAGNER: Effects of common dead space on inert gas exchange in mathematical models of the lung. *J.Appl.Physiol.* 47: 896–906 (1979)

[10] FOWLER, W.S.: Lung function studies. II. The respiratory dead space. *Am.J.Physiol.* 154: 405–416 (1948)

[11] GOMEZ, D.M.: A mathematical treatment of the distribution of tidal volume throughout the lung. *Proc.Nat.Acad.Sci.U.S.* 49: 312–319 (1963)

[12] GOMEZ, D.M., W.A. BRISCOE, G. GORDON: Continuous distribution of specific tidal volume throughout the lung. *J.Appl.Physiol.* 19: 683–692 (1964)

[13] HEISE, M., H. KOPPE, K. SCHMIDT: Mathematical treatment of inert gas clearence curve as a method for studying regional inhomogeneity of alveolar ventilation in the lung. *Respiration* 31: 310–317 (1974)

[14] HLASTALA, M.P., B. WRANNE, C.J. LENFANT: Cyclical variations in FRC and other respiratory variables in resting man. *J.Appl.Physiol.* 34: 670–676 (1973)

[15] LENFANT, C.: Time-dependent variations of pulmonary gas exchange in normal man at rest. *J.Appl.Physiol.* 22: 675–684 (1967)

[16] LEWIS, S.M., J.W. EVANS, A.A. JALOWAYSKI: Continuous distributions of specific ventilation recovered from inert gas washout. *J.Appl.Physiol.* 44: 416–423 (1978)

[17] LEWIS, S.M.: Emptying patterns of the lung studied by multiple-breath N_2 washout. *J.Appl.Physiol.* 44: 424–430 (1978)

[18] LIEW, H.D. VAN: Semilogarithmic plots of data which reflect a continuum of exponential processes. *Science* 138: 682–683 (1962)

[19] MILIC-EMILI, J., J.A.M. HENDERSON, M.B. DOLOVICH, D. TROP, K. KANEKO: Regional distribution of inspired gas in the lung. *J.Appl.Physiol.* 21: 749–759 (1966)

[20] NAKAMURA, T., T. TAKISHIMA, T. OKUBO, T. SASAKI, H. TAKAHASHI: Distribution function of the clearence time constant in the lung. *J.Appl.Physiol.* 21: 227–232 (1966)

[21] NAKAMURA, T., T. TAKISHIMA, Y. SAGI, T. SASAKI, T. OKUBO: A new method of analyzing the distribution of mechanical time constants in the lung. *J.Appl.Physiol.* 21: 265–270 (1966)

[22] NUNN, J.F., E.J.M. CAMPBELL, B.W. PECKETT: Anatomical subdivisions of the volume of respiratory dead space and effect of position of the jaw. *J.Appl.Physiol.* 14: 174–176 (1959)

[23] OKUBO, T., C. LENFANT: Distribution function of lung volume and ventilation determined by lung N_2 washout. *J.Appl.Physiol.* 24: 658–667 (1968)

[24] PAIVA, M., L.A. ENGEL: Theoretical studies of gas mixing and ventilation distribution in the lung. *Physiol.Rev.* 67: 750–796 (1987)

[25] PESLIN, R., S. DAWSON, J. MEAD: Analysis of multicomponent exponential curves by the Post-Widder's equation. *J.Appl.Physiol.* 30: 462–472 (1971)

[26] SHEPARD, R.H., E.J.M. CAMPBELL, H.B. MARTIN, T. ENNS: Factors affecting the pulmonary dead space as determined by single breath analysis. *J.Appl.Physiol.* 11: 241–244 (1957)

[27] WAGNER, P.D.: Information content of the multibreath nitrogen washout. *J.Appl.Physiol.* 46: 579–587 (1979)

[28] WESSEL, H.U., R.L. STOUT, C.K. BASTANIER, M.H. PAUL: Breath-by-breath variation of FRC: effect on V_{O_2} and V_{CO_2} measured at the mouth. *J.Appl.Physiol.* 46: 1122–1126 (1979)

[29] WEST, J.B., C.T. DOLLERY: Ventilation-perfusion ratio in the lung, measured with radioactive CO_2. *J.Appl.Physiol.* 15: 405–410 (1960)

CONTINUOUS DISTRIBUTIONS OF VENTILATION AND GAS CONDUCTANCE TO PERFUSION IN THE LUNGS

Kazuhiro Yamaguchi, Akira Kawai, Masaaki Mori, Kohichiro Asano, Tomoaki Takasugi, Akira Umeda and Tetsuro Yokoyama

Department of Medicine, School of Medicine, Keio University, Tokyo 160, Japan

INTRODUCTION

Overall gas transfer in the lung under a steady state is commonly considered to be limited mainly by uneven distribution of ventilation-perfusion ratios ($\dot{V}_A/\dot{Q}$) and by diffusion impairment (Wagner, 1977). Although remarkable methods allowing one to know the distribution of $\dot{V}_A/\dot{Q}$ in the lung have recently been developed (Yokoyama and Farhi, 1967; Wagner et al., 1974; Evans and Wagner, 1977), no reliable tool has been introduced for directly solving the issue of whether diffusion impairment across the blood-gas barrier contributes significantly to determining the efficiency of pulmonary gas exchange. In 1961, Piiper theoretically analyzed the importance of $\dot{V}_A/\dot{Q}$ and diffusion impairment in terms of "$\dot{V}_A/\dot{Q}$-D/$\dot{Q}$ field", but rigorous method to realize his theory has not been accomplished. The present study was, therefore, undertaken to develop a new method for detection of distribution of $\dot{V}_A/\dot{Q}$ and diffusing capacity in the lung using nine gases as indicator gases. Applying the present procedure to patients with interstitial lung disease, the importance of $\dot{V}_A/\dot{Q}$ inequality and of diffusion impairment will be discussed.

METHODS

Indicator gases. In order to assess a quantitative significance of maldistribution of $\dot{V}_A/\dot{Q}$ as well as of diffusion impairment for the gas exchange in the lung, nine gases with varied diffusivity and solubility were used for analysis. Among them, sulfur hexafluoride (SF_6), ethane, cyclopropane, halothane, diethyl ether and acetone are physiologically inert gases. The remainder are O_2, CO_2 and CO, all of which combine chemically with hemoglobin molecules (Hb) in red cells. The reason why these gases are used in the present study will be discussed in detail later (see below).

Diffusive conductance. The gas transfer efficiency at each lung unit was taken to be limited both by $\dot{V}_A/\dot{Q}$ and by $G/\dot{Q}$, G is diffusive conductance for a given indicator gas (i.e. diffusing capacity) defined in a certain lung unit. Gas transfer between alveolar gas and pulmonary capillary blood may be perturbed by several factors such as diffusion limitation in gas phase, in alveolar-capillary membrane including plasma layer and in red cells as well as chemical reactions (Scheid and Piiper, 1989). Since the relative contribution of these individual components in determining the efficiency of gas transfer might not be quantified by any physiological approach, use was made of a simple model in which two capacitors, for the gas storage in alveolar space and in capillary blood, were connected by a resistive membrane characterized by its transfer conductance, G. G at each gas exchange unit is defined as:

$$G = \dot{M}/(P_A - P\bar{c}) \tag{1}$$

where $\dot{M}$ and P_A are, respectively, transfer rate and alveolar partial pressure of a given gas. $P\bar{c}$ is mean partial pressure of the gas in the pulmonary capillary. In case of inert gases for which a diffusive process is predominant, G values are mutually related through their Krogh diffusion constants, K:

$$G_1/G_2 = K_1/K_2 \tag{2}$$

K is expressed in terms of the product of the diffusion coefficient, d and the solubility, α of the gas in the resistive membrane. Although limiting roles imposed by diffusion in the resistor are not uniform but consist of at least three major processes such as gas-phase diffusion disequilibrium within the alveolar region (i.e. stratified inhomogeneity), diffusion which occurs in aqueous media including plasma layer and inside the cells and that across the cell membranes (mainly composed of lipids), their effects were quantitatively estimated, as a whole, converting any diffusive resistance into that caused by diffusion in aqueous media (see discussion). This consideration led one to practically assume that the solubility of the inert gas within the hypothetical membrane, α was identical to that in the blood, β. Diffusion coefficient, d in the membrane was taken to be inversely proportional to the square root of the molecular weight of the gas (Wagner, 1977).

Among the gases combining with Hb, O_2 is least affected by a limiting role imposed by chemical reaction kinetics so that diffusive process, especially in aqueous media, may be the prevalent determinant for O_2 transfer within the resistive membrane (Yamaguchi et al., 1987). In contrast, CO_2 transfer may be significantly limited by reactions (Klocke, 1987) so that G_{CO_2}/G_{O_2} of the resistive membrane was taken to be 3.3 in place of 20. The former value, 3.3 was reported by Piiper et al. (1980) who determined G_{CO_2} and G_{O_2} simultaneously by rebreathing in healthy adults, while the latter was predicted from diffusion in aqueous media.

CO transfer between alveolar space and capillary blood may be limited by both diffusive processes and reaction kinetics possessed of a relatively slow rate of CO combining with Hb in red cells (cf. Scheid and Piiper, 1989). Although a general agreement regarding the relative significance of diffusion and reaction accounting for CO gas exchange has not been reached, the analysis was made assuming the value of G_{O_2}/G_{CO} in a homogeneous lung unit to be 1.18, the value experimentally obtained by Meyer et al. (1981).

Partial pressures in pulmonary capillary. To estimate effective partial pressure of the gas along the pulmonary capillary, the solubility of a given inert gas, β was taken to be equal in plasma and interior of the red cell.

Because of nonlinear dissociation curves, the effective solubility, β (=slope of dissociation curves in terms of concentration-partial pressure plots) for O_2, CO_2 and CO are not constant but vary depending on their partial pressures surrounding Hb molecules. Furthermore, there exist allosteric interactions among three gases in combining with Hb in

626

red cells. To estimate the changes of P_{O2} and P_{CO} in the pulmonary capillary, Hb dissociation curves for O_2 and CO established in our laboratory were applied. In them, quantitative influences of allosteric interactions mediated by the hydration of CO_2, i.e. CO_2 Bohr effect and by intracellular 2,3-diphosphoglycerate (DPG) were taken into account in addition to the competitive interactions between O_2 and CO. Detailed descriptions on Hb dissociation curves in the presence of O_2 and CO were given elsewhere (Yamaguchi et al., 1981; 1988).

To obtain a relationship between content and partial pressure of CO_2 in pulmonary capillary, the CO_2 dissociation curve originally reported by Kelman (1967) was combined with the empirical expression on the interrelation of P_{CO2} and pH given by Mengden et al. (1969). The ratio determining the distribution of CO_2 in red cells and in plasma was assumed to be a function both of pH and total saturation of Hb with O_2 and CO.

Lung model and principle of analysis. A lung model consisting of a large number of compartments ventilated and perfused in parallel, each characterized by its own value of $\dot{V}_A/\dot{Q}$ and of $G/\dot{Q}$, was used for analysis.

The units with infinitely large resistance to gas transfer ($G/\dot{Q}=0$) do not contribute to gas exchange irrespective of the $\dot{V}_A/\dot{Q}$ ratio, constituting alveolar dead space and blood shunt concurrently. On the other hand, gas exchange efficiency of the units having no resistance to gas transfer ($G/\dot{Q}=\infty$) is determined solely by their $\dot{V}_A/\dot{Q}$. The lung units which are perfused but not ventilated ($\dot{V}_A/\dot{Q}=0$) may show a small amount of mass flow unless their $G/\dot{Q}$ is definitely zero. However, gas exchange provided by such a mechanism may be nearly negligible as compared to that provided by convective flow. Therefore, unventilated units were designated as blood shunt. The units which are not perfused but ventilated ($\dot{V}_A/\dot{Q}=\infty$) may not significantly contribute to gas exchange regardless of the $G/\dot{Q}$ ratio. Their ventilation, thereby, constitutes alveolar dead space. There may exist a number of lung units simultaneously exhibiting the effect of alveolar dead space and of blood shunt (i.e. units with $G/\dot{Q}=0$). However, these might not be practically specified in terms of any physiological procedure so that two particular compartments were assumed for representing either dead space or blood shunt.

Generally the conservation of mass for a given indicator gas, i, at a lung unit, j, yields the following relation:

$$\dot{V}_{AIj} \cdot P_{Ii} - \dot{V}_{Aj} \cdot P_{Ai,j} = \dot{Q}_j \cdot (Cc'_{i,j} - C\bar{v}_i)/Bg \tag{3}$$

where $\dot{V}_{AI}$ is inspired alveolar ventilation, while P_I denotes inspired partial pressure. Cc' and $C\bar{v}$ are, respectively, end-capillary and mixed venous contents. The term Bg is the capacitance coefficient in a gaseous medium (Piiper et al, 1971).
Equilibration kinetics across the resistive membrane can be described by the differential equation:

$$(P_{Ai,j} - Pc_{i,j}) \cdot dG_{i,j} = \dot{Q}_j \cdot dCc_{i,j} \tag{4}$$

where the increment of the blood content of the gas is expressed by dCc which is brought about due to gas transfer through an element of the membrane with diffusive conductance, dG.

Since the sum of all pressures within alveolar space should be equal to barometric pressure, P_B, the following relation holds:

$$\sum_{i=1}^{9} P_{Ai,j} + P_{AN2j} + P_{H2O} = P_B \tag{5}$$

in which P_{AN2} indicates alveolar N_2 pressure existing as the resident gas, while P_{H2O} is water vapor pressure at body temperature. Combining equations (3) and (4) with (5), alveolar partial pressure and end-capillary content of the indicator gas can be expressed either explicitly (for inert gases) or implicitly (for O_2, CO_2 and CO) as a function of $\dot{V}_A/\dot{Q}$ and $G/\dot{Q}$. The integration of equation (4), i.e. well-known Bohr integration, can readily be performed for the inert gas, resulting in an exponential partial pressure equilibration between alveolar space and capillary blood. On the other hand, numerical procedures are of necessity to perform the Bohr integration for O_2, CO_2 or CO owing to the nonlinearity of the Hb dissociation curves as well as the allosteric interaction among them. The numerical computations were carried out using the effective dissociation curves for O_2, CO_2 and CO as described above.

Based on mixing equations, the concentration of any gas in mixed arterial blood can be described as the perfusion-weighted average of the concentrations of the gas in the individual lung units:

$$\sum_{j=1}^{N} Q_j \cdot Cc'_{i,j} = Q_T \cdot Ca_i \tag{6}$$

where $\dot{Q}_T$ is total blood flow, while Ca is arterial content of the gas. N is the number of gas exchange units assumed in the lung model. To obtain a representative distribution of $\dot{V}_A/\dot{Q}$ and of $G/\dot{Q}$ by virtue of perfusion, a sum of squares of deviations, S, between the measured arterial content of the gas and its predicted value from the model was minimized with respect to the fractional perfusion q_j defined by $\dot{Q}_j/\dot{Q}_T$. S is given as:

$$S = \sum_{i=1}^{9} W_i \cdot \left(\sum_{j=1}^{N} Cc'_{i,j} \cdot q_j - Ca_i \right)^2 \tag{7}$$

in which weights W_i are the reciprocals of the estimated variance of the arterial contents Ca_i. Since the lung model consists of parameters whose number greatly exceeds the number of measured variables, unknowns in a system are not statistically orthogonal to one another and are very sensitive to experimental errors (Evans and Wagner, 1977). To overcome this difficulty, the ridge regression (Hoerl and Kennard, 1970) was introduced incorporated with the equality constraint that in the solution fractional perfusion q_j sums to 1.0:

$$L = \sum_{i=1}^{9} W_i \cdot \left(\sum_{j=1}^{N} Cc'_{i,j} \cdot q_j - Ca_i \right)^2 + \mu_1 \cdot \left(1 - \sum_{j=1}^{N} q_j \right) + \sum_{j=1}^{N} \mu_{2j} \cdot q_j^2 \tag{8}$$

where L is the function leading to optimization. The second and last terms of the equation designate the equality constraint and the revision in terms of ridge regression, respectively. μ_1 and μ_{2j} are Lagrange multipliers for enforcing these constraints. Modifying the numerical techniques developed by Wagner et al. (1977; 1980) for processing inert gas data to find a representative distribution of $\dot{V}_A/\dot{Q}$ in the lung, the function L was minimized with respect to q_j. In the analysis, a Lagrange multiplier μ_1 was readily determined and replaced with an adequate value in the midst of computation (Evans and Wagner, 1977). Choice of optimal values of μ_{2j} stabilizing the resulting distribution was made empirically by recovering a wide variety of known distributions of $\dot{V}_A/\dot{Q}$ and $G/\dot{Q}$ from error-perturbed data and comparing them to original distributions (see results). A physiologically important requirement that all perfusions should be positive, i.e. nonnegativity constraint

628

was settled by a trial and error procedure outlined by Olszowka and Wagner (1980).

Clinical application. 15 patients with interstitial pneumonia of unknown etiology were studied. They were placed in the supine position and a balloon-tipped Swan-Ganz catheter was introduced into the femoral vein and advanced into the pulmonary artery. An arterial cannula was inserted into the femoral artery. The patients were given a mixture of 21% O_2 and 0.1% CO in N_2 as the inspired gas and a normal saline containing a small quantity of SF_6, ethane, cyclopropane, halothane, diethyl ether and acetone via the antecubital vein. After a steady state was established, the expired gas and the samples of both arterial and mixed venous blood were simultaneously taken slowly at a uniform rate. Concentrations of O_2 and CO_2 in the expired gas were measured with the Scholander gas analyzer, P_{O_2} and P_{CO_2} in the blood sample with the electrodes. Measurements on inert gases were made using a gas chromatograph equipped either with a flame ionization detector (FID) or with an electron capture detector (ECD). Concentration of CO in the sample was examined with FID after converting CO into methane by means of nickel as catalyst.

RESULTS

Detectable range of $\dot{V}_A/\dot{Q}$ and of $G/\dot{Q}$. The $\dot{V}_A/\dot{Q}$ ratio is dimensionless, but the $G/\dot{Q}$ is given in the unit of ml(STPD)/(ml·Torr) in the present study. The relationship between $\dot{V}_A/\dot{Q}$ and retention, R of the inert gas is depicted in Figure 1. R was defined as $Cc'/C\bar{v}$. As far as $G/\dot{Q}$ was constant, the position of R-$\dot{V}_A/\dot{Q}$ curve given by each inert gas differed significantly from the other. Thus, R of the least soluble gas, SF_6 fell away steeply at $\dot{V}_A/\dot{Q}$ areas as low as 0.005, while that of the most soluble gas, acetone at $\dot{V}_A/\dot{Q}$ as high as 100. The R-$\dot{V}_A/\dot{Q}$ curves were also affected by $G/\dot{Q}$ especially for the range less than 10^{-4} where a marked diffusion limitation might exist (see below). However, the shift caused by $G/\dot{Q}$ was not distinct at both ends of the R-$\dot{V}_A/\dot{Q}$ curve. Thus, the $\dot{V}_A/\dot{Q}$ spectrum detectable by the inert gases used would range between 0.005 and 100 regardless of $G/\dot{Q}$.

R values for O_2, CO_2 and CO were defined as $Cc'_{O_2}/(Hb)$, $Cc'_{CO_2}/C\bar{v}_{CO_2}$ and $Cc'_{CO}/(Hb)$, respectively, where (Hb) denoted an effective concentration of Hb. R-$\dot{V}_A/\dot{Q}$ curves of O_2, CO_2 and CO showed that these gases would detect the lung units with the $\dot{V}_A/\dot{Q}$ ranging from 0.1 to 10, not exceeding the range provided by inert gases.

R value of a given inert gas changed significantly at the $G/\dot{Q}$ below 10^{-4} due to diffusion limitation but not at the $G/\dot{Q}$ over 10^{-3} where the gas exchange was solely governed by $\dot{V}_A/\dot{Q}$ (Fig. 2).

Variation of R for CO_2 was noticeable in the range of $G/\dot{Q}$ between 10^{-4} and 10^{-3}, whereas that for CO at the $G/\dot{Q}$ more than 0.1 where the most precipitous change was found at the $G/\dot{Q}$ of around 1 (Fig. 2). R for O_2 increased at the lung units having the $G/\dot{Q}$ ranging between 10^{-4} and 10^{-3} mainly due to the diminution in a limiting role imposed by diffusion. On the other hand, R for O_2 decreased considerably at the lung units with the $G/\dot{Q}$ more than 0.1 because of concomitantly increasing C_{CO} at a higher $G/\dot{Q}$ unit in which equilibration between alveolar gas and capillary blood was practically attained for O_2 but not for CO (Fig. 2).

It is, thus, clear that nine gases used as for the indicator may allow one to characterize the behavior of gas exchange attributed to the lung units having the $\dot{V}_A/\dot{Q}$ ranging from 0.005 to 100 and $G/\dot{Q}$ from 10^{-5} to 1 with considerable reliability.

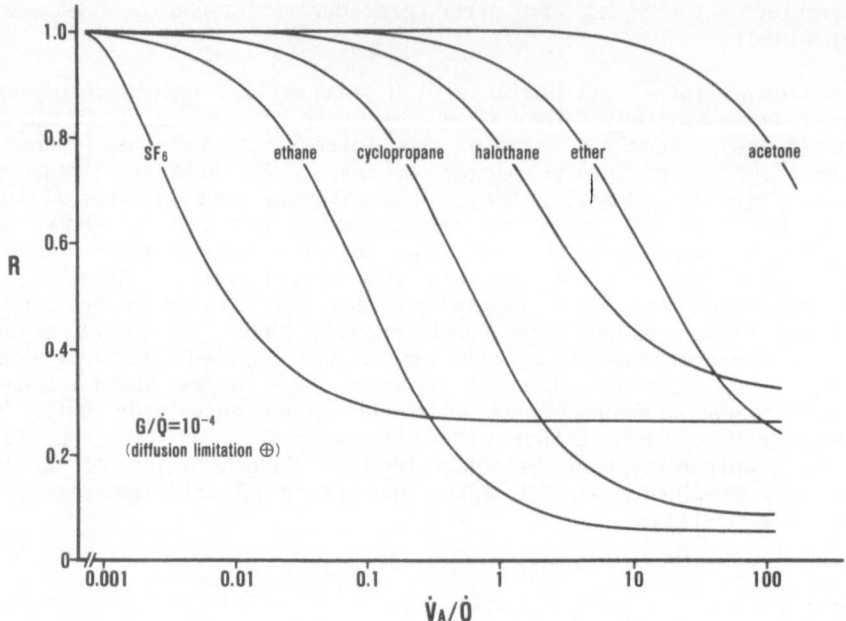

Fig.1.   Effect of $\dot{V}_A/\dot{Q}$ on inert gas exchange. $R:Cc'/C\bar{v}$. $G/\dot{Q}$ is given in the unit of ml(STPD)/(ml·Torr).

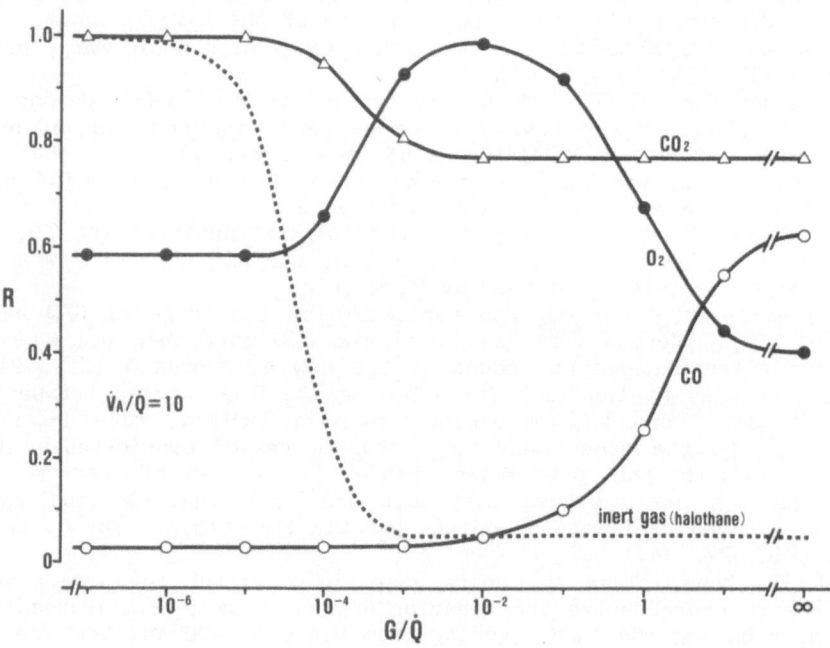

Fig.2.   Effect of $G/\dot{Q}$ on gas exchange of indicator gases. See text for further explanation.

The number of lung units. As long as there were sufficient units more than 20 along either $\dot{V}_A/\dot{Q}$ or $G/\dot{Q}$ axis, the discrepancy between the original and recovered distributions, quantified by the root-mean-square difference (RMS), did not improve. Adding more lung units simply resulted in a scaling change on both axes without significant improvement in RMS.

Lagrange multipliers, μ_2. Choice of adequate values for μ_2 in eq. (8) was empirically made by a trial and error procedure comparing original distribution and those recovered by error-perturbed data which were constructed by addition of 3% random errors, the magnitude encountered in practice, to a set of ideal retentions. For this 20 different distributions of $\dot{Q}$ up to quadrimodal patterns on $\dot{V}_A/\dot{Q}-G/\dot{Q}$ field were originally assumed and the retentions not containing any appreciable error were calculated. Subsequently these ideal retentions were perturbed by adding the small numbers selected randomly by computer and thus 30 sets of retentions for each distribution, mimicking realistic data, were generated. Thereafter resulting sets of retentions with substantial errors were used to compute the distributions predicted with various μ_2. Finally the agreement between original and recovered distributions was examined in terms of the RMS difference to choose the optimal μ_2 describing a reliable distribution of $\dot{V}_A/\dot{Q}$ and of $G/\dot{Q}$ in the lung. The following μ_2 was, thus, preferred:

$$\mu_{2j} = \left[20 \cdot \left\{ 1 + \left(\frac{\dot{Q}_T}{\dot{V}_E} \right)^2 \cdot \left(\frac{\dot{V}_A}{\dot{Q}} \right)^2_j \right\} \cdot \left\{ 1 + 10^n \cdot \left(\frac{\dot{Q}_T}{\dot{V}_E} \right)^2 \cdot \left(\frac{G}{\dot{Q}} \right)^2_j \right\} \cdot \left\{ 1 + \frac{10^{-4}}{(G/\dot{Q})_j} \right\}^2 \right]^{\frac{1}{2}} \qquad (9)$$

where n=3 when $G/\dot{Q}<=0.1$, while n=5 when $G/\dot{Q}>0.1$.

$\dot{V}_A/\dot{Q}$ and $G/\dot{Q}$ distributions in patients with interstitial pneumonia. Availing of the method developed in this study, distributions of $\dot{V}_A/\dot{Q}$ and $G/\dot{Q}$ in patients with interstitial pneumonia of unknown etiology (IIP) were examined. In Figure 3 is depicted the representative distribution of $\dot{Q}$ on $\dot{V}_A/\dot{Q}-G/\dot{Q}$ field obtained from the typical case with active inflammation in the interstitium which is pathologically consistent with desquamative interstitial pneumonia. $\dot{Q}$ distribution along $G/\dot{Q}$ axis was bimodal and peak flows were observed at $G/\dot{Q}$ of $5 \cdot 10^{-5}$ and 10^{-2}. Extremely low $G/\dot{Q}$ regions less than 10^{-4} did exist and received 5% of the cardiac output. On the other hand, most of the lung was operating in the normal $\dot{V}_A/\dot{Q}$ around 0.9 but 9.5% of the total blood flow was associated with essentially unventilated units.

The mean values of fractional perfusions in the lung units obtained from taking the average of all patients are shown in Table 1. The lung units with $G/\dot{Q}$ below 10^{-4} and those between 10^{-4} and 10^{-3} received, respectively, 0.8% and 9.9% of the cardiac output, the remainder being associated with units having $G/\dot{Q}$ ranging from 10^{-3} to 10^{-1}. Most of the blood flow along the $\dot{V}_A/\dot{Q}$ axis was found in the areas with $\dot{V}_A/\dot{Q}$ between 0.1 and 1.0. Exceedingly low $\dot{V}_A/\dot{Q}$ areas ($\dot{V}_A/\dot{Q}<0.1$) received the blood flow less than 10% of the cardiac output whereas 25% of the total flow was associated with relatively high $\dot{V}_A/\dot{Q}$ areas ($1<\dot{V}_A/\dot{Q}<10$).

DISCUSSION

Critique of methods. The crucial assumption in the present analysis is that on the mechanism limiting gas transfer across the blood-gas barrier in a homogeneous lung unit. O_2 transfer was presumed to be predominantly limited by diffusion in aqueous media. This is true for O_2 movement in alveolar-capillary membrane but may not hold strictly in the blood particularly in red cells where limitation by finite reaction

631

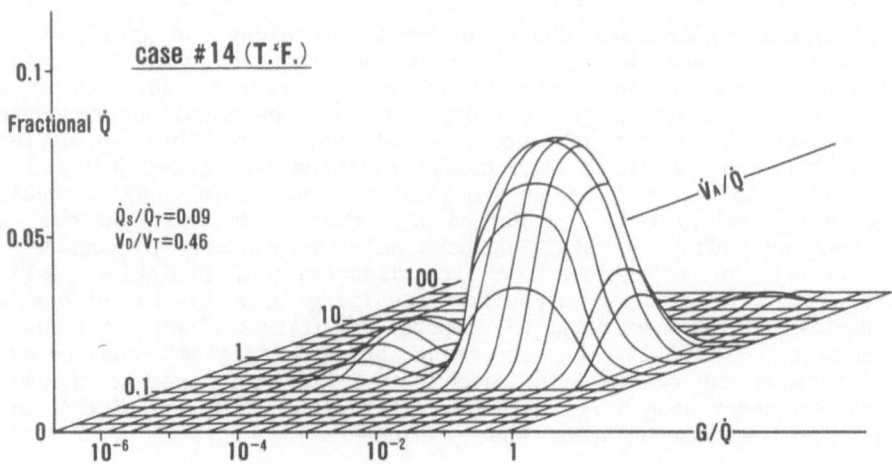

Fig.3. $\dot{Q}$ distribution on $\dot{V}_A/\dot{Q}$-$G/\dot{Q}$ field in patient with desquamative interstitial pneumonia.

Table 1. Fractional perfusion along $\dot{V}_A/\dot{Q}$ and $G/\dot{Q}$ axes in patients with interstitial pneumonia (mean, n=15).

$\dot{V}_A/\dot{Q}$	$\dot{Q}\,(\%)$	$G/\dot{Q}$	$\dot{Q}\,(\%)$
0 - 0.01	0.7	10^{-5}-10^{-4}	0.8
0.01 - 0.1	9.4	10^{-4}-10^{-3}	9.9
0.1 - 1	64.5	10^{-3}-10^{-2}	54.9
1 - 10	25.4	10^{-2}-10^{-1}	34.4
10 - 100	0.6	10^{-1}-1	0.02

rate of O_2 combining with Hb plays a role. However, recent studies on O_2 kinetics through red cells have confirmed that O_2 diffusion inside the red cell, i.e. diffusion in aqueous media is mainly responsible for the limitation but reaction of O_2 with Hb appears to be fast enough to prevent any major limitation of O_2 exchange (Yamaguchi et al., 1985; Hook et al., 1988). This may lend support to the appropriateness of the assumption that O_2 transfer through the blood-gas barrier can be treated in qualitatively the same manner as the inert gas excepting the varying effective solubility of O_2 in the blood.

Since the Krogh diffusion constant of CO_2 in tissues, K_{CO2} is larger by a factor of 20 than K_{O2}, diffusion equilibration for CO_2 is much more rapid than for O_2 (Wagner, 1977). However, a high diffusivity of CO_2 may be canceled out to a significant extent by the associated chemical reactions in the blood. Because of finite rates of reactions including carbamate formation, kinetics of the Haldane effect (i.e. interaction of O_2 and CO_2 via Hb) and trans-membrane ionic movement (i.e. exchange of chloride with bicarbonate), G values of CO_2, G_{CO2} at a given lung unit may be a considerably smaller than that predicted solely from diffusion limitation. In the present analysis, therefore, G_{CO2}/G_{O2} in each lung unit was taken to be 3.3 according to the experimental results observed by Piiper et al. (1980) for healthy adults in virtue of a rebreathing technique having an important advantage of diminishing any kind of inhomogeneity.

The rate of CO transfer is expected to be limited both by diffusion and reaction, however the relative importance of the individual components has not been identified with certainty. Although there has been much debate on the ratio of G_{O2} to G_{CO} in the lung (cf. Scheid and Piiper, 1989), the most appealing value has been proposed by Meyer et al. (1981) who simultaneously determined G_{O2} and G_{CO} in humans with a rebreathing technique using isotopic O_2 and CO, yielding G_{O2}/G_{CO} averaging 1.18. As this ratio is very close to that of the Krogh diffusion constants in tissues, one might interpret their results as indicative of the dominant role of diffusion in aqueous media, as compared to chemical reaction, in transfer of O_2 and of CO in the lung. Since the issue of whether or not G_{CO2}/G_{O2} and G_{O2}/G_{CO} vary along the capillary has not yet settled, the fixed ratios were simply used as a first approximation.

Our recent studies on O_2 transfer kinetics of red cells (Yamaguchi et al., 1985; 1987; 1988) have failed to provide any evidence for a limiting role of red cell membrane. Thus, diffusion limitation in red cell membrane, i.e. in nonaqueous phase, was considered to be of minor importance. Since the cells constituting the blood-gas barrier possess qualitatively the same cell membrane as the red cell, the assumption as described above may hold true in the whole process of gas exchange between alveolar gas and capillary blood in each lung unit, as well.

Any gas may be more or less affected by diffusion in gas phase depending on its diffusivity in the gaseous medium, the process being of much importance particularly in inert gas exchange (cf. Scheid and Piiper, 1989). The difficult problem arises as to discriminating the effect of diffusion limitation in gas phase from that in aqueous media. Since both effects might not be separated by means of the present method, a limiting role of diffusion in alveolar gas, if any, was evaluated converting it into the equivalent resistance imposed by diffusion in aqueous media.

The present study extended the numerical procedures elaborated by Wagner et al. (1977; 1980) for assessing a representative distribution of $\dot{V}_A/\dot{Q}$ from the multiple inert gas technique. Their method is characterized by introducing the idea of ridge regression originally proposed by Hoerl and Kennard (1970). The reliability of their numerical procedures has extensively been discussed by themselves on the basis of linear programming (Evans and Wagner, 1977; Olszowka and

Wagner, 1980). To examine the limits of the analytical technique of least-squares approach incorporated with ridge regression, recovery of $\dot{V}_A/\dot{Q}$ and $G/\dot{Q}$ distributions was tested based on the data for retentions containing substantial errors (see results). The current approach permitted to describe the location and magnitude of the distribution consisting of up to two modes on $\dot{V}_A/\dot{Q}$-$G/\dot{Q}$ field with considerable accuracy. Trimodal distributions were occasionally predicted but none of quadrimodal patterns were recovered.

Diffusion impairment in clinical cases. It has generally been believed that every inert gas does attain a complete equilibration of partial pressure between alveolar gas and capillary blood in a functionally homogeneous lung unit (cf. Wagner and West, 1980). The theoretical calculations, however, revealed that disequilibrium of an inert gas may appreciably take place in the lung unit with $G/\dot{Q}$ less than 10^{-4} (Fig. 2). Since experimental verification on this issue has not been made, we have refrained from assuming the completeness in diffusion equilibration of the inert gas across the blood-gas barrier. Thus, in two patients out of fifteen with IIP were detected such extremely low $G/\dot{Q}$ areas receiving 3-5% of total blood flow (Fig. 3). One is the patient of desquamative interstitial pneumonia with highly active inflammation and the other is the case with far advanced fibrosis. The results may indicate that diffusion impairment occurring either in alveolar gas space or in alveolar-capillary membrane somewhat interferes with inert gas transfer in certain cases having diffuse interstitial involvement. However, since remaining thirteen patients showed no areas of extremely low $G/\dot{Q}$ less than 10^{-4}, relative importance of diffusion limiting inert gas transfer is uncertain and requires further investigation.

O_2 exchange may fairly be perturbed in a lung unit with $G/\dot{Q}$ less than 10^{-3} owing mainly to diffusion limitation (Fig. 2). Thirteen patients manifested such relatively low $G/\dot{Q}$ areas where, on the average, about 10% of total blood flow was observed (Table 1). On the other hand, two patients at an early stage of the disease exhibited neither systemic hypoxemia nor lung units with $G/\dot{Q}$ below 10^{-3}. The findings may be consistent with the idea that failure of diffusion equilibration across the blood-gas barrier plays a role partly in the mechanism of hypoxemia in advanced disease of interstitial pneumonia. However, fractional perfusion flowing into the low $G/\dot{Q}$ areas is so small, i.e. 10%, that the drop in arterial P_{O_2} attributable to diffusion impairment may be difficult to detect by indirect classical methods. In fact, Wagner et al. (1976) failed to show a contribution of diffusion limitation to the hypoxemia observed at rest in patients with interstitial lung disease by means of the method based on multiple inert gas elimination, whereas they found distinct evidence for an important role of diffusion dominating a considerable portion of the hypoxemia in the same patients on exercise which might make $G/\dot{Q}$ distribution worsen further.

SUMMARY

Theoretical analysis and experimental observations were conducted to establish a method allowing to demonstrate the characteristics of distribution of ventilation ($\dot{V}_A$) as well as of diffusive conductance (G) to perfusion ($\dot{Q}$) in the lungs. O_2, CO_2 and CO binding to hemoglobin molecules within the erythrocyte together with six inert gases including SF_6, ethane, cyclopropane, halothane, diethyl ether and acetone, of varied solubility in blood and different diffusivity in lung tissue, were used as indicator gases. 15 patients with interstitial pneumonia of unknown etiology, placed in the supine position, were given a mixture of 21% O_2 and 0.1% CO in N_2 as the inspired gas and saline containing appropriate amount of the six inert gases was infused via an antecubital vein. After a steady state was established, the expired gas was collected and the samples of both arterial and mixed venous blood were simultaneously taken through catheters inserted into the femoral and pulmonary artery. The concentrations of the indicator gases in the samples were measured by gas chromatography, with electrodes or with the Scholander gas analyzer.

Assuming that the mass transfer efficiency of a given indicator gas at each gas exchange unit would be limited by $\dot{V}_A/\dot{Q}$ and $G/\dot{Q}$ ratios, the data obtained from the human subjects were analyzed in terms of a lung model having 20 units along the $\dot{V}_A/\dot{Q}$ and $G/\dot{Q}$ axes, respectively. The numerical analysis including the procedure of simultaneous Bohr integration for O_2, CO_2 and CO in a pulmonary capillary and the method of weighted least-squares combined with constrained optimization permitted the data to be transformed into a virtually continuous distribution of $\dot{Q}$ against $\dot{V}_A/\dot{Q}$ and $G/\dot{Q}$ axes. The numerical procedure was strictly tested using various artificial distributions of $\dot{V}_A/\dot{Q}$ and $G/\dot{Q}$ ratios, showing that it could characterize the distributions containing up to at least two modes on $\dot{V}_A/\dot{Q}$-$G/\dot{Q}$ field with a substantial accuracy.

Analytical results estimated from the patients with interstitial lung disease revealed the following features. (1) There appears to be bimodal distribution of $\dot{Q}$ along $G/\dot{Q}$ axis extending to relatively low $G/\dot{Q}$ less than 10^{-3} ml(STPD)/(ml·Torr), which may limit O_2 exchange between alveolar gas and capillary blood. This area of low $G/\dot{Q}$ receives 10% of total $\dot{Q}$. (2) Severe diffusion limitation causing disequilibrium of the inert gas across the blood-gas barrier is solely observed in 2 out of 15 patients and an amount of $\dot{Q}$ associated with this phenomenon is very small (below 1%). (3) These findings consistently suggest that inhomogeneity of $G/\dot{Q}$ does exist and may play an appreciable role in causing impairment of gas exchange in patients with interstitial pneumonia.

REFERENCES

Evans, J.W., and Wagner, P.D., 1977, Limits on $\dot{V}_A/\dot{Q}$ distributions from analysis of experimental inert gas elimination, J. Appl. Physiol., 42: 889.

Hoerl, A.E., and Kennard, R., 1970, Ridge regression: biased estimation for nonorthogonal problems, Technometrics, 12: 55.

Hook, C., Yamaguchi, K., Scheid, P., and Piiper, J., 1988, Oxygen transfer of red blood cells: experimental data and model analysis, Respir. Physiol., 72: 65.

Kelman, G.R., 1967, Digital computer procedure for the conversion of P_{CO_2} in blood CO_2 content, Respir. Physiol., 3: 111.

Klocke, R.A., 1987, Carbon dioxide transport, in: Handbook of Physiology, A.P. Fishman ed., Am Physiol. Soc., Bethesda, Maryland, sect. 3, vol. IV, p. 173.

Mengden, H-J., Schultehinrichs, D., and Thews, G., 1969, Dependence of plasma pH on oxygen saturation, Respir. Physiol., 6: 151.

Meyer, M., Scheid, P., Riepl, G., Wagner, H-J., and Piiper, J., 1981, Pulmonary diffusing capacities for O_2 and CO measured by a rebreathing technique, J. Appl. Physiol., 51: 1643.

Olszowka, A.J., and Wagner, P.D., 1980, Numerical analysis of gas exchange, in: Pulmonary gas exchange, J.B. West ed., Academic Press, N.Y., vol. I, p. 263.

Piiper, J. 1961, Variations of ventilation and diffusing capacity to perfusion determining the alveolar-arterial O_2 difference: theory, J. Appl. Physiol., 16: 507.

Piiper, J., Dejours, P., Haab, P., and Rahn, H., 1971, Concepts and basic quantities in gas exchange physiology, Respir. Physiol., 13: 292.

Piiper, J., Meyer, M., Marconi, C., and Scheid, P., 1980, Alveolar-capillary equilibration kinetics of $^{13}CO_2$ in human lungs studied by rebreathing, Respir. Physiol., 47: 29.

Scheid, P., and Piiper, J., 1989, Blood gas equilibration in lungs and pulmonary diffusing capacity, in: Respiratory Physiology, H.K. Chang and M. Paiva ed., Marcel Dekker, INC., N.Y. p.453.

Wagner, P.D., Saltzman, H.A., and West, J.B., 1974, Measurement of continuous distributions of ventilation-perfusion ratios: theory, J. Appl. Physiol., 36: 588.

Wagner, P.D., Dantzker, D.R., Dueck, R., dePolo, J.L., Wasserman, K., and West, J.B., 1976, Distribution of ventilation-perfusion ratios in patients with interstitial lung disease, Chest, 69: 256.

Wagner, P.D., 1977, Diffusion and chemical reaction in pulmonary gas exchange, Physiol. Review, 57: 257.

Wagner, P.D., and West, J.B., 1980, Ventilation-perfusion relationships, in: Pulmonary gas exchange, J.B., West, ed., Academic Press, N.Y., vol. I, p.219.

Yamaguchi, K., Kawashiro, T., and Yokoyama, T., 1981, Mathematical representation of the CO-Hb dissociation curves of whole blood under various O_2 tensions, Prog. Respir. Res., 16: 158.

Yamaguchi, K., Nguyen-Phu, D., Scheid, P., and Piiper, J., 1985, kinetics of O_2 uptake and release by human erythrocytes studied by a stopped-flow technique, J. Appl. Physiol., 58: 1215.

Yamaguchi, K, Glahn, J., Scheid, P., and Piiper, J., 1987, Oxygen transfer conductance of human red blood cells at varied pH and temperature, Respir. Physiol., 67: 209.

Yamaguchi, K., Mori, M., Kawai, A., and Yokoyama, T., 1988, influences of carbon monoxide on the binding of oxygen, carbon dioxide, proton and 2,3-diphosphoglycerate to human hemoglobin, Adv. exp. Med. Biol., 222: 299.

Yokoyama, T., and Farhi L.E., 1967, The study of ventilation-perfusion ratio distribution in the anesthetized dog by multiple inert gas washout, Respir. Physiol., 3: 166.

ASSESSMENT OF STRATIFIED INHOMOGENEITY WITHIN DISTAL ALVEOLAR SPACE WITH

RESPECT TO OXYGEN UPTAKE

Klaus-Dieter Schuster and Hartmut Heller

Institute of Physiology I
University of Bonn
5300 Bonn 1, FRG

INTRODUCTION

The persistence of longitudinal partial pressure gradients inside the alveolar space during the respiratory cycle, i.e. axial gas mixing deficit, is usually referred to as stratified inhomogeneity or stratification. The question whether or not there is a limitation of oxygen transport by stratified inhomogeneity in distal alveolar space has been studied in several experimental and theoretical approaches, but remains a controversial issue. Whereas the results of most theoretical studies (Rauwerda, 1946; Cumming et al., 1971; Paiva, 1973; Pack et al., 1977) suggest that the influence of stratification on gas exchange in lungs can be considered negligible, in some experimental investigations (Sikand et al., 1976; Hlastala et al., 1982), partial pressure gradients within alveolar space were estimated to be significantly higher than zero, indicating a relative importance of stratified inhomogeneity. For quantifying stratificational effects, Okubo and Piiper (1974) introduced a compartment model of the lung including a "stratificational conductance". There is evidence from all investigations performed up to now that such a stratificational conductance is considerably higher than the pulmonary diffusing capacity of oxygen. A major problem in determining a stratificational conductance is seperating it from the much lower conductances brought about by ventilation, diffusion and blood flow. This difficulty can be considerably reduced by applying a rapid single breath procedure to increase the conductance of ventilation, and by using oxygen-labeled carbon dioxide, $C^{18}O_2$. This isotopic species undergoes instantaneous removal by isotopic exchange when in contact with blood, so that limitations caused by diffusion and blood flow are lower by more than an order of magnitude for this gas compared with oxygen.

METHODS

Experimental procedure

Single breath manoeuvres were performed, as schematically shown in Fig. 1. These consisted of the following steps: (i) rapid inspiration of a

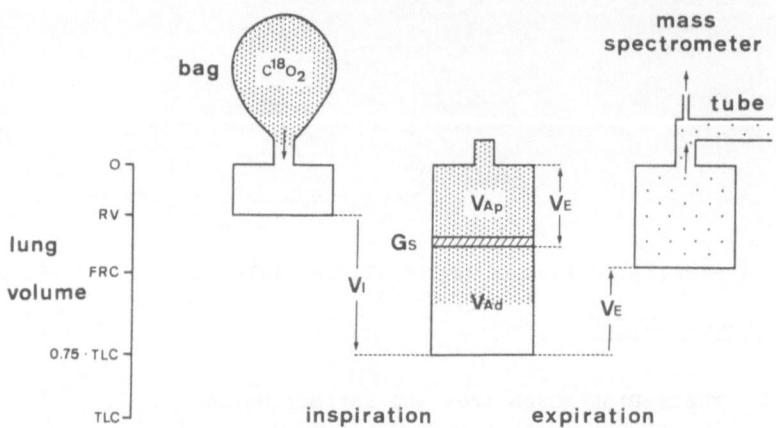

Fig. 1   Lung volume before inspiration, during breath holding and after expiration. The alveolar space is divided into a proximal and distal compartment V_{Ap} and V_{Ad}, which are separated by the stratificational conductance Gs. It has been arbitrarily assumed that Gs is located at the level at which the end-expired gas analysed comes from. RV: residual volume, FRC: functional residual capacity, TLC: total lung capacity, V_i, V_E: inspiratory and expiratory volume.

gas mixture containing a small amount of $C^{18}O_2$ from residual volume to about 75% of total lung capacity, (ii) breath holding (with varying duration in different experiments), (iii) rapid expiration down to a level slightly below functional residual capacity. Inspired and expired volume as well as $C^{18}O_2$ partial pressures of inspiratory and end-expiratory gas were measured.

Model analysis

The model as shown in Fig. 2 primarily consists of a proximal and distal alveolar compartment V_{Ap} and V_{Ad} as well as a compartment with flowing capillary blood, and these are separated from each other by boundaries representing conductances of gas transport. On account of serial arrangement, the overall conductance G_0 of this system can be calculated from

$$\frac{1}{G_0} = \frac{1}{G_s} + \frac{1}{D} + \frac{1}{\beta_B \dot{Q}} \tag{1}$$

where $\beta_B \dot{Q}$ is the conductance of blood flow $\dot{Q}$, D the pulmonary diffusing capacity and Gs the stratificational conductance. The latter is arbitrarily assumed to be located at the alveolar level at which the end-expired gas analysed came from (s. Fig. 1). Diffusing capacity D and blood capacitance coefficient β_B have been shown to be very high (Schuster,1987; Nanassy,1986).

When assuming the extreme case $D = \infty$ and $ß_B = \infty$, it follows that

$$G_O = G_S \tag{2}$$

and the model of Fig. 2 is reduced to only one compartment, that of the proximal alveolar volume V_{Ap}. Mathematical analysis of such a model shows that partial pressure P_{Ap} at the end of expiration can be described as the

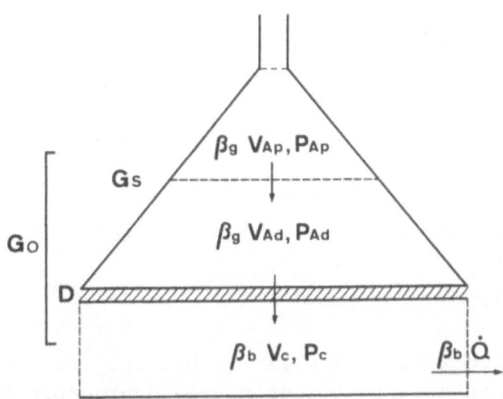

Fig. 2. Compartment model of the lung. V_{Ap}, V_{Ad}: volumes of proximal and distal alveolar compartment, V_c: capillary blood volume, P_{Ap}, P_{Ad}, P_c: partial pressures within the respective compartments. $ß_g$, $ß_B$ capacitance coefficients of gas phase and blood, G_s: stratificational conductance (s.Fig.1), D: diffusing capacity, $ß_b\dot{Q}$: blood flow conductance, G_O: overall conductance. For $C^{18}O_2$, D and $ß_B$ are very high. When assuming infinity for these parameters, the model is reduced to one compartment and G_O becomes a lower limit of G_s.

product of three functions

$$P_{Ap} = fi \cdot f_{BH} \cdot f_E \tag{3}$$

where fi, f_{BH} and f_E are partial pressure-time functions during inspiration, breath holding and expiration, respectively. Due to constant alveolar volume during breath holding, f_{BH} should follow the monoexponential term

$$f_{BH} = \exp \left(- \frac{G_O}{ß_g V_{Ap}} \cdot t_{BH} \right) \tag{4}$$

639

where β_g is the capacitance coefficient of gas phase. During inspiration and expiration, anatomical proportions within the lung vary in a complicated manner so that f_i and f_E are considered to be unknown. This difficulty can be circumvented by performing a set of experiments with very similar phases of inspiration and expiration but varying breath-hold time t_{BH}. Under these conditions, f_i and f_E should be the same for every experiment i.e. constant for the whole set. This can be formulated as follows:

$$c = f_i \cdot f_E \quad . \tag{5}$$

Therefore end-expired partial pressures P_{Ap} are expected to vary only with breath-hold function f_{BH}. Combining equations (3),(4) and (5) leads to

$$P_{Ap} = C \cdot \exp\left(- \frac{G_0}{\beta_g \, V_{Ap}} \cdot t_{BH}\right) \quad . \tag{6}$$

For estimating G_0, the rate constant λ was determined from semi-logarithmic plots of P_{Ap} against t_{BH}, and G_0 was calculated from

$$G_0 = \lambda \cdot \beta_g \cdot V_{Ap} \quad . \tag{7}$$

With respect to equation (2), G_0 equals G_s, but only for infinite values of D and β_B. In reality, these assumptions may be approached but not fulfilled. For finite values of D and β_B, it follows from equation (1) that

$$G_0 < G_s \tag{8}$$

i.e. G_0 can be regarded as a lower limit of G_s.

RESULTS

Experiments were performed on two healthy resting men, 36 and 37 years of age. End-expiratory partial pressures as a function of breath-hold time measured on subject A are given in Fig. 3. $C^{18}O_2$ disappearance from alveolar gas shows a biexponential time course. The slow phase of $C^{18}O_2$ removal may be caused by releasing the label from various pulmonary dead spaces as discussed in a previous paper (Schuster, 1985). For the points of the fast phase of $C^{18}O_2$ disappearance, the rate constant $\lambda = 3.9$ s^{-1} was evaluated for subject A. Table 1 contains data obtained for both subjects. Overall conductance G_0 amounts to 641 and 553 ml·mmHg^{-1}·min^{-1}. Diffusing capacity of oxygen, D_{O_2}, was determined by multiplying single breath carbon monoxide diffusing capacity with 1.2, according to Haab and Geiser (1986). The lowest value found for G_0 is considered to be a lower limit of G_s, so that stratificational conductance within distal alveolar gas should be higher than 553 ml·mmHg^{-1}·min^{-1}. For assessing this value with respect to oxygen uptake performance of lungs, the D_{O_2}/G_s ratio is given, and this turns out to be lower than 0.076.

640

Table 1. Data of 2 subjects for assessing stratificational conductance G_s.

S	λ s^{-1}	V_{Ap} l BTPS	G_o	D_{O_2} ml$\cdot$mmHg$^{-1}\cdot$min^{-1}	$\dfrac{D_{O_2}}{G_{O_2}}$
A	3.9	2.37	641	40	0.063
E	2.4	3.31	553	42	0.076

$$G_s > 553 \qquad \frac{D_{O_2}}{G_s} < 0.076$$

λ : rate constant of $C^{18}O_2$ disappearance, V_{Ap}: volume of proximal alveolar compartment, G_o: overall conductance of $C^{18}O_2$ uptake, D_{O_2}: oxygen diffusing capacity. The lowest value of G_o can be regarded as a lower limit of G_s.

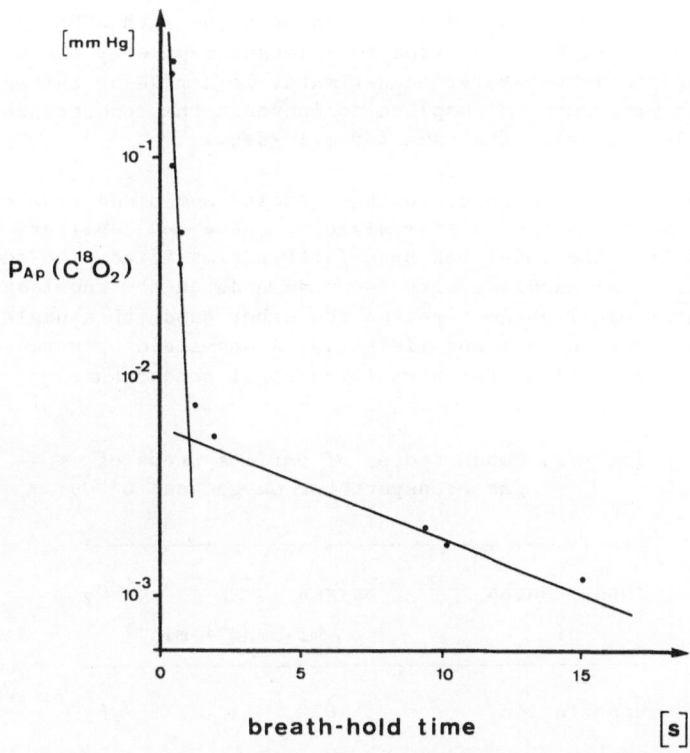

Fig. 3. Semi-logarithmic plot of end-expiratory $C^{18}O_2$ partial pressure against breath-hold time for subject A. The rate constant for $C^{18}O_2$ disappearance from proximal compartment V_{Ap} amounts to 3.9 s^{-1}.

641

DISCUSSION

Methodical aspects

An extremely simplified compartment model similar to that introduced by Okubo and Piiper (1974), was applied for assessing quantitatively the effects of stratified inhomogeneity. Advantages and disadvantages of such a model have been discussed in detail elsewhere (Okubo and Piiper,1974; Scheid and Piiper, 1980; Paiva and Engel, 1987). Therefore only those differences between the models and experimental approaches will be discussed which are expected to be of significance.

Whereas poorly soluble test gases such as helium and SF_6 (Okubo and Piiper, 1974; Kawashiro et al., 1976) or gases of medium solubility such as acetylen and freon (Adaro and Farhi, 1971; Hlastala et al., 1982) were used by other groups, the highly soluble species $C^{18}O_2$ was applied in our experiments. For the latter, a virtual capacitance coefficient of blood was calculated to be 125 $ml \cdot mmHg^{-1} \cdot ml^{-1}$, due to isotopic exchange (Nanassy, 1986). As a consequence, diffusing capacity as well as blood flow conductance are very high for this species, which is illustrated by comparing conductances of the various steps of gas transport between oxygen and $C^{18}O_2$ (Table 2). Diffusing capacity was found to be greater than 1100 $ml \cdot mmHg^{-1} \cdot min^{-1}$ (Schuster, 1987) and blood flow conductance is expected to be as high as 625 000 $ml \cdot mmHg^{-1} \cdot min^{-1}$. The consequence of such high values is that $C^{18}O_2$ uptake is far less limited by diffusion and blood flow when compared with other gases i.e. the stratificational effect is covered to a lesser degree by these steps of gas transport and should be easier to estimate. To arrive at this goal, a rapid single breath procedure was applied to increase the conductance of ventilation, which is primarily the same for all gases.

On account of the high diffusing capacity and blood flow conductance of $C^{18}O_2$, the compartments of distal alveolar space and capillary blood were neglected so that the model has been further simplified. The advantage is an easier mathematical handling with fewer demands on the knowledge of physiological or anatomical parameters. On the other hand, this neglection leads to a loss of information: a lower limit i.e. a one-sided open range instead of a single value was obtained for stratificational conductance.

Table 2. Conductances of various steps of gas transport for oxygen and $C^{18}O_2$

conductances of	oxygen	$C^{18}O_2$
	$ml \cdot mmHg^{-1} \cdot min^{-1}$	
ventilation	6	6
diffusion	40	>1100
blood flow	6	625 000

During phases of inspiration and expiration, anatomical proportions of the lung as well as gas transport performance vary in a complicated manner and modelling of these phases without introducing unprovable assumptions seems to be an unsolvable problem. Other groups have applied the simplified approach of assuming inspiration and expiration to be instantaneous so that half the ventilatory cycle is considered to consist only of breath holding at constant conditions (Okubo and Piiper, 1974; Adaro and Piiper, 1976). The consequences of this simplification are not easily assessable. Attempts were made to circumvent this problem in the present paper by estimating the stratificational conductance from a set of experiments with equal phases of inspiration and expiration but varying breath-hold time. It was rather difficult to fulfill the above conditions. In spite of a lot of training, only two of seven persons succeeded satisfactorily.

Assessment of the result and conclusions reached

Attempts to estimate effects due to stratified inhomogeneity are complicated by the fact that admixtures of dead spaces as well as of parallel inhomogeneities may also contribute to end-expiratory gas composition, and it turns out to be rather difficult to separate these different influences from each other. When attributing contribution of dead spaces and parallel inhomogeneities to the stratificational effect, stratificational conductance is underestimated. Supposing this case for our experiments, the "true stratificational conductance" should be greater than the estimated lower limit i.e. it is within the range found.

The result obtained in this paper represents a situation of gas transfer for a fixed state of lung expansion, as shown in Fig. 1, and is expected to depend on respiratory pattern to a lower extent than in other experimental approaches due to elimination of the varying phases of inspiration and expiration. Inspiratory volume was as great as 50 to 60 % of total lung capacity so as to guarantee the loading of the proximal compartment with inspiratory gas at the beginning of the breath-hold phase. On account of these conditions it is expected that the result obtained here represents gas mixing performance within distal alveolar space.

Although diffusing capacity and blood flow conductance is much higher for $C^{18}O_2$ than for O_2, stratificational conductances should be of similar size because of similar behaviour of both species within the gas phase. Fig.4 shows the result of the present paper as well as values of D_{O_2}/G_s ratios estimated from literature. The lower limit of D_{O_2}/G_s ratio determined in this paper is close to experimental results found by Okubo and Piiper (1974) on isolated dog lung lobes, but considerably lower than comparable findings of other authors (Fig. 4). Several causes may account for these differences, such as expansion status of lung and breathing pattern as well as methodical difficulties of estimating the true stratificational effect as discussed above.

Here G_s in distal alveolar space is found to be more than an order of magnitude greater compared to D_{O_2}. When assuming normal data of oxygen uptake at rest, an oxygen partial pressure drop within distal alveolar gas can be calculated to be lower than 0.5 mmHg, which is less than 1 % of the gradient between inspiratory gas and arterial blood. Therefore stratificational conductance within distal alveolar gas does not exhibit a limiting factor of oxygen uptake in lungs.

A well-known phenomenon of respiratory physiology is the sloping alveolar plateau. Several mechanisms are expected to contribute to alveolar slope, such as continuing gas exchange, sequential emptying of areas underlying unequal ventilation/volume or ventilation/perfusion ratios and, finally, the stratificational effect (Scheid and Piiper, 1980). With respect to the small partial pressure drop given above, it can be concluded that only a very small contribution of stratification to alveolar slope is expected.

SUMMARY

Investigations were made as to whether or not there is a limitation of oxygen transport by stratified inhomogeneity in distal alveolar gas. Experiments were performed on 2 subjects. Single breath manoeuvres were carried

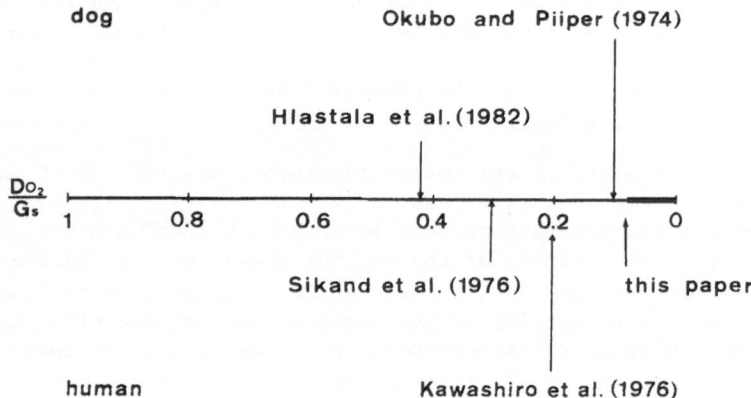

Fig. 4. D_{O_2}/G_s ratios as determined from literature data experimentally obtained on dogs and humans

out with similar phases of inspiration and expiration but varying breath-hold times. The inspiratory gas contained a small amount of oxygen labeled carbon dioxide, $C^{18}O_2$. End-expiratory gas was analysed on its residual $C^{18}O_2$ partial pressure by mass spectrometry. For evaluating a stratificational conductance, compartment model analysis was applied on the breath-hold data. Stratificational conductance has been found to be higher than 553 ml·mmHg^{-1}·min^{-1}. When transfered to oxygen transport, this means that stratificational conductance is more than 10 times higher than oxygen diffusing capacity. It can be concluded that (i) stratified inhomogeneity in distal alveolar space does not exhibit a limiting factor of oxygen uptake in lungs, (ii) a contribution of stratificational effects to sloping alveolar plateau is expected to be of minor importance.

644

REFERENCES

Adaro, F., and Farhi, L.E., 1971, Effects of intralobular gas diffusion on alveolar gas exchange, Fed. Proc., Fed. Am. Soc. Exp. Biol. 30:437.

Adaro, F., and Piiper, J., 1976, Limiting role of stratification in alveolar exchange of oxygen, Respir. Physiol., 26:195-206.

Cumming, G., Horsfield, K., and Preston, S., 1971, Diffusion equilibrium in the lungs examined by nodal analysis, Respir. Physiol., 12:329-345.

Haab, P., and Geiser, J., 1986, Relationship between O_2 and CO diffusing capacities, Prog. Resp. Res., 21:56-59.

Hlastala, M.P., McKenna, H.P., Middaugh, M., and Robertson, H.T., 1982, Role of diffusion-dependent gas inhomogeneity in gas exchange in the dog, Bull. Eur. Physiopathol. Respir., 18:373-380.

Kawashiro, T., Sikand, R.S., Adaro, F., Takahashi, H., and Piiper, J., 1976, Study of intrapulmonary gas mixing in man by simultaneous wash-out of helium and sulfur hexafluoride, Respir. Physiol., 28:261-275.

Nanassy, E., 1986, Der Transfer von ^{18}O-markiertem Kohlendioxid ($C^{18}O_2$) und Kohlenmonoxid (CO) zwischen Alveolarraum und Blut - Vergleich der Diffusionskapazitäten beider Gase, Inaugural-Dissertation, Universität Bonn, 1-85.

Okubo, T., and Piiper, J., 1974, Intrapulmonary gas mixing in excised dog lung lobes studied by simultaneous wash-out of two inert gases, Respir. Physiol., 21:223-239.

Pack, A., Hooper, M.B., Nixon, W., and Taylor, J.C., 1977, A computational model of pulmonary gas transport incorporating effective diffusion, Respir. Physiol. 29:101-124.

Paiva, M., 1973, Gas transport in the human lung, J. Appl. Physiol. 35:401-410.

Paiva, M., and Engel, L.A., 1987, Theoretical studies of gas mixing and ventilation distribution in the lung, Physiol. Rev., 76:750-796.

Piiper, J., 1979, Series ventilation, diffusion in airways, and stratified inhomogeneity, Fed. Proc., 38:17-21.

Rauwerda, P.E., 1946, Unequal ventilation of different parts of the lung and the determination of cardiac output, Ph. D. Thesis, State Univ. of Groningen, The Netherlands.

Scheid, P., and Piiper, J., 1980, Intrapulmonary gas mixing and stratification, in: "Pulmonary Gas Exchange", J. B. West, ed., Academic Press, New York, 1:87-130.

Schuster, K.-D., 1985, Kinetics of pulmonary CO_2 transfer studied by using labeled carbon dioxide $C^{16}O^{18}O$, Respir. Physiol, 60:21-37.

Schuster, K.-D., 1987, Diffusion limitation and limitation by chemical reactions during alveolar-capillary transfer of oxygen-labeled CO_2, Respir. Physiol., 67:13-22.

Sikand, R.S., Magnussen, H., Scheid, P., and Piiper, J., 1976, Convective and diffusive gas mixing in human lungs: experiments and model analysis, J. Appl. Physiol., 40:362-371.

Computer Modeling of Gas Phase O_2

Airway Transport

P.W. Scherer, G.R. Neufeld, S.J. Aukberg and S. Gobran

Dept. of Bioengineering and Dept. of Anesthesia
University of Pennsylvania, Philadelphia, PA   19104

INTRODUCTION

The first step in the exchange of O_2 between the atmosphere and the alveolar capillary blood is the exchange between the inhaled tidal volume and the functional residual capacity. This exchange accurs completely in the gas phase by molecular diffusion across the moving gaseous interface that separates these two volumes during inspiration and expiration. On this moving interface gas transport is governed by Fick's first law of diffusion which can be written for O_2 as

$$\dot{V}_{O_2} = D_{O_2-N_2} \cdot \frac{A}{P} \cdot \frac{dP_{O_2}}{dx} \tag{1}$$

where P is total gas pressure in the acinar airways.

Utilizing the relation dV = Adx between cumulative airway volume V, total airway cross section A, and airway path length x in a single path trumpet bell model gives

$$\dot{V}_{O_2} = D_{O_2-N_2} \cdot \frac{A^2}{P} \cdot \frac{dP_{O_2}}{dV} = D_{O_2-N_2} \cdot \frac{A^2}{P} \cdot \frac{\Delta P_{O_2}}{\Delta V} \tag{2}$$

The resistance to this diffusive gas exchange between V_T and FRC can be found by rearranging eqn. (2) as

$$R_{O_2}^{gas} = \frac{P\Delta V}{D_{O_2-N_2}A^2} = \frac{\Delta P_{O_2}}{\dot{V}_{O_2}} \tag{3}$$

where ΔP_{O_2} represents the moving air phase O_2 partial pressure difference along acinar airway volume ΔV.

Oxygen Transport to Tissue XII, Edited by J. Piiper *et al.*
Plenum Press, New York, 1990

METHODS AND RESULTS

We used a single path (Hansen et al; 1975) numerical trumpet bell model (Scherer et al., 1988) to investigate the relationship between ΔPo_2 as measured by the slope of phase III of the single breath O_2 washout curve and the values of ΔPo_2 that exist along the airways during inspiration when most of the O_2 exchange between the V_T and the FRC occurs.

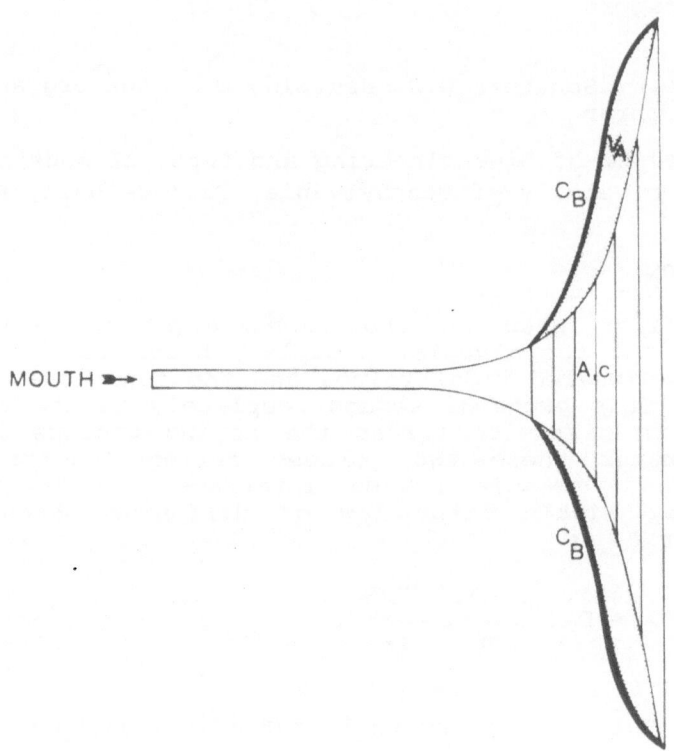

Figure 1. Schematic diagram of trumpet bell model used in numerical calculations.

Figure 1 is a schematic diagram of the single path trumpet bell model used in the calculations. The airways of cross sectional area A are assumed rigid while the alveolar volume V_A which is distributed over the acinar generations expands around them to inhale and exhale the tidal volume.

The governing equation for the intra airway O_2 concentration c is

$$\left(1 + \frac{V_A}{V}\right) \frac{\partial c}{\partial t} + \frac{\dot{Q}}{V} \frac{\partial c}{\partial z} = \frac{1}{V} \frac{\partial}{\partial z}\left(\frac{DA}{\ell} \frac{\partial c}{\partial z}\right) + \frac{1}{V} \alpha \dot{Q}_B \left(Cont._{art} - Cont._{ven}\right) \quad (4)$$

648

which along with boundary and initial conditions has been derived previously in the literature (Scherer et al., 1988) by the authors for CO_2. In equation 4, $D_{O_2-N_2}$ is the molecular diffusivity of O_2 in alveolar gas, z is the dimensionless generational coordinate, $\dot{Q}$ is airway gas volume flowrate per generation, α is the fraction of total alveoli per generation, ℓ is airway length per generation, and $\dot{Q}_B$ is total cardiac output. The last term on the right is a source term which represents the uptake of O_2 from the airway gas phase into the alveolar capillary blood. Cont. stands for the O_2 content bound to hemoglobin in arterial and venous blood.

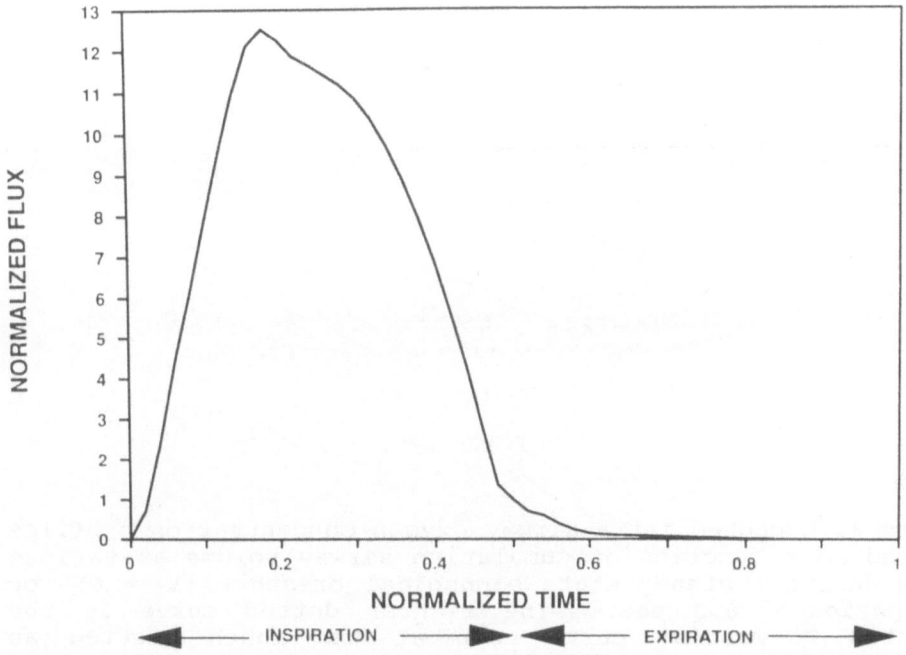

Figure 2. Computed normalized O_2 flux out of the moving tidal volume over a single sinusoidal breath of period 6.0 sec.

Figure 2 shows a numerical model computation of the O_2 flux out of the moving tidal volume into the FRC during sinusoidal breathing at 10BPM with a tidal volume of 650 cc. The O_2 flux (vol O_2/sec) is normalized by dividing by the average $\dot{V}_{O_2}$ uptake over the breath. It can be seen that most of the O_2 flux out of the tidal volume occurs during inspiration with a peak flux occurring about half way through inspiration.

To investigate the relationship between the intra airway O_2 concentration gradients and the O_2 concentration gradient (phase III) seen at the mouth on washout, we computed (see Fig. 3) intra airway P_{O_2} at various times and airway locations during inspiration and expiration and plotted it as a function of cumulative airway volume. We also superimposed on this plot the computed expired O_2 washout curve seen at the mouth (dotted curve).

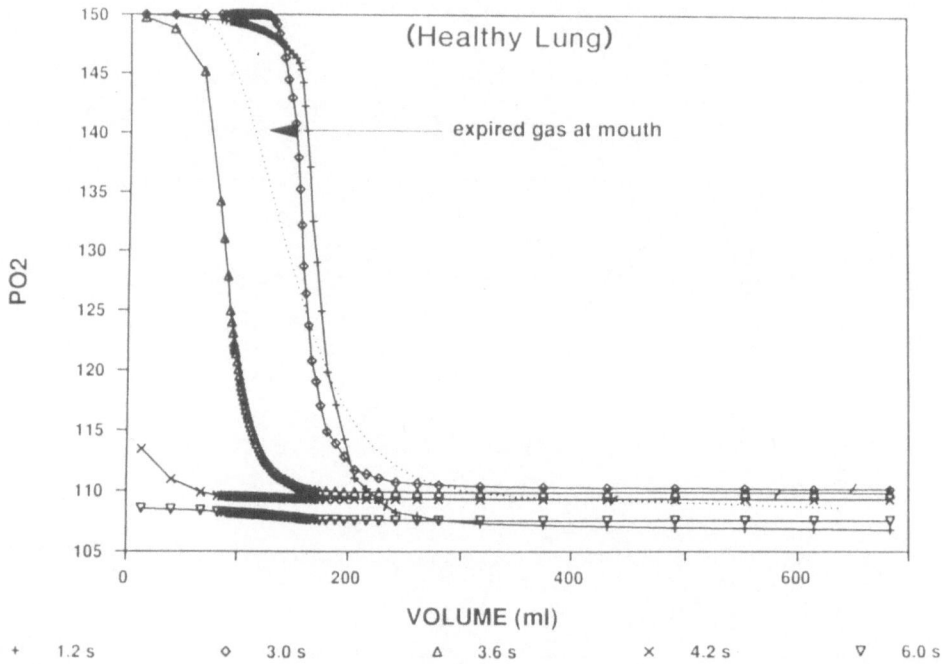

Figure 3. Computed intra-airway oxygen concentration profiles plotted as a function of cumulative airway volume at various times during a steady state sinusoidal breath of V_T = 650 cc and period = 6.0 sec. The lighter dotted curve is the computed O_2 washout curve seen at the mouth plotted as function of volume exhaled.

If we compare in figure 3 the slope of phase III of the mouth washout curve with the acinar airway P_{O_2} concentration profile curve at 1.2 seconds after the start of inspiration (lowest curve in fig. 3) it can be seen that the slopes are equal to each other within about 10% derivation. The intra airway O_2 concentration profile at 1.2 seconds on figure 3 corresponds to the time of peak O_2 flux out of the tidal volume shown in figure 2.

The agreement of the phase III slope of the O_2 mouth washout curve with the intra airway P_{O_2} concentration profile at the time of peak flux suggests that the slope of phase III of the washout curve can be used to obtain an estimate of

650

ΔP_{O_2} in equation 3 and thereby an estimate also of the gas phase resistance to O_2 transport from the V_T to the FRC once $\dot{V}_{O_2}$ is also known from the washout curve.

DISCUSSION AND SUMMARY

Using $\dot{V}_{O_2}$ and ΔP_{O_2} obtained from the phase III slope (over the entire alveolar plateau) of experimental O_2 washout curves measured by mass spectrometry and substituting these values in equation 3 gives

$$R_{O_2gas} \simeq 0.0104 \text{ mmHg} \cdot \text{min/ml} \tag{5}$$

This can be compared to total (gas + membrane + blood) O_2 alveolar transport resistance for a subject at rest (Weibel, 1989) as

$$R_{O_2tot} = \frac{1}{D_{LO_2}} = 0.0333 \text{ mmHg} \cdot \text{min/ml} \tag{6}$$

Therefore the fraction of total alveolar O_2 transport resistance due to air phase diffusion between the V_T and the FRC would be

$$\frac{R_{O_2gas}}{R_{O_2tot}} \simeq 0.31 \tag{7}$$

This estimate indicates that in the healthy state, the gas phase resistance to alveolar O_2 exchange is significant and

accounts for at least 31% of the total resistance measured by $\frac{1}{D_{LO_2}}$. Since the ΔP_{O_2} driving the O_2 from the V_T into the FRC can be even larger than that given by the phase III slope of the O_2 washout curve and could approach that of phase II of the curve at some times during inspiration, the estimate of R_{O_2gas} found above may be low.

The numerical model predicts that the major mechanism of O_2 transport between the V_T and the blood in a healthy lung is first by longitudinal gas phase diffusion of O_2 from the V_T into the FRC along the concentration gradients shown in figure 3 followed by gas phase and membrane diffusion from the FRC into the alveolar capillary blood.

On the basis of this study, we conclude that airphase resistance to O_2 transport between the V_T and the FRC is significant and may be as high as 50% of the total resistance to O_2 transport between the V_T and the alveolar capillary blood in a healthy person.

651

REFERENCES

HANSEN JE, AMPAYA EP, BRYANT GH, and NAVIND JJ, 1975, Human Air Space Shapes, Sizes, Areas, and Volumes, J. Appl. Physiol. 38(6):990-995.

SCHERER PW, GOBRAN S, AUKBERG SJ, BAUMGARDNER JE, BARTKOWSKI R, and NEUFELD GR, 1988,Numerical and Experimental Study of Steady-State CO_2 and Inert Gas WAshout, J. Appl. Physiol. 64(3):1022-1029.

WEIBEL, ER, 1989, Lung Morphomethy and Models in Respiratory Physiology in: "Respiratory Physiology An Analytical Approach," Chang, H.K., Paiva, M, Eds. Marcel Dekker, Inc., New York.

DISTRIBUTION OF VENTILATION AND DIFFUSION WITH PERFUSION IN A

TWO-COMPARTMENT MODEL OF GAS EXCHANGE

Vidal Melo, M.F.*, Caprihan, A., Luft, U.C., Loeppky, J.A.

Lovelace Medical Foundation - Albuquerque - New Mexico - USA
Abteilung für Experimentelle Chirurgie
Im Neuenheimer Feld 347, D-6900 Heidelberg 1, FRG

INTRODUCTION

Since the 1940's mathematical models have been applied to the study of gas exchange (Riley and Cournand, 1949). This is because it is possible to describe the lung, in respiratory physiology, as a set of units with qualitatively similar mathematical description, which can be added for an overall result. As chemical processes are involved in the relationships between gas partial pressures and contents in the blood, such a description involves the solution of non-linear, interdependent and differential equations. Due to the difficulties of such calculations at those early times, most of the work was restricted to graphical analyses with several simplifying assumptions.

With the availability of computational facilities, models applying numerical techniques allowed for more complex theoretical analyses (Chinet et al., 1971; Wagner et al., 1974). These models have been mainly related to the study of the alveolar ventilation to perfusion ratio ($\dot{V}_A/\dot{Q}$), i.e., how the inspired air matches the pulmonary capillary blood flow. This ratio is topographically distributed in the lung and varies in different lung diseases. There are, however, other important features involved in the gas exchange process, such as gas mixing in the alveolar phase and alveolar-capillary diffusion. This was cited as playing a role in diseases such as emphysema, chronic bronchitis and the syndrome of alveolar capillary block (Briscoe, 1980) and can be the key factor to explain some experimental and clinical observations in states of lung edema (Granger et al., 1987).

King and Briscoe (1967a, 1967b) suggested a set of graphical procedures to estimate $\dot{V}_A/\dot{Q}$ - diffusion interaction. However, the method was difficult to understand and apply. An aspect of great interest in this subject is to estimate the relative influence of $\dot{V}_A/\dot{Q}$ and diffusion (D_L) on the alveolar-arterial oxygen and carbon dioxide differences, i.e., the difference between the partial pressures of each of these gases in the alveoli and that in the arterial blood. Ideally, such differences would equal zero signifying no barrier to exchange between gas and blood. The estimation of the relative importance of components causing the overall difference would be helpful to understand pathophysiological processes which often affect each of them differently. The use of models based on only one component, usually the $\dot{V}_A/\dot{Q}$ ratio, attributes only this influence to the differences and so does not allow for a comprehensive analysis.

Oxygen Transport to Tissue XII, Edited by J. Piiper *et al.*
Plenum Press, New York, 1990

A numerical model has been proposed to estimate $\dot{V}_A/\dot{Q}$ distribution from experimental data (Wagner et al., 1974). Gases with linear blood pressure versus content relationships (inert gases) are used, simplifying the computations. This model was used to estimate $D_L O_2$ from measured $\dot{V}_A/\dot{Q}$ distributions and gas exchange data (Hammond and Hempleman, 1987). However, gas chromatography and elaborate procedures are involved, making the routine use of these precise methods impractical in a clinical setting. It would be desirable to have a simpler methodology to estimate the relevant parameters of gas exchange.

A two-compartment model of gas exchange was recently proposed (Loeppky et al., 1987) which uses as input data routine measurements obtained in the pulmonary function laboratory. It allows for the estimation of $\dot{V}_A/\dot{Q}$ and was useful in describing data of patients suffering from chronic obstructive pulmonary disease (COPD).

This work describes a two-compartment mathematical model derived from this previous one where diffusion impairment, in addition to $\dot{V}_A/\dot{Q}$ inequality, is taken into account. This is done by considering a fixed diffusion to perfusion ratio homogeneous to the whole lung. A method to estimate the partitioning of alveolar-arterial oxygen and carbon dioxide differences into $\dot{V}_A/\dot{Q}$ and diffusion components is described and the model is applied in the analysis of data of COPD patients.

METHODS

Model Description

The model proposed in this work is shown in fig. 1. It is represented by a lung composed of two gas exchange units plus a true shunt (not shown). The alveolar ventilation ($\dot{V}_A$) is distributed to each of the two compartments according to f_v, the fraction of ventilation. The same principle is applied to blood flow (f_q), including the consideration of the true shunt fraction (f_s).

A membrane with a diffusing capacity, D, exists between the alveolar gas and blood. The graphic at the bottom displays the time course of the O_2 partial pressure of a slug of blood as it passes by an alveolus. Depending on the diffusion impairment, alveolar and end capillary O_2 and CO_2 pressures may or may not attain an equilibrium after the transit time T.

The diffusing capacity to perfusion ratio, $D/\dot{Q}$, is considered equal in both compartments, i.e., diffusion is assumed to be distributed relative to perfusion.

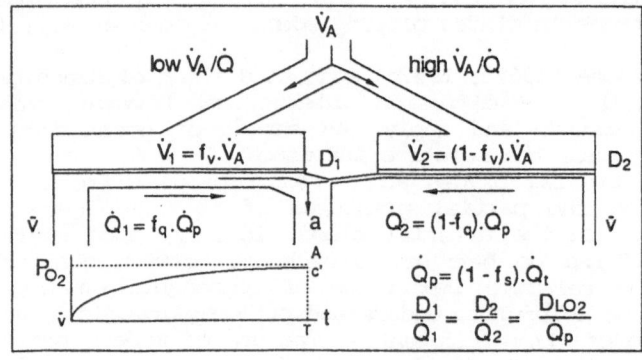

Fig. 1.   Schematic of 2-compartment lung model.

The gas exchange features of each compartment are essentially defined by its $\dot{V}_A/\dot{Q}$ and $D/\dot{Q}$ characteristics. If $\dot{V}_A$ and $\dot{Q}_p$ are known, the $\dot{V}_A/\dot{Q}$ ratios are determined by f_v an f_q:

$$(\dot{V}_A/\dot{Q})_1 = f_v \cdot \dot{V}_A/f_q \cdot \dot{Q}_p \qquad (1) \qquad (\dot{V}_A/\dot{Q})_2 = (1-f_v) \cdot \dot{V}_A/(1-f_q) \cdot \dot{Q}_p \qquad (2)$$

Total alveolar ventilation is calculated using the alveolar ventilation equation as expressed by Rahn and Fenn (1955):

$$\dot{V}_A = 0.863 \cdot \dot{V}_{O_2}(R+F_I CO_2 \cdot (1-R))/(P_A CO_2 - P_I CO_2) \qquad (3)$$

where, $\dot{V}_{O_2}$ = O_2 uptake (ml/min)
$\quad\quad\quad$ R $\quad$ = respiratory quotient
$\quad\quad\quad$ $F_I CO_2$ = CO_2 fraction in inspired air
$\quad\quad\quad$ $P_A CO_2$ = alveolar CO_2 partial pressure (mmHg)
$\quad\quad\quad$ $P_I CO_2$ = inspiratory CO_2 partial pressure (mmHg)

For the calculations presented here, cardiac output, $\dot{Q}_t$, was estimated from body surface area calculated from height and weight according to DuBois and DuBois (1916). In resting supine subjects with normal cardiac function, the mean value for cardiac index is 3.5 L/min/m^2 (Wade and Bishop, 1962). This value was used both for normal and for chronic obstructive patients since none of the latter were in cardiac failure at the time the measurements were performed. $\dot{Q}_p$ (1/min) is calculated as:

$$\dot{Q}_p = \dot{Q}_t \cdot (1-f_s) \qquad (4)$$

where f_s is computed with the equation:

$$f_s = (P_A O_2 - P_a O_2) \cdot 0.0031/[(P_A O_2 - P_a O_2) \cdot 0.0031 + (a-v)O_2] \qquad (5)$$

$P_A O_2$ $\quad$ = alveolar O_2 partial pressure
$P_a O_2$ $\quad$ = arterial O_2 partial pressure
$(a-v)O_2$ = O_2 content difference between arterial and mixed
$\quad\quad\quad\quad$ venous blood.

The diffusion values for O_2 and CO_2 are estimated from steady state measurements done with carbon monoxide, using generally accepted relations. The effect of chemical reaction rate (θ) was not taken into account.

$$D_L O_2 = 1.23 \cdot D_L CO \qquad (6) \qquad D_L CO_2 = 20.0 \cdot D_L O_2 \qquad (7)$$

where $D_L CO$, $D_L O_2$ and $D_L CO_2$ are the lung diffusion capacities for CO, O_2 and CO_2, respectively.

Assuming a log normal $\dot{V}_A/\dot{Q}$ distribution, some indices can be calculated from $(\dot{V}_A/\dot{Q})_1$ and $(\dot{V}_A/\dot{Q})_2$ to numerically characterize the extent of $\dot{V}_A/\dot{Q}$ inequality. These are the mean $\ln(\dot{V}_A/\dot{Q})$ and ln standard deviation, (SD_{VQ}) which are calculated by weighting the compartment $\dot{V}_A/\dot{Q}$ ratios with their respective ventilation and perfusion.

$$\text{Mean } \ln(\dot{V}_A/\dot{Q}) = \overline{VQ} = f_g{}^1[\ln(\dot{V}_A/\dot{Q})_1] + f_g{}^2[\ln(\dot{V}_A/\dot{Q})_2] \qquad (8)$$

$$SD_{VQ} = [f_g{}^1(\ln(\dot{V}_A/\dot{Q})_1 - \overline{VQ})^2 + f_g{}^2(\ln(\dot{V}_A/\dot{Q})_2 - \overline{VQ})^2]^{1/2} \qquad (9)$$

$$\text{where } f_g{}^1 = (\dot{V}_A + \dot{Q}_p)_1/(\dot{V}_A + \dot{Q}_p) \qquad (10) \qquad f_g{}^2 = (\dot{V}_A + \dot{Q}_p)_2/(\dot{V}_A + \dot{Q}_p) \qquad (11)$$

Fitting the Model to Patient Data

The procedure is sketched in the flowchart of fig. 2. The input data are: blood hemoglobin concentration (Hb), arterial O_2 and CO_2 partial

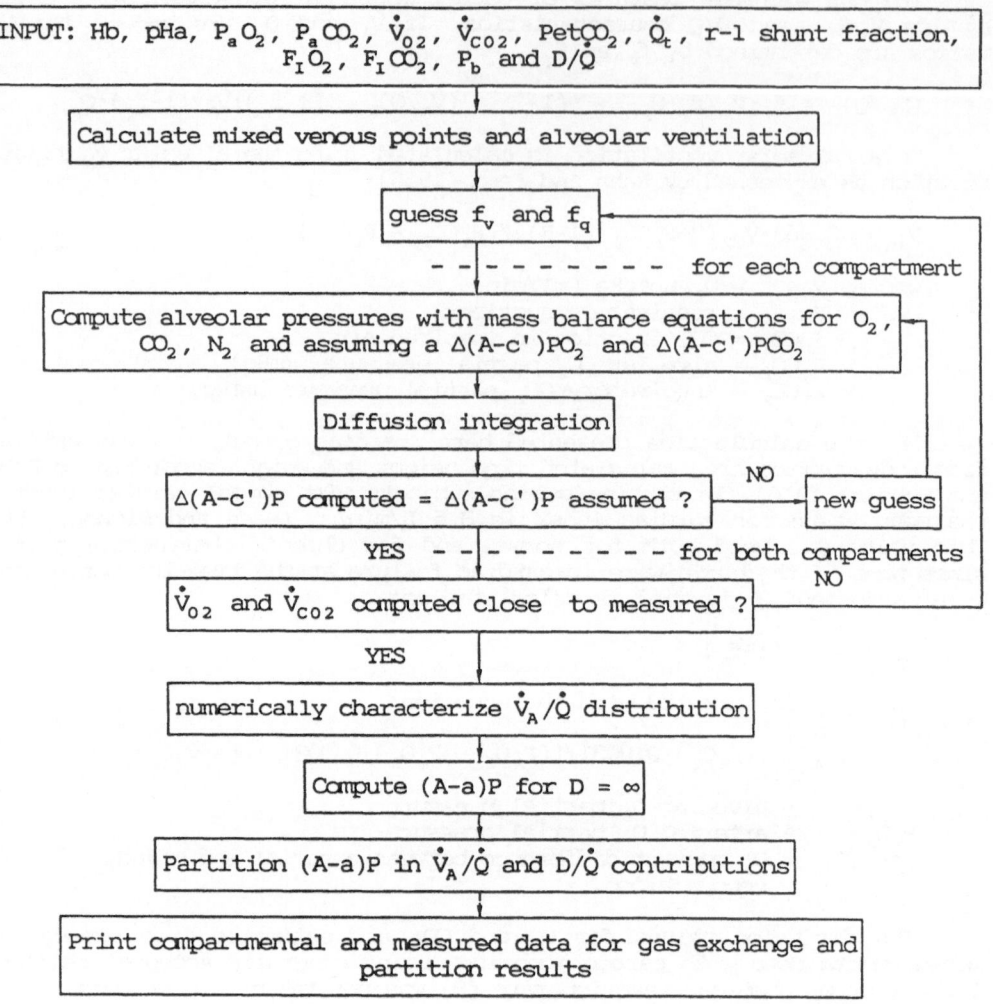

Fig. 2. Flowchart of the 2-compartment model.

pressures (P_aO_2, P_aCO_2) and pH, end tidal CO_2 partial pressure ($P_{et}CO_2$, assumed = P_ACO_2), $\dot{Q}_t$, f_s, O_2 uptake ($\dot{V}_{O2}$) and CO_2 production ($\dot{V}_{CO2}$), fractions of O_2 and CO_2 in the inspired air (F_IO_2, F_ICO_2), $D/\dot{Q}$ and the barometric pressure (P_b).

The fitting of the model to patient data aims to find the parameters f_v and f_q that lead to $\dot{V}_{O2}$ and $\dot{V}_{CO2}$ calculated from the model ($\dot{V}_{O2}{}^m$, $\dot{V}_{CO2}{}^m$) being equal to those measured in a subject. Because of the complexity of the problem, expressed mathematically as a set of non-linear, interdependent, differential equations, a closed formula to calculate such compartments, i.e., to calculate f_v and f_q is not feasible. So, (f_v, f_q) are determined by computer through methodic guesses to arrive at a unique solution. The used technique is the downhill simplex method (Press et al., 1986), which finds the minimum of a multidimensional function. Here the function (F) to be minimized depends on the difference between measured and computed $\dot{V}_{O2}$ and $\dot{V}_{CO2}$ and has two dimensions: f_v and f_q.

$$F(\dot{V}_{O2}, \dot{V}_{CO2}, \dot{V}_{O2}{}^m, \dot{V}_{CO2}{}^m) = [(\dot{V}_{O2} - \dot{V}_{O2}{}^m)^2 + (\dot{V}_{CO2} - \dot{V}_{CO2}{}^m)^2]^{-1/2} \qquad (12)$$

656

Given the input data, $\dot{V}_A$ and $\dot{Q}_p$ are calculated as in equations (3) and (4). Mixed venous O_2 and CO_2 contents (C_vO_2 and C_vCO_2) are determined through Fick's equation using O_2 and CO_2 dissociation curves (see below) to calculate arterial O_2 and CO_2 contents (C_aO_2 and C_aCO_2):

$$C_vO_2 = C_aO_2 - \dot{V}_{O2}/\dot{Q}_t \qquad (13) \qquad C_vO_2 = C_aCO_2 + \dot{V}_{CO2}/\dot{Q}_t \qquad (14)$$

Once a pair (f_v, f_q), and therefore, $(\dot{V}_A/\dot{Q})_1$ and $(\dot{V}_A/\dot{Q})_2$ is chosen, and knowing mixed venous points, inspired gas fractions and the dissociation curves for O_2 and CO_2, it is possible to calculate the alveolar partial pressures P_AO_2, P_ACO_2 and P_AN_2 for each $\dot{V}_A/\dot{Q}$ ratio. This is done by solving mass conservation equations for each respiratory gas, considering their fraction sum equal to 1. The equations are recent variations of classical equations (Olszowka and Wagner, 1980) taking into account N_2 inequality between alveolar gas and blood:

$$P_AN_2 = \frac{P_IN_2[(\dot{V}_A/\dot{Q})P_AO_2 + 8.63(C_aO_2 - C_vO_2) + 0.0017 \cdot 8.63 \cdot P_IO_2]}{P_IO_2[(\dot{V}_A/\dot{Q}) + 0.0017 \cdot 8.63]} \qquad (15)$$

$$(\dot{V}_A/\dot{Q})P_ACO_2 = 8.63(C_vCO_2 - C_aCO_2) \qquad (16)$$

$$P_b - 47.1 - P_AO_2 - P_ACO_2 - P_AN_2 = 0 \qquad (17)$$

where P_IN_2 (assumed equal to P_vN_2) and P_IO_2 are inspired values. Since the blood contents are a function of the gas partial pressures, the solution of the set of equations involves the handling of implicit functions.

The O_2 and CO_2 dissociation curves are required. For CO_2, the relationship described by Loeppky et al. (1983) was utilized. In this procedure, the curve relating the blood PCO_2 to the whole blood C_{CO2} is an exponential. A Haldane factor of 0.28 (unit increase in C_{CO2} per unit decrease in oxygenated Hb) was incorporated, denoted as ΔH, and the exponent b is obtained by an iterative procedure from the knowledge of Hb and a single coordinate on the curve (PCO_2 and C_{CO2}):

$$C_{CO2} = K \cdot PCO_2{}^b + \Delta H \qquad (18)$$

The description of the O_2 dissociation curve and Bohr effect given by Severinghaus (1979) was employed in the form of two equations:

$$PO_2(s) = PO_2 \exp(-[(PO_2/26.6)^{0.184} - 2.2](pH-7.4)) \qquad (19)$$

where PO_2 is at 37°C and actual pH and $PO_2(s)$ is PO_2 corrected to a pH of 7.4 and

$$SO_2 = (((PO_2(s)^3 + 150 PO_2(s))^{-1} \cdot 23,400)+1)^{-1} \qquad (20)$$

In order to calculate C_{O2}, the dissolved O_2 was obtained by multiplying PO_2 by 0.003 and adding this amount to the product of $SO_2 \cdot Hb \cdot 1.36$:

$$C_{O2} = 0.003 \cdot PO_2 + SO_2 \cdot Hb \cdot 1.36 \qquad (21)$$

Note that C_{CO2} is a function of O_2 saturation (and so PO_2) and that PO_2 is a function of pH which depends on PCO_2 (as shown below), i.e., the functions are not only non-linear, but interdependent.

Diffusion Limitation

If alveolar-arterial diffusion equilibrium is assumed, $P_AO_2 = P_aO_2$ and

657

$P_A CO_2 = P_a CO_2$. The solution of equations (15) to (17) provides, then, the final blood and gas values.

As diffusion impairment is considered in the present model through attribution of a fixed $D/\dot{Q}$ to both compartments, the assumption of alveolar-end capillary ($P_{c'}$) partial pressures equality is no longer valid. Thus, two more equations must be introduced for determination of alveolar and blood pressures:

$$P_A O_2 = P_{c'} O_2 + \Delta PO_2 \qquad (22) \qquad\qquad P_A CO_2 = P_{c'} CO_2 - \Delta PCO_2 \qquad (23)$$

where ΔPO_2 and ΔPCO_2 are the alveolar-end capillary differences, not known initially.

The computation of ΔPO_2 and ΔPCO_2 is done assuming that for each of these gases crossing the alveolo-capillary membrane its instantaneous mass flow rate across that membrane is directly proportional to the partial pressure difference across the membrane. This is a one-dimensional expression of Fick's first law of diffusion (Kellogg,1987). The application of this law, assuming transit time equal to 1 sec, leads to the following differential equations, describing gas changes in a slug of blood during pulmonary transit (Wagner, 1977):

$$dC(t)/dt = 0.1 \cdot (D/\dot{Q}_p) \cdot (P_A - Pc(t)) \qquad (24)$$

where $C(t)$ and $Pc(t)$ are the gas instantaneous blood content and partial pressure at time t, P_A is the alveolar partial pressure, D (ml/min/mmHg) is the diffusing capacity, representing the ability of the gas to diffuse across the blood-gas barrier. It is directly proportional to the cross-sectional area over which diffusion occurs and the solubility of the gas in the blood-gas barrier and indirectly related to the thickness of the barrier and the square root of the molecular weight of the gas.

This differential equation is integrated for O_2 and CO_2 with a 4th order Runge-Kutta method in the interval 0 to T with $\Delta t = 0.01$ sec, the interval for optimal integration in the work of Wagner and West (1972), starting at the mixed venous point. Because of time constraints, a simplification procedure (described below) was used to compute the O_2 and CO_2 dissociation curves in this step.

At the end of each integration a value of ΔPO_2 and ΔPCO_2 is determined and can be used in equations (22) and (23). These, together with equations (15) to (17), are used to calculate alveolar and arterial pressures up to convergence of ΔPO_2 and ΔPCO_2. At the beginning of each iteration ΔPO_2 and ΔPCO_2 are taken to equal zero, i.e., the situation of no diffusion impairment.

Once blood and alveolar points are determined for the compartments, $\dot{V}_{O2}$ and $\dot{V}_{CO2}$ are calculated for each model compartment as:

$$\dot{V}_{O2}{}^m = f_q \cdot \dot{Q}_p (C_{c'} O_2 - C_v O_2) + (1 - f_q) \cdot \dot{Q}_t (C_{c'} O_2 - C_v O_2) \qquad (25)$$

$$\dot{V}_{CO2}{}^m = f_q \cdot \dot{Q}_p (C_v CO_2 - C_{c'} CO_2) + (1 - f_q) \cdot \dot{Q}_t (C_v CO_2 - C_{c'} CO_2) \qquad (26)$$

and compared with measured values.

Mean alveolar pressures for the model are calculated by weighting compartments with their ventilation fraction:

$$P_A O_2 = f_v \cdot P_A O_2{}^1 + (1 - f_v) \cdot P_A O_2{}^2 \qquad (27)$$

$$P_A CO_2 = f_v \cdot P_A CO_2{}^1 + (1 - f_v) \cdot P_A CO_2{}^2 \qquad (28)$$

658

Mean end capillary contents are computed by weighting compartment contents with its perfusion. The gas partial pressures are obtained iteratively as described below:

$$C_c'O_2 = f_q \cdot C_a O_2^1 + (1 - f_q) \cdot C_a O_2^2 \tag{29}$$

$$C_c'CO_2 = f_q \cdot C_a CO_2^1 + (1 - f_q) \cdot C_a CO_2^2 \tag{30}$$

At this step the alveolar-end capillary difference for O_2 and CO_2 can be computed. This difference is related to $\dot{V}_A/\dot{Q}$ and $D/\dot{Q}$ effects:

$$(A-c')PO_2 = P_A O_2 - P_c \cdot O_2 \tag{31} \qquad (c'-A)PCO_2 = P_c \cdot CO_2 - P_A CO_2 \tag{32}$$

By using measured arterial values instead of computed end capillary, the total alveolar arterial difference is computed, taking the shunt into account.

<u>Approximate procedure to compute diffusion integrals</u>

<u>Gas contents for given partial pressures.</u> Given a pair (PO_2, PCO_2), a pair (C_{O2}, C_{CO2}) can only be calculated through an iterative procedure, since C_{O2} can not be written as a function of PO_2 in equations (19) to (21). During this procedure one must take into account both the influence of CO_2 levels on the O_2 curve (Bohr effect) as well as the influence of O_2 saturation on the CO_2 dissociation curve (Haldane effect).

Assuming a pH value, O_2 saturation is calculated through (19) and (20). C_{CO2} can be calculated by equation (18) and also by (Loeppky et al., 1983):

$$C_{CO2} = [1 - \frac{0.02924Hb}{(2.244-0.422SO_2)(8.74-pH)}] \cdot 2.226 \cdot 0.0301 PCO_2 (1+10^{pH-6.1}) \tag{33}$$

Both values of C_{CO2} are computed and iterated up to convergence by varying pH using Brent's method (Press et al., 1986). It is therefore apparent that pH is one of the determinants of the position of the O_2 dissociation curve.

<u>Approximate procedure.</u> The integration of equation (24) through the Runge-Kutta method implies the calculation of O_2 and CO_2 pressures from given contents, done by using the procedure above. However, the superposition of these loops leads to prohibitive computation times, as already noticed by Hammond and Hempleman (1987).

The calculation of dissociation curves is, therefore, simplified by taking a fixed O_2 dissociation curve computed from a pH equal to the average of arterial and mixed venous pH, corresponding to a mean Bohr effect. A mean Haldane effect is also used as stated above. For the first iteration of a pair (f_v, f_q), given ΔPO_2 and ΔPCO_2, the value of pH is fixed at the average pH = $(pH_a + pH_v)/2$. After this iteration, a vector of pH values (pH(i)) is calculated from C_{CO2} at each step of integration $(C_{CO2}(i))$, and mixed venous (v) and end capillary (c') pH and C_{CO2} values and applied on next iterations when convergence is not attained at the first step:

$$pH(i) = pH_v + (pH_c \cdot - pH_v)(C_{CO2}(i)-C_v CO_2)/(C_c \cdot CO_2 - C_v CO_2) \tag{34}$$

Thus, once guesses C'_{O2} and C'_{CO2} are determined, the corresponding pressures are calculated as:

$$P'CO_2 = (C'_{CO2} + \Delta H)^{1/b}/K \tag{35}$$

and the PO_2 is determined using equations (19)-(21) as the value that leads to a $C_{O2} = C'_{O2}$, with pH calculated as above.

This approximation was evaluated by plotting the time course of PO_2, PCO_2, C_{O2} and C_{CO2} for situations of low (= 0.5) and high (= 10.0) $\dot{V}_A/\dot{Q}$ and low (= 1.0) and high (= 10.0) $D/\dot{Q}$ with mean data of normal and most compromised patients (group 3 below). The final differences between the more precise (std) and the approximate (app) procedures were then calculated and expressed as:

$$\% \text{ error} = (Xstd-Xapp)/(Xstd-Xv) \quad (36) \quad \text{where } X = PO_2, PCO_2, CO_2 \text{ and } CO_2$$
$$v = \text{mixed venous value}$$

<u>Partitioning of ventilation/perfusion and diffusion/perfusion effects on alveolar-end capillary partial pressure differences</u>

Once final estimates $(\dot{V}_A/\dot{Q})_1$ and $(\dot{V}_A/\dot{Q})_2$ for the compartments are known, the partitioning of $(A-c')PO_2$ and $(A-c')PCO_2$ due to $\dot{V}_A/\dot{Q}$ inequality and diffusion can be attempted. Two assumptions are made for these calculations: (1) the observed $(A-c')P$ are dependent only on these two factors and (2) $\dot{V}_A/\dot{Q}$ and $D/\dot{Q}$ are independent and additive. As a consequence, the effect of $\dot{V}_A/\dot{Q}$ inequality is expressed by the $(A-c')P$ computed using the two calculated compartments, when $D/\dot{Q} = \infty$. This is done by solving equations (15) to (17) for each compartment considering equality between arterial and alveolar values, i.e., in equations (22) and (23) ΔPO_2 and $\Delta PCO_2 = 0$. The previously inspired and computed mixed venous points, as well as the determined values for $(\dot{V}_A/\dot{Q})_1$ and $(\dot{V}_A/\dot{Q})_2$ are used as inputs. The overall alveolar and end capillary pressures can be calculated as stated in equations (27) to (32). Assuming P^* as the pressures so calculated, i.e., those due only to $\dot{V}_A/\dot{Q}$ inequality, the percent contribution of $\dot{V}_A/\dot{Q}$ inequality to the overall $(A-c')PO_2$ and $(c'-A)PCO_2$ are:

$$\Delta_{VQ}O_2 = (P^*_A O_2 - P^*_{c'} O_2)/(A-c')PO_2 \tag{37}$$

$$\Delta_{VQ}CO_2 = (P^*_{c'} CO_2 - P^*_A CO_2)/(c'-A)PCO_2 \tag{38}$$

The Δ_{VQ} so calculated are subtracted from 100 % to compute the fraction due to diffusion.

$$\Delta_{DQ}O_2 = 1 - \Delta_{VQ}O_2 \tag{39}$$

$$\Delta_{DQ}CO_2 = 1 - \Delta_{VQ}CO_2 \tag{40}$$

The calculations for the alveolar-arterial differences would be similar, taking arterial instead of end capillary points. The arterial values can be calculated from end capillary plus shunt component, weighting the contents with the perfusion. These computations will produce values very close to measured ones.

<u>Patients</u>

Mean data from 53 patients is used in this study. Ten are normal (N) subjects and 43 are patients suffering from chronic obstructive pulmonary disease (COPD), in this case emphysema of varying degrees of severity. These patients are divided into three groups according to their P_aO_2 values breathing air. These classification values together with D and $D/\dot{Q}$ mean values for each group are shown in table 1. Physical characteristics and pulmonary function tests of normal and diseased patients are presented in a previous publication (Loeppky et al., 1987).

660

Table 1. P_aO_2 (mmHg) classification values for COPD groups and group D_LO_2 (ml/min/mmHg) and D/$\dot{Q}$ mean values (measurements made at 1646 m).

	N	G1	G2	G3
P_aO_2	> 68	56 - 64	45 - 55	< 45
D_LO_2	47.20	14.65	9.83	8.14
D/$\dot{Q}$	7.44	2.42	1.76	1.48

RESULTS

Tables 2(a) and 2(b) present the errors due to the approximation procedure in the integration of equation (24) for O_2 and CO_2 using data of normals and group 3, respectively. It can be seen that the errors are always smaller than 1 % except in the case of low $\dot{V}_A/\dot{Q}$ and low D/$\dot{Q}$ when they range from 1 to 3 % and the PO_2 at high $\dot{V}_A/\dot{Q}$ and low D/$\dot{Q}$ values.

Table 2. Errors associated with approximate (app) procedure in calculations of diffusion integrals for different $\dot{V}_A/\dot{Q}$-D/$\dot{Q}$ combinations. Mean data from (a) normals and (b) severe COPD patients. Error = [Xstd-Xapp]·100/[Xstd-Xv], X= PO_2, PCO_2, CO_2, CCO_2. std=standard method.

(a)Normals:

$\dot{V}_A/\dot{Q}$	D/$\dot{Q}$	PO_2	PCO_2	C_{O2}	C_{CO2}
low	low	-1.2	-2.9	-1.0	-2.4
low	high	0.3	-0.4	0.1	-0.1
high	low	-0.3	0.0	0.0	0.0
high	high	0.0	0.1	0.0	0.2

(b)Group 3:

$\dot{V}_A/\dot{Q}$	D/$\dot{Q}$	PO_2	PCO_2	C_{O2}	C_{CO2}
low	low	0.8	-1.5	-0.1	-1.3
low	high	-0.2	0.2	-0.1	0.0
high	low	-2.0	0.0	-0.4	0.0
high	high	0.0	0.2	0.0	0.2

The order of magnitude of the error does not increase when the calculations are based not on one step but on the final values after a convergence procedure to ΔPO_2 and ΔPCO_2. Table 3 presents ventilation and perfusion fractions f_v and f_q, respectively, calculated for mean values of normals and group 3 with approximate and standard methods. The similarity of the results is apparent, with the computing time in the order of 1/25 the one needed for the std procedure.

Table 3. Ventilation (f_v) and perfusion (f_q) fractions calculated with approximate (app) and standard (std) methods for mean data of normals and group 3.

	Normals		Group3	
	app	std	app	std
f_v	0.550	0.550	0.234	0.236
f_q	0.833	0.833	0.001	0.002

Figure 3 shows log means and standard deviations of the $\dot{V}_A/\dot{Q}$ using mean data of each group. The values were estimated for the case of no diffusion impairment and with the measured diffusion impairment. The effect of D/$\dot{Q}$ consideration in reducing the values of standard deviations is observed in each case, except for normals.

661

Figures 4(a) and 4(b) show the results of the partitioning procedure when applied on mean data of each group. It can be seen that, as the degree of diffusion impairment increases, its relative contribution to the overall (A-c')P increases as well. This contribution being more important for O_2 than CO_2 exchange.

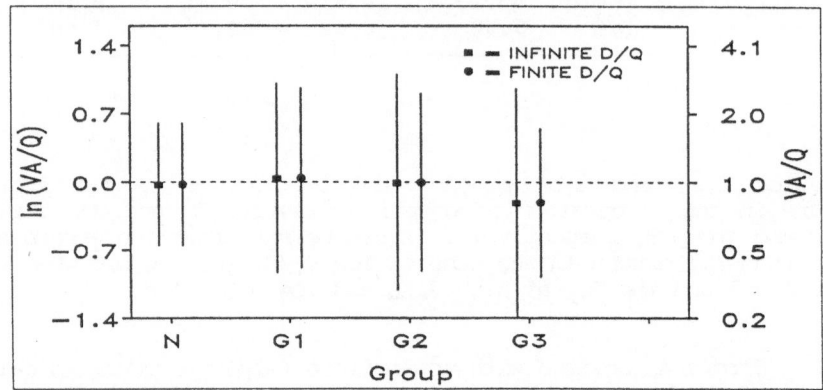

Fig. 3. Estimated $\dot{V}_A/\dot{Q}$ log means and standard deviations calculated for mean data of each group assuming $D/\dot{Q}$ measured and infinite.

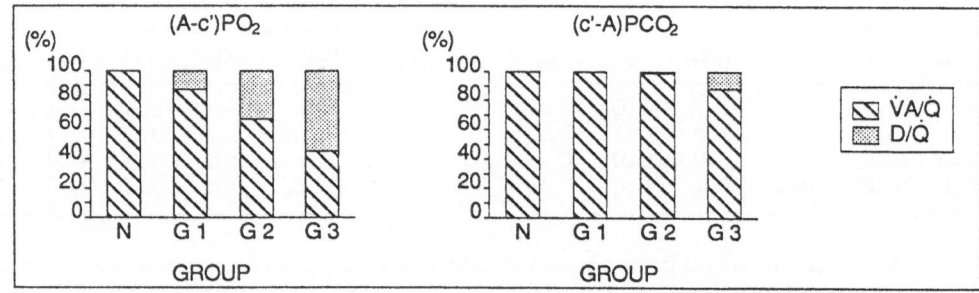

Fig. 4. Partitioning of (A-c') (a) O_2 and (b) CO_2 partial pressure differences into $\dot{V}_A/\dot{Q}$ and $D/\dot{Q}$ components.

DISCUSSION

The effect of chemical reaction rate (θ) on the lung diffusing capacity was not considered in this work. This factor has shown to be of much less importance than the membrane diffusing capacity (Wagner and West, 1972) in computations of (A-a)PO_2 and (a-A)PCO_2.

The clinical and physiological interests of analyzing $D/\dot{Q}$-$\dot{V}_A/\dot{Q}$ interactions have been apparent for many years. Hammond and Hempleman (1987) studied the subject using the compartments calculated through the inert gases technique. However, the consideration of diffusion was done after the definition of $\dot{V}_A/\dot{Q}$ distribution.

The introduction of diffusion into the procedures to determine the compartments can be excessively time consuming if the interdependence between O_2 and CO_2 dissociation curves is computed continuously. Those authors cited this time demand as a limiting factor in their work, where the integration was used only for the calculation of final results. As the present work aims to apply the model to patient data, this is indeed more critical since the integration is to be used during the iterations.

662

When O_2 values are in the linear part of the O_2 dissociation curve, contents and pressures are related through a constant factor, ß, and the use of Piiper's exponential relation (1969) to compute the $(A-a)PO_2$ is possible. However several patients have values out of the hypoxic range and even for hypoxic patients, the resolution of 2 compartments leads to one with high $\dot{V}_A/\dot{Q}$. Thus, it was necessary to search for an alternative approach which would improve computation time without compromising the precision in the range of studied values. As shown by the simulations, the fixation of pH values seems to satisfy these conditions even in a wide range of $\dot{V}_A/\dot{Q}$ and $D/\dot{Q}$ combinations.

The effect of diffusion consideration in the estimates of SD_{VQ} was to reduce SD_{VQ} values. This would be expected since when $\dot{V}_A/\dot{Q}$ is the unique factor to explain gas exchange characteristics, its inequality, i.e., SD_{VQ}, must account for the whole $(A-a)P$. When diffusion is considered, a part of $(A-a)P$ is assigned to $D/\dot{Q}$, reducing SD_{VQ}. The absence of SD_{VQ} modifications in normals with diffusion limitation imposed demonstrates that diffusion doesn't represent a significant impairment factor for gas exchange in these subjects.

The assumption of a constant $D/\dot{Q}$ is a simplification procedure which corresponds to the assumption of diffusion and perfusion being equally distributed in the lungs. This hypothesis has been stated as the most probable by Hammond and Hempleman (1987). On the other hand some experimental evidence suggests a $D/\dot{Q}$ heterogeneity. The presented model could be used to study this factor by assigning different $D/\dot{Q}$ values to the compartments while keeping a constant overall $D/\dot{Q}$. It shall be noted that by computing the diffusion integrations, ß was allowed to vary. So, although $D/\dot{Q}$ was constant, a degree of $D/\dot{Q}$ß heterogeneity was in fact considered.

The participation of diffusion impairment as a factor to compromise gas exchange in addition to $\dot{V}_A/\dot{Q}$ in COPD patients has already been demonstrated (King and Briscoe, 1967b). However, mathematical model approaches to these cases have concentrated mainly on $\dot{V}_A/\dot{Q}$ or alveolar ventilation to volume ratio $(\dot{V}_A/V)$ inequalities. The results of this model suggest that diffusion plays an increasing role in $(A-a)PO_2$ and $(a-A)PCO_2$ the more severe is the disease, even in face of large SD_{VQ} values.

The percent diffusion impairment contribution was greater for O_2 than CO_2 in all cases when $\Delta_{DQ}O_2 \neq 0$. This can be explained by the fact that the largest of the 2 compartments, i.e., that receiving the largest sum $(\dot{V}_A+\dot{Q}_p)$, has $\dot{V}_A/\dot{Q}$ ratios around the normal range and it was demonstrated by Wagner and West (1972) that, given a diffusion impairment, $(A-a)PO_2$ varies with $\dot{V}_A/\dot{Q}$, attaining a maximum for values of $\dot{V}_A/\dot{Q}$ around 1. In the case of CO_2, this maximum is attained at high $\dot{V}_A/\dot{Q}$ values. The computation of an increasing diffusion impairment contribution to $(a-A)PCO_2$ for the most compromised patients illustrates this fact, as for such patients the high $\dot{V}_A/\dot{Q}$ compartment accounts for a relatively greater percentage of gas exchange.

Measurements of diffusion are known to be underestimated in the presence of $\dot{V}_A/\dot{Q}$ heterogeneity. The interaction of $\dot{V}_A/\dot{Q}$ heterogeneity with different diffusion heterogeneities was studied theoretically by Chinet et al. (1971), who studied the error for different combinations of these factors. Since the present model allows for an estimate of $\dot{V}_A/\dot{Q}$ inequality in the presence of $D/\dot{Q}$ limitation, it could be associated with this previous analysis to correct the estimate of $D_L O_2$.

In conclusion, a two-compartment model of gas exchange was presented which considers the effects of $\dot{V}_A/\dot{Q}$ and $D/\dot{Q}$. The model allows for partitioning of $\dot{V}_A/\dot{Q}$ and $D/\dot{Q}$ influences on alveolar-arterial differences of O_2 and CO_2 using generally available gas exchange data. This possibility was demonstrated to be useful in the description of gas exchange impairment in COPD patients. Other applications are, in conjunction with previous works, the analysis of diffusion measurements and the study of diffusion heterogeneity.

REFERENCES

Briscoe, W.A., 1980, Gas Transfer in Diseased Lungs, in:"Pulmonary Gas Exchange"- vol.II, J.B. West, ed., Academic Press, New York, 275-314.

Chinet, A., J.L. Micheli, P. Haab, 1971, Inhomogeneity effects on O_2 and CO_2 pulmonary diffusion capacity estimates by steady-state methods. Theory, Resp. Physiol., 13:1.

DuBois, D., DuBois, E.F., 1916, Clinical calorimetry. A formula to estimate the approximate surface area if height and weight be known, Arch. Internal Med., 17:863.

Fick, A., 1855, Üver diffusion, Pogg. Ann., 94:59.

Granger, W.M., Miller, D.A., Ehrhart, I.C., Hofman, W.F., 1987, The effect of blood flow and diffusion impairment on pulmonary gas exchange: a computer model, Comp. Biom. Res., 20:497.

Hammond, M.D., Hempleman, S.C., 1987, Oxygen diffusing capacity estimates derived from measured $\dot{V}_A/\dot{Q}$ distributions in man, Respir. Physiol., 69:129.

Kellogg,R.H., 1987, Laws of physics pertaining to gas exchange, in "Handbook of Physiology - Section 3 - The Respiratory System". Fishman A.P., ed., Amer. Physiol. Society, Bethesda, Maryland.

King, T.K.C., Briscoe, W.A., 1967a, Bohr integral isopleths in the study of blood gas exchange in the lung, J. Appl. Physiol., 22:659.

King, T.K.C., Briscoe, W.A., 1967b, Blood gas exchange in emphysema: an example illustrating method of calculation, J. Appl. Physiol., 23:672.

Loeppky, J.A., Luft, U.C., Fletcher, E.R., 1983, Quantitative description of whole blood CO_2 dissociation curve and Haldane effect, Respir. Physiol., 51:167.

Loeppky, J., Caprihan, A., Luft, U.C., 1987, $\dot{V}_A/\dot{Q}$ inequality during clinical hypoxemia and its alterations, in: "Man in Stressful Environments", K. Shiraki and M.K. Yousef, ed., C.C. Thomas, Springfield, Il, 199-232.

Olszowka, A.J., Wagner, P.D., 1980, Numerical analysis of gas exchange, in: "Pulmonary Gas Exchange" - vol. I, J.B. West, Academic Press, New York, 33-66.

Press, W.H., Flannery, B.P., Teukolsky, S.A., Vetterling, W.T., 1986, "Numerical Recipes", Cambridge University Press.

Piiper, J., 1969, Apparent increase in O_2 diffusing capacity with increased O_2 uptake in inhomogeneous lungs: Theory, Respir. Physiol., 6:209.

Rahn, H., Fenn, W.O., 1955, A Graphical Analysis of the Respiratory Gas Exchange: The O_2-CO_2 Diagram, Amer.Physiol.Society, Washington D.C..

Riley, R.L. and Cournand, A., 1949, "Ideal" alveolar air and the analysis of the ventilation-perfusion relationships in the lungs, J. Appl. Physiol., 1: 825.

Severinghaus, J.W., 1979, Simple, accurate equations for human blood O_2 dissociation computations, J. Appl. Physiol, 46:599.

Wade, O.L., Bishop, J.M., 1962, Cardiac output and regional blood flow, F.A. Davas, Philadelphia, 26-38.

Wagner, P.D., West, J.B., 1972, Effects of diffusion impairment on O_2 and CO_2 time courses in pulmonary capillaries, J. Appl. Physiol., 36:62.

Wagner, P.D., Salzman, H.A., West, J.B., 1974, Measurement of continuous distributions of ventilation perfusion ratios: Theory, J. Appl. Physiol., 36:588.

Wagner, P.D., 1977, Diffusion and chemical reaction in pulmonary gas exchange, Physiol. Rev., 57:257.

Dr. Vidal Melo was supported by a grant of the Rotary Foundation

664

OXYGEN TRANSPORT THROUGH LUNG SURFACTANT

AND THE SURFACTANT SPECIFIC PROTEINS

Erna Ladanyi and Karlheinz Stalder

Department of Occupational Health
University of Göttingen
Federal Republic of Germany
D-3400 Göttingen, Windausweg 2

INTRODUCTION

Inhaled oxygen can reach alveolar and capillary cell walls only after having crossed both the so-called lung surfactant surface layer (LSSL) lining the alveoles at the air / water interface, and the underlying aqueous hypophase. The hypophase contains the LSSL precursors: a great variety of phospholipids and neutral lipids organized in different morphological forms, and three specific proteins called Sp-A, Sp-B and Sp-C. Therefore the transport of oxygen to the lung tissue is a rather complicated process, including such steps as the penetration through the air / aqueous interface, diffusion through the obviously viscous subface, and possibly an interaction with some of the LSSL or subphase components.

In a previous paper one of us described in detail the structure and role of the lung surfactant system in general and that of the LSSL in particular (Ladanyi, 1988). Electrochemical methods such as d.c. and a.c. polarography using dropping or hanging mercury electrods proved to be valuable tools in studying such properties of the LSSL in vitro as surface activity (Ladanyi, Zugravu and Tomoaia, 1974; Ladanyi and Stalder, 1979), composition (Ladanyi, 1980), double-layer capacitance (Ladanyi et al., 1988) and finally the influence of environmental and occupational noxious agents on different LSSL-properties (Ladanyi et al., 1974; Ladanyi, 1980; Stalder and Ladanyi, 1980; Ladanyi and Stalder, 1983; Ladanyi, 1986; Ladanyi, 1987; Ladanyi and Stalder, 1987).

Concerning the oxygen transport to tissues it was reported that LSSL represents a certain barrier and a place of storage for the crossing oxygen (Ladanyi, 1988). It could be shown that LSSL-lipids extracted from lung lavages / amniotic fluid increased while Sp-A diminished the energetic barrier the crossing oxygen had to overcome. An accumulation of oxygen in lung lavages of experimental animals was also observed.

The aim of the work we describe in this paper was to study the kinetics of oxygen accumulation in human broncho-alveolar lavages BAL, to investigate any differences between healthy subjects and those complaining of pulmonary disorders. Additionally, we wished to localize the oxygen storage within the lavage components.

EXPERIMENTAL

The measurements were carried out with both Differential Pulse Polarographic(DPP) and Alternating Current Tast-Polarographic (AC1$_t$) techniques. A dropping mercury electrode (DME) was used as a working electrode and two KCl saturated Ag/AgCl microelectrodes as reference and auxilliary electrodes, respectively. The polarograph was a Polarecord E 506 (Metrohm, Herisau, Switzerland) equipped with an x-y recorder. The dropping time was mechanically adjusted to 1.4 sec.

The biological material consisted of human natural BAL obtained by rinsing with saline the lung (one lobe) of a volunteer (a gift from Prof.Akino, Sapporo, Japan) and the lung of patients complaining about one or the other pulmonary disorder (Prof.Hüttemann, Lenglern, FRG).

The influence of some individual lavage components and their mixture on oxygen transport from the air to the air / aqueous interface onto the DME was studied on synthetic lipids and specific proteins. The lipids used were: diphosphatidylcholine (DPPC) and diphosphatidylethanolamine (DPE) (Sigma, Germany). The surfactant specific proteins used were a gift from Prof.Schäfer, Byk Gulden Pharmaceuticals, Konstanz, Germany and included: 34 kd recombinant Sp-A, 3.8 kd recombinant Sp-C and 17 kd bovine Sp-B. The molecular ratio of DPPC:DPE:Sp-A:Sp-B:Sp-C was 20:20:5:1:1. The absolute concentrations of the proteins in saline were 0.16; 0.54; 0.22 mg/ml, respectively. Saline was prepared of p.a. grade NaCl (Merck Germany) in deionized, double-destilled water. Nitrogen of 99.999 % purity was used for de-aeration without any additional purification. All mesurements were carried out at room temperature.

The experiment consisted basically in recording the oxygen reduction peaks of non-de-aerated samples (in order to see maximal oxygen saturation current and saturation potential values), in de-aerating the samples and in recording the time-dependence of the oxygen concentration during their re-aeration.

To evaluate adsorption / desorption potentials and surface activity, alternating-current tast-polarographic curves were also obtained. Water insoluble or hydrophobic components, which were to be dispersed in saline, were first transferred from their solutions in organic solvents onto the aqueous surface, then the organic solvents were evaporated in the presence of nitrogen and the hydrophobic components were finally dispersed in saline by sonication.

RESULTS AND DISCUSSION

The components of the LSSL behave not only in vivo but also in vitro as typical surface-active agents: they adsorb at interfaces changing the properties of the given interface. The adsorption of LSSL and hypophase components at the mercury / saline interface results in easily measurable changes of some

666

electrochemical parameters, for example the double-layer capacitance, the appearance of typical capacitance maxima on the adsorption current-potential curve of the saline at a well-defined potential.

Oxygen dissolved in saline is reduced at the mercury electrode (Fig.1a) in two steps at very well defined potential values:

I. $O_2 + 2 H^+ + 2 e^-$ -----------> H_2O_2

the peak potential E_P^1 in DPP is equal to - 0.06 V

II. $H_2O_2 + 2 H^+ + 2 e^-$ ----------> $2 H_2O$

the peak potential E_P^2 in DPP is equal to - 1.012 V

The resulting reduction current is proportional to the oxygen concentration.

Therefore, any change in the current intensity will reflect changes in the dissolved oxygen concentration. Furthermore, any change of the peak-potential value of the oxygen reduction in the presence of lung surfactant components will show surfactant-determined changes in the ability of the reducing species to reach the electrode.

In Fig.1 are shown the oxygen reduction peaks in saline (a), in healthy BAL (b), in BAL of patients (c and d), as well as the way the reduction current and peak potentials were measured (a). As it can be seen, much more oxygen is stored in the BAL than in the saline. The healthy lavage stores more oxygen than the pathologic samples. Furthermore, the peak potentials are shifted with lavages towards more negative values compared with saline. The largest peak potential shift was obtained with the healthy lavage.

It is obvious that in the presence of lung surfactant components in saline, different processes influence the two oxygen reduction steps. To reach the electrode for step I oxygen has to cross the surface active LSSL at the air / water interface, has to diffuse through the viscous aqueous subphase and finally, has to cross the surfactant layer adsorbed at the water / mercury interface. So the potential shift ΔE_P^1 of step I will be the sum of all these processes. For step II, the oxygen peroxide already formed in step I and dissolved in the saline in the neighbourhood of the electrode has only to cross the surfactant layer adsorbed at the water / mercury interface. Therefore ΔE_P^2 of step II will mirror solely the hindering effect of the adsorbed surfactant components.

As it can be seen, both ΔE_P^1 and ΔE_P^2 are shifted with BAL towards more negative potential values, the shift ΔE_P^2 with healthy BAL reaching even 252 mV. This indicates that this BAL gives the most tightly adsorbed layer.

The re-aeration time dependent changes in the height of the peaks obtained with previously de-aerated samples could be used to study the kinetics of oxygen transport through the system. Considering the solubility of oxygen in saline to be 8 µg/ml, peak height could be converted into oxygen concentration.

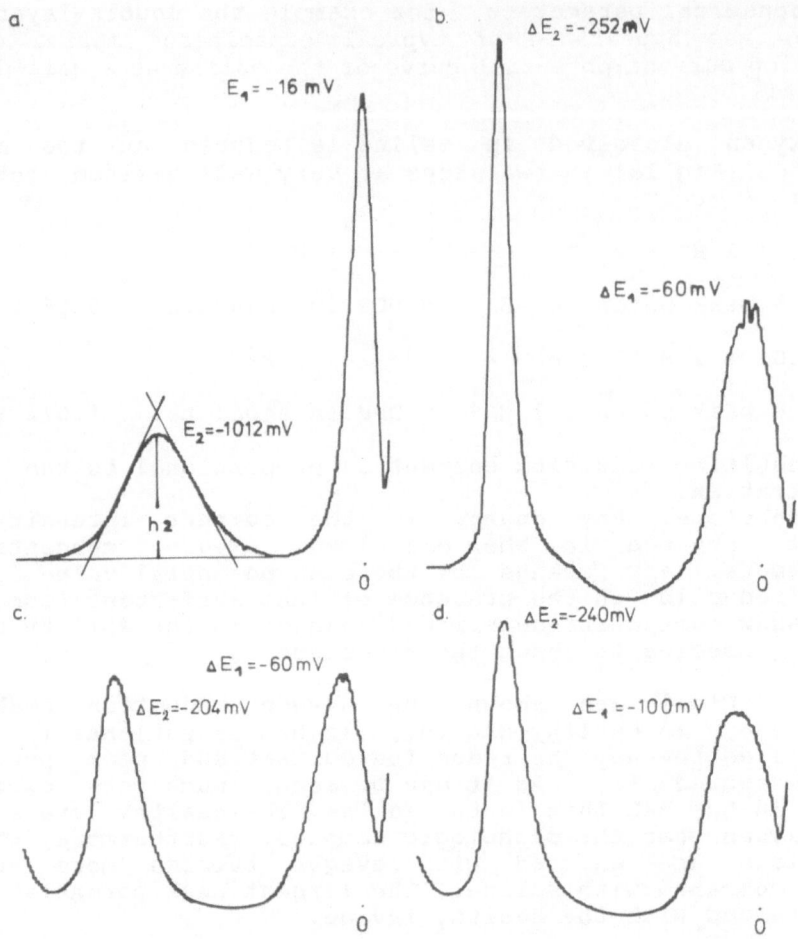

Fig.1   Current-potential curves of oxygen reduction in BAL
        and saline. a: saline, b: BAL of healthy volunteer, c
        and d: BAL of patients.

Fig.2 evaluates the changes in peak height due to step I,
and Fig.3 the changes due to step. II. According to both
reduction steps the redissolution kinetics of oxygen are
different in saline (curve 1), in the BAL of a patient (curve
2), in the BAL of a healthy volunteer (curve 3) and in the
saline dispersion of surfactant components. BAL of the healthy
person (Fig.2 and 3, curves 3) stored not only more oxygen but
also stored it more rapidly than did the BAL of patients.
However, according to the $\Delta E_P{}^1$ and $\Delta E_P{}^2$ values, LSSL structure
is more tight with healthy BAL than with any other studied
samples. Fig.3 illustrating the transport kinetics of oxygen
through this layer shows that BAL components in low
concentration (for values see experimental part) are not able
to influence the kinetics and therefore curve 4 overlaps

668

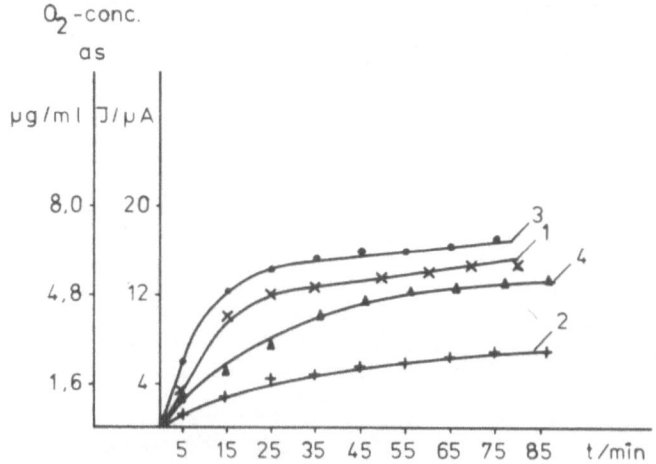

Fig.2   Kinetics of oxygen transport through the lung surfactant measured as oxygen reduction current. Reduction step I.

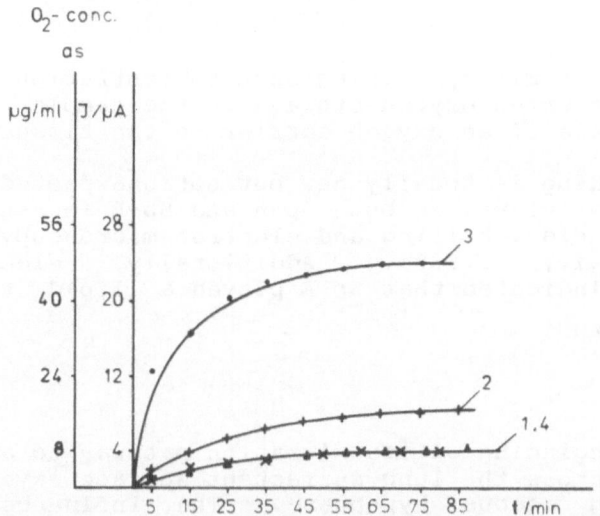

Fig.3   Kinetics of oxygen transport through the lung surfactant measured as oxygen reduction current. Reduction step II.

curve 1 of the saline. This figure shows that for the five to six times larger oxygen storage in BAL as opposed to saline, one or some of the BAL components have to be responsible.

By studying the maximal adsorption current and peak potential shift produced by two lipid and three specific protein components of the BAL, we found that the water soluble major surfactant protein called Sp-A could be the binding site of the oxygen.

669

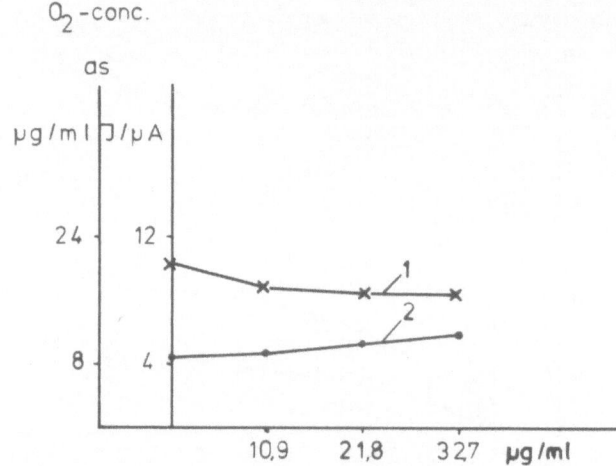

Fig.4   Augmentation of stored oxygen in saline in the
         presence of the major surfactant specific protein
         Sp-A. Curve 1: peak current of the reduction step I;
         curve 2: pek current of the reduction step II.

As seen in Fig.4, rising Sp-A concentration in saline resulted in augmented oxygen storage in the sample. Thus, Sp-A may play the role of an oxygen carrier to the tissue.

This finding is totally new but not unexpected since we could recently visualize both Sp-A and Sp-B in the LSSL by using immuno-gold-labelling and electron microscopy (Ladanyi, Suzuki et al., 1989). Additionally, electrochemical measurements indicated that Sp-A played a lipid transporting role.

SUMMARY

At the very beginning of its migration pathway to any tissue, oxygen has to cross the lung surfactant surface layer (LSSL) and the underlying aqeous hypophase. The influence of human broncho-alveolar lavage and its lipid and specific protein components on oxygen transport were studied in vitro using sensitive electrochemical techniques. LSSL adsorbing from BAL at the dropping mercury electrode / saline interface shifted the peak potentials of oxygen reduction E_P towards more negative values. The magnitude of ΔE_P was dependant on the quality of BAL. The kinetics of the oxygen transport were evaluated by measuring changes in the intensities of the reduction current. Storage of oxygen in BAL was several times higher than in saline and indicated the presence of one or more binding sites or promoters for oxygen among BAL components. The surfactant specific protein was found to be one of the binding sites (or promoters) for the oxygen. Since electron microscopic immuno-gold labelling demonstrated the presence of this protein in the LSSL, and electrochemically it proved to transport lipids from the surface layer to the subphase, it was suggested that Sp-A plays the role of an oxygen carrier.

670

ACKNOWLEDGMENT

We thank Mrs. Dina von Pape von Beyme for her skilful technical assistance.

REFERENCES

Ladanyi, E., Zugravu, E., Tomoaia,M.,1974, Electrochemical Methods in Surface-Activity Studies of Lung Surfactant.I. Polarographic Maximum Suppressing Ability of Lung Surfactant, Int. Arch. Arbeitsmed., 33:245

Ladanyi, E., Stalder,K., 1979, Alternating current-tast-polarographic determination of surface activity of lung surfactant, J. Electroanal.Chem., 99:321

Ladanyi, E., 1980, Polarographische Elektrosorptionsanalyse des oberflächenaktiven Systems der lunge (Lung surfactant), Dissertation, Technische Universität Clausthal

Ladanyi, E., Stalder, K., 1980, Changes occuring in the lung surfactant under the action of inhalative occupational substances, Verh. Dtsch. Ges. Arbeitsmed., 20:519

Ladanyi, E., Stalder, K.,1983, Contribution to the medical importance of dusts resulting by the use of agricultural machines, Verh.Dtsch.Ges.Arbeitsmed., 23:523

Ladanyi, E., 1986, Inhalative Noxen und das Surfactant-System der Lunge, Prax.Klin.Pneumol., 40:465

Ladanyi, E., Möbius, D., Stalder, K., von Wichert, P., 1987, Structure of isolated lung surfactant monolayer, Symposium on Membrane Lipids, 20-21 March,Sintra, Portugal, ACTAS do INSTITUTO de BIOQIMICA (in press)

Ladanyi, E., Stalder, K., 1987, Modellversuche zum Einfluß von Formaldehyd auf das Lungensurfactant, Verh.Dtsch.Ges.Arbeitsmed.,27:545

Ladanyi, E., 1987, Present knowledge in the field of lung surfactant electrochemistry, J.Bioel.Bioenerg., (in Press)

Ladanyi, E., Miller, I., Popovitz-Biro, R., Marikovsky, J., von Wichert, P., Müller, B., Stalder, K., 1988, Molecular structure of the extracellular surface-layer of the human lung surfactant, 3d International Symposium, Basic Research on Lung Surfactant, Marburg, 12-14 September

Ladanyi, E., 1988, "Oxygen Transport to Tissue XI", Plenum, New York

Ladanyi, E., Suzuki, Möbius,D., Schäfer, K., Stalder, K., Presence, localization and possible role of Sp-A and Sp-B in the human lung surfactant surface layer. An electronmicroscopic study using immuno-gold double labelling, 60 Years of Surfactant Research, Floating congress on the river Rhine, 11-17 November 1989.

Stalder, K., Ladanyi, E., 1980, Changes occuring in the lung surfactant under the action of inhalative occupational substances, Verh. Dtsch. ges. Arbeitsmed., 20:519

AN ULTRASTRUCTURAL STUDY OF PULMONARY CAPILLARY VESSELS IN BLOOD VOLUME-OVERLOADED RAT

Masato Sageshima*, Koichi Kawamura, Kohei Toda, Hirotake Masuda and Takeshi Shozawa

Department of Laboratory Medicine*, Second Department of Pathology
Akita University, School of Medicine
Akita, Japan

INTRODUCTION

Luminal dilatation and endothelial proliferation of arteries can be induced by an overload of blood flow as shown by Kamiya and Togawa (1980) and Masuda et al. (in press). Blood flow often increases physiologically and pathologically. In contrast to chronic congestion, morphological changes of the pulmonary vessels induced by an overload of pulmonary blood flow are not well understood.

In this paper, we describe changes of the capillary endothelial ultrastructure, capillary surface area and volume in the volume overloaded rat lung induced by arterio-venous fistula and discuss the adaptation of the pulmonary capillary to the blood volume overload.

MATERIALS AND METHODS

Sixty-two Sprague-Dawley rats (8 week-old, body weight 282 ± 30g) were used. Fourty-two rats were anesthesized with sodium pentobarbital (50mg/kg) intraperitoneally for shunt operation. A side-to-side anastomosis was constructed between the left common carotid artery and the left external jugular vein at the level of about 2 cm from aortic arch by microsurgery. Twenty rats were used for controls. Blood flow rate of both common carotid arteries was measured before anastomosis, just after anastomosis and at the final measurement with electromagnetic flow-meter (Vl6-A, Nihon Koden, Japan) 1.5 cm distal to the aortic arch. Measurements of the controls was done before sacrifice. The animals were sacrificed at 0 week (control, n = 4), 1 week (shunt, n = 10; control, n = 2), 2 weeks (shunt, n = 8; control, n = 4), 4 weeks (shunt, n = 8; control, n = 4), 8 weeks (shunt, n = 10; control, n = 3) and 24 weeks (shunt, n = 6; control, n = 3). A cannula was placed into the abdominal aorta and 75 ml of oxygen-saturated heparinized Ringer solution was injected and drained via renal vein for 5 minutes. After whole blood was washed out, 50 ml of 3% glutaraldehyde solution in cacodylate buffer (pH 7.4) was injected from the abdominal aorta under 100 mmHg. Pulmonary vessels were dilated and fixed better than by tracheal instillation fixation. Lung tissue samples were obtained from the upper lobe and fixed in 1% osmic acid solution in cacodylate buffer (pH 7.4) for 2 hours at 4°C, dehydrated and embedded in Epon 812 resin. The ultrathin sections were observed by transmission electron microscope (TEM) (LEM 2000, Akashi, Japan). For morphometrical study, panorama views composed of 2000x magnified photomicrographs of peripheral lung area, 0.48 mm x 0.32

mm in size, were prepared. The air ducts proximal to respiratory bronchiole, pulmonary arteriole and interlobular and interlobar connective tissue were excluded for this study.

Electron microscopic and morphometrical studies were carried out in 12 shunted cases and 8 controls of 2, 4 and 24 weeks (Shunt, n = 4 for each shunt time group. Control, n = 3 in 2 weeks, n = 3 in 4 weeks, n = 2 in 24 weeks).

The number of the endothelial nuclei was counted on each composite panorama view and was expressed by number per cm. The capillary surface area and volume were estimated by linear integration and point counting method on the same composite panorama view, respectively (Weibel, 1969).

For the indication of the grade of volume overload, the weight of both ventricles, right ventricular free wall, and left ventricle including ventricular septum were measured. The grade of right ventricular hypertrophy was expressed by relative weight of right ventricle (weight of right ventricular free wall/body weight). We used the sum of the common carotid artery flows for calculation of blood flow index. Blood flow index was given by blood flow rate/body weight (ml/min/kg).

Statistical analysis was performed by means of the student t-test. Differences were considered significant if p values were less than 0.05. The results are given as mean values ± SD.

RESULTS

Blood flow index was 21.28 ± 6.18 ml/min/kg before anastomosis and 85.00 ± 19.72 ml/min/kg just after anastomosis. The final measurement yielded: 83.02 ± 16.01 ml/min/kg at 1 week, 86.20 ± 21.56 ml/min/kg at 2 weeks, 91.65 ± 14.18 ml/min/kg at 4 weeks and 77.24 ± 10.89 ml/min/kg at 24 weeks. The blood flow indexes after the anastomosis were significantly higher than those of the controls throughout the experiment. Relative weight of right ventricle of the shunted rats was about 1.5 times larger than that of the controls, the difference being significant ($p < 0.05$).

Morphometrical results (Fig.1)

In the shunted rats, the average number of the endothelial nuclei was (96.5 ± 18.0) x $10^3/cm^2$ in 2 weeks, (86.8 ± 7.1) x $10^3/cm^2$ in 4 weeks and (56.0 ± 16.5) x $10^3/cm^2$ in 24 weeks. The controls showed (64.5 ± 13.7) x $10^3/cm^2$ in 2 weeks, (74.2 ± 3.8) x $10^3/cm^2$ in 4 weeks and (60.3 ± 2.8) x $10^3/cm^2$ in 24 weeks. The numbers were significantly larger in 2 and 4 weeks of the shunted rats than in the controls ($p < 0.05$). The average capillary surface was significantly larger in 2 weeks $(1776.6 \pm 65.9$ cm^2/cm^3) and 4 weeks $(1825.0 \pm 135.0$ cm^2/cm^3) than that of the controls in 2 weeks $(1411.8 \pm 76.0$ cm^2/cm^3) and 4 weeks $(1562.7 \pm 165.1$ cm^2/cm^3), respectively ($p < 0.05$). There was no significant difference between the shunted $(1630.0 \pm 208.3$ cm^2/cm^3) and the control $(1535.7 \pm 106.9$ cm^2/cm^3) in 24 weeks. The average capillary volume was significantly larger in 2 weeks $(0.340 \pm 0.043$ cm^3/cm^3) and 4 weeks $(0.340 \pm 0.052$ cm^3/cm^3) than that of controls in 2 weeks $(0.260 \pm 0.014$ cm^3/cm^3) and 4 weeks $(0.225 \pm 0.021$ cm^3/cm^3), respectively ($p < 0.05$). There was no significant difference between the shunted $(0.331 \pm 0.063$ cm^3/cm^3) and the control $(0.255 \pm 0.015$ cm^3/cm^3) in 24 weeks.

Electron microscopic study

1) 2 weeks. In the capillary endothelium, the cytoplasm was irregularly thickened and the nucleus enlarged. The capillary lumen was often stenotic (Fig. 2). Pinocytotic vesicles

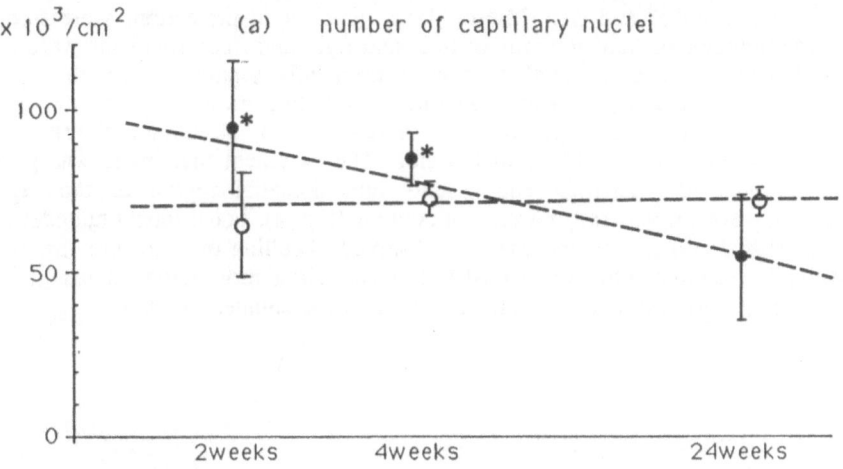

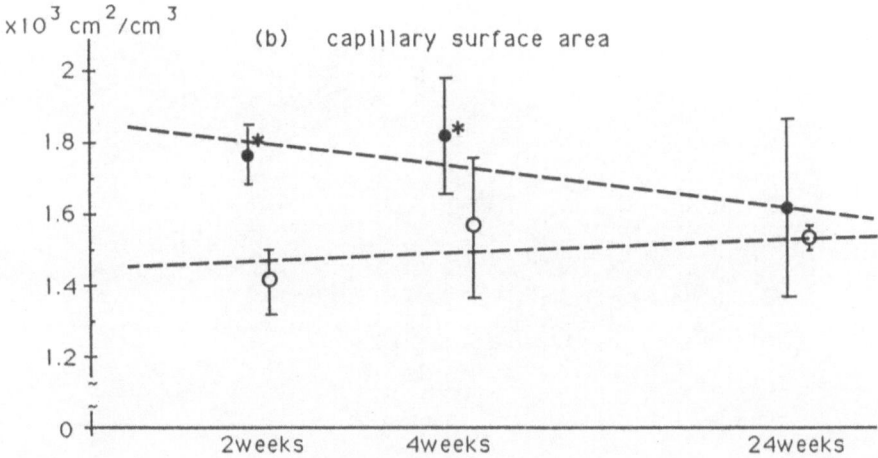

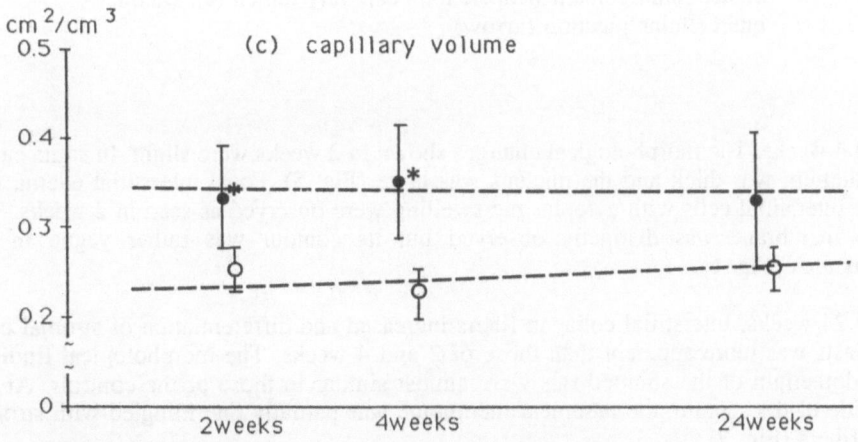

Fig. 1. Morphometric results of pulmonary capillary vessels.
Closed circles, shunted; open circles, controls; asterisk, difference between shunted and controls statistically significant.

were increased in number and size. Micropillous cytoplasmic projections were frequently observed. The intercellular junction was distinct and tight and microfilaments were distinct around the junction. The endothelial cells were frequently separated from the basement membrane to various degree, showing sometimes bulla-like protrusion into the capillary lumen (Fig. 3). This endothelial separation was more frequently observed at the stromal side of the alveolar septum than at the epithelial side. The basement membrane was generally thin. In the area of interstitial edema and subendothelial region of the separated endothelium, however, it was rather vague in contour (Fig. 4). Focal interstitial edema with dispersed interstitial collagen fibers was also observed. Swelling of immature stromal cells was occasionally seen in addition to stromal fibroblasts. These morphological changes varied in degree from case to case but were clearly found in the shunted animals.

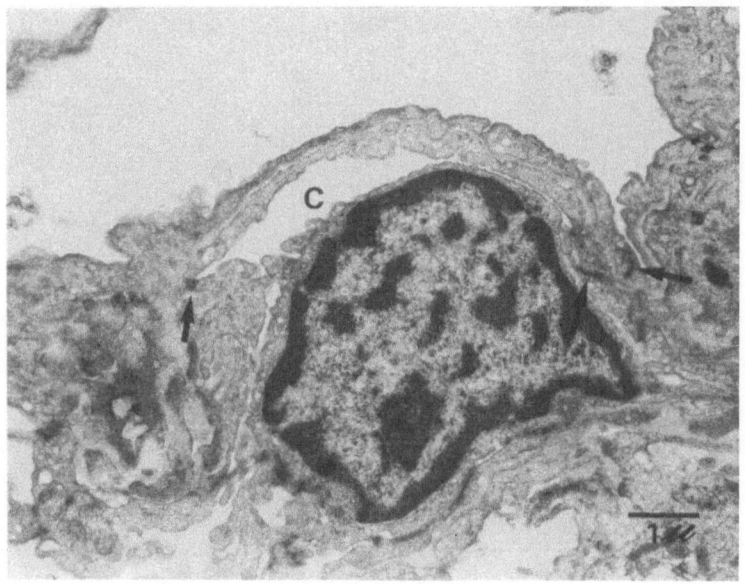

Fig. 2. Shunt, 2 weeks. Swelling of the endothelium and marked nuclear enlargement with stenotic capillary lumen (c). Distinct intercellular junction (arrow)

2) 4 weeks. The morphological changes shown in 2 weeks were slight. In some cases, the endothelium was thick and the nucleus was large (Fig. 5). Focal interstitial edema was observed. Interstitial cells with cytoplasmic swelling were observed as seen in 2 weeks. The basement membrane was distinctly observed but its contour was rather vague in the edematous area (Fig. 6).

3) 24 weeks. Interstitial collagen fibers increased and differentiation of stromal cells to fibroblasts was more apparent than those of 2 and 4 weeks. The morphological findings of the endothelium of the shunted rats were almost similar to those of the controls. At the stromal side of the septum, the basement membrane was partially intermingled with stromal collagen fibers (Fig. 7).

In the shunted rats of every groups, there was no evidence of edema or hemorrhage in the alveolar air spaces.

676

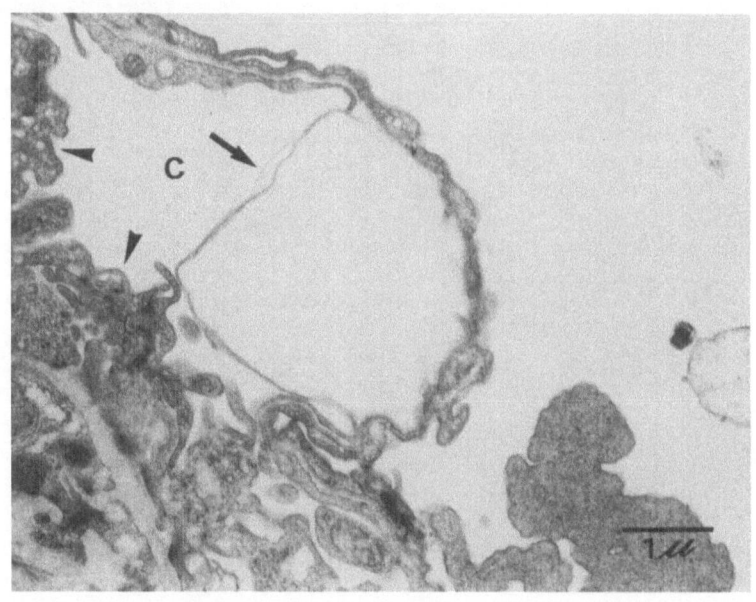

Fig. 3. Shunt, 2 weeks. Severe endothelial separation showing bulla-like change (arrow). Microvillous projection of the cytoplasm (arrow head). C: Capillary lumen

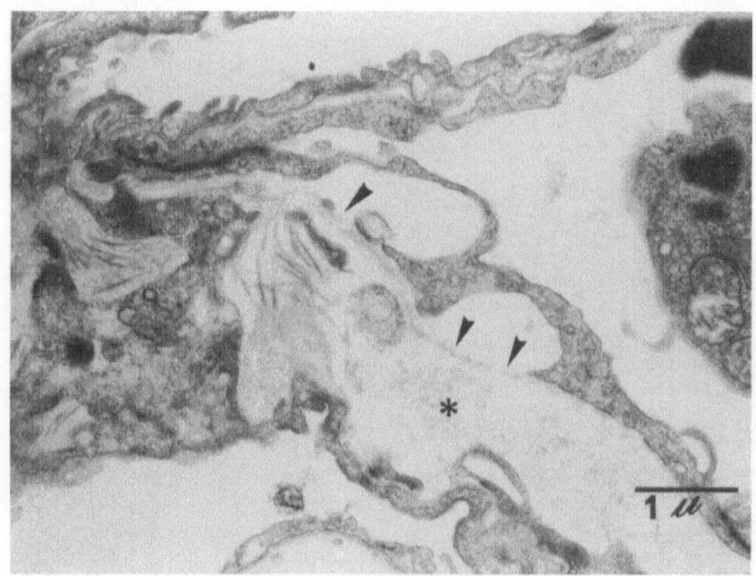

Fig. 4. Shunt, 2 weeks. The basement membrane was vague in contour at the area of endothelial separation (arrow head). Dispersed collagen fibers in the interstitial edema (asterisk)

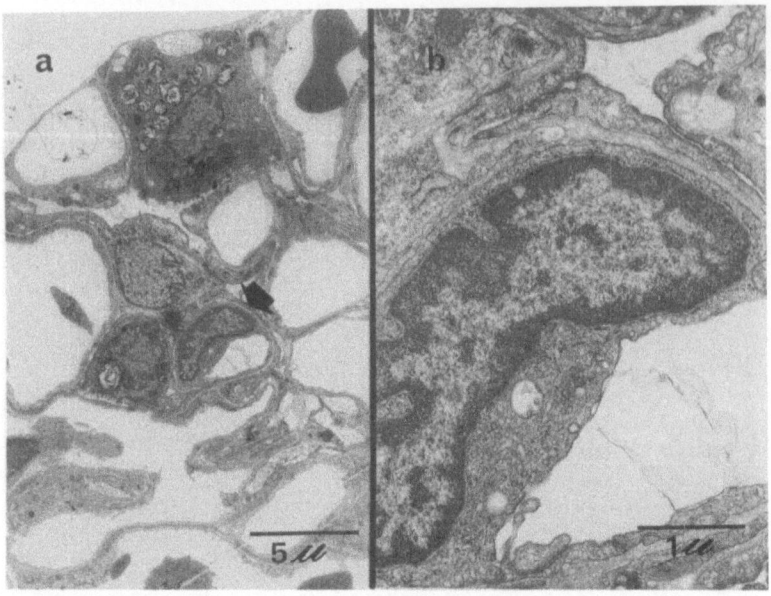

Fig. 5. Shunt, 4 weeks. (a) Swelling of the endothelium with nuclear enlargement (arrow). (b) High power magnification of the endothelium

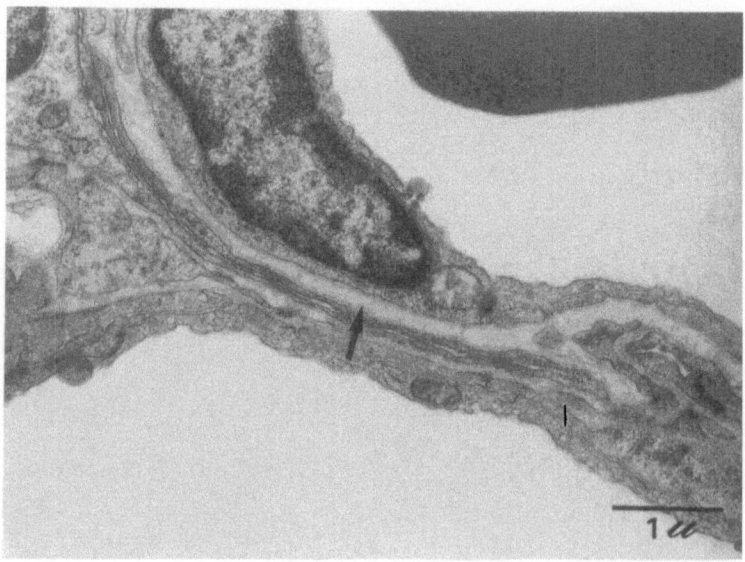

Fig. 6. Shunt, 4 weeks. Mild interstitial edema and the basement membrane was vague in contour (arrow).

DISCUSSION

In order to increase pulmonary blood volume, injection of plasma expander or mechanical obstruction of venous drainage has been introduced experimentally (Hovig et al., 1971). These experimental procedures were inadequate for a long-term study because of hypoxemia. But arterio-venous fistula formation between the common carotid artery and the external jugular vein were useful for a long-term study. In our experiments, an overload of pulmonary blood flow was suggested by a constant increase of venous return, because the blood flow index of the common carotid arteries and the relative weight of right ventricle increased significantly.

Nuclear enlargement, swelling of the cytoplasm with an increase of pinocytotic vesicles and numerical increase of the nuclei of endothelium were most prominent in 2 weeks and indistinct in 24 weeks. The capillary surface area and the capillary volume increased significantly in 2 and 4 weeks. Kamiya and Togawa (1980) showed that high shear stress in canine common carotid artery induced by an overload of blood flow was normalized by luminal dilatation. Recently, using the same model, we have shown that in the early phase of the experiment, proliferation of endothelial cells is observed in the canine common carotid artery and in chronic phase, it is indistinct due to decrease of shear stress (Masuda et al., in press). We also showed proliferation of capillary endothelial cells in the myocardium of rats in the early phase of experimental volume hypertrophy (Kawamura et al., this volume). In experimental pulmonary edema induced by injection of plasma expander, any ultrastructural changes except for interstitial edema have not been reported (Hovig et al., 1971). We think that the morphological changes of the capillary endothelium and increases of the capillary surface area and volume are responses of capillary endothelium to an overload of blood flow in the early phase. Alteration of the morphological changes of the endothelium in the chronic phase might be adaptation to an overload of blood flow.

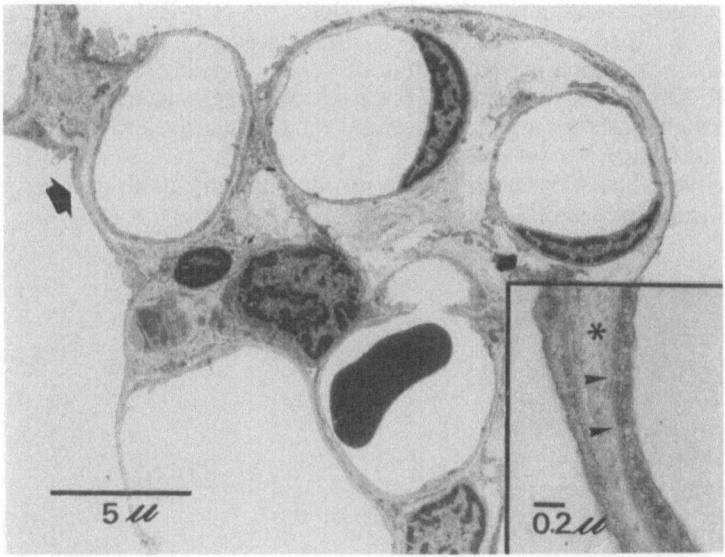

Fig. 7. Shunt, 24 weeks. The morphological changes of the endothelium are unnoticeable as compared to 2 weeks. Inset: Increase of stromal collagen fibers (asterisk) associated with indistinct basement membrane (arrow heads)

At the stromal side of the septum, the basement membrane was vague in contour in the shunted rats of 2 and 4 weeks. This might be induced by stromal edema. The vagueness of the basement membrane seen in both groups of 24 weeks might be due to stromal collagenization in the course of organization of edema.

SUMMARY

Blood volume overload to pulmonary capillaries was experimentally induced by arterio-venous fistula between the left common carotid artery and the left external jugular vein in 42 S-D rats. The animals were sacrificed 1, 2, 4, 8 and 24 weeks after the operation. Just after anastomosis, blood flow index of the shunted rats was about 4 times larger than in the controls and the high value was kept throughout the experiment ($p < 0.05$). The relative right ventricular weight was about 1.5 times larger than in the controls ($p < 0.05$). The capillary endothelium of the alveolar septum showed nuclear enlargement, swelling of the cytoplasm with microvillous projection, and increases of pinocytotic vesicles and microfilaments. In some cases, focal edema was seen accompanied with an increase of immature stromal cells. The basement membrane showed vague in contour in edematous area, especially in the epithelial side of the alveolar septum. These changes appeared most predominantly in 2 weeks and subsided almost completely in 24 weeks. Number of endothelial nucleus of the shunted rats increased significantly in 2 and 4 weeks and was normalized in 24 weeks. The surface area and volume of the capillaries also increased significantly in 2 and 4 weeks of the shunted animals. These findings showed that the capillary endothelium reacted and then was adapted to the blood volume overload.

REFERENCES

Hovig, T., A. Nicolaysen and G. Nicolaysen, 1971, Ultrastructural studies of the alveolar-capillary barrier in plasma perfused rabbit lungs. Effect of EDTA and increased capillary pressures, Acta Physiol. Scand., 82: 417-432.

Kamiya, A. and T. Togawa, 1980, Adaptive regulation of wall shear stress to flow change in the canine carotid artery, Am. J. Physiol., 239: H14-H21.

Kawamura, K., K. Tohda, M. Kobayashi, H. Masuda and T. Shozawa, Fine structure of capillary proliferation in myocardium of volume overloaded rats (this volume).

Masuda, H., K. Kawamura, K. Tohda, T. Shozawa, M. Sageshima and A. Kamiya, Increase in endothelial cell density prior to artery enlargement in flow loaded canine carotid artery arteriosclerosis (in press).

Weibel, E.R., 1969, Stereological principles for morphometry in electron microscopy, Int. Rev. Cytol., 26: 235-302.

OTHER ORGANS AND TISSUES

SEVERITY OF OXYGEN FREE RADICAL EFFECTS AFTER ISCHEMIA AND REPERFUSION IN INTESTINAL TISSUE AND THE INFLUENCE OF DIFFERENT DRUGS

J. Lutz, A. Augustin and E. Friedrich

Physiologisches Institut der Universitaet
Roentgenring 9, D-8700 Wuerzburg, F.R.G.

INTRODUCTION

The intestinal tissue reacts to a temporary total occlusion of the superior mesenteric artery with a marked increase of lipid peroxidation as shown in previous experiments (Augustin and Lutz, 1988, Lutz and Augustin, 1989). This increase, as measured by the content of thiobarbituric acid reactive substances (TBARS), exceeds the reaction of the kidney to similar manipulations (e.g. Bird et al., 1988, Paller, 1988) severalfold: Whereas the increase of TBARS in the kidney is described as about 2-3 fold, we observed a more than 10 fold increase in intestinal tissue. Thus, the ischemic intestine is well suited as a model to test the effect of various drugs. According to different types and sources of oxygen free radicals in ischemic tissue, several therapeutic substances were chosen. Superoxide dismutase and catalase are scavengers of O_2 and H_2O_2. Allopurinol is believed to act chiefly as xanthine oxidase inhibitor. An antibiotic therapy was chosen to restrain the inflammatory response and production of radicals via myeloperoxidase by neutrophiles. Deferoxamine, an iron chelator, was given to prevent the iron dependent reaction of superoxide anion radical and hydrogen peroxide to the most aggressive hydroxyl radical by the Haber Weiss and Fenton mechanism. An artificial oxygen carrier (perfluorochemical) has been shown by us (Lutz et al., 1985, Hamar et al., 1987) to extend survival of rats after a temporary mesenteric artery occlusion. In the experiments presented here, polymerized stromafree hemoglobin was used and tested alone as well as in combination with the above mentioned substances.

MATERIALS AND METHODS

Experiments were performed in male Wistar rats by reversibly occluding the superior mesenteric artery under ether anesthesia. This was done by pulling the vessel against a thread, which was lead via a tube through the abdominal wall and fixed until removal. The rats were given an analgesic (tramadol, 2.5 mg/kg b.wt.) immediately after the initial

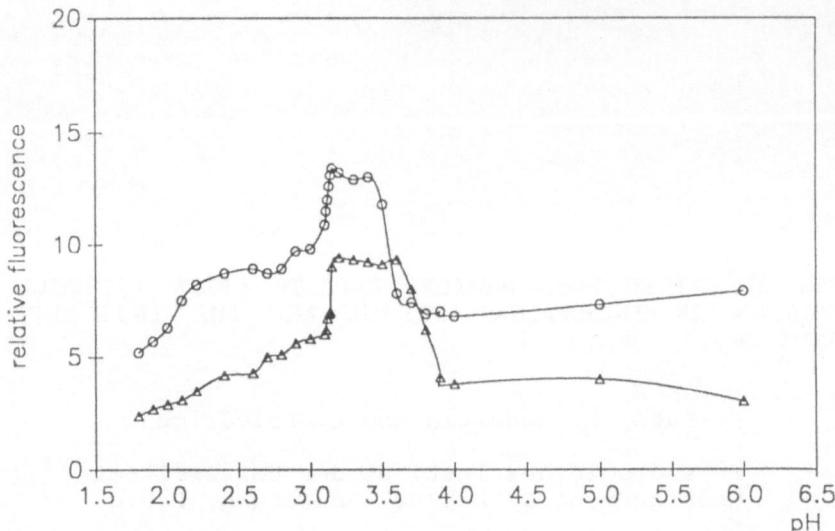

Figure 1a. pH-dependence of the reaction of lipid peroxides from plasma with TBA. Open circles: human plasma, triangles: rat plasma. The curves show a broad plateau, but have very steep shoulders.

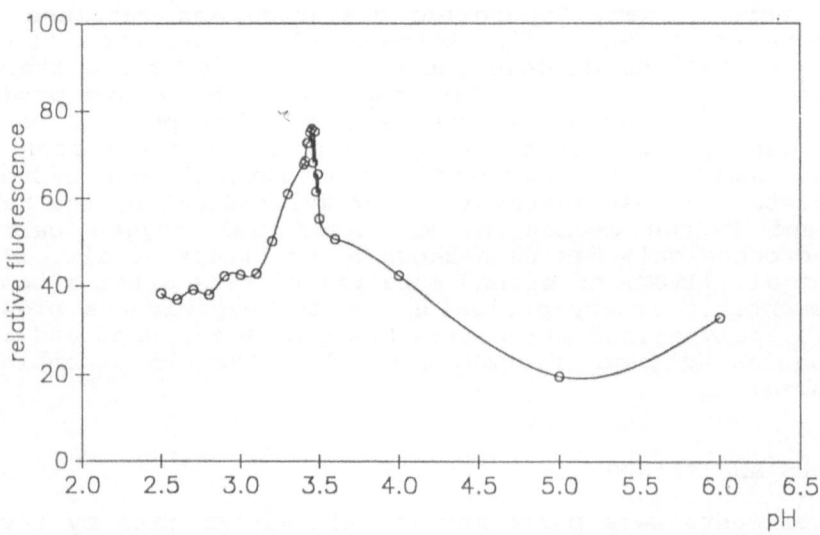

Figure 1b. The same curve for the reaction of lipid peroxides from intestinal tissue with TBA. The optimum of the reaction has a sharper peak as compared to plasma probes. In each case a strict maintenance of the pH in samples is required.

operation and again when the occlusion was reopened under ether anesthesia. The occlusion period lasted 90 min, the reperfusion time (till sacrificing the animals) 2.5 hours. To determine the level of lipid peroxides, a modification of the methods of Ohkawa et al. (1979) and Yagi (1976) was applied, based on the reaction with thiobarbituric acid. Samples were spectrofluorometrically measured (excitation 515 nm, emission 553 nm); all determinations were performed in duplicate.

The results are expressed as the tissue content of TBARS. The expression TBARS was chosen for malondialdehyde equivalent substances, since mostly malondialdehyde precursors and relative substances are determined by this reaction. As methodical questions arouse concerning disturbance of the TBA-reaction by hemoglobin (Fuld et al., 1986, Gamelin and Zager, 1988), some important details of our method are given here. Of great importance is the observation of a definite pH-value for plasma and tissue probes, as Figure 1a and b demonstrate. Furthermore, the influence of hemoglobin in the samples is greatly diminished if a spectrofluorometric instead of a spectrophotometric method is used (Figure 2a and b).

The following drugs were given in comparison to an untreated group (n.th.). If not stated otherwise they were given at the end of the occlusion period: superoxide dismutase (SOD), 10 mg/kg bwt. = 30.000 U/kg, given together with CAT, half the dose i.p. and i.v.; catalase (CAT), 10 mg/kg =340.000 U/kg; allopurinol (ALLO), 100 mg/kg iv.; ciprofloxacin (CF), 20 mg/kg given one hour before occlusion, 60 mg/kg at the end of the occlusion period; deferoxamine (DFO), 30 mg/kg sc. one hour before the occlusion and 30 mg/kg iv. at the end of occlusion. A polymerized stromafree hemoglobin solution (PHb) with a hemoglobin content of 10 g/dl was given alone or in combination with the above mentioned drugs in a dose of 2 g PHb/kg bwt. Statistical analysis was performed by Mann-Whitney's U-test. Error bars in the figures correspond to SEM. The groups consisted of 4 to 8 animals, in the n.th. group of 13 animals.

RESULTS

Like in previous experiments the animals without therapy revealed very high tissue levels of TBARS (Figure 3). SOD in combination with CAT caused a depression to less than half of these control values. A still larger effect was seen after administration of ALLO, which is believed attacking the major points of the oxygen free radical chain. From several antibiotics tested, CF, a fluorinated quinolone was most effective, depressing the increased TBARS content in intestinal tissue to about 10 % of the n.th. group. DFO, the iron chelator, showed also a significantly depressant effect on the production of lipid peroxides, even though variances were higher. Thus, all the drugs used protected the rats to a considerable degree from the formation of lipid peroxides measured as TBARS; the decrease in TBARS was statistically significant at levels of p<0.01 (**) to p<0.001 (***).

Treatment of the animals at the end of the occlusion period with PHb did not cause a further increase in TBARS (Figure 4). This finding is compatible with the previously

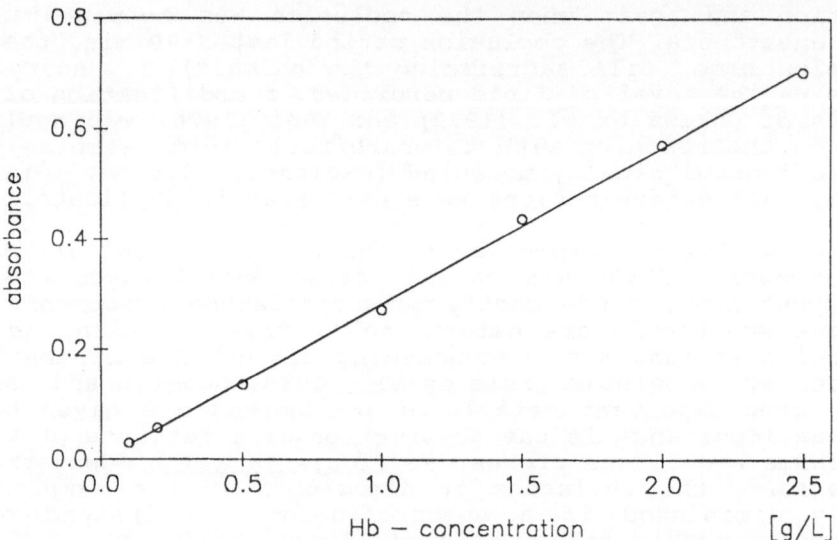

Figure 2a. Spectrophotometric calibration curve for hemoglobin solutions, measured at 532 nm. As small a concentration as 0.25 g Hb/l will considerably influence the spectrophotometric result of the TBA reaction.

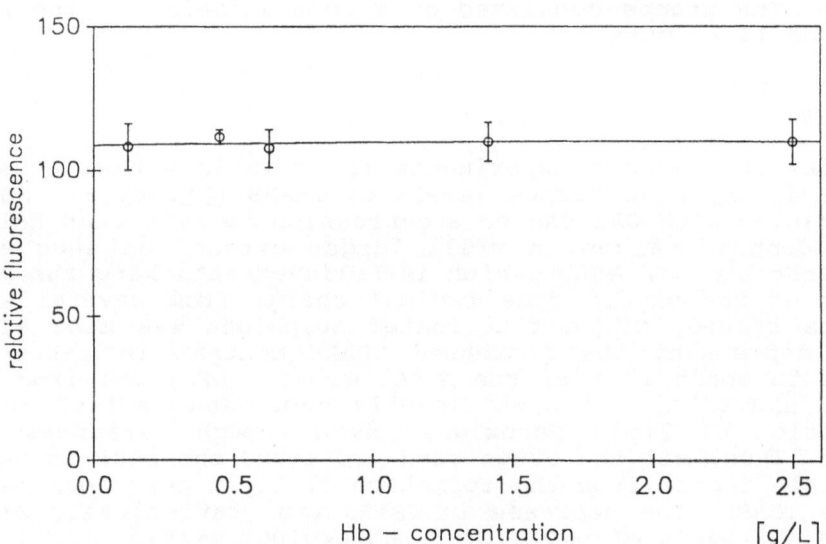

Figure 2b. Relative fluorescence of plasma samples with increasing Hb concentrations. The fluorescence is not influenced by concentrations up to 2.5 g Hb/l.

686

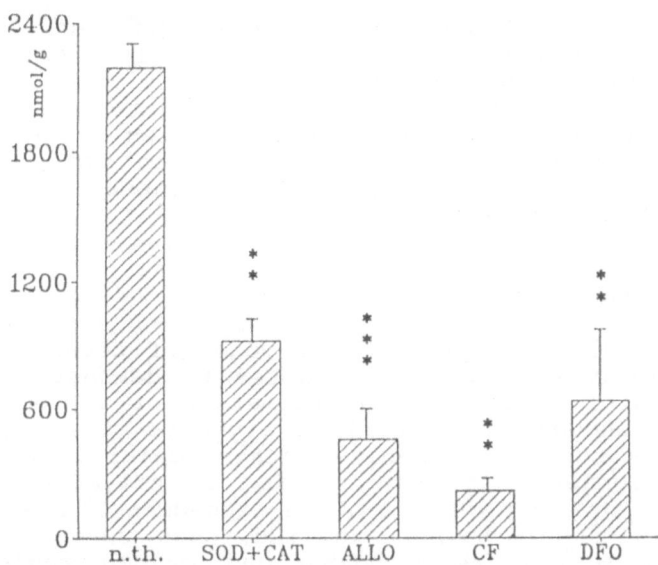

Figure 3. Content of lipid peroxides, expressed in terms of TBA reactive substances in intestinal tissue. All values are highly significant (p< 0.01) as compared to the n.th. group. Abreviations and doses see methods.

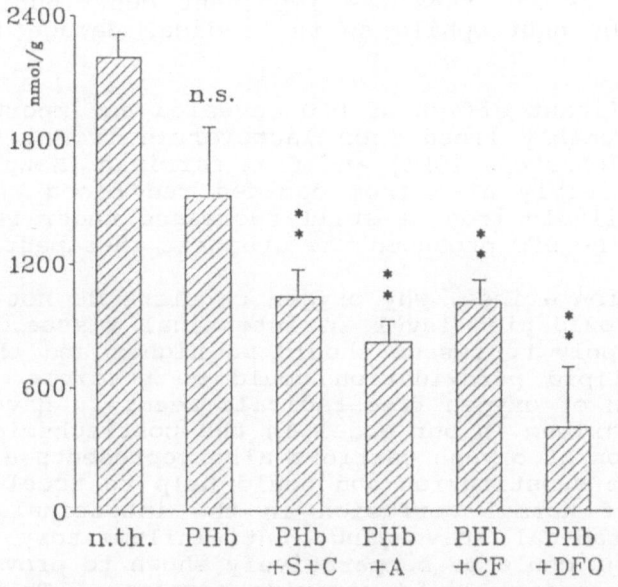

Figure 4. Content of lipid peroxides, expressed in terms of TBA reactive substances in intestinal tissue. Polymerized Hb, given as an additive oxygen carrier influences the production of TBARS, but this effect can be reduced by the used drugs. S = SOD, C = CAT, A = ALLO.

reported effects of perfluorochemicals (Augustin and Lutz, 1988, Lutz and Augustin, 1989); they revealed that a further increase of TBARS following the administration of such oxygen carrying substances did not occur in the intestinal tissue. A highly protective effect from the production of lipid peroxides could also be shown for drugs used in combination with PHb, though to a lesser degree. Under these conditions DFO was the most effective drug.

DISCUSSION AND CONCLUSION

These results reveal that the model of mesenteric artery occlusion in rats is superbly suited to demonstrate protective effects of drugs from the formation of lipid peroxides. SOD and CAT showed an effect that might be still further increased if a sufficiently high intravascular level could be maintained longer, e.g. by coupling to polyethylene glycol (Liu et al. 1989) or entrapment in liposomes (Turrens et al., 1984), but it should be born in mind, that these enzymes only act against precursors of dangerous oxygen radicals like hydroxyl radical and OCl^-, which can be produced also without the participation of O_2-radical and H_2O_2. ALLO was very effective; this doesn't exclude the possibility that it acted as an oxygen free radical scavenger rather than by inhibition of the ischemia produced conversion of xanthine dehydrogenase to an oxidase: allopurinol, especially when converted to oxypurinol has been shown to be a very potent scavenger itself (Grootveld et al., 1987, Moorhouse et al., 1987). The strong effect of the antibiotic CF also highlights the fact that it is not mainly the role of xanthine oxidase alone that leads to the post-occlusive elevation of TBARS in the intestine. The development of an additional peritonitis is apparently suppressed by CF and thus the important contribution of an infiltration by neutrophils to the radical damage is emphasized.

The significant effect of DFO revealed an important role of iron, presumably freed from lactoferrin from neutrophils (Ambruso and Johnston, 1981) and from ferritin (Mazur et al., 1955), and possibly also from damaged red blood cells. The amount of available iron is still increased under PHb therapy. In this case DFO produced the greatest therapeutic effect.

The question arises, why oxygen carriers do not increase the lipid peroxidation level in intestinal tissue, although the oxygen supply to tissue should be higher and thereby an increase of lipid peroxidation could be possible. However, the production of oxygen free radicals seems to have already reached its maximum in our model in the postischemic period. The application of oxygen carriers also represents a very effective volume substitution and could help to accelerate the restoration of normal perfusion in the intestinal circulation. Thus, radical scavenging, anti-inflammatory and iron chelating drugs could be comparatively shown to provide protective action against lipid peroxide formation. This appears to be additive to the increased survival, facilitated by oxygen carriers in the mesenteric occlusion shock.

688

SUMMARY

The influence of different drugs on ischemia induced oxygen free radical damage was examined in intestinal tissue of rats by determination of thiobarbituric acid reactive substances (TBARS). Some methodical aspects of this method were considered. Experiments were done with and without the use of polymerized stromafree hemoglobin (PHb) as an additional oxygen carrier. Reversible total occlusion of the superior mesenteric artery was performed for 90 min, reperfusion time was 2.5 hours. Despite higher O_2 availability PHb did not increase the TBARS level any further. Superoxide dismutase with catalase; allopurinol; ciprofloxacin; and deferoxamine produced a highly significant reduction of TBARS, even if used together with PHb.

REFERENCES

Ambruso D.R., Johnston R.B. (1981). Lactoferrin enhances hydroxyl radical production by human neutrophils, neutrophil particulate fractions and an enzymatic generating system. J. Clin. Inv. 67, 352-360.

Augustin A., Lutz J. (1988). The effect of a temporary occlusion of the superior mesenteric artery on the level of lipid peroxides in plasma and intestinal tissue. Europ. J. Physiol. 11, R 51

Bird J.E., Milhoan K., Wilson C.B., Young S.G., Mundy C.A., Parthasarathy S., Blantz R.C. (1988). Ischemic acute failure and antioxidant therapy in the rat. J. Clin. Inv. 81, 1630-1638

Fuld R., Spar B., Urbaitis B.K. (1986). Mesurement of malonaldehyde in kidney following ischemia and reflow. Kidney Int. 29, 301 A

Gamelin L.M., Zager R.A. (1988). Evidence against oxidant injury as a critical mediator of postischemic acute renale failure. Am. J. Physiol. 255, F 450-460

Grootveld M., Halliwell B., Moorhouse C.P. (1987). Action of uric acid, allopurinol and oxypurinol on the myeloperoxidase-derived oxidant hypochlorous acid. Free Rad. Res. Comms. 4, 69-76

Hamar J., Dezsi L., Adam E., Netzer K.O., Stark M., Lutz J. (1987). Role of fluid replacement, increased oxygen availability by perfluorchemicals and enhanced RES function. Res. Exp. Med. 187, 451-459

Liu T.H., Beckman J.S., Freeman B.A., Hogan E.L., Hsu C.Y. (1989). Polyethylene glycol-conjugated superoxide dismutase and catalase reduce ischemic brain injury. Am. J. Physiol. 256, H589-593 (1989)

Lutz J., Hamar J., Netzer K.O., Stark M. (1985). Survival from mesenteric occlusion shock influenced by different treatment in rats. Int. J. Microcirc. Clin. Exp. 4, 103

Lutz J., Augustin A. (1989). The influence of a temporary cessation and reperfusion of intestinal blood flow on the level of hepatic lipid peroxides. In: Oxygen Transport to Tissue XI (Eds. K. Rakusan et al.) Plenum Pub. N.Y. p. 803-808

Mazur A., Baez S., Shorr E. (1955). The mechanism of iron release from ferritin as related to its biological properties. J. Biol. Chem. 213, 147-160

Moorhouse P.C., Grootveld M., Halliwell J.G., Quinlan G., Gutteridge J.M.C. (1987). Allopurinol and oxypurinol are hydroxyl radical scavengers. FEBS Lett. 213, 23-28

Ohkawa H., Ohishi N., Yagi K. (1979). Assay for lipid peroxides in animal tissues by thiobarbituric acid reaction. Anal. Biochem. 95, 351-358

Paller M.S. (1988). Hemoglobin- and myoglobin-induced acute renal failure in rats: role of iron in nephrotoxicity. Am. J. Physiol. 255, F 539-544

Yagi K. (1976). A simple fluorimetric assay for lipoperoxide in blood plasma. Biochemical Med. 15, 212-216

EFFECT OF ETHANOL ON HEPATIC OXYGENATION: EVIDENCE OF HEPATIC HYPOXIA

Hiroyuki Fukui, Nobuhiro Sato, Sunao Kawano, Harumasa Yoshihara, Taizo Hijioka, Hiroshi Eguchi, Moritaka Goto, Takashi Matsunaga, Shinji Kubota-Kashio, Takenobu Kamada and Hermann Metzger*

The First Dept. of Medicine, Osaka University Medical School, Fukushima- ku, Osaka, Japan and *Dept. of Physiology, Medizinische Hochschule Hannover, FRG

INTRODUCTION

It is well established that oxygen is required for the metabolism of ethanol in the liver and that hepatic oxygen consumption is stimulated by acute or chronic administration of ethanol. Our recent study in the perfused liver clearly demonstrated a vasoconstrictive effect of ethanol in the presinusoidal region, which led to hepatic tissue hypoxia due to a reduction in oxygen supply to hepatocytes (Hijioka et al., 1988). This evidence suggests that ethanol could damage the liver via mechanisms involving tissue hypoxia. However, a recent study in baboons has revealed an impaired hepatic oxygen utilization after large doses of ethanol (40-70mM) (Lieber et al., 1989). The mechanism may be a direct toxic effect of ethanol on oxygen utilization in mitochondria, possibly via accumulation of aldehyde, or by a microcirculatory disturbance which reduces delivery of oxygen to the regional hepatic tissue. We have determined the sinusoidal blood hemoglobin oxygenation using a sensitive spectrophotometer, coupled with an in vivo microscope, and demonstrated a clear oxygenation gradient in the sinusoidal blood from periportal to pericentral regions. Ethanol exaggerated this oxygen gradient (Sato et al., 1987). The purpose of the present study is to evaluate the effect of ethanol on hepatic oxygenation by means of that spectrophotometric method and also a polarographic method and thereby to clarify whether ethanol produces hepatic hypoxia.

MATERIALS AND METHODS

Male Sprague-Dawley rats weighing 100-250 g were fasted for 24 hours. Rats were anesthetized with ketamine hydrochloride (100 mg/kg body wt) or sodium pentobarbital (35 mg/kg body wt), and the liver was exposed by making a left subcostal incision and was positioned on a specially designed stage. Ethanol (1 g/kg or 4 g/kg body wt, 30% V/V in saline solution) was ingested by a gastric tube.

1) Analysis of hepatic oxygenation by in vivo microscopy system

Transmitted light through specific areas (window 87 µm) in the periportal and pericentral regions of the hepatic lobule was detected by a sensitive spectrophotometer (UNISOKU, USP-430B) coupled with an in vivo microscope via an optical fiber bundle (Eguchi et al., 1987). The index of sinusoidal blood hemoglobin oxygenation (ISO_2) was calculated from the absorption spectra of hemoglobin at those regions using the equation:

$$ISO_2 = 0.673[Er_{577} - (9\ Er_{569} + 8\ Er_{586})/17]/(Er_{569} - Er_{586})$$

where Er_{569}, Er_{577} and Er_{586} indicate intensities of the absorption at those wavelengths.

2) <u>Measurement of hepatic surface tissue oxygen tension</u>

A micro-O_2 electrode (diameter 15 μm) (Metzger and Schywalsky, 1988) was gently contacted with the surface of the liver and the hepatic surface Po_2 was detected by this polarographic method.

3) <u>Measurement of blood ethanol concentration</u>

Ethanol concentrations in the peripheral and portal blood, 15 minutes after the ethanol load, were measured by gas chromatography.

RESULTS

At the basal state, before administration of ethanol, ISO_2 at the periportal and pericentral regions of several hepatic lobules showed a large variation: in the former it was 25-40 (average 35) while in the latter it was 20-40 (average 30). A low dose of ethanol (1 g/kg body wt) produced a significant increase in ISO_2 at the periportal regions of the liver lobule as shown by Sato et al. (1987) (Fig. 1). A high dose of ethanol (4 g/kg body wt) significantly increased ISO_2 at the periportal regions, similar to the low dose, but significantly reduced ISO_2 at the pericentral regions (Fig. 2).

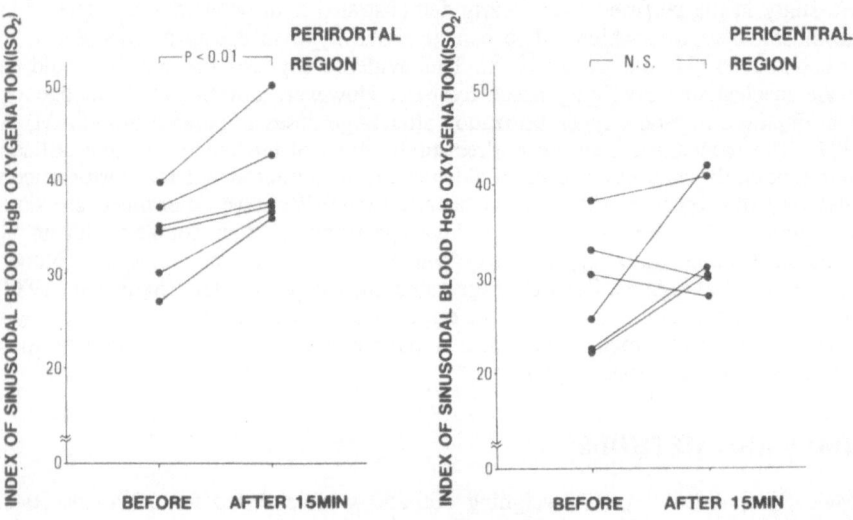

Fig. 1. Effect of acute ethanol (1 g/kg body wt) ingestion on the index of sinusoidal blood hemoglobin oxygenation in the periportal and pericentral regions of hepatic lobules in rats.

At the basal state, before administration of ethanol, hepatic surface Po_2 measured by the micro-O_2 electrode was 20-50 mmHg. It was significantly elevated (before 38.3 ± 4.2 mmHg; after 15 min. 50.4 ± 5.6 mmHg) by the low dose of ethanol (1 g/kg body wt) (Fig. 3). In contrast, it was significantly reduced (before 40.8 ± 7.8 mmHg; after 15 min. 31.1 ± 10.7 mmHg) by the high dose of ethanol (4 g/kg body wt). At that time, some of hepatic surface regions showed high Po_2, but other hepatic surface regions showed extremely low Po_2 (< 20 mmHg) (Fig. 4).

692

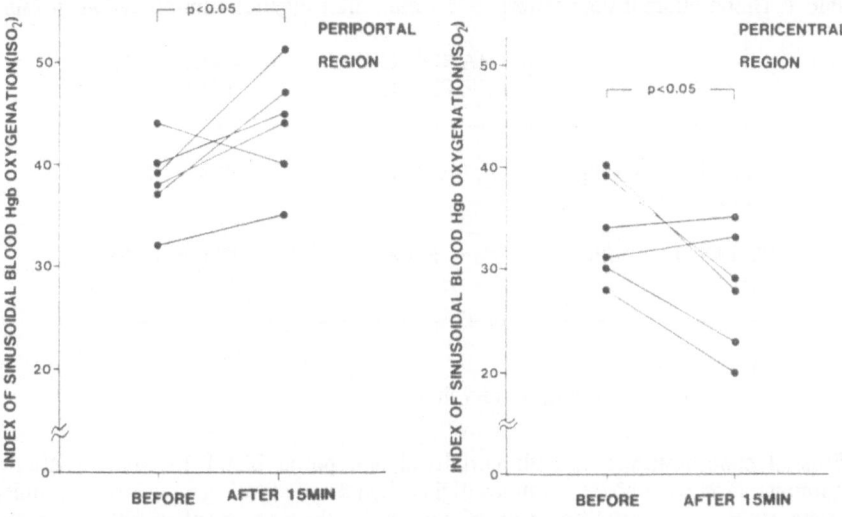

Fig. 2. Effect of acute ethanol (4 g/kg body wt) ingestion on the index of sinusoidal blood hemoglobin oxygenation in the periportal and pericentral regions of hepatic lobules in rats.

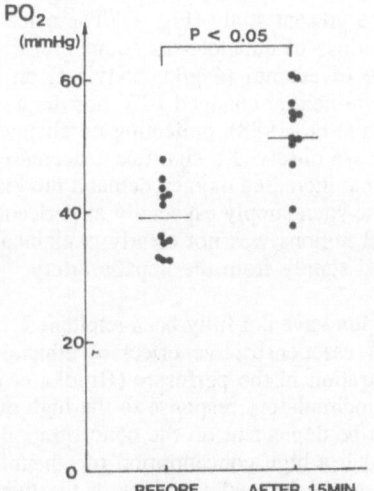

Fig. 3. Effect of acute ethanol ingestion (1 g/kg body wt) on hepatic surface tissue oxygen tension in rats.

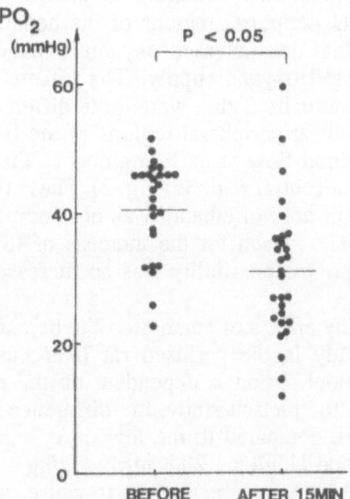

Fig. 4. Effect of acute ethanol ingestion (4 g/kg body wt) on hepatic surface tissue oxygen tension in rats.

Table 1. Blood ethanol concentrations 15 min. after ethanol administration in rats.

	ETHANOL CONCENTRATION (mM)	
	PERIPHERAL	PORTAL
ETHANOL 1 g/kg n=3	10.0 ± 6.3	20.8 ± 7.2
ETHANOL 4 g/kg n=4	61.0 ± 7.8	112.6 ± 11.5*

Mean $\pm$ S.E.M.
*p < 0.05 versus peripheral value

Ethanol concentrations in both peripheral and portal blood 15 minutes after 4 g/kg ethanol administration were about six times higher than those after 1 g/kg ethanol administration. Ethanol concentrations in portal blood were about twice those in peripheral blood after either 1 g/kg or 4 g/kg ethanol administration (Table 1).

DISCUSSION

The present study demonstrated, using both an in vivo microscopic and spectrophotometric system and also a micro-O_2 electrode, that administration of a high dose of ethanol to rats reduced hepatic tissue oxygenation, especially at pericentral regions of the hepatic lobules, and resulted in focal hypoxia in surface hepatic tissue.

We have reported (Sato et al., 1987) that, after administration of a low dose of ethanol (1 g/kg body wt) to rats, sinusoidal blood erythrocyte flow velocity (EFV) increases significantly while the width of sinusoids shows no significant change. Since the width of sinusoids did not change, the increase in EFV after administration of the low dose of ethanol indicated enhanced hepatic blood flow, resulting in increased oxygen supply. In fact ISO_2 showed a significant increase at periportal regions of the hepatic lobules in the present study (Fig. 1). These results indicate that the increased oxygen demand after the low dose of ethanol was compensated for by increased oxygen supply. The effects of a high dose of ethanol (4 g/kg body wt) on the hepatic microcirculation were quite different. The high dose neither changed EFV nor the width of sinusoids at peripheral regions of the liver lobule (Sato et al., 1988), indicating no change in hepatic blood flow at these regions. In the present study, we observed a significant decrease in ISO_2 at pericentral regions (Fig. 2). These results suggest that increased oxygen demand produced by the high dose of ethanol was not compensated for by oxygen supply especially at pericentral regions. The reason for the increase of ISO_2 at periportal regions was not clearly explained in this study. One possibility was an increased arterial blood supply from the hepatic artery.

The effects of ethanol on the hepatic microcirculation have not fully been explained. Our recent study in the perfused rat liver has shown that a vasoconstrictive effect of ethanol at presinusoidal region is dependent on the ethanol concentration in the perfusate (Hijioka et al., 1988). In the present study, the difference in hepatic microcirculatory response to the high dose of ethanol, compared to the low dose, was considered to be dependent on the concentration in portal blood (1 g/kg : 20.8 mM; 4 g/kg : 112.6 mM). Such a high concentration of ethanol in the portal vein may be enough to cause vasoconstriction at presinusoidal regions of the liver.

The ISO_2 measurements indicate that oxygenation of hemoglobin in sinusoids would be influenced by other factors such as pH and carbon dioxide (CO_2) in the blood. On the other hand, Po_2 is considered to indicate the oxygen tension of hepatic surface tissue including hepatocytes. The area of interest of Po_2 measured by our micro-O_2 electrode (diameter 15 μm)

694

is smaller than that of ISO_2 (diameter 87 μm). Measurement of Po_2, a direct index of local hepatic tissue oxygenation, shed further light on the local hepatic tissue oxygenation. The significant increase in Po_2 observed after administration of the low dose of ethanol (1 g/kg body wt) (Fig. 3) confirmed the compensation for oxygen demand by the increased oxygen supply, even in such a small region. On the contrary, an interesting variation in Po_2 was observed after administration of the high dose of ethanol (4 g/kg body wt) (Fig. 4). Some areas showed an increase while others showed a significant decrease. This suggested that some of the hepatocytes did not compensate for the increase in oxygen demand. Surprisingly, there appeared hepatic surface areas where Po_2 was extremely low (< 20 mmHg) after the high dose of ethanol. This evidence suggested the occurrence of local hypoxia in surface hepatic tissue. The areas showing an increased Po_2 after ethanol must have been in periportal regions because at some periportal regions ISO_2 increased after administration of the high dose (Fig. 2). These data indicate that local hepatic oxygenation showed a heterogenous response to ethanol and that an imbalance of hepatic oxygen demand and supply, at some hepatic regions, caused by the high dose of ethanol may be associated with local hepatic hypoxia.

There is a controversy about hypoxia due to ethanol administration. Shaw et al. (1977) reported that ethanol administration increases hepatic blood flow and thereby compensates for the increase of hepatic oxygen demand. Iturriaga et al. (1980) reported a Po_2 increase in the hepatic vein after administration of ethanol. On the other hand there is a report that ethanol reduces hepatic blood flow in the human liver (Lundquist et al., 1962). These conflicting reports may stem from differnces in methods and doses of ethanol. In the present study, the higher dose of ethanol was administered and the concentration of ethanol in the peripheral blood was very high, as is often found in alcoholics. Lieber and we have reported that a high concentration of ethanol (peripheral 40-70 mM) causes the reduction of regional hepatic blood volume, hepatic oxygen consumption and release of enzymes from the liver (Lieber et al., 1988). Tsukamoto has reported that chronically alcohol treated rats show high plasma ethanol levels (40-70 mM in peripheral blood) and hepatic fibrosis and necrosis (Tsukamoto and Ping, 1989). This evidence supports our idea that a continuous high plasma level of ethanol would cause hepatocyte injury and hepatic functional disorder via local hepatic hypoxia.

In conclusion, ingestion of a high dose of ethanol reduces local hepatic tissue oxygenation via an imbalance between delivery and demand of oxygen in pericentral regions of the liver lobule, resulting in local hypoxia especially in surface hepatic tissue. This mechanism might be associated with the pathogenesis of alcoholic liver injury.

SUMMARY

The effects of low (1 g/kg body wt) and high doses (4 g/kg body wt) of ethanol on hepatic oxygenation in rats was investigated employing an in vivo microscopic and spectrophotometric system and also a micro oxygen electrode. The low dose of ethanol increased sinusoidal blood hemoglobin oxygenation (ISO_2) at periportal regions in the liver lobule. The high dose of ethanol increased ISO_2 at periportal regions, but decreased ISO_2 at pericentral regions. The low dose of ethanol increased hepatic surface tissue Po_2, but the high dose decreased this Po_2.

From these data it is concluded that the increase of hepatic blood flow, after administration of a low dose of ethanol, compensated for the increase of oxygen demand in hepatocytes but that administration of a high dose of ethanol reduced hepatic tissue oxygenation via an imbalance between delivery and demand of oxygen in pericentral regions of the liver lobule, resulting in local hypoxia in surface hepatic tissue.

REFERENCES

Eguchi, H., N. Sato, T. Matsumura, et al., 1987, In vivo estimation of oxygen saturation of hemoglobin in hepatic lobules in rats, Adv. Exp. Med. Biol., 222: 591-596.

Hijioka, T., N. Sato, H. Yoshihara, et al., 1988, A new mechanism for alcoholic liver damage: hepatic hypoxia due to ethanol-induced presinusoidal vasoconstriction, Hepatology, 8: 1422.

Israel, Y., L. Videla, and J. Bernstein, 1975, Liver hypermetabolic state after chronic ethanol consumption: hormonal interrelations and pathogenic implications, Federation Proc., 34: 2052-2059.

Iturriaga, H., G. Ugarte, and Y. Israel, 1980, Hepatic vein oxygenation, liver blood flow and the rate of ethanol metabolism in recently abstinent alcoholic patients, Eur. J. Clin. Invest., 10: 211-218.

Lieber, C.S., E. Baraona, N. Sato, et al., 1989, Impaired oxygen utilization: A new mechanism for the hepatotoxicity of ethanol in sub-human primates, J. Clin. Invest., 83: 1682-1690.

Lundquist, F., N. Tygstrup, K. Winkler, et al., 1962, Ethanol metabolism and production of free acetate in the human liver, J. Clin. Invest., 41: 955-961.

Metzger, H.P. and M. Schywalsky, 1988, Observation of microcirculation disorders of the fluorescence-stained gamma globulins, in: "Microcirculation in Circulatory Disorders," H. Manabe, B. W. Zweifach and K. Messmer, eds., Springer-Verlag, Tokyo.

Sato, N., H. Eguchi, Y. Takei, et al., 1987, Microcirculatory aspects for the mechanism of alcoholic liver disease - Sinusoidal blood flow and oxygenation at periportal and pericentral regions of hepatic lobules in rats -, in: "Microcirculation - an Update", vol. 2, M. Tsuchiya et al., eds., Excerpta Medica, Amsterdam.

Sato, N., S. Kawano, T. Matsumura, et al., 1988, Abnormal heterogenic distribution of hepatic sinusoidal blood flow and oxygenation after acute ethanol consumption in rats: Direct evidence for pericentral hypoxia, in: "Biomedical and Social Aspects of Alcohol and Alcoholism," K. Kuriyama, et al., eds., Excerpta Medica, Amsterdam.

Shaw, S., E. Heller, H. Friedman, et al., 1977, Increased hepatic oxygenation following ethanol administration in the baboon, Proc. Soc. Exp. Biol. Med., 156: 509-513.

Tsukamoto, H. and Xiao Ping Xi, 1989, Incomplete compensation of enhanced hepatic oxygen consumption in rats with alcoholic centrilobular liver necrosis, Hepatology, 9: 302-306.

REDISTRIBUTION OF LOCAL HEPATIC BLOOD FLOW DURING ACUTE BLEEDING AND PROLONGED HEMORRHAGIC HYPOTENSION STUDIED USING FLUOROCHROMED PLASMA PROTEINS AND SURFACE Po_2 MEASUREMENTS

Michael Schywalsky and Hermann P. Metzger

Department of Physiology, Medical School Hannover
Postbox 610 180
3000 Hannover 61, Federal Republic of Germany

INTRODUCTION

The degree and local distribution of morphological alterations occurring in the rat liver following hemorrhagic hypotension is influenced by a sequence of microcirculatory changes resulting in an increased heterogeneity of the microcirculatory perfusion pattern (Vanecko et al., 1969; Koo and Liang, 1977). On the other hand, morphological alterations develop during prolonged hemorrhagic hypotension which force changes in the hepatic microcirculation. Thus there exists a strong mutual relationship between changes in hepatic microcirculatory function and changes in hepatic parenchymal structure (Rappaport, 1976; Sherman and Fisher, 1987). Therefore it is necessary to distinguish between acute microcirculatory changes during the bleeding phase of hemorrhagic hypotension, and time dependent microcirculatory changes during prolonged hemorrhagic hypotension. In order to study the development and kinetics of sinusoidal flow heterogeneity during stepwise bleeding and during prolonged hemorrhagic hypotension, hepatic surface Po_2 was measured and microcirculatory perfusion patterns were demonstrated by use of fluorochromed plasma proteins (Metzger and Schywalsky, 1988).

MATERIALS AND METHODS

Wistar rats (200-300 g) were anaesthetized with Ketamine (15 mg/100 g b.w.) and Xylazine (0.2 mg/100 g b.w.). The mean arterial blood pressure (MAP) was measured via the a. carotis. The animals were allowed to breathe spontaneously. During the experiments the animals were monitored by controlling respiration and heart rate. Body temperature was maintained between 36.5 and 37.5°C. Liver surface Po_2 (sPo_2) was measured with a specially developed multi-gold-wire-cathode placed in riding position on the surface of the lobus sinister (Metzger and Schywalsky, 1988).

Microcirculatory perfusion patterns were demonstrated by use of the fluorescent dyes fluoresceine isothiocanate (FITC) and lissamine rhodamine B 200 (RB 200) conjugated with plasma proteins (Vetterlein et al., 1982; Metzger and Schywalsky, 1988). Five minutes before ending the experiment RB 200 stained protein was infused (0.3 ml of a 6% solution) followed by a FITC protein bolus injected intraarterially in the arc of the aorta at the end of the experiment. Tissue samples were taken very quickly 10 seconds after and rapidly frozen in liquid nitrogen. Fluorescence microscopic analysis of cryofixed tissue slices showed the green stained convective front spreading from the portal field to the hepatic vein on the background of the red stained sinusoids representing the whole sinusoidal network. Two experimemtal groups of rats were investigated:

Group I: Stepwise bleeding and lowering of the MAP were induced. As a function of MAP, hepatic sPo_2 was registered until a MAP of 40 mmHg was reached. Following these measurements perfusion patterns were analyzed.

Oxygen Transport to Tissue XII, Edited by J. Piiper *et al.*
Plenum Press, New York, 1990

Group II: Hemorrhage to MAP of 40 mmHg was induced. The time dependent course of hepatic sPo_2 was registered every twenty minutes during short reinfusion intervals (5 min) to MAP of 60 mmHg. Following sPo_2 registration MAP was lowered again to 40 mmHg. Additionally the blood volume needed for the adjustment of the MAP to the mentioned values was measured. After a period of five hours, the perfusion pattern was observed.

RESULTS

The relative decrease of individual sPo_2 curves in group I expressed as the percentage of their initial state showed different slopes (Fig. 1a,b). The difference between the highest and lowest sPo_2 reached a maximum value of about 50 % in a MAP range between 50 and 80 mmHg (Fig. 2). The convective front of the fluorescent labelled proteins showed an irregular front spreading from the portal field into the sinusoidal network (Fig. 3a,b).

In group II, a time dependent sPo_2 decrease to 0 mmHg occurred in one group of the curves during prolonged hemorrhagic hypotension while in others sPo_2 values close to the initial state were reached (Fig. 4). In many of the fluorochrome stained slices single predominantly perfused sinusoids were observed (Fig. 5). The reinfused and withdrawn blood volumes required for adjusting the MAP between 60 and 40 mmHg were nearly equal for the first four sPo_2 measuring intervals. Subsequently the volumes differed progressively. Increasing blood volumes were needed to raise MAP to 60 mmHg, less volume was withdrawn to lower MAP to 40 mmHg.

DISCUSSION

The results indicate reductions in hepatic sPo_2 and sinusoidal blood flow in response to stepwise bleeding are not a passive reaction closely related to MAP. These findings are in general agreement with intravital microscopic studies on sinusoidal blood flow following acute hepatic denervation (Koo, 1987). Distribution of local hepatic blood flow during stepwise bleeding is characterized by the maximum differences between the highest and lowest relative sPo_2 values. In

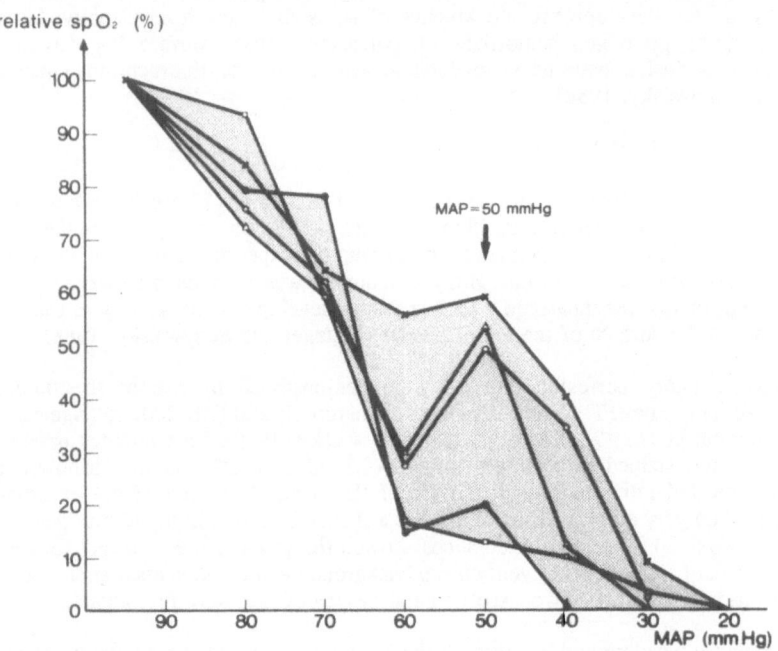

Fig. 1a. Relative hepatic surface Po_2 for the five individual, simultaneous measurements in response to stepwise blood withdrawal (rat A).

698

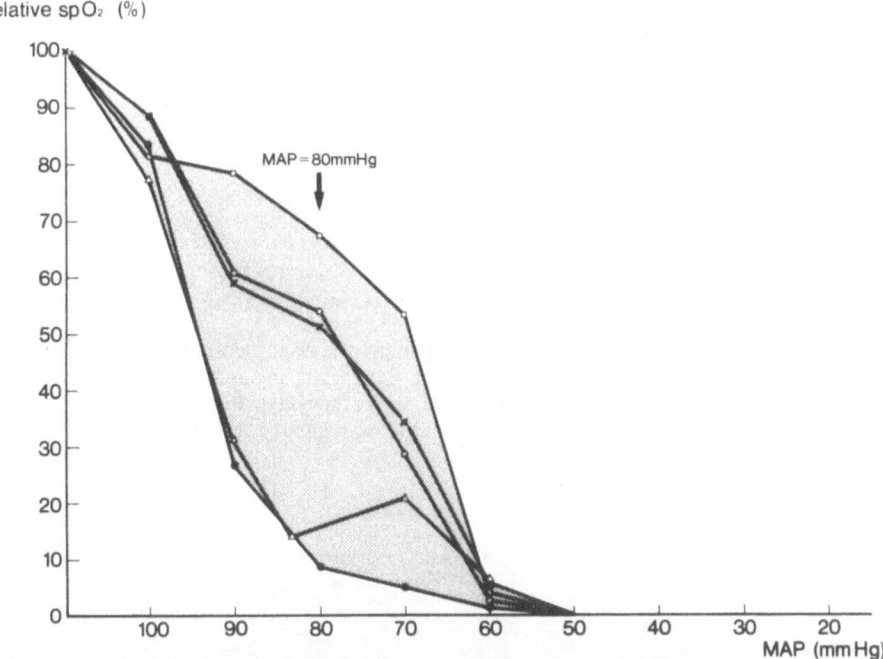

Fig. 1b. Relative hepatic surface Po_2 for the five individual, simultaneous measurements in response to stepwise blood withdrawal (rat B).

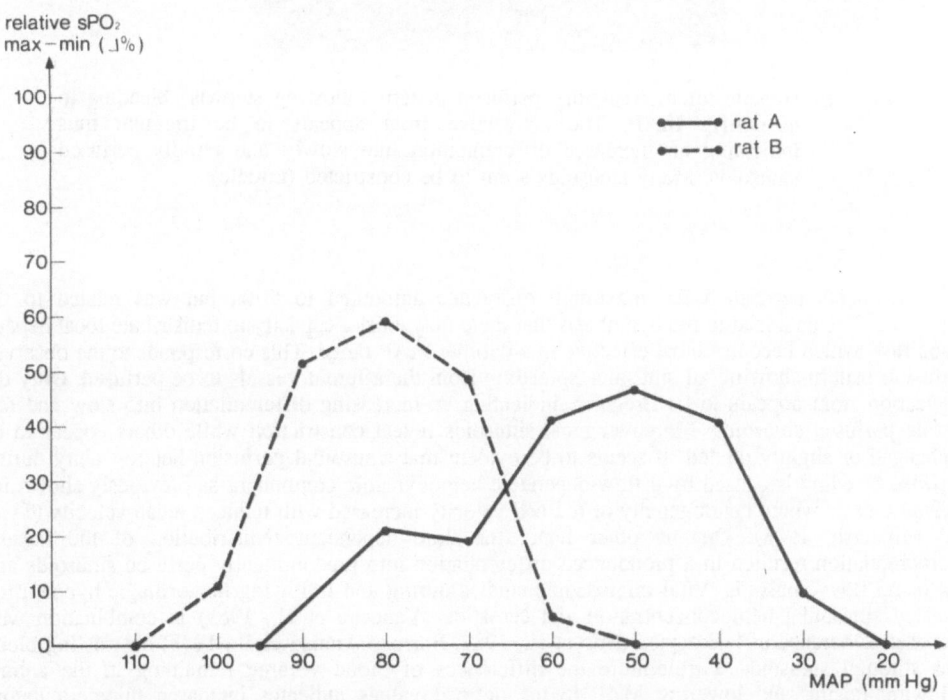

Fig. 2. Differences between highest and lowest relative hepatic sPo_2 vs. MAP during stepwise blood withdrawal. Maximum heterogeneity appears at different MAP values depending on the individual.

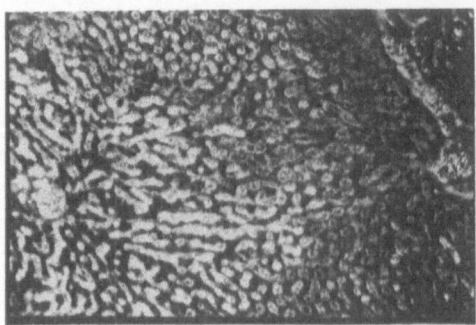

Fig. 3a. Hepatic microcirculatory perfusion pattern of a normotensive control rat. All sinusoids spreading from the portal field (left) towards the central vein (right) are homogeneously perfused. The convective front, appearing 10 s after bolus injection of FITC-globuline, is regularly formed.

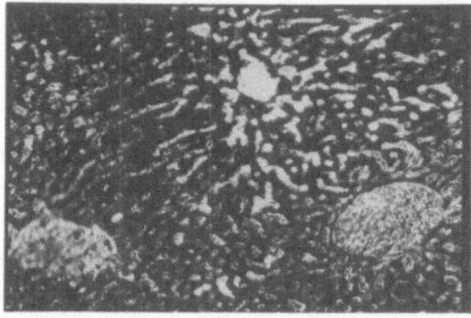

Fig. 3b. Hepatic microcirculatory perfuson pattern following stepwise bleeding to 40 mmHg MAP. The convective front appears to be irregular thus indicating an increased differentiation into slowly and rapidly perfused sinusoids. Many sinusoids seem to be constricted (middle).

all experiments performed the maximum difference amounted to 50%, but was related to the individual. The data lead to the hypothesis that there may exist a capacity to redistribute local hepatic blood flow which becomes most effective in a definite MAP range. This corresponds to the observed perfusion pattern showing all sinusoids spreading from the afferent vessels to be perfused. Only the convection front appears to be irregular, indicating an increasing differentiation into slow and fast plasma perfused sinusoids. Moreover most sinusoids reveal constriction while others appear to be unchanged or slightly dilated. It seems to be evident that sinusoidal perfusion heterogeneity during stepwise bleeding is caused by a flow-dependent hemodynamic component as previously shown for skeletal muscle where heterogeneity of red cell velocity increased with reduced mean velocity (Tyml and Mikulash, 1988). On the other hand, the time dependent redistribution of the hepatic microcirculation resulted in a pronounced differentiation into predominantly perfused sinusoids and low or no flow sinusoids. Vital microscopic studies during and following hemorrhagic hypotension revealed sinusoidal hemoconcentration and cessation (Vanecko et al., 1969) in combination with increased adherence of leukocytes (Koo et al., 1987; Barroso-Aranda et al., 1988) obstructing blood flow through sinusoids. Furthermore the differences of blood volume remaining in the animal following raising and lowering MAP to the defined values indicates increased microcirculatory decompensation (Zweifach, 1974).

The phenomenon of hepatic sPo_2 values increasing close to initial state during reinfusion intervals may be attributed to the predominantly perfused sinusoids shown in the corresponding perfusion pattern. However, these "fast" sinusoids might be taken as sinusoidal shunts having no nutritional effect. These observations suggest that damage to the hepatic microcirculation has developed, finally resulting in hepatic parenchymal necrosis.

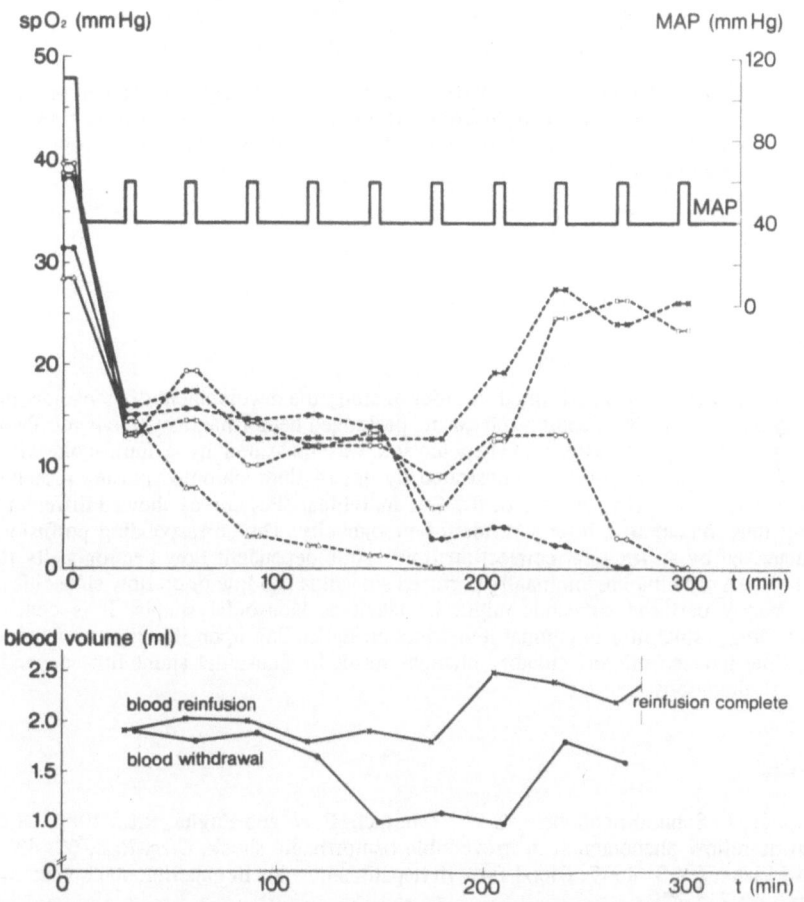

Fig. 4. Hepatic surface Po_2 of hemorrhaged rats during short reinfusion periods. Time dependent decrease occurred in one group of the curves while in the other, values close to the initial state and above those of the first reinfusion interval were reached. The lower diagram shows the blood volumes needed for adjusting MAP to 60 mmHg for the purpose of measuring surface Po_2 and for lowering MAP again to 40 mmHg. These volumes differed progressively indicating microcirculatory decompensation.

It is concluded that changes in microcirculatory pattern during stepwise bleeding might be regarded as functional redistribution, while time dependent changes are mainly caused by morphological alterations which result in sinusoidal shunt flow.

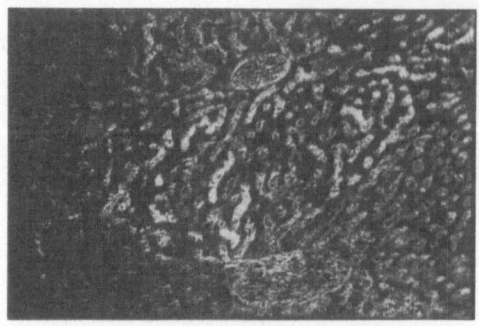

Fig. 5. Hepatic microcirculatory perfusion pattern following prolonged hemorrhagic hypotension. A marked differentiation into predominantly perfused sinusoids and low or no flow sinusoids can be observed. No convection front could clearly be indentified.

SUMMARY

The present study was performed in order to study the development of sinusoidal blood flow heterogeneity during stepwise bleeding and during prolonged hemorrhagic hypotension. Two methods have been applied: Hepatic surface oxygen tension was measured by a multi-gold-wire-cathode. Sinusoidal perfusion patterns were demonstrated by use of fluorochromed plasma proteins. During stepwise bleeding the relative decrease of the five individual sPo_2 curves showed different slopes in the same rat thus indicating a flow dependent heterogeneity. The corresponding perfusion patterns were characterized by an irregular convection front. Time dependent flow heterogeneity resulted in a perfusion pattern showing predominantly perfused sinusoids and low or no flow sinusoids, whereby the predominantly perfused sinusoids might be taken as sinusoidal shunts. It is concluded that stepwise bleeding results in a functional redistribution depending upon the degree of reduced flow, while long time induced microcirculatory changes result in sinusoidal shunt flow depending upon morphological alterations.

REFERENCES

Barroso-Aranda, J., Schmid-Schönbein, G.W., Zweifach, B.W. and Engler, R.L., 1988, Granulocytes and no-reflow phenomenon in irreversible hemorrhagic shock, *Circ. Res.*, 63: 437-447.

Koo, A. and Liang, I.Y.S., 1977, Blood flow in hepatic sinusoids in experimental shock, *Microvasc. Res.*, 13: 315-325.

Koo, A., 1987, Nervous control of the hepatic microcirculation, in: "Microcirculation - an update, vol. 2", M. Tsuchiya et al., eds., Elsevier Sciences Publishers B.V., Amsterdam.

Koo, A., Breit, G. and Intaglietta, M., 1988, Leukocyte adherence in hepatic microcirculation in ischemia reperfusion, in: "Microcirculation in circulatory disorders", H. Manabe et al., eds., Springer, Tokyo.

Metzger, H.P. and Schywalsky, M., 1988, Observation of microcirculatory disorders of the hemorrhagic rat liver by use of fluorescence stained gamma globulines, in: "Microcirculation in circulatory disorders", H. Manabe et al., eds., Springer, Tokyo.

Rappaport, A.M., 1976, The microcirculatory acinar concept of normal and pathological hepatic structure, *Beitr. Path.*, 157: 215-243.

Sherman, I.A. and Fisher, M.M., 1987, Hepatic microvascular patterns in normal and diseased liver, in: "Microcirculation - an update, vol. 2", M. Tsuchiya et al., eds., Elsevier Sciences Publishers B.V., Amsterdam.

Tyml, K. and Mikulash, K., 1988, Evidence for increased perfusion heterogeneity in skeletal muscle during reduced flow, *Microvasc. Res.*, 35: 316-324.

Vanecko, R.M., Szanto, P.B. and Shoemaker, W.C., 1969, Microcirculatory changes in primate liver during shock, *Surg. Gynecol. Obstet.*, 129: 995-1004.

Vetterlein, F., dal Ri, H. and Schmidt, G., 1982, Capillary density in rat myocardium during timed plasma staining, *Am. J. Physiol.*, 242: H133-H141.

Zweifach, B.W., 1974, Mechanisms of blood flow and fluid exchange in micro-vessels: Hemorrhagic hypotension model, *Anesthesiology*, 41: 157-168.

SUPPORT OF HYPOXIC RENAL CELL VOLUME REGULATION BY GLYCINE

G. Gronow, N. Klause and M. Mályusz

Department of Physiology
University of Kiel
D-2300 Kiel, Fed. Rep. of Germany

INTRODUCTION

In vivo measurements of frequency distributions of extracellular PO_2 indicated local regions of tissue hypoxia in the renal cortex. Thus, almost 50% of the values ranged between 24 and 40 mmHg, about 10% were in the range of 10 to 20 mm Hg, and in 3% of the measurements even values between 1 to 10 mm Hg were obtained (Baumgärtl et al., 1972). The marginal oxygen supply of cortical cells was interpreted to be the result of a) the known high metabolic activity of renal cortical cells and b) a reduced vascular oxygen supply due to O_2-shunting from descending arterial vasa recta into closely arranged ascending renal veins. Accordingly, at an insufficient arterial oxygen supply local regions of tissue hypoxia became the predominant sites of hypoxic cellular damage (Alcorn et al., 1981).

Amino acids seem to modify the hypoxic tolerance of the kidney. In the cell-free perfused rat kidney, for example, renal function became insufficient if no amino acid mixtures were added to the perfusate (Brezis et al., 1984). In contrast, a mixture of 8 amino acids significantly supported renal function in this preparation even in hypoxia (Gronow et al., 1986). A mixture of 3 amino acids (proline, glycine, aspartic acid) proved to be as protective as 8 amino acids in the intact kidney (Mályusz and Gronow, 1986) as well as in isolated renal cells (Gronow et al., 1988). In studies on the protective role of glutathione it was suggested that a metabolic cleavage product of glutathione, the amino acid glycine, may have increased the hypoxic tolerance of renal tubular cells (Weinberg et al., 1987).

Therefore it was the aim of the present study to test the effect of glycine on renal cellular hypoxic tolerance. We isolated tubules of the rat kidney cortex by collagenase treatment and measured cellular function at different levels of extracellular oxygen tensions (1, 2.5, 5, 10, 40, and 100 mm Hg) both with and without glycine in the incubation medium. Functional parameters under study were a) the tubular diameter as an index of cellular volume regulation, b) intracellular K^+ homeostasis, c) cellular aerobic (gluconeogenetic) and hypoxic (glycolytic) metabolic activity, and, as an index of disruptive cell swelling, c) the hypoxic loss of 4 marker enzymes.

MATERIAL AND METHODS

Details of the method have been reported elsewhere (Gronow et al., 1984). Briefly, kidneys of starved male Sprague-Dawley rats (350-450g) were cooled and rinsed in situ with ice-cold Krebs-Ringer-Bicarbonate solution (KRB). Isolated tubular segments (ITS) were then prepared by mechanical and collagenase treatment (0.2g /100 ml CLS II, Worthington, Freehold, N.J., USA), washed twice, incubated at 37°C in basic KRB medium (pH 7.4, 10 g /l bovine albumine, 10 mmol/l glucose). In glycine experiments the amino acid was added to the basic medium in a concentration of 5 mmol /l.

Different levels of extracellular oxygen tension were adjusted by gassing the surface of incubation media with a mixture of 95% O_2 : 5% CO_2 and 95% N_2 : 5% CO_2. Extracellular oxygen tension in the incubation medium was registrated polarographically with a Clark - type oxygen electrode. Measurements of tubular diameter were performed under a light microscope (Olympus BH-2, PM-10 AK, Polzin, Kiel, FRG). In each series of experiment bath PO_2 in the incubation medium was kept constant for 45 min at either 1.0, 2.5, 5.0, 10, 40, or 100 mm Hg, respectively. Gluconeogenesis of isolated tubular cells was tested in an additional, second incubation period: tubules were washed twice in glucose-free medium and 30 min reincubated in an oxygenated basic medium containing 10 mmol/l lactate instead of glucose.

Parameters measured at different levels of oxygen tension were a) the tubular diameter as an index of hypoxic volume regulation, b) the formation of lactate from 10 mmol /l glucose, c) intracellular K^+ accumulation (K^+) and d) the loss of 4 marker enzymes: cytoplasmatic lactate dehydrogenase (LDH); brush border τglutamyltransferase (τGT); mitochondrial glutamate dehydrogenase (GlDH); and lysosomal acid phosphatase (APase). Assays of enzyme activities, glucose, and lactate content in the incubation medium as well as of intracellular K^+ were performed according to standard procedures and previously described methods (Gronow et al., 1984, 1986). A P-value of 0.05 or less (Student's t-Test for paired observations, n = 16) was regarded to indicate a statistically significant difference.

RESULTS AND DISCUSSION

Hypoxic Volume Regulation: At an extracellular PO_2 of 100 mm Hg the addition of glycine had no significant effect on mean tubular diameter of isolated tubular cells (37.8 ± 4.5 µm, Fig.1, left panel, triangles). With no glycine in the incubation medium (circles), however, the tubular diameter rose with decreasing oxygen tension, at 10 mm Hg by about 40%, and at 1 mm Hg by about 170%, respectively. In the presence of glycine (triangles) no significant swelling was observed until the extracellular PO_2 was adjusted to 1 mm Hg. At this extracellular oxygen tension tubular diameter rose in the presence of glycine by about 25%.

In the K^+ content of isolated tubular cells (Fig. 1, right panel) no significant effects of glycine could be observed at a PO_2 between 100 - 10 mm Hg. This range might reflect the physiological gradient from arterial vessels to cell surfaces (Jones and Kennedy, 1982). It included 97% of the PO_2-frequency distribution in the renal cortex in vivo (Baumgärtl et al., 1972). Below 10 mm Hg, however, K^+ decreased to about one third of the value observed at 100 mm Hg both in presence (triangles) and absence (circles) of glycine. The effect of glycine on intracellular K^+ was insignificant despite the observed concomitant support of hypoxic volume regu-

706

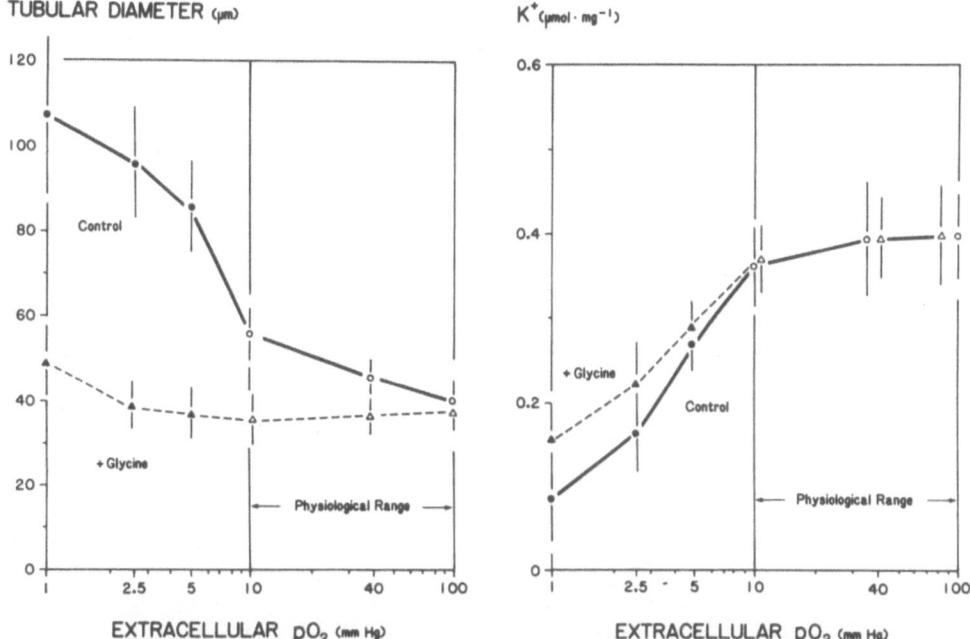

Fig.1 Volume regulation of renal cells in isolated tubular segments of rat kidney cortex (37°C). Left panel: mean tubular diameter at different levels of extracellular oxygen tension. Right panel: intracellular K^+ content at different levels of extracellular oxygen tension. Incubation medium: Ringer solution containing 10 mmol /l glucose (control observations = circles) plus additional 5 mmol /l glycine (+ Glycine = triangles). Mean ± SD (n = 16)

lation (left panel). The hypoxic release of intracellular K^+ was similar to K^+ losses observed previously in anoxic tubular cells (Gronow et al., 1984). It might reflect active volume regulation more than cell swelling or cell death (Lang et al., 1988).

Metabolic Activity: Under aerobic conditions, glucose is not a preferred substrate in the renal cortex. In extreme hypoxia, however, it may well support basic renal functions (Gronow and Cohen, 1984). Isolated tubular segments preincubated without glucose at a PO_2 of 1 mm Hg produced 77 ± 21 nmol lactate /mg protein ·60 min (100% in Fig.2, left panel). In the presence of glucose (circles), glycolysis rose by about 100%. Addition of glycine did not increase the rate of lactate formation. Instead, a small but significant suppression of glycolysis could be measured in the range of PO_2 = 1-5 mm Hg. Thus, an increased rate of glycolytic generation of metabolic energy did not contribute to the observed support of hypoxic volume regulation by glycine (Fig. 1, left panel). In the "physiological range", at higher extracellular oxygen tensions (10 - 100 mm Hg), no significant formation of lactate occurred.

According to our earlier observations (Gronow et al., 1984), gluconeogenesis (GNG) of isolated tubular segments which had been incubated in a

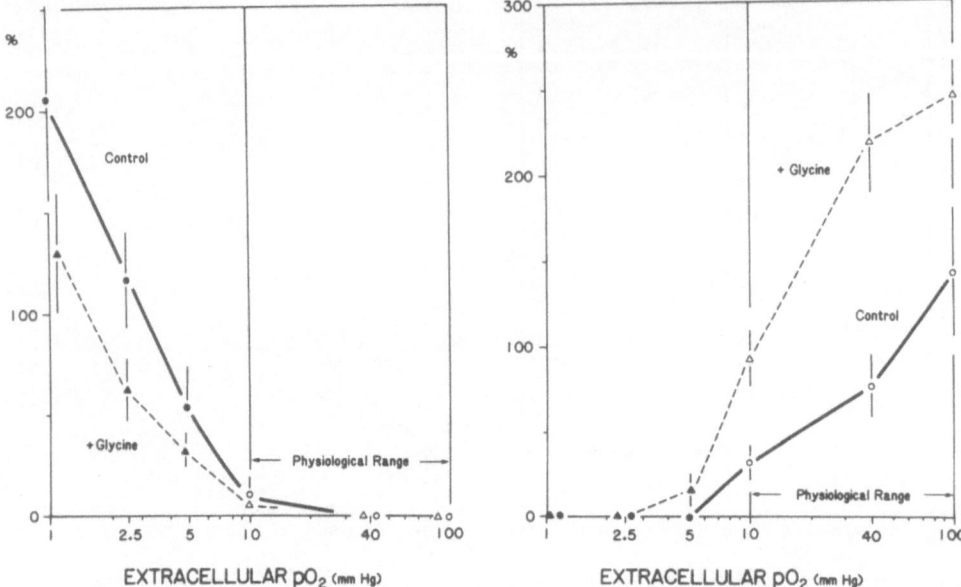

LACTATE FORMATION   GLUCOSE FORMATION

EXTRACELLULAR pO₂ (mm Hg)   EXTRACELLULAR pO₂ (mm Hg)

Fig.2 Metabolic activity of renal cells in isolated tubular segments of rat kidney cortex (37° C). Left panel: glycolysis at different levels of extracellular oxygen tension. Right panel: gluconeogenesis after incubation of tubular cells at different levels of extracellular oxygen tension and subsequent reoxygenation in the presence of 10 mmol /l lactate. Incubation medium: Ringer solution containing 10 mmol /l glucose at PO_2 = 1 - 10 mm Hg, or 10 mmol /l lactate at PO_2 = 10 - 100 mm Hg (control observations = circles) plus additional 5 mmol /l glycine (+ Glycine = triangles). Mean ± SD (n = 16)

preceding test period in standard medium without glucose amounted to 225 ± 34 nmol glucose /mg protein·60 min in the presence of 10 mmol /l lactate (100% in Fig. 2, right panel). After incubation in a the basic medium containing 10 mmol /l glucose, however, intracellular glucose probably leaked out of the washed cells. GNG rose, as expected, above control values (at PO_2 = 100 mm Hg by about 45%, Fig. 2, right panel, circles).

This "increased" rate of gluconeogenesis at 100 mm Hg, however, fell by about 30% when tubular cells were suspended at a "mixed venous" PO_2 of 40 mm Hg. If one takes into consideration the transfer of "non-metabolic" glucose from the glucose-containing basic medium, the liberation of glucose at lower oxygen tensions (PO_2 = 10 mm Hg) does not necessarily represent gluconeogenetic activity of the tubular cells. At lower oxygen tensions (5, 2.5, and 1 mm Hg) no measurable GNG could be detected in the reoxygenation period. The accelerated conversion of glucose to lactate (left panel) may have reduced its liberation in the reoxygenation period (right panel).

708

Glucose formation does not only include the consumption of metabolic energy but also requires the transport of metabolites across intact mitochondrial membranes. The observed irreversible reduction in the ability of reoxygenated tubular cells to form glucose in the presence of lactate (Fig. 2, right panel) may indicate irreversible mitochondrial dysfunction, as has been observed in the reoxygenated rat kidney (Gronow et al., 1986). According to our expectations, the addition of a glucoplastic amino acid increased glucose formation in the presence of lactate (Fig.2, right panel, triangles).

In the presence of glycine GNG rose at PO_2 =100 mm Hg significantly by about 155%. Compared to control observations (circles) GNG remained elevated even at a PO_2 of 10 mm Hg. When tubuli were incubated at lower oxygen tensions ($PO_2 <$ 10 mm Hg), no significant reversibility of GNG could be observed in the reoxygenation period. Thus, in the preceding hypoxic period (PO_2 = 2.5 mm Hg) cellular functions, which generated metabolic energy, may already have become insufficient.

Hypoxic Enzyme Losses: Lack of metabolic energy reduces transmembraneous ion pumping in renal tubular cells. Thus oncotic forces of intracellular macromolecules would have attracted extracellular fluid. Accordingly, without glycine in the incubation medium, renal cellular volume rose at already PO_2=10 mm Hg, indicated by the 40% increase in mean tubular diameter (Fig.1, left panel, circles). Extracellular Cl^-, Ca^{++}, and more water moved probably into the cells (Hochachka, 1986), and the liberation of lysosomal enzymes as well as disruptive swelling of renal cells and cell organelles led to a marked loss of cellular constituents (Fig. 3):

Under aerobic control conditions (100% level in Fig. 3) renal tubular cells released only small amounts of intracellular enzymes (in mU /mg protein $\cdot$60 min): cytoplasmatic LDH = 66.1 ± 12.4., brush border τGT = 48.2 ± 13, lysosomal APase = 4.87 ± 1.1, and mitochondrial GlDH = 6.0 ± 1.3. This enzyme leakage remained unchanged down to an extracellular PO_2 of 10 mm Hg, both with (triangles) or without (circles) glycine in the incubation medium. With no glycine in the medium (circles), a significant increase in the loss of cytoplasm (LDH) and of brush border microvilli (τGT) occurred, indicating hypoxic membrane deformation at PO_2= 5 mm Hg (Fig.3, left panel, circles). This loss of cell constituents may have been transient as long as mitochondrial structure was maintained. This view was supported by the finding that at PO_2= 5 mm Hg no significant losses of GlDH could be seen. However, in the presence of glycine (triangles) no significant enzyme loss occurred at all.

At lower oxygen tensions (2.5 - 1 mm Hg), however, the liberation of lysosomal APase and mitochondrial GlDH was no longer suppressed by glycine; under both experimental conditions a significant increase in enzyme activities could be observed in the incubation medium (Fig. 3, right panel). This enzyme release probably indicated irreversible changes in cellular structure, resulting in uncoupling of mitochondrial respiratory function in both the reoxygenated kidney and in isolated tubular cells (Gronow et al., 1986, 1988). Thus, Ca^{++}-activated lysosomal enzymes, especially phospholipases, may have attacked mitochondrial membranes at this low PO_2 (Hochachka, 1986), as indicated in the present experiments by a significant loss of a mitochondrial matrix enzyme (GlDH) and lysosomal APase. Finally, at a PO_2 of 1 mm Hg, cell death was obvious, indicated by the high activity of all 4 marker enzymes in the incubation medium. At this low PO_2 an insufficient driving force for oxygen transport along the diffusion gradient from cellular surface into mitochondrial space limited any effective oxidative phosphorylation (Jones and Kennedy, 1982) .

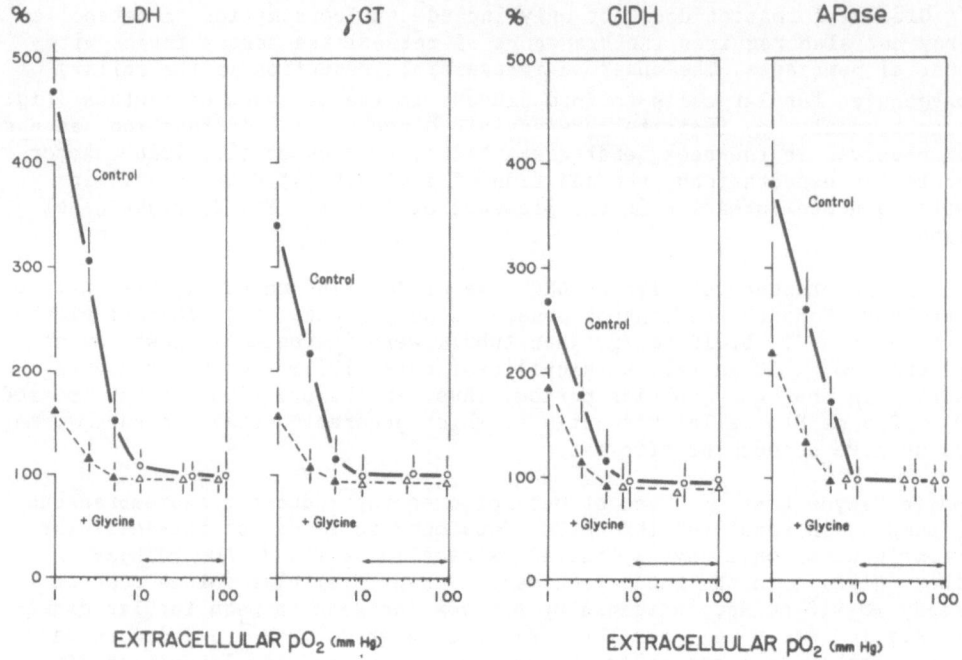

EXTRACELLULAR pO$_2$ (mm Hg)　　　　　EXTRACELLULAR pO$_2$ (mm Hg)

Fig.3　Enzyme loss of renal cells in isolated tubular segments of rat kidney cortex (37° C). Left panels: loss of cytoplasmatic lactate dehydrogenase (LDH) and of brush border γglutamyltransferase (γGT) at different levels of extracellular oxygen tension. Right panel: loss of mitochondrial glutamate dehydrogenase (GlDH) and of lysosomal acid phosphatase (APase) at different levels of extracellular oxygen tension. Incubation media: Ringer solution containing 10 mmol /l glucose (control observations = circles) plus additional 5 mmol /l glycine (+ Glycine = triangles). Mean ± SD (n = 16)

The observed protective effect of glycine in the range of marginal oxygen tensions (5 - 10 mm Hg), however, remains unclarified. Some authors suggest a direct stimulatory substrate effect on renal aerobic energy metabolism (Brezis et al., 1984). However, in view of a reduced oxygen consumption of isolated renal tubules at an extracellular PO_2 of 5 mm Hg and of a parallel increase in the redox state of mitochondrial cytochromes (Balaban et al., 1980) a stimulation of aerobic energy metabolism at this low oxygen tension is unlikely. Inhibition by amino acids of lysosomal enzyme liberation, thus of the uncoupling of mitochondrial respiration, seems to be a more probable explanation (Mellors et al., 1967; Gronow et al., 1986).

In contrast to the known inhibiting effect of several amino acids on lysosomal autophagy in liver cells, little is known about the specific effects of amino acids in renal tubular cells. Glycine, for example, reduced lysosomal enzyme liberation in the present experiments significantly, although it had little effect on autophagy in liver cells. Lysine, on the other hand, caused severe cellular destruction in the kidney, but inhibited lysosomal autophagy in the liver (Seglen et al., 1980; Racusen et al.,

1985). Obviously, further investigations are necessary to clarify the mechanism by which specific amino acids support hypoxic volume regulation in renal cortical cells.

SUMMARY

PO_2 declines to less than 10 mm Hg in local regions of the renal cortex. Amino acids seem to modify the hypoxic tolerance of renal cells. It was suggested that glycine may support renal function in hypoxia. Aim of the present study was to test the effect of glycine on renal cellular hypoxic tolerance. We isolated tubules of the rat kidney cortex (ITS) by collagenase treatment and measured cellular function at different levels of extracellular oxygen tensions (1, 2.5, 5, 10, 40, and 100 mm Hg) both with and without glycine in the incubation medium. No significant effects were observed in the "physiological" range at an extracellular $PO_2 = 100 - 10$ mm Hg. With no glycine in the incubation medium, the outer tubular diameter of ITS rose at lower oxygen tensions, at PO_2 of 1 mm Hg by about 170%, and the loss of 4 marker enzymes increased about 2 - 4fold. Hypoxic lactate formation increased at extracellular oxygen tensions < 10 mm Hg. Intracellular K^+ fell in parallel to about one third of the aerobic control values. Addition of glycine to the incubation medium did not significantly change intracellular K^+ or anaerobic lactate formation. In contrast, the loss of marker enzymes was significantly suppressed by glycine, lysosomal APase and mitochondrial GlDH by about 30%, cytoplasmatic LDH and brush border τGT by about 50%. Accordingly, at $PO_2 = 1$ mm Hg the hypoxic swelling of renal cells was suppressed in the presence of glycine by about 50%. It is concluded that glycine may support integrity of cellular and mitochondrial membranes by an inhibitory effect on lysosomal enzyme liberation, which in turn may also reduce the uncoupling of mitochondrial respiration.

REFERENCES

Alcorn, D., Emslie, K. R., Ross, B. D., Ryan, G. B., and Tange, J. D., 1981, Selective distal nephron damage during isolated kidney perfusion. Kidney Int., 19:638.

Balaban, R. S., Soltoff, S. P., Storey, J. M., and Mandel, L. J., 1980, Improved renal cortical tubule suspension: spectrophotometric study of O_2 delivery. Am. J. Physiol., 238:F50.

Baumgärtl, H., Leichtweiss, H. P., Lübbers, D. W., Weiss, Ch., and Huland, H., 1972, Microvasc.Res. 4:247.

Brezis, M., Silva, P., and Epstein, F. H., 1984, Amino acids induce renal vasodilatation in isolated perfused kidney: coupling to oxidative metabolism, Am. J. Physiol. 247:H999.

Gronow, G. H. J., Benk, P., and Franke, H., 1984, Effect of anaerobic substrates on post-anoxic cellular functions in isolated tubular segments of rat kidney cortex, Adv. Exp. Med. Biol. 180:403.

Gronow, G. H. J., and Cohen, J. J., 1984, Substrate support for renal functions during hypoxia in the perfused rat kidney, Am. J. Physiol. 247:F618.

Gronow, G., Skrezek, Ch., and Kossmann, H., Correlation between mitochondrial respiratory dysfunction and Na^+-reabsorption in the reoxygenated rat kidney, Adv. Exp. Med. Biol. 200:515.

Gronow, G., Klause, N., and Mályusz, M., 1988, Amino acid - mediated reduction of hypoxic uncoupling of mitochondrial respiration in isolated kidney tubules, Pflügers Arch. 411:R91.

Hochachka, P. W., Defense strategies against hypoxia and hypothermia, 1986, Science 231:234.

Jones, D. P., and Kennedy, F. G., 1982, Intracellular oxygen supply during hypoxia, Am. J. Physiol. 243:C247.

Lang, F., Völkl, P., and Paulmichl, M., 1988, How do cells regulate their volume, Pflügers Arch. 411:R4

Mályusz, M., and Gronow, G., 1987, Contrasting effects of amino acid mixtures on hypoxic dysfunction in the rat kidney, in: "Molecular Nephrology. Biochemical aspects of kidney function", Z. Kovacevic and W. G. Guder, eds., Walter de Gruyter, Berlin.

Mellors, A., Tappel, L., Sawant, P. L., Desai, I. D., 1967, Mitochondrial swelling and uncoupling of oxidative phosphorylation by lysosomes, Biochim. Biophys. Acta 143:299.

Racusen, C. R., Finn, W. F., Whelton, A., and Solez, K., 1985, Mechanisms of lysine-induced acute renail failure in rats, Kidney Int. 27:517.

Seglen., P. O., Gordon, P. B., and Poli, A., 1980, Amino acid inhibition of the autophagic/lysosomal pathway of protein degradation in isolated rat hepatocytes, Biochem. Biophys. Acta 630:103.

Weinberg, J. M., Davis, J. A., Abarzua, M., and Rajan, T., 1987, Cytoprotective effect of glycine and glutathione against hypoxic injury to renal tubules, J. Clin. Invest. 80:1446.

AEROBIC GLYCOLYSIS IN THE RETINA OF THE CRAB OCYPODE RYDERI

U. Knollmann[2], H. Acker[1], H. Langer[2], M. A. Delpiano[1]

[1]Max-Planck-Institut für Systemphysiologie, 4600 Dortmund
Rheinlanddamm 201, F.R.G.
[2]Institut für Tierphysiologie, Ruhr-Universität Bochum
4630 Bochum, F.R.G.

INTRODUCTION

Higher crustaceans (Malacostracan) possess compound eyes of a similar type as insects (apposition eyes). They consist of several thousands of ommatidia, each composed of a cornea followed by the dioptric apparatus and the long visual receptor cells (500 μm), which contain the light-sensitive rhabdoms and are enveloped by the pigment cells and separated from each other by a large extracellular space. The axons of the receptor cells penetrate the basal lamina to contact the optical ganglia. Since insect eyes are supplied with sufficient oxygen by tracheols, their metabolism is exclusively aerobic (Tsacopoulos et al., 1981). In contrast, the retina of crustaceans is supplied with oxygen by haemolymph in a similar way as the mammalian retina by blood, which is known to perform aerobic glycolysis (Warburg et al., 1924). Therefore, to investigate the metabolism of the crab retina and see whether aerobic glycolysis also exists, we measured tissue Po_2 (Pgo_2) and extracellular pH (pHe) under normoxic and hypoxic conditions and determined lactate production.

METHODS

The isolated retina of the terrestrial shore-crab, Ocypode ryderi, was superfused in a lucite chamber (A, Fig. 1) with a physiological solution (Cole, 1941) (in mM): NaCl 460, KCl 15, $MgCl_2$ 3.8, $MgSO_4$ 5.2, $CaSO_4$ 25, HEPES 10, $NaHCO_3$ 6, glucose 5.5, pH 7.3 at 25°C. The medium was equili-brated with different gas mixtures using two gas-mixing pumps (B) in separate vessels (C). A peristaltic pump (D) drained the superfusate and returned it to the reservoir (C). In some experiments using drugs, the stopcock (E) was open to avoid recirculation. The Po_2 in the medium (Pmo_2) of 280 or 150 Torr was referred to as normoxia or lowered to 27 Torr in experiments referred to as hypoxia. Pco_2 and pH in the medium were kept constant. Pgo_2 and pHe were measured with two channel-barrelled microelectrodes (Acker et al., 1983; Delpiano and Acker, 1985). Lactate in tissue and superfusate were continuously determined in fluorometric assays according to Lowry and Passonneau (1972).

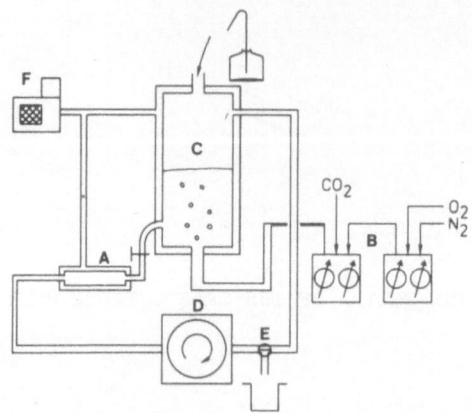

Fig. 1. Schematic representation of the superfusion system: A, lucite chamber containing the crab retina; B, gas-mixing pumps to equilibrate Cole's solution in reservoir C; D, peristaltic pump promotes circulation of the solution; E, stopcock interrupts recirculation; F, thermostat.

RESULTS AND DISCUSSION

In a previous comparative study the specific activity of several enzymes in the eyes of the crab (Ocypode) and the blowfly Calliphora erythrocephala were measured. The results showed that Ocypode eyes have a very high glycolytic capacity (LDH) and a relatively low capacity for oxydative degradation (citrate synthase, cytochrome c oxydase) as compared to the eye of Calliphora (Rivera and Langer, 1983). Since this could be due to insufficient oxygen supply from the superfusion medium to the distal part of the receptor cells, we measured the Pg_{O_2} in different depth of the retina.

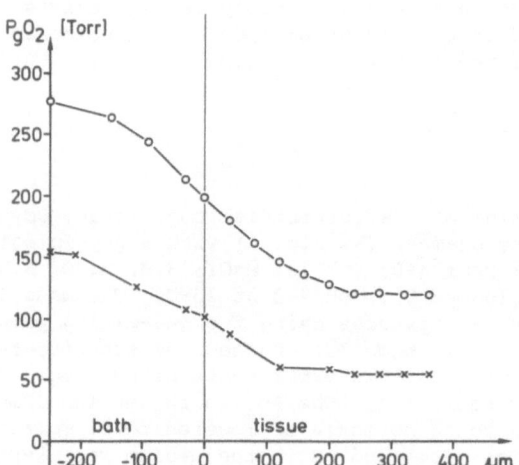

Fig. 2. Pg_{O_2} profile above and within the retina of the crab measured with Po_2 microelectrodes under different Po_2 in the medium (O, 280 Torr, X, 150 Torr). The abscissa gives depth of puncture, the ordinate the Po_2 above and inside the tissue. $0\ \mu m$ represents the tissue surface or the level corresponding to the basal lamina.

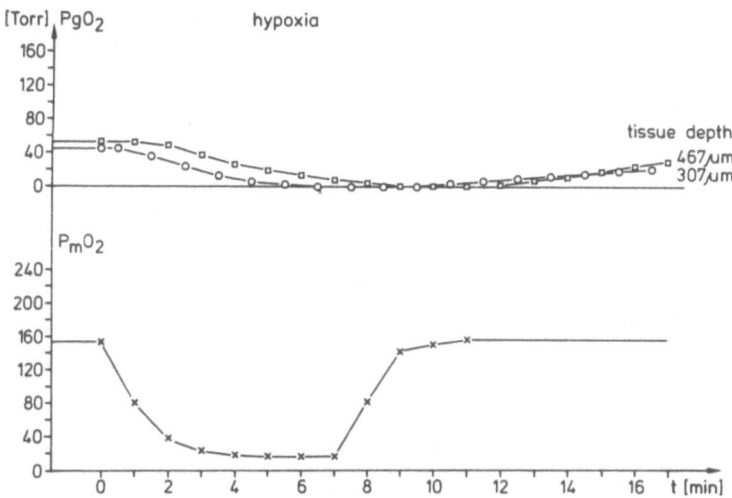

Fig. 3. Time course of the PgO_2 and PmO_2 at the ordinate during hypoxia created by lowering PmO_2 in the superfusion medium. The upper panel of the figure shows PgO_2 values measured with PO_2 micro-electrodes at two different tissue depth, as indicated on the right. The lower panel gives the PmO_2 (X) measured by a catheter PO_2 electrode. The abscissa gives the experimental time (min).

For this purpose the PO_2 microelectrode was inserted step by step into the tissue by impailing the basal lamina (tissue surface orientated to the bath) by means of a microdrive manipulator. As illustrated in Fig. 2, a downward slope in oxygen partial pressure could be found starting already at a distance of about 200 μm above the tissue surface (indicated by the vertical line). This PmO_2 gradient continued inside the tissue as a PgO_2 downward gradient over the next 200 μm to reach steady levels depending on the PmO_2 at about 300 μm, i.e. for a PmO_2 of 280 and 150 Torr the PgO_2 was 138 and 50 Torr, respectively (Fig. 2).

When hypoxia was produced by lowering the PmO_2, as shown in the lower panel of Fig. 3, the PgO_2 in the retina went down in a delayed fashion to reach values near to zero Torr at the end of the hypoxic period. In this experiment the oxygen partial pressure in the medium was adjusted to 152 Torr, nearer to physiological conditions, but still higher than in the haemolymph of the crab (Morris and Bridges, 1985). It was interesting to observe that the delayed decrease in PgO_2 at the onset of hypoxia was more pronounced the deeper the tip of the microelectrode was inside the tissue and that the recovery time of the PgO_2 after hypoxia was extremely slow. Such behaviour as well as the PmO_2-dependent PgO_2 gradient (Fig. 2) indicate an oxygen consumption in the retina. This finding is supported by the fact that antimycin A applied to the medium abolished both, the PgO_2 gradient and the delayed PgO_2 recovery after hypoxia (Delpiano et al., in preparation). Thus, the high glycolytic capacity (LDH) found in the crab retina (see above) cannot only be regarded as an energy preserving mechanism in case of an insufficient oxygen supply. However, pH profiles inside the retina could be measured with a pH gradient starting already 200 μm above the tissue surface (Fig. 4), similar as in the case of the PgO_2 profile (Fig. 2). This pH gradient is obviously built up by a steady lactate pro-duction and release by the crab retina, which could be directly measured during superfusion in previous experiments (Langer et al., 1988). Moreover, during hypoxia, the pH level inside the tissue was not changed (not shown), which can also be explained by a permanent lactate production and release.

715

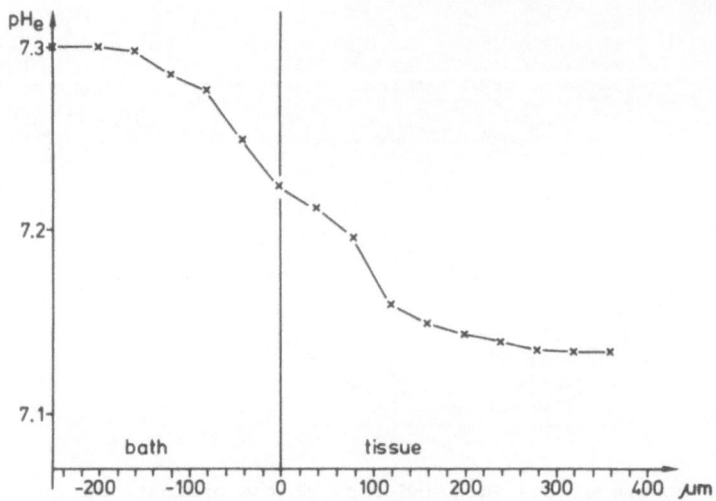

Fig. 4. pH profile outside and inside the tissue of the superfused retina of Ocypode ryderi measured with a double-barrelled pH-sensitive microelectrode at a Pm_{O_2} of 280 Torr. Details of diagram as in Fig. 2.

From these measurements of the Pg_{O_2}, pHe, and lactate determination it is assumed that the retina of the crab Ocypode ryderi performs aerobic glycolysis.

SUMMARY

Our experiments on the isolated and superfused crab retina reveal that pronounced gradients for P_{O_2} and pH exist in this tissue. The Pg_{O_2} profiles and a delayed recovery of the Pg_{O_2} after hypoxia seem to be a consequence of the oxygen consumption inside the tissue, as much as both characteristics can be abolished by impairment of electron transport in the respiratory chain after application of antimycin A. The pH profile is obviously created by a production and steady release of lactate, which could be measured in the superfusate. As this lactate release is occurring inspite of a sufficient oxygen supply and consumption, it can be concluded that this tissue performs aerobic glycolysis.

REFERNCES

Acker, H., Holtermann, G., Carlsson, J., Nedermann, T., 1983, Methodological aspects of microelectrode measurements in cellular spheroids, Adv. Exp. Med. Biol., 159:445.

Cole, W. H., 1941, A perfusing solution for the lobster (Homarus) heart and the effect of its constituent ions on the heart, J. Gen. Physiol. 25:1.

Delpiano, M. A., and Acker, H., 1985, Extracellular pH changes in the superfused cat carotid body during hypoxia and hypercapnia, Brain Res., 342:273.

Delpiano, M. A., Knollmann, U., Acker, H., and Langer, H., 1989, Po_2 and pH in the retina of the crab Ocypode ryderi - Evidence for aerobic glycolysis (in preparation).

716

Langer, H., Delpiano, M. A., Knollmann, U., and Acker, H., 1988, Oxygen and and glycolysis in the retina of the compound eye of a crab, in: "Oxygen Sensing in Tissues," H. Acker, ed., Springer-Verlag, Berlin.

Lowry, O. H., and Passoneau, J. V., 1972, "A Flexible System of Enzymatic Analysis," Academic Press, New York.

Morris, S., and Bridges, C. R., 1985, An investigation of haemocyanin oxygen affinity in the semi-terrestrial crab Ocypode saratan Forsk, J. Exp. Biol., 117:119.

Rivera, M. E., and Langer, H., 1983, Enzyme pattern of energy releasing metabolism in eyes, optical ganglia of the blowfly Calliphora erythrocephala and the crab Ocypode ryderi, Mol. Physiol., 4:265.

Tsacopoulos, M., Poitry, S., and Borsellino, A., 1981, Diffusion and consumption of oxygen in the superfused retina of the drone (Apis mellifera) in darkness, J. Gen. Physiol., 77:601.

Warburg, O., Posener, K., Negelein, E., 1924, Über den Stoffwechsel der Carcinomzelle, Biochem. Z., 152:309.

AN ISOLATED PERFUSED FROG SKIN PREPARATION FOR THE STUDY OF GAS EXCHANGE

Alan Pinder*, Daniel Clemens†, and Martin Feder†

*Biology Department
Dalhousie University
Halifax, Nova Scotia
Canada B3H 4J1

†Department of Anatomy
University of Chicago
1025 E. 57 St.
Chicago, IL 60637, U.S.A.

INTRODUCTION

The vascular arrangement of frog skin offers two advantages for studying gas exchange within the microcirculation, both due to the relatively simple two-dimensional geometry of the skin: 1) all vessels are visible, thus amenable to techniques being developed for analysing blood flow from video images, and 2) oxygen uptake, gas partial pressure gradients, perfusion rate, perfusate capacitance etc. can be easily manipulated. Other tissues have much more complex three dimensional vasculatures in which it is difficult to observe an entire capillary bed supplied by a particular vessel. Tissues that are thin enough to observe red cell flow through the entire capillary bed (mesentery, thin muscles) have too low an oxygen consumption to simultaneously measure oxygen delivery. Thus, it is difficult to test models of oxygen delivery being developed to predict the effects of microcirculatory red cell flow patterns on gas exchange. Although the isolated perfused frog skin preparation was developed specifically for studying cutaneous oxygen uptake, it has promise for studying more general features of gas exchange in capillary beds, such as the effects of blood flow heterogeneity and red cell spacing on oxygen uptake and delivery.

VASCULAR ANATOMY OF FROG SKIN

Most of the skin on the body of a frog is supplied by the cutaneous artery, a branch of the pulmocutaneous artery, which arrives at the skin just behind the tympanum. The same section of skin is drained by the cutaneous vein, which runs in a loop from behind the mouth posteriorly along the lateral body wall, turns 180° in front of the hind limb, and returns ventrolaterally to below the forelimb to join the brachial vein. Small arteries and veins distributing blood to the skin are on the inside surface of the dermis. Terminal arterioles and venules run perpendicular the skin to take the blood to a dense layer of capillaries about 50 μm below the surface of the epidermis. There is also a very much more diffuse capillary network in the dermis itself, presumably supplying oxygen and nutrients to the dermis. A more detailed description is available from de Saint-Aubain (1982).

A very simple model is thus appropriate (Figure 1). Virtually all gas exchange is between the external medium and the superficial capillary bed across the diffusion barrier of the epidermis. Nutritive capillaries in the dermis consitute a functional shunt around the gas exchanger, adding deoxygenated blood to the oxygenated blood returning from the superficial capillaries. There may also be anastomoses between cutaneous arteries and veins, although these have not been demonstrated. Only a small proportion of cutaneous blood flow appears to travel through shunt vessels.

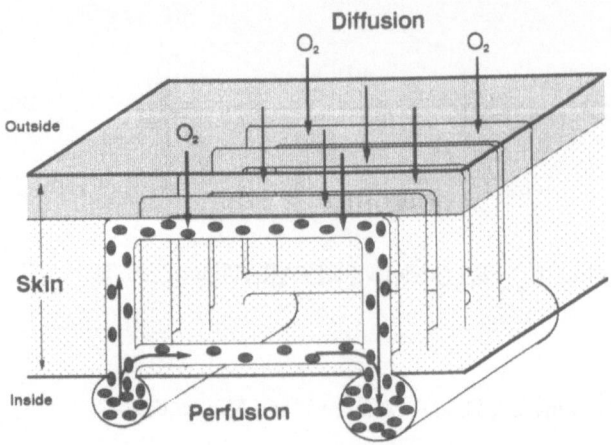

Figure 1. Simplified model of frog cutaneous circulation. See explanation in text

The vascular bed, although it forms a single layer beneath the skin, of course is not as geometrically regular as diagrammed; in fact the capillaries form a highly irregular two dimensional network. The capillaries are close enough to the surface to see individual red cells, and red cell flow is highly heterogenous when observed *in vivo*.

METHODS

The skin is removed from 250 to 350 g bullfrogs half immersed in MS222 (tricaine methanesulfonate, 0.5 g/l, adjusted to pH 7 with NaOH) anaesthetic dissolved in protein-containing Ringer's solution, with the frogs on ice to slow metabolism. Bovine serum albumin is added to all superfusates (0.1%) and perfusates (0.2%) to prevent the microvasculature from becoming excessively leaky (Mason et al. 1977, Curry et al. 1983). Bullfrog skin is only loosely attached to the body wall, with large lymph spaces underlying most of the surface removed. The lymph spaces are separated by septa of connective tissue through which sparse blood vessels and nerves run. The ventrolateral section of the cutaneous vein runs in one of these septa and must be carefully separated from the body wall with a fine-tipped cauterizing tool to seal all small veins returning from the body wall, freeing the vein almost up to the junction with the brachial vein. All minor blood vessels in other septa are also cauterized as the skin is being removed. The cutaneous artery and vein are left intact until the separation of the skin from the body wall is complete. The cutaneous artery is cannulated with the shank of a 23g needle and the skin is flushed with perfusate before removing it from the frog; the cutaneous vein is very large and can easily be cannulated with PE 60 - 90 tubing after the skin is removed.

Red cells are drained from the same animal from which the skin was taken, washed in three changes of saline and resuspended in an artificial plasma consisting of 111mM NaCl, 2.4 mM KCl, 1.0 mM $MgSO_4$, 1.1 mM $CaCl_2$, 6 mM glucose, 2 mM HEPES buffer, 0.2% BSA, and 500 units/ml heparin, adjusted to pH 7.6.

The cut edges of the skin skin are sealed by clamping the skin between two halves of a plexiglas chamber; grooves are cut in the plexiglass for the arterial and venous cannulae. 23g needle tips set in sockets around the edge of the chamber hold the skin in place before clamping. Since distribution of blood is from the middle of the skin patch toward the edges and drainage is from the edge toward the middle, clamping does not interfere with perfusion of the skin patch. The bottom of the chamber is filled with protein-containing saline and the top is ventilated with gas mixtures or water.

The skin is perfused in a recirculating system (Figure 2) to conserve red cells. Because there is a continuous return of oxygenated blood to the tonometer, the tonometer must be quite efficient. In our system we used an intermittently spinning tonometer produced by Walter

Nüsse at the Max Planck Institut für experimentelle Medizin (Göttingen). In order to draw a continuous stream of perfusate we added a 6 mm diameter bulb to the bottom of the tonometer chamber; the bulb retains a sufficient pool of blood to draw from when the tonometer is spinning. Flow rates were varied from 27 µl min^{-1} to 280 µl min^{-1}.

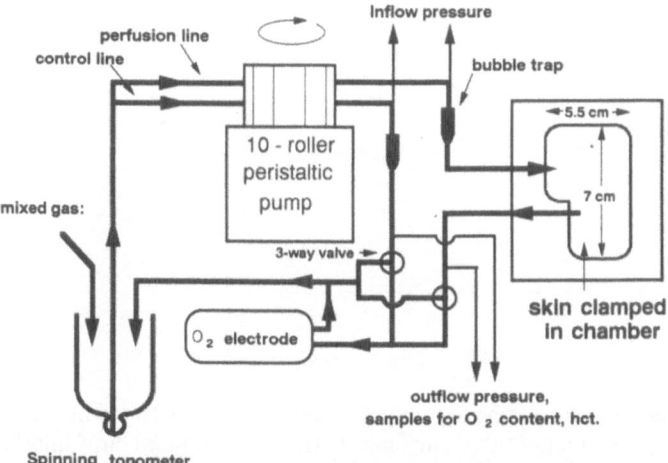

Figure 2. Perfusion apparatus. Perfusate is pumped by a peristaltic pump (Gilson Minipuls 3) through a bubble trap to the skin. An identical parallel perfusion line is used to control for resistance and gas exchange in the tubing. Flow from either line can be diverted past an oxygen electrode (Radiometer E5046 in D616 cuvette). Outflow P$_{O2}$ and input and output pressures are continuously monitored and spot samples are taken for input-output oxygen content differences for calculating oxygen uptake. Return of perfusate to the tonometer is by gravity. The skin chamber is set about 20 cm higher than the tonometer to maintain a negative pressure in the return lines collapsing the relatively large capacitance cutaneous vein and decreasing response time to experimental manipulations. Total circulating volume of the system is 8 to 10 ml, with 6 to 8 ml in the tonometer.

PRELIMINARY RESULTS

This preparation can now be done routinely; data from the first three successful preparations are presented here. All experiments were done at 20°C, the perfusate equilibrated with 2% CO$_2$ and 4% O$_2$, hematocrit maintained between 10 and 15% (all within the physiological range for amphibians) and the external respiratory medium was humidified air.

Cutaneous diffusing capacity was comparable to that measured in vivo at 5°C (0.018 nmol mmHg^{-1} min^{-1} cm^{-2}, Pinder 1987) and decreases slowly over the course of the experiment (Figure 3). Two of the preparations showed a more rapid decrease at the start of perfusion then stabilized for several hours. The preparations gradually took on a pink tinge, perhaps the result of a buildup of stagnant red cells in the skin.

Oxygen uptake was proportional to perfusion rate (perfusion limited) at the lowest perfusion rates (Figure 4). At progressively higher perfusion rates oxygen uptake approached a maximum set by the rate of diffusion through the epidermis, so that oxygen uptake as a function of perfusion approximated a rectangular hyperbola. Even when the skin patch became primarily diffusion limited, oxygen uptake increased with increasing perfusion rate (Figure 5). This was due primarily to an increase in diffusing capacity, perhaps caused by the recruitment of additional capillaries by increased perfusion pressure.

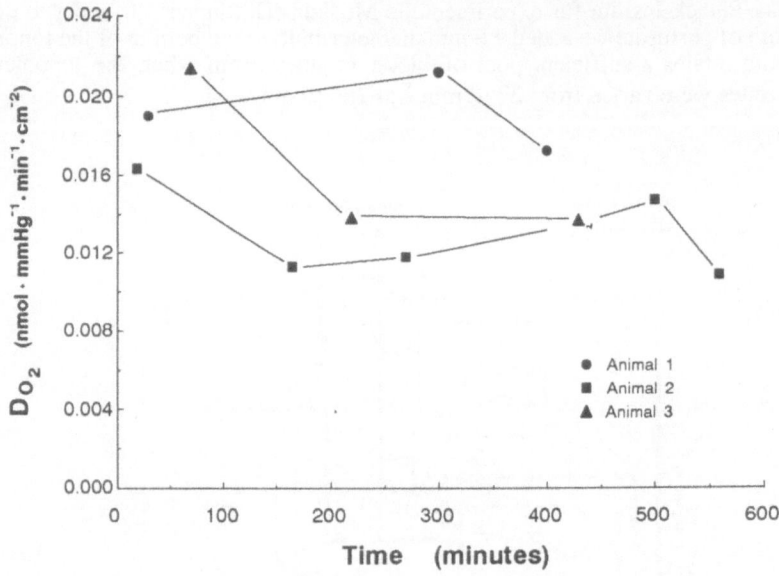

Figure 3. changes in diffusing capacity over time. All measurements were done at the same blood flow rate (3.9 μl cm^{-2} min^{-1}; Hb flow 2.0 nmol cm^{-2} min^{-1}).

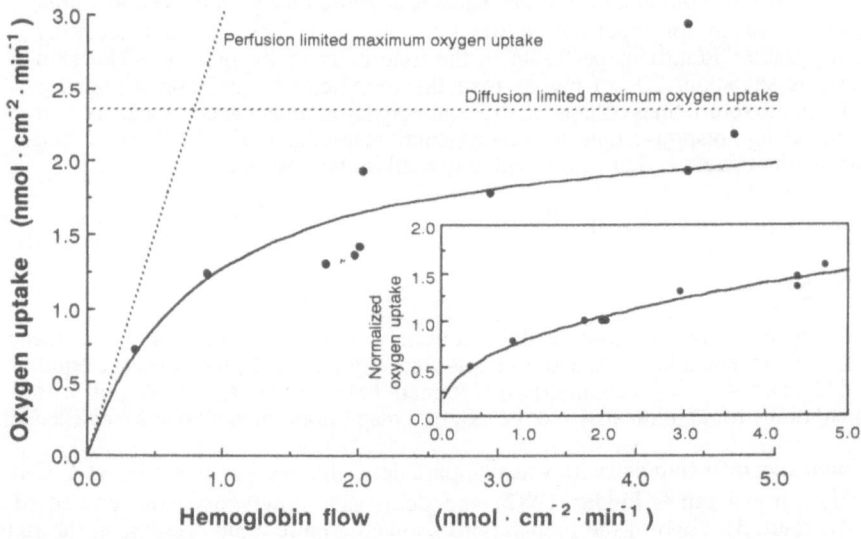

Figure 4. Oxygen uptake as a function of hemoglobin flow in a single preparation. "Perfusion limited" maximum oxygen uptake was calculated as Hb flow x 4 O_2/Hb x (1 - proportion of Hb-O_2 in perfusate). "Diffusion limited" maximum oxygen uptake was calculated from the y-intercept of a double-reciprocal (Lineweaver-Burk) plot. This is an underestimate, since diffusing capacity increases with perfusion rate or pressure (see below). The inset graph is the same data set after normalization: all oxygen uptakes are compared to oxygen uptakes measured over the course of the experiment at an Hb flow of 2 nmol cm^{-2} min^{-1} to control for time-dependent changes in diffusing capacity.

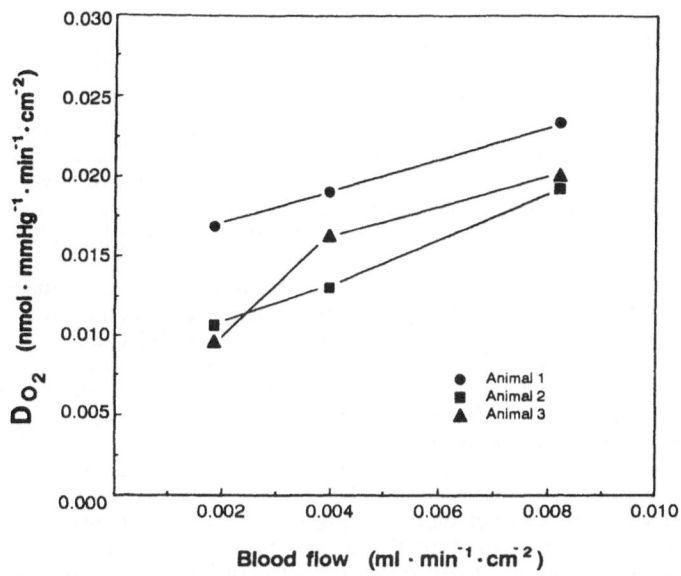

Figure 5. Diffusing capacity as as function of perfusion rate. The lowest perfusion rates are not included because the calculation of the mean gradient across the skin is inaccurate without a Bohr integration when the exiting blood is almost fully saturated, resulting in a gross underestimate of diffusing capacity.

DISCUSSION

The reduction in diffusing capacity over the course of the experiment creates difficulty in comparing early measurements of diffusing capacity to later measurements. It is possible to normalize the data by comparing all measurements to a frequently repeated standard perfusion rate (inset, Figure 4), but this considerably slows the progress of the experiment.

Decreased diffusing capacity is associated with increased vascular resistance. The reasons for these changes are not clear but may be related to 1) progressive blocking of some vascular pathways by particles in the perfusate or by clumped red cells or 2) increased vasoconstriction. Addition of a filter to the perfusion line would reduce problems associated with particulates. In recent experiments, addition of ortho- and paracresol (6 mg/l), used in other perfusion preparations to prevent vasoconstriction (e.g. Geiser and Betticher, 1989) reversed the increased resistance. Whether it can also prevent the reduction in diffusing capacity is not yet known.

A difficulty with this preparation, as with all isolated perfused preparations, is that normal neural input to blood vessels is interrupted with unknown results on the distribution of blood flow. It is possible that heterogeneity of flow increases or that shunt pathways dilate, decreasing efficiency of gas exchange. Unlike most preparations, distribution of blood flow in isolated frog skin can be compared to the organ *in vivo* because the capillaries are visible from the outside of the animal. In any case, a major use for this preparation is in testing the effects of manipulations such as varying the heterogeneity of blood flow or red cell spacing; it is not as important that the isolated skin reflect *in vivo* responses as that changes in blood flow and capillary hematocrit can be quantified in the isolated skin. Other experiments can then be done to test the *in vivo* significance of such changes.

There are potential problems associated with the use of a recirculating perfusate. The red cells may be damaged or hemoglobin effectors may change over the course of the experiment; some hemolysis is usually visible in the perfusate after one or two hours of recirculation. Capillaries in frog skin are fairly leaky to fluid *in vivo* so "plasma" is gradually lost from the perfusate and hematocrit increases (*in vivo* lymph is collected in sacs just under the skin; in the isolated preparation the "lymph" flows into the saline bath under the skin. There is no evidence of edema during perfusion). Hematocrit must thus be checked frequently and adjusted as necessary. Finally, frog skin contains a wide range of peptides, some of which may be vasoactive and which may be released into the perfusate. A non-recirculating system would avoid these difficulties, but would necessitate the use of several donor animals. More easily available mammalian blood cells might be used, but mammalian blood cells are much smaller than amphibian cells and may not flow similarly through amphibian capillaries.

The value of this preparation is in the potential to simultaneously quantify gas exchange and patterns of blood flow, including capillary recruitment, tube hematocrit, red cell velocity, transit time, flux, and spacing. Thus for the preparation to be fully useful it must be possible to 1) record high quality video images of red cells in cutaneous capillaries for computer analysis of red cell movement and 2) manipulate blood flow heterogeneity, tube hematocrit, capillary recruitment etc. Although red cells are visible from the outside of the skin, they are obscured in heavily pigmented areas. In order to improve imaging of red cells sufficiently for computer analysis it may be necessary to use albino animals (both bullfrog and *Xenopus laevis* albinos are available) or use fluorescently labelled red cells. We have not yet attempted to manipulate blood flow except to change the bulk perfusion rate. It may be possible to alter capillary recruitment and the distribution of red cell flow by stimulating nerves removed along with the skin, adding vasoactive agents to the perfusate, or by changing the P_{O_2}, P_{CO_2}, or pH of the perfusate (Malvin and Hlastala, 1986).

CONCLUSIONS

Gas exchange and vascular resistance in isolated perfused frog skin can be monitored for several hours. This preparation has the potential to be much more widely useful than just to characterize cutaneous gas exchange in amphibians if 1) video microscopy and computer analysis of the movements of individual red cells can be applied and 2) ways are found to manipulate capillary recruitment, red cell spacing, and heterogeneity of blood flow. No other preparation has the potential for simultaneously measuring gas exchange and flow distribution and heterogeneity in an entire vascular network as easily as in frog skin.

REFERENCES

Curry, F.E., Huxley, V.H. and Sarelius I.H., 1983, Techniques in the microcirculation: Measurement of permeability, pressure and flow, in: "Techniques in the Life Sciences P3/I, Cardiovascular Physiology" P309:1-34, Elsevier Scientific Publishers Ireland Ltd., County Clare.

de Saint-Aubain, M.L., 1982, The morphology of amphibian skin before and after metamorphosis, Zoomorph. Berl. 100:55-63.

Geiser, J. and Betticher D.C., 1989, Gas transfer in isolated lungs perfused with red cell suspension or hemoglobin solution, Respir. Physiol. 77:31-40.

Malvin, G.M. and Hlastala M.P., 1986, Regulation of cutaneous gas exchange by environmental O_2 and CO_2 in the frog, Respir. Physiol. 65:99-111.

Mason J.C., Curry, F.E. and Michel, C.C., 1977, The effects of proteins upon the filtration coefficient of individually perfused frog mesenteric capillaries, Microvasc. Res. 13:185-202.

Pinder, A.W., 1987, Cutaneous diffusing capacity increases during hypoxia in cold submerged bullfrogs (Rana catesbeiana), Respir. Physiol. 70:85-95.

REDUCTION OF GAS SOLUBILITY IN THE FISH SWIMBLADDER

Bernd Pelster, Hirosuke Kobayashi and Peter Scheid

Institut für Physiologie
Ruhr-Universität Bochum
4630 Bochum

INTRODUCTION

The gas filled swimbladder serves many fish as a hydro-static organ to achieve neutral buoyancy. The gas enters the bladder by diffusion from the swimbladder vessels. The high gas partial pressures necessary for establishing a diffusion gradient from the vessels to the swimbladder is achieved by reducing the solubility of gases in the swimbladder blood. This 'single concentrating effect' (Kuhn et al., 1963), the increase in gas partial pressure induced by a change in solubility, is then multiplied by countercurrent multiplication in the rete mirabile (Steen, 1970; Fänge, 1983).

The initial decrease in gas solubility includes a reduction in physical solubility, for example via the salting-out effect, but also a liberation of gases from a chemical bond, as for O_2 and CO_2. The present study analyses the reduction in gas solubilities that can be expected on the basis of the changes in blood metabolite concentrations measured in swimbladder vessels.

MATERIAL AND METHODS

Specimens of the European eel, Anguilla anguilla, were purchased from a local supplier and kept in a freshwater aquarium at 12 - 14 °C until the experiment. Blood was obtained via a catheter inserted in the bulbus arteriosus. Glassware and syringes were heparinized (120 IU heparin/ml saline) prior to use to prevent coagulation. The saline solution consisted of (in $mmol \cdot l^{-1}$): NaCl, 124; KCl, 5; $MgSO_4$, 0.9; $CaCl_2$, 1.1; $NaHCO_3^-$, 10; glucose, 5.

Measurement of inert gas solubility

The variation of argon and nitrogen solubility with changes in pH and lactate concentration have been measured using a mass spectrometer according to the method of Meyer and Scheid (1980) with the modifications of Pelster et al. (1988).

Oxygen Transport to Tissue XII, Edited by J. Piiper et al.
Plenum Press, New York, 1990

725

Blood samples with different pH and lactate concentrations were equilibrated for 20 minutes with known gas mixtures (20% O_2, $0.1 - 10\%$ CO_2, rest N_2 or Ar) provided by precision gas mixing pumps (Wösthoff, Bochum, FRG). An aliquot of the equilibrated blood was then transferred to a gas tight tonometer vessel containing the same gas mixtures, the test gas being replaced by a different inert gas. At the end of the second equilibration the fraction of the test gas was measured with a mass spectrometer (Scheid, 1983). For calibration the test gas mixture was analysed in a similar way using a larger gas tight vessel for the second equilibration. The solubility of gas in the blood sample could be calculated as

$$\alpha = \beta_g \left[1 - \frac{V}{V_2} \right] \bigg/ \left[\frac{P_2}{P_1} \cdot \frac{V_2}{V_1} - \frac{V}{V_1} \right],$$

in which β_g is the capacitance coefficient (55.6 $\mu mol \cdot l^{-1} \cdot mmHg^{-1}$ at $15°C$; Piiper et al., 1971), V is the blood sample volume (1 ml), V_1 is the volume of the extraction vessel for blood, V_2 is the volume of the vessel used for the gas sample and P_1 and P_2 are the partial pressures of test gas in the blood or gas vessel after re-equilibration.

Measurement of the Root effect

Blood samples were equilibrated for 25 minutes in a tonometer (model 273, IL, Paderno Dugano, Italy) with gas mixtures containing 50% O_2, $0.1 - 10\%$ CO_2 and N_2, provided by precision gas mixing pumps (Wösthoff, Bochum, FRG). Plasma pH of the equilibrated sample (pH_e) was measured with a capillary pH electrode (G 299, Radiometer, Copenhagen, Denmark) and the intracellular pH (pH_i) was measured using the freeze thaw method (Zeidler and Kim, 1977). The oxygen content was determined with a Tucker chamber (Tucker, 1967) equipped with a P_{O_2} electrode (E 5047, Radiometer).

In vivo preparation

Eels were anaesthetised with 0.1 $g \cdot l^{-1}$ MS 222 and the swimbladder was exposed as described earlier (Kobayashi et al., 1989a; Pelster et al., 1989). Blood samples from swimbladder vessels entering and leaving the gas gland were obtained by micropuncture. pH and P_{CO_2} of these samples were measured using Radiometer electrodes (G 299 and E 5037, Radiometer), CO_2 content, C_{CO_2}, was determined using the method of Cameron (Cameron, 1971).

RESULTS

In the pH range of 5.5 to 8.4 the solubility of N_2 and of Ar did not depend on the pH (Fig. 1), and this was true irrespective of the mechanisms, by which the pH of the blood samples was changed (P_{CO_2}, HCl or lactic acid). Furthermore, increasing the lactate concentration to as much as 21 $mmol \cdot l^{-1}$ did not induce a significant reduction in physical solubility (Fig. 2).

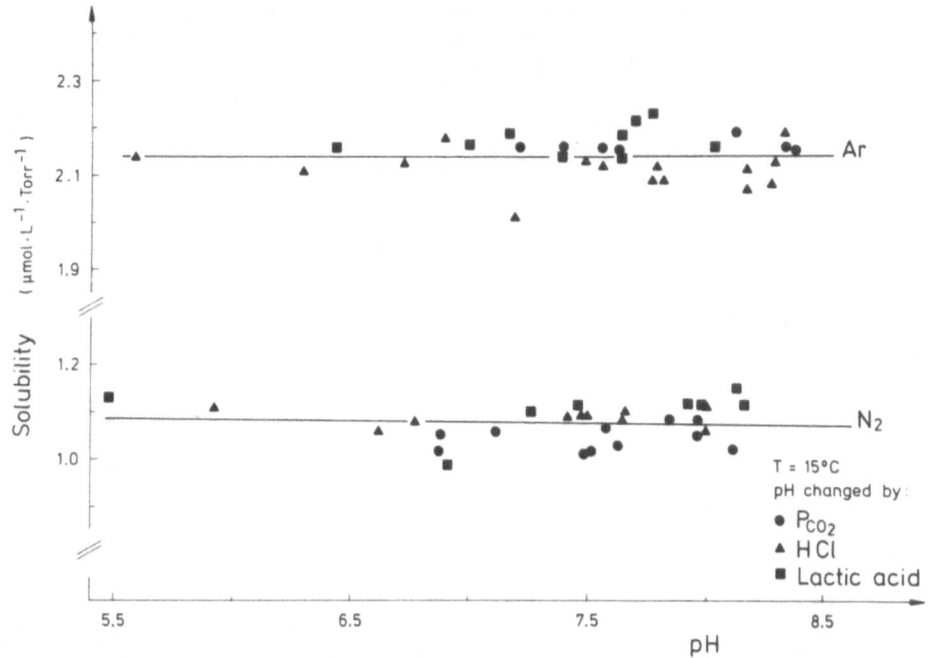

Fig. 1 Solubility of nitrogen (N_2) and argon (Ar) in whole blood of the eel at varying pH. The pH was varied with CO_2 (●), HCl (▲) or lactic acid (■).

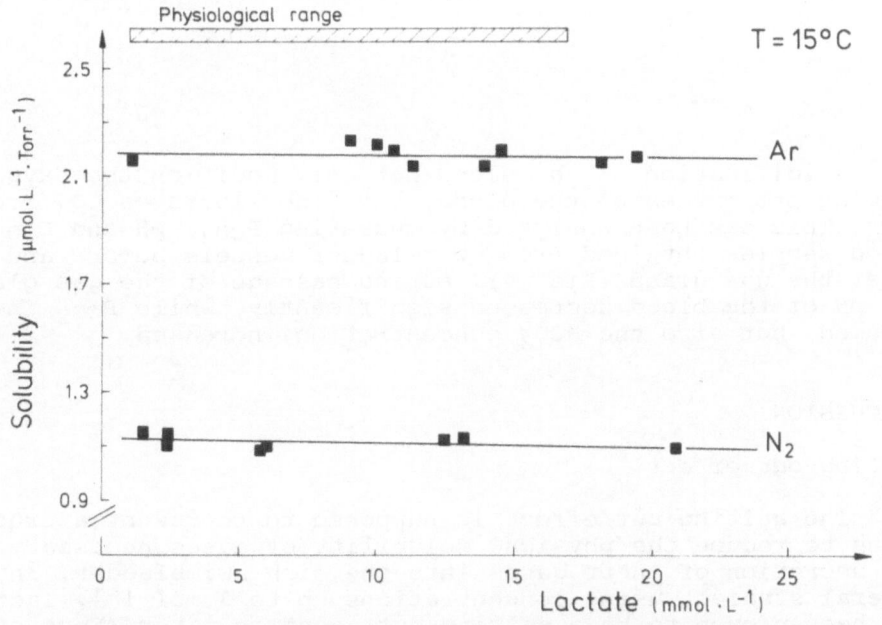

Fig. 2 Solubility of nitrogen (N_2) and argon (Ar) at varying lactate concentrations. The bar indicates lactate concentration range measured in swimbladder vessels.

The influence of adding CO_2 to the blood on the oxygen carrying capacity is illustrated in Fig. 3. The resulting acidification significantly reduced the oxygen content of the blood samples, and at pH 7.1, a value which has been measured in the swimbladder vessels, the oxygen carrying capacity was reduced by about 40%. The relation of pH_e and pH_i is shown in the inset.

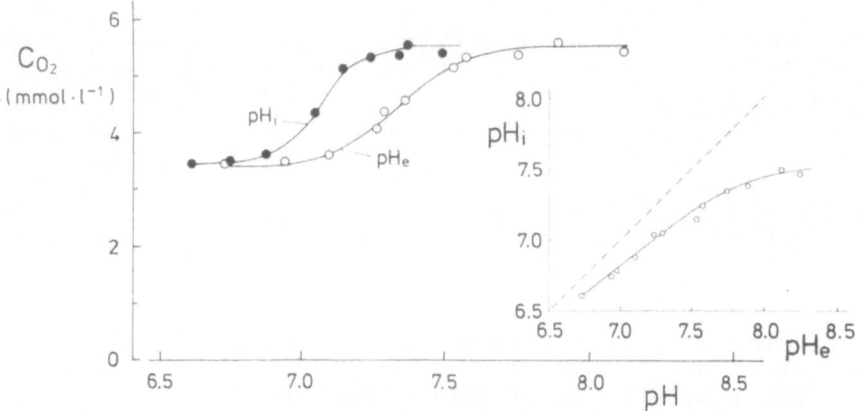

Fig. 3 Oxygen content (C_{O_2}) of whole blood at high P_{O_2} (350 Torr) and different pH values (●, pH_i; o, pH_e) attained by titration with CO_2. Inset shows the pH_i/pH_e relation.

Acidification of the blood not only modifies the oxygen binding properties of the blood, but also liberates CO_2 from HCO_3^-. This has been analysed by measuring P_{CO_2}, pH and C_{CO_2} in blood samples obtained from swimbladder vessels before and after the gas gland (Fig. 4). During passage of the gas gland the pH of the blood decreased significantly, while P_{CO_2} increased, but also the HCO_3^- concentration increased.

DISCUSSION

Salting-out effect

The salting-out effect is supposed to represent a crucial means to reduce the physical solubility of gases necessary for the secretion of inert gases into the fish swimbladder. In several studies, using concentrations up to 3 mol·l^{-1}, lactate has been proven to be a salting-out agent in water (Enns *et al.*, 1967; Gerth and Hemmingsen, 1982). Variation in salt concentration of 20 mmol·l^{-1}, which seems to be a reasonable physiological value, causes only changes in solubility of 1%

or even less. Therefore, the physiological importance of this effect for the secretion of inert gases has been questioned (Gerth and Hemmingsen, 1982). The present study clearly shows that varying the blood lactate concentration in the physiological range does not result in significant changes in physical solubility of gases and thus presents no evidence that the salting effect of lactate in whole blood is significantly larger compared to aqueous solutions.

Beside lactate also protons have been assumed to act as a salting-out agent (Steen, 1963a). This observation, however, could not be confirmed in the present study (see Fig. 1). Based on the metabolite concentrations measured *in vivo* in swimbladder vessels we can therefore expect only very small changes in physical solubility of gases as a result of changes in pH (Fig. 5B).

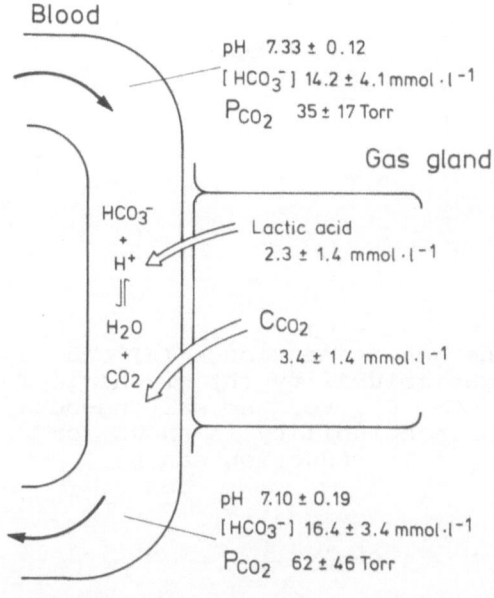

Fig. 4
In vivo pH, P_{CO_2} and HCO_3^- concentrations (calculated from C_{CO_2} and $\alpha \cdot P_{CO_2}$) of the blood before and after passage of the gas gland and the production of lactic acid and CO_2 (C_{CO_2}) by the gas gland tissue.

We have, however, shown that even a one percent reduction in physical solubility may be sufficient to generate very high inert gas partial pressures by countercurrent multiplication in the rete mirabile, if the permeability of the rete mirabile for gases and solutes is high enough (Kobayashi *et al.*, 1989b). But very little is known about permeabilities in the rete and the importance of the salting-out effect can not be assessed.

Root effect

The P_{O_2} is increased by a significant decrease in the oxygen carrying capacity of the hemoglobin (Root, 1931; Bridges et al., 1983; Brittain, 1987). Depending on the hemoglobin concentration, a reduction of the oxygen capacity by about 40% (see Fig. 3) allows a large increase in the oxygen partial pressure. If we neglect the diminution of oxygen content by the secretion of oxygen, all the oxygen molecules liberated from the hemoglobin stay in solution thus giving a very high P_{O_2} value (Fig. 5C).

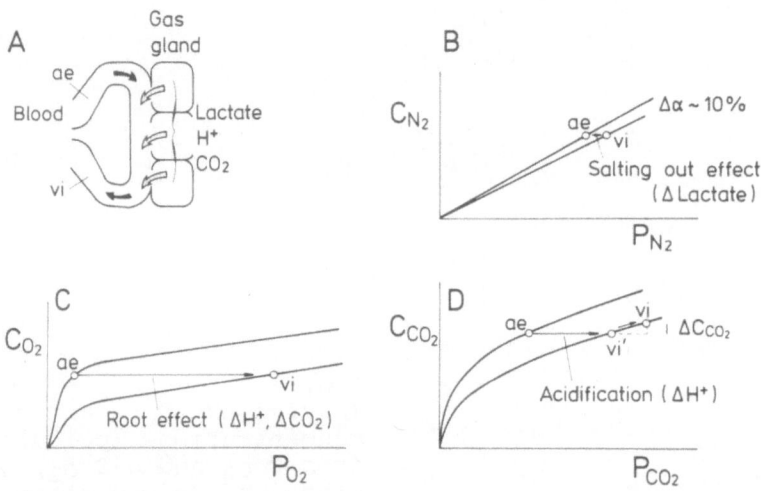

Fig. 5 Metabolites produced by the gas gland tissue (A) and their effect on physical gas solubility (B) or chemical binding of oxygen (C) and CO_2 (D). For the salting-out effect (B) a 10% reduction in solubility is shown for illustration, although only a 1% reduction can be expected.

The acidification of blood necessary to initiate the Root effect is brought about by the production of lactic acid and probably of CO_2 (Pelster et al., 1989) in the gas gland cells. Our data show a drop in pH of more than 0.2 pH units in the blood during passage of the gas gland which allows a reduction in oxygen carrying capacity of about 15%. The pH of the blood entering the gas gland, however, is far below the arterial value of about 7.8 indicating that the blood has already been acidified during passage of the rete mirabile. The acidification of the arterial blood in the rete is caused by a significant veno-arterial shift of acid, mainly CO_2, in the rete mirabile (Kobayashi et al., in prep.). Comparing the in vivo pH values in the swimbladder vessels and the relation

730

of pH_e and $O_2cap/(O_2cap)_{max}$ measured *in vitro* elucidates that the two step acidification of the blood by acid shift in the rete and acid production in the gas gland brings about an almost maximal reduction of the oxygen carrying capacity and accordingly a significant increase in P_{O_2}.

Increase in P_{CO_2}

Freshly secreted swimbladder gas usually contains quite high fractions of CO_2 (Jacobs, 1932; Wittenberg *et al.*, 1964). The initial increase in P_{CO_2} is brought about by acidification of the blood in the gas gland. The vessels of the swimbladder tissue represent a closed system and any addition of protons must, *via* the bicarbonate buffer, result in an increase in P_{CO_2} (Fig. 5D). In mammals, approximately 2/3 of the protons gained from the metabolism combine with the bicarbonate buffer. If this is similar in fishes, the addition of more than 2 mmol·l^{-1} lactic acid by the gas gland will result in a considerable increase in P_{CO_2} on the expenses of the HCO_3^- concentration. Our results, however, show an increase in P_{CO_2} as well as an increase in HCO_3^- concentration, indicating a release of CO_2 from the tissue into the blood. Pelster *et al.* (1989) assume anaerobic decarboxylation reactions to be responsible for the observed increase in C_{CO_2} of the blood during passage of the gas gland. As shown in Fig. 5D, the addition of CO_2 to the blood will, on top of the acidification by the release of lactic acid from the gas gland, further enhance the P_{CO_2}.

Conclusion

The swimbladder tissue has adopted two different strategies to reduce the solubility of gases: general reduction of the physical solubility of any gas, and release of a certain gas from a chemical binding site. Due to the small changes in metabolite concentrations that have been found in the swimbladder capillaries we can expect only a very small increase in the partial pressure of gases caused by the salting-out effect and indeed, the physiological significance of this effect still remains to be shown. The release of gases from a chemical binding site, on the other hand, has been proven to be a very efficient means to increase the gas partial pressure. Accordingly, significant increases in P_{CO_2} and P_{O_2} of the blood have been observed during passage of the gas gland (Steen, 1963b; Kobayashi *et al.*, in prep.). The composition of freshly secreted swimbladder gas - mainly oxygen, carbon dioxide usually ranging between 10% and 40%, rest inert gases (Fänge, 1983) - thus correlates with the magnitude of the single concentrating effect which we expect on the basis of changes in blood metabolite concentrations measured in swimbladder vessels.

ACKNOWLEDGEMENTS

We thank Mrs. G. Ryfa and Mr. S. Röhr for skillful technical assistance. Financial support by the Ministerium für Wissenschaft und Forschung des Landes NRW (Grant No. IVB4-10200687) is gratefully acknowledged.

REFERENCES

Bridges, C. R., Hlastala, M. P., Riepl, G., and Scheid, P., 1983, Root effect induced by CO_2 and by fixed acid in the blood of the eel, Anguilla anguilla., *Respir. Physiol.*, 51: 275-286.

Brittain, T., 1987, The Root effect, *Comp. Biochem. Physiol.*, 86B: 473-481.

Cameron, J. N., 1971, Rapid method for determination of total carbon dioxide in small blood samples, *J. Appl. Physiol.*, 31: 632-634.

Enns, T., Douglas, E., and Scholander, P. F., 1967, Role of the swimbladder rete of fish in secretion in inert gas and oxygen, *Adv. Biol. Med. Phys.*, 11: 231-244.

Fänge, R., 1983, Gas exchange in fish swim bladder, *Rev. Physiol. Biochem. Pharmacol.*, 97: 111-158.

Gerth, W. A., and Hemmingsen, E. A., 1982, Limits of gas secretion by the salting-out effect in the fish swimbladder rete, *J. Comp. Physiol.*, 146: 129-136.

Jacobs, W., 1932, Untersuchungen zur Physiologie der Schwimmblase der Fische. II. Die Volumregulation in der Schwimmblase des Flussbarsches, *Z. vergl. Physiol.*, 18: 125-156.

Kobayashi, H., Pelster, B., and Scheid, P., 1989a, Water and lactate movement in the swimbladder of the eel, Anguilla anguilla, *Respir. Physiol.*, in press.

Kobayashi, H., Pelster, B., and Scheid, P., 1989b, Solute back-diffusion raises the gas concentrating efficiency in counter-current flow, *Respir. Physiol.*, in press.

Kuhn, W., Ramel, A., Kuhn, H.J., and Marti, E., 1963, The filling mechanism of the swimbladder generation of high gas pressures through hairpin countercurrent multiplication. *Experientia*, 19: 497-511.

Meyer, M., and Scheid, P., 1980, Solubility of acetylene in human blood determined by mass spectrometry, *J. Appl. Physiol.*, 48: 1035-1037.

Pelster, B., Kobayashi, H., and Scheid, P., 1988, Solubility of nitrogen and argon in eel whole blood and its relationship to pH, *J. Exp. Biol.*, 135: 243-252.

Pelster, B., Kobayashi, H., and Scheid, P., 1989, Metabolism of the perfused swimbladder of European eel: O_2, CO_2, glucose and lactate balance, *J. Exp. Biol.*, in press.

Piiper, J., Dejours, P., Haab, P., and Rahn, H., 1971, Concepts and basic quantities in gas exchange physiology, *Respir. Physiol.*, 13: 292-304.

Root, R. W., 1931, The respiratory function of the blood of marine fishes, *Biol. Bull.*, 61: 427-456.

Scheid, P., 1983, Respiratory mass spectrometry, *in*: "Measurement in clinical respiratory physiology", G. Laszlo and M. F. Sudlow, eds., London, New York: Academic Press, pp. 131-166.

Steen, J. B., 1963a, The physiology of the swimbladder of the eel Anguilla vulgaris. I. The solubility of gases and the buffer capacity of the blood, *Acta Physiol. Scand.*, 58: 124-137.

Steen, J. B., 1963b, The physiology of the swimbladder in the eel Anguilla vulgaris. III. The mechanism of gas secretion, *Acta Physiol. Scand.*, 59: 221-241.

Steen, J. B., 1970, The swim bladder as a hydrostatic organ, *in*: "Fish Physiology", Vol. IV, W. S. Hoar and D. J. Randall, eds., New York: Academic Press, pp. 413-443.

Tucker, V. A., 1967, Method for oxygen content and dissociation curves on microliter blood samples, *J. Appl. Physiol.*, 23: 410-414.

Wittenberg, J. B., Schwend, M. J., and Wittenberg, B. A., 1964, The secretion of oxygen into the swim-bladder of fish. III The role of carbon dioxide, *J. Gen. Physiol.*, 48: 337-355.

Zeidler, R., and Kim, H. D., 1977, Preferential hemolysis of postnatal calf red cells induced by internal alkalinization, *J. Gen. Physiol.*, 70: 385-401.

GAS EXCHANGE IN THE FISH SWIMBLADDER

Peter SCHEID, Bernd PELSTER, Hirosuke KOBAYASHI

Institut für Physiologie
Ruhr-Universität Bochum
4630 Bochum, F.R.G.

Introduction

Neutral buoyancy is of great advantage to any object in water, a submarine like an animal, since it will help to save energy to avoid sinking or rising. Depending on their composition, fish are usually somewhat denser than the surrounding water, largely because of the skeletal and protein masses (densities around 2-3 and 1.3 g/cm^3, respectively) which override the "floating" mass of fat (density 0.9). Thus their overall density is some 5% larger than that of water, and this results in a sinking force of about 5% of their weight, which to overcome in water can be shown to be a formidable energy expenditure (Denton, 1961). There are several ways in which animals have solved this problem to become neutrally buoyant (Denton, 1961), and the swimbladder is one such floating device.

Swimbladder as a float

An air-filled swimbladder reduces the density of fish without adding much weight, since the density of air is only 0.00125 g/cm^3 at 1 atm, and still only 0.7 g/cm^3 at 700 atm, a pressure experienced in the sea at a depth of 7000 m below the surface. There are, however, problems with such a floating device.

For one, a given swimbladder volume will give the fish neutral buoyancy only at one particular depth, provided the gas-filled swimbladder volume obeys Boyle's law, which it apparently does (Denton, 1961). Secondly, elevation results in increased swimbladder volume and thus in a lifting force, and *vice versa*. Since it is well-known that fish vary their depth, partly in a diurnal rhythm, there must be provision for adjustments of swimbladder volume, and these are deposition or resorption of gas in the swimbladder. But aside from these adjustments there must also be provision for keeping the gases inside the swimbladder. Let us briefly look at the anatomy of the swimbladder and review the mechanisms by which the swimbladder is capable of depositing, resorbing, and keeping gas in it.

Swimbladder architecture

A swimbladder occurs in many marine fish, mostly in inhabitants of the upper 200 m in free ocean water and in species living near the bottom down to about 2000 m, the depth record being 7000 m.

The swimbladder may be viewed as a gas-containing sac. Despite differences in details, it is functionally very similar in most fishes. There are two main types, known as physostome and physoclist bladders, the former with, the latter without an open connection to the digestive tract. Let us from hereon consider the eel, since this has served as an experimental animal in many studies on the swimbladder function.

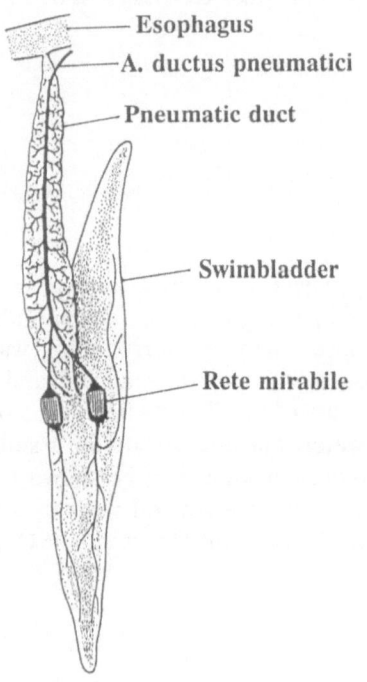

The swimbladder of the eel (fig. 1) constitutes an unpair, elongated sac which is connected to the esophagus through the pneumatic duct. The opening to the esophagus is, however, too small in this animal to be of significance for gas filling or emptying of the bladder by this route. Swimbladder and pneumatic duct serve quite different functions, the former as the site for gas deposition, the latter for gas resorption. The opening between both can be occluded.

The swimbladder is lined by a cuboidal epithelium , whereas the epithelium of the pneumatic duct is flat. Both structures are supplied with blood through a common artery. But, whereas

Fig. 1. Schematic drawing of the swimbladder and pneumatic duct in the eel. There are two retia mirabilia, arranged in parallel. Only the arterial vascular supply is shown. After Dorn (1961).

the capillaries of the pneumatic duct are drained by veins that enter directly into the large veins, there is the counter-current capillary arrangement in the red bodies of the swimbladder, the retia mirabilia. Thus arterial blood enters the arterial capillary supply of the rete, an arrangement of some 100,000 arterial capillaries, before it re-assembles into (post-rete) arterial vessels that feed the capillaries of the swimbladder wall. Blood draining from them into veins in the swimbladder wall feed, in turn, the venous capillaries in the rete, again about 100,000 capillaries that run strictly parallel and in checker-board arrangement with the arterial capillaries. Venous blood continues on to the liver (again a capillary network) before it returns to the main circulation. The rete thus presents the basis for a hairpin counter-current system, similar to that of the kidney, and this will be of functional significance for the gas deposition. A schema of the swimbladder and rete is shown in fig. 2.

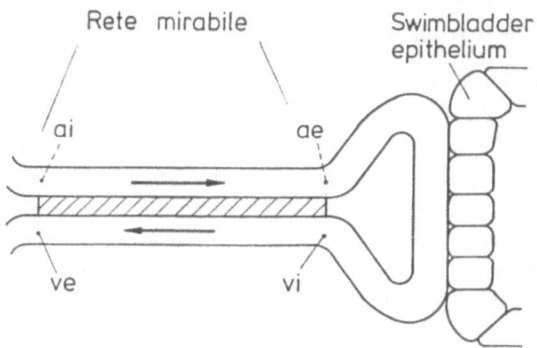

Fig. 2. Schema of the rete mirabile, with its counter-current capillaries and the swimbladder epithelium. The symbols ai, ae, vi, ve denote arterial influx and efflux, and venous influx and efflux of blood to the rete.

Composition of swimbladder gas

The total pressure of the swimbladder gas at any depth is in equilibrium with the hydrostatic pressure. Thus at 2000 m this pressure would be about 200 atm. Analysis of swimbladder gas from fish at various depths has shown that O_2, CO_2, N_2, and other inert gases occur in varying fractions: In shallow water, the fractional composition resembles that of air, whereas in deeper regions, O_2 becomes the major component. There is, furthermore, a time variation in fraction: CO_2 is enriched in freshly deposited gas, but is later replaced by O_2 and by inert gases (Steen, 1970).

What is important, though, is that at any depth the partial pressure of any gas component exceeds that in the surrounding water. Indeed the partial pressures of dissolved gases hardly increase with depth in the oceans. At a depth of 2000 m, P_{O_2} in the swimbladder may thus be close to 180 atm and P_{N_2} around 20 atm, while in arterial blood, P_{O_2} is below 0.1 atm and P_{N_2}, 0.8 atm.

What prevents gases from leaving the swimbladder?

Irrespective of how the gases have entered the swimbladder in the first place, there is a tremendous pressure head for their leaving, into the blood or the surrounding tissues. What keeps gases inside the swimbladder? We mention three factors of importance.

(1) The entrance to the pneumatic duct may be closed off by muscles in the wall.

(2) The swimbladder wall is covered by a silvery layer that is impregnated with crystals of guanine and some hypoxanthine, constituting an impermeable coat. Removing this silvery layer increases the gas permeability of the swimbladder wall some 100 times (Denton et al., 1972).

(3) The rete counter-current system acts as a barrier in that gas is shunted from the rete venous capillaries to the rete arterial capillaries. This is analogous to the diffusion shunt that has been assumed for muscle tissue in mammals (Piiper et al., 1984; Piiper, 1987; Kobayashi et al., 1989a). The rete, thus, does not only play a key role in gas deposition (see below) but also in its maintenance. It should, however, be noted that complete cessation of blood flow to the swimbladder epithelium would constitute an even more effective way of preventing gas loss.

How can gas be released from the swimbladder?

When fish decide to float at lesser depth, the gas volume of the swimbladder must be reduced. In physostome fish, gas can exit to the gut, and bubbles can be observed to leave from the fish's mouth. In physoclist, where there is no such connection, gas is resorbed in structures like the pneumatic duct of the eel, or in resorptive parts of the swimbladder wall normally separated from the main swimbladder by muscular activity. These resorptive epithelia are well vascularized, and the resorptive process apparently employs the same mechanisms as, for example, in the alveolar lung (Denton, 1961).

Mechanisms for deposition of gas

The most interesting and least completely understood mechanism is the deposition of gas into the swimbladder. This was earlier thought to occur as an active gas secretion. Although it is now generally accepted that passive diffusion governs gas deposition, the term gas secretion has been retained (Fänge, 1983). For gases to be deposited by passive diffusion from the blood in the capillaries of the swimbladder into its gas phase, high partial pressures must be created in this blood, at least as high as in the gas phase. Two independent mechanisms are thought to be responsible for the formation of high gas partial pressures in the swimbladder blood: (1) Reduction of gas solubility in blood by the metabolic action of the swimbladder epithelium; (2) Counter-current concentration of gases in the rete mirabile.

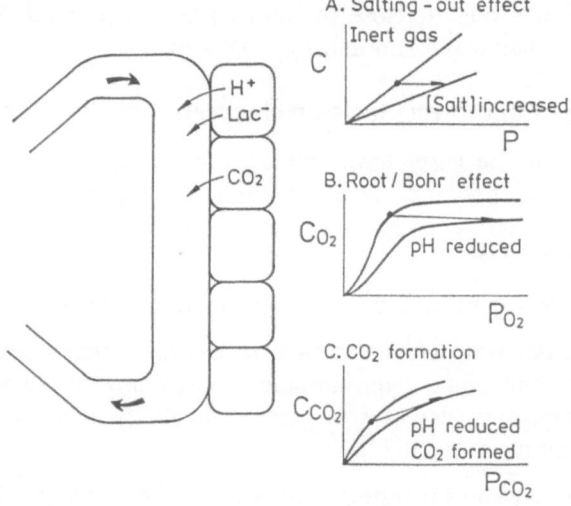

Fig. 3. Synopsis of the mechanisms that lead to increased partial pressure in blood flowing through the capillaries of the swimbladder epithelium. Release of H^+ and lactate leads to increased inert gas partial pressure, due to the salting-out-effect (A); to increased P_{O_2}, due to Bohr and Root effects (B); and to increased P_{CO_2} due to conversion from HCO_3^- (C). CO_2 release increases P_{CO_2} further (C). This schema is based on the assumption that gas deposition into the swimbladder gas phase is small compared with the gas supply by the inflowing blood, so that the content is not notably reduced on the passage through the epithelium.

Reduction of gas solubility

The swimbladder epithelium forms lactic acid, even in the presence of high O_2 concentrations. As a result, the pH of the capillary blood is lowered, and its salt (lactate) concentration is increased. This acts on the solubility of inert gases, O_2 and CO_2.

Inert gases are known to display a salting-out effect, *i.e.* a reduction of solubility with increased salt concentration. This reduction may be very small indeed under the physiological conditions encountered in the swimbladder (Pelster *et al.*, 1988), probably only a few percent; this will, nonetheless, be enough for an efficient concentrating action in the rete (see below).

Reduction in pH leads to *reduced O_2 affinity* via the Bohr effect, and thus to increased P_{O_2} (provided O_2 content is not lowered too much by O_2 diffusion into the swimbladder). In some fish, there is, in addition to the Bohr effect, the Root effect (Root, 1931; Bridges *et al.*, 1983), *i.e.*, a reduction in O_2-*capacity* when pH is reduced. This leads to a particularly large increase in P_{O_2} with acidification, when the O_2 saturation is high. Most fish equipped with a swimbladder show a Root effect in their blood (Farmer *et al.*, 1979).

Acidification converts HCO_3^- *into* CO_2, and the P_{CO_2} is thus also increased. We have recently shown that a further substantial increase in P_{CO_2} of swimbladder blood is achieved by *anoxidative CO_2 formation* from glucose in the pentose phosphate cycle, the levels of the key enzymes of this pathway being fairly high in the swimbladder epithelium (Pelster *et al.*, 1989).

Figure 3 summarizes the mechanisms for increasing partial pressures in the swimbladder blood.

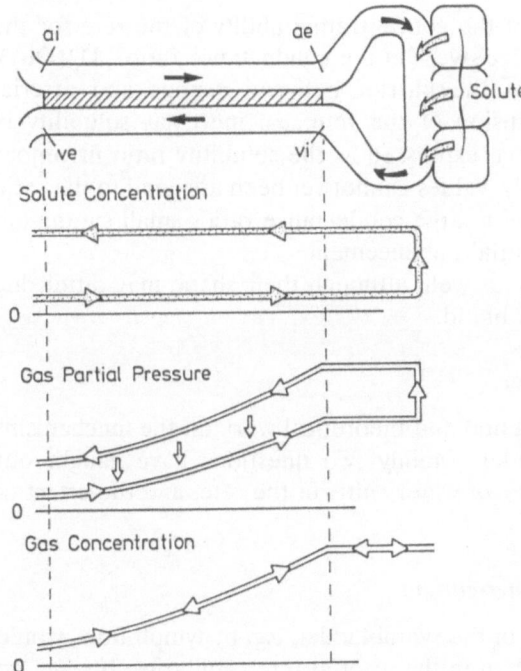

Counter-current enhancement in the rete

When blood with increased gas partial pressure enters the rete, there will be a partial pressure head from the venous to the arterial capillaries, and gas will diffuse back. Thereby, the partial pressure will be reduced in the rete venous blood, on its way leaving the swimbladder, but will be enhanced in the rete arterial blood, on its way towards the swimbladder (fig. 4).

Fig. 4. Schema of the counter-curent concentration of gas in the rete capillaries, represented for an inert gas (Salting-out effect, cf. fig. 3). Note that the salt is assumed not to diffuse back in the rete. For details see text. Symbols as in fig. 2.

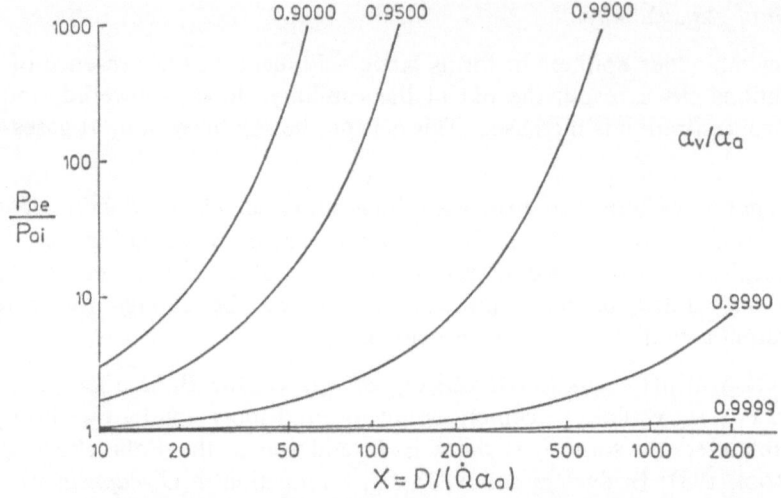

Fig.5. Concentrating efficiency of the rete counter-current flow system is calculated as the ratio of (inert gas) partial pressures in arterial efflux and influx, Pae/Pai (ordinate; symbols as in fig.2). X is the dimensionless ratio of diffusing capacity (D) and perfusive conductance ($\dot{Q}\alpha_a$; $\dot{Q}$, rete blood flow; α_a, inert gas solubility in arterial capillary blood in the rete). Various curves for the solubility ratio of venous to arterial rete blood (α_v/α_a), which is below unity by virtue of the salting-out effect (fig. 3).

It was mainly Kuhn and his colleagues who have worked out the details of how this hairpin counter-current concentration operates in the swimbladder as in the kidney (cf. Fänge, 1983).

Figure 5 shows a theoretical plot of the concentrating ability of the rete for the case of an inert gas. Two factors are decisive: (1) the conductance ratio, $D/(\dot{Q}\alpha_a)$, where D is the diffusing capacity of the barrier between venous and arterial capillaries in the rete; $\dot{Q}$, blood perfusion of the rete; α_a, inert gas solubility in arterial blood. (2) The salting-out effect, expressed as the solubility ratio in venous and arterial blood, α_v/α_a. Unfortunately, values cannot yet been assigned to the ratio $D/(\dot{Q}\alpha_a)$; it is evident, however, that with large conductance ratios small salting-out effects, 5% or less, can lead to substantial enhancement.

Similar plots apply to O_2 and CO_2 as well, although their shape may differ due to the alinear binding charateristics in blood.

Further addition to the classical model

We have recently resumed experimental and theoretical work on the mechanisms of gas secretion in the eel swimbladder. Mainly two questions have caught our interest, the occurrence and significance of water shifts in the rete, and the effect of salt back-diffusion in the rete.

Flow must be balanced in the rete counter-current

Loss of fluid from the vascular bed in the swimbladder, *e.g.* by lymph flow, would yield a lower flow rate in the venous than in the arterial rete capillaries. It was first Niesel and Röskenbleck (1963) who showed that such a flow imbalance in a hairpin counter-current system would largely compromise its concentrating efficiency, and this holds true for the kidney medulla as for the rete.

We have recently measured the water shift in the eel from the capillaries to the swimbladder epithelium, and have found this indeed to be very small, at most 5% of the total flow (Kobayashi et al., 1989b). This indicates virtually balanced rete flows.

We have theoretically investigated the problem of flow mismatch, and our results show that for all practical possibilities flow mismatch largely destroys the rete concentrating ability. There is, however, one exception, and that is O_2. If plasma water is drained in the swimbladder wall, its O_2 concentration will be much lower than that of whole blood. Under this condition, the concentrating efficiency can be strongly enhanced by the reduction of venous compared with arterial rete flow. This mechanism would not work for other gases, since the draining plasma water would contain gases at about the same concentration as that of whole blood (Kobayashi et al., 1989b). This analysis shows that, except perhaps for O_2, flow match may be of vital importance for the rete function.

Salt back-diffusion is advantageous

Most studies into the rete function have concentrated on gas back-diffusion while solute back-diffusion was assumed to exert negligible effects (Kuhn et al., 1963) or be detrimental (Steen, 1970). Since there is undebatable evidence for lactate back-diffusion from venous to arterial capillaries in the rete (Steen, 1963; Kobayashi et al., 1989b), we have theoretically re-investigated its implications (Kobayashi et al., 1989c). The analysis shows that salt back-diffusion significantly enhances the concentrating efficiency for inert gases in the rete. This can easily be understood, since the salt concentration will there exert an enhancement on gas partial pressure by the salting-out effect.

Kobayashi et al. (1989a) have applied this model to other tissues, e.g. mammalian muscle, where a counter-current arrangement between pre- and post-capillary vessels would largely enhance tissue P_{O_2} levels. In fact, back-diffusion of CO_2 would in this case increase the P_{O_2} via the Bohr effect which would then yield P_{O_2} enhancement in a hairpin counter-current system. This would be of significant advantage for tissue O_2 supply in hypoxia. In hyperoxia, on the other hand, this counter-current arrangement would act as a barrier, much like that discussed above, to prevent O_2 toxicity in the tissue.

Summary

The fish swimbladder acts as a device to adjust for neutral buoyancy at various depths. High gas pressures, corresponding to the ambient hydrostatic pressure, are encountered, most of which is made up by O_2 and N_2. To prevent gas loss, the swimbladder wall is made impermeable by guanine crystals in its wall. Gas deposition is made possible by lactic acid production in the swimbladder epithelium, which increases blood gas partial pressures of inert gases (salting-out effect), O_2 (Bohr and Root effects) and CO_2 (conversion from HCO_3^-). The hairpin counter-current blood flow in the rete mirabile enhances this partial pressure increase to the tremendous values, up to several 100 atm, encountered in deep sea dwellers. Flow balance in the rete capillaries is found to be crucial, and salt back-diffusion to be advantageous, for the concentrating efficiency in the rete mirabile.

References

Bridges, C.R., Hlastala, M.P., Riepl, G., and Scheid, P., 1983, Root effect induced by CO_2 and by fixed acid in the blood of the eel *Anguilla anguilla, Respir. Physiol.*, 51: 275-286.

Denton, E.J., 1961, The buoyancy of fish and cephalopods, *Progr. Biophys. Biophys. Chem.*, 11: 177-234.

Denton, E.J., Liddicoat, J.D., and Taylor, D.W., 1972, The permeability to gases of the swimbladder of the conger eel (*Conger conger*), *J. Mar. Biol. Ass. U.K.*, 52: 727-746.

Fänge, R., 1983, Gas exchange in fish swimbladder, *Rev. Physiol. Biochem. Pharmacol.*, 97: 111-158.

Farmer, M., Fyhn, H.J., Fyhn, U.E.H., and Noble, R.W., 1979, Occurrence of Root effect hemoglobins in Amazonian fishes, *Comp. Biochem. Physiol.*, 62A, 115-124.

Kobayashi, H., Pelster, B., Piiper, J., and Scheid, P., 1989a, Significance of the Bohr effect for tissue oxygenation in a model with counter-current blood flow, *Respir. Physiol.* 76: 277-288.

Kobayashi, H., Pelster, B., and Scheid, P., 1989b, Water and lactate movement in the swimbladder of the eel, *Anguilla anguilla, Respir. Physiol.* (in press).

Kobayashi, H., Pelster, B., and Scheid, P., 1989c, Solute back-diffusion raises the gas concentrating efficiency in counter-current flow, *Respir. Physiol.* (in press).

Kuhn, W., Ramel, A., Kuhn, H.J., and Marti, E., 1963, The filling mechanism of the swimbladder. Generation of high gas pressures through hairpin counter-current multiplication, *Experientia*, 19: 497-552.

Niesel, W., and Röskenbleck, H., 1963, Die Bedeutung der Stromgeschwindigkeiten in den Gefäßsystemen der Niere und der Schwimmblase für die Aufrechterhaltung von Konzentrationsgradienten, *Pflügers Arch.*, 277, 302-315.

Pelster, B., Kobayashi, H., and Scheid, P., 1988, Solubility of nitrogen and argon in eel whole blood and its relationship to pH, *J. exp. Biol.*, 135: 243-252.

Pelster, B., Kobayashi, H., and Scheid, P., 1989, Metabolism of the perfused swimbladder of European eel: O_2, CO_2, glucose and lactate balance, *J. exp. Biol.* (in press).

Piiper, J., Meyer, M., and Scheid, P., 1984, Dual role of diffusion in tissue gas exchange: blood-tissue equilibration and diffusion shunt, *Respir. Physiol.*, 56: 131-144.

Piiper, J., 1987, Role of diffusion shunt in transfer of inert gases and O_2 in muscle, in: "Oxygen Transport to Tissue", Vol. X, M. Mochizuki, C.R. Honig, T. Koyama, T.K. Golstick and D.F. Bruley, eds., Plenum Press, New York and London, pp. 55-61.

Root, R.W., 1931, The respiratory function of the blood of marine fishes, *Biol. Bull.*, 61: 427-546.

Steen, J.B., 1963, The physiology of the swimbladder in the eel *Anguilla vulgaris*, III. The mechanism of gas secretion, *Acta Physiol. Scand.*, 59: 221-241.

Steen, J.B., 1970, The swimbladder as a hydrostatic organ, in: "Fish Physiology", Vol. IV, W.S. Hoar, and D.J. Randall, eds., Academic Press, New York, pp. 413-443.

FETUS AND NEONATE

ALTERATIONS IN INTRAUTERINE OXYGEN TENSION DURING THE ESTROUS CYCLE
IN THE RAT AND HAMSTER AND ITS REGULATION BY OVARIAN STEROID
HORMONES: A COMPARATIVE STUDY

David L. Kaufman and Jerald A. Mitchell

Department of Anatomy/Cell Biology
Wayne State University School of Medicine
Detroit, Michigan

INTRODUCTION

Previous studies have established that oxygen tension within the
lumen of the uterus of the rat (Mitchell and Yochim, 1968a) and guinea
pig (Garris and Mitchell, 1979) undergoes marked variation during the
course of the normal estrous cycle. Such alterations are thought to
result from the complex interplay of a number of physiologic and anatomic
factors, including uterine blood flow, endometrial metabolism, and uterine
structure. All of these determinants of intrauterine pO_2 are directly
influenced by fluctuations in endogenous ovarian steroid hormone levels.
In order to gain further insight into the factors which regulate intrauterine oxygen tension, comparative studies have been undertaken in various species. The present report concerns observations made in the rat
and hamster, specifically, the patterns of alteration in intrauterine
pO_2 observed during the normal estrous cycle and the effects of ovarian
steroid deprivation and stimulation on oxygen tension within the lumen
of the uterus.

MATERIALS AND METHODS

Animals: Adult virgin rats (Rattus norvegicus), 250-300 g, of the Holtzman strain (Spraque-Dawley derived) were housed 1/cage in a 14 hour light/
10 hour dark photoperiod. Vaginal smears were recorded daily for at
least 2 complete estrous cycles prior to use. Adult virgin female golden
hamsters (Mesocricetus auratus), 70-200 g, were housed 2-5/cage in a
12 h light/12 h dark photoperiod (lights on 0700). Animals were examined
daily for vaginal discharge, and only regularly cycling animals were
used. Temperature was maintained at 20-24°C; Purina Lab Chow and water
were available ad libitum to all animals.

Measurement of Intrauterine Oxygen Tension: Intrauterine oxygen tension
(pO_2) was measured in vivo in the hamster with a polarographic needle
electrode (Model 760) and chemical microsensor (Model 1201) manufactured
by Diamond General Corp., Ann Arbor, MI. The microelectrode consisted
of a glass insulated platinum sensor mounted in a 22 gauge stainless
steel tube. A silver/silver chloride reference electrode was used. The
sensor was calibrated in physiological saline maintained at animal body
temperature (38°C) and gassed with 0% O_2:100% N_2, 5% O_2: 95% N_2 or 21%

O_2 (air). Corrections were made for prevailing atmospheric pressure. All measurements were conducted at 0.8 volts within a calibration range of 10^{-7} to 10^{-11} amperes. The measurement of intrauterine oxygen tension in the rat has been previously reported (Mitchell and Yochim, 1968 a,b)

Animals were anesthetized with Nembutal (rats 50 mg/kg; hamsters 90 mg/kg BW) and the tubal end of the uterine cornu was exposed by a flank incision and secured by passing a loop of suture around the utero-tubal junction. The calibrated sensor was gently inserted into the uterine lumen through a small incision at the uterotubal junction and held stationary by a clamp assembly. Rectal temperature was monitored and normal body temperature maintained with the aid of an incandescent lamp. The entire operative area was protected by surgical gauze moistened with warm physiological saline. During the experiment, gentle traction was applied to the uterus by the securing ligature, thus ensuring extension of the horn throughout its length. Intrauterine pO_2 was recorded for 15 minutes following a 5 minute equilibration period. The electrode calibration was then checked. If deviations in calibration exceeded +/-5 mmHg, the recording was rejected. Records were analyzed for minute-to-minute changes in pO_2 as previously described (Mitchell and Yochim, 1968a). From such recordings the maxima and minima were determined and the mean pO_2 for the 15 min recording interval was computed (mean maxima + mean mimina/2). Intrauterine oxygen tension was measured during proestrus, estrus, metestrus, and diestrus of the estrous cycle and during days 1, 3 and 5 following ovariectomy with and without ovarian steroid hormone replacement therapy. All recordings were made between 1030 h and 1500 hours.

Ovarian Steroid Hormone Treatments: Randomly selected animals were ovariectomized bilaterally during estrus, and hormone treatment was initiated. Each animal received daily s.c. injections of either vehicle (0.1 ml sesame oil), 1 µg estrone (E), or 2.0 mg progesterone (P) or E + P (100 µg + 2.0 mg).

Measurement of arterial and venous pO_2: Oxygen tension was measured in vivo in the aorta and vena cava with the oxygen sensor in rats (Mitchell and Yochim, 1968a). Arterial and venous pO_2 were measured in heparinized samples (100 µl) of whole blood taken from the left ventricle and jugular vein respectively of anesthetized hamsters by mean of a Radiometric ABL-30 semi-automatic blood gas analyzer.

RESULTS

Changes in intrauterine oxygen tension during the estrous cycle in the rat and hamster: Intrauterine pO_2 was observed to fluctuate from moment-to-moment in both species; peak and trough values were averaged to determine mean oxygen levels. Mean pO_2 varied during the normal estrous cycle in both rats and hamsters. In the rat (Table 1), mean pO_2 increased from a postovulatory (estrus) level of 25.7 ± 4.2 mmHg to 32.5 ± 4.2 during metestrus to reach a peak of 47.6 ± 3.3 mmHg (p< 0.01 vs estrus) during diestrus-1 before declining to 33.8 ± 6.4 mmHg during proestrus. In the hamster (Table 2), mean pO_2 increased from a post-ovulatory level of 19.0 ± 1.3 mmHg, (estrus) to 33.3 ± 4.6 during metestrus, peaked between metestrus and diestrus (46.5 mmHg) and declined to 5.5 ± 1.1 mmHg (p< 0.001 vs estrus) during proestrus.

Arterial and Venous Oxygen Tension in the Rat and Hamster: In the rat, mean arterial (aortic) and venous (caval) pO_2 were 78.5 ± 4.5 (5 animals) and 29.3 ± 0.6 (3) mmHg, respectively. In the hamster, mean pO_2 was 89.9 ± 4.8 (3 animals) in arterial blood and 38.9 ± 1.76 (3) mmHg in venous blood.

746

Table 1. Changes in Intrauterine Oxygen Tension During the Estrous Cycle in the Rat

Stage of Cycle	Number	Mean pO_2 (mmHg)
Estrus	8	25.7 ± 4.2
Metestrus	9	32.5 ± 4.2
Diestrus -1	6	47.6 ± 3.3*
Diestrus -2	7	36.4 ± 3.4
Proestrus	8	33.8 ± 6.4

Values: mean $\pm$ s.e.m.
* Value significantly different from that of estrus; $p < 0.01$.

Table 2. Changes in Intrauterine Oxygen Tension During the Estrous Cycle in the Hamster

Stage of Cycle	Number	Mean pO_2 (mmHg)
Estrus	5	19.0 ± 1.3
Metestrus	5	33.3 ± 4.6*
Metestrus + 12 hrs.	2	46.5
Diestrus	3	24.8 ± 1.5*
Proestrus	5	5.5 ± 1.0*

Values: mean $\pm$ s.e.m.
* value significantly different from that of estrus; $p < 0.001$.

Effects of Ovariectomy and Ovarian Steroid Hormone Treatment on Intrauterine Oxygen Tension: The effects of ovariectomy and of ovarian steroid hormone replacement on intraluminal pO_2 were determined. In the rat, (Table 3), ovariectomy increased mean pO_2 from a pre-ovariectomy level of 25.7 ± 4.2 mmHg to 61.4 ± 2.7 mmHg at 120 hrs. The post-ovariectomy rise in pO_2 was reduced by steroid treatment (120 hrs): estrone (29.1 ± 3.8), progesterone (47.9 ± 5.9) and estrone plus progesterone (41.2 ± 4.2 mmHg). In the hamster (Table 4), ovariectomy resulted in a rapid rise in mean pO_2: from a pre-ovariectomy level of 19.0 ± 3.1 mmHg to 46.1 ± 4.5; ($p < 0.001$) at 72 hrs. and remained elevated through 120 hrs. (35.7 ± 3.4 mmHg). Daily subcutaneous injections of ovarian steroids modified the post-ovariectomy increase in pO_2. The 72 hr. peak in pO_2 (46.1 mmHg) was markedly suppressed by estrone (25.6 ± 2.8 mmHg; p 0.001) and inhibited by progesterone (18.8 ± 4.2 mmHg; $p < 0.001$). Simultaneous injection of both hormones also prevented the post-ovariectomy increase in luminal pO_2 (46.1 ± 4.5 vs 22.7 ± 4.7 mmHg). Daily injection of hormone vehicle (sesame oil) did not alter the post-ovariectomy increase in luminal pO_2 in either species.

Table 3. Changes in Intrauterine Oxygen Tension in the Rat following Ovariectomy and Treatment with Ovarian Steroid Hormones

		Ovariectomy		
Day	Vehicle	Estrone	Progesterone	Estrone + Progesterone
0	25.7 ± 4.2 (9)	25.7 ± 4.2 (9)	25.7 ± 4.2 (9)	25.7 ± 4.2 (9)
1	33.3 ± 4.8 (6)	39.6 ± 6.1 (5)	34.5 ± 6.1 (6)	33.9 ± 2.4 (6)
3	48.5 ± 6.2 (7)	30.3 ± 4.1 (5)	42.5 ± 4.6 (6)	33.0 ± 3.7 (7)
5	61.4 ± 2.7 (6)	29.1 ± 3.8*(6)	47.9 ± 5.9 (5)	41.2 ± 4.2*(6)

Values: mean $pO_2 \pm$ s.e.m., ( ) number of animals, Day 0: Estrus.
* Value significantly different from that of vehicle treated group on Day 5; p<0.01.

Table 4. Changes in Intrauterine Oxygen Tension in the Hamster following Ovariectomy and Treatment with Ovarian Steroid Hormones

		Ovariectomy		
Day	Vehicle	Estrone	Progesterone	Estrone + Progesterone
0	19.0 ± 3.1 (6)	19.0 ± 3.1 (6)	19.0 ± 3.1 (6)	19.0 ± 3.1 (6)
1	30.9 ± 3.2 (5)	34.2 ± 6.9 (6)	22.6 ± 2.0 (6)	15.0 ± 4.0 (5)
3	46.1 ± 4.6 (6)	25.6 ± 2.8*(5)	18.8 ± 4.2*(4)	22.7 ± 4.7*(3)
5	35.7 ± 3.4 (9)	28.4 ± 3.2 (5)	35.2 ± 2.3 (7)	19.8 ± 0.8 (3)

a Mean $\pm$ s.e.m., ( ) number of animals, Day 0: Estrus.
* Value significantly different from that of vehicle treated control Day 3; p<0.001.

DISCUSSION

The regulation of oxygen tension within the lumen of the uterus is a complex process. In previous studies, we have established that intrauterine pO_2 varies according to reproductive state (Mitchell and Yochim, 1968b; Yochim and Mitchell, 1968; Garris and Mitchell, 1978, 1979), is dependent on the metabolic status of the endometrium (Mitchell and Yochim, 1968; Simpson and Mitchell, 1981; Yochim and Mitchell, 1968), correlates with uterine blood flow (Hammer, Goldman and Mitchell, 1981; Mitchell and Hammer, 1983; Mitchell, Hammer and Goldman, 1983) and is influenced by ovarian steroid hormones (Mitchell and Yochim, 1968a; Garris and Mitchell, 1978, 1979). The present study indicates that ovarian steroid hormones are also potent regulators of intrauterine oxygen tension in the hamster.

SUMMARY

The results indicate that marked fluctuations in oxygen availability occur during the estrous cycle in both rats and hamsters. Patterns of luminal pO_2 were similar in that levels were intermediate during metestrus and maximal during diestrus. By contrast, minimal pO_2 levels occurred during estrus in the rat vs. during proestrus in the hamster. The species also differed in the range of mean pO_2 occurring during the cycle: approximately 5-50mmHg in the hamster vs. 25-50 mmHg in the rat. Ovariectomy results in marked increases in luminal pO_2 in both species. The increase is reduced by hormone replacement.

REFERENCES

Garris, D.R., and Mitchell, J.A., 1978, Temporal correlations between plasma ovarian steroid hormone levels and intrauterine oxygen tension in the guinea pig in: "Oxygen Transport to Tissue III." eds. I.A. Silver, M. Erecinska and H.I. Bicher, Plenum Press, New York, p. 473.

Garris, D.R. and Mitchell, J.A., 1979, Intrauterine oxygen tension during the estrous cycle of the guinea pig: Its relation to uterine blood volume and plasma estrogen and progesterone levels, Biol. Reprod., 21: 149.

Hammer, R.E., Goldman, H. and Mitchell, J.A., 1981, The effects of nicotine on uterine blood flow and intrauterine oxygen tension in the rat, J. Reprod. Fertil., 63: 163.

Mitchell, J.A. and Hammer, R.E., 1983, Serotonin-induced disruption of implantation in the rat: I. Effects on serum progesterone, implantation site blood flow and intrauterine pO_2, Biol. Reprod., 28: 830.

Mitchell, J.A., Hammer, R.E. and Goldman, H., 1983, Concomitant reduction in uterine blood flow and intrauterine oxygen tension in the rat following nicotine administration, in: "Oxygen Transport to Tissue IV." eds., H. Bicher and D. Bruley, Plenum Press, New York, p. 231.

Mitchell, J.A. and Yochim, J.M., 1968a, Measurement of intrauterine oxygen tension in the rat and its regulation by ovarian steroid hormones. Endoc., 83: 691.

Mitchell, J.A. and Yochim, J.M., 1968b Intrauterine oxygen tension during the estrous cycle in the rat: Its relation to uterine respiration and vascular activity. Endoc., 83: 701.

Simpson, D.A. and Mitchell, J.A., 1983, Temporal correlations between intrauterine oxygen tension and endometrial H4-Lactic Dehydrogenase in the guinea pig, in: "Oxygen Transport to Tissue IV," eds., H. Bicher and D. Bruley, Plenum Press, New York, p. 587.

Yochim, J.M. and Mitchell, J.A. 1968, Intrauterine oxygen tension in the rat during progestation: Its possible relation to carbohydrate metabolism and the regulation of nidation. Endoc., 83: 706.

Participation in the meeting of the International Society on Oxygen Transport to Tissue was made possible by a travel grant to J.A.M. by the Alexander von Humboldt Stiftung; D.L.K. is a Wayne State University Medical Alumni Association Medical Student Summer Fellowship Recipient, 1988.

CHANGES IN BLOOD PCO$_2$ AND ACID-BASE STATUS IN CHICK EMBRYO BETWEEN DAY 4 AND 6 OF INCUBATION

Hans-Jürgen Meuer and Petra Tietke

Zentrum Physiologie
Medizinische Hochschule Hannover
D-3000 Hannover 61, FRG

INTRODUCTION

Warm blooded embryos grow rapidly during the first third of the developmental period. Embryonic mass can multiply five-fold and more within one day. To meet the metabolic demands of the growing organism adequate adaptations of the transport function of the blood are necessary. It has been shown that blood volume, hematocrit and cardiac output increase, and that specific properties of the blood change (e.g. oxygen affinity of the hemoglobin), but the few available data are not sufficient to describe even the basic physiological functions during the early stages of development. On the other hand, there is a high rate of abortions during early human pregnancy which is due to other than genetic abnormalities and only poorly understood.

Our aim was to gain insight in the developmental changes of embryonic respiration shortly after the onset of circulation. Studies of physiological functions in early mammalian embryos in vivo are handicapped by the difficulties in keeping normal conditions during the experiments. This does in general not apply to birds, because the avian embryo develops separately from the mother. Therefore we use the chick embryo for our experiments.

The hatching time of the chick is 21 days. We investigated 4- and 6-days-old embryos which are in terms of developmental stage comparable to human embryos of 5 to 6 weeks. In this age respiratory gas exchange with the environment takes place via the vitelline blood vessels of the yolk sac membrane. The vitelline arteries carry the blood with low oxygen tension to the extraembryonic capillaries, and the oxygenated blood is returned to the embryo by the vitelline veins (figs. 1 & 7).

Recently we measured PO$_2$ and pH of the blood of vitelline and intraembryonic arteries and veins in ovo (Meuer and Baumann, 1988; Meuer et al., 1989) and evaluated the oxygen saturation of the blood using oxygen hemoglobin equilibrium curves. In the present study we determined the PCO_2 in extra-embryonic blood vessels under normocapnic and increasing hypercapnic conditions in order to investigate CO_2 transport and acid base balance of the blood.

METHODS

PCO_2 was registered with ion-selective microelectrodes (fig. 2) according to the method described by Bomsztyk and Calalb (1986). The tips of the two microelectrodes (tip diameter 3 μm) were filled with a liquid ion exchanger (Fluka) which forms a proton selective membrane (Ammann et al., 1981). Therefore, the electrode potential is defined by the ratio of the proton concentrations at both sides of the membrane. The pH of the electrode buffer solution of the PCO_2 electrode (bicarbonate solution) changes with environmental PCO_2, since the membrane is permeable to CO_2, whereas the buffer solution of the reference (pH) electrode (sodium citrate solution) is insensitive PCO_2 changes. The PCO_2 signal is obtained from the difference between both electrode potentials. Amplifier ground is connected to the circuit via an agar-KCl-bridge.

Bomsztyk and Calalb (1986) used this method for PCO_2 measurements above 20 torr. To adapt the setup to lower PCO_2 values which are characteristic for early embryonic blood the composition of the electrode buffer solution for the PCO_2 electrode had to be modified. The best performance was obtained with a solution containing 20 mmol/l $NaHCO_3$, 20 mmol/l NaCl and 0.1 mg/l carbonic anhydrase. The slope of the calibration curve at 38°C was about 60 mv/decade above a PCO_2 of 20 torr and about 55 mv/decade below this value (fig. 3).

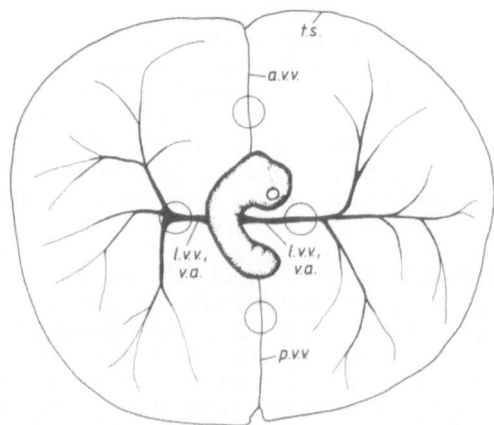

Fig. 1. Schematic of the yolk sac circulation of a 4-day-old chick embryo. The vitelline arteries run mostly parallel to the lateral vitelline veins. The approximate sites of PCO_2 measurements are indicated by circles. Abbreviations: v.a., vitelline arteries; l.v.v., lateral vitelline veins; a.v.v., anterior vitelline vein; p.v.v., posterior vitelline vein; t.s., terminal sinus.

752

Measurements were performed in ovo after removing part of the eggshell covering the embryo and the area vasculosa of the yolk sac. The egg was placed in a thermostated (38°C) and humidity (80%) controlled chamber. Immediately before and after each experiment the electrode setup was calibrated in the experimental chamber with citrate buffer solution equilibrated with 1% or 3% CO_2 obtained from gas mixing pumps (Wösthoff, Bochum, W.-Germany). For the PCO_2 measurements the two microelectrodes were inserted as close as possible (distance less than 1 mm) into the same blood vessel. The approximate sites of measurement are indicated in fig. 1.

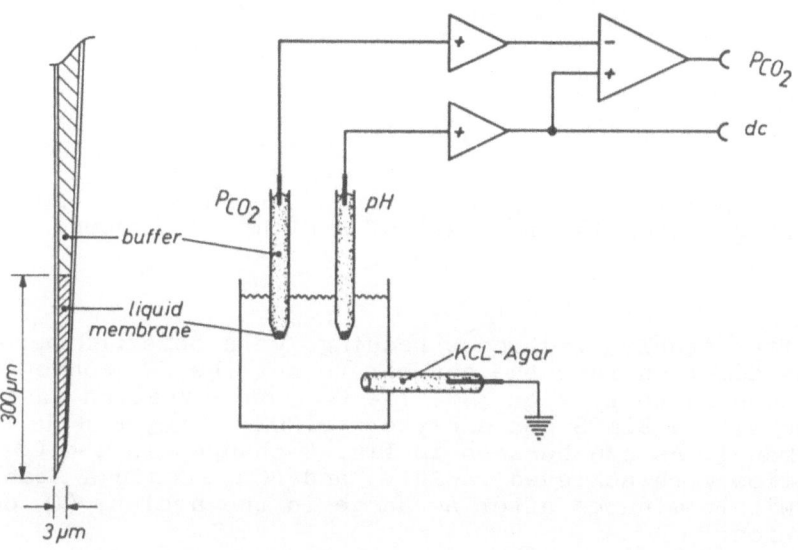

Fig. 2. Schema of the microelectrode tips and electronic circuit for determining PCO_2 in embryonic blood vessels.

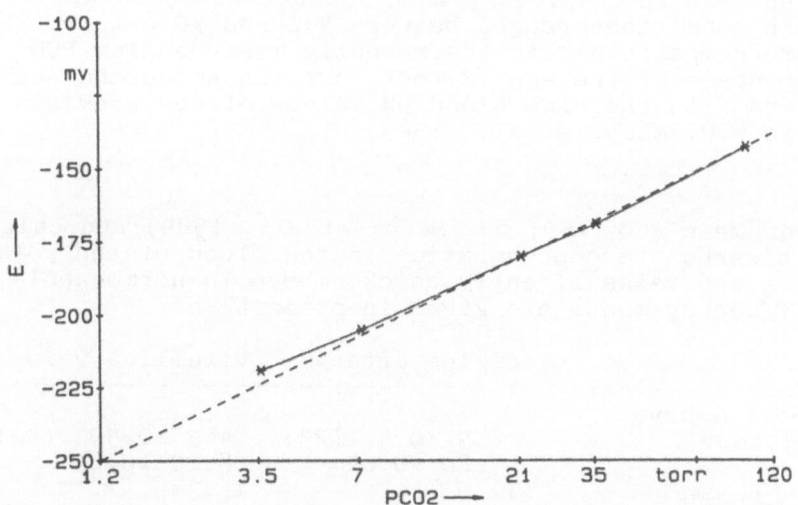

Fig. 3. Typical calibration curve of a PCO_2 microelectrode circuit. Broken line: Ideal curve with a slope of 62 mv/decade (38°C).

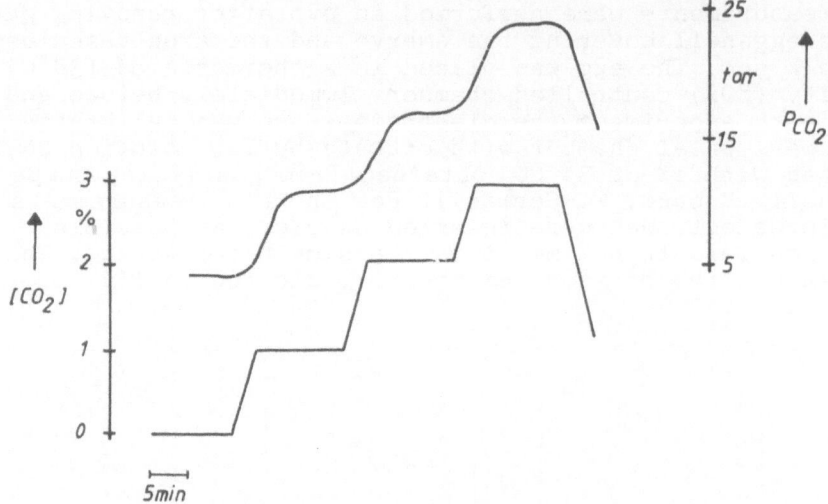

Fig. 4. Typical PCO_2 signals obtained in embryonic blood in ovo during a stepwise increase of ambient CO_2 concentration.

After stable electrodes readings were obtained pure CO_2 gas was admitted into the chamber to set the CO_2 concentration stepwise to 1, 2 or 3 %. The CO_2 concentration was controlled with a BINOS gas analyzer (BINOS 1, Leybold-Heraeus, W.-Germany). As can be seen in fig. 4 changes in the CO_2 concentration were achieved rapidly, and PCO_2 readings stabilized within minutes after a change in the ambient CO_2 concentration.

RESULTS

The mean PCO_2 values (table 1) determined under normocapnic conditions ranged between 4.2 and 10.6 torr. These data are compatible with the recently measured low PCO_2 in the air space of the egg (2 to 3 torr, Lapennas and Reeves, 1983) and with the high blood pH values of our previous study (7.66 to 8.0, Meuer et al., 1989).

Table 1. Mean PCO_2 ±SD, pH (Meuer et al., 1989) and calculated bicarbonate concentration in the blood of the yolk sac arteries and veins of early chick embryo in normocapnia. Numbers of measurements are given in parentheses.

	vitelline artery	vitelline vein
4-day-old embryo		
PCO_2, torr	6.9 ±0.4 (23)	4.2 ±0.49 (40)
pH	7.80 ±0.04	8.00 ±0.05
$[HCO_3^-]$, mmol/l	11.9	12.3
6-day-old embryo		
PCO_2, torr	10.6 ±1.0 (13)	7.0 ±0.9 (13)
pH	7.66 ±0.07	7.89 ±0.08
$[HCO_3^-]$, mmol/l	12.9	15.3

Increasing environmental CO_2 concentration raised blood PCO_2 concomitantly. Using the mean pH values which were determined under the same conditions (Meuer et al., 1989) the in-ovo-relationship between log PCO_2 and pH was obtained (fig. 5). From these data linear regressions were calculated:

4 days: $\log PCO_2 = 15.1 - 1.81 * pH$
6 days: $\log PCO_2 = 13.1 - 1.56 * pH$

From the measured mean PCO_2 and pH values the bicarbonate concentration was calculated using the Henderson-Hasselbalch equation. For the CO_2 solubility in plasma the value for cerebrospinal fluid at 37°C (0.0318 mmol/1/torr, Altman and Dittmer, 1974) was taken, because the protein concentration of the plasma (11 g/1, Romanoff, 1967) is low. For the evaluation of pK' we used the equation given by Siggaard-Andersen (1974) for whole blood at 37°C:

$pK' = 6.125 - \log(1 + 10^{pH-8.7}) - 0.01$

Linear regression gave the following results for the relationship between the bicarbonate concentration and the pH:

4 days: $[HCO_3^-] = 241 - 28.8 * pH$
6 days: $[HCO_3^-] = 182 - 21.4 * pH$

Estimates of the CO_2 dissociation curves of the blood (fig. 6) were obtained from the sum of the bicarbonate concentration and the respective concentration of physically dissolved CO_2. CO_2 bound to hemoglobin was neglected, because the hemoglobin concentration is low (20 g/1 at day 4 and 35 g/1 at day 6; Romanoff, 1967; Baumann et al., 1983).

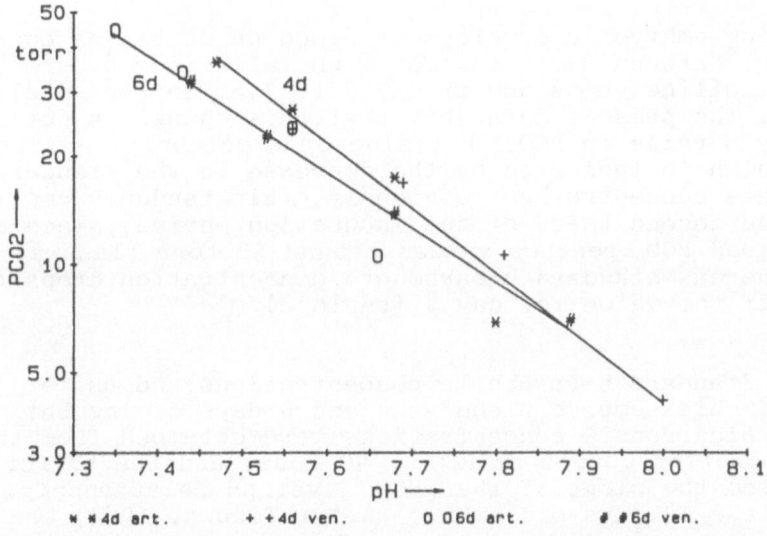

Fig. 5. CO_2 titration curves of the blood of the vitelline arteries and veins of 4- and 6-days-old chick embryos obtained by stepwise increase of the ambient CO_2 concentration. Symbols: Measured mean values, solid lines: calculated by linear regression.

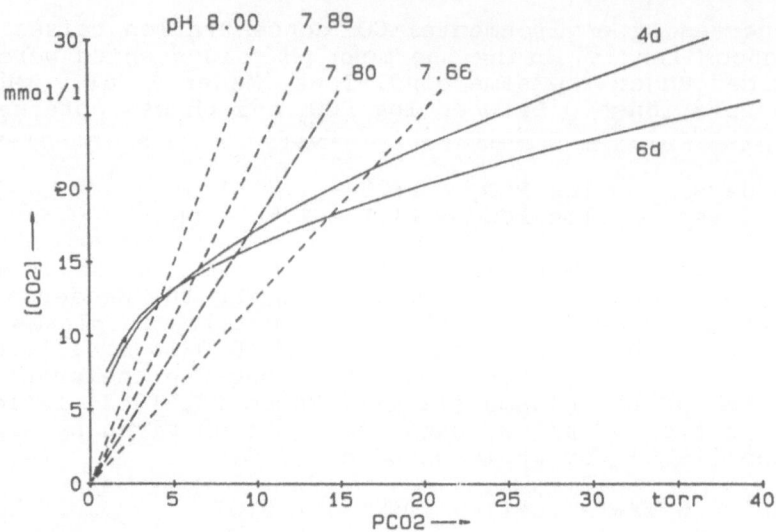

Fig. 6. Calculated CO_2 dissociation curves for chick embryonic blood at 4 and 6 days of incubation.
Broken lines: Iso-pH-curves.

DISCUSSION

The present study investigates for the first time parameters concerning CO_2 transport by the blood and acid-base status during the early developmental stages of the chick embryo. Taken together the present data and the recently determined PO_2 and pH values we can now describe the average blood gas status in extra- and intraembryonic arteries and veins (fig. 7).

During embryonic development blood pH decreases continuously. Between day 4 and day 6 pH falls from 8.0 to 7.89 in the vitelline veins and from 7.8 to 7.66 in the vitelline arteries. The present data show that this change is not only caused by a raise in PCO_2, but also by a metabolic acidification which is indicated by the decrease in the standard bicarbonate concentration. Obviously, this tendency continues during the second third of the incubation period, since at day 14 blood PCO_2 reaches values around 32 torr (Tazawa, 1984), whereas standard bicarbonate concentration drops to about half the value for day 4 (table 2).

Table 2. Standard bicarbonate concentrations and buffer values of chick embryo blood at 4 and 6 days of incubation. Standard bicarbonate concentrations were obtained from the CO_2 dissociation curves at $PCO_2 = 40$ torr, and the buffer values from the slope of the $[HCO_3^-]$ vs. pH relationship. The data for the 14-days-old embryo (after Tazawa, 1984) are given for comparison.

	4 days	6 days	14 days
Standard $[HCO_3^-]$, mmol/l	29.4	24.4	14
Buffer value, mmol/l	28.8	21.4	18

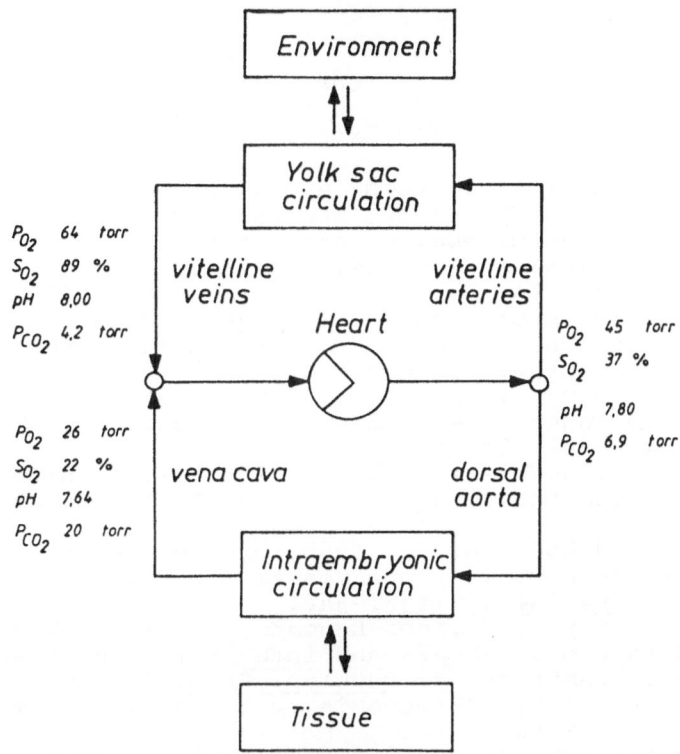

Fig. 7. Circulation block diagram and schematic of respiratory gas transport in the 4-days-old chick embryo. Mean PO_2 and mean pH were determined recently (Meuer and Baumann, 1988; Meuer et al., 1989). Oxygen saturation (SO_2) was obtained from O_2-Hb-equilibrium curves at actual pH. Mean PCO_2 in vitelline vessels was determined in this study. Intraembryonic venous PCO_2 was calculated from the CO_2 titration curve.

SUMMARY

PCO_2 was measured in extraembryonic blood vessels of early chick embryos (first third of incubation period) using PCO_2 microelectrodes with liquid ion exchanger. Between day 4 and day 6 of incubation mean PCO_2 in the vitelline veins increased from 4.2 to 7.0 torr and in the vitelline arteries from 6.9 to 10.6 torr. From the CO_2 titration curves obtained in ovo by measurements in hypercapnic environment buffer values of the blood were evaluated (28.8 mmol/l at day 4 and 21.4 mmol/l at day 6). Calculated standard bicarbonate concentration was 29.4 mmol/l at day 4 and decreased to 24.4 mmol/l by day 6 indicating metabolic acidification. These results are compatible to the data previously determined for older embryos.

REFERENCES

Altman, P.L., Dittmer, D.S., eds., 1971, "Biological handbooks: Respiration and circulation", Federation of American Societies for Experimental Biology, Bethesda, MD.

Ammann, D., Canter, F., Stein, R. A., Schulthess, P., Shijo, Y., and Simon, W., 1981, Neutral carrier based hydrogen-ion selective microelectrode for extra- and intracellular studies, Anal. Chem., 53:2267-2269.

Baumann, R., Padeken, S., Haller, E. A., and Brilmayer, T., 1983, Effects of hypoxia on oxygen affinity, hemoglobin pattern and blood volume of early chicken embryos, American Journal of Physiology, 244:R733-R741.

Bomsztyk, K., and Calalb, M.B., 1986, A new microelectrode for simultaneous measurement of pH and PCO_2, Am. J. Physiol., 251:F933-F937.

Lapennas, G.N., and Reeves, R.B., 1983, Oxygen affinity and equilibrium curve shape in blood of chicken embryos, Respir. Physiol., 52:13-26.

Meuer, H.J. and Baumann, R., 1988, Oxygen pressure in intra- and extraembryonic blood vessels of early chick embryo, Respir. Physiol., 71:331-342.

Meuer, H.J., Sieger, U., and Baumann, R., 1989, Measurement of pH in blood vessels and interstitium of 4 and 6 day old chick embryos, J. Develop. Physiol., in press.

Romanoff, A.L., 1967, "Biochemistry of the Avian Embryo", Wiley, New York.

Siggaard-Andersen, O., 1974, "The acid-base status of the blood", Munksgaard, Copenhagen.

Tazawa, H., 1984, Carbon dioxide transport and acid-base balance in chickens before and after hatching, in: "Respiration and Metabolism of Embryonic Vertebrates," R.S. Seymour, ed., Dr. W. Junk Publishers, Dordrecht, Boston, London.

TWO-DIMENSIONAL MODEL OF TISSUE OXYGEN GRADIENTS IN AVIAN GROWTH CARTILAGE

S.F. Silverton, M. Pacifici, J.C. Haselgrove,
S.H. Colodny, and R.E. Forster

Dept. of Physiology, School of Medicine and Dept.
of Biochemistry, School of Dental Medicine,
University of Pennsylvania, Philadelphia, PA and
ASTRO Space division of GE, East Windsor, N.J.

INTRODUCTION

The avian growth plate cartilage provides an appropriate model for the study of oxygen supply to tissue. Unlike cardiac or skeletal muscle, which have been the classical organs to study oxygen diffusion (Krogh), the cartilage is non-contractile and vessels are confined to sparsely spaced channels which traverse the growth plate longitudinally. Thus, because of the tissue constraints, this organ is incapable of opening new vascular channels during adaptation to oxygen stress. Oxygen stress is a probable concomitant of the growth plate environment. First, because of the low priority of blood flow to the bone (current estimates are that 3% of total body blood flow traverses the entire skeleton) (Tothill). Secondly, cartilage cells are far removed from vascular channels compared to cells in other organs. The average intervascular distance in hypertrophic cartilage (near the mineralization front) is 318 μm, and in resting cartilage (near the articular side of the growth plate) this distance is increased to 404 μm (Silverton 1989b). Measurements of oxygen pressures by glass microelectrode in growth plates are recorded at 20-25 mm Hg (Brighton).

We have been interested in the redox status of cartilage cells, and in particular in the availability of oxygen for cellular metabolism in chondrocytes, which generate the ground substance for the total linear growth of the organism over a period of 4-6 weeks (Shapiro 1983, Silverton 1989a). Our studies have focused on the oxygen consumption characteristics of these cells and on a model of oxygen supply built up from the anatomic constraints of the growth plate.

The oxygen utilization of cells isolated from avian growth plates (chondrocytes) as measured by oxygen quenching of phosphorescence of Palladium-coproporphyrin is unusual. Unlike hepatocytes, neuroblastoma cells or osteoclasts, which demonstrate constant oxygen consumption over the physiological range of oxygen pressures with an abrupt decrease below 5 μM [O_2] (Silverton 1989a), chondrocytes showed a gradual decrease of oxygen utilization starting at oxygen pressures as high as 20 μM. To rule out cell damage as the cause of the anomalous oxygen uptake curves, we have measured oxygen usage in cultured chondrocytes. We were able to harvest populations of cells resembling resting or hypertrophic chondrocytes by varying the origin (upper or lower sternum) and culture time of embryonic chondrocytes. These populations have been previously characterized by morphology, enzyme expression and collagen subtypes and populations can be selected which resemble the resting or hypertrophic cells found in vivo in the growth plate.

The oxygen uptake characteristics of these cells and electron micrographs of growth plate cells in situ were used in concert with anatomic data from our previous studies to calculate oxygen gradients using the Krogh-Erlang equation modified for a variable oxygen consumption. Using the gradients calculated by this method, we have also constructed a model of oxygen distribution in the growth plate which allows 90% of oxygen consumption to occur at locations which comprise only a small percentage of the tissue area. The model utilizes a spread sheet program in reiterative mode to solve a two-dimensional matrix constructed from anatomic data. The simulation incorporates non-linear oxygen consumptions operative at matrix points where "mitochondria" are localized according to a cell and tissue construct. The theoretical basis of the simulation is derived from an application of spread sheet mathematics to solve complex thermal and electrical resistance problems.

METHODS

Preparation of Two Regions of Avian Growth Plate for Electron Microscopy

Growth plates from 6 week old chicks were divided into hypertrophic and resting regions under a dissecting microscope, fixed with diluted Karnovsky's fixative at 350 mosm (0.9% gluteraldehyde and 1% paraformaldehyde) in 0.1 M cacodylate buffer (pH 7.4) overnight. After rinsing, the small blocks were post-fixed with 1% osmium tetroxide for 60 minutes at 4°C and stained with 1% uranyl acetate in 0.5 M maleate buffer (pH 4.6). Specimens were dehydrated with increasing concentrations of ethanol, and embedded in Epon. Thin sections of resting and hypertrophic regions were stained with 25% uranyl acetate in methanol and 0.5% lead citrate in distilled water and examined with a JEOL Model JEM 100-CX II, transmission electron microscope.

Preparation of Cultured Chondrocytes

Upper third and lower third portions of sterna are dissected from 19 day chick embryos, digested with bacterial collagenase/trypsin and plated in Dulbecco's high glucose modified Eagle's medium (Gibco Laboratories) containing 10% fetal calf serum (Hyclone Laboratories), 2 mM L-glutamine, and 50 units penicillin/streptomycin (Pacifici 1988). Cells were maintained in culture for 10 days at which time, as shown by

760

others (Gibson, 1985), upper sternal cells have become hypertrophic as indicated by cell size, and Type X collagen formation. Lower sternal cells, on the other hand, synthesize only Type II collagen and resemble growth plate resting cells.

Oxygen Uptake by Pd-coproporphyrin Phosphorescence Quenching

Oxygen consumption by cultured cartilage cells was quantitated at physiological oxygen pressures using the technique of Vanderkooi (1987). Briefly, cells were suspended in HBSS (Hank's Balanced Salt Solution) containing 20 mM HEPES, 0.2% bovine serum albumin, and 10 μl of a solution containing 1 μM Pd-coproporphyrin in dimethylformamide (final concentration of the porphyrin probe was 4.4 pmoles/ml) in an airtight glass cuvette. The buffer was gassed with nitrogen to reduce the O_2 pressure to the physiological range and the cuvette placed in an adapted fluorometer chamber at 25°C. The phosphorescence quenching constant, τ, of the heavy metal-heme compound was sampled repetitively and the corresponding O_2 concentration calculated. Protein concentrations were measured with the Lowry (1951) method.

Two Dimensional Model of Oxygen Supply

Composition of Physical Matrix. Electron micrographs of resting and hypertrophic chondrocyte profiles were analyzed for cytoplasmic area, number of mitochondria, and nuclear area. The cells were ranked for these features. Three profiles representing the median rank of hypertrophic chondrocytes and 6 profiles of resting chondrocytes were selected as the sample cells. These profiles were used to form an array of cells extending from the theoretical blood vessel boundary to midway between two theoretical vessels.

Composition of Spread Sheet Matrix. The correspondence between the physical matrix and the spread sheet matrix was 1:1, with each 1 μm square from the physical matrix represented by a spread sheet cell. The exception for this correspondence was that an extra line containing the information for calculation of non-linear oxygen consumption was added in the x-direction for each mitochondrion in the physical matrix. Thus, the spread-sheet contained additional lines per cell which were not present in the physical matrix. Each cell in the spread-sheet was then letter coded to represent the character of the spread sheet cell (to avoid confusion, we will term these "cells", pixels), mitochondrion or not, and to represent the character of the four neighboring pixels. Thus, a pixel which was non-mitochondrial and was surrounded by non-mitochondrial, non-vessel border pixels was "A". A pixel at the "vessel"-"tissue" boundary was "B", etc. Formulas representing the flux of oxygen expected for each of these pixel types were substituted into the spread-sheet matrix.

Formula Sets for Oxygen Supply in Growth Plate Cartilage. Oxygen, at a concentration of 20 μM (at 37°C, 1.2 μM is equivalent to 1 mm Hg), was presumed to diffuse from an infinite source at the "vessel" boundary. A mitochondrion was treated as an oxygen sink with oxygen consumption characteristics dependent on the local concentration. The curves for oxygen consumption of hypertrophic and resting chondrocytes were taken from measurements of the cultured chondrocytes described above. Instead of curve fitting, the experimental data were broken up into four linear slopes each of which was operative within a particular range of oxygen concentrations. Additional oxygen consumption of 10% of total consumption was calculated as a shunt and represents non-mitochondrial oxygen metabolism of the cell. The oxygen concentration at each pixel of the matrix was derived from the product of the average of the oxygen concentrations of its four neighbors times a conductivity value. The conductivity value was derived from the value of oxygen diffusibility as used in Krogh cylinder calculations (1.5×10^{-5} μmoles/cm^2/sec).

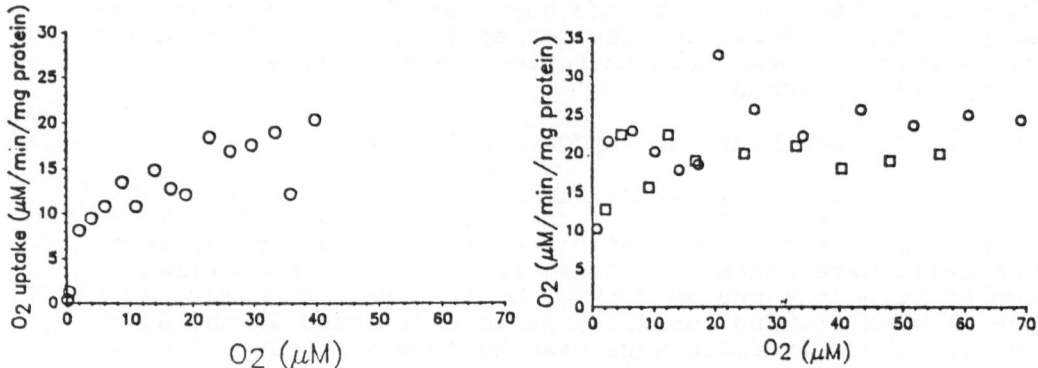

Figure 1. Oxygen Consumption of Cultured Chondrocytes. Oxygen uptake of cultured chondrocytes with characteristics of resting (left) and hypertrophic cells (right) was measured by phosphorescence quenching. x-axis shows ambient [O$_2$]. y-axis is rate of oxygen uptake at 25°C.

RESULTS

The oxygen dependency of oxygen consumption in cultured chondrocytes varied with cell phenotype. Cells which resembled hypertrophic chondrocytes and were derived from the upper sternum of embryonic chicks demonstrated an oxygen uptake similar to most other cells, with a constant consumption throughout most of the physiological range (Figure 1, right). In contrast, cultured cells which were derived from the lower sternum and resembled resting chondrocytes showed an oxygen consumption similar that previously reported from isolated growth plate chondrocytes, e.g. decreasing oxygen consumption commencing at 20 μM [O$_2$] (Figure 1, left).

A simulation of oxygen gradients in resting zone of the cartilage growth plate based on the Krogh-Erlang equation and using a variable oxygen consumption was carried out. Gradients which were originally calculated by Krogh-Erlang Equation to reach zero oxygen concentration half way through the tissue cylinder showed a decrease in oxygen only to 1 μM when a modified oxygen dependency was utilized.

Simulation of oxygen gradients in the hypertrophic and resting zones using the model constructed from electron micrographs of resting and hypertrophic chondrocytes and carried out with a spread-sheet program showed contours of oxygen concentrations dependent on the distance from the "vessel" and the extent of localized oxygen consumption around mitochondria. Oxygen concentrations fell 3-6 μM around mitochondria located near the "vessel". In more distant areas with lower oxygen pressures, the perimitochondrial drop in oxygen concentration was 2-4 μM. In the resting zone, the furthest regions from the "vessel" demonstrated increasing anoxia. Also, mitochondria far removed from the "vessel" were subject to oxygen pressures lower than those seen in the tissue furthest from the vessel.

DISCUSSION

Our primary goal in these studies was to further define the oxygen gradients which may be present in the avian growth cartilage. We have shown that although hypertrophic chondrocytes resemble other cell types and demonstrate constant oxygen consumption throughout the physiological range of oxygen pressures, resting cartilage cells show a pattern of variable oxygen uptake. Thus, refining our original calculation of oxygen gradients through the hypertrophic zone, using the Krogh cylinder model, we estimate that the oxygen pressure drop from the vessel to the edge of the cylinder is 17 mm Hg.

The problem of oxygen supply to the resting zone is further complicated by the variable oxygen uptake in these cells. To explore this problem, we have used two models. The first is an extension of the Krogh-Erlang equation with a variable oxygen consumption. This simulation predicts an oxygen pressure of 1 μM at the boundary of the tissue cylinder if vessel oxygen pressure is 20 μM. This compares favorably to the previous zero levels calculated with the unmodified Krogh-Erlang equation. Oxygen pressures near 1 μM have been presumed to be adequate for mitochondrial function providing no other oxygen gradients exist between the tissue and the mitochondria.

To initiate an investigation of cellular and intercellular oxygen gradients in the avian growth plate, we employed a non-homogeneous model of oxygen usage, based on the paradigm originally utilized for solving electrical and thermal flows. Our first solutions with this model utilized the underlying gradients calculated with a modified Krogh equation and looked for local gradients around the "mitochondria". In the hypertrophic zone, gradients of oxygen around mitochondria were from 3 to 7 μM, depended on the total area of the mitochondrion, and were relatively insensitive to distance from the blood vessel. In contrast, gradients of oxygen around the resting region mitochondria were more pronounced near the vessel and decreased with distance from the oxygen supply. This difference is a consequence of the decreasing oxygen consumption of these cells with decreasing oxygen supply. Of interest, in both regions, was the effect of the localized consumption of oxygen on the perimitochondrial space. Thus, regions surrounding a mitochondrion demonstrated lower oxygen pressures than the edge of the tissue cylinder. The effect of this very local decrease of $[O_2]$ on adjacent cellular elements is unknown.

The diffusibility coefficient is an important determinant of oxygen flow in tissue. Unfortunately, this parameter is only an assumed value in our calculations. Also, ease of oxygen diffusion could be expected to vary through the different elements of the tissue which include: extracellular collagen and proteoglycans, cell membranes, and cell organelles. An improvement in the predictions of this model could be obtained through better measurements of oxygen diffusion. Another troublesome unknown is the average vessel oxygen concentration. Further use of the model would be to determine the minimum oxygen at the vessel needed to feed the most distant mitochondrion at 1 μM $[O_2]$. This could be achieved by allowing the flow equations to reach equilibrium and testing increasing oxygen concentrations at the vessel. Ideally, this information would be correlated with oxygen electrode micropuncture studies of the region.

Finally, the anomalous oxygen uptake characteristics of resting chondrocytes pose the question of biochemical control of mitochondria in anoxic environments. The mechanism of oxygen uptake regulation by these cells requires further investigation.

REFERENCES

Brighton, C.T. and Heppenstall, R.B., 1971, Oxygen tension in zones of the epiphyseal plate, the metaphysis and diaphysis. J. Bone and Joint Surgery, 53-A:719-728.

Gibson, G.J., Flint, M.H., 1985, Type X collagen synthesis by chick sternal cartilage and its relationship to endochondral development. J. Cell Biol. 101:277-284.

Lowry, O.H., Rosebrough, W.J., Farr, A.L., Randall, R.J. 1951, Protein measurement with folin phenol reagent. J. Biol. Chem. 193:265-275.

Pacifici, M. Iozzo, R.V. 1988 Remodeling of the rough endoplasmic reticulum during stimulation of procollagen secretion by ascorbic acid in cultured chondrocytes. J. Biol. Chem. 263:2483-2492.

Shapiro, I.M., Golub, E.E., May, M. and Rabinowitz, J.S., 1983, Studies of nucleotides of growth plate cartilage: Evidence linking changes in cellular metabolism with cartilage calcification. Bioscience Rep. 3:345-351.

Silverton, S.F., Matsumoto, H., DeBolt, K., Reginato, A., and Shapiro, I.M., 1989a, Pentose phosphate shunt metabolism by cells of the chick growth cartilage. Bone 10:45-51,.

Silverton, S.F., Wagerle, L.C., Haselgrove, J.C., Forster, R.E., 1989b, Oxygen supply to the cartilage growth plate: an estimation of oxygen gradients in two regions of the epiphyseal growth plate. in Oxygen Transport to Tissue XI ed. K. Rakusan (in press).

Tothill, P., 1984, Bone blood flow measurement. J. Biomed. Eng. 6:251-256.

Vanderkooi, J.M., Grzegroz, M., Green, T.J. and Wilson, D.F., 1987, An optical method for measurement of dioxygen concentration based upon quenching of phosphorescence. J. Biol. Chem. 262:5476-5482.

EFFECT OF LONG-TERM HYPOXIA ON OXYGEN TRANSPORT PROPERTIES OF

BLOOD IN PREGNANT GUINEA PIGS

C. Geisen[1] , K. Mottaghy[1] , I. Scheffen[2]
and P. Kaufmann[2]

Depts. of Physiology[1] and Anatomy[2] ,
Medical Faculty, Technical University of Aachen
D-5100 Aachen, F.R.G.

INTRODUCTION

Chronic hypoxia is associated with diverse and complex adaption mechanisms, including increases in haemoglobin and 2,3-diphosphoglycerate (2,3-DPG) concentrations and tissue capillary density, as well as changes in cardiac output and other compensatory mechanisms (Kitanaka et al., 1988). During pregnancy adaptive physiological changes include increases in cardiac output, ventilation and oxygen consumption (Gilbert et al., 1979). Adjustment to hypoxia during pregnancy is complicated by the fact, that compensatory mechanisms have to insure a sufficient O_2 supply to the maternal as well as the fetal organism. Placental O_2 transfer depends on a number of factors, including the diffusion capacity of the placenta, the O_2 status, the perfusion rate and morphology of uterine and umbilical vessels, furthermore the O_2 affinity and O_2 capacity of maternal as well as fetal blood (Christensen et al., 1986; Longo et al., 1972; Moll and Kastendieck, 1977)

Examining only the gas exchange properties of blood, one should consider two different aspects: First, biochemical aspects, such as variations of 2,3-DPG concentrations or pH values and second, rheological aspects, e.g. changes in red blood cell (RBC) deformability, haematocrit, plasma protein concentrations as well as plasma viscosity. Model rheological investigations may deliver useful contribution to the complex situation of hypoxia during pregnancy. One should remember, that during usual analyses of blood gas status the RBC's are at rest; the effects of shear flows are not considered. It is a known fact, that the kinetics of oxygen uptake are enhanced bv intraerythrocyte convection (Mottaghy et al., 1982; Zander and Schmid-Schönbein, 1972),

The technique of rheo-oxymetry (Mottaghy et al., 1987) allows a continuous analysis of the oxygen uptake of blood samples under shear flow conditions. By varying the shear rate or contact time of RBC`s with the gas compartment, different rheological situations can be simulated. We used this method together with a rapid O_2 dissociation curve (ODC) measurement technique (Sick and Gersonde, 1980), to examine blood samples of chronically hypoxic pregnant animals.

The present study was undertaken in order to establish the value of these techniques in obtaining information concerning the kinetics of oxygen exchange during hypoxia. This report provides preliminary data for pregnant guinea pigs. In a parallel investigation, the effects of long term hypoxia on the placental morphology of the same animals were additionally studied (a report can be found in this volume by Scheffen et al.).

METHODS

Seven pregnant guinea pigs (Pirbright white) were kept in an isobar hypoxic environment (12% oxygen) from approx. the 15th day of gestation on until near term. The cages were sealed and ventilated with an oxygen-nitrogen gas mixture, whereby the oxygen concentration was monitored by an Oxytest S (Hartmann und Braun, Frankfurt, FRG) and the partial pressure of oxygen with a PO_2 electrode (IL 227/213, Instr.Lab.Inc., Lexington, Mass., USA). The animals were handled and nursed using integrated non-gasdiffusive gloves, thus avoiding an interruption of the hypoxic conditions.

After 5 (n=2), 25 (n=2) and 45 (n=3) days of hypoxia, i.e. after each third of pregnancy, blood samples were taken and analyzed. Seven pregnant animals in a normal atmosphere acted as a control group. Additional blood samples of non-pregnant and non-hypoxic guinea pigs (n=2) were taken for base values. Figure 1 illustrates the time-flow diagram of the study.

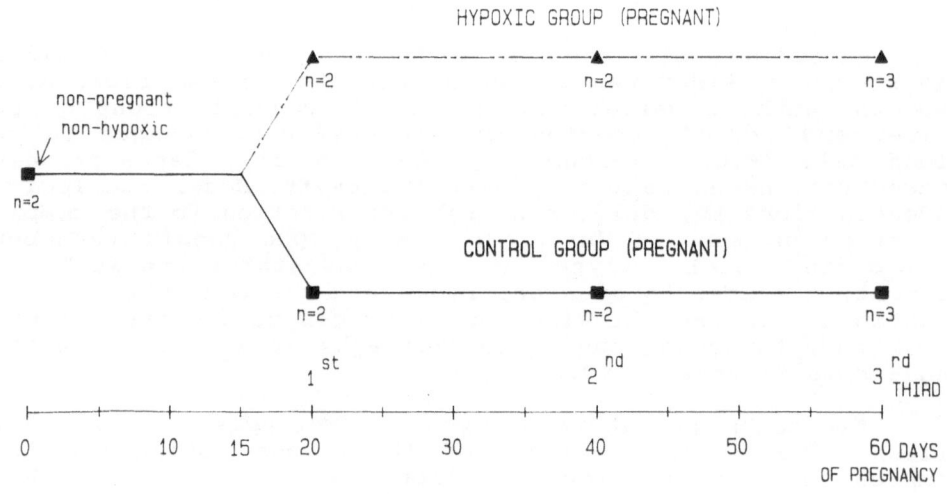

Fig.1. Time-flow diagram of the experiments

Blood samples were immediately obtained from the right ventricle after thoracotomy under general anaesthesia with ketamine-HCl and pentobarbital. Further details concerning the technical procedure of the experiments and the preparation for morphological investigations are reported by Scheffen et al.(1989). Blood gas values and acid base parameters (AVL 947, Bad Homburg, FRG) of central venous blood were measured immediately after aspiration in addition to haemotological parameters, such as haematocrit (HCT), haemoglobin (Hb) (photometrically, PM 2A, Zeiss, FRG) and also 2,3-DPG concentrations (photometrically, UV-Kit, Sigma Diagnostics, St.Louis, USA).

A rheo-oxymeter was used for shear modeling and simultaneous oxygen uptake measurements of RBC`s under defined shear flow conditions. The technical principle of this method has previously been described in detail (Mottaghy et al. 1984; Mottaghy and Hanse 1985), and will only be briefly mentioned here (see Figure 2). The blood is sheared, in a Couette-flow-model, in the gap between two coaxial cylinders. The inner cylinder wall is gas permeable allowing the sheared blood film to be gassed. Shearing is produced by rotating the inner cylinder. In analogy to a fluid droplet (Schmid-Schönbein, and Wells 1969) the RBC`s are sheared inducing intracellular convection facillating the simultaneous oxygenation. The shear rate in the present study was set to 800 s^{-1} by adjusting the rotation frequency of the inner cylinder. Taylor- vortices generated inside the gap enhance the diffusion of oxygen through the plasma.

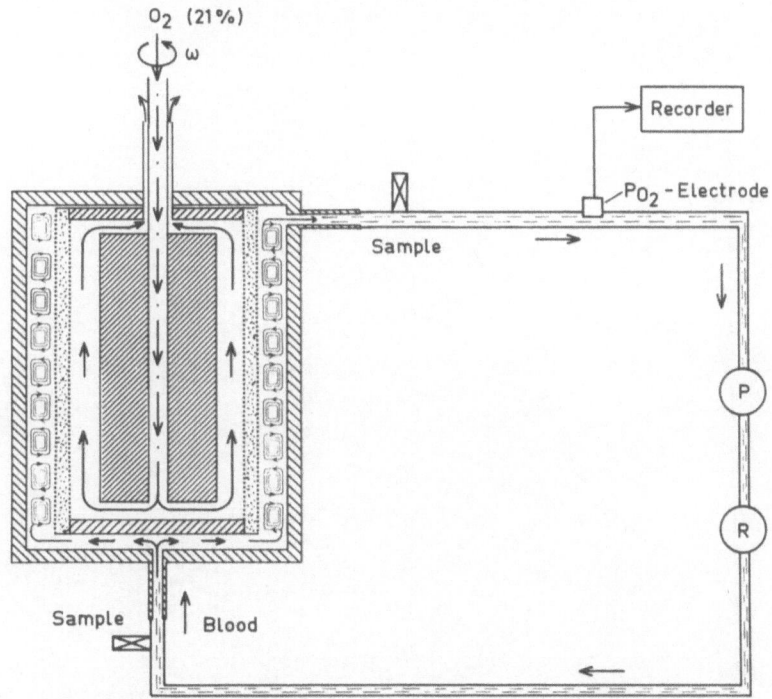

Fig.2. Schematic drawing of the Rheo-oxymeter and the experimental circuit: P = roller pump; R = reservoir; ω = rotation frequency; ◎ indicate Taylor-vortices (see Mottaghy et al. 1987).

769

Blood is led by a roller pump through the apparatus from the base to the top. The pump velocity determines the contact time of the blood with the gas compartment. During the experiments the blood flow rate was held at a constant 5 ml/min. A P_{O_2} electrode (IL 227/213) at the outlet of the rheo-oxymeter continuously measured the P_{O_2}. The blood was recirculated through the device until a full oxygenation was achieved. All measurements were carried out at room temperature (21°C). Figure 3 shows an original registration of the oxygen uptake kinetics of a venous blood sample. The rheo-oxymetry time (RT) is defined here as the time required for achieving a P_{O_2} of 120 mm Hg.

The second method used is the O_2 depletion technique according to Sick and Gersonde (1985) which allows a rapid and continuous measurement of the ODC. The instrument consists of a gas reaction cell and a photometer for the detection of changes in the O_2 saturation. A thin layer of RBC`s (thickness approx. 20 µm) suspended in a buffer solution is smeared on a cuvette, which is placed in the light beam of the photometer. The measurement (at 37°C) starts after a complete saturation of the RBC`s has been achieved by filling the reaction cell with oxygen. The oxygen is then depleted by a nitrogen flow with a constant flow rate through the reaction cell. The actual O_2 partial pressure for each photometrically measured saturation is calculated and graphed automatically by an integrated computer system. For more details see Sick and Gersonde (1980).

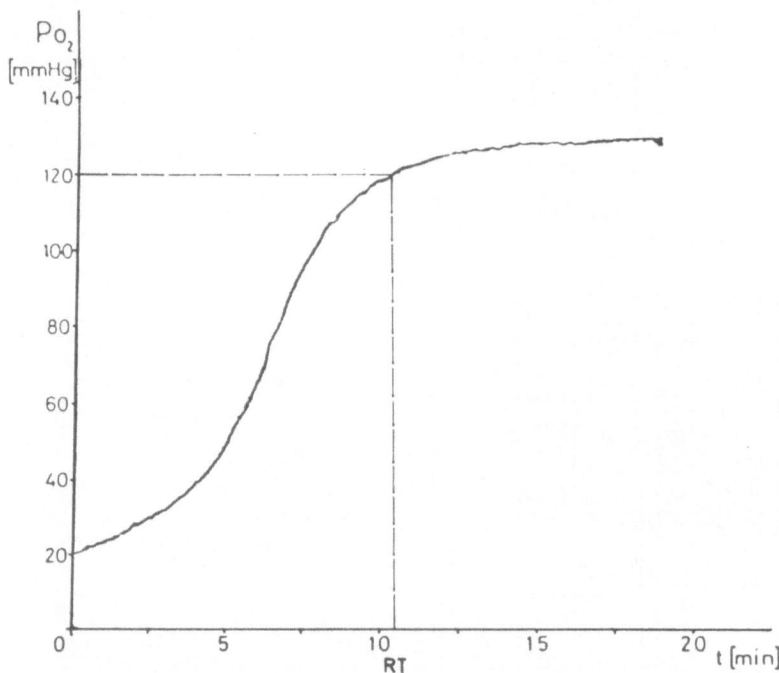

Fig.3. Course of P_{O_2} during Rheo-oxymetry in normal blood as a function of the contact time: RT = Rheo-oxymetry time (HCT = 0.43; P_{50} = 41 mmHg; shear rate = 800 s⁻¹; temperature = 21°C)

770

RESULTS

Changes in central venous P_{O_2} ($P\bar{v}_{O_2}$) during pregnancy are shown in Figure 4. The base $P\bar{v}_{O_2}$ value, as observed in non-pregnant non-hypoxic animals was approx. 19 mmHg. In the control animals, the mean $P\bar{v}_{O_2}$ is 17 mmHg after the first, and 19 mmHg after the second third of pregnancy. During the final third, the mean $P\bar{v}_{O_2}$ decreases to 12 mmHg. In hypoxic animals, however, the mean $P\bar{v}_{O_2}$ decreases to 10 mmHg after 5 days of hypoxia; the mean $P\bar{v}_{O_2}$ is 16 mmHg after the second third of pregnancy and decreases to 13 mmHg during the final third. The difference between $P\bar{v}_{O_2}$ in hypoxic vs. control animals therefore decreased near term.

Changes in haematocrit and haemoglobin concentrations in the two groups are shown in Figures 5 and 6. In non-pregnant non-hypoxic animals the respective mean values are 0.43 and 129 g/l. The mean haematocrit in the control group varies from 0.35 after the first third of pregnancy to 0.41 after the second, and 0.35 after the final third of pregnancy. In contrast, the mean haematocrit values in hypoxic animals are respectively 0.40, 0.42 and 0.49 during pregnancy. Similar changes also occur in mean haemoglobin concentrations. The mean cellular haemoglobin concentration (MCHC) remains stable around 300 g/l in all animals, with one exception, the average MCHC in hypoxic animals is 329 g/l after 40 days of pregnancy.

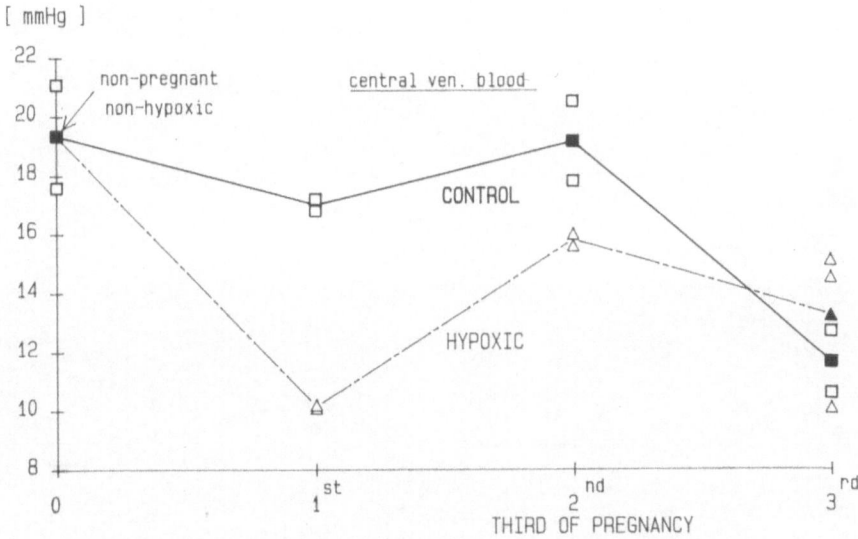

Fig.4. Changes in central venous P_{O_2} ($P\bar{v}_{O_2}$) during different pregnancy periods. Here and in the following figures the dashed lines indicate the hypoxic and the solid lines the control group. Time "0" indicates data of the non-pregnant and non-hypoxic animals. The solid symbols indicate the mean values of the corresponding measurments (open symbols).

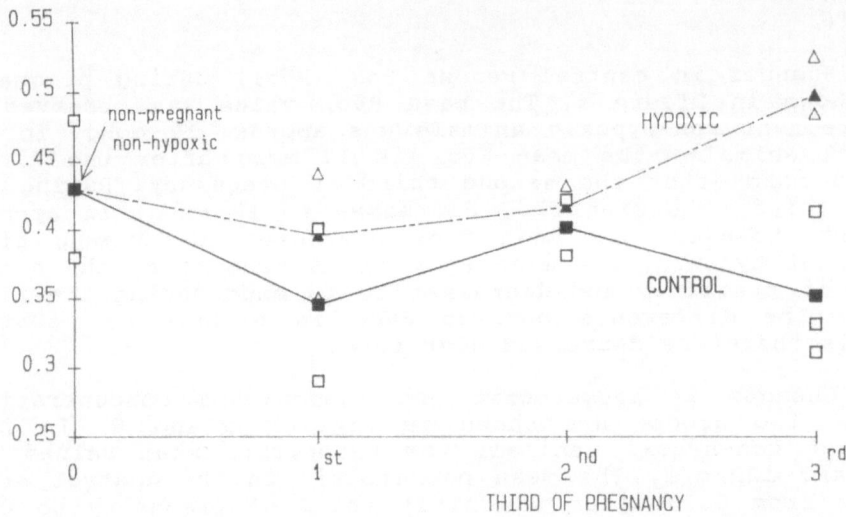

Fig.5. Alterations in haematocrit values during pregnancy in both groups.

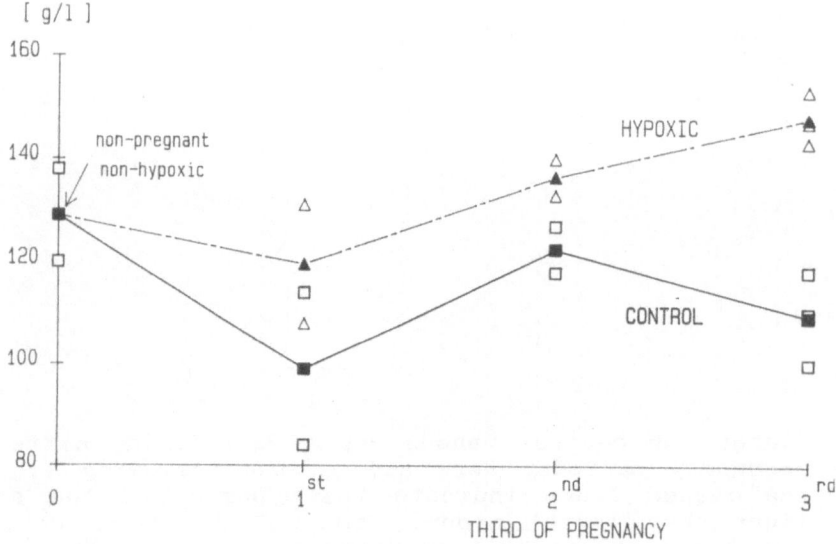

Fig.6. Changes in haemoglobin concentrations in the different stages of pregnancy.

Figure 7 illustrates the changes in P_{50} occurring during pregnancy. This data was derived from the ODC of each blood sample obtained by O_2-depletion technique. The base value of non-pregnant and non-hypoxic guinea pigs is 41 mmHg. During the first third of pregnancy a marked decrease of up to 6.5 mmHg (-16%) is observed in control animals and of up to 7.5 mmHg (-18%) in hypoxic animals. The mean P_{50} stays almost stable in control animals (34.7 mmHg after the second and 35.5 mmHg after the final third of pregnancy). In contrast, the mean P_{50} increases in hypoxic animals to 34.5 mmHg after the second and 38 mmHg after the last third of pregnancy.

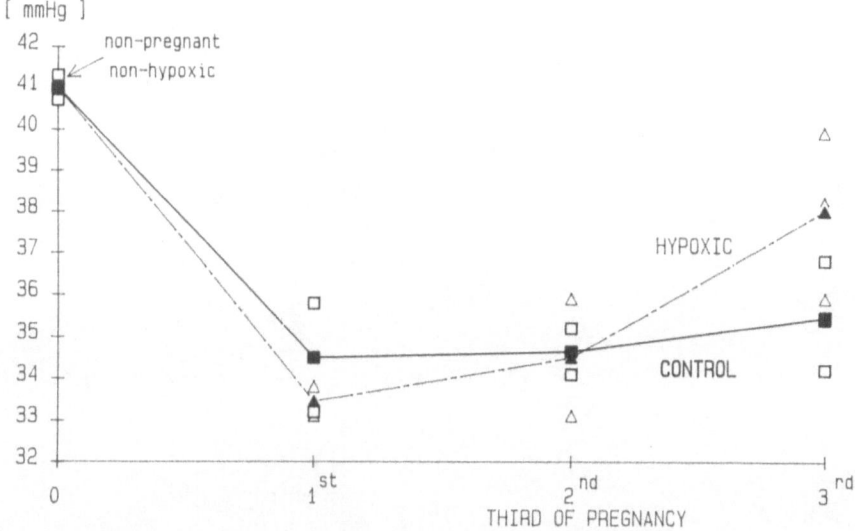

Fig.7. Changes in P_{50} values during pregnancy in the hypoxic and the control group

Alterations in the oxygen uptake of animal blood under defined shear rates measured by rheo-oxymetry are depicted in Figure 8. As already mentioned, the blood samples were immediately brought into the rheo-oxymeter, subjected to a constant shear rate of 800 s^{-1} during which the time for full saturation was measured (Rheo-oxymetry time). This parameter is shown for hypoxic vs. control animals. The base value RT (non-pregnant, non-hypoxic animals) is approx. 10.2 minutes. In control animals the RT drops to 7.1 minutes by the 20th day of pregnancy, and after increasing to 9.1 minutes during the second third of pregnancy, remains stable at 9 minutes in the final third. Hypoxic animals also show a primary decrease of the mean RT during the first third of pregnancy up to 8.5 min. During the final two third's of pregnancy, however, an increase of 72% (up to 14.6 minutes) was registered.

DISCUSSION

No differences were observed in control vs. hypoxic guinea pigs in regard to fetal weight, number of fetuses or abortion rate in the present study. This indicates that hypoxic animals develop various compensatory mechanisms in order to provide a sufficient oxygen supply to fetal tissue.

As shown in the early stages of pregnancy, a decrease in $P\bar{v}_{O_2}$ occurs in hypoxic animals after 5 days of exposure to hypoxia. This is an immediate response to hypoxia in connection with hyperventilation and an increased cardiac output (Kitanaka et. al. 1988). The difference in $P\bar{v}_{O_2}$ between hyp-

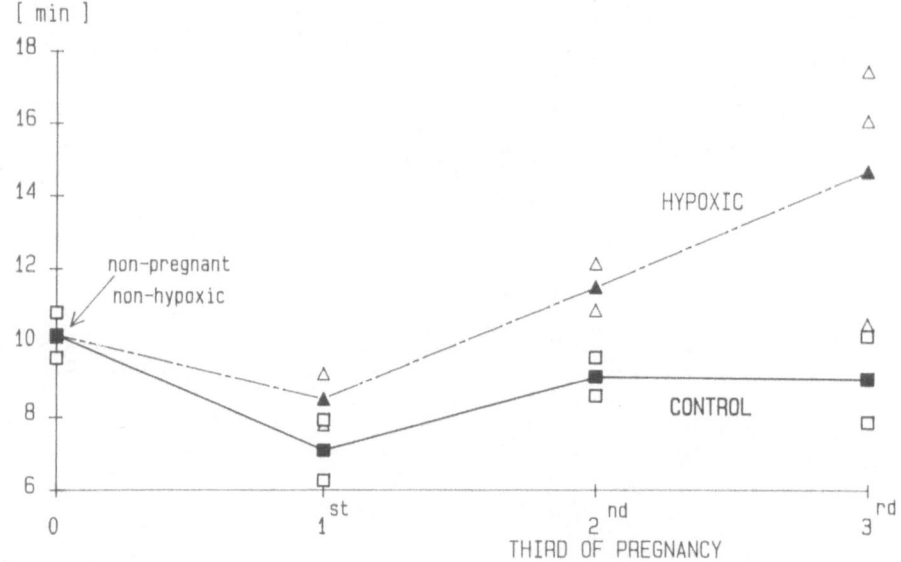

Fig.8. Alterations in rheo-oxymetry time of sheared blood (shear rate = 800 s^{-1}) during the different pregnancy periods in the hypoxic and the control group.

oxic and control animals (Figure 4) decreases progressively during pregnancy, suggesting that other compensatory mechanisms come into effect e.g., an increase of Hb concentration and haematocrit. Even though the oxygen transport capacity is enhanced by 35% due to this increase (hypoxic vs. control animals), haemorheological consequences arise by the final third of pregnancy. This is reflected in the rheo-oxymetry time results (Figure 8). In the first third of pregnancy, the RT is reduced. This is in accordance with a reduced HCT and is supported by a decreased P_{50}. In contrast, this situation is reversed during the following pregnancy periods; here the RT is prolonged. Thus, a longer contact time of the RBC's with the gas compartment is necessary for full oxygenation of the increased haemoglobin amount. Since, obviously, a

774

prolonged contact time, or in other words a lowering of the blood flow in the tissue is not a biologically optimal response, other compensatory mechanisms have to be established.

Changes in the placental morphology of the hypoxic animals were found in the course of the following investigations: scanning electron microscopic examination of placentae from hypoxic animals showed, in comparison to control animals, a less orientated capillary bed with more branchings and capillary loops. Light and transmission electron microscopic examinations of placentae from hypoxic animals showed a thinning of placental trophoblast and a reduction in capillary diameter, while the total number of capillaries per crossection was enhanced (the complete results are reported by Scheffen et al. in this volume).

Obviously, the necessity of prolonged contact times can be by-passed by morphological adaption mechanisms e.g. increasing the gas exchange area and reducing diffusion distance. This would explain the changes described above. Similar examples are known in comparative physiology, where animals with nucleated erythrocytes or with less deformable RBC`s possess an increased tissue capillary diameter and density in addition to a higher heart-body weight ratio (Schmid-Schönbein, 1981).

The biochemical changes already mentioned must still be taken into consideration. We measered a slight increase of 2,3-DPG from the first third of pregnancy on in hypoxic animals, which would explain the increase of P_{50} (Figure 7). It should be mentioned, that the values obtained here by the ODC device used differ from those reported in literature (see Bartels and Harms, 1959). The present development level of the ODC unit used only allows measurements of RBC suspensions in the absence of CO_2. The measurements, however, are reproducible and should be regarded as relative. The ODC of hypoxic animals are right-shifted (-> 2,3-DPG level rise), which, of course has a positive effect on the oxygen delivery to tissue. This fact should not be overestimated, especially when compared to compensation mechanisms related to haemo-rheological alterations. A recent report (Schmid-Schönbein, 1988), describes the significance of rheological parameters for the blood flow in haemochorial, multivillous placentae e.g. the human placenta. The morphology of the guinea pig placenta, however, differs from that of the human placenta, exhibiting countercurrent flow conditions.

It can be concluded, that shear model studies deliver information which may help to explain the complex subject of tissue gas exchange. The purpose of this study was to examine the capability of blood shear model studies in a limited number of animal experiments. The data presented should be regarded as preliminary results. Further studies should be combined with e.g. in-vivo blood flow measurements of the umbilical and uterine vessels, the cardiac performance and ventilation. Also, methodical improvements, e.g. simultaneous continuous oxygen saturation measurements with rheo-oxymeters and the reconstruction of ODC-devices for whole blood measurements are necessary.

SUMMARY

Pregnant guinea pigs undergoing long-term hypoxia were studied and the results compared with those of control animals (pregnant, but non-hypoxic). Hypoxic animals demonstrated a decrease of O_2 affinity (-7%) and an increase of O_2 capacity (+35%). In addition, the HCT was found to be higher in the hypoxic group (+41%), causing haemorheological disadvantages; in a shear model study the blood of hypoxic animals had to be exposed to the gas compartment of the rheooxymeter up to 62% longer than that of the control group. We have postulated, that this rheological impairement is compensated, since no abnormalities in number and abortion rate of fetuses (due to a possible O_2 delivery impairment) were found. Our morphological studies in fact support this opinion, showing e.g. more capillary branchings and loops and a reduction of diffusion distances between maternal and fetal blood in hypoxic guinea pig placentae. The results emphasize the importance of more detailed rheological studies in connection with other investigations for a complete description of compensatory mechanisms.

ACKNOWLEDGEMENTS

The authors appreciate the cooperation of Prof. K. Gersonde, (St. Ingbert, FRG) and the excellent technical assistance of M. Wiesen, M. Nittritz, F.J. Kaiser, B. Hoffmann and S. Böhm. We thank also Prof. J.A. Mitchell (Detroit, USA), B. Oedekoven and J. Beckman for helpful discussions during the preparation of the manuscript.

REFERENCES

Bartels, H. and Harms, H., 1959, Sauerstoffdissoziationskurven des Blutes von Säugetieren, Pflügers Arch. 268: 334-365.

Christensen, P., Grønlund, J. and Carter A.M., 1986, Placental gas exchange in the guinea-pig: fetal blood gas tensions following the reduction of maternal oxygen capacity with carbon monoxide, J. Develop. Physiol. 8: 1-9.

Gilbert, R.G., Cummings, L.A., Juchau, M.R. and Longo, L.D., 1979, Placental diffusing capacity and fetal development in exercising or hypoxic guinea pigs, J. Appl. Physiol.: Respirat. Environ Exercise Physiol. 46 (4): 828-834.

Kitanaka, T., Gilbert, R.D. and Longo, L.D., 1988, Maternal and Fetal Responses to Long-term Hypoxemia in Sheep, in: "The Endocrine Control of the Fetus," W. Künzel and A.Jensen, eds.38-63, Springer-Verlag Berlin Heidelberg.

Longo, L. D., Hill, E. P. and Power, G. G., 1972, Theoretical analysis of factors affecting placental O_2 transfer, Am. J. Physiol., 222 (3): 730-739.

Moll, W. and Kastendieck, E., 1977, Transfer of N_2O, CO and HTO in the artificially perfused guine-pig placenta, Resp. Physiol. 29: 283-302.

Mottaghy, K., Cremer, J., and Pescarmona, J. P., 1987, Rheo-oxymetrie ein neues Verfahren zur Bestimmung der O_2-Transporteigenschaften der Erythrozyten unter Scherbedingungen, in: "Fortschritte in der kardiovaskulären Hämorheologie," B. E. Strauer, A. M. Ehrly, M. Leschke, eds., Münchner Wissenschaftliche Publikationen, München.

Mottaghy, K., Haest, C.W.M., Cremer, J. and Derissen, W., 1984, Oxygen Uptake Into The Sheared Flowing Blood: Effects Of Red Cell Membranes And Haematocrit, in: "Oxygen Transport To Tissue, Vol. V," D. W. Lübbers, H. Acker, E. Leniger-Follert and T. K. Goldstick, eds., Plenum Pub. Corp., New York.

Mottaghy, K., Haest, C.W.M. and Schleuter, H.J., 1982, Effect of red cell rigidity on gas transport by sheared flowing blood, Chem. Eng. Commun. 15: 157–167.

Mottaghy, K. and Hanse, H.J., 1985, Effect of combined shear, secondary flow and axial flow of blood on oxygen uptake, Chem. Eng. Commun. 36: 269–279.

Scheffen, I., Kaufmann, P., Philippens, L., Leiser, R., Geisen, C. and Mottaghy, K., 1989, Alterations of the fetal capillary bed in the guinea pig placenta following long-term hypoxia: a morphometrical study, at: "Meeting of the International Society on Oxygen Transport to Tissue", Göttingen, FRG, July 21–24, 1989.

Schmid-Schönbein, H., 1981, Blood rheologicy and oxygen transport to tissues, in: "Adv. Physiol. Sci. Vol. 25 Oxygen Transport to Tissue," A. G. B. Kovach, E. Dora, M. Kessler, I. A. Silver, eds., Akademiai Kiado, Budapest.

Schmid-Schönbein, H., 1988, Conceptional proposition for a specific microcirculatory problem: maternal blood flow in hemochorial multivillous placentae as percolation of a "Porpous medium", in: "Throphoblast Research, Vol. 3," P. Kaufmann and K. Miller, eds., Plenum Publishing Corporation.

Schmid-Schönbein, H. and Wells, R.E., 1969, Fluid drop-like transition of erythrocytes under shear, Science 165: 288.

Sick, H. and Gersonde, K., 1980, Rapid Measurement and Computer Analysis of Complete Oxygen Dissociation Curves of Red Blood Cells, J. Clin. Chem. Biochem., 18, 10: 689.

Sick, H. and Gersonde, K., 1985, Continuous Gas-Depletion Technique for Measuring O_2-Dissociation Curves of High-Affinity Hemoglobins, Analyt. Biochem. Vol. 146: 277–280.

Zander, R. and Schmid-Schönbein, H., 1972, Influence of intracellular convection on the oxygen release by human erythrocytes. Pflügers Arch. 335: 58–73.

ALTERATIONS OF THE FETAL CAPILLARY BED IN THE GUINEA PIG PLACENTA FOLLOWING LONG-TERM HYPOXIA

I. Scheffen[1], P. Kaufmann[1], L. Philippens[1],
R. Leiser[2], C. Geisen[3], and K. Mottaghy[3]

Depts. of Anatomy[1] and Physiology[3]
Medical Faculty, Technical University of Aachen
D-5100 Aachen, F.R.G.
Dept. of Veterinary Anatomy[2], Berne, CH

INTRODUCTION

Several pathohistological studies (Chabes et al., 1967; Salvatore, 1968; Hölzl et al., 1974; Kaufmann, 1982; Jackson et al., 1988; Kaufmann et al., 1988) concerning the maturation of the human placental villi provide evidence that there is a mutual relation between terminal villus development and capillary growth. Two types of fetal villous hyper-capillarisation have been described (Kaufmann et al., 1988) which are combined with typical patterns of terminal villus malformations:
Type a: abnormally long, largely unbranched, highly dilated capillary loops, and
Type b: dense, highly branched capillary networks, composed of short, narrow capillaries.

It is well known that hypoxia stimulates capillary growth even in the placenta (Bacon et al., 1984; Jackson et al., 1988). However, so far it has proven impossible to determine which of the above morphological reaction patterns correlates with a decreased oxygen supply.

Recently, it was demonstrated that long-term hypoxia in the pregnant guinea pig simultaneously produces an increased fetal capillary growth and a reduced capillary diameter (Bacon et al., 1984). From a physiological point of view, it is very unlikely that longitudinal capillary growth without capillary branching (corresponding to type a, cf. above) can be accompanied by a reduction of the mean capillary diameter, since this would greatly increase blood flow resistance. Because of this, capillary growth by increased branching (type b, cf. above) is rather likely to be the placental response to hypoxia. However, up to now this assumption lacked proof. The aim of the present study was to determine which type of three-dimensional capillary arrangement is induced by hypoxia. The physiological regulator mechanisms adjusting the placental transfer or exchange capacity to guarantee appropriate fetal development were examined in a parallel study by Geisen et al. (1989).

MATERIAL AND METHODS

30 pregnant guinea pigs were kept in an isobar hypoxic environment (12% O_2) from about day 15 to about day 60 of gestation. The control group consisted of 15 animals, exposed to normal atmospheric pressure. The hypoxic animals were kept in sealed acryl glass boxes divided into two compartments being connected by a lock. The boxes were ventilated with an oxygen-nitrogen mixture, controlled by two flowmeters (Fischer/Porter), an oxymeter (Oxytest S, H+B, Frankfurt, F.R.G.), and a Po_2 electrode (IL 227/213, Instr. Lab. Inc., Lexington, Mass., USA). This design (fig. 1) enabled us to keep the guinea pigs under uninterrupted hypoxia througout the entire experimental period (45 days). Near term, respective specimens were obtained by supravital perfusion of the in-situ placentae under general anaesthesia with Ketamine-HCL and Pentobarbital. The morphology of the fetal placental capillaries was examined by means of scanning electron microscopy of vessel casts, semithin histology, and transmission electron microscopy.

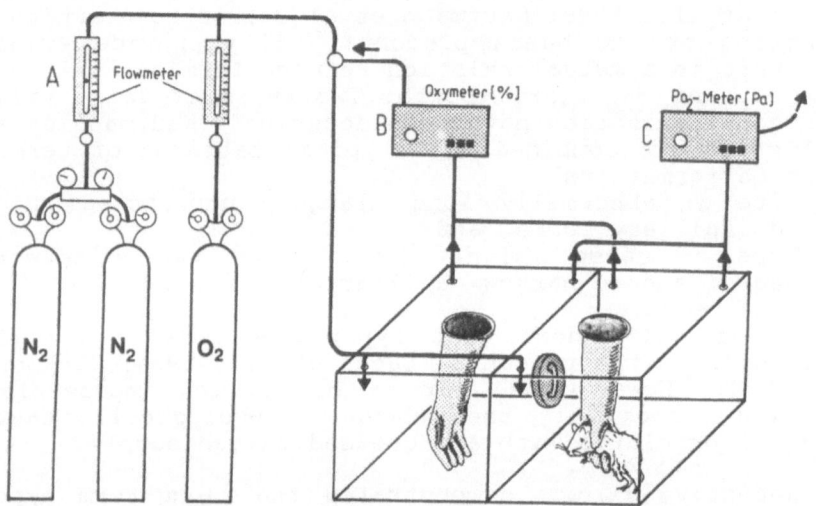

Fig. 1. Scheme of the expertimental design. Details see text

The vessel casts were prepared according to Leiser and Kohler (1983) and Leiser (1985). The fetal placental vessels were rinsed with warm (37°C) physiological salt solution to remove blood. The injection was performed with cooled and freshly prepared plastic components Batson No. 17^R corrosion compound (Polysciences) mixed with Sevriton^R. After the start of polymerization, the placentae were excised. Final hardening of the plastic was achieved in water baths at 25°C for 2 hours, followed by heating to

80°C for 12 hours or overnight. Placental tissue was removed from the vessel casts by alternating immersion in 40% KOH and in water at 60°C. To obtain suitable pieces for the scanning electron microscopy, the specimens were embedded in fluid gelatin (50°C, 20%) and frozen to -5°C for cutting with a knife. The gelatin was removed by KOH as described above for the placental tissue. The cleaned and dried specimens were mounted and sputter-coated with gold, prior to scanning electron microscopical investigation.

For semithin histology and ultrastructural studies the placentae were perfused with 2.2% phosphate buffered glutaraldehyde (340 mosmol). Small pieces of placental tissue were cut and postfixed with 1% phophate buffered OsO_4. Following dehydration in a graded series of ethanol, the material was embedded in Araldit and cut on a ultra-microtome.

To quantify the data obtained from the vessel casts, the following parameters were measured:
A. as a measure of capillary looping: the mean degree of deviation of the individual capillary longitudinal axis from the axis connecting fetal arteries and venules.
B. as a measure of capillary branching: the mean distance between two points of branching (in μm).
From these data we computed mean values, standard deviations and used analysis of variance and Student`s test.

For both purposes a three-dimensional system had to be projected on a two-dimensional screen, thus producing errors by hidden branching, wrong projection of angles, etc. Since there was no other possibility to establish more reliable quantitative results from three-dimensional structures, we tried to minimize the errors by studying the specimens in various tilting angles (details are reported by Scheffen 1989).

RESULTS

Differences in the vessel casts in the experimental group as compared to the controls, were found only in the fetal capillary system. The fetal capillaries in the periphery of the placental lobes of the control group are characterized by few loops and branches. They are largely oriented in parallel. Towards the center of the lobe the fetal capillaries are nearly straight and unbranched (figs. 2, 4, 9). In contrast, the vessel casts of the hypoxic animals showed a considerably higher degree of branching and coiling in the periphery . The transition into straight and uncoiled fetal capillaries takes place much closer to the labyrinthine center than described for the normoxic controls (figs. 3, 5). Quantitative analysis of the control and experimental group showed a statistically significant difference in the two measured parameters (A, B) in the central as well as in the peripheral part of the lobe (fig. 10).

The placental semithin sections from the hypoxic group, have an increased number of fetal capillary cross-sections, a reduced mean capillary diameter and thinner

781

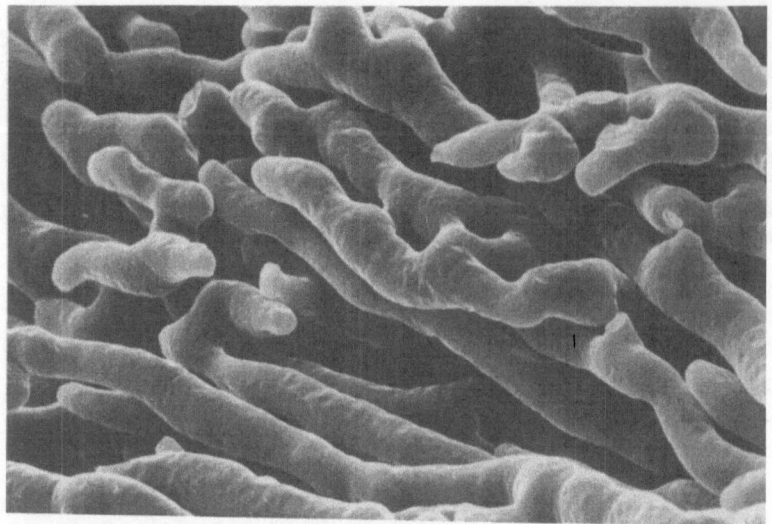

Fig. 2. Scanning electron micrograph (2500X) of the vessel cast of the peripheral labyrinth of the normal guinea pig placenta. The picture shows fetal capillaries with a parallel orientation, some loops and a low degree of branching.

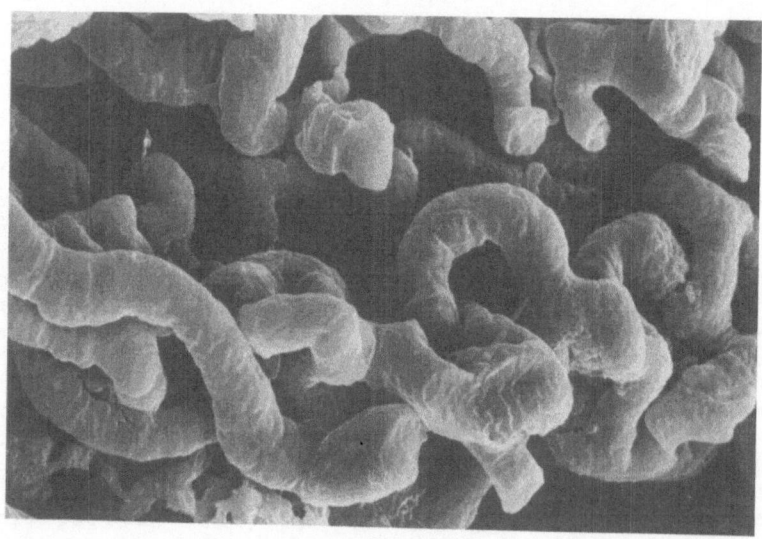

Fig. 3. Scanning electron micrograph (2600X) of the vessel cast of the peripheral labyrinth of the hypoxic group. In contrast to the control the hypoxic group shows less orientated capillaries with a higher degree of branching and coiling.

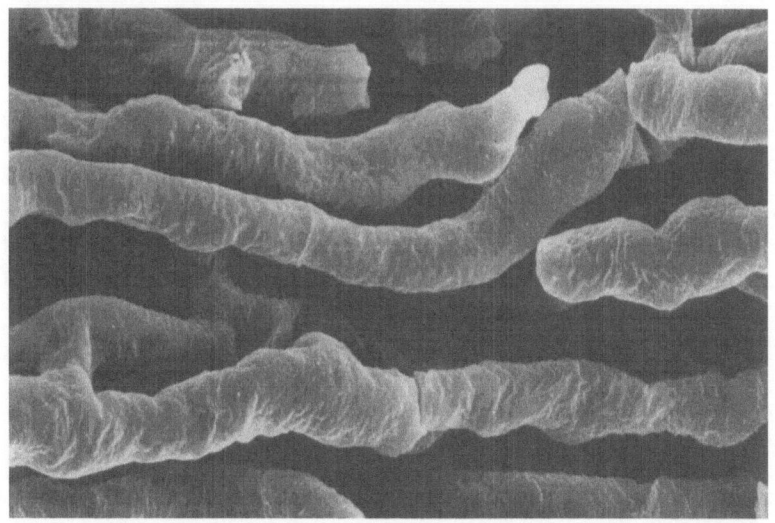

Fig. 4. Scanning electron micrograph (2450X) of the vessel cast of the central zone of the control group. Compared to the peripheral part of the lobe, the capillaries are mainly orientated in parallel and poorly branched.

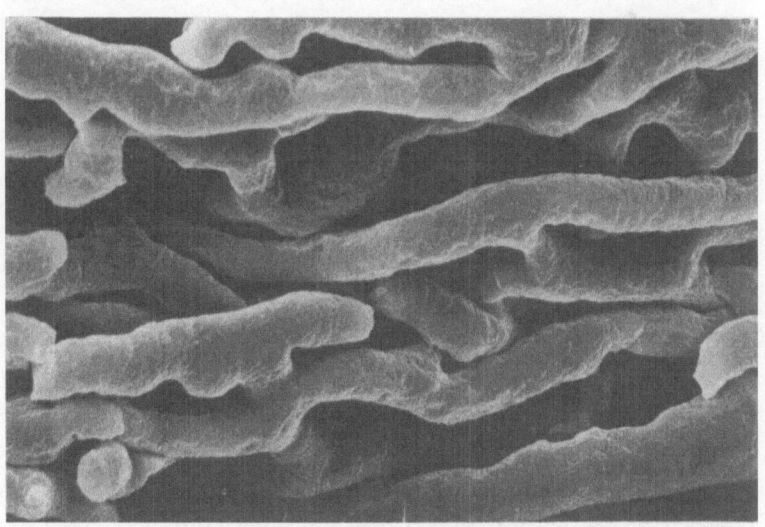

Fig. 5. Scanning electron micrograph (2450X) of the vessel cast of the central zone of the hypoxic group. The capillaries show an obviously greater number of branches and a higher degree of coiling.

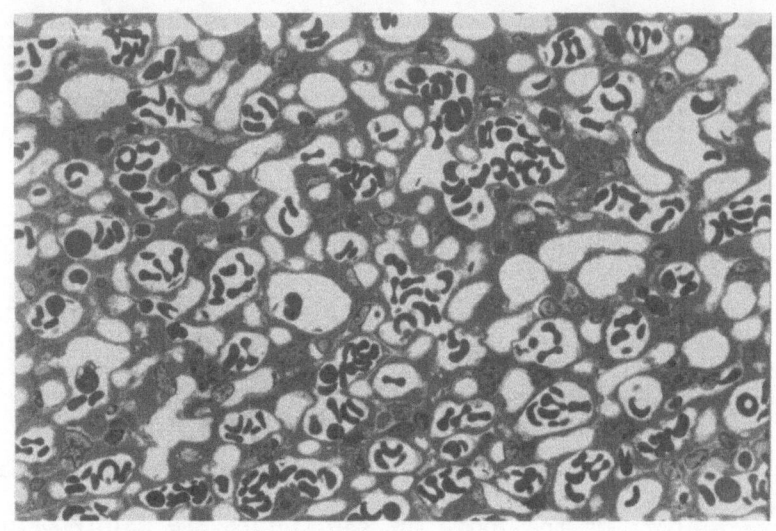

Fig. 6

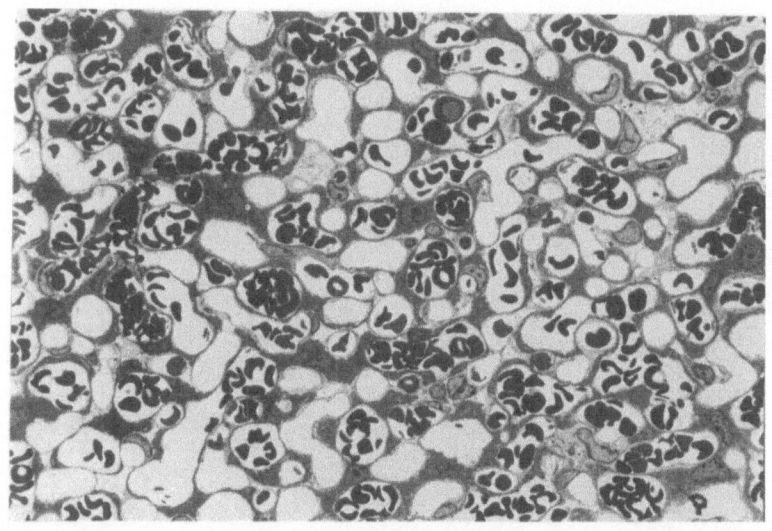

Fig. 7

Fig. 6 and fig. 7. Comparison of the semithin sections of the periphery of the lobe from the control (fig. 6) and the hypoxic groups (fig. 7):
Fig. 6 shows thick syncytioplasmic lamellae, which on one side come in contact with varying sized maternal blood spaces (l) and on the other side with the fetal vessels (c).
In contrast the hypoxic animals (fig. 7) show an increased number of capillary cross-sections per unit of sectional area and thinner tissue lamellae, separating maternal and fetal blood. (fig.6+7: 512X)

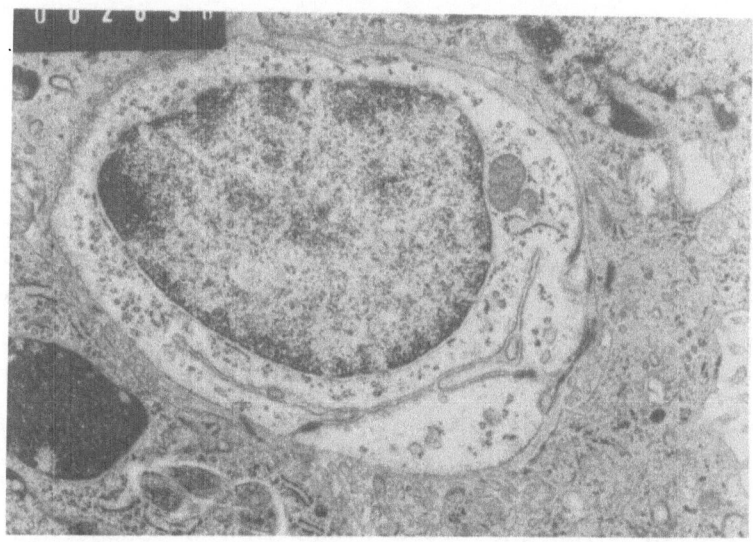

Fig. 8. Transmission electron micrograph (18050X) of the peripheral part of the lobe . This picture shows a sprouting capillary. Endothelial cells border a cleft-like lumen. This characteristical form is more often seen in the hypoxic group as compared to the control.

syncytioplasmic lamellae separating maternal and fetal circulations compared to controls (figs. 6, 7).

The ultrastructural pictures confirm the findings of the light-microscopical studies. In addition, most of the hypoxic placentae revealed characteristical features of sprouting fetal capillaries (fig. 8). No endothelial mitoses were seen. In contrast to the controls, the trophoblastic and endothelial mitochondria were hydropic and swollen.

Placental and fetal weights are not reported here, since inter-individual variations caused by varying litter size and variations of gestational age don't permit a statistical calculation.

DISCUSSION

For our study an experimental design was developed which avoided interruptions in the hypoxic conditions for the entire experimental period. As a result death of the dams (0/30), as well as a high frequency of abortion (3/30) were avoided, both typical complications of the previous study by Bacon et al. (1984). This standardization of the conditions should have minimized other experimental variability as well. We believe that the results visible in our morphological material represent typical, unselected, placental response to hypoxia.

The reaction patterns of the mammalian placenta under hypoxic conditions are composed of a variety of factors, such as hemoglobin concentration, hematocrit, arterial oxygen pressure, placental blood flow, structural adaptation of the vessel system, changes of the barrier thickness, etc. Many of these factors have been studied in-vivo (Delaquerriere-Richardson et al., 1967; Power, 1968; Longo et al., 1969, 1972; Bacon et al., 1984; Christensen et al., 1986; Jackson et al., 1988) as well as in-vitro (Tominaga and Page, 1966; Fox , 1970; Mac Lennan et al., 1972; Amaladoss and Burton, 1985). This study, however, a part of a larger collaborative investigation, focuses on adaptative mechanisms related to the fetal vessel system.

Like the human placenta, the guinea pig placenta represents the hemomonochorial type. Different from the human placenta, however, which is villous in type, the guinea pig has a labyrinthine placenta: the maternal blood flows in lacunae which are more or less parallel channels, passing the syncytiotrophoblast from the center of a lobe to the periphery. Between the lacunae lies a network of fetal capillaries in which the blood flows in the opposite direction namely from the periphery of the lobe towards the center. The resulting counter-current flow system of maternal and fetal blood streams is the optimal condition for oxygen diffusion (Faber, 1973). Furthermore, this type of exchanger causes a larger difference in oxygen concentration between the maternal and fetal plasma in the center of the lobes of the guinea pig placenta than in the periphery. This feature could explain why there is a difference between the arrangement of the fetal capillaries in the two zones (fig. 9).

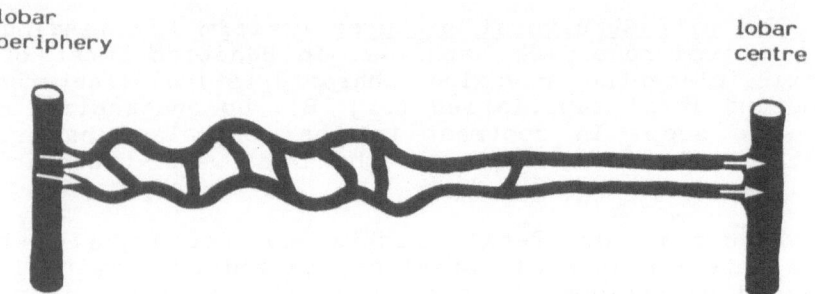

lobar periphery

lobar centre

Fig. 9. Schematic simplified representation of the fetal vessels in the lobe of the normal guinea pig placenta at term. The fetal blood flows from the periphery of the lobe to the center, i. e. centripetally, whereas the maternal blood flows centrifugally.

Placental diffusing capacity depends upon several factors: 1) membrane diffusing capacity (exchange area, barrier thickness, gas solubility in tissue fluids and membranes), 2) capillary blood volume, and 3) diffusing capacity of blood (O_2 capacity, hemoglobin contents).

Oxygen transfer is regulated by placental diffusing capacity, by the rates of maternal and fetal blood flow, by arterial oxygen pressure, and by the vascular geometry of the placenta.

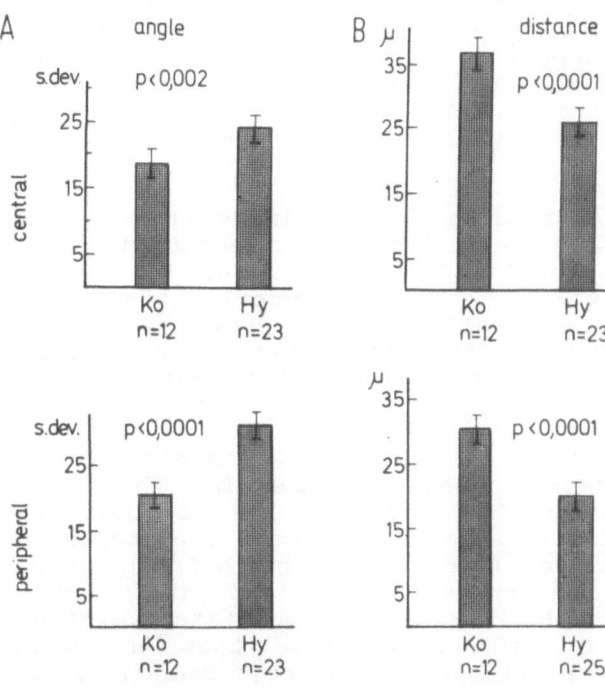

A angle
s.dev. p<0,002
central
Ko n=12   Hy n=23

B μ distance
p<0,0001
Ko n=12   Hy n=23

s.dev. p<0,0001
peripheral
Ko n=12   Hy n=23

μ p<0,0001
Ko n=12   Hy n=25

Fig. 10.
Graphic illustration of the morphometrical parameters A and B. We used the mean values of the standard deviations and of the distances between two points of branching for the Student's test. There is a significant difference for both parameters between the control group (Ko) and the hypoxic group (Hy) in the central (above) as well as in the peripheral part (below) of the lobe.

lobar periphery   lobar centre

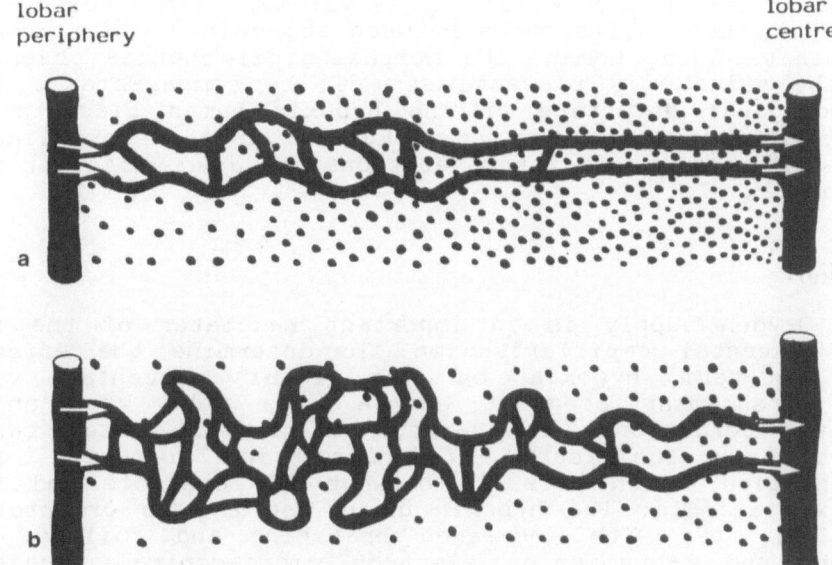

Fig. 11 a, b. Greatly simplified representations of the lobar arrangement of the fetal vessels in relation to the concentration of oxygen in the vessels and tissues a: the fetal vessel arrangement in a normoxic environment b: the fetal vessel arrangement in a hypoxic environment (12% O_2)

From these theoretical considerations, we hypothesize that the typical arrangement of fetal placental vessels under hypoxic conditions should resemble type b of hyper-capillarisation (dense, highly branched capillary networks, composed of short, narrow capillaries). Our results support this hypothesis.

The morphological finding of 1) increasing number of capillary cross-sections per unit of sectional area, 2) a decreased mean capillary diameter, and 3) a reduced trophoblastic thickness under hypoxic conditions are consistant with adaptation to improve placental diffusing capacity and O_2 transfer (fig. 6,7). These results ar consistent with those published by Bacon et al. (1984).

However, such results obtained from tissue sections cannot answer our initial question concerning the exact type of hyper-capillarisation. This can only be addressed from the three-dimensional vessel casts data. The increased coiling of the fetal capillaries in the hypoxic group (fig., 3) might be interpreted as an indicator for a simple longitudinal growth. However, the quantitative studies of the casts (fig. 10) demonstrate that the mean distance between two points of capillary branching is significantly reduced. Therefore branching with establishment of complex capillary nets must be the prevailing mode. These results are in agreement with our initial hypothesis, and demonstrate the dependence of oxygen exchange upon placental vessel structure (fig. 11a, b). The parallel physiological study shows additional adaptative mechanisms concerning the diffusion capacity of blood as well as oxygen transfer (cf. Geisen et al., this volume). In spite of the structural differences between the guinea pig placenta and that of the human, the morphological changes observed in the guinea pig placenta seem to be comparable to the histological structure of the hypoxic human placenta as described by Jackson et al. (1988). It is therefore likely that our results and interpretations are also valid for the human.

SUMMARY

Oxygen supply is an important regulator of the fetal placental capillarisation. To determine the effects of long-term hypoxia on the fetal placental vessel arrangement pregnant guinea pigs were kept under hypoxic conditions (12% O_2 for 45 days). Vessel casts showed a significant difference in branching and orientation of the vessels between the controls and the hypoxic animals. The hypoxic group had a less orientated capillary bed with increased branching and coiling. By light- and transmission electron microscopical studies, there was a decreased diffusion distance, a decreased diameter of the fetal capillaries, and an increased number of capillary cross-sections. These results are consistent with the hypothesis that chronic hypoxia is responsible for increased branching and coiling of the capillaries resulting in a dense network of short and narrow capillaries in the placenta.

REFERENCES

Amaladoss, A. S. P., and Burton, G. J., 1985, Organ culture of human placental villi in hypoxic and hyperoxic conditions: a morphometric study, J. Develop. Physiol., 7, 13-118

Bacon, B. J., Gilbert, R. D., Kaufmann, P., Smith, A. D., Trevino, F. T., and Longo, L. D., 1984, Placental anatomy and diffusing capacity in guinea pigs following long-term maternal hypoxia, Placenta, 5, 475-488.

Chabes, A., Peroda, J., and Perez, J., 1967, Morphometry of human placenta at high altitude, Abstract, Am. J. Pathol., 50, 14a-15a.

Christensen, P., Grønlund, J., and Carter, A. M., 1986, Placental gas exchange in the guinea pig: fetal blood gas tensions following the reduction of maternal oxygen capacity with carbon monoxide, J. Develop. Physiol., 8, 1-9.

Delaquerriere-Richardson, L., and Valdivia, E., 1967, Effects of simulated high altitude on pregnancy, Arch. Path., 84, 405-417.

Faber, J. J., 1973, Diffusional exchange between foetus and mother as a function of the physical properties of the diffusing materials, in: "Foetal and neonatal physiology," Combin, R. S., Gross, K. W., Dawes, G. S., and Nathaniels, P. W., eds., 306-327, London: Cambridge University Press.

Fox, H., 1970, Effect of hypoxia on trophoblast in organ culture. A morphologic and autoradiographic study, Am. J. Obstet. Gynecol., 107, 1058-1064

Geisen, C., Mottaghy, K., Scheffen, I., and Kaufmann, P., 1989, Effect of long-term hypoxia on oxygen transport properties of blood in pregnant guinea pigs, at: "Meeting of the Internat. Society on Oxygen Transport to Tissue", Göttingen, F.R.G., July 21-24, 1989

Hölzl, M., Lüthje, D., and Seck-Ebersbach, K., 1974, Placentaveränderungen bei EPH-Gestose, Arch. Gynec., 217, 315-334.

Jackson, M. R., Mayhew, T. M., and Haas, J. D., 1988, On the factors which contribute to thinning of the villous membrane in human placentae at high altitude. II. An increase in the degree of peripheralization of fetal capillaries, Placenta, 9, 9-18.

Jackson, M. R., Mayhew, T. M., and Haas, J. D., 1988, Effects of high altitude on the vascularization of terminal villi in human placentae, Trophoblast Research, 3, 351-360.

Kaufmann, P., 1982, Development and differentation of the human placental villous tree, Bibltheca. Anat. 22, 29-39.

Kaufmann, P., Luckhardt, M., and Leiser, R., 1988, Three-dimensional representation of the fetal vessel system in the human placenta, Trophoblast Research, Vol.3, 113-137.

Leiser, R., 1985, Fetal vasculature of the human placenta: scanning electron microscopy of microvascular casts, Contr. Gynecol. Obstet. 13, 27-31.

Leiser, R., and Kohler, T., 1984, The blood vessels of the cat girdle placenta. Observations on corrosion casts, scanning electron microscopical and histological studies. II. Fetal vasculature, Anat. Embryol. 170, 209-216.

Longo, L. D., Hill, E. P., and Power, G.G., 1972, Theoretical analysis of factors affecting placental O_2 transfer, Am. J. Physiol. 222, 730-739

Longo, L. D., Power, G. G., and Forster, R. E., 1969, Placental diffusing capacity for carbon monoxide at varying partial pressures of oxygen, J. Appl. Physiol., 26, 360-370.

Mac Lennan, A. H., Sharp, F., and Shaw-Dum, J., 1972, The ultrastructure of human trophoblast in spontaneous and induced hypoxia using a sytem of organ culture: a comparison with ultrastructural changes in pre-eclampsia and placental insufficiency, J. Obstet. Gynecol. Br. Commonw. 79, 113-121.

Power, G. G., 1968, Solubility of oxygen and carbon monoxid in blood and pulmonary and placental tissue, J. Appl. Physiol., 24, 468-474.

Scheffen, I., 1989, Veränderungen des fetalen Gefäßbettes in der Plazenta durch chronischen Sauerstoffmangel, Dissertation, Tech. Univ. of Aachen, F.R.G.

Tominaga, T.,and Page, E. W., 1966, Accomodation of the human placenta to hypoxia, Am. J. Obstet. Gynecol. 94, 679-691.

Van der Heijden, F. L., 1981, Compensation mechanisms for experimental reduction of the functional capacity in the guinea pig placenta I. Changes in the maternal and fetal placenta vascularization, Acta. anat. 111, 352-358.

PERINATAL CHANGES IN HEMOGLOBIN CONCENTRATION IN RATS

F.L.Ubels*, W.P.Meeuwsen*, S.Wijkstra*, B.Oeseburg**

*Dept. of Physiology, University of Groningen and **Dept. of Physiology, University of Nijmegen, The Netherlands

INTRODUCTION

During experiments on oxygen availability in fetal rats using pulse oximetry as presented by Wijkstra et al.(1989), we were struck by the low total hemoglobin concentration (c_{Hb}) in the fetal blood. Since hemoglobin serves as the main oxygen carrier, the course of c_{Hb} during fetal life should be investigated before using the rat as an animal model in oxygen transport research.

In human fetuses, the c_{Hb} value is higher than the maternal one. Prenatal fetal human c_{Hb} is presented by Falkner and Tanner (1978) as being approximately 20 g/L higher than the maternal c_{Hb}. Guyton (1986) describes prenatal c_{Hb} as being 150% of the maternal c_{Hb}. In human newborns, c_{Hb} usually exceeds the normal adult value, as described by Oski and Naiman (1982), Dacie and Lewis (1963), Forestier et al. (1986) and Nelson (1987).

For rats only scanty data are available in the literature. Prenatal data were presented by Nicholas and Bosworth (1928), Kindred and Corey (1931) and Wintrobe and Shumacker (1936). Wintrobe and Shumacker (1936) showed that c_{Hb} in newborn rats was approximately 60% of the adult value, already indicating the immaturity of newborn rats in comparison to humans. This finding was confirmed by Bruner et al. (1938) and by Garcia (1957). The postnatal c_{Hb} showed a decrease during the first 20 days after birth. Thereafter, c_{Hb} slightly increased until the adult value was reached after about 2 months (Williamson and Ets (1926), Bruner et al. (1938) and Garcia (1957)). However, since standardization of the measurement of hemoglobin concentration (Van Kampen and Zijlstra (1961)) became internationally accepted only in the mid-sixties, the absolute values of c_{Hb} presented before that time are uncertain. For this reason and since only fragmentary data on the time course of c_{Hb} in the fetal rat are available, we decided to go more systematically into this matter.

MATERIALS AND METHODS

Fertile female Wistar rats were kept with fertile males overnight until conception had taken place. The day of conception was indicated day zero for fetal age. Pregnancy in the rat is of 21 days duration. From day 18 onward we were able to reliably obtain fetal blood samples for the determination of c_{Hb}. To this end, the pregnant mothers were anaesthetized with ether. The fetuses were removed from the uterus by Caesarian section. All animals were sacrificed after the experiment. The fetuses were dried using a gauze so that all amniotic fluid was removed, in order that dilution of the blood sample was prevented. A blood sample of 10 μL was obtained by an incision of the carotid artery. Blood was directly sampled from the artery into a HemoCue microcuvette (HemoCue AB, Helsingborg, Sweden). These cuvettes fill by capillary suction only. The postnatal c_{Hb} was obtained from spontaneously born offspring immediately after birth as well as during the following four days. c_{Hb} was determined by means of the HemoCue system (HemoCue AB, Helsingborg, Sweden), which involves the following procedure. The inside of the disposable HemoCue microcuvette is coated with a reagent mixture of sodium deoxycholate, sodium nitrite and azide. The erythrocytes are lysed by the sodium deoxycholate and all hemoglobin is converted to hemiglobin by the sodium nitrite. The azide then converts the hemiglobin to hemiglobinazide, which is a stable compound with a distinct optical spectrum, displaying a rather flat part around 565 nm. c_{Hb} is actually measured at a wavelength of 565 nm, while a second measurement at 880 nm is used for compensating the influence of any residual turbidity. Within one minute one can read c_{Hb} from the photometer display. The HemoCue is calibrated so that its results are in agreement with those of the standardized methemoglobin cyanide method (Kwant et al. (1987)).

To determine the total amount of Hb in mg per g of fetal body mass we weighed all the fetuses from one litter. These were then homogenized (Braun MR100) and the c_{Hb} of the homogenate was determined using a special HemoCue microcuvette (TUR cuvette). This cuvette allows measurements to be performed on samples with higher turbidity, even on those which contain tissue fragments.

To determine the plasma protein concentration (c_{Pr}) we obtained blood samples as described above. The blood samples were centrifuged, the hematocrit was measured and 4 μl of plasma was used for the determination of the total protein concentration (Sopar-Biochem, Brussels). In this method proteins react with a pyrogallol-molybdate complex. The optical density of the sample-reagent mixture is proportional to its protein concentration at 598 nm.

RESULTS

The perinatal hemoglobin concentration values are presented in figure 1 and table I. The great number of fetuses at day 20 resulted from a series of pulse oximetry experiments performed at that age only (Wijkstra et al., 1989).

Figure 2 shows the time course of the fetal body mass and the Hb content of 21, 22, 20 and 16 fetuses from 2 litters each, at day 18, 19, 20 and 21 after conception, respectively. Hematocrit and hemoglobin concentration of these fetuses are presented in figure 3.

792

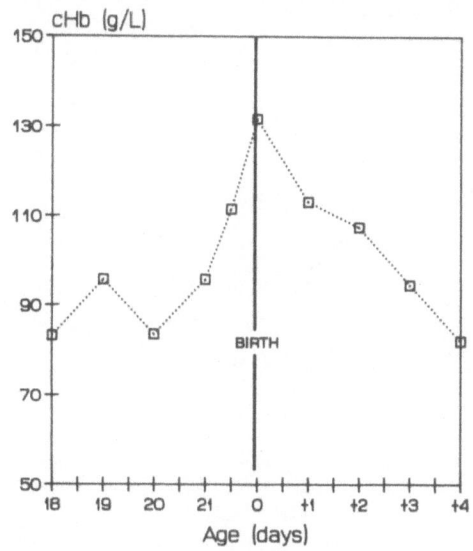

Figure 1. Course of perinatal hemoglobin concentration (c_{Hb}).

Table I. Perinatal hemoglobin concentration (g/L).

Day	18	19	20	21	21.5	P	1	2	3	4
c_{Hb}	83.3	95.7	83.6	95.7	111.5	131.6	113.1	107.5	94.5	82.1
SEM	2.5	2.6	1.1	1.4	2.4	3.9	3.5	3.2	2.8	2.4
N_F	20	21	63	24	19	14	12	12	12	12
N_L	2	2	6	2	3	3	3	3	3	2

Day = days of gestation and after parturition (P); c_{Hb} = average total hemoglobin concentration; SEM = standard error of the mean; N_F = number of fetuses; N_L = number of litters.

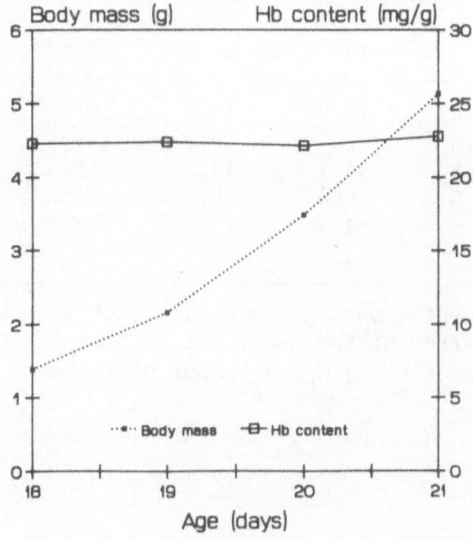

Figure 2. Course of prenatal body mass and hemoglobin content.

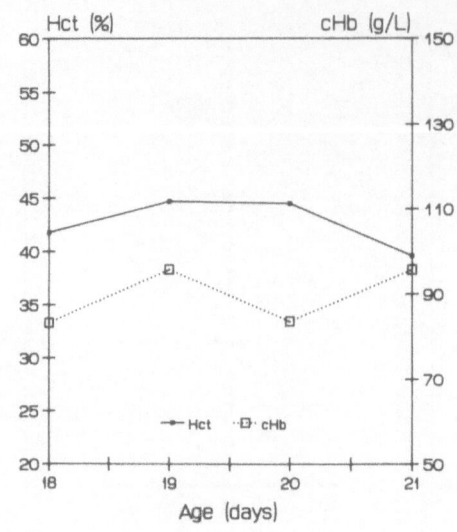

Figure 3. Course of prenatal hemoglobin concentration(c_{Hb}) and hematocrit (Hct).

Using the hematocrit values and the corresponding values of c_{Hb}, the mean corpuscular hemoglobin concentration (MCHC) was calculated to be 199, 214, 188 and 242 g/L from day 18 to day 21. In table II prenatal values of the plasma protein concentration and hematocrit are given, obtained from another series of experiments.

Table II. Prenatal c_{Pr} (g/L) and Hct (%)

Day	18	19	20	21
c_{Pr}	12.3	12.8	15.0	16.5
SEM	0.3	0.2	0.3	0.2
Hct	41.8	44.7	44.5	39.6
SEM	0.6	0.6	0.5	0.6
N_F	10	23	21	15
N_L	2	3	2	2

Day = days of gestation; c_{Pr} = plasma protein concentration; SEM = standard error of the mean; Hct = hematocrit; N_F = number of fetuses ; N_L = number of litters.

The prenatal course of c_{Pr}, together with the prenatal course of c_{Hb} of the experiments given in Table I, is depicted in figure 4.

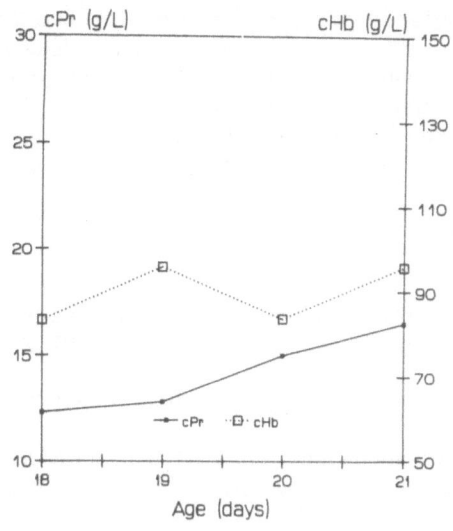

Figure 4. Course of prenatal plasma protein concentration (c_{Pr}) and hemoglobin concentration (c_{Hb}).

DISCUSSION

In rats in utero c_{Hb} is much lower than the adult (maternal) value, which is in contrast to the situation in humans. This has implications for the trans-placental oxygen transport and should be taken into account when results of perinatal physiological research obtained in rats are extrapolated to the human situation.

Although the erythropoietic system keeps pace with the enormous growth of over 50% per day during the last three days in utero, as is demonstrated by the results shown in figure 2, a quickly developing 'physiological anaemia' is observed in the early postnatal period (figure 1).

The slight increase in c_{Hb} from the 20th to the 21st day, although highly significant (table I), together with the concomitant fall in hematocrit (table II), indicates a fast maturation of the erythropoietic system. Nevertheless mean corpuscular hemoglobin concentration is still low, indicating that the whole system is still quite immature.

The consistent increase in c_{Pr} from the 18th day on (figure 4), together with the time course of the hematocrit (figure 3) imply that the changes in c_{Hb} are caused by fluctuations in the amount of hemoglobin produced in the erythropoietic system. From day 19 to 20 the hematocrit remains constant whereas c_{Hb} decreases, while from day 20 to 21 c_{Hb} increases with a decreasing hematocrit. These rapid changes of hematocrit and c_{Hb}, sometimes in opposite direction, thus cannot be explained by hemodilution or hemoconcentration only. If one calculates the total blood volume from c_{Hb} and the body content of hemoglobin (figure 2), assuming all hemoglobin to be present in the vascular system, mean values of about 25% of the body mass are obtained. This indicates that a substantial part of the hemoglobin present in the body is still in the extravascular compartment, i.e. in the tissue spaces where the erythrocytes are formed.

795

These erythrocytes presumably are responsible for the fast rise in c_{Hb} on the last day before birth. The statement that the sharp rise in c_{Hb} directly after birth is caused by hemoconcentration during parturition is not supported by our findings. Since the rise in c_{Hb} is already found before parturition started (day 21.5: table I) the higher values directly after birth can only partly be explained by hemoconcentration.

CONCLUSIONS

The use of HemoCue cuvettes, demanding a filling volume of 10 µL only, made reliable c_{Hb} measurements in these small fetuses possible.

Also with respect to hematological values rats are born very immature.

The described differences should be born in mind when extrapolating perinatal oxygen transport phenomena found in rats to the human situation.

REFERENCES

Bruner, H. D., Erve, J. van de and Carlson, A. J., 1938, The blood picture of rats from birth to twenty-four days of age, Am J Physiol., 124:620.

Dacie, J. V. and Lewis, S. M., 1963, p 12, in: "Practical Haematology," 3rd ed., J. and A. Churchill Ltd. London.

Falkner, F. and Tanner, J. M., 1978, p 359, in: "Human Growth. 1. Principles and Prenatal Growth," Plenum Press, New York.

Forestier, F., Daffos, F., Galacteros, F., Bardakjian, J., Rainaut, M. and Beuzard, Y., 1986, Hematological values of 163 normal fetuses between 18 and 30 weeks of gestation, Pediatr Res.,20 (4):342.

Garcia, J. F., 1957, Changes in blood, plasma and red cell volume in the male rat, as a function of age, Am J Physiol., 190 (1):19.

Guyton, A.C., 1986, p 986, in: "Textbook of medical physiology," 7th ed., W. B. Saunders Company, Philadelphia.

Kindred, J. E. and Corey, E. L., 1931, Studies on the blood of the fetal albino rat. II. The hemoglobin content of the blood as related to the differential erythrocyte count, Physiol Zool., 4:294.

Kwant, G., Oeseburg, B., Zwart, A. and Zijlstra, W.G., 1987, Calibration of a practical haemoglobinometer, Clin lab Haemat., 9:387.

Nelson, J., 1987, p 1035, in: "Textbook of Pediatrics," 13th ed., W. B. Saunders Company, Philadelphia.

Nicholas, J. S. and Bosworth, E. B., 1928, The determination of the amount of hemoglobin present in rat fetuses during development, Am J Physiol., 83:499.

Oski, F. A. and Naiman, J. L., 1982, Normal blood values in the newborn period, in: "Hematologic problems in the newborn," 3rd ed., W. B. Saunders Company, Philadelphia.

Van Kampen, E. J. and Zijlstra, W. G., 1961, Standardization of hemoglobinometry. II. The hemiglobincyanide method, Clin Chem Acta., 6:538.

Williamson, C. S. and Ets, H. N., 1926, The effect of age on the hemoglobin of the rat, Am J Physiol., 77:480.

Wintrobe, M. M. and Shumacker, H. B., Jr., 1936, Erythrocyte studies in the mammalian fetus and newborn, Am J Anat., 58:313.

Wijkstra, S., Schuiling, G., Kwant, G. and Oeseburg, B., 1989 in press, Pulse-oximetry in fetal rats, in: " Proceedings of the III conference on: Fetal and Neonatal Physiological measurements."

METABOLIC AND DEVELOPMENTAL RESPONSES OF THE CALF TO A CHRONIC HYPOXIC EPISODE IN THE IMMEDIATE NEWBORN PERIOD

Howard D. Tyler and Harold A. Ramsey

Department of Animal Science
North Carolina State University
Raleigh, North Carolina 27695

INTRODUCTION

Oxygen consumption of the newborn increases 3-fold above fetal levels in the first 2 days (Dawes and Mott, 1959), with much of this increase occurring in the first 12 hours (Acheson et al., 1957). This is due in large part to energy expenditures for maintenance of thermal neutrality (Mount, 1958; Dawes and Mott, 1959; Alexander, 1975). Activity of the gastrointestinal tract also contributes to a significant extent, with oxygen consumption rising 3.5-fold in postnatal intestinal tissue at rest (Reeves et al., 1972; Edelstone and Holtzman, 1981b). Oxygen consumption increases an additional 65-72% during digestion (Brodie et al., 1910; Edelstone and Holtzman, 1981a).

The change in oxygen availability normally occurring at birth has been shown to play a regulatory role in the induction of certain liver enzymes (Ballard, 1971; Warnes et al., 1977), and has been suggested to be involved in other adaptive changes occurring during the newborn period (Villee and Hagerman, 1958). The primary objective of this experiment, therefore, was to determine if hypoxia in the newborn period (resulting in maintenance of P_{O2} near fetal levels) would delay developmental changes in gastrointestinal tissue normally occurring in the first 24 hours of life. Specifically, the time of cessation of macromolecular transport in the small intestine (closure) was compared between hypoxic and normoxic groups. Secondary objectives were to observe the effects of dietary treatments in modulating this response and to document the response of various metabolic parameters to the experimental conditions imposed.

METHODS

Twelve Holstein calves were obtained at birth and assigned within 2 minutes postpartum to a primary treatment group (level of inspired oxygen) and secondary treatment group (dietary regimen). Hypoxia was imposed on selected calves by introduction of inspired gas mixtures to the animal via a face

Oxygen Transport to Tissue XII, Edited by J. Piiper *et al.*
Plenum Press, New York, 1990

mask. Oxygen content of the inspired air was limited by introduction of nitrogen. Flow rate was controlled and adjusted to the needs of the calf. A 30-liter neoprene gas bag placed within the gas delivery system allowed mixing of the gases and facilitated visualization of gas consumption by the calf. The duration of the hypoxic treatment was 24 hours.

Dietary regimens superimposed upon the primary treatment included colostrum feeding, milk feeding or fasting. Colostrum fed calves recieved 1 kg colostrum at birth and every 12 hours thereafter through 48 hours. Milk-fed calves received milk for the first 2 feedings after which their diet was identical to colostrum-fed calves. Fasted calves received no feed until 24 hours after which their diet was identical to the other two groups. All calves received 2 kg milk at 60 and 72 hours. To insure complete feed delivery and uniformity between groups, an esophageal feeder was utilized for all calves at all scheduled feedings and frozen pooled colostrum was the source for all colostrum fed.

Arterial blood samples were drawn several times daily. Most samples were drawn from either the ventral coccygeal or carotid arteries, although the dorsal metatarsal was more accessible in some situations. Samples were analyzed on an Instrumentation Laboratories System 1302 Blood Gas System and subsequently on an Instrumentation Laboratories 482 Co-oximeter System.

Venous blood was drawn from the jugular vein at birth and every 6 hours thereafter up to 48 hours, always prior to feeding. Samples were also drawn at 60 and 72 hours. Plasma was separated and stored for later analysis at -20°C.

Plasma samples were analyzed for fructose, glucose, lactate and IgG. Fructose was determined spectrophotometrically by the method of Roe (1934). Glucose was determined by the oxygen rate method with a Beckman oxygen electrode (Beckman Instruments, Inc., Brea, CA). Lactate was determined enzymatically using a commercial kit (Boehringer Mannheim, Indianapolis, IN). Concentrations of IgG were quantitated via radial immunodiffusion with commercial gels (ICN Immunobiologicals, Lisle, IL).

Time of closure was estimated by calculation of the join point (J) as described by Hudson (1966) and modified by Stott et al. (1979). All data were analyzed using the General Linear Models Procedure of SAS (SAS Institute, Cary, NC). The statistical model included treatment, time and calf effects. In all cases, probabilities greater than 0.05 were considered not significant and the results are reported accordingly.

RESULTS AND DISCUSSION

A decreased rate of absorption was observed in hypoxic calves, although absorptive capacity was unchanged. Time of closure was significantly delayed in colostrum-fed calves (p=0.02) from 20 hours in normoxic calves to 40.5 hours in hypoxic calves. This developmental delay was not seen in fasted or milk-fed calves.

798

Glucose values were not significantly different at any time period although the rate of the postnatal increase in plasma glucose was slower in hypoxic calves. Hypoxic calves peaked at 42 hours with values averaging 110 mg/dl, while normoxic calves peaked at 24 hours averaging 105 mg/dl. Activity of glucose-6-phosphatase, induced postnatally by increased oxygenation of the liver (Dawkins, 1961; Ballard, 1971) may have been inhibited to some degree by hypoxia since variation among hypoxic calves was high.

Plasma fructose was monitored to observe its possible interaction with glucose. Fructose levels were unchanged by hypoxia in this study. Values averaged 60-65 mg/dl at birth. From 12-72 hours, levels ranged between 0-10 mg/dl.

Hypoxia significantly increased plasma lactate values between 6 hours and 42 hours (p<0.035). Levels at 24 hours were 90 mg/dl and 23 mg/dl in hypoxic and normoxic calves, respectively. This contributed to a progressive primary metabolic acidosis developing in hypoxic calves. Blood pH decreased postnatally in hypoxic calves, while in normoxic calves a gradual increase in pH was observed throughout the duration of the experiment. Post-hypoxic calves experienced a dramatic resolution of their acidosis, with pH values equivalent to normoxic calves within a few hours of cessation of hypoxia.

In normoxic calves P_{CO2} was unchanged throughout the experimental period (Figure 1). In hypoxic calves, arterial P_{CO2} was similar to normoxic calves. Post-hypoxic calves had lower P_{CO2} values than normoxic calves. Similar phenomena have

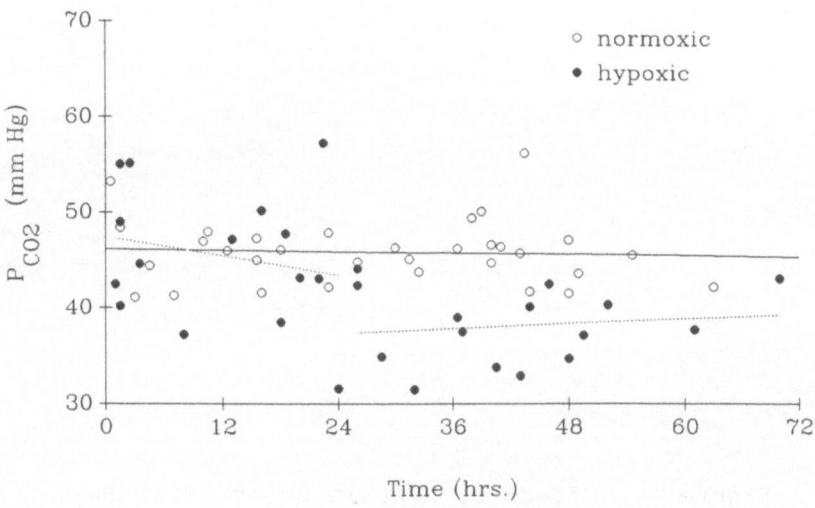

Figure 1. The effect of hypoxia on arterial P_{CO2} levels in newborn calves.

799

been reported in newborn rats subjected to chronic hypoxia (Mortola, 1986). Newborn animals respond to extended periods of hypoxia with a biphasic ventilatory response (Cross and Malcolm, 1952), although the magnitude of response is variable during the first day of life (Blanco et al., 1984). This response is evidenced by an immediate increase in ventilation followed by a gradual decrease to near control levels (Eden and Hanson, 1987). If hypoxia is extended for a long enough period, ventilation gradually rises again and persists beyond the period of hypoxia (Mortola et al., 1986). This increased ventilation is responsible for the lowered P_{CO2} levels observed in post-hypoxic calves in this experiment.

The primary goal of the hypoxic treatment was to maintain arterial P_{O2} near prenatal levels to prevent postnatal changes triggered by increased oxygen tension occurring at parturition. Published values for arterial P_{O2} in the fetal calf vary from 19.5 mm Hg (Gahlenbeck et al., 1968) to 29.5 mm Hg (Reeves et al., 1972) depending on sampling site and use of anaesthesia. Therefore, values obtained in this experiment ($P_{O2} \approx 25$ mm Hg) (Figure 2) can be considered successful. Arterial P_{O2} remained fairly constant over time in hypoxic calves, with minimal variation between calves. Cessation of hypoxia resulted in a rapid rise in arterial oxygen tension, with an apparent compensation to levels above control calves. This may have been due to a combination of factors including the increased ventilation described previously and some left-right shunting through the ductus arteriosus. Closure of the ductus is dependent in part

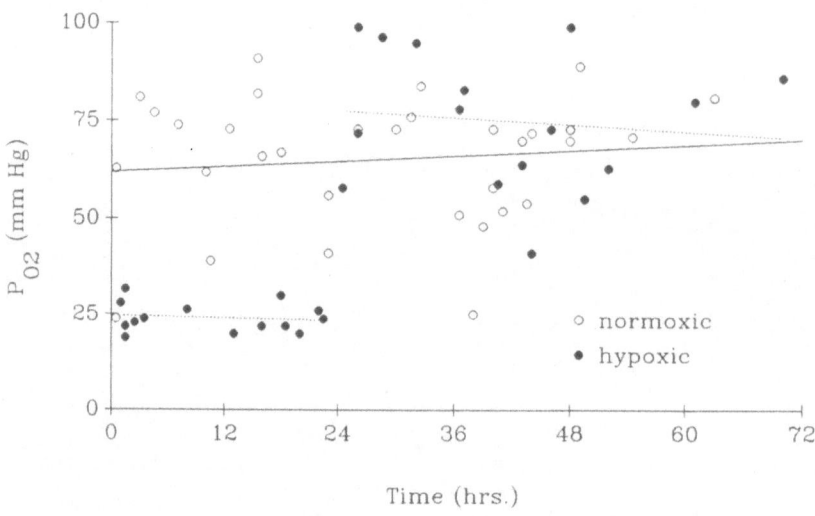

Figure 2. Effect of hypoxia on arterial P_{O2} in newborn calves.

800

on the postnatal increase in P_{O2} (Assali and Morris, 1964; Detweiler, 1984) and the critical P_{O2} for induction of ductus closure is $\approx$ 55 mm Hg. Therefore, left-right shunting may have contributed to blood oxygenation in post-hypoxic calves in this study.

Levels of HCO_3^- in arterial blood decreased to between 13-18 mmol/L during the period of hypoxia , while values for normoxic calves gradually rose from 25 mmol/L to 30 mmol/L during the same time period. Since P_{CO2} remained stable throughout this period, the decrease in HCO_3^- would indicate a shift in the ratio of $HCO_3^-:H_2CO_3$. Carbonic acid levels in blood are directly proportional to P_{CO2} as per the equation:

P_{CO2} (mm Hg) x 0.03 = H_2CO_3 (mmol/L)

Incomplete resolution of HCO_3^- in post-hypoxic calves reflects the lowered P_{CO2} in those calves.

Base excess is a calculated value representing deviation of anionic buffering components of the blood from normal. It includes buffering capacity of hemoglobin and therefore is useful in situations, such as the newborn period, where hemoglobin values vary. Reference values reflect normal adult human values and therefore are of questionable value in calves. Data collected from normoxic calves in this experiment suggest that a calf born with a pH of 7.35, a P_{CO2} of 46 and an HCO_3^- of 26 is normal, while a 48-hour-old calf would have a pH of 7.45, a P_{CO2} of 45 and an HCO_3^- of 30. A newborn calf born with 48-hour values is therefore not normal, but rather in metabolic alkalosis with respect to normal newborn values. With this in mind, base excess for clinically normal unstressed calves should be zero by definition. Only in this manner can deviations from normal be readily observed and their causes elucidated. Thus, the regression line for normoxic calves in this experiment becomes the definition of zero for all age groups, with all values being adjusted accordingly (Figure 3).

Levels of total hemoglobin were relatively unaffected by hypoxia. Birth values ranged from 6.5-13.5 mg/dl. Although variability between calves was high, concentrations within calves were stable over time, decreasing only 1-2 g/dl over the experimental period. Measurements of oxygen capacity of the blood, which varied directly with level of hemoglobin, ranged from 8 to 18 vol% O_2 at birth. Oxygen content, however, reflected the influence of hypoxia, although variability was greater between calves than was seen with oxygen tension. This variability was in part due to the variability in oxyhemoglobin levels between calves. Given the steep slope of the oxygen dissociation curve at 25 mm Hg, even slight differences in oxygen tension between calves would translate into large differences in percent oxyhemoglobin. The stability of values in normoxic calves suggests that the large fluctuations in P_{O2} in these calves were all at or near the upper plateau of the oxygen dissociation curve. Oxygen content in hypoxic calves averaged 7.5 vol% O_2 while normoxic calves had values averaging 14 vol% O_2. Since oxygen capacity between groups was not different, any differences in oxygen content can be considered treatment effects.

801

SUMMARY

While there is an extensive bank of literature on hypoxia and anoxia in the neonate, the primary focus of previous research has been pulmonary responses to acute hypoxia. Furthermore, no research on the metabolic effects of chronic hypoxia in the immediate newborn period has been published to our knowledge. Given the dynamic state of the newborn in the first hours of life, the relevence of previous observations to the results of the present study is questionable.

Previous studies on calves subjected to periods of acute hypoxia have reported decreased P_{O2}, decreased P_{CO2} and increased pH (Reeves and Leathers, 1964). Thus, it is clear that newborn calves respond differently to chronic hypoxia than they do to acute hypoxia. Much of this difference is due to the biphasic ventilatory response previously discussed.

Past reports of acid-base changes in newborn calves utilized venous blood (Moore, 1969; Schlerka et al., 1979; Eigenmann et al., 1981; Maurer-Schweitzer et al., 1977) and are therefore subject to criticism given the variability inherent in venous blood values. Waizenhoffer and Mulling (1978) compared arterial and venous blood gases, but only drew a limited number of arterial samples at 12 and 24 hours. These values were comparable to 12- and 24-hour values in the present experiment.

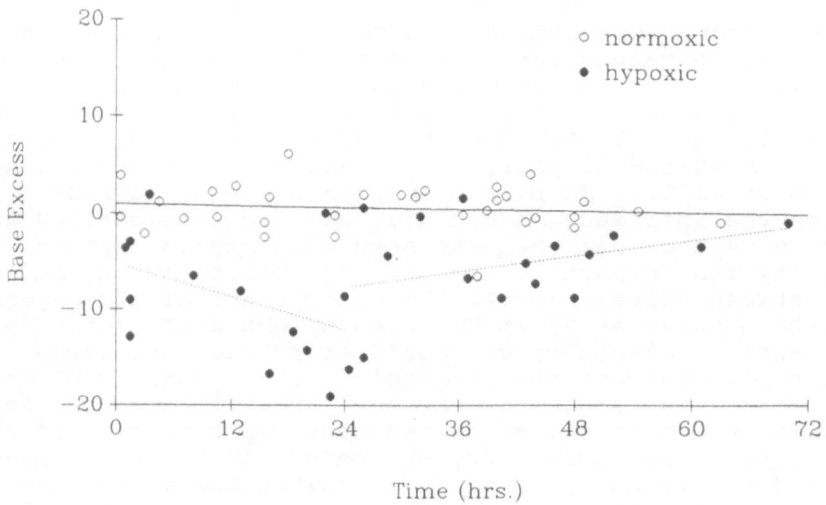

Figure 3. Effect of hypoxia on base excess values in newborn calves.

802

In conclusion, the metabolic effects of a 24-hour hypoxic episode in the newborn calf are relatively mild, consisting primarily of the development of a primary metabolic acidosis due in large part to accummulation of lactate. A relatively large base deficit was incurred concurrently. Most values returned to normal following return of P_{O2} to normal levels at 24 hours. The exceptions were values for arterial P_{CO2} which dropped below normoxic values during the post-hypoxic period. Developmental delays due to hypoxia were noted in colostrum-fed calves, but not in fasted or milk-fed calves.

REFERENCES

Alexander, G. 1975. Body temperature control in mammalian young. Br. Med. Bull. 31:62-68.

Ballard, F.J. 1971. The development of gluconeogenesis in rat liver. Biochem. J. 124:265-274.

Blanco, C.E., G.S. Dawes, M.A. Hanson and H.B. McCooke. 1984. The response to hypoxia of arterial chemoreceptors in fetal sheep and new-born lambs. J. Physiol. 351:25-37.

Brodie, T.G., W.C. Cullis and W.D. Halliburton. 1910. The gaseous metabolism of the small intestine. Part II. The gaseous exchanges during the absorption of Witte's peptone. J. Physiol. 40:173-189.

Cross, K.W. and J.L. Malcolm. 1952. Evidence of carotid body and sinus activity in new-born and foetal animals. J. Physiol. 118:10P-11P.

Dawes, G.S. and J.C. Mott. 1959. The increase in oxygen consumption of the lamb after birth. J. Physiol. 146:295-315

Dawkins, M.J.R. 1961. Changes in glucose-6-phosphatase activity in liver and kidney at birth. Nature 191:72-73.

Detweiler, D.K. 1984. Regional and fetal circulations. In: M. J. Swenson (Ed): Duke's Physiology of Domestic Animals, pp 192-206. Comstock Publishing Associates Ithaca.

Edelstone, D.I. and I.R.Holtzmann.1981a. Oxygen consumption by the gastrointestinal tract and liver in conscious newborn lambs. Am. J. Physiol. 240:G297-G304.

Edelstone, D.I. and I.R.Holtzmann.1981b. Gastrointestinal tract O_2 uptake and regional blood flows during digestion in conscious newborn lambs. Am. J. Physiol. 241:G289-G293.

Eden, G.J. and M.A. Hanson. 1987. Maturation of the respiratory response to acute hypoxia in the newborn rat. J. Physiol. 392:1-9.

Eigenmann, V.U.J.E., E. Grunert and U. Koppe. 1981. Zur Spätasphyxie des Kalbes. Berl. Münch. Tierarztl. Wschr. 94:249-254.

Gahlenbeck, H., H. Frerking, A.M. Rathschlag-Schaefer and H. Bartels. 1968. Oxygen and carbon dioxide exchange across the cow placenta during the second part of pregnancy. Resp. Physiol. 4:119-131.

Hudson, D.J. 1966. Fitting segmented curves whose join points have to be estimated. J. Am. Stat. Assoc. 61: 1097-1129.

Maurer-Schweitzer, V.H., U. Wilhelm and K. Walser. 1977. Blutgas-und Säure-Basen-Verhältnisse bei lebensfrischen Kaiserschnittkälbern in den ersten 24 Lebensstunden. Berl. Munch. Tierarztl. Wschr. 90:215-218.

Moore, W.E. 1969. Acid-base and electrolyte changes in normal calves during the neonatal period. Am. J. Vet. Res. 30:1133-1138.

Mortola, J.P. 1987. Dynamics of breathing in newborn mammals. Physiol. Rev. 67:187-243.

Mortola, J.P., C.A. Morgan and V. Virgona. 1986. Respiratory adaptation to chronic hypoxia in newborn rats. J. Appl. Physiol. 61:1329-1336.

Reeves, J.T., F.S. Daoud and M. Gentry. 1972. Growth of the fetal calf and its arterial pressure, blood gases, and hematologic data. J. Appl. Physiol. 32:240-244.

Reeves, J.T. and J.E. Leathers. 1964. Circulatory changes following birth of the calf and the effect of hypoxia. Circ. Res. 15:343-354.

Schlerka, V.G., W. Petschenig and J. Jahn. 1979. Untersuchungen über die Blutgase, den Säure-Basen-Haushalt, Elektrolytgehalt, einige Enzyme und Inhaltsstoffe im Blut neugeborener Kälber. Dtsch. Tierarztl. Wschr. 86:95-100.

Stott, G.H., D.B. Marx, B.E. Menefee and G.T. Nightengale. 1979. Colostral immunoglobulin transfer in calves. I. Period of absorption. J. Dairy Sci. 62:1632-1638.

Villee, C.A. and D.D. Hagerman. 1958. Effect of oxygen deprivation on the metabolism of fetal and adult tissues. Am. J. Physiol. 194:457-464.

Waizenhofer, V.H. and M. Mulling. 1978. Untersuchungen über das Verhalten von pH_{akt}, P_{O2} und P_{CO2} im venösen, kapillaren und arteriellen Blut neugeborener Kälber. Berl. Münch. Tierarztl. Wschr. 91:173-176.

Warnes, D.M., R.F. Seamark and F.J. Ballard. 1977. The appearance of gluconeogenesis at birth in sheep. Biochem. J. 162:627-634.

WHOLE BODY

EFFECTS OF NORMOBARIC HYPEROXIA ON HEMODYNAMICS AND O_2 UTILIZATION IN CONSCIOUS DOGS

R. F. Lodato

Division of Pulmonary Medicine and Critical Care
University of Texas Health Science Center
Houston, Texas, USA

INTRODUCTION

It seems reasonable to speculate, as Priestley did in 1775 shortly after his discovery of oxygen, that breathing oxygen-enriched air at normal atmospheric pressure, or normobaric hyperoxia, might increase the overall rate of oxidative metabolism, according to the law of mass action. Indeed, the opposite may occur under resting conditions. Chapler, Cain, and Stainsby (1984) presented intriguing data in anesthetized dogs showing that hyperoxia can paradoxically decrease O_2 consumption. More recently, Lodato (1989) has reported that whole-body O_2 consumption is decreased during hyperoxia in intact conscious dogs. This paper briefly summarizes this latter work and, in addition, presents a new analysis of the relationship between O_2 consumption and O_2 transport during normobaric hyperoxia.

METHODS

The experimental methods have been previously detailed by Lodato (1989). In brief, six conditioned mongrel dogs, 4 males and 2 females, weighing 28 to 30 kg, were instrumented with chronically indwelling, fluid-filled, femoral arterial catheters, using sterile technique under general anesthesia. Care of the animals was in accordance with the recommendations of the American Association for Accreditation of Laboratory Animal Care and met all standards prescribed by the "Guide for the Care and Use of Laboratory Animals" [DHEW (DHHS) Publication No. (NIH) 78-23, revised 1978]. The dogs were studied only after they had fully recovered from surgery and had, in addition, a minimum of two weeks training to become accustomed to the laboratory and its personnel and to accept breathing from a mask (see below).

They were studied fasted and lying on their side, unsedated and fully conscious, in a quiet room thermostatically controlled to 23°C. A 7.5F flow-directed pulmonary artery oximetry/thermodilution catheter (American Edwards Laboratories) was inserted sterilely under local anesthesia. Stability of the dog's body temperature throughout the experiment was documented by continuously monitoring the catheter thermistor. The dogs were continuously attended to assure that they never fell asleep during the experiment.

The dogs breathed through a standard loose-fitting cone-shaped veterinary mask, which delivered air or oxygen-enriched air with a nominal $F_IO_2 = 0.75$, at 15 l/min.

Mean and phasic systemic arterial (SAP), pulmonary arterial (PAP), and right atrial (RAP) pressures were continuously recorded on a Hewlett-Packard recording system (model 7758B) using strain gauge manometers (Hewlett-Packard model 1290A), hydrostatically referenced to midchest level. The pulmonary arterial wedge pressure (PAWP) was determined by balloon inflation. Heart rate (HR) was determined from the arterial pressure waveform. Cardiac output (CO) was determined by averaging at least 6 measurements (3 before and 3 after each blood gas sample) obtained by thermodilution using an American Edwards Laboratories SAT-1 Oximeter Cardiac Output Computer and injectates of 5 ml iced physiologic saline. Cardiac output values were then normalized by dividing by the dog's body weight in kg. The pulmonary vascular pressure gradient was calculated as: PAP - PAWP. Systemic (SVR) and pulmonary (PVR) vascular resistances were calculated as: SVR = (SAP - RAP)/CO, and PVR = (PAP - PAWP)/CO. Stroke volume (SV) was calculated as: SV = CO/HR. Right (RVWR) and left (LVWR) ventricular steady-flow external work rates, or hydraulic power outputs (Milnor, 1982), were calculated as: RVWR = 0.0136 x (PAP - RAP) x CO, and LVWR = 0.0136 x (SAP - PAWP) x CO, where the constant 0.0136 converts to units of g·m/(min·kg). Right (RVSW) and left (LVSW) ventricular stroke works were calculated as: RVSW = RVWR/HR, and LVSW = LVWR/HR.

Systemic arterial (a) and mixed-venous (v) blood pH, CO_2 and O_2 tensions (PCO_2 and PO_2, torr), oxyhemoglobin saturation (SO_2, %), and hemoglobin concentration (Hb, g/dl) were determined from blood samples drawn anaerobically from the arterial line and distal port of the pulmonary artery catheter. The samples were analyzed immediately with an Instrumentation Laboratory System 1302 pH/Blood Gas Analyzer with a bath controlled to 37.0 ± 0.1 °C, and a Model IL 282 CO-Oximeter. The pH/Blood Gas Analyzer was calibrated with whole blood using an Instrumentation Laboratory 237 Tonometer (blood-gas factor of 1.016).

Systemic arterial oxygen content (CaO_2, ml O_2/dl) was calculated as: $CaO_2 = 1.34$ x Hb x SaO_2/100 + 0.003 x PaO_2. Mixed venous O_2 content (CvO_2) was likewise calculated. The arterial-venous difference in O_2 content (ml O_2/dl) was calculated as $CaO_2 - CvO_2$; O_2 delivery (ml O_2/min/kg) as CO x CaO_2; O_2 consumption (ml O_2/min/kg) as CO x ($CaO_2 - CvO_2$); and O_2 extraction ratio as O_2 consumption/ O_2 delivery.

Because of the critical importance of SvO_2 to the results and conclusions, several steps were taken to assure the validity of SvO_2 as measured by blood sampling. First, proximal positioning of the catheter tip in the pulmonary artery was confirmed from the continuous recording of the PAP waveform and the requirement that at least 0.9 ml inflation of the catheter balloon was needed to achieve wedging. Additionally, in one experiment, dorsal-ventral and lateral chest radiographs were obtained repeatedly throughout the experiment to confirm proximal positioning of the catheter tip in the pulmonary artery whenever the balloon was fully deflated (as for blood gas sampling). On deflation of the balloon from its wedge position, the radiographs (Fig. 1) showed that the tip typically recoiled at least 5 cm more proximally in the pulmonary artery. Second, the oximetry catheter-computer system was calibrated in vivo against the CO-Oximeter and was used to continuously monitor and record SvO_2 to assure stability of the SvO_2 values, particularly during blood gas sampling, and to corroborate the magnitude of the change in SvO_2 on switching between air and O_2. Third, the paired mixed-venous and arterial blood samples were drawn simultaneously, slowly, and continuously over 30 sec. The protocol

808

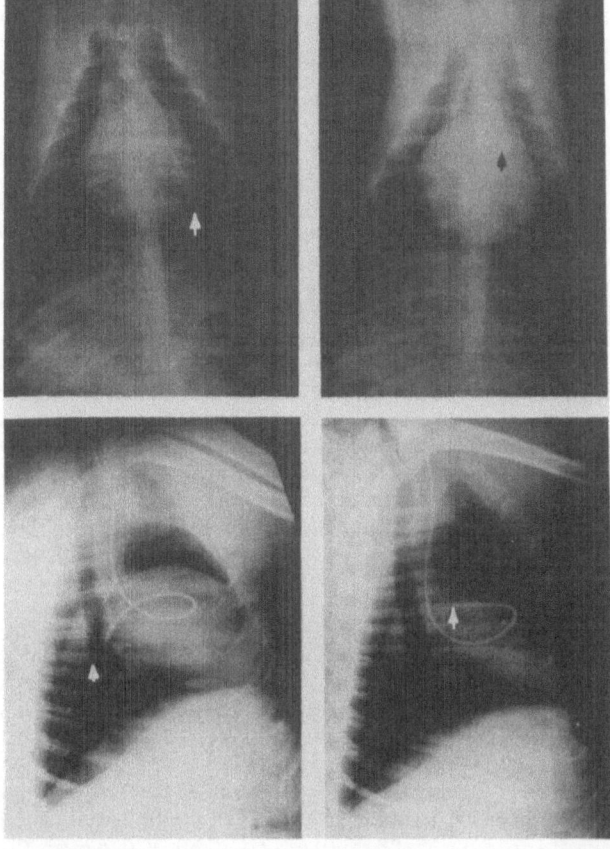

Fig. 1. Dorsoventral (upper panel) and lateral (lower panel) chest radiographs in a conscious dog showing the position of the tip (arrows) of the pulmonary artery catheter in the "wedged" position with the balloon inflated (left panel) and on subsequent deflation (right panel). Deflation caused the catheter tip to recoil about 5 cm into the more proximal pulmonary artery, precluding the possibility of contaminating a mixed-venous blood sample by "arterialized" blood.

was as follows. The initial set of hemodynamic and blood gas measurements was obtained while the dog breathed room air without a mask. Then the mask was applied, and for the next 5 h the inspired gas was alternated hourly between air and O_2, beginning with air. During the entire 5 h period, sets of hemodynamics and blood gas data were obtained every 20 min.

For statistical analysis the effect of hyperoxia on each parameter was assessed using a paired 2-tailed Student's t test to compare the mean value for the group of 6 dogs during air breathing with the group mean during hyperoxia. To enable comparisons both among and within the five 1-hour periods, one-way analysis of variance with repeated measures and, when F was significant, orthogonal contrasts or modified-t tests (Winer,

1971) were done. Least-squares linear regression analysis was used to study the relationship between O_2 consumption and O_2 delivery among the dogs; the slopes and intercepts were compared by t tests. All values are reported as mean $\pm$ SEM.

RESULTS

The basic hemodynamics and gas exchange data have been reported elsewhere (Lodato, 1989) previously and are summarized here before presenting the results on the O_2 consumption vs. O_2 delivery relationship.

Hyperoxia increased PaO_2 from 95 ± 2 to 475 ± 18 torr and PvO_2 from 43 ± 1 to 56 ± 3 torr. Hyperoxia decreased both the heart rate (P=0.0018) and the cardiac output (P=0.0024). Right atrial pressure rose slightly, but systemic arterial, pulmonary arterial, and pulmonary arterial wedge pressures did not change. Stroke volume also did not change with hyperoxia, but systemic vascular resistance increased (P=0.0167), while both the pulmonary vascular pressure gradient (PAP - PAWP) and pulmonary vascular resistance fell. Hyperoxia also decreased right ventricular stroke work (P=0.0149), right ventricular work rate (P=0.0058), and left ventricular work rate (P<0.0001), but did not change left ventricular stroke work.

The group mean Hb was 11.6 ± 0.7 g/dl. Hyperoxia decreased mixed venous pH (P=0.0166) and increased mixed venous PCO_2 (P=0.0478), but did not change arterial pH or PCO_2. As expected, hyperoxia increased (P<0.0001) the PO_2, SO_2, and O_2 contents of both the mixed venous and the arterial blood. Unexpectedly, however, hyperoxia decreased the arterial-venous difference in O_2 content, from 3.96 ± 0.22 to 3.53 ± 0.16 ml O_2/dl (P=0.0025). Paradoxically, O_2 delivery did not change (from 23.63 ± 4.63 to 23.40 ± 4.78 ml O_2/min/kg; P=0.51). Most notably, hyperoxia decreased O_2 consumption from 5.88 ± 0.68 to 4.80 ± 0.62 ml O_2/min/kg (P=0.0002). It is important to note that even within each dog statistically significant decreases in O_2 consumption were also observed, with P values for individual dogs ranging from P=0.0345 to P=0.0002. O_2 extraction ratio for the group also decreased, from 0.268 ± 0.022 to 0.217 ± 0.016 (P=0.0008). Fig. 2 shows the time course for O_2 consumption for the group of dogs as the inspired gas was alternated hourly between air and O_2. Each point is expressed as a percent of the mean of the first 1-hour control period. Following each switch in inspired gas, a new steady state level of O_2 consumption was already fully developed by 20 min into the hour and was maintained for the remainder of the hour (F not significant within any of the 1-hour periods). The mean within each 1-hour period was significantly different from that of the immediately preceding hour. These results indicate that the hyperoxia-induced decrease in O_2 consumption was both reproducible and reversible. Although there was no significant difference between the means of the two 1-hour periods of hyperoxia (P=0.77), there were differences among the means of the three 1-hour periods of air breathing (P=0.0097). Considering only the three 1-hour periods of breathing air, the first period was greater than both the second (P=0.0037) and the third (P=0.0205), but there was no difference between the second and third periods (P=0.45). This failure of O_2 consumption to completely return to the first-hour control level suggests a persistence or "lingering" effect of hyperoxia. In a separate preliminary study on the time course of the heart rate response to hyperoxia, a similar persistence-like effect was noted for heart rate. Fig. 3 shows an example of this pattern of heart rate response in a single conscious dog. It shows a continuous time series recording of arterial (by pulse oximetry) and mixed venous oxyhemoglobin saturation and heart rate (by cardiotachometer) as the inspired gas was alternated every

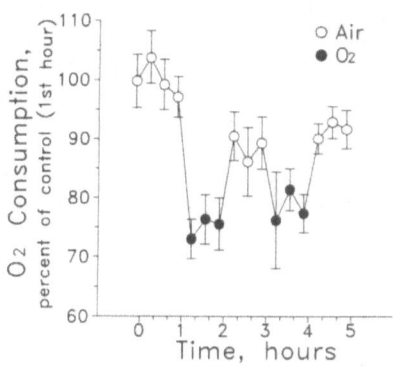

Fig. 2. Time course of resting whole-body O_2 consumption
as the inspired gas was alternated hourly between
air and O_2 (hyperoxia) in 6 conscious dogs.
Hyperoxia reversibly and reproducibly decreased
O_2 consumption. See text for details. (From
Lodato, 1989)

20 minutes between air and O_2; with each period of hyperoxia the heart rate
decreased, but on return to normoxia, the heart rate returned only
partially toward its normoxia baseline. It should be emphasized that
these latter results are preliminary and require control experiments to
differentiate a true persistence effect of hyperoxia from a simple time-
dependent effect.

Figure 4 shows the relationship between O_2 consumption and O_2 delivery
among all the dogs for air versus hyperoxia. Each dog is represented by
its 10 determinations (i.e., 10 points) on air and its 6 determinations
during hyperoxia. The slopes of the two regression lines shown are
not different (P=0.525), but the intercept for hyperoxia is decreased
(P=0.030), suggesting that hyperoxia causes a parallel shift downward in
the O_2 consumption-O_2 delivery relationship. Figure 5 shows the O_2
extraction ratios corresponding to the data points in Figure 4; the
smooth-curve fits to the data were determined directly from the regression
line parameters in Figure 4. Again, hyperoxia appears to depress O_2
extraction ratio at all levels of O_2 delivery.

DISCUSSION

The principal findings that normobaric hyperoxia decreased resting
whole-body O_2 consumption and produced associated hemodynamic changes in
conscious dogs has been recently reported in full elsewhere (Lodato,
1989). This report extends this work by presenting a simple analysis of
the effect of hyperoxia on the relationship between O_2 consumption and O_2
delivery. The nature of the relationship between O_2 consumption and O_2
delivery has received considerable attention recently (Kreuzer and Cain,
1985). Some generalizations have emerged, including that for most tissues
(the highly metabolic heart is an important exception), O_2 delivery is
sufficiently greater than the needs of the tissue. This excess
availability of O_2 apparently allows the tissues to regulate their
individual O_2 consumption according to their metabolic demand for O_2 and
relatively independently of O_2 delivery. As perturbations in O_2 delivery

occur within the physiological range, the tissues alter their O_2 extraction ratios, largely by capillary recruitment and consequent decrease in diffusion distance, in the direction opposite to the perturbation in O_2 delivery. A natural limit to the effectiveness of this autoregulation of O_2 consumption occurs at very low levels of O_2 delivery, below which level O_2 consumption also falls. These concepts have provided a ready explanation for much of the experimental data which has been generated. When plotted using O_2 consumption as the ordinate and O_2 delivery as the

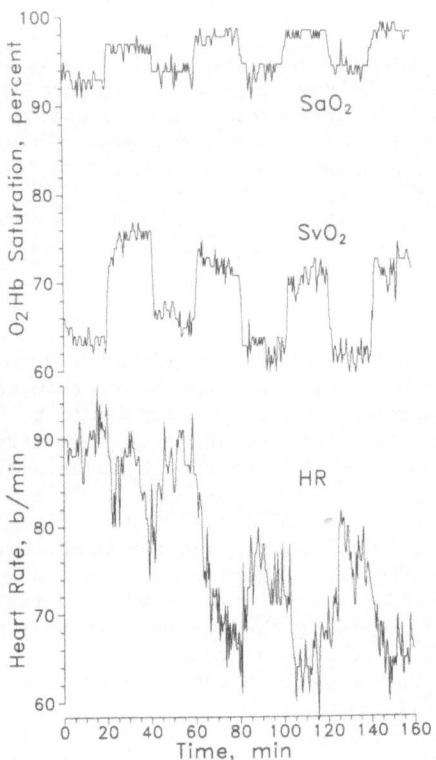

Fig. 3. Continuous time series recordings of arterial (SaO_2) and mixed-venous oxyhemoglobin saturation (SvO_2) and heart rate (HR) in a single conscious dog as the inspired gas was alternated every 20 min between air and O_2 (hyperoxia). Heart rate decreased during hyperoxia but remained below baseline on return to air.

abscissa, a _horizontal_ trajectory is observed for modest perturbations in O_2 delivery over the physiological range. In contrast, in critically ill patients, most notably those with the adult respiratory distress syndrome, attention has been focused on a pathological dependence of O_2 consumption on O_2 delivery (Kreuzer and Cain, 1985). When plotted, an _oblique_ trajectory with a positive slope is noted over much of the physiological range of O_2 delivery, suggesting inadequate autoregulatory reserves at the microcirculatory level. In contrast to the two situations above, the data

812

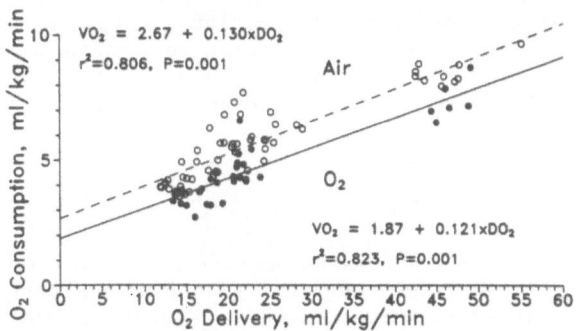

Fig. 4. Relationship between O_2 consumption and O_2 delivery among all 6 dogs for air versus hyperoxia. Hyperoxia produced a unique parallel shift downward in the O_2 consumption-O_2 delivery relationship. See text for details.

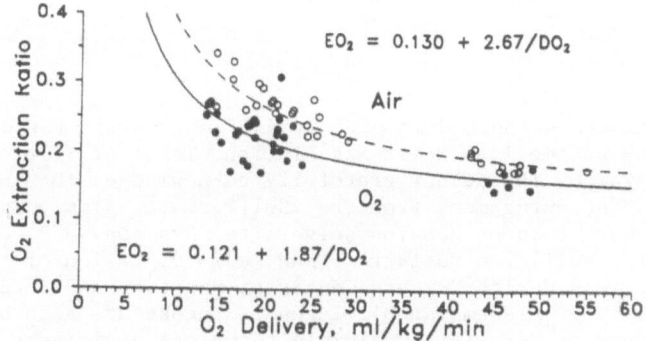

Fig. 5. Oxygen extraction ratios corresponding to the data points in Fig. 4. See text for details.

of the present study when plotted (Fig. 4) show two features. First, there are modest oblique (positive) slopes, of a common magnitude, for both normoxia and hyperoxia. Second, the challenge with hyperoxia appears to produce a <u>vertical</u> downward displacement of O_2 consumption with respect to O_2 delivery. This vertical trajectory was seen within dogs as well as among dogs, accounting for the parallel shift downward in O_2 consumption. This unexpected pattern in the O_2 consumption-delivery relationship induced by hyperoxia appears to represent a unique physiological phenomenon.

The mechanisms by which hyperoxia could induce such a decrease in O_2 consumption, in the face of an unchanged O_2 supply, are obscure. The possibilities would include that hyperoxia may produce: 1) systemic cellular O_2 toxicity (Lambertsen, 1978); 2) a paradoxically inadequate O_2 supply at the microcirculatory level, as described by Bourdeau-Martini et al (1974) and Lund et al (1980); or 3) a facultative decrease in O_2 demand, as has been demonstrated in mammals under certain physiological challenges unrelated to hyperoxia (Cain, 1987). Each of these potential mechanisms, acting alone, would be expected to produce an oblique orientation to the O_2 consumption-delivery relationship, and therefore, do not readily predict the vertical displacement in the relationship observed in this study.

SUMMARY

Normobaric hyperoxia decreased resting whole-body O_2 consumption in conscious dogs by equal (Fick) contributions from decreases in cardiac output and in the arterial-venous difference in O_2 content. The decrease in O_2 consumption was fully developed by 20 min, was maintained for at least 1 h, and was both reversible and reproducible. Hyperoxia also decreased heart rate, right and left ventricular work rates (and there-fore, presumably myocardial O_2 consumption), and pulmonary vascular resis-tance; and increased systemic vascular resistance and right atrial pres-sure. Paradoxically, hyperoxia did not change O_2 delivery. This latter observation together with the decrease in O_2 consumption produced a unique vertical orientation to the O_2 consumption-delivery relationship induced by hyperoxia. It is concluded that hyperoxia may decrease metabolic rate and substantially alter hemodynamics, which may have important implications for understanding the metabolic regulation of oxygen utilization and for the medical and nonmedical uses of oxygen.

ACKNOWLEDGEMENTS

This work was supported in part by National Institutes of Health PHS HL-06815 and by an American Heart Association, Texas Affiliate, Grant-in-Aid Award. The author gratefully acknowledges the invaluable discussions and encouragement from Drs. Sol Permutt, Jimmie T. Sylvester, and Roy G. Brower of Johns Hopkins University; Drs. David R. Dantzker, Martin J. Tobin, Guillermo Gutierrez, and Kevin D. Fallon of the University of Texas Health Science Center at Houston; and Stephen M. Cain of the University of Alabama at Birmingham. Thanks are also owed to Rosie Cousins and Kevin B. Kern for invaluable technical assistance, and to Jeri Lynn Bastin for secretarial and other support.

REFERENCES

Bourdeau-Martini, J., Odoroff, C.L., and Honig, C.R., 1974, Dual effect of oxygen on magnitude and uniformity of coronary intercapillary distance, Am. J. Physiol., 226(4):800-810.
Cain, S. M., 1987, Gas Exchange in Hypoxia, Apnea, and Hyperoxia, Chapter 19 in: "Handbook of Physiology", Sec. 3 "Respiration," Vol. 4., "Gas Exchange," Fishman, A. (ed), Am. Physiol. Soc., Maryland.
Chapler, C. K., Cain, S.M., and Stainsby, W. N., 1984, The effects of hyperoxia on oxygen uptake during acute anemia, Can. J. Physiol. Pharmacol., 62:809-814.
Kreuzer, F., and Cain, S. M., 1985, Regulation of the peripheral vasculature and tissue oxygenation in health and disease, Crit. Care Clin., 1(3):453-470.
Lambertsen, C. J., 1978, Effects of hyperoxia on organs and their tissues, in: "Extrapulmonary Manifestations of Respiratory Disease", Robin E. D., ed, "Lung Biology in Health and Disease", New York, Marcel Dekker, Vol. 8, p 239-303.
Lodato, R. F., 1989, Decreased O_2 consumption and cardiac output during normobaric hyperoxia in conscious dogs, J. Appl. Physiol., 67(4):1551-1559.

Lund, N., Jorfeldt, L., and Lewis, D. H., 1980, Skeletal muscle oxygen pressure fields in healthy human volunteers, Acta Anaesth. Scand., 24:272-278.

Milnor, W. R., 1982, "Hemodynamics," Baltimore, MD: Williams & Wilkins.

Winer, B. J., 1971, "Statistical Principles in Experimental Design," 2nd ed., McGraw-Hill Book Co., New York.

ACID-BASE CHARACTERISTICS OF STEADY-STATE EXERCISE IN RATS ADAPTED TO SIMULATED ALTITUDE

Norberto C. Gonzalez, Susan Dolezal and Richard L. Clancy

Department of Physiology
University of Kansas Medical Center
Kansas City, KS 66103, USA

INTRODUCTION

Acclimation to altitude hypoxia results in adaptive changes which ultimately lead to an improvement in tissue oxygen delivery (for review, see Bouverot, 1985). Some of these adaptations produce changes in systems not directly related to oxygen transport. An example of this are the changes in the acid-base balance that follow prolonged exposure to altitude. The hyperventilation leads to hypocapnia and extra- and intracellular bicarbonate depletion (Freeman and Fenn, 1953, Olson and Dempsey, 1979, Bouverot, 1985, Gonzalez and Clancy, 1986 a,b) which tends to lower the buffer value of extra- and intracellular fluids. On the other hand, the increased hemoglobin concentration associated with hypoxia results in an increase in the non-bicarbonate buffer value of blood (Gonzalez and Clancy, 1986a). These features, coupled with differences in the rate of renal excretion of acid-base equivalents (Widener et al., 1986), modify the responses of hypoxia-adapted animals to challenges in the acid-base balance.

Exercise represents an interesting model for the study of the regulation of acid-base balance since it involves an increase in the generation of both volatile and non-volatile acid equivalents. Secondly, the changes in the oxygen transport system that occur in exercise influence the acid-base balance, and vice-versa. While the acid-base consequences of exercise in normoxia have been studied extensively (Parkhouse and McKenzie, 1984), less is known concerning this subject on animals acclimated to altitude hypoxia. The rat is frequently used in studies of exercise, and the cardiovascular (Flaim et al., 1979, Laughlin and Armstrong, 1982) respiratory and acid-base changes (Fregosi and Dempsey, 1984) that occur during steady-state locomotory exercise have been described. On the other hand, information on the effect of exercise on circulation, respiration and acid-base balance of altitude-adapted rats is scarce. The objective of these experiments was to describe some of the respiratory and acid-base characteristics of steady-state exercise in rats adapted to simulated altitude.

METHODS

Male Sprague-Dawley rats, 200-225 g, were maintained for 3 weeks in a chamber where barometric pressure was kept at 370-380 mmHg (3WHx). This resulted in a P_{O_2} of moist inspired air of 68-70 mmHg. Hypobaria was interrupted 3 times a week for approximately 30 min. each time in order to clean the cages and feed and water the animals. At two weeks of hypoxia, the animals were removed from the chamber, anesthetized by methoxyflurane inhalation and a PE50 catheter introduced in the right carotid artery. The catheter was filled with a solution of heparine in 50% glucose, flame-sealed, and exteriorized at the back of the neck. After recovery from anesthesia the animals were returned to the hypobaric chamber.

Oxygen Transport to Tissue XII, Edited by J. Piiper *et al.*
Plenum Press, New York, 1990

At 3 weeks, the animals were removed from the hypobaric chamber and placed in a sampling chamber which could hold up to six rats. PIo_2 of the sampling chamber was maintained at 68-70 mmHg by mixing air and nitrogen at ambient barometric pressure. Controls (Nx) were pair-fed littermates maintained at ambient barometric pressure throughout, but otherwise treated in the same manner.

The animals were removed one by one from the sampling chamber and placed in a treadmill. The treadmill consists of a 37 cm long track enclosed in a plexiglass box with a volume of 6 L. Gas enters and leaves the treadmill through separate inflow and outflow tubes. A fan insures rapid mixing of the gas in the treadmill. At a flow of 18 l/min, a change in FIo_2 of 0.01 is completed with a half time of 18 sec. Ports in the inflow and outflow tubes allow sampling of the gas entering and leaving the treadmill. The Po_2 of the inflowing gas can be adjusted by mixing O_2 and N_2 from CO_2-free gas cylinders; for 3WHx animals Po_2 was 68-70 mmHg; for the Nx animals it was 140-145 mmHg. The treadmill was tested for leaks by filling it with 100% nitrogen at a pressure of 5-10 cmH_2O and monitoring the O_2 concentration of the air in the immediate vicinity of the treadmill. Oxygen consumption ($\dot{V}o_2$) and CO_2 production ($\dot{V}co_2$, ml STPD/min·kg) were calculated from the measurement of the O_2 concentrations of the inflowing and the outflowing gas, the CO_2 concentration of the outflowing gas, and the volume flow of the outflowing gas, using standard steady-state gas exchange equations (Otis, 1964). O_2 concentration of gas entering and leaving the treadmill was measured continuously and simultaneously with a 2-channel Applied Electrochemistry O_2 analyzer. CO_2 concentration of the effluent gas was monitored continuously with a Beckman LB2 CO_2 analyzer. The inflowing gas contained no CO_2. For the measurement of effluent gas flow, the gas sampling ports were disconnected and all the effluent gas was collected in a calibrated spirometer during 1 minute. Airflow measurements were performed while the animal was at rest and immediately after a run was terminated. Since the flow of gas entering the treadmill was maintained constant, the only variation in the effluent gas flow was determined by a difference between the $\dot{V}o_2$ and $\dot{V}co_2$ of the animal. At flow rates of 18 l/min, a variation in respiratory exchange ratio from 0.5 to 1.5 at the highest level of $\dot{V}o_2$ measured would result in changes of flow of less than 0.1%. Measurements for $\dot{V}o_2$ and $\dot{V}co_2$ were made when a steady state was observed in the concentration of O_2 and CO_2 of the outflowing gas.

In most animals, two running protocols were followed: one for the determination of maximal $\dot{V}o_2$ ($\dot{V}o_2(max)$); another for the determination of acid-base and blood gas composition of arterial blood during submaximal steady-state exercise. For the determination of $\dot{V}o_2(max)$ the treadmill was set at a 10° angle. After a 90 sec warm-up period at a speed of 14 m/min, treadmill speed was increased by 7 m/min every 2 min. This time allowed stabilization of the O_2 and CO_2 concentration of the effluent gas. Speed was increased in this manner until the rat could not keep pace with the treadmill. $\dot{V}o_2(max)$ was defined as the $\dot{V}o_2$ at which an increment in work rate did not result in an increase in oxygen consumption. In this protocol no blood samples were taken. In the second protocol, after a 1 min warm-up at a speed of 14 m/min, treadmill speed and angle were set to provide a work rate between 60 and 95% of $\dot{V}o_2(max)$; this work rate was then maintained for 5-10 min. The indwelling catheter was connected to sampling ports placed on the top of the treadmill and arterial blood samples were obtained immediately before, and during the last minute of the run. Measurements for $\dot{V}o_2$ and $\dot{V}co_2$ were carried out simultaneously with blood sampling. The volume of blood withdrawn was replaced immediately after sampling with fresh blood obtained from donors of the corresponding experimental group. Each rat was run twice; first to establish $\dot{V}o_2(max)$, and the second time, usually 3 to 5 days later, at one level of submaximal exercise. After the second run the rats were used as blood donors.

Arterial blood samples were analyzed for pH, Po_2 and Pco_2 in a Radiometer BMS blood gas analyzer, and for Hb concentration and O_2 saturation in an OSM2 Radiometer Hemoxymeter. Blood lactate concentration was measured using the Sigma enzymatic assay. Blood gas measured values were corrected for the rectal temperature of the animal (Bradley et al., 1956). Rectal temperature was measured before and immediately (30-45 sec) after the run. Plasma HCO_3^- concentration was calculated from measured pH and Pco_2 values using pK of 6.1 and sCO_2 of 0.03 mmol/(L·mmHg).

Results are expressed as means ± SEM. The data were analyzed using analysis of variance. The F test was used to determine the existence of a significant difference between means. The means were ranked using Tukey test. A value of p < 0.05 was considered to indicate a significant difference.

RESULTS AND DISCUSSION

$\dot{V}o_2(max)$ (ml STPD/min·kg) was 70.3 ± 1.3 in Nx (N = 18), and 53.6 ± 1.7 in 3WHx (N = 19), (p < 0.01). $\dot{V}co_2(max)$ was 78.5 ± 1.9 and 58.3 ± 1.9 ml STPD/min·kg in Nx and 3WHx, respectively (p < 0.01). Although maximal oxygen uptake was decreased in the altitude-adapted rats, the $\dot{V}o_2$ observed for common levels of work rate was the same in both groups. This agrees with previous findings (Pugh et al., 1964) which show that, for a given level of work rate, $\dot{V}o_2$ is independent of inspired Po_2.

Fig. 1 shows arterial plasma pH plotted as a function of $\dot{V}o_2$. The data corresponds to the animals which were run at submaximal levels of exercise. The highest values of $\dot{V}o_2$ shown correspond to 97 and 95% of $\dot{V}o_2(max)$ of the 3WHx and Nx rats, respectively. No significant difference in resting arterial plasma pH was observed between 3WHx and Nx. As exercise intensity increased, pHa decreased in both groups. For a given $\dot{V}o_2$ (and work rate), pH was lower in 3WHx than in Nx. However, the pH observed at the highest $\dot{V}o_2$ was essentially the same in both groups.

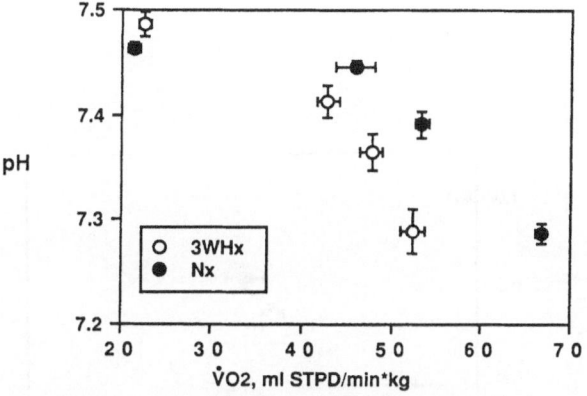

Fig. 1. Arterial plasma pH as a function of oxygen consumption during steady-state exercise in simulated altitude-adapted rats (3WHx) and normoxic controls (Nx). Each point in these and the following figures represents the mean of 18 rats for the resting values of Nx, and 19 rats for the resting values of 3WHx. The exercise values represent an average of 6-7 rats. The bars represent 1 SEM on either side of the mean.

Fig. 2 shows arterial blood lactate concentration as a function of $\dot{V}o_2$. Maximal lactate concentration, as well as maximal $\dot{V}o_2$, were higher in Nx than in 3WHx. However, no difference in blood lactate concentration was observed between Nx and 3WHx either in resting conditions, or during exercise at comparable $\dot{V}o_2$ levels. Previous determinations in other altitude-acclimated animals, including humans, have also shown that although maximal $\dot{V}o_2$ and blood lactate concentration are higher in normoxic subjects, blood lactate concentration during submaximal exercise is not different in sea level and altitude-acclimated subjects at comparable $\dot{V}o_2$ values (Hansen et. al, 1967, Cerretelli, 1980).

Since, for a given $\dot{V}o_2$, lactate concentration is the same in Nx and 3WHx, the lower plasma pH observed in 3WHx at comparable $\dot{V}o_2$ suggests that, when challenged with an acid load of comparable magnitude, the altitude-adapted rats have a lower ability to regulate their plasma pH than the Nx controls.

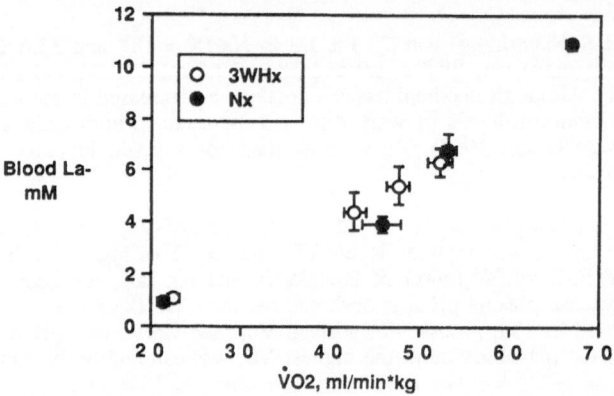

Fig. 2. Blood lactate concentration as a function of oxygen uptake during steady-state exercise.

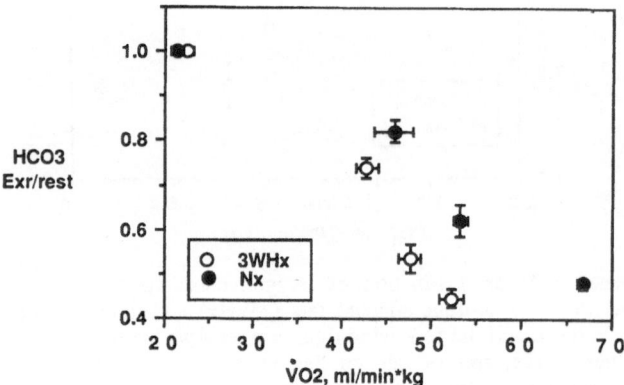

Fig. 3. Ratio of exercise (exr) to resting (rest) plasma bicarbonate concentration as a function of oxygen uptake during steady-state exercise.

Fig. 3 shows that the changes in plasma HCO_3^- concentration (expressed as the ratio of exercise/rest) mirror the changes in plasma pH. As exercise intensity increased, the relative decrease in plasma HCO_3^- concentration was larger in the 3WHx than in the Nx rats. Resting plasma HCO_3^- concentration was 24.2 ± 0.4 and 15.0 ± 0.5 mM in Nx and 3WHx, respectively. The HCO_3^- concentration at the highest exercise level was 11.8 ± 0.2 mM in Nx and 7.6 ± 0.9 mM in 3WHx.

Fig. 4 shows that exercise is associated with a decrease in $Paco_2$ below resting values in Nx as well as in 3WHx. Exercise hyperventilation has been previously shown to occur in normoxic rats (Fregosi and Dempsey, 1984). Our data indicate that hyperventilation also occurs in altitude-adapted rats exercising in hypoxia. The pattern of $Paco_2$ decrease was similar in both groups: at the lowest

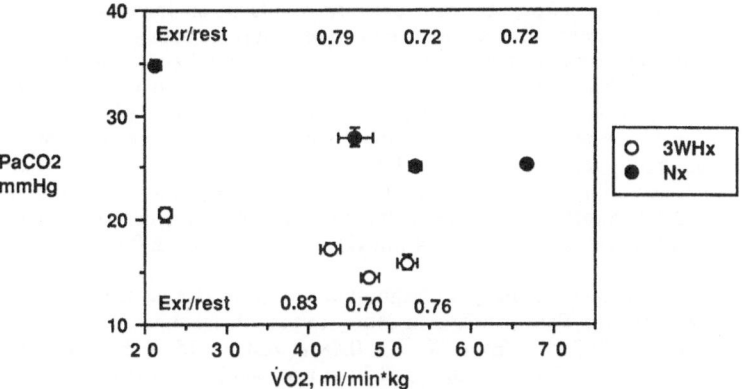

Fig. 4. Arterial Pco$_2$ as a function of oxygen uptake. The numbers indicate the average ratio exercise/resting (exr/rest) values at the various levels of exercise.

level of exercise, Paco$_2$ decreased to approximately 80% of the resting value; further increase in exercise intensity was accompanied by relatively smaller changes in Paco$_2$. No difference in Paco$_2$ was observed between the two highest levels of exercise in either group. These data indicate that the lower pH values observed in 3WHx at comparable exercise levels can not be explained on the basis of a less effective ventilatory response to exercise in this group.

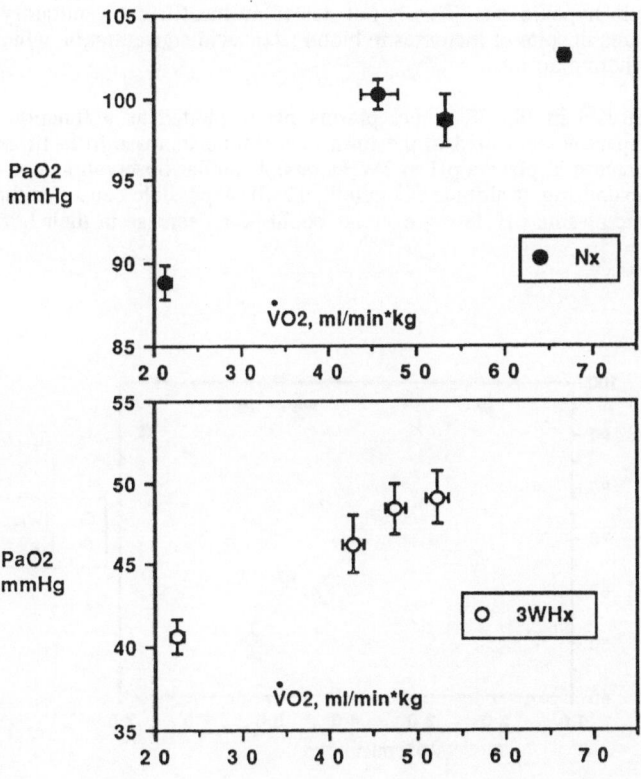

Fig. 5. Arterial Po$_2$ as a function of oxygen uptake during steady-state exercise. Upper panel: normoxic controls; lower panel: altitude-adapted rats.

The hyperventilation associated with exercise results in an increase in Pao_2 as well as a decrease in $Paco_2$. Fig. 5 shows that Pao_2 increases significantly in both groups. Exercise at altitude is sometimes accompanied by a decrease in Pao_2, which is more marked at higher levels of exercise and at highest altitudes. This decrease in Pao_2 is thought to be due to the combined effect of diffusion impairment and increased heterogeneity of $\dot{V}_A/\dot{Q}$ distribution (West et al., 1962, Wagner et al., 1987). However, our data do not support the presence of a major impairment of gas exchange in the 3WHx rats during exercise at this altitude. Alveolar Po_2, calculated from the alveolar ventilation equation assuming arterial and alveolar Pco_2 to be equal, increased from 44.4 ± 0.7 mm Hg at rest to 53.2 ± 0.9 mmHg at the highest level of exercise. Alveolar-to-arterial Po_2 difference did not change significantly: it was 3.9 ± 0.4 mmHg at rest, and 4.1 ± 0.8 at the highest $\dot{V}o_2$.

In spite of the increase in Pao_2, O_2 saturation of Hb decreased in the 3WHx group at the highest $\dot{V}o_2$ levels. (Fig. 6). The decrease in HbO_2 saturation resulted in a decrease in total O_2 concentration (Cao_2, ml STPD/dL) from 18.39 ± 0.48 at rest, to 15.54 ± 0.87 at the highest $\dot{V}o_2$ ($p < 0.01$). In contrast, Cao_2 in Nx was 19.64 ± 0.22 at rest, and 20.27 ± 0.15 at the highest exercise level (N.S.). The decrease in HbO_2 saturation seen in the 3WHx rats was probably due, at least in part, to the decrease in pH associated with exercise. Although 3WHx and Nx rats showed essentially the same arterial blood pH at the highest levels of exercise, the effect of low pH on oxygen saturation of hemoglobin should be more marked at the Po_2 values prevailing in the 3WHx rats, due to the shape of the oxyhemoglobin dissociaton curve. An additonal factor that may have contributed to lower the HbO_2 saturation is the increase in temperature that occured during exercise. The rectal temperature measured immediately after the run was completed increased from 38.1 ± 0.1 °C at rest to 40.9 ± 0.2 °C at the highest $\dot{V}o_2$ in the 3WHx.

In summary, rats adapted to simulated altitude exercising in hypoxia show a lower $\dot{V}o_2(max)$ than normoxic rats exercising at ambient PO_2. Both normoxic and hypoxic rats show hyperventilation, increase in blood lactate concentration and a decrease in pH. Although the hyperventilation results in an increase in Pao_2, arterial O_2 concentration actually decreases at the highest $\dot{V}o_2$ in the 3WHx rats. While plasma pH decreases as exercise intensity increases in both groups, at comparable $\dot{V}o_2$ plasma pH is lower in hypoxic rats. This is not due to an insufficient ventilatory response by the hypoxic rats, and occurs in spite of increases in blood lactic acid concentration which are comparable to those seen in the normoxic rats.

This is illustrated in Fig. 7, where plasma pH is plotted as a function of blood lactate concentration: a comparable acid load in the form of a similar increase in lactic acid concentration results in a larger decrease in plasma pH in 3WHx rats. A similar observation was made in altitude-acclimated humans exercising at altitude (Cerretelli, 1980). A possible cause for the lower ability of 3WHx rats to regulate plasma pH during exercise could be a decrease in their buffer capacity. The

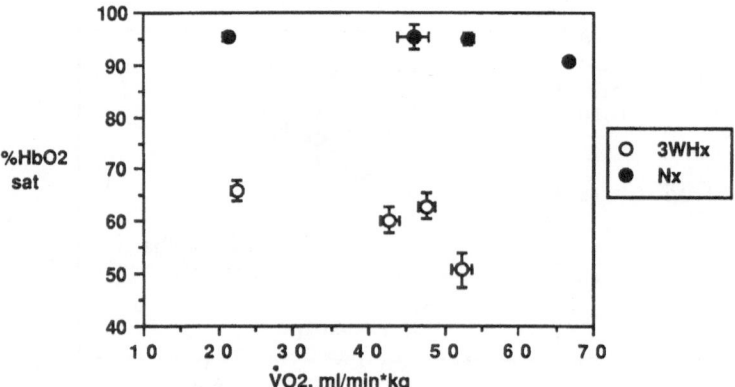

Fig. 6. Percent oxygen saturation of hemoglobin of arterial blood as a function of oxygen uptake during steady-state exercise.

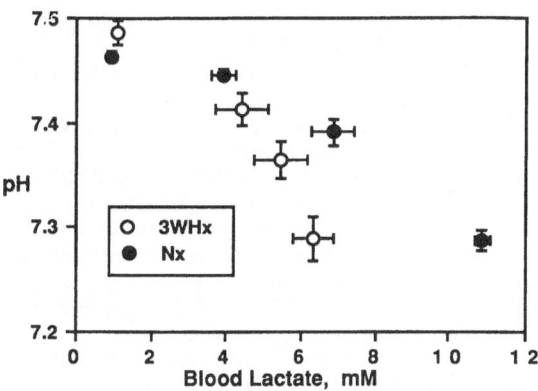

Fig. 7. Plasma pH of arterial blood as a function of blood lactate concentration during steady-state exercise.

decrease in blood bicarbonate concentration is only partially compensated by the increase in non-bicarbonate buffer value secondary to the elevated hemoglobin concentration (Gonzalez and Clancy, 1986a). Whether the decrease in cell bicarbonate of prolonged hypoxia (Freeman and Fenn, 1953, Gonzalez and Clancy, 1986b) is compensated by an increase in the concentration of non-bicarbonate buffers is not known. Simultaneous measurements of pH and lactate of intra- and extracellular fluids in exercise should help clarify this point.

Finally, although for comparable $\dot{V}_{O_2}$ plasma pH is lower in 3WHx, the pH reached at the highest level of exercise is the same as that of Nx rats. Plasma pH of arterial blood during exercise is a complex function of the magnitude of the acid load generated by the exercising muscles, and the effectiveness of buffering and removal of the proton equivalents generated. The fact that arterial plasma pH at maximal exercise is the same at two widely different absolute levels of $\dot{V}_{O_2}$ may be a coincidence. On the other hand, it may reflect a relationship between maximal oxygen uptake and acid-base balance; whether this is the case should be determined by further research.

ACKNOWLEDGEMENTS

This research was supported by NIH grant HL39443.

REFERENCES

Bouverot, P., 1985, Adaptation to Altitude Hypoxia in Vertebrates, Springer Verlag, Berlin.
Bradley, A.F., M. Stupfel and J.W. Severinghaus, 1956, Effect of temperature on P_{CO_2} and P_{O_2} of blood in vitro, J. Appl. Physiol., 9: 201-204.
Cerretelli, P., 1980, Gas exchange in altitude, in: "Pulmonary Gas Exchange," J.B. West, ed., Vol. II: Organism and Environment, pp. 97-147, Academic Press, New York.
Flaim, S.F., Minteer, W.J., Clark, D.P. and Zelis, R., 1979, Cardiovascular response to acute aquatic and treadmill exercise in the untrained rat, J. Appl. Physiol.: Respir. Environm. Exercise Physiol., 46: 302-308.
Freeman, J.W. and Fenn, W.O., 1953, Changes in carbon dioxide stores of rats due to atmospheres low in oxygen or rich in carbon dioxide, Am. J. Physiol., 174: 422-430.
Fregosi, R.F. and Dempsey, J.A., 1984, Arterial blood acid-base regulation during exercise in rats, J. Appl. Physiol.: Respir. Environm. Exercise Physiol., 57: 396-402.

Gonzalez, N.C. and Clancy, R.L., 1986a, Acid-base regulation in prolonged hypoxia: Effects of increased Pco_2, *Respir. Physiol.*, 64: 213-22.

Gonzalez, N.C. and Clancy, R.L., 1986b, Intracellular pH regulation during prolonged hypoxia in rats, *Respir. Physiol.*, 65: 331-339.

Hansen, J.E., Stelten, G.P. and Vogel, J.A., 1967, Arterial pyruvate, lactate, pH and Pco_2 during work at sea level and at high altitude, *J. Appl. Physiol.*, 23: 523-530.

Laughlin, M.H. and Armstrong, R.B., 1982, Muscular blood flow distribution patterns as a function of running speed in rats, *Am. J. Physiol.* 243 (Heart Circ. Physiol. 12): H296-H306.

Olson, E.B. and Dempsey, J.A., 1978, Rat as a model of humanlike ventilatory adaptation to chronic hypoxia, *J. Appl. Physiol.: Respir. Environm. Exercise Physiol.*, 44: 763-769.

Otis, A.B., 1964, Quantitative relationships in steady-state gas exchange, in: "Handbook of Physiology," Section 3: Respiration, Vol. I., W.O. Fenn and H. Rahn, ed., pp. 681-698, American Physiological Society, Washington, D.C.

Parkhouse, W.S. and McKenzie, D.C., 1984, Possible contribution of skeletal muscle buffers to enhanced anaerobic performance: a brief review. *Med. Sci. Sports Exerc.*, 16: 328-338.

Pugh, L.G.C.E., Gill, M.B., Lahiri, S., Milledge, J.S., Ward, M.P. and West, J.B., 1964, Muscular exercise at great altitudes, *J. Appl. Physiol.*, 19: 431-440.

Wagner, P.D., Sutton, J.R., Reeves, J.T., Cymerman, A., Groves, B.M. and Malconian, M.K., 1987, Operation Everest II: Pulmonary gas exchange during a simulated ascent to Mount Everest, *J. Appl. Physiol.: Respir. Environm. Exercise Physiol.*, 63: 2348-2359.

West, J.B., Lahiri, S., Gill, M.B., Milledge, J.S., Pugh, L.G.C.E. and Ward, M.P., 1962, Arterial oxygen saturation during exercise at high altitude, *J. Appl. Physiol.*, 17: 617-621.

Widener, G., Sullivan, L.P., Clancy, R.L. and Gonzalez, N.C., 1986, Renal compensation of hypercapnia during prolonged hypoxia, *Respir. Physiol.*, 65: 341-350.

EXPERIMENTAL SUPPORT FOR THE THEORY OF DIFFUSION LIMITATION
OF MAXIMUM OXYGEN UPTAKE

Peter D. Wagner,[1] Josep Roca,[2] Michael C. Hogan,[1] David C.
Poole[1], D.C. Bebout[1], and Pierre Haab[3]

[1]Department of Medicine, M-023A, University of California San
Diego, La Jolla, CA 92093, USA
[2]Servei de Pneumologia, Hospital Clinic, Villarroel, 170, 08036
Barcelona, Spain
[3]Institut de Physiologie, Université de Fribourg Suisse, Rue du
Musée 5, CH-1700 Fribourg, Switzerland

INTRODUCTION

In spite of many years of investigation, the question of what
determines maximum oxygen uptake ($\dot{V}O_2$max) remains controversial. While it
has been suggested that $\dot{V}O_2$max is not limited by oxygen supply (Jobsis and
Stainsby, 1986), the preponderance of experimental evidence suggests the
contrary. Thus, increasing oxygen supply by breathing hyperoxic gas
mixtures produces small but measurable increments in maximum $\dot{V}O_2$ (Welch,
1983). Other forms of increasing convective oxygen delivery to muscle
also result in increased $\dot{V}O_2$max. Thus, as reviewed by Gledhill (1982),
blood transfusion acutely increases $\dot{V}O_2$max. In the opposite direction, it
is well-known that $\dot{V}O_2$max can be acutely reduced by breathing hypoxic gas
mixtures, and in the extreme with chronic hypoxia, $\dot{V}O_2$max can be reduced
to some 25% of sea level values when the inspired PO_2 is equivalent to
that of the summit of Mt. Everest (Ward et al., 1989). It would thus seem
reasonable to conclude that oxygen supply is in fact limiting maximum
$\dot{V}O_2$, at least within an experimentally feasible range of investigation.
The question then becomes: where in the oxygen transport pathway from the
atmosphere to the mitochondria within the muscle are these limits set?
Based presumably upon the strong correlations observed repeatedly between
maximum oxygen uptake and convective oxygen delivery to the muscle by the
arterial blood (Pirnay et al., 1972; Horstman et al., 1976; Saltin, 1985),
it has frequently been suggested that maximum $\dot{V}O_2$ is limited by oxygen
delivery. This hypothesis cannot account for the residual oxygen left in
effluent muscle venous blood. Under normoxic conditions, it is common to
see femoral venous PO_2 in the range of 20-30 torr in normal individuals
cycling maximally (Roca et al., 1989). What is more, we recently found
both in intact normal human subjects and in the electrically stimulated,
isolated canine gastrocnemius that there was a strong linear relationship
between maximum oxygen uptake and either measured muscle venous PO_2 or
calculated mean muscle capillary PO_2 (Hogan et al., 1988; Hogan et al.,
1989; Roca et al., 1989). Based upon the data of the Rochester group
suggesting very low cytoplasmic PO_2 values at $\dot{V}O_2$max in normal muscle (1-3
torr, Honig et al., 1984; Gayeski and Honig, 1988), we have suggested that

the linear relationship, which passes through the origin, is a manifestation of the laws of diffusion in that if the rate of diffusion of oxygen from hemoglobin to the mitochondria is not sufficiently rapid, the flux of oxygen and hence $\dot{V}O_2max$ will be limited by the diffusive properties. In such a system, the flux of oxygen would be linearly related to the partial pressure driving gradient which is essentially equal to the mean capillary PO_2 since, according to Honig et al. above, intramyocyte PO_2 is so low.

The above collection of findings therefore suggested that convective oxygen delivery was only one part of the system that limited maximum $\dot{V}O_2$, and that subsequent peripheral muscle tissue diffusion of oxygen was an important additional constraining factor. This led to the hypothesis of diffusion limitation of $\dot{V}O_2max$ (Wagner, 1988a; Wagner, 1988b) which we proposed could account for current observations on how maximum $\dot{V}O_2$ is set. The hypothesis simply states that for a given oxygen delivery to the muscle blood vessels, the amount which can be extracted (by diffusion) is constrained by the diffusing capacity of the muscle. In this manner, increasing oxygen delivery increases $\dot{V}O_2max$ only by increasing the partial pressure of oxygen in the capillary, which when coupled to a given diffusing capacity (presumably essentially structurally determined), will account for the increase in $\dot{V}O_2max$.

In this poster, we bring together four experiments that support this integrative theory in which the diffusion of oxygen within muscle plays a critical role in setting maximum $\dot{V}O_2$.

Experiment 1

Figure 1 shows the relationships between maximum $\dot{V}O_2$ and muscle effluent venous PO_2, mean capillary PO_2, and convective oxygen delivery in electrically stimulated, isolated canine gastrocnemius performing at peak $\dot{V}O_2$. These are data taken from Hogan et al. (1988). The linear relationship between $\dot{V}O_2max$ and mean capillary PO_2 demonstrates consistency with the diffusion limitation hypothesis and therefore supports it. It does not exclude competing hypotheses, however. Notice that there is an equally good relationship between $\dot{V}O_2max$ and convective oxygen delivery in the same experiment (lower panel) as would be predicted from the diffusion limitation hypothesis (Wagner, 1988a; Wagner, 1988b). In other words, the data in Figure 1 are not only consistent with the basic hypothesis advanced in this presentation, but also with the observations of other authors on the linear relationship between delivery and uptake as expressed in the lower panel.

Experiment 2

In this study recently reported by Roca et al. (1989), normal human volunteers were requested to exercise on a cycle ergometer to maximum VO_2, breathing three different gas mixtures (21% oxygen, 15% oxygen and 12% oxygen) administered in random order to each subject. Maximum whole body $\dot{V}O_2$ was measured at each inspired concentration along with arterial and femoral venous PO_2 values. In this experiment we also had access to mixed venous blood, and data from the pulmonary artery are therefore shown as well. Figures 2 and 3 show data in a form analogous to that of Figure 1. Both individual subject data (Fig. 2) and group data are shown. The individual subject data are those obtained at maximal $\dot{V}O_2$ conditions (established prior to the study by a conventional ramp test), whereas the group data (Fig. 3) reflect both maximal conditions and a work load 10% greater than maximal in order to be sure of attaining $\dot{V}O_2max$. Both group and individual data show remarkably linear relationships between flux and

826

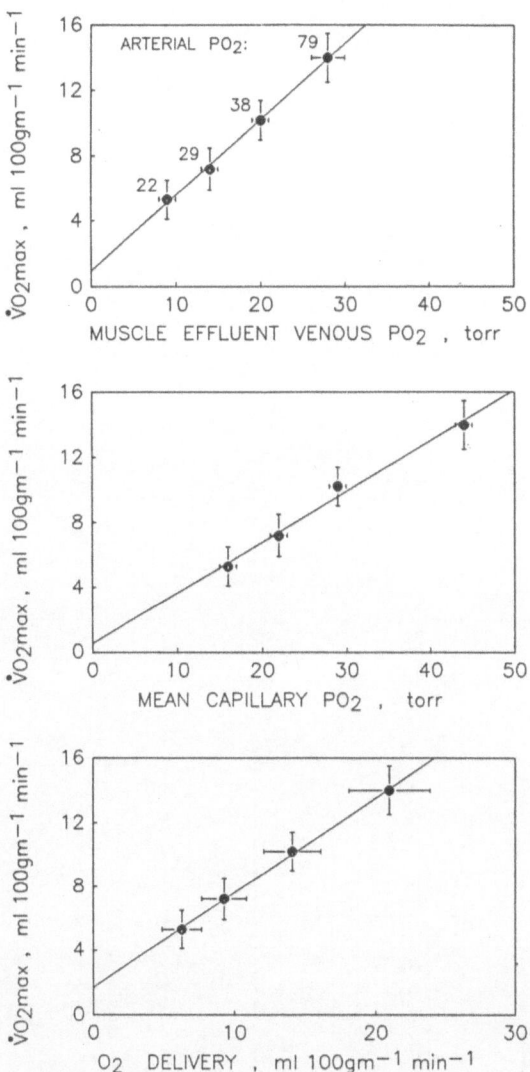

Fig. 1. Reduction with PaO_2 in maximum $\dot{V}O_2$ and in measured muscle effluent venous PO_2, calculated mean capillary PO_2, and measured arterial O_2 delivery in isolated canine gastrocnemius during maximal contractions produced by electrical stimulation. Group data are shown at four arterial PO_2 values ranging from about 80 to about 20 torr, applied in random sequence.

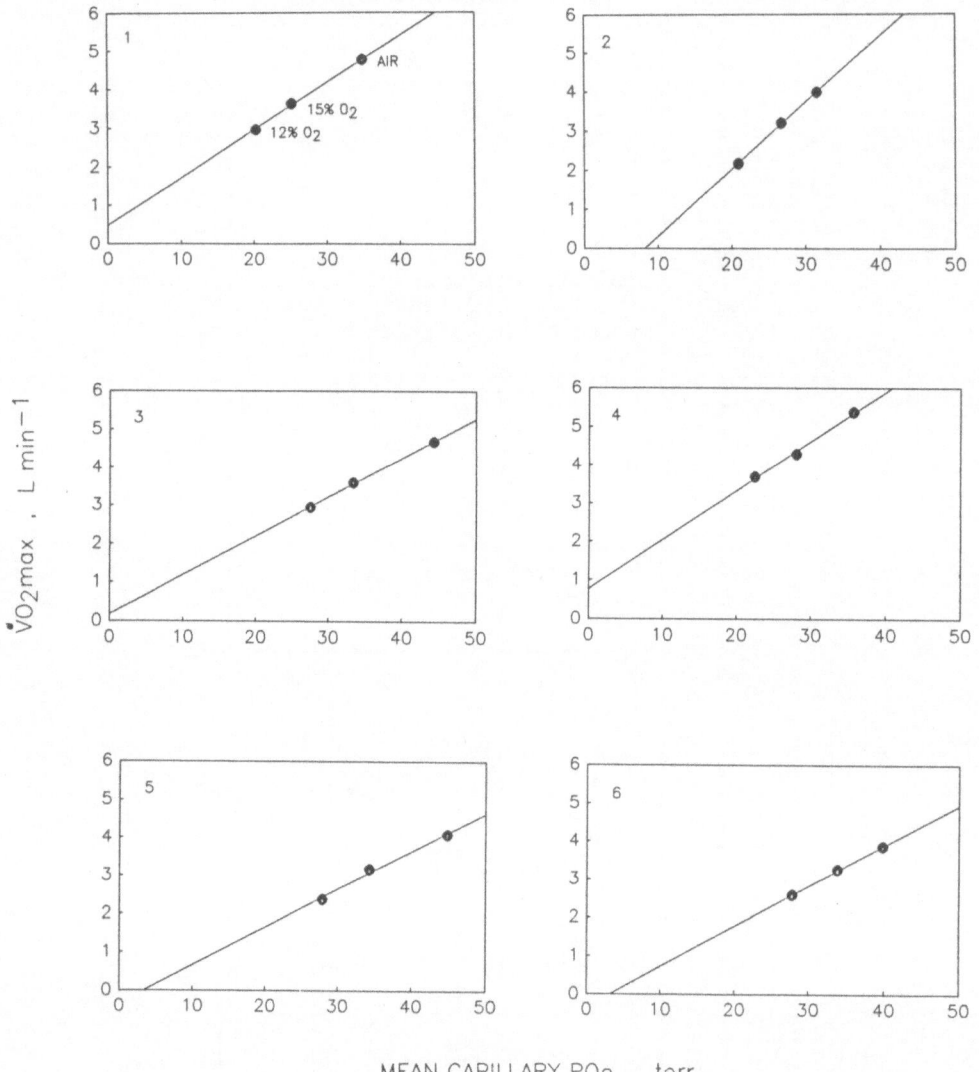

Y-axis label: $\dot{V}O_2max$, L min^{-1}

X-axis label: MEAN CAPILLARY PO$_2$, torr

Panel 1 labels: AIR, 15% O$_2$, 12% O$_2$

Fig. 2. Individual (n=6) normal human subject data showing the relationship between whole body maximum $\dot{V}O_2$ and calculated mean muscle capillary PO$_2$. Data are for maximal exercise and reflect three different inspired oxygen concentrations breathed in random order.

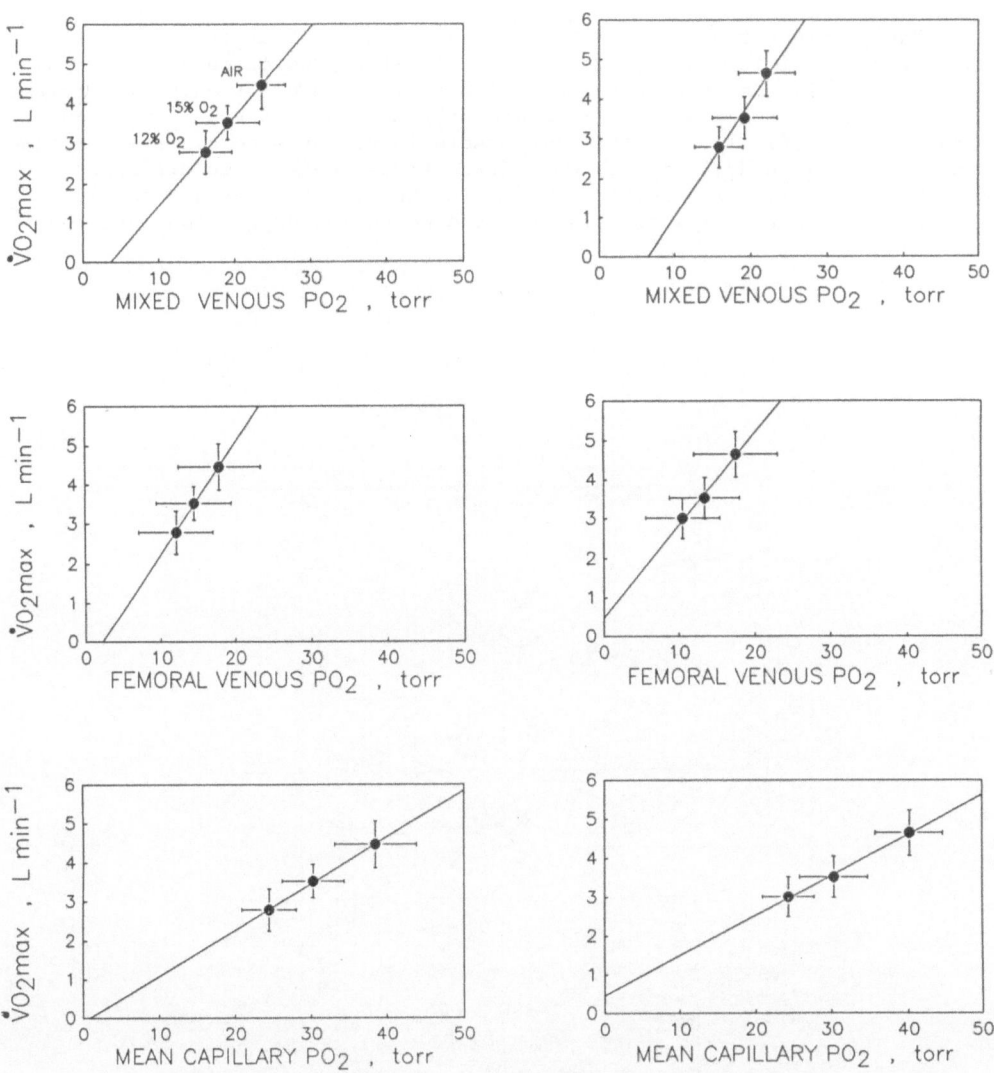

Fig. 3. Mean (n=6) human data at maximum $\dot{V}O_2$ (left) and also at a work load 10% greater (right). The linear relationships observed, with intercepts that as a group are not statistically different from zero, are consistent with the hypothesis of diffusive/convective interaction determining maximum $\dot{V}O_2$.

driving gradient as shown. Again, our interpretation is that these data are consistent with the importance of diffusion of oxygen as a limiting process in the setting of maximum $\dot{V}O_2$. Although not shown, these results also demonstrated as good correlations between maximum $\dot{V}O_2$ and oxygen delivery as in the canine experiments of Figure 1.

Experiment 3

The previous two experiments represent the earliest explorations of the diffusion limitation hypothesis. A specific prediction outlined in Experiment 3 is that the manner in which a given level of oxygen delivery is attained will play a role in determining the maximum $\dot{V}O_2$ associated with it. Specifically, if the same muscle is perfused to achieve the same convective oxygen delivery by two different strategies, two different maximum $\dot{V}O_2$'s will be observed. Thus, in the situation of a high blood flow combined with a low arterial PO_2 and hence low oxygen concentration,

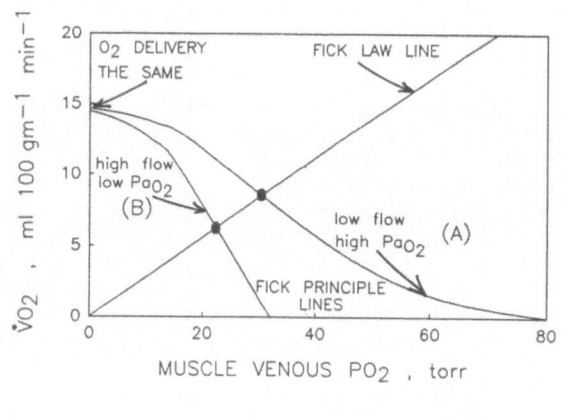

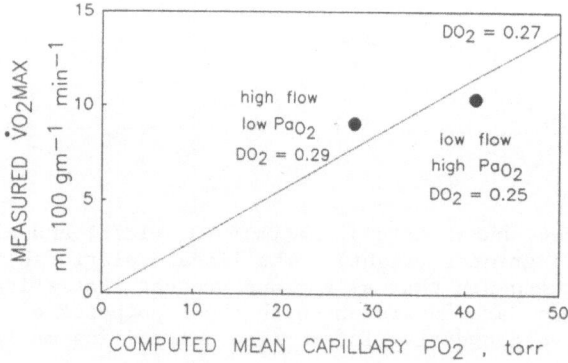

Fig. 4. Theory (top) and experimental results (bottom) showing how maximum $\dot{V}O_2$ and associated mean capillary PO_2 can be different at the same level of arterial oxygen delivery according to how the oxygen delivery was produced. The experimental data show concurrence with theoretical predictions, although in the low flow state, the results may have been affected by perfusion heterogeneity.

830

we predict maximum $\dot{V}O_2$ to be less than when oxygen delivery is at the same level but achieved by a low blood flow combined with a high arterial PO_2, and hence oxygen content. This prediction is presented in Figure 4 (upper panel) showing three lines intersecting at two points. The straight line through the origin reflects Fick's law of diffusion, namely, that maximum $\dot{V}O_2$ will be a linear function of the driving gradient, here assumed proportional to muscle effluent venous PO_2. For a given structural situation (i.e., a given specimen of muscle), the diffusing capacity would be constant as oxygen delivery is altered, such that $\dot{V}O_2max$ should lie somewhere along the straight line under the different conditions referred to above. The two curved lines represent graphical solutions to the Fick principle equation for given values of blood flow and arterial oxygen content as venous effluent PO_2 is increased from zero to the arterial value. Notice that when venous effluent PO_2 is zero, the value indicated by the Fick principle would be equivalent to oxygen delivery and be visualized as the Y intercept at the extreme lefthand edge of the plot.

As can be seen from the upper panel of Figure 4, the two curved lines representing graphical solutions to the Fick principle, meet with a common intercept, expressing the intended experimental design of achieving the same convective oxygen delivery. However, because the slope of the curved line is in part dependent upon blood flow, that associated with the high flow, low PO_2 state has a more negative (steeper slope) than in the other situation. Consequently, the points of intersection as shown by the solid circles will lie at different points along the line describing Fick's law of diffusion. As argued previously (Wagner, 1988a; Wagner, 1988b), the points of intersection reflect the predicted values of maximum $\dot{V}O_2$ (and the associated effluent muscle venous PO_2 values). The lower panel of Figure 4 gives the experimental results. In isolated, maximally stimulated canine gastrocnemius, the high flow, low PO_2 condition did in fact produce a lower maximum $\dot{V}O_2$ than the low flow high PO_2 condition. Similarly, effluent venous PO_2 as well as computed mean capillary PO_2 (shown in the figure) was less in the former condition as well. While the points did not lie precisely on a single straight line as predicted in the upper panel, the data certainly demonstrate that maximum $\dot{V}O_2$ cannot be a unique function of convective oxygen delivery since delivery was the same under both conditions. That the $\dot{V}O_2max$ for the low flow condition is a little less than would be predicted from that under the high flow condition and a constant diffusing capacity may well be explained by the development of perfusion heterogeneity under low flow conditions, but this postulate remains to be evaluated.

Experiment 4

Experiment 2 referred to above in normal human subjects explored the diffusion limitation hypothesis by comparing whole body $\dot{V}O_2$ measured from expired gas collection to PO_2 measured in the femoral vein during cycle exercise. In Experiment 4, by virtue of thermodilution blood flow measurements, it was possible to measure not only whole body $\dot{V}O_2$ but also, simultaneously, oxygen uptake in the exercising leg from the product of measured leg blood flow and arteriovenous oxygen concentration difference (Andersen and Saltin, 1985). This was done both before and after 18 days of high altitude exposure in a group of 7 volunteers. Measurements were made breathing sea level room air, and hypoxic gas equivalent to that breathed normally at the summit of Pike's Peak where the subjects were kept for the 18 days (altitude 4300 meters). The relationship between whole body $\dot{V}O_2max$ and femoral venous PO_2 under all conditions were again consistent with the linear relationships shown in Figure 2. This was also found to be the case when leg $\dot{V}O_2$ computed as above was related to femoral venous PO_2, and 18 days of residence on Pike's Peak had no effect on the slope of the line.

SUMMARY

The four experiments summarized above demonstrate that there is a strong relationship between both measured muscle venous PO_2 and calculated mean muscle capillary PO_2 and $\dot{V}O_2max$. This is true for whole body or exercising muscle $\dot{V}O_2max$, and is seen both in isolated canine gastrocnemius and intact man. This behavior is exactly what would be expected if the diffusing properties for oxygen in skeletal muscle play a constraining role in setting maximum $\dot{V}O_2$. These data therefore support the hypothesis we advanced (Wagner, 1988a; Wagner, 1988b), that it is a quantitative integrative relationship between convective and diffusive phenomena that combine to set maximum $\dot{V}O_2$. A specific prediction of this integrative hypothesis (i.e., the non-uniqueness of $\dot{V}O_2max$ as a function of convective oxygen delivery) was confirmed (Experiment 3). While at this point in time phenomena such as perfusion heterogeneity and muscle shunts cannot be quantitatively taken into account in such analyses, the remarkable concurrence between expectations of the hypothesis and experimental data continue to lend support to the basic idea that maximum $\dot{V}O_2$ is not limited by any single step of the oxygen transport pathway from atmosphere to mitochondria, but rather by the way in which each and every step combines with every other step to determine oxygen supply.

ACKNOWLEDGEMENTS

This study was funded by NIH grant HL 17731. The authors wish to thank all collaborators in the acquisition of experimental data, and also Tania Davisson for secretarial assistance in preparation of this manuscript.

REFERENCES

Andersen, P., and Saltin B., 1985, Maximum perfusion of skeletal muscle in man. <u>J. Physiol. Lond.</u>, 366:233-249.

Gayeski, T. E. J., and Honig, C. R., 1988, Intracellular PO_2 in long axis of individual fibers in working dog gracilis muscle. <u>Am. J. Physiol. (Heart Circ. Physiol.)</u>, 254:H1179-H1186.

Gledhill, N., 1982, Blood doping and related issues: a brief review. <u>Med. Sci. Sports Exerc.</u>, 14(3):183-189.

Hogan, M. C., Roca, J., Wagner, P. D., and West, J. B., 1988, Limitation of maximal O_2 uptake and performance by acute hypoxia in dog muscle in situ. <u>J. Appl. Physiol.</u>, 65(2):815-821.

Hogan, M. C., Roca, J., Wagner, P. D., and West, J. B., 1989, Dissociation of maximal O_2 uptake from O_2 delivery in canine gastrocnemius in situ. <u>J. Appl. Physiol.</u>, 66(3):1919-1226.

Honig, C. R., Gayeski, T. E. J., Federspiel, W., Clark Jr., A., and Clark, P., 1984, Muscle O_2 gradients from hemoglobin to cytochrome: new concepts, new complexities. <u>Adv. Exp. Med. Biol.</u>, 169:23-38.

Horstman, D. H., M. Gleser, and Delehunt, J., 1976, Effects of altering O_2 delivery on $\dot{V}O_2$ of isolated, in situ working muscle. <u>Am. J. Physiol.</u>, 230:327-334.

Jobsis, F. F., and Stainsby, W. N., 1986, Oxidation of NADH during contractions of circulated mammalian skeletal muscle. <u>Respir. Physiol.</u>, 45:2937-2941.

Pirnay, F., Lamy, M., Dujardin, J., Deroanne, R., and Petit, J. M., 1972, Analysis of femoral venous blood during maximum muscular exercise. <u>J. Appl. Physiol.</u>, 33(3):289-292.

Roca, J., Hogan, M. C., Story, D., Bebout, D. E., Haab, P., Gonzalez, R., Ueno, O., and Wagner, P. D., 1989, Evidence for tissue diffusion limitation of $\dot{V}O_2max$ in normal humans. <u>J. Appl. Physiol.</u>, 67(1):291-299.

Saltin, B., 1985, Hemodynamic adaptations to exercise. Am. J. Cardiol., 55:42D-47D.

Wagner, P. D., 1988a, An integrated view of the determinants of maximum oxygen uptake, in: "Oxygen Transfer from Atmosphere to Tissues", N. C. Gonzalez, and M. R. Fedde, eds., vol. 227, pp. 245-256. Plenum Press, New York.

Wagner, P. D., 1988b, The determinants of $\dot{V}O_2$max, in: "Annals of Sports Medicine", 4(4), pp. 196-212. Oxford Univ. Press.

Ward, M. P., Milledge, J. S., and West, J. B., eds., 1989, in: "High Altitude Medicine and Physiology". Univ. of Pennsylvania Press, Philadelphia.

Welch, H. G., 1983, Hyperoxia and human performance: a brief review. Med. Sci. Sports Med., 14(4):253-262.

EVIDENCE SUPPORTING THE EXISTENCE OF AN EXERCISE ANAEROBIC THRESHOLD

Akira Koike , Karlman Wasserman, William L. Beaver, Daniel Weiler-Ravell, David K. McKenzie, and Stephania Zanconato

Department of Medicine, University of California
Harbor-UCLA Medical Center
1000 W. Carson St., Torrance, CA 90509

INTRODUCTION

Since 5 liters of blood with normal arterial oxygen (O_2) concentration contains only 1 liter of O_2, the increase in muscle blood flow to support an O_2 requirement of 1 L/min, aerobically, must be greater than 5 L/min. The slope for the increase in cardiac output for an increase in oxygen uptake ($\dot{V}O_2$) is normally approximately 6 (Faulkner et al., 1977; Yamaguchi et al., 1986), suggesting that approximately 18% of the O_2 delivered to the exercising muscles remains at the venous end of the muscle capillary. This would result in an O_2 tension of approximately 15 mmHg. This is consistent with the 8 mmHg O_2 tension in exercising muscle below which the muscle lactate/pyruvate ratio was observed to abruptly increase by Bylund-Fellenius et al (1981). To maintain total aerobic metabolism, there must be little inhomogeneity in the muscle blood flow/O_2 consumption ratio ($Qm/\dot{V}O_2m$) since a ratio of 5.5 or less would obligate anaerobic metabolism (Figure 1). After maximal hyperemia, the $Qm/\dot{V}O_2m$ ratio must increase to obtain an increased capillary to mitochondrial PO_2 difference ($P(c-m)O_2$) necessary to achieve the mass flow of O_2 to avoid anaerobiosis for increasing work rate. An additional mechanism regulating the decrease in capillary PO_2 is the shift to the right of the oxyhemoglobin dissociation curve accompanying the local acidosis of anaerobic metabolism.

It had been hypothesized that the development of a lactic acidosis during exercise was in response to anaerobic metabolism (Hill et al., 1924) and the $\dot{V}O_2$ at which the lactic acidosis developed during a progressively increasing work rate test was the threshold of anaerobic metabolism

Oxygen Transport to Tissue XII, Edited by J. Piiper *et al.*
Plenum Press, New York, 1990

(Wasserman and McIlroy, 1964). The anaerobic threshold concept postulates that as work rate increases, the O_2 consumed is equal to the O_2 required from external sources. However in the mid-range of the subject's exercise capacity profile, O_2 transport is inadequate to provide the O_2 partial pressure gradient between blood and mitochondria for the metabolic

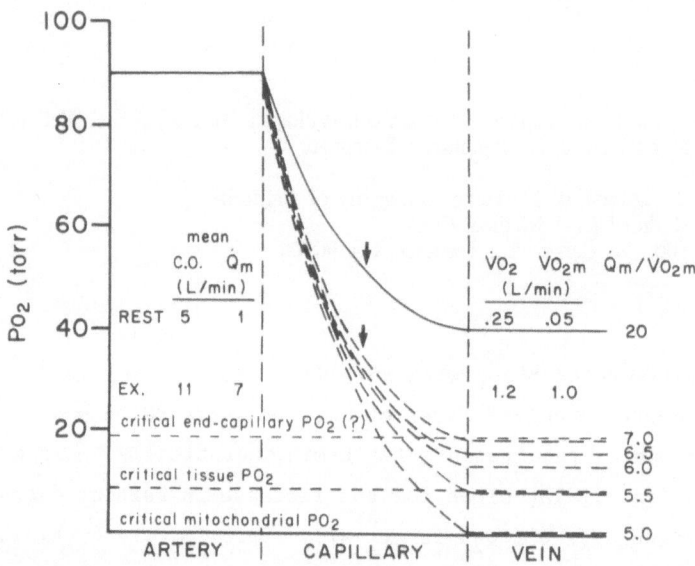

Figure 1. Conceptual model of muscle capillary O_2 partial pressure relationships along a representative muscle capillary during exercise. Assumes a hemoglobin concentration of 15 gm/dl, arterial PO_2 of 90 mmHg and muscle $\dot{V}O_2$ ($\dot{V}O_2m$) of 1 L/min (corresponding to an adult walking at 3 miles/hour). The O_2 diffused out of the capillary at a given point is assumed to be proportional to the capillary-mitochondrial partial pressure gradient. Curves are contours of PO_2 along the capillary for various muscle perfusion (Qm) to $\dot{V}O_2m$ ratios. This rate of O_2 transfer to the mitochondria must satisfy the cellular energy requirement, and is governed by the law of diffusion. The critical PO_2, which defines the adequacy of O_2 supply to the mitochondrion varies with the site (capillary, tissue fluid, mitochondrion). Not only is the partial-pressure gradient between the O_2 source (capillary) and O_2 sink (mitochondrion) a critical determinant of the mass transfer, but the distance between them is equally critical. (Reprinted from Wasserman et al. 1989 with permission of authors and American Heart Association).

stress as illustrated in Figure 2. Thus O_2 consumption is less than the O_2 required, the difference being anaerobic and reflected in the development of lactic acidosis. This study describes our test of the hypothesis that O_2 flow to the mitochondria is limiting during exercise above but not below the anaerobic threshold (AT), using gas exchange evidence of the development of metabolic acidosis.

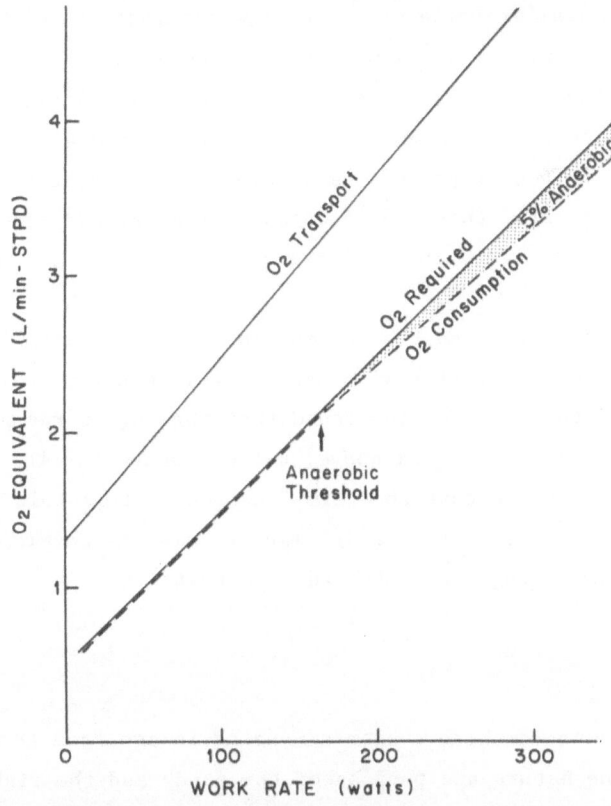

Figure 2. Oxygen transport (cardiac output x arterial O_2 content), the O_2 required by the tissues to perform work totally aerobically and the O_2 consumed as related to work rate. The difference between O_2 required and O_2 consumed is the anaerobic contribution to metabolism. The point where O_2 consumption is below O_2 required is the anaerobic threshold.

To reduce O_2 transport to the exercising muscles, we had subjects breathe low concentrations of carbon monoxide (CO) thereby increasing the carboxyhemoglobin (COHb) concentration to approximately 20%. This should affect $\dot{V}O_2$ at work levels for which energy generation is partially anaerobic, but not at work levels for which O_2 flow to the mitochondria was sufficient to support work totally aerobically. In this study, $\dot{V}O_2$ was studied breath-by-breath, in man, in response to a progressively increasing (ramp pattern) work rate test (Whipp et al., 1981) while breathing air and air with supplemental CO.

The AT was determined by gas exchange evidence of developing metabolic acidosis using the V-slope analysis of Beaver et al (1986). In this method, carbon dioxide output ($\dot{V}CO_2$) is plotted as a function of $\dot{V}O_2$

during an incremental exercise test. At low and moderate work rates, $\dot{V}CO_2$ increases linearly with $\dot{V}O_2$ with a slope between 0.9 and 1.0. But at the $\dot{V}O_2$ above which lactic acidosis occurs, intracellular bicarbonate buffers the hydrogen ion, releasing additional CO_2 and increasing the $\dot{V}CO_2-\dot{V}O_2$ slope to a value clearly steeper than 1.0. The slope is usually greater than 1.3 in normal subjects, and may be considerably higher in patients with cardiovascular disease (Wasserman, 1988).

We found that decreasing the oxygen content of the arterial blood by displacing oxygen on hemoglobin with non-toxic levels (at rest) of carbon monoxide, reduced the AT. We also found that the $\dot{V}O_2$ in response to incremental exercise was reduced above, but not below the AT. These findings provide evidence that the term "anaerobic threshold" correctly describes the exercise $\dot{V}O_2$ above which lactic acidosis develops in response to an increasing work rate exercise test.

METHODS

Eight normal non-smoking subjects ranging in age from 18 to 45 years were studied. The nature and purpose of the study and the risks involved were explained. Each subject voluntarily consented to participate in the study. The protocol and procedures for this study were reviewed and approved by the institution's Human Subjects Committee.

Carbon Monoxide Loading

Carbon monoxide (CO) loading was accomplished using a procedure similar to that of Vogel and Gleser (1972). This technique involved establishing a titration curve by monitoring venous COHb levels after breathing successive levels of 5-10 liters of 1% CO in air using an Instrumentation Laboratories Co-oximeter to measure COHb. This titration was used to estimate the volume of 1% CO that the subject needed to breathe in order to achieve the 20% COHb level on the study days. No subjects experienced symptoms from breathing the gas. Resting minute ventilation and heart rate were not affected after the 20% COHb level had been established.

Exercise Protocol

Each subject performed incremental and constant work rate exercise while breathing either air (control) or air supplemented with carbon

838

monoxide (20% COHb) on different days. The order of testing was randomized. The incremental exercise test consisted of 3 minutes of unloaded
cycling followed by a progressively increasing work rate (ramp pattern)
test (Whipp et al, 1981) using an electromagnetically-braked cycle
ergometer (Lanooy). The rate of increase was designed to reach the maximum
tolerated work rate in approximately 8-15 minutes. The same rate of
increase in work rate was used for the CO study.

During the exercise test on the CO study day, following loading of CO
into the blood, the subjects breathed 0.0225% CO in air in order to
maintain the COHb level at approximately 20% during exercise.

Expired Gas Analysis

During the test, subjects breathed through a mouthpiece attached to a
turbine volume transducer (Alpha Technologies) for measurement of
ventilatory volumes. Oxygen, carbon dioxide and nitrogen tensions were
measured continuously in respired gas, which was sampled at a rate of 1
ml/s from the mouthpiece, by mass spectrometry (Perkin-Elmer). $\dot{V}O_2$ and
$\dot{V}CO_2$ were calculated breath-by-breath, as previously described using a
computer (Hewlett-Packard 1000) (Beaver et al., 1981).

Anaerobic Threshold Determination

$\dot{V}CO_2$ was plotted against $\dot{V}O_2$ on equal arithmetic coordinates for
each subject during the air and CO plus air breathing study days
(Figure 3). The AT was measured by the V-slope method of Beaver et al.
(1986) as modified by Sue et al. (1988), using a clear plastic 45° right
triangle. The break-point in the $\dot{V}CO_2$-$\dot{V}O_2$ relationship became evident
when $\dot{V}CO_2$ increased more steeply than $\dot{V}O_2$ as evidenced by the data
points increasing in a systematic fashion at an angle greater than 45°,
as shown in Figure 3. The $\dot{V}O_2$ at the break-point was read from the $\dot{V}O_2$
axis and defined as the AT.

Effect of Increased COHb on Exercise O_2 Consumption

The difference in $\dot{V}O_2$ at each work rate for the control and
increased COHb studies was determined every 10 seconds for each subject.
The AT of the increased COHb studies for all subjects were time-aligned, in
order to determine if there was a systematic difference in $\dot{V}O_2$ between
the two tests.

839

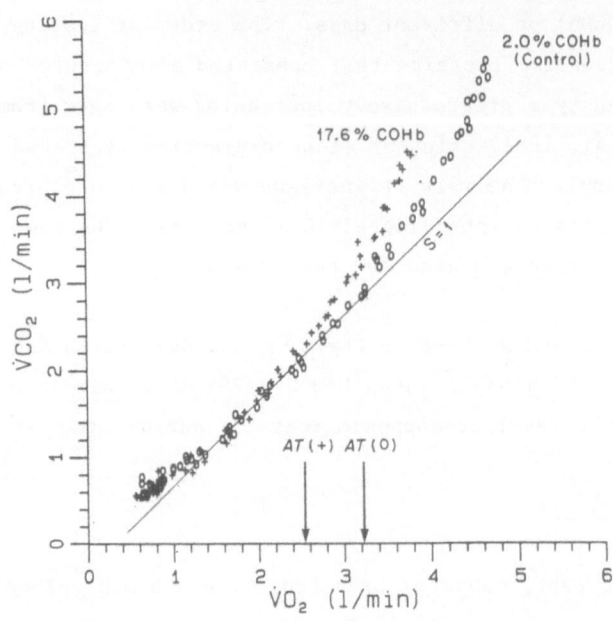

Figure 3. Plot of carbon dioxide output ($\dot{V}CO_2$) L/min-STPD against oxygen uptake ($\dot{V}O_2$) L/min-STPD. Except for the lowest points when the slope is quite shallow because of increasing CO_2 stores, the points run along a slope of approximately 1 (indicated by the line labelled S = 1) as work rate is increased. In the mid-work region, the slope becomes steeper than 1 without evidence of hyperventilation (lowering of $PETCO_2$). This is interpreted to indicate a reduction in bicarbonate stores secondary to buffering of lactic acid. The break-point reflecting the buffering of lactic acid or anaerobic threshold (AT) could be readily determined using a $45°$ right triangle or by a computer using the principles described by Beaver et al (1986). It is seen that the AT (0) for the control study (air breathing) has a value of approximately 3.2 L/min-STPD, whereby the study in which the carboxyhemoglobin (COHb) was increased to 17.6% resulted in a reduction in AT (+) to approximately 2.5 L/min-STPD.

RESULTS

Actual COHb levels for the control and air plus CO study were 1.7 $\pm$ 0.8 (mean $\pm$ SD), and 19.8 $\pm$ 1.4%, respectively. Figure 3 is a plot of $\dot{V}CO_2$ vs $\dot{V}O_2$ for the air-breathing and air plus CO-breathing study for one representative subject. The point of steepening of the $\dot{V}CO_2$-$\dot{V}O_2$ relationship to a slope greater than one is the AT (right arrow). When COHb was increased to approximately 20%, the same relationship was found below the AT except that the AT was reduced (left arrow). The AT and $\dot{V}O_2$ max were similarly reduced in all subjects when COHb was increased.

Figure 4 shows a plot of the $\dot{V}O_2$ vs time from the start of the work rate increase in a typical subject. Since work rate was continuously increased with time (40 W/min for this subject), this is also a plot of $\dot{V}O_2$ vs work rate. It is noted that the $\dot{V}O_2$-time ($\dot{V}O_2$-work rate) relationships for the two studies were essentially superimposable until the subject reached a time (work rate) in the vicinity of the AT of the air plus CO-breathing study. Above the AT, the $\dot{V}O_2$ was lower in the air plus CO-breathing study than during the air-breathing study.

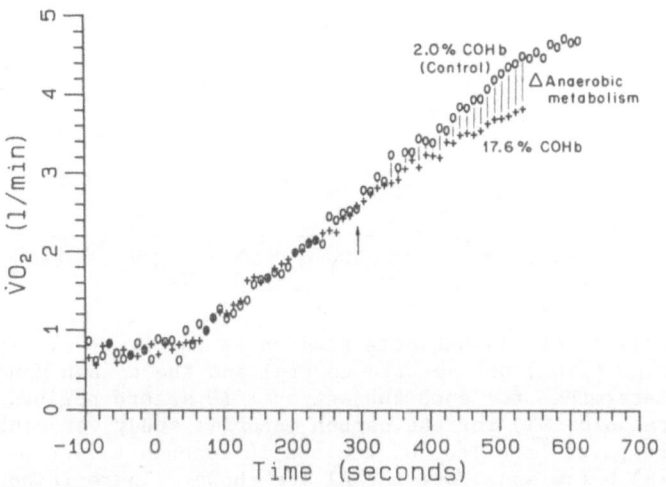

Figure 4. Oxygen uptake ($\dot{V}O_2$) as a function of time in response to progressively increasing work rate tests from unloaded cycling. The points to the left of "0" time are measured during unloaded cycling. At "0" time the work rate was increased at 40 W/min to the subject's maximum tolerated work level during both the control and carbon monoxide (CO) breathing study. In the latter, the carboxyhemoglobin (COHb) increased to 17.6%. The arrow depicts the subject's anaerobic threshold (AT) for the CO breathing study. It is noted that the $\dot{V}O_2$-time or work rate relationship is the same during the control and CO studies until the AT is surpassed. Thereafter, $\dot{V}O_2$ is reduced in the CO study. The shaded area reflects the additional anaerobic metabolism in the CO study. Note a delay time (lower left) between the start of the progressively increasing work rate and the linear increase in $\dot{V}O_2$ of about 40 seconds due to the time constant of $\dot{V}O_2$ kinetics.

Figure 5 shows a composite plot of the differences in $\dot{V}O_2$ at each work rate during the two studies for all 8 subjects. The AT work rates of the CO study for all subjects were time-aligned to determine the effect of the reduced blood O_2 content caused by CO breathing on $\dot{V}O_2$ above and below the AT. There was no systematic difference in $\dot{V}O_2$ as related to

841

work rate below the subjects' $\underline{AT}$. However above their $\underline{AT}$, all subjects demonstrated a reduction in $\dot{V}O_2$ as compared to the air-breathing control study, when the COHb level was increased to approximately 20%, similar to that shown for the subject in Figure 4.

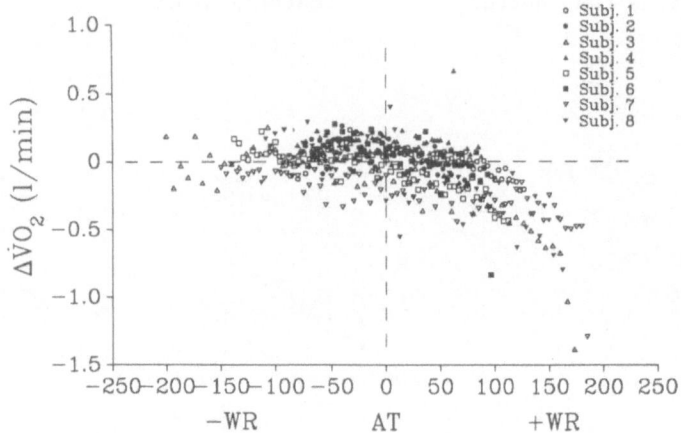

Figure 5. Analysis of all subjects studied as in Figure 4. The difference in oxygen uptake ($\Delta\dot{V}O_2$) between the control and the carbon monoxide (CO) studies are determined for each subject over 10 second periods. The anaerobic threshold ($\underline{AT}$) for the carbon monoxide study for each subject was aligned at the point labelled "0" and the 10 seconds values of $\Delta\dot{V}O_2$ for the work rates (WR) below and above the $\underline{AT}$ are shown. There is no systematic difference in $\dot{V}O_2$ below the $\underline{AT}$. Above the $\underline{AT}$, the $\dot{V}O_2$ is less for the study with raised carboxyhemoglobin as compared to the control.

DISCUSSION

The present study addresses two misconceptions prevalent in the exercise physiology literature. It is assumed that if $\dot{V}O_2$ increases linearly with work rate below the $\dot{V}O_2$ max, anaerobiosis cannot exist. But linearity is only relative, depending on the ability of the investigator to identify the small change in slope expected when aerobic energy generation is supplemented by anaerobic mechanisms. From the stoichiometry for O_2 equivalents of lactate, it can be calculated that 5% of the total energy obtained from anaerobic mechanisms at an oxygen consumption of 2 L/min will result in an increase in lactate production of 10 meq/min. While lactate will increase detectibly in the blood and create a significant metabolic acidosis, the 5% change in the VO_2-work rate slope

842

would be very difficult to detect unless a very clear reference slope for the condition of non-anaerobic metabolism were available. By reducing the O_2 content of arterial blood by approximately 18%, as was done in this study, it was possible to compare the differences in $\dot{V}O_2$ as work rate was progressively increased. It was only above the AT (identified by the V-slope method) that $\dot{V}O_2$ as a function of work rate was reduced when carboxyhemoglobin (COHb) was increased (Figures 4 and 5). Thus below the AT, reduced oxygen flow to the tissues caused by reduced arterial oxygen content (surmised by the increase in COHb) did not affect the $\dot{V}O_2$ increase with work rate. This suggests that, in this work rate domain, O_2 flow is not critically limited. In contrast, for work rates above the AT, the reduced arterial O_2 content was accompanied by a reduction in $\dot{V}O_2$. Thus in this work rate domain, O_2 flow to the muscles is critically limited. Consequently, the transitional $\dot{V}O_2$ demarcates the exercise metabolic stress above which O_2 consumption is sensitively influenced by O_2 available to the tissues. This, we believe, is good evidence that the anaerobic threshold is an appropriate term applied to this level of work.

The second misconception which this paper addresses is that as long as oxygen remains in the venous blood leaving the muscle, the muscle O_2 supply is adequate to avoid anaerobiosis. But the law of diffusion states that the mass transfer of a substance from one site to another depends on an activity difference or diffusion gradient. In this case, the activity difference is the PO_2 difference between capillary and mitochondria. The mass transfer of oxygen depends on this partial pressure difference as well as the diffusion distance and the surface area for diffusion. The partial pressure of O_2 in the capillary (critical PO_2) below which the mass flow of O_2 into the mitochondria is inadequate to maintain the mitochondrial PO_2 at a level adequate to produce sufficient high energy PO_4 for all the ADP available is not constant. Assuming maximal hyperemia (maximal surface area for diffusion and shortest diffusion distance from the blood to the mitochondria), the critical capillary PO_2 (Figure 1) must increase as work rate ($\dot{V}O_2$) increases for metabolism to be 100% aerobic. From measurements of venous oxyhemoglobin saturation from exercising extremities, this does not appear to be evident (Andersen and Saltin, 1985). Thus partial anaerobiosis might be expected to be present at high levels of work, after maximal exercise hyperemia has taken place, despite the presence of a PO_2 in the muscle venous effluent which did not cause muscle lactic acidosis at lower work rates. The lactic acidosis, locally, can benefit the exercising muscle by unloading O_2 from hemoglobin at a higher PO_2 thereby facilitating O_2 diffusion under conditions of high O_2 need.

A final point should be addressed with respect to the $\dot{V}O_2$-work rate relationship. Some investigators assume that the $\dot{V}O_2$-exercise intensity relationship is constant from the lowest work level to $\dot{V}O_2$ max at which point anaerobiosis abruptly occurs (Medbo et al., 1988). This concept is inconsistent with experimental data. As Hansen et al (1988) have shown, in contrast to work rates below the AT, the $\dot{V}O_2$-work rate slope above the AT is dependent on the rate of work rate increase (Figure 6). Thus when the rate of increase is relatively slow with respect to the subject's fitness, e.g., a one minute incremental exercise test lasting approximately 20 minutes from the lowest work rate to max, the slope of the $\dot{V}O_2$-work rate relationship steepens above the AT. The precise mechanism for this steepening is unknown, but reflects a relative inefficiency of energy

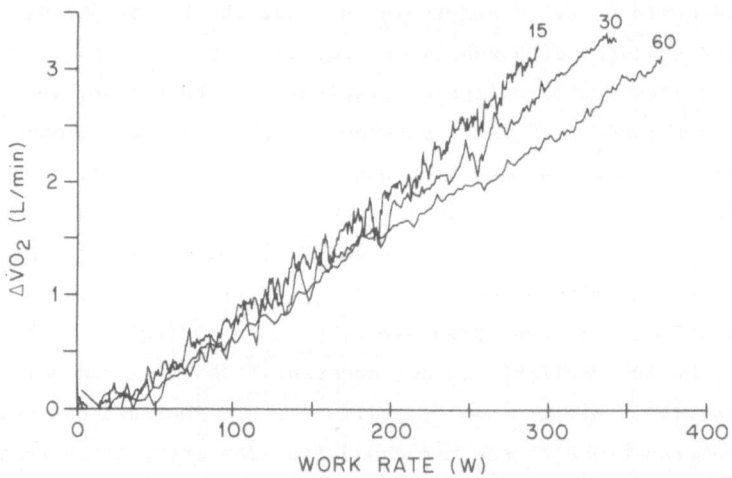

Figure 6. Oxygen uptake above that of unloaded pedalling ($\Delta\dot{V}O_2$) as a function of work rate for a normal subject when the work rate is increased progressively in ramp pattern at 15 W/min, 30 W/min and 60 W/min. The incremental exercise period lasted 1185, 682 and 381 seconds for the 15 W/min, 30 W/min and 60 W/min exercise studies, respectively. Below the anaerobic threshold (AT), the slopes are the same for the $\dot{V}O_2$-work rate relationship. Above the AT, the slope for the 15 W/min is steeper, while that for the 60 W/min is more shallow than that for the work rate increments below the AT. The slope above the AT was intermediate for the 30 W/min ramp study. (Republished with permission from Hansen et al).

844

expenditure during anaerobic work and appears to be related to the lactate increase (Roston et al., 1987). In contrast, when the work rate is increased relatively rapidly to max, e.g., approximately 6 minutes, the rate of increase in $\dot{V}O_2$ relative to work rate is slowed compared to that below the AT. Work rate protocols between these two extremes result in an increase in $\dot{V}O_2$ relative to work rate above the AT which is similar to that below the AT (Figure 6). This appears to be consequent to the offsetting effects above the AT of the mechanism to cause the $\dot{V}O_2$-work rate relationship to steepen as work rate increases slowly (15 W/min ramp in Figure 6) and the mechanism to cause the $\dot{V}O_2$ to become more shallow with respect to work rate (60 W/min ramp in Figure 6).

We believe that the increased O_2 cost of work above the anaerobic threshold of slowly increasing work rate protocols reflect the increased oxygen cost of lactate catabolism for studies with sustained high blood lactate levels with sufficient tissue for lactate to diffuse into tissues where it can undergo oxidative metabolism or gluconeogenesis. When the work rate protocol is rapid, the ratio of anaerobic to aerobic energy-generating mechanisms is relatively great above the anaerobic threshold. Thus the increased O_2 cost of lactate catabolism might be offset by the high rate of anaerobic metabolism. In this case, the onset of fatigue might stop the exercise before the high blood lactate could be catabolized rapidly enough to mask the anaerobic contribution.

SUMMARY

In this paper, we provided evidence to support the concept that work above the anaerobic threshold, measured by the V-slope method, is, in fact, performed partially anaerobically. In contrast, work performed below the anaerobic threshold is totally aerobic (Figures 2, 4 and 5). The $\dot{V}O_2$ at the transition from aerobic to partial anaerobic metabolism must depend on cardiovascular performance since it regulates the capillary PO_2 level needed for O_2 diffusion transport into the mitochondria (Figure 1). At high work rates, the capillary PO_2 needed for the oxygen requirement might not be met by the cardiovascular oxygen supply. This would result in the oxygen consumed being less than the oxygen required by the working tissue (Figure 2), with the oxygen equivalent difference necessarily coming from anaerobic metabolism. The consequences are increased lactate formation and metabolic acidosis, and the physiological and biochemical disturbances which result from the latter.

Supported by Public Health Service Grant No. HL11907. Dr. A. Koike was supported by funds from Otsuka Pharmaceutical Company.

REFERENCES

Andersen P, Saltin B., 1985, Maximal perfusion of skeletal muscle in man. J Physiol 366:233-249.

Beaver WL, Lamarra N, Wasserman K., 1981, Breath-by-breath measurement of true alveolar gas exchange. J Appl Physiol:Respirat Environ Exercise Physiol 51:1662-1675.

Beaver WL, Wasserman K, Whipp BJ., 1986, A new method for detecting anaerobic threshold by gas exchange. J Appl Physiol 60:2020-2027.

Bylund-Fellenius AC, Walker PM, Elander A, Holm S, Holm J, Schersten T., 1981, Energy metabolism in relation to oxygen partial pressure in human skeletal muscle during exercise. Biochem J 200:247-255.

Faulkner JA, Heigenhauser GF, and Schork MA., 1977, The cardiac output-oxygen uptake relationship of men during graded bicycle ergometry. Med Sci Sports 9:148-154.

Hansen JE, Casaburi R, Cooper DM, Wasserman K., 1988, Oxygen uptake as related to work rate increment during cycle ergometer exercise. Eur J Appl Physiol 57:140-145.

Hill AV, Long CNH, Lupton H., 1924, Muscular exercise, lactic acid, and the supply and utilization of oxygen. VI. The oxygen debt at the end of exercise. Proc R Soc Lond 97:127-137.

Medbo JI, Mohn A-C, Tabata I, Bahr R, Vaage O, Sejersted OM., 1988, Anaerobic capacity determined by maximal accumulated O_2 deficit. J Appl Physiol 64:50-60.

Roston WL, Whipp BJ, Davis JA, Cunningham DA, Effros RM, Wasserman K., 1987, Oxygen uptake kinetics and lactate concentration during exercise in humans. Am Rev Respir Dis 135:1080-1084.

Sue DY, Wasserman K, Moricca RB, Casaburi R., 1988, Metabolic acidosis during exercise in patients with chronic obstructive pulmonary disease: Use of the V-slope method for anaerobic threshold determination. Chest 94:931-938.

Vogel JA, Gleser MA., 1972, Effect of carbon monoxide on oxygen transport during exercise. J Appl Physiol 32:234-239.

Wasserman K., 1988, New concepts in assessing cardiovascular function. Circulation 78:1060-1071.

Wasserman K, Beaver WL, Whipp BJ., 1989, Gas exchange theory and the anaerobic (lactate) threshold. Circulation (Suppl): (in press).

Wasserman K, McIlroy MB., 1964, Detecting the threshold of anaerobic metabolism in cardiac patients during exercise. Am J Cardiol 14:844-852.

Whipp BJ, Davis JA, Torres F, Wasserman K., 1981, A test to determine parameters of aerobic function during exercise. J Appl Physiol:Respirat Environ Exercise Physiol;50:217-221.

Yamaguchi I, Komatsu E, Miyazawa K., 1986, Intersubject variability in cardiac output-O_2 uptake relation of men during exercise. J Appl Physiol 61:2168-2174.

O_2 SUPPLY DEPENDENCY IN PATIENTS WITHOUT HYPERLACTEMIA

L. Hannemann, K. Reinhart, Ch. Conrad, O. Grenzer, B. Kuss, K. Eyrich

Dept. Anesthesiology and Intensive Care Medicine, Free University of Berlin
Klinikum Steglitz, West-Berlin

INTRODUCTION

Oxygen consumption (VO_2) may be limited by oxygen delivery (DO_2) despite a normal cardiac output in critically ill patients with sepsis and acute respiratory failure (Bihari et al., 1987). However, some controversy exists whether this phenomenon can only be seen in patients with elevated lactate levels or also in hyperdynamic patients without signs of anaerobic metabolism (Haupt et al., 1985, Gilbert et al., 1986). Therefore, it was our aim to test the hypothesis that inadequate tissue oxygenation may occur in hyperdynamic patients with DO_2 values about 600 ml/min/m^2 and without hyperlactemia.

PATIENTS AND METHODS

Twenty patients of a surgical intensive care unit were investigated after informed consent and approval by our local research committee had been obtained. All patients required ventilatory support and pulmonary artery catheterization due to systemic sepsis or acute respiratory failure.

The patients were stabilized by fluid administration to keep pulmonary capillary wedge pressure above 15 mmHg. In addition, all patients needed inotropic support to maintain mean arterial blood pressure (MAP) above 70 mmHg. Thus dobutamine (DOB) was administered at a dosage range of up to 16 μg/kg/min. When this regimen was insufficient to achieve an adequate MAP, norepinephrine (NOR) was added. The first measurement was taken after the patients had been hemodynamically stable for at least 30 min under this therapy.

Then dobutamine or the combination of DOB and NOR were replaced by dopamine alone at a dosage titrated to achieve a similar MAP as observed with the catecholamines. The second measurement was taken when the patients had been hemodynamically stable under the new regimen for 30 min.

Measurements included a complete hemodynamic profile. Cardiac output was measured in triplicate by thermodilution. VO_2 was calculated as the product of cardiac index and arterial venous oxygen concentration difference. DO_2 was calculated as the product of cardiac index and arterial oxygen concentration.

RESULTS

In 14 patients, the treatment with catecholamines resulted in a significant increase in DO_2, accompanied by a significant increase in oxygen consumption. In contrast, 6 patients exhibited a fall in DO_2 which was parallelled in 4 cases by a decrease in VO_2. In both subgroups, the serum lactate levels were within normal ranges (Table 1).

Oxygen Transport to Tissue XII, Edited by J. Piiper *et al.*
Plenum Press, New York, 1990

Table 1. mean ± SD, * difference between (1) and (2) significant at p < 0,05

Measurements	Increase in DO_2(n = 14)		Decrease in DO_2(n = 6)	
	1	2	1	2
DO_2 (ml·min^{-1}·m^{-2})	730 ± 207 *	930 ± 276	725 ± 142 *	654 ± 165
VO_2 (ml·min^{-1}·m^{-2})	157 ± 34 *	175 ± 39	166 ± 26	134 ± 27
PCWP (mmHg)	17 ± 3	17 ± 5	15 ± 4 *	19 ± 4
MAP (mmHg)	85 ± 13	82 ± 16	78 ± 6	71 ± 12
Lactate (mmol/l)	1.75 ± 1.13		1.55 ± 1.00	

DISCUSSION

We conclude that a tissue oxygen debt may occur in the hyperdynamic critically ill patient even in the absence of hyperlactemia. A possible explanation for this observation could be the fact, that a considerable amount (10-20%) of the basal oxygen consumption is used for reactions that are largely extramitochondrial (Robin, 1980). These reactions occur in various intracellular compartments and involve the biosynthesis, degradation and detoxification of a number of molecular species that are critical for normal cell function (Block, 1962). Some of these extramitochondrial reactions have relatively low affinities for oxygen. As a result, these reactions would be impaired with even moderate degrees of oxygen lack on the tissue level despite adequate O_2 tensions on the mitochondrial level and without signs of anaerobic metabolism. One example is the biosynthesis of some neurotransmitters that appear to have a high Michaelis constant for oxygen. It has been assumed that some of the neurological manifestations of dysoxia are caused by "transmitter failure" rather than ATP depletion (Siesjo, 1978).

Thus, a covered tissue oxygen debt may exist even in hyperdynamic patients with adequate fluid loading, inotropic support and normal arterial oxygen tension due to maldistribution of blood flow at the microcirculatory level. A tissue O_2 debt in the critically ill can only be detected by increasing DO_2 and subsequently measuring the VO_2. Tissue hypoxia is considered to be an important mechanism in the development of multiple organ failure (Bihari et al., 1987, Sibbald et al., 1989). Under these conditions it may be useful trying to increase DO_2 in these patients until no further increase in VO_2 can be achieved.

However, the potential side effects of therapeutic efforts to increase DO_2 should be considered. In this study dopamine, which had to be administered in a dosage up to 50 g/kg/min, resulted in a decrease in oxygen delivery in 6 patients, due to a decrease in cardiac index. This fall in cardiac index may have occurred because of considerable increase in cardiac filling pressures well above normal, which may have compromised cardiac work due to decrease in myocardial compliance (Chernow, 1982). Furthermore, one has to consider metabolic side effects of the drug as well as possible influences on nutritive blood flow within the microcirculation (Reinhart, 1989).

Further investigations in hyperdynamic septic patients or appropriate animal models are needed to clarify the possible influences and side effects of catecholamines and other therapeutic measures, that are chosen to increase DO_2 in already hyperdynamic patients.

REFERENCES

Bihari, D., Smithies, M., Gimson, A., Tinker, J., 1987, The effects of vasodilatation with prostacyclin on oxygen delivery and uptake in critically ill patients, New Engl. J. Med., 317: 397-403.

Block, K., 1962, Oxygen and biosynthetic patterns, Fed. Proc., 21: 1058-63.

Chernow, B., Rainey. T.G., Lake, R., 1982, Endogenous and exogenous catecholamines in critical care medicine, Crit. Care Med., 10:6, 409-416.

Gilbert, E.M., Haupt, M.T., Mandanas, R.J., Huaringa, A.J., Carlson, R.W., 1986, The effect of fluid loading, blood transfusion, and catecholamine infusion on oxygen delivery and consumption patients sepsis, Ann. Rev. Resp. Dis., 134: 873-878.

Haupt, M.T., Gilbert, E.M., Carlson, R.W., 1985, Fluid loading increases oxygen consumption in septic patients with lactic acidosis, Am. Rev. Respir. Dis., 131: 912-916.

Reinhart, K., 1989, Oxygen transport and tissue oxygenation in sepsis and septic shock, in: "Sepsis - an interdisciplinary challenge", K. Reinhart, K. Eyrich eds., Springer Berlin, Heidelberg, New York.

Robin, E.D., 1980, Of men and mitochondria, Coping with dysoxia, Am. Rev. Respir. Dis., 122:517-531.

Sibbald, W.J., Bersten, A., Rutledge, F.S., 1989, The role of tissue hypoxia in multiple organ failure in: "Clinical aspects of O_2 transport and tissue oxygenation", Reinhart K, Eyrich K eds. Springer Berlin, Heidelberg, New York.

Siesjo, B.K., 1978, Brain energy metabolism, New York: John Wiley & Sons, 398-526.

THE EFFECT OF EPINEPHRINE ON OXYGEN CONSUMPTION, OVERALL ENERGY METABOLISM, AND SUBSTRATE UTILIZATION IN RATS

L. Benthem, J. van der Leest, W.P. Meeuwsen, H. van der Molen, J.P. Zock, W.G. Zijlstra and A.B. Steffens°

Department of Physiology, and Department of Animal Physiology°
University of Groningen, The Netherlands

INTRODUCTION

In the past 25 years the relationship between plasma free fatty acids (FFA) and glucose metabolism has been studied extensively (Wolfe et al., 1988). In several studies an invers relationship between the availability of FFA and the rate of glucose oxidation has been recognized (Randle et al., 1963; Jansson, 1980). The underlying mechanism, however, has not yet been completely clarified. Several hypotheses have been proposed. Jansson (1980) suggests that FFAs entering the muscle cell by diffusion, a concentration-dependent process, influence glucose oxidation by inhibiting two enzyms, Pyruvate Dehydrogenase (PDH) and Phosphofructokinase (PFK). PDH is inhibited by acetyl-CoA from the breakdown of FFA, PFK by citrate formed from acetyl-CoA in the first step of the citric acid cycle. The rise in concentration of these metabolites resulting from the increased breakdown of FFA thus leads to an enhanced inhibition of the degradation of glucose.

The ratio between glucose and FFA utilization can be determined by measuring oxygen consumption (VO_2) and carbondioxide production (VCO_2). Their ratio, the respiratory quotient ($RQ = VCO_2/VO_2$), depends on the ratio in which glucose and FFAs are utilized. When only glucose is oxidised, the RQ will be 1.00, when only FFAs are used, the RQ will be 0.70. The degradation of proteins is not taken into account because this parameter may be assumed to be constant (Westerterp, 1976). Under circumstances of stress, like increased physical activity, an increase in RQ can be recognized (Galbo, 1983, Musch et al., 1988), indicating that relatively more glucose is oxidized. This is favourable, because less oxygen is needed for the release of energy. Simultaneous with the change in RQ, a rise in the plasma epinephrine (E) concentration is observed (Galbo, 1983). Circulating E, originating from the adrenal medulla seems primarily to be related with emotional stress (Scheurink et al., 1989 a&b) and has a stimulating effect on the release of glucose from hepatocytes (Steffens et al., 1984).

The aim of this study was to clarify the influence of circulating E upon VO_2 and RQ in relation to blood glucose, plasma FFA, and plasma insulin in unrestrained, freely moving rats. Therefore E was intravenously administred in three doses; 20, 35 and 50 ng /min for 40 min. Infusion of saline served as control. Before, during, and after the infusion, VO_2, VCO_2, and the concentrations of the above-mentioned substances were measured.

Oxygen Transport to Tissue XII, Edited by J. Piiper *et al.*
Plenum Press, New York, 1990

MATERIALS AND METHODS

Gas Analysis. Metabolic rate and RQ were determinated with an open air system (fig 1). The rat was placed in a metabolic chamber (5.5 l) through which a flow of air (2.5 l/min), free from water (silicagel) and CO_2 (sodalime), was drawn by an electric membrane pump (WISA). After desiccation ($CaCl_2$) of the effluent gas from the box, temperature and flow rate were determined, using a Fleisch pneumatograph. Thereafter, the gas flow was diverted into three channels. Two of them led to O_2 and CO_2 analysers at a flow rate of 200 and 20 ml/min, respectively. The third channel was an exit for the rest of the gas. CO_2 was measured with a Balzers QMG511 mass spectrometer. For determination of O_2 a Beckmann OM11 oxygen analyser was used. To correct for drift in the signal of the oxygen analyser, dried and CO_2 free air was analysed for a period of 20 s, after measuring for 100 s. Before each experiment, the CO_2 and O_2 analysers were calibrated using a certified gas (HOEKLOOS, Schiedam), containing approximately 20% O_2 and 1% CO_2, and dried, CO_2-free air (20.95% O_2).

The analog outputs from flowmeter, thermometer, O_2 and CO_2 analysers were fed into a HP9200 computer system. VO_2 and VCO_2 were calculated using the measured flow through the metabolic box, the expired fractions of O_2 and CO_2, and the fractions of O_2 and CO_2 in room air (Sonne and Galbo, 1980). The respiratory quotient (RQ) was calculated as VCO_2/VO_2.

Animals. Male wistar rats weighing 300-350 g at the beginning of the experiments were used. They were kept separately in Plexiglas cages (25 * 25 * 30 cm) at room temperature (22 ± 2 °C) and had free access to standard carbohydrate-rich food (HOPE FARMS) and water unless otherwise stated. The rats were maintained on a 12-12 h light-dark cycle (7-19 h light), and were handled and weighed daily at 9:00 h.

Surgery. All surgery was performed under ether anesthesia. The experiments started as soon as the rats had regained their preoperative body weight. There were no significant differences in body weight at the beginning of the experiments. All animals were provided with two silicon vascular catheters with their tips at the entrance of the right atrium, one for blood sampling

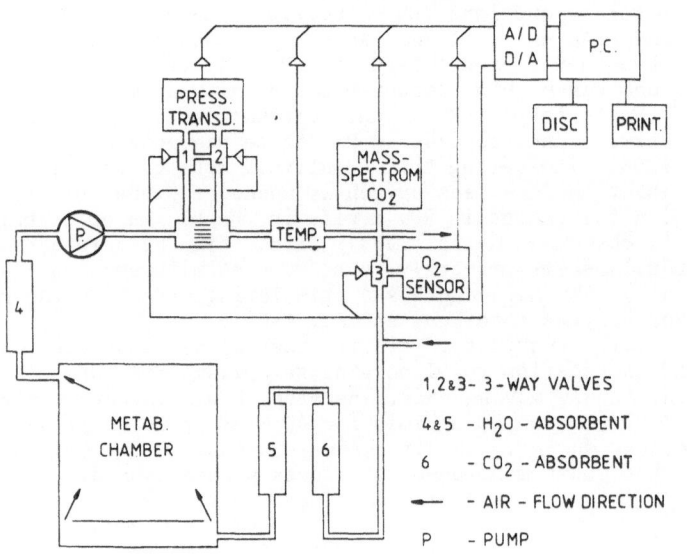

Fig. 1.   Schematic diagram of metabolic chamber and auxiliary equipment.

852

and one for infusion of fluids. The catheters were introduced through the jugular veins and the distal ends were externalized at the top of the skull according to Steffens (1969b). This method allows repeated blood sampling and continuous infusion of fluids in unanesthetized, unrestrained rats, without disturbing the animals (Steffens, 1969a).

Blood Sampling Procedure and Chemical Determinations. After the animal had been placed in the metabolic box, it was connected to polyethylene tubes [lenght 450 mm, OD 1.25 mm, ID 0.75 mm] through which blood was sampled or fluids were infused (Steffens, 1969b). During the entire experiment, 20 blood samples of 0.35 ml each were withdrawn for the determination of blood glucose, plasma FFA, and insulin. After each sample was withdrawn, a transfusion of 0.35 ml citrated donorblood was given in return. This blood was obtained from donor rats provided with similar catheters. Between the withdrawals of blood, the tip of the catheter was filled with 6% citrate solution as anticoagulant. Citrate was used instead of heparin to avoid activation of endothelial lipase. The blood samples were immediately transferred to chilled (0°C) centrifuge tubes containing 10 μl EDTA solution as anticoagulant. Blood glucose was determined by an enzymatic colorimetric method (Sigma). The remaining blood was centrifuged at 2600 G for 10 min. Plasma was stored at -30 °C until determination of FFA and insulin. Plasma FFA was determined by an enzymatic colorimetric method (Mulder et al., 1983). Rat specific plasma immunoreactive insulin (IRI) was determined by means of a radioimmunoassay (NOVO, Copenhagen). Guinea pig serum M8309 served as antiserum. Duplicate assays were performed on 25-μl plasma samples. Bound and free 125I-labeled insulin were separated by means of polyethylene glycol solution (23.75% wt/wt) as suggested by Henquin et al. (1974) and developed by Steffens et al. (1984).

Experimental Procedure. All experiments were performed between 1200 and 1600 h, i.e. in the light period. Food was removed about 3 h before the start of the experiment, and the animal transferred to the experimental room. After calibration of the gas-analysers, the animal was placed into the metabolic chamber and attached to the blood sampling and infusion tubes. Then we waited till the rat metabolized at a low level. At that time the experiment was started and two blood samples (t = -11 and t = -1 min) were taken to measure the basal levels of the blood components. Subsequently, the pump was connected to the infusion tube and the infusion was started (t = 0) at a rate of 0.05 ml/min. During the infusion-period (40 min) blood samples were withdrawn at t = 2, 7, 12, 17, 22, 27, 32, & 37 min. After the infusion period blood samples were taken at t = 42, 47, 52, 57, 62, 67, 77, 87, 97, & 107 min. The animals received infusions of either 20, 35 & 50 ng E per min or saline (0.05 ml/min), in a random order with at least one day between consecutive experiments.

TABLE 1.Basal values of VO_2, RQ, blood glucose, plasma FFA, and plasma insulin in each of the four experiments.

	n	SAL	n	E20	n	E35	n	E50
VO_2	9	.737±.021	9	.724±.026	8	.767±.035	8	.790±.032
RQ	9	.820±.020	9	.871±.015	8	.808±.009	8	.849±.016
[GLU]	11	6.18±.07	11	6.11±.01	12	6.05±.04	11	6.02±.10
[FFA]	7	.275±.010	7	.205±.011	7	.303±.018	7	.244±.030
[INS]	6	48.5±.50	6	52.0±4.0	9	49.5±2.56	7	62.5±2.5

Values are means ± SEM; n, no. of rats; SAL, SALINE; E20, E35, E50: EPINEPHRINE 20, 35, 50 ng/min, respectively; VO_2, Oxygen consumption (mmol/kg*min); [GLU], blood GLUCOSE (mmol/l); [FFA], plasma FREE FATTY ACIDS (mmol/l); [ins], plasma INSULIN (μU/ml).

Statistics. The total time of an experiment was divided into five periods: $t = -10 - 0$, $t = 1 - 20$, $t = 21 - 40$, $t = 41 - 75$ and $t = 76 - 110$ min. The results for VO_2, blood glucose, plasma FFA, and plasma insulin obtained in the respective periods were pooled and the means, expressed as a fraction of means of the period $t = -10 - 0$, and SEM are presented. The results for RQ are expressed as mean absolute change related to the period from $t = -10 - 0$, and SEM. The Student t-test was used to compare VO_2 and RQ values obtained in experiments with infusion of epinephrine with values obtained in experiments in which only the solvent, saline, was administered. Two way ANOVA followed by a Mann-Whitney U test was used to compare the results for blood glucose, plasma FFA and plasma insulin, obtained in experiments in which E was administered with those in which only saline was infused. The level of significance was set at $p < 0.05$.

RESULTS

Table 1 presents the basal levels of oxygen consumption, RQ, and the concentrations of blood glucose, plasma free fatty acids, and plasma insulin. VO_2 and RQ represent the mean values over the period of $t = -10$ to 0 before the start of the infusion. Blood glucose, plasma FFA and insulin represent the mean value of the blood samples taken at $t = -10$ and $t = -1$ min. There were no initial differences between the four groups, except for the basal RQ value of the 20 ng/min experiments and the basal plasma insulin level of the 50 ng/min experiments. Figures 2-6 show the relative values for VO_2 (fig 2), blood glucose (fig 4), plasma FFA (fig 5), and plasma insulin (fig 6) and the absolute changes in RQ (fig 3). The figures show the mean values ± SEM over the periods $t = 1-20$ and $t = 21-40$ during infusion, and $t = 41-75$ and $t = 76-110$ post-infusion.

With regard to VO_2 (fig 2), we observed a slight elevation of the metabolic rate during the infusion of saline (8% in the second infusion period). In the post-infusion periods metabolic rate stayed at this level. Infusion of epinephrine caused increases in VO_2 of 20, 32, and 23% for 20, 35, and 50 ng/min, respectively. These elevations differ significantly from that in the control experiment. In the post-infusion periods of the 20 and 50 ng/min experiments, VO_2 returned to pre-infusion value. In the 35 ng/min experiments, VO_2 stayed at a significantly elevated level.

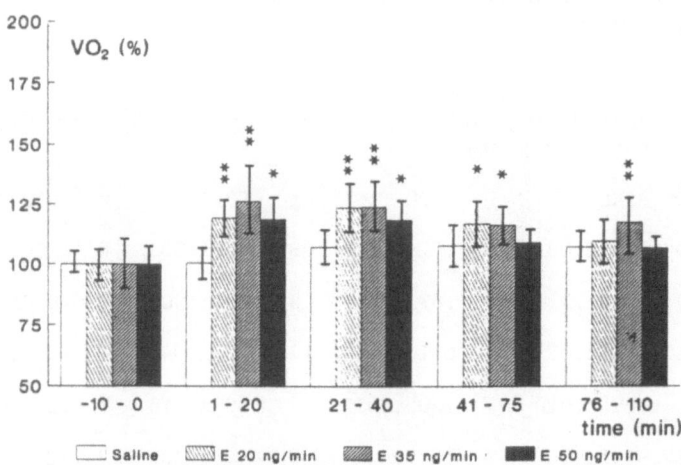

Fig. 2. Effect of epinephrine on oxygen consumption. Data are expressed as mean changes relative to pre-infusion values. Statistical differences are reported in respect to saline. * $p < .05$; ** $p < .01$

854

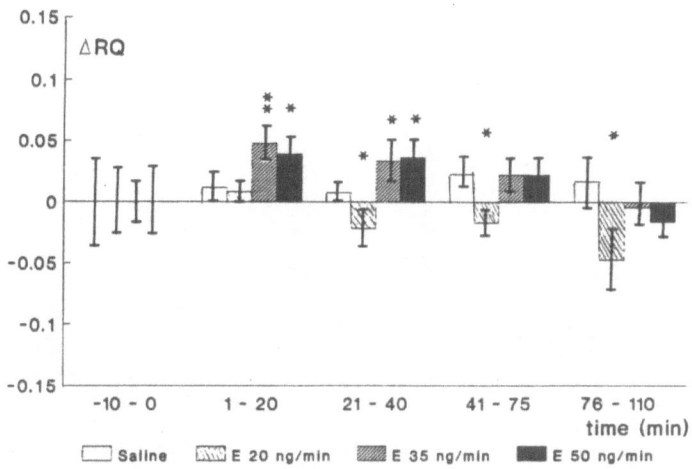

Fig. 3. Effect of epinephrine on RQ. Data are expressed as average changes from pre-infusion level. Statistical differences are expressed as in Fig. 2.

During infusion of saline, mean RQ values stayed at their pre-infusion value. In the post-infusion periods RQ was slightly elevated (0.02). When E was administered at a rate of 35 or 50 ng/min, mean RQ values rose significantly in the first infusion period, reaching 0.05 and 0.04 above pre-infusion values, respectively. They stayed at this level in the second infusion period. In the first post-infusion period they did not differ from control values. In the second post-infusion period they were lower than control as well as pre-infusion levels. This fall was not significant. When E was infused at a rate of 20 ng/min, RQ values were significantly reduced in the second period of the infusion and in both the post-infusion periods.

Only minor changes in blood glucose concentrations were observed when saline was infused (fig4). In the second post-infusion period of the saline experiments, the mean blood glucose level declined to 4% below the pre-infusion level. Infusion of E had a dose-related effect upon blood glucose. In the

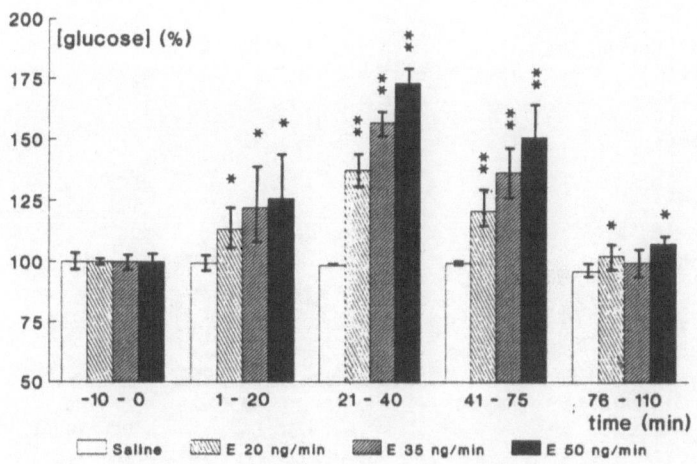

Fig. 4. Effect of epinephrine on blood glucose concentration. Data are expressed as in Fig. 2.

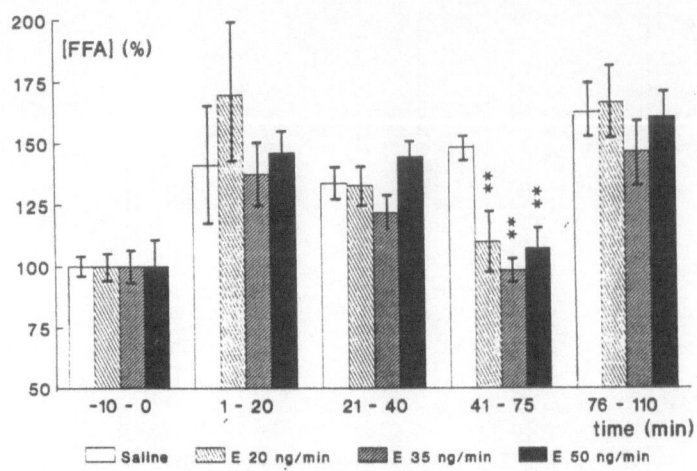

Fig. 5.   Effect of epinephrine on plasma FFA concentration. Data are expressed as in Fig. 2.

second infusion period significant increases were found of 37, 57, and 73%, when E was infused at rates of 20, 35, and 50 ng/min, respectively. In the post-infusion periods blood glucose concentrations decreased, but a significant elevation was still present in the second post-infusion period of the 20 and 50 ng/min E experiments.

The plasma FFA concentration showed a strong increase, 40 and 37% in the first and second infusion period, respectively, when saline was administered. After termination of the infusion, FFA stayed at this elevated level (47% in the first post-infusion period, 61% in the second). Compared to the control experiment, infusion of E caused no significant differences in plasma FFA. Only in the first post-infusion period, plasma FFA levels were significantly lower than in the control experiments. In the second post-infusion period FFA concentrations were back on control level.

Plasma insulin concentrations did not change when saline was infused

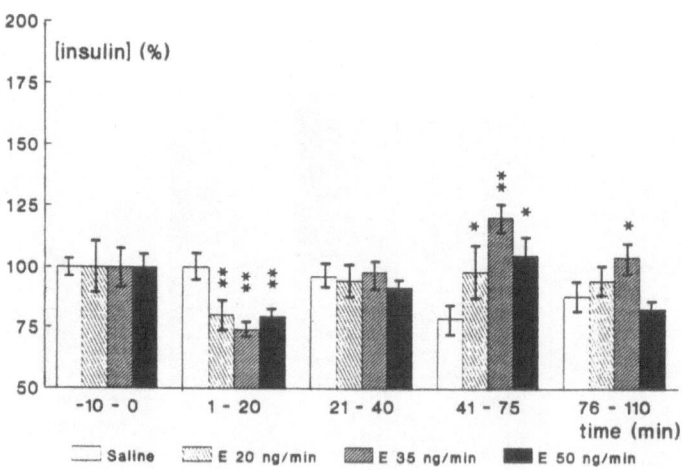

Fig. 6.   Effect of epinephrine on plasma insulin concentration. Data are expressed as in Fig. 2.

856

(fig 6). Only in the first post-infusion period a reduction in the insulin concentration was observed (23%). In the second post-infusion period insulin was only slightly decreased. When E was infused, plasma insulin showed a significant decrease during the first infusion period, in the second period it had returned to the control value. In the first period after termination of the E infusion, the plasma insulin concentration was significantly above the control level. In the second post-infusion period only the insulin concentration of the 35 ng/min experiment differed significantly from the control value.

DISCUSSION

The most conspicuous results of this study are 1) the elevated VO_2 levels during and after infusion of E; 2) the rise in RQ when E is infused at rates of 35 and 50 ng/min; 3) the fall in RQ when E is infused at a rate of 20 ng/min; 4) the dose related way in which the blood glucose concentration increases during infusion of E; 5) the rise in FFA concentration even when only saline was infused; and 6) the unchanged insulin concentrations in the second infusion period.

The blood concentrations of nutrients are the net result of the uptake of these substances from the blood and their release from stores, like hepatocytes (glucose) and lipocytes (FFA), into the blood. The increase in glucose concentration during infusion of E is due to an enhanced release from the liver rather than to a decreased uptake, because total metabolism is increased (VO_2, RQ). The increased release will be caused by an elevated stimulation of α-adrenoceptors on hepatocytes (Scheurink et al., 1988), and a simultaneous decrease in insulin concentration. After termination of the E infusion, blood glucose levels decline almost to pre-infusion values. This may be explained by a decreased release from the hepatocytes and an increased uptake by hepatocytes and muscle cells, stimulated by the elevated insulin level.

The decreased plasma insulin levels in the first infusion period of the E experiments will be the result of an active inhibition of insulin release, by stimulation of the α-adrenoceptors on the pancreas β-cell. This mechanism was first recognised by Porte (1967). In the second infusion period of the E experiments, the plasma insulin levels have returned to their pre-infusion values. Blood glucose levels are still elevated in this period, but have reached a stable value, as shown by the low SEM-values. This may be due to blood glucose having reached a new set-point, induced directly by elevated concentrations of E, or indirectly by other circulating hormones stimulated by the elevated E levels. After termination of the E infusion, an increase in plasma insulin levels is found, which may be explained by a decreased inhibition of insulin release.

Plasma FFA-levels rise during infusion of saline as well as during infusion of E. A tentative explanation for this phenomenon may be found in the infusion and the increased blood sampling frequency disturbing the animals. Normally, rats are not disturbed by sampling and infusion (Steffens, 1969a). In our study, however, the circumstances were somewhat different. Steffens (1969a) took the blood samples in the homecage of the rat. We took the blood samples after transfer of the animals from their homecages to the metabolic chamber. This might have induced elevated plasma E levels as seen in previous studies (Scheurink et al., 1989a). Although we waited till the animals metabolised at a low rate, expecting that then they were well-habituated to the new environment and had low rates of E release, it may be that the transfer of the animals induces a lowered threshold for E release from the adrenal medulla. This would mean that after transfer a smaller disturbance would, nevertheless, result in elevated E levels. Hence, that the infusion of only saline, which had no effect in the homecage (Steffens, 1969a), may now be able to induce an increase in E concentration. These slightly elevated E levels may have induced an increased release of FFA. There is no difference in FFA response between infusion of E and saline, nor between the used doses of E. Thus it seems that FFA release is the result of a triggered process rather than an dose dependent-reaction, as are the blood glucose concentrations. The threshold for FFA release will in that case be at a lower set-point than that for glucose release, because blood glucose is not influenced by the saline infusion. The decrease in the FFA levels found in the first post-infusion period is caused by the elevated insulin levels, since insulin stimulates the re-esterifacation of FFA. The rise in FFA concentration in the second post-infusion period will be a consequence of the above-mentioned mechanism.

Compared to values known from the literature, the VO_2 levels found in the pre-infusion periods of this study are rather low (Sonne and Galbo, 1980; Musch et al., 1988). For example Musch et al. (1988) presented resting VO_2 levels of 0.9 - 1.0 mmol/ kg·min. In this study the mean pre-infusion VO_2 level was 0.75 mmol/ kg·min. VO_2 rose only slightly during infusion of saline, and reached values as found by Musch et al. (1988) to be resting levels. During infusion of E, significantly elevated VO_2 levels, as compared to saline, were found. After termination of the saline infusion, VO_2 stayed at a slightly elevated level. Except for the 35 ng/min E experiments, VO_2 in the post-infusion periods declined to a level as found in the control experiment. The results presented for VO_2 confirm our hypothesis that the animals were not stressed or excited during the pre-infusion period, but that the infusion and the more frequent blood sampling cause some stress under the circumstances of our experiment. The stimulatory effect of E upon metabolism has been recognized for many years (Juschtschenko, 1909). However, it is remarkable that E does not influence total metabolism in a dose-related way.

The RQ values obtained in this study seem to be conflicting. In the control experiments the FFA levels were increased, while blood glucose stayed at its pre-infusion level. According to Jansson (1980) a fall in RQ would be expected under these circumstances. However, it should be considered that such a shift in substrate utilization is thought to take place in metabolising muscle cells. In this study only a slight increase in VO_2 was found, when saline was infused. Hence metabolism was only slightly above basal conditions and it may be that only a few muscle cells were activated.

When E was infused at rates of 35 or 50 ng/min, there was no difference in FFA response, between the two doses. The glucose response in the 50 ng/min experiments appeared to be larger than in the 35 ng/min experiments. In both the experiments VO_2 was increased to nearly the same level, significantly above that in the control experiment. The RQ values of both the experiments show an equal rise in the infusion period. In the second post-infusion period RQ was not significantly different in comparison with the control level, nor were the concentrations of the blood components. These results suggest that infusion of E causes a shift towards glucose utilization, probably by influencing the glucose-FFA ratio in the blood.

When E was infused at a rate of 20 ng/min, VO_2 and FFA rose to the same level as reached in the 35 and 50 ng/min experiments, but the blood glucose concentration increased less. The RQ value in the second infusion period appeared to be significantly lower than in the control experiment. In the post-infusion periods, when VO_2, FFA and glucose were at control levels, RQ was still lower than in the control experiment. A possible explanation for this divergent result may be found in the high pre-infusion value for the RQ in the 20 ng/min experiments. A spontaneous decline in RQ thus may have thwarted the effectof the E infusion. The absolute RQ levels in the second post-infusion periods of the 20 ng/min experiments are at the same level as those of the control experiment.

The overall results for RQ in this study seem to justify the conclusion that E can influence the ratio in which glucose and FFA are utilized. This influence is exerted indirectly, by influencing the availability of the substrates, rather than directly, by influencing utilization. The results further suggest that E has a stimulating effect on total metabolism and on the release of glucose to the blood. The release of insulin, however, seems to be inhibited by E. Finally, it should be stressed that in experiments of

858

this kind due attention should be given to the influence the experimental conditions may exert on the experimental animal.

SUMMARY

In this study the influence of epinephrine (E) on oxygen consumption, overall energy metabolism, and substrate utilization in rats has been investigated. Therefore E was infused at rates of 20, 35, and 50 ng/min for 40 min. Infusion of the solvent, saline, served as control experiment. Before, during, and after the infusion, VO_2, as parameter for total metabolism, and RQ, as parameter for substrate utilization, were determined using an open circuit. In addition blood samples were taken for determination of blood glucose, plasma free fatty acids (FFA) and plasma insulin concentrations. The results show a rise in VO_2 and blood glucose during infusion of E. Plasma FFA concentrations were elevated during infusion of E and of saline. Plasma insulin decreased when E was administered. RQ values were increased when E was infused at rates of 35 and 50 ng/min. The results suggest that E can influence the ratio in which glucose and FFA are utilized. This influence seems to be excerted indirectly by influencing the availability of the substrates, rather than directly, by influencing utilization.

REFERENCES

Galbo, H., 1980, "Hormonal and Metabolic Adaptation to Exercise", Thieme, Stuttgart.

Henquin, J.C., Malvaux, P., and Lambert, A.E., 1974, Glucagon Immunoassay Using Polyethylene Glycol to Precitipate Antibodybound Hormone, Diabetologica, 10: 61.

Jansson, E., 1980, Diet and Muscle Metabolism in Man, CATA Physiol. Scand., Suppl. 487.

Juschtschenko, A.J., 1909, Der Einfluss des Thyreoidins, Spermins und Adrenalins, sowie der Entfernung der Schilddrüse und der Testikeln auf die Oxydationsprozesse, den Atmungsgasaustausch und die Giftigkeit des Harns bei Tieren, Biochem. Z., 15: 365.

Mulder, C., Schouten, J.A., and Popp-Snijders, C., Determination of Free Fatty Acids: A Comparative Study of the Enzymatic Versus the Gas Chromatografic and the Colorimetric Method, J. Clin. Chem. Clin. Biochem., 21: 823.

Musch, T.I., Bruno, A., Bradford, G.E., Vayonis, A., and Moore, R.L., 1988, Measurements of Metabolic Rates in Rats: a Comparison of Techniques, J. Appl. Physiol., 65: 964.

Porte, D. jr., 1967, A Receptor Mechanism for the Inhibition of Insulin Release in Man, J. Clin. Invest., 46: 86.

Randle, P.J., Garland, P.B., Hales, C.N., et al., 1963, The Glucose and Fatty Acid Cycle: Its Role in Insulin Sensitivity and the Metabolic Disturbance of Diabetes Mellitus, Lancet 1: 785.

Scheurink, A.J.W., Steffens, A.B., Benthem, L., 1988, Central and Peripheral Adrenoceptors Affect Glucose, Free Fatty Acids, and Insulin in Exercising Rats, Am. J. Physiol., 255: R255.

Scheurink, A.J.W., Steffens, A.B., Dreteler, G.H., Benthem, L., and Bruntink, R., 1989a, Experience Affects Exercise-Induced Changes in Catecholamines, Glucose, and FFA, Am. J. Physiol., 256: R169.

Scheurink, A.J.W., Steffens, A.B., Bouritius, H., Dreteler, G.H., Bruntink, R., Remie, R., and Zaagsma, J., 1989b, Adrenal and Sympathetic Catecholamines in Exercising Rats, Am. J. Physiol., 256: R155.

Sonne, B., and Galbo, H., 1980, Simultaneous Determinations of Metabolic and Hormonal Responses, Heart Rate, Temperature and Oxygen Uptake in Running Rats, Acta Physiol. Scand., 109: 201.

Steffens, A.B., 1969a, Blood Glucose and FFA Levels in Relation to Meal Pattern in the Normal Rat and the Ventromedial Hypothalamic Lesioned Rat, Physiol. Behav., 4: 215.

Steffens, A.B., 1969b, A Method for Frequent Sampling of Blood and Infusion of Fluids in the Rat Without Disturbing the Animal, Physiol. Behav.4: 833.

Steffens, A.B., Damsma, G., van der Gugten, J., and Luiten, P.G.M., 1984, Circulating Free Fatty Acids, Insulin, and Glucose During Chemical Stimulation of Hypothalamus in Rats, Am. J. Physiol., 247: E765.

Westerterp, K.R., 1976, "How Rats Economize", Thesis, Groningen.

Wolfe, B.M., Klein, S., Peters, E.J., Schmidt, B.F., and Wolfe, R.R., 1988, Effect of Elevated Free Fatty Acids on Glucose Oxidation in Normal Humans, Metabolism, 37: 323.

OXYGEN TRANSPORT RELATED VARIABLES AND MUSCLE TISSUE OXYGENATION IN CRITICALLY ILL PATIENTS WITH AND WITHOUT SEPSIS

Konrad Reinhart, Frank Bloos, Frank König, Lutz Hannemann, Brigitte Kuss

Dept. Anesthesiology and Intensive Care Medicine, Free University of Berlin
Klinikum Steglitz, West-Berlin

INTRODUCTION

Impaired tissue oxygenation and tissue hypoxia are considered to be the final common pathway of the various clinical insults that are responsible for the development of multiple systems organ failure (Sibbald et al., 1989). It was hypothesized that, in sepsis and ARDS patients tissue hypoxia may exist despite normal or even supranormal convective O_2-transport to tissues by reduced O_2 extraction capacity (Schumaker, Cain, 1987). In various clinical studies supply dependency of O_2 uptake (VO_2) has been demonstrated in patients with sepsis and ARDS (Danek et al., 1980, Mohsenifar et al., 1983, Bihari et al. 1987) which was interpreted as the indirect proof for the existance of a covert tissue hypoxia (Bihari et al., 1987). To test this hypothesis we compared variables related to hemodynamics- and oxygen transport - in two groups of critically ill patients.

PATIENTS AND METHODS

All patients that were investigated required pulmonary artery catheterization due to septic shock (n = 10) or cardiorespiratory failure without shock (n = 10). Diagnosis of sepsis was based on the evaluation of a mosaic of factors from the patients' medical history, symptoms, clinical and laboratory findings (Gramm et al., 1989). Despite adequate volume therapy, all patients with septic shock required a combination of dobutamine and norepinephrine for maintenance of adequate mean arterial pressure (MAP) > 70 mmHg. The patients without sepsis only required inotropic support in form of dobutamine. All patients from both groups were mechanically ventilated.

Patients were studied only during hemodynamically stable periods in which no volume replacement, or change in inotropic support or body temperature took place.

The paired arterial and mixed venous blood samples were drawn simultaneously and slowly from the radial and pulmonary artery, respectively. Oxygen content for arterial (CaO_2) and mixed venous (CvO_2) blood was derived from immediate measurements of hemoglobin concentration and O_2 saturation (IL-282 Co-oximeter) and PO_2 (Radiometer ABL-2, Copenhagen). Cardiac output was measured in triplicate by thermodilution. O_2 delivery was calculated as the product of cardiac index and CaO_2.

Tissue PO_2 values were obtained from the quadriceps femoris muscle using a fast responding polarographic-type hypodermic needle probe (Sigma-PO_2 Histograph/KIMOC, Eppendorf Instruments, FRG) as described in detail earlier (Kersting et al.). The probe has a stainless metal shaft with a diameter of 350 µm and a response time of < 500 ms. The probe was inserted in a stepwise manner 20 - 30 mm into the muscle. Rapid forward movement was immediately followed by a shorter backward movement to minimize the mechanical effects of adhesive forces between the probe surface and the tissue. 200 measurements at each study stage were displayed as a PO_2 histogram. Data were pooled for all subjects and separately for septic and non-septic patients for composite histograms.

Statistical analysis was performed using Student's two-tailed t-test for independent data.

RESULTS

Compared to the control group, septic patients had significantly higher O_2 uptake, O_2 delivery and SvO_2. Heart rate (HR), mean arterial pressure (MAP) and pulmonary capillary wedge pressure (PCWP) were not different between the two groups, whereas cardiac index (CI) was significantly higher and systemic vascular resistance (SVR) was significantly lower in the septic shock patients (Table 1).

Table 1: Hemodynamic and O_2 transport related variables of the two study groups.
Mean values $\pm$ SD; Asterisk: difference statistically significant at $p < 0.05$

		Patients with septic shock (n = 10)		Patients without septic shock (n = 10)
HR	(b/min)	101 ± 21		100 ± 15
MAP	(mm Hg)	75 ± 25		87 ± 19
CI	($l \cdot min^{-1} \cdot m^{-2}$)	5.19 ± 1.44	*	3.8 ± 0.8
SVR	($dyne \cdot sec \cdot cm^{-5}$)	581 ± 303	*	815 ± 250
PCWP	(mm Hg)	16 ± 2.5		15 ± 3
DO_2	($ml \cdot min^{-1} \cdot m^{-2}$)	770 ± 220	*	550 ± 180
VO_2	($ml \cdot min^{-1} \cdot m^{-2}$)	183 ± 34	*	166 ± 33
SvO_2	(%)	76 ± 4	*	69 ± 3
ptO_2	(mm Hg)	23.3 ± 22		30.6 ± 23

The tissue PO_2 histograms revealed left shift with a higher number of low tissue PO_2 classes in the latter group and a mean value (pt O_2) 7 mmHg below that of the non-sepitc patients (Fig. 1).

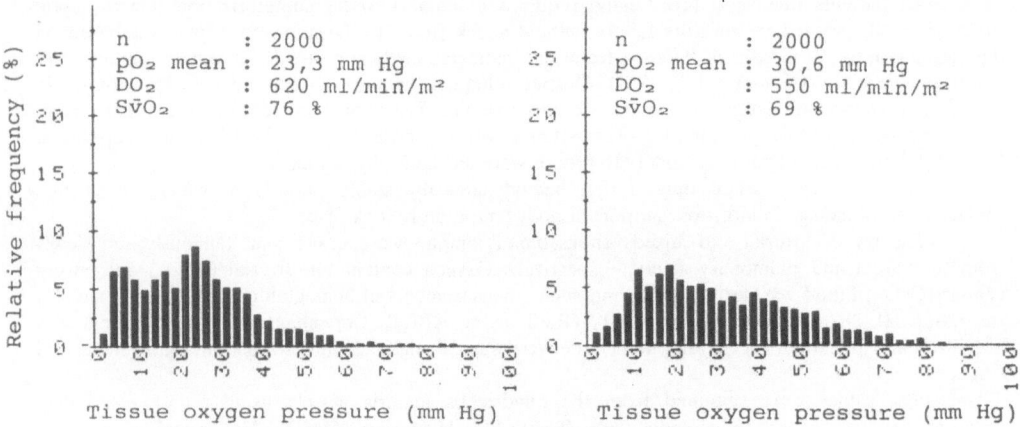

Fig. 1: Tissue O_2 pressure determined in the quadriceps
femoris muscle in patients with and without sepsis.

7 of these 10 patients died from multiple systems organ failure, whereas only 2 patients died in the control group.

DISCUSSION

Human skeletal muscle is an organ with great capillary reserves. The occurence of impaired oxygenation in such an organ in hyperdynamic septic shock and despite O_2 delivery well above normal is surprising. This can only be explained by impaired tissue O_2 flux due to reduced capillary surface area, maldistribution of nutritive blood flow, endothelial cell damage, interstitial edema and increased arterio-venous shunting. All those alterations may occur in sepsis and septic shock (Schumaker, Cain, 1987, Bihari et al., 1987). Also abnormalities in regional vascular tone due to vasoactive mediators may contribute to reduced muscle blood flow. Both increases and descreases of muscle blood flow have been demonstrated in sepsis (Finley et al., 1975, Lang et al., 1984). It cannot be ruled out that in our septic shock patients, the high dosage of the vasopressor norepinephrine additionally impaired nutritive blood flow. We have previously observed a decrease of muscle PO_2 after augmenting the dosage of norepinephrine in septic shock patients (Reinhart, 1989).

Tissue hypoxia may play the major role for the abnormal dependence of O_2 uptake on delivery in the critically ill patient with sepsis and ARDS (Sibbald et al., 1989, Bihari et al., 1987). One might assume that tissue hypoxia in other organ system with less capillary reserves may even be more severe than that in skeletal muscle. In critically ill patients (Liaw, 1985, Grum et al., 1985) tissue hypoxia is in accordance with findings of decreased tissue high energy phosphate stores and elevated ATP degradation products.

It may be aggravated by the increased tissue O_2 needs. Elevated systemic O_2 uptake has been demonstrated in the sepsis patients in this as well as in other studies (Reinhart, 1989). Though, multiple organ failure is the result of various clinical insults, where many toxic mediators are involved, impaired tissue oxygenation, due to a failure of metabolic autoregulation at the microcirculatory level is likely to be an important cofactor in the pathogenesis of this syndrome (Sibbald et al., 1989).

In the clinical setting a covert O_2 debt cannot easily be detected by monitoring of the usual hemodynamic or oxygen-transport-related variables. It can be only quantified after increasing DO_2 and/or improving nutritive blood flow and measurement of the consequent changes in VO_2. Unfortunately up to now there are no monitoring techniques for routine use at the bedside that allow direct evaluation of the tissue bioenergetic status such as NMR spectroscopy and near infrared spectroscopy.

REFERENCES

Bihari,D., Smithies,M., Gimson,A., et al., 1987, The effects of vasodilation with prostacyclin on oxygen delivery and uptake in critically ill patients, N.E.J.M, 317:397.

Danek, S.J., Lynch, J.P., Weg, J.G., Dantzker, D.R., 1980, The dependence of oxygen uptake on oxygen delivery in the adult respiratory distress syndrome, Am. Rev. Respir. Dis., 122:387.

Finley, R.J., Holliday, R.L., Lefcoe, M., et al., 1975, Capillary muscle blood flow in human sepsis. Surgery ,78:87.

Gramm, H.-J., Reinhart, K., Goecke, J., et al., 1989, Early clinical, laboratory and hemodynamic indicators of sepsis and septic shock, in: " Sepsis - an interdisciplinary challenge", K. Reinhart, K. Eyrich eds. Springer Berlin, Heidelberg, New York.

Grum, C.M., Simon, R.H., Dantzker, D.R., et al., 1985, Evidence for adenosine triphosphate degradation in critically ill patients. Chest, 88:763.

Kersting, Th., Reinhart, R., Dennhardt, R., et al., Effect of low dose dopamine on muscle PO_2 in healthy volunteers and intensive care patients. Europ. J. Anaesth., 2:143

Lang, C.H., Gregory, J., Bagby, G.J., et al., 1984, Cardiac output and redistribution of organ blood flow in hypermetabolic sepsis. Am. J. Physiol., 246:R331.

Liaw, K.Y., 1985, Effect of injury, sepsis, and parenteral nutrition on high-energy phosphate in human liver and muscle. J.P.E.N, 9:28.

Mohsenifar, Z., Goldbach, P., Tashkin, D.P., Campisi, D.J., 1983, Relationship between O_2 delivery and O_2 consumption in the adult respiratory distress syndrome. Chest, 84:267.

Reinhart, K., 1989, Oxygen transport and tissue oxygenation in sepsis and septic shock. In: "Sepsis - an interdisciplinary challenge" K. Reinhart, K. Eyrich eds. Springer Berlin, Heidelberg, New York.

Schumaker, P.T., Cain, S.M., 1987, The concept of critical oxygen delivery. Intensive Care Med. 13:223.

Sibbald, W.J., Berten, A., Rutledge, F.S., 1989, The role of tissue hypoxia in multiple organ failure in: "Clinical aspects of O2 transport and tissue oxygenation", K. Reinhart, K. Eyrich, eds. Springer Berlin, Heidelberg, New York.

AUTOREGULATION REMAINS INTACT DURING STABLE XENON INHALATION IN THE BABOON

SK Wolfson Jr^{1,2}, H Yonas^2, D Gur^3, EE Cook^1, J Greenberg^4, RP Brenner^5

^{1}Surgical Research Laboratory, Montefiore Hospital
Departments of Neurological Surgery
and ^{3}Radiology, University of Pittsburgh
^{4}Department of Neurology, University of Pennsylvania, Philadelphia
^{5}Department of Psychiatry and Neurology, University of Pittsburgh
3459 Fifth Avenue, Pittsburgh, PA 15213, USA

Introduction

For the measurement of cerebral blood flow (CBF) using the stable xenon/CT method to be recognized as a valuable adjunct to clinical management of ischemic brain syndromes and to both clinical and animal research, it must provide assessment of perfusion reserve. Does blood flow remain under control by metabolic need (autoregulation) or has it become pressure dependent, that is, on the elbow of the autoregulation curve? Usually, resting gray matter flow below 40 to 50 mL/(100 cm$^3\cdot$min) indicates that autoregulation is exceeded or impaired, and levels of 20 to 25 mL/(100 cm$^3\cdot$min) imply exhaustion of reserve with imminent danger of ischemic injury. Since xenon, in concentrations of 30% to 35% as used for clinical CBF determination, is known to alter blood flow with reports of 15% to 20% flow increase toward the end of the 5 min period of Xe breathing, it is important to be certain that autoregulation is not disturbed by the methodology. It has been shown that agents which are known to alter cerebrovascular resistance (CVR) and CBF do not necessarily affect the ability of the brain to regulate flow in the face of varying cerebral perfusion pressure (CPP). Examples are hypocapnea, a potent cerebral vasoconstrictor that does profoundly alter CBF in dogs but does not substantially affect the ability to autoregulate CBF [1], and Mannitol, which improves O_2 transport by lowering viscosity but does not raise CBF due to a compensatory elevation in CVR in cats and humans [2,3]. By contrast, the calcium entry blocker, nimodipine, has been shown to impair CBF autoregulation in rats [4]. The present work was an attempt to determine if stable Xe, as used clinically in the Xe/CT local CBF method, has a significant effect upon the autoregulation of CBF in a primate.

Oxygen Transport to Tissue XII, Edited by J. Piiper *et al.*
Plenum Press, New York, 1990

Methods

All experimental procedures were carried out in accordance with institutional guidelines for the care and maintenance of laboratory animals.

Five baboons were anesthetized with Halothane (Ayerst Laboratories, Inc, New York, NY), intubated, provided with a 7 f abdominal aortic bleeding/reinfusion catheter and other smaller lines for central arterial and venous blood pressure monitoring, and for IV fluid and drug administration (Fig 1). The animal was placed on the CT table, heparinized (150 u/kg), the lines attached to appropriate monitoring devices, and the bleeding/reinfusion line attached to a servo-controlled withdrawal/infusion pump (Fig 2). The animal was switched to an anesthetic "cocktail" designed to provide minimal alteration of CBF: Morphine SO_4, 0.2 mg/kg (Eli Lilly, Indianapolis, IN), diazepam, 0.2 mg/kg (Valium^R, Hoffman LaRoche, Nutley, NJ), both every 2 h, propranolol HCL, 0.02 mg (Inderal^R, Ayerst Laboratories, Inc, New York, NY), pancuronium bromide, 0.2 mg/kg (Pavulon^R, Organon Pharmaceuticals,

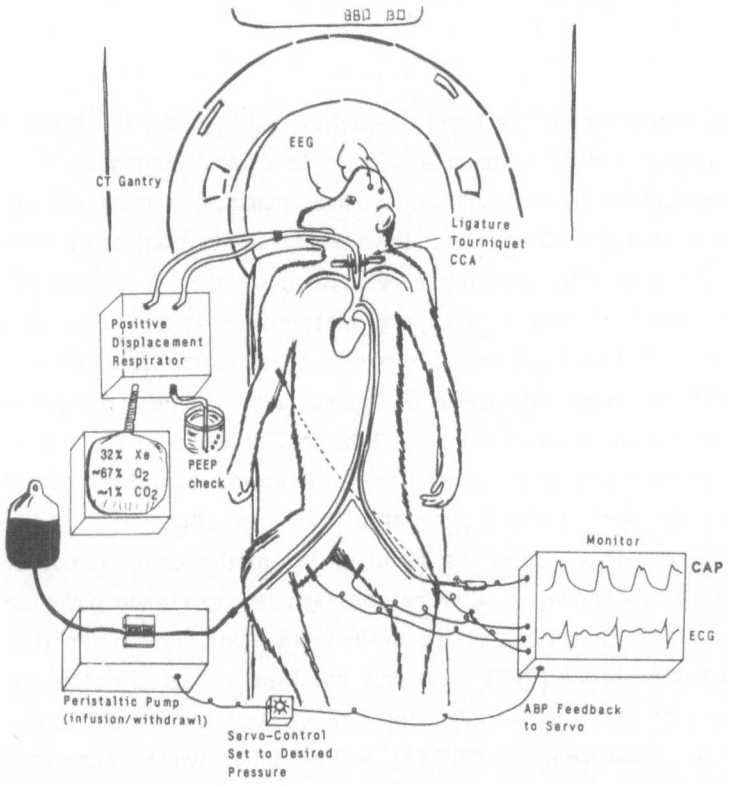

Fig 1. Diagram of the experimental arrangement. The immobilized baboon is placed on the CT table after insertion of a 7 FR catheter into the abdominal aorta via the right femoral artery and a 1.5 mm line into the aortic arch via the left femoral artery. A venous infusion line and CVP line were also inserted via femoral veins. Arterial blood pressure was regulated by setting the servo to the desired level so that blood infusion or withdrawal was effected as needed.

866

West Orange, NJ) both every hour. This regimen provided sedation and analgesia to minimize pain and stress–related cardiovascular effects and paralysis for total immobilization during Xe/CT CBF measurements. $PaCO_2$ was maintained at 30 to 35 torr by controlled ventilation using a positive displacement respirator with 5 cm PEEP. CAB was varied by setting the servo control to the desired level so as to control the withdrawal or reinfusion of blood, automatically. Partial or complete common carotid ligation was added when the lowest pressures were desired. EEG electrodes were applied with collodion in accordance with the international 10–20 system [10]. Those electrodes used were ones that did not interfere with CBF measurement by virtue of placement within CT scan plane: frontal (F3,FZ,F4), central (C3,CZ,C4), parietal (P3,PZ,P4) and ear (A1,A2). Both referential and bipolar derivations were used.

Local blood flow was measured using the stable Xe/CT method [5–7]. This noninvasive method has been shown to provide reproducible CBF measurements with excellent resolution (125 mm^3) and to correlate well with other (invasive) CBF techniques such as tissue autoradiography using labeled microspheres [8] and ^{14}C iodoantipyrine (IAP) [9]. We used a standard GE 9800 scanner with an add–on

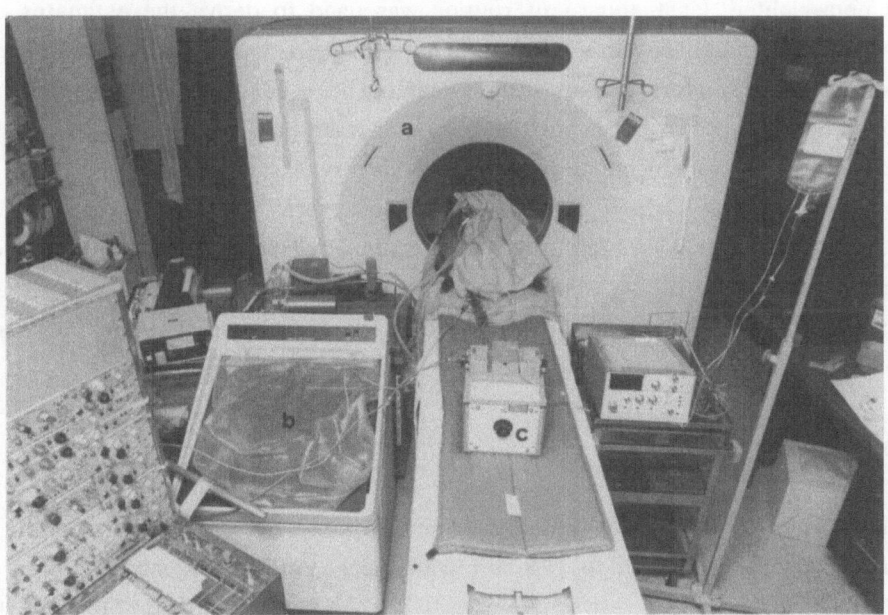

Fig 2. Photograph of the experiment carried out in the computed tomography suite. a) CT gantry; b) Xe administration device; c) Servo-controlled aortic blood infusion/withdrawal pump.

inhalation system designed specifically for this purpose. Details of the methodology and its limitations are described elsewhere [7,11]. Similar add-on systems are currently available as a commercial product on several CT scanners by several manufacturers (for example: Siemens, Picker, Philips, and Toshiba). After the animal was positioned in the scanner, two brain levels were selected for each study, and two nonenhanced baseline scans were obtained at each level. Then the inhaled gas was switched to a 32% Xe/68% oxygen mixture. Rapid serial scanning was initiated at 0.3 min after beginning Xe inhalation. Six enhanced scans were obtained at each level. The Xe/CT studies lasted 6.5 min. After completion of the scanning routine, the animal was returned to room air for a period of 1 h before the next experiment was begun. This allowed for complete Xe washout and for establishment and stabilization of the next blood flow state.

The computational methodology was as follows: the averaged baseline images were subtracted from the enhanced images, and each voxel was subsequently defined by a series of enhancement values as a function of time $[\Delta C_t(t)]$. This series was used in conjunction with end-tidal measurements assumed to be proportional to Xe concentrations in arterial blood $[\Delta C_a]$, and to solve for a monocompartmental Kety equation in which $C_a(u)$ and $C_t(t)$ are used as input data:

$$\Delta C_t(t) = \lambda k \int_0^t \Delta C_a(u) e^{-k(t-u)} du$$

A nonweighted, least-square fit routine was used to derive the estimates of two parameters, λ and f (where $f = \lambda k$). The derivation of the various independent and dependent parameters for each voxel resulted in a set of flow values that was used to generate a flow image. In addition, moderate preanalysis and postanalysis smoothing routines were used to reduce pixel-to-pixel variation (using a 3 pixel x 3 pixel bell-shaped filter). Each animal was studied several times in a preset protocol using the Xe/CT method in each series of studies. Preliminary Xe/CT measurements were made before and during hemorrhagic shock as a means of establishing the desired low flow state. Changes in EEG were used as evidence that the flow was low enough to affect function (Fig 3).

Results

In 4 animals the CAP was varied between 18 and 150 torr. In the 5th, the pressure range was extended to 196 torr with the IV infusion of phenylephrine. Fig 4 illustrates flow maps derived from Xe/CT measurements at 38, 115 and 165 torr. For this study, the entire slice was considered a region of interest and the average flow from two levels was calculated and plotted. The values contain both gray and white matter including a wide variety of cortical, central ganglion, and central white matter tracts. The mean flow of all 5 animals are plotted together in Fig 5. This graph resembles the familiar autoregulation curve for mammalian brain. Flow remained relatively constant between 40 and 150 torr. Significant deviations of flow were restricted to CAPs below 40 torr and above 150 torr.

Ordinarily, autoregulation of CBF is studied in relationship to cerebral perfusion pressure (CPP, often equated to intraventricular cerebrospinal fluid, CSP, or intracranial pressure, ICP) which were not measured in these already complex and very invasive experiments where we also compared tissue autoradiography (IAP) with Xe/CT [9]. Thus we substitute CAP for CPP. We had no evidence of brain edema (from histology of autoradiography sections) or other cause of significant variations of ICP. In general, the EEG results were in agreement with the measured flow

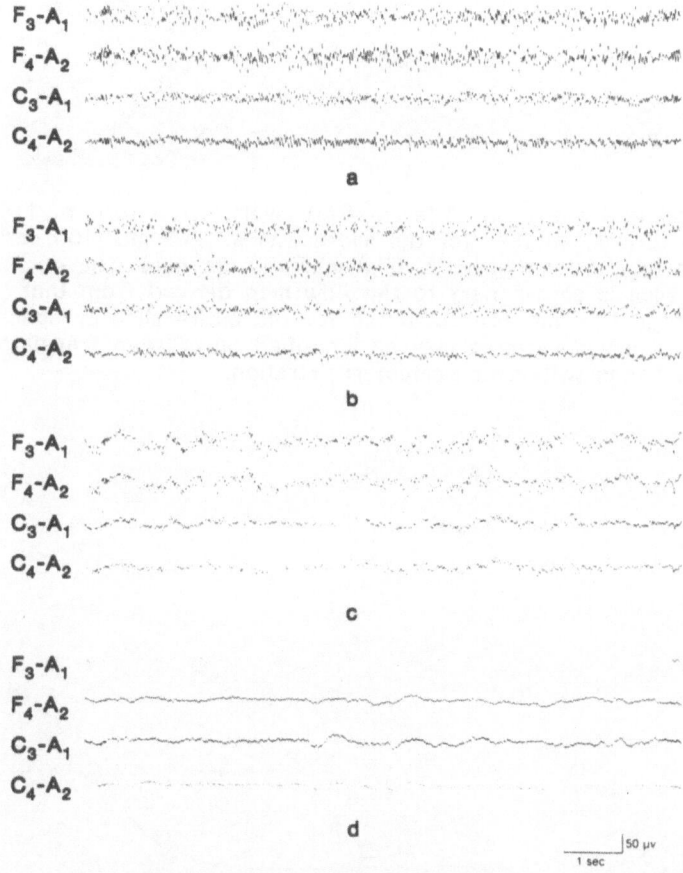

Fig 3. Representative EEG samples from the animal whose flow is shown in Fig 4: (a) baseline tracing (anesthesia only) -- activity primarily in the beta frequency; (b) increase in theta activity following the administration of 32% Xe ($P_aXe = 27\%$); (c) further decrease of faster frequency activity with an increase of delta activity during low flow state; (d) suppression of activity when Xe ($P_aXe = 29\%$) is present in addition to low flow.

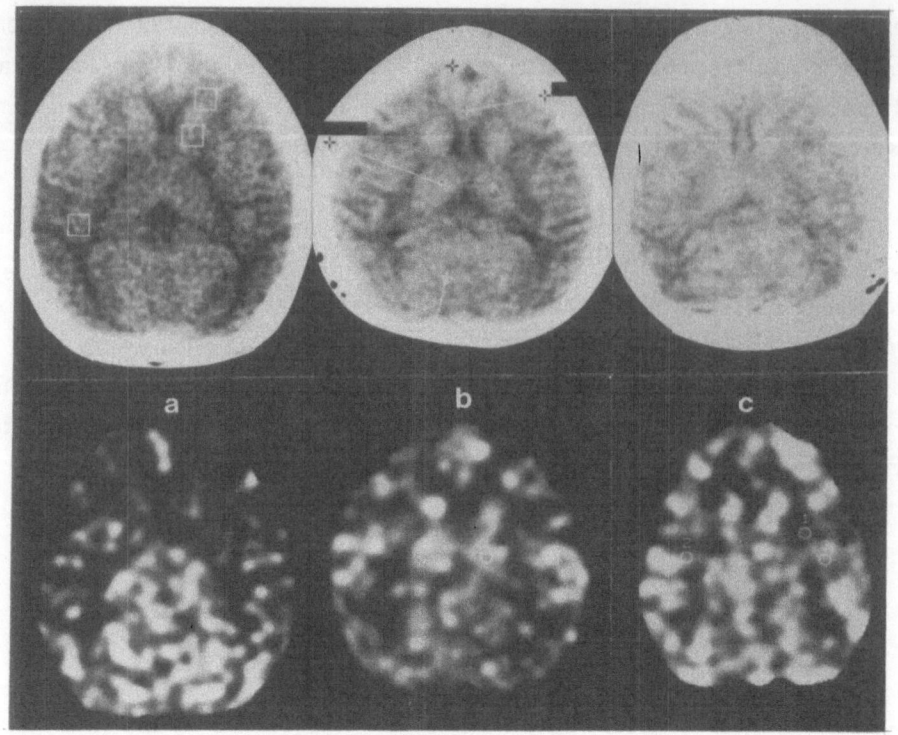

Fig 4. Blood flow maps for different CAB levels. a. 38 torr, b. 115 torr, c. 165 torr. Mean flow for the slices was a. 25.3 mL/(100 cm^3·min) b. 40.4 mL/(100 cm^3·min), c. 57.8 mL/(100 cm^3·min). The corresponding CT slice is shown next to the flow map derived from that level. The average flow may be computed for the entire slice or specific anatomical structures may be identified on CT and translated to the flow image with near perfect registration.

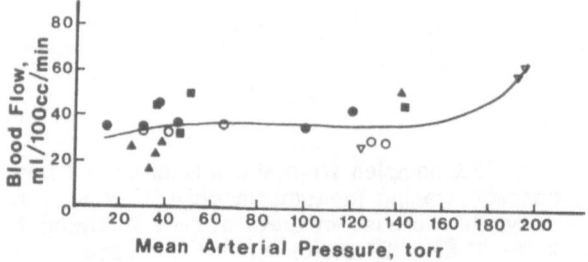

Fig 5. The CAP and Xe/Ct CBF values are plotted for the 5 baboons. Each different symbol represents a different animal experiment.

values. The expected effects of low perfusion were associated only with the lowest pressures encountered. Since Xe is an anesthetic in doses above 50% with known EEG changes in that concentration, it is not unexpected that increased theta activity was encountered with 32% Xe administration. Reduced blood flow alone also caused EEG changes, but the combination Xe administration and low flow (below the elbow of the autoregulation curve) produced the most profound effects. The profound EEG slowing was not seen while the CAB was above 40 to 50 torr and with autoregulation intact. We conclude that normal flow regulatory responses are operative during 32% Xe breathing, permitting both reasonable clinical inferences to be made and physiologic mechanisms to be studied.

Acknowledgements

This work is supported in part by USPHS Grant No HL27208. D Gur is the recipient of an Established Investigator Award from the American Heart Association. Funds are provided by the AHA, Pennsylvania Affiliate.

Summary

To test the possible effect of 32% end-tidal Xe concentration upon autoregulation, 5 baboons, Papio anubis/cynocephalus, were anesthetized/paralyzed with proprandol 0.02, diazepam 0.1, morphine sulfate 0.1, and pancuronium 0.2 (mg/(h·kg)). The animals were subjected to a servocontrolled blood infusion-withdrawal program to control central aortic blood pressure (CAP). $PaCO_2$ was held to 30 to 35 torr, with individual variation <3 torr by control of ventilation and by including CO_2 in the Xe/O_2 mixture. Three to six CBF measurements were made in each subject over the above range. In four animals the CAP was varied between 18 and 150 torr, with corresponding CBF measurements. The CAP range was extended to 196 torr in the 5th animal by IV administration of phenylephrine. Significant lowering of global blood flow did not occur above 40 torr mean CAP. While regulated flow persists to about 150 torr at the high end, there is a breakaway between 150 and 190 torr where flow increased 90%. A 4th order polynomial fit of the data has the characteristic appearance of the familiar autoregulation curve. We conclude that autoregulation is preserved even in the presence of FI_{Xe} of 32% in the breathing mixture.

References

1. Artru AA, Katz RA, Colley PS. Autoregulation of cerebral blood flow during normocapnia and hypocapnia in dogs. Anesthesiology. 1989; 70:288-292.

2.   Muizelaar JP, Lutz HA, Becker DP. Effect of mannitol on ICP and CBF and correlation with pressure autoregulation in severely head-injured patients. J. Neurosurg. 1984; 61:700-706.

3.   Muizelaar JP, Wei EP, Kontos HA, et al. Mannitol causes compensatory cerebral vasoconstriction and vasodilation in response to blood viscosity changes. J. Neurosurg. 1983; 59:822-828.

4.   Hollerhage HJ, Gaab MR, Zumkeller M, Walter GF. The influence of nimodipine on cerebral blood flow autoregulation and blood-brain barrier. J. Neurosurg. 1988; 69:919-922.

5.   Drayer BP, Wolfson Jr SK, Reinmuth OM, Dujovny M, Boehkne M, Cook EE. Xenon enhanced computed tomography for the analysis of cerebral integrity, perfusion, and blood flow. Stroke. 1978; 9:123-130.

6.   Yonas H, Gur D, Claasen D, Wolfson Jr SK, Moossy J. Stable xenon enhanced computed tomography in the study of clinical and pathologic correlates of focal ischemia in baboons. Stroke. 1982; 13:750-758.

7.   Gur D, Wolfson Jr SK, Yonas H, Good WF, Shabason L, Latchaw RE, Miller DM, Cook EE. Progress in cerebrovascular disease: local cerebral blood flow by xenon enhanced CT. Stroke. 1982; 13:750-758.

8.   Gur D, Yonas H, Jackson DL, Wolfson Jr SK, Rockette H, Good WF, Cook EE, Arena VC, Willy JA, Maitz GS. Simultaneous measurements of cerebral blood flow by the xenon/CT method and the microsphere method: a comparison. Invest Radiol. 1985a; 20(5):672-677.

9.   Wolfson Jr SK, Clark J, Greenberg JH, Gur D, Yonas H, Brenner RP, Cook EE, Lordeon PA. Simultaneous measurements of normal and low cerebral blood flow states by the Xe/CT method and the [^{14}C]iodoantipyrine method. Submitted for publication.

10.   Jasper HH. Report of committee on methods of clinical examination in electroencephalography. Electroencephalogr Clin Neurophysiol. 1958; 10:370-375.

11.   Yonas H, Good WF, Gur D, Wolfson Jr SK, Latchaw RE, Good B, Leanza R, Miller DM. Mapping cerebral blood flow by xenon-enhanced computerized tomography: clinical experience. Radiol. 1984; 19:228-238.

CARDIOVASCULAR RESPONSES, HEMODYNAMICS AND OXYGEN
TRANSPORT TO TISSUE DURING MODERATE ISOVOLEMIC
HEMODILUTION IN PIGS

A. Trouwborst, R. Tenbrinck, M. Fennema, M. Bucx,
W.G.M. v.d. Broek and B.K. Trouwborst-Weber

Department of Anesthesiology, Erasmus University
Rotterdam, The Netherlands

INTRODUCTION

As postulated, acute isovolemic hemodilution induces a decrease in systemic vascular resistance (SVR) almost parallel to the decrease in blood viscosity, whereas cardiac index (CI) increases significantly without increase in myocardial contractility, while oxygen affinity of hemoglobin is unaffected (Messmer et al., 1973). Over a wide range of hematocrit (Hct) levels, the rise in CI compensates for the decreased oxygen transport capacity, thereby maintaining oxygen transport to the tissue unaltered. Nearly all data in the literature about the effects of hemodilution are obtained from experiments in dogs. Dogs, however, differ from humans in anatomy, distribution of coronary arteries and in sympathetic responses (Weaver et al., 1986). Therefore in this study cardiovascular responses, hemodynamics, oxygen transport to tissue and the oxygen affinity of hemoglobin during normoxic acute isovolemic moderate hemodilution were studied in pigs. Ample evidence exist to demonstrate that the pig is closely related to the human both anatomically and physiologically. The cardiovascular system and metabolism show similarities with respect to the size and distribution of coronary vessels, blood pressure, heart rate, cardiac index, regional distribution of cardiac output and maximum oxygen consumption (Swindle, 1984; Mc Krinan et al., 1986; Weaver et al., 1986).

MATERIAL AND METHODS

Twelve male Yorkshire pigs (10.2 - 12.0 kg) were used. After giving 0.3 mg/kg midazolam (Dormicum®) i.m. a catheter was introduced into one of the ear veins and the trachea was cannulated below the larynx. Throughout the experimental procedure anesthesia was maintained with a continuous i.v. infusion of 0.2 mg/kg/h midazolam (Dormicum®). After cannulation of the trachea 0.1 mg/kg pancuronium was given with an additional continuous i.v. infusion of 0.3 mg/kg/h. The pigs were ventilated with air with a tidal volume adjusted to keep end-tidal CO_2 between 34 and 38 mm Hg. Catheters (Cook Europe BP8) were placed in the left femoral artery, in the right femoral vein and in the right femoral artery (the latter for arterial blood pressure monitoring). Via the left femoral vein a thermodilution catheter (Swan Ganz® 93A-095-7F, Am. Edwards Lab., U.S.A.) was introduced into a pulmonary artery. In all animals the injectate port was located in the right atrium as was proven by postmortem examination. The urine bladder was cannulated. Body temperature (blood temperature) was measured with the thermistor electrode of the thermodilution catheter and was kept stable throughout the procedure by means of a heating pad. After all preparations were completed, the anesthetized paralyzed animals were ventilated until blood gas tensions, pH and hemodynamic parameters were stabilized (average time: 30 min). Pulse rate (HR), arterial blood pressure, pulmonary arterial pressure and the right atrium pressure were monitored continuously (Horizon 2000®, Mennen Medical, Israel). After the stabilization period baseline measurements were performed, including: pulse rate (HR), mean arterial blood pressure (MAP), mean pulmonary arterial pressure (mPAP), pulmonary wedge pressure (PWP), right atrial pressure (RAP) and cardiac output (CO). From these data pulmonary vascular resistance (PVR) and systemic vascular resistance (SVR) as well as left ventricular stroke work index (LVSWI) were calculated. In addition, arterial and mixed venous blood samples were taken for measurements of PO_2, pH and PCO_2 (ABL 330, Radiometer, Copenhagen) and for measurement of hemoglobin and oxyhemoglobin content (spectrophotometer OSM3, Radiometer, Copenhagen). Oxygen flux (O_2 flux) was calculated as the product of arterial oxygen content and CO. Oxygen uptake (VO_2) was defined as the product of CO and the arteriovenous oxygen content difference. From the mixed venous blood sample, the P_{50} (PO_2 at which hemoglobin is 50% saturated), corrected to 38°C and pH = 7.40, was calculated from a single measurement of venous pH, PO_2 and oxygen saturation of the hemoglobin (Siggaard-Andersen et al., 1988), assuming a value for n_{HILL} of 2.83, a value previously established as a standard for pig blood (Trouwborst et al.,1989).

The first step of isovolemic hemodilution with dextran 40 (iso-oncotic, 50 g/l in 0.9% salt solution; Isodex®, N.P.B.I., Holland) began after baseline measurements were completed. The dextran solution (warmed to 38°C) was instilled slowly into the right femoral vein at the same time and at the same rate that blood was removed from the left femoral artery. Stepwise isovolemic hemodilution was induced by steps of 10 ml/kg bodyweight until a total exchange of 40 ml/kg bodyweight. New sets of data were obtained 5 min after each step of isovolemic hemodilution. The time between each step of blood exchange was 15 min.

Data are presented as means ± S.D. The accepted probability for a statistical significance between means was $P < 0.05$. Differences were tested by the Wilcoxon signed-rank test. Regression lines were estimated by methods of least squares. For regression analysis the Spearman rank correlation coefficient was used.

RESULTS

Hematocrit (Hct) decreased stepwise from 30.5% to 15.9%. Cardiovascular and hemodynamic responses are summarized in table 1. During isovolemic hemodilution the HR increased by 20%. A slight but not significant increase in MAP could be observed. CI rose by 39% accompanied by a significant rise in LVSWI of 29%. An initial statistically significant increase in PWP up to 24% of baseline value was observed, presumably due to an increase in venous return of the blood volume to the heart. The SVR declined by only 16%.

The parameters of oxygen transport and oxygen uptake are summarized in table 2. During hemodilution VO_2 increased statistically significantly by nearly 15% in spite of a gradual decrease in O_2 flux of 27%, resulting in a higher oxygen extraction ratio (ER). The P_{50} value rose gradually from 32.3 mm Hg to 34.3 mm Hg. Throughout the whole experimental procedure, arterial and mixed venous pH and PCO_2 remained unchanged.

874

Table 1.

Hemodynamics during moderate normovolemic hemodilution in twelve pigs.

blood-exchange ml/kg bw	Hct %	HR b/min	MAP mmHg	PWP mmHg	CI l/min/m²	LVSWI g.cm/m²	SVR dyns/cm⁻⁵
0	30.5 3.0	175 29	106 17	6.8 2.5	5.4 1.1	42 11.6	3425 1013
10	24.8 2.7 #	186 29	106 15	8.5 3.6 #	6.0 0.9 #	44.5 12.5 #	3254 1020
20	21.2 2.6 #	194 28 #	115 14	7.8 3.4	7.0 1.1 #	51.4 12 #	2933 805 #
30	18.4 2.0 #	207 26 #	115 10	7.1 3.8	7.4 1.4 #	53.2 10.4 #	2819 977 #
40	15.9 1.5 #	210 20 #	118 11	7.5 2.8	7.5 1.4 #	54 11.7 #	2863 947 #

Values are expressed as mean and standard deviation. # mean significantly different from initial value (p<0.05). Hct = hematocrit; HR = heartrate; MAP = mean arterial bloodpressure; PWP = pulmonary wedge pressure; CI = cardiac index; LVSWI = left ventricular stroke work index; SVR = systemic vascular resistance.

Table 2.

Data of systemic oxygenation during moderate normovolemic hemodilution in twelve pigs.

blood-exchange ml/kg bw	Hb gm%	O_2 fl ml/min/m²	VO_2 ml/min/m²	ER	P_{50} mmHg	pvO_2 mmHg
0	10.1 1.2	791 168	215 55	0.28 0.08	32.3 1.6	45.0 5.0
10	8.3 0.8 #	712 146	227 58 #	0.32 0.06 #	32.7 1.6 #	43.4 3.8 #
20	7.0 0.7 #	695 135 #	227 54 #	0.33 0.07 #	33.0 1.4 #	42.9 4.0 #
30	6.2 0.7 #	651 114 #	242 67 #	0.37 0.08 #	33.5 1.8 #	40.6 3.5 #
40	5.3 0.5 #	579 96 #	240 72 #	0.41 0.08 #	34.3 1.7 #	39.6 4.6 #

Values are expressed as mean and standard deviation. # mean significantly different from initial value ($p<0.05$). Hb = hemoglobin; O_2 fl = oxygen flux; VO_2 = oxygen uptake; ER = oxygen extraction ratio; P_{50} = oxygen tension at 50% saturation; PvO_2 = mixed venous oxygen tension (not corrected for pH)

DISCUSSION

As reported, hemodilution induces a decrease in SVR almost parallel to the decrease in blood viscosity (Murray and Escotar, 1968; Fowler and Holmes, 1975), whereas CI increases significantly without an increase in myocardial contractility (Messmer et al., 1973). However, an increase in oxygen uptake by the heart during hemodilution has also been reported (Von Restorff et al., 1975). Furthermore it has been observed that both anemic and hypoxic hypoxia in dogs increases total catecholamine levels with consequent stimulation of the cardiovascular system (Sylvester et al., 1979). In our study a 39% increase in CI was observed. The percentage increase in cardiac output over pre-anemic values reported by Cain (1977) was more pronounced (130%). Others found an increase up to 70% in comparison to baseline values (Lundsgaard-Hansen, 1979). In humans at Hct of 20%, an increase in cardiac output of about 35% has been reported (Laks et al., 1974). There are possible explanations for the differences between data of our study and that of others, including species difference. In most studies dogs were used, with the spleen excluded from the circulation, while in our study the intact pig model was chosen. Differences in response may be due to differences in the sympathetic responses and to differences in the anatomy and distribution of coronary arteries in both animals. Our CI response findings were more in accordance with data obtained in humans, probably due to similarities in the cardiovascular system and metabolism of man and swine. The estimated decrease in SVR in our pig study differs from the observed decrease in SVR in the dog studies. Cain reports a decrease in SVR of 60% at the initial stage of anemic hypoxia and of 74% at the final stage (Cain, 1977). In our study the decrease in SVR was not more than 16% of the baseline value. The observed increase in CI in our study was more related to the concomitant 30% increase in LVSWI, which was accompanied by an increase of 20% in heart rate. This increase in workload of the heart could explain the reported increase in myocardial oxygen consumption during normovolemic hemodilution (Von Restorff et al., 1975), while the increase in heart rate might be due to the increase in total circulating catecholamine levels as has been observed in anemic hypoxia (Sylvester et al., 1979). In the case of a decreased venous return of blood, as in hypovolemia, an increase in heart rate can follow. In our study, however, this reason does not apply because the estimated increase in PWP suggests, in contrast, an improved venous return. This is in accordance with the observation that venous return can be significantly improved in the presence of decreased viscosity such as with hemodilution (Messmer et al., 1986). In return, the increase in venous return could have contributed to some extent to the increase in CI.

In our study in pigs, in contrast to studies in dogs, the increase in CI did not compensate for the decreased oxygen transport capacity. However, despite a decrease in O_2flux, the VO_2 rose, resulting in a higher ER. Also in contrast to reports in dogs (Messmer et al., 1973), in our study an immediate and acute shift to the right of the oxyhemoglobin dissociation curve (ODC), as expressed by a change in the P_{50} (corrected to pH = 7.40), could be observed. No correlation was found between the change in P_{50} and changes in the arterial or mixed venous actual pH, PCO_2 or temperature of the blood. It has been reported that a chronically reduced hemoglobin concentration in man initiates an improved oxygen unloading from the hemoglobin, afforded by an increased P_{50} (Blumberg and Marti, 1972; Edwards and Canon, 1972). Blumberg and Marti (1972) reported that in chronically anemic patients on dialysis, 2,3 -diphosphoglycerate (2,3 - DPG) correlated with decreased oxygen affinity of hemoglobin. Edwards et al removed blood over 1-3 weeks from humans and found an increased P_{50}. value but with no change in 2,3 -DPG. They discussed an intrinsic decrease in oxygen affinity of young erythrocytes that has not been satisfactorily explained in chemical terms. These data, however, were obtained in chronic anemia whereas our study deals with the effects of acute hemodilution. It is unlikely that the observed change during acute hemodilution can be explained by the factors discussed in relation to chronic anemia. Messmer et al., (1973) studied in dogs the actual in vivo effects on the P_{50} of stepwise induced hemodilution with a dextran solution. An acute increase in P_{50}. was observed but not before Hct decreased to 10%. The acute significant increase in the P_{50} was even more enhanced 24 h after the induction of acute hemodilution, suggesting the existence of a rapid adaptation mechanism to anemia besides a later response (Messmer et al., 1973). Acute changes in P_{50} are reported of coronary sinus, venous and arterial blood during signs of hypoxia of the heart in man (Colvard and Longmuir, 1973; Kostuk et al., 1973; Shappell et al., 1970). From the reported data of Messmer no information about PO_2 of the mixed venous blood is available. The mean PO_2 values in various organs in his dog model however did not decrease during hemodilution, indicating the absence of hypoxia. In our study however a gradual decrease in mixed venous PO_2 could be observed during isovolemic hemodilution.

SUMMARY

In conclusion, in contrast to many reports obtained from dogs, in pigs the rise in CO during moderate isovolemic hemodilution is, besides the increased venous return, more induced by increased work performance of the heart and less by a decreased SVR. The rise in CO did not compensate for the decrease in oxygen transport capacity. Our results confirm most of the reported findings in humans at the same stage of hemodilution. Besides changes in hemodynamics, in our study a gradual decreased oxygen affinity of hemoglobin could be observed. That the pig animal model for studying hemodilution and oxygen transport to the tissue is more appropriate than the dog model is open to discussion.

ACKNOWLEDGMENT

This work was supported by Foundation for Fundamental Medical Research (SFMO, the Netherlands). The authors would like to thank Mr. A. Kok, Mrs. L. Visser and Ms. D. Haas for technical and secretarial assistance.

REFERENCES

Blumberg, A., and Marti, H.R., 1972, Adaptation to anemia by decreased oxygen affinity of hemoglobin in patients on dialysis, *Kidn. Int. 1:263.*

Cain, S.M., 1977, Oxygen delivery and uptake in dogs during anemic and hypoxic hypoxia,*J. Appl. Physiol., 42:228.*

Colvard, M.L. and Longmuir, J.S., 1973, The effects of pacing on oxygen hemoglobin dissociation and oxygen carrying capacity in patients suspected of coronary artery, disease, *Am. Heart J.,85: 662*

Edwards, M.J., and Canon B., 1972, Oxygen transport during erythropoietic response to moderate blood loss, *New Eng. J. of Med., 287:115.*

Fowler, N.O., and Holmes, J.C., 1975, Blood viscosity and cardiac output in acute experimental anemia, *J.Appl. Physiol., 39:453.*

Kostuk, W.J., Sierra, K., Bernstein, E.P.,and Sebel, B.E., 1973, Altered hemoglobin oxygen affinity in patients with acute myocardial infarction, *Am. J. Cardiol., 31: 295.*

Laks, H., Pilon, R.N., Klovekorn, W.P., et al., 1974, Acute hemodilution: its effect on hemodynamics and oxygen transport in anesthetized man, *Ann. Surg., 180:103.*

Lundsgaard-Hansen, P., 1979, Hemodilution- New clothes for an Anemic Emperor, *Vox Sang., 36:321.*

Messmer, K., Görnandt, L., Jesch, E., Sinagowitz, E., Sunder-Plassmann, L., and Kessler, M., 1973, Oxygen transport and tissue oxygenation during hemodilution with dextran, *Adv. Exp. Med. Biol.*, 37:669.

Messmer, K., Kreimeier, U., and Intaglietta, M., 1986, Present state of intentional hemodilution, *Eur. Surg. Res.*, 18:254.

Mc Krinan, M.D., White, B.D. Guth, and Bloor, C.M., 1986, Cardiovascular and Metabolic responses to acute and chronic exercise in swine, in : *"Swine in Biomedical research", M.E. Tumbleson, ed., Plenum Press, New York and London.*

Murray, J.F., and Escobar, E., 1968, Circulatory effects of blood viscosity: comparison of methemoglobinemia and anemia, *J. Appl. Physiol.*, 25:594.

Shappell, S., Murray, J.A., Nasser, M.G., Willis, R.E., Torrance, J.O., Lenfant, C.J.M., 1970, Acute changes in hemoglobin affinity for oxygen during angina pectoris, *N. Eng. J. Med.*, 282: 1219.

Siggaard -Andersen, O., Wimberley, P.D., Fogh-Andersen, N., Cothgen, J.H., 1988, Measured and derived quantities with modern pH and bloodgas equipment: calculation algorithms with 54 equations, *Scand. J. Clin. Lab. Invest.*, 48:S189:7.

Swindle, M.M., 1984, Swine as replacement for dogs in the surgical teaching and research laboratory, *Lab. Anim. Sci.*, 34:383.

Sylvester, J.G., Scharf, S.M., and Gilbert, R.D., 1979, Hypoxic and CO hypoxia in dogs: hemodynamics, carotic reflexes and catecholamines, *Am. J. Physiol.*, 236:H22.

Trouwborst, A., van den Broek, W.G.M., Tenbrinck, R., Groenland, T.H.N., Bucx, M., Faithfull, N.S., 1989, Alterations in oxyhemoglobin dissociation curve during normoxic acute normovolemic hemodilution, *Adv. Exp. Med. Biol.*, 248:419.

Von Restroff, W.B., Höfling, J., and Bassenge, E., 1975, Effect of increased blood fluidity through hemodilution on coronary circulation at rest and during exercise, *Pluegers Arch.*, 357:15.

Weaver, M.E., Pantely, G.A., Bristow, J.D., and Ladley, H.D., 1986, A quantitive study of the anatomy and distribution of coronary arteries in swine in comparison with other animals and man, *Cardiovasc. Res.*, 20:907.

ANALYSIS OF OXYGEN TRANSPORT TO TISSUE

DURING EXTREME HEMODILUTION

Amy G. Tsai, Karl-E. Arfors* and Marcos Intaglietta

Department of AMES-Bioengineering; R-012
University of California, San Diego; La Jolla, California, 92093

*Pharmacia Experimental Medicine - La Jolla, California, 92037

INTRODUCTION

The condition of reduced systemic hematocrit is frequently encountered in the clinical setting. Intentional dilution is employed therapeutically as well as prophylactically and is implemented by partially replacing blood with RBC free plasma expanders.

Extreme hemodilution, i.e. the reduction of systemic hematocrit to levels as low as 7%, is commonly employed during open-heart and open-cranial surgeries (Chapler and Cain, 1986; Laver *et al.*, 1975; Niinikoski *et al.*, 1981; Kim *et al.*, 1989) with no apparent adverse effect to tissue function. This procedure is typically performed at hypothermic conditions to slow the metabolism of the tissue; however, successful application at normal body temperature has also been reported (Lilleaasen, 1981). During moderate reductions in systemic hematocrit, several compensatory adjustments by the body are evoked to contend with the reduction of the oxygen content of the blood; these include an increase in cardiac output, the redistribution of blood flow to some tissues, and an increase in the whole body oxygen extraction ratio (Chapler and Cain, 1986; Messmer, 1975). The main mechanism for the noted beneficial effect is proponed to be the increased fluidity of blood as a consequence of the reduced viscosity (Messmer *et al.*, 1972) thus allowing the delivery of oxygen to regions previously afflicted with reduced or no flow and bordering on anoxia.

During extreme hemodilution, when the oxygen content of blood is severely curtailed, the deficit cannot simply be offset by maintaining a constant flux of RBCs to the capillaries (Tsai and Intaglietta, 1989). The subsequent reduced oxygen content of blood, leads to an increasingly discontinuous supply of oxygen to the tissue amplifying the effects due to the particulate nature of blood at the capillary level. Under normal conditions the resulting discontinuity may jeopardize tissue function. A mathematical model previously developed and which assumes that tissue is oxygenated by the transit of discrete sources of oxygen was used to determine how RBC spacing effects tissue oxygenation. Simulations of synthetic spacing patterns were made to elucidate the effect of spacing on tissue oxygenation.

MATHEMATICAL MODEL

The mathematical model which is fully described elsewhere, was previously used to compare tissue oxygenation at different hematocrits during constant flux conditions (Tsai and Intaglietta, 1989). Briefly, the representation of blood in the mathematical model is based on the observation that capillary hematocrit is on the order of half the systemic. Therefore, as a consequence of the spacing between RBCs, these RBCs at times may act as a discrete source of oxygen to the tissue. Figure 1 is a schematic of the underlying difference in characterizing the oxygenation potential of blood when studying low hematocrit conditions. The classical concept of blood as a continuous source of oxygen is shown in the upper panel; however a more appropriate description, especially during extreme hemodilution, is when the RBCs should be considered individual sources of oxygen to the tissue (lower panel).

The particulate nature of blood is introduced into the model as cylinders of well-mixed oxygen sources separated by segments of plasma (whose length defines RBC separation or spacing). The design of the model accounts for hemodilution by designating the RBC as the sole source of oxygen to the tissue. This configuration, allows for the study of changes in RBC spacing as well as hematocrit and RBC velocity. The diffusion of oxygen from the cells is dependent on the pO_2 gradients within the tissue where the consumption rate is tension dependent. The pO_2 levels within the tissue are governed by Fick's law of diffusion. The RBC velocity determines the amount of time the RBC resides within the tissue volume, during

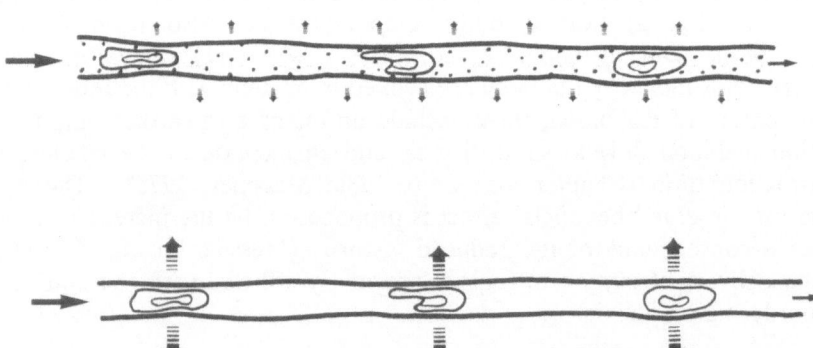

Figure 1. Schematic of RBC as discrete sources of oxygen illustrating the underlying basis of the mathematical model. Upper panel depicts the classical view that blood in capillaries presents the tissue with a constant flux of oxygen. The lower panel in which RBCs act as point sources of oxygen to the tissue is appropriate for characterizing the conditions during extreme hemodilution.

which time oxygen is transferred from the RBC to the tissue depending on the oxygen gradient in the tissue as defined by the governing equation. Capillary hematocrit is inversely related to the distance between RBCs. The effect of each individual RBC is assessed, as its transfer of oxygen to the tissue is dependent on the past events which occurred in the tissue.

The model was used to ascertain the effects of different patterns of RBC spacing on tissue oxygenation. In these spacing pattern simulations, the potential oxygen to the tissue is maintained by delivering an identical number of RBC per unit time to the capillary, i.e. a constant RBC flux condition. Both velocity and time averaged capillary hematocrit are constant, while the relative spacing between RBC is varied. Corresponding velocity and capillary hematocrit values were used to mimic normal and extreme hemodilution conditions. The values for velocity and capillary hematocrit, representative of normal conditions, are 500 μm/sec and 20% respectively. Extreme hemodilution conditions are simulated by doubling the velocity and halving the hematocrit, in this manner the flux is kept constant. These changes in hemodynamics are commiserate with those that have been observed to occur during isovolemic hemodilution (Lindbom *et al.*, 1988; Mirhashemi *et al.*, 1987, Tigno *et al.*, 1986).

Two oxygenation indices are used to compare the effects of each pattern, time averaged tissue pO_2 and the amount of oxygen transferred by each cell into the tissue during transit. The first index gives an indication as to the overall tissue pO_2 levels over time, while the second index serves to give a more detail understanding of this resultant pO_2 by assessing the contribution of each cell in the pattern. The time averaged pO_2 is the average tension within the volume of tissue understudy over the period of time, dictated by the RBC velocity, to pass a pattern of cell. The amount of oxygen loss by each cell in terms of saturation drop after its transit is used as an indicator of the amount of oxygen transferred by each cell. Assessment of these parameters are made when the oxygen tension within the tissue converges to a quasi-steady state. This point in time is reached when the amount of oxygen lost by the first cell in the pattern converges to a steady state. To maintain computational levels within limits, the length of capillary through which the train of RBC passes is set equal to the length of a RBC.

RESULTS

Figure 2 shows the four patterns of spacing considered in the analysis and the values of the oxygenation indices relative to those obtained with an even pattern. The spacing pattern is in units of RBC length, thus a pattern of 25 - 1 - 1 translates into a repetitive pattern of 4 cells, separated by 25, 1, and 1 cell length(s) respectively. The time averaged tissue pO_2 and the amount of oxygen loss by each cell was assessed for an inlet RBC saturation of 50% into a tissue slice of RBC length under extreme hemodilution conditions.

The homogenous spacing pattern resulting in optimized time average tissue pO_2 within the tissue, despite the constant delivery rates among the patterns. The

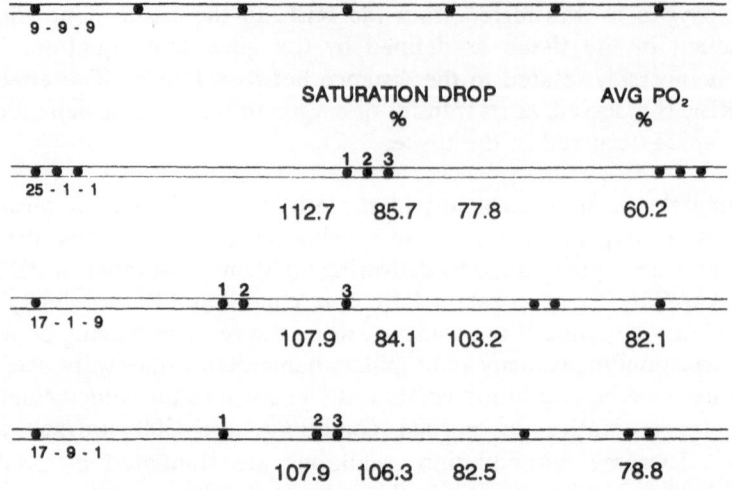

	SATURATION DROP %			AVG PO₂ %
9 - 9 - 9				
	1 2 3			
25 - 1 - 1	112.7	85.7	77.8	60.2
17 - 1 - 9	1 2 3			
	107.9	84.1	103.2	82.1
17 - 9 - 1	1 2 3			
	107.9	106.3	82.5	78.8

Figure 2. Comparison of average tissue pO_2 and amount of oxygen transferred by each cell in the spacing for different spacing patterns. Inlet saturation of cells is 50% under hemodilution conditions. Percent values for the parameters are relative to that obtained during even RBC configurations. Calculations are made across a finite length of capillary.

25 - 1 - 1 pattern resulted in a reduction of average tissue pO_2 by 39.8%. The amount of oxygen transferred by each cell with even spacing is constant. A higher average reduction in saturation drop was found for the non-uniform spacing when compared to the uniform spacing situation. The most deviant pattern from an even distribution, 25 - 1 - 1, resulted in a maximum saturation drop of 22.2% in the lead cell in the pattern. Simulations of spacing patterns under normal conditions resulted in no significant changes in the two oxygenation indices relative to even spaced cells.

DISCUSSION

The results of this analysis shows that during extreme hematocrit reduction conditions, the oxygenation levels within the tissue fluctuate as a result of a passing cells to the point of preventing optimal tissue oxygenation given an identical RBC delivery rate. Under normal conditions, the oxygenation levels immediately adjacent to the capillary were sensitive to cell passage, but spacing pattern was not a factor on the overall tissue oxygenation as assessed by the oxygen indices. This result is similar to the findings of Federspiel and Sarelius (1984), which show that the limit to the assumption that blood is a continuous source of oxygen to the tissue is a capillary hematocrit of 20%.

The variability observed in the amount of oxygen transferred from the cell to the tissue in Figure 2 leads to several interesting ideas. It can be concluded that the non-uniformity of the spacing results in the cells being exposed to unfavorable gradients for optimal oxygen release with closely spaced cells and excessive gradients for widely spaced cells. Closely spaced cells are prevented from releasing oxygen due to exposure to relatively lower gradients, while greatly spaced cells are exposed to high gradients. Evenly spaced cells minimize these deviations, thus maximizing tissue oxygenation. Contributions to the variability in single RBC saturations found experimentally (Ellsworth *et al.*, 1988) may in part be due to the effect of cell spacing, which is more significant during reduced hematocrit conditions.

CONCLUSIONS

Model simulations suggest that RBC spacing has no appreciable effect on tissue oxygenation during normal conditions; however as the oxygen content of blood is reduced during hemodilution, the tissue becomes increasingly sensitive to the passage of each source.

ACKNOWLEDGEMENT

Supported by USPHS Grants HL 12493, HL 17421, and HL 07089-15.

REFERENCES

Chapler, C.K. and S.M. Cain, 1986, The physiologic reserve in oxygen carrying capacity: studies in experimental hemodilution, *Can. J. Physiol. Pharmacol.*, **64**:7-12.

Federspiel, W.J. and I.H. Sarelius, 1984, An examination of the contribution of red cell spacing to the uniformity of oxygen flux at the capillary wall, *Microvasc. Res.*, **27**:273-285.

Kim, Y.D., Katz, N.M., Ng, L., Nancherla, A., Ahmed, S.W., and Wallace, R.B., 1989, Effects of hypothermia and hemodilution on oxygen metabolism and hemodynamics in patients recovering from coronary artery bypass operations, *J. Thorac. Cardiovas. Surg.*, **97**:36-42.

Laver, M.B., Buckley, M.I. and W.G. Austen, 1975, Extreme hemodilution in man; its use in conjunction with profound hypothermia and circulatory arrest in man, *in*: "Intentional Hemodilution," K. Messmer and H. Schmid-Schoenbein, eds., *Bibl. Haematolog.*, **41**:225-232.

Lilleaasen, P., 1981, Haemodilution in open-heart surgery. Volume and acid base studies, *Ann. Clinc. Res.*, **13**, Suppl. **33**:72-77.

Lindbom, L., Mirhashemi, S., Intaglietta, M. and K.-E. Arfors, 1988, Increase in capillary blood flow and relative haematocrit in rabbit skeletal muscle following acute normovolaemic anaemia, *Acta Physiol. Scand.*, **134**: 503-512.

Messmer, K., 1975, Hemodilution, *Surg. Clinics of N. Amer.*, **55**(3):659-678.

Messmer, K., Sunder-Plassmann, L., Klovekorn, W.P. and K. Holper, 1972, Circulatory significance of hemodilution: Rheological changes and limitations, *Adv. Microcirc.*, **4**:1-77.

Mirhashemi, S., Messmer, K., Arfors, K.-E. and M. Intaglietta, 1987, Microcirculatory effects of normovolemic hemodilution in skeletal muscle, *Int. J. Microcirc.: Clin. Exp.*, **6**:359-369.

Microcirc.: Clin. Exp., 6:359-369.

Niinikoski, J. Laaksonen, V., Meretoja, O., Jalonen, J. and M.V. Inberg, 1981, Oxygen transport to tissue under normovolemic moderate and extreme hemodilution during coronary bypass operation, *Ann. of Thorac. Surg.*, 31(2):134-143.

Tigno, X.T. and H. Henrich, 1986, Flow characteristics of the microcirculation following intentional hemodilution, *Acta Med. Phillippina*, 22:5-12.

Tsai, A.G. and M. Intaglietta, 1989, Local tissue oxygenation during constant red blood cell flux: A discrete source analysis of velocity and hematocrit changes, *Microvasc. Res.*, 37:308-322.

TUMORS

SIZE-DEPENDENT OXYGENATION AND ENERGY STATUS IN MULTICELLULAR TUMOR SPHEROIDS

Stefan Walenta, Jörg Dötsch, Beatrice Bourrat-Flöck, and Wolfgang Mueller-Klieser

Institute of Physiology and Pathophysiology, University of Mainz, D-6500 Mainz, FRG

INTRODUCTION

Multicellular tumor spheroids show numerous analogies to tumor microregions in vivo, such as the development of central necrosis at a certain spheroid size (for reviews see: Mueller-Klieser, 1987; Sutherland, 1988). The histological structure of the cell aggregates suggests that diffusion limitation of oxygen or nutrients in spheroids may cause cell death in the innermost parts of the spheroids. However, measurements with oxygen-sensitive microelectrodes are indicative of necrosis arising in the presence of relatively high oxygen tension (PO_2) values, as they were found in normal tissue (Carlsson and Acker, 1985; Mueller-Klieser et al., 1986). Although still controversial in literature, recent determinations of glucose diffusion coefficients allow for the calculation of penetration depths for this metabolite that exclude any diffusion limitation for glucose in spheroids with early necrosis (Casciari et al., 1988; Doerschel et al., 1989). To investigate whether the cells in the spheroid center die from the lack of ATP, and whether the ATP content of spheroids is related to the spheroid oxygenation, PO_2 measurements in spheroids were compared with ATP determinations using a novel technique for metabolic imaging.

MATERIALS AND METHODS

Measurements were carried out on EMT6/Ro spheroids that were derived from a mouse mammary sarcoma and that were cultured under standard cell culturing conditions. The oxygen tension (PO_2) distribution in these spheroids was investigated with oxygen-sensitive microelectrodes as a function of spheroid size. The histological structure of spheroids at various stages of growth was analyzed in h. e. stained standard paraffin or cryostat sections through the spheroid centers. Details on spheroid culturing, PO_2 measurements, and histological investigations in spheroids are published elsewhere (Mueller-Klieser and Sutherland, 1982a,b; Mueller-Klieser et al., 1986).

The distribution of ATP within spheroids of various sizes was assessed by a high resolution imaging technique with quantitative bioluminescence (Mueller-Klieser et al., 1988). This technique allows for the detection of ATP concentrations in cryostat sections of tissues or spheroids using the ATP-dependent luciferase of fireflies. The biochemical reaction is illustrated in Fig. 1. For imaging, the frozen spheroid section is covered with a cryostat section of a frozen cocktail containing gelatine and luciferase. The luciferase reaction with light emission is then started by thawing this sandwich. The spatial distribution of the light intensity is measured directly using an appropriate microscope (Axiophot, Zeiss, Oberkochen, FRG) and an imaging photon counting system (ARGUS 100, Hamamatsu, Herrsching, FRG).

The intensity of the bioluminescence can be calibrated in absolute terms using heat-inactivated tissue homogenates with different ATP concentrations that were determined with HPLC. The frozen homogenates were processed for metabolic imaging in the way described for tissue biopsies or spheroids. A linear correlation has been obtained between the bioluminescence intensity and the ATP concentration of the homogenates within a physiologically relevant range which is demonstrated in Fig. 2.

$$ATP + O_2 + luciferin$$

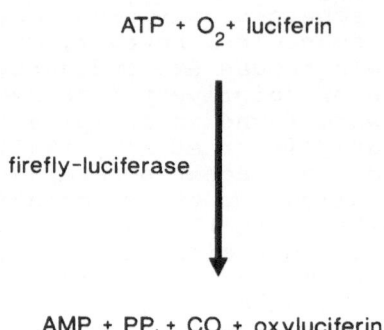

firefly-luciferase

$$AMP + PP_i + CO_2 + oxyluciferin + h\upsilon$$

Fig. 1. Enzymatic reaction for quantitative registration of ATP.

RESULTS AND DISCUSSION

To compare the emergence and growth of necrosis in the center of spheroids with the local metabolic milieu of these cellular areas, central oxygen tensions and ATP concentrations in EMT6 spheroids were determined as a function of spheroid size (see Fig. 3). The respective values decrease from rather high initial levels to very low signals close to or at

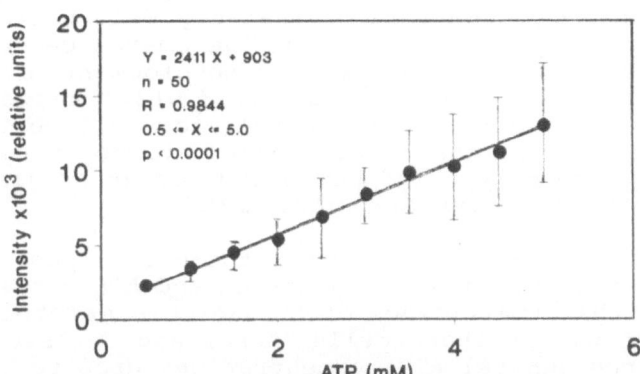

Fig. 2. Bioluminescence intensity as a function of ATP concentration in tissue homogenates investigated with the high resolution imaging technique described in the text.

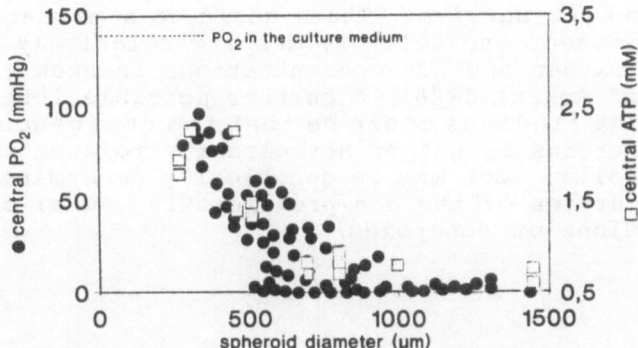

Fig. 3. Central PO_2 (dots) and central ATP concentration (squares) in multicellular EMT6 spheroids as a function of spheroid size.

the background level. The data in Fig. 3 indicate that the average diameters at which the background level is reached are similar for PO_2 and ATP being 700 - 900 µm. On the other hand, central necrosis were found in essentially all spheroids that were larger than 500 µm in diameter under these conditions (for further details on the histological analysis see: Mueller-Klieser et al., 1986; Bourrat-Flöck, 1988). It can be concluded from this comparison that necroses arise in the centers of these cell aggregates in the presence of oxygen tensions and ATP concentrations that may well sustain cellular viability in normal tissues. Thus, the emergence of necrosis during spheroid growth precedes the development of hypoxia with PO_2 values of a few mm Hg and less which suggests that hypoxia is unlikely to be responsible for cell death in spheroids. This is in accordance with observations in a number of different spheroid types of human and rodent origin (Carlsson and Acker, 1985; Sutherland et a., 1986).

The data in Fig. 3 show that there is a positive correlation between spheroid oxygenation and ATP content which may reflect the significance of the aerobic metabolism for the energetic status of these cells (Acker and Carlsson, 1988). However, since central ATP concentrations drop to the background level only after the development of massive central necroses during spheroid culturing, low ATP levels are presumably the consequence of ATPases released upon cell death in the spheroid center and may not be caused by diffusion limitation of oxygen and/or substrates.

One possible interpretation of the data obtained with regard to the development of necrosis in spheroids would be the hypothesis that cells in the center of spheroids are confronted with a hostile environment due to the accumulation of protons and metabolic wastes leading to an extremely high energy demand for survival. There could be a mismatch between energy requirement and delivery despite relatively high environmental oxygen and ATP concentrations in such a case (Carlsson and Acker, 1988). A further possible interpretation of the present findings could be that the emergence of cell death in spheroids is not or not directly related to the energy metabolism, but may be genetically determined by intrinsic properties of the non-proliferating tumor cells in central portions of spheroids.

ABSTRACT

To evaluate interrelationships among the oxygenation, the energy status, and the development of necrosis in tumor microregions, oxygen tensions were measured with microelectrodes, and ATP distributions were determined with a quantitative imaging technique using multicellular spheroids as in vitro tumor models. The results obtained show a positive correlation between central oxygen tensions and ATP concentrations in spheroids. During spheroid growth, both quantities decrease from rather high values to recordings close to or at the background level within similar ranges of spheroid size. Since the emergence of central necroses precedes this drop in energy-rich phosphate, the data may suggest that energy metabolism is not   directly involved in the development of necrosis in the spheroids investigated.

ACKNOWLEDGEMENTS

This work was supported by the Deutsche Forschungsgemeinschaft (Mu576/2-3, Mu5762-4) and by the Bundesministerium für Forschung und Technologie (01 ZO 8801).

REFERENCES

Bourrat-Flöck, B., 1988," Bedeutung des metabolischen Milieus für das Wachstumsverhalten multizelulärer Sphäroide", Thesis, University of Mainz; FRG.

Carlsson, J., and Acker, H., 1985, Influence of the oxygen pressure in the culture medium on the oxygenation of different types of multicellular spheroids, Int. J. Radiat. Oncol. Biol. Phys., 11:535.

Carlsson, J., and Acker, H., 1988, Relations between pH, oxygen partial pressure and growth in cultured cell spheroids, Int. J. Cancer, 42:715.

Casciari, J.J., Sotirchos, S. V., and Sutherland, R. M., 1988, Glucose diffusivity in multicellular tumor spheroids, Cancer Res., 48:3905.

Doerschel, D., Karbach, U., Groebe, K. F., and Mueller-Klieser, 1989, Assessment of glucose penetration into tumor microregions by high resolution autoradiography, Cytotechnology, 2(Suppl.):36.

Mueller-Klieser, W., 1987, Multicellular spheroids. A review on cellular aggregates in cancer research, J. Cancer Res. Clin. Oncol., 113:101.

Mueller-Klieser, W., and Sutherland, R. M., 1982a, Influence of convection in the growth medium on oxygen tensions in multicellular tumor spheroids, Cancer Res., 42:237.

Mueller-Klieser, W., and Sutherland, R. M., 1982b, Oxygen tensions in multicell spheroids of two cell lines, Br. J. Cancer, 45:256.

Mueller-Klieser, W., Freyer, J. P., and Sutherland, R. M., 1986, Influence of glucose and oxygen supply conditions on the oxygenation of multicellular spheroids, Br. J. Cancer, 53:345.

Sutherland, R. M., 1988, Cell and environment interactions in tumor microregions: The multicell spheroid model, Science, 240: 177.

Sutherland, R. M., Sordat B., Bamat, J., Gabbert, H., Bourrat, B., and Mueller-Klieser, W., 1986, Oxygenation and differentiation in multicellular spheroids of human colon carcinoma, Cancer Res., 46:5320.

BLOOD FLOW, OXYGEN CONSUMPTION AND TISSUE OXYGENATION OF HUMAN TUMORS

P. Vaupel, F. Kallinowski, P. Okunieff

Dept. Radiation Medicine, Massachusetts General Hospital
Cancer Center, Harvard Medical School
Boston, MA 02114, USA

INTRODUCTION

A great number of malignancies are relatively resistant to radio-therapy, chemotherapy and other non-surgical treatment modalities. A variety of factors are involved in the lack of responsiveness of these neoplasms including an intrinsic, genetically determined resistance and physiological, extrinsic (epigenetic, environmental) factors primarily created by inadequate and heterogeneous vascular networks[1-3]. Thus, properties such as tumor blood flow and tissue oxygen supply, factors which usually go hand in hand, can markedly influence the therapeutic response. Data on these parameters are mostly derived from rodent tumors. However, fast-growing rodent tumors might not adequately re-present the multitude of neoplastic growths encountered in patients. Unfortunately, data on human tumors in situ are scarce and there may be significant errors associated with the techniques used for measurements. This should be kept in mind when comparing available results from the literature.

In order to create a factual basis for further research, the currently available information on tumor blood flow, O_2 supply, and tissue oxygen distribution is compiled in this article. This synopsis attempts to identify areas in which future research might be most beneficial (e.g., for designing specifically tailored treatment protocols for individual subjects, for assessing early tumor response to treatment and/or for examining potentially useful tools for prediction of long-term tumor response).

BLOOD FLOW OF HUMAN TUMORS

Whereas blood flow rates of most rodent tumors decrease with increasing tumor size, a similar relationship has not been found to be valid for all human tumors[4-9]. Whether the latter indicates a lack of sufficient data or a fundamental deviation from animal tumors is unclear at this time. Considering the pathogenetic mechanisms directly responsible for the weight-adjusted flow decline that occurs with tumor growth, (i) a progressive rarefaction of the vascular bed, i.e., a decrease in the number of patent vessels per gram tumor, (ii) severe structural and functional abnormalities of the tumor microcirculation, and (iii) the development of necrosis have to be taken into account. These disturbances of tumor microcirculation are apparent even at early growth stages and, at least in iso- or xenografted rodent tumor models, become more pronounced with tumor growth. There are some indications that considerable spatial and temporal heterogeneity of the microcirculation exists amongst individual vessels and different microareas within one tumor, in the same tumor line at different sizes or at different growth sites, between different tumor lines with the same grading and staging as well as between tumors of different histologies.

Global tumor perfusion (blood flow) has been measured in patients using various techniques. To date, none of the approaches employed to measure tumor perfusion provide the information needed on the events occurring at the microscopic level. Furthermore they cannot give detailed data on arterio-venous shunt perfusion.

For comparison the blood supply to normal organs, and of tumors is given in Fig. 1 using specific perfusion rate units ($ml \cdot g^{-1} \cdot min^{-1}$). Flow values through normal organs are averages from which there may be departures in individual cases. Moreover, the specific perfusion may vary considerably in different parts of an organ (e.g., renal cortex vs. medulla, subendocardial vs. subepicardial regions of the myocardium,

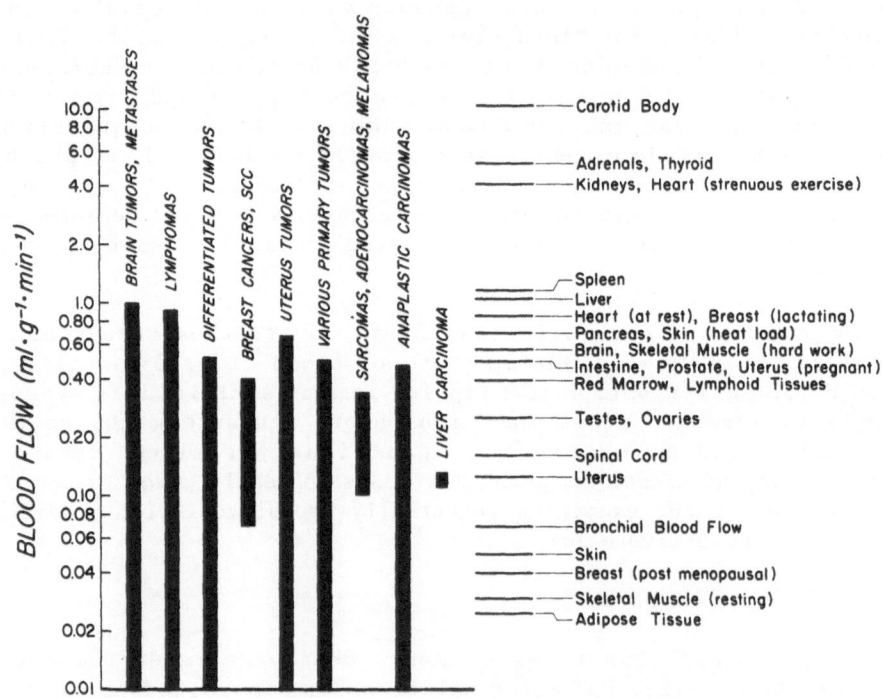

Fig.1. Variability of blood flow in human malignancies (vertical, black bars) and mean flow values of normal tissues (horizontal lines). Pooled data are given for the various tumors[6-9].

896

white matter vs. gray matter of the brain), and under different conditions of activity (e.g., heart at rest or during strenuous exercise, skeletal muscle at rest or during hard work, skin in a cold environment and during heat load).

A number of studies on blood flow through human tumors have been reported. Most of them are more or less casuistic reports rather than systematic investigations and therefore definite conclusions cannot be drawn. Flow studies were performed on brain tumors, breast cancers, metastatic lesions, anaplastic carcinomas, differentiated tumors and lymphomas, carcinomas of the uterus, adenocarcinomas and squamous cell cancers, melanomas, liver tumors, lymphangiomas, and osteosarcomas[6-9].

Considering all presently available data on human tumors in situ, the following (preliminary) conclusions can be drawn if flow data for the different tumor types are pooled:
- blood flow can vary considerably despite similar histological classification and primary site,
- tumors can have flow rates which are similar to those measured in organs with a high metabolic rate such as liver, heart or brain,
- some tumors exhibit flow values which are even lower than those of tissues with a low metabolic rate such as skin, resting skeletal muscle or adipose tissue,
- blood flow in human tumors can be higher or lower than that of the tissue of origin, depending in the functional state of the latter tissue (e.g., average blood flow in breast cancers is substantially higher than that of post-menopausal breast and significantly lower than flow data obtained in the lactating, parenchymal breast),
- the average perfusion rate of carcinomas does not deviate substantially from that of sarcomas, and
- metastatic lesions exhibit a blood supply which is comparable to that of the primary tumors.

Due to methodological uncertainties, these preliminary conclusions should be confirmed in future studies. Blood flow measurement techniques of the future should be non-invasive and should allow for serial investigations. Current promising methodologies employing radiation activation (a washout technique), NMR and computerized body tomography (CBT) angiographic and cinematographic techniques (flow reading techniques) and other washout procedures (e.g., ^{201}Tl and D_2O) may achieve this goal.

OXYGEN CONSUMPTION RATES OF HUMAN TUMORS IN SITU

Cellular respiration was among the first processes to be investigated in early cancer biochemistry. The studies of Warburg[10] resulted in his formulation of a pattern essential for malignancies, namely, an impaired respiratory rate of the tumor tissue, a partial failure of the Pasteur effect, and an excessive rate of aerobic glycolysis, i.e., the formation of lactic acid from glucose in the presence of oxygen. Subsequent studies indicated, however, that these notions were neither characteristic nor unique to malignant tumors (e.g., aerobic glycolysis is also found in the renal medulla, in the retina, in leukocytes and in other phagocytic cells). Oxygen consumption rates of tumors in vivo are inter-

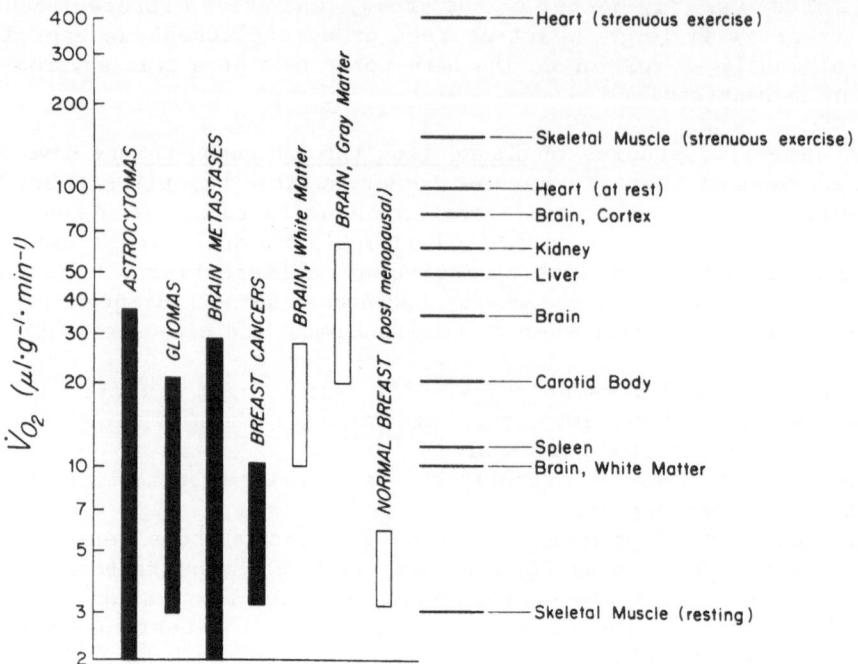

Fig.2. Range of oxygen consumption rates ($\dot{V}O_2$) of various human malignancies (black bars) and of the respective normal tissues (white bars: values obtained with the $^{15}O_2$ inhalation technique and positron emission tomography; horizontal lines: data obtained with other techniques). Pooled data are given[8,9].

mediate between normal tissues with low metabolic rates and normal tissues with quite high activities (see Fig. 2). The same holds true for the oxygen utilization, e.g., O_2 extraction is not impaired and respiratory function of cancer cells is not deficient (see Table I).

TUMOR TISSUE OXYGENATION

Besides mathematical evaluations of the pO_2 distribution in tissues, polarographic and cryospectrophotometric microtechniques have been used, to gain an insight into the oxygenation of tissue microareas. Utilizing O_2-sensitive, invasive microelectrode techniques, the pO_2 distribution has been measured in normal tissues and tumors of patients. In contrast, the spectrophotometric microtechnique enables the measurement of the oxyhemoglobin saturation (HbO_2) of individual red blood cells within microvessels ($\emptyset < 12$ um) of cryobiopsies taken from patient malignancies and from the normal tissue of origin.

OXYGEN PARTIAL PRESSURE DISTRIBUTION

Frequency distributions of measured pO_2 values (pO_2 histograms) for various normal human tissues are given in Fig. 3. As expected, there is a scattering of the tissue pO_2 values between 1 mmHg and values typical for arterial blood (80-100 mmHg). The median pO_2 values for some normal tissues are given in Table II. Whereas in these normal tissues the medians range from 24 to 66 mmHg, the respective values in the malignan-

898

Table I. Oxygen utilization (O_2 extraction ratio, O_2 extraction fraction) of human tumors and normal tissues

Tissue	O_2-utilization (v/v)
Astrocytomas	$0.03 - 0.52^*$
Brain metastases	$0.17 - 0.55^*$
Breast cancer	$0.10 - 0.35^*$
Brain (whole)	$0.30 - 0.47^*$
Brain gray matter	$0.30 - 0.68^*$
Brain white matter	$0.25 - 0.67^*$
Skeletal muscle	$0.45 - 0.75$
Spleen	0.05
Kidney	$0.008 - 0.10$
Heart	
at rest	$0.50 - 0.75$
strenuous exercise	0.85
Liver	
portal vein	$0.20 - 0.25$
hepatic artery	$0.40 - 0.50$
Carotid body	0.01
Breast (postmenopausal)	$0.50 - 0.75^*$

*PET study ($^{15}O_2$)

cies analyzed so far are $\leqslant 20$ mmHg. Comparing the pO_2 histograms of normal tissues with those of squamous cell carcinomas or breast cancers (see Fig. 4) there is clear evidence that in tumors
- there is a distinct shift of the distribution curve to the left,
- on the average, the mean pO_2 values are lower in malignancies than in the surrounding normal tissues (see Table III),
- there is less scattering of the pO_2 values due to lacking high pO_2 values, and
- there is an accumulation of pO_2 values in the lower pO_2 classes indicating tissue hypoxia and, thus, reduced radiosensitivity in tumors.

Data obtained so far are indicative of an inadequate O_2 supply to the tumor tissue, most probably due to a restriction of the microcirculation and thus of the O_2 availability to the cancer cells in vivo. As the tumors increase in size, this situation is significantly aggravated. Besides intra-tumor heterogeneities, there is a marked tumor-to-tumor variability, even if tumors of the same size, at the same growth site and of the same cell line are compared.

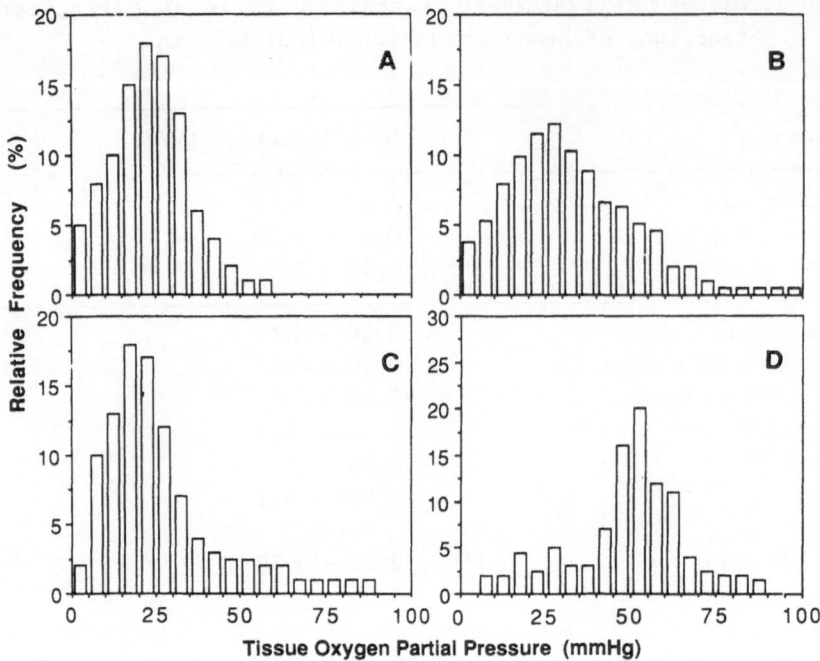

Fig.3. Frequency distributions of measured oxygen partial pressures (pO_2 histograms) for various normal tissues. A: liver, B: skeletal muscle, C: brain, D: subcutis[8-9].

Table II. Median pO_2 values in various normal tissues and in tumors of patients.

Tissue	Median pO_2 (mmHg)
Spleen	66*
Subcutis	50
Gastric mucosa	47
Uterine cervix	36
Skeletal muscle	28
Myocardium	25
Liver	24
Brain	24
Cervix cancers	
stage 0	20
stage I	13
stage II	5
Adenocarcinomas	10 – 12
Squamous cell carcinomas	15
Breast cancers	17

*Arterial pO_2 = 100 mmHg

As already mentioned, characterization of the oxygen status is also possible using a cryospectrophotometric ex vivo technique which allows for the measurements of the HbO_2 saturation of individual red blood cells in tumor microvessels.

Characterization of the oxygen status of <u>human tumors</u> in situ was also possible using this technique. As a typical example, in Fig. 5 the oxyhemoglobin saturation (HbO_2) frequency distribution is shown for differentiated adenocarcinomas of the rectum and for the normal rectal

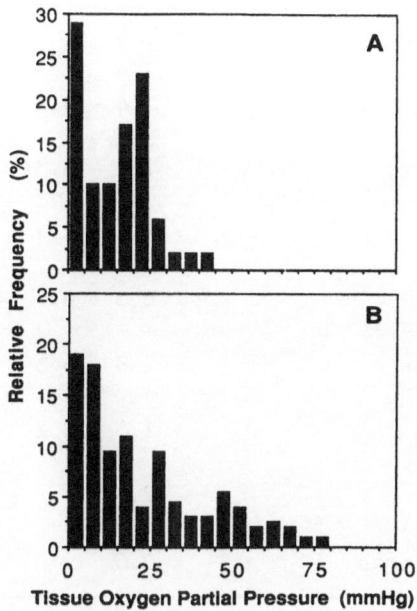

Fig.4. pO_2 histograms for squamous cell carcinomas (A) and breast cancers (B)[8-9].

mucosa[11,12]. As a rule, the mean HbO_2 saturation values observed in the tumors are distinctly lower than those found in the normal tissue at the site of tumor growth. The same holds for squamous cell carcinomas of the oral cavity[12,13]. The medians of the HbO_2 frequency distributions of the normal oral mucosa and of the tumors decreased from 80 to 49 sat.% (see Fig. 6) and correlated with changes in vascular density. Here again, in various malignant tumors considerable inter- and intra-individual differences were observed, even when tumors of the same clinical stage and grade were investigated.

Considering pooled data for the oxygen partial pressure distributions in various human malignancies and taking into account the available information on the HbO_2 saturation in human tumors there is experimental evidence for the existence of hypoxia in human tumors especially at advanced growth stages, i.e., in bulky tumors. In the case of the oxyhemoglobin saturation measurements, hypoxia is to be expected in

Table III. Comparison between the mean pO_2 values in normal tissues and in human malignancies.

Tumor type	$\dfrac{pO_2 \text{ (normal tissue)}^*}{pO_2 \text{ (tumor)}}$
Cervix cancer (stage 0)	1.6
(stage 1)	2.4
(stage 2)	1.4 – 3.2
Squamous cell cancers	1.7 – 6.3
Breast cancer	1.4 – 4.4
Melanomas	6.3 – 6.7
Soft tissue sarcomas	2.8 – 6.3
Malignant lymphomas	1.5 – 2.2
Adenocarcinomas	7.1
Basal cell epitheliomas	5.6

*Ratio of mean O_2 tension in normal tissue to mean O_2 tension in tumors

central portions of the intercapillary space if HbO_2 values fall below 30 sat.%. Both, in diffusion-limited hypoxia ("chronic hypoxia") and in ischemic hypoxia ("acute hypoxia") O_2 deficiency starts to develop first in tissue areas far away from a tumor microvessel and, most seriously, at the venous end of the microvessel.

SUMMARY

The objective of this article was to summarize current knowledge of blood flow and oxygen supply to human tumors, parameters which go hand in hand, and in turn critically determine the cellular metabolic micro-environment of human malignancies. A compilation of available data on blood flow, oxygen supply, and tissue oxygen distribution in human tumors is presented. Though data on human tumors in situ are scarce and there may be significant errors associated with the techniques used for

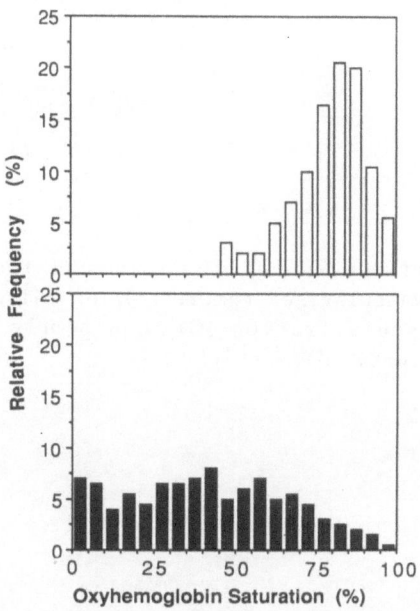

Fig.5. Frequency distributions of measured oxyhemoglobin (HbO_2) satura-tion values of individual red blood cells within microvessels of the normal rectal mucosa (top) and of differentiated adenocar-cinomas of the rectum (bottom)[8,9,11,12].

measurements, experimental evidence is provided for the existence of a compromised and anisotropic blood supply to many tumors. Comparable to rodent tumors, O_2-depleted areas develop in human malignancies which coincide with nutrient and energy deprivation, and with a hostile meta-bolic microenvironment. Significant variations in these relevant parame-ters have to be expected between different locations within the same tumor, at the same location at different times, and between individual tumors of the same grading and staging.

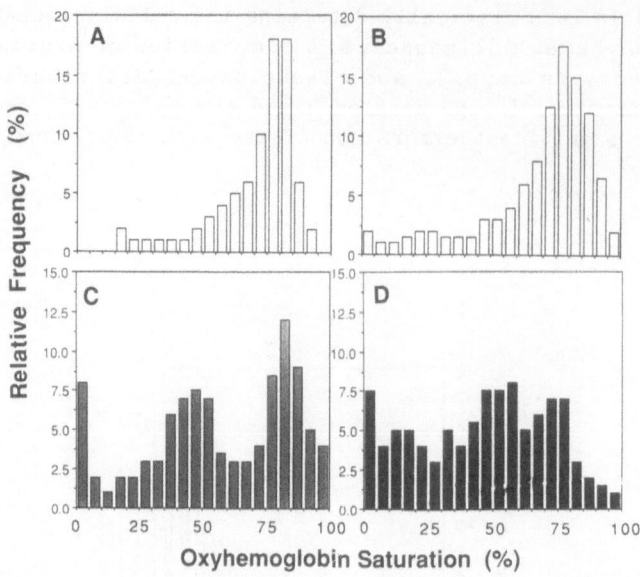

Fig.6. Frequency distributions of HbO_2 values of the normal oral mucosa (A), of well-vascularized tumors (B), of malignancies with medium quality of vascularization (C), and poorly vascularized cancers of the oral cavity (D)[8,9,12,13].

REFERENCES

1. P. Vaupel and W. Müller-Klieser, Interstitieller Raum und Mikromilieu in malignen Tumoren. Progr. Appl. Microcirc. 2:78 (1983).

2. P. Vaupel and F. Kallinowski, Microcirculation and metabolic micromilieu in malignant tumors. Funktionsanalyse biolog. Systeme 18: 265 (1988).

3. R. M. Sutherland, Cell and environment interactions in tumor microregions. The multicell spheroid model. Science 240:177 (1988).

4. R. K. Jain and K. Ward-Hartley, Tumor blood flow – Characterization, modifications, and role in hyperthermia. IEEE Trans. Sonics Ultrasonics SU-31:504 (1984).

5. P. Vaupel, Pathophysiologie der Durchblutung maligner Tumoren. Funktionsanalyse biolog. Systeme 8:155 (1982).

6. M. Mäntylä, J. Heikkonen, and J. Perkkiö, Regional blood flow in human tumours measured with argon, krypton and xenon. Brit. J. Radiol. 61:379 (1988).

7. H. S. Reinhold, Physiological effects of hyperthermia. Rec. Res. Cancer Res. 107:32 (1988).

8. P. Vaupel, F. Kallinowski, and P. Okunieff, Blood flow, oxygen and nutrient supply, and metabolic microenvironment of human tumors. A review. Cancer Res. (in press).

9. P. Vaupel, F. Kallinowski, and P. Okunieff, Blood flow, metabolic micromilieu and bioenergetics of human tumors: Critical parameters in hyperthermic treatment. Progr. Hyperthermia 1 (in press).

10. O. Warburg, On respiratory impairment in cancer cells. Science 124: 269 (1956).

11. P. Wendling, R. Manz, G. Thews, and P. Vaupel, Inhomogeneous oxygenation of rectal carcinomas in humans. A critical parameter for preoperative irradiation? Adv. Exp. Med. Biol. 180: 293 (1984).

12. P. Vaupel and F. Kallinowski, Tissue oxygenation of primary and xenotransplanted human tumours. In: "Radiation Research", E.M. Fielden, J.F. Fowler, J.H. Hendry, and D. Scott (eds.), London, New York, Philadelphia, Taylor & Francis (1987).

13. W. Mueller-Klieser, P. Vaupel, R. Manz, and R. Schmidseder, Intracapillary oxyhemoglobin saturation of malignant tumors in humans. Int. J. Radiat. Oncol. Biol. Phys. 7:1397 (1981).

OXYGENATION OF TUMORS DERIVED FROM *ras* TRANSFORMED CELLS

F. Kallinowski[1], R.R. Friis[2], F. Van Roy[3], P. Vaupel[1]

[1]Dept.of Radiation Medicine, Mass. General Hospital, Harvard Medical School, Boston, MA 02114, USA

[2]Institute of Clinical-Experimental Cancer Research, Tiefenau Hospital, CH-3004 Bern, Switzerland

[3]Laboratory of Molecular Biology, State University, B-9000 Ghent, Belgium

INTRODUCTION

Malignant transformation involves the activation of growth-promoting and/or the loss of growth suppressing genes. Activated ras genes are frequently found in malignant tumors (Barbacid, 1987). Considering tumor treatment, malignant transformation by ras genes has been reported to increase the cellular resistance to ionizing radiation and chemotherapy (Sklar, 1988 a,b). Besides these genetic factors, epigenetic influences such as the tumor oxygenation can modulate the effectiveness of non-surgical treatment modalities (Teicher et al., 1981; Sutherland, 1988). Since there are only insufficient data on possible interrelationships between ras activation and changes of the tumor micromilieu the tissue oxygenation of tumors derived from ras transformed fibroblast-like cells was investigated.

MATERIALS AND METHODS

Tumors and Animals. Spontaneously tumorigenic Fischer 344 rat embryo cells were chosen as a ras-free cell line (Rat1, Freeman et al., 1973). The effect of ras transformation was evaluated with Rat1 cells transfected with copies of a variant human c-Ha-ras1 protooncogene (Rat1pEJ6.6, Land et al., 1983). These cells were received from two different laboratories in order to assess the stability of the expressed phenotype. The influence of a cooperating oncogene was determined with cells obtained from low-passage Fischer 344 rat embryonic fibroblast-like cultures and transfected with plasmids pT24, pSVvmyc and pHSG272, encoding, respectively, a mutated c-Ha-ras1 gene as well, a viral myc oncogene and a neomycin-resistance gene (REFpneoMYCrasEpool). In order to have another species, immortalized 3T3 cells from Balb/c mice transfected with a mutant cHa-ras1 gene and a neomycin-resistance gene were studied (3T3pEJ6.6, Dr. Rollins, cell line previously not described). The constancy of the ras-transformed phenotype and possible differences of the host strain were evaluated using primary Sprague-Dawley rat embryonic cells infected with a retroviral vector expressing the Adenovirus E1A protein (Dr. B. Vennström, Stockholm, Sweden). Infected cells were selected using the neomycin-resistance marker coexpressed from the virus. Three separate superinfections with HaMuSV (Harvey murine sarcoma virus) were performed from each of which a single colony growing in soft agar suspension was randomly isolated (E1AHaMuSV1-3).

All cell lines were maintained in conventional culture media (Dulbecco's minimal essential medium, glucose 1 g/l, Sigma, St. Louis, MO, USA; 10% fetal calf serum, JR Scientific, Woodland, CA, USA; penicillin 50 U/ml, streptomycin 50 µg/ml, neomycin 100 µg/ml, Sigma, St. Louis, MO, USA; 5 % CO_2 and air). The cultures were passaged once to twice a week. Exponentially growing cells were harvested by trypsinization at 37°C for 5-10 min (0.05% trypsin, 0.53 mM EDTA, Gibco Laboratories, Grand Island, NY, USA).

Nude mice were chosen as tumor hosts in order to exclude possible artifacts by comparing tumors grown in different species. The mice were bred and maintained in a defined flora colony (Sedlacek and Mason, 1977). Tumors grew subcutaneously in the hind foot dorsum of 8 week old animals after the injection of approximately 10^6 cells. The mice received 6 Gy whole body irradiation about 24 hrs prior to tumor implantation. Tumor growth was assessed by serial caliper measurements of three orthogonal diameters and subsequent calculation of a rotation ellipsoid. Volume doubling times were determined according to Steel (1977).

Southern Blot Analysis. Gene analysis was performed as described by Ausubel et al. (1987). High molecular weight DNA was prepared from cell cultures by phenol extraction and ethanol precipitation. 10 µg samples of genomic DNA were restricted with an excess (160-240 units) of *Kpn*I restriction enzyme (Boehringer Mannheim, Mannheim, FRG). Fragments were electrophoretically separated in a 0.7% (w/v) agarose gel and blotted to GeneScreen® hybridization transfer membrane (NEN Products, Boston, MA, USA).^{32}P-labelled probes with a specific activity of 5 • 10^7 cpm/µg were obtained by a random heximer priming procedure using the 3 kB *Sac*I fragment of plasmid pEJ6.6 containing coding regions of the biologically active mutant cHa-ras1 gene. Hybridization was performed at 42°C with 50% formamide and 10% dextran sulfate. The final wash solution contained 0.1% SSC and 0.1% SDS at 65°C. Autoradiographs were obtained by exposing standard X-ray film for 72 hrs at -70°C using an intensifier screen. A similarly prepared blot was hybridized with a nick translated v-Ha-ras probe (1 • 10^7 cpm/µg). Hybridization conditions included 50% formamide at 37°C. The final wash was performed at 55°C with 0.1% SSC and 0.1% SDS. An autoradiograph was obtained by overnight exposure using an intensifier screen at - 70°C. The results are shown in Fig. 1.

Oxygenation measurements. The measurement system was originally described by Weiss and Fleckenstein (1986) and was recently applied to oxygen measurements in animal tumors (Vaupel et al., 1989). Briefly, tissue oxygen levels were determined with needle electrodes featuring a gold disc with a radius of 6 μm recessed within a steel shaft of 350 μm outer diameter. The needles were sharply ground and contained a teflon filled recess of up to 20 μm. The electrodes were

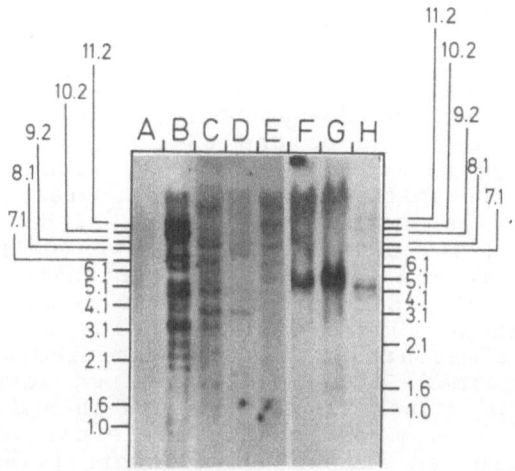

Fig. 1. Autoradiographs of Southern blots from Rat1 (A), REFpneoMYCrasEpool (B), Rat1pEJ6.6 (C from Dr. Mareel, D from Dr. Weinberg), 3T3pEJ6.6 (E), and E1AHaMuSV1-3 cells (F,G,H). Lanes A-E are from the same membrane and probed with the 3kB *Sac*I fragment of plasmid pEJ6.6. Lanes F-H are from another membrane and probed with a v-Ha-ras probe. Numbers indicate the sizes of the DNA fragments (in kb).

connected to an amplifier unit which provided the polarization voltage of -700 mV against a Ag/AgCl reference electrode, controlled the electrode movement and performed the data evaluation (KIMOC 6650, Eppendorf, Hamburg, FRG). The electrodes were calibrated in 0.9% saline equilibrated with air or pure nitrogen gas (pO_2 = 0 mmHg). Typical electrode characteristics: oxygen sensitivity = 2-3 pA/mmHg pO_2; response times (T_{90}) around 600 ms; drift < 1% between pre- and poststudy calibration (T = 37°C). Tissue measurements typically took less than 5 min. For pO_2 measurements, an electrode was inserted through a skin incision (< 1 mm) and advanced about 1 mm. After an initial equilibration period < 2 min, electrodes were moved through the tissues with a step length of maximally 1 mm. Each forward movement was immediately followed by a withdrawal motion of 0.3 mm in order to minimize tissue compression. Data were obtained 1.4 sec after completion of movements. In this way, 20 - 60 values were measured in tumors and, as a control, in untreated subcutaneous tissue of the hind foot dorsum. Temperature differences between the calibration vessel, tumors and subcutis were taken into account in the evaluation of local pO_2 values. From these data, frequency distributions were compiled. Median pO_2 values and 10th and 90th percentiles were used to characterize the histogram obtained.

RESULTS

Median pO_2 values around 47 mmHg were measured in the subcutis of the hind foot dorsum of 10 tumor-free mice the 10th and 90th percentile being 33 and 63 mmHg, respectively. As a rule, lower oxygenation levels were found in tumors as evidenced by a left-shift of the pO_2 histogram, a reduction of median oxygen levels and an accumulation of data in lower pO_2 classes (Figs. 2-4). Considering sizes < 100 µl, median pO_2 values of 27 mmHg were detected in spontaneously tumorigenic Rat1 tumors (Fig. 2A). Six percent of all values ranged between 0 and 5 mmHg. These tumors exhibited slow growth rates with volume doubling times around 28 days. Compared with this tumor line, ras transformed Rat1 cells exhibited lower oxygenation levels with median values ranging between 18 and 23 mmHg (Figs. 2C and D). An even more pronounced left-shift of the pO_2 histogram was observed in the v-myc immortalized and cHa-ras1 transformed Fischer rat tumor line (Fig. 2B; median pO_2 value = 9 mmHg). Comparable pO_2 values were detected in tumors derived from cHa-ras1 transformed 3T3 cells (Fig. 2E; median pO_2 value = 7 mmHg). The frequency distributions obtained in tumors formed by ras transformed Sprague-Dawley rat cells were similar to those previously described with median values ranging between 9 and 17 mmHg (Figs. 2F-H). All ras transformed tumors grew much more rapidly than the spontaneously tumorigenic Rat1 line with volume doubling times ranging between 2.5 and 4 days. With increasing sizes, a marked left-shift of the pO_2 histograms was observed indicating a further reduction of the tissue oxygenation (Figs. 3 and 4). As already described for sizes < 100 µl, higher pO_2 values were generally observed in slowly growing, spontaneously tumorigenic Rat1 tumors as compared with rapidly doubling ras-transformed lines.

DISCUSSION

Formation of new blood vessels is a necessary step for malignant growth to clinically relevant sizes (Gimbrone et al., 1972). Angiogenesis is initiated by several angiogenic factors (Folkman and Klagsbrun, 1987) and can first be detected during the transition from hyperplasia to neoplasia (Gimbrone and Gullino, 1976; Folkman et al., 1989). The newly formed

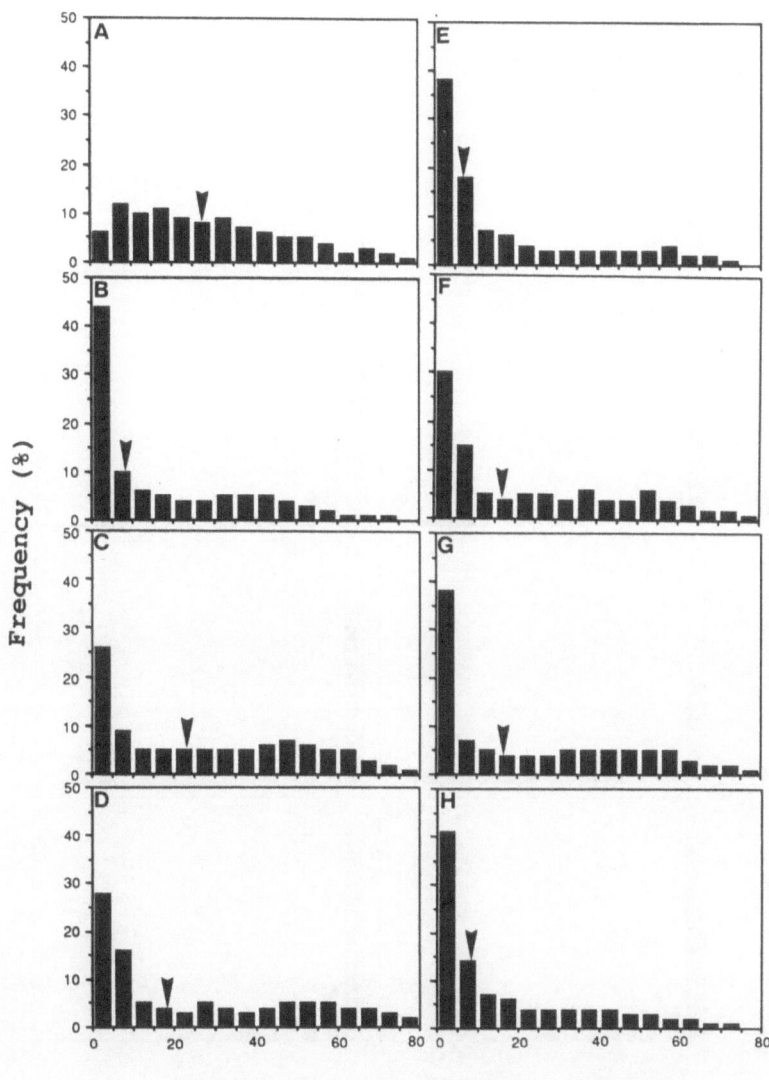

Fig. 2. Frequency distributions of pO_2 values measured in Rat1 (A), REFpneoMYCrasEpool (B), Rat1pEJ6.6 (C from Dr. Mareel, D from Dr. Weinberg), 3T3pEJ6.6 (E), and E1AHaMuSV1-3 (F,G,H) tumors with sizes < 100 µl. The arrows indicate the respective median pO_2 values.

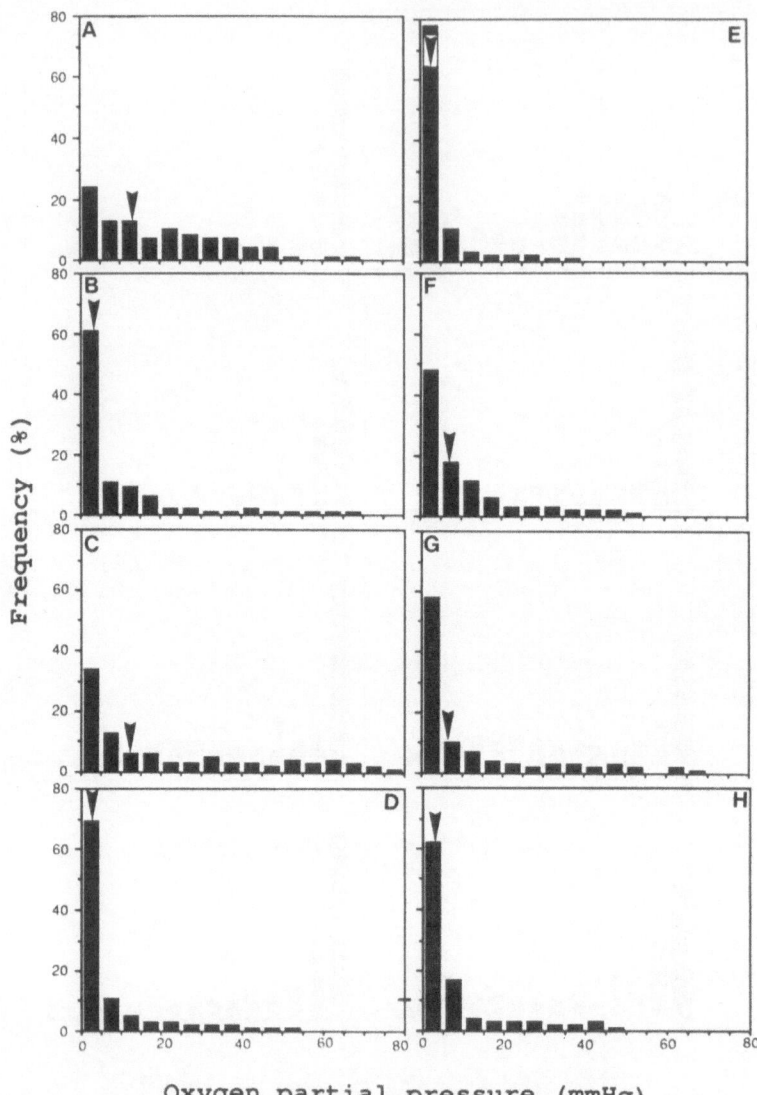

Fig. 3. Histograms of pO$_2$ values measured in Rat1 (A), REFpneoMYCrasEpool (B), Rat1pEJ6.6 (C from Dr. Mareel, D from Dr. Weinberg), 3T3pEJ6.6 (E), and E1AHaMuSV1-3 (F,G,H) tumors with sizes between 200 - 300 μl. The arrows indicate the respective median pO$_2$ values.

912

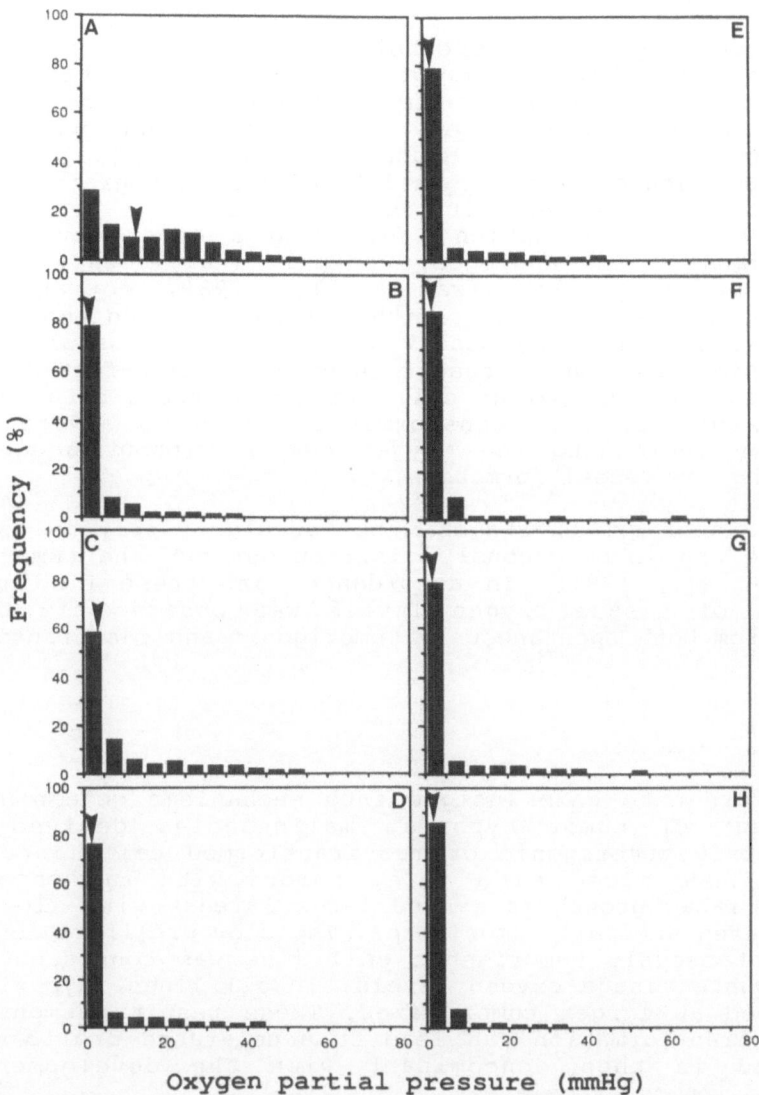

Fig. 4. pO_2 histograms obtained in Rat1 (A), REFpneoMYCrasEpool (B), Rat1pEJ6.6 (C from Dr. Mareel, D from Dr. Weinberg), 3T3pEJ6.6 (E), and E1AHaMuSV1-3 (F,G,H) tumors with sizes between 400 and 500 µl. The arrows indicate the respective median pO_2 values.

vasculature exhibits severe structural and functional abnormalities (Peterson, 1979). As a consequence, low oxygenation values have generally been found in advanced tumors in experimental animals and in patients (Vaupel, 1977; Vaupel et al., 1989). In accordance with these findings, pO_2 levels lower than those in normal subcutis were measured in all tumors investigated here.

Tissue oxygenation is determined by the relationship of oxygen supply and consumption. In this study, ras transformation accelerated tumor growth thus necessitating elevated consumption rates for oxygen and nutrients. Concomitantly, severe tumor hypoxia was observed indicating an inadequate vascular supply. In human tumor xenografts, rapid tumor growth coincided with high perfusion rates and an adequate tissue oxygenation indicating a close coupling of growth-promoting and angiogenic stimuli (Kallinowski et al., 1989). Recently, Ishikawa et al. (1989) cotransfected ras-transformed NIH 3T3 cells with platelet-derived endothelial cell growth factor. They observed an increased vascularity in vivo but similar growth rates of secondarily transfected tumors as compared with the parent cell line. From these data, it may be concluded that ras transformation leads to severe tumor hypoxia by increasing the oxygen demand without adequately stimulating new vessel formation.

At later growth stages, the vascular supply is usually inadequate and inhomogenously distributed over the tumor mass (Vaupel et al., 1981). In accordance with these findings, a reduction of tissue oxygen levels was observed in tumors derived from both spontaneously tumorigenic and ras transformed cells.

SUMMARY

In order to gain insight into mechanisms governing the development of tumor hypoxia, malignancies derived from spontaneously tumorigenic or ras-transformed cell lines were grown in nude mice. As a rule, tumors with ras oncogenes exhibited rapid growth rates and large areas with low pO_2 readings even at small tumor sizes. The slow proliferation rate of a spontaneously tumorigenic cell line was consistent with more adequate tissue oxygen levels. In all lines, hypoxia was accentuated at larger tumor sizes. These results demonstrate that ras transformation can lead to accelerated proliferation rates and is then concomitant with the development of pronounced tumor hypoxia.

ACKNOWLEDGEMENTS

Rat1, Rat1pEJ6.6 and REFpneoMYCrasEpool cells were provided by Dr. Mareel, Dept. of Experimental Cancerology, Ghent, Belgium. Rat1pEJ6.6 cells were additionally obtained from Dr. Weinberg, Whitehead Institute for Biomedical Research, Cambrigde, MA, USA. Ras-transformed Balb/c 3T3 cells were received from Dr. Rollins, Dana Farber Cancer Institute, Boston, MA, USA. The murine retroviral vector expressing the E1A protein was a gift of Dr.

914

B. Vennström, Stockholm, Sweden. The authors gratefully acknowledge the excellent quality of the nude mice contributed by R.S. Sedlacek, Edwin L. Steele Laboratory for Radiation Biology, Mass. General Hospital, Harvard Medical School, Boston, MA, USA. The electrode measurement device was generously supplied by Eppendorf Geraetebau, Hamburg, FRG.

REFERENCES

Ausubel, F.M., Brent, R., Kingston, R.E., Moore, D.D., Seidman, J.G., Smith, J.A., and Struhl, K. (eds.), 1987, Current protocols in molecular biology. Wiley, New York.

Barbacid, M., 1987, ras genes. Ann. Rev. Biochem., 56: 779.

Folkman, J., and Klagsbrun, M., 1987, Angiogenic factors. Science (Wash., DC), 235: 442.

Folkman, J., Watson, K., Ingber, D., and Hanahan, D., 1989, Induction of angiogenesis during the transition from hyperplasia to neoplasia. Nature, 339: 58.

Freeman, A.E., Gilden, R.V., Vernon, M.N., Wolford, R.G., Hugunin, P.E., and Huebner, R.J., 1973, 5-Bromo-2'-deoxy-uridine potentiation of transformation of rat embryo cells induced in vitro by 3-methylcholanthrene: induction of rat leukemia virus gs antigen in transformed cells. Proc. Nat. Acad. Sci. USA, 70: 2415.

Gimbrone, M.A., and Gullino, P.M., 1976, Angiogenic capacity of preneoplastic lesions of the murine mammary gland as a marker of neoplastic transformation. Cancer Res., 36: 2611.

Gimbrone, M.A., Leapman, S.B., Cotran, R.S., and Folkman, J., 1972, Tumor dormancy in vivo by prevention of neovascularization. J. Exptl. Med., 136: 261.

Ishikawa, F., Miyazono, K., Hellman, U., Drexler, H., Wernstedt, C., Hagiwara, K., Usuki, K,. Takaku, F., Risau, W., and Heldin, C.H., 1989, Identification of angiogenic activity and the cloning and expression of platelet derived endothelial cell growth factor. Nature, 338: 557.

Kallinowski, F., Schlenger, K.H., Runkel, S., Kloes, M., Stohrer, M., Okunieff, P., and Vaupel, P., 1989, Blood flow, metabolism, cellular microenvironment, and growth rate of human tumor xenografts. Cancer Res., 49: in press.

Land, H., Parada, L.F., and Weinberg, R.A., 1983, Tumorigenic conversion of primary embryo fibroblasts requires at least two cooperating oncogenes. Nature, 304: 596.

Peterson, H.-I., 1979, Tumor blood circulation: angiogenesis, vascular morphology and blood flow of experimental and human tumors. Boca Raton, CRC Press.

Sedlacek, R.S., and Mason, K.S., 1977, A simple and inexpensive method for maintaining a defined flora mouse colony. Lab. Animal Sci., 27: 667.

Sklar, M. D., 1988a, The ras oncogenes increase the intrinsic resistance of NIH 3T3 cells to ionizing radiation. Science (Wash., DC), 239: 645.

Sklar, M. D., 1988b, Increased resistance to cis-diammine-dichloroplatinum(II) in NIH 3T3 cells transformed by ras oncogenes. Cancer Res., 48: 793.

Steel, G. G., 1977, Growth kinetics of tumours. Clarendon Press, Oxford.

Sutherland, R. M., 1988, Cell and environment interactions in tumor microregions: the multicell spheroid model. Science (Wash., DC), 240: 177.

Teicher, B. A., Lazo, J. S., and Sartorelli, A. C., 1981, Classification of antineoplastic agents by their selective toxicities towards oxygenated and hypoxic tumor cells. Cancer Res., 41: 73.

Van Roy, F. M, Messiaen, L., Liebaut, G.T., Gao, J., Dragonetti, C.H., Fiers, W.C., and Mareel, M., 1986, Invasiveness and metastatic capability of rat fibroblast-like cells before and after transfection with immortalizing and transforming genes. Cancer Res., 46: 4787.

Vaupel, P., 1977, Hypoxia in malignant tumors. Microvasc. Res., 13: 399.

Vaupel, P., Frinak, S., and Bicher, H.I., 1981, Heterogeneous oxygen partial pressure and pH distribution in C3H mouse mammary adenocarcinoma. Cancer Res., 41: 2008.

Vaupel, P., Kallinowski, F., and Okunieff, P., 1989, Blood flow, oxygen and nutrient supply, and metabolic microenvironment of human tumors: a review. Cancer Res., 49: in press.

Vaupel, P., Okunieff, P., Kallinowski, F., and Neuringer, L. J., 1989, Correlations between ^{31}P-NMR spectroscopy and tissue O tension measurements in a murine fibrosarcoma. Radiat. Res., in press.

Weiss, C., and Fleckenstein, W., 1986, Local tissue pO measured with "thick" needle probes. Funktionsanalyse biolog. Systeme, 15, 155.

TUMOUR RADIOSENSITIZATION BY CLOFIBRATE AND

ITS ANALOGS: POSSIBLE MECHANISMS

David G. Hirst

CRC Gray Laboratory, P.O. Box 100
Mount Vernon Hospital
Northwood, Middx HA6 2JR
England

INTRODUCTION

The oxygenation of radioresistant hypoxic cells in malignant tumours has been a major goal of radiation therapists and biologists for many years. Increased release of bound oxygen from blood to tissues was recognised as a possible way of achieving this (Siemann et al., 1979), and several other studies since then (Hirst et al, 1987 a, b; Siemann et al, 1986; Siemann and Macler, 1986) have supported this approach, showing that the affinity with which haemoglobin binds oxygen has an effect on the radiosensitivity of several experimental mouse tumours. Facilitated release of oxygen from blood to tissues can be achieved by reducing the binding affinity of haemoglobin for oxygen. Several techniques can be used <u>in vivo</u> to achieve this but they fall into two basic categories: the alteration of the intraerythrocytic concentration of the naturally occurring allosteric modifier, 2,3-diphosphoglycerate (2,3-DPG) (Siemann et al, 1979; Siemann et al, 1986; Hirst and Wood, 1987), or the administration of compounds which are direct modifiers of haemoglobin/oxygen affinity, such as inositol hexaphosphate (Teissiere et al, 1985) or derivatives of chlorophenoxyacetic acid (Hirst and Wood, 1987; 1989 a,b).

Our own recent studies have focused on the last of these methods but although some sensitization of mouse tumours was achieved, we were reluctant to attribute the effects entirely to changes in haemoglobin affinity. A large number of structurally-related analogs of the lead compound clofibrate were tested, and several of these were shown to have the clinically favourable properties of potent haemoglobin/oxygen affinity reduction and very low toxicity (Hirst and Wood, 1989b). Unfortunately, these compounds were not impressive sensitizers of one of our mouse tumours (SCCVII/St) to radiation. This discrepancy lead us to conclude that these compounds must have properties other than their ability to bind to haemoglobin which influence tumour radiosensitivity, probably through changes in tissue oxygenation. A major determinant of oxygenation is blood flow, so this was an obvious area to investigate. The present study reports our observations on blood flow changes in SCCVII/St tumours (using two different methods) after administration of bezafibrate, clofibrate and ML1024 (etophylline clofibrate).

MATERIALS AND METHODS

Mice, tumour systems and irradiations

Experiments were carried out exclusively with SCCVII/St carcinomas implanted intradermally on the backs of C3H/Km female mice, 11-13 weeks old. These were housed six to a cage under defined flora conditions with free access to food and water. The derivation, passage and maintenance of this tumour line has been previously described (Hirst et al, 1982). Tumours were implanted as a suspension of 2×10^5 cells in medium and took about 12 days to reach the treatment size of 150-300 mg. Irradiations with 250 kVp X-rays were carried out whole body at a dose rate of 2.85 Gy/min.

Measurement of tumour blood flow

The ^{86}Rb Cl uptake method was used to measure the perfusion of the tumours - the technique is basically that first described by Sapirstein (1958). Mice were given a bolus injection of 3 μ Ci/mouse ^{86}Rb Cl (specific activity 13 m Ci/m mole) in 0.1 ml of saline via a tail vein. The mice were killed by cervical dislocation 2 minutes later and tissues of interest excised. In most cases these included tumour, muscle (quadriceps), kidney and tail, the last to confirm that there was not an excessive amount of activity left at the injection site. In some experiments, the entire animal was dissected so that the fate of all the injected activity could be ascertained.

^{14} SR 2508 binding

The use of nitroimidazole binding to estimate the amount of hypoxia in tumours in vivo was first described by Garrecht and Chapman (1983) and has since been used by others (Hirst et al., 1986). In these studies ^{14}C misonidazole was used but as it was unavailable, the analog ^{14}C SR 2508 was substituted in the present experiments. It has been shown to have similar, though not identical, preferential binding characteristics under hypoxia. Immediately before irradiation, ^{14}C SR 2508 was injected i.p. at a dose of 0.2 μ Ci/g in saline (0.01 ml/g). Mice were killed 24 hours later and their tumours weighed and solubilized in Protosol (~10 ml/g of tissue) at 50°C overnight. Liquid scintillation fluid (15 ml Aquasol, universal LSC cocktail; NEM, Boston, MA) was added to the vials and samples were counted in a scintillation counter. Results were expressed as: cpm (sample) - cpm (tissue blank) per gram of tissue divided by the injected cpm. This number was then compared for control and drug-treated animals.

Clonogenic excision assay

Mice were killed 24 hours after irradiation and their tumours excised. Single cell suspensions were prepared by enzymic digestion, as previously described (Hirst et al., 1982). Cells were plated in Waymouth's medium + 15% fetal calf serum in tissue culture dishes, and the number of colonies arising after 12 days of incubation at 37°C was counted. Plating efficiencies and surviving fractions were then calculated.

RESULTS

The effects of three chlorophenoxyacetic acid derivatives, clofibrate, bezafibrate and ML1024 (etophylline clofibrate) on tumour perfusion and the binding of ^{14}C SR 2508 were compared with surviving fraction data for radiation sensitivity. We have found previously in this tumour system that clofibrate is the most potent radiosensitizer, followed by ML1024, which had a small but significant effect (Hirst and Wood, 1988b). High dose

918

bezafibrate had not previously been tested in this tumour. It had no significant effect in the present study at a dose of 2 mmol/kg on the radiation sensitivity of the SCCVII/St carcinoma as measured by an in vivo/in vitro excision assay (data not shown). These results do not correlate well with the potency of these three compounds as inhibitors of Hb/O_2 binding (Hirst and Wood, 1987; Hirst and Wood, 1988a).

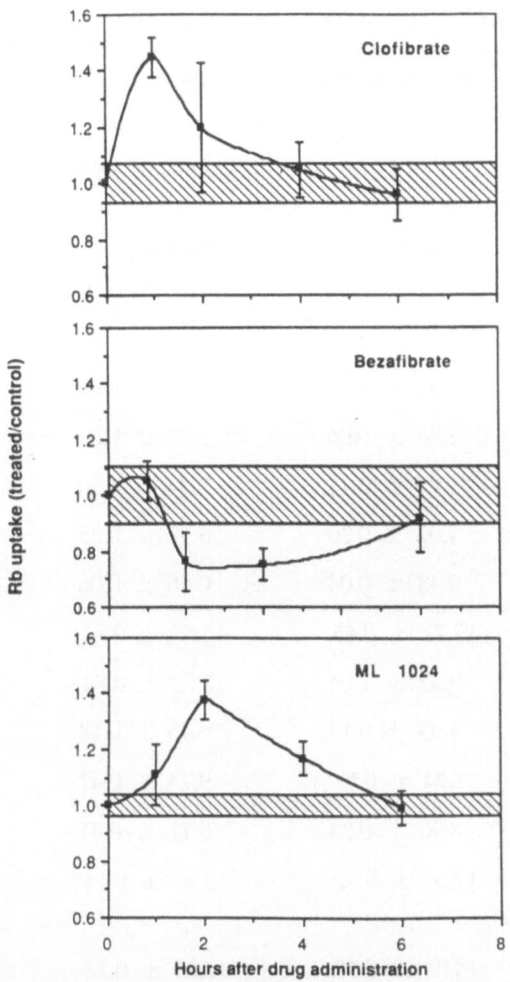

Fig.1 ^{86}Rb uptake in the SCCVII/St tumour at different times after 3 mmole/Kg clofibrate, bezafibrate or ML1024. Data points are means (± 1 s.e.) of 6-8 tumours in two separate experiments. Hatched areas are ± 1 s.e. for untreated tumours.

Relative tumour blood flow after administration of these three compounds is shown in Fig.1. ^{86}Rb extraction by tumours was increased significantly, but only transiently, by clofibrate, returning to control levels within 2 hours. ^{86}Rb extraction was significantly decreased 4 hours after bezafibrate, while ML1024 caused a significant increase in extraction at 2 and 4 hours after administration.

We wished to study these blood flow changes in more detail. If clofibrate increases the relative perfusion of the tumour (Fig.1) then do other organs also show changes? Table I shows the relative distribution of the cardiac output as measured by ^{86}Rb uptake to dissected parts of mice (15) without, or 1 hour after, administration of 3 mmole/Kg clofibrate. While the blood flow changes for the individual tissues were not all statistically significant, blood flow to the visceral organs as a group was significantly increased (p <0.05) while that to the skeleton and muscle was significantly reduced. Furthermore, the relative change in tumour perfusion shown in Table 1 actually understates the magnitude of the effect because no account was taken of tumour size. When this correction is made, relative perfusion per gram of tumour was increased by 33 ± 7.5% (p <0.05) 1 hour after clofibrate, a result which is consistent with those in Fig.1.

Table I.   The effect of clofibrate on the distribution of injected ^{86}Rb in the mouse (%).

Tissue	Control	Clofibrate 3 m mole/Kg	% change	P
Head	9.58 ± 0.19	8.23 ± 0.46	- 14	0.024
Heart	2.84 ± 0.06	3.02 ± 0.06	+ 6	0.061
Lungs	1.91 ± 0.15	2.34 ± 0.10	+ 23	0.047
Liver	4.02 ± 0.20	5.32 ± 0.25	+ 32	0.001
Spleen	0.48 ± 0.04	0.56 ± 0.06	+ 17	0.311
Gut	13.72 ± 0.65	16.50 ± 0.84	+ 20	0.024
Kidneys	9.82 ± 1.04	12.12 ± 0.84	+ 23	0.104
Skin	4.93 ± 0.19	4.46 ± 0.18	- 10	0.095
Fore limbs	11.24 ± 0.57	9.78 ± 0.47	- 13	0.066
Hind limbs	8.90 ± 0.35	8.11 ± 0.41	- 9	0.169
Thorax	17.65 ± 0.52	15.89 ± 0.64	- 10	0.056
Lower back and pelvis	14.04 ± 0.78	12.65 ± 0.28	- 10	0.100
Tumour	0.85 ± 0.04	1.02 ± 0.09	+ 20	0.140

We have previously used ^{14}C-misonidazole binding to measure differences in the number of hypoxic cells in tumours after manipulations of the oxygen transport characteristics of the blood (Hirst and Wood, 1987). A similar procedure using the binding of another more readily available labeled nitroimidazole, ^{14}C SR 2508, was used to obtain the results given in Fig.2. Binding is shown as a function of time between administration of the test drug and injection of the ^{14}C SR 2508. Clofibrate and ML1024 both produced rapid and sustained reductions in binding which were statistically significant; bezafibrate, on the other hand, gave highly variable results but with a clear trend towards <u>increased</u> binding.

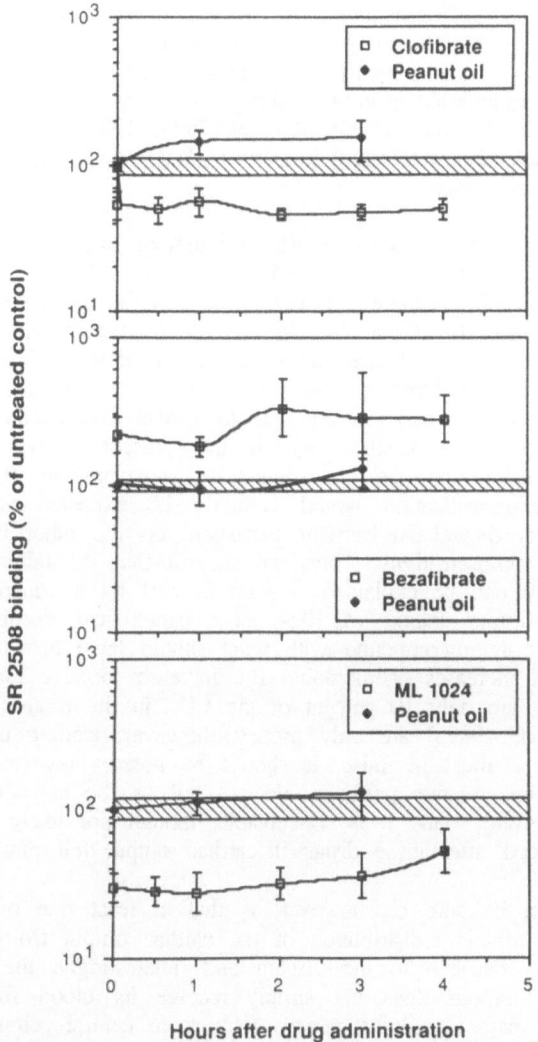

Hours after drug administration

Fig.2 Binding of ^{14}C-SR2508 to SCCVII/St tumours at different times after clofibrate, bezafibrate, ML1024, or the peanut oil vehicle. Data was obtained in three separate experiments giving 6-11 tumours per point. Error bars are $\pm$ 1 s.e. Hatched areas are range ($\pm$ 1 s.e.) in untreated tumours.

DISCUSSION

We have previously reported on the radiosensitizing effects of some derivatives of chlorophenoxyacetic acid (Hirst et al, 1987; Hirst and Wood. 1988 a,b). Our results have been disappointing for two main reasons: First, although we have identified new compounds that are potent modifiers of haemoglobin affinity for oxygen and of low

921

toxicity, they have been poor or totally ineffective radiosensitizers of mouse tumours. Second, where radiosensitization was seen, it seemed to be unrelated to the reduction in haemoglobin affinity achieved with a given compound. There could be two possible explanations for these anomalies: The simplest is that haemoglobin affinity has no effect on tumour oxygenation and radiosensitivity. Alternatively, these compounds could have other effects which enhance or negate the changes in affinity. We feel that the former is unlikely in view of our own observations and those of others, albeit in different tumour systems, that reduced haemoglobin affinity resulting from elevated 2,3-DPG levels does lead to increased tumour radiosensitivity (Siemann et al, 1979; 1985; Hirst et al, 1987). This leaves the possibility that other interesting properties of this class of compound could have radiobiologically-significant effects.

Perhaps the most important parameter affecting tumour oxygenation is tissue perfusion (Degner and Sutherland, 1988). Our "blood flow" data (Fig.1) suggest a plausible hypothesis for the action of clofibrate and bezafibrate. Clofibrate causes a small reduction in haemoglobin/oxygen affinity (Hirst and Wood, 1987), which would be expected to increase tumour oxygenation and hence radiosensitivity modestly (Siemann, 1979; Hirst and Wood, 1987). It also produces a substantial increase in the fraction of the cardiac output perfusing the tumour, which would lead to further oxygenation and sensitization. Bezafibrate at the toxic dose levels used in the present study, gives haemoglobin modification that we have previously concluded is super-optimal (Hirst et al, 1987), although significant radiosensitization would certainly be expected at some time after administration. Its effects on relative tumour perfusion, on the other hand, would reduce tissue oxygenation and radiosensitivity. Thus, we suggest that the failure of bezafibrate to radiosensitize consistently can be explained at least in part by a concomitant reduction in "blood flow". The clofibrate analog ML1024 gave paradoxical results which defy any simple explanation. Here is a compound with what should have been ideal properties for tumour oxygenation: it increases haemoglobin P_{50}, increases relative tumour perfusion and does so at doses which are only 10 percent of the LD_{50} in the mouse. Its radiosensitizing effects (Hirst and Wood, 1989b) are only modest, however, leading us to conclude that some of our assumptions must be false. It should be noted, however, that the methods we have used to measure relative perfusion do not tell us the absolute blood flow (mls of blood/100 mg tissue/min); and it is conceivable, though not likely, that absolute flow might actually be reduced after these drugs if cardiac output fell dramatically.

One thing we can be sure of, however, is that at least one of these compounds, clofibrate, produces a dramatic redistribution of the cardiac output from the major masses of skeletal muscle, and the skin to the viscera and, interestingly, the tumour (Table I). This suggests that the tumour does not simply receive its blood from the local skin vessels, but must have major supplying vessels with more central origins lying in parallel with those of the normal skin.

A recurring question in this study has been why the radiosensitization achieved with these drugs is less impressive than can be produced by modification of 2,3-DPG in the red blood cell (Siemann, 1979; Hirst and Wood, 1987; Siemann, 1987). An obvious difference is that the improved oxygenation achieved either by restoration of ambient oxygen tension or transfusion of 2,3-DPG enriched blood was almost instantaneous in these studies, whereas the administration of the clofibrate analogs gives a gradual increase in P_{50} over 1-3 hr (Hirst and Wood, 1987). This effect would only be important if there was some form of physiological adaptation to the improvement in oxygen delivery, such as reduction in cardiac output (not detectable with the assays used in this study). Adaptation to changes in oxygen delivery can be a powerful mechanism, as we have previously shown for haematocrit alterations (Hirst and Wood, 1986).

922

In previous studies, we have used nitroimidazole binding as an end point to support the concept of an oxygen-dependent mechanism for the radiosensitivity changes, and in each case a close correlation was found (Hirst, Hazlehurst and Brown, 1985; Hirst and Wood, 1987a). That was not the case in the present study. While clofibrate decreased SR 2508 binding, the effect was rather modest and more rapid in onset than the radiosensitization, although it should be noted that time zero is not very meaningful as the SR 2508 has a half life of about 50 minutes in the mouse (Brown and Workman, 1980). ML1024, on the other hand, caused an abrupt and sustained reduction in binding which was larger than would have been predicted on the basis of its effects on radiosensitivity. Finally, bezafibrate caused increased binding which, at certain times after administration (1 hour), was highly significant. A closer examination of the individual animals revealed, however, that the dramatic increases occurred only in those few animals which had been noted as reaching a semi-comatose state at the time of sacrifice. It should be noted that the dose used was close to the LD_{50} of this drug in our animals.

Our results are difficult to interpret in a way which offers a satisfactory explanation for the activity of this class of drugs in vivo. Their original attraction was that they offered a novel means of oxygenating tumours, but we must now conclude that a major part of their effectiveness is attributable to a phenomenon which is becoming quite frequently reported: alterations in tumour blood flow. Do they, therefore, have any clear advantages over other recently studied drugs shown to have similar properties, such as nicotinamide (Horsman et al, 1988), or the calcium channel blockers (Hill and Stirling, 1987; Wood and Hirst, 1988). Our conclusion, based on the compounds tested to date, is that they do not. For example, we have shown (Wood and Hirst, 1988) that a 5-10-fold reduction in hypoxic fraction of the SCCVII/St tumour can be achieved by doses of 0.05-1 mg/Kg of several calcium channel blockers (Wood and Hirst, 1989 a,b) an effect requiring a 100-1000X higher dose of clofibrate.

We do not feel that the concept of haemoglobin affinity modification for tumour radiosensitization should be abandoned; rather, we must examine other ways of applying it clinically or else develop other drugs with different characteristics from the clofibrate analogs.

ACKNOWLEDGEMENTS

We are grateful to Doug Menke and Nixy Kaul for their excellent technical assistance, and to Ann Hanford for help in preparing the manuscript.

REFERENCES

Brown, J.M., and Workman, P., 1980, Partition coefficient as a guide to the development of radiosensitizers which are better than misonidazole. Radiation Research, 82, 171-190.

Chaplin, D.J., Durand, R.E., and Olive, P.L., 1986, Acute hypoxia in tumours: implication for modifiers of radiation effects. International Journal of Radiation Oncology, Biology and Physics, 12, 1279-1282.

Degner, F.L., and Sutherland, R.M., 1988, Mathematical modelling of oxygen supply and oxygenation in tumour tissues: prognostic, therapeutic and experimental implications. International Journal of Radiation Oncology, Biology and Physics, 15, 391-397.

Garrecht, B.M., and Chapman, J.D., 1983, The labelling of EMT-6 tumours in BALB/c mice with ^{14}C misonidazole. British Journal of Radiology, 56, 745-753.

Hill, R.P., and Stirling, D., 1987, Oxygen delivery and tumour response. Radiation Research, Proceedings of the Eighth International Congress of Radiation Research, edited by E.M. Fielden, J.F. Fowler, J.H. Hendry, and D. Scott (London: Taylor & Francis Ltd.), pp. 725-730.

Hirst, D.G., Brown, J.M., and Hazlehurst, J.L., 1982, Enhancement of CCNU cytotoxicity by misonidazole: possible therapeutic gain. British Journal of Cancer, 46, 109-116.

Hirst, D.G., Hazlehurst, J.L., and Brown, J.M., 1985, Changes in misonidazole binding with hypoxic fraction in mouse tumours. International Journal of Radiation Oncology, Biology and Physics, 11, 1349-1355.

Hirst, D.G., and Wood, P.J., 1987a, The adaptive response of mouse tumours to anaemia and retransfusion. International Journal of Radiation Biology, 51, 597-609.

Hirst, D.G., and Wood, P.J., 1987b, The influence of haemoglobin affinity for oxygen on tumour radiosensitivity. British Journal of Cancer, 55, 487-491.

Hirst, D.G., Wood, P.J., and Schwartz, H.C., 1987, The modification of haemoglobin affinity for oxygen and tumour radiosensitivity by antilipidemic drugs. Radiation Research, 112, 164-172.

Hirst, D.G., and Wood, P.J., 1989a, Altered radiosensitivity in a mouse carcinoma after administration of clofibrate and bezafibrate. Radiotherapy and Oncology, 15, 55-61.

Hirst, D.G., and Wood, P.J., 1989b, Chlorophenoxyacetic acid derivatives as haemoglobin modifiers and tumour radiosensitizers. International Journal of Radiation Oncology, Biology and Physics, 16, 1183-1186.

Horsman, M.R., Brown, J.M., Hirst, V.K., Lemmon, M.J., Wood, P.J., Dunphy, P.D., and Overgaard, J., 1988, Mechanism of action of the selective tumour radiosensitizer nicotinamide. International Journal of Radiation Oncology, Biology and Physics, 15, 685-690.

Sapirstein, L.A., 1958, Regional blood flow by fractional distribution indicators. American Journal of Physiology, 193, 161-168.

Siemann, D.W., 1987, New trends in improving oxygen delivery to tumour tissues. Radiation Research, Proceedings of the Eighth International Congress of Radiation Research, edited by E.M. Fielden, J.F. Fowler, J.H. Hendry and D. Scott (London: Taylor & Francis Ltd.), pp. 713-718.

Siemann, D.W., Hill, R.P., Bush, R.S., and Chhabra, P., 1979, The in vivo radiation response of an experimental tumour: The effect of exposing tumour-bearing mice to a reduced oxygen environment prior to but not during irradiation. International Journal of Radiation Oncology, Biology and Physics, 5, 61-68.

Siemann, D.W., and Macler, L.M., 1986, Tumour radiosensitization through reduction in haemoglobin affinity. International Journal of Radiation Oncology, Biology and Physics, 12, 1295-1297.

Teisseire, B.P., Ropars, C., Vallez, M.O., Herigault, R.A., and Nicolau, C., 1985, Physiological effects of high - P_{50} erythrocyte transfusion on piglets. Journal of Applied Physiology, 38, 1820.

Wood, P.J., and Hirst, D.G., 1989a, Cinnarizine and flunarizine as radiation sensitizers in two murine tumours. British Journal of Cancer, 58, 742-745.

Wood, P.J., and Hirst, D.G., 1989b, Calcium antagonists as radiation modifiers: site specificity in relation to tumour response. International Journal of Radiation Oncology, Biology and Physics, 16, 1141-1144.

GROUP PHOTO

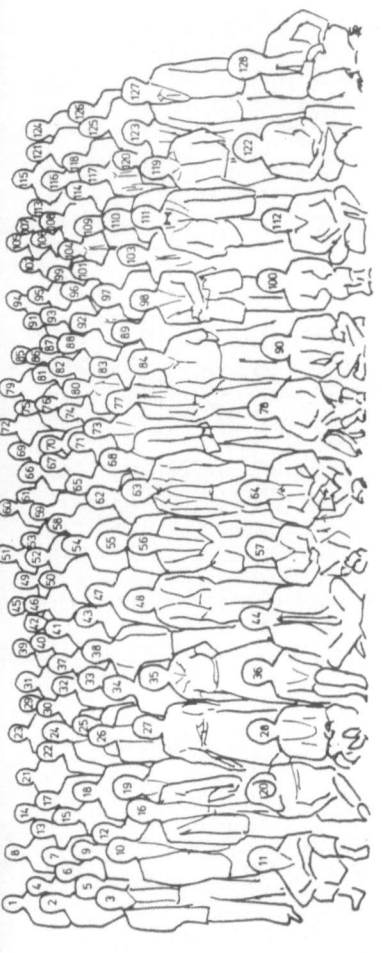

1. EIJKING, E.P.
2. HAGENDORFF, A.
3. MEUER, H.J.
4. HANNEMANN, L.
5. SHOZAWA, T.
6. MAGUIRE, D.
7. EIRING, P.
8. SCHEID, P.
9. HOOFD, L.
10. YOKOYAMA, T.
11. DEGENS, H.
12. GOLDSTICK, T.
13. FERRARI, M.
14. GROS, G.
15. BEBOUT, D.
16. COLODNY, S.
17. TYLER, H.
18. CAIN, S.M.
19. KOBAYASHI, H.
20. KING, C.E.
21. SCHERER, P.W.
22. MEYER, M.
23. AR, A.
24. BOS, J.A.H.
25. TUREK, Z.
26. JÜRGENS, K.
27. LIVERA, N.
28. SILVERTON, S.F.
29. PAGANELLI, C.H.
30. TATEISHI, N.
31. GEISEN, C.
32. POSS, M.

33. JAKOBSEN, K.
34. YAMAGUCHI, K.
35. LADANYI, E.
36. KANG, K.A.
37. GREGERSEN, O.
38. REINHART, K.
39. KUWAHIRA, I.
40. BRULEY, D.F.
41. REIJNEN, J.A.M.
42. HASHIMOTO, M.
43. LAMANNA, J.C.
44. FAITHFULL, S.
45. OKADA, Y.
46. GROTE, J.
47. GUTIERREZ, G.
48. METZGER, S.
49. CALZIA, E.
50. ARNZ, I.E.
51. RÜTTEN, H.
52. GROEBE, K.
53. SCHEFFEN, I.
54. STOKES, B.T.
55. LONGMUIR, I.S.
56. MOCHIZUKI, M.
57. McCABE, M.
58. OKUNIEFF, P.
59. PHILIPPENS, L.
60. HOLST, H.
61. SCHWALEN, A.
62. MARINI, C.
63. LÜBBERS, D.W.
64. HOLLAND, R.A.B.

65. CAMPBELL, S.E.
66. OLDERS, J.
67. HIRST, D.
68. MAYEVSKY, A.
69. RAHMEL, A.
70. RAYNAUD, J.
71. BIRO, G.P.
72. BENTHEM, L.
73. LODATO, R.F.
74. HONIG, C.R.
75. HAK, J.B.
76. KREUZER, F.
77. TROUWBORST, A.
78. SEIYAMA, A.
79. MOHR, M.
80. RAKUSAN, K.
81. OESEBURG, B.
82. GAYESKI, T.
83. BRAUN, R.D.
84. VIDAL MELO, M.F.
85. BASSINGTHWAIGHTE, J.
86. INCE, C.
87. HAAB, P.
88. BATRA, S.
89. TOMITA, M.
90. ITO, K.
91. KISLYAKOV, Y.U.
92. PINDER, A.W.
93. KALLINOWSKI, F.
94. LOUTIS
95. ZOCK, J.P.
96. NELSON, J.A.

97. NEMOTO, E.M.
98. VAUPEL, P.
99. ROGERS, G.
100. NIOKA, S.
101. GONZALEZ, N.C.
102. VAN BEEK, JHGM
103. WALENTA, S.
104. MORNEBURG, H.
105. GRONOW, G.
106. ESSENPREIS, M.
107. RINNERT, H.
108. SCHYWALSKY, M.
109. HEISLER, N.
110. THIMM, F.
111. FARHI, L.
112. DELPY, D.F.
113. SHAMS, H.
114. MERTZLUFT, F.
115. RAHN, H
116. METZGER, H.P.
117. HUDETZ, A.G.
118. VAN DER ZEE, P.
119. SUGIOKA, K.
120. COLANTUONI, A.
121. PELSTER, B.
122. GOTO, M.
123. BERTUGLIA, S.
124. BARTECZKO, I.
125. EKE, A.
126. HILLEBRECHT, A.
127. PIIPER, J.
128. FUKUI, H.

929

AUTHOR INDEX

SUBJECT INDEX